Handbook of
Materials for
Product Design

HANDBOOK OF MATERIALS FOR PRODUCT DESIGN

Charles A. Harper   Editor in Chief

Technology Seminars, Inc., Lutherville, Maryland

Third Edition

McGRAW-HILL

New York   Chicago   San Francisco   Lisbon   London   Madrid
Mexico City   Milan   New Delhi   San Juan   Seoul
Singapore   Sydney   Toronto

Library of Congress Cataloging-in-Publication Data

Handbook of materials and product design / Charles A. Harper, editor in chief.
    p. cm.
   ISBN 0-07-135406-9
    1. Materials—Handbooks, manuals, etc. 2. Engineering design—Handbooks, manuals,
etc.   I. Harper, Charles A.

  TA403.4.H365 2001
  620.1'1—dc21   2001030028
                                    CIP

McGraw-Hill

A Division of The McGraw·Hill Companies

ISBN 0-07-135406-9

*The sponsoring editor for this book was Kenneth McCombs and the
production supervisor was Pamela A. Pelton. It was set in Century
Schoolbook by J. K. Eckert & Company, Inc.*

Printed and bound by R. R. Donnelley & Sons Company.

This book is printed on acid-free paper.

CONTENTS

Chapter 10. Metallic Finishes and Processes   **10.1**

Chapter 11. Plastics Joining Materials and Processes   **11.1**

Chapter 12. Plastics and Elastomers as Adhesives   **12.1**

Chapter 13. Testing of Materials   **13.1**

Chapter 14. Materials Recycling   14.1

CONTRIBUTORS

Thomas A. Andersen   *Northrop-Grumman Corporation, Baltimore, Maryland* (CHAP. 10)

R. J. Del Vecchio   *Technical Consulting Services, Fuquay-Varina, North Carolina* (CHAP. 6)

Simon Durham   *Pratt & Whitney Canada, Longuevil, Quebec, Canada* (CHAP. 3)

J. Donald Gardner   *Northrup Grumman Electronic Sensors and Systems Sector, Columbia, Maryland* (CHAP. 1)

Carl P. Izzo   *Industrial Paint Consultant, Export, Pennsylvania* (CHAP. 9)

J. Randolph Kissell   *TGB Partnership, Hillsborough, North Carolina* (CHAP. 2)

James Margolis   *Consultant, Montreal, Quebec, Canada* (CHAP. 6)

Perry L. Martin   *Martin Testing Laboratories, Yuba City, California, www.martintesting.com* (CHAP. 13)

Robert Ohm   *Uniroyal Chemical, Naugatuck, Connecticut* (CHAP. 6)

Stanley T. Peters   *Process Research, Mountain View, California, www.process-research.com* (CHAP. 5)

Edward M. Petrie   *ABB Power T & D Company, Inc., Raleigh, North Carolina* (CHAPS. 11, 12)

Jordon I. Rotheiser   *Rotheiser Design, Inc., Highland Park, Illinois* (CHAP. 4)

Susan E.M. Selke   *School of Packaging, Michigan State University, East Lansing, Michigan* (CHAP. 14)

Jerry E. Sergent   *TCA Inc., Corbin, Kentucky* (CHAP. 7)

Thomas P. Seward III   *New York State College of Ceramics, Alfred University, Alfred, New York* (CHAP. 8)

Arun Varshneya   *New York State College of Ceramics, Alfred University, Alfred, New York* (CHAP. 8)

Steven Yue   *McGill University, Montreal, Quebec, Canada* (CHAP. 3)

PREFACE

While the role of materials has always been important in product design, materials are now often the keystones for successful products in our modern world of high technology. In fact, it might even be said that materials are the critical limiting factor for achieving the high performance and reliability demanded of today's products. Next generation's products usually require new or improved materials, and necessity often becomes the mother of invention. Materials scientists always rise to meet the need.

Success in achieving outstanding materials is not adequate, however. Since most product designers are mechanical or electrical engineers, and since materials are chemical, these significantly different technical languages lead to a critical knowledge and understanding gap. Successful product design requires, first, bridging this technical language barrier gap and, second, providing the product designer with the information, data, and guidelines necessary to select the optimum material for a given product design. It is the purpose of this *Handbook of Materials for Product Design* to provide both an understanding of the many classes of materials that the product designer has available to him, and the information, data, and guidelines that will lead the product designer to the best choice of materials for his specific product. Toward this end, this book has been prepared as a thorough sourcebook of practical data for all ranges of interests. It contains an extensive array of materials properties and performance data, presented as a function of the most important product variables. In addition, it contains very useful reference lists at the end of each chapter and a thorough, easy-to-use index.

The chapter organization of this *Handbook of Materials for Product Design* is well suited for reader convenience. The initial three chapters deal with metal materials—first, the important ferrous metals, then second, the broadly used aluminum metals and alloys, and third, metals other than those covered in the first two chapters. The second set of three chapters covers polymeric materials first, the all-important group of plastic materials, then, second, that specially reinforced group of plastics known as *composites,* and third, that important group of rubbery polymeric materials known as *elastomers*. Next come two chapters on the two major groups of nonmetallic, inorganic materials, namely, ceramics and glasses. These are followed by two chapters on finishes, first organic finishes and paints, and second, electrodeposited or electroplated metallic finishes.

Following all of the above chapters on specific groups of materials are two chapters on the always critical and often difficult areas of bonding materials. First is a chapter on the joining of plastics, with explanations of the various processes and their trade-offs. Next comes a very practical and useful chapter on the many adhesive bonding materials, techniques, and processes, along with their trade-offs.

The final two chapters in the book are both increasingly important and critical in modern product design applications. First is a chapter on materials testing and reliability, and second is a chapter on material recycling. These are especially important, since they affect not only optimum product design but also environmental and even legal issues.

The result of these presentations is an extremely complete and comprehensive single reference text—a must for the desk of anyone involved in product design, development, and application. This *Handbook of Materials for Product Design* will also be invaluable for every reference library.

As will be evident from a review of the subject and author listings, I have had the good fortune to be able to bring together a team of outstanding chapter authors, each with a great depth of experience in his or her field. Together, they offer the reader a base of knowledge as perhaps no other group could. Hence, I would like to give special credit to these authors in this preface.

It is my hope and expectation that this *Handbook of Materials for Product Design* will serve its readers well. Any comments or suggestions will be welcomed.

Charles A. Harper
Technology Seminars, Inc.
Lutherville, Maryland

1

Ferrous Metals[*]

J. Donald Gardner
Northrup Grumman Electronic Sensors and Systems Sector
Columbia, Maryland

1.1 Introduction

One major technical advancement of the early Greek period was the widespread use of iron. Furnaces were developed that could reach the high melting temperature of that metal. Iron technology had spread throughout the classical world by about 500 B.C. Adding small amounts of carbon to iron as it was hammered over a charcoal fire inaugurated early steels. Mining became well developed and included the use of pumps to keep mines from flooding. Technology also advanced weaponry with the development of catapults, better swords, and body armor.

The main reasons for the popularity of steel are the relatively low cost of making, forming, and processing it; the abundance of its two raw materials (iron ore and scrap); and its unparalleled range of mechanical properties. More than any other material, the quality of life, in many respects, has improved on the planet as the quality of steel has improved.

Initially, tools of iron to were used to form many of the other needed goods. Eventually, this was followed by the Industrial Revolution and the mechanization of farms. Machine tools and other equipment made of iron and steel changed the economy of both city and farm.

Steel is the most widely used material for building the world's infrastructure. It is used to fabricate everything from pins to skyscrapers. In addition, the tools required to build and manufacture such articles are also made of steel.

---

* With special credit to **Stephen G. Konsowski,** Consultant, Glen Burnie, Maryland.

Today, most of the finished steel produced in the United States is shipped to five domestic markets. The automotive industry takes the greatest share, about 20 percent. Almost 20 percent goes into the warehouses of steel service centers, where the metal is sold as bar stock or plate, or it is processed to order for specific industrial applications. Some 15 percent is used, either directly or indirectly, for construction. Cans and other containers take about six percent of the production of finished steel. Another six percent goes directly from the mill to farm and electrical equipment manufacturers.

The steel industry, like many of the metal industries, offers a high degree of standardization. Because of the degree of standardization, a designer need only be concerned with specifying the proper alloy, product form, and heat treatment and can be less concerned with the ultimate supplier of the steel. Specifications for steel products, as well as testing procedures, are normally included in the general standard systems of most industrial countries. Such standardization still does not exist for nearly all forms of organic materials.

1.2   The Structure of Iron

1.2.1   Phase Diagrams[1]

The term *phase* is variously defined in the dictionary. The chemical definition, which applies for this chapter, is *a solid, liquid, or gaseous homogeneous form of matter existing as a distinct part of a heterogeneous system.*

One of the simplest and most common examples of this definition is water (H_2O). The H_2O chemical compound is a solid phase (ice) at temperatures at or below 0°C (32°F). It is a liquid phase (water) between the temperatures of 0°C (32°F) and 100°C (212°F) at sea level pressure, and a gaseous phase (steam) at a temperature of 100°C (212°F) or above.

Metals and alloys may also exist as solid, liquid, and gaseous phases. With one notable exception (mercury), metals exist in their solid phase at room temperatures. A few metals and alloys do change from solid to liquid.

1.2.2   Phase Changes

In most simple substances, the phases are very straightforward: solid, liquid and gas, as described above. This is less likely to be true for all metals and alloys. Some pure metals can exist as more than one phase within their solid state, depending on temperature. Alloys vary widely and may contain several phases within their solid state.

The commonly accepted system of designating the various phases utilizes the chemical symbol of the element and letters of the Greek alphabet. See Fig. 1.1 for the iron-carbon phase diagram.

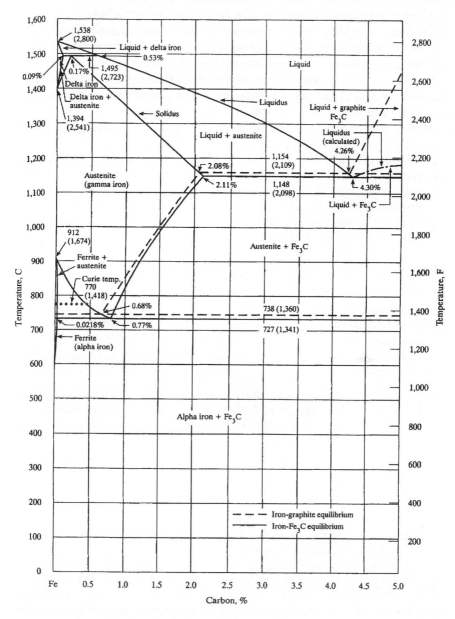

Figure 1.1 Major groups in the AISI-SAE designation system.[2]

Fluid at temperatures near that of boiling water, most metals and alloys change from their solid to their liquid phases at considerably elevated temperatures. Many metals undergo this change within the range of 540 to 1540°C (1000 to 2800°F). At still higher temperatures, metals and alloys are converted into gaseous phases.

In heat processing of metals, the transition from the solid to the gaseous phase (sublimation) sometimes must be considered—most notably in vacuum processing of metals that have relatively low vapor pressures. An outstanding example of sublimation is heat treating of stainless steels in a vacuum furnace, under which condition the loss of chromium can be excessive when elevated temperature and a hard vacuum (a pressure of 10^{-6} torr) are used.

Definitions of terms related to metallurgical phase changes are:

1. *Eutectic.* (1) An isothermal (constant temperature) reversible reaction in which a liquid solution is converted into two or more intimately mixed solids on cooling, the number of solids formed being the same as the number of components in the system. (2) An alloy having the composition indicated by the eutectic point on an equilibrium diagram. (3) An alloy structure of intermixed solid constituents formed by a eutectic reaction.

2. *Eutectoid.* (1) An isothermal reversible reaction in which a solid solution is converted into two or more intimately mixed solids on cooling, the number of solids formed being the same as or the number of components in the system. (2) An alloy having the composition indicated by the eutectoid point on an equilibrium diagram. (3) An alloy structure of intermixed solid constituents formed by a eutectoid reaction.

3. *Intermetallic compound.* An intermediate phase in an alloy system having a narrow range of homogeneity and relatively simple stoichiometric proportions. The nature of the atomic binding can be of various types, ranging from metallic to ionic.

4. *Peritectic.* An isothermal reversible reaction in which a liquid phase reacts with a solid phase to produce a single (and different) solid phase on cooling.

5. *Peritectoid.* An isothermal reversible reaction in which a solid phase reacts with a second solid phase to produce a single (and different) solid phase on cooling.

1.2.3 Crystallography of Iron

Allotropy is the nature of a solid metal to assume more than one lattice arrangement (crystallographic structure), depending on the temperature and energy level. Several of the pure metals are allotropic. In some instances, the allotropic characteristic is of minor importance in relation to heat processing. For iron and titanium, however, their allotropic characteristics are extremely important, serving as the mechanism that allows development of specific properties by means of heat treatment.

The simplest forms of phase diagrams are those that represent pure metals. These are straight lines, usually beginning at room temperature and extending up to or above the metals' melting temperatures. The phase diagram for pure iron generally is considered the most important, because iron is the major constituent of all steels. At room temperature, the iron atoms are body-centered cubic (bcc) structure. This phase is designated *alpha iron*. Iron in this phase is attracted to a magnet. When the temperature reaches 770°C (1418°F) on heating, the bcc alpha iron structure becomes nonmagnetic; this is not considered to be a phase change.

When the temperature reaches 912°C (1674°F), the atom arrangement changes to the face-centered cubic (fcc) phase, which is designated *gamma iron*. This phase change is accompanied by a decrease in volume and absorption of heat (heat of transformation). The fcc phase prevails until the temperature reaches 1394°C (2541°F), at which point a third modification takes place—the atom arrangement reverts back to bcc and is denoted as *delta iron*. Although the alpha and delta phases are similar, both being bcc, they are separately termed, because one exists below the gamma range, while the other exists above the gamma range. The delta phase melts at 1538°C (2800°F).

For most heat processing operations of ferrous alloys, the portion above the gamma range is not important. The lattice modifications of pure iron are reversible, because they are temperature dependent; that is, when the liquid phase is cooled to 1538°C (2800°F), it first freezes as delta iron (bcc) phase, then changes to gamma iron (fcc) at 1394°C (2541°F).

When two or more metals are combined in their liquid states then allowed to solidify, a wide array of events may take place. In many instances, because two metals are totally insoluble in each other, the solidified mass is not an alloy but simply a mixture. A notable example of a metal mixture is lead in iron (or steel). Lead is completely insoluble in iron; if the lead is added to molten iron, it solidifies as a dispersion of lead particles in the solidified iron and is thus a true mixture.

At the opposite extreme, two metals may be completely soluble in each other in the liquid phase as well as in the solid phase. A notable example of such an alloy system is copper and nickel. In other instances, the solid solubility of one metal in another may vary greatly with temperature. Numerous phases may be formed, depending to a great extent on cooling rate. In these cases, when one metal dissolves in another, the result is an alloy that is sharply segregated from a mixture, such as lead in iron as described above. Nonmetals dissolved in metals may form alloys such as steel, an alloy of iron and carbon.

Also, when iron cools from the gamma phase (fcc) to the alpha phase (bcc), there is an evolution of heat. The alpha phase is constant to room temperature and below. An isotherm (constant temperature) occurs on the phase diagram, because heat is evolved on cooling and absorbed on heating during phase changes and at the magnetic change.

1.3   Steelmaking[2]

1.3.1   Process Metallurgy of Steelmaking

Steel is by definition an alloy of iron and carbon, but this statement must be qualified by placing limits on carbon content. When iron-carbon alloys have less than 0.005% carbon present at room temperature, they are considered to be pure iron. Pure iron is soft, ductile, and relatively weak. It is not normally used as an engineering material because of its low strength, but it is used for special applications such as magnetic devices and enameling steels (steels that are glass coated, like bathtubs). From the commercial standpoint, steels have a low carbon limit of approximately 0.06% carbon. On the other end of the carbon-content scale, iron-carbon alloys with more than approximately 2% by weight of carbon are considered to be cast irons. Above this carbon level, casting is about the only way that a useful shape can be made from the alloy, since the high carbon makes the iron alloy too brittle for rolling, forming, shearing, or other fabrication techniques. Thus, steels are alloys of iron and carbon with carbon limits between approximately 0.06 and 2.0%.

Typically, the carbon content in pig iron may be 4 or 5%, which is too high to use as a steel. In addition to the high carbon content, the pig iron may contain high amounts of silicon, sulfur, phosphorus, and manganese, as well as physical inclusions of nonmetallic materials from the ore.

1.3.1.1   Refining processes.   The early Bessemer furnaces used in the purification of pig iron to steel employed a sustained blast of air in the molten iron. Today, the purification of pig iron into steel is accomplished in three basic processes.

1. Basic oxygen furnace

2. Open hearth furnace

3. Electric furnace

Basic oxygen furnace. The basic oxygen furnace was introduced into the United States in the 1960s. This process for making steel involves charging a large ladle-shaped vessel with about 25% scrap and 75% molten pig iron. Oxygen lances are lowered into the melt, and blowing is initiated. Large amounts of pressurized oxygen are introduced into the molten charge. The heat of this reaction provides the heat for the process; the furnace is not externally heated. When the charge is sufficiently blown (the carbon content of the charge is reduced to the desired level), the oxygen lances are withdrawn, and the furnace is tipped to pour the molten charge. Refining times are very short, and sophisticated computer equipment is used to analyze the composition of the charge; 300 tons of steel can be refined in 25 min.

Obviously, the short refining time of the basic oxygen furnace (BOF) makes it the process of choice and economic necessity for competitive steel mills. Since the 1960s, there has been an evolutionary change in the U.S. steel industry to make all large-tonnage steel products in basic oxygen furnaces

Open hearth furnaces. Open hearth furnaces have been the workhorse of the steel industry since the turn of this century. They can produce as much as 450 tons of steel in a single batch, and they have been used for the bulk of steel production in the United States since the phaseout of the Bessemer process. The starting material for the open hearth furnace is molten pig iron that is transported to the furnace in hot metal cars and scrap (possibly equal amounts). These starting materials are charged into the hearth, which is really nothing but a shallow rectangular pool of molten metal. Large quantities of air are supplied to the area over the molten pool to produce oxidation of the carbon in the material to be refined. Oxygen lances are also placed into the pool to help reduce the carbon level. When the charge has been refined to the desired carbon level, the slag that forms on the top of the pool is tapped, and the refined steel is tapped into a ladle. Subsequently, the metal in the ladle is poured into ingot molds. A charge of 300 tons may take 8 hr of refining before it is ready to pour.

Electric furnaces. Electric furnaces are usually charged with scrap, and melting is accomplished by establishing an electric arc between carbon electrodes that are lowered in proximity to the charge. These furnaces take massive amounts of energy, but they do not compare in complexity and enormity with the open hearth equipment. Refining times are shorter, and the furnace is pivoted for pouring. Electric furnaces tend to produce cleaner steels than open hearth furnaces. Until the 1970s, these furnaces were mostly used for specialty alloys, such as tool steels, bearing steels, and stainless steels. These furnaces are also used in mini-mills that are often dedicated to limited groups of steel products.

1.3.1.2   Clean steels.   With increased emphasis on reliability, many industries are specifying that steels for critical parts be made by special melting practices. Many strength-of-materials studies have shown, for example, that the fatigue life and toughness of steels are directly proportional to the size and volume fraction of nonmetallic inclusions in the steel. These inclusions are usually oxides, silicates, sulfides, or aluminas that form during conventional melting and refining. The inclusion rating of a piece of steel can be measured by sawing a thin slice from the end of a steel shape and etching it in an acid. A dirty steel will show pits when the inclusions are etched away by the acid. There are also micro-cleanliness standards for steels that measure inclusion ratings by microscopic examination of a polished sample from a steel shape. These inclusion ratings can be added to a steel purchasing specification so that a designer has the option of specifying steel of special cleanliness.

There are two major techniques and a number of options in each category:

Vacuum melting or remelting

- Vacuum degassing (VD)

- Vacuum arc remelting (VAR)

- Vacuum induction melting (VIM)

- Electron beam refining (EBR)

Chemical reaction

- Argon oxygen decarburization (AOD)

- Electroslag remelting (ESR)

- Ladle refining (LR)

Vacuum refining.   In vacuum refining, steel is usually melted in a basic oxygen or electric furnace and, by some secondary technique, the steel is exposed to a vacuum while it is molten. Dissolved gases rise to the surface of the melt and go off in the vacuum. With sufficient stirring and mixing, solid inclusions can also be vaporized and drawn away by the vacuum. The differences between the various processes involve how melting and stirring are achieved. Vacuum degassing (VD) is usu-

ally accomplished by lance stirring molten steel and pouring it into ingots in a large evacuated enclosure. Vacuum arc remelting (VAR), the
most popular technique, involves casting of steel from the BOF or electric furnace into cylindrical ingots. A stub shaft is welded to these ingots, and the ingot is remelted in a vacuum by establishing an arc
between the ingot (electrode) and a water-cooled copper mold. The ingot becomes a giant welding electrode. This process is very effective in
removal of inclusions because it is very energetic. Every drop of the ingot is exposed to vacuum as the droplets transfer in the arc.

Vacuum induction melting (VIM) is a process that is used to melt
solid scrap or liquid charges. The charge is placed in a crucible and
heated by high-frequency induced currents. This produces convection
current mixing of the melt. The entire crucible is in the vacuum, and
ingots are also cast in the vacuum.

In electron beam refining, molten metal is poured down a tundish
(chute) into an ingot mold. The tundish and mold are in a vacuum. As
the metal flows down the tundish, it is subjected to an electron beam
that vaporizes impurities so that they can be removed as vapors in the
vacuum.

The processes grouped under chemical reaction are different from
the vacuum refining mechanism in that impurities are removed by reaction with some species introduced into the melt; they are not in vacuum. In the AOD process, argon and oxygen are introduced into a
crucible containing a molten heat from an electric furnace or BOF. The
oxygen reduces carbon level (decarburization), sulfides, and other impurities. The argon causes significant stirring to disperse oxides and
make them smaller. The argon also promotes removal of dissolved
gases.

Electroslag refining. Electroslag refining is similar to VAR without the
vacuum. A VIM or electric furnace melt is cast into remelt ingots.
These ingots have a stub welded to them, and they are made into electrodes for arc remelting in a water-cooled copper mold. Purification is
accomplished when the melting metal from the ingot passes through a
molten flux that acts like a welding electrode slag to remove impurities. Shrinkage voids are minimal in ESR ingots.

1.3.1.3 Steel terminology. The selection of steels requires consultation
on property information and supplier information on availability. If
the designer is to make any sense out of handbook information, it is
necessary to become familiar with the terms used to describe mill processing operations. There are so many terms that it can be very confusing. The following is a tabulation of steel product terms and what
they mean.

Carbon steel   Steels with carbon as the principal hardening agent. All other alloying elements are present in small percentages, with manganese being limited to 1.65% maximum, silicon to 0.60% maximum, copper to 0.60% maximum, and 0.05% maximum for sulfur and phosphorus.

Alloy steels   This term can refer to any steel that has significant additions of any element other than carbon but, in general usage, alloy steels are steels with total alloy additions of less than about 5%. They are used primarily for structural applications.

Rimmed   Slightly deoxidized steels that solidify with an outer shell on the ingot that is low in impurities and very sound. These steels can retain a good finish even after severe forming because of the surface cleanliness.

Killed   Strongly deoxidized, usually by chemical additions to the melt. These steels have less segregation than rimmed steels. The mechanical properties are more predictable than commercial and merchant-grade steels.

Concast   Steel produced in a continuous steel casting facility. These steels are deoxidized (usually with aluminum additions).

Galvanized   Zinc-coated steel products. The zinc is applied by hot dipping.

Galvannealed   Zinc-coated and heat-treated steel. There are usually paint-adhesion problems with galvanized steels. The heat treatment given to galvannealed steels creates an oxide layer that allows better paint adhesion.

Sheet   Rolled steel primarily in the thickness range of 0.010 to 0.250 in. (0.25 to 6.4 mm) thick and with a width of 24 in. (610 mm) or more.

Bar   Hot- or cold-rolled rounds, squares, hexes, rectangles, and small shapes. Round bars can be as small as 0.25 in. (6.4 mm); flats can have a minimum thickness of 0.203 in. (5.0 mm); shapes have a maximum dimension less than 3 in. (76 mm).

Coil   Rolled steel in the thickness range of sheet or strip.

Flat wire   Small hot- or cold-rolled rectangles often made by cold-reducing rounds to rectangular shape.

Wire   Hot- or cold-drawn coiled rounds in varying diameters, usually not exceeding 0.25 in. (6.4 mm).

Shapes   Hot-rolled I-beams, channels, angles, wide-flange beams, and other structural shapes. At least one dimension of the cross section is greater than 3 in. (76 mm).

Tin plate   Cold-rolled steel with a usual thickness range of 0.005 to 0.014 in. (0.1 to 0.4 mm). It may or may not be tin coated.

Free machining   Steels with additions of sulfur, lead, selenium, or other elements in sufficient quantity that they machine more easily than untreated grades.

Drawing quality Hot- or cold-rolled steel specially produced or selected to satisfy the elongation requirements of deep drawing operations.

Merchant quality Steels with an M suffix on the designation intended for nonstructural applications. A low-quality material.

Commercial quality Steels produced from standard rimmed, capped, concast, or semikilled steel. These steels may have significant segregation and variation in composition, and they are not made to guaranteed mechanical property requirements (most widely used grade).

H Steels Steels identified by an H suffix on the designation and made to a guaranteed ability to harden to a certain depth in heat treatment.

B Steels Steels with small boron additions as a hardening agent. These steels are identified by a B inserted between the first two and last two digits in the four-digit identification number (xx B xx).

PicklingUse of acids to remove oxides and scale on hot-worked steels.

Temper rolling Many steels will exhibit objectionable strain lines when drawn or formed. Temper rolling involves a small amount of roll reduction as a final operation on annealed material to eliminate stretcher strains. This process is sometimes used to improve the surface finish on a steel product.

TemperThe amount of cold reduction in rolled sheet and strip.

E Steels Steels with an E prefix on the four-digit designation are melted by electric furnace.

1.4   Carbon and Alloy Steels[2]

1.4.1   Basic Definitions of Carbon and Alloy Steels

Carbon steels are simply alloys of iron and carbon, with carbon as the major strengthening agent. The American Iron and Steel Institute (AISI) defines carbon steels as steels with up to 2% carbon and only residual amounts of other elements except those added for deoxidation (for example, aluminum), with silicon limited to 0.6%, copper to 0.6%, and manganese to 1.65%. Other terms applied to this class of steels are plain carbon steels, mild steels, low-carbon steels, and straight carbon steels. These steels make up the largest fraction of steel production. They are available in almost all product forms: sheet, strip, bar, plates, tube, pipe, wire. They are used for high-production items such as automobiles and appliances, but they also play a major role in machine design for base plates, housings, chutes, structural members, and literally hundreds of different parts.

Alloy steel is not a precise term. It could mean any steel other than carbon steels, but accepted application of the term is for a group of steels with varying carbon contents up to about 1% and with total alloy content below 5%. The AISI defines alloy steels as steels that exceed one or more of the following limits: manganese, 1.65%; silicon, 0.60%; copper, 0.60%. A steel is also an alloy steel if a definite concentration of various other elements is specified: aluminum, chromium (to 3.99%), cobalt, molybdenum, nickel, titanium, and others. These steels are widely used for structural components that are heat treated for wear, strength, and toughness. They are the types of steels used for axle shafts, gears, and hand tools such as hammers and chisels.

A family of steels related to, but different from, alloy steels is high-strength, low-alloy (HSLA) steels. This term is used to describe a specific group of steels that have chemical compositions balanced to produce a desired range of mechanical properties. Some of these steels also have alloy additions to improve their atmospheric corrosion resistance. They are available in various products, but usage centers about sheet, bar, plate, and structural shapes. The yield strength is usually in the range of 42 to 70 ksi (289 to 482 MPa), with the tensile strength 60 to 90 ksi (414 to 621 MPa). The primary purpose of these steels is weight reduction through increased strength. Smaller section sizes are possible.

In addition to carbon, alloy, and high-strength, low-alloy steels, there are tool steels, steels for special applications such as pressure vessels and boilers, mill-heat-treated (quenched and tempered or normalized) steels, and ultrahigh-strength steels. All these steels can be useful in engineering design, but the most important are undoubtedly the carbon and low-alloy steels and tool steels.

1.4.1.1 Alloy designation. The last 30 years in the steel industry have seen a great deal of activity in the area of alloy development. The high-strength, low-alloy, quenched and tempered, and some of the ultrahigh-strength steels were developed in this period. They meet industry needs for weight reduction, higher performance, and, in many cases, lower costs. The disadvantage from the designer's standpoint is that it is becoming difficult to categorize steels in an orderly fashion to aid selection. The common denominator for the steel systems is use. They are the types of steels that would be used for structural components. Figure 1.2 outlines the categories. Even with the abundance of special-purpose steels, the workhorses are (and will continue to be for some time) the wrought ASTM, AISI-SAE carbon and alloy steels. Fortunately, these steels have an understandable and orderly designation system. We shall describe this system in detail and make some general comments on the other steel systems shown in Fig. 1.2.

1.4.2   Carbon and Alloy Steels

The most important identification system for carbon and alloy steels in the U.S. is the system adopted by the American Iron and Steel Institute (AISI) and the Society of Automotive Engineers (SAE). This system usually employs only four digits. The first digit indicates the grouping by major alloying elements. For example, a first digit of 1 indicates that carbon is the major alloying element. The second digit in some instances suggests the relative percentage of a primary alloying element in a given series. The 2xxx series of steels has nickel as the primary alloying element. A 23xx steel has approximately 3% nickel; a 25xx steel has approximately 5% nickel. The last two digits (sometimes the last 3) indicate median carbon content in hundreds of a percent. A 1040 steel will have a normal carbon concentration of 0.40%. The classes of steels in this system are shown in Table 1.1.

In addition to the four digits, various letter, prefixes, and suffixes provide additional information on particular steels.

AISI 1020 steel is covered by UNS 610200. The first four digits come from the AISI system, and a 0 is added as the last digit. An outline of the entire Unified Numbering System is shown in Table 1.2.

The UNS is described in detail in ASTM E527 specification. The system is discussed in this text, because some property handbooks have adopted this system in identifying alloys, and at least one trade organization, the Copper Development Association, has adopted UNS numbers as the official identification system for all copper alloys. Since AISI and most of the steel industry does not utilize the UNS designation, it will be noted only in passing in this chapter.

1.4.3   High-Strength, Low-Alloy Steel

There is a significant amount of commercial competition in this family of steels, and trade names are often used to designate high-strength, low-alloy steels, but it is wise to avoid this practice. The most accepted practice is to separate these alloys by minimum tensile properties for a given section thickness.

Chemical compositions are published, but the ranges are wide, and alloy additions are balanced to meet mechanical properties rather than composition limits. The preferred system to use in specifying one of these alloys on an engineering drawing is to use an American Society for Testing Materials (ASTM) designation number followed by the strength grade desired. ASTM specifications A 242, A 440, A 441, A

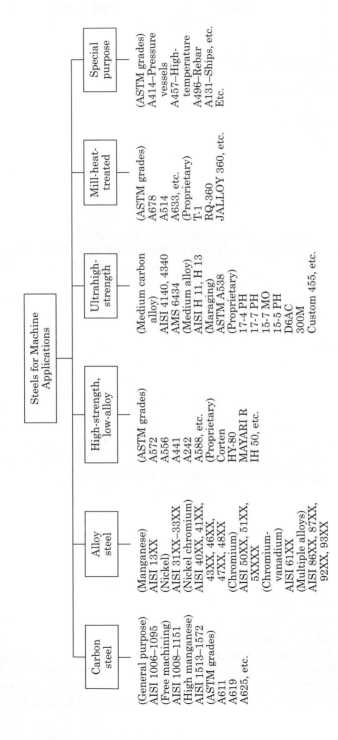

Figure 1.2 Steel types used for machine applications.[2]

TABLE 1.1   Major Groups in the AISI–SAE Designation System[2]

Class	AISI series	Major constituents
Carbon steels	10xx	Carbon steel
	11xx	Resulfurized carbon steel
Alloy steels		
Manganese	13xx	Manganese 1.75%
	15xx	Manganese 1.00%
Nickel	23xx	Nickel 3.50%
	25xx	Nickel 5.00%
Nickel-chromium	31xx	Nickel 1.25%, chromium 0.65 or 0.80%
	33xx	Nickel 3.50%, chromium 1.55%
Molybdenum	40xx	Molybdenum 0.25%
	41xx	Chromium 0.95%, molybdenum 0.20%
	43xx	Nickel 1.80%, chromium 0.50 or 0.80%, molybdenum 0.25%
	46xx	Nickel 1.80%, molybdenum 0.25%
	48xx	Nickel 3.50%, molybdenum 0;25%
Chromium	50xx	Chromium 0.30 or 0.60%
	51xx	Chromium 0.80, 0.95 or 1.05%
	5xxxx	Carbon 1.00%, chromium 0.50, 1.00, or 1.45%
Chromium-vanadium	61xx	Chromium 0.80 or 95%, vanadium 0.10 or 0.15% min.
Multiple alloy	86xx	Nickel 0.55%, chromium 0.50%, molybdenum 0.20%
	87xx	Nickel 0.55%, chromium 0.50%, molybdenum 0.25%
	92xx	Manganese 0.85%, silicon 2.00%
	93xx	Nickel 3.25%, chromium 1.20%, molybdenum 0.12%
	94xx	Manganese 1.00%, nickel 0.45%, chromium 0.40%, molybdenum 0.12%
	97xx	Nickel 0.55%, chromium 0.17%, molybdenum 0.20%
	98xx	Nickel 1.00%, chromium 0.80%, molybdenum 0.25%

588, and A 572 cover some of these steels in structural shapes. ASTM specifications A 606, A 607, A 715, A 568, A 656, A 633, A 714, and A 749 cover some of these steels in sheet and strip form.

The discriminating characteristic of this class of steels is high strength. Yield strengths usually exceed 175 ksi (1206 MPa). Certain of the alloy steels are considered to be ultrahigh strength (e.g., 4140,

TABLE 1.2   Outline of the Unified Numbering System for Metals[3]

UNS number	Alloy system
Axxxxx	Aluminum and aluminum alloys
Cxxxxx	Cooper and copper alloys
Exxxxx	Rare earth and rare element-like metals and alloys
Fxxxxx	Cast irons
Gxxxxx	AISI and SAE carbon and alloy steels
Hxxxxx	AISI and SAE H steels
Jxxxxx	Cast steels (except tool steels)
Kxxxxx	Miscellaneous steels and ferrous alloys
Lxxxxx	Low-melting metals and alloys
Mxxxxx	Miscellaneous nonferrous metals and alloys
Nxxxxx	Nickel and nickel alloys
Pxxxxx	Precious metals and alloys
Rxxxxx	Reactive and refractory metals and alloys
Sxxxxx	Heat and corrosion resistant (stainless steels)
Txxxxx	Tool steels, wrought and cast
Zxxxxx	Zinc and zinc alloys

4340), as are some of the AISI tool steels (H11 and H13). In these cases, proper designation is achieved by using the AISI system. Some steels are strictly proprietary (for example, 17-4PH). In this instance, there is no recourse but to use the trade name on the drawing along with the name and address of the manufacturer. A very useful class of ultrahigh-strength structural steels is the 18% nickel maraging steels. These are covered by ASTM specification A 538, and this specification can be used for alloy designation.

1.4.4   Mill-Heat-Treated Steel

Proprietary and ASTM specifications cover these alloys. The ASTM specifications again are the preferred method of designation. For example, ASTM specification A 678 covers quenched and tempered carbon steel plates for structural applications. ASTM A 663 covers "Normalized High-Strength Low-Alloy Structural Steel," and A 514 covers "High Yield Strength, Quenched and Tempered Alloy Steel Plate Suitable for Welding." There are abrasion-resistant grades with hardnesses between about 300 and 400 HB.

1.4.5   Special-Purpose Steel

Many steel alloys have been developed for special applications. There are steels intended for high-temperature and low-temperature service, for springs, for pressure vessels, for boilers, for use in concrete, for railroads, and for all sorts of service conditions. In addition to steels for specific types of service, there are steels with various coatings (e.g., tin plate). These can be applied to any type of service, and specifications center on base metal and coating characteristics. The AISI has product specifications on coated steels, rail steels, and steels for various building applications. There is no numbering system on these steels, and product information is best obtained from the AISI steel products handbooks. The ASTM has specifications on many coated products, rail products, and steels intended for particular types of service. The ASTM specification number can be used for specifying these steels. There is an ASTM index that supplies information on the availability of specifications on these types of steels. Special-purpose steels and many steels from the other categories mentioned are covered by competitive alloy-designation systems developed by various government regulatory bodies. These alloy-designation systems are usually mandated for government work but, in private industry, the predominating alloy-designation systems for steels used in machines are the AISI and ASTM systems.

1.4.6   Carbon Steels

Carbon steel is an alloy of iron with carbon as the major strengthening element. The carbon strengthens by solid solution strengthening and, if the carbon level is high enough, the alloy can be quench-hardened. All metals, in addition to strengthening by alloying, can be strengthened by mechanical working, or *cold finishing,* as it is more appropriately termed. Carbon steels are available in all the mill forms (bar, strip, sheet shapes), and an important selection factor pertaining to carbon steels is whether to use a cold- or hot-finished product. Similarly, the designer must make a decision about whether to use a hardenable or nonhardenable alloy.

1.4.6.1   Cold-finished.   Cold finishing causes work hardening by grain reduction and buildup of dislocation density. This phenomenon occurs in most ductile metals. The strength of a metal can be increased by as much as a factor of 100 by simply reducing the cross section of a shape by rolling, drawing, swaging, or some related process. The mechanism by which this strengthening occurs is not as simple as we have implied. It is the complex atomic interactions that occur in the crystalline structure of metals.

1.4.7 Behavior of Carbon and Alloy Steels in Tension

Figure 1.3 shows some idealized stress-strain diagrams for a metal and a ceramic. A typical stress-strain diagram as one would generate in tensile testing a piece of low-carbon steel would look something like *a-b-g* in Fig. 1.3. There is an anomaly in traditional stress-strain curves that may not be apparent. The stress at failure (point *g*) is lower than the stress reached when the sample is stretching. This is because the stress is calculated on the original cross-sectional area of the test sample, but the sample is really necking down, so its area is smaller than when the test started. People who perform tensile tests of metals have learned to accept this, and since everybody tests this way, there is no problem. However, if we were able to measure the instantaneous area of a tensile sample in testing, the curve generated

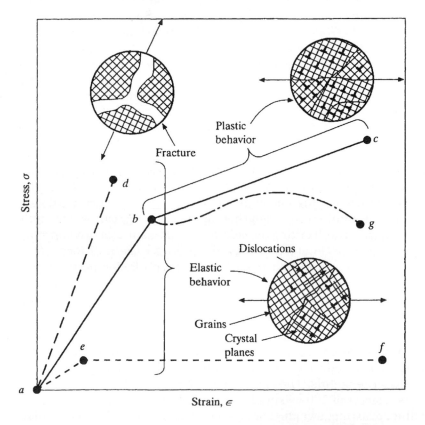

Figure 1.3 Idealized stress-strain diagrams for a metal and a ceramic. Curve *a-b-g* is the room-temperature stress-strain curve for a ductile steel. Curve *a-d* is the true stress-strain curve for a ceramic. Curve *a-e-f* is the true stress-strain curve for a steel at hot-working temperatures.[2]

would be the true stress-strain diagram. The part fails at the highest stress. The area used in the calculation of true stress is the instantaneous area, and the true strain is calculated as the natural log (ln) of the instantaneous length divided by the original sample length. Tensile testing machines that use computers to process the test data now frequently generate these curves.

A stress-strain curve for red-hot steel may look something like *a-e-f*. The stress required for plastic flow is very low, and there is no work hardening. The material is completely plastic. It is obvious that plastic flow is much easier at elevated temperatures, and this is why much of the shaping of steel products is done when the steel is red hot; this is hot working. Dynamic recrystallization occurs, and dislocations and other defects are annihilated.

Other dislocation phenomena happen in cold work, but it is important to remember that the ability to work harden is a very special property that belongs only to certain metals, and carbon steels are so useful from the materials standpoint because they have favorable plastic flow characteristics.

Steel sheet, bar, strip, and special shapes may be purchased from mills in various degrees of cold-work strengthening. This is accomplished by rolling sheet cold through large mills or by drawing through dies that reduce their cross-sectional area. The thickness or diameter goes down, the sheet or strip gets longer, and the strength goes up, depending on how much reduction is produced in rolling. Most carbon steel products are available as cold-rolled product. Typically, cold finished products do not receive reductions greater than about 10%, but the strengthening effect can yield as much as a 20% increase in tensile strength over the annealed condition. The yield strength (at 10% reduction) may be increased as much as 50%. If a one-of-a-kind part is being designed with cold-finished material, and it is overdesigned to the point that strength is not critical, it is probably adequate to specify cold-rolled steel. If strength is critical, it may be necessary to specify a desired degree of cold finishing. In bars and sheet, the temper designation system is one-quarter, one-half, three-quarters, full-hard, and skin-rolled temper. In strip and tin mill products, a number system is used to designate temper or degree of cold work. Table 1.3 illustrates the difference in mechanical properties for these various strip tempers. The numbering system for tin mill products is different from the strip system; T1 temper is the softest temper. It is uncommon to get sheet materials in the harder tempers, since the rolling of large widths would require extraordinary rolling mills.

From the standpoint of formability, the harder the temper, the less the ability to cold form into the desired shape. In general, full-hard

TABLE 1.3   Mechanical Properties of Carbon Steel Strip with a Maximum 0.25% Carbon[2]

Temper	Rockwell hardness	Nominal tensile strength [psi (MPa)]
No. 1 hard temper	B–85 minimum	90,000 (620)
No. 2 half-hard temper	B–70 to 85	65,000 (450)
No. 3 quarter-hard temper	B–60 to 75	55,000 (380)
No. 4 skin-rolled temper	B–65 maximum	50,000 (350)
No. 5 dead-soft temper	B–55 maximum	50,000 (350)

materials cannot be bent even 90° without a generous bend radius. Skin rolling, besides providing a good finish, eliminates stretcher strains. Stretcher strains are roadmap-type lines on a severely deep drawn sheet. They come from nonuniform atomic slip in the area of the yield point. By yielding the surface of a sheet before forming, this objectionable defect is eliminated. Figure 1.4 illustrates the effect of temper rolling on formability.

All steels can be cold finished, but it is not common to severely cold work high-carbon or alloy steels, since it is difficult to do; the material has poor formability, the strengthening effect is not as pronounced, and it will be lost if the steel is subsequently heat treated. The most commonly used cold-finished carbon steels are AISI grades 1006 through 1050, 1112, 1117, and other free-machining steels. The benefits are closer size tolerances, better mechanical properties, and improved surface finish. The disadvantages are lower ductility, greater instability during machining operations, and possibly higher price. Weighing these factors, it is usually desirable to make general machine parts such as brackets, chutes, guards, and mounting plates from cold-finished steels unless the part requires significant machining of surfaces.

Many sheet steels are used for items such as appliance housings and automobile panels. In these types of applications, a prime material requirement is the ability to be formed or drawn into shapes. Thus, cold-rolled sheet steels are available in four basic types that vary in forming characteristics and strength characteristics. Only the structural quality (SQ) has definite mechanical property requirements. The other grades have varying degrees of formability, and it has been common practice to use the lowest (and lowest-cost) grade that will do the job. High-strength sheet steels are available for appli-

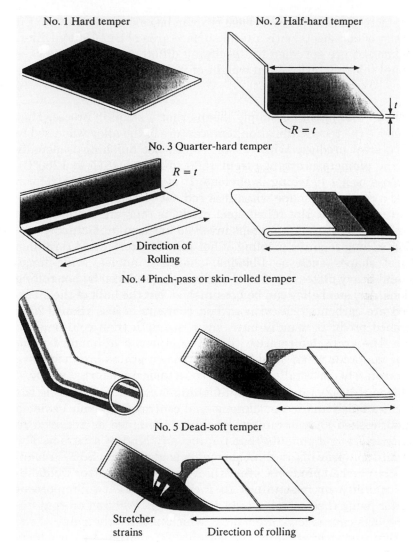

No. 1 Hard temper

No. 2 Half-hard temper

$R = t$

t

No. 3 Quarter-hard temper

$R = t$

Direction of
Rolling

No. 4 Pinch-pass or skin-rolled temper

No. 5 Dead-soft temper

Stretcher
strains

Direction of rolling

Figure 1.4 Formability of various tempers of 0.25% max. carbon steel strip.[2]

cations where weight reduction is a concern. ASTM specifications can
be used for drawing designation.

Black plate.   Tin mill steel without a coating is called *black plate*. The
plated coatings are applied by continuous web electrodeposition. These
steels are available in large coils (up to 15,000 lb) or in cut sheet form.
The largest volume use of these steels is for beverage cans, and for this
application cut sheets are usually supplied. The unit of measure for

steel quantity is the *base box*. A base box can be various weights, but it will consist of cut sheets with a total surface area of 217.78 ft^2. Different thicknesses are specified by specifying different weights per base box (10-mil thick material has a weight of 90 lb per base).

Hot-finished. *Hot finishing* simply means that a shaping process that is done above the recrystallization temperature of the alloy was used to produce a steel product. With carbon steels, hot finishing is usually done in the temperature range from 1500 to 2300°F (815 to 1260°C). The purpose of hot finishing is obvious. It is simply easier to deform steel into a particular shape when it is red hot compared to when it is at room temperature. Hot-rolled steel products thus are cheaper than cold-finished. There are fewer steps involved in the manufacturing process. Cold rolling requires pickling of hot-rolled material and rerolling. Structural shapes such as I-beams, channels, angles, wide-flange beams, and heavy plates are almost exclusively produced by hot-rolling operations. Any steel alloy can be hot finished, but the bulk of the steels produced are carbon steels with carbon contents of less than 0.25%. Hot-finished products usually have lower strength than cold-finished products. The grain deformation that strengthens cold-finished products does not occur when steel is red hot. When grains are deformed, they immediately recrystallize and return to their undistorted shape.

The biggest disadvantage of hot finishing over cold finishing is the poor surface finish and looser dimensional control. The oxide scale on some hot-finished shapes almost precludes their use as received in making machine components that require any kind of accuracy. The same limitation prevails in dimensional considerations. Some advantages of hot-finished products, other than low cost, are better weldability and stability in machining. In fact, these are two important reasons for using these steels in machine design. Welding on cold-finished products causes local annealing, which negates the value of the cold working in strengthening. Machining of cold-finished products unbalances the cold-working stresses, and the part is likely to distort badly. Since there are low residual stresses from hot working, hot-finished products are preferred over cold-finished for parts requiring stability during and after machining operations.

Typical carbon steel grades used for machine bases, frames, and structural components are AISI 1020, 1025, and 1030. Normal (merchant) grades of hot-finished steel do not have rigorous control on composition or properties. If the mechanical properties of the steel must be guaranteed, it is best to specify a hot-finished shape to ASTM specifications. ASTM specification A 283 covers several strength grades of structural-quality steel. A 284 covers machine steels, A 36 covers bridge and building steel, and A 285 covers steels for flanges and fire-boxes. The proper drawing specification would show the ASTM specification number and the strength grade.

Hardening. A requirement for hardening is sufficient carbon content. With carbon steels, to get 100% martensite in moderate sections, a carbon content of about 0.6 wt% is desirable. This does not mean, however, that any size part can be made from 1060 steel and that it will harden to 60 HRC. Carbon steels have poor hardenability, and it is difficult to meet quenching requirements. Each steel alloy has certain time requirements on quenching if hardened structure is to be obtained. Time-temperature-transformation diagrams, for example, tell us that a 1080 steel must be quenched from its hardening temperature of about 1600°F (870°C) to a temperature below about 1000°F (530°C) in less than 1 s. This rapid cooling rate can be achieved in water with thin section, sheet metal, and bars up to 1 in. (25 mm) dia., but, on heavier sections, only the surface may harden. As bar diameter or section size increases, even the surface will not harden. As shown in Fig. 1.5, hardenability increases with carbon content. Maximum hardenability is achieved at about 0.8% carbon. Hardenability decreases somewhat as carbon content is increased over 1%, since carbon tends to promote the formation of ferrite.

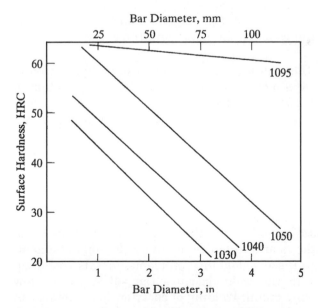

Figure 1.5 Approximate maximum surface hardness of carbon steel of varying bar diameter (water quenched).[2]

The hardening situation with carbon steels is even worse than implied by Fig. 1.5. If hardness readings were taken in the center of the bar, these data would show that even the 1095 steel reached a hardness of only about 40 HRC. Thus, plain carbon steels have low hardenability, and large parts will harden only on the surface, if at all. Rapid heating and quenching techniques, such as flame or induction, work very well on these steels and overcome hardenability limitations. Thin sections, such as flat springs and wire springs, are well suited for manufacture from carbon steels. AISI 1080 to 1090 steels are commonly used for dowel pins, springs, knife blades, doctor blades, and the like. Large parts made from 1040 to 1060, or their free-machining counterparts 1140 to 1151, are commonly flame or induction hardened. All the low-carbon (less than 0.3% carbon) grades can be carburized and quench hardened to obtain hardened surfaces.

Weldability.   All metals can be welded to themselves by at least one commonly used process. Similarly, any metal can be welded to another metal, but this does not mean that the weld will have usable properties. It is a design rule that titanium cannot be welded to other metals (except for several exotic metals). Titanium can be fusion welded to steel, but the weld will be like glass. Tool steels can be welded but, unless special precautions are taken, nine times out of ten, the weld will crack. The *weldability* of metal refers to these kinds of welding effects or aftereffects. The weldability of a particular metal combination is the ease with which the weld can be made and the soundness of the weld after it is made.

The primary factor that controls the weldability of metals is chemical composition, the basis metal composition, and the composition of the filler metal, if any. A fusion weld cannot be made between dissimilar metals unless they have metallurgical solubility. Titanium cannot be welded to most other metals, because it has limited solubility and tends to form brittle compounds. Phase diagrams can be consulted to determine if dissimilar metals can be welded. If the phase diagrams show low solid solubility, the weld cannot be made. The combination will have poor weldability.

With metals that have adequate solubility to be fusion welded, poor weldability will be manifested by such things as cracking of the weld, porosity, cracking of the heat-affected zone, and weld embrittlement. Many things can cause these weldability problems. Metals with high sulfur contents crack in the weld, because the sulfur causes low strength in the solidifying metal. Welding of rusty or dirty metals can cause weld porosity, and welding of hardenable steels can lead to cracking in the heat-affected zone.

Of all the potential weldability problems that can occur, two stand out as the most common and most troublesome:

1. Arc welding resulfurized steels: 11xx carbon steels
2. Arc welding hardenable steels: 1030 to 1090 carbon steels, alloy steels with C > 0.3%

High-sulfur steels.   High-sulfur steels (>0.1%) are widely used for parts that require significant machining. They save money; they are a real aid in lowering machining costs. However, there is no way to avoid weld-cracking problems with these steels other than to establish a hard and fast rule never to arc weld them. Brazing and soldering are acceptable, but not arc welding to themselves or to other metals.

Welding hardenable steels causes a weldability problem, because there is a high risk of cracking in the weld or adjacent to the weld. When a weld is made, it and the heat-affected zone go through a thermal cycle not unlike a quench-hardening cycle. It makes no difference whether the parts are hardened or soft before welding. Melting of the steel requires a temperature of about 3000°F (1650°C). The hardening (austenitizing) temperature range for most steels is 1500 to 2000°F (815 to 1093°C). Obviously, the weld is going to be cycled in this temperature range, as is some of the metal adjacent to the weld. The mass of the metal being welded serves as a quenching medium, and hard martensite will form either in the weld, the heat-affected zone, or both. This structure will be brittle, and if the weld is under restraint, the brittle structure will crack.

An obvious way to prevent this cracking is to prevent the quench that leads to martensite formation. Preheating the parts helps to accomplish this. If the part survives the welding operation without cracking, it may be prone to cracking in service if any brittle martensite formed. The solution to this is to postheat or temper the weld. A 1100 to 1200°F (593 to 648°C) stress relieve is even more desirable.

Thus, there are ways to prevent cracking in welding hardenable steels, but the risk is so high that the designer is wise to prevent fusion welds on hardenable steels if at all possible. The steels that have a high risk of cracking in welding are alloy steels, tool steels, and plain carbon steels with carbon contents greater than 0.3%. Since alloy steels have higher hardenability than carbon steels, weldability can be a problem when the carbon equivalent is greater than about 0.4%.

1.4.8 Physical Properties

Other physical properties of steel must be noted. One of the most useful physical properties of carbon and most alloy steels is ferromagnetism—attraction to a magnet. This allows carbon and alloy steels to be used in magnetic devices such as motors and solenoids, and they can be held in machining with magnetic chucks.

The physical property that makes steel the most useful structural metal is the modulus of elasticity. Carbon and alloy steels have a value of 30 million psi (207 GPa), making them on of the stiffest common engineering metals. The only engineering metals that have higher stiffness are some nickel alloys, beryllium alloys, tungsten, and molybdenum.

The thermal conductivity of carbon and alloy steels is about 27 Btu-ft/hr-ft^2-°F (47 W/mK). Electrical conductivity is about 15% of pure copper; they are not good conductors of heat or electricity, but they are better conductors than stainless steels and some of the high-alloy steels. The coefficient of thermal expansion is about 7×10^{-6} in/in/°F (12.6×10^{-6} m/m/K). All these physical properties are about the same for high-strength, low-alloy steels, mill-heat-treated steels, and the low-alloy ultrahigh-strength steels.

1.4.9 Alloy Steels

We have defined alloy steels as a group of mill products that meet certain composition limits. They are the 13xx, 4xxx, 5xxx, 6xxx, 8xxx, and 9xxx series in the AISI designation system. The same AISI alloys are specified in ASTM A 29 and A 29M (M specs are metric).

There are several things that can happen when alloy atoms are added to a pure metal. The atoms can go into a random solid solution either substitutionally (Fig. 1.6a) or interstitially (Fig. 1.6b), or the atoms may want to pair with the host atoms in some definite proportion (Fig. 1.6c) or stay by themselves (Fig. 1.6d). We have already discussed the strengthening effects of alloying elements going into solution. Their presence impedes dislocation motion. Ordered arrangements of alloy atoms with some stoichiometric relationship of host atoms to alloy atoms lead to the formation of compounds with

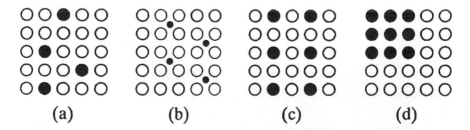

Figure 1.6 Distribution of alloy atoms (black) in a pure metal atomic lattice. (a) Substitutional solution (random), (b) interstitial solution (random), (c) ordered substitutional solution, and (c) clustering of like-alloy atoms.[2]

ionic or covalent-type bonds rather than metallic electron bonding. The formation of compounds in metals is common, and their occurrence can have a strengthening effect, as does a pearlite structure in steel (pearlite is a compound of Fe_3C layered with ferrite), or it can sometimes cause brittleness. Some intermetallic compounds cause alloys to become useless as structural materials. Finally, as in the solute atoms, they may give rise to the formation of separate phases, rich in solute. The phases can have a good or bad effect on properties, depending on the nature of the phase.

Which of these things happens when alloy is added depends on a number of thermodynamic factors, such as the effect on free energy. Other factors controlling the effect of alloying are the relative size of the solute atoms compared with the host atoms, the electrochemical nature of solute atoms, and even the valence of the alloy atoms. Alloy steels are really quite complex in that commercial alloys invariably contain more than one major alloying element. The exact effects of adding, for example, manganese, molybdenum, and chromium to make a 4000 series alloy steel are not fully understood, but the net effect of these additions on phase equilibrium, transition temperatures, and properties is known. Isothermal transformation diagrams have been established on most alloy steels, and these can be used to determine the net effect of alloying on the quenching requirements of a steel. Figure 1.7 shows time-temperature-transformation (TTT) diagrams for two steels with the same carbon content but different contents of major alloying elements. The diagram for 1340 steel indicates that the heat treater has to quench the steel very rapidly to get it to harden (less than 1 s). The effect of the nickel, chromium, and molybdenum combination is to lessen the quenching requirements. An oil quench can probably meet the 10-s requirement indicated on the TIT curve for the 4340 steel.

The phase diagrams would similarly be changed by alloy addition. Iron-carbon alloys (carbon steels) are binary alloys. When an alloy steel contains four major alloy additions, a quaternary phase diagram is needed to determine phase relationships, the austenite transition temperature changes, as do the phases that can be present. These equilibrium diagrams most design engineers will probably never use, but it is important to realize that heat-treating temperatures for alloy steels will be different from those for carbon steels.

Probably the most important effect of alloying steels is to alter hardenability. The TTT diagrams are an indicator of the hardenability of steel, but an equally useful tool is the Jominy end-quench test. In this test, a specified size bar is heated to a prescribed austenitizing temperature and water quenched on one end in a special fixture under prescribed conditions. The quenched end cools rapidly, and the rate of

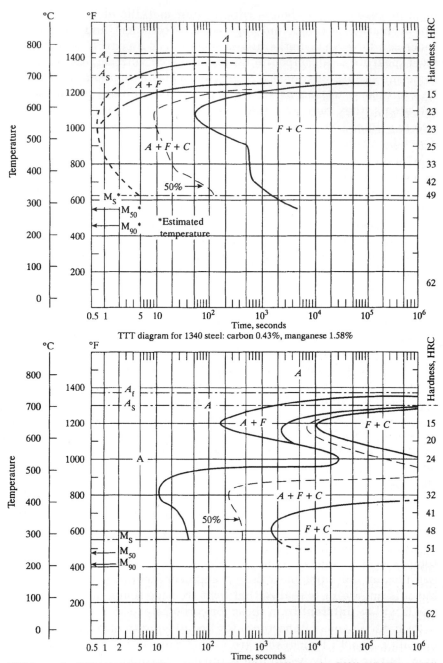

TTT diagram for 1340 steel: carbon 0.43%, manganese 1.58%

TTT diagram for 4340 steel: carbon 0.42%, manganese 0.78%, nickel 1.79%, chromium 0.80%, molybdenum 0.33%

Figure 1.7 Time-temperature-transition diagrams for two steel types with the same carbon content but different alloy content.[4]

cooling is progressively less as cooling progresses toward the other end of the bar. After the bar cools, a small flat is ground onto it, and the hardness is measured at various distances from the quenched end. These data are plotted on a curve of hardness versus distance in Fig. 1.8, and the curve serves as a measure of hardenability. It can also be used to calculate the section size that will through harden for a given alloy. The factors that control the hardenability are the alloy content, cooling rate, and the grain size of the steel. Thus, if a steel has only 0.20% carbon but also contains 1% molybdenum, the carbon equivalent will be 0.45%, and the material will have the hardenability of a steel with 0.45% carbon rather than 0.2% carbon. This equation is normally used in welding to determine if a particular steel requires preheating, but it shows which elements are more important in improving hardenability.

There are other reasons for adding alloying elements to steels besides improving hardenability. Sometimes they improve corrosion characteristics, physical properties, or machinability. Copper improves corrosion resistance. Sulfur and phosphorus additions improve machinability by producing small inclusions in the structure that aid chip formation. Sulfur contents over approx. 0.06%, on the other hand, make steels unweldable due to hot shortness. Thus, there are good and bad effects of adding alloying elements. Table 1.4 summarizes the effects of common alloying elements in alloy steels.

Most of the common alloy elements used in alloy steels tend to increase hardenability. They do so by altering transformation temperatures and by reducing the severity of the quench required to get transformation to hardened structure. If a steel has very low hardenability, it is possible that it will not harden at all in heavy sections. Also, a severe quenching rate leads to distortion and cracking tendencies.

1.5   Selection of Alloy Steels

As shown in Table 1.5, numerous alloy steels are commercially available. Added to this list are H controlled hardenability steels, boron steels, nitrogenated steels, and free machining grades. How does one select one of these for a machine application? One step that can be taken in categorizing these steels is to separate them into grades intended for through hardening and grades intended for carburizing. All the grades with carbon contents of about 0.2% or less are intended for carburizing. The remainder of the alloys through harden, but they do so to different degrees; they have differing hardenabilities. The carburizing grades also have different hardenabilities and differing core strengths after carburizing. There are also some special grades, such

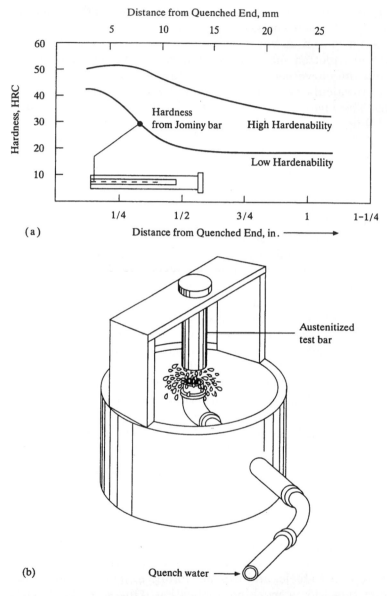

Figure 1.8 (a) Use of Jominy data to measure hardenability and (b) Jominy end-quenched fixture.[2]

as the aluminum-bearing grades, that are intended for nitriding applications.

The physical properties of most grades do not vary significantly from those of carbon steels; thus, the selection procedure can be re-

TABLE 1.4   Effects of Alloying Elements in Alloy Steel[2]

	Typical ranges in alloy steels (%)	Principal effects
Aluminum	<2	Aids nitriding Restricts grain growth
Sulfur	<0.5	Adds machinability Reduces weldability and ductility
Chromium	0.3–4	Increases resistance to corrosion and oxidation Increases hardenability Increases high-temperature strength Can combine with carbon to form hard, wear-resistant microconstituents
Nickel	0.3–5	Promotes an austenitic structure Increases hardenability Increases toughness
Copper	0.2–0.5	Promotes tenacious oxide film to aid atmospheric corrosion resistance
Manganese	0.3–2	Increases hardenability Promotes an austenitic structure Combines with sulfur to reduce its adverse effects
Silicon	0.2–2.5	Removes oxygen in steel making Improves toughness Increases hardenability
Molybdenum	0.1–0.5	Promotes grain refinement Increases hardenability Improves high-temperature strength
Vanadium	0.1–0.3	Promotes grain refinement Increases hardenability Will combine with carbon to form wear-resistant microconstituents

duced to weighing mechanical properties and hardening characteristics. If an alloy steel is to be used for a load-bearing member, the properties of prime importance may be yield strength and toughness. If the part is intended for a wear application, an alloy might be desired that can harden to 60 HRC on the surface with a particular section thickness. Keeping these things in mind, let us categorize alloy steels into through-hardening and carburizing grades and discuss the relative merits of some of the more common grades.

1.5.1   Through Hardening

At least 40 AISI alloy steels are considered to be through hardening. They are those with a carbon content greater than about 0.3%. The

TABLE 1.5   Chemical Composition of AISI–SAE Steels[5]

AISI/SAE[b] number	C	Mn	Ni	Cr	Mo	Other elements
1330	0.28/0.33	1.60/1.90	–	–	–	–
1335	0.33/0.38	1.60/1.90	–	–	–	–
1340	0.38/0.43	1.60/1.90	–	–	–	–
1345	0.43/0.48	1.60/1.90	–	–	–	–
4012	0.09/0.14	0.75/1.00	–	–	0.15/0.25	–
4023	0.20/0.25	0.70/0.90	–	–	0.20/0.30	–
						S
4024	0.20/0.25	0.70/0.90	–		0.20/0.30	0.035/0.050
4027	0.25/0.30	0.70/0.90	–	–	0.20/0.30	–
4028	0.25/0.30	0.70/0.90	–	–	0.20/0.30	0.035/0.050
4032[a]	0.30/0.35	0.70/0.90	–	–	0.20/0.30	–
4037	0.35/0.40	0.70/0.90	–	–	0.20/0.30	–
4042[a]	0.40/0.45	0.70/0.90	–	–	0.20/0.30	–
4047	0.45/0.50	0.70/0.90	–	–	0.20/0.30	–
4118	0.18/0.23	0.70/0.90	–	0.40/0.60	0.08/0.15	–
4130	0.28/0.33	0.40/0.60	–	0.80/1.10	0.15/0.25	–
4135[a]	0.33/0.38	0.70/0.90	–	0.30/1.10	0.15/0.25	–
4137	0.35/0.40	0.70/0.90	–	0.80/1.10	0.15/0.25	–
4140	0.38/0.43	0.75/1.00	–	0.80/1.10	0.15/0.25	–
4142	0.40/0.45	0.75/1.00	–	0.80/1.10	0.15/0.25	–
4145	0.43/0.48	0.75/1.00	–	0.80/1.10	0.15/0.25	–
4147	0.45/0.50	0.75/1.00	–	0.80/1.10	0.15/0.25	–
4150	0.48/0.53	0.75/1.00	–	0.80/1.10	0.15/0.25	–
4161	0.56/0.64	0.75/1.00	–	0.70/0.90	0.25/0.35	–
4320	0.17/0.22	0.45/0.65	1.65/2.00	0.40/0.60	0.20/0.30	–
4340	0.38/0.43	0.60/0.80	1.65/2.00	0.70/0.90	0.20/0.30	–
E4340	0.38/0.43	0.65/0.85	1.65/2.00	0.70/0.90	0.20/0.30	–
4419	0.18/0.23	0.45/0.65	–	–	0.45/0.60	–
4422[a]	0.20/0.25	0.70/0.90	–	–	0.35/0.45	–

TABLE 1.5   Chemical Composition of AISI–SAE Steels[5] *(continued)*

AISI/SAE[b] number	C	Mn	Ni	Cr	Mo	Other elements
4427[a]	0.24/0.29	0.70/0.90	–	–	0.35/0.45	–
4615	0.13/0.18	0.45/0.65	1.65/2.00	–	0.20/0.30	–
4617[a]	0.15/0.20	0.45/0.65	1.65/2.00	–	0.20/0.30	–
4620	0.17/0.22	0.45/0.65	1.65/2.00	–	0.20/0.30	–
4621	0.18/0.23	0.70/0.90	1.65/2.00	–	0.20/0.30	–
4626	0.24/0.29	0.45/0.65	0.70/1.00	–	0.15/0.25	–
4718	0.16/0.21	0.70/0.90	0.90/1.20	–	0.30/0.40	–
4720	0.17/0.22	0.50/0.70	0.90/1.20	0.35/0.55	0.15/0.25	–
4815	0.13/0.18	0.40/0.60	3.25/3.75	–	0.20/0.30	–
4820	0.18/0.23	0.50/0.70	3.25/3.75	–	0.20/0.30	–
5015	0.12/0.17	0.30/0.50	–	0.30/0.50	–	–
5046[a]	0.43/0.48	0.75/1.00	–	0.20/0.35	–	–
5060[a]	0.56/1.00	0.75/1.00	–	0.40/0.60	–	–
5115[a]	0.13/0.18	0.70/0.90	–	0.70/0.90	–	–
5120	0.17/0.22	0.7/0.90	–	0.70/0.90	–	–
5130	0.28/0.33	0.70/0.90		0.80/1.10	–	–
5132	0.30/0.35	0.60/0.80	–	0.75/1.00	–	–
5135	0.33/0.38	0.60/0.80	–	0.80/1.05	–	–
5140	0.38/0.43	0.70/0.90	–	0.70/0.90	–	–
5145	0.43/0.48	0.70/0.90	–	0.70/0.90	–	–
5147	0.46/0.51	0.70/0.95	–	0.85/1.15	–	–
5150	0.48/0.53	0.70/0.90	–	0.70/0.90	–	–
5155	0.51/0.59	0.70/0.90	–	0.70/0.90	–	–
6160	0.56/0.64	0.75/1.00	–	0.70/0.90	–	–
50100[a]	0.98/1.10	0.25/0.45	–	0.40/0.60	–	–
E51100	0.98/1.10	0.25/0.45	–	0.90/1.15	–	–
E52100[c]	0.98/1.10	0.25/0.45	–	1.30/1.60	–	–
						V
6118	0.16/0.21	0.50/0.70	–	0.50/0.70	–	0.10/0.15

TABLE 1.5   Chemical Composition of AISI–SAE Steels[5] (continued)

AISI/SAE[b] number	C	Mn	Ni	Cr	Mo	Other elements
6150	0.48/0.53	0.70/0.90	–	0.80/1.10	–	0.15 min
8115[a]	0.13/0.18	0.70/0.90	0.20/0.40	0.30/0.50	0.08/0.15	–
8615	0.13/0.18	0.70/0.90	0.40/0.70	0.40/0.60	0.15/0.25	–
8617	0.15/0.20	0.70/0.90	0.40/0.70	0.40/0.60	0.15/0.25	–
8620	0.18/0.23	0.70/0.90	0.40/0.70	0.40/0.60	0.15/0.25	–
8622	0.20/0.25	0.70/0.90	0.40/0.70	0.40/0.60	0.15/0.25	–
8625	0.23/0.28	0.70/0.90	0.40/0.70	0.40/0.60	0.15/0.25	–
8627	0.25/0.30	0.70/0.90	0.40/0.70	0.40/0.60	0.15/0.25	–
8630	0.28/0.33	0.70/0.90	0.40/0.70	0.40/0.60	0.15/0.25	–
8637	0.35/0.40	0.75/1.00	0.40/0.70	0.40/0.60	0.15/0.25	–
8640	0.38/0.43	0.75/1.00	0.40/0.70	0.40/0.60	0.15/0.25	–
8642	0.40/0.45	0.75/1.00	0.40/0.70	0.40/0.60	0.15/0.25	–
8645	0.43/0.48	0.75/1.00	0.40/0.70	0.40/0.60	0.15/0.25	–
8650[a]	0.48/0.53	0.75/1.00	0.40/0.70	0.40/0.60	0.15/0.25	–
8655	0.51/0.59	0.75/1.00	0.40/0.70	0.40/0.60	0.15/0.25	–
8660[a]	0.56/0.64	0.75/1.00	0.40/0.70	0.40/0.60	0.15/0.25	–
8720	0.18/0.23	0.70/0.90	0.40/0.70	0.40/0.60	0.20/0.30	–
8740	0.38/0.43	0.75/1.00	0.40/0.70	0.40/0.60	0.20/0.30	–
8822	0.20/0.25	0.75/1.00	0.40/0.70	0.40/0.60	0.30/0.40	–
						Si
9254	0.52/0.59	0.60/0.80	–	0.60/0.80	–	1.20/0.60
9255	0.51/0.59	0.70/0.95	–	–	–	1.80/1.20
9260	0.56/0.64	0.75/1.00	–	–	–	1.80/1.20
9310[a]	0.08/0.13	0.45/0.65	3.00/3.50	1.00/1.40	0.08/0.15	–

Source: Modern Steels, Bethlehem Steel Corp., Bethlehem, PA 1980

[a]SAE only.
[b]AISI designation number coincides with ASTM A 29 grades.
[c]Also covered by ASTM A 295.

differences in hardenability on some of the more popular alloys can be seen in Fig. 1.9. From the use standpoint, if large section parts are required, an alloy such as 4340 should be used, because it will harden through in heavy sections. All the alloys shown would develop comparable hardness and strength in small sections. A carbon steel with the same carbon content as the alloy steels will not through harden with an oil quench even in thin sections. All the hardness-versus-bar diameters in Fig. 1.9 would shift upward with a water quench.

Processes are required to reduce the sulfur content to improve toughness in these steels. Calcium argon degassing (CAD) is one such process. Electric furnace-melted steels are transferred to a degassing vessel. Calcium is added to the melt to scavenge sulfur and make it float up to the slag. Argon is then blown into the charge to mix, agitate, and further complete the sulfide control. Rare earth elements can also be used in this type of process, but the net effect in all cases is a greatly reduced volume fraction of sulfides. The ones that remain are globular rather than stringers: the objectives of these inclusion control processes are elimination of anisotropy and improved toughness.

One application of high-strength sheet steels is usually on structural components where weight and material cost savings can be realized by reducing part thickness without lowering strength. This can be done with the higher strengths of these steels. If weight reduction is desirable, then the use of these stronger steels should be investigated through steel mills.

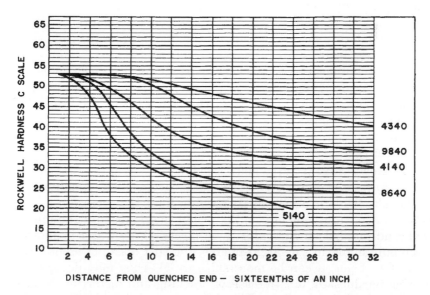

Figure 1.9 Hardening characteristics of some alloy steels, oil quenched.[4]

1.5.2 High-Strength, Low-Alloy Steels

These steels are defined by AISI as a *group of steels with chemical composition specially developed to impart higher mechanical properties and in certain of these steels materially greater resistance to atmospheric corrosion than is obtainable from conventional carbon steels.* Unfortunately, how higher mechanical properties are produced varies significantly with the manufacturer. Chemical composition and processing are the usual factors that are varied to increase mechanical properties, but these alloys are made primarily to a designated strength level rather than to rigorous chemical composition specification.

In general, high-strength, low-alloy steels (HSLA) are not hardened by heat treatments. They are supplied and used in the hot-finished condition; they have low carbon contents (less than 0.2%). They have a microstructure that consists of ferrite and pearlite, they all contain about 1% manganese to produce solid solution strengthening of the ferrite, and they usually contain small concentrations (less than 0.5%) of other elements for additional strengthening or other effects.

HSLA steels are primarily intended for structural-type applications where weldability is a prime selection requirement. This is why the carbon and alloy content of these steels is low. They do not have the hardenability of alloy steels and, even though they contain significant amounts of alloy (Mn), they are not considered to be alloy steels. Table 1.6 is a tabulation of some HSLA steels showing some of the alloying elements that are used for strengthening.

Copper is one of the most important alloying elements in HSLA steels. In small concentrations (less than 0.5%), it produces solid solution strengthening of ferrite, as well as a side benefit that is even more important: the atmospheric corrosion resistance of the steel can be improved by a factor of four. Chromium, nickel, and phosphorus are usually used in these alloys to assist in the formation of the protective oxide film. If the copper content is over about 0.75%, these alloys can be precipitation hardened by heating in the range from 1000 to 1100°F (540 to 600°C).

These steels have been widely used for structural applications (bridges, towers, railings, and stairs), and they are left unpainted. It was thought that the adherent rust would limit the corrosion to just this layer. After a decade of use, the use of these steels in the unprotected condition is losing popularity. In places where water is allowed to accumulate and in salt atmospheres, the rust does not stop at the thin surface layer; it can be more substantial. They may develop respectable hardnesses, but the risk of cracking and distortion is such that water quenching is not advisable unless economics dictate the

TABLE 1.6   ASTM Specifications for High-Strength, Low-Alloy Steels

ASTM specification	Title
A20	Standard specification for general requirements for steel plates for pressure vessels
A203/A203M	Pressure vessel plates, alloy steel, nickel
A204/A204M	Pressure vessel plates, alloy steel, molybdenum
A225/A225M	Pressure vessel plates, alloy steel, manganese-vanadium
A285/A285M	Pressure vessel plates, carbon steel, low- and intermediate-tensile strength
A299/A299M	Pressure vessel plates, carbon steel, manganese-silicon
A302/A302M	Pressure vessel plates, alloy steel, manganese-molybdenum and manganese-molybdenum-nickel
A322	Standard specification for steel bars, alloy, standard grades
A353/A353M	Pressure vessel plates, alloy steel, 9% nickel double-normalized and tempered
A387/A387M	Pressure vessel plates, alloy steel, chromium-molybdenum
A455/A455M	Pressure vessel plates, carbon steel, high-strength manganese
A515/A515M	Pressure vessel plates, carbon steel, for intermediate- and higher-temperature service
A516/A516M	Pressure vessel plates, carbon steel, moderate- and lower- temperature service
A517/A517M	Pressure vessel plates, alloy steel, high-strength, quenched and tempered
A533/A533M	Pressure vessel plates, alloy steel, quenched and tempered manganese-molybdenum and manganese-molybdenum-nickel
A537/A537M	Pressure vessel plates, heat-treated, carbon-manganese-silicon steel
A542/A542M	Pressure vessel plates, alloy steel, quenched and tempered chromium-molybdenum
A543/A543M	Pressure vessel plates, alloy steel, quenched and tempered nickel-chromium-molybdenum
A562/A562M	Pressure vessel plates, carbon steel, manganese-titanium for glass or diffused metallic coatings
A612/A612M	Pressure vessel plates, carbon steel, high-strength, for moderate- and lower-temperature service

TABLE 1.6   ASTM Specifications for High-Strength, Low-Alloy Steels *(continued)*

ASTM specification	Title
A645/A645M	Pressure vessel plates, 5% nickel alloy steel, specially heat treated
A662/A662M	Pressure vessel plates, carbon-manganese, for moderate- and lower-temperature service
A681	Standard specification for tool steels alloy
A724/A724M	Pressure vessel plates, carbon steel, quenched and tempered, for welded layered pressure vessels
A734/A734M	Pressure vessel plates, alloy steel and high-strength, low-alloy steel, quenched and tempered
A735/A735M	Pressure vessel plates, low-carbon manganese-molybdenum-columbium alloy steel, for moderate and lower temperature service
A736/A736M	Pressure vessel plates, low-carbon age-hardening nickel-copper-chromium-molybdenum-columbium alloy steel
A737/A737M	Pressure vessel plates, high-strength, low-alloy steel
A738/A738M	Pressure vessel plates, heat-treated, carbon-manganese-silicon steel, for moderate- and lower-temperature service
A782/A782M	Pressure-vessel plates, quenched and tempered, manganese-chromium-molybdenum-silicon-zirconium alloy steel
A832/A832M	Pressure vessel plates, alloy steel, chromium-molybdenum-vanadium

use of a lean alloy and water quench. When serviceability rather than cost is the prime selection consideration, the designer can probably get along nicely using only one or two alloys: AISI 4140 and 4340. The mechanical properties of all the alloy steels depend on the condition of heat treat (the tempered hardness). The mechanical properties vary slightly with alloy at a given hardness, but these two alloys have excellent strength and toughness at most hardness levels, and at about 30 HRC. The toughness is comparable to that of a low-carbon steel, but the yield strength is still four or five times as high as that of mild steel. These two alloys are widely used for shafting, gears, and other power transmission components hardened and tempered to 30 to 32 HRC. They can be final machined after hardening, and this provides another advantage. The distortion in heat treatment is eliminated. The reason for two alloys is that AISI 4340 has better hardenability than 4140. It is true that 4340 would suffice for all applications, but it is common warehouse practice to stock 4140 in small sizes where high hardenability is not essential. These steels should be used over the standard alloy steel grades when repeatable hardening characteristics are essential on high-production parts.

1.5.3 B Steels

Boron and nitrogen are next to carbon in the periodic table and, as we might expect, they behave like carbon as hardening agents in steel.

There is a family of about 10 AISI steels that contain small amounts of boron (0.0005 to 0.003%) as an alloying element. They are identified by the standard AISI designation, with a B in the center of the alloy number; e.g., 50B44 is an AISI 5044 steel with boron.

The incentive for producing boron steels is that with low- to medium-carbon contents, an extremely small amount of boron can conceivably provide the same hardenability improvement as adding, for example, 1% of an expensive alloying element such as molybdenum or chromium. This can provide a significant cost reduction on high-production parts. There are many metallurgical considerations associated with the use of boron. Boron lowers hardenability when added to high-carbon steels; too much boron can cause formation of Fell compounds, which lower toughness; nitrogen must be kept low, or it will combine with boron to form boron nitride compounds. Boron strengthening is not widely used because of these factors but, for suitable applications, it can provide a hardenable steel with good formability, machinability, and heat-treating characteristics at a significantly lower material cost as compared to standard grades.

1.5.4 High-Strength Sheet Steels

There has been a continuing trend toward weight reduction in all forms of vehicles for fuel savings. Since a significant amount of the weight of automobiles and trucks is in sheet steel components, various steel producers started commercial production of sheet steels with minimum yield strengths higher than the 30 ksi (206 MPa) that is typical on low-carbon sheet steel. Many of these steels are proprietary in nature, but alloying, melting practice, heat treatment, or a combination of these usually achieves higher strengths. ASTM specifications cover a number of these steels: ASTM A 568, A 572, A 606, A 607, A 715, and A 816.

The structural-quality, low-alloy, and weathering grades of these steels have solid solution strengthening by alloy additions as a part of their strengthening mechanism. Dual-phase steels usually contain some alloy additions such as manganese or silicon, but the strengthening effect employed in these steels is the formation of martensite or bainite in the ferrite matrix. After cold rolling, dual-phase sheet steels are continuously heated to the temperature region at which the structure is part austenite and part ferrite. A temperature in the range from 1360 to 1400°F (730 to 760°C) is typical. The steel is then cooled at a rate sufficient to cause the austenite to transform to martensite or lower bainite. This results in soft, ductile ferrite matrix containing islands of quench-hardened martensite. The strengthening effect is almost directly proportional to the volume percentage of hard structure. Yield strength can be increased to as high as 100 ksi (700 MPa) while still maintaining formability. These steels can be further strengthened by a post-forming age hardening.

Bake hardening sheet steel is steel that increases in strength when heated for short times (as short as 30 s) at elevated temperatures [in the range of 930 to 1300°F (500 to 700°C)]. These steels have very low carbon content (interstitial free) and copper contents up to 1.5%. These steels are continuously annealed and subsequently age hardened to get microscopic copper precipitates (3 nm) that strengthen the steel. Tensile strength can be increased from 60 ksi (420 MPa) to as high as 95 ksi (660 MPa). Most ingot-cast sheet steels age harden during paint baking due to reversion of carbon atoms to dislocations, but the strengthening effect was not as pronounced as in these copper-bearing steels.

Deoxidation practices are used to enhance the properties of some grades of HSLA steels. As mentioned in our discussion of melting practice, killed steels have finer grain size and less chemical segregation than nonkilled steel.

Niobium, titanium, vanadium, and nitrogen are used as microalloying additions in a significant number of HSLA steel (A 575, A 607). Their role is to form precipitates that restrict grain growth during hot rolling. Concentrations of these elements are usually less than about 0.2%. Thus, the use of these elements for strengthening does not significantly increase manufacturing costs.

Rare earth elements such as cerium, lanthanum, and praseodymium are used in HSLA steels to obtain shape control of sulfide and oxide inclusions. These elements tend to combine with sulfides and oxides and essentially reduce their plasticity so that they do not elongate to form stringers during hot rolling. Elongated sulfides and oxides lead to tearing in high-restraint welds. The globular sulfides and oxides that are produced in shape controlled steels are less detrimental.

Aluminum and silicon are used in HSLA steels for deoxidation. They are used to kill steels and reduce chemical segregation. Silicon also contributes to solid solution strengthening of ferrite. Calcium additions such as in the CAD process are also used for control of sulfides and oxides.

The hot-rolled HSLA steels have minimum yield strengths in the 40 to 80 ksi (275 to 550 MPa) range and tensile strengths in the range from 55 to 90 ksi (379 to 620 MPa) with good formability and acceptable weldability. These strength levels are achieved by major alloy additions of manganese, microalloying, inclusions shape control, grain refinement, deoxidation, and controlled rolling. The mechanisms of strengthening vary from simple solid solution strengthening of ferrite to complex reactions between microalloying additions and dislocations. As a class of steels, they should be used wherever their high strength or their corrosion resistance is an advantage. Many proprietary alloys are considered to be in the HSLA category but, wherever possible, these alloys should be specified on engineering drawings by ASTM number and strength grade.

1.5.5   Special Steels

1.5.5.1   Austenitic manganese steels. This family of steels was introduced in the United States in 1892, and they continue to be used for special applications. Alloy additions can cause substantial changes in phase equilibrium. Austenitic manganese steels contain from 1 to 1.4% carbon and from 10 to 14% manganese. The high manganese content causes the austenite phase that is normally stable in steels at temperatures in excess of about 1400°F (760°C) to be the stable phase at room temperature.

Austenitic manganese steels are available as castings, forgings, and hot-rolled shapes. Their yield and tensile strengths are comparable to the HSLA steels, but their unique use characteristic is their ability to work harden when impacted. They are supplied with an annealed hardness of about 200 HB. With impacting, the surface hardness for a depth of 1 or 2 mm can increase to 500 HB.

They have application in design for wear plates on earth-moving equipment, rock crushers, ball mills, and the like where the predominating wear mode is gouging, abrasion, and battering. They are weldable, which allows easy application in high-wear areas. There are a number of grades of these steels (ASTM A 128), but their use characteristics are all similar.

1.5.5.2   Mill-heat-treated steels. Some mill-heat-treated steels are approximately high strength, low alloy in composition, but they are supplied from the mill in other than hot-rolled or hot-rolled and annealed condition. They are either normalized or quenched and tempered. Yield strengths can be as high as 110 ksi (758 MPa). Some of these steels are alloyed with manganese and silicon; some are plain carbon in composition. They are available in most forms suitable for structural use. They are usually proprietary, but there are some ASTM specifications that can be used for designation (ASTM A 321, A 514, A 577, A 633, and A 678).

These steels have good toughness and formability, but welding heavy sections requires special welding precautions. A particularly useful group of these alloys is the abrasion-resistant grades. They have a hardness in the range of 250 to 400 HB. This hardness is not high enough to resist scratching from hard particles, but the high yield strength (about 100 ksi, or 689 MPa) minimizes denting and gouging in handling rocks, coal, and similarly aggressive media. These steels are widely used in truck bodies and ore-handling equipment. In machine design, they find application where the high strengths can be used for weight reductions and for making hoppers, chutes, and loading platforms where the abrasion-resistant grades will provide resistance to abuse. Three mill-heat-treated steels that are widely used for very high-strength structural fabrications are HY-80, HY-100, and HY-130. The numbers stand for the minimum yield strength of these alloys, in ksi.

As in the HSLA steels, the carbon is kept low to achieve weldability. They have a predominantly martensitic structure, and special welding techniques are required. They are used for high-strength, high-toughness plates for pressure vessels, nuclear reactors, and submersible vessels.

1.5.5.3 Ultrahigh-strength steels. When all the steels that we have previously discussed still do not meet desired strength requirements, the ultrahigh-strength steels can be employed. This includes alloy steels, tool steels, steels with a high concentration of alloying elements, and stainless steels. The through-hardening alloy steels that we recommended for general use, AISI 4140 and 4340, when heat treated to high strength levels, are considered to be ultrahigh-strength steels. Two hot-work tool steels, AISI H11 and H13, have sufficient strength and toughness to be used for structural applications. The stainless steels will similarly be discussed in subsequent chapters. In the high-alloy category, a particularly useful group of steels in design is the 18% nickel maraging steels.

All these steels are available in some of the standard wrought forms, but the bulk of these steels are supplied as forged shapes. Many of these alloys are also melted using vacuum, VAR, and other special techniques aimed at improved cleanliness. The common denominator for ultrahigh-strength steels is high strength with measurable toughness. Most tool steels have high strengths, but they are so brittle that they are unusable except where bending is not expected. Their elongation in a tensile test is almost immeasurable. The strength and toughness characteristics of some ultrahigh-strength steels are shown in maraging nickel and stainless steels. These latter steels are usually supplied from mills in the martensitic or semiaustenitic condition, and hardening to maximum strength is accomplished on most grades by a low-temperature aging process (900 to 1150°F; 428 to 620°C). As shown in Table 1.7, the maraging steels come in different strength grades. It is common to designate these alloys by tensile strength; a 300-ksi maraging steel is an 18% nickel steel with a minimum tensile strength of 300,000 psi. The more appropriate designation would be the ASTM specifications and grade: ASTM A 538, grade A, B, or C.

Experience in design shows that three alloys, 4340 and the 250- and 300-ksi grades of the 18% nickel maraging steels, will meet most needs. The 250-ksi maraging steel has solved innumerable breakage problems in injection-molding cavities. It also provides an extra measure of toughness for high-loaded shafts with unavoidable stress concentrations. It has solved problems on components with operating stresses over 100 ksi (689 MPa). The alloy steel 4340 is more available and cheaper to use when the service is not quite as severe.

1.6 Selection and Specification

How is the designer to cope with the breath of the number and variety of steels? We have discussed a wide range of steels, but familiarization with a few specific steels from these various categories is generally enough. If a primary design criterion is strength, as shown in Fig. 1.6, the strength ranges overlap. If a design calls for yield strength of 100 ksi (68.9 MPa), the job could be done with a hardenable carbon steel. If lowest cost is a second criterion, this may be the proper choice. If the design is a large structure such as a vessel or a tank, a mill-heat-treated steel will eliminate heat-treatment difficulties. If the design is for a few power transmission shafts, alloy steels would be a better choice. They cost more than carbon steels, but they have better safety in hardening because of the oil quench. This weighing of relative merits must be done to arrive at a single material choice.

Some of the more commonly used alloys are listed below.

- AISI 1015: The best forming and welding characteristics. It is good general purpose plate.

- AISI 1020: The higher carbon range improves machinability while keeping good formability and weldability.

TABLE 1.7   Chemistry and Mechanical Properties for Maraging Steels[4]

Chemical requirements (%)	Grade A (200)	Grade B (250)	Grade C (300)
Carbon, maximum	0.03	0.3	0.03
Nickel	17.0–19.0	17.0–19.0	18.0–19.0
Cobalt	7.0–8.5	7.0–8.5	8.0–19.0
Molybdenum	4.0–4.5	4.6–5.1	4.6–5.2
Titanium	0.10–0.25	0.30–0.50	0.55–0.80
Silicon, maximum	0.10	0.10	0.10
Manganese, maximum	0.10	0.10	0.10
Sulfur, maximum	0.010	0.010	0.010
Phosphorus, maximum	0.010	0.010	0.010
Aluminum	0.05–0.15	0.05–1.15	0.05–0.15
Boron (added)	0.003	0.003	0.003
Zirconium (added)	0.02	0.05	0.05
Calcium (added)	0.05	0.05	0.05
Tensile requirements			
Tensile strength, ksi, minimum	210	240	280
	(1448 MPa)	(1655 MPa)	(1930 MPa)
Yield strength, 0.2% offset, ksi	200–235	230–260	275–305
	(1379–1619 MPa)	(1586–1793 MPa)	(1891–2100 MPa)
Elongation in 2 in., minimum	8	6	6
Reduction in area for round specimens, % minimum	40	35	30

- 1025: Plates >1 in. thick in this category are silicon killed along with having the higher carbon range for greater strength. This produces the optimum combination of strength, weldability, and good surface quality after machining, plus internal soundness.

- 1035: An intermediate-carbon, special-quality machinery steel, higher in strength and hardness than low-carbon steel. Used for studs, bolls, etc.

- 1040, 1045: Medium-carbon steels, used when greater strength and hardness is desired in the as-rolled condition. Can be hammer forged and responds to heat treatment. Suitable for flame and induction hardening. Uses include gears, shafts, axles, bolts and studs, machine parts, etc.

- 1045: Silicon killed with higher carbon content for greater strength. In the lighter and medium thicknesses, heat treatment will provide still higher strength. Machinability is good. Forming and welding opportunities are limited.

- 1060 ANNEALED: A nominal 0.60-carbon, silicon-killed, fine-grain steel designed primarily for surface flame hardening. Hardness to 62–65 Rockwell C in depths up to 4 in. with excellent uniformity of case depths and little distortion while retaining a soft, ductile core. Used for machine ways and wear strips on machine tools, cams and cam followers, unusual and intricate shapes, conveyor sprockets, sheave wheels, automotive dies.

- 1117: A widely used special-quality manganese steel. Develops a uniformly hard case supported by a strong, tough core. Machinability is increased over 1018. Used for medium-duty shafts, studs, pins, and other machine parts.

- 11L17: An addition of 0.15 to 0.35% lead to the basic 1117 analysis provides greatly improved machinability without changing the excellent case hardening characteristics and core properties of 1117.

- 1141: A medium-carbon, special-quality, manganese steel with improved machinability and better heat treatment response (surface hardness is deeper and more uniform) than plain carbon steels. Good as-rolled strength and toughness. Uses include gears, shafts, axles, bolts, studs, pins, etc.

- 11L41: Has greatly superior machinability due to 0.15/0.35% lead addition.

- 1144: Similar to 1141 with slightly higher carbon and sulfur content, resulting in superior machinability and improved response to heat treating. Often used for induction-hardened parts requiring 55 RC surface.

- 1215: A resulfurized and rephosphorized low-carbon steel for use in screw machines. An excellent carburizing grade.

- 4140 annealed: A superior free-machining, direct-hardening alloy. A chrome moly alloy with 0.40 carbon and 0.15/0.35 lead addition can be quenched and tempered to a broad range of strength levels.

- 4150 annealed: Has a higher carbon content to provide deeper hardenability for uniform through hardening in larger sections with little loss in machinability.

- 4140/42/47 quenched and tempered: A medium-carbon, chrome-moly alloy with optimal strength and hardness already in the bar. You save the cost and delay of heat treating to eliminate double handling for rough machining and finishing. In addition, the carbon content is matched to bar size, giving uniform properties through-out each section and from bar to bar across the broad size range. Surface hardness ranges between 269 and 321 Brinell. All bars are stress relieved.

- 8620: Most widely used of all case hardening alloys; truly general-purpose, oil-hardening steels with good core properties. Low in cost. Composition is balanced, nominally 0.55% nickel, 0.50% chromium, 0.20% molybdenum. Minimum distortion and growth characteristics.

- 4340: A highly alloyed steel, nominally 1.80% nickel, 0.80% chromium, 0.25% molybdenum, assuring deep hardness when oil quenched, with high-strength characteristics attained throughout the section. Used for heavily stressed parts operating under heavy-duty conditions.

- 6150. An electric furnace alloy, nominally 0.95% chromium, 0.15% vanadium content. Has high surface hardness combined with low distortion characteristics and good impact resistance.

1.6.1 Stainless and Speciality Steels[6]

Stainless steels are an important class of engineering alloys used for a wide range of applications and in many environments. Stainless steels are used extensively in the power generation, pulp and paper, and chemical processing industries, but they are also chosen for use in many everyday household and commercial products. They are iron-base alloys that contain a minimum of approximately 11 Cr, the amount needed to prevent the formation of rust in unpolluted atmospheres (hence, the designation *stainless*). Few stainless steels contain more than 30% Cr or less than 50% Fe. They achieve their stainless characteristics through the formation of an invisible and adherent chromium-rich oxide surface film. This oxide forms and heals itself in the presence of oxygen. Other elements added to improve particular characteristics include nickel, molybdenum, copper, titanium, aluminum, silicon, niobium, nitrogen, sulfur, and selenium. Carbon is normally present in amounts ranging from less than 0.03% to over 1.0% in certain martensitic grades. Figure 1.10 provides a useful summary of some of the compositional and property linkages in the stainless steel family.

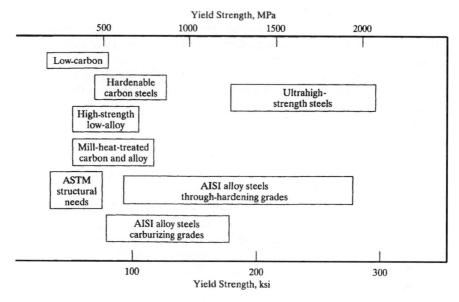

Figure 1.10   Yield strength ranges for various types of steel.[2–4]

1.6.2   Early History of Stainless Steels

Stainless steels were examined by a number of people before anyone observed that they resisted corrosion. The earliest published record of corrosion tests on iron-chromium alloys describes the work that was carried out by a Frenchman, Berthier, who examined a series of such alloys around 1820. Berthier noted that acids less readily attacked these alloys than iron.

Between this first experiment in 1820 and the early 1900s, there was a lack of progress toward the development of stainless steel. This lack of appreciation of the potential benefit to corrosion resistance that could be realized by alloying chromium and iron can be explained in two ways. The first reason was that practically all of the scientists prepared their alloys by using carbon to reduce the chromium ore to chromium. As a result, their alloys were laden with carbon. The detrimental effects of high carbon contents in stainless steels will be discussed more thoroughly later, but this effect was responsible for many erroneous conclusions in early experiments. The second reason was the belief that resistance to sulfuric acid was a good measure of resistance to all forms of corrosion. Occasionally, an investigator did depart from the pattern of evaluating corrosion resistance by using sulfuric acid, but invariably these investigators then chose hydrochloric acid. It is a well known fact today that stainless steels do show corrosion in these strong reducing acids. Laboratory tests made in these solutions are not a true indication of corrosion resistance in other environments. There are two common characteristics of most experiments conducted in the 1800s:

1. Most chromium alloys had a large amount of carbon in them.
2. Most corrosion testing was done in sulfuric acid.

These are the primary reasons for the lack of development of the stainless steels prior to 1900.

The first important breakthrough in this pattern occurred in 1892, when Goldschmidt developed his *aluminothermic process,* which utilized aluminum rather than carbon in the reduction of chromium ore. Utilization of this process permitted investigation of chromium-iron steels with lower carbon contents.

The years between 1910 and 1915 may be considered the *golden years* for inventions of stainless steels. The three basic classes of stainless steels were all developed in this period, and in three different countries. The man commonly referred to as the inventor of stainless steels is Harry Brearly, an Englishman, who developed the 0.35% carbon, 14% chromium alloy that is known today as Type 420. He was searching for a better alloy for making gun barrels at the time and was trying unsuccessfully to etch a sample of this composition for metallographic examination. Others had experienced similar trouble, but the spark of genius in Mr. Brearly enabled him to turn this annoyance into a discovery. At the same time, in America, Mr. Dantsizen, working for General Electric Company, was investigating similar alloys primarily for lead-in wires in incandescent lamps. Certain properties of these alloys led him to develop a lower-carbon variation, similar to AISI Type 410, for use as turbine blades. The application is still popular. Concurrently, in Germany, two men named Maurer and Strauss were studying iron-nickel alloys for thermocouple protection tubes. They eventually added some chromium, which led to an alloy whose composition was very similar to AISI Type 302, 18-8.

As the properties of the different types of stainless steels were established more completely, it became apparent that no one or two grades of stainless steel would satisfy all requirements. General corrosion resistance was the initial advantage found in these alloys, specific applications often required improved characteristics such as better machinability, better pitting resistance, increased resistance to high-temperature scaling, less work hardening, and corrosion resistance of welds. This led to modification of the basic alloys and to an ever-increasing family of stainless steels. There are almost 50 different steels that are recognized by the American Iron and Steel Institute as distinct. Each has an AISI type number that defines the chemical analysis. To this list we must add the many proprietary and patented alloys that fall in the stainless steel family but are not given AISI type numbers until they cease to be proprietary, and until they are sold in sufficient tonnage.

What makes stainless steel stainless? The simple answer is chromium. The addition of chromium to iron markedly reduces atmospheric corrosion. At a level of approximately 11% chromium, the steel will exhibit no apparent corrosion beyond superficial rusting without weight loss in the atmosphere. Actually, stainless characteristics can be observed in a 10% chromium alloy on a farm far from industrial areas, but in the atmospheres of Pittsburgh and New York the extra 1% chromium is important. This addition of 11% chromium to iron the is the definition of stainless steel, imparting atmospheric corrosion resistance to the alloys. All stainless steels have this minimum of 11% chromium and all are resistant to atmospheric corrosion. This is the only feature that they have in common. For ease of discussion, the entire group of steels has been divided into three classes. These classes are named for the metallurgical condition of the steel in the form in which it is used, as follows:

Class I—Martensitic stainless steels

Straight chromium, magnetic and heat treatable. Heat treatment produces martensite that is hard, strong, and has acicular microstructure.

Class II—Ferritic stainless steels

Straight chromium, magnetic and not heat treatable. The ferrite phase is always present

Class III— Austenitic stainless steels

Chromium nickel, nonmagnetic, not heat treatable. The austenite is very tough, ductile, and quite strong as annealed and can develop high-strength only by cold work.

These three basic classes have been used for many years. The picture has been changed, however, in the past few years with the increasing popularity of a relatively new group of steels, and these are now included in most classifications as follows:

Class IV—Precipitation hardening stainless steels

This group contains steels that may be either martensitic or austenitic as used, but they develop strength during heat treatment by utilizing precipitation hardening reactions rather than phase transformations such as employed in heat treating steel.

These descriptive terms of the four classes are very clear and understandable to a metallurgist, but, for most people, a simple definition of these metallurgical terms helps to understand the use, utility, and restrictions of the steels in each case.

1.6.2.1 Class I: martensitic, chromium, heat treatable. The basic alloy in Class I corresponds to Type 420. This alloy is referred to as *martensitic* by reasons of its structure in the hardened condition.

This needle-like structure is known as *martensite*. The term, incidentally, is derived simply from the name of the man, Martens, who first examined metals under the microscope. The structure is needle-like or acicular and is present in all hardened steels, whether they be stainless, carbon, or alloy. High hardness and relatively low ductility characterize it. All other hardenable stainless steels in Class I may be regarded as modifications of this basic alloy.

Type 420. Type 420 is corrosion resistant only in the hardened condition. The reason is that, in the annealed condition, the carbon in chromium stainless steels combines with the chromium to form a chromium carbide having a formula of $Cr_{23}C_6$; that is, about four parts of chromium are combined with one part carbon. Carbon has an atomic weight only one-fourth that of chromium so, in percent weight, 1 part of carbon combines with close to 16 parts by weight of chromium. If we take an alloy which has 0.35% carbon, and all the carbon is combined with chromium to form chromium carbide, roughly 5.6% ($16 \times 0.35\%$) of the chromium in the alloy is tied up with the carbon. If there is 13% chromium to start with, and 5.6% is removed, only 7.4% remains in the matrix of the alloy. This is insufficient to give the alloy desired corrosion resistance. It takes about 10.5 to 11% chromium to give reasonable resistance to atmospheric corrosion. When Type 420 is hardened, the chromium carbides are removed from the structure, and the chromium is all available to provide corrosion resistance. For this reason, the alloy is corrosion resistant in the hardened condition but not in the annealed condition.

Type 410. To get around this limitation, it would be logical to reduce the carbon content somewhat, and that is exactly what was done in the case of Type 410. In this type, the carbon content is dropped from 0.35% down to about 0.10%. This latter amount of carbon does not take too much chromium out of the lattice in the annealed condition, and therefore this alloy is resistant to corrosion in the annealed condition. It can develop a very wide range of mechanical properties by different heat treatments and still be resistant to corrosion. Naturally, this alloy soon became a popular standard stainless grade.

Type 440C. Increasing the carbon content also modifies Type 420. The objective in this case is to get an alloy that is very hard and abrasion resistant, comparable to a tool steel. The demand for such an alloy resulted in Type 440C, which has 1% carbon and 17% chromium. You will notice that the chromium content has been pushed up some 4%. The reason is that there is a point in carbon content above which carbides no longer are removed in the hardened condition. The rest of the carbon remains in the structure as free chromium carbides, even in the hardened condition. To compensate for the chromium that is tied up in those free chromium carbides, the chromium content is raised to 17%. This alloy hardens to around 600 Brinell and has remarkable resistance to abrasion, but it is limited in application because of its brittleness in the hardened condition.

Types 440A and 440B. Attempts have been made to arrive at a compromise between ability to harden and brittleness by backing off in the carbon contents, and there are two variations of Type 440C, now designated as "440A and 440B," which are a 0.60% carbon, and 0.80% carbon variety, respectively. These alloys harden to a few fewer Rockwell points than Type 440 C and are a little tougher in the hardened condition.

Early experience with stainless steels showed that they were quite different in machining properties from carbon steels. Efforts were made to improve these properties, and it was not long until it was found that additions of sulfur or selenium would appreciably improve machinability.

Type 416. A free-machining variation of Type 410 appeared, now known as Type 416, which was simply AISI 410 with an addition of about 0.30% sulfur.

Types 420F and 440F. Types 420 and 440C also have a free-machining counterpart in Type 420F and Type 440F. In these types, selenium is most frequently employed as the free-machining element.

Type 403. There are two other alloys that might be considered as modifications of Type 410 alloy. The first one is Type 403. The usefulness of the alloy is limited for, while it can be tempered or annealed to give quite a range of hardness and tensile strength, it should not be used in those conditions.

Type 430F. Even though Type 430 is considered soft in the family of stainless steels, there remains room for improvement in machinability. Sulfur is used as the addition in obtaining the free machining Type 430F.

Type 430 Ti. Class II alloys, like Type 430, are substantially nonhardenable. It is true that, if they are quenched from high temperatures, they may increase somewhat in hardness. They are never used in this condition, because they have poor toughness and corrosion resistance. Where the higher hardness is a requisite, some other steel, such as Type 410 or 17-4PH, is used. There are instances in which manufacturing methods demand short-time exposure to elevated temperature and where Type 430 satisfies all other requirements. Glass sealing operations in the making of television tubes and all welding applications are examples. In these applications, Type 430 Ti is used. An addition of a small amount of titanium to Type 430 is sufficient to suppress all hardening during exposure to elevated temperature.

1.6.2.2   Class III: austenitic, chromium nickel, nonhardenable by heat treatment.

The addition of nickel to the chromium-iron steels is the common feature of the Class III alloys. Alloys of this class have a characteristic grain structure that first was observed by Sir W. C. Roberts-Austen. Austenite is nonmagnetic, and its microstructure contains parallel bands inside the grain. They are referred to as *twins* and are caused by a certain type of shearing that occurs inside the grain. Austenite is a soft and extremely ductile and tough structure even at very low subzero temperatures.

Type 302. The basic alloy of Class III is Type 302. It has 18% chromium and 8% nickel as the major alloying additions. It performs exceedingly well in a multitude of diverse applications and is the most popular grade of stainless steel.

The best-known problem encountered with Type 302 is the phenomenon of *sensitization*. When these steels are heated at temperatures ranging from about 900 to 1600°F, chromium carbides form and precipitate at the grain boundaries rather than being randomly dispersed throughout the structure as they are in the Class I and II alloys.

In this case, an 18-8 alloy was heated for 100 hr at 1200°F. The grain boundaries are loaded with chromium carbides. This type of carbide precipitation is bad because, when the carbides form, they take chromium from a band immediately adjacent to the grain boundary. The chromium is reduced to a very low level at that point.

Much of the carbon migrates to the grain boundary and reacts with the chromium it extracts from a narrow band. The net result is that a fairly continuous band of low chromium content is formed around the boundaries, and this band is very low in resistance to corrosion. If such material were exposed to fairly corrosive medium, corrosion would follow along these bands, penetrating through the alloy. In very severe cases, the material could disintegrate into a powder of individual grains. This is termed *sensitization,* because the metal is now sensitive to attack from acids.

If carbide precipitation does occur, it can be remedied, of course, by solution annealing the material to redissolve the carbides. In some cases, it is impractical to consider such reannealing treatments. Examples are very large welded structures and parts that will be in service for long periods in the sensitizing temperature range. In such cases, it is necessary to resort to some device to alleviate or prevent harmful carbide precipitation.

Type 304.   One obvious method to help the situation is to lower the carbon content and thereby limit the amount of carbides that can precipitate. This approach is employed in the case of Type 304, which is a modified Type 302 with the carbon held to a maximum of 0.08%. This grade is somewhat less susceptible to carbide precipitation and can be welded in certain sections and exposed to mild corrosive conditions without harm.

Types 321, 347, and 348.   Where a service environment is more corrosive, the slightly lower carbon content of Type 304 compared to Type 302 is not sufficient to prevent harmful sensitization in weldments. Another solution is necessary. Types 321, 347, and 348 are modifications of Type 302 to which an addition of titanium or columbium has been made. The function of these added elements is to form stable and insoluble carbides of titanium or columbium to tie up all the carbon in the alloy so it cannot precipitate as chromium carbide. The columbium carbide is more stable than the titanium carbide (that is, more insoluble at higher temperatures), and for that reason the columbium alloy will withstand slightly more severe conditions than Type 321.

Tantalum occurs in nature in conjunction with columbium. The two have very similar chemical characteristics, and tantalum acts very much like columbium in reacting with the carbon preferentially. There is, however, a large difference in their response to radiation. Radioactive tantalum has a much longer half-life than columbium and is therefore objectionable as construction material for radioactive service. For atomic energy applications, Type 348 is specified, and it differs from 347 only in the maximum permissible tantalum content.

All three of these grades have 10% nickel rather than the 8% found in Type 304. Carbon acts like nickel in making the steels austenitic, and when, it is essentially removed by reacting with columbium or titanium, the chemical balance of the steel is upset. Proper balance for regaining a completely austenitic structure is obtained by adding 2% more nickel.

Type 304 L.   Type 304 L represents a lower-carbon version of Type 302, but it is still susceptible to sensitization. Even lower-carbon contents were necessary to avoid harmful carbide precipitation during welding. The *rustless* process of melting and the development of oxygen blowing during melting were two steps that enabled the making of Type 304 L. This is the extra-low-carbon version of 18-8 in which carbon is held to a maximum of 0.03%.

Type 304 L is naturally lower in price than Types 321 and 347, which have the cost of the added element plus the extra nickel. It has proved to be the number one choice among stainless steels where sensitization or welding is a consideration.

Types 303 and 303 Se.   Type 303 is a modification of Type 302, 18-8, to which sulfur or a combination of phosphorus and selenium are added to improve the machinability. We designate the steels as Type 303 and Type 303 Se. They are not made in flat-rolled products such as sheet or strip. There is a difference in the machining properties of the sulfur-containing Type 303 and the selenium-containing Type 303.

The general distinction between them is this: The sulfur-bearing grade is the more free-cutting grade. It can be machined at the highest rate. The selenium-bearing grade gives the best surface finish. It is not quite as free cutting. Sulfur in Type 303, which makes it more free-cutting, makes for a poorer surface finish on turning and forming and other machining operations.

Consider Type 302 and Type 303 together. Type 302 has the lowest free-cutting properties and has by far the best surface finish. The selenium is next in line. It has better free-cutting properties and a little poorer surface than 18-8, but better than the sulfur-bearing alloy.

Type 302B.   The addition of 2% silicon adds appreciably to the scaling resistance. Type 302B has a maximum operating temperature of 1800°F as opposed to 1650°F for Type 302, where destructive scaling is the important factor. These temperature limits apply to heating in an air atmosphere, and they vary considerable if the atmosphere contains certain detrimental products of combustion such as sulfur dioxide or mixtures of sodium and vanadium oxides.

Types 308, 309, and 310.   In the discussion of Class II alloys, it was shown how scaling resistance was improved as chromium content was raised. This advantage was utilized in developing ferritic grades with higher chromium contents and is also employed in the austenitic grades. Type 308 is not significantly better than 18-8. The major use of this grade is in welding wire that will be used to deposit an 18-8 analysis in the weld, allowing for some dilution during welding.

It is largely the chromium content that determines the scaling resistance of the alloy. The nickel has to be added with the added chromium to get steels that are fully austenitic and that will hot work satisfactorily. In this class we have Type 309.

Type 310 is the grade with the highest chromium content and is chiefly known for its high scaling resistance. It is superior in resistance to scaling to all grades of stainless steel except Type 446. It should be pointed out that there is one important difference between the chromium and the chromium-nickel grades with respect to their high-temperature strength. The difference is brought out in Fig. 5.11, which compares the tensile strength of alloys of all three classes. Generally speaking, at temperatures up to 900–1000°F, the chromium grades such as Type 410 have very high creel strengths and are commonly used for applications such as steam turbine blading. At higher temperatures, however, the chromium nickel grades show a decided superiority over chromium grades, and they are most commonly used for high-temperature applications where strength is a factor. Where it is simply a matter of scaling resistance without any stress being applied to the material, such as a heater baffle, a chromium grade will work very satisfactorily. If stresses are present, the chromium-nickel grades are a more logical choice.

Types 316, 316L, 317, and 317L. Further modifications of Type 302 include alloys with molybdenum added. These are called Type 316, which has 2% molybdenum, and Type 317, which has 3% molybde-

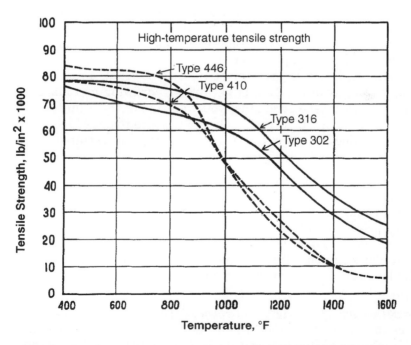

Figure 1.11 Tensile strength for various types of stainless steels at elevated temperatures.[6]

num. Adding molybdenum to 18-8 helps in two ways: corrosion resistance and high-temperature strength.

The most important benefit is in corrosion resistance to reducing acids and to pitting resistance. Type 316 has been particularly popular in corrosion environments such as chemical and paper industry plants where such media as sulfurous acid and similar reagents are used. Type 316 is generally considered to be the most corrosion resistant of all stainless steels. However, even Type 316 is considered borderline as to corrosion resistance in certain substances in the chemical industry. An increase in the molybdenum content of approximately 1% is believed to provide pitting and general corrosion resistance. The grade is Type 317. The molybdenum content of Type 316 also contributes appreciably to elevated temperature creep and rupture strength. It is as strong as Type 310 up to the point at which scaling resistance becomes a factor, and then the increased chromium present in Type 310 makes this grade preferable.

Type 301. Another modification of Type 302 is Type 301. In this case, the chromium and nickel contents have been lowered. This is done primarily to increase the cold-work hardening rate of the alloy. It is well known that the chromium-nickel alloys all have a rather high cold-work hardening rate. In a word, when the metal is cold-drawn, cold-rolled, or worked at room temperature in any way, it tends to increase in tensile strength and hardness rapidly. At the same time, the ductility falls off. This property, which is illustrated in Fig. 1.12, is an important factor in many applications.

In making high-tensile sheet or strip, it is desirable to use a fast work-hardening analysis. This is because, in drawing or rolling the material to high tensile strengths, you want to get them up to those values without having to do too much cold-work on them, which would exhaust their ductility. It is this latter property that resulted in the use of Type 301 alloy for high-tensile applications.

Type 305. High work-hardening rates are valuable in making high-tensile strip. On the other hand, in wire intended for cold heading or in sheet intended for multiple drawing operations or for free spinning, high work-hardening rates are obviously undesirable. Here, you want to use a minimal amount of work to end up with a product that is not too hard. For such jobs, Type 305 is used. It is a free-spinning alloy, and the low work-hardening rate is obtained by increasing the nickel content.

Types 201 and 202. The original purpose of the 200 series alloys was to conserve nickel during a nickel shortage. By substituting increased amounts of nitrogen and manganese for some of the nickel, it is possible to arrive at an analysis that has engineering properties very simi-

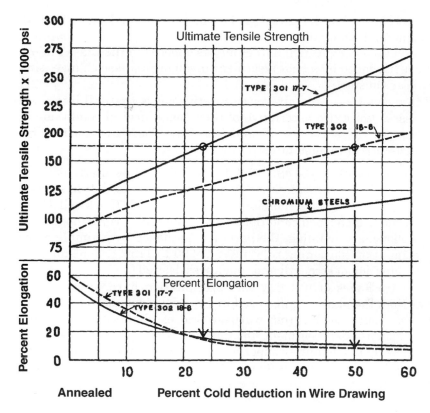

Figure 1.12 Tensile strength for various types of stainless steels at elevated temperatures.[6]

lar to those of the 300 series alloys, but which uses less nickel. Type 201, containing 17Cr-4Ni-7Mn, and Type 202, with 18Cr-5Ni-8Mn, have properties that are very similar to Types 301 and 302, respectively. There are some differences. The 200 series steels have slightly higher high-temperature strengths and room temperature hardness due to the nitrogen. They also have higher work-hardening rates and are slightly less corrosion resistant.

That completes the list of the Class III alloys, all of which resemble 18-8 in that they are austenitic. Each of the other steels has some modification in chemical analysis that produces some improved characteristic.

1.6.2.3   Class IV: precipitation hardening, chromium nickel.   Only recently have metallurgists begun speaking of a fourth class of stainless steels. Three classes were sufficient for many years, but the increased popularity of these relatively new grades of precipitation hardening stainless steels has led to the grouping of them into a special class. They all have one common quality: they can be hardened by a heat treatment that produces precipitation hardening, as opposed to the Class I alloys that are hardened by a heat treatment involving a phase change (similar to carbon steel).

Within this class, we have representative steels that fall into three subclasses, which are descriptive of their metallurgical conditions as was true for Class I, Class II, and Class III. These are martensitic, semiaustenitic, and austenitic.

Martensitic: AISI 630 or 17-4 PH. 17-4 PH (AISI 630) is martensitic at room temperature regardless of any heat treatment used. It has a very low carbon content and therefore is relatively soft (Rockwell C 30) even in the transformed condition. This grade contains a copper addition. Following machining, the parts are hardened by a simple aging or precipitation hardening treatment at 900 to 1150°F, depending on the combination of strength and toughness desired.

The advantages of utilizing this precipitation hardening method, rather than the phase change used in Type 410, include freedom from heavy scaling, freedom from warpage and excellent surface quality undamaged by heat treatment. In addition, 17-4 PH exhibits corrosion resistance that is definitely superior to the Class I alloys and is equivalent to 18-8 in most corrosion environments. Also, a final machining operation to size is usually unnecessary with 17-4PH but is practically always necessary with Class I steels.

Semiaustenitic: 17-7 PH and PH 15-7 Mo. Class I alloys are always martensitic, and Class III alloys are always austenitic at room temperature. The semiaustenitic steels are austenitic at room temperature when they have been solution annealed at 1900–2000°F and rapidly cooled. As such, they are soft and ductile and can be formed or drawn. This is followed by a conditioning heat treatment at 1400 to 1750°F that changes the balance so that, during cooling, they transform to martensite. However, this cooling does not have to be drastic, and warpage can be minimized. As with 17-4 PH, the hardness immediately after transformation is relatively low: Rockwell C 30. The final heat treatment once again consists of an aging or precipitation hardening heat treatment at 950 to 1150°F and hardness from Rockwell C 38 to C 49 is possible. The precipitation hardening is achieved by the addition of roughly 1% aluminum.

The first steel of this type was grade 17-7 PH, and it still serves in many applications. A modification, PH 15-7 Mo, was then made in which part of the chromium was replaced with molybdenum to achieve higher creep and rupture strengths in the temperature range of 700–1000°F.

In both 17-7 PH and PH 15-7 Mo, cold work can be used as a means of obtaining the transformation rather than using the conditioning heat treatment. The cold-worked strip or wire is then quite hard and cannot be severely formed, but it can be utilized as springs or saw blades, and the aging treatment produces a very high level of hardness and proportional yield strength in conjunction with the excellent corrosion resistance.

TABLE 1.8   Martensitic, Austenitic, and Precipitation Hardening Alloys [6]

Martensitic alloys	
420	The basic martensitic alloy; corrosion resistant only in the hardened condition
410	Lower carbon content version of 420
440C	Higher carbon content version of 420; very hard and abrasion resistant
440A and 440B	Lower carbon content version of 440C; trades off some hardness to reduce brittleness
416	A free-machining variation of Type 410
420F and 440F	Free-machining variations of 420 and 440 C
430F	Free-machining variations of 430
430Ti	A small amount of titanium to is added to 430 to suppress all hardening during exposure to elevated temperature
Austenitic alloys	
302	The basic austenitic alloy containing 18% Cr and 8% Ni
304	Lower carbon content version of 302; better weldability and corrosion resistance
321, 347, 348	Type 302 with additions of titanium or columbium. The function of these added elements is to tie up all the carbon in the alloy so it cannot precipitate as chromium carbide. The columbium carbide is more stable than the titanium carbide (that is, more insoluble at higher temperatures), and for that reason the columbium alloy will withstand more severe conditions than Type 321.
304L	Even lower carbon version of Type 302, but still susceptible to sensitization; 0.03% C max
303 and 303Se	Modifications of 302, to which sulfur or a combination of phosphorus and selenium are added to improve the machinability
302B	Type 302 with an addition 2% silicon; has a maximum operating temperature of 1800°F

TABLE 1.8   Martensitic, Austenitic, and Precipitation Hardening Alloys *(continued)*[6]

308	Used mainly as a welding wire used to deposit an 18-8 analysis in the weld, allowing for some dilution during welding
310	The grade with the highest chromium content, chiefly known for its high scaling resistance
316, 316L, 317, 317L	Type 316 is generally considered to be the most corrosion resistant of all stainless steels. However, even Type 316 is considered borderline corrosion resistant when exposed to certain chemicals. An increase in the molybdenum content of approximately 1% (317) is believed to provide pitting and general corrosion resistance. Both of these molybdenum-bearing stainless steels find many applications requiring welding. 316L and 317L are the extra low carbon grades produced for these applications.
301	The corrosion-resistant grade includes a lower carbon level and a small increase in columbium to eliminate sensitization. The presence of copper improves the corrosion resistance in phosphoric acid solutions especially.
305	Low work-hardening rate alloy for cold-heading applications
201 and 202	200 series alloys were developed to conserve nickel during a time of nickel shortage. By substituting increased amounts of nitrogen and manganese for some of the nickel, it is possible to arrive at an analysis that has engineering properties very similar to the 300 series. The 200 series steels have slightly higher high-temperature strengths and room-temperature hardness due to the nitrogen. They also have higher work-hardening rates and are slightly less corrosion resistant.

Precipitation hardening alloys

17-4PH	The advantages of utilizing this precipitation hardening method rather than the phase change used in Type 410 include freedom from heavy scaling, freedom from warpage, and excellent surface quality, undamaged by heat treatment. In addition 17-4 PH exhibits corrosion resistance that is definitely superior to the Class I alloys and is equivalent to 18-8 in most corrosion environments.
17-7PH and PH 15-7 Mo	These semiaustenitic steels are austenitic at room temperature when they have been solution annealed at 1900–2000°F and rapidly cooled. As such, they are soft and ductile and can be formed or drawn. As with 17-4 PH, the hardness immediately after transformation is relatively low: Rockwell C 30. The final heat treatment once again consists of an aging or precipitation hardening heat treatment at 950–1150°F, and hardness from Rockwell C 38 to C 49 is possible. The precipitation hardening is achieved by the addition of about 1% aluminum. 17-7 PH serves in many general applications. In PH 15-7 Mo, the chromium was replaced with molybdenum to achieve higher creep and rupture strengths in the temperature range of 700–1000°F.

1.7 Welding Ferrous Metals[7]

1.7.1 Welding Carbon Steel

Carbon steels are defined as alloys of iron and carbon in which the manganese content did not exceed 1.65%, and the silicon and copper contents did not exceed 0.60% each. Specified limits are not to be set for any other alloying elements. Carbon steels sometimes contain additions of elements like aluminum or titanium for purposes of deoxidation or to secure a fine austenitic grain size. Small amounts of needling agents containing boron or vanadium may be added to increase hardening capability. Grain size inhibitors like columbium or nitrogen are added at times to increase strength—particularly yield strength. Carbon steels may be rephosphorized to levels somewhat above the usual residual level to achieve a modest increase in strength. Sulfur or lead may be added to improve machinability. Obviously, the "carbon steels" encountered in welding are more than simple iron-carbon alloys with restricted manganese, silicon, and copper contents. Small changes in the chemical composition of carbon steel can have a profound influence on a welding operation. A small addition of an element to improve a particular quality or property may change for better or worse in the steel's weldability. Although carbon is the element wielding the greatest influence over the welding behavior of carbon steel, the remainder of the composition deserves as close study as might be given another alloy steel.

A sound weld in carbon steels is determined to a large degree by familiarity with the microstructural characteristics of each particular type and our ability to avoid the development of an unsuitable structure in or adjacent to the weld joint. The importance of cooling rate on the final structure of a particular steel in its heat-affected zone adjacent to the weld deposit cannot be overemphasized. With increasing carbon content, faster rates of cooling tend to produce a martensitic structure whose properties are not desirable for weld joints. Microstructural changes involving transformations of austenite, ferrite, and martensite are a most important part of welding carbon steels and must be given attention before dealing with any other metallurgical aspect.

1.7.2 Cooling Rates in Welds

The fastest cooling rates occur in the metal that has been heated to the highest temperature during welding. Most carbon steels have so low a carbon content that the carbon does not change its properties to any great extent under the most rapid cooling likely to occur in welding. In multilayer welding, the next bead always refines the preceding bead. Instead of the weld metal, it is primarily the base metal in the immediate vicinity of the weld that required most attention, because it is the metal closest to the weld that cools rapidly and is most likely to exhibit important changes in the structure and properties as a result of welding.

1.7.3 Quenching

Quenching is rapid cooling by immersion in liquids or gases or by contact with metal. The term is commonly applied to the heat treating in which a medium-carbon or high-carbon steel is heated above the upper critical temperature and, after removal from the furnace, is immersed in a quenching medium at or near room temperature. Immersing the steel in cold water is known as *water quenching,* in oil as *oil quenching,* in air as *air quenching.* The steel cools extremely quickly in water quenching, slowly in oil quenching, and still more slowly in air quenching.

During water quenching of a thin section of steel, about 1/2 in. dia., the austenite is supercooled to 200 to 400°F. It then changes to martensite and appears to our microscope at room temperature as martensite. The steel is said to be *fully hardened.* The maximum hardness that can be obtained in any carbon steel by a full hardening treatment like this is governed by the carbon content. Oil quenching usually is too slow to fully harden plain carbon steel. Water quenching cools a thin piece of steel at the rate of over 900°F/s while the steel is in the temperature range of 800–1200°F, in which the austenite transforms most rapidly. In other words, the steel remains in the range less than 1/2 second, which is insufficient time for the first transformed crystal to appear. The austenite continues to cool until, at 200–400°F, it is changed to martensite. During oil quenching, on the other hand, the rate of cooling of a thin bar of steel in the temperature range of 800–1200°F is only about 100°F/s. The steel remains in the range about 4 s, which is sufficient to permit the austenite to change partly or wholly to fine pearlite. When this transformation starts, heat is evolved, which tends to reduce the cooling rate still further.

Air quenching is usually so slow that the steel transforms entirely to pearlite during cooling in the range 1200–1100°F. It must be remembered that the surface of a part cools more rapidly than the interior, and that thin sections of uniformly heated metal cool more rapidly than thick. In that respect, air cooling may harden a wire, yet it may produce a soft pearlite in a 2-in. thick plate. Indeed, air cooling of steel 1/2 in. and thicker from 100°F above the critical range is the customary means of creating a strong, ductile structure in steel.

In welding, the rate of cooling after the weld melt has passed by is governed by contact with adjacent steel rather than by immersion in air.

1.7.4   Microstructural Changes

The actual carbon content will be a major factor in determining the exact nature and importance of the microstructural changes. For purposes of evaluating welding processes, it is better to group the carbon steels by particular ranges of carbon content. As each group is discussed, we examine the metallurgical features of specific types of steels that require consideration for welding.

1.7.4.1   Low-carbon steel (0.04 to 0.15% carbon).   A large amount of the steel used today is of the low-carbon variety. The carbon range of 0.04 to 0.15% is intended as a guide for a group of steels that give practically no difficulty with hardening during welding. Low-carbon steel is found in auto bodies, sheet and strip used for stamped and formed articles, drums, cans, electric motor shells and frames, and thousands of other applications. Many of these mass-produced articles call for steel within a narrower carbon range than 0.04 to 0.15%, and with various restrictions on manganese and silicon contents. If steel were needed for a welded motor frame that would be subjected to severe forming and bending after welding, the material might be ordered as AISI 1006. The carbon maximum of 0.08% would ensure low hardness. If steel were needed for an easily formed and welded tank support on a tractor where somewhat greater strength would be desirable, AISI 1010 might be specified. Composition requirements are needed to tailor the steel for both fabricating operations and end use. To obtain the best formability, the steel may be made by any one of a number of steelmaking practices. To obtain steel of lowest cost and with excellent surface for forming, the heat of steel may be finished as rimmed steel. Where best drawing quality is needed, a special killed steel may be preferred. Deoxidation is accomplished with aluminum rather than silicon to obtain the cleanest steel for this particular practice. The aluminum in excess of that required for deoxidation ties up the nitrogen in the steel and makes the material nonaging. Flat-rolled steel does not undergo an increase in strength and a decrease in ductility with lapse of time after the final cold rolling or flattening operation. This steel does not display a return of pronounced yield-point elongation that results in localized stretching and breakage during drawing operations. From a welding standpoint, these steelmaking practices introduce certain metallurgical conditions in the steel that require consideration when applying many of the welding processes.

Rimmed low-carbon steel is not deoxidized and tends to have a concentration of carbon and certain other elements in the central core area. It is important to recognize that, during any fusion joining operation, a rimmed steel will attempt to continue with the chemical reaction between the contained carbon and oxygen just as occurred during the freezing of the ingot. In the case of fusion joining, the carbon monoxide gas formed by the reaction bubbles out of the weld pool. When oxyacetylene welding is performed, the rate of solidification usually is sufficiently slow to allow the gas to escape, and a reasonably sound weld can be expected. However, if the rimmed steel is fused and solidified rather quickly, as occurs during mechanized gas tungsten arc welding, then the solid weld metal will contain many gas pockets.

A practical means of avoiding weld metal porosity in joining a rimmed steel by arc welding is simply to deoxidize the fused metal. This can be accomplished by a number of techniques. In gas tungsten arc welding, a welding rod containing sufficient deoxidizers such as manganese, silicon, or aluminum can be added to the pool, or a thin coating of powdered aluminum can be applied to the edges of the steel to be fused. In metal arc welding with covered electrodes, the covering usually contains ample amounts of silicon and possibly other deoxidizing agents to tie up the oxygen in the rimmed steel core wire and rimmed steel base metal as well. Submerged arc welding may, or may not be capable of accommodating a rimmed steel base metal, depending on the kind of electrode employed and the nature of the flux composition. It is a simple matter to select an electrode or a flux with sufficient deoxidizer, usually silicon, to produce a sound weld. Gas metal arc welding requires a little more attention when welding a rimmed steel, because all of the deoxidizer required must be in the consumable electrode. If the shielding gas contains an appreciable percentage of carbon dioxide, a gas that permits some oxidation of manganese and silicon, then the electrode must be quite high in deoxidation ability. An electrode for gas metal arc welding a rimmed steel often contains about 0.60% silicon, which is about twice the amount ordinarily found in a low-carbon steel weld deposit. During deposition, approximately half of the silicon in the electrode is lost through oxidation.

When fusion joining a low-carbon killed steel, no particular problem arises in obtaining a sound deposit, unless perhaps the operation is conducted in some manner that permits unnecessary exposure of the molten metal to the air, and an excessive quantity of oxygen is absorbed. A problem that sometimes arises with killed steel is the accumulation of slag on the surface of the weld, which hampers the proper flow of molten metal. This slag is formed when the deoxidizers employed in the steel tend to form a refractory oxide with a relatively high melting point. Aluminum, titanium, and zirconium are examples of excellent deoxidizers, but they tend to produce refractory oxides. When welding a steel containing these elements, the weld slag is formed by the oxide inclusions originally trapped in the steel ingot and by any additional oxide that may form during the welding operation if the oxidation shielding is less than perfect. These refractory slags are apt to be most troublesome in oxyacetylene welding, where the lower temperature heat source does not make the slag fluid by virtue of temperature. In arc welding operations, the weld slag is made quite fluid by the high temperature of the arc or is assimilated by a flux employed in the process. In any fusion joining operation, including the oxyacetylene process, a refractory slag, formed by deoxidizers or deoxidation products in the base metal, can be altered by introducing a filler metal containing generous amounts of manganese and silicon. The oxides of these elements mix with the refractory oxides and lower the melting point of the slag, which in turn tends to increase fluidity. For example, an aluminum-killed low-carbon steel being oxyacetylene welded might produce an objectionable amount of viscous slag consisting mostly of aluminum oxide. The use of a low-carbon steel welding rod containing about 1% manganese and 0.5% silicon would facilitate the operation. In most applications, there should be no objection to adding filler metal of this kind to the weld, except that its cost will be a little higher that of than steel with lower manganese and silicon.

One instance in which the hardness of weld metal or the base metal heat-affected zone in low-carbon steel may be of concern is a weldment that is subjected to severe cold forming in the weld area. The rapidly cooled weld-affected structure may not have adequate ductility to accommodate the cold-forming operation, particularly if the carbon is at the high side of the range, that is, 0.10–0.15%. Weld joints in sheet subjected to sharp bends or lock seaming operations have been observed to crack when the carbon content was high. Lower carbon content, perhaps 0.06% or less, greatly improves the ability of the weld joint to undergo cold-forming operations.

1.7.4.2 Very low-carbon steel (0.03% max. carbon). A number of unique properties are found in steel containing 0.03% carbon or less. The steel has a very high, narrow transformation range. The lower the carbon, the closer the range approaches the transformation point of iron at 1670°F. If the steel also is very low in manganese and silicon contents, the magnetic properties will be comparable to those of pure iron, and therefore the metal will be very useful for electromagnetic equipment. Also, the corrosion resistance of this very low-carbon, low-alloy-content steel will be unusually good. This does not imply a degree of corrosion resistance like that of stainless steel; the metal will rust at a slower rate than ordinary steel and will have substantially longer life in painted and coated products.

Metals and alloys have a very, very low yield strength as they undergo an allotropic transformation. In addition, the volume change, which occurs during these transformations, plays a part in the distortion or warpage that occurs. Many of the articles made of very low-carbon steel are welded, and while no difficulty could possibly arise with hardening in heat-affected zones, any high-speed welds made by fusion joining in steel of the rimmed type may require deoxidation of the weld metal to avoid porosity. The techniques for handling the deoxidation of the weld can be about the same as described for the low-carbon steel containing about 0.04 to 0.15% carbon (that is, the use of aluminum and silicon as deoxidizers).

1.7.4.3 Decarburized steel (less than 0.005% carbon).

There is another class of steels, known as *decarburized steels,* containing as little as 0.001% carbon. These steels give absolutely no difficulty with weld hardening, but, as might be expected, the tailoring of the steel for best single-coat enameling capability does not result in best weldability under all welding processes.

Open coil annealing can be applied to steel of any reasonably low-carbon content and virtually eliminate this alloying element. There appears to be an advantage in starting the decarburizing treatment with a steel that is not too low in carbon. If a steel containing up to approximately 0.15% carbon can be processed to form massive carbides in the microstructure, these large particles will serve a very useful purpose. When the steel is cold reduced by rolling, the massive carbide particles develop microscopic cracks or voids. During the decarburization treatment, the carbide particles are removed, but many microscopic voids remain. These voids serve as cavities into which hydrogen may diffuse during cooling and thus relieve the pressure for the gas to escape from the surface of the sheet by spalling off small flakes of porcelain coating.

There are few pertinent facts about the welding characteristics of decarburized single-coat enameling steel. The microstructure of the steel is composed entirely of ferrite grains, there being no carbon beyond the alpha iron solubility limit to form carbides. The steel for this almost carbon-free product may have been originally produced as a rimmed steel or as a killed steel—usually using aluminum as the deoxidizing element in the killed type. Either type can be decarburized to the 0.001% carbon level by the open coil annealing operation.

In welding decarburized steel of the rimmed type, a remarkable change in behavior will be noted. Because the carbon has been almost completely removed, the contained oxygen can create no boiling or rimming action. A sound weld metal is produced, provided, of course, that no filler metal of higher carbon content has been introduced and that no carbon pickup has occurred from other sources during the welding operation. Precautions against carbon pickup are of no concern.

1.7.4.4   Mild steel (0.15 to 0.30% carbon).   Mild steel has a carbon content in the range of 0.10 to 0.25%. For purposes of welding, we will consider it to be 0.15 to 0.30% carbon, because this is the range within which falls a large amount of the steel fabricated by welding. Many weldment designers and welding engineers consider carbon steel in this range to be readily weldable and use the term *mild steel* in everyday discussion of this material. Mild steels are used in products and weldments in many fields. Buildings, bridges, piping, castings, pressure vessels, and machine bases are typical examples. The fabricators and users of these products in many cases have generated specifications to cover the steel employed in their particular commodity. While these specifications are necessary for reasons of surface finish, dimensional tolerances, and quality standards required by each particular industry, it is unfortunate that they also present a very large number of seemingly different steel compositions. Close study will show, however, that the analysis requirements differ mostly in the numbers used to define the limits. Much of the carbon ranges overlap so frequently that the composition of a single type of steel will meet the requirements of many specifications. Sometimes the designation of the steel is the only difference between specifications. The whole lot could be reduced to a reasonable number of standardized steels by some coordinated effort. Rather than attempt a complete list of all the carbon steels in the 0.15 to 0.30% carbon range, it should suffice to discuss steels that are representative of the various types within this category.

Hot-rolled carbon steels containing 0.15 to 0.30% carbon will show that a modest change occurs in properties as the three major elements (carbon, manganese, and silicon) are varied from low levels to the maximum permissible limits for these elements in a carbon steel. These changes apparently are sufficient to make the types of steels better suited for particular kinds of weldments. At the low side of the ranges for C, Mn, and Si, steel exhibits a tensile strength of about 45 to 65 ksi and a yield strength of about 25 to 30 ksi. These strength properties serve very well for a great number of articles, but, in many cases, a higher strength could be used to advantage to reduce the weight of the article. By increasing the carbon, manganese, and silicon to the top of our *mild steel* ranges, a level of 50 tensile strength approaching 90 ksi can be secured. To gain higher strength in a hot-rolled carbon steel, it would be necessary to increase the carbon content. The adverse influence of this kind of alloying on weldability will be explained in the next section, on *medium-carbon* steels. To gain higher strength by alloying and also retain good weldability, the kind of alloying required elevates the steel into a *high-strength, low-alloy* category. Even the modest increases in carbon, manganese, and silicon employed in the mild steels will increase the cost of the steel. If the desired higher strength can be achieved with a minimal alloy increase, it is possible to lower the total material cost for the article. This simple relationship between the alloy content of steel, its strength, and its cost often is a foremost consideration in selecting carbon and alloy steels. Another important consideration at the moment, however, is whether a variation in alloy content from the low side to the high side of the mild steel range in any way affects the reputed good weldability of this class of steel.

A critical examination of the behavior of mild steels during welding and the performance of the weldment in service shows that this class of steels varies in weldability over its strength range. Not only is there a gradual increase in the need for welding precautions as the tensile strength is shifted from a low of 45 ksi to a high of 90 ksi, but certain mild steels, depending on the steelmaking practice, have other characteristics that exert a profound influence on weldability. In evaluating each of these mild steels, it is necessary to consider much more than just the chemical composition. The thickness of the sections, the form, the steelmaking practice, and the condition of heat treatment will be important factors. The weldability probably will vary with the different welding processes that are applicable. The final estimation of weldability must take into account the service for which the weldment is intended—particularly the temperature range over which the weldment must exhibit satisfactory performance.

1.7.4.5 Medium-carbon steels (0.30 to 0.60% carbon). Carbon steels containing about 0.30% carbon have generally good weldability, with some need for caution if the carbon and the manganese simultaneously border at the maxilla arbitrarily set for mild steel. However, in the medium-carbon steels, over a small spread of 30 points of carbon (i.e., from 0.30 to 0.60%), a pronounced change occurs in weldability. The steel containing 0.60% carbon when welded by ordinary procedures easily may be fully hardened (martensitic) in the heat-affected zones. This level of carbon not only provides the hardenability to achieve a fully martensitic structure, but the hardness of the martensitic zone will be about Rockwell 63 C, which is almost the maximum hardness that can be attained in a carbon or an alloy steel. These heat-affected zones will be low in toughness and ductility. They have a strong propensity to develop cracks when cooling to room temperature, and they may fracture in service, particularly if subjected to impact loading. If the weld metal is substantially increased in carbon content by virtue of dilution with the 0.60% carbon base metal, the weld also will be likely to display high hardness, cracking susceptibility, and a tendency toward brittle failure. The procedure for welding a steel containing 0.60% carbon might well include (1) preheating to about 500°F, (2) welding with a low-hydrogen process, (3) avoiding carbon pickup in the weld deposit by minimizing dilution with melted base metal, (4) maintaining a 500°F minimum interpass temperature, and (5) transferring the weldment without dropping below the interpass temperature to a postweld heat treating furnace for annealing or an equally suitable heat treatment. Whether all of these precautions are mandatory in the procedure will depend on the actual article to be welded, its size, joint design, joining process, kind of filler metal, etc. If the steel being welded contained less than 0.60% carbon, less need would exist for the precautionary measures outlined. It is quite difficult to predict the minimum precautionary measures required in the welding procedure for a medium carbon steel. Attempts have been made by a large number of investigators to provide empirical formulas for calculating the required preheat temperature, etc. from the known base metal composition, thickness, and process to be applied. In lieu of any guidance from a weldability prediction system, or experience with a comparable weldment, every available precaution would have to be taken to insure against an unsound or unreliable weldment. Despite the unfavorable response of these steels to the thermal cycle of fusion joining, many machinery parts of medium-carbon steel are assembled without difficulty by welding through the use of a properly planned procedure.

Steels containing 0.30 to 0.60% carbon are extensively used in machinery. Tractors, earth-moving vehicles, mining equipment, power shovels, derricks, and pumps are examples of the many kinds of machinery that have components made of medium-carbon steels. This kind of steel often is selected for wear resistance rather than its greater strength. The parts frequently are used in the heat-treated condition to ensure strength or hardness in the required range. Assembly by welding may be performed before or following final heat treatment, depending on the nature of the weldment. Medium-carbon steels usually are produced as fully killed steels using hot-topped ingots to avoid pipe in the product. Defects of this nature are highly undesirable in the higher carbon steels, because they often produce splitting during hot rolling or forging, and cracking during welding and during heat treatment. These steels often are produced with a controlled austenitic grain size. Fine-grain steels usually help gain better notched bar impact strengths. Coarse-grain steels generally display greater hardenability and sometimes are preferred for heavier sections to be heat treated. While carbon steels have a relatively low hardenability as compared to alloy steels, this feature often is evaluated and considered very carefully before proceeding with the production of heat treated parts. The actual hardenability of a particular steel may determine whether production parts are to be quenched in water or in oil for hardening, and this in turn may call for some adjustment of welding procedure if the weld metal also must meet specified mechanical property limits. Carbon steels cannot be purchased to standardized hardenability limits (H-bands), as can the alloy steels.

1.7.4.6 High-carbon steels (above 0.60% carbon). Steel containing carbon in the range of about 0.60 to 1.00% usually is pictured in springs, cutting tools, gripper jaws, mill rolls, crane and railroad car wheels, and other articles that seldom call for assembly by welding. More often, welding is applied as a maintenance or repair operation. This alone would justify attention being given to the metallurgy of welding high-carbon steels. However, a much greater amount of welding is being performed on high-carbon steels than might be imagined, and this arises because of an interesting case of economical salvage.

Welding engineers differ on the required procedures for joining high-carbon steel. One procedure obtained by extrapolation from the medium-carbon steels would entail, of course, preheat, low-hydrogen conditions during fusion, maintaining of high interpass temperature, and postweld heat treatment. Is is thought that similar high-carbon steels can be successfully welded for many applications without preheat and postweld heat treatment. For the most part, high heat input is advocated, along with good protection of the molten metal, and use of a low-hydrogen type welding electrode. This practice may produce joints that are free of underbead cracking, because avoiding hydrogen pickup in the base metal heat-affected zone eliminates the strongest promoter of this defect. The final microstructure of the heat-affected zone still is a matter deserving of careful consideration. Many weldments can be devised to make maximum use of (a) retarded cooling rates from high heat input, (b) multilayer welds to secure the tempering effect from each pass, and (c) tempering beads deposited atop the weld reinforcement for the restricted heat effect. Yet, our knowledge of the limited toughness in a weld affected zone of 0.80% carbon steel suggests that as-welded joints in steel of this kind be employed with the greatest of caution. A safer approach is to use a postweld heat treatment to reduce the hardness of the heat-affected areas and increase their toughness and ductility.

1.7.4.7   Cast iron (above 1.7% carbon).   *Cast iron* generally covers iron in cast form that contains a very high carbon content—perhaps 1.7 to 4.5%. This carbon may be varied in the mode of distribution in the microstructure, and this gives rise to a number of different forms of cast iron that differ to a surprising extent in mechanical properties—and in weldability.

Grey cast iron.   This is the most widely used form, so named because of the dull grey color on fractured surfaces. By adding approximately 1 to 3% silicon, the cast alloy, on slowly cooling, will precipitate its carbon as flakes of free graphite in the microstructure. It has the unique properties of a ferrite matrix with numerous soft flake-like inclusions (of graphite) dispersed throughout. Grey cast iron has a tensile strength of 25 to 50 ksi and displays no yield strength. It has a very high compressive yield strength, very good damping capacity, and excellent machinability. The toughness and ductility of gray cast iron can vary considerably, depending on the exact size and shape of the graphite flakes and whether any combined carbon remained in the alloy to form some pearlitic microstructure during cooling. Gray cast iron has, at best, modest toughness.

White cast iron.   White cast iron is not widely used, because of extreme brittleness. By control of chemical composition, the structure of white cast iron is kept free of graphitic carbon. A fractured surface will appear white, as contrasted with the grey-colored fracture of grey cast iron. The microstructure of white cast iron consists of primary carbides in a fine dendritic formation. The matrix may be either martensite or a fine pearlite, depending on the composition and the cooling rate. While the hardness of martensitic white cast irons may be as high as 600 BHN, the material may exhibit only 20 ksi in a tensile test because of its very low ductility. Through use of the chill plate in the mold, only a skin of white cast iron is produced on a casting to gain this high hardness for abrasion resistance.

Malleable iron.   Malleable iron is made in two types: (1) ferritic malleable iron and (2) pearlitic malleable iron. Both types are made from essentially the same iron-carbon alloy composition, but different heat treatments are employed to obtain the particular microstructures that distinguish the two types. Ferritic malleable iron consists of a matrix of ferrite grains in which all of the carbon is dispersed as tiny patches of temper carbon (graphite). Pearlitic malleable iron contains patches of temper carbon, but some of the carbon is dispersed in the matrix as cementite. Depending on the heat treatment, this combined carbon may appear in pearlite, tempered martensite, or spherodized carbide. Malleable iron, especially the ferritic hype, exhibits a higher tensile strength and better ductility than gray cast iron simply because of the mechanical effect of patches of graphite as compared with flakes of graphite. Malleable iron castings are used, therefore, in a wider variety of articles. Good machinability still is one of the chief advantages of the material.

Nodular cast iron. Nodular iron is cast iron in which free carbon or graphite is dispersed as tiny balls or spherulites instead of flakes as found in grey iron, or patches as found in malleable iron. The composition of nodular iron is similar to that of grey iron except for a small addition of a nodularizing agent, which may be cerium, calcium, lithium, magnesium, sodium, or a number of other elements. Magnesium is commonly used for this purpose. The nodularizing treatment is so effective in causing the carbon contained in the molten iron alloy to form spheroids of graphite that some castings are used in the as-cast condition. More often, the castings are heat treated much in the same manner as malleable iron castings to produce a matrix that is ferritic, pearlitic, or tempered martensite.

Weldability of cast iron. All cast irons, whether grey, white, malleable or nodular, suffer from essentially the same handicap in fusion joining: too much carbon. While the manufacturing process (i.e., casting and possibly heat treating) may be capable of producing a microstructure that possesses useful mechanical properties, the thermal cycle of fusion joining ordinarily does not produce a desirable microstructural condition. The temperature immediately adjacent to the weld becomes too high, and the cooling rate of the entire heat-affected zone is too rapid. Massive carbides tend to form in the zone immediately adjacent to the weld, while the remainder of the heat-affected zone tends to form a high-carbon martensite. Both of these microstructural conditions are very brittle and are subject to cracking, either spontaneously or from service applied loads. The degree of brittleness and propensity to cracking will depend to some extent on the kind of cast iron, its condition of heat treatment, and the welding procedure.

Fusion joining, because of its localized nature, produces stress in the weld area. The base metal must be capable of some plastic deformation on a localized scale to accommodate these stresses, or else cracking will result. Nodular iron and malleable iron treated to a ferritic matrix are better suited to absorb the stresses from welding than are grey or white cast iron. Arc welding exaggerates weld stress and is more likely to cause cracking than gas welding.

The composition and structure of the cast iron play a part in determining the brittleness and cracking susceptibility of a weld joint by affecting the amount of carbon that goes into solution during the austenization of the heat-affected zone. To minimize the formation of massive carbides and high-carbon martensite, it is most helpful to have all carbon present as free carbon (graphite) and in the form of not-too-small spheroids. The smaller the surface area of graphite in contact with the hot austenitic matrix, the less carbon that will be dissolved to appear later as combined carbon in the structure at room temperature. Flakes of graphite, as are present in grey iron, display the greatest tendency to enter solution because of their greater surface area. The graphite is rather slow to dissolve, and free graphite often remains in the weld melt. The process of fusion joining is a reversal of the solidification process. Those areas last to solidify are the first areas to melt. The composition of a typical cast iron might be 3.5% carbon, 0.5% manganese, 0.04% phosphorus, 0.06% sulfur, and 2.5% silicon. The addition of 0.07% magnesium to this composition would promote the formation of nodular graphite. Increasing the manganese content would act to decrease graphitization. Higher phosphorus encourages embrittlement. Higher sulfur also acts to decrease graphitization. Silicon, it will be remembered, promotes graphitization of the carbon.

Avoidance of hydrogen pickup during any arc-welding on cast iron reduces the likelihood of cracking on cooling. This factor is of lesser importance than in the welding of hardenable steels, and it must not be assumed that the use of a low-hydrogen flux covering on a mild steel arc-welding electrode spells success in welding cast iron.

The mechanical properties of weld metal employed on cast iron can play a major part in the success of the operation. If the yield strength is held quite low, the weld metal imposes stresses of lower intensity during cooling, which reduces the likelihood of cracking. During service, the weld metal deforms easily to minimize stress concentrations on the brittle base metal. This sacrificial action by weld metal can be seen to a degree when using austenitic stainless steel weld deposits. Weld metals of nickel, or an alloy of approximately 50% nickel and 50% iron, are so effective in providing this kind of relief that considerable use is made of nickel and nickel-iron alloy filler metals in arc-welding cast iron. Ordinary low-carbon steel electrodes are not satisfactory for welding cast iron, because the carbon picked up by the weld deposit quickly increases the yield. Another advantage of the austenitic-like weld deposits of stainless steel or nickel alloy is the ease with which they can be machined in the as-welded condition.

Preheating cast iron to modest temperatures (up to 600°F) does not ensure success in an arc-welding operation with mild steel filler metal, as often is the case with hardenable steel. Positive benefit from preheating is secured in gas welding cast iron with a cast iron filler rod. In this case, the entire joint area of the cast iron article is preheated to almost a red heat (900°F or higher) and is slowly cooled after fusion joining has been completed. This procedure produces a weld with a microstructure of graphitic carbon in a matrix of ferrite and pearlite. A preheat of 300 to 400°F often is applied when arc welding cast iron with nickel or nickel-iron alloy electrodes (although a temperature in the range of about 400 to 600°F is to be strongly recommended).

Postweld heat treatment of cast iron weldments can be performed to relieve residual stresses and to improve the microstructure in the area of the weld joint. One practice is to heat slowly to about 1150°F immediately upon completion of welding, and to slowly cool after soaking at temperature for about one hour. A more thorough postweld anneal, often called a *graphitizing-ferritizing* treatment, requires heating to soak at 1650°F for four hours, furnace cooling at 60°F per hour to 1000°F or lower, and cooling in air to room temperature.

A novel procedure, recommended for welding nodular iron, that does not require a postweld heat treatment to obtain optimum weld joint ductility is based upon a surfacing or *buttering* technique. The procedure requires advance knowledge of the surfaces of the casting to be joined. A thick layer (about 5/16 in. thick) of weld metal is deposited on these surfaces prior to assembly into a weldment and at a time when the cast components can be conveniently annealed immediately after the surfacing or buttering operation. The weld metal employed for surfacing does not necessarily have to be the same as subsequently used for joining the cast pieces together; however, it must be a weld metal that is suitable to serve as part of the base metal. This surfacing-annealing-welding procedure has been successfully demonstrated with shielded metal arc welding (employing a preheat of 600°F) and covered electrodes of ENiFe, E307-15, and E6016. These electrodes represent nickel base alloy, austenitic Cr-Ni stainless steel, and a mild steel (low-hydrogen covering), respectively. The object of the surfacing weld is to arrange for the heat-affected zone of the final assembly weld to fall within the surfacing weld, rather than the cast iron base metal.

1.7.5 Estimating the Weldability of Carbon Steels

Our discussion of carbon steels has been carried from "steel" containing less than 0.005% carbon to cast iron, which may contain as much as 5% of this alloying element. Although we probed unusual aspects of steel (e.g., degree of deoxidation) in assessing weldability, the property that obviously exerted the greatest influence was the propensity to harden when heated to a high temperature and quickly cooled. The manner in which the hardness of the heat-affected structure was controlled by the carbon content, and its ability to harden on cooling was controlled by the carbon, manganese, and silicon contents, was explained by describing the formation of martensite and its properties. The carbon range over which the greatest change occurred in the weldability of steel appeared to be about 0.30 to 0.50%. Below this range, there appeared to be little cause for concern that the hardenability of the steel might produce underbead cracking or brittle heat-affected zones. Above this range, there was little doubt that precautions had to be taken in planning the welding procedure to avoid underbead cracking or brittle heat-affected zones. Within the 0.30 to 0.50% range, steels responded according to the amounts of carbon, manganese, and silicon present. Because of the demand for strength, welding engineers are continually seeking ways of welding steel in the 0.30 to 0.50% carbon range without risk of cracking, without serious impairment of toughness or ductility, and without costly or inconvenient innovations in the procedure. It does not appear possible to develop a simple system for precisely predicting the entire behavior of a particular steel during a welding operation, or the performance of welded joints in the steel in service. The features embodied in an actual weldment and the conditions of service are much too diverse to be represented in a reasonable number of practicable weldability test specimens. Progress has been made, however, on simple evaluations of a number of the major individual features involved in a welding procedure that affect weldability. The welding engineer, in developing a satisfactory procedure, can use these pieces of information as guideposts.

1.7.6   Filler Metals for Joining Iron and Steel

The base metal and filler metal are the two components that determine the composition of the weld metal. Together they are important factors in establishing the final properties of the solidified weld. The base metal commonly is a fixed component, because it is presented to the welding engineer as "the material to be joined." The filler metal, however, plays a more complex role. Filler metals offer the welding engineer an area of choice that can be effectively utilized to control the final chemical composition and the mechanical properties of the weld.

Many of the welding processes involve the deposition of filler metal. Some arc welding processes employ a consumable electrode that is deposited as filler metal, while other processes may use a supplementary rod or wire that is melted into the joint by a heat source, such as an arc supported by a nonconsumable electrode or a gas flame. Brazing and soldering make use of filler metals, even though only a thin film of filler metal is left between the workpieces. Filler metals are often employed in the form of cast rods, flat strip, thin foil, square bars, powdered metal, and even precipitated metal from aqueous solutions or gaseous compounds in addition to the traditional wire form.

Filler metals are a special category of materials. They have a higher cost relative to equivalent base metal cost. Design engineers should be aware of special standards establishing their various classifications. Filler metals are not generally the same materials as the base metals they are designed to join. It must be recognized that it is the weld metal that, in the end, bonds the workpieces together.

1.7.6.1 Important facts about weld metal. The differences between the base metal and the filler metal are quite marked when the weld metal is in the as-deposited condition. Where the weld metal has been reheated, such as the first bead of a two-pass weld, the differences can still be seen. Postweld heat treatment, such as normalizing, usually does not completely eradicate the microstructural differences. The unique features found in the weld metal microstructure arise from the unusual conditions under which solidification has taken place. Weld metal will display a microstructure and properties that are not exactly like those of wrought metal, or even a casting, of the same chemical composition. Sometimes certain properties of the weld metal may be regarded as inferior; sometimes they may be considered superior. A given base metal type may not represent the optimal chemical composition for weld metal. For virtually all metals and alloys used in wrought or cast form, modification in chemical composition will improve their properties in weld metal form. This is the principal reason why welding rods and electrodes have evolved as a separate class of materials. A second reason is the influence that filler metal composition exerts on the mechanics of deposition. The effects observed in this area of filler metal technology will be highly dependent, of course, on the particular welding process employed. Deposition characteristics will be touched on later as the various kinds of filler metal are reviewed.

Simply melting the tightly abutting edges of base metal workpieces together can form weld metal, in which case the joint is called an *autogenous* weld, meaning that the weld metal was produced entirely from the base metal. For the majority of weld joints, however, some filler metal is added during the formation of the weld metal. For a complete appraisal of the weld metal origin, we must look to three possible contributing sources: (1) the base metal, (2) filler metal, which may be a welding rod or a consumable electrode, and (3) metal carried in a flux or slag. In much of the fusion joining, the major percentage of the weld metal is derived from filler metal in the form of a consumable electrode or a supplementary rod. Not as much use is made of slag or flux as the primary source or carrier of metal for the weld deposit. The base metal that is melted and thus mixes or alloys with any deposited filler metal is a component to be considered for two reasons.

First, the filler metal ordinarily is of a composition that has been carefully designed to produce satisfactory weld metal. If this optimal composition is adulterated with an excess of the base metal composition, the properties of the weld metal may be less than satisfactory. The percentage of base metal that represents an excess in the weld metal naturally will depend on the steels involved and many factors concerning the weldment.

Second, if the alloy composition of the filler metal and the base metal are quite dissimilar, it remains to be seen whether the resultant weld metal alloy composition will be satisfactory for the application. As the requirements for weld joints in alloy steels become more stringent, circumstances arise in which the welding engineer must do more than merely select a classification of filler metal reputed to be compatible with the type of steel base metal to be joined. It may be necessary to specify composition requirements for the weld metal, in situ. Consequently, the filler metal composition can be chosen only after the base metal composition and the percent base metal that will enter the weld metal are known. This admixture of base metal into the weld metal is called *dilution*. A simple technique for coordinating filler metal with dissimilar composition base metal at different levels of dilution to secure a particular weld metal composition will be illustrated in this chapter.

The homogeneity of weld metal deposits often has been questioned because of alloys being contributed by as many as three separate sources. Chemical analyses have been made of drillings from very small holes positioned on the cross-section of weld metal deposited by the shielded metal arc process in a joint. The results showed that electromagnetic stirring of the molten weld melt had accomplished remarkable uniformity of chemical composition from side to side and from top to bottom in each bead. More recent studies, however, utilizing metallographic examination and the electron microprobe analyzer, have shown that, under certain welding conditions, the final weld deposit can be heterogeneous in nature to some degree. The principal conditions that encourage heterogeneity are (1) very high weld travel speed, (2) very large additions of alloy in an adjuvant material, and a variable arc length, and (3) an arc that produces deep penetration in a central area and secondary melting. Of course, the degree of heterogeneity observed would also depend on the amount and kind of alloys involved, their sources, and many aspects of the welding conditions. Most weld deposits, however, can be regarded as being essentially homogeneous both over their cross-section and along their length, providing welding conditions have been held constant. Homogeneity on a microscopic scale in the weld metal structure is a basic matter that has been given scant attention. Only recently has the partitioning of elements in the dendritic structure of certain weld metals been analyzed with the electron microprobe analyzer. Information on microstructural heterogeneity may be useful in determining how the properties of weld metal can be improved.

1.7.6.2 Mechanical properties of weld metal. Some very helpful general remarks can be made about the mechanical properties of steel weld metals. The welding engineer has been aware for a long time that most weld metals as deposited display an unusually high yield strength as compared with the same composition steel in the cast or in the wrought conditions. For example, low-carbon steel weld metal regularly has a yield strength of at least 50 ksi, whereas a wrought steel of this same composition would possess a yield strength of only about 30 ksi. The tensile strength of the weld metal is somewhat higher than its wrought or cast counterparts. These facts regarding strength often are discussed in terms of yield strength/tensile strength ratio. In low-carbon steel, weld metal has a YS-UTS ratio of about 0.75. Cast and wrought steels of this same composition ordinarily have a ratio of about 0.50; that is, the yield strength is about one-half of the tensile strength. Because of this unusual inherent strength of weld metal, it is not necessary to employ as much carbon or other alloying elements in the filler metals for many of the steels as compared to that present in the base metal. The higher strength of weld metal is a peculiarity deserving of study. We should determine the reasons for this difference in strength and ascertain whether any circumstances arise in which weld metal does not exhibit this strength advantage.

Little difference exists in the strength of weld metal deposited by any of the fusion joining processes. Shielded metal arc, submerged arc, gas metal arc, gas tungsten arc, atomic-hydrogen arc, and the oxyacetylene gas welding processes have been compared, both in making single-bead deposits and in making multilayer welds. In comparing processes, those that accomplish welding with lowest heat input, and are characterized by more rapid heating and cooling rates, tend to produce a finer-grain, acicular microstructure. In the arc-welding processes, shielded metal arc and gas metal arc welding tend to produce the fine-grain, acicular structure, and the YS-UTS ratio of their weld deposits may range as high as 0.90. Processes that involve slower rates of heating and cooling, like atomic-hydrogen arc and oxyacetylene gas welding, produce weld metal with slightly larger grains. The strength and the YS-UTS ratio is correspondingly lower but usually not less than about 0.60. It should be noted that, as the rate of cooling increases with the different processes, a finer grain size is produced, and the yield strength is raised.

In the past, the remarkable strength of weld metal was attributed to its fine grain size. The cooling rate of the weld metal also affects the distribution of carbide particles that form in the microstructure. As expected, faster cooling results in finer carbides or pearlite lamellae, and this also increases strength. Some evidence has been obtained through careful examination of carbon replicas and electro-thinned specimens of weld metal that extremely small, elongated areas of retained austenite exist along the ferrite boundaries. This information, at first thought, may seem to be of little importance, but it helps explain the unusual resistance of the fine grains of weld metal to recrystallization. The retention of these small areas of austenite, although quite surprising in view of the low- carbon and low alloy content, is thought to be attributable to stabilization through plastic deformation during rapid cooling under restraint.

Reheating of weld metal by multipass deposition does little to change the grain size and alter the dislocation density. Multipass weld metal is virtually as strong (both UTS and YS) as single-bead weld metal. In metal arc deposited weld metal, the small degree of recrystallization that occurs from deposition of the multiple passes tends to produce a heterogeneous, duplex grain pattern of the original fine acicular grains and a small number of larger equiaxed grains. Weld metal from the atomic hydrogen arc and the oxyacetylene gas welding processes is more equiaxed in the as-deposited condition and undergoes even less change during multipass welding.

When weld metal is postweld heated, no significant change occurs in room-temperature strength on exposure to reheating temperatures as high as 1200°F and for times as long as 5 hr. At a temperature of about 700°F, the very small areas of retained austenite at the ferrite grain boundaries are believed to undergo transformation to ferrite. Extremely small carbides are precipitated in the newly formed ferrite. These areas then appear to serve very effectively to prevent recrystallization. The very fine ferrite grain size is preserved, along with its inherent high strength, until the metal is heated to the point where austenite begins to form (eutectoid temperature). At temperatures above 1200°F, the number of dislocations in the lattice begin to diminish, and this acts to lower the yield strength. Temperatures about 1300°F and higher are above the eutectoid point and cause some austenite to form. This results in the formation of equiaxed ferrite grains when transformation occurs on cooling. Therefore, temperatures above 1300°F reduce dislocation density and produce recrystallization. With microstructural changes of this kind, the weld metal strength (and the YS-UTS ratio) will decrease to that normally found in cast and wrought steel of the same composition. Annealing at 1750°F reduces the dislocation density to the low level found in annealed wrought steel. However, the grain structure of weld metal heated to this temperature, though equiaxed, still is finer than regular wrought steel and is reflected in somewhat higher strength in the weld metal. Heating to temperatures above approximately 1750°F is required to increase the grain size of the weld metal to equal that of wrought metal.

Weldments would be much less complicated if weld metal could be readily secured that possessed mechanical properties and physical characteristics matching those of the base metal. This seemingly simple objective is difficult to attain, as we now understand, because the base metal composition, when fused to form weld metals, invariably offers a uniquely different set of properties. Often, the properties of the weld would not be entirely satisfactory. Base metal composition may be quite unsuitable for undergoing the conditions of droplet transfer, exposure to oxidizing conditions, rapid freezing, and the many other unusual conditions to which a steel filler metal is subjected during deposition. Therefore, the welding engineer, in planning practically all weldments, must look for a new composition of steel that will serve as filler metal. There will be circumstances, of course, where a nonferrous filler metal will offer a better solution to the weld metal problem. Before this search for a filler metal can be started, the engineer must know what properties are deemed important in the weld metal, and the required levels of test values for these properties. If the specific levels needed are not known, we should at least give some thought as to how closely the properties of the weld metal must match those of the base metal.

Tensile strength is usually the first property that receives attention in considering the kind of weld metal needed. For the majority of weldments, the designer's goal is to just have the weld metal match the base metal in strength. It would appear to be desirable to have the weld metal in a butt joint equal in strength to the base metal. There are instances, however, where a somewhat lower strength can be tolerated in the weld metal. This is often true of fillet welds where a relatively large cross section of weld metal easily can be deposited to compensate for lower strength, and where the greater toughness and ductility that normally go with lower strength could be an attribute. Fillet welded joints often contain points of stress concentration, and greater demands sometimes are made of the weld metal to exhibit toughness and ductility. It is a rare case in which the weld metal is required to be substantially stronger than the base metal. Weld metal of significantly higher strength is likely to be cause for concern. If the weldment containing extra-strong weld metal was accidentally overloaded beyond its yield point, a weld joint being subjected to transverse plastic bending might cripple or buckle in the heat-affected zone adjacent to the weld metal. If the weld joint was being forced to elongate longitudinally along with the base metal, the extra-strong weld metal might have inadequate ductility to accompany the base metal through the deformation. For the majority of weldments, therefore, it is considered desirable to have the weld metal strength match that of the base metal. For this reason, many specifications for welded joints require the weld metal to achieve a level near the minimum. This discussion of weld strength also provides some explanation for the classification of most steel welding rods and electrodes on the basis of strength. Furthermore, these standardized steel filler metals are designed to deposit weld metals that possess adequate ductility and toughness for most services.

Toughness is the property that appears to be next to strength in terms of importance in weld metal. Of course, there may be exceptional weldments where toughness is of primary importance. Again, we find that weld metal toughness is controlled through chemical composition, and the alloying that promotes greater toughness in base metal is not necessarily the best condition for weld metal. A particularly difficult problem is to secure weld metal that is comparable in toughness to a quenched and tempered high-strength steel base metal with the condition of the welded joint being restricted to the as-welded, or the welded and stress-relief heat treated conditions. This problem of providing weld metal with comparable toughness to a specially heat treated base metal becomes a real challenge when the weldment is to be used at cryogenic temperatures.

Many other properties can be of special importance in the weld metal, depending on the nature of the weldment and its intended service. It may be imperative that the corrosion resistance of the weld metal in atmospheric exposure equal that of the base metal. This requirement may appear to call for the filler metal to have at least the same amount and kind of alloy content as is present in the base metal. This yields the fact that less alloy is required in weld metal to produce corrosion resistance equal to that of a low-alloy wrought steel. Often, high-strength, low-alloy wrought steels that have been selected for their corrosion resistance are welded with unalloyed mild steel filler metal. As will be explained shortly, enough alloy is picked up by the weld metal to increase its corrosion resistance to an adequate level. Of course, where unusual service conditions promote corrosion, such as elevated temperature oxidation or scaling, then the weld metal probably will be required to have an alloy composition somewhat similar to the base metal so as to exhibit comparable resistance. In this case, the element chromium probably would be employed, and the amount required in the weld metal and the base metal would depend on the nature of the environment to which the weldment will be exposed.

A weld metal may be required to exhibit good machinability, a property that often is secured in wrought steel by additions of sulfur during steelmaking. Weld metal machinability must be improved via another more complex alloying system because of the harmful effects that a high sulfur content would have on weld metal soundness. Weld metal in an article to be coated with vitreous or porcelain enamel is expected to undergo this operation as readily as the base metal, which, in many cases, is an enameling iron. The weld metal composition required will be highly dependent on both the type of iron or steel in the base metal and the exact nature of the enameling technique. Weld metal sometimes is required to be capable of extensive, uniform tensile elongation, so a welded article can be subjected to severe cold forming operations and not exhibit susceptibility to weld joint breakage. Although many additional examples of special requirements for weld metal can be cited, the aforementioned should serve to emphasize that, when specific properties are demanded in the weld metal of a weldment, the chemical composition of the weld metal must be designed to provide these properties.

With strength and toughness over a relatively narrow range of temperature commonly being the only requirements, the welding engineer usually can find the standardized welding rods or electrodes satisfactory for the great majority of applications. We now find more often that the performance demanded of weld joints calls for a more detailed study of the weld metal to be certain that this portion of the weld joint possesses all the properties needed to ensure satisfactory service performance. The modern engineering approach to providing weld metal that is best suited for a particular weldment is to formulate its composition on the basis of test data and experience with weld metal. While the amount of such information available is only a mere shadow of that accumulated for wrought and cast steels, the data being reported in the literature grow steadily in volume and in completeness as their importance is recognized.

When weld metal composition limits are firmly fixed, the welding engineer easily can determine in a quantitative manner how the base metal will affect the filler metal composition requirements. As mentioned earlier, fusion welding invariably involves some melting of the base metal, and this impending diluent requires recognition in anticipating the weld metal composition. Ordinarily, a welding engineer who selects the edge preparation, joint geometry, penetration and weld metal area will be able to predetermine with sketches of the joint cross-section, or with welded test coupons the percent dilution of the weld metal by the base metal. There can be some uncertainty about the exact amount of dilution that will occur. Various weld beads deposited in the joints may undergo different amounts of dilution. A root bead, for example, deposited with a technique designed for deep penetration, may encounter very heavy dilution—perhaps 80%. That is, the weld metal is made up of 80% base metal and only 20% filler metal. The final weld beads in the joint will not penetrate the base metal as extensively and will require a greater proportion of filler metal to fill the joint and complete the weld. The dilution in such beads may be only on the order of 20% (i.e., 20% base metal and 80% filler metal). Coordination will be needed among (a) base metal composition, (b) percent dilution, and (c) weld metal composition to project the desired filler metal composition.

1.7.7   Filler Metals for Joining

A wide variety of metals and alloys are used as filler metal in joining operations on carbon and alloy steels. They range from the nonferrous metals and alloys employed in soldering, brazing, and braze welding to the high-alloy steeled in welding processes. To cover all welding, brazing, and soldering, include the following:

1. Ingot iron or decarburized steel

2. Carbon steel

3. Low-alloy steel

4. High-alloy (stainless) steel

Filler metals are employed in the joining processes in a number of different forms. One of the earliest forms of filler metal was a shearing taken from the edge of thin base metal. Although shearings still are occasionally used for some operations, we now know that this practice is questionable, because base metal seldom represents the optimal filler metal composition. As welders called for more convenient forms of filler metal, material was supplied as thin cast bars and then as smooth, round wire. Filler metal is also produced in the form of tubular powder-filled rods and wire, thin flat strip, pellets, and powdered metal. There are certain soldering operations in which the metal for joining is chemically precipitated from an aqueous flux solution. New brazing operations are reported in which the required filler metal is produced from a mixture of gases in a controlled atmosphere.

There is good reason, however, for starting this review of filler metals with a discussion of the two most widely used forms; namely, electrodes and welding rods. Because careless use of the terms *rods* and *wires* in place of *electrodes* and *welding rods* often causes confusion in discussions of welding procedures, it should be worthwhile to explain the correct terminology in the AWS-ASTM classification system for filler metals.

1.7.7.1 Designation system. The designation system used for filler metals begins with the initial letters of the designations to indicate the basic process categories by which the filler metals are intended to be deposited. The letter E stands for electrode, R for welding rod, and B for brazing filler metal. Combinations of ER and RB indicate suitability for either of the process categories designated. Therefore, some filler metals cannot be identified as electrodes or welding rods until they have been put to use. This may account, in part, for the looseness with which these two terms are commonly used. Furthermore, the various shapes in which these filler metals are commonly supplied are so similar to the usual concepts of rods, wires, sticks, etc., that the ordinary use of such terms is natural and expressive. Be that as it may, some filler metals are immediately identifiable as electrodes or welding rods, and, as to those which are not, welding procedures are quite specific in regard to the process used, so the technical language can and should be quite exact.

1.7.7.2 Electrodes. An electrode, in general, is a terminal that serves to conduct current to or from an element in an electrical circuit. In welding, a filler metal electrode serves as the terminal of an arc, the heat of which progressively melts the electrode as it is advanced to maintain an approximately constant arc length. Whenever the term *electrode* is used alone in this chapter and elsewhere, it should be clear from the context whether its use as a filler metal is intended. A welding rod, on the other hand, carries no current. It is advanced at a suitable rate from an external position into the heat source, which may be an arc or a gas flame, and is melted approximately as it advances. Even though there should be no problem in understanding the operational difference between electrodes and welding rods, these filler metals are supplied in such great variety of forms, shapes, and sizes that some attempt at further description is warranted.

Welding rod ordinarily is a bare rod or wire that is employed as filler metal in any fusion joining process, and that does not act as an electrode if used in an arc welding process. One exception to the bare condition is the flux-covered bronze rod, which is sometimes used for braze welding. Welding rod may be either solid or composite. Solid products are made as a cast rod or as drawn wire. The solid wrought wire is available in straightened and cut lengths or in coils. The cast rod is marketed in straight lengths. Solid welding rod may be used with any of the many fusion joining processes. For this reason, the chemical composition is governed by analysis requirements on the actual rod. Composite welding rod is manufactured in several different kinds of construction. The rod may be a tube filled with any desired combination of flux and powdered metal, or it may be a folded length of strip in which the folds have been filled with powdered ingredients and then closed at the surface by crimping. Composite rods are manufactured to secure an overall composition that sometimes is difficult to produce as wrought solid wire. Occasionally, a number of fine, solid wires of different metals and alloys are braided together so that their overall composition fills the need for a particular alloy. Composite welding rod usually is subject to analysis requirements based on the composition of undiluted weld metal deposited by a prescribed process and procedure. This practice is followed, because the recovery of alloying elements in the weld metal deposit will depend to some extent on their form in the composite rod.

Filler metal electrodes, often called *consumable electrodes,* may be in the form of straight or coiled wire, either solid or composite. The solid electrode may be bare, lightly coated with a flux or an emissive material, or heavily covered with fluxing and slag-forming ingredients. If the solid electrode bears a coating or a covering, the heart of the electrode is the core wire. However, the flux may not necessarily be present as a surface covering. Sometimes the flux is included as a core material in a tubular wire or enfolded in a crimped or wrapped electrode. A braided electrode of fine wires may be impregnated with a flux. A solid core wire in coils may have a fine wire spirally wrapped on the surface and a flux covering applied after wrapping. The flux covering is then lightly brushed or sandblasted to expose a portion of the surface of the spirally wrapped wire. This wire permits electrical contact through the flux covering from the contact jaws (of a continuous type welding head) to the core wire. Another covered electrode using solid, coiled wire makes use of a wire mesh sleeve that is imbedded in the flux (but in contact with the core wire and exposed at the surface) to pass current for welding.

Composite electrodes are those in which two or more metal components are combined mechanically. As another example of this type, a tube filled with powdered metal, may be used instead of solid wire. These tubular electrodes permit the formulation of alloys that are difficult to produce or to utilize in the form of coils of solid drawn wire. Chemical analysis determinations concerning composite electrodes are made on undiluted weld metal deposited by the electrode using the process for which the product was designed.

Knowledge of the construction and formulation of an electrode can be of considerable help in avoiding difficulties. This is particularly true in the case of composite electrodes where the components have been proportioned by the manufacturer to provide the required alloy composition in the weld deposit. In using tubular powder-filled wire, care must be taken to avoid loss of the metal powder from the core. This may occur if the electrode is bent awkwardly or crushed and the seam is opened sufficiently for the powder to sift out. If the powder is not bonded in the core, a portion may run out when the tubular electrode end is cut off. Loss of metal powder in any manner results in a weld deposit that is deficient in the alloying elements contained in the powder. This loss may not be detectable by the appearance of the deposit, but it is likely to become apparent later. This portion of the welded joint probably will show a deficiency in mechanical properties, corrosion resistance, or whatever properties were to be gained from the missing alloy content.

Covered electrodes that contain large amounts of powdered metal in the covering also must be given similar consideration. Even the method of preparing the striking end of any covered electrode can be very important. It will be recalled that a covered electrode in supporting the welding arc melts with a conical sheath. If the covering is chamfered excessively at the striking end, a smaller-than-normal amount of covering is melted with the initially deposited metal. If an electrode is used part way, and the welder in restriking the arc dislodges a large fragment of covering from the end, then the deposit will be deficient in alloy at the start of the bead. Whether the smaller amount of covering and its contained alloy will be significant depends, of course, on the amount of alloy normally secured via the covering, the nature of the alloying elements, and their role in the deposit. An alloy deficiency, even in a small portion of a weld bead, can be metallurgically significant.

Finally, those electrodes that contain greater amounts of easily oxidized alloying elements require more care during deposition. Electrodes that depend upon elements like chromium, molybdenum, vanadium, or columbium to secure particular weld metal properties should be deposited with a short arc length and with as little weaving as possible. This technique is intended to minimize exposure of the metal droplets being transferred and the weld melt surface to any oxygen and nitrogen from the atmosphere that may have infiltrated the arc.

1.7.8 Specifications for Welding Rods and Electrodes

Specifications for welding rods and electrodes have been issued by a dozen or more domestic organizations, including the following:

American Welding Society

American Society for Testing and Materials

American Society of Mechanical Engineers

American Bureau of Shipping

Society of Automotive Engineers

U. S. Army Ordnance

U. S. Navy, Bureau of Ships

U. S. Federal Specifications

U. S. Navy, Bureau of Aeronautics

U. S. Coast Guard

These organizations have done much over the years to standardize the composition and construction of welding rods and electrodes, and to secure consistency in performance from lot to lot. Presently, most welding rod and electrodes for fabricating military equipment are procured against the military specifications of the "MIL-" series. Commercial users naturally turn to the specifications developed jointly by the AWS and the ASTM. Much attention has been given to having common requirements in the specifications of the military and the AWS-ASTM, but a complete merging or interchange has yet to be achieved. Specifications issued by AWS-ASTM are widely recognized and serve as good examples for discussion. Their specifications are quite complete, and each includes an appendix of helpful information. Rather than reproduce these specifications here, even in abbreviated form, the reader is urged to study complete copies. The development and standardization of filler metals is a never-ending activity, and new and novel welding rods and electrodes periodically appear on the market. Those that have yet to be included in specifications, but that have achieved significant use, are discussed after the standard classes in each kind of alloy.

Most of the AWS-ASTM specifications, it may be recalled, deal with a single kind of alloy and either the bare rod or covered electrodes. Because covered electrodes have been used in much greater quantities than bare electrodes in recent years, more attention was given to the preparation of specifications for the former. Specifications have been issued by the AWS-ASTM for the bare solid and composite electrodes employed in gas metal arc and submerged arc welding of carbon and low-alloy steels.

Virtually all covered electrodes of the carbon and low-alloy steels are made from a single kind of steel core wire; namely, a low-carbon, rimmed steel. Therefore, the flux covering on an electrode is a most important factor in determining operating characteristics, and the classification of covered electrodes is determined to a large extent by the nature of the covering. The addition of deoxidizing agents and alloying elements to the deposit is accomplished by incorporating suitable powdered materials in the flux covering. As will be shown, a remarkably wide array of filler metal compositions are produced with electrodes that employ this covering technique. While covering formulas with respect to alloy content are held as proprietary information by the electrode manufacturer, the coverings on carbon and low-alloy steel electrodes are identified by a unique numbering system that employs four or five digits following the E prefix. The first two (and sometimes three) digits indicate the approximate minimum tensile strength expected of the weld metal in a certain condition; that is, plain steel weld metal is tested as deposited, while the majority of the low-alloy steel weld metals are tested in the stress-relieved condition. The next-to-last digit in the classification number indicates the position in which the electrode is capable of making satisfactory welds. Only three numbers are employed, and they indicate the following:

The last digit in the classification number indicates the kind of current to be used with the electrode and the kind of covering; however, the significance of a zero as the last digit will depend to some extent on the character of the electrode covering. Not all the coverings are available on the more highly alloyed steels. For example, the EXX10 covering, which contains a high cellulose content (and therefore is hydrogen bearing), is not employed when strength above approximately 100,000 psi UTS is required. High-strength filler metals ordinarily are employed to join hardenable steels that are susceptible to cracking from hydrogen in the heat-affected zones. Furthermore, the mechanical properties of the high-strength weld metal also would be adversely affected by hydrogen picked up in the deposit from the covering.

1.7.9 Iron and Carbon Steel Filler Metals

The number of iron and carbon steel welding rods and electrodes of different chemical analyses does not approach, of course, the great variety of alloy steel welding rods and electrodes. Nevertheless, in addition to the dozen or more flux coverings on carbon steel electrodes, several different steelmaking practices may be employed in making carbon steel welding rods, and the products differ sufficiently in welding properties to justify a distinct class identification. It is well to keep in mind that the large number of electrodes and welding rods developed by demand; that is, each is designed to best fill a particular set of needs, which may involve mechanical properties, operating characteristics, weld appearance, and so forth. Because the details of electrode construction can influence the composition and properties of the weld deposit, some time is taken here to discuss features like the kind of core wire, the nature of electrode coverings, and their influence on deposit properties.

Information on covered electrodes and their deposits is presented in specification AWS A5.5. Both carbon and low-alloy steel electrodes are included in these tables, although, for the moment, we will direct our attention only to the carbon steel classifications.

1.7.9.1 Carbon steel covered arc welding electrodes. AWS A5.1 is a specification titled *Mild Steel Covered Arc-Welding Electrodes*. The majority of covered electrodes used in the United States are manufactured to comply with this specification, even though only two modest levels of strength presently are provided. The electrodes are classified on the basis of (1) mechanical properties of deposited metal, (2) type of covering and its operating characteristics, and (3) kind of current with which the electrode is usable. The level of minimum tensile strength in the as-welded condition is the first distinguishing feature, namely, 62 ksi and 67 ksi (for the E60 series) and 72 ksi (for the E70 series). These levels of strength in the weld metal are achieved by regulation of the carbon and manganese contents. A single kind of core wire generally is employed in making all mild steel electrodes. This is a rimmed steel containing approximately 0.06 to 0.15% carbon, 0.30 to 0.60% manganese, residual amounts of phosphorus and sulfur, and, of course, very low silicon content—which is characteristic of a rimmed steel. The use of steel of this character plays an important part in the operating performance of the electrode, particularly in aiding the deposition of a weld metal in the overhead position. It is believed that the expansion gases contained in this steel at the rapidly melting tip of the electrode acts to propel minute droplets of metal away from the molten end. As droplets of metal enter the weld pool, the deoxidizing elements (previously contained in the electrode covering, but now transferred to the weld metal) quickly take up the oxygen and change the deposit to a killed steel. Because of this desirable operating behavior, rimmed steel core wire is used even in the majority of alloy-steel covered welding electrodes. Where the amount of alloy required in the weld deposit cannot be conveniently carried in the flux covering, the only alternative is to employ an alloy-steel core wire that contains all, or a major portion, of the needed alloy. Because alloy-steel wire usually is a killed steel, its use as the electrode core wire generally will detract from the all-position operating capability of the electrode.

If the testing requirements of the AWS-ASTM filler metal specifications are examined, it will be seen that the welding procedures are very much like those used in good shop practice, but many pertinent details are stipulated. The purpose of this close control of welding procedure is to ensure that a valid comparison can be made of results from repeated tests, or possibly from different testing facilities. In fact, every effort is made to employ similar procedures in the filler metal specifications for the different welding processes to permit direct comparison of property values obtained. In all cases, an interpass temperature is specified, which is intended to minimize the most potent variables that affect properties, namely, interpass temperature and bead size. Also, an artificial aging treatment consisting of heating to 200 to 220°F for 48 hr is applied to the welded test plates made with all electrodes, except the low-hydrogen classifications, to accelerate the effusion of hydrogen and secure the level of ductility characteristic of the weld metal under test.

In studying the weld deposit analyses for the various classes of carbon steel electrodes, note that small variations in composition seem to be related to the kinds of covering on each electrode. These composition variations, while not large, are sufficient to cause differences in mechanical properties, particularly when the composition changes are accompanied by different degrees of soundness (porosity) and by variations in hydrogen content. Although the tensile strength and ductility do not show marked changes, notch toughness is particularly sensitive to chemical composition and is discussed in some detail elsewhere in this chapter. Charpy V-notch impact test properties are a recently added requirement to the AWS-ASTM specification for certain of the mild steel arc welding electrodes. A minimum requirement of 20 ft-lb at −20°F is expected of weld metal deposited from the E6010, E6011, E6027, E7015, E7016, and E7018 class electrodes. A minimum requirement of 20 ft-lb at 0°F is expected of the E7028 electrodes. No impact requirements are set for any of the remaining electrode classes in the AWS A5.1 specification.

1.7.10   Other Filler Metals

1.7.10.1   E45 series of coated electrodes.   A thinly coated E45 series of electrodes were included as standard classes in the AWS-ASTM specification, but these were dropped long ago because of limited usage. They present an interesting aspect of the metallurgy of electrodes.

Because the thin coating on the E45XX electrodes allows a significant loss of carbon and manganese and does little to avoid porosity, the strength of weld metal from these electrodes may vary from 45 to 65 ksi UTS. The light coatings on these electrodes originated during the early days of arc welding when a surface film of powdered lime, sulcoat (controlled rusting), or other arc-stabilizing compounds was found to improve the operational characteristics of the electrode. However, these light coatings did little to improve the soundness and mechanical properties of the deposited weld metal, and so the more heavily coated or covered electrodes soon became the mainstay for the metal arc welding process.

Yet, E4510 and E4520 electrodes continue to be used to a limited extent on certain noncritical articles where electrode cost is a major consideration. E45 series electrodes are manufactured from rimmed steel wire. No deoxidizers are contained in the electrode coatings. Therefore, the deposited metal regularly contains considerable porosity caused by the oxygen in the steel and the oxygen and nitrogen picked up from the air, and any hydrogen that may have been held in some form in the light coating. The deposit is not required to meet any particular chemical requirements, but the deposited metal is expected to have sufficient strength ductility to display 45,000 psi min UTS and 5% min elongation in EP inches. The E4510 and E4520 electrodes usually are operated on direct current-straight polarity.

E60 series of covered electrodes. These *mild steel* electrodes are the most widely used for arc welding. Consequently, they are produced with the greatest number of electrode coverings having special operating characteristics. The following paragraph gives a brief insight into the metallurgical relationship between covering formulation and such aspects as operating behavior, weld composition, soundness, mechanical properties and deposit shape.

To achieve weld metal strengths required in the classifications of the E60 series, weld metal carbon content of about 0.06 to 0.09% is sought, along with manganese content of about 0.30 to 0.75%. To raise the weld metal strength sufficiently to qualify for the E70XX classifications, small increases in carbon (0.08 to 0.12%) and manganese (0.40 to 1.00%) are required.

1.8 Summary

In summary, ferrous metals through the years have been the most tested and well characterized materials that exist. With hundreds of alloys readily available, often with a variety of heat treatments, they will continue to be a primary structural metal of choice for the foreseeable future.

References

1. Boyer, H., and Devis, J. *Phase Diagrams: Interpretation and Application to Commercial Alloys.* Metals Engineering Institute, American Society of Metals, 1983.
2. Budinski, K. G. *Engineering Materials—Properties and Selection,* 5/e, Prentice Hall, 1996.
3. ASTM E527, Standard Practice for Numbering Metals and Alloys (UNS). American Society for Testing and Materials.
4. Herbick and Held. *The Making, Shaping, and Treating of Steel.* United States Steel Corp., 1971.
5. *Metals Handbook,* 10/e, vol. 1. ASM International, 1990.
6. Bloom, F. K., and Waxweiller, J. H. *Development of Stainless Steels.* Armco Research and Technology.
7. Linnert, G. E. *Welding Metallurgy, Carbon and Alloy Steels,* 4/e. GLM Publications.

2

Aluminum and Its Alloys

J. Randolph Kissell
TGB Partnership
Hillsborough, North Carolina

2.1 Introduction

This chapter describes aluminum and its alloys and their mechanical, physical, and corrosion resistance properties. Information is also provided on aluminum product forms and their fabrication, joining, and finishing. A glossary of terms used in this chapter is given in Section 2.10, and useful references on aluminum are listed at the end of the chapter.

2.1.1 History

When a six-pound aluminum cap was placed at the top of the Washington Monument upon its completion in 1884, aluminum was so rare that it was considered a precious metal and a novelty. In less than 100 years, however, aluminum became the most widely used metal after iron. This meteoric rise to prominence is a result of the qualities of the metal and its alloys as well as its economic advantages.

In nature, aluminum is found tightly combined with other elements, mainly oxygen and silicon, in reddish, clay-like deposits of bauxite near the Earth's surface. Of the 92 elements that occur naturally in the Earth's crust, aluminum is the third most abundant at 8%, surpassed only by oxygen (47%) and silicon (28%). Because it is so difficult to extract pure aluminum from its natural state, however, it wasn't until 1807 that it was identified by Sir Humphry Davy of England, who named it aluminum after alumine, the name the Romans gave the metal they believed was present in clay. Davy successfully produced small, relatively pure amounts of potassium but failed to isolate aluminum.

In 1825, Hans Oersted of Denmark finally produced a small lump of aluminum by heating potassium amalgam with aluminum chloride.

Napoleon III of France, intrigued with possible military applications of the metal, promoted research leading to Sainte-Claire Deville's improved production method in 1854, which used less costly sodium in place of potassium. Deville named the aluminum-rich deposits near Les Baux in southern France *bauxite* and changed Davy's spelling to "aluminium." Probably because of the leading role played by France in the metal's early development, Deville's spelling was adopted around the world, including Davy's home country; only in the U.S.A. and Canada is the metal called "aluminum" today.

These chemical reaction recovery processes remained too expensive for widespread practical application, however. In 1886, Charles Martin Hall of Oberlin, Ohio, and Paul L. T. Héroult in Paris, working independently, discovered virtually simultaneously the electrolytic process now used for the commercial production of aluminum. The Hall-Héroult process begins with aluminum oxide (Al_2O_3), a fine white material known as alumina, produced by chemically refining bauxite. The alumina is dissolved in a molten salt called cryolite in large, carbon-lined cells. A battery is set up by passing direct electrical current from the cell lining acting as the cathode and a carbon anode suspended at the center of the cell, separating the aluminum and oxygen. The molten aluminum produced is drawn off and cooled into large bars, called ingots. Hall went on to patent this process and to help found, in nearby Pittsburgh in 1888, what became the Aluminum Company of America, now called Alcoa. The success of this venture was aided by the discovery of Germany's Karl Joseph Bayer about this time of a practical process that bears his name for refining bauxite into alumina.

2.1.2 Attributes

Aluminum is a silvery metallic chemical element with the symbol Al, atomic number 13, atomic weight 26.98 based on ^{12}C, and valence +3. There are eight isotopes of aluminum, but by far the most common is aluminum-27, a stable isotope with 13 protons and 14 neutrons in its nucleus. Aluminum, in the solid state, has a face-centered crystal structure.

Although aluminum is the most abundant metal in the Earth's crust, it costs more than some less plentiful metals because of the cost to extract the metal from natural deposits. Its widespread use is due to the unique characteristics of aluminum and its alloys. The most significant of these properties are:

High strength-to-weight ratio. Aluminum is the lightest metal other than magnesium, with a density about one-third that of steel. The strength of aluminum alloys, however, rivals that of mild carbon steel and can approach 100 ksi (700 MPa). This combination of high strength and light weight makes aluminum especially well suited to transportation vehicles such as ships, rail cars, aircraft, trucks, and, increasingly, automobiles, as well as portable structures such as ladders, scaffolding, and gangways.

Ready fabrication. Aluminum is one of the easiest metals to form and fabricate, including operations such as extruding, bending, roll-forming, drawing, forging, casting, spinning, and machining. In fact, all methods used to form other metals can be used to form aluminum. Aluminum is the metal most suited to extruding. This process (by which solid metal is pushed through an opening outlining the shape of the resulting part, like squeezing toothpaste from the tube) is especially useful, since it can produce parts with complex cross sections in one operation. Examples include aluminum fenestration products such as window frames and door thresholds, and mullions and framing members used in curtainwalls, the outside envelope of many buildings.

Corrosion resistance. The aluminum cap placed at the top of the Washington Monument in 1884 is still there today. Aluminum reacts with oxygen very rapidly, but the formation of this tough oxide skin prevents further oxidation of the metal. This thin, hard, colorless oxide film tightly bonds to the aluminum surface and quickly reforms when damaged.

High electrical conductivity. Aluminum conducts twice as much electricity as an equal weight of copper, making it ideal for use in electrical transmission cables.

High thermal conductivity. Aluminum conducts heat three times as well as iron, benefitting both heating and cooling applications, including automobile radiators, refrigerator evaporator coils, heat exchangers, cooking utensils, and engine components.

High toughness at cryogenic temperatures. Aluminum is not prone to brittle fracture at low temperatures and has a higher strength and toughness at low temperatures, making it useful for cryogenic vessels.

Reflectivity. Aluminum is an excellent reflector of radiant energy; hence its use for heat and lamp reflectors and in insulation.

Non-toxicity.   Because aluminum is non-toxic, it is widely used in the packaging industry for food and beverages, as well as cooking utensils and piping and vessels used in food processing.

Recyclability.   Aluminum is readily recycled; about 30% of U.S. aluminum production is from recycled material. Aluminum made from recycled material requires only 5% of the energy needed to produce aluminum from bauxite.

Often, a combination of the properties of aluminum plays a role in its selection for a given application. An example is gutters and other rain-carrying goods, made of aluminum because it can easily be roll-formed with portable equipment on site, and it is so resistant to corrosion from exposure to the elements. Another is beverage cans, which benefit from aluminum's light weight for shipping purposes, and its recyclability.

2.1.3   Applications

In the U.S.A., about 21 billion pounds of aluminum worth $30 billion was produced in 1995, about 23% of the world's production. (To put this in perspective, about $62 billion of steel is shipped each year). Of this, about 25% is consumed in transportation applications, 25% in packaging, 15% in the building and construction market, and 13% in electrical products. Other markets include consumer durables such as appliances and furniture; machinery and equipment for use in petrochemical, textile, mining, and tool industries; reflectors; and powders and pastes used for paint, explosives, and other products.

The current markets for aluminum have developed over the relatively brief history of industrial production of the metal. Commercial production became practical with the invention of the Hall-Héroult process in 1886 and the birth of the electric power industry, a requisite because of the energy required by this smelting process. The first uses of aluminum were for cooking utensils in the 1890s, followed by electrical cable shortly thereafter. Shortly after 1900, methods to make aluminum stronger by alloying it with other elements (such as copper) and by heat treatment were discovered, opening new possibilities. Although the Wright brothers used aluminum in their airplane engines, it wasn't until the second world war that dramatic growth in aluminum use occurred, driven largely by the use of aluminum in aircraft. Following the war, building and construction applications of aluminum boomed due to growth in demand and the commercial advent of the extrusion process, an extremely versatile way to fabricate prismatic members. Then, between the late 1960s and the 1980s, the aluminum share of the U.S. beverage can market went from zero to nearly 100%. The most recent growth in aluminum use has been in automobiles and light trucks; over 220 pounds of aluminum were used, on average, in each car produced in North America in 1996. In the 1990s, aluminum use grew at a mean rate of about 3% annually in the U.S.A.

2.1.4 The Aluminum Association

The aluminum industry association in the United States is the Aluminum Association, founded in 1933 and composed of the primary American aluminum producers. The Aluminum Association is the main source of information, standards, and statistics concerning the U.S. aluminum industry. Contacts for the Association are:

Mail: 900 19th Street, N.W., Suite 300, Washington, DC, 20006

Phone: (202) 862-5100

Fax: (202) 862-5164

Internet: www.aluminum.org

The Aluminum Association is the secretariat for the American National Standards Institute (ANSI) for standards on aluminum alloy and temper designations and tolerances for aluminum mill products. Publications offered by the Association also provide information on applications of aluminum such as automotive body sheet and electrical conductors. Other parts of the world are served by similar organizations, including the European Aluminum Association in Brussels, the Aluminum Association of Canada in Montreal, and the Japan Aluminum Association in Tokyo.

2.2 Alloy and Temper Designation System

Metals enjoy relatively little use in their pure state. The addition of one or more elements to a metal results in an alloy, which often has significantly different properties from those of the unalloyed material. While the addition of alloying elements to aluminum sometimes degrades certain characteristics of the pure metal (such as corrosion resistance or electrical conductivity), this is acceptable for certain applications, because other properties (such as strength) can be so markedly enhanced. While the approximately 15 alloying elements used with aluminum are often called *hardeners*, they serve purposes in addition to increasing strength; even though alloying elements usually constitute less than 10% of the alloy by weight, they can dramatically affect many material properties.

Aluminum alloys are divided into two categories: *wrought alloys*, those that are worked to shape; and *cast alloys*, those that are poured in a molten state into a mold that determines their shape. The Aluminum Association maintains a widely recognized designation system for each category, described in ANSI H35.1, *Alloy and Temper Designations for Aluminum*, and discussed below. The Unified Numbering System (UNS), developed by the Society of Automotive Engineers and ASTM in conjunction with other technical societies, U.S. government agencies, and trade associations to identify metals and alloys, includes aluminum alloys. The UNS number for wrought aluminum alloys uses the same number as the Aluminum Association designation but precedes it with "A9" (for example, UNS A95052 for 5052). The UNS number for cast aluminum alloys also uses the same number as the Aluminum Association designation but precedes it with A and a number 0 or higher (for example, UNS A14440 for A444.0).

2.2.1 Wrought Alloys

The Aluminum Association's designation system for aluminum alloys was introduced in 1954. Under this system, a four-digit number is assigned to each alloy registered with the Association. The first number of the alloy designates the primary alloying element, which produces a group of alloys with similar properties. The last two digits are assigned sequentially by the Association. The second digit denotes a modification of an alloy. For example, 6463 is a modification of 6063 with slightly more restrictive limits on certain alloying elements such as iron, manganese, and chromium to obtain better finishing characteristics. The primary alloying elements and the properties of the resulting alloys are listed below and summarized in Table 2.1.

1xxx. This series is for *commercially pure aluminum*, defined in the industry as being at least 99% aluminum. Alloy numbers are assigned within the 1xxx series for variations in purity and which elements compose the impurities, the main ones being iron and silicon. The primary uses for alloys of this series are electrical conductors and chemical storage or processing, because the best properties of the alloys of this series are electrical conductivity and corrosion resistance. The last two digits of the alloy number denote the two digits to the right of the decimal point of the percentage of the material that is aluminum. For example, 1060 denotes an alloy that is 99.60% aluminum. The strength of pure aluminum is relatively low.

2xxx. The primary alloying element for this group is *copper*, which produces high strength but reduced corrosion resistance. These alloys

TABLE 2.1   Wrought Alloy Designation System and Characteristics

Series number	Primary alloying element	Relative corrosion resistance	Relative strength	Heat treatment
1xxx	none	excellent	fair	not heat treatable
2xxx	copper	fair	excellent	heat treatable
3xxx	manganese	good	fair	not heat treatable
4xxx	silicon	—	—	not heat treatable
5xxx	magnesium	good	good	not heat treatable
6xxx	magnesium and silicon	good	good	heat treatable
7xxx	zinc	fair	excellent	heat treatable

were among the first aluminum alloys developed and were originally called *duralumin.* Alloy 2024 is perhaps the best known and most widely used alloy in aircraft. The aluminum-copper alloys have fallen out of favor, though, in most applications that are to be welded or exposed to the weather for long periods of time.

3xxx.   *Manganese* is the main alloying element for the 3xxx series, increasing the strength of unalloyed aluminum by about 20%. The corrosion resistance and workability of alloys in this group, which primarily consists of alloys 3003, 3004, and 3105, are good. The 3xxx series alloys are well suited to architectural products such as rain-carrying goods and roofing and siding.

4xxx.   *Silicon* is added to alloys of the 4xxx series to reduce the melting point for welding and brazing applications. Silicon also provides good flow characteristics, which in the case of forgings provide more complete filling of complex die shapes. Alloy 4043 is commonly used for weld filler wire.

5xxx.   The 5xxx series is produced by adding *magnesium*, resulting in strong, corrosion-resistant, high-welded-strength alloys. Alloys of this group are used in ship hulls and other marine applications, weld wire, and welded storage vessels. The strength of alloys in this series is directly proportional to the magnesium content, which ranges up to about 6%.

6xxx. Alloys in this group contain *magnesium and silicon* in proportions that form magnesium silicide (Mg_2Si). These alloys have a good balance of corrosion resistance and strength. 6061 is one of the most popular of all aluminum alloys, and it has a yield strength comparable to mild carbon steel. The 6xxx series alloys are also very readily extruded, so they compose the majority of extrusions produced and are used extensively in building, construction, and other structural applications.

7xxx. The primary alloying element of this series is *zinc*. The 7xxx series includes two types of alloys—the aluminum-zinc-magnesium alloys (such as 7005) and the aluminum-zinc-magnesium-copper alloys (such as 7075 and 7178). The alloys of this group include the strongest aluminum alloy, 7178, which has a minimum tensile ultimate strength of 84 ksi (580 MPa), and are used in aircraft frames and structural components. The corrosion resistance of those 7xxx series alloys alloyed with copper is less, however, than the 1xxx, 3xxx, 5xxx, or 6xxx series. Some 7xxx alloys without copper (such as 7008 and 7072) are used as cladding to cathodically protect less corrosion-resistant alloys.

8xxx. The 8xxx series is reserved for alloying elements other than those used for series 2xxx through 7xxx. Iron and nickel are used to increase strength without significant loss in electrical conductivity and so are useful in conductor alloys like 8017. Aluminum-lithium alloy 8090, which has exceptionally high strength and stiffness, was developed for aerospace applications.

2.2.1.1   9xxx.   This series is not currently used.

Experimental alloys are designated in accordance with the above system, but with the prefix X until they are no longer experimental. Producers may also offer proprietary alloys to which they assign their own designation numbers.

The chemical composition limits in percent by weight for common wrought alloys are given in Table 2.2. Wrought aluminum alloys are sometimes identified by a color code using tags or paint on the product. Colors have been established for the alloys given in Table 2.3. Table 2.4 correlates current alloy designations with designations used prior to the current system.

National variations of these alloys may be registered by other countries under this system. Such variations are assigned a capital letter following the numerical designation (for example, 6005A, used in Europe and a variation on 6005). The chemical composition limits for national variations are similar to the Aluminum Association limits but vary slightly. Some standards-writing organizations of other countries have their own designation systems that are different from the Aluminum Association system. A comparison of some alloy designations is given in Table 2.5.

The 2xxx and 7xxx series are sometimes referred to as aircraft alloys, but they are also used in other applications, including bolts and screws used in buildings. The 1xxx, 3xxx, and 6xxx series alloys are sometimes referred to as "soft," while the 2xxx, 5xxx, and 7xxx series alloys are called "hard." This description refers to the ease of extruding the alloys—hard alloys are more difficult to extrude, requiring higher-capacity presses and are thus more expensive.

2.2.2 Cast Alloys

Casting alloys contain larger proportions of alloying elements than wrought alloys. This results in a heterogeneous structure, which is generally less ductile than the more homogeneous structure of the wrought alloys. Cast alloys also contain more silicon than wrought alloys to provide the fluidity necessary to make a casting.

While the Aluminum Association cast alloy designation system uses four digits like the wrought alloy system, most similarities end there. The cast alloy designation system has three digits, followed by a decimal point, followed by another digit. The first digit indicates the primary alloying element. The second two digits designate the alloy or, in the case of commercially pure casting alloys, the level of purity. The last digit indicates the product form—1 or 2 for ingot (depending on impurity levels) and 0 for castings. A modification of the original alloy is designated by a letter prefix (A, B, C, etc.) to the alloy number. The primary alloying elements are:

1xx.x. These are the *commercially pure aluminum* cast alloys; an example of their use is cast motor rotors.

2xx.x. The use of *copper* as the primary alloying element produces the strongest cast alloys. Alloys of this group are used for machine tools, aircraft, and engine parts. Alloy 203.0 has the highest strength at elevated temperatures and is suitable for service at 400°F (200°C).

TABLE 2.2 Chemical Composition Limits of Wrought Aluminum Alloys [1] [2]

AA DESIG-NATION	SILICON	IRON	COPPER	MAN-GANESE	MAG-NESIUM	CHROM-IUM	NICKEL	ZINC	TITAN-IUM	OTHERS[20] Each [20]	OTHERS[20] Total [3]	ALUMI-NUM Min. [4]
1050	0.25	0.40	0.05	0.05	0.05	…	…	0.05	0.03	0.03[9]	…	99.50
1060	0.25	0.35	0.05	0.03	0.03	…	…	0.05	0.03	0.03[9]	…	99.60
1100	0.95 Si + Fe		0.05–0.20	0.05	…	…	…	0.10	…	0.05[5][10]	0.15	99.00
1145[8]	0.55 Si + Fe		0.05	0.05	…	…	…	0.05	0.03	0.03[10]	…	99.45
1175[7]	0.15 Si + Fe		0.10	0.02	0.02	…	…	0.04	0.02	0.02[10]	…	99.75
1200	1.00 Si + Fe		0.05	0.05	…	…	…	0.10	0.05	0.05	0.15	99.00
1230[7]	0.70 Si + Fe		0.10	0.05	0.05	…	…	0.10	0.03	0.03[9]	…	99.30
1235	0.65 Si + Fe		0.05	0.05	0.05	…	…	0.05	0.06	0.03[9]	…	99.35
1345	0.30	0.40	0.10	0.05	0.05	…	…	0.05	0.03	0.03[9]	…	99.45
1350[6]	0.10	0.40	0.05	0.01	…	0.01	…	0.05	…	0.03[13]	0.10	99.50
2011	0.40	0.7	5.0–6.0					0.30		0.05[11]	0.15	Remainder
2014	0.50–1.2	0.7	3.9–5.0	0.40–1.2	0.20–0.8	0.10		0.25	0.15	0.05	0.15	Remainder
2017	0.20–0.8	0.7	3.5–4.5	0.40–1.0	0.40–0.8	0.10		0.25	0.15	0.05	0.15	Remainder
2018	0.9	1.0	3.5–4.5	0.20	0.45–0.9	0.10	1.7–2.3	0.25		0.05	0.15	Remainder
2024	0.50	0.50	3.8–4.9	0.30–0.9	1.2–1.8	0.10		0.25	0.15	0.05	0.15	Remainder
2025	0.50–1.2	1.0	3.9–5.0	0.40–1.2	0.05	0.10		0.25	0.15	0.05	0.15	Remainder
2036	0.50	0.50	2.2–3.0	0.10–0.40	0.30–0.6	0.10		0.25	0.15	0.05	0.15	Remainder
2117	0.8	0.7	2.2–3.0	0.20	0.20–0.50	0.10		0.25		0.05	0.15	Remainder
2124	0.20	0.30	3.8–4.9	0.30–0.9	1.2–1.8	0.10		0.25	0.15	0.05	0.15	Remainder
2218	0.9	1.0	3.5–4.5	0.20	1.2–1.8	0.10	1.7–2.3	0.25		0.05	0.15	Remainder
2219	0.20	0.30	5.8–6.8	0.20–0.40	0.02			0.10	0.02–0.10	0.05[12]	0.15	Remainder
2319	0.20	0.30	5.8–6.8	0.20–0.40	0.02			0.10	0.10–0.20	0.05[12]	0.15	Remainder
2618	0.10–0.25	0.9–1.3	1.9–2.7	…	1.3–1.8		0.9–1.2	0.10	0.04–0.10	0.05	0.15	Remainder
3003	0.6	0.7	0.05–0.20	1.0–1.5				0.10		0.05	0.15	Remainder
3004	0.30	0.7	0.25	1.0–1.5	0.8–1.3			0.25		0.05	0.15	Remainder
3005	0.6	0.7	0.30	1.0–1.5	0.20–0.6	0.10		0.25	0.10	0.05	0.15	Remainder
3105	0.6	0.7	0.30	0.30–0.8	0.20–0.8	0.20		0.40	0.10	0.05	0.15	Remainder
4032	11.0–13.5	1.0	0.50–1.3	…	0.8–1.3	0.10	0.50–1.3	0.25		0.05	0.15	Remainder
4043	4.5–6.0	0.8	0.30	0.05	0.05			0.10	0.20	0.05[12]	0.15	Remainder
4045[11]	9.0–11.0	0.8	0.30	0.05	0.05			0.10	0.20	0.05	0.15	Remainder
4047[11]	11.0–13.0	0.8	0.30	0.15	0.10			0.20		0.05[12]	0.15	Remainder
4145[11]	9.3–10.7	0.8	3.3–4.7	0.15	0.15	0.15		0.20		0.05[12]	0.15	Remainder
4343[11]	6.8–8.2	0.8	0.25	0.10	…			0.20		0.05	0.15	Remainder
4643	3.6–4.6	0.8	0.10	0.05	0.10–0.30			0.10	0.15	0.05[12]	0.15	Remainder

Alloy												Remainder
5005	0.30	0.7	0.20	0.20	0.50–1.1	0.10	…	0.25	…	0.05	0.15	Remainder
5050	0.40	0.7	0.20	0.10	1.1–1.8	0.10	…	0.25	…	0.05	0.15	Remainder
5052	0.25	0.40	0.10	0.10	2.2–2.8	0.15–0.35	…	0.10	…	0.05	0.15	Remainder
5056	0.30	0.40	0.10	0.05–0.20	4.5–5.6	0.05–0.20	…	0.25	0.15	0.05	0.15	Remainder
5083	0.40	0.40	0.10	0.40–1.0	4.0–4.9	0.05–0.25	…	0.25	0.15	0.05	0.15	Remainder
5086	0.40	0.50	0.10	0.20–0.7	3.5–4.5	0.05–0.25	…	0.20	0.20	0.05	0.15	Remainder
5154	0.25	0.40	0.10	0.10	3.1–3.9	0.15–0.35	…	0.25	0.15	0.05①	0.10	Remainder
5183	0.40	0.40	0.10	0.50–1.0	4.3–5.2	0.05–0.25	…	0.25	…	0.03⑨	0.15	Remainder
5252	0.08	0.10	0.05	0.10	2.2–2.8	…	…	0.05	0.05	0.05	0.15	Remainder
5254	0.45 Si + Fe		0.10	0.01	3.1–3.9	0.15–0.35	…	0.20	0.05	0.05①⑩	0.15	Remainder
5356	0.25	0.40	0.10	0.05–0.20	4.5–5.5	0.05–0.20	…	0.10	0.06–0.20	0.05	0.15	Remainder
5454	0.25	0.40	0.10	0.50–1.0	2.4–3.0	0.05–0.20	…	0.25	0.20	0.05	0.15	Remainder
5456	0.25	0.40	0.20	0.50–1.0	4.7–5.5	0.05–0.20	…	0.25	0.20	0.03⑨	0.10	Remainder
5457	0.08	0.10	0.20	0.15–0.45	0.8–1.2	…	…	0.05	…	0.05①⑩	0.15	Remainder
5554	0.25	0.40	0.10	0.50–1.0	2.4–3.0	0.05–0.20	…	0.25	0.05–0.20	0.05①⑩	0.15	Remainder
5556	0.25	0.40	0.10	0.50–1.0	4.7–5.5	0.05–0.20	…	0.25	0.05–0.20	0.05	0.15	Remainder
5652	0.40 Si + Fe		0.04	0.01	2.2–2.8	0.15–0.35	…	0.10	…	0.05①⑩	0.15	Remainder
5654	0.45 Si + Fe		0.05	0.01	3.1–3.9	0.15–0.35	…	0.20	0.05–0.20	0.05①⑩	0.15	Remainder
5657	0.08	0.10	0.10	0.03	0.6–1.0	…	…	0.05	0.05–0.15	0.02⑨	0.05	Remainder
6003⑦	0.35–1.0	0.6	0.10	0.8	0.8–1.5	0.35	…	0.20	0.10	0.05	0.15	Remainder
6005	0.6–0.9	0.35	0.10	0.10	0.40–0.6	0.10	…	0.10	0.10	0.05	0.15	Remainder
6053	⑬	0.35	0.10	…	1.1–1.4	0.15–0.35	…	0.10	…	0.05	0.15	Remainder
6061	0.40–0.8	0.7	0.15–0.40	0.15	0.8–1.2	0.04–0.35	…	0.25	0.15	0.05	0.15	Remainder
6063	0.20–0.6	0.35	0.10	0.10	0.45–0.9	0.10	…	0.10	0.10	0.05	0.15	Remainder
6066	0.9–1.8	0.50	0.7–1.2	0.6–1.1	0.8–1.4	0.40	…	0.25	0.20	0.05	0.15	Remainder
6070	1.0–1.7	0.50	0.15–0.40	0.40–1.0	0.50–1.2	0.10	…	0.25	0.15	0.03⑰	0.10	Remainder
6101⑫	0.30–0.7	0.50	0.10	0.03	0.35–0.8	0.03	…	0.10	…	0.05	0.15	Remainder
6105	0.6–1.0	0.35	0.10	0.15	0.45–0.8	0.10	…	0.10	0.10	0.05	0.15	Remainder
6151	0.6–1.2	1.0	0.35	0.20	0.45–0.8	0.15–0.35	…	0.25	0.15	0.05	0.15	Remainder
6162	0.40–0.8	0.50	0.20	0.10	0.7–1.1	0.10	…	0.25	0.10	0.03⑰	0.10	Remainder
6201	0.50–0.9	0.50	0.10	0.03	0.6–0.9	0.03	…	0.10	…	0.05⑤	0.15	Remainder
6253⑦	⑮	0.50	0.10	…	1.0–1.5	…	…	1.6–2.4	…	0.05	0.15	Remainder
6262	0.40–0.8	0.7	0.15–0.40	0.15	0.8–1.2	0.04–0.35	…	0.15	0.15	0.05	0.15	Remainder
6351	0.7–1.3	0.50	0.10	0.40–0.8	0.40–0.8	0.04–0.14	…	0.20	0.20	0.05	0.15	Remainder
6463	0.20–0.6	0.15	0.20	0.05	0.45–0.9	…	…	0.05	…	0.05	0.15	Remainder
6951	0.20–0.50	0.8	0.15–0.40	0.10	0.40–0.8	…	…	0.20	…	0.05	0.15	Remainder

TABLE 2.2 Chemical Composition Limits of Wrought Aluminum Alloys (Concluded) ① ②

AA DESIG-NATION	SILICON	IRON	COPPER	MAN-GANESE	MAG-NESIUM	CHROM-IUM	NICKEL	ZINC	TITAN-IUM	OTHERS② Each ⑳	OTHERS② Total ③	ALUMI-NUM Min. ④
7005	0.35	0.40	0.10	0.20–0.7	1.0–1.8	0.06–0.20	...	4.5–5.0	0.01–0.06	0.05⑪	0.15	Remainder
7008⑦	0.10	0.10	0.05	0.05	0.7–1.4	0.12–0.25	...	4.5–5.5	0.05	0.05	0.10	Remainder
7049	0.25	0.35	1.2–1.9	0.20	2.0–2.9	0.10–0.22	...	7.2–8.2	0.10	0.05	0.15	Remainder
7050	0.12	0.15	2.0–2.6	0.10	1.9–2.6	0.04	...	5.7–6.7	0.06	0.05⑫	0.15	Remainder
7072①	0.7 Si + Fe		0.10	0.10	0.10		...	0.8–1.3	...	0.05	0.15	Remainder
7075	0.40	0.50	1.2–2.0	0.30	2.1–2.9	0.18–0.28	...	5.1–6.1	0.20	0.05	0.15	Remainder
7175	0.15	0.20	1.2–2.0	0.10	2.1–2.9	0.18–0.28	...	5.1–6.1	0.10	0.05	0.15	Remainder
7178	0.40	0.50	1.6–2.4	0.30	2.4–3.1	0.18–0.28	...	6.3–7.3	0.20	0.05	0.15	Remainder
7475	0.10	0.12	1.2–1.9	0.06	1.9–2.6	0.18–0.25	...	5.2–6.2	0.06	0.05	0.15	Remainder
8017	0.10	0.55–0.8	0.10–0.20	...	0.01–0.05	...	...	0.05	...	0.03㉓	0.10	Remainder
8030	0.10	0.30–0.8	0.15–0.30	...	0.05	...	...	0.05	...	0.03㉔	0.10	Remainder
8176	0.03–015	0.40–1.0	...	...		...	...	0.10	...	0.05㉕	0.15	Remainder
8177	0.10	0.25–0.45	0.04	...	0.04–0.12	...	...	0.05	...	0.03㉖	0.10	Remainder

Note: Listed herein are designations and chemical composition limits for some wrought unalloyed aluminum and for wrought aluminum alloys registered with The Aluminum Association. This list does not include all alloys registered with The Aluminum Association. A complete list of registered designations is contained in the "Registration Record of International Alloy Designations and Chemical Composition Limits for Wrought Aluminum and Wrought Aluminum Alloys." These lists are maintained by the Technical Committee on Product Standards of the Aluminum Association.

① Composition in percent by weight maximum unless shown as a range or a minimum.

② Except for "Aluminum" and "Others," analysis normally is made for elements for which specific limits are shown. For purposes of determining conformance to these limits, an observed value or a calculated value obtained from analysis is rounded off to the nearest unit in the last right-hand place of figures used in expressing the specified limit, in accordance with ASTM Recommended Practice E 29.

③ The sum of those "Other" metallic elements 0.010 percent or more each, expressed to the second decimal before determining the sum.

④ The aluminum content for unalloyed aluminum not made by a refining process is the difference between 100.00 percent and the sum of all the other metallic elements present in amounts of 0.010 percent or more each, expressed to the second decimal before determining the sum.

⑤ Also contains 0.40–0.7 percent each of lead and bismuth.

⑥ Electric conductor. Formerly designated EC.

⑦ Cladding alloy. See Table 6.1.

⑧ Foil.

⑨ Vanadium 0.05 percent maximum.

⑩ Also contains 0.20–0.6 percent each of lead and bismuth.

⑪ Brazing alloy.

⑫ Bus conductor.

⑬ Vanadium plus titanium 0.02 percent maximum; boron 0.05 percent maximum; gallium 0.03 percent maximum.

⑭ Zirconium 0.08–0.20.

⑮ Silicon 45 to 65 percent of actual magnesium content.

⑯ Beryllium 0.0008 maximum for welding electrode and welding rod only.

⑰ Boron 0.06 percent maximum.

⑱ Vanadium 0.05–0.15; zirconium 0.10–0.25.

⑲ Gallium 0.03 percent maximum; vanadium 0.05 percent maximum.

⑳ In addition to those alloys referencing footnote ⑯, a 0.0008 weight percent maximum beryllium is applicable to any alloy to be used as welding electrode or welding rod.

㉑ Zirconium 0.08–0.15.

㉒ "Others" includes unlisted elements for which no specific limit is shown as well as unlisted metallic elements. The producer may analyze samples for trace elements not specified in the registration or specification. However, such analysis is not required and may not cover all metallic "Other" elements. Should any analysis by the producer or the purchaser establish that an "Others" elements exceeds the limit of "Total," the material shall be considered nonconforming.

㉓ Boron 0.04 percent maximum; lithium 0.003 percent maximum.

㉔ Boron 0.001–0.04.

㉕ Gallium 0.03 percent maximum.

㉖ Boron 0.04 percent maximum.

TABLE 2.3 Wrought Alloy Color Code[1]

Alloy	Color	Alloy	Color
1100	White	5154	Blue and green
1350	Unmarked	5183	Orange and brown
2011	Brown	5356	Blue and brown
2012	Yellow and white	5456	Gray and purple
2014	Gray		
2017	Yellow	5554	Red and brown
2018	White and green	5556	Black and gray
2024	Red	6013	Red and blue
2025	White and red	6053	Purple and black
2111	Black and green	6061	Blue
2117	Yellow and black	6063	Yellow and green
2214	White and gray	6066	Red and green
2218	White and purple	6070	Blue and gray
2219	Yellow and blue	6101	Red and black
2618	Brown and black	6151	White and blue
3003	Green	6262	Orange
		6351	Purple and orange
4032	White and orange	7005	Brown and purple
4043	White and brown	7049	Blue and purple
5052	Purple	7050	Yellow and orange
		7075	Black
		7076	White and black
5056	Yellow and brown	7149	Orange and black
Alclad 5056	Orange and gray	7150	Yellow and purple
5083	Red and gray	7175	Green and Brown
5086	Red and orange	7178	Orange and blue

[1]Wrought aluminum mill products are sometimes identified as to alloy by the use of a color code; for example, tags or paint on the end of rod and bar. Colors have been established for the alloys listed in the following table and chart. *Note: thee colors do not apply to ink used for identification marking.*

Color	Orange	Gray	Purple	Brown	Green	Glue	Yellow	Red	Black	White
White	4032	2214	2218	4043	2018	6151	2012*	2025	7076*	1100
Black	7149*	5556	6053	2618	2111*	–	2117	6101	7075	
Red	5086	5083	–	5554	6066	6013*	–	2024		
Yellow	7050	–	7150	5056	6063	2219	2017			
Blue	7178	6076	7049	5356	5154	6061				
Green	–	–	–	7175	3003					
Brown	5183	–	7005	2011						
Purple	6351	5456	5052							
Gray	Alclad 5056	2014								
Orange	6262									

TABLE 2.4   Wrought Alloy Designations, Old and New

New designation	Old designation	New designation	Old designation
1100	2S	5056	56S
1350	EC	5086	K186
2014	14S	5154	A54S
2017	17S	6051	51S
2024	24S	6053	53S
2025	25S	6061	61S
2027	27S	6063	63S
2117	A17S	6066	66S
3003	3S	6101	No. 2 EC
3004	4S	6951	J51S
4043	43S	7072	72S
5050	50S	7075	75S
5052	52S	7076	76S

3xx.x. *Silicon*, with *copper* and/or *magnesium*, is used in this series. These alloys have excellent fluidity and strength and are the most widely used aluminum cast alloys. Alloy 356.0 and its modifications are very popular and used in many different applications. High-silicon alloys have good wear resistance and are used for automotive engine blocks and pistons.

4xx.x. The use of *silicon* in this series provides excellent fluidity in cast alloys as it does for wrought alloys, and so these are well suited to intricate castings such as typewriter frames and they have good general corrosion resistance. Alloy A444.0 has modest strength but good ductility.

5xx.x. Cast alloys with *magnesium* have good corrosion resistance, especially in marine environments (for example, 514.0), good machinability, and can be attractively finished. They are more difficult to cast than the 200, 300, and 400 series, however.

6xx.x. This series is unused.

TABLE 2.5 Foreign Alloy Designations and Similar AA Alloys

Alloy designation	Designating country	Equivalent or similar AA alloy	Alloy designation	Designating country	Equivalent or similar AA alloy
Al99	Austria	1200	E-A1995[4]	Germany	1350
Al99,5	(Önorm)[1]	1050	3.0257[5]		"
E-Al		1350	AlCuBiPb[4]		2011
AlCuMg1		2017	3.1655[5]		"
AlCuMg2		2024	AlCuMg0.5[4]		2117
AlCuMg0,5		2117	3.1305[5]		"
AlMg5		5056	AlCuMg1[4]		2017
AlMgSi0,5		6063	3.1325[5]		"
E-AlMgSi		6101	AlCuMg2[4]		2024
AlZnMgCu1,5		7075	3.1355[5]		"
			AlCuSiMn[4]		2014
990C	Canada	1100	3.1255[5]		"
CB60	(CSA)[2]	2011	AlMg4.5Mn[4]		5083
CG30		2117	3.3547[5]		"
CG42		2024	AlMgSi0.5[4]		6063
CG42 Alclad		Alclad 2024	3.3206[5]		"
CM41		2017	AlSi5[4]		4043
CN42		2018	3.2245[5]		"
CS41N		2014	E-AlMgSi0.5[4]		6101
CS41N Alclad		Alclad 2014	3.3207[5]		"
CS41P		2025	AlZnMgCu1.5[4]		7075
GM31N		5454	3.4365[5]		"
GM41		5083			
GM50P		5356	1E	Great	1350
GM50R		5056	91E	Britain	6101
GR20		5052	H14	(BS)[6]	2017
GS10		6063	H19		6063
GS11N		6061	H20		6061
GS11P		6053	L.80, L.81		5052
MC10		3003	L.86		2117
S5		4043	L.87		2017
SG11P		6151	L.93, L.94		2014A
SG121		4032	L.95, L.96		7075
ZG62		7075	L.97, L.98		2024
ZG62 Alclad		Alclad 7075	2L.55, 2L.56		5052
			1L.58		5056
A5/L	France	1350	3L.44		5050
A45	(NF)[3]	1100	5L.37		2017
A-G1		5050	6L.25		2218
A-G0.6		5005	N8		5083
A-G4MC		5086	N21		4043
A-GS		6063			
A-GS/L		6101	150A	Great	2017
A-M1		3003	324A	Britain	4032
A-M1G		3004	372B	(DTD)[7]	6063
A-U4G		2017	717, 724, 731A		2618
A-U2G		2117	745, 5014, 5084		"
A-U2GN		2618	5090		2024
A-U4G1		2024	5100		Alclad 2024
A-U4N		2218			
A-U4SG		2014			
A-S12UN		4032			
A-Z5GU		7075			

TABLE 2.5   Foreign Alloy Designations and Similar AA Alloys

Alloy designation	Designating country	Equivalent or similar AA alloy	Alloy designation	Designating country	Equivalent or similar AA alloy
P-AlCu4MgMn	Italy	2017	Al-Mg-Si	Switzerland	6101
P-AlCu4.5MgMn	(UNI)[8]	2024	Al1.5Mg	(VSM)[10]	5050
P-AlCu4.5MgMnplacc.		Alclad 2024	Al-Cu-Ni		2018
P-AlCu2.5MgSi		2117	Al3.5Cu0.5Mg		2017
P-AlCu4.4SiMnMg		2014	Al4Cu1.2Mg		2027
P-AlCu4.4SiMnMgplacc.		Alclad 2014	Al-Zn-Mg-Cu		7075
P-AlMg0.9		5657	Al-Zn-Mg-Cu-pl		Alclad 7075
P-AlMg1.5		5050			
P-AlMg2.5		5052	Al99.0Cu	ISO[11]	1100
P-AlSi0.4Mg		6063	AlCu2Mg		2117
P-AlSi0.5Mg		6101	AlCu4Mg1		2024
			AlCu4SiMg		2014
Al99.5E	Spain	1350	AlCu4MgSi		2017
L-313	(UNE)[9]	2014	AlMg1		5005
L-314		2024	AlMg1.5		5050
L-315		2218	AlMg2.5		5052
L-371		7075	AlMg3.5		5154
			AlMg4		5086
			AlMg5		5056
			AlMn1Cu		3003
			AlMg3Mn		5454
			AlMg4.5Mn		5083
			AlMgSi		6063
			AlMg1SiCu		6061
			AlZn6MgCu		7075

[1]Austrian Standard M3430
[2]Canadian Standards Association
[3]Normes Françaises
[4]Deutsche Industrie-Norm.
[5]Werkstoff-Nr.
[6]British Standard
[7]Directorate of Technical Development
[8]Unificazione Nazionale Italiana
[9]Una Norma Español
[10]Verein Schweizerischer Maschinenindustrieller
[11]International Organization for Standardization

7xx.x. Primarily alloyed with *zinc*, this series is difficult to cast and so is used where its finishing characteristics or machinability is important. These alloys have moderate or better strengths and good general corrosion resistance but are not suitable for elevated temperatures.

8xx.x. This series is alloyed with about 6% *tin* and primarily used for bearings, being superior to most other materials for this purpose. These alloys are used for large rolling mill bearings and connecting rods and crankcase bearings for diesel engines.

9xx.x. This series is reserved for castings alloyed with elements other than those used in the other series.

The chemical composition limits for common cast alloys are given in Table 2.6.

Other standards-writing organizations such as the federal government and previous ASTM specifications have assigned different designations to cast alloys. A cross-reference chart is provided in Table 2.7.

2.2.3 Tempers

Aluminum alloys are tempered by *heat treating* or *strain hardening* to further increase strength beyond the strengthening effect of adding alloying elements. Alloys are divided into two groups based on whether their strengths can be increased by heat treating. Both *heat-treatable* and *non-heat-treatable* alloys can be strengthened by strain hardening, also called cold-working. The alloys that are not heat treatable may be strengthened only by cold working. Whether an alloy is heat treatable depends on its alloying elements. Alloys in which the amount of alloying element in solid solution in aluminum increases with temperature are heat treatable. In general, the 1xxx, 3xxx, 4xxx, and 5xxx series wrought alloys are not heat treatable, while the 2xxx, 6xxx, and 7xxx wrought series are, but there are exceptions to this rule. Strengthening methods are summarized in Table 2.8.

Non-heat-treatable alloys may also undergo a heat treatment, but this heat treatment is used only to stabilize properties so that strengths don't decrease over time (behavior called *age softening*) and is only required for alloys with an appreciable amount of magnesium (the 5xxx series). Heating to 225°F to 350°F (110°C to 180°C) causes all the softening to occur at once and thus is used as the stabilization heat treatment.

Before tempering, alloys begin in the annealed condition, the weakest but most ductile condition. Tempering, while increasing the strength, decreases ductility and therefore decreases workability. To reduce material to the annealed condition, the typical annealing treatments given in Table 2.9 can be used.

Strain hardening is achieved by mechanical deformation of the material at ambient temperature. In the case of sheet and plate, this is done by reducing its thickness by rolling. As the material is worked, it becomes resistant to further deformation, and its strength increases. The effect of this work on the yield strength of some common non-heat-treatable alloys is shown in Figure 2.1.

Two heat treatments can be applied to annealed condition heat-treatable alloys. First, the material can be *solution heat treated*. This allows soluble alloying elements to enter into solid solution; they are retained in a supersaturated state upon *quenching*, a controlled rapid cooling usually performed using air or water. Next, the material may

TABLE 2.6 Chemical Composition Limits for Cast Alloys

Alloy	Product[c]	Silicon	Iron	Copper	Manganese	Magnesium	Chromium	Nickel	Zinc	Titanium	Others Each	Others Total[a]
201.0	S	0.10	0.15	4.0–5.2	0.20–0.50	0.15–0.55	—	—	—	0.15–0.35	0.05[b]	0.10
204.0	S&P	0.20	0.35	4.2–5.0	0.10	0.15–0.35	—	0.05	0.10	0.15–0.30	0.05[f]	0.15
208.0	S&P	2.5–3.5	1.2	3.5–4.5	0.50	0.10	—	0.35	1.0	0.25	—	0.50
222.0	S&P	2.0	1.5	9.2–10.7	0.50	0.15–0.35	—	0.50	0.8	0.25	—	0.35
242.0	S&P	0.7	1.0	3.5–4.5	0.35	1.2–1.8	0.25	1.7–2.3	0.35	0.25	0.05	0.15
295.0	S	0.7–1.5	1.0	4.0–5.0	0.35	0.03	—	—	0.35	0.25	0.05	0.15
296.0	P	2.0–3.0	1.2	4.0–5.0	0.35	0.05	—	0.35	0.50	0.25	—	0.35
308.0	P	5.0–6.0	1.0	4.0–5.0	0.50	0.10	—	—	1.0	0.25	—	0.50
319.0	S&P	5.5–6.5	1.0	3.0–4.0	0.50	0.10	—	0.35	1.0	0.25	—	0.50
328.0	S	7.5–8.5	1.0	1.0–2.0	0.20–0.6	0.20–0.6	0.35	0.25	1.5	0.25	—	0.50
332.0	P	8.5–10.5	1.2	2.0–4.0	0.50	0.50–1.5	—	0.50	1.0	0.25	—	0.50
333.0	P	8.0–10.0	1.0	3.0–4.0	0.50	0.05–0.50	—	0.50	1.0	0.25	—	0.50
336.0	P	11.0–13.0	1.2	0.50–1.5	0.35	0.7–1.3	—	2.0–3.0	0.35	0.25	0.05	—
354.0	S&P	8.6–9.4	0.20	1.6–2.0	0.10	0.40–0.6	—	—	0.10	0.20	0.05	0.15
355.0	S&P	4.5–5.5	0.6[d]	1.0–1.5	0.50[d]	0.40–0.6	0.25	—	0.35	0.25	0.05	0.15
C355.0	S&P	4.5–5.5	0.20	1.0–1.5	0.10	0.40–0.6	—	—	0.10	0.20	0.05	0.15
356.0	S&P	6.5–7.5	0.6[d]	0.25	0.35[d]	0.20–0.45	—	—	0.35	0.25	0.05	0.15
A356.0	S&P	6.5–7.5	0.20	0.20	0.10	0.25–0.45	—	—	0.10	0.20	0.05	0.15
357.0	S&P	6.5–7.5	0.15	0.05	0.03	0.45–0.6	—	—	0.05	0.20	0.05	0.15
A357.0	S&P	6.5–7.5	0.20	0.20	0.10	0.40–0.7	—	—	0.10	0.04–0.20	0.05[e]	0.15
359.0	S&P	8.5–9.5	0.20	0.20	0.10	0.50–0.7	—	—	0.10	0.20	0.05	0.15
443.0	S&P	4.5–6.0	0.8	0.6	0.50	0.05	0.25	—	0.50	0.25	—	0.35
B443.0	S&P	4.5–6.0	0.8	0.15	0.35	0.05	—	—	0.35	0.25	0.05	0.15
A444.0	P	6.5–7.5	0.20	0.10	0.10	0.05	—	—	0.10	0.20	0.05	0.15
512.0	S	1.4–2.2	0.6	0.35	0.8	3.5–4.5	0.25	—	0.35	0.25	0.05	0.15
513.0	P	0.30	0.40	0.10	0.30	3.5–4.5	—	—	1.4–2.2	0.20	0.05	0.15
514.0	S	0.35	0.50	0.15	0.35	3.5–4.5	—	—	0.15	0.25	0.05	0.15
520.0	S	0.25	0.30	0.25	0.15	9.5–10.6	—	—	0.15	0.25	0.05[f]	0.15
535.0	S&P	0.15	0.15	0.05	0.10–0.25	6.2–7.5	—	—	—	0.10–0.25	0.05	0.15

705.0	S&P	0.20	0.8	0.20	0.40-0.6	1.4-1.8	0.20-0.40	—	2.7-3.3	0.25	0.05	0.15
707.0	S&P	0.20	0.8	0.20	0.40-0.6	1.8-2.4	0.20-0.40	—	4.0-4.5	0.25	0.05	0.15
710.0	S	0.15	0.50	0.35-0.65	0.05	0.6-0.8	—	—	6.0-7.0	0.25	0.05	0.15
711.0	P	0.30	0.7-1.4	0.35-0.65	0.05	0.25-0.45	—	—	6.0-7.0	0.20	0.05	0.15
712.0	S	0.30	0.50	0.25	0.10	0.50-0.65	0.40-0.6	—	5.0-6.5	0.15-0.25	0.05	0.20
713.0	S&P	0.25	1.1	0.40-1.0	0.6	0.20-0.50	0.35	0.15	7.0-8.0	0.25	0.10	0.25
771.0	S	0.15	0.15	0.10	0.10	0.8-1.0	0.06-0.20	—	6.5-7.5	0.10-0.20	0.05	0.15
850.0	S&P	0.7	0.7	0.7-1.3	0.10	0.10	—	0.7-1.3	—	0.20	—[j]	0.30
851.0	S&P	2.0-3.0	0.7	0.7-1.3	0.10	0.10	—	0.30-0.7	—	0.20	—[j]	0.30
852.0	S&P	0.40	0.7	1.7-2.3	0.10	0.6-0.9	—	0.9-1.5	—	0.20	—[j]	0.30

[a] The alloys listed are those which have been included in Federal Specifications QQ-A-596d, ALUMINUM ALLOYS PERMANENT AND SEMI-PERMANENT MOLD CASTINGS, QQ-A-601E, ALUMINUM ALLOY SAND CASTINGS, and Military Specification MIL-A-21180c, ALUMINUM ALLOY CASTINGS, HIGH STRENGTH. Other alloys are registered with The Aluminum Association and are available. Information on these should be requested from individual foundries or ingot suppliers.

[b] Except for "Aluminum" and "Others," analysis normally is made for elements for which specific limits are shown. For purposes of determining conformance to these limits, an observed value or calculated value obtained from analysis is rounded off to the nearest unit in the last right hand place of figures used in expressing the specified limit, in accordance with the following:

When the figure next beyond the last figure or place to be retained is less than 5, the figure in the last place retained should be kept unchanged.

When the figure next beyond the last figure or place to be retained is greater than 5, the figure in the last place retained should be increased by 1.

When the figure next beyond the last figure or place to be retained is 5 and

(1) there are no figures or only zeros, beyond this 5, if the figure in the last place to be retained is odd, it should be increased by 1; if even, it should be kept unchanged;

(2) if the 5 next beyond the figure in the last place to be retained is followed by any figures other than zero, the figure in the last place retained should be increased by 1; whether odd or even.

[c] S = Sand Cast   [d] P = Permanent Mold Cast

[e] If iron exceeds 0.45 percent, manganese content shall not be less than one-half the iron content.

[f] Also contains 0.04-0.07 percent beryllium.

[g] Also contains 0.003-0.007 percent beryllium, boron 0.005 percent maximum.

[h] Also contains 5.5-7.0 percent tin.

[i] Also contains 0.40-1.0 percent silver.

[j] Also contains 0.05 max. percent tin.

[k] The sum of those "Others" metallic elements 0.010 percent or more each, expressed to the second decimal before determining the sum.

TABLE 2.7   Cast Alloy Cross-Reference Chart

ANSI AA	Former designation	UNS	Federal (QQ-A-596) (QQ-A-601)	Former ASTM (B26) (B108)	Former SAE (J453c)	Military (MIL-A-21180)
201.0	—	A02010	—	CQ51A	382	—
204.0	—	A02040	—	—	—	—
208.0	108	A02080	108	CS43A	—	—
213.0	C113	A02130	113	CS74A	33	—
222.0	122	A02220	122	CG100A	34	—
242.0	142	A02420	142	CN42A	39	—
295.0	195	A02950	195	C4A	38	—
296.0	B295.0	A02960	B195	—	380	—
308.0	A108	A03080	A108	—	—	—
319.0	319, Allcast	A03190	319	SC64D	326	—
328.0	Red X-8	A03280	Red X-8	SC82A	327	—
332.0	F332.0	A03320	F132	SC103A	332	—
333.0	333	A03330	333	SC94A	331	—
336.0	A332.0	A03360	A132	SN122A	321	—
354.0	354	A03540	—	—	—	C354
355.0	355	A03550	355	SC51A	322	—
C355.0	C355	A33550	C355	SC51B	335	C355
356.0	356	A03560	356	SG70A	323	—
A356.0	A356	A13560	A356	SG70B	336	A356
357.0	357	A03570	357	—	—	—
A357.0	A357	A13570	—	—	—	A357
359.0	359	A03590	—	—	—	359
443.0	43	A04430	—	S5B	35	—
B443.0	43	A24430	43	S5A	—	—
A444.0	—	A14440	—	—	—	—
512.0	B514.0	A05120	B214	GS42A	—	—
513.0	A514.0	A05130	A214	GZ42A	—	—
514.0	214	A05140	214	G4A	320	—
520.0	220	A05200	220	G10A	324	—
535.0	Almag 35	A05350	Almag35	GM70B	—	—
705.0	603, Ternalloy 5	A07050	Ternalloy 5	ZG32A	311	—
707.0	607, Ternalloy 7	A07070	Ternalloy 7	ZG42A	312	—
710.0	A712.0	A07100	A612	ZG61B	313	—
711.0	C721.0	A07110	—	ZC60A	314	—
712.0	D712.0	A07120	40E	ZG61A	310	—
713.0	613, Tenzaloy	A07130	Tenzaloy	ZC81A	315	—
771.0	Precedent 71A	A07710	Precedent 71A	—	—	—
850.0	750	A08500	750	—	—	—
851.0	A850.0	A08510	A750	—	—	—
852.0	B850.0	A08520	B750	—	—	—

undergo a *precipitation heat treatment,* also called *artificial aging,* by which constituents are precipitated from solid solution to increase the strength. An example of this process is the production of 6061-T6 sheet. From its initial condition, 6061-O annealed material is heat

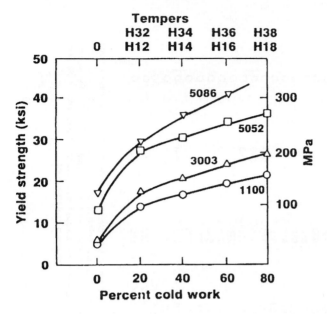

Figure 2.1 Effect of cold work on yield strength of several work-hardening alloys.

TABLE 2.8   Strengthening Methods

Pure aluminum 1xxx	Alloying 2xxx—Cu 6xxx—Mg, Si 7xxx—Zn	Heat treatment Solution heat treatment; Natural aging or artificial aging	Strain hardening (cold working)	-T tempers
	Alloying 3xxx—Mn 5xxx—Mg	Strain hardening (cold working)		-H tempers

treated to 990°F (530°C) as rapidly as possible (solution heat treated), then cooled as rapidly as possible (quenched), which renders the temper T4. Then the material is heated to 320°F (160°C) and held for 18 hours (precipitation heat treated); upon cooling to room temperature, the temper is T6.

Solution heat treated aluminum may also undergo *natural aging*. Natural aging, like artificial aging, is a precipitation of alloying elements from solid solution but, because it occurs at room temperature, it occurs much more slowly (over a period of days and months rather than hours) than artificial aging. Both aging processes result in an in-

TABLE 2.9 Typical Annealing Treatments for Aluminum Alloy Mill Products

The treatments listed in this table are typical for various sizes and methods of manufacture and may not exactly describe the optimum treatment for a specific item.

ALLOY	METAL TEMPERATURE °F	APPROX. TIME AT TEMPERATURE Hours	TEMPER DESIGNATION
1060	650	①	O
1100	650	①	O
1145	650	①	O
1235	650	①	O
1345	650	①	O
1350	650	①	O
2014	775②	2–3	O
2017	775②	2–3	O
2024	775②	2–3	O
2117	775②	2–3	O
2219	775②	2–3	O
3003	775	①	O
3004	650	①	O
3005	775	①	O
3105	650	①	O
5005	650	①	O
5050	650	①	O
5052	650	①	O
5056	650	①	O
5083	650	①	O
5086	650	①	O
5154	650	①	O
5254	650	①	O
5454	650	①	O
5456	650	①	O
5457	650	①	O
5652	650	①	O
6005	775②	2–3	O
6053	775②	2–3	O
6061	775②	2–3	O
6063	775②	2–3	O
6066	775②	2–3	O
7072	650	①	O
7075	775③	2–3	O
7175	775③	2–3	O
7178	775③	2–3	O
7475	775③	2–3	O
Brazing Sheet:			
Nos. 11 & 12	650	①	O
Nos. 23 & 24	650	①	O

① Time in the furnace need not be longer than necessary to bring all parts of load to annealing temperature. Rate of cooling is unimportant.

② These treatments are intended to remove effects of solution heat treatment and include cooling at a rate of about 50° F per hour from the annealing temperature to 500° F. The rate of subsequent cooling is unimportant. Treatment at 650° F, followed by uncontrolled cooling, may be used to remove the effects of cold work, or to partially remove the effects of heat treatment.

③ This treatment is intended to remove the effects of solution heat treatment and includes cooling at an uncontrolled rate to 400° F or less, followed by reheating to 450° F for 4 hours. Treatment at 650° F, followed by uncontrolled cooling, may be used to remove the effects of cold work, or to partially remove the effects of heat treatment.

crease in strength and a corresponding decrease in ductility. Material that will be subjected to severe forming operations (like cold heading wire to make rivets or bolts) is often purchased in a T4 temper, formed, and then artificially aged or allowed to naturally age. Care must be taken to perform the forming operation before too long a period of time elapses, or natural aging of the material will cause it to harden and decrease its workability. Sometimes T4 material is refrigerated to prevent natural aging if cold forming required for fabrication into a product such as a fastener or a tapered pole won't be performed soon after solution heat treatment.

The temper designation system is the same for both wrought and cast alloys, although cast alloys are only heat treated and not strain hardened, with the exception of some 85x.0 casting alloys. The temper designation follows the alloy designation, the two being separated by a hyphen (for example, 5052-H32). Basic temper designations are letters. Subdivisions of the basic tempers are given by one or more numbers following the letter.

The basic temper designations are as listed below:

F—as fabricated. Applies to the products of shaping processes in which no special control over thermal conditions or strain hardening is employed. For wrought products, there are no mechanical property limits.

O—annealed. Applies to wrought products that are annealed to obtain the lowest strength temper, and to cast products that are annealed to improve ductility and dimensional stability. The O may be followed by a number other than zero.

H—strain hardened (wrought products only). Applies to products that have their strength increased by strain hardening, with or without supplementary thermal treatments to produce some reduction in strength. The H is always followed by two or more numbers.

W—solution heat treated. An unstable temper applicable only to alloys that spontaneously age at room temperature after solution heat-treatment. This designation is specific only when the period of natural aging is indicated; for example, W 1/2 hour.

T—thermally treated to produce stable tempers other than F, O, or H. Applies to products that are thermally treated, with or without supplementary strain-hardening, to produce stable tempers. The T is always followed by one or more numbers.

2.2.3.1 Strain-hardened tempers. For strain-hardened tempers, the first digit of the number following the H denotes the following:

H1—strain hardened only. Applies to products that are strain hardened to obtain the desired strength without supplementary thermal treatment. The number following this designation indicates the degree of strain hardening. (Example: 1100-H14)

H2—strain hardened and partially annealed. Applies to products that are strain hardened more than the desired final amount and then reduced in strength to the desired level by partial annealing. For alloys that age soften at room temperature, the H2 tempers have the same minimum ultimate tensile strength as the corresponding H3 tempers. For other alloys, the H2 tempers have the same minimum ultimate tensile strength as the corresponding H1 tempers and slightly higher elongation. The number following this designation indicates the strain hardening remaining after the product has been partially annealed. (Example: 3005-H25)

H3—strain hardened and stabilized. Applies to products that are strain hardened and whose mechanical properties are stabilized either by a low temperature thermal treatment or as a result of heat introduced during fabrication. Stabilization usually improves ductility. This designation is applicable only to those alloys that, unless stabilized, gradually age soften at room temperature. The number following this designation indicates the degree of strain hardening remaining after the stabilization has occurred. (Example: 5005-H34)

H4—strain hardened and lacquered or painted. Applies to products that are strain hardened and subjected to some thermal operation during subsequent painting or lacquering. The number following this designation indicates the degree of strain hardening remaining after the product has been thermally treated as part of the painting or lacquering curing. The corresponding H2X or H3X mechanical property limits apply.

The digit following the designation H1, H2, H3, or H4 indicates the degree of strain hardening. Number 8 is for the tempers with the highest ultimate tensile strength normally produced. Number 4 is for tempers whose ultimate strength is approximately midway between that of the O temper and the HX8 temper. Number 2 is for tempers whose ultimate strength is approximately midway between that of the O temper and the HX4 temper. Number 6 is for tempers whose ultimate strength is approximately midway between that of the HX4 temper and the HX8 temper. Numbers 1, 3, 5, and 7 similarly designate intermediate tempers between those defined above. Number 9 designates tempers whose minimum ultimate tensile strength exceeds that of the HX8 tempers by 2 ksi (15 MPa) or more.

The third digit, when used, indicates a variation in the degree of temper or the mechanical properties of a two-digit temper. An example is pattern or embossed sheet made from the H12, H22, or H32 tempers; these are assigned H124, H224, or H324 tempers, respectively, since the additional strain hardening from embossing causes a slight change in the mechanical properties.

2.2.3.2 Heat-treated tempers.   For heat-treated tempers, the numbers 1 through 10 following the T denote the following:

T1—cooled from an elevated temperature shaping process and naturally aged to a substantially stable condition. Applies to products that are not cold worked after cooling from an elevated temperature shaping process, or in which the effect of cold work in flattening or straightening may not be recognized in mechanical property limits. (Example: 6005-T1 extrusions)

T2—cooled from an elevated temperature shaping process, cold worked, and naturally aged to a substantially stable condition. Applies to products that are cold worked to improve strength after cooling from an elevated temperature shaping process, or in which the effect of cold work in flattening or straightening is recognized in mechanical property limits.

T3—solution heat treated, cold worked, and naturally aged to a substantially stable condition. Applies to products that are cold worked to improve strength after solution heat treatment, or in which the effect of cold work in flattening or straightening is recognized in mechanical property limits. (Example: 2024-T3 sheet)

T4—solution heat treated and naturally aged to a substantially stable condition. Applies to products that are not cold worked after solution heat treatment, or in which the effect of cold work in flattening or straightening may not be recognized in mechanical property limits. (Example: 2014-T4 sheet)

T5—cooled from an elevated temperature shaping process and then artificially aged. Applies to products that are not cold worked after cooling from an elevated temperature shaping process, or in which the effect of cold work in flattening or straightening may not be recognized in mechanical property limits. (Example: 6063-T5 extrusions)

T6—solution heat treated and then artificially aged. Applies to products that are not cold worked after solution heat treatment, or in which the effect of cold work in flattening or straightening may not be recognized in mechanical property limits. (Example: 6063-T6 extrusions)

T7—solution heat treated and then overaged/stabilized. Applies to wrought products that are artificially aged after solution heat treatment to carry them beyond a point of maximum strength to provide control of some significant characteristic. Applies to cast products that are artificially aged after solution heat treatment to provide dimensional and strength stability. (Example: 7050-T7 rivet and cold heading wire and rod)

T8—solution heat treated, cold worked, and then artificially aged. Applies to products that are cold worked to improve strength, or in which the effect of cold work in flattening or straightening is recognized in mechanical property limits. (Example: 2024-T81 sheet)

T9—solution heat-treated, artificially aged, and then cold worked. Applies to products that are cold worked to improve strength after artificial aging. (Example: 6262-T9 nuts)

T10—cooled from an elevated temperature shaping process, cold worked, and then artificially aged. Applies to products that are cold worked to improve strength, or in which the effect of cold work in flattening or straightening is recognized in mechanical property limits.

Additional digits may be added to designations T1 through T10 for variations in treatment. Stress relieved tempers follow the conventions listed below.

Stress relieved by stretching:

T_51—Applies to plate and rolled or cold-finished rod or bar, die or ring forgings, and rolled rings when stretched after solution heat treatment or after cooling from an elevated temperature shaping process. The products receive no further straightening after stretching. (Example: 6061-T651)

T_510—Applies to extruded rod, bar, profiles, and tubes and to drawn tube when stretched after solution heat treatment or after cooling from an elevated temperature shaping process.

T_511—Applies to extruded rod, bar, profiles, and tubes and to drawn tube when stretched after solution heat treatment or after cooling from an elevated temperature shaping process. These products may receive minor straightening after stretching to comply with standard tolerances.

These stress-relieved temper products usually have larger tolerances on dimensions than products of other tempers.

Stress relieved by compressing:

T_52—Applies to products that are stress relieved by compressing after solution heat treatment or cooling from an elevated temperature shaping process to produce a permanent set of 1 to 5%.

Stress relieved by combined stretching and compressing:

T_54—Applies to die forgings that are stress relieved by restriking cold in the finish die.

For wrought products heat treated from annealed or F temper (or other temper when such heat treatments result in the mechanical properties assigned to these tempers):

T42—Solution heat treated from annealed or F temper and naturally aged to a substantially stable condition (Example: 2024-T42)

T62—Solution heat treated from annealed or F temper and artificially aged (Example: 6066-T62)

Typical heat treatments for wrought alloys are given in Table 2.10. Heat treatments for cast alloys are given in Table 2.11.

2.2.4   Alloy Selection

Tables for wrought (Table 2.12) and cast alloys (Table 2.13), rating properties of interest such as corrosion resistance and weldability, are useful for selecting an alloy for a given application. The ratings are relative and should be considered in that light.

2.3   Physical Properties

Physical properties include all properties other than mechanical ones. The physical properties of most interest to material designers include density, melting point, electrical conductivity, thermal conductivity, and coefficient of thermal expansion. While these properties vary among alloys and tempers, average values can be useful to the designer.

Density doesn't vary much by alloy (since alloying elements make up such a small portion of the composition), ranging from 0.095 to 0.103 lb/in^3 and averaging around 0.1 lb/in^3 (2700 kg/m^3). This compares to 0.065 for magnesium, 0.16 for titanium, and 0.283 lb/in^3 for steel. Density is calculated as the weighted average of the densities of the el-

TABLE 2.10 Typical Heat Treatments for Wrought Alloys ①

The typical treatments listed in this table are for furnaces operating to instructions (mill practices) given in Fahrenheit. For furnaces operating to instructions given in Celsius, see Table 3.4 in the metric unit edition of ALUMINUM STANDARDS AND DATA.

ALLOY	PRODUCT	SOLUTION HEAT TREATMENT ②		PRECIPITATION HEAT TREATMENT		
		METAL TEMPERATURE ③ °F	TEMPER DESIGNATION	METAL TEMPERATURE ③ °F	APPROX. TIME AT TEMPERATURE ④ Hours	TEMPER DESIGNATION
2011	Rolled or Cold Finished Wire, Rod & Bar	975	T3⑤	320	14	T8⑤
			T4	–	–	–
			T451⑥	–	–	T8⑤
	Drawn Tube	960	T3⑤	310	14	T8⑤
			T4511⑥	–	–	–
2014⑦	Flat Sheet	935	T3⑤	–	–	–
			T42	320	18	T62
	Coiled Sheet	935	T4	320	18	T6
			T42	320	18	T62
	Plate	935	T451⑥	320	18	T651⑥
			T42	320	18	T62
	Rolled or Cold Finished Wire, Rod & Bar	935	T4	320⑧	18	T6
			T451⑥	320⑧	18	T651⑥
			T42	320⑧	18	T62
	Extruded Wire, Rod, Bar, Profiles (Shapes) & Tube	935	T4	320⑧	18	T6
			T4510⑥	320⑧	18	T6510⑥
			T4511⑥	320⑧	18	T6511⑥
			T42	320⑧	18	T62
	Drawn Tube	935	T4	320⑧	18	T6
			T42	320⑧	18	T62

Alloy	Product form	Solution temp	Temper	Aging temp	Time (hr)	Temper
	Die Forgings	935⑨	T4	340	10	T6
	Hand Forgings and Rolled Rings	935⑨	T4㉙	340	10	T6
			T452⑩㉙	340	10	T652⑩
2017	Rolled or Cold Finished Wire, Rod & Bar	935	T4	—	—	—
			T451⑥	—	—	—
2018	Die Forgings	950⑪	T42㉙	340	10	T61
2024⑦	Flat Sheet	920	T3⑤	375	12	T81⑤
			T361⑤	375	8	T861⑤
			T42	375	9	T62
			T42	375	16	T72
	Coiled Sheet	920	T4	—	—	—
	Plate	920	T351⑥	375	12	T851⑥
			T361⑤	375	8	T861⑤
			T42	375	9	T62
	Rolled or Cold Finished Wire, Rod & Bar	920	T351⑥	375	12	T851⑥
			T36⑤	—	—	—
			T4	375	12	T6
			T42	375	16	T62
	Extruded Wire, Rod, Bar Profiles (Shapes) & Tube	920	T3⑤	375	12	T81⑤
			T3510⑥	375	12	T8510⑥
			T3511⑥	375	12	T8511⑥
	Drawn Tube	920	T3⑤	—	—	—
			T42	—	—	—
2025	Die Forgings	960	T4㉙	340	10	T6
2036	Sheet	930	T4	—	—	—

2.29

TABLE 2.10 Typical Heat Treatments for Wrought Alloys (Continued)[1]

ALLOY	PRODUCT	SOLUTION HEAT TREATMENT[2]		PRECIPITATION HEAT TREATMENT		
		METAL TEMPERATURE[3] °F	TEMPER DESIGNATION	METAL TEMPERATURE[3] °F	APPROX. TIME AT TEMPERATURE[4] Hours	TEMPER DESIGNATION
2117	Rolled or Cold Finished Wire and Rod	935	T4	—	—	—
2124	Plate	920	T42	—	—	—
			T351[6]	375	12	T851[6]
2218	Die Forgings	950[11] 950[12]	T4[9] T41[9]	340 460	10 6	T61 T72
2219[7]	Flat Sheet	995	T31[5] T37[5]	350 325	18 24	T81[5] T87[5]
			T42[9]	375	36	T62
	Plate	995	T37[5] T351[6]	350 350	18 18	T87[5] T851[6]
			T42[9]	375	36	T62
	Rolled or Cold Finished Wire, Road & Bar	995	T4[9] T351[9]	375 375	36 18	T6 T851[6]
	Extruded Rod, Bar, Profiles (Shapes) & Tube	995	T31[5] T3510[6] T3511[6]	375 375 375	18 18 18	T81[5] T8510[6] T8511[6]
			T42[9]	375	36	T62
	Die Forgings and Rolled Rings	995	T4[9]	375	26	T6
	Hand Forgings	995	T4[9] T352[10][9]	375 350	26 18	T6 T852[10]
2618	Forgings and Rolled Rings	985[11]	T4[9]	390	20	T61
4032	Die Forgings	950[9]	T4[9]	340	10	T6

6005	Extruded Rod, Bar, Profiles (Shapes) & Tube	[30]	T1	350	8	T5
6053	Rolled or Cold Finished Wire and Rod	945	T4[29]	355	8	T61
6061 [1]	Die Forgings	970	T4[29]	340	10	T6
	Sheet	990	T4	320	18	T6
			T42	320	18	T62
	Plate	990	T4[27] T451[6]	320 320	18 18	T6[27] T651[6]
			T42	320	18	T62
	Rolled or Cold finished Wire, Rod & Bar	990	T4 T3[30] T4 T4 T451[6]	320[13] 320[13] 320[13] 320[13] 320[13]	18 18 18 18 18	T6 T89[5] T913[22] T94[22] T651[6]
			T42	320[13]	18	T62
	Extruded Rod, Bar, Profiles (Shapes) and Tube	[30]	T1	350	8	T51
		990[15]	T4 T4510[6] T4511[6]	350 350 350	8 8 8	T6 T6510[6] T6511[6]
		990	T42	350	8	T62
	Structural Profiles (Shapes)	990[15]	T4[29]	350	8	T6
	Pipe	990[15]	T4[29]	350	8	T6
	Drawn Pipe	990	T4	320[13]	18	T6
			T42	320[13]	18	T62
	Die and Hand Forgings	990	T4[29]	350	8	T6
	Rolled Rings	990	T4[29] T452[10][29]	350 350	8 8	T6 T652[10]

TABLE 2.10   Typical Heat Treatments for Wrought Alloys   *(Continued)* ①

ALLOY	PRODUCT	SOLUTION HEAT TREATMENT ②		PRECIPITATION HEAT TREATMENT		
		METAL TEMPERATURE ③ °F	TEMPER DESIGNATION	METAL TEMPERATURE ③ °F	APPROX. TIME AT TEMPERATURE ④ Hours	TEMPER DESIGNATION
6063	Extruded Rod, Bar, Profiles (Shapes) & Tube	㉚	T1 T1	360 ⑯ 360	3 3	T5 T52
		950 ⑯	T4	350 ⑰	8	T6
		970	T42	350 ⑫	8	T62
	Drawn Tube	970	T4 T3 ⑤⑩㉒ T3 ⑤⑩㉒ T3 ⑤⑩㉒	350 350 350 350	8 8 8 8	T6 T83 ⑤ T831 ⑤ T832 ⑤
			T42	350	8	T62
	Pipe	970 ⑯	T4 ㉓	350 ⑰	8	T6
6066	Extruded Rod, Bar, Profiles (Shapes) & Tube	990	T4 T4510 ⑥ T4511 ⑥	350 350 350	8 8 8	T6 T6510 ⑥ T6511 ⑥
			T42	350	8	T62
	Drawn Tube	990	T4	350	8	T6
			T42	350	8	T62
	Die Forgings	990	T4 ㉓	350	8	T6
6070	Extruded Rod, Bar Profiles (Shapes) & Tube	1015 ⑲	T4 ㉓	320	18	T6
		1015	T42 ㉓	320	18	T62
6101	Extruded Rod, Bar, Tube, Pipe and Structural Profiles (Shapes)	970 ⑱	T4 ㉓ T4 ㉓ T4 ㉓ T4 ㉓ T4 ㉓	390 440 410 535 430	10 5 9 7 3	T6 T61 T63 T64 T65

6105	Extruded Rod, Bar Profiles (Shapes) and Tube	⑳	T1	350	8	T5
6151	Die Forgings	960	T4⑳	340	10	T6
	Rolled Rings	960	T4⑳	340	10	T6
			T452⑩⑳	340	10	T652⑩
6162	Extruded Rod, Bar, Profiles (Shapes) & Tube	⑳	T1⑳	350	8	T5
			T1510⑥⑳	350	8	T5510⑥
			T1511⑥⑳	350	8	T5511⑥
		980⑲	T4⑳	350	8	T6
			T4510⑥⑳	350	8	T6510⑥
			T4511⑥⑳	350	8	T6511⑥
6201	Wire	950	T3⑤⑳	320	4	T81⑤
6262	Rolled or Cold Finished Wire, Rod and Bar	1000	T4⑳	340	8	T6
			T4⑳	340	12	T9⑭
			T451⑥⑳	340	8	T651⑥
			T42⑳	340	8	T62
	Extruded Rod, Bar, Profiles (Shapes) and Tube	1000⑮	T4⑳	350	12	T6
			T4510⑥⑳	350	12	T6510⑥
			T4511⑥⑳	350	12	T6511⑥
		1000	T42⑳	350	12	T62
	Drawn Tube	1000	T4⑳	340	8	T6
			T4⑳	340	8	T9⑭
			T42⑳	340	8	T62
6351	Extruded Rod, Bar and Profiles (Shapes)	⑳	T1	250	10	T54
		⑳	T1	350	8	T5
		985	T4	350	8	T6
6463	Extruded Rod, Bar and Profiles (Shapes)	⑳	T1	400	1	T5
		970⑯	T4⑳	350⑰	8	T6
		970	T42⑳	350⑰	8	T62
6951⑳	Sheet	985	T42	320	18	T62

TABLE 2.10 Typical Heat Treatments for Wrought Alloys *(Continued)* ①

ALLOY	PRODUCT	SOLUTION HEAT TREATMENT ②		PRECIPITATION HEAT TREATMENT		
		METAL TEMPERATURE ③ °F	TEMPER DESIGNATION	METAL TEMPERATURE ③ °F	APPROX. TIME AT TEMPERATURE ④ Hours	TEMPER DESIGNATION
7005	Extruded Rod, Bar and Profiles (Shapes)	㉘	T1 ㉙	㉒	㉒	T53
7049	Die Forgings	875 ⑨	W	⑧	⑧	T73
	Hand Forgings	875 ⑨	W	⑧	⑧	T73
7050	Plate	890	W52 ⑩	⑧	⑧	T7352 ⑩
	Plate	890	W51 ⑥ W51 ⑥	㉛ ㉗	㉛ ㉜	T7451 ⑥ T7651 ⑥
	Rolled or Cold Finished Wire and Rod	890	W	㉘	㉘	T7
	Extruded Rod, Bar and Profiles (Shapes)	890	W510 ⑥ W510 ⑥ W510 ⑥	㉘ ㉘ ㉝	㉘ ㉘ ㉝	T73510 ⑥ T74510 ⑥ T76510 ⑥
			W511 ⑥ W511 ⑥ W511 ⑥	㉘ ㉗ ㉝	㉘ ㉗ ㉝	T73511 ⑥ T74511 ⑥ T76511 ⑥
	Die Forgings	890	W	㉘	㉙	T74
	Hand Forgings	890	W52 ㉞	㉘	㉙	T7452 ⑩
7075 ①	Sheet	900 ㉓	W W W	250 ⑱ ㉚ ㉚ ㉘	24 ㉚ ㉚ ㉘	T6 T73 ㉗ T76 ㉗
			W	250 ⑱	24	T62
	Plate	900 ㉓ ㉔	W51 ⑥ W51 ⑥ W51 ⑥	250 ⑱ ㉚ ㉚ ㉘	24 ㉚ ㉚ ㉘	T651 ⑥ T7351 ⑥ ㉗ T7651 ⑥ ㉗
			W	250 ⑱	24	T62

Product	Temp (°C)	Cold water quench / W designation	Time (h)	Temp (°C)	Temper
Rolled or Cold finished Wire, Rod and Bar	915 ㉓㉔	W / W	250 ⑲㉒	24 ⑲㉒	T6 / T73㉗
		W	250	24	T62
Extruded Rod, Bar and Profiles (Shapes)	870	W51⑥ / W51⑥	250 ⑲㉒	24 ⑲㉒	T651⑥ / T7351⑥㉗
		W / W / W	250⑲ / ㉒㉒ / ㉘	24 / ㉒㉒ / ㉘	T6 / T73㉗ / T76㉗
		W	250⑲	24	T62
		W510⑥ / W510⑥ / W510⑥	250⑲ / ㉒㉒ / ㉘	24 / ㉒㉒ / ㉘	T6510⑥ / T73510⑥㉗ / T76510⑥㉗
		W511⑥ / W511⑥ / W511⑥	250⑲ / ㉒㉒ / ㉘	24 / ㉒㉒ / ㉘	T6511⑥ / T73511⑥㉗ / T76511⑥㉗
Extruded Tube	870	W / W	250⑲ / ㉒㉒	24 / ㉒㉒	T6 / T73㉗
		W	250⑲	24	T62
		W510⑥ / W510⑥	250⑲ / ㉒㉒	24 / ㉒㉒	T6510⑥ / T73510⑥㉗
		W511⑥ / W511⑥	250⑲ / ㉒㉒	24 / ㉒㉒	T6511⑥ / T73511⑥㉗
Drawn Tube	870	W / W	250 / ㉒㉒	24 / ㉒㉒	T6 / T73㉗
		W	250	24	T62
Die Forgings	880⑨	W / W	250 / ㉒	24 / ㉒	T6 / T73㉗
		W52⑩	㉘	㉘	T7352⑩㉒

TABLE 2.10 Typical Heat Treatments for Wrought Alloys *(Continued)* ①

ALLOY	PRODUCT	SOLUTION HEAT TREATMENT ②		PRECIPITATION HEAT TREATMENT		
		METAL TEMPERATURE ③ °F	TEMPER DESIGNATION	METAL TEMPERATURE ③ °F	APPROX. TIME AT TEMPERATURE ④ Hours	TEMPER DESIGNATION
7075 ⑦	Hand Forgings	880 ⑨	W W	250 ⑧	24 ⑧	T6 T73 ⑫
			W52 ⑩ W52 ⑩	250 ⑧	24 ⑧	T652 ⑩ T7352 ⑩ ⑫
	Rolled Rings	880	W	250	24	T6
7178 ⑦	Sheet	875	W W	250 ⑧	24 ⑧	T6 T76 ⑫
			W	250	24	T62
	Plate	875	W51 ⑥ W51 ⑥	250 ⑧	24 ⑧	T651 ⑥ T7651 ⑥ ⑫
			W	250	24	T62
	Rolled or Cold Finished Wire and Rod	870	W	250	24	T6
	Extruded Rod, Bar and Profiles (Shapes)	870	W W	250 ⑧	24 ⑧	T6 T76 ⑫
			W	250	24	T62
			W510 ⑥ W510 ⑥	250 ⑧	24 ⑧	T6510 ⑥ T76510 ⑥ ⑫
			W511 ⑥ W511 ⑥	250 ⑧	24 ⑧	T6511 ⑥ T76511 ⑥ ⑫
7475	Sheet	900 ⑬	W W	⑧ ⑭	⑭ ⑭	T61 T761
	Plate	900 ⑬	W51 ⑥ W51 ⑥ W51 ⑥	240 ⑪ ⑭	24 ⑪ ⑭	T651 T7351 T7651
	Rod	900	W	⑭	⑭	T62

① The times and temperatures shown are typical for various forms, sizes and methods of manufacture and may not exactly describe the optimum treatment for a specific item.

② Material should be quenched from the solution heat-treating temperature as rapidly as possible and with minimum delay after removal from the furnace. Unless otherwise indicated, when material is quenched by total immersion in water, the water should be at room temperature and suitably cooled to remain below 100° F during the quenching cycle. The use of high-velocity high-volume jets of cold water is also effective for some materials. For additional details on aluminum alloy heat treatment and for recommendations on such specifics as furnace solution heat treat soak time see military Specification MIL-H-6088 or ASTM B597.

③ The nominal metal temperatures should be attained as rapidly as possible and maintained ±10° F of nominal during the time at temperature.

④ The time at temperature will depend on time required for load to reach temperature. The times shown are based on rapid heating, with soaking time measured from the time the load reaches within 10° F of the applicable temperature.

⑤ Cold work subsequent to solution heat treatment and, where applicable, prior to any precipitation heat treatment is required to attain the specified mechanical properties for these tempers.

⑥ Stress-relieved by stretching. Required to produce a specified amount of permanent set subsequent to solution heat treatment and, where applicable, prior to any precipitation heat treatment.

⑦ These heat treatments also apply to alclad sheet and plate in these alloys.

⑧ An alternative treatment comprised of 8 hours at 350° F also may be used.

⑨ Quench after solution treatment in water at 140° F to 180° F.

⑩ Stress-relieved by 1–5 percent cold reduction subsequent to solution heat treatment and prior to precipitation heat treatment.

⑪ Quench after solution heat treatment in water at 212° F.

⑫ Quench after solution heat treatment in air blast at room temperature.

⑬ An alternative treatment comprised of 8 hours at 340° F also may be used.

⑭ Cold working subsequent to precipitation heat treatment is necessary to secure the specified properties for this temper.

⑮ By suitable control of extrusion temperature, product may be quenched directly from extrusion press to provide specified properties for this temper. Some products may be adequately quenched in air blast at room temperature.

⑯ An alternate treatment comprised of 1–2 hours at 400° F also may be used.

⑰ An alternate treatment comprised of 6 hours at 360° F also may be used.

⑱ An alternate two-stage treatment comprised of 4 hours at 205° F followed by 8 hours at 315° F also may be used.

⑲ An alternate three-stage treatment comprised of 5 hours at 210° F followed by 4 hours at 250° F followed by 4 hours at 300° F also may be used.

⑳ Two-stage treatment comprised of 6 to 8 hours at 225° F followed by a second-stage of:

(a) 24-30 hours at 325° F for sheet and plate
(b) 8-10 hours at 350° F for rolled or cold-finished rod and bar.
(c) 6-8 hours at 350° F for extrusions and tube.
(d) 8-10 hours at 350° F for forgings in T73 temper and 6-8 hours at 350° F for forgings in T7352 temper.

㉑ Applies to tread plate only.

㉒ Held at room temperature for 72 hours followed by two stage precipitation heat-treatment of 8 hours at 225° F plus 16 hours at 300° F.

㉓ With optimum ingot homogenization, heat treating temperatures as high as 928° F are sometimes acceptable.

㉔ An alternate two-stage treatment for sheet, plate, tube and extrusions comprised of 6 to 8 hours at 225° F followed by a second stage of 14-18 hours at 335° F may be used providing a heating-up rate of 25° F per hour is used. For rolled or cold-finished rod and bar the alternate treatment is 10 hours at 350° F.

㉕ A two-stage treatment comprised of 3-5 hours at 250° F followed by 15-18 hours at 325° F.

㉖ A two-stage treatment comprised of 3-5 hours at 250° F followed by 18-21 hours at 320° F.

㉗ The aging of aluminum alloys 7075 and 7178 from any temper to the T73 (applicable to alloy 7075 only) or T76 temper series requires closer than normal controls on aging practice variables such as time, temperature, heating-up rates, etc., for any given item. In addition to the above, when re-aging material in the T6 temper series to the T73 or T76 temper series, the specific condition of the T6 temper material (such as its property level and

TABLE 2.11   Recommended Times and Temperatures for Heat Treating Commonly Used Aluminum Sand and Permanent Mold Castings[1]

Alloy	Temper	Product[3]	Solution heat treatment[2] Metal temps. ±10°F[4]	Time, hr.	Aging treatment Metal temps. ±10°F[4]	Time, hr.
201.0	T6	S	950–960 then 980–990	2 14–20	Room temp. then 310	14–24 20
201.0	T7	S	950–960 then 980–990	2 14–20	Room temp. then 370	12–14 5
204.0	T4	S or P	970	10	Room temp.	5 days
208.0	T4	P	940	4–12	—	—
208.0	T6	P	940	4–12	310	2–5
208.0	T7	P	940	4–12	500	4–6
222.0	O[5]	S	—	—	600	3
222.0	T61	P	950	12	310	11
222.0	T551	P	—	—	340	16–22
222.0	T65	S	950	4–12	340	7–9
242.0	O[6]	S	—	—	650	3
242.0	T571	S	—	—	400	8
242.0	T77	S	960	5[7]	625–675	2 min
242.0	T571	S or P	—	—	400	7–9
242.0	T61	S or P	960	4–12[7]	400–450	3–5
295.0	T4	S	960	12	—	—
295.0	T6	S	960	12	310	3–6
295.0	T62	S	960	12	310	12–24
295.0	T7	S	960	12	500	4–6
296.0	T6	P	950	8	310	1–8
319.0	T5	S	—	—	400	8
319.0	T6	S	940	12	310	2–5
319.0	T6	P	940	4–12	310	2–5
328.0	T6	S	960	12	310	2–5
332.0	T5	P	—	—	400	7–9
333.0	T5	P	—	—	400	7–9
333.0	T6	P	940	6–12	310	2–5
333.0	T7	P	940	6–12	500	4–6
336.0	T551	P	—	—	400	7–9
336.0	T65	P	960	8	400	7–9
354.0	—	See note[8]	980–995	10–12	See note[9]	See note[9]
355.0	T51	S or P	—	—	440	7–9
355.0	T6	S	980	12	310	3–5
355.0	T7	S	980	12	440	3–5
355.0	T71	S	980	12	475	4–6
355.0	T6	P	980	4–12	310	2–5
355.0	T62	P	980	4–12	340	14–18
355.0	T7	P	980	4–12	440	3–9
355.0	T71	P	980	4–12	475	3–6

TABLE 2.11 Recommended Times and Temperatures for Heat Treating Commonly Used Aluminum Sand and Permanent Mold Castings[1] *(Continued)*

Alloy	Temper	Product[3]	Solution heat treatment[2]		Aging treatment	
			Metal temps. ±10°F[4]	Time, hr.	Metal temps. ±10°F[4]	Time, hr.
C355.0	T6	S	980	12	310	3–5
C355.0	T61	P	980	6–12	Room temp. then 310	8 min 10–12
356.0	T51	S or P	—	—	440	7–9
356.0	T6	S	1000	12	310	3–5
356.0	T7	S	1000	12	400	3–5
356.0	T71	S	1000	10–12	475	3
356.0	T6	P	1000	4–12	310	2–5
356.0	T7	P	1000	4–12	440	7–9
356.0	T71	P	1000	4–12	475	3–6
A356.0	T6	S	1000	12	310	3–5
A356.0	T61	P	1000	6–12	Room temp. then 310	8 min 6–12
357.0	T6	P	1000	8	330	6–12
A357.0	—	See note[8]	1000	8–12	See note[9]	See note[9]
359.0	—	See note[8]	1000	10–14	See note[9]	See note[9]
A444.0	T4	P	1000	8–12	—	—
520.0	T4	S	810	18[10]	—	—
535.0	T5[6]	S	—	—	750	5
705.0	T5	S	—	—	Room temp. 210	21 days, or 8
705.0	T5	P	—	—	Room temp. 210	21 days, or 10
707.0	T7	S	990	8–16	350	4–10
707.0	T7	P	990	4–8	350	4–10
710.0	T5	S	—	—	Room temp.	21 days
711.0	T1	P	—	—	Room temp.	21 days
712.0	T5	S	—	—	Room temp. 315	21 days, or 6–8
713.0	T5	S or P	—	—	Room temp.	21 days
771.0	T5	S	—	—	355	3–5[11]
771.0	T51	S	—	—	405	6
771.0	T52[5]	S	—	—	330	6–12[11]
771.0	T53[5]	S	—	—	360	4[11]
771.0	T6	S	1090	6[11]	265	3
771.0	T71	S	1090	6[5]	430	15
850.0	T5	S or P	—	—	430	7–9
851.0	T5	S or P	—	—	430	7–9
851.0	T6	P	900	6	430	4

TABLE 2.11 Recommended Times and Temperatures for Heat Treating Commonly Used Aluminum Sand and Permanent Mold Castings[1] *(Continued)*

Alloy	Temper	Product[3]	Solution heat treatment[2]		Aging treatment	
			Metal temps. ±10°F[4]	Time, hr.	Metal temps. ±10°F[4]	Time, hr.
852.0	T5	S or P	—	—	430	7–9

[1]The heat treat times and temperatures given in this standard are those in general use in the industry. The times and temperatures shown for solution heat treatment are critical. Quenching must be accomplished by complete immersion of the castings with a minimum delay after the castings are removed from the furnace.

Under certain conditions, complex castings that might crack or distort in the water quench can be oil or air blast quenched. When this is done, the purchaser and the foundry must agree to the procedure and also agree on the level of mechanical properties that will be acceptable. Aging treatments can be varied slightly to attain the optimum treatment for a specific casting or to give agreed upon slightly different levels of mechanical properties.

Temper designations for castings are as follows:

F   As cast—cooled naturally from the mold in room temperature air with no further heat treatment.

O   Annealed. Usually the weakest, softest, most ductile, and most dimensionally stable condition.

T4   Solution heat treated and naturally aged to substantially stable condition. Mechanical properties and stability may change over a long period of time.

T5   Naturally cooled from the mold and then artificially aged to attain improved mechanical properties and dimensional stability.

T6   Solution heat treated and artificially aged to attain optimum mechanical properties and generally good dimensional stability.

T7   Solution heat treated and over-aged for improved dimensional stability, but usually with some reduction from the optimum mechanical properties.
The T5, T6, and T7 designations are sometimes followed by one or more numbers that indicate changes from the originally developed treatment.
[2]Unless otherwise noted, quench in water at 150–212°F.
[3]S = sand cast, P = permanent mold cast.
[4]Temperature range unless otherwise noted.
[5]Stress relieve for dimensional stability in the following manner: (1) Hold at 775 —25°F for 5 hr. Then (2) furnace cool to 650°F for 2 or more hr. Then (3) furnace cool to 450°F for not more than 1/2 hr. Then (4) furnace cool to 250°F for approximately 2 hr. Then (5) cool to room temperature in still air outside the furnace.
[6]No quench required. Cool in still air outside furnace.
[7]Use air blast quench.
[8]Casting process varies; sand, permanent mold, or composite to obtain desired mechanical properties.
[9]Solution heat treat as indicated then artificially age by heating uniformly for the time and temperature necessary to obtain the desired mechanical properties.
[10]Quench in water at 150–212°F for 10–20 seconds only.
[11]Cool in still air outside furnace to room temperature.

ements comprising the alloy; the 5xxx and 6xxx series alloys are the lightest since magnesium is the lightest of the main alloying elements. Densities for wrought aluminum alloys are listed in Table 2.14. Densities for cast alloys are given in Table 2.15. The density of a casting is less than that of the cast alloy, because some porosity cannot be avoided in producing castings. The density of castings is usually about 95 to 100% of the theoretical density of the cast alloy.

The *melting point* also varies by alloy. While pure aluminum melts at about 1220°F (660°C), the addition of alloying elements depresses the melting point to between about 900 and 1200°F (500 to 650°C) and produces a melting range, since the different alloying elements melt at different temperatures. Most aluminum alloys' mechanical properties are significantly degraded well below their melting point. Few alloys are used above 400°F (200°C), although some, like 2219, have applications in engines up to about 600°F (300°C)

Thermal and electrical conductivity also vary widely by alloy. The purer grades of aluminum have the highest conductivities, up to a *thermal conductivity* of about 1625 Btu-in./ft^{2}hr°F (234 W/m-K) and an *electrical conductivity* of 62% of the International Annealed Copper Standard (IACS) at 68°F (20°C) for equal volume, or 204% of IACS for equal weight.

The *coefficient of thermal expansion*, the rate at which material expands as its temperature increases, is itself a function of temperature, being slightly higher at greater temperatures. Average values are used for a range of temperature, usually from room temperature [68°F (20°C)] to water's boiling temperature [212°F (100°C)]. A commonly used number for this range is 13×10^{-6}/°F (23×10^{-6}/°C). This compares to 18 for copper, 15 for magnesium, 9.6 for stainless steel, and 6.5×10^{-6}/°F for carbon steel.

For wrought alloys, typical physical properties are given in Table 2.16. Typical physical properties of cast alloys are given in Table 2.15.

2.4 Mechanical Properties

Mechanical properties are properties related to the behavior of material when subjected to force. Most are measured according to standard test methods provided by the American Society for Testing and Materials (ASTM). The mechanical properties of interest for aluminum and ASTM test methods by which they are measured are:

Strength

   Tensile yield strength (F_{ty})—B557
   Tensile ultimate strength (F_{tu})—B557
   Compressive yield strength (F_{cy})—E9
   Shear ultimate strength (F_{su})—B565 (fastener material), B769, B831 (thin material)

Modulus of elasticity (E)—E111

Modulus of rigidity (G)

Table 2.12 Comparative Characteristics and Applications of Wrought Alloys

ALLOY AND TEMPER	RESISTANCE TO CORROSION		Workability (Cold) ③	Machinability ④	Brazeability ⑤	WELDABILITY ⑥			SOME APPLICATIONS OF ALLOYS
	General ①	Stress-Corrosion Cracking ②				Gas	Arc	Resistance Spot and Seam	
1060-O	A	A	A	E	A	A	A	B	Chemical equipment, railroad tank cars
H12	A	A	A	E	A	A	A	A	
H14	A	A	A	D	A	A	A	A	
H16	A	A	B	D	A	A	A	A	
H18	A	A	B	D	A	A	A	A	
1100-O	A	A	A	E	A	A	A	B	Sheet metal work, spun hollowware, fin stock
H12	A	A	A	E	A	A	A	A	
H14	A	A	A	D	A	A	A	A	
H16	A	A	B	D	A	A	A	A	
H18	A	A	C	D	A	A	A	A	
1350-O	A	A	A	E	A	A	A	B	Electrical conductors
H12, H111	A	A	A	E	A	A	A	A	
H14, H24	A	A	A	D	A	A	A	A	
H16, H26	A	A	B	D	A	A	A	A	
H18	A	A	B	D	A	A	A	A	
2011-T3	D③	D	C	A	D	D	D	D	Screw machine products
T4, T451	D③	D	B	A	D	D	D	D	
T8	D	B	D	A	D	D	D	D	
2014-O	:	:	:	D	D	D	D	B	Truck frames, aircraft structures
T3, T4, T451	D③	C	C	B	D	B	B	B	
T6, T651, T6510, T6511	D	C	D	B	D	B	B	B	
2017-T4, T451	D③	C	C	B	D	B	B	B	Screw machine products, fittings
2018-T61	:	:	C	B	D	D	C	B	Aircraft engine cylinders, heads and pistons

Alloy & Temper									Applications
2024-O	D③	...	...	D	D	D	D	D	Truck wheels, screw machine products, aircraft structures
T4, T3, T351, T3510, T3511	D③	C	C	B	D	D	B	B	
T361	D	C	D	B	D	D	C	B	
T6	D	B	C	B	D	D	C	B	
T861, T81, T851, T8510, T8511	D	B	D	B	D	D	C	B	
T72	...	...	...	B	D	D	C	B	
2025-T6	D	C	B	B	D	D	B	B	Forgings, aircraft propellers
2036-T4	C	...	B	C	C	C	B	B	Auto body panel sheet
2117-T4	C	A	D	C	D	D	B	B	Rivets
2124-T851	D	B	...	B	D	D	C	B	Aircraft structures
2218-T61	D	C	...	...	D	D	C	B	Jet engine impellers and rings
T72	D	C	...	B	D	D	C	B	
2219-O	...	...	...	...	D	D	A	B	Structural uses at high temperatures (to 600 F) / High strength weldments
T31, T351, T3510, T3511	D③	C	C	B	A	A	A	A	
T37	D③	C	D	B	A	A	A	A	
T81, T851, T8510, T8511	D	B	D	B	A	A	A	A	
T87	D	B	D	B	A	A	A	A	
2618-T61	D	C	...	B	D	D	C	B	Aircraft engines
3003-O	A	A	A	E	A	A	A	B	Cooking utensils, chemical equipment, pressure vessels, sheet metal work, builder's hardware, storage tanks
H12	A	A	A	E	A	A	A	A	
H14	A	A	B	D	A	A	A	A	
H16	A	A	C	D	A	A	A	A	
H18	A	A	C	D	A	A	A	A	
H25	A	A	B	D	A	A	A	A	
3004-O	A	A	A	D	B	A	A	B	Sheet metal work, storage tanks
H32	A	A	B	C	B	A	A	A	
H34	A	A	B	C	B	A	A	A	
H36	A	A	C	C	B	A	A	A	
H38	A	A	C	C	B	A	A	A	
3105-O	A	A	A	E	A	A	A	B	Residential siding, mobile homes, rain carrying goods, sheet metal work
H12	A	A	B	E	A	A	A	A	
H14	A	A	B	D	A	A	A	A	
H16	A	A	C	D	A	A	A	A	
H18	A	A	C	D	A	A	A	A	
H25	A	A	B	D	A	A	A	A	

Table 2.12 Comparative Characteristics and Applications of Wrought Alloys (Continued)

ALLOY AND TEMPER	RESISTANCE TO CORROSION General (1)	RESISTANCE TO CORROSION Stress-Corrosion Cracking (2)	Workability (Cold) (3)	Machinability (4)	Brazeability (5)	WELDABILITY (6) Gas	WELDABILITY (6) Arc	WELDABILITY (6) Resistance Spot and Seam	SOME APPLICATIONS OF ALLOYS
4032-T6	C	B	...	B	D	D	B	C	Pistons
5005-O	A	A	A	E	B	A	A	B	Appliances, utensils, architectural, electrical conductor
H12	A	A	A	E	B	A	A	A	
H14	A	A	B	E	B	A	A	A	
H16	A	A	C	D	B	A	A	A	
H18	A	A	C	D	B	A	A	A	
H32	A	A	A	E	B	A	A	A	
H34	A	A	B	D	B	A	A	A	
H36	A	A	C	D	B	A	A	A	
H38	A	A	C	D	B	A	A	A	
5050-O	A	A	A	E	B	A	A	B	Builder's hardware, refrigerator trim, coiled tubes
H32	A	A	A	D	B	A	A	A	
H34	A	A	B	D	B	A	A	A	
H36	A	A	C	C	B	A	A	A	
H38	A	A	C	C	B	A	A	A	
5052-O	A	A	A	D	C	A	A	B	Sheet metal work, hydraulic tube, appliances
H32	A	A	B	D	C	A	A	A	
H34	A	A	B	C	C	A	A	A	
H36	A	A	C	C	C	A	A	A	
H38	A	A	C	C	C	A	A	A	
5056-O	A[4]	B[4]	A	D	D	C	C	B	Cable sheathing, rivets for magnesium, screen wire, zipper
H111	A[4]	B[4]	A	D	D	C	C	A	
H12, H32	A[4]	B[4]	B	D	D	C	C	A	
H14, H34	A[4]	B[4]	B	C	D	C	C	A	
H18, H38	A[4]	C[4]	C	C	D	C	C	A	
H192	B[4]	D[4]	D	B	D	C	C	A	
H392	B[4]	D[4]	D	B	D	C	C	A	

Alloy-Temper									Applications
5083-O	A④	A④	B	D	D	C	A	B	
H321, H116	A④	A④	C	D	D	C	A	A	
H111	A④	B④	C	D	D	C	A	A	
5086-O	A④	A④	A	D	D	C	A	B	Unfired, welded pressure vessels, marine, auto aircraft cryogenics, TV towers, drilling rigs, transportation equipment, missile components
H32, H116	A④	B④	B	D	D	C	A	A	
H34	A④	B④	B	D	C	C	A	A	
H36	A④	B④	C	D	C	C	A	A	
H38	A④	B④	C	D	C	C	A	A	
H111	A④	A④	B	D	D	C	A	A	
5154-O	A④	A④	A	D	D	C	A	B	Welded structures, storage tanks, pressure vessels, salt water service
H32	A④	A④	B	D	D	C	A	A	
H34	A④	A④	B	D	C	C	A	A	
H36	A④	A④	C	D	C	C	A	A	
H38	A④	A④	C	D	C	C	A	A	
5252-H24	A	A	B	C	D	B	A	A	Automotive and appliance trim
H25	A	A	B	C	D	B	A	A	
H28	A	A	C	C	C	C	A	A	
5254-O	A④	A④	A	D	D	C	A	B	Hydrogen peroxide and chemical storage vessels
H32	A④	A④	B	D	D	C	A	A	
H34	A④	A④	B	D	C	C	A	A	
H36	A④	A④	C	D	C	C	A	A	
H38	A④	A④	C	D	C	C	A	A	
5454-O	A	A	A	D	D	C	A	B	Welded structures, pressure vessels, marine service
H32	A	A	B	D	D	C	A	A	
H34	A	A	B	D	C	C	A	A	
H111	A	A	B	D	D	C	A	A	
5456-O	A④	B④	B	D	D	C	A	B	High strength welded structures, pressure vessels, marine applications, storage tanks
H321, H116	A④	B④	C	D	D	C	A	A	
5457-O	A	A	A	B	E	B	A	B	
5652-O	A	A	A	C	D	C	A	B	Hydrogen peroxide and chemical storage vessels
H32	A	A	B	C	D	C	A	A	
H34	A	A	B	C	D	C	A	A	
H36	A	A	C	C	C	A	A	A	
H38	A	A	C	C	C	A	A	A	

Table 2.12 Comparative Characteristics and Applications of Wrought Alloys *(Continued)*

ALLOY AND TEMPER	RESISTANCE TO CORROSION		Workability (Cold) ③	Machinability ④	Brazeability ⑤	WELDABILITY ⑥			SOME APPLICATIONS OF ALLOYS
	General ①	Stress-Corrosion Cracking ②				Gas	Arc	Resistance Spot and Seam	
5657-H241	A	A	A	D	B	A	A	A	Anodized auto and appliance trim
H25	A	A	B	D	B	A	A	A	
H26	A	A	B	D	B	A	A	A	
H28	A	A	C	D	B	A	A	A	
6005-T1, T5	:	:	:	:	A	A	A	A	
6053-O	:	:	:	E	B	A	A	B	Wire and rod for rivets
T6, T61	A	A	:	C	B	A	A	A	
6061-O	B	A	A	D	A	A	A	B	Heavy-duty structures requiring good corrosion resistance, truck and marine, railroad cars, furniture, pipelines
T4, T451, T4510, T4511	B	B	B	C	A	A	A	A	
T6, T651, T652, T6510, T6511	B	A	C	C	A	A	A	A	
6063-T1	A	A	B	D	A	A	A	A	Pipe railing, furniture, architectural extrusions
T4	A	A	B	D	A	A	A	A	
T5, T452	A	A	B	C	A	A	A	A	
T6	A	A	C	C	A	A	A	A	
T83, T831, T832	A	A	C	C	A	A	A	A	
6066-O	C	A	B	D	D	D	B	B	Forgings and extrusion for welded structures
T4, T4510, T4511	C	B	C	C	D	D	B	B	
T6, T6510, T6511	C	B	C	B	D	D	B	B	
6070-T4, T4511	B	B	B	C	D	A	A	A	Heavy duty welded structures, pipelines
T6	B	B	C	C	D	A	A	A	
6101-T6, T63	A	A	C	C	A	A	A	A	High strength bus conductors
T61, T64	A	A	B	D	A	A	A	A	
6151-T6, T652	:	:	:	:	B	:	:	:	Moderate strength, intricate forgings for machine and auto parts
6201-T81	A	A	:	C	A	A	A	A	High strength electric conductor wire

Designation									Some applications
6262-T6, T651, T6510, T6511	B	A	C	B	B	B	B	A	Screw machine products
T9	B	A	D	B	B	B	B	A	
6351-T1	..	..	C	C	C	B	A	B	Extruded shapes, structurals, pipe and tube
T4	A	..	C	C	C	B	A	B	
T5	A	..	C	C	C	B	A	A	
T6	A	..	C	C	C	B	A	A	
6463-T1	A	A	B	D	A	A	A	A	Extruded architectural and trim sections
T5	A	A	B	C	A	A	A	A	
T6	A	A	C	C	A	A	A	A	
6951-T42, T62	..	..	..	..	A	A	A	A	
7005-T53	..	..	..	B	B	C	A	A	
7049-T73, T7352	C	B	D	B	D	D	D	B	Aircraft forgings
7050-T73510, T73511 T74[1], T7451[1], T74510[1], T74511[1], T7452[1], T7651, T76510, T76511	C	B	D	B	D	D	D	B	Aircraft and other structures
7075-O	..	..	..	..	D	D	D	B	Aircraft and other structures
T6, T651, T652, T6510, T6511	C[3]	C	D	B	D	D	D	B	
T73, T7351	C	B	D	B	D	D	D	B	
7175-T74, T7452, T7454	C	B	D	B	D	D	C	B	
7178-O	..	..	..	..	D	D	D	B	Aircraft and other structures
T6, T651, T6510, T6511	C[3]	C	D	B	D	D	D	B	
7475-O	..	..	..	..	D	D	D	B	Shell Casings Aircraft & Other Structures
7475-T61, -T651	C	C	D	B	D	B	B	B	
7475-T761, T7351	C	B	D	B	D	D	D	B	
8017-H12, H22, H221	A	A	A	D	A	A	A	A	Electrical conductors
8030-H12, H221	A	A	A	E	A	A	A	A	Electrical conductors
8176-H14, H24	A	A	A	D	A	A	A	A	Electrical conductors
8177-H13, H23, H221	A	A	A	E	A	A	A	A	Electrical conductors

Notes for Table 2.12.

① Ratings A through E are relative ratings in decreasing order of merit, based on exposures to sodium chloride solution by intermittent spraying or immersion. Alloys with A and B ratings can be used in industrial and seacoast atmospheres without protection. Alloys with C, D and E ratings generally should be protected at least on faying surfaces.

② Stress-corrosion cracking ratings are based on service experience and on laboratory tests of specimens exposed to the 3.5% sodium chloride alternate immersion test.

A = No known instance of failure in service or in laboratory tests.

B = No known instance of failure in service; limited failures in laboratory tests of short transverse specimens.

C = Service failures with sustained tension stress acting in short transverse direction relative to grain structure; limited failures in laboratory tests of long transverse specimens.

D = Limited service failures with sustained longitudinal or long transverse areas.

These ratings are neither product specific nor test direction specific and therefore indicate only the general level of stress-corrosion cracking resistance. For more specific information on certain alloys, see ASTM G64.

③ In relatively thick sections the rating would be E.

④ This rating may be different for material held at elevated temperature for long periods.

⑤ Ratings A through D for Workability (cold), and A through E for Machinability, are relative ratings in decreasing order of merit.

⑥ Ratings A through D for Weldability and Brazeability are relative ratings defined as follows:

A = Generally weldable by all commercial procedures and methods.

B = Weldable with special techniques or for specific applications that justify preliminary trials or testing to develop welding procedure and weld performance.

C = Limited weldability because of crack sensitivity or loss in resistance to corrosion and mechanical properties.

⑦ T74 type tempers, although not previously registered, have appeared in various literature and specifications as T736 type tempers.

TABLE 2.13 Characteristics of Aluminum Sand and Permanent Mold Castings [1]

Alloy	Product	Fluidity	Resistance to hot cracking	Pressure tightness	Normally heat treated?	Strength at elevated temps.	Corrosion resistance	Machinability	Polishing	Anodizing appearance	Weldability
201.0	S	3	4	3	Yes	1	4	1	1	2	4
204.0	S&P	3	4	3	Yes	1	4	1	2	3	4
208.0	S	2	2	2	Optional	3	4	3	3	3	2
222.0	S&P	3	3	3	Yes	1	4	1	2	3	3
242.0	S&P	3	4	4	Yes	1	4	2	2	3	4
295.0	S	3	4	4	Yes	3	4	2	1	2	3
296.0	P	3	4	3	Yes	2	4	3	3	3	3
308.0	P	2	2	2	No	3	3	3	1	4	2
319.0	S&P	2	2	2	Optional	3	3	3	3	4	2
328.0	S	1	1	2	Optional	2	3	3	3	4	1
332.0	P	1	2	2	Yes	1	3	4	4	4	2
333.0	P	1	2	2	Optional	2	3	3	3	4	3
336.0	P	1	2	2	Yes	1	3	4	4	4	3
354.0	P	1	1	1	Yes	2	3	4	4	4	3
355.0	S&P	1	1	1	Yes	2	3	3	3	4	1
C355.0	S&P	1	1	1	Yes	2	3	3	3	4	1
356.0	S&P	1	1	1	Yes	3	2	3	4	4	1
A356.0	S&P	1	1	1	Yes	3	2	3	4	4	1
357.0	S&P	1	1	1	Yes	3	2	3	4	4	1
A357.0	S&P	1	1	1	Yes	2	2	3	4	4	1
359.0	S&P	1	2	2	Yes	2	2	4	4	4	1
443.0	S&P	1	1	1	No	4	3	5	4	4	1
B443.0	S&P	1	1	1	No	4	2	5	4	4	1
A444.0	P	1	1	1	Optional	4	2	5	4	4	1
512.0	S	3	3	4	No	3	1	2	2	2	3
513.0	P	4	4	4	No	3	1	1	1	1	3

TABLE 2.13 Characteristics of Aluminum Sand and Permanent Mold Castings (Continued)[1]

Alloy	Product	Fluidity	Resistance to hot cracking	Pressure tightness	Normally heat treated?	Strength at elevated temps.	Corrosion resistance	Machinability	Polishing	Anodizing appearance	Weldability
514.0	S	4	4	5	No	3	1	1	1	1	3
520.0	S	4	4	5	Yes	5	1	1	1	1	4
535.0	S	5	4	5	Optional	3	1	1	1	1	4
705.0	S&P	4	4	4	No	4	2	1	2	2	4
707.0	S&P	4	4	4	No	4	2	1	2	2	4
710.0	S	4	5	4	No	4	4	1	2	2	4
711.0	P	4	5	4	Yes	5	2	1	1	1	4
712.0	S	3	5	4	No	4	3	1	2	2	4
713.0	S&P	3	4	4	No	4	3	1	1	1	4
771.0	S	3	4	4	Yes	4	3	1	1	1	4
850.0	S&P	4	5	5	Yes	5	4	1	3	n/a[2]	5
751.0	S&P	4	5	5	Yes	5	4	1	3	n/a[2]	5
852.0	S&P	4	5	5	Yes	5	4	1	3	n/a[2]	5

[1]Selection of an alloy for a particular application requires consideration not only of mechanical properties but also of numerous other characteristics such as behavior in the casting process or subsequent treatments in the course of manufacture, and response to the environmental conditions of service. This table includes several significant characteristics that deserve consideration in the selection of an alloy. The characteristics are comparatively rated from 1 to 5 in decreasing order of performance.
[2]Information not available.

TABLE 2.14   Nominal Densities of Wrought Aluminum and Aluminum Alloys[1]

Alloy	Density (lb/in^3)	Specific gravity	Alloy	Density (lb/in^3)	Specific gravity
1050	.0975	2.705	5252	.096	2.67
1060	.0975	2.705	5254	.096	2.66
1100	.098	2.71	5356	.096	2.64
1145	.0975	2.700	5454	.097	2.69
1175	.0975	2.700	5456	.096	2.66
1200	.098	2.70	5457	.097	2.69
1230	.098	2.70	5554	.097	2.69
1235	.0975	2.705	5556	.096	2.66
1345	.0975	2.705	5652	.097	2.67
1350	.0975	2.705	5654	.096	2.66
2011	.102	2.83	5657	.097	2.69
2014	.101	2.80	6003	.097	2.70
2017	.101	2.79	6005	.097	2.70
2018	.102	2.82	6053	.097	2.69
2024	.100	2.78	6061	.098	2.70
2025	.101	2.81	6063	.097	2.70
2036	.100	2.75	6066	.098	2.72
2117	.099	2.75	6070	.098	2.71
2124	.100	2.78	6101	.097	2.70
2218	.101	2.81	6105	.097	2.69
2219	.103	2.84	6151	.098	2.71
2618	.100	2.76	6162	.097	2.70
3003	.099	2.73	6201	.097	2.69
3004	.098	2.72	6262	.098	2.72
3005	.098	2.73	6351	.098	2.71
3105	.098	2.72	6463	.097	2.69
4032	.097	2.68	6951	.098	2.70
4043	.097	2.69	7005	.100	2.78
4045	.096	2.67	7008	.100	2.78
4047	.096	2.66	7049	.103	2.84
4145	.099	2.74	7050	.102	2.83
4343	.097	2.68	7072	.098	2.72
4643	.097	2.69	7075	.101	2.81
5005	.098	2.70	7175	.101	2.80
5050	.097	2.69	7178	.102	2.83
5052	.097	2.68	7475	.101	2.81
5056	.095	2.64	8017	.098	2.71
5083	.096	2.66	8030	.098	2.71
5086	.096	2.66	8176	.098	2.71
5154	.096	2.66	8177	.098	2.70
5183	.096	2.66			

[1]Density and specific gravity depend on composition, and variations are discernible from one cast to another for most alloys. The nominal values shown above should not be specified as engineering requirements but are used in calculating typical values for weight per unit length, weight per unit area, covering area, etc. The density values are derived from the metric and subsequently rounded. These values are not to be converted to the metric. X.XXX0 and X.XXX5 density values and X.XX0 and X.XX5 specific gravity values are limited to 99.35 percent or higher purity aluminum.

TABLE 2.15 Typical Physical Properties of Commonly Used Sand and Permanent Mold Casting Alloys

These typical properties are not guaranteed, and should not be used for design purposes but only as a basis for general comparison of alloys and tempers with respect to any given characteristic.

Alloy	Temper	Specific Gravity [a]	Density [a] lb. per cu. in.	Approximate Melting Range °F	Electrical Conductivity % IACS	Thermal Conductivity at 25°C, CGS [b]	Coeff. of Thermal Expansion, per °F × 10⁻⁶	
							68–212°F	68–572°F
201.0	T6	2.80	0.101	1060–1200	27–32	0.29	19.3	24.7
	T7	2.80	0.101	1060–1200	32–34	0.29	—	—
204.0	T4	—	—	985–1200	—	—	—	—
208.0	F	2.79	0.101	970–1160	31	0.30	12.4	13.4
222.0	T61	2.95	0.107	965–1155	33	0.31	12.3	13.1
242.0	T571 [c]	2.81	0.102	990–1175	34	0.32	12.6	13.6
	T77	2.81	0.102	990–1175	38	0.36	12.6	13.6
295.0	T6	2.81	0.102	970–1190	35	0.33	12.7	13.8
296.0	T6 [c]	2.80	0.101	970–1170	33	0.31	12.2	13.3
308.0	F	2.79	0.101	970–1135	37	0.35	11.9	12.9
319.0	F	2.79	0.101	960–1120	27	0.26	11.9	12.7
328.0	F	2.70	0.098	1025–1105	30	0.29	11.9	12.9
332.0	T5 [c]	2.76	0.100	970–1080	26	0.25	11.5	12.4
333.0	F [c]	2.77	0.100	960–1085	26	0.25	11.4	12.4
	T5 [c]	2.77	0.100	960–1085	29	0.28	11.4	12.4
	T6 [c]	2.77	0.100	960–1085	29	0.28	11.4	12.4
	T7 [c]	2.77	0.100	960–1085	35	0.33	11.4	12.4
336.0	T551 [c]	2.72	0.098	1000–1050	29	0.28	11.0	12.0
354.0	T61	2.71	0.098	1000–1105	32	0.30	11.6	12.7
355.0	T51	2.71	0.098	1015–1150	43	0.40	12.4	13.7
	T6	2.71	0.098	1015–1150	36	0.34	12.4	13.7
	T6 [c]	2.71	0.098	1015–1150	39	0.36	12.4	13.7
	T61	2.71	0.098	1015–1150	37	0.35	12.4	13.7
	T62 [c]	2.71	0.098	1015–1150	38	0.35	12.4	13.7
	T7	2.71	0.098	1015–1150	42	0.39	12.4	13.7
	T71	2.71	0.098	1015–1150	39	0.36	12.4	13.7
C355.0	T61	2.71	0.098	1015–1150	39	0.36	12.4	13.7
356.0	T51	2.68	0.097	1035–1135	43	0.40	11.9	12.9

Alloy	Temper	Specific gravity[a]	Specific heat	Approx. melting range (°F)	Electrical conductivity	Thermal conductivity[b]	Coefficient of thermal expansion	
A356.0	T6	2.68	0.097	1035–1135	39	0.36	11.9	12.9
	T6[c]	2.68	0.097	1035–1135	41	0.38	11.9	12.9
	T7	2.68	0.097	1035–1135	40	0.37	11.9	12.9
	T7[c]	2.68	0.097	1035–1135	43	0.40	11.9	12.9
357.0	T61	2.67	0.097	1035–1135	39	0.36	11.9	12.9
A357.0	F	2.67	0.097	1035–1135	39	0.36	11.9	12.9
359.0	T61	2.67	0.097	1045–1115	39	0.36	11.9	12.9
443.0	T6	2.69	0.097	1065–1170	35	0.33	11.6	12.7
B443.0	F	2.69	0.097	1065–1170	37	0.35	12.3	13.4
A444.0	F	2.68	0.097	1070–1170	37	0.35	12.3	13.4
512.0	F	2.65	0.097	1090–1170	41	0.38	12.1	13.2
513.0	F[c]	2.68	0.096	1075–1180	38	0.35	12.7	13.8
514.0	F	2.65	0.097	1110–1185	34	0.32	13.4	14.5
520.0	T4	2.57	0.093	840–1120	21	0.21	13.4	14.5
535.0	F	2.62	0.095	1020–1165	23	0.23	13.7	14.8
705.0	F	2.76	0.100	1105–1180	25	0.25	13.1	14.8
707.0	F	2.77	0.100	1085–1165	25	0.25	13.1	14.3
710.0	F[c]	2.81	0.102	1105–1195	35	0.33	13.2	14.4
711.0	F	2.84	0.103	1120–1200	40	0.37	13.4	14.6
712.0	F	2.81	0.101	1135–1200	35	0.33	13.1	14.2
713.0	F	2.81	0.100	1100–1180	30	0.29	13.7	14.8[d]
771.0	F[c]	2.81	0.102	1120–1190	37	0.33	13.4[d]	14.6[d]
850.0	T5[c]	2.88	0.104	435–1200	47	0.43	13.0	[e]
851.0	T5[c]	2.83	0.103	440–1165	43	0.40	12.6	[e]
852.0	T5[c]	2.88	0.104	400–1175	45	0.41	12.9	[e]

[a] Assuming solid (void-free) metal. Since some porosity cannot be avoided in commercial castings, the actual values will be slightly less than those given.

[b] Cgs units equals calories per second per square centimeter per centimeter of thickness per degree centigrade.

[c] Chill cast samples; all other samples cast in green sand mold.

[d] Estimated value.

[e] Exceeds operating temperature.

Reference: Aluminum, Volume I. Properties, Physical Metallurgy and Phase Diagrams, American Society for Metals, Metals Park, Ohio (1967). Data for alloy 771.0 supplied by the U.S. Reduction Company, East Chicago, Indiana.

TABLE 2.16 Typical Physical Properties of Wrought Alloys[1]

Alloy	Average[2] coefficient of thermal expansion 68° to 212°F per °F.	Melting range,[3,4] approx. °F	Temper	Thermal conductivity at 77°F English units[5]	Electrical conductivity at 68°F, percent of International Annealed Copper Standard		Electrical resistivity at 68°F Ohm-cir. Mil/foot
					Equal volume	Equal weight	
1060	13.1	1195–1215	O	1625	62	204	17
			H18	1600	61	201	17
1100	13.1	1190–1215	O	1540	59	194	18
			H18	1510	57	187	18
1350	13.2	1195–1215	All	1625	62	204	17
2011	12.7	1005–1190[6]	T3	1050	39	123	27
			T8	1190	45	142	23
2014	12.8	945–1180[7]	O	1340	50	159	21
			T4	930	34	108	31
			T6	1070	40	127	26
2017	13.1	955–1185[7]	O	1340	50	159	21
			T4	930	34	108	31
2018	12.4	945–1180[6]	T61	1070	40	127	26
2024	12.9	935–1180[7]	O	1340	50	160	21
			T3, T4, T361	840	30	96	35
			T6, F81, T861	1050	38	122	27
2025	12.6	970–1185[7]	T6	1070	40	128	26
2036	13.0	1030–1200[6]	T4	1100	41	135	25

2117	13.2	1030–1200[6]	T4	1070	40	130	26
2124	12.7	935–1180[7]	T851	1055	38	122	27
2218	12.4	940–1175[7]	T72	1070	40	126	26
2219	12.4	1010–1190[7]	O	1190	44	138	24
			T31, T38	780	28	88	37
			T6, T81, T87	840	30	94	35
2618	12.4	1020–1180	T6	1020	37	120	28
3003	12.9	1190–1210	O	1340	50	163	21
			H12	1130	42	137	25
			H14	1100	41	134	25
			H18	1070	40	130	26
3004	13.3	1165–1210	All	1130	42	137	25
3105	13.1	1175–1210	All	1190	45	148	23
4032	10.8	990–1060[7]	O	1070	40	132	26
			T6	960	35	116	30
4043	12.3	1065–1170	O	1130	42	140	25
4045	11.7	1065–1110	All	1190	45	151	23
4343	12.0	1070–1135	All	1250	47	158	25
5005	13.2	1170–1210	All	1390	52	172	20
5050	13.2	1155–1205	All	1340	50	165	21
5052	13.2	1125–1200	All	960	35	116	30
5056	13.4	1055–1180	O	810	29	98	36
			H38	750	27	91	38
5083	13.2	1095–1180	O	810	29	98	36
5086	13.2	1085–1185	All	870	31	104	33
5154	13.3	1100–1190	All	870	32	107	32
5252	13.2	1125–1200	All	960	35	116	30
5254	13.3	1100–1190	All	870	32	107	32
5356	13.4	1060–1175	O	810	29	98	36

TABLE 2.16 Typical Physical Properties of Wrought Alloys[1] (Continued)

Alloy	Average[2] coefficient of thermal expansion 68° to 212°F per °F.	Melting range,[3,4] approx. °F	Temper	Thermal conductivity at 77°F English units[5]	Electrical conductivity at 68°F, percent of International Annealed Copper Standard		Electrical resistivity at 68°F Ohm–cir. Mil/foot
					Equal volume	Equal weight	
5454	13.1	1115–1195	O	930	34	113	31
			H38	930	34	113	31
5456	13.3	1055–1180	O	810	29	98	36
5457	13.2	1165–1210	All	1220	46	153	23
5652	13.2	1125–1200	All	960	35	116	30
5657	13.2	1180–1215	All	1420	54	180	19
6005	13.0	1125–1210[6]	T1	1250	47	155	22
			T5	1310	49	161	21
6053	12.8	1070–1205[6]	O	1190	45	148	23
			T4	1070	40	132	26
			T6	1130	42	139	25
6061	13.1	1080–1205[6]	O	1250	47	155	22
			T4	1070	40	132	26
			T6	1160	43	142	24
6063	13.0	1140–1210	O	1510	58	191	18
			T1	1340	50	165	21
			T5	1450	55	181	19
			T6, T83	1390	53	175	20

Alloy							
6066	12.0	1045–1195[7]	O	1070	40	132	26
			T6	1020	37	122	28
6070	..	1050–1200[7]	T6	1190	44	145	24
6101	13.0	1150–1210	T6	1510	57	188	18
			T61	1540	59	194	18
			T63	1520	58	191	18
			T64	1570	60	198	17
			T65	1510	58	191	18
6105	13.0	1110–1200[6]	T1	1220	46	151	23
			T5	1340	50	165	21
6151	12.9	1090–1200[6]	O	1420	54	178	19
			T4	1130	42	138	25
			T6	1190	45	148	23
6201	13.0	1125–1210[6]	T81	1420	54	180	19
6253	..	1100–1205	..	..	..	..	..
6262	13.0	1080–1205[6]	T9	1190	44	145	24
6351	13.0	1030–1200	T6	1220	46	151	23
6463	13.0	1140–1210	T1	1340	50	165	21
			T5	1450	55	181	19
			T6	1390	53	175	20
6951	13.0	1140–1210	O	1480	56	186	19
			T6	1370	52	172	20
7049	13.0	890–1175	T73	1070	40	132	26
7050	12.8	910–1165	T74[9]	1090	41	135	25
7072	13.1	1185–1215	O	1540	59	193	18
7075	13.1	890–1175[8]	T6	900	33	105	31

TABLE 2.16 Typical Physical Properties of Wrought Alloys[1] (Continued)

Alloy	Average[2] coefficient of thermal expansion 68° to 212°F per °F.	Melting range,[3,4] approx. °F	Temper	Thermal conductivity at 77°F English units[5]	Electrical conductivity at 68°F, percent of International Annealed Copper Standard — Equal volume	Equal weight	Electrical resistivity at 68°F Ohm–cir. Mil/foot
7175	13.0	890–1175[8]	T74	1080	39	124	26
7178	13.0	890–1165[8]	T6	870	31	98	33
7475	12.9	890–1175	T61, T651	960	35	116	30
			T76, T651	1020	40	132	26
			T7351	1130	42	139	25
8017	13.1	1190–1215	H12, H22	..	59	193	18
			H212	..	61	200	17
8030	13.1	1190–1215	H221	1600	61	201	17
8176	13.1	1190–1215	H24	1600	61	201	17

[1]The following typical properties are not guaranteed, since in most cases they are averages for various sizes, product forms, and methods of manufacture and may not be exactly representative of any particular product or size. These data are intended only as a basis for comparing alloys and tempers and **should not be specified as engineering requirements or used for design purposes.**
[2]Coefficient to be multiplied by 10^{-6}. Example: $12.2 \times 10^{-6} = 0.0000122$.
[3]Melting ranges shown apply to wrought products of 1/4 inch thickness or greater.
[4]Based on typical composition of the indicated alloys.
[5]English units = but-in./ft²hr°F
[6]Eutectic melting can be completely eliminated by homogenization.
[7]Eutectic melting is not eliminated by homogenization.
[8]Homogenization may raise eutectic melting temperature 20–40°F but usually does not eliminate eutectic melting.
[9]Although not formerly registered, the literature and some specifications have used T736 as the designation for this temper.

Poisson's ratio (v)

Fracture toughness—B645, B646

Elongation—B557

Hardness—B647, B648, B724, E10, E18

Fatigue strength

Mechanical properties are a function of the alloy and temper as well as, in some cases, product form. For example, 6061-T6 extrusions have a minimum tensile ultimate strength of 38 ksi (260 MPa), while 6061-T6 sheet and plate have a minimum tensile ultimate strength of 42 ksi (290 MPa).

2.4.1 Minimum and Typical Mechanical Properties

There are several bases for mechanical properties. A *typical property* is an average property; if you test enough samples, the average of the test results will equal the typical property. A *minimum property* is defined by the aluminum industry as the value that 99% of samples will equal or exceed with a probability of 95%. (The U.S. military calls such minimum values "A" values and also defines "B" values as those for which 90% of samples will equal or exceed with a probability of 95%, a slightly less stringent criterion that yields higher values). Typical mechanical properties are given in Table 2.17. Some minimum mechanical properties are given in ASTM and other specifications; more are given in Table 2.18 for wrought alloys and Table 2.19 for cast alloys. Minimum mechanical properties are called "guaranteed" when product specifications require them to be met, and they are called "expected" when they are not required by product specifications.

Structural design of aluminum components is usually based on minimum strengths. The rules for such design are given in the Aluminum Association's *Specification for Aluminum Structures*, part of the *Aluminum Design Manual*. Safety factors given there, varying from 1.65 to 2.64 by type of structure, type of failure (yielding or fracture), and type of component (member or connection), are applied to the minimum strengths to determine the safe capacity of a component. Typical strengths should be used to determine the capacity of fabrication equipment (for example, the force required to shear a piece) or the strength of parts designed to fail at a given force to preclude failure of an entire structure. (Pressure relieving panels are an example of this, called frangible design). Maximum ultimate strengths are specified for some aluminum products (usually in softer tempers), but these materials are usually intended to be cold worked into final use products, changing their strength.

TABLE 2.17 Typical Mechanical Properties of Wrought Alloys[1,2]

Alloy and temper	Tension Strength, ksi		Tension Elongation, % in 2 in.		Hardness Brinnell number 500 kg load 10 mm ball	Shear Ultimate shearing strength, ksi	Fatigue Endurance[3] limit, ksi	Modulus Modulus[4] of elasticity, ksi × 10³
	Ultimate	Yield	1/16 in. thick specimen	1/2 in. dia. specimen				
1060-O	10	4	43	..	19	7	3	10.0
1060-H12	12	11	16	..	23	8	4	10.0
1060-H14	14	13	12	..	26	9	5	10.0
1060-H16	16	15	8	..	30	10	6.5	10.0
1060-H18	19	18	6	..	35	11	6.5	10.0
1100-O	13	5	35	45	23	9	5	10.0
1100-H12	16	15	12	25	28	10	6	10.0
1100-H14	18	17	9	20	32	11	7	10.0
1100-H16	21	20	6	17	38	12	9	10.0
1100-H18	24	22	5	15	44	13	9	10.0
1350-O	12	4	..	..[5]	..	8	..	10.0
1350-H12	14	12	..	..	..	9	..	10.0
1350-H14	16	14	..	..	..	10	..	10.0
1350-H16	18	16	..	..[6]	..	11	..	10.0
1350-H19	27	24	..	..	..	15	7	10.0
2011-T3	55	43	..	15	95	32	18	10.2
2011-T8	59	45	..	12	100	35	18	10.2

2014-O	27	12	..	18	45	18	13	10.6
2014-T4, T451	62	42	..	20	105	38	20	10.6
2014-T6, T651	70	60	..	13	135	42	18	10.6
Alclad 2014-O	25	10	21	..	..	18	..	10.5
Alclad 2014-T3	63	40	20	..	..	37	..	10.5
Alclad 2014-T4, T451	61	37	22	..	..	37	..	10.5
Alclad 2014-T6, T651	68	60	10	..	..	41	..	10.5
2017-O	26	10	..	22	47	18	13	10.5
2017-T4, T451	62	40	..	22	105	38	18	10.5
2018-T61	61	46	..	12	120	39	17	10.8
2024-O	27	11	20	22	47	18	13	10.6
2024-T3	70	50	18	..	120	41	20	10.6
2024-T4, T351	68	47	20	19	120	41	20	10.6
2024-T361[7]	72	57	13	..	130	42	18	10.6
Alclad 2025-O	26	11	20	..	..	18	..	10.6
Alclad 2024-T3	65	45	18	..	..	40	..	10.6
Alclad 2024-T4, T351	64	42	19	..	..	40	..	10.6
Alclad 2024-T361[7]	67	63	11	..	..	41	..	10.6
Alclad 2024-T81, T851	65	60	6	..	..	40	..	10.6
Alclad 2024-T861[7]	70	66	6	..	..	42	..	10.6
2025-T6	58	37	..	19	110	35	18	10.4
2036-T4	49	28	24	..	..	..	18[8]	10.3
2117-T4	43	24	..	27	70	28	14	10.3
2124-T851	70	64	..	8	..	..	..	10.6

TABLE 2.17 Typical Mechanical Properties of Wrought Alloys[1,2] (Continued)

Alloy and temper	Tension				Hardness	Shear	Fatigue	Modulus
	Strength, ksi		Elongation, % in 2 in.		Brinnell number 500 kg load 10 mm ball	Ultimate shearing strength, ksi	Endurance[3] limit, ksi	Modulus[4] of elasticity, ksi × 10³
	Ultimate	Yield	1/16 in. thick specimen	1/2 in. dia. specimen				
2218-T72	48	37	...	11	95	30	...	10.8
2219-O	25	11	18	...	...	...	...	10.6
2219-T42	52	27	20	...	...	...	...	10.6
2219-T31, T351	52	36	17	...	...	...	...	10.6
2219-T37	57	46	11	...	...	...	...	10.6
2219-T62	60	42	10	...	...	...	15	10.6
2219-T81, T851	66	51	10	...	...	...	15	10.6
2219-T87	69	57	10	...	...	...	15	10.6
2618-T61	64	54	...	10	115	38	18	10.8
3003-O	16	6	30	40	28	11	7	10.0
3003-H12	19	18	10	20	35	12	8	10.0
3003-H14	22	21	8	16	40	14	9	10.0
3003-H16	26	25	5	14	47	15	10	10.0
3003-H18	29	27	4	10	55	16	10	10.0
Alclad 3003-O	16	6	30	40	...	11	...	10.0
Alclad 3003-H12	19	18	10	20	...	12	...	10.0
Alclad 3003-H14	22	21	8	16	...	14	...	10.0
Alclad 3003-H16	26	25	5	14	...	15	...	10.0
Alclad 3003-H18	29	27	4	10	...	16	...	10.0

3004-O	26	10	20	25	45	16	14	10.0
3004-H32	31	25	10	17	52	17	15	10.0
3004-H34	35	29	9	12	63	18	15	10.0
3004-H36	38	33	5	9	70	20	16	10.0
3004-H38	41	36	5	6	77	21	16	10.0
Alclad 3004-O	26	10	20	25	...	16	...	10.0
Alclad 3004-H32	31	25	10	17	...	17	...	10.0
Alclad 3004-H34	35	29	9	12	...	18	...	10.0
Alclad 3004-H36	38	33	5	9	...	20	...	10.0
Alclad 3004-H38	41	36	5	6	...	21	...	10.0
3105-O	17	8	24	...	...	12	...	10.0
3105-H12	22	19	7	...	...	14	...	10.0
3105-H14	25	22	5	...	...	15	...	10.0
3105-H16	28	25	4	...	...	16	...	10.0
3105-H18	31	28	3	...	...	17	...	10.0
3105-H25	26	23	8	...	...	15	...	10.0
4032-T6	55	46	...	9	120	38	16	11.4
5005-O	18	6	25	...	28	11	...	10.0
5005-H12	20	19	10	...	...	14	...	10.0
5005-H14	23	22	6	...	...	14	...	10.0
5005-H16	26	25	5	...	...	15	...	10.0
5005-H18	29	28	4	...	...	16	...	10.0
5005-H32	20	17	11	...	36	14	...	10.0
5005-H34	23	20	8	...	41	14	...	10.0
5005-H36	26	24	6	...	46	15	...	10.0
5005-H38	29	27	5	...	51	16	...	10.0

TABLE 2.17 Typical Mechanical Properties of Wrought Alloys[1,2] *(Continued)*

Alloy and temper	Tension				Hardness	Shear	Fatigue	Modulus
	Strength, ksi		Elongation, % in 2 in.					
			1/16 in. thick specimen	1/2 in. dia. specimen	Brinnell number 500 kg load 10 mm ball	Ultimate shearing strength, ksi	Endurance[3] limit, ksi	Modulus[4] of elasticity, ksi $\times 10^3$
	Ultimate	Yield						
5050-O	21	8	24	..	36	15	12	10.0
5050-H32	25	21	9	..	46	17	13	10.0
5050-H34	28	24	8	..	53	18	13	10.0
5050-H36	30	26	7	..	58	19	14	10.0
5050-H38	32	29	6	..	63	20	14	10.0
5052-O	28	13	25	30	47	18	16	10.2
5052-H32	33	28	12	18	60	20	17	10.2
5052-H34	38	31	10	14	68	21	18	10.2
5052-H36	40	35	8	10	73	23	19	10.2
5052-H38	42	37	7	8	77	24	20	10.2
5056-O	42	2	..	35	65	26	20	10.3
5056-H18	63	59	..	10	105	34	22	10.3
5056-H38	60	50	..	15	100	32	22	10.3
5083-O	42	21	..	22	..	25	..	10.3
5083-H321, H116	46	33	..	16	..	..	23	10.3
5086-O	38	17	22	..	..	23	..	10.3
5086-H32, H116	42	30	12	..	..	..	..	10.3
5086-H34	47	37	10	..	..	27	..	10.3
5086-H112	39	19	14	..	..	..	..	10.3

5154-O	35	17	27	.	58	22	17	10.2
5154-H32	39	30	15	.	67	22	18	10.2
5154-H34	42	33	13	.	73	24	19	10.2
5154-H36	45	36	12	.	78	26	20	10.2
5154-H38	48	39	10	.	80	28	21	10.2
5154-H112	35	17	25	.	63	.	17	10.2
5252-H25	34	25	11	.	68	21	.	10.0
5252-H38, H28	41	35	5	.	75	23	.	10.0
5254-O	35	17	27	.	58	22	17	10.2
5254-H32	39	30	15	.	67	22	18	10.2
5254-H34	42	33	13	.	73	24	19	10.2
5254-H36	45	36	12	.	78	26	20	10.2
5254-H38	48	39	10	.	80	28	21	10.2
5254-H112	35	17	25	.	63	.	17	10.2
5454-O	36	17	22	.	62	23	.	10.2
5454-H32	40	30	10	.	73	24	.	10.2
5454-H34	44	35	10	.	81	26	.	10.2
5454-H111	38	26	14	.	70	23	.	10.2
5454-H112	36	18	18	.	62	23	.	10.2
5456-H	45	23	.	24	.	.	.	10.3
5456-H25	45	24	.	22	.	.	.	10.3
5456-H321, H116	51	37	.	16	90	30	.	10.3
5457-O	19	7	22	.	32	12	.	10.0
5457-H25	26	23	12	.	48	16	.	10.0
5457-H38, H28	30	27	6	.	55	18	.	10.0

TABLE 2.17 Typical Mechanical Properties of Wrought Alloys[1,2] (Continued)

Alloy and temper	Tension Strength, ksi Ultimate	Yield	Elongation, % in 2 in. 1/16 in. thick specimen	1/2 in. dia. specimen	Hardness Brinnell number 500 kg load 10 mm ball	Shear Ultimate shearing strength, ksi	Fatigue Endurance[3] limit, ksi	Modulus Modulus[4] of elasticity, ksi × 10³
5652-HO	28	13	25	30	47	18	16	10.2
5652-H32	33	28	12	18	60	20	17	10.2
5652-H34	38	31	10	14	68	21	18	10.2
5652-H36	40	35	8	10	73	23	19	10.2
5652-H38	42	37	7	8	77	24	20	10.2
5657-H25	23	20	12	..	40	12	..	10.0
5657-H38, H28	28	24	7	..	50	15	..	10.0
6061-O	18	8	25	30	30	12	9	10.0
6061-T4, T451	35	21	22	25	65	24	14	10.0
6061-T6, T651	45	40	12	17	95	30	14	10.0
Alclad 6061-O	17	7	25	..	..	11	..	10.0
Alclad 6061-T4, T451	33	19	22	..	..	22	..	10.0
Alclad 6061-T6, T651	42	37	12	..	..	27	..	10.0
6063-O	13	7	..	..	25	10	8	10.0
6063-T1	22	13	20	..	42	14	9	10.0
6063-T4	25	13	22	..	..	..	..	10.0
6063-T5	27	21	12	..	60	17	10	10.0
6063-T6	35	31	12	..	73	22	10	10.0

Alloy and temper							
6063-T83	37	35	9		82	22		10.0
6063-T831	30	27	10		70	18		10.0
6063-T832	42	39	12		95	27		10.0
6066-O	22	12		18	43	14		10.0
6066-T4, T451	52	30		18	90	29		10.0
6066-T6, T651	57	52		12	120	34	16	10.0
6070-T6	55	51	10			34	14	10.0
6101-H111	14	11						10.0
6101-T6	32	28	15[9]		71	20		10.0
6262-T9	58	55		10	120	35	13	10.0
6351-T4	36	22	20		42	14		10.0
6351-T6	45	41	14		95	29	13	10.0
6463-T1	22	13	20		42	14	10	10.0
6463-T5	27	21	12		60	17	10	10.0
6463-T6	35	31	12		74	22	10	10.0
7049-T73	75	65		12	135	44		10.4
7049-T7352	75	63		11	135	43		10.4
7050-T73510, T73511	72	63		12				10.4
7050-T7451[10]	76	68		11		44		10.4
7050-T7651	80	71		11		47		10.4
7075-O	33	15	17	16	60	22		10.4
7075-T6, T651	83	73	11	11	150	48	23	10.4
Alclad 7075-O	32	14	17			22		10.4
Alclad 7075-T6, T651	76	67	11			46		10.4

TABLE 2.17 Typical Mechanical Properties of Wrought Alloys[1,2] (Continued)

| Alloy and temper | Tension | | | | Hardness | Shear | Fatigue | Modulus |
| | Strength, ksi | | Elongation, % in 2 in. | | Brinnell number 500 kg load 10 mm ball | Ultimate shearing strength, ksi | Endurance[3] limit, ksi | Modulus[4] of elasticity, ksi $\times 10^3$ |
	Ultimate	Yield	1/16 in. thick specimen	1/2 in. dia. specimen				
7175-T74	76	66	..	11	135	42	23	10.4
7178-O	33	15	15	16	..	..	..	
7178-T6, T651	88	78	10	11	..	..	..	10.4
7178-T76, T7651	83	73	..	11	..	..	..	10.4
Alclad 7178-O	32	14	16	..	..	..	..	10.4
Alclad 7178-T6, T651	81	71	10	..	..	..	..	10.4

7475-T61	82	71	11	..	..	..	10.2
7475-T651	85	74	..	13	..	..	10.4
7475-T7351	72	61	..	13	..	..	10.4
7475-T761	75	65	12	..	..	..	10.2
7475-T7651	77	67	..	12	..	..	10.4
Alclad 7475-T61	75	66	11	..	..	..	10.2
Alclad 7475-T761	71	61	12	..	..	..	10.2
8176-H24	17	14	15	..	10	..	10.0

[1] The typical properties listed in this table are not guaranteed, since in most cases they are averages for various sizes, product forms, and methods of manufacture and may not be exactly representative of any particular product or size. These data are intended only as a basis for comparing alloys and tempers and should not be used for design purposes.

[2] The indicated typical mechanical properties for all except 0 temper material are higher than the specified minimum properties. For 0 temper products typical ultimate and yield values are slightly lower than specified (maximum) values.

[3] Based on 500,000,000 cycles of completely reversed stress using the R.R. Moore type of machine and specimen.

[4] Average of tension and compression moduli. Compression modulus is about 2% greater than tension modulus.

[5] 1350-O wire will have an elongation of approximately 23% in 10 inches.

[6] 1350-H19 wire will have an elongation of approximately 1.5% in 10 inches.

[7] Tempers T361 and T861 were formerly designated T36 and T86, respectively.

[8] Based on 10^7 cycles using flexural type testing of sheet specimens.

[9] Based on 1/4 in thick specimen.

[10] T7451, although not previously registered, has appeared in literature and in some specifications such as T73651.

TABLE 2.18 Minimum Mechanical Properties for Wrought Aluminum Alloys

ALLOY AND TEMPER	PRODUCT	THICKNESS RANGE IN.	TENSION		COMPRESSION	SHEAR		COMPRESSIVE MODULUS OF ELASTICITY‡
			F_{tu} ksi	F_{ty} ksi	F_{cy} ksi	F_{su} ksi	F_{sy} ksi	E ksi
1100-H12	Sheet, Plate (Rolled Rod & Bar)	All	14	11	10	9	6.5	10,100
-H14		All	16	14	13	10	8	10,100
2014-T6	Sheet	0.040-0.249	66	58	59	40	33	10,900
-T651	Plate	0.250-2.000	67	59	58	40	34	10,900
-T6, T6510, T6511	Extrusions	All	60	53	52	35	31	10,900
-T6, T651	Cold Finished Rod & Bar, Drawn Tube	All	65	55	53	38	32	10,900
Alclad 2014-T6	Sheet	0.025-0.039	63	55	56	38	32	10,800
-T6	Sheet	0.040-0.249	64	57	58	39	33	10,800
-T651	Plate	0.250-0.499	64	57	56	39	33	10,800
3003-H12	Sheet & Plate	0.017-2.000	17	12	10	11	7	10,100
-H14	Sheet & Plate	0.009-1.000	20	17	14	12	10	10,100
-H16	Sheet	0.006-0.162	24	21	18	14	12	10,100
-H18	Sheet	0.006-0.128	27	24	20	15	14	10,100
-H12	Drawn Tube	All	17	12	11	11	7	10,100
-H14	Drawn Tube	All	20	17	16	12	10	10,100
-H16	Drawn Tube	All	24	21	19	14	12	10,100
-H18	Drawn Tube	All	27	24	21	15	14	10,100
Alclad 3003-H12	Sheet & Plate	0.017-2.000	16	11	9	10	6.5	10,100
-H14	Sheet & Plate	0.009-1.000	19	16	13	12	9	10,100
-H16	Sheet	0.006-0.162	23	20	17	14	12	10,100
-H18	Sheet	0.006-0.128	26	23	19	15	13	10,100
Alclad 3003-H14	Drawn Tube	0.025-0.259	19	16	15	12	9	10,100
-H18	Drawn Tube	0.010-0.500	26	23	20	15	13	10,100

3004-H32	Sheet & Plate	0.017-2.000	28	21	18	17	12	10,100
-H34	Sheet & Plate	0.009-1.000	32	25	22	19	14	10,100
-H36	Sheet	0.006-0.162	35	28	25	20	16	10,100
-H38	Sheet	0.006-0.128	38	31	29	21	18	10,100
3004-H34	Drawn Tube	0.018-0.450	32	25	24	19	14	10,100
-H36	Drawn Tube	0.018-0.450	35	28	27	20	16	10,100
Alclad								
3004-H32	Sheet	0.017-0.249	27	20	17	16	12	10,100
-H34	Sheet	0.009-0.249	31	24	21	18	14	10,100
-H36	Sheet	0.006-0.162	34	27	24	19	16	10,100
-H38	Sheet	0.006-0.128	37	30	28	21	17	10,100
-H131, H241, H341	Sheet	0.024-0.050	31	26	22	18	15	10,100
-H151, H261, H361	Sheet	0.024-0.050	34	30	28	19	17	10,100
3005-H25	Sheet	0.013-0.050	26	22	20	15	13	10,100
-H28	Sheet	0.006-0.080	31	27	25	17	16	10,100
3105-H25	Sheet	0.013-0.080	23	19	17	14	11	10,100
5005-H12	Sheet & Plate	0.017-2.000	18	14	13	11	8	10,100
-H14	Sheet & Plate	0.009-1.000	21	17	15	12	10	10,100
-H16	Sheet	0.006-0.162	24	20	18	14	12	10,100
-H32	Sheet & Plate	0.017-2.000	17	12	11	11	7	10,100
-H34	Sheet & Plate	0.009-1.000	20	15	14	12	8.5	10,100
-H36	Sheet	0.006-0.162	23	18	16	13	11	10,100
5050-H32	Sheet	0.017-0.249	22	16	14	14	9	10,100
-H34	Sheet	0.009-0.249	25	20	18	15	12	10,100
-H32	Cold Fin. Rod & Bar Drawn Tube	All	22	16	15	13	9	10,100
-H34	Cold Fin. Rod & Bar Drawn Tube	All	25	20	19	15	12	10,100

TABLE 2.18 Minimum Mechanical Properties for Wrought Aluminum Alloys (Continued)

ALLOY AND TEMPER	PRODUCT	THICKNESS RANGE IN.	TENSION		COMPRESSION	SHEAR		COMPRESSIVE MODULUS OF ELASTICITY‡
			F_{tu} ksi	F_{ty} ksi	F_{cy} ksi	F_{su} ksi	F_{sy} ksi	E ksi
5052-O	Sheet & Plate	0.006-3.000	25	9.5	9.5	16	5.5	10,200
-H32	Sheet & Plate	All	31	23	21	19	13	10,200
-H34	Cold Fin. Rod & Bar / Drawn Tube	All	34	26	24	20	15	10,200
-H36	Sheet	0.006-0.162	37	29	26	22	17	10,200
5083-O	Extrusions	up thru 5.000	39	16	16	24	9	10,400
-H111	Extrusions	up thru 0.500	40	24	21	24	14	10,400
-H111	Extrusions	0.501-5.000	40	24	21	23	14	10,400
-O	Sheet & Plate	0.051-1.500	40	18	18	25	10	10,400
-H116	Sheet & Plate	0.188-1.500	44	31	26	26	18	10,400
-H321	Sheet & Plate	0.188-1.500	44	31	26	26	18	10,400
-H116	Plate	1.501-3.000	41	29	24	24	17	10,400
-H321	Plate	1.501-3.000	41	29	24	24	17	10,400
5086-O	Extrusions	up thru 5.000	35	14	14	21	8	10,400
-H111	Extrusions	up thru 0.500	36	21	18	21	12	10,400
-H111	Extrusions	0.501-5.000	36	21	18	21	12	10,400
-O	Sheet & Plate	0.020-2.000	35	14	14	21	8	10,400
-H112	Plate	0.250-0.499	36	18	17	22	10	10,400
-H112	Plate	0.500-1.000	35	16	16	21	9	10,400
-H112	Plate	1.001-2.000	35	14	15	21	8	10,400
-H112	Plate	2.001-3.000	34	14	15	21	8	10,400
-H116	Sheet & Plate	All	40	28	26	24	16	10,400
-H32	Sheet & Plate	All	40	28	26	24	16	10,400
-H34	Sheet & Plate / Drawn Tube	All	44	34	32	26	20	10,400
5154-H38	Sheet	0.006-0.128	45	35	33	24	20	10,300

Alloy-Temper	Product	Thickness						
5454-O	Extrusions	up thru 5.000	31	12	12	19	7	10,400
-H111	Extrusions	up thru 0.500	33	19	16	20	11	10,400
-H111	Extrusions	0.501-5.000	33	19	16	19	11	10,400
-H112	Extrusions	up thru 5.000	31	12	13	19	7	10,400
-O	Sheet & Plate	0.020-3.000	31	12	12	19	7	10,400
-H32	Sheet & Plate	0.020-2.000	36	26	24	21	15	10,400
-H34	Sheet & Plate	0.020-1.000	39	29	27	23	17	10,400
5456-O	Sheet & Plate	0.051-1.500	42	19	19	26	11	10,400
-H116	Sheet & Plate	0.188-1.250	46	33	27	27	19	10,400
-H321	Sheet & Plate	0.188-1.250	46	33	27	27	19	10,400
-H116	Plate	1.251-1.500	44	31	25	25	18	10,400
-H321	Plate	1.251-1.500	44	31	25	25	18	10,400
-H116	Plate	1.501-3.000	41	29	25	25	17	10,400
-H321	Plate	1.501-3.000	41	29	25	25	17	10,400
6005-T5	Extrusions	up thru 1.000	38	35	35	24	20	10,100
6061-T6, T651	Sheet & Plate	0.010-4.000	42	35	35	27	20	10,100
-T6, T6510, T6511	Extrusions	All	38	35	35	24	20	10,100
-T6, T651	Cold Fin. Rod & Bar	up thru 8.000	42	35	35	25	20	10,100
-T6	Drawn Tube	0.025-0.500	42	35	35	27	20	10,100
-T6	Pipe	All	38	35	35	24	20	10,100
6063-T5	Extrusions	up thru 0.500	22	16	16	13	9	10,100
-T5	Extrusions	0.500-1.000	21	15	15	12	8.5	10,100
-T6	Extrusions & Pipe	All	30	25	25	19	14	10,100
6066-T6, T6510, T6511	Extrusions	All	50	45	45	27	26	10,100
6070-T6, T62	Extrusions	up thru 2.999	48	45	45	29	26	10,100
6105-T5	Extrusions	up thru 0.500	38	35	35	24	20	10,100
6351-T5	Extrusions	up thru 1.000	38	35	35	24	20	10,100
6463-T6	Extrusions	up thru 0.500	30	25	25	19	14	10,100

† F_{tu} and F_{ty} are minimum specified values (except F_{ty} for 1100-H12, -H14 Cold Finished Rod and Bar and Drawn Tube, Alclad 3003-H18 Sheet and 5050-H32, -H34 Cold Finished Rod and Bar which are minimum expected values); other strength properties are corresponding minimum expected values.

TABLE 2.18a Minimum Mechanical Properties for Wrought Aluminum Alloys (SI)

ALLOY AND TEMPER	PRODUCT	THICKNESS RANGE mm	TENSION		COMPRESSION	SHEAR		COMPRESSIVE MODULUS OF ELASTICITY‡
			F_{tu}† MPa	F_{ty}† MPa	F_{cy} MPa	F_{su} MPa	F_{sy} MPa	E MPa
1100-H12	Sheet, Plate	All	95	75	70	62	45	69,600
-H14	Rolled Rod & Bar	All	110	95	90	70	55	69,600
2014-T6	Sheet	1.00-6.30	455	400	405	275	230	75,200
-T651	Plate	6.30-50.00	460	405	400	275	235	75,200
-T6, T6510, T6511	Extrusions	All	415	365	360	240	215	75,200
-T6, T651	Cold Finished Rod & Bar, Drawn Tube	All	450	380	365	260	220	75,200
Alclad 2014-T6	Sheet	0.63-1.00	435	380	385	260	220	74,500
-T6	Sheet	1.00-6.30	440	395	400	270	230	74,500
-T651	Plate	6.30-12.50	440	395	385	270	230	74,500
3003-H12	Sheet & Plate	0.40-50.00	120	85	70	75	48	69,600
-H14	Sheet & Plate	0.20-25.00	140	115	95	85	70	69,600
-H16	Sheet	0.15-4.00	165	145	125	95	85	69,600
-H18	Sheet	0.15-3.20	185	165	140	105	95	69,600
-H12	Drawn Tube	All	120	85	75	75	48	69,600
-H14	Drawn Tube	All	140	115	110	85	70	69,600
-H16	Drawn Tube	All	165	145	130	95	85	69,600
-H18	Drawn Tube	All	185	165	145	105	95	69,600
Alclad 3003-H12	Sheet & Plate	0.40-50.00	115	80	62	70	45	69,600
-H14	Sheet & Plate	0.20-25.00	135	110	90	85	62	69,600
-H16	Sheet	0.15-4.00	160	140	115	95	85	69,600
-H18	Sheet	0.15-3.20	180	160	130	105	90	69,600
Alclad 3003-H14	Drawn Tube	0.63-6.30	135	110	105	85	62	69,600
-H18	Drawn Tube	0.25-12.50	180	160	140	105	90	69,600

Designation	Product	Thickness						Modulus
3004-H32	Sheet & Plate	0.40-50.00	190	145	125	115	85	69,600
-H34	Sheet & Plate	0.20-25.00	220	170	150	130	95	69,600
-H36	Sheet	0.15-4.00	240	190	170	140	110	69,600
-H38	Sheet	0.15-3.20	260	215	200	145	125	69,600
3004-H34	Drawn Tube	0.45-11.50	220	170	165	130	95	69,600
-H36	Drawn Tube	0.45-11.50	240	190	185	140	110	69,600
Alclad								
3004-H32	Sheet	0.40-6.30	185	140	115	110	85	69,600
-H34	Sheet	0.20-6.30	215	165	145	125	95	69,600
-H36	Sheet	0.15-4.00	235	185	165	130	110	69,600
-H38	Sheet	0.15-3.20	255	205	195	145	115	69,600
-H131, H241, -H341	Sheet	0.60-1.20	215	180	150	125	105	69,600
-H151, H261, -H361	Sheet	0.60-1.20	235	205	195	130	115	69,600
3005-H25	Sheet	0.32-1.20	180	150	140	105	90	69,600
-H28	Sheet	0.15-2.00	215	185	170	115	110	69,600
3105-H25	Sheet	0.32-2.00	160	130	115	95	75	69,600
5005-H12	Sheet & Plate	0.40-50.00	125	95	90	75	55	69,600
-H14	Sheet & Plate	0.20-25.00	145	115	105	85	70	69,600
-H16	Sheet	0.15-4.00	165	135	125	95	85	69,600
-H32	Sheet & Plate	0.40-50.00	120	85	75	75	48	69,600
-H34	Sheet & Plate	0.20-25.00	140	105	95	85	59	69,600
-H36	Sheet	0.15-4.00	160	125	110	90	75	69,600
5050-H32	Sheet	0.40-6.30	150	110	95	95	62	69,600
-H34	Sheet	0.20-6.30	170	140	125	105	85	69,600
-H32	Cold Fin. Rod & Bar† Drawn Tube	All	150	110	105	90	62	69,600
-H34	Cold Fin. Rod & Bar† Drawn Tube	All	170	140	130	105	85	69,600

TABLE 2.18a Minimum Mechanical Properties for Wrought Aluminum Alloys (SI) (Continued)

ALLOY AND TEMPER	PRODUCT	THICKNESS RANGE mm	TENSION		COMPRESSION	SHEAR		COMPRESSIVE MODULUS OF ELASTICITY[‡]
			F_{tu}[†] MPa	F_{ty}[†] MPa	F_{cy} MPa	F_{su} MPa	F_{sy} MPa	E MPa
5052-O	Sheet & Plate	0.15-80.00	170	65	66	110	38	70,300
-H32	Sheet & Plate	All	215	160	145	130	90	70,300
-H34	Cold Fin. Rod & Bar / Drawn Tube	All	235	180	165	140	105	70,300
-H36	Sheet	0.15-4.00	255	200	180	150	115	70,300
5083-O	Extrusions	up thru 130.00	270	110	110	165	62	71,700
-H111	Extrusions	up thru 12.70	275	165	145	165	95	71,700
-H111	Extrusions	12.70-130.00	275	165	145	160	95	71,700
-O	Sheet & Plate	1.20-6.30	275	125	125	170	70	71,700
-H116	Sheet & Plate	4.00-40.00	305	215	180	180	125	71,700
-H321	Sheet & Plate	4.00-40.00	305	215	180	180	125	71,700
-H116	Plate	40.00-80.00	285	200	165	165	115	71,700
-H321	Plate	40.00-80.00	285	200	165	165	115	71,700
5086-O	Extrusions	up thru 130.00	240	95	95	145	55	71,700
-H111	Extrusions	up thru 12.70	250	145	125	145	85	71,700
-H111	Extrusions	12.70-130.00	250	145	125	145	85	71,700
-O	Sheet & Plate	0.50-50.00	240	95	95	145	55	71,700
-H112	Sheet & Plate	4.00-12.50	250	125	115	150	70	71,700
-H112	Plate	12.50-40.00	240	105	110	145	62	71,700
-H112	Plate	40.00-80.00	235	95	105	145	55	71,700
-H116	Sheet & Plate	1.60-50.00	275	195	180	165	110	71,700
-H32	Sheet & Plate / Drawn Tube	All	275	195	180	165	110	71,700
-H34	Sheet & Plate / Drawn Tube	All	300	235	220	180	140	71,700
5154-H38	Sheet	0.15-3.20	310	240	230	165	140	71,000
5454-O	Extrusions	up thru 130.00	215	85	85	130	48	71,700
-H111	Extrusions	up thru 12.70	230	130	110	140	75	71,700

Temper	Product	Thickness (mm)						E (MPa)
-H111	Extrusions	12.70-130.00	230	130	110	130	75	71,700
-H112	Extrusions	up thru 130.00	215	85	90	130	48	71,700
-O	Sheet & Plate	0.50-80.00	215	85	85	130	48	71,700
-H32	Sheet & Plate	0.50-50.00	250	180	165	145	105	71,700
-H34	Sheet & Plate	0.50-25.00	270	200	185	160	115	71,700
5456-O	Sheet & Plate	1.20-6.30	290	130	130	180	75	71,700
-H116	Sheet & Plate	4.00-12.50	315	230	185	185	130	71,700
-H321	Sheet & Plate	4.00-12.50	315	230	185	185	130	71,700
-H116	Plate	12.50-40.00	305	215	170	170	125	71,700
-H321	Plate	12.50-40.00	305	215	170	170	125	71,700
-H116	Plate	40.00-80.00	285	200	170	170	115	71,700
-H321	Plate	40.00-80.00	285	200	170	170	115	71,700
6005-T5	Extrusions	up thru 25	260	240	240	165	140	69,600
6061-T6, T651	Sheet & Plate	0.25-100.00	290	240	240	185	140	69,600
-T6, T6510, T6511	Extrusions	All	260	240	240	165	140	69,600
-T6, T651	Cold Fin. Rod & Bar	up thru 200	290	240	240	170	140	69,600
-T6	Drawn Tube	0.63-12.50	290	240	240	185	140	69,600
-T6	Pipe	All	260	240	240	165	140	69,600
6063-T5	Extrusions	up thru 12.50	150	110	110	90	62	69,600
-T5	Extrusions	12.50-25.00	145	105	105	85	59	69,600
-T6	Extrusions & Pipe	All	205	170	170	130	95	69,600
6066-T6, T6510, T6511	Extrusions	All	345	310	310	185	180	69,600
6070-T6, T62	Extrusions	up thru 80.00	330	310	310	200	180	69,600
6105-T5	Extrusions	up thru 12.50	260	240	240	165	140	69,600
6351-T5	Extrusions	up thru 25.00	260	240	240	165	140	69,600
6463-T6	Extrusions	up thru 12.50	205	170	170	130	95	69,600

† F_{tu} and F_{ty} are minimum specified values (except F_{ty} for 1100-H12, -H14 Cold Finished Rod and Bar and Drawn Tube, Alclad 3003-H18 Sheet and 5050-H32, -H34 Cold Finished Rod and Bar which are minimum expected values); other strength properties are corresponding minimum expected values.

‡Typical values. For deflection calculations an average modulus of elasticity is used; this is 700 MPa lower than values in this column.

TABLE 2.19 Mechanical Property Limits for Commonly Used Aluminum Sand Casting Alloys[1]

| Alloy | Temper[3] | Tensile strength | | | | Percent elongation in 2 in. or 4 × dia. | Typical Brinell[2] hardness 500-kgf load 10-mm ball |
| | | Ultimate | | Yield (0.% offset) | | | |
		ksi	MPa	ksi	MPa		
201.0	T7	60.0	414	50.0	345	3.0	110–140
204.0	T4	45.0	310	28.0	193	6.0	—
208.0	F	19.0	131	12.0	83	1.5	40–70
222.0	O	23.0	159	—	—	—	65–95
222.0	T61	30.0	207	—	—	—	100–130
242.0	O	23.0	159	—	—	—	55–85
242.0	T571	29.0	200	—	—	—	70–100
242.0	T61	32.0	221	20.0	138	—	90–120
242.0	T77	24.0	165	13.0	90	1.0	60–90
295.0	T4	29.0	200	13.0	90	6.0	45–75
295.0	T6	32.0	221	20.0	138	3.0	60–90
295.0	T62	36.0	248	28.0	193	—	80–110
295.0	T7	29.0	200	16.0	110	3.0	55–85
319.0	F	23.0	159	13.0	90	1.5	55–85
319.0	T5	25.0	172	—	—	—	65–95
319.0	T6	31.0	214	20.0	138	1.5	65–95
328.0	F	25.0	172	14.0	97	1.0	45–75
328.0	T6	34.0	234	21.0	145	1.0	65–95
354.0	see note[4]	—	—	—	—	—	—
355.0	T51	25.0	172	18.0	124	—	50–80
355.0	T6	32.0	221	20.0	138	2.0	70–105

355.0	T7	35.0	241	—	—	—	70–100
355.0	T71	30.0	207	22.0	152	—	60–95
C355.0	T6	36.0	248	25.0	172	2.5	75–105
356.0	F	19.0	131	—	—	2.0	40–70
356.0	T51	23.0	159	16.0	110	—	45–75
356.0	T6	30.0	207	20.0	138	3.0	55–90
356.0	T7	31.0	214	29.0	200	—	60–90
356.0	T71	25.0	172	18.0	124	3.0	45–75
A356.0	T6	34.0	234	24.0	165	3.5	70–105
357.0	see note[4]	—	—	—	—	—	—
A357.0	see note[4]	—	—	—	—	—	—
359.0	see note[4]	—	—	—	—	—	—
443.0	F	17.0	117	7.0	49	3.0	25–55
B433.0	F	17.0	117	6.0	41	3.0	25–55
512.0	F	17.0	117	10.0	69	—	35–65
514.0	F	22.0	152	9.0	62	6.0	35–65
520.0	T4[5]	42.0	209	22.0	152	12.0	60–90
535.0	F or T5	35.0	241	18.0	124	9.0	60–80
705.0	F or T5	30.0	207	17.0	117	5.0	50–80
707.0	T5	33.0	228	22.0	152	2.0	70–100
707.0	T7	37.0	255	30.0	207	1.0	65–95
710.0	F or T5	32.0	221	20.0	138	2.0	60–90
712.0	F or T5	34.0	234	25.0	172	4.0	60–90
713.0	F or T5	32.0	221	22.0	152	3.0	60–90
771.0	T5	52.0	290	38.0	262	1.5	85–115
771.0	T51	32.0	221	27.0	186	3.0	70–100
771.0	T52	36.0	248	30.0	207	1.5	70–100
771.0	T53	36.0	248	27.0	186	1.5	—
771.0	T6	42.0	290	35.0	241	5.0	75–105
771.0	T71	48.0	331	45.0	310	2.0	105–35
850.0	T5	16.0	110	—	—	5.0	30–60
851.0	T5	17.0	117	—	—	3.0	30–60
852.0	T5	24.0	165	18.0	124	—	45–75

TABLE 2.19 Mechanical Property Limits for Commonly Used Aluminum Permanent Mold Casting Alloys[1] (Continued)

Alloy	Temper[3]	Minimum properties				Percent elongation in 2 in. or 4 × dia.	Typical Brinell[2] hardness 500-kgf load 10-mm ball
		Tensile strength					
		Ultimate		Yield (0.% offset)			
		ksi	MPa	ksi	MPa		
204.0	T4	48.0	331	29.0	200	8.0	—
208.0	T4	33.0	228	15.0	103	4.5	60–90
208.0	T6	35.0	241	22.0	152	2.0	75–105
208.0	T7	33.0	228	16.0	110	3.0	65–95
222.0	T551	30.0	207	—	—	—	100–130
222.0	T65	40.0	276	—	—	—	125–155
242.0	T571	34.0	234	—	—	—	90–120
242.0	T61	40.0	276	—	—	—	95–125
296.0	T6	35.0	241	—	—	2.0	75–105
308.0	F	24.0	165	—	—	—	55–85
319.0	F	28.0	193	14.0	97	1.5	70–100
319.0	T6	34.0	234	—	—	2.0	75–105
332.0	T5	31.0	214	—	—	—	90–120
333.0	F	28.0	193	—	—	—	65–100
333.0	T5	30.0	207	—	—	—	70–105
333.0	T6	35.0	241	—	—	—	85–115
333.0	T7	31.0	214	—	—	—	75–105
336.0	T551	31.0	214	—	—	—	90–120
336.0	T65	40.0	276	—	—	—	110–140
354.0	T61	48.0	331	37.0	255	3.0	—
354.0	T62	52.0	359	42.0	290	2.0	—

355.0	T51	27.0	186	—	—	60–90
355.0	T6	37.0	255	—	1.5	75–105
355.0	T62	42.0	290	—	—	90–120
355.0	T7	36.0	248	—	—	70–100
355.0	T71	34.0	234	27.0	186	—	65–95
C355.0	T61	40.0	276	30.0	207	3.0	75–105
356.0	F	21.0	145	—	—	3.0	40–70
356.0	T51	25.0	172	—	—	—	55–85
356.0	T6	33.0	228	22.0	152	3.0	65–95
356.0	T7	25.0	172	—	—	3.0	60–90
356.0	T71	25.0	172	—	—	3.0	60–90
A356.0	T61	37.0	255	26.0	179	5.0	70–100
357.0	T6	45.0	310	—	—	3.0	75–105
A357.0	T61	45.0	310	36.0	248	3.0	85–115
359.0	T61	45.0	310	34.0	234	4.0	75–105
359.0	T62	7.0	324	38.0	262	3.0	85–115
443.0	F	21.0	145	7.0	49	2.0	30–60
B443.0	F	21.0	145	6.0	41	2.5	30–60
A444.0	T4	20.0	138	—	—	20.0	—
513.0	F	22.0	152	12.0	83	2.5	45–75
535.0	F	35.0	241	18.0	124	8.0	60–90
705.0	T5	37.0	255	17.0	117	10.0	55–85
707.0	T7	45.0	310	35.0	241	3.0	80–110
711.0	T1	28.0	193	18.0	124	7.0	55–85
713.0	T5	32.0	221	22.0	152	4.0	60–90
850.0	T5	18.0	124	—	—	8.0	30–60
851.0	T5	17.0	117	—	—	3.0	30–60
851.0	T6	18.0	124	—	—	8.0	—
852.0	T5	27.0	186	—	—	3.0	55–85

Notes for Table 2.19

[1]Values represent properties obtained from separately cast test bars and are derived from ASTM B-26, Standard Specification for Aluminum-Alloy Sand Castings; Federal Specification QQ-A-601e, Aluminum Alloy Sand Castings; and Military Specification MIL-A-21180c, Aluminum Alloy Castings, High Strength. Unless otherwise specified, the average tensile strength, average yield strength, and average elongation values of specimens cut from castings shall be not less than 75 percent of the tensile and yield strength values and not less than 25 percent of the elongation values given above. The customer should keep in mind that (1) some foundries may offer additional tempers for the above alloys, and (2) foundries are constantly improving casting techniques and, as a result, some may offer minimum properties in excess of the above.

[2]Hardness values are given for information only; not required for acceptance.

[3]F indicates "as cast" condition; refer to AA-CS-M11 for recommended times and temperatures of heat treatment for other tempers to achieve properties specified.

[4]Mechanical properties for these alloys depend on the casting process. For further information, consult the individual foundries.

[5]The T4 temper of Alloy 520.0 is unstable; significant room temperature aging occurs within life expectancy of most castings. Elongation may decrease by as much as 80 percent.

2.4.2 Strengths

While the stress-strain curve of aluminum is approximately linear in the elastic region, aluminum alloys do not exhibit a pronounced yield point like mild carbon steels. Therefore, an arbitrary definition for the yield strength has been adopted by the aluminum industry: a line parallel to a tangent to the stress-strain curve at its initial point is drawn, passing through the 0.2% strain intercept on the x (strain) axis. The stress where this line intersects the stress-strain curve is defined as the *yield stress*. The shape of the stress-strain curve for H, O, T1, T2, T3, and T4 tempers has a less pronounced knee at yield when compared to the shape of the curve for the T5, T6, T7, T8, and T9 tempers. (This causes the inelastic buckling strengths of these two groups of tempers to differ, since inelastic buckling strength is a function of the shape of the stress-strain curve after yield.)

Ultimate strength is the maximum stress the material can sustain. All stresses given in aluminum product specifications are engineering stresses; that is, they are calculated by dividing the force by the original cross sectional area of the specimen rather than the actual cross sectional area under stress. The actual area is less than the original area, since necking occurs after yielding; thus the engineering stress is slightly less than the actual stress.

When strengths are not available, relationships between the unknown strength and known properties may be used. The tensile ultimate strength (F_{tu}) is almost always known, and the tensile yield strength (F_{ty}) is usually known, so other properties are related to these:

$F_{cy} = 0.9\,F_{ty}$ for cold-worked tempers

$F_{cy} = F_{ty}$ for heat-treatable alloys and annealed tempers

$F_{sy} = 0.6\,F_{ty}$

$F_{su} = 0.6\,F_{tu}$

These relationships are approximate, but usually accurate enough for design purposes.

Tensile ultimate strengths vary widely among alloys and tempers, from a minimum of 8 ksi (55 MPa) for 1060-O and 1350-O to a maximum of 84 ksi (580 MPa) for 7178-T62. For some tempers (usually the annealed temper) of certain alloys, strengths are also limited to a maximum value to ensure workability without cracking.

The strength of aluminum alloys is a function of temperature. Most alloys have a plateau of strength between roughly −150°F (−100°C) and 200°F (100°C), with higher strengths below this range, and lower strengths above it. Ultimate strength increases 30 to 50% below this range, while the yield strength increase at low temperatures is not so dramatic, being on the order of 10%. Both ultimate and yield strengths drop rapidly above 200°F, dropping to nearly zero at 750°F (400°C). Some alloys (such as 2219) retain useful (albeit lower) strengths as high as 600°F (300°C). Figure 2.2 shows the effect of temperature on strength for various alloys.

Heating tempered alloys also has an effect on strength. Heating for a long enough period of time reduces the condition of the material to the annealed state, which is the weakest temper for the material. The higher the temperature, the briefer the period of time required produce annealing. The length of time of high temperature exposure causing no more than a 5% reduction in strength is given in Table 2.20 for 6061-T6. Since welding introduces heat to the parts being welded, welding reduces their strength. This effect is discussed in Section 2.8.1 below, and minimum reduced strengths for various alloys are given there.

TABLE 2.20   Maximum Time at Elevated Temperatures, 6061-T6

Elevated temperature		
°F	°C	Maximum time[1]
800	430	not recommended
500	260	not recommended
450	230	5 minutes
425	220	15 minutes
400	200	30 minutes
375	190	1 to 2 hours
350	180	8 to 10 hours
230–325	110–165	50 hours

[1]Loss of strength will not exceed 5% at these times.

Under a constant stress, the deformation of an aluminum part may increase over time, behavior known as *creep*. Creep effects increase as the temperature increases. At room temperature, very little creep occurs unless stresses are near the tensile strength. Creep is usually not a factor unless stresses are sustained at temperatures over about 200°F (95°C).

2.4.3   Modulus of Elasticity, Modulus of Rigidity, and Poisson's Ratio

The *modulus of elasticity (E)* (also called Young's modulus) is the slope of the stress-strain curve in its initial, elastic region prior to yield. The

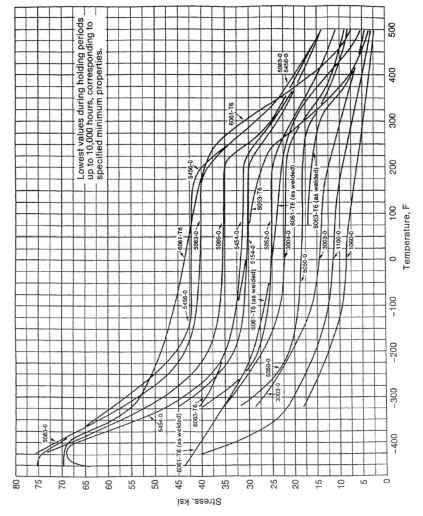

Figure 2.2 Typical tensile strengths of some aluminum alloys at various temperatures.

modulus is significant, because it is a measure of a material's stiffness, or resistance to elastic deformation, and its buckling strength. The modulus varies slightly by alloy, since it is a function of the alloying elements. It can be estimated by averaging the moduli of the alloying elements according to their proportion in the alloy, although magnesium and lithium tend to have a disproportionate effect. An approximate value of 10,000 ksi (69,000 MPa) is sometimes used, but moduli range from 10,000 ksi for pure aluminum (1xxx series), manganese (3xxx series), and magnesium-silicon alloys (6xxx series) to 10,800 ksi (75,000 MPa) for the aluminum-copper alloys and 11,200 ksi (77,200 MPa) for 8090, an aluminum-lithium alloy. Moduli of elasticity for various alloys are given in Table 2.18. This compares to 29,000 ksi (200,000 MPa) for steel alloys (about three times that of aluminum) and to 6,500 ksi (45,000 MPa) for magnesium.

For aluminum, the tensile modulus is about 2% less than the compressive modulus. An average of tensile and compressive moduli is used to calculate bending deflections; the compressive modulus is used to calculate buckling strength.

Aluminum's modulus of elasticity is a function of temperature, increasing about 10% around −300°F (−200°C) and decreasing about 30% at 600°F (300°C).

At strains beyond yield, the slope of the stress-strain curve is called the *tangent modulus* and is a function of stress, decreasing as the stress increases. Values for the tangent modulus or the Ramberg-Osgood parameter n define the shape of the stress-strain curve in this inelastic region and are given in the *U.S. Military Handbook on Metallic Materials and Elements for Aerospace Structures (MIL HDBK 5)* for many aluminum alloys. The Ramberg-Osgood equation is

$$\varepsilon = \frac{\sigma}{E} + 0.002\left(\frac{\sigma}{F_y}\right)^n \tag{2.1}$$

where ε = strain
σ = stress
F_y = yield strength

The *modulus of rigidity* (G) is the ratio of shear stress in a torsion test to shear strain in the elastic range. The modulus of rigidity is also called the *shear modulus*. An average value for aluminum alloys is 3,800 ksi (26,000 MPa).

Poisson's ratio (ν) is the negative of the ratio of transverse strain that accompanies longitudinal strain caused by axial load in the elastic range. Poisson's ratio is approximately 0.33 for aluminum alloys, similar to the ratio for steel. While the ratio varies slightly by alloy and decreases slightly as temperature decreases, such variations are insignificant for most applications. Poisson's ratio can be used to relate the modulus of rigidity (G) and the modulus of elasticity (E) through the formula

$$G = \frac{E}{2(1 + \nu)}$$
(2.2)

2.4.4 Fracture Toughness and Elongation

Fracture toughness is a measure of a material's resistance to the extension of a crack. Aluminum has a face-centered cubic crystal structure and so does not exhibit a transition temperature (like steel) below which the material suffers a significant loss in fracture toughness. Furthermore, alloys of the 1xxx, 3xxx, 4xxx, 5xxx, and 6xxx series are so tough that their fracture toughness cannot be readily measured by the methods commonly used for less tough materials and is rarely of concern. Alloys of the 2xxx and 7xxx series are less tough, and when they are used in fracture critical applications such as aircraft, their fracture toughness is of interest to the designer.

The plane strain fracture toughness (K_{Ic}) for some products of the 2xxx and 7xxx alloys can be measured by ASTM B645. For those products whose fracture toughness cannot be measured by this method (such as sheet, which is too thin for applying B645), nonplane strain fracture toughness (K_c) may be measured by ASTM B646. Fracture toughness limits established by the Aluminum Association are given in Table 2.21. Fracture toughness is a function of the orientation of the specimen and the notch relative to the part, and so toughness is identified by two letters: L for the length direction, T for the width (long transverse) direction, and S for the thickness (short transverse) direction. The first letter denotes the specimen direction perpendicular to the crack, and the second letter the direction of the notch.

Ductility, the ability of a material to absorb plastic strain before fracture, is related to elongation. *Elongation* is the percentage increase in the distance between two gage marks of a specimen tensile tested to fracture. All other things being equal, the greater the elongation, the greater the ductility. The elongation of aluminum alloys tends to be less than mild carbon steels; while A36 steel has a minimum elongation of 20%, the comparable aluminum alloy, 6061-T6, has a minimum elongation requirement of 8 or 10%, depending on the product form. An alloy that is not ductile may fracture at a lower tensile stress than its minimum ultimate tensile stress because it is unable to deform plastically at local stress concentrations. Instead, brittle fracture occurs at a stress riser, leading to premature failure of the part.

TABLE 2.21   Fracture Toughness Limits

Alloy and temper	Thickness (in.)	K_{Ic}, $ksi\sqrt{in.}$ min.		
		L-T	T-L	S-L
Limits for Plate[1]				
2124-T851	1.500–6.000	24	20	18
7050-T7451[2,3]	1.000–2.000	29	25	—
	2.001–3.000	27	24	21
	3.001–4.000	26	26	21
	4.001–5.000	25	22	21
	5.001–6.000	24	22	21
7050-T7651[2]	1.000–2.000	36	24	—
	2.001–3.000	34	23	20
7475-T651	1.250–1.500	30	28	—
7475-T7351	1.250–2.499	40	33	—
	2.500–4.000	40	33	25
7475–T7651	1.250–1.500	33	30	—
Limits for Sheet[4]				
7475-T61	0.040–0.125	—	75	—
	0.126–0.249	60	60	—
7475-T761	0.040–0.125	—	87	—
	0.126–0.249	—	80	—

[1]When tested per ASTM Test Method E399 and ASTM Practice B645.
[2]Thickness for K_{Ic} specimens in the T-L and L-T test orientations: up through 2 in. (ordered, nominal thickness), use full thickness; over 2 through 4 in., use 2-in. specimen thiccentered at T/2; over 4 in., use 2-in. specimen thickness centered at T/4. Test location for K_{Ic} specimensn the S-L test orientation: locate crack at T/2.
[3]T74 type tempers, although not previously registered, have appeared in the literature and in some specifications as T736 type tempers.
[4]When tested per ASTM Practice B646 and ASTM Practice E561.

The elongation of annealed tempers is greater than that of strain hardened or heat treated tempers, while the strength of annealed tempers is less. Therefore, annealed material is more workable and able to undergo more severe forming operations without cracking.

Elongation values are affected by the thickness of the specimen, being higher for thicker specimens. For example, typical elongation values for 1100-O material are 35% for a 1/16 in. thick specimen, and 45% for a 1/2 in. diameter specimen. For this reason, it is important to specify the type of specimen used to obtain the elongation value. Elongation is also very much a function of temperature, being lowest at room temperature and increasing at both lower and higher temperatures.

2.4.5 Hardness

The hardness of aluminum alloys can be measured by several methods, including Webster hardness (ASTM B647), Barcol hardness (ASTM B648), Newage hardness (ASTM B724), and Rockwell hardness (ASTM E18). The Brinnell hardness (ASTM E10)for a 500 kg load on a 10 mm ball is used most often and is given in Tables 2.17 and 2.19. Hardness measurements are sometimes used for quality assurance purposes on temper. The Brinnell hardness number (BHN) is approximately related to minimum ultimate tensile strength: BHN = $0.556 \, F_{tu}$; this relationship can be useful to help identify material or estimate its strength based on a simple hardness test. The relationship between hardness and strength is not as dependable for aluminum as for steel, however, so this equation is only approximate.

2.4.6 Fatigue Strength

Tensile strengths established for metals are based on a single application of load at a rate slow enough to be considered static. The repeated application of loads causing tensile stress in a part may result in fracture at a stress less than the static tensile strength. This behavior is called *fatigue*. While the fatigue strength of aluminum alloys varies by alloy and temper, it does not vary as much as the static strength (Figure 2.3). For this reason, designers often consider fatigue strength to be independent of alloy and temper, especially when the number of load cycles is high.

The fatigue strengths of the various aluminum alloys can be compared based on the endurance limits given in Table 2.17. These endurance limits are the stress range required to fail an R. R. Moore specimen in 500 million cycles of completely reversed stress. Endurance limits are not useful for designing components, however, because the conditions of the test by which endurance limits are established are rarely duplicated in actual applications. Also, endurance limit test specimens are small compared to actual components, and fatigue strength is a function of size, being lower for larger components. This is because fatigue failure initiates at local discontinuities such as scratches or weld inclusions and the probability that a discontinuity will be present is greater the larger the component.

Fatigue strength is strongly influenced by the number of cycles of load and the geometry of the part. Geometries such as connections

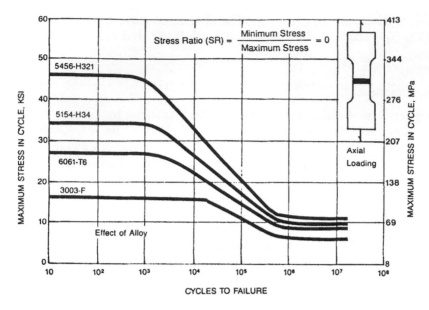

Figure 2.3   Fatigue strengths of MIG welded butt joints.

that result in stress concentrations due to abrupt transitions such as
sharp corners or holes have lower fatigue strengths than plain metal
without such details. Therefore, for design purposes, applications are
categorized by the severity of the detail, from A (being least severe,
such as base metal in plain components) to F (being most severe, such
as fillet weld metal). Design strengths in fatigue can be found in Table
2.22 by substituting parameters given there into the equation

$$ S_{rd} = \frac{C_f}{N^{1/m}} \tag{2.3} $$

where S_{rd} = allowable stress range, which is the algebraic difference
between the minimum and maximum stress (tension is
positive, compression is negative)
C_f = constant from Table 2.22
N = number of cycles of load
m = constant from Table 2.22

This equation is set so that there is a 95% probability that 97.7% of
components subjected to fatigue will be strong enough to withstand
the stress range given by the equation.

This equation shows that fatigue strength decreases rapidly as the
number of load cycles increases. For loads of constant amplitude, how-

TABLE 2.22   Fatigue Strengths of Aluminum Alloys

Category	C_f (ksi)	m	Fatigue limit (ksi)	Category examples
A	96.5	6.85	10.2	Plain metal
B	130	4.84	5.4	Members with groove welds parallel to the direction of stress
C	278	3.64	4.0	Groove welded transverse attachments with transition radius 24 in. $\geq R \geq 6$ in. (610 mm $\geq R \geq 150$ mm)
D	157	3.73	2.5	Groove welded transverse attachments with transition radius 6 in. $\geq R \geq 2$ in. (150 mm $\geq R \geq 50$ mm)
E	160	3.45	1.8	Base metal at fillet welds
F	174	3.42	1.9	Fillet weld metal

ever, it is believed that the fatigue strength of aluminum alloys does not decrease once the number of cycles reaches approximately 5 million. The fatigue strength predicted by the above equation for $N = 5$ million is called the *constant amplitude fatigue limit* (CAFL, or simply fatigue limit) and is given in Table 2.22. Loads may also have variable amplitudes, such as the loads on a beam in a bridge carrying traffic composed of cars and trucks of various weights. For variable amplitude loads, no lower bound on the fatigue strength is believed to exist, but some design codes use one-half of the constant amplitude fatigue limit as the fatigue limit for variable amplitude loading. Fatigue strengths of aluminum alloys are 30 to 40% of those of steel under similar circumstances of loading and severity of the detail.

Fatigue is also affected by environmental conditions. The fatigue strength of aluminum in corrosive environments such as salt spray can be considerably less than the fatigue strength in laboratory air. This may be because corrosion sites such as pits act as points of initiation for cracks, much like flaws such as dents or scratches. The more corrosion resistant alloys of the 5xxx and 6xxx series suffer less reduction in fatigue strength in corrosive environments than the less corrosion resistant alloys such as those of the 2xxx and 7xxx series. On the other hand, fatigue strengths are higher at cryogenic temperature than at room temperature. There isn't enough data on these effects to establish design rules, so designers must test specific applications to determine the magnitude of environmental factors on fatigue strength.

The fatigue strength of castings is less than that of wrought products, and no fatigue design strengths are available for castings. While castings are less notch sensitive than wrought products, they, like wrought products, should be designed with as few stress concentrations as possible to improve fatigue life.

2.5   Corrosion Resistance

As mentioned above, aluminum is resistant to corrosion from many agents. The hard aluminum oxide skin that forms on the surface in the presence of oxygen discourages further oxidation of the metal. Thus aluminum is often used without any protective coating. For some applications, the metal may be protected with a coating. An example is anodizing, a process that accelerates the formation of the protective oxide layer, as discussed in Section 2.9.4.1 below.

2.5.1   General Corrosion Resistance

Most, but not all, aluminum alloys are less corrosion resistant than pure aluminum. General corrosion resistance of aluminum alloys is usually an inverse function of the amount of copper used in the alloy. Thus the 2xxx series alloys are the least corrosion resistant alloys, since copper is their primary alloying element and all have appreciable (around 4%) levels of copper. (The only alloy that the Aluminum Association *Specification for Aluminum Structures* requires to be painted for atmospheric exposure applications is 2014-T6). Some 7xxx series alloys contain about 2% copper in combination with magnesium and zinc to develop strength. Such 7xxx series alloys (such as 7049, 7050, 7075, 7175, 7178, and 7475) are the strongest but least corrosion resistant of their series. Low copper aluminum-zinc alloys, such as 7005, are also available, and have become more popular recently. Copper does have a beneficial effect in 7xxx series alloys' resistance to stress corrosion cracking (discussed further in Section 2.5.5 below), however, by allowing them to be precipitated at higher temperatures without loss in strength in the T73 temper, which has good strength and good stress corrosion cracking resistance. Among the 6xxx series alloys, higher copper content (1% in 6066) generally decreases corrosion resistance, but most 6xxx series alloys contain little copper.

Some other alloying elements also decrease corrosion resistance. Lead (added to 2011 and 6262 for machining characteristics), nickel (added to 2018, 2218, and 2618 for elevated temperature service), and tin (used in 8xx castings) all tend to decrease the corrosion resistance, but not enough to matter in most applications.

Many of the 5xxx series alloys have general corrosion resistance as good as commercially pure aluminum and are more resistant to salt water, and so are useful in marine applications.

Heat treatment parameters, such as the rate of quenching and the temperature and time of artificial aging, can also affect general corrosion resistance. The relative general corrosion resistance of the various alloys is given in Tables 2.12 and 2.13.

2.5.2 Galvanic Corrosion

Galvanic corrosion occurs when two conductors with different electric potentials are electrically connected by an electrolyte, which is provided by the moisture present in most applications. The conductor that is more anodic is corroded, and the conductor that is more cathodic is protected from corrosion. In a saltwater solution, the order of most anodic to most cathodic is

magnesium

zinc

aluminum

cadmium

mild steel

lead

tin

brasses

copper

bronzes

stainless steels

While this list applies only when the electrolyte is a saline solution, it is useful, because such a solution is similar to marine environments. The list shows that aluminum tends to be corroded in the presence of steel but protected when in electrical contact with zinc. An exception to the series is austenitic stainless steel, which, despite its place at the end of the list, tends to be polarized and therefore doesn't corrode aluminum. The electric potential of different aluminum alloys varies slightly, so galvanic corrosion can also occur when different alloys of aluminum are in contact, such as where 5xxx series filler wires are used to weld 6xxx series base metal. The electric potentials of many common alloys of aluminum and other metals are listed in Table 2.23 for a salt water electrolyte. The electric potential of non-heat-treatable alloys is not a function of the temper, but the potential for heat-treatable alloys varies slightly by temper and even for a given temper depending on the quenching rate.

TABLE 2.23   Relative Electrical Potential of Metals in Saltwater[1]

Alloy-temper	Potential (V)	Alloy-temper	Potential (V)
Magnesium	−1.73	6063-T5, T6	−0.83
Zinc	−1.10	6061-T6, 6351-T5	−0.83
7072	−0.96	2024-T8, 2219-T8	−0.82
7005-T6	−0.94	Cadmium	−0.82
5056, 5083, 5182, 5456	−0.87	2024-T6	−0.81
5154, 5454	−0.86	2219-T6	−0.80
5052, 5086	−0.85	2014-T6	−0.78
7075-T73, T76; 7475-T73, T76	−0.84	2014-T4	−0.69
7049-T73, T76; 7050-T73, T76	−0.84	Mild carbon steel	−0.58
1060, 3004, 5050	−0.84	Lead	−0.55
1100, 3003	−0.83	Copper	−0.20
7075-T6, 7178-T6, 7475-T6	−0.83		

[1]Where temper is not given, potential is for all tempers of that alloy.

The corrosion of aluminum in contact with more cathodic metals can be minimized by maximizing the ratio of exposed surface area of the aluminum to that of the other metal. This minimizes the current density and thereby slows the rate of corrosion. Coating the cathodic metal is an effective way of doing this; coating the aluminum only is not and can actually have the opposite effect. Because of this principle, the use of steel fasteners in an aluminum part usually will not cause significant corrosion of the aluminum, because the total surface area of the part is typically much greater than that of the fasteners, the part being much larger. Aluminum fasteners in a steel component, on the other hand, will tend to corrode rapidly.

Some products can be provided with a thin coating of pure aluminum or corrosion resistant aluminum alloy (such as 7072); the resulting product is called *alclad*. This cladding is metallurgically bonded to one or both sides of sheet, plate, 3003 tube, or 5056 wire, and may be 1.5 to 10% of the overall thickness. The cladding alloy is chosen so it is anodic to the core alloy and so protects it from corrosion. Any corrosion that occurs proceeds only to the cladding-core interface and then spreads laterally, making cladding very effective in protecting thin materials. Because the coating generally has a lower strength than the base metal, alclad alloys have slightly lower strengths than non-alclad alloys of the same thickness. Alloys used in clad products are given in Table 2.24.

Aluminum alloys can also be cathodically protected with zinc or magnesium sacrificial anodes or with impressed currents. Buried aluminum pipelines are typically protected with anodes.

2.5.3 Pitting Corrosion

Pitting corrosion is localized, resulting in small, randomly located pits roughly hemispherical in shape, and it most commonly occurs in the presence of chloride ions. The pits tend to be covered with corrosion product, and so the rate of growth of the pit depth tends to diminish over time. Pit depth has been observed to be approximately a function of the cube root of time. Therefore, doubling the thickness of a part increases the time to perforation eightfold. If the part thickness is sufficient and appearance is not an issue, pitting corrosion can be tolerated.

2.5.4 Deposition Corrosion

The ions of heavy metals such as copper, mercury, lead, nickel, cobalt, and tin can cause severe localized corrosion of aluminum, especially in acidic solutions where they are most soluble. This action is called *deposition corrosion,* because the metal reduced from these ions plates onto the aluminum, setting up galvanic cells, usually resulting in pitting. A copper concentration of 0.02 to 0.05 ppm in neutral or acidic solutions is the approximate threshold above which pitting may occur. The exact threshold is a function of the particular aluminum alloy, the pH of the solution, the concentration of other ions (especially bicarbonate, chloride, and calcium), and whether the pits that develop are open or occluded. Significantly higher copper concentrations may be acceptable, depending on these factors and the duration of exposure, since pitting penetration rates decrease over time.

Mercury is among the most corrosive heavy metals to aluminum. Where aluminum's oxide coat is damaged by chemical or mechanical attack, mercury can corrode aluminum rapidly, especially in the presence of moisture. For this reason, a 0.005 ppm limit on the mercury content of water is suggested. This compares to the U.S. Environmental Protection Agency maximum contaminant level goal for mercury in drinking water of 0.002 ppm.

2.5.5 Stress Corrosion Cracking

Stress corrosion cracking (SCC) is localized directional cracking caused by a combination of tensile stress and corrosive environment.

TABLE 2.24 Components of Clad Products

DESIGNATION	COMPONENT ALLOYS①		TOTAL SPECIFIED THICKNESS OF COMPOSITE PRODUCT in.	SIDES CLAD	CLADDING THICKNESS PER SIDE Percent of Composite Thickness		
	CORE	CLADDING			NOMINAL	AVERAGE② min.	AVERAGE② max.
Alclad 2014 Sheet and Plate	2014	6003	Up thru 0.024	Both	10	8	::
			0.025–0.039	Both	7.5	6	::
			0.040–0.099	Both	5	4	::
			0.100 and over	Both	2.5	2	3③
Alclad 2024 Sheet and Plate	2024	1230	Up thru 0.062	Both	5	4	::
			0.063 and over	Both	2.5	2	3③
1½% Alclad 2024 Sheet and Plate	2024	1230	0.188 and over	Both	1.5	1.2	3③
Alclad One Side 2024 Sheet and Plate	2024	1230	Up thru 0.062	One	5	4	::
			0.063 and over	One	2.5	2	3③
1½% Alclad One Side 2024 Sheet and Plate	2024	1230	0.188 and over	One	1.5	1.2	3③
Alclad 2219 Sheet and Plate	2219	7072	Up thru 0.039	Both	10	8	::
			0.040–0.099	Both	5	4	::
			0.100 and over	Both	2.5	2	3③
Alclad 3003 Sheet and Plate	3003	7072	All	Both	5	4	6③
Alclad 3003 Tube	3003	7072	All	Inside	10	::	::
			All	Outside	7	::	::
Alclad 3004 Sheet and Plate	3004	7072	All	Both	5	4	6③
Alclad 5056 Rod and Wire	5056	6253	Up thru 0.375	Outside	20	16 (of total cross-sectional area)	6③
Alclad 6061 Sheet and Plate	6061	7072	All	Both	5	4	6③
Alclad 7050 Sheet and Plate	7050	7072	Up thru 0.062	Both	4	3.2	6③
			0.063 and over	Both	2.5	2	::

Product	Alloy	Cladding	Thickness	Sides			
7108 Alclad 7050 Sheet and Plate	7050	7108	Up thru 0.062	Both	4	3.2	..
			0.063 and over	Both	2.5	2	..
Alclad 7075 Sheet and Plate	7075	7072	Up thru 0.062	Both	4	3.2	..
			0.063-0.187	Both	2.5	2	..
			0.188 and over	Both	1.5	1.2	3③
2 1/2% Alclad 7075 Sheet and Plate	7075	7072	0.188 and over	Both	2.5	2	4③
Alclad One Side 7075 Sheet and Plate	7075	7072	Up thru 0.062	One	4	3.2	..
			0.063-0.187	One	2.5	2	..
			0.188 and over	One	1.5	1.2	3③
2 1/2% Alclad One Side 7075 Sheet and Plate	7075	7072	0.188 and over	One	2.5	2	4③
7008 Alclad 7075 Sheet and Plate	7075	7008	Up thru 0.062	Both	4	3.2	..
			0.063-0.187	Both	2.5	2	..
			0.188 and over	Both	1.5	1.2	3③
7011 Alclad 7075 Sheet and Plate	7075	7011	Up thru 0.062	Both	4	3.2	..
			0.063-0.187	Both	2.5	2	..
			0.188 and over	Both	1.5	1.2	3③
Alclad 7178 Sheet and Plate	7178	7072	Up thru 0.062	Both	4	3.2	..
			0.063-0.187	Both	2.5	2	..
			0.188 and over	Both	1.5	1.2	3③
Alclad 7475 Sheet	7475	7072	Up thru 0.062	Both	4	3.2	..
			0.063-0.187	Both	2.5	2	..
			0.188-0.249	Both	1.5	1.2	..
No. 7 Brazing Sheet	3003	4004	Up thru 0.024	One	15	12	18
			0.025-0.062	One	10	8	12
			0.063 and over	One	7.5	6	9
No. 8 Brazing Sheet	3003	4004	Up thru 0.024	Both	15	12	18
			0.025-0.062	Both	10	8	12
			0.063 and ove	Both	7.5	6	9
No. 11 Brazing Sheet	3003	4343④	Up thru 0.063	One	10	8	12
			0.064 and over	One	5	4	6

TABLE 2.24 Components of Clad Products *(Continued)*

DESIGNATION	COMPONENT ALLOYS①		TOTAL SPECIFIED THICKNESS OF COMPOSITE PRODUCT in.	SIDES CLAD	CLADDING THICKNESS PER SIDE Percent of Composite Thickness		
	CORE	CLADDING			NOMINAL	AVERAGE②	
						min.	max.
No. 12 Brazing Sheet	3003	4343④	Up thru 0.063	Both	10	8	12
			0.064 and over	Both	5	4	6
No. 23 Brazing Sheet	6951	4045	Up thru 0.090	One	10	8	12
			0.091 and over	One	5	4	6
No. 24 Brazing Sheet	6951	4045	Up thru 0.090	Both	10	8	12
			0.091 and over	Both	5	4	6
Clad 1100 Reflector Sheet	1100	1175	Up thru 0.064	Both	15	12	18
			0.065 and over	Both	7.5	6	9
Clad 3003 Reflector Sheet	3003	1175	Up thru 0.064	Both	15	12	18
			0.065 and over	Both	7.5	6	9

NOTE: This table does not include all clad products registered with The Aluminum Association.
① Cladding composition is applicable only to the aluminum or aluminum alloy bonded to the alloy ingot or slab preparatory to processing to the specified composite product. The composition of the cladding may be subsequently altered by diffusion between the core and cladding due to thermal treatment.
② Average thickness per side as determined by averaging cladding thickness measurements taken at a magnification of 100 diameters on the cross-section of a transverse sample polished and etched for microscopic examination.
③ Applicable for thicknesses of 0.500 in. and greater.
④ The cladding component, in lieu of 4343 alloy, may be 5% 1xxx Clad 4343.

SCC only affects aluminum alloys with appreciable amounts of alloying elements such as copper, magnesium, silicon, and zinc, and it is usually only of concern among the 2xxx and 7xxx alloys. Because failure is sudden and fracture of the parts occurs, designers are especially wary of this type of corrosion. Any source of tensile stress can cause SCC—not only externally applied service loads, but also residual stresses from fabrication and stresses induced by slight misfits in assembled parts.

The susceptibility to SCC is a function of both the magnitude and duration of the tensile stress. Design against SCC is conducted on the basis of testing that establishes a threshold stress below which, for a given environment and duration of stress, fracture will not occur. Since water or water vapor is a key environmental factor producing SCC and chloride ions present in marine applications greatly accelerate attack, salt solutions are frequently used as the test environment. The SCC strengths of certain high strength 2xxx and 7xxx alloys guaranteed by aluminum producers on the basis of ASTM G47, *Test Method for Determining Susceptibility to Stress Corrosion Cracking of High Strength Aluminum Products,* are given in Table 2.25.

When an aluminum product is quenched as part of its heat treatment, the rapid cooling at the surface of the part and slower cooling at the core produce residual stresses in the part. The surface, which is cooled first, is in compression, and the center is in tension. The resulting product is resistant to SCC because external forces must first overcome the initial compressive stresses before the surface experiences tension, a requisite for SCC to occur. If, however, the compressive stress regions are removed by fabrication processes such as machining, the exposed tensile stress regions near the center of the part become susceptible to SCC. To overcome this, aluminum products of constant cross section (such as sheet and plate and extrusions) are often stress relieved by stretching to improve their SCC performance. Such products are stretched from 1/2 to 5% and designated by a temper Tx5x or Tx5xx (such as 2014-T6510, and as described in Section 2.2.3.2 above).

Cold working (such as rolling) of wrought products often results in highly directional grain structures, which causes anisotropic resistance to stress corrosion cracking. Usually, the longitudinal resistance is highest, the short transverse resistance is lowest, and the long transverse resistance in between. The directional differences in resistance to cracking in tempers produced by extended precipitation treatments such as T6 and T8 for the 2xxx alloys and T73, T736 and T76 tempers for the 7xxx alloys is usually much less than the directional differences in other tempers. So applications using thick parts, machined parts, and components subject to significant through-

TABLE 2.25 Corrosion Resistance Test Criteria

PRODUCT	ALLOY AND TEMPER	STRESS CORROSION RESISTANCE TEST ① STRESS LEVEL	EXFOLIATION CORROSION RESISTANCE TEST ② SAMPLE LOCATION
Sheet	7075-T76	—	③
	7178-T76	—	③
	7475-T761	—	③
Plate	2124-T851⑥	50% RLTYS③	—
	2219-T851⑥	75% RLTYS⑤	—
	2219-T87⑥	75% RLTYS⑤	—
	7050-T7451④	35 ksi	③
	7050-T7651	25 ksi	③
	7075-T7351	75% RLTYS⑤	—
	7075-T7651	25 ksi	③
	7178-T7651	25 ksi	③
	7475-T7351	40 ksi	—
	7475-T7651	25 ksi	③
Extruded, or Cold Finished from Extruded, Wire, Rod, and Bar, and Extruded Profiles (Shapes) and Tube	2219-T8510, T8511	30 ksi	—
	7050-T73510, T73511	75% RLYS⑤	—
	-T74510④, T74511④	35 ksi	③
	-T7651, T76511	17 ksi	③
	7075-T73, T7351	75% RLYS⑤	—

Product form	Temper	Criterion	Note
Rolled, or Cold Finished from Rolled Wire, Rod and Bar	-T73510, T73511	75% RLYS⑤	—
	-T76	25 ksi	③
	-T76510, T76511	25 ksi	③
	7178-T76	25 ksi	③
	-T76510, T76511	25 ksi	③
Drawn Tube	7075-T73, T7351	75% RLYS⑤	—
Hand Forgings	7049-T73, T7352	75% RLYS⑤	③
	7050-T74④	35 ksi	—
	7075-T73, T7352	75% RLYS⑤	③
	7075-T74, T7452	35 ksi for thickness 3 in. and less and less 50% RLYS for thickness over 3 in.	—
Die Forgings	7049-T73, T7352	75% RLYS⑤	—
	7050-T74④	35 ksi	③
	7075-T73, T7352	75% RLYS⑤	—

① Tested in accordance with ASTM G47.
② Tested in accordance with ASTM G34-72 and displays corrosion less than that pictured by Photo B, Fig. 2.
③ Sample location to be the same as the electrical conductivity test location per Table 6.5.

④ T74 type tempers, although not previously registered, have appeared in literature and in some specifications as T736 type tempers.
⑤ RLTYS—Registered long transverse yield strength.
⑥ These 2xxx alloys do not routinely require SCC lot release testing. The criteria shown are in accordance with certain government and customer specifications, but apply only when specified.

thickness tensile stresses benefit from using these tempers, while thin parts and other parts not subject to significant through-thickness tensile stresses may be of the T3 or T4 tempers of the 2xxx alloys or the T6 tempers of the 7xxx alloys. The effectiveness of the extended precipitation treatments is demonstrated by 7075-T73, which resists SCC in the least favorable direction at stresses up to 44 ksi (300 MPa), while, under similar conditions, 7075-T6 resists cracking only up to 7 ksi (50 MPa). The performance of some 7xxx alloys promoted as corrosion resistant in T73, T74, and T76 tempers is demonstrated by testing the material against combination criteria of yield strength and electrical conductivity. Although these are indirect measures of SCC performance, they correlate well with actual SCC resistance.

5xxx series alloys with more than about 3% magnesium and in strain hardened tempers may become prone to SCC after exposure to elevated temperatures. The service record of the 6xxx series alloys, on the other hand, has no reports of SCC. Casting alloys other than the 7xx series and 520.0-T4 have not been reported to experience SCC in service.

The resistance of wrought alloys to stress corrosion cracking is ranked in Table 2.12.

2.5.6   Exfoliation Corrosion

Exfoliation corrosion is a delamination parallel to the metal surface caused by the formation of corrosion product, and is also called *lamellar, layer,* or *stratified* corrosion. The corrosion product causes the material to swell, pushing flakes of metal up at the surface. This kind of corrosion is accelerated in slightly acidic environments and when aluminum is electrically coupled with a cathodic material. Unlike stress corrosion cracking, it is unaffected by the presence of stress. Exfoliation is greatly increased by exposure to salts such as those found in de-icing salts and sea water.

Alloys most prone to exfoliation are the 2xxx and 7xxx series and some cold-worked 5xxx series alloys such as 5456-H321. Aluminum-magnesium alloys with more than 3% magnesium (such as 5056, 5083, 5086, 5154, 5254, and 5456) held at temperatures above 150°F (65°C) are also susceptible. Exfoliation of 5456-H321 boat hull plate led to the development of special tempers for this application (H116 and H117, which have longer precipitation heat treating times) for alloys 5083, 5086, and 5456.

Resistance to exfoliation may be measured by immersion tests such as ASTM G66, *Test Method for Visual Assessment of Exfoliation Corrosion Susceptibility of 5XXX Series Aluminum Alloys (ASSET Test),* and ASTM G34, *Test Method for Exfoliation Corrosion Susceptibility in 2XXX and 7XXX Series Aluminum Alloys (EXCO Test)*. Specimens are compared to reference photographs so they can be ranked according to the following code:

P—pitting

EA—superficial corrosion

EB—moderate corrosion

EC—severe corrosion

ED—very severe corrosion

Aluminum producers use ASTM G34 rating EB an acceptance criterion for certain 2xxx and 7xxx alloys as stated in Table 2.25.

2.5.7   Atmospheric Corrosion

Many aluminum alloys are used in products exposed to the elements and without any protective coatings. The rate of corrosion is mainly a function of the concentration of pollutants in the air and the proximity to bodies of water. The appearance of bare aluminum gradually changes over time from an initial shiny silver color to dull gray. Mild pitting occurs, which results in a general roughening of the surface. All other factors being equal, aluminum corrodes at a rate about one-tenth that of mild carbon steel under atmospheric exposure.

Laboratory exposure tests such as salt-spray tests have not been good predictors of atmospheric corrosion rates. Instead, long-term outdoor exposure tests are used. Environments are categorized as seacoast, industrial, rural, or desert, listed in the approximate order of severity as observed in tests. Pitting depth, weight loss, and loss of tensile strength have been used as measures of the amount of corrosion in tests. Since most atmospheric corrosion manifests as pitting, loss of tensile strength is the best indicator of the structural consequences of corrosion, since this accounts for the effects of the distribution, size, number, depth, and shape of the pitting. The average strength loss for 1100, 3003, and 3004 alloys all in the H14 temper has been measured as about 12% over 30 years in a seacoast environment and 7.5% in an industrial exposure. Weight loss is used to calculate a theoretical corrosion rate (in units such as mils per year), as if the corrosion were uniform over the entire surface of the specimen.

For atmospheric exposure, corrosion is initially (the first six months to two years) relatively rapid, followed by a very slow, steady state. This behavior is independent of alloy and type of environment. The linear rate after the initial period varies by environment, ranging from about 0.03 mils per year in rural or desert environments to 0.11 mils per year for seacoast exposures. The 1xxx, 3xxx, 5xxx, and 6xxx alloys fare better than non-alclad 2xxx and 7xxx alloys.

2.5.8   Corrosion in Waters

2.5.8.1   High-purity water.   Only a small amount of aluminum is dissolved by high-purity water at room temperatures, and, after a few days, an oxide film on the aluminum prevents further oxidation. Also, aluminum is not affected by the various chemicals added to water in the steam power industry to inhibit the corrosion of steel. Aluminum is resistant to steam below about 300°F (150°C), but high-purity aluminum sheet disintegrates in a few days when exposed to higher temperatures. Ironically, alloys that are more corrosion resistant to water at room temperature are less corrosion resistant at elevated temperatures. Alloying elements that decrease resistance at room temperature increase resistance at elevated temperatures.

2.5.8.2   Natural fresh waters.   The corrosion resistant aluminum alloys (1xxx, 3xxx, 5xxx, and 6xxx series) are compatible with exposure to most natural fresh waters and have been widely used in boats. The pitting that may occur is a function of the presence of dissolved minerals in the water, such as chlorides, sulfates, bicarbonates, and heavy metals, and the water temperature and pH. The effect of heavy metals is lessened in more alkaline waters. Chlorine, in the levels used for potable water, has a negligible effect on aluminum.

2.5.8.3   Seawater.   Extensive experience with the 1xxx, 3xxx, 5xxx, and 6xxx wrought alloy series and 356.0 and 514.0 castings in marine environments has been good, including partial, intermittent, and total immersion conditions. The 5xxx and 3xxx alloys are preferred; corrosion is of the pitting type and the corrosion rate based on weight loss is less than 0.2 mils/yr, about 5% that of uncoated carbon steel. The 6xxx series alloys are slightly less resistant and suffer a corrosion rate about two to three times that of the 3xxx and 5xxx series. Alloys of the 2xxx and 7xxx series should be alclad or painted for marine applications.

2.5.9   Corrosion in Soils

Soils vary greatly in pH, mineral content, moisture, organic content, resistivity, aeration, and other parameters, so their corrosion effects can vary significantly. Most soils cause pitting on aluminum that can range from negligible to severe. For example, cinders have caused corrosion of aluminum pipe but, in the proper soil conditions, alclad 3004-H34, 5052-H141, and 6061-T6 have been used without coatings for corrugated pipe used in sewers, drains, and culverts since 1960. Extrapolating this experience, these aluminum alloys are estimated to have a 50-year life. Alloys 3003, 3004, 5050, 5052, 6061, and 6063 have been used widely in on the surface and buried in irrigation, petroleum, light pole, and mining applications. Guidelines for the quality of soil used as backfill can be found in ASTM B789, *Standard Practice for Installing Corrugated Aluminum Structural Plate Pipe for Culverts and Sewers.* The Aluminum Association also provides recommendations for soil conditions for uncoated culvert. They suggest the soil and water pH be between 4 and 9 and that the soil and water resistivities by 500 Ω-cm or greater (except in seawater, where it can be as low as 35 Ω-cm). Others have suggested soil resistivity be 1500 Ω-cm or greater for buried pipelines to avoid pitting.

Because of the great variability in soil conditions, aluminum pipelines are often protected with bituminous coating or tape wrap or by cathodic protection.

2.5.10 Water Staining

Aluminum can be stained (Figure 2.4) by water that is drawn by capillary action between adjacent pieces of aluminum that are tightly packed together, such as stacks of sheet or plate or rolled coils. While all alloys can suffer water stains, the 5xxx series is slightly more susceptible than the others. The resulting stains vary from white to gray to nearly black and may exhibit an iridescent appearance. The stains are a roughening of the surface due to the formation of aluminum oxide corrosion product, but the corrosion is superficial, and mechanical properties are not affected. In fact, the thickened oxide skin in the stained areas is more resistant to subsequent corrosion. Water stains can be removed mechanically or chemically, but the process is often difficult, and it may be impossible to restore the original surface brightness. Therefore, the best policy is prevention.

The moisture that causes water staining often comes from condensation. This can occur when cold metal is brought indoors to a heated environment, where the temperature of the metal is below the dew point of the air. To avoid this, cold metal should be warmed in a dry place before being brought into a heated area with high humidity. If wet metal is dried within a short time, staining will not occur. Moisture can also come from exposure to the elements, and so aluminum products that are stacked should be protected from the weather. Strippable plastic film attached to sheets at the mill is effective.

Figure 2.4 Water stains.

2.5.11 Compatibility with Various Substances

Aluminum tends to be more resistant to acids than alkalis, but compatibility with various substances must be evaluated individually. Compatibility of aluminum with an extensive list of substances is tabulated in *Guidelines for the Use of Aluminum with Food and Chemicals,* published by the Aluminum Association. The suitability of aluminum for use with many compounds is based on laboratory data with high-purity chemicals. In actual practice, impurities may be present. Their combined effect can be greater than the sum of their individual effects, and most materials are degraded over time when exposed to other substances in service conditions. For example, aluminum is very resistant to gasoline, but over time it can be corroded by water or other impurities that are sometimes present in gasoline storage tanks.

Nonetheless, aluminum has been found to be the best material for many specific applications. Examples are alloys 5254 and 5652 for hydrogen peroxide storage vessels, 5454 for gasoline tank trucks, 1100 for nitric acid, 1100, 3003, 5052, and 6061 for acetic acid, and 3003 for service with ammonium nitrate and urea. In specific instances, a small amount of an inhibitor may be added to the substance aluminum is exposed to for the purpose of reducing or preventing corrosion that would otherwise occur. A list of inhibitors is given in *Aluminum with Food and Chemicals*. Sodium compounds are frequently used with aluminum alloys.

2.6 Product Forms

Ingot (large, unfinished bars of aluminum) is the result of primary aluminum production. This material is not in a useful form for purchasers, so it is wrought into semifabricated mill products that include flat rolled products (foil, sheet, and plate, depending on thickness), rolled elongated products (wire, rod, and bar, depending on dimensions), drawn tube, extrusions, and forgings. Ingot is also furnished to foundries that remelt it to produce castings.

2.6.1 Wrought Products

Aluminum mills produce wrought products. ASTM specifications provide minimum requirements for wrought aluminum alloys by product (for example, sheet and plate) rather than by alloy. Each ASTM specification addresses all the alloys that may be used to make that product. Some ASTM specifications for wrought products are

B209 *Sheet and Plate*

B210 *Drawn Seamless Tubes*

B211 *Bar, Rod, and Wire*

B221 *Extruded Bars, Rods, Wire, Shapes, and Tubes*

B234 *Drawn Seamless Tubes for Condensers and Heat Exchangers*

B236 *Bars for Electrical Purposes (Bus Bars)*

B241 *Seamless Pipe and Seamless Extruded Tube*

B247 *Die Forgings, Hand Forgings, and Rolled Ring Forgings*

B308 *6061-T6 Standard Structural Shapes*

B313 *Round Welded Tubes*

B316 *Rivet and Cold-Heading Wire and Rods*

B317 *Extruded Bar, Rod, Pipe, and Structural Shapes for Electrical Purposes (Bus Conductors)*

B345 *Seamless Pipe and Seamless Extruded Tube for Gas and Oil Transmission and Distribution Piping Systems*

B361 *Welding Fittings*

B373 *Foil for Capacitors*

B404 *Seamless Condenser and Heat-Exchanger Tubes with Integral Fins*

B429 *Extruded Structural Pipe and Tube*

B479 *Foil for Flexible Barrier Applications*

B483 *Drawn Tubes for General Purpose Applications*

B491 *Extruded Round Tubes for General Purpose Applications*

B547 *Formed and Arc-Welded Round Tube*

B632 *Rolled Tread Plate*

Tolerances on the dimensional, mechanical, and other properties of wrought alloys are given in the Aluminum Association's *Aluminum Standards and Data* and ANSI H35.2 *Dimensional Tolerances for Aluminum Mill Products*. The two are the same, although they are revised at different dates and so may not match exactly until the next revision. The tolerances are standard tolerances (also called commercial or published tolerances). Special (either more strict or less strict than standard) tolerances may be met by agreement between the purchaser and the supplier. Tolerances approximately one-half of standard tolerances can usually be provided if the purchaser so specifies.

2.6.1.1 Flat rolled products (foil, sheet, and plate). Flat rolled products are produced in rolling mills, where cylindrical rolls reduce the thickness and increase the length of ingot. Initially, huge heated ingots six feet wide, 20 feet long, and more than two feet thick and weighing over 20 tons are rolled back and forth in a breakdown mill to reduce the thickness to a few inches. Plate of this thickness can be heated, quenched, stretched (to straighten and relieve residual stresses), aged at temperature, and shipped, or it may be coiled and sent on to a continuous mill to further reduce its thickness. Before further rolling at the continuous mill, the material is heated to soften it for cold rolling. Heat treatments and stretching are also applied there after rolling to the desired thickness. Alternatively, sheet can be produced directly from molten metal rather than ingot in the continuous casting process, by which molten aluminum passes through water-cooled casting rollers and is solidified. The thickness is then further reduced by cold rolling.

The resulting flat rolled products are rectangular in cross section and are called *foil, sheet,* or *plate,* depending on their thickness (Table 2.26). *Foil* has a thickness less than 0.006 in. (up through 0.15 mm). *Sheet* has a thickness less than 0.25 in., but not less than 0.006 in. (over 0.15 mm through 6.3 mm). *Plate* is 0.25 in. (over 6.3 mm) thick or more. The use of the term *flat* in flat rolled products is to distinguish these from rolled wire, rod, or bar, which are discussed in Section 2.6.1.3.

TABLE 2.26   Flat Rolled Products

	Thickness			
	Minimum		Maximum	
Product	in.	mm	in.	mm
Foil			<0.006	0.15
Sheet	0.006	<0.15	<0.25	6.3
Plate	0.25	<6.3		

Foil is produced as thin as 0.00017 in. (0.004318 mm) thick, in alloys 1100, 1145, 1235, 2024, 3003, 5052, 5056, 6061, 8079, and 8111, in both rolls and sheets, and in the annealed and H19 tempers. The H19 variety is called hard foil because it is fully strain hardened.

Uses for foil include the cores of aluminum honeycomb panels used in aircraft, capacitors (ASTM B373 *Foil for Capacitors*), and for packaging (ASTM B479 *Foil for Flexible Barrier Applications*). Standard household foil is approximately 0.0006 in. thick, or about 1/4 the thickness of the paper used in this book, while capacitor foil is four times thinner. In many packaging applications, foil is laminated to paper or plastic films for strength.

Foil finishes include:

- *Mill finish (MF),* a nonuniform finish as produced by the mill and which may vary from coil to coil and within a coil
- *Chemically cleaned,* washed to remove lubricant and foreign material
- *Embossed,* a impressed pattern made by an engraved roll or plate
- *Etched,* roughened mechanically or electrochemically to provide an increased surface area

- *Mechanically grained*, roughened for applications such as lithographing
- *Scratch brushed*, abraded, usually with wire brushes, to produce a roughened surface
- *Matte, one side (M1S) or two sides (M2S)*, a diffuse reflecting finish
- *Bright two sides (B2S)*, a uniform bright specular finish
- *Extra bright two sides (EB2S)*, a uniform extra bright specular finish

Fin stock is coiled sheet or foil in specific alloys (1100, 1145, 3003, and 7072), tempers, and thicknesses (0.004 to 0.030 in.) used for the manufacture of fins for heat exchangers.

Sheet is one of the most widely used aluminum products and is produced in more alloys than any other aluminum product. Sheet is available rolled into coils with slit edges or in flat sheets with sheared, slit, or sawed edges. Circular blanks from coil or flat sheet are available. Panel flat sheet is flat sheet with a tighter tolerance on flatness than flat sheet.

Sheet is available in a number of finishes:

- *Mill finish (MF)*, nonuniform as produced by the mill
- *One side bright mill finish (1SBMF)*, with a moderate degree of brightness on one side and mill finish on the other
- *Standard one side bright finish (S1SBF)*, with a uniform bright finish on one side and mill finish on the other
- *Standard two sides bright finish (S2SBF)*, with uniform bright finish on both sides
- *Painted*, with a factory applied paint on one or both sides

The maximum weight of a coil is about 9900 lb (4500 kg). Coils are available in standard widths such as 24, 30, 36, 48, 60, and 72 in., and up to about 108 in. (2740 mm). Commonly available widths for flat sheet are 96, 120, and 144 in. Aluminum sheet gauges are different from steel sheet gauges. Therefore, it's better to identify aluminum sheet by thickness rather than gage number to avoid confusion. Table 2.27 lists some common aluminum sheet thicknesses, corresponding gage numbers, and weights per unit area based on 0.100 lb/in^3 density.

Plate is available in thicknesses up to about 8 in. (200 mm) in certain alloys; single plates may weigh up to 7900 lb (3600 kg). Circular plate blanks are available. Common widths and lengths for plate are similar to those for sheet. Tapered thickness plate is produced for storage vessels because the pressure at the bottom of such vessels is greater than nearer to the top.

TABLE 2.27   Common Aluminum Sheet Thicknesses

Thickness (in.)	Gage no.	Weight (lb/ft^2)	Thickness (in.)	Gage no.	Weight (lb/ft^2)
0.006	—	0.086	0.063	14	0.907
0.008	—	0.115	0.071	—	1.02
0.012	—	0.173	0.080	12	1.15
0.019	—	0.274	0.090	11	1.30
0.020	24	0.288	0.100	10	1.44
0.024	—	0.346	0.125	—	1.80
0.025	22	0.360	0.160	—	2.30
0.032	20	0.461	0.180	—	2.59
0.040	18	0.576	0.190	—	2.74
0.050	16	0.720			

Sheet and plate tolerances have been established for width, length, deviation of edges from straight, squareness, and flatness. Thickness tolerances are different for alloys specified for aerospace applications (including 2014, 2024, 2124, 2219, 2324, 2419, 7050, 7150, 7178, and 7475) than for other alloys. The ASTM specification for sheet and plate is B209, *Sheet and Plate.*

A number of sheet and plate products are made for specific applications by the primary producers. These include tread plate (ASTM B632 *Rolled Tread Plate*) made with a raised diamond pattern on one side to provide improve traction, duct sheet, and roofing and siding, available in corrugated and other shapes.

2.6.1.2 Extrusions. *Extrusions* are products formed by pushing heated metal, in a log-shaped form called a billet, through an opening called a die, the outline of which defines the cross-sectional shape of the extrusion. Some examples are shown in Figure 2.5. Thousands of pounds of pressure are exerted by the extrusion press as it forces the aluminum through the die and onto a runoff table, where the extrusion is straightened by stretching and cut to length. Artificial aging heat treatment may then be applied to extrusions made of heat-treatable alloys.

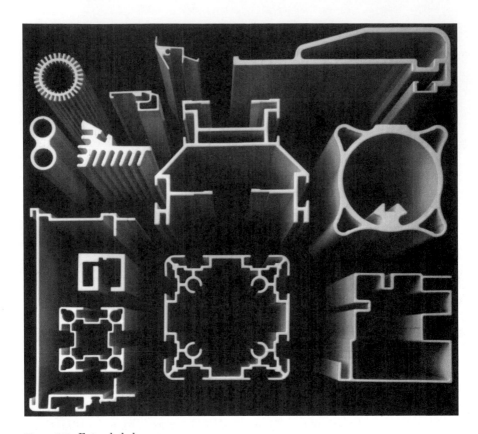

Figure 2.5 Extruded shapes.

Extruding was developed in the 1920s and replaced rolling around 1950 for producing standard shapes such as I beams and angles. Today it is used for virtually all aluminum shapes. Shapes, called *profiles* in the aluminum industry, are all products that are long relative to their cross-sectional dimensions and are not sheet, plate, tube, wire, rod, or bar. Commonly available shapes include I beams, tees, zees, channels, and angles; a great variety of custom shapes can be produced with modest die costs and short lead times. In the late 1960s, the Aluminum Association developed standard I beam and channel sizes that are structurally efficient and have flat (as opposed to tapered) flanges for convenient connections (ASTM B308 *6061-T6 Standard Structural Shapes*).

Profiles are sized by the smallest diameter circle than encloses their cross-sectional shape, called the *circumscribing circle size* (Figure 2.6). The size of an extruded shape is limited by the capacity and size of the extrusion press used to produce the shape. Presses that are used for

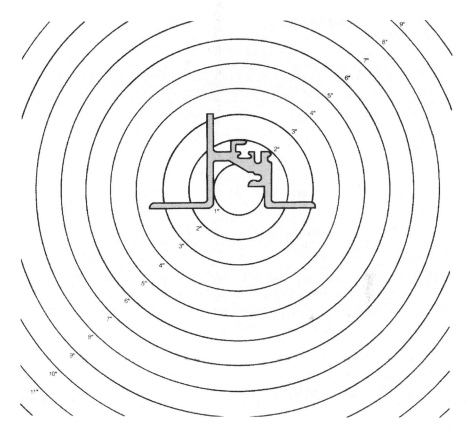

Figure 2.6 Extrusion circumscribing circle size.

most practical applications are limited to those profiles that fit within about a 15 in. (375 mm) diameter circle, although recently some larger presses have come on line; presses for military and aerospace applications can handle shapes as large as 31 in. (790 mm) in circle size. Standard I beams range up to 12 in. (305 mm) in depth, because the entire shape can fit in a 15 in. diameter circle. Standard channels are as large as 15 in. deep.

Some alloys are more difficult to extrude than others; generally, the stronger the alloy, the more difficult it is to extrude. The relative extrudability of some alloys is given in Table 2.28. Alloy 6063 is the benchmark for extrudability, because it's easy to extrude and widely used; about 75% of all extrusions are 6xxx series alloys. The 1xxx, 3xxx, and 6xxx series alloys are called *soft alloys,* while the 2xxx, 5xxx, and 7xxx are called *hard alloys,* based on the relative difficulty of extruding.

TABLE 2.28   Relative Hot Extrudability

Alloy	Extrudability, % of 6063	Alloy	Extrudability, % of 6063
1060	135	5456	20
1100	135	6061	60
1350	160	6063	100
2011	35	6066	40
2014	20	6101	100
2024	15	6151	70
3003	120	6253	80
5052	80	6351	60
5083	20	6463	100
5086	25	7001	7
5154	50	7075	10
5254	50	7079	10
5454	50	7178	7

Hollow shapes are extruded with hollow billets or with solid billets through porthole or bridge dies. The billet metal must divide and flow around the support for the die outlining the inside surface of the hollow extrusion and then weld itself back together as it exits the die. Such dies are more expensive than those for solid shapes. Hollow shapes produced with porthole or bridge dies are not considered seamless and are not used for parts designed to hold internal pressure. (See Section 2.6.1.4 below for information on tubes used to contain internal pressure.)

A profile's shape factor (the ratio of the perimeter of the profile to its area) is an approximate indication of its extrudability. The higher the ratio, the more difficult the profile is to extrude. In spite of this, profiles that are thin and wide are extrudable, but the trade-off between the higher cost of extruding and the additional cost of using a thicker section should be considered. Sometimes a thicker part is less expensive in spite of its greater weight, because it is more readily extruded. Parts as thin as 0.04 in. (1 mm) can be extruded but, as the circle size increases, the minimum extrudable thickness also increases.

There are also other factors at play in cost. Unsymmetric shapes are more difficult to extrude, as are shapes with sharp corners. Providing a generous fillet or rounded corner decreases extrusion cost. Profiles with large differences in thickness across the cross section or with fine details are also more difficult to extrude. Consulting extruders when designing a shape can help reduce costs.

Metal flows fastest at the center of the die, so as shapes become larger, it becomes more difficult to design and construct dies that keep the metal flowing at a uniform rate across the cross section. For this reason, larger shapes have larger dimensional tolerances. Tolerances on cross section dimensions, length, straightness, twist, flatness, and other parameters are given in *Aluminum Standards and Data.* Tolerances on cross sectional dimensions are also given here in Table 2.29.

Extrusions can incorporate a number of useful details, including:

1. *Screw chases.* A ribbed slot, called a screw chase or boss, can be extruded so as to receive a screw, eliminating the need for drilling and tapping a hole in the extrusion for attachments.

2. *Indexing marks.* A shallow groove extruded in a part can locate where a line of holes are to be punched or drilled. Marks (either raised or grooved) can be used to identify the extruder if more than one supplier is used, to orient a part, or to distinguish parts otherwise very similar in appearance.

3. *Gasket retaining grooves or guides.* In curtainwall, fenestration, and other applications, elastomeric gaskets are often required to seal between parts. By extruding a groove in the metal that matches a protrusion on a gasket, the need for adhesives to attach the gasket to the extrusion can be avoided.

4. *Interlocks.* Extrusions can be designed with interlocks of several kinds: those that serve as hinges, permitting rotation of the parts relative to each other; those that serve to align parts relative to each other, such as tongue and groove; and those that act as snap or friction fit connections to hold the parts together without the need for additional fastening.

5. *Fastener retaining grooves.* Grooves can be provided to prevent a bolt head or nut from rotating, or to permit fastener heads to be held flush without the need for countersinking.

6. *Weld details.* Edge preparation such as V, J, or U joint details or integral backing bars can be extruded along the edges of parts to be welded.

Cold extrusions (also called *impact extrusions* or *impacts*) are also produced, an effective method for making tubes or cup-like pieces that are hollow with one end partially or totally closed. The metal is formed at room temperature, and any heating of the metal is a consequence of the conversion of deformation energy to heat. The slug or preform is struck by a punch and deformed into the shape of a die, resulting in a wrought product with tight tolerances and zero draft and no parting lines. The five most commonly cold extruded alloys and their relative cold extrudability are shown in Table 2.30. An example of a cold extrusion is irrigation tubing, which can be produced up to 6 in. in diameter, 0.058 in. wall thickness, and 40 ft in length.

TABLE 2.29 Cross-Sectional Dimension Tolerances—Profiles (Shapes) ①
EXCEPT FOR T3510, T4510, T6510, T73510, T76510, T77510 AND T8510 TEMPERS ⑦

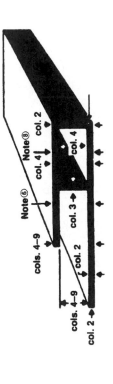

SPECIFIED DIMENSION	METAL DIMENSIONS				TOLERANCE ② ③—in. plus and minus									
	ALLOWABLE DEVIATION FROM SPECIFIED DIMENSION WHERE 75 PERCENT OR MORE OF THE DIMENSION IS METAL ⑨ ⑩				SPACE DIMENSIONS									
	All Except Those Covered by Column 3		Wall Thickness ④ Completely ⑤ Enclosing Space 0.11 sq. in. and Over (Eccentricity)		ALLOWABLE DEVIATION FROM SPECIFIED DIMENSION WHERE MORE THAN 25 PERCENT OF THE DIMENSION IS SPACE ⑥ ⑧									
					At Dimensioned Points 0.250–0.624 inches from Base of Leg		At Dimensioned Points 0.625–1.249 inches from Base of Leg		At Dimensioned Points 1.250–2.499 inches from Base of Leg		At Dimensioned Points 2.500–3.999 inches from Base of Leg		At Dimensioned Points 4.000–5.999 inches from Base of Leg		At Dimensioned Points 6.000–8.000 inches from Base of Leg	
in.	Col. 2		Col. 3		Col. 4		Col. 5		Col. 6		Col. 7		Col. 8		Col. 9
	Alloys 5083 5086 5454	Other Alloys ⑪	Alloys 5083 5086 5454	Other Alloys ⑪	Alloys 5083 5086 5454	Other Alloys ⑪	Alloys 5083 5086 5454	Other Alloys ⑪	Alloys 5083 5086 5454	Other Alloys ⑪	Alloys 5083 5086 5454	Other Alloys ⑪	Alloys 5083 5086 5454	Other Alloys ⑪	Alloys 5083 5086 5454	Other Alloys ⑪
Col. 1															

Tolerance notes: ±15% of specified dimension; ±.090 max, ±.015 min. · ±10% of specified dimension; ±.060 max, ±.010 min.

Size Range	1	2	3	4	5	6	7	8	9	10	11	12	13	14
Up thru 0.124	.009	.006	.013	.010	.015	.012								
0.125–0.249	.011	.007	.016	.012	.018	.014	.020	.016						
0.250–0.499	.012	.008	.018	.014	.020	.016	.022	.018	.024	.020				
0.500–0.749	.014	.009	.021	.016	.023	.018	.025	.020	.027	.022				
0.750–0.999	.015	.010	.023	.018	.025	.020	.027	.022	.030	.025	.035	.030		
1.000–1.499	.018	.012	.027	.021	.029	.023	.032	.026	.036	.030	.041	.035		
1.500–1.999	.021	.014	.031	.024	.033	.026	.038	.031	.043	.036	.049	.042	.057	.050
2.000–3.999	.036	.024	.046	.034	.050	.038	.060	.048	.069	.057	.080	.068	.092	.080
4.000–5.999	.051	.034	.061	.044	.067	.050	.081	.064	.095	.078	.111	.094	.127	.110
6.000–7.999	.066	.044	.076	.054	.084	.062	.104	.082	.121	.099	.142	.120	.162	.140
8.000–9.999	.081	.054	.091	.064	.101	.074	.127	.100	.147	.120	.182	.145	.197	.170

Tolerance notes: ±15% of specified dimension; ±.090 max, ±.025 min. · ±15% of specified dimension; ±.090 max, ±.015 min.

Size Range	1	2	3	4	5	6	7	8	9	10	11	12	13	14
Up thru 0.124	.021	.014	.025	.018	.027	.020								
0.125–0.249	.022	.015	.026	.019	.029	.022	.035	.028						
0.250–0.499	.024	.016	.028	.020	.032	.024	.038	.030	.058	.050				
0.500–0.749	.025	.017	.030	.022	.035	.027	.049	.040	.068	.060				
0.750–0.999	.027	.018	.031	.023	.039	.030	.057	.050	.079	.070	.099	.090		
1.000–1.499	.028	.019	.033	.024	.043	.034	.069	.060	.089	.080	.109	.100		
1.500–1.999	.036	.024	.046	.034	.056	.044	.082	.070	.102	.090	.122	.110	.182	.170
2.000–3.999	.051	.034	.061	.044	.071	.054	.097	.080	.117	.100	.137	.120	.197	.180
4.000–5.999	.066	.044	.076	.054	.086	.064	.112	.090	.132	.110	.152	.130	.212	.190
6.000–7.999	.081	.054	.091	.064	.101	.074	.127	.100	.147	.120	.167	.140	.227	.200
8.000–9.999	.096	.064	.106	.074	.116	.084	.142	.110	.162	.130	.182	.150	.242	.210
10.000–11.999	.111	.074	.121	.084	.131	.094	.157	.120	.177	.140	.197	.160	.257	.220
12.000–13.999	.126	.084	.136	.094	.146	.104	.172	.130	.192	.150	.212	.170	.272	.230
14.000–15.999	.141	.094	.151	.104	.161	.114	.187	.140	.207	.160	.227	.180	.287	.240
16.000–17.999	.156	.104	.166	.114	.176	.124	.202	.150	.222	.170	.242	.190	.302	.250
18.000–19.999	.171	.114	.181	.124	.191	.134	.217	.160	.237	.180	.257	.200	.317	.260
20.000–21.999	.186	.124	.196	.134	.206	.144	.232	.170	.252	.190	.272	.210	.332	.270
22.000–24.000	.201	.134	.211	.144	.221	.154	.247	.180	.267	.200	.287	.220	.347	.280

TABLE 2.29 Notes

① These Standard Tolerances are applicable to the average profile (shape); wider tolerances may be required for some profiles (shapes) and closer tolerances may be possible for others.

② The tolerance applicable to a dimension composed of two or more component dimensions is the sum of the tolerances of the component dimensions if all of the component dimensions are indicated.

③ When a dimension tolerance is specified other than as an equal bilateral tolerance, the value of the standard tolerance is that which applies to the mean of the maximum and minimum dimensions permissible under the tolerance for the dimension under consideration.

④ Where dimensions specified are outside and inside, rather than wall thickness itself, the allowable deviation (eccentricity) given in Column 3 applies to mean wall thickness. (Mean wall thickness is the average of two wall thickness measurements taken at opposite sides of the void.)

⑤ In the case of Class 1 Hollow Profiles (Shapes) the standard wall thickness tolerance for extruded round tube is applicable. (A Class 1 Hollow Profile (Shape) is one whose void is round and one inch or more in diameter and whose weight is equally distributed on opposite sides of two or more equally spaced axes.)

⑥ At points less than 0.250 inch from base of leg the tolerances in Col. 2 are applicable.

⑦ Tolerances for extruded profiles (shapes) in T3510, T4510, T6510, T73510, T76510 and T8510 tempers shall be as agreed upon between purchaser and vendor at the time the contract or order is entered.

⑧ The following tolerances apply where the space is completely enclosed (hollow profiles (shapes)): For the width (A), the balance is the value shown in Col. 4 for the depth dimension (D). For the depth (D), the tolerance is the value shown in Col. 4 for the width dimension (A). In no case is the tolerance for either width or depth less than the metal dimensions (Col. 2) at the corners.

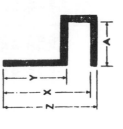

Example—Alloy 6061 hollow profile (shape) having 1 × 3 rectangular outside dimensions; width tolerance is ±0.021 inch and depth tolerance ±.034 inch. (Tolerances at corners, Col. 2, metal dimensions, are ±0.024 inch for the width and ±0.012 inch for the depth.) Note that the Col. 4 tolerance of 0.021 inch must be adjusted to 0.024 inch so that it is not less than the Col. 2 tolerance.

⑨ These tolerances do not apply to space dimensions such as dimensions "X" and "Z" of the example (right), even when "Y" is 75 percent or more of "X." For the tolerance applicable to dimensions "X" and "Z," use Col. 4, 5, 6, 7, 8 or 9, dependent on distance "A."

3t or Greater

3t or Greater

⑩ The wall thickness tolerance for hollow or semihollow profiles (shapes) shall be as agreed upon between purchaser and vendor at the time the contract or order is entered when the nominal thickness of one wall is three times or greater than that of the opposite wall.

⑪ Limited to those alloys listed in Table 11.1

TABLE 2.30   Relative Cold Extrudability of Annealed Alloys

Alloy	Relative cold extrusion pressure
1100	1.0
3003	1.2
6061	1.6
2014	1.8
7075	2.3

2.6.1.3 Wire, rod, and bar. Wire, rod, and bar are defined as products that have much greater lengths than cross-section dimensions (ASTM B211 *Bar, Rod, and Wire*). *Wire* is rectangular (with or without rounded corners), round, or a regular hexagon or octagon in cross section and with one perpendicular distance between parallel faces less than 0.375 in. (10 mm or less). Material that has such a dimension of 0.375 in. or greater (greater than 10 mm) is *bar*, if the section is rectangular or a regular hexagon or octagon, and *rod*, if the section is round (see Table 2.31).

TABLE 2.31   Wire, Rod, and Bar

Width or dia.	Square, rectangular, hexagon, or octagon	Circular
<0.375 in. (≤10 mm)	Wire	Wire
≥0.375 in. (>10 mm)	Bar	Rod

Rod and bar can be produced by hot rolling long, square ingot or hot extruding; the product may also be subsequently cold finished by drawing through a die. Wire can be hot extruded or drawn (pulled through a die or series of dies that define its cross-sectional shape) or flattened by roll flattening round wire into a rectangular shape with rounded corners. Drawing and cold finishing result in much tighter tolerances on thickness than rolling, which in turn produces more precise dimensions than extruding (ASTM B221, *Extruded Bars, Rods, Wire, Shapes, and Tubes*).

The round products (wire and rod), when produced for subsequent forming into fasteners such as rivets or bolts, are called *rivet and cold-heading wire and rod* (ASTM B316). Rivet and cold-heading wire and rod are produced in the alloys and with the strengths shown in Table 2.32; these alloys have good machinability for threading. Rivet and cold heading wire and rod, and the fasteners produced from it, shall upon proper heat treatment be capable of developing the properties presented in Table 2.32. Tensile tests are preferred for the rivet and cold heading wire and rod, and shear tests for the fasteners made from it. Section 2.8.2 below furnishes additional information on aluminum rivets and bolts.

TABLE 2.32   Minimum Strengths of Rivet and Cold Heading Wire and Rod

Alloy-temper	Diameter (in.)	Tensile strength		Shear ultimate strength (ksi)
		Ultimate (ksi)	Yield (ksi)	
2017-T4	0.063–1.000	55	32	33
2024-T42	0.163–0.124	62	—	37
2024-T42	0.125–1.000	62	40	37
2117-T4	0.063–1.000	38	18	26
2219-T6	0.063–1.000	55	35	30
6053-T61	0.063–1.000	30	20	20
6061-T6	0.063–1.000	42	35	25
7050-T7	0.063–1.000	70	58	39
7075-T6	0.063–1.000	77	66	42
7075-T73	0.063–1.000	68	56	41
7178-T6	0.063–1.000	84	73	46
7277-T62	0.500–1.250	60	—	35

2.6.1.4   Tubes.   Tube is a product that is hollow and long in relation to its cross section, which may be round, a regular hexagon, a regular octagon, an ellipse, or a rectangle, and tube has a uniform wall thickness. Seamless tubes, common in pressure applications, are made without a metallurgical weld resulting from the method of manufacture; such tubes can be produced from a hollow ingot or by piercing a solid ingot. Pipe is tube that is made in standardized diameter and wall thickness combinations. Tube is produced by several different methods.

Drawn tube is made by pulling material through a die (ASTM B210, *Drawn Seamless Tubes* and B483, *Drawn Tubes for General Purpose Applications*). Drawn tube is available in straight lengths or coils, but coils are generally available only as round tubes with a wall thickness of 0.083 in. (2 mm) or less and only in non-heat-treatable alloys. Drawn seamless tubes are used in surface condensers, evaporators, and heat exchangers, in wall thicknesses up to 0.200 in. and diameters up to 2.00 in., and in alloys 1060, 3003, 5052, 5454, and 6061 (ASTM B234, *Drawn Seamless Tubes for Condensers and Heat Ex-*

changers and B404, *Seamless Condenser and Heat-Exchanger Tubes with Integral Fins*). Heat exchanger tube is very workable, and is tested for leak tightness and marked "HE."

Welded tube is produced from sheet or plate that is rolled into a circular shape and then longitudinally welded by gas tungsten or gas metal arc welding. ASTM B547, *Formed and Arc-Welded Round Tube*, is available in diameters from 9 to 60 in. (230 to 1520 mm) in wall thicknesses from 0.125 to 0.500 in. (3.15 to 12.5 mm). ASTM B313, *Round Welded Tubes*, are made in wall thicknesses from 0.032 in. (0.80 mm) to 0.125 in. (3.20 mm).

Extruded tube (ASTM B241, *Seamless Pipe and Seamless Extruded Tube*, B345, *Seamless Pipe and Seamless Extruded Tube for Gas and Oil Transmission and Distribution Piping Systems*, B429, *Extruded Structural Pipe and Tube*, and B491, *Extruded Round Tubes for General Purpose Applications*) is made by the extrusion process, discussed above. Tube may be extruded and then drawn to minimize ovality, a process sometimes called *sizing.*

2.6.1.5 Forgings. Forgings are one of the oldest wrought products, since they can be produced by simply hammering a hot lump of metal into the desired shape. A hammer, hydraulic press, mechanical press, upsetter, or ring roller is used to form the metal. Both castings and forgings can be used to produce parts with complex shapes; forgings are more expensive than castings but have more uniform properties and better ductility.

There are two types of forgings: open die forgings and closed die forgings (Table 2.33). *Open die forgings* (also called *hand forgings*) are produced without lateral confinement of the material during the forging operation. Minimum mechanical properties are not guaranteed for open die forgings unless specified by the customer, so they don't tend to be used for applications where structural integrity is critical.

Closed die forgings (also called *die forgings*) are more common and are produced by pressing the forging stock (made of ingot, plate, or extrusion) between a counterpart set of dies. Popular uses of closed die forgings are automotive and aerospace applications; they have been made up to 23 feet (7 m) long and 3100 lb (1400 kg) in weight. Die forgings are divided into four categories described below, from the least intricate, lowest quantity forgings, and lowest cost, to the most sharply detailed, highest quantity type, and highest cost. Less intricate forgings are used when quantities are small, because it is more economical to incur machining costs on each of a few pieces than to incur higher one-time die costs. The most economical forging for a par-

TABLE 2.33   Forging Types

	Closed die				
	Blocker type	Finish only	Conventional	Precision	
Typical production quantity	<200	500	>500		
No. of sets of dies	1	1	2–4		Open die
Fillet radius	2	1.5	1.0	<1.0	
Corner radius	1.5	1.0	1.0	<1.0	

ticular application depends on the dimensional tolerances and quantities required.

Blocker-type forgings have large fillet and corner radii and thick webs and ribs so that only one set of dies is needed; generally, two squeezes of the dies are applied to the stock. Fillets are about two times the radius of conventional forgings, and corner radii are about 1.5 times that of conventional forgings. Usually, all surfaces must be machined after forging. Blocker-type forgings may be selected if tolerances are so tight that machining would be required in any event, or if the quantity to be produced is small (typically up to 200 units). Blocker-type forgings can range in size from small to very large.

Finish only forgings also use only one set of dies, like blocker-type forgings, but typically one more squeeze than with blocker-type forgings is applied to the part. Because of the additional squeezes, the die experiences more wear than for other forging types, but the part can be forged with tighter tolerances and reduced fillet and corner radii and web thickness. Fillets are about 1.5 times the radius of conventional forgings, and corner radii about the same as that of conventional forgings. The average production quantity for finish only forgings is 500 units.

Conventional forgings are the most common of all die forging types. Conventional forgings require two to four sets of dies; the first set produces a blocker forging that is subsequently forged in finishing dies. Fillet and corner radii and web and rib thicknesses are smaller than for blocker-type or finish only forgings. Average production quantities are 500 or more.

Precision forgings, as the name implies, are made to closer than standard tolerances and include forgings with smaller fillet and corner radii and thinner webs and ribs.

There are other ways to categorize forgings. *Can* and *tube forgings* are cylindrical shapes that are open at one or both ends; these are also called *extruded forgings*. The walls may have longitudinal ribs or be flanged at one open end. *No-draft forgings* require no slope on vertical walls and are the most difficult to make. *Rolled ring forgings* are short cylinders circumferentially rolled from a hollow section.

Die forging alloys and their mechanical properties are listed in Table 2.34. Alloys 2014, 2219, 2618, 5083, 6061, 7050, 7075, and 7175 are the most commonly used. The ASTM specification for forgings is B247 *Die Forgings, Hand Forgings, and Rolled Ring Forgings*.

2.6.1.6 Electrical conductors. Aluminum is used as a conductor because of its excellent electrical conductivity. Alloys 1350 [formerly known as EC (electrical conductor) grade] 5005, 6201, 8017, 8030, 8176, and 8177 are used in the form of wire, and alloys 1350 and 6101 are produced as bus bar (ASTM B317 *Extruded Bar, Rod, Pipe, and Structural Shapes for Electrical Purposes (Bus Conductors)* and ASTM B236 *Bars for Electrical Purposes (Bus Bars)*), made by extruding, rolling, or sawing from plate or sheet. The minimum conductivity of aluminum conductors is about 60% of the International Annealed Copper Standard (IACS).

In power transmission lines, the necessary strength for long spans is obtained by stranding aluminum wire around a high strength galvanized or aluminized steel core. This product is called *aluminum conductor, steel reinforced (ACSR)*. The resulting strength-to-weight ratio is about twice that of copper of equal conductivity.

2.6.2 Cast Products

The first aluminum products, including the first commercial application (a tea kettle), were castings, made by pouring molten aluminum into a mold. They are useful for making complex shapes and are produced by a number of methods. Common methods and their ASTM specifications are

B26   *Sand Castings*

B85   *Die Castings*

B108   *Permanent Mold Castings*

B618   *Investment Castings*

Castings are made in foundries. Usually, the aluminum to be cast is received in ingot form, but foundries located next to a smelter may receive molten aluminum directly from the reduction plant, and some foundries use recycled material. Castings make up about one-half the aluminum used in automotive applications.

TABLE 2.34 Mechanical Property Limits—Die Forgings ⑥

ALLOY AND TEMPER ②	SPECIFIED THICKNESS ② in.	SPECIMEN AXIS PARALLEL TO DIRECTION OF GRAIN FLOW				SPECIMEN AXIS NOT PARALLEL TO DIRECTION OF GRAIN FLOW			BRINELL HARDNESS ⑤ 500 Kg Load—10 mm ball min.
		TENSILE STRENGTH ksi min.		ELONGATION percent min. in 2 in. or 4D ③		TENSILE STRENGTH ksi min.		ELONGATION percent min. in 2 in. or 4D ③	
		ULTIMATE	YIELD	COUPON	FORGING	ULTIMATE	YIELD	FORGING	
1100-H112 ④	Up thru 4.000	11.0	4.0	25	18	...	...	...	20
2014-T4	Up thru 4.000	55.0	30.0	16	11	...	...	...	100
2014-T6	Up thru 1.000	65.0	56.0	8	6	64.0	55.0	3	125
	1.001–2.000	65.0	56.0	⊖	6	64.0	55.0	2	125
	2.001–3.000	65.0	55.0	⊖	6	63.0	54.0	2	125
	3.001–4.000	63.0	55.0	⊖	6	63.0	54.0	2	125
2018-T61	Up thru 4.000	55.0	40.0	10	7	...	...	...	100
2025-T6	Up thru 4.000	52.0	33.0	16	11	...	...	...	100
2218-T61	Up thru 4.000	55.0	40.0	10	7	...	...	...	100
2218-T72	Up thru 4.000	38.0	29.0	8	5	...	...	...	85
2219-T6	Up thru 4.000	58.0	38.0	10	8	56.0	36.0	4	100
2618-T61	Up thru 4.000	58.0	45.0	6	4	55.0	42.0	4	115
3003-H112 ④	Up thru 4.000	14.0	5.0	25	18	...	...	...	25
4032-T6	Up thru 4.000	52.0	42.0	5	3	...	...	...	115
5083-H111 ④	Up thru 4.000	42.0	22.0	...	14	39.0	20.0	12	...
5083-H112 ④	Up thru 4.000	40.0	18.0	...	16	39.0	16.0	14	...
5456-H112	Up thru 4.000	44.0	20.0	...	16	...	...	...	...
6053-T6	Up thru 4.000	36.0	30.0	16	11	...	...	...	75
6061-T6	Up thru 4.000	38.0	35.0	10	7	38.0	35.0	5	80
6066-T6	Up thru 4.000	50.0	45.0	12	8	...	...	...	100
6151-T6	Up thru 4.000	44.0	37.0	14	10	44.0	37.0	6	90
7049-T73 ⑦	Up thru 1.000	72.0	62.0	10	7	71.0	61.0	3	135
	1.001–2.000	72.0	62.0	10	7	70.0	60.0	3	135
	2.001–3.000	71.0	61.0	10	7	70.0	60.0	3	135
	3.001–4.000	71.0	61.0	10	7	70.0	60.0	2	135
	4.001–5.000	70.0	60.0	10	7	68.0	58.0	2	135
7050-T74 ⑧⑨	Up thru 2.000	72.0	62.0	...	7	68.0	56.0	5	...
	2.001–4.000	71.0	61.0	...	7	67.0	55.0	4	...
	4.001–5.000	70.0	60.0	...	7	66.0	54.0	3	...
	5.001–6.000	71.0	59.0	...	7	66.0	54.0	3	...

Temper	Thickness								
7050-T74 ⑧⑨	Up thru 2.000	7	72.0	62.0	...	68.0	56.0	5	...
	2.001–4.000	7	71.0	61.0	...	67.0	55.0	4	...
	4.001–5.000	7	70.0	60.0	...	66.0	54.0	3	...
	5.001–6.000	7	71.0	59.0	...	66.0	54.0	3	...
7075-T6	Up thru 1.000	7	75.0	64.0	10	71.0	61.0	3	135
	1.001–2.000	7	74.0	63.0	⊖	71.0	61.0	3	135
	2.001–3.000	7	74.0	63.0	⊖	70.0	60.0	3	135
	3.001–4.000	7	73.0	62.0	⊖	70.0	60.0	2	135
7075-T73 ⑦	Up thru 3.000	7	66.0	56.0	...	62.0	53.0	3	125
	3.001–4.000	7	64.0	55.0	...	61.0	52.0	2	125
7075-T7352 ⑦	Up thru 3.000	7	66.0	56.0	...	62.0	51.0	3	125
	3.001–4.000	7	64.0	53.0	...	61.0	49.0	2	125
7175-T74 ⑧⑩	Up thru 3.000	7	76.0	66.0	...	71.0	62.0	4	...
7175-T7452 ⑧⑩	Up thru 3.000	7	73.0	63.0	...	68.0	55.0	4	...
7175-T7454 ⑧⑩	Up thru 3.000	7	75.0	65.0	...	70.0	61.0	4	...

① When separately forged coupons are used to verify acceptability of forgings in the indicated thicknesses, the properties shown for thicknesses "Up thru 1 inch," including the test coupon elongation, apply.

② As-forged thickness. When forgings are machined prior to heat treatment, the properties will also apply to the machined heat treat thickness, provided the machined thickness is not less than one-half the original (as-forged thickness.

③ D equals specimen diameter.

④ Properties of H111 and H112 temper forgings are dependent on the equivalent cold work in the forgings. The properties listed should be attainable in any forging within the prescribed thickness range and may be considerably exceeded in some cases.

⑤ For information only: The Brinell Hardness is usually measured on the surface of a heat-treated forging using a 500 kg load and a 10 mm penetrator ball.

⑥ The data base and criteria upon which these mechanical property limits are established are outlined on page 6-1 under "Mechanical Properties."

⑦ Material in this temper, 0.750 inch and thicker, when tested in accordance with ASTM G47 in the short transverse direction at a stress level of 75 percent of the specified minimum yield strength, will exhibit no evidence of stress corrosion cracking. Capability of individual lots to resist stress corrosion is determined by testing the previously selected tensile test sample in accordance with the applicable lot acceptance criteria outlined on pages 6-7 through 6-10.

⑧ T74 type tempers, although not previously registered, have appeared in the literature and in some specifications as T736 type tempers.

⑨ Material in this temper when tested at any plane in accordance with ASTM G34-72 will exhibit exfoliation less than that shown in Category B, Figure 2 of ASTM G34-72. Also, material, 0.750 inch and thicker, when tested in accordance with ASTM G47 in the short transverse direction at a stress level of 35 ksi, will exhibit no evidence of stress corrosion cracking. Capability of individual lots to resist exfoliation corrosion and stress corrosion cracking is determined by testing the previously selected tensile test sample in accordance with the applicable lot acceptance criteria outlined on pages 6-7 through 6-10.

⑩ Material in this temper, 0.750 inch and thicker, when tested in accordance with ASTM G47 in the short transverse direction at a stress level of 35 Ksi, will exhibit no evidence of stress corrosion cracking. Capability of individual lots to resist stress corrosion cracking is determined by testing the previously selected tensile test sample in accordance with the applicable lot acceptance criteria outlined on pages 6-7 through 6-10.

The minimum mechanical properties of separately cast test bars of cast alloys are given in Table 2.19. The average tensile ultimate strength and tensile yield strength of specimens cut from castings need only be 75% of the minimum strengths given in Table 2.19, and 25% of the minimum elongation values given in Table 2.19. The values for specimens cut from castings should be used in design, because they are more representative of the actual strength of the casting.

2.6.2.1   Casting types

Sand castings are made with a sand mold that is used only once. This method is used for larger castings that have no intricate details and are produced in small quantities. The mold material is sometimes referred to as *green sand* or *dry sand*. Aluminum sand castings as large as 7000 lb (3000 kg) have been produced.

Permanent mold castings are made in reusable molds; sometimes the flow is assisted by a small vacuum but otherwise is gravity induced. Permanent mold castings are more expensive than sand castings but can be held to tighter tolerances and finer details, including wall thicknesses as small as 0.09 in. (2 mm). Semi-permanent molds made of sand or other material are used when the geometry of the casting makes it impossible to remove the mold in one piece from the solidified part.

Die castings are made by injecting the molten metal under pressure into a reusable steel die at high velocity. Solidification is rapid, so high production rates are possible. Die castings are usually smaller and may have thinner wall thicknesses and tighter tolerances than either sand or permanent mold castings.

Investment castings are made by surrounding (investing) an expendable pattern (usually wax or plastic) with a refractory slurry that sets at room temperature. The pattern is then removed by heating, and the resulting cavity is filled with molten metal.

Not all casting alloys are appropriate for all production methods, but some may be produced by multiple methods. Fewer alloys are suitable for die casting than the other methods.

New methods, such as squeeze casting and thixocasting, are showing promise in producing high strength, ductile castings but have not been proven in aluminum yet. Thixocasting has more recently been called *semi-solid forming* and may be thought of as a cross between casting and forging. Semi-solid forming stock has a special globular crystal structure that behaves as a solid until sufficient shearing forces are applied during forming, upon which the material flows like a viscous liquid.

2.6.2.2 Casting quality. Foundries will hold only those tolerances that are specified by the purchaser. This is unlike the case for wrought products, for which mills will meet standard mill tolerances as a minimum. The dimensions of castings can be difficult to control, because it is sometimes difficult to predict the shrinkage during solidification and the warping that may be produced by nonuniform cooling.

The quality of cast material may also vary widely, and any inspection methods must be specified by the purchaser. The most commonly used inspection techniques are radiography and penetrant methods. Radiography is performed by X-raying the part to show discontinuities such as gas holes, shrinkage, and foreign material. These discontinuities are then rated by comparing them to reference radiographs shown in ASTM E155. The ratings are then compared to inspection criteria agreed to beforehand by the customer and the foundry. The inspection criteria for quality and frequency of inspection can be selected and then specified from the Aluminum Association's casting quality standard AA-CS-M5-85, which provides seven quality levels and four frequency levels from which to choose. The penetrant inspection method is only useful for detecting surface defects. Two techniques are available. The fluorescent penetrant procedure is to apply penetrating oil to the part, remove the oil, apply developer to absorbed oil bleeding out of surface discontinuities, and then inspect the casting under ultraviolet light. The dye penetrant method uses a color penetrant, enabling inspection in normal light. Frequency levels are given in AA-CS-M5-85 for penetrant testing also.

A test bar cast with each heat is also useful. It can be tested and the results compared directly to minimum mechanical properties listed for the alloy in Table 2.19.

2.6.3 Metal Matrix Composites

A relatively new product, aluminum metal matrix composite (MMC), consists of an aluminum alloy matrix with carbon, metallic, or, most commonly, ceramic reinforcement. Of all metals, aluminum is the most commonly used matrix material in MMCs. MMCs combine the low density of aluminum with the benefits of ceramics such as strength, stiffness (by increasing the modulus of elasticity), wear resistance, and high temperature properties. They can be formed from both solid and molten states into forgings, extrusions, sheet and plate, and castings. Disadvantages include decreased ductility and higher cost; MMCs cost about three times more than conventional aluminum alloys. Yet, even though they're still being developed, MMCs have been applied in automotive parts such as diesel engine pistons, cylinder liners, drive shafts, and brake components such as rotors.

Reinforcements are characterized as continuous or discontinuous, depending on their shape, and make up 10 to 70% of the composite by volume. Continuous fiber or filament reinforcements (designated f) include graphite, silicon carbide (SiC), boron, and aluminum oxide (Al_2O_3). Discontinuous reinforcements include SiC whiskers (designated w), SiC or Al_2O_3 particles (designated p), or short or chopped (designated c) Al_2O_3 or graphite fibers. The Aluminum Association standard designation system for aluminum MMCs identifies each in the following form:

matrix material/reinforcement material/reinforcement volume %, form

For example, 2124/SiC/25w is aluminum alloy 2124 reinforced with 25% by volume of silicon carbide whiskers; 6061/Al_2O_3/10p is aluminum alloy 6061reinforced with 10% by volume of aluminum oxide particles.

2.6.4 Aluminum Powder

There are many uses for aluminum powder particles, which can be as small as a few microns thick. Larger particles are used in the chemical and metal production industries; one of the first uses of aluminum was as particles to remove oxygen from molten steel during its production. Finer particles are used as an explosive in fireworks and flares and as a solid fuel for rockets. Each launch of the space shuttle uses 350,000 pounds (160,000 kg) of aluminum powder. Powder is also flattened into flakes in a rotating mill and used as a constituent for paints to provide a metallic finish. Finally, aluminum powder may be pressed into parts, referred to as powder metallurgy, competing with conventionally cast aluminum parts.

2.6.5 Aluminum Foam

Closed cell aluminum foam is made by bubbling gas or air through aluminum alloys or aluminum metal matrix composites (see Section 2.6.3) to create a strong but lightweight product. The foam density is 2 to 20% that of solid aluminum. Foamed aluminum's advantages include low weight, fire retardant properties, a high strength-to-weight ratio, and good rigidity and energy absorbency. Current applications include sound insulation panels. Standard size blocks as well as parts with complex shapes can be cast.

2.7   Fabrication

Fabrication of mill products into end use products includes a number of processes; these are divided into *forming* and *machining*.

2.7.1   Forming

All of the forming methods used for other metals may be used to form aluminum. The relative formability of various wrought alloys is given under the heading "workability" in Table 2.12.

Some issues are common to more than one forming method. Mechanical properties (such as ductility) of aluminum mill products of medium and hard tempers are not isotropic; that is, they are not the same in all directions. Therefore, the orientation of bends with respect to the grain of a part will affect how severe the forming operation may be without cracking. Heat treatable alloys are less likely to crack when bends are made perpendicular to the rolling or extruding direction; non-heat treatable alloys are less likely to crack when bends are made parallel to the rolling or extruding direction.

Lubricants useful in forming aluminum, in order of increasing effectiveness, are:

Kerosene

Mineral oil [viscosity 40 to 300 SUS at 100°F (40°C)]

Petroleum jelly

Mineral oil plus 10 to 20% fatty oil

Tallow plus 50% paraffin

Tallow plus 70% paraffin

Mineral oil plus 10 to 15% sulfurized fatty oil and 10% fatty oil

Dried soap or wax films

Fat emulsions in aqueous soap solutions with finely divided fillers

Mineral oil with sulfurized fatty oil, fatty oil, and finely divided fillers

2.7.1.1   Blanking and piercing.   *Blanking* produces a piece by cutting it from a larger piece, such as cutting a circular blank from a flat sheet or plate. *Piercing* produces a piece with holes by cutting out material (called a *slug*), such as punching holes in an extrusion. Both are usually performed in punch presses. The force required for the blanking or piercing operation can be calculated using the typical shear strengths given in Table 2.17. The clearance between the mating surfaces of the punch and die (per side) are expressed in terms of a percentage of the work thickness. The greater the strength of the part, the greater the clearance required. Values range from 5% for 1100-O, 3003-O, 5005-O, and 5050-O to 8% for 2014-T4, -T6, 2024-T3, -T4, 7075-T6, and 7178-T6. The walls of the die opening are usually tapered 1.2° from the axis of the force to minimize sticking of the blank or slug in the die. Lubricants also help minimize sticking and aid stripping from the punch without buckling.

2.7.1.2 Press-brake forming. In press-brake forming, the work is placed over a die and pressed down by a punch that is actuated by the ram of a press brake. The minimum bend radius for 90° cold bends of various aluminum alloys is given in Table 2.35; using smaller radii can cause cracking. ASTM B209 also gives a bend diameter factor N for sheet and plate. Such material is capable of being bent cold through an angle of 180° around a pin with a diameter equal to N times the material thickness without cracking. The maximum thickness that can undergo a 180° bend without cracking according to the Aluminum Association is given for common alloys in Table 2.36. Heating the work permits smaller bend radii but will result in a loss in strength unless the original material is O temper.

Often, the oils present on sheet from rolling at the mill are sufficient lubrication for press-brake forming.

Parts that are bent undergo *springback*, a partial return of the part to its original shape upon removal of the bending forces. The amount of springback is proportional to the yield strength and bend radius and inversely proportional to the part thickness and is usually determined by trial and error. Springback is offset by bending beyond the angle desired in the finished shape.

2.7.1.3 Roll forming. Roll forming makes parts with a constant cross section and longitudinal bends using a series of cylindrical dies in male-female pairs to progressively form a sheet or plate to a final shape in a continuous operation. This method is well suited to aluminum and is used to make roofing and siding, standing seam roofing, automotive trim, gutters and downspouts, structural shapes, furniture parts, and picture frames. The speed of the part through the rolls can be as high as hundreds of feet per minute. Because roll forming pressures are low, and some lubrication may already be present (in the case of painted sheet or mill oil on bare sheet), lubrication is not always necessary.

Roll forming can be combined with high-frequency resistance welding to produce tubular parts such as irrigation pipe, condenser tubing,

TABLE 2.35   Recommended Minimum Bend Radii for 90° Cold Forming of Sheet and Plate[1,2,3,4,5]

Alloy	Temper	Radii for various thicknesses expressed in terms of thickness "t" (in.)							
		$\frac{1}{64}$	$\frac{1}{32}$	$\frac{1}{16}$	$\frac{1}{8}$	$\frac{3}{16}$	$\frac{1}{4}$	$\frac{3}{8}$	$\frac{1}{2}$
1100	O	0	0	0	0	½t	1t	1t	1½t
	H12	0	0	0	½t	1t	1t	1½t	2t
	H14	0	0	0	1t	1t	1½t	2t	2½t
	H16	0	½t	1t	1½t	1½t	2½t	3t	4t
	H18	1t	1t	1½t	2½t	3t	3½t	4t	4½t
2014	O	0	0	0	½t	1t	1t	2½t	4t
	T3	1½t	2½t	3t	4t	5t	5t	6t	7t
	T4	1½t	2½t	3t	4t	5t	5t	6t	7t
	T6	3t	4t	4t	5t	6t	8t	8½t	9½t
2024	O	0	0	0	½t	1t	1t	2½t	4t
	T3	2½t	3t	4t	5t	5t	6t	7t	7½t
	T361[6]	3t	4t	5t	6t	6t	8t	8½t	9½t
	T4	2½t	3t	4t	5t	5t	6t	7t	7½t
	T81	4½t	5½t	6t	7½t	7½t	9t	10t	10½t
	T861[6]	t5	6t	7t	8½t	8½t	10t	11½t	11½t
2036	T4	. .	1t	1t	. .	. .	. .	. .	. .
3003	O	0	0	0	0	½t	1t	1t	1½t
	H12	0	0	0	½t	1t	1t	1½t	2t
	H14	0	0	0	1t	1t	1½t	2t	2½t
	H16	½t	1t	1t	1½t	2½t	3t	3½t	4t
	H18	1t	1½t	2t	2½t	3½t	4½t	5½t	6½t
3004	O	0	0	0	½t	1t	1t	1t	1½t
	H32	0	0	½t	1t	1t	1½t	1½t	2t
	H34	0	1t	1t	1½t	1½t	2½t	2½t	3t
	H36	1t	1t	1½t	2½t	3t	3½t	4t	4½t
	H38	1t	1½t	2½t	3t	4t	5t	5½t	6½t
3105	H25	½t	½t	½t	. .	. .	. .	. .	. .
5005	O	0	0	0	0	½t	1t	1t	1½t
	H12	0	0	0	½t	1t	1t	1½t	2t
	H14	0	0	0	1t	1½t	1½t	2t	2½t
	H16	½t	1t	1t	1½t	2½t	3t	3½t	4t
	H18	1t	1½t	2t	2½t	3½t	4½t	5½t	6½t
	H32	0	0	0	½t	1t	1t	1½t	2t
	H34	0	0	0	1t	1½t	1½t	2t	2½t
	H36	½t	1t	1t	1½t	2½t	3t	3½t	4t
	H38	1t	1½t	2t	2½t	3½t	4½t	5½t	6½t
5050	O	0	0	0	½t	1t	1t	1½t	1½t
	H32	0	0	0	1t	1t	1½t	. .	. .
	H34	0	0	1t	1½t	1½t	2t	. .	. .
	H36	1t	1t	1½t	2t	2½t	3t	. .	. .
	H38	1t	1½t	2½t	3t	4t	5t	. .	. .

TABLE 2.35 Recommended Minimum Bend Radii for 90° Cold Forming of Sheet and Plate[1,2,3,4,5] (Continued)

Alloy	Temper	Radii for various thicknesses expressed in terms of thickness "t" (in.)							
		1/64	1/32	1/16	1/8	3/16	1/4	3/8	1/2
5052	O	0	0	0	½t	1t	1t	1½t	1½t
	H32	0	0	1t	1½t	1½t	1½t	1½t	2t
	H34	0	1t	1½t	2t	2t	2½t	2½t	3t
	H36	1t	1t	1½t	2½t	3t	3½t	4t	4½t
	H38	1t	1½t	2½t	3t	4t	5t	5½t	6½t
5083	O	. .	. .	½t	1t	1t	1t	1½t	1
	H321	. .	. .	1t	1½t	1½t	1½t	2t	2½t
5086	O	0	0	½t	1t	1t	1t	1½t	1½t
	H32	0	½t	1t	1½t	1½t	2t	2½t	3t
	H34	½t	1t	1½t	2t	2½t	3t	3½t	4t
	H36	1½t	2t	2½t	3t	3½t	4t	4½t	5t
5154	O	0	0	½t	1t	1t	1t	1½t	1½t
	H32	0	½t	1t	1½t	1½t	2t	2½t	3½t
	H34	½t	1t	1½t	2t	2½t	3t	3½t	4t
	H36	1t	1½t	2t	3t	3½t	4t	4½t	5t
	H38	1½t	2½t	3t	4t	5t	5t	6½t	6½t
5252	H25	0	0	1t	2t	. .	. .	. .	. .
	H28	1t	1½t	2½t	3t	. .	. .	. .	. .
5254	O	0	0	½t	1t	1t	1t	1½t	1½t
	H32	0	½t	1t	1½t	1½t	2t	2½t	3½t
	H34	½t	1t	1½t	2t	2½t	3t	3½t	4t
	H36	1t	1½t	2t	3t	3½t	4t	4½t	5t
	H38	1½t	2½t	3t	4t	5t	5t	6½t	6½t
5454	O	O	½t	1t	1t	1t	1½t	1½t	2t
	H32	½t	½t	1t	2t	2t	2½t	3t	4t
	H34	½t	1t	1½t	2t	2½t	3t	3½t	4t
5456	O	. .	. .	1t	1t	1½t	1½t	2t	2t
	H321	. .	. .	. .	2t	2t	2½t	3t	3½t
5457	O	0	0	0	0	. .	. .	. .	. .
5652	O	0	0	0	½t	1t	1t	1½t	1½t
	H32	0	0	1t	1½t	1½t	1½t	1½t	2t
	H34	0	1t	1½t	2t	2t	2½t	2½t	3t
	H36	1t	1t	1½t	2½t	3t	3½t	4t	4½t
	H38	1t	1½t	2½t	3t	4t	5t	5½t	6½t
5657	H25	0	0	0	1t	. .	. .	. .	. .
	H28	1t	1½t	2½t	3t	. .	. .	. .	. .
6061	O	0	0	0	1t	1t	1t	1½t	2t
	T4	0	0	1t	1½t	2½t	3t	3½t	4t
	T6	1t	1t	1½t	2½t	3t	3½t	4½t	5t

TABLE 2.35   Recommended Minimum Bend Radii for 90° Cold Forming of Sheet and Plate[1,2,3,4,5] *(Continued)*

		Radii for various thicknesses expressed in terms of thickness "t" (in.)							
Alloy	Temper	1/64	1/32	1/16	1/8	3/16	1/4	3/8	1/2
7050	T7	. .	. .	. .	. .	. .	8t	9t	9½t
7072	O	0	0	. .	. .	. .	. .	. .	. .
	H14	0	0	. .	. .	. .	. .	. .	. .
	H18	1t	1t	. .	. .	. .	. .	. .	. .
7075	O	0	0	1t	1t	1½t	2½t	3½t	4t
	T6	3t	4t	5t	6t	6t	8t	9t	9½t
7178	O	0	0	1t	1½t	1½t	2½t	3½t	4t
	T6	3t	4t	5t	6t	6t	8t	9t	9½t

[1]The radii listed are the minimum recommended for bending sheets and plates without fracturing in a standard press brake with air band dies. Other types of bending operations may require larger radii or permit smaller radii. The minimum permissible radii will also vary with the design and condition of the tooling.
[2]Alclad sheet in the heat-treatable alloys can be bent over slightly smaller radii than the corresponding tempers of the bare alloy.
[3]Heat-treatable alloys can be formed over appreciably smaller radii immediately after solution heat treatment.
[4]The H112 temper (applicable to non-heat-treatable alloys) is supplied in the as-fabricated condition without special property control but usually can be formed over radii applicable to the H14 (or H34) temper or smaller.
[5]The reference test method is ASTM E290.
[6]Tempers T361 and T861 were formerly designated T36 and T86, respectively.

TABLE 2.36   Sheet Thickness for 180° Cold Bending

Alloy-temper	Maximum metal thickness (in.)	Alloy-temper	Maximum metal thickness (in.)
1100-O	0.125	5005-H14	0.031
1100-H14	0.063	5005-H34	0.031
2014-O	0.063	5050-O	0.063
2024-O	0.063	5050-H34	0.031
3003-O	0.125	5052-O	0.063
3003-H14	0.063	5052-H32	0.031
3004-O	0.063	5052-H34	0.016
3004-H32	0.031	5086-O	0.031
3004-H34	0.016	5154-H32	0.016
5005-O	0.125	6061-O	0.063
5005-H12	0.063	6061-T4	0.031
5005-H32	0.063	7075-O	0.031

and furniture tubing. Portable roll forming machines are used to make aluminum gutters on the site where they will be installed.

2.7.1.4 Roll bending. Roll bending is used to curve tube, rod, bar, and shapes to a radius. The stock is fed into a set of grooved forming rolls that apply opposing compressive forces to the part to produce the curvature. Annealed or intermediate tempers are generally used, because full hard or fully heat treated tempers suffer excessive springback unless the radius of curvature is very large.

2.7.1.5 Drawing. In drawing, the center portion of a flat, circular blank is pressed into a die opening to form sheet into cylindrical, cuplike containers. Drawing aluminum usually does not reduce the thickness of the work, but it does reduce its diameter in plan view. Maximum recommended diameter reductions, expressed as a fraction of the diameter of the blank for the first draw and of the prior punch diameter for subsequent draws, are given in Table 2.37 for annealed ma-

TABLE 2.37   Diameter Reductions for Deep Drawing

Operation	Blank or punch diameter	1100, 3003, 3004, 3005, 5005, 5050, 5052, 5457, 6061	2014, 2024, 5083, 5086, 5154, 5456
Blank	D		
1st draw	D_1	0.40D	0.30D
2nd draw	D_2	$0.20D_1$	$0.15D_1$
3rd draw	D_3	$0.15D_2$	$0.15D_2$
4th draw	D_4	$0.15D_3$	

terial. An example for a 3003-O 25 in. diameter flat blank would be to reduce its diameter to 15 in. in the first draw [a reduction of 0.40×25 in $= 10$ in. $= (25-15)$ in]; to reduce it to 12 in. in the second draw [a reduction of 0.20×15 in. $= 3$ in. $= (15-12)$ in.]; and to reduce it to 10 in. in the third and final draw [a reduction of 0.167×12 in.) $= 2$ in. $= (12-10)$ in.]. The diameter reduction, depth of draw, and blank diameter to thickness ratio are larger for the lower strength alloys than for higher strength alloys.

Drawing can be combined with ironing; this method is used to make beverage cans. Ironing squeezes the walls, making them thinner, and it can reduce the wall thickness by up to 40% in the first pass while increasing the wall height. Multiple ironing steps may be used, and the work may be annealed between ironing passes. Drawing may also be done at elevated temperatures 350°F to 600°F (175°C to 315°C); this approach is useful for drawing parts thicker than 0.125 in. (3 mm) and for stronger alloys such as 2024, 2219, 5083, 5086, 5456, 6061, 7075, and 7178.

2.7.1.6 Stretch forming. In stretch forming, the work is stressed beyond yield and stretched over a form to achieve the desired shape. Stretch forming can produce sheets or plates with compound curvature and compound bends in extrusions. Equipment costs are high, so production quantities must be high enough to justify this fabrication method. Applications of stretch forming include hull panels for boats and wing skin and fuselage panels for aircraft.

Material to be stretch formed needs to have a wide forming range (defined as the difference between yield strength and ultimate strength) and high elongation. The heat-treatable alloys such as 7075, 2024, and 6061 are most suitable for stretch forming in the W temper immediately after quenching after solution heat treatment but may also be formed in the annealed condition. Non-heat-treatable alloys such as 1100 and 3003 are formed in the annealed temper. The stretch forming rating of several alloys and tempers are given in Table 2.38.

2.7.1.7 Spinning. Spinning forms seamless axisymmetric shapes using a combination of rotation and force. Usually, a flat blank is forced by a blunt rounded tool against a rotating mandrel that has the desired shape, but hollow shapes like tube can also be spun. The resulting shapes are surfaces of revolution. Applications include bowls, pitchers, and other hollowware; kettles, heads for tanks and other vessels; and tapered tubes used for light poles.

Thin stock [up to about 0.08 in. (2 mm) thick] can be spun on hand-spinning lathes. Thicker, larger blanks require more powerful equipment to apply pressure to the stock. Blanks up to about 1/2 in. (13 mm) thick can be spun at room temperature; hot spinning is used for heavier gauges. Parts as wide as 16 ft (5 m) can be spun. Both heat-treatable and non-heat-treatable alloys are spun, including 1100, 2014, 2024, 3003, 3004, 5052, 5086, 5154, and 6061. Most are spun in the annealed temper and may require annealing at various times during spinning if forming is severe enough. Heat-treatable alloys can

TABLE 2.38 Stretch Forming Ratings

Alloy-temper	Forming range (ksi)	Elongation	Stretch forming rating
7075-W	28	19	100
2024-W	28	20	98
2024-T3	20	18	95
6061-W	14	22	90
7075-O	18	17	80
2024-O	16	19	80
6061-O	10	22	75
3003-O	10	30	75
1100-O	8	35	70
7075-T6	9	11	10

then be solution heat treated, quenched, and immediately spun again to attain their final shape before hardening occurs.

Spinning usually has high unit labor costs because production rates are low, but spinning is useful for parts that are too large to be formed in presses and for parts to be produced in relatively small quantities.

2.7.1.8 Other forming methods. Numerous other forming methods can be applied to aluminum. These include high energy rate forming (HERF) methods applied in the aerospace industry, during which deformation rates are on the order of several hundred feet per second versus conventional forming rates of 0.5 to 20 ft/sec. HERF methods include (1) explosive forming, where a shock wave, usually in water, acts as the punch, and only a die is needed, and (2) electrohydraulic forming, where a spark gap or an exploding bridgewire discharges electrical energy in water to generate a shock wave.

2.7.2 Machining

Aluminum is more machinable than most other metals. Its high thermal conductivity conducts heat away from the cutting tool, prolonging tool life, and the ease in cutting aluminum permits high-speed machining. The main difficulty in machining aluminum is the tendency for long continuous chips to form. Alloys with small amounts of low melting point metals [e.g., lead (like 2011 and 6262), bismuth, and tin] are the most machinable. Lead-free variations on 2011 and 6262 with good machinability are also available from some producers. The relative machinability of various wrought alloys, given in Tables 2.12 and 2.13, is based on chip characteristics and surface finish rather than tool life, since tool life is usually very long when machining any of the aluminum alloys.

Layout lines are usually scribed on the stock and measured at room temperature, out of direct exposure to sunlight, because of aluminum's high coefficient of thermal expansion.

Aluminum can be cut by thermal and mechanical methods. Thermal cutting methods are discussed in Section 2.8.1.4 below; mechanical cutting methods, which are probably the more commonly used of the two, are addressed here. The advantage of mechanical cutting methods is that they do not have an adverse effect on mechanical properties as does the heat induced by thermal methods. Aluminum is readily cut using the same types of tools and cutting speeds used to cut wood. The main difference is that more power is needed to cut aluminum than wood in order to maintain those speeds.

Circular saws, floor-mounted band saws, and saber saws all cut aluminum effectively. Recommended blade speeds are:

- 8000 ft/min (2400 m/min) for high-speed steel circular blades hardened to Rockwell C65

- 12,000 ft/min (3700 m/min) for tungsten carbide tipped circular blades

- 5000 ft/min (1500 m/min) for band saw blades

Tooth shapes developed by blade manufacturers for cutting aluminum have more rake and clearance than those used to cut steel. Tooth spacing on ordinary circular saw blades is adequate for aluminum, but band saw blades for aluminum should have greater spacing than those for other metals and not have more than 3 or 4 teeth per inch (1 to 1.5 teeth per cm).

Aluminum parts up to about 1/4 in. (6 mm) may also be sheared.

An alternative to thermal cutting methods used for cutting thick aluminum stock is cutting with water at high pressure with abrasives such as garnet.

Grinding aluminum is used to bevel edges prior to welding or to remove weld reinforcement to blend the weld smoothly into base metal. Much higher speeds are used than for ferrous materials. High-speed, semiflexible grinding disks used for aluminum are usually 7 in. (180 mm) diameter with grit sizes from 24 to 120 and maximum rated rotation speeds of 8600 rpm. Grinding machines must have sufficient power to operate at these high speeds so the disks do not clog.

Mechanical gouging can be done with rotary cutter machines designed for this purpose. Routers developed from woodworking tools are also used to shape aluminum. Aluminum can also be chemically milled, usually with sodium hydroxide based or other alkaline solutions. A typical removal rate is 0.0001 in. (0.0025 mm) per minute. Metal removal is controlled by masking, duration of immersion, and composition of the bath.

2.8   Joining

2.8.1   Welding

Welding is the process of uniting parts by either heating, applying pressure, or both. Welding is like the little girl who, when she was good, was very, very good and, when she was bad, was horrid. Improper welding can be awful, while correctly designed and executed welds can solve problems intractable by other means. When heat is used to weld aluminum (as is usually the case), it reduces the strength of all tempers other than annealed material, and this must be taken into account where strength is a consideration. Also, welding aluminum is different from welding steel, and most steel welding techniques are not transferable to aluminum.

Aluminum's affinity for oxygen, which quickly forms a thin, hard oxide surface film, has much to do with the welding process. This oxide is nearly as hard as diamonds, attested to by the fact that aluminum oxide grit is often used for grinding. It has a much higher melting point than aluminum itself [3725°F (2050°C), versus 1220°F (660°C)], so trying to weld aluminum without first removing the oxide melts the base metal long before the oxide. The oxide is also chemically stable; fluxes to remove it require corrosive substances that can damage the base metal unless they are fully removed after welding. Finally, the oxide is an electrical insulator and porous enough to retain moisture. For all these reasons, the base metal must be carefully cleaned and wire brushed immediately before welding, and the welding process must remove and prevent reformation of the oxide film during welding.

The metal in the vicinity of a weld can be considered as two zones: the weld bead itself, a casting composed of a mixture of the filler and the base metal, and the heat affected zone (HAZ) in the base metal outside the weld bead. The extent of the HAZ is a function of the thickness and geometry of the joint, the welding process, the welding procedure, and preheat and interpass temperatures, but it rarely exceeds 1 in. (25 mm) from the centerline of the weld. The strength of the metal near a weld is graphed in Figure 2.7. Smaller welds and higher

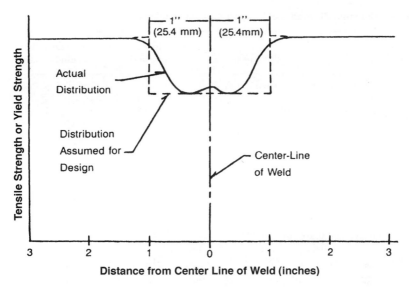

Figure 2.7 Strength near a weld.

welding speeds tend to have a smaller HAZ. As the base metal and filler metal cool after freezing, if the joint is restrained from contracting and its strength at the elevated temperature is insufficient, hot cracking may occur.

The magnitude of the strength reduction from welding varies: for non-heat-treatable alloys, welding reduces the strength to that of the annealed (O) temper of the alloy; for heat-treatable alloys, the reduced strength is slightly greater than that of the solution heat treated but not artificially aged temper (T4) of the alloy. Minimum tensile strengths across groove welded aluminum alloys are given in Table 2.39. These strengths are the same as those required to qualify a welder or weld procedure in accordance with the American Welding Society (AWS) D1.2 *Structural Welding Code—Aluminum* and the American Society of Mechanical Engineers (ASME) *Boiler and Pressure Vessel Code*, Section IX. They are based on the most common type of welding (gas-shielded arc, discussed next) and, as long as a recommended filler alloy is used, they are independent of filler. Yield strengths for welded material are also given in the Aluminum Association's *Aluminum Design Manual,* but they must be multiplied by 0.75 to obtain the yield strength of the weld-affected metal, because the Association's yield strengths are based on a 10 in. (250 mm) long gage length, and only about 2 in. (50 mm) of that length is heat affected metal.

Fillet weld shear strengths *are* a function of the filler used; minimum shear strengths for the popular filler alloys are given in

TABLE 2.39   Minimum Strengths of Welded Aluminum Alloys

Alloy	Product	Thickness (in.)	Tensile ultimate strength (ksi)	Tensile yield strength (ksi)
1060	sheet and plate	up thru 3.000	8	2.5
1060	extrusion	all	8.5	2.5
1100	all	up thru 3.000	11	3.5
2219	all	all	35	–
3003	all	up thru 3.000	14	5
Alclad 3003	tube	all	13	4.5
Alclad 3003	sheet and plate	up to 0.500	13	4.5
Alclad 3003	plate	0.500 to 3.000	14	5
3004	all	up thru 3.000	22	8.5
Alclad 3004	sheet and plate	up to 0.500	21	8
Alclad 3004	plate	0.500 to 3.000	22	8.5
5005	all	up thru 3.000	15	5
5050	all	up thru 3.000	18	6
5052	all	up thru 3.000	25	9.5
5083	forging	all	39	16
5083	extrusion	all	39	16
5083	sheet and plate	up thru 1.500	40	18
5083	plate	> 1.500, thru 3.000	39	17
5083	plate	> 3.000, thru 5.000	38	16
5083	plate	> 5.000, thru 7.000	37	15
5083	plate	> 7.000, thru 8.000	36	14
5086	all	up thru 2.000	35	14
5086	extrusion	> 2.000, thru 5.000	35	14
5086	plate	> 2.000, thru 3.000	34	14
5154	all	up thru 3.000	30	11
5254	all	up thru 3.000	30	11

TABLE 2.39   Minimum Strengths of Welded Aluminum Alloys *(Continued)*

Alloy	Product	Thickness (in.)	Tensile ultimate strength (ksi)	Tensile yield strength (ksi)
5454	all	up thru 3.000	31	12
5456	extrusion	up thru 5.000	41	19
5456	sheet and plate	up thru 1.500	42	19
5456	plate	> 1.500, thru 3.000	41	18
5456	plate	> 3.000, thru 5.000	40	17
5456	plate	> 5.000, thru 7.000	39	16
5456	plate	> 7.000, thru 8.000	38	15
5652	all	up thru 3.000	25	9.5
6005	extrusion	up thru 1.000	24	–
6061	all	all	24	–
Alclad 6061	all	all	24	–
6063	extrusion	up thru 1.000	17	–
6351	extrusion	up thru 1.000	24	–
7005	extrusion	up thru 1.000	40	–
356.0	casting	all	23	–
443.0	casting	all	17	7
A444.0	casting	all	17	–
514.0	casting	all	22	9
535.0	casting	all	35	18

Table 2.40. Fillet welds transverse (perpendicular) to the direction of force are generally stronger than fillet welds longitudinal (parallel) to the direction of force. This is because transverse welds are in a state of combined shear and tension, and longitudinal welds are in shear, and tension strength is greater than shear strength.

Heat-treatable base metal alloys welded with heat-treatable fillers can be heat treated after welding to recover strength lost by heat of welding. This post-weld heat treatment can be a solution heat treatment and aging or just aging (see Section 2.2.3). While solution heat treating and aging will recover more strength than aging alone, the

TABLE 2.40   Minimum Shear Strengths of Filler Alloys

Filler alloy	Longitudinal shear strength (ksi)	Transverse shear strength (ksi)
1100	7.5	7.5
2319	16	16
4043	11.5	15
4643	13.5	20
5183	18.5	–
5356	17	26
5554	17	23
5556	20	30
5654	12	–

rapid quenching required in solution heat treating can cause distortion of the weldment because of the residual stresses that are introduced. Natural aging will also recover some of the strength; the period of time required is a function of the alloy. The fillet weld strengths for 4043 and 4643 in Table 2.40 are based on 2 to 3 months of natural aging.

Prior to 1983, the ASME *Boiler and Pressure Vessel Code, Section IX, Welding and Brazing Qualifications* was the only widely available standard for aluminum welding. Many aluminum structures other than pressure vessels were welded in accordance with the provisions of the *Boiler and Pressure Vessel Code*, therefore, due to the lack of an alternative standard. In 1983, the American Welding Society's (AWS) D1.2 *Structural Welding Code—Aluminum* was introduced as a general standard for welding any type of aluminum structure (e.g., light poles, space frames, etc.). In addition to rules for qualifying aluminum welders and weld procedures, D1.2 includes design, fabrication, and inspection requirements. There are other standards that address specific types of welded aluminum structures, such as ASME B96.1 *Welded Aluminum-Alloy Storage Tanks*, AWS D15.1 *Railroad Welding Specification—Cars and Locomotives*, and AWS D3.7 *Guide for Aluminum Hull Welding*.

2.8.1.1 Gas-shielded arc welding. Before World War II, shielded metal arc welding (SMAW) using a flux coated electrode was one of the few ways aluminum could be welded. This process, however, was inefficient and often produced poor welds. In the 1940s, inert gas-shielded arc welding processes were developed that used argon and helium instead of flux to remove the oxide, and they quickly became more popular. Other methods of welding aluminum are used (and will be discussed below), but today most aluminum welding is by the gas-shielded arc processes.

There are two gas-shielded arc methods: gas metal arc welding (GMAW), also called metal inert gas welding or MIG, and gas tungsten metal arc welding (GTAW), also called tungsten inert gas welding or TIG. MIG welding uses an electric arc between the base metal being welded and an electrode filler wire. The electrode wire is pulled from a spool by a wire-feed mechanism and delivered to the arc through a gun. In TIG welding, the base metal and, if used, the filler metal are melted by an arc between the base metal and a nonconsumable tungsten electrode in a holder. Tungsten is used because it has the highest melting point of any metal [6170°F (3410°C)] and reasonably good conductivity—about one-third that of copper. In each case, the inert gas removes the oxide from the aluminum surface and protects the molten metal from oxidation, allowing coalescence of the base and filler metals.

TIG welding was developed before MIG welding and was originally used for all metal thicknesses. Today, however, TIG is usually limited to material 1/4 in. (6 mm) thick or less. TIG welding is slower and does not penetrate as well as MIG welding. In MIG welding, the electrode wire speed is controlled by the welding machine and, once adjusted to a particular welding procedure, does not require readjustment, so even manual MIG welding is considered to be semiautomatic. MIG welding is suitable for all aluminum material thicknesses.

The weldability of wrought alloys depends primarily on the alloying elements, discussed below for the various alloy series:

1xxx: Pure aluminum has a narrower melting range than alloyed aluminum. This can cause a lack of fusion when welding, but generally the 1xxx alloys are very weldable. The strength of pure aluminum is low, and welding decreases the strength effect of any strain hardening, so welded applications of the 1xxx series are used mostly for their corrosion resistance.

2xxx: The 2xxx alloys are usually considered poor for arc welding, being sensitive to hot cracking, and their use in the aircraft typically has not required welding. However, alloy 2219 is readily weldable, and 2014 is welded in certain applications.

3xxx: The 3xxx alloys are readily weldable but have low strength and so are not used in structural applications unless their corrosion resistance is needed.

5xxx: The 5xxx alloys retain high strengths, even when welded, are free from hot cracking, and are very popular in welded plate structures such as ship hulls and storage vessels.

6xxx: The 6xxx alloys can be prone to hot cracking if improperly designed and lose a significant amount of strength due to the heat of welding, but they are successfully welded in many applications. Postweld heat treatments can be applied to increase the strength of 6xxx weldments. The 6xxx series alloys (like 6061 and 6063) are often extruded and combined with the sheet and plate products of the 5xxx series in weldments.

7xxx: The low-copper-content alloys (such as 7004, 7005, and 7039) of this series are weldable; the others are not, losing considerable strength and suffering hot cracking when welded.

Some cast alloys are readily welded, and some are postweld heat-treated, because they are usually small enough to be easily placed in a furnace. The condition of the cast surface is key to the weldability of castings; grinding and machining are often needed to remove contaminants prior to welding. The weldability of the 355.0, 356.0, 357.0, 443.0, and A444.0 alloys is considered excellent.

Filler alloys can be selected based on different criteria, including resistance to hot cracking, strength, ductility, corrosion resistance, elevated temperature performance, MIG electrode wire feedability, and color match for anodizing. Recommended selections are given in Table 2.41, and a discussion of some fillers is given below. Material specifications for these fillers are given in AWS A5.10, *Specification for Bare Aluminum and Aluminum Alloy Welding Electrodes and Rods*. There is no ASTM specification for aluminum weld filler.

Filler alloys 5356, 5183, and 5556 were developed to weld the 5xxx series alloys, but they have also become useful for welding 6xxx and 7xxx alloys. Alloy 5356 is the most commonly used filler due to its good strength, compatibility with many base metals, and good MIG electrode wire feedability. Alloy 5356 also is used to weld 6xxx series alloys, because it provides a better color match with the base metal than 4043 when anodized. Alloy 5183 has slightly higher strength than 5356, and 5556 higher still. Because these alloys contain more than 3% magnesium and are not heat treatable, however, they are not suitable for elevated temperature service or postweld heat treating. Alloy 5554 was developed to weld alloy 5454, which contains less than 3% magnesium so as to be suitable for service over 150°F (66°C).

Alloy 5654 was developed as a high-purity, corrosion-resistant alloy for welding 5652, 5154, and 5254 components used for hydrogen peroxide service. Its magnesium content exceeds 3%, so it is not used at elevated temperatures.

Alloy 4043 was developed for welding the heat-treatable alloys, especially those of the 6xxx series. Its has a lower melting point than the 5xxx fillers and so flows better and is less sensitive to cracking. Alloy 4643 is for welding 6xxx base metal parts over 0.375 in. (10 mm) to 0.5 in. (13 mm) thick that will be heat treated after welding. Alloys 4047 and 4145 have low melting points and were developed for brazing but are also used for some welds; 4145 is used for welding 2xxx alloys, and 4047 is used instead of 4043 in some instances to minimize hot cracking and increase fillet weld strengths.

Alloy 2319 is used for welding 2219; it's heat treatable and has higher strength and ductility than 4043 when used to weld 2xxx alloys that are postweld heat treated.

Pure aluminum alloy fillers are often needed in electrical or chemical industry applications for conductivity or corrosion resistance. Alloy 1100 is usually satisfactory, but for even better corrosion resistance (due to its lower copper level), 1188 may be used. These alloys are soft and sometimes have difficulty feeding through MIG conduit.

The filler alloys used to weld castings are castings themselves (C355.0, A356.0, 357.0, and A357.0), usually 1/4 in. (6 mm) rod used for TIG welding. They are mainly used to repair casting defects. More recently, wrought versions of C355.0 (4009), A356.0 (4010), and A357.0 (4011) have been produced so that they can be produced as MIG electrode wire. (Alloy 4011 is only available as rod for GTAW, however, since its beryllium content produces fumes too dangerous for MIG welding.) Like 4643, 4010 can be used for postweld heat treated 6xxx weldments.

Weld quality may be determined by several methods. *Visual inspection* detects incorrect weld sizes and shapes (such as excessive concavity of fillet welds), inadequate penetration on butt welds made from one side, undercutting, overlapping, and surface cracks in the weld or base metal. *Dye penetrant inspection* uses a penetrating dye and a color developer and is useful in detecting defects with access to the surface. *Radiography* (making X-ray pictures of the weld) can detect defects as small as 2% of the thickness of the weldment, including porosity, internal cracks, lack of fusion, inadequate penetration, and inclusions. *Ultrasonic inspection* uses high-frequency sound waves to detect similar flaws, but it is expensive and requires trained personnel to interpret the results. Its advantage over radiography is that it is better suited to detecting thin planar defects parallel to the X-ray beam. Destructive tests, such as bend tests, fracture (or nick break) tests, and tensile tests are usually reserved for qualifying a welder or a weld procedure. Acceptance criteria for the various methods of inspection and tests are given in AWS D1.2 and other standards for specific welded aluminum components or structures.

TABLE 2.41 Guide to the Choice of Filler Metal for General Purpose Welding

Base Metal	319.0,333.0 354.0,355.0 C355.0,380.0	356.0,A356.0 A357.0,359.0 413.0,A444.0 443.0	511.0,512.0 513.0,514.0	7005[k],7039 710.0,711.0 712.0	6070	6061 6063 6101 6201 6151 6351 6951	5456	5454	5154 5254[a]	5086	5083	5052 5652[a]	5005 5050	3004 Alc.3004	2219 2519	2014 2036	1100 3003 Alc.3003	1060 1070,1080 1350
1060,1070, 1080,1350	4145[c,j]	4145[j,f]	4043[e,i]	4043[j]	4043[j]	4043[j]	5356[c]	4043[j]	4043[e,i]	5356[c]	5356[c]	4043[j]	1100[c]	4043	4145	4145	1100[c]	1188[j]
1100,3003 Alclad 3003	4145[c,j]	4145[j,f]	4043[e,i]	4043[j]	4043[j]	4043	5356[c]	4043[e,i]	4043[e,i]	5356[c]	5356[c]	4043[e,i]	4043[e]	4043[e]	4145	4145	1100[c]	
2014,2036	4145[g,c,j]	4145			4145	4145									4145[g]	4145[g]		
2219,2519	4043[j]	4043[j]	4043[j]	4043[j]	4043[f,i,j]	4043[f,i,j]	4043	4043[j]	4043	4043	4043	4043[i]	4043	4043	2319[c,f,i]			
3004 Alclad 3004	4043[j]	4043[j]	5654[b]	5356[e]	4043[e]	4043[b]	5356[e]	5654[b]	5654[e]	5356[e]	5356[e]	4043[i]	4043	4043[e]				
5005,5050	4043[j]	4043[b]	5654[b]	5356[e]	4043[e]	4043[b]	5356[e]	5654[b]	5654[b]	5356[e]	5356[e]	4043[e,i]	4043[d,e]					
5052,5652[a]	4043[j]	4043[b,i]	5654[b]	5356[e]	5356[b,c]	5356[b,c]	5356[e]	5654[b]	5654[b]	5356[e]	5356[e]	5654[a,b,c]						
5083	4043[b,i]	5356[c,e,i]	5356[e]	5183[e]	5356[e]	5356[e]	5183[e]	5356[e]	5356[e]	5356[e]	5183[e]							
5086	5356[c,e,i]	5356[c,e,i]	5356[e]	5356[e]	5356[e]	5356[e]	5356[e]	5356[e]	5356[e]	5356[e]								
5154,5254[a]	4043[b,i]	4043[b,i]	5654[b]	5356[e]	5356[b,c]	5356[b,c]	5356[b]	5654[b]	5654[a,b]									
5454	4043[b,i]	4043[b,i]	5654[b]	5356[e]	5356[b,c]	5356[b,c]	5356[b,c]	5554[c,e]										
5456	5356[c,e,i]	5356[c,e,i]	5356[e]	5556[e]	5356[e]	5356[e]	5556[e]											

Base metal						
6061,6063, 6351,6101, 6201,6151, 6951	4145^{c,j}	4043^{f,j}	5356^{b,c}	5356^{b,c,i}	4043^{b,j}	4043^{b,i}
6070	4145^{c,j}	4043^{f,j}	5356^{c,e}	5356^{c,e,i}	4043^{b,j}	4043^{e,i}
7005^{k},7039, 710.0,711.0, 712.0	4043^{j}	4043^{b,j}	5356^{b}	5356^{e}		
511.0,512.0 513.0,514.0		4043^{b,j}	5654^{b,d}			
356.0,A356.0 A357.0,359.0 413.0						
A444.0,443.0	4145^{c,j}	4043^{d,j}				
319.0,333.0, 354.0,355.0, C355.0,380.0	4145^{d,c,i}					

2.8.1.2 Other arc welding processes. *Stud welding* (SW) is a process used to attach studs to a part. Two methods are used for aluminum: arc stud welding, which uses a conventional welding arc over a timed interval, and capacitor discharge stud welding, which uses an energy discharge from a capacitor. Arc stud welding is used to attach studs ranging from 1/4 in. (6 mm) to 1/2 in. (13 mm) in diameter, while capacitor discharge stud welding uses studs 1/16 in. (1.6 mm) to 1/4 in. (6 mm) in diameter. Capacitor discharge stud welding is very effective for thin sheet [as thin as 0.040 in. (1.0 mm)], because it uses much less heat than arc stud welding and does not mar the appearance of the sheet on the opposite side from the stud. Studs are inspected using bend, torque, or tension tests. Stud alloys are the common filler alloys. Stud welding requirements are included in AWS D1.2.

Plasma arc welding with variable polarity (PAW-VP) [also called variable polarity plasma arc (VPPA) welding] is an outgrowth of TIG welding and uses a direct current between a tungsten electrode and either the workpiece or the gas nozzle. Polarity is constantly switched from welding to oxide cleaning modes at intervals tailored to the joint being welded. Two gases, a plasma gas and a shielding gas, are provided to the arc. Welding speed is slower than MIG welding, but often fewer passes are needed; single pass welds in metal up to 5/8 in. (16 mm) thick have been made. The main disadvantage is the cost of the required equipment.

Plasma-MIG welding is a combination of plasma arc and MIG welding, by which the MIG electrode is fed through the plasma coaxially, superimposing the arcs of each process. Higher deposition rates are possible, but equipment costs are also higher than for conventional MIG welding.

Arc spot welding uses a stationary MIG arc on a thin sheet held against a part below, fusing the sheet to the part. The advantage over resistance welding (discussed below) is that access to both sides of the work is unnecessary. Problems with gaps between the parts, overpenetration, annular cracking, and distortion have limited the application of this method. It has been used to fuse aluminum to other metals such as copper, aluminized steel, and titanium for electrical connections.

Shielded metal arc welding (SMAW) is an outdated, manual process that uses a flux-coated filler rod, the flux taking the place of the shielding gas in removing oxide. Its only advantage is that it can be performed with commonly used shielded metal arc steel welding equipment. Shielded metal arc welding is slow, prone to porosity [especially in metal less than 3/8 in. (10 mm) thick], and susceptible to corrosion if the slightest flux residue is not removed, and it produces spatter (especially if rods are exposed to moisture) and requires preheating for metal 0.10 in. thick and thicker. Only 1100, 3003, and 4043 filler alloys are available for this process; see AWS A5.3, *Specification for Aluminum and Aluminum Alloy Electrodes for Shielded Metal Arc Welding* for more information. For these reasons, gas-shielded arc welding is preferred.

2.8.1.3 Other fusion welding processes.

Fusion welding is any welding method that is performed by melting of the base metal or base and filler metal. It includes the arc welding processes mentioned above and several others discussed below as they apply to aluminum.

Oxyfuel gas welding (OFW), or oxygas welding, was used to weld aluminum prior to development of gas-shielded arc welding. The fuel gas, which provides the heat to achieve coalescence, can be acetylene or hydrogen, but hydrogen gives better results for aluminum. The flux can be mixed and applied to the work prior to welding, or flux-coated rods used for shielded metal arc welding can be used to remove the oxide. Oxyfuel gas welding is usually confined to sheet metal of the 1xxx and 3xxx alloys. Preheating is needed for parts over 3/16 in. (5 mm) thick. Problems include large heat affected zones, distortion, flux residue removal labor and corrosion, and the high degree of skill required. The only advantage is the low cost of equipment; so oxyfuel gas welding of aluminum is generally limited to less developed countries where labor is inexpensive and capital is lacking.

Electrogas welding (EGW) is a variation on automatic MIG welding for single-pass, vertical square butt joints such as in ship hulls and storage vessels. It has not been widely applied for aluminum, because the sliding shoes needed to contain the weld pool at the root and face of the joint have tended to fuse to the molten aluminum and tear the weld bead.

Electroslag welding uses electric current through a flux without a shielding gas; the flux removes the oxide and provides the welding heat. This method has been only experimentally applied to aluminum for vertical welds in plate.

Electron beam welding (EBW) uses the heat from a narrow beam of high-velocity electrons to fuse plate. The result is a very narrow heat affected zone and suitability for welding closely fitted, thick parts [even 6 in. (150 mm) thick] in one pass. A vacuum is needed, or the electron beam is diffused; also, workers must be protected from X-rays resulting from the electrons colliding with the work. Thus, electron beam welding must be done in a vacuum chamber or with a sliding seal vacuum and a lead-lined enclosure.

Laser beam welding (LBW) is an automatic welding process that uses a light beam for heat; for aluminum, a shielding gas is also used. Equipment is costly.

Thermit welding uses an exothermic chemical reaction to heat the metal and provide the filler; the process is contained in a graphite mold. Its application to aluminum is for splicing high-voltage aluminum conductors. These conductors must be kept dry, because the copper and tin used in the filler have poor corrosion resistance when exposed to moisture.

2.8.1.4 Arc cutting. Arc cutting is not a joining process but, rather, a cutting process. However, it is included in this section on joining, because it is similar to welding in that an arc from an electrode is used. Plasma arc cutting is the most common arc cutting process used for aluminum. It takes the place of flame cutting (such as oxy-fuel gas cutting) used for steel, a method unsuited to aluminum, because aluminum's oxide has such a high melting point relative to the base metal that flame cutting produces a very rough severing.

In plasma arc cutting, an arc is drawn from a tungsten electrode, and ionized gas is forced through a small orifice at high velocity and temperature, melting the metal and expelling it and, in so doing, cutting through the metal. To cut thin material, a single gas (air, nitrogen, or argon) may act as both the cutting plasma and to shield the arc, but to cut thick material, two separate gas flows (nitrogen, argon, or, for the thickest cuts, an argon-hydrogen mix) are used. Cutting can be done manually, usually on thicknesses from 0.040 in. to 2 in. (1 to 50 mm), or by machine, more appropriate for material 1/4 to 5 in. (6 to 125 mm) thick.

Arc cutting leaves a heat-affected zone and microcracks along the edge of the cut. Thicker material is more prone to cracking, since thick metal provides more restraint during cooling. The cut may also have some roughness and may not be perfectly square in the through thickness direction. The *Specification for Aluminum Structures* therefore requires that plasma-cut edges be machined to a depth of 1/8 in. (3 mm). The quality of the cut is a function of alloy (6xxx series alloys cut better than 5xxx), cutting speed, arc voltage, and gas flow rates.

Plasma arc gouging, used to remove metal to form a bevel or groove, is also performed on aluminum. It can be performed manually or by machine and leaves a clean surface that clearly indicates where the gouging has reached sound base metal. The orifice in the gun is larger than for plasma cutting and a longer arc is used. Groove depths up to 1/4 in. (6 mm) per pass can be achieved, and multiple passes may be made.

2.8.1.5 Resistance welding. Resistance welding is a group of processes that use the electrical resistance of an assembly of parts for the heat required to weld them together. Resistance welding includes both fusion and solid state welding (see Section 2.8.1.6), but it's useful to consider the resistance welding methods as their own group. Because aluminum's electrical conductivity is higher than steel's, it takes more current to produce enough heat to fuse aluminum by resistance welding than for steel.

Resistance spot welding (RSW) produces a spot weld between two or more parts that are held tightly together by briefly passing a current between them. It is useful for joining aluminum sheet and can be used on almost every aluminum alloy, although annealed tempers may suffer from excessive indentation due to their softness. Its advantages are that it is fast, automatic, uniform in appearance, not dependent on operator skill, and strong, and it minimizes distortion of the parts. Its disadvantages are that it applies only to lap joints, is limited to parts no thicker than 1/8 in. (3 mm), requires access to both sides of the work, and requires equipment that is costly and not readily portable. Tables are available that provide the minimum weld diameter, minimum spacing, minimum edge distance, minimum overlap, and shear strengths as a function of the thickness of the parts joined. Proper cleaning of the surface by etching or degreasing and mechanical cleaning is needed for uniform quality.

Weld bonding is a variation on resistance spot welding in which adhesive is added at the weld to increase the bond strength.

Resistance roll spot welding is similar to resistance spot welding except that the electrodes are replaced by rotating wheel electrodes. Intermittent seam welding has spaced welds; seam welding has overlapped welds and is used to make liquid or vapor tight joints.

Flash welding (FW) is a two-step process: heat is generated by arcing between two parts, and then the parts are abruptly forced together. The process is automatically performed in special-purpose machines, producing very narrow welds. It has been used to make miter and butt joints in extrusions used for architectural applications and to join aluminum to copper in electrical components.

High-frequency resistance welding uses high-frequency welding current to concentrate welding heat at the desired location and for aluminum is used for longitudinal butt joints in tubular products. The current is supplied by induction for small diameter aluminum tubing, and through contacts for larger tubes.

2.8.1.6   Solid state welding.   Solid state welding encompasses a group of welding processes that produce bonding by the application of pressure at a temperature below the melting temperatures of the base metal and filler.

Explosion welding (EXW) uses a controlled detonation to force parts together at such high pressure that they coalesce. Explosion welding has two applications for aluminum: it has been used to splice natural gas distribution piping in rural areas where welding equipment and skilled labor are scarce, and to bond aluminum to other metals like copper, steel, and stainless steel to make bimetallic plates.

Ultrasonic welding (USW) produces coalescence by pressing overlapping parts together and applying high-frequency vibrations that disperse the oxide films at the interface. Ultrasonic welding is very well suited to aluminum: spot welds join aluminum wires to themselves or to terminals, ring welds are used to seal containers, line and area welds are used to attach mesh, and seam welds are used to join coils for the manufacture of aluminum foil. Welds between aluminum and copper are readily made for solid state ignition systems, automotive starters, and small electric motors. The advantages of the process are that it requires less surface preparation than other methods, is automatic, fast (usually requiring less than a second), and produces joint strengths that approach that of parent material. Joint designs are similar to resistance spot welds, but edge distance and spot spacing requirements are much less restrictive.

Diffusion welding uses pressure, heat, and time to cause atomic diffusion across the joint and produce bonding, usually in a vacuum or inert gas environment. Pressures can reach the yield strength of the alloys, and times may be in the range of a minute. Sometimes a diffusion aid such as aluminum foil is inserted in the joint. Diffusion welding has been useful to join aluminum to other metals or to join dissimilar aluminum alloys. Welds are of high quality and leak tightness.

Pressure welding uses pressure to cause localized plastic flow that disperses the oxide films at the interface and causes coalescence. When performed at room temperature, it is called cold welding (CW); when at elevated temperature, it is termed hot pressure welding (HPW). Cold welding is used for lap or butt joints. Butt welds are made in wire from 0.015 in. (0.4 mm) to 3/8 in. (10 mm) in diameter, rod, tubing, and simple extruded shapes. Lap welds can be made in thicknesses from foil to 1/4 in. (6 mm). 5xxx alloys with more than 3% magnesium, 2xxx and 7xxx alloys, and castings fracture before a pressure weld can be made and so are not suitable for this process. Hot pressure welding is used to make alclad sheet.

Friction stir welding (FSW) is a new technique by which a nonconsumable tool is rotated and plunged into the joint made by abutting parts. The tool then moves along the joint, plasticizing the material to join it. No filler or shielding gas is needed, nor is there any need for current or voltage controls. It has been applied to 2xxx, 5xxx, 6xxx, and 7xxx alloys, in thicknesses up to 1 in. (25 mm). Friction stir welding produces uniform welds with little heat input and attendant distortion and loss of strength. The disadvantage is that high pressures must be brought to bear on the work and equipment costs are high.

2.8.1.7 Brazing. *Brazing* is the process of joining metals by fusion using filler metals with a melting point above 840°F (450°C), but lower than the melting point of the base metals being joined. *Soldering* also joins metals by fusion, but filler metals for soldering have a melting point below 840°F (450°C). Brazing and soldering differ from welding in that no significant amount of base metal is melted during the fusion process. Ranking the temperature of the process and the strength and the corrosion resistance of the assembly, from highest to lowest, are welding, brazing, and then soldering.

Brazing's advantage is that it is very useful for making complex and smoothly blended joints, using capillary action to draw the filler into the joint. A disadvantage is that it requires that the base metal be heated to a temperature near the melting point; since yield strength decreases drastically at such temperatures, parts must often be supported to prevent sagging under their own weight. Another disadvantage is the corrosive effect of flux residues, which can be overcome by using vacuum brazing or chloride-free fluxes.

Brazing can be used on lap, flange, lock-seam, and tee joints to form smooth fillets on both sides of the joint. Joint clearances are small, ranging from 0.003 in. (0.08 mm) to 0.025 in. (0.6 mm), and depend on the type of joint and the brazing process.

Non-heat-treatable alloys 1100, 3003, 3004, and 5005; heat-treatable alloys 6061, 6063, and 6951; and casting alloys 356.0, A356.0, 357.0, 359.0, 443.0, 710.0, 711.0, and 712.0 are most the commonly brazed of their respective categories. The melting points of 2011, 2014, 2017, 2024, and 7075 alloys are too low to be brazed, and 5xxx alloys with more than 2% magnesium are not very practically brazed, because fluxes are ineffective in removing their tightly adhering oxides. Brazing alloys are shown in Table 2.42, and brazing sheet (cladding on sheet) parameters are given in Table 2.43.

Brazing fluxes are powders that are mixed with water or alcohol to make a paste that removes the oxide film from the base metal upon heating. Chloride fluxes have traditionally been used, but their residue is corrosive to aluminum. More recently, fluoride fluxes, which are not corrosive and thus do not require removal, have come into use. They are useful where flux removal is difficult, such as in automobile radiators.

Brazing can be done by several processes. Torch brazing uses heat from an oxyfuel flame and can be manual or automatic. Furnace braz-

Table 2.42 Common Brazing Filler Alloys and Forms

Brazing alloy designation	AWS classification number	Nominal composition % Si	Cu	Mg	Melting range °F (°C)	Normal brazing °F (°C)	Rod	Sheet	Clad[1]	Powder	Torch	Furnace	Dip	Remarks
4343	BA1Si-2	7.5	—	—	1070–1135 (517–613)	1110–1150 (599–621)		X	X			X	X	
4145	BA1Si-3	10	4	—	970–1085 (521–585)	1060–1120 (571–604)	X	X	X		X	X	X	Desirable where control of fluidity is necessary
4047	BA1Si-4	12	—	—	1070–1080 (577–582)	1080–1120 (582–604)	X	X		X	X	X	X	Fluid in entire brazing range
4045	BA1Si-5	10	—	—	1070–1095 (577–591)	1090–1120 (588–604)		X	X			X	X	
4004	BA1Si-7	10	—	1.5	1030–1105[2] (554–596)[2]	1090–1120 (588–604)			X			X		Vacuum furnace brazing
4147	BA1Si-9	12	—	2.5	1044–1080[2] (562–582)[2]	1080–1120 (582–604)			X			X		Vacuum furnace brazing
4104[3]	BA1Si-11	10	—	1.5	1030–1105[2] (554–596)[2]	1090–1120 (588–604)			X			X		Vacuum furnace brazing
4044	– – –	8.5	—	—	1070–1115[2] (577–602)[2]	1100–1135 (593–613)			X			X	X	

[1]As a cladding on aluminum brazing sheet (Table 15.2).
[2]The melting range temperatures shown for this filler were obtained in air. These temperatures are different in vacuum.
[3]Also contains 0.10 Bi.

TABLE 2.43 Some Standard Brazing Sheet Products

Commercial	Number of sides cladding	Core alloy	Cladding composition	Thickness		% Cladding on each side	Brazing range °F (°C)
				Sheet in.	mm		
No. 7	1	3003	4004	0.024 and less	0.61 and less	15	1090–1120
No. 8	2			0.025 to 0.062	0.62 to 1.59	10	(588–604)
				0.063 and over	1.60 and over	7.5	
No. 11	1	3003	4343	0.063 and less	1.60 and less	10	1100–1150
No. 12	2			0.064 and over	1.62 and over	5	(593–621)
No. 13	1	6951	4004	0.024 and less	0.61 and less	15	1090–1120
No. 14	2			0.025 to 0.062	0.62 to 1.59	10	(588–604)
				0.063 and over	1.60 and over	7.5	
No. 21	1	6951	4343	0.090 and less	2.29 and less	10	1100–1150
No. 22	2			0.091 and over	2.3 and over	5	(593–621)
No. 23	1	6951	4045	0.090 and less	2.29 and less	10	1090–1120
No. 24	2			0.091 and over	2.3 and over	5	(588–604)
No. 33	1	6951	4044	All	All	10	1100–1135
No. 34	2						(593–613)
No. 44	see note[1]	6951	4044/7072	All	All	15/5	1100–1135
							(593–613)

[1]This product is Clad with 4044 on one side and 7072 on the other side for resistance to corrosion.

ing is most common and is used for complex parts like heat exchangers where torch access is difficult. Assemblies are cleaned, fluxed, and sent through a furnace on a conveyor. Dip brazing is used for complicated assemblies with internal joints. The assemblies are immersed in molten chloride flux; the coating on brazing sheet or preplaced brazing wire, shims, or powder supply the filler. Vacuum brazing does not require fluxes and is done in a furnace; it's especially useful for small matrix heat exchangers, which are difficult to clean after fluxing.

Upon completion of brazing, the assembly is usually water quenched to provide the equivalent of solution heat treatment and to assist in flux removal. The work may subsequently be naturally or artificially aged to gain strength.

Minimum requirements for fabrication, equipment, material, procedure, and quality for brazing aluminum are given in the American Welding Society's publication C3.7 *Specification for Aluminum Brazing*.

2.8.1.8 Soldering. *Soldering* is the process of joining metals by fusion with filler metals that have a melting point below 840°F (450°C). [*Brazing*, described in Section 2.8.1.6, uses filler metals with a melting point above 840°F (450°C), but lower than the melting point of the base metals being joined.]

Soldering is much like brazing but conducted at lower temperatures. Soldering is limited to aluminum alloys with no more than 1% magnesium or 4% silicon, because higher levels produce alloys that have poor flux wetting characteristics. Alloys 1100 and 3003 are suitable for soldering, as are clad alloys of the 2xxx and 7xxx series. Alloys of zinc, tin, cadmium, and lead are used to solder aluminum; they are classified by melting temperature and described in Table 2.44.

TABLE 2.44 Classification of Aluminum Solders

Type	Melting range °F (°C)	Common constituents	Ease of application	Wetting of aluminum	Relative strength	Relative corrosion resistance
Low temp.	300–500 (149–260)	Tin or lead plus zinc and/or cadmium	Best	Poor to fair	Low	Low
Intermediate temp.	500–700 (260–371)	Zinc base plus cadmium or zinc–tin	Moderate	Good to excellent	Moderate	Moderate
High temp.	700–840 (371–449)	Zinc base plus aluminum, copper, etc.	Most difficult	Good to excellent	High	Good

Soldering fluxes are classified as organic and inorganic. Organic fluxes are used for low temperature [300 to 500°F (150 to 260°C)] soldering and usually need not be removed, being only mildly corrosive. Inorganic fluxes are used for intermediate [500 to 700°F (260 to 370°C)] and high temperature [700 to 840°F (370 to 450°C)] soldering. Inorganic flux must be removed, since it is very corrosive to aluminum. Both fluxes produce noxious fumes that must be properly ventilated.

Like brazing, soldering can be performed by several processes. Soldering with a hot iron can be done on small wires and sheet less than 1/16 in. (1.6 mm) thick. Torch soldering can be performed in a much wider variety of cases, including automatic processes used to make automobile air conditioning condensers. Torch soldering can also be done without flux by removing the aluminum oxide from the work by rubbing with the solder rod, called *abrasion soldering*. Abrasion soldering can also be performed by ultrasonic means. Furnace and dip soldering are much like their brazing counterparts. Resistance soldering is well suited to spot or tack soldering; flux is painted on the base metal, the solder is placed, and current is passed through the joint to melt the solder.

Soldered joint shear strengths vary from 6 to 40 ksi (40 to 280 MPa) depending on the solder used. Corrosion resistance is poor if chloride containing flux residue remains and the joint is exposed to moisture. Zinc solders have demonstrated good corrosion resistance, even for outdoor exposure.

2.8.2 Fastening

The types of fasteners used to connect aluminum parts are bolts, rivets, screws, nails, and special-purpose fasteners. Where holes are required, they may be punched, drilled, or punched or drilled and then reamed. If holes are punched and then enlarged, the amount by which the diameter of hole is enlarged should be at least 1/4 of the thickness of the piece and no less than 1/32 in. (0.8 mm). Punching should be limited to material that is no thicker than the diameter of the hole to avoid tear out at the back side of the work. For design purposes such as the determination of the net cross-sectional area of the part at a hole, the size of punched holes is taken as the nominal hole diameter plus 1/32 in. (0.8 mm).

Aluminum sheet may also be fastened by mechanical clinches that locally deform the material on both sides of the joint to hold it together.

2.8.2.1 Bolts. Aluminum *bolts* are made of 2024-T4, 6061-T6, and 7075-T73 material conforming to ASTM B316 in. diameters from 1/4 in.

(6 mm) to 1 in. (25 mm) with the finished product conforming to ASTM F468, *Nonferrous Bolts, Hex Cap Screws, and Studs for General Use.* Minimum ultimate tensile and shear strengths are given in Table 2.45. Bolts should be spaced no closer together than 2.5 times the bolt diameter measured center to center, no closer than two bolt diameters from the center of the bolt to the edge of the part, and in holes no larger than 1/16 in. (1.6 mm) larger than the nominal bolt diameter. The effective area of the bolt resisting shear loads is based on the diameter of the bolt in the shear plane.

TABLE 2.45   Minimum Strengths of Aluminum Bolts

Alloy-temper	Minimum tensile strength (ksi)	Minimum shear strength (ksi)
2024-T4	62	37
6061-T6	42	25
7075-T73	68	41

Aluminum structural bolts for use in aluminum transmission towers, substations, and similar aluminum structures are made of 2024-T4 with 6061-T6 or 6262-T9 nuts in 5/8, 3/4, and 7/8 in. diameters to ASTM F901.

Aluminum *nuts* are made of ASTM B211 material and are available in 2024-T4, 6061-T6, and 6262-T9 with properties conforming with ASTM F467, *Nonferrous Nuts for General Use.* Full thickness nuts of 6262-T9 are strong enough to develop the full strength of bolts made of 2024-T4, 6061-T6, or 7075-T73; nuts of 6061-T6 are strong enough to develop the full strength of 2024-T4 and 6061-T6 bolts. Machine screw nuts and other styles of small nuts [1/4 in. (6 mm) and smaller] are usually made of 2024-T4. Flat *washers* are usually made of alclad 2024-T4 and helical spring washers of 7075-T73.

Galvanized and plated steel and austenitic stainless steel bolts are also used to fasten aluminum parts. Galvanized, high-strength (ASTM A325) steel bolts can be used in joints that are designed to prevent slip of the connected parts relative to each other, because A325 bolts are strong enough to apply compression to the joint to develop the necessary friction between the faying surfaces. Such joints are called *slip critical joints* and have greater fatigue strengths than other bolted joints. To resist slip, the surfaces of the aluminum parts that will be in contact must be roughened before the parts are assembled. Roughening aluminum by abrasion blasting to an average substrate profile of 2.0 mils (0.05 mm) in contact with similar aluminum surfaces or with zinc painted steel surfaces with a maximum dry film paint thickness of 4 mils (0.1 mm) will achieve a friction coefficient of 0.5. Turning the nut a prescribed rotation (for example, 2/3 of a complete rotation) is commonly used to tighten such connections for steel assemblies. Using the same number of turns as used for turn-of-nut methods to tighten steel assemblies on aluminum assemblies produces the same pretension in the bolt.

2.8.2.2 Rivets. *Rivets* are used to resist shear loads only; they cannot be relied on to resist tensile loads. Usually, the rivet alloy is similar to the base metal. Table 2.46 lists common aluminum rivet alloys and their minimum ultimate shear strengths. Many different head types are available, including countersunk styles. Hole diameters for cold-driven rivets should not exceed 4% more than the nominal rivet diameter; hole diameters for hot-driven rivets should not exceed 7% more than the nominal rivet diameter. The effective area of the rivet resisting shear is based on the hole diameter, since the rivet is designed to completely fill the hole when properly installed. Rivets should be spaced no closer together than three times the rivet diameter measured center to center. Specifications for rivets are given in Table 2.47, and identification markings are shown in Figure 2.8 for the various rivet alloys.

TABLE 2.46   Minimum Expected Shear Strengths of Aluminum Rivets

Designation before driving	Minimum expected ultimate shear strength (ksi)
1100-H14	9.5
2017-T4	33
2117-T4	26
5056-H32	25
6053-T61	20
6061-T6	25
7050-T7	39

2.8.2.3 Screws. Wood and sheet metal *screws* are made of 2024-T4 or 7075-T73 aluminum; austenitic stainless steel screws may also be used to connect aluminum parts. Equations for the shear and tensile strengths of tapping screw connections in aluminum parts can be found in the *Specification for Aluminum Structures*, Section 5.3.

TABLE 2.47 Rivet Specifications

Alloy and temper	Specification number	Grade or code
1100-F	MIL-R-5674	A
2017-T4	MIL-R-5674	D
2117-T4	MIL-R-5674	AD
2024-T4	MIL-R-5674	DD
5056-H32	MIL-R-5674	B
6053-T61	MIL-R-1150	E
6061-T6	MIL-R-1150	F
1100-F	AMS 7220	99A1
2024-T4	AMS 7223	4.5 Cu, 1.5 Mg, 0.6 Mn
2117-T4	AMS 7222	2.5 Cu, 0.3 Mg
2017-T4	FF-R-556	B

2.8.2.4 Other Fasteners. Aluminum *nails* (screw shank and ring shank) and *staples* are made of 5056-H19 or 6061-T6 wire and are used in building construction to fasten aluminum attachment clips. These clips in turn fasten sheet metal to substrate or to attach wall or roof covering materials.

There are also many proprietary fasteners made of aluminum and designed to serve a particular purpose. One example is the lockbolt, which consists of a pin with concentric grooves onto which a collar is swaged, forming a permanently fastened joint. Others include blind rivets that can be installed with access to only one side of a joint. These rivets form their own head on the back side of the joint during installation.

2.8.3 Adhesives

Adhesive bonding is a process of joining materials with an adhesive placed between the faying surfaces. The suitability of adhesives for aluminum is demonstrated by their successful use in aircraft since the 1950s. Examples include the adhesive bonding of aluminum face sheets to honeycomb cores to make honeycomb panels, and bonding aluminum face sheets to plastic cores to make sandwich panels, sometimes called aluminum composite material (ACM). Helicopter rotor blades are now joined only by adhesives, since adhesives have proven more durable than their mechanical fastener predecessors.

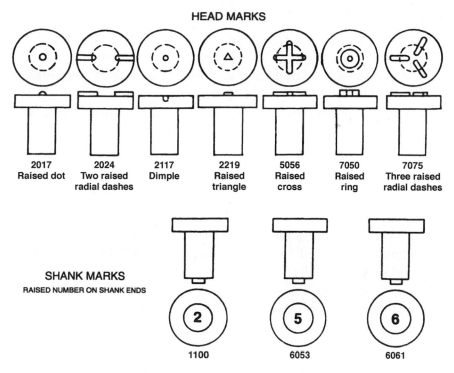

Figure 2.8 Rivet identification markings.

The advantages of adhesives are:

- Joints are sealed, improving corrosion resistance.
- Stress concentrations inherent in mechanically fastened joints are avoided, allowing a more uniform transfer of stress through the joint and improving fatigue performance.
- Bonds can provide electrical and thermal insulation between parts joined.
- Bonds can act as vibration dampers.
- The clean appearance and aerodynamic streamlining of joints is afforded without fasteners.
- Aluminum can be joined to dissimilar materials.

Disadvantages are:

- Adhesively bonded joints tend to have low peel strengths. For this reason, they are often used in conjunction with fasteners or welds that resist the peeling, while the adhesive resists shearing forces.

- Adhesive shelf life can be short.

- Surface preparation is critical to the strength and durability of the joint.

- Most adhesives lose strength at elevated temperatures more rapidly than the aluminum parts they are joining.

- Skill and care are required to properly make adhesively bonded joints, and verification of joint integrity is difficult.

Surface pretreatment is by degreasing and mechanical abrasion for less critical applications, and by etching or anodizing in acid solutions for more rigorous service such as in aircraft. The four most common preparations are:

1. The Forest Products Laboratory (FPL) chromic-sulfuric acid etching procedure, which may also be used as the first step of the anodizing pretreatments

2. The P2 etch, which uses ferric sulfate, and so is a less hazardous treatment than FPL

3. Phosphoric acid anodization (PAA), a popular method in the U.S. aerospace industry

4. Chromic acid anodization (CAA), often used in European aerospace applications

Adhesives are classified as *thermoplastic resins*, which can be repeatedly softened by heat and hardened by cooling to ambient temperature; and *thermosetting resins*, which cannot be resoftened by heating. Thermoplastic resins are generally less durable, less rigid, and less solvent resistant than thermosetting resins, and they have a lower modulus of elasticity and will creep under load. Thermoplastics are usually not used for structural applications but may be blended with thermosetting resins for such cases; examples include vinyls (thermoplastic) combined with epoxy resins (thermosetting), a combination particularly well suited to aluminum. Thermosetting resins are usually cured with chemical hardeners, heat, or both. Care must be taken to account for the effect of any heat applied for curing adhesive on the strength of tempered aluminum products.

2.9   Finishes

2.9.1   General

Although many proprietary designations have been used for aluminum finishes, almost all can be placed in one of three categories: mechanical finishes, chemical finishes, and coatings. The Aluminum Association has adopted a designation system based on these three categories: M for mechanical finishes, C for chemical finishes, and for coatings: A for anodic, R for resinous and other organic coatings, V for vitreous (porcelain and ceramic), E for electroplated and other metal coatings, and L for laminated coatings, including veneers, plastic coatings, and films. A finish may include all three categories; for example, an architectural building panel may receive a directionally textured, medium satin mechanical finish, and a chemical cleaning treatment followed by an anodic coating, and be identified as AA-M32C12A31. Whenever the Aluminum Association designation system is used, the designation is preceded by "AA." Aluminum Association designations for the finishes described below are given in parentheses.

2.9.2 Mechanical Finishes

Mechanical finishes include as-fabricated (M1x), buffed (M2x), directional textured (M3x), and nondirectional textured (M4x), as shown in Table 2.48. As-fabricated finishes have no mechanical finishing other than that produced by the fabrication methods used to make the part. Several mechanical finishing methods are used as described below.

Abrasion blasting (M4x) is used to clean aluminum surfaces (especially castings), prepare surfaces for subsequent finishes such as organic coatings, and to produce a decorative, nondirectional textured matte finish. Blasting should not be used on material less than about 1/8 in. (3 mm) thick, because the residual stresses induced can curl thin material. A number of different substances are used to blast aluminum, including washed silica sand (from 20 to over 200 grit size), aluminum oxide, coal slag, steel grit, steel or stainless steel shot, plastic pellets, and crushed walnut shells. When steel is used, it can become embedded in the aluminum and subsequently stains the surface when it rusts, so steel grit-blasted aluminum is usually cleaned in a 50% nitric acid solution at ambient temperature for 20 minutes after blasting. Fine abrasives (that may also be wet blasted with water) produce a fine-grain matte finish.

Barrel finishing, in which parts are tumbled in a barrel with either a wet or a dry medium, is used to smooth sharp edges, give a matte finish, and prepare the surface for coatings. Barrel deburring is usually done wet with synthetic detergents mixed with granite fines or limestone chips. Barrel burnishing, which produces a smooth, mirror-like finish, is done wet with steel balls. Care must be taken to keep the media pH near neutral to avoid chemical attack on the aluminum parts that can generate explosive gases. Barrels must be vented, and steel drums should be rubber lined to prevent rust particles from staining the aluminum parts.

TABLE 2.48 Mechanical Finishes (M)

Type of finish	Designation[1]	Description	Examples of methods of finishing[2]
As-fabricated	M10	Unspecified	To be specified.
	M11	Specular as fabricated	
	M12	Nonspecular as fabricated	
	M1X	Other	
Buffed	M20	Unspecified	
	M21	Smooth specular	Polished with grits coarser than 320. Final polishing with a 320 grit using peripheral wheel speed of 30 m/s (6,000 ft/min). Polishing followed by buffing, using tripoli based buffing compound and peripheral wheel speed of 36 to 41 m/s (7,000 to 8,000 ft/min).
	M22	Specular	Buffed with tripoli compound using peripheral wheel speed 36 to 41 m/s (7,000 to 8,000 ft/min).
	M2X	Other	To be specified.
Directional textured	M30	Unspecified	
	M31	Fine satin	Wheel or belt polished with aluminum oxide grit of 320 to 400 size; peripheral wheel speed 30 m/s (6,000 ft/min).
	M32	Medium satin	Wheel or belt polished with aluminum oxide grit of 180 to 220 size; peripheral wheel speed 30 m/s (6,000 ft/min).
	M33	Coarse satin	Wheel or belt polished with aluminum oxide grit of 80 to 100 size; peripheral wheel speed 30 m/s (6,000 ft/min).
	M34	Hand rubbed	Hand rubbed with stainless steel wool lubricated with neutral soap solution. Final rubbing with No. 00 steel wood.
	M35	Brushed	Brushed with rotary stainless steel wire brush, wire diameter 0.24 mm (0.0095 in.); peripheral wheel speed 30 m/s (6,000 ft/min.); or various proprietary satin finishing wheels or satin finishing compounds with buffs.
	M3X	Other	To be specified.

Nondirectional textured	M40	Unspecified	
	M41	Extra fine matte	Air blasted with finer than 200 mesh washed silica or aluminum oxide. Air pressure 310 kPa (45psi); gun distance 203–305 mm (8–12 in.) from work at 90° angle.
	M42	Fine matte	Air blasted with 100 to 200 mesh silica sand if darkening is not a problem; otherwise aluminum oxide type abrasive. Air pressure 207 to 621 kPa (30 to 90 psi) (depending upon thickness of material); gun distance 305 mm (12 in.) from work at angle of 60° to 90°.
	M43	Medium matte	Air blasted with 40 to 50 mesh silica sand if darkening is not a problem; otherwise aluminum oxide type abrasive. Air pressure 207 to 621 kPa (30 to 90 psi) (depending upon thickness of material); gun distance 305 mm (12 in.) from work at angle of 60° to 90°.
	M44	Coarse matte	Air blasted with 16 to 20 mesh silica sand if darkening is not a problem; otherwise aluminum oxide type abrasive. Air pressure 207 to 621 kPa (30 to 90 psi) (depending upon thickness of material); gun distance 305 mm (12 in.) from work at angle of 60° to 90°.
	M45	Fine shot blast	Shot blasted with cast steel shot of ASTM size 70–170 applied by air blast or centrifugal force. To some degree, selection of shot size is dependent on thickness of material since warping can occur.
	M46	Medium shot blast	Shot blasted with cast steel shot of ASTM size 23–550 applied by air blast or centrifugal force. To some degree, selection of shot size is dependent on thickness of material since warping can occur.
	M47	Coarse shot blast	Shot blasted with cast steel shot of ASTM size 660–1320 applied by air blast or centrifugal force. to some degree, selection of shot size is dependent on thickness of material since warping can occur.
	M4X	Other	To be specified.

[1]The complete designation must be preceded by AA–signifying Aluminum Association.
[2]Examples of methods of finishing are intended for illustrative purposes only.

Polishing with either belt or setup wheel polishers and aluminum oxide or silicon carbide abrasives with a lubricant is used to produce directional textured satin finishes. The finer the grit size, the finer the finish. *Brushing* with rotary stainless steel wire brushes can also be used to produce a directional textured satin finish. *Buffing* with a buffing wheel cloth and buffing compound produces high luster and specular (mirrorlike) finishes.

2.9.3 Chemical Finishes

Chemical finishes include nonetching cleaning treatments (C1x), etching (C2x), chemical or electrochemical brightening (C3x), and chemical coatings (C4x), as shown in Table 2.49.

Nonetching cleaning treatments (C1x) are used to remove oils, grease, and dirt, a process called *degreasing,* and to prepare the surface for subsequent finishes such as anodizing or the application of a chemical conversion coating. One method (C11) uses organic solvents such as kerosene, Stoddard solvent, and mineral spirits, with small amounts of emulsifiers and surfactants; the solution is applied to the material and then sprayed off with water. Because of environmental concerns regarding the use of such solvents, inhibited chemical cleaning (C12) has become more common. Alkaline cleaners can be inhibited with sodium silicates and are based on aqueous solutions of sodium carbonate, trisodium phosphate, or other alkalis to which small amounts of sodium silicate are added to inhibit etching. The bath is held at 140 to 160°F (60 to 70°C), and the material is immersed for 2 to 5 minutes.

Chemical etching (C2x) uses either an alkaline or acid solution at elevated temperatures to produce a matte finish on aluminum, usually prior to chemical brightening or coating. Etching prior to anodizing removes surface contaminants that can cause discolorations. Alkaline solutions used include sodium hydroxide, potassium hydroxide, trisodium phosphate, sodium fluoride, and sodium carbonate. Acid solutions are preferred for castings with high silicon content.

Chemical brightening (C3x), also called *bright dipping* or *chemical polishing,* smooths and brightens the surface by immersion and agitation in an acid bath (usually sulfuric, nitric, phosphoric, or acetic) with oxidizing agents. *Electrolytic brightening*, also called *electrobrightening* or *electropolishing,* produces similar surfaces but uses a different process. The part is first buffed, cleaned, and rinsed and then immersed in an acid or alkaline bath through which direct current is passed with the part as the anode. Both chemical and electrolytic brightening work by selectively dissolving the high points of a rough surface.

TABLE 2.49 Chemical Finishes (C)

Type of finish	Designation[1]	Description	Examples of methods of finishing[2]
Nonetched cleaned	C10	Unspecified	Organic solvent treated.
	C11	Degreased	Inhibited chemical type cleaner used
	C12	Inhibited chemical cleaned	To be specified.
	C1X	Other	
Etched	C20	Unspecified	
	C21	Fine matte	Trisodium phosphate, 22–45 g/l (3–6 oz per gal) used at 60–71°C (140–160°F) for 3 to 5 min. Sodium hydroxide, 30–45 g/l (4–6 oz per gal) used at 49–66°C (120–150°F) for 5 to 10 min.
	C22	Medium matte	Sodium fluoride, 11g/l (1.5oz) plus sodium hydroxide 30–45 g/l (4–6 oz per gal) used at 54–66°C (130–150°F) for 5 to 10 min.
	C23	Coarse matte	To be specified.
	C2X	Other	
Brightened	C30	Unspecified	
	C31	Highly specular	Chemical bright dip solution of the proprietary phosphoric–nitric acid type used, or proprietary electrobrightening or electropolishing treatment.
	C32	Diffuse bright	Etched finish C22 followed by brightened finish C31.
	C3X	Other	To be specified.
Chemical coatings[3]	C40	Unspecified	
	C41	Acid chromate–fluoride	Proprietary chemical treatments used producing clear to typically yellow colored surfaces.
	C42	Acid chromate–fluoride–phosphate	Proprietary chemical treatments used producing clear to typically green colored surfaces. Proprietary chemical treatments used producing clear to typically gray colored surfaces.
	C43	Alkaline chromate	Proprietary chemical coating treatment employing no chromates.
	C44	Non-chromate	Proprietary chemical coating treatment in which coating liquid is dried on the work with no subsequent water rinsing.
	C45	Non-rinsed chromate	To be specified.
	C4X	Other	

[1]The complete designation must be preceded by AA–signifying Aluminum Association.
[2]Examples of methods of finishing are intended for illustrative purposes only.
[3]Includes chemical conversion coatings.

Chemical coatings (C4x) include chemical conversion coatings. Chemical conversion coatings are low-solubility oxide, phosphate, or chromate compounds that are formed when agents react with the metal surface and adhere to it. They differ from anodic coatings (described below), because they are formed by a chemical reaction rather than an electrochemical reaction. Chemical conversion coatings are used to improve adhesion for subsequent organic finishes, to provide corrosion protection without decreasing electrical conductivity, to spot treat damaged anodic coatings, and for decorative purposes. Examples include baking pans, storm doors, beverage cans, aircraft fuselage skins, and electronic components. Chemical conversion coatings cost less than anodizing, which is discussed below, but are not as tenacious.

The chemical conversion process steps are: removing organic contaminants, removing the surface aluminum oxide from the part, conditioning to accept the coating, applying the coating, rinsing, and drying. Clear, yellow, green, or gray colors can be produced.

2.9.4   Coatings

2.9.4.1   Anodizing.   Anodizing is the conversion of the aluminum surface to aluminum oxide while the part is the anode in an electrolytic cell. During anodizing, the part is immersed in an acid solution that serves as the electrolyte at a controlled temperature and time while electric current is introduced. The main reasons for anodizing are as follows.

- It increases corrosion resistance. The aluminum oxide coating produced by anodizing is thicker than the natural oxide that occurs without anodizing. This coating protects the underlying metal from corrosion.

- It prepares the surface for subsequent painting, adhesives, or electroplating

- It increases abrasion resistance. So-called hard anodizing coatings are used on parts exposed to wear.

- It provides electrical insulation.

- It provides a lustrous, decorative appearance: clear and colored anodizing is available.

Anodizing designations are shown in Table 2.50.

Anodizing steps include: solvent cleaning to degrease the part, chemical cleaning, etching (or brightening, if desired), anodizing, and finally, sealing of the anodized coating, usually in slightly acidified hot water. *Anodizing processes* include chromic acid (A12), sulfuric acid (A2x, A3x, A4x), hard anodizing (A13), and other specialized processes. Chromic acid treatments are used on assemblies with recesses such as lapped joints where it is difficult to remove all of the electrolyte. The chromic acid produces colors ranging from yellow to dark olive, depending on anodizing thickness. Sulfuric acid treatments cannot be used when the electrolyte is difficult to remove. Hard anodizing treatments use sulfuric acid alone or with additives, or other acids, and vary from conventional sulfuric acid processes in the operating temperature and current density used.

Anodizing *thicknesses* range from 0.2 mils (5 µm) to 0.7 mils (18 µm), except for hard anodizing, which is from 1 mil (25µm) to over 4 mils (100 µm) thick. Thicknesses less than 0.4 mils (10 µm) are called *protective and decorative coatings* (A2x) and are not recommended for outdoor exposure. Thicknesses between 0.4 (10 µm) and 0.7 mils (18 µm) are called *architectural Class II coatings* (A3x), and those 0.7 mils (18 µm) and thicker are called *architectural Class I coatings* (A4x).

Coloring processes include:

1. *Impregnated coloring.* This process uses dyes before the coating is sealed. An example is a gold color produced by precipitating iron oxide from an aqueous solution of ferric ammonium oxalate. Care must be taken to obtain uniform and colorfast results.

2. *Integral color.* In integral color anodizing the color is inherent in the oxide itself, usually producing an earth tone. The color produced is a function of the alloy and temper; copper alloys anodize to a yellow or green in sulfuric acid, while manganese and silicon alloys anodize gray to black in this process. Integral coloring has been largely replaced by the electrolytic coloring.

3. *Two-step electrolytic coloring.* The first step is conventional anodizing in sulfuric acid, followed by immersion in an electrolyte with a dissolved metal salt. Tin (producing bronze shades), nickel, cobalt, and copper (producing burgundy and blue) are used.

Anodizing different alloys by the same process can produce different colors, so filler alloys for weldments to be anodized should be chosen with color compatibility in mind. An example is 6061; when welded with 4043 filler, the filler anodizes much darker than the base metal; 5356 filler welded 6061 more closely matches the color of the base metal.

Alloys containing more than 5% copper or 7.5% total alloying elements are not suited to chromic acid, because excessive pitting occurs.

TABLE 2.50 Anodic Coatings (A)

Type of Finish	Designation[1]	Description	Examples of Methods of Finishing[2]
General	A10	Unspecified	
	A11	Preparation for other applied coatings	$3\mu m$ (0.1 mil) anodic coating produced in 15% H_2SO_4 at 21° ±1°C (70°F ± 2°F) at 129 A/m^2 (12 A/ft^2) for 7 min, or equivalent.
	A12	Chromic acid anodic coatings	To be specified.
	A13	Hard, wear and abrasion resistant coatings	To be specified.
	A1X	Other	To be specified.
Protective and decorative coatings less than 10 μm (0.4 mil) thick	A21	Clear coating	Coating thickness to be specified. 15% H_2SO_4 used at 21° ±1°C (70°F ± 2°F) at 129 A/m^2 (12A/ft^2).
	A211	Clear coating	Coating thickness–3 μm (0.1 mil) minimum. Coating weight–6.2 g/m^2 (4 mg/in^2) minimum.
	A212	Clear coating	Coating thickness–5 μm (0.2 mil) minimum. Coating weight–12.4 g/m^2 (8 mg/in^2) minimum.
	A213	Clear coating	Coating thickness–8 μm (0.3 mil) minimum. Coating weight–18.6 g/m^2 (12 mg/in^2) minimum.
	A22	Coating with integral color	Coating thickness to be specified. Color dependent on alloy and process methods.
	A221	Coating with integral color	Coating thickness–3 μm (0.1 mil) minimum. Coating weight–6.2 g/m^2 (4 mg/in^2) minimum.
	A222	Coating with integral color	Coating thickness–5 μm (0.2 mil) minimum. Coating weight–12.4 g/m^2 (8 mg/in^2) minimum.
	A223	Coating with integral color	Coating thickness–8 μm (0.3 mil) minimum. Coating weight–18.6 g/m^2 (12 mg/in^2) minimum.
	A23	Coating with impregnated color	Coating thickness to be specified. 15% H_2S_4 used at 27°C ± °C (80°F ± 2°F) at 129 A/m^2 (12 A/ft^2) followed by dyeing with organic or inorganic colors.

Category	Code	Description	Specification
	A231	Coating with impregnated color	Coating thickness–3 μm (0.1 mil) minimum. Coating weight–6.2 g/m² (4mg/in²) minimum.
	A232	Coating with impregnated color	Coating thickness–5 μm (0.2 mil) minimum. Coating weight–12.4 g/m² (8 mg/in²) minimum.
	A233	Coating with impregnated color	Coating thickness–8 μm (0.3 mil) minimum. Coating weight–18.6 g/m² (12 mg/in²) minimum.
	A24	Coating with electrolytically deposited color	Coating thickness to be specified. Application of the anodic coating, followed by electrolytic deposition of inorganic pigment in the coating.
	A2X	Other	To be specified.
Architectural Class II[3] 10 to 18 μm (0.4 to 0.7 mil) coating	A31	Clear coating	15% H_2SO_4 used at 21°C ± 1°C (70°F ± 2°F) at 129 A/m² (12 A/ft²) for 30 min, or equivalent.
	A32	Coating with integral color	Color dependent on alloy and anodic process.
	A33	Coating with impregnated color	15% H_2SO_4 used at 21°C ± 1°C (70°F ± 2°F) at 129 A/m² (12 A/ft²) for 30 min, followed by dyeing with organic or inorganic colors.
	A34	Coating with electrolytically deposited color	Application of the anodic coating followed by electrolytic deposition of inorganic pigment in the coating.
	A3X	Other	To be specified.
Architectural Class I[3] 18 μm (0.8 mil) and thicker coatings	A41	Clear coating	15% H_2SO_4 used at 21°C ± 1°C (70°F ± 2°F) at 129 A/m² (12 A/ft²) for 60 min, or equivalent.
	A42	Coating with integral color	Color dependent on alloy and anodic process.
	A43	Coating with impregnated color	15% H_2SO_4 used at 21°C ± 1°C (70°F ± 2°F) at 129 A/m² (12 A/ft²) for 60 min, followed by dyeing with organic or inorganic colors, or equivalent.
	A44	Coating with electrolytically deposited color	Application of the anodic coating followed by electrolytic deposition of inorganic pigment in the coating.
	A4X	Other	To be specified.

[1]The complete designation must be preceded by AA—signifying Aluminum Association.
[2] Examples of methods of finishing are intended for illustrative purposes only.
[3]Aluminum Association Standards for Anodized Architectural Aluminum. (No longer in print)

Alloys that are produced because they are especially suited to bright anodizing are 1100, 3002, 5252, 5657, 6463, 7016, and 7029. Other alloys, like 3003, are not specifically produced for this purpose but also lend themselves to bright anodized finishes.

Anodizing reduces the light reflectance of aluminum alloys. It also reduces the fatigue strength of wrought alloys by introducing a brittle surface where fatigue cracks can more readily initiate.

2.9.4.2   Resinous and other organic coatings. Organic coatings work well with aluminum when the surface is properly cleaned and prepared to accept such coatings. A wash primer or zinc chromate primer is usually applied before the finish coat of paint. The alternatives to primers are conversion coatings or anodizing. Usually, only thin anodic coatings are required if paint will subsequently be applied; sulfuric or chromate acid electrolyte anodizings are acceptable.

A popular finish coat paint is polyvinylidene fluoride (PVDF) resin (commonly known as Kynar or Hylar), which provides good weatherability and a wide choice of colors. The disadvantage is that this paint is more expensive than anodizing.

Painted sheet may be subsequently formed without cracking the coating, within the limitations given in Table 2.51. When paint is baked on, mechanical properties for the material must be obtained from the supplier, since heat can reduce the strength of tempered aluminum alloys. The temper designation H4x applies to products that are strain hardened and then subjected to some heat during the painting process.

2.9.4.3   Vitreous coatings. Vitreous coatings include porcelain and ceramics. Porcelain enamels are glass coatings that enhance appearance and protect the metal. They are distinguished from organic coatings such as paint, because they are inorganic and are fused to the metal substrate. Tanks, vessels, cookware, and signs are examples of aluminum products that have been porcelain enameled.

Porcelain enamels include lead base, barium, and phosphate enamels; they may be colored with pigments. After preparation, aluminum parts are coated with enamel, usually by spraying, and fired at temperatures from 980 to 1020°F (525 to 550°C) (nearly the melting point of the aluminum being coated) for 5 to 15 minutes. The high temperatures are needed to fuse the coating, but they also reduce the strength of the aluminum to essentially that of annealed material. The resulting coatings are not as hard or durable as those produced on cast iron, because only enamels with low melting temperatures can be used on aluminum alloys due to aluminum's relatively low melting point.

TABLE 2.51   Recommended[1] Minimum Bend Radii[2] for Painted Sheet[3]

Alloy	Temper before film application	Thickness of base sheet (in.)					
		0.016	0.025	0.032	0.040	0.050	0.064
1100	0	1T	1T	1T	1T	1T	1T
	H12	1T	1T	1T	1T	1T	1T
	H14	1T	1T	1T	1T	1T	1T
	H16	1T	1T	1T	1T	2T	3T
	H18	2T	2T	3T	3T	4T	5T
3003	0	1T	1T	1T	1T	1T	1T
	H12	1T	1T	1T	1T	1T	1T
	H14	1T	1T	1T	1T	1T	1T
	H16	1T	1T	2T	3T	3T	4T
	H18	2T	3T	4T	5T	6T	7T
3105	0	1T	1T	1T	1T	1T	1T
	H12	1T	1T	1T	1T	1T	1T
	H14	1T	1T	1T	1T	1T	1T
	H16	1T	1T	2T	3T	3T	4T
	H18	2T	3T	4T	5T	6T	7T
5005	0	1T	1T	1T	1T	1T	1T
	H32	1T	1T	1T	1T	1T	1T
	H34	1T	1T	1T	1T	1T	1T
	H36	1T	1T	2T	3T	3T	4T
	H38	2T	3T	4T	5T	6T	7T
5052	0	1T	1T	1T	1T	1T	1T
	H32	1T	1T	1T	1T	1T	1T
	H34	1T	1T	1T	2T	2T	3T
	H36	2T	3T	3T	3T	4T	5T
	H38	2T	3T	4T	5T	6T	7T

[1]90° bends for high gloss alkyd, acrylic, siliconized acrylic, polyester, or siliconized polyester films recommended for moderate forming; and 180° bends for high gloss vinyl and medium gloss fluoropolymer films recommended for severe forming. For sheet painted with medium gloss paints other than fluoropolymers or with low gloss paints, minimum bend raidus usually must be greater than shown in the table to prevent or minimize paint microcracking.
[2]Minimum radius over which painted sheet may be bent varies with type and gloss of paint, nature of forming operation, type of forming equipment, and design and condition of tools. Minimum radius for a specific material, or hardest alloy and temper for a specific radius, can be closely determined only by actual trial under contemplated conditions of fabrication.
[3]The reference test method is ASTM E290.

2.9.4.4   Electroplating and other metal coatings.   Electroplating is the deposition of a metal (usually chromium, nickel, cadmium, copper, tin, zinc, gold, or silver) on an aluminum surface by immersing the aluminum in an electrolyte through which a current is passed. Electroplating is used for decorative or functional purposes such as enhancing corrosion resistance and is often less than 1 mil (25 μm) thick. An example is aluminum automotive bumpers with copper, nickel, and chromium coatings.

Aluminum's rapidly forming oxide skin, and the fact that aluminum is anodic to most plating metals, make plating aluminum more difficult than other metals. Any discontinuity in the metal coating when the coating is cathodic to aluminum (which includes all those mentioned above except zinc and cadmium) causes corrosion of the aluminum at the discontinuity rather than protecting it. Alloys with more than 3% magnesium are generally not electroplated.

Surface preparations used for electroplating are surface roughening, anodizing, or immersion coating. Immersion coating is usually done with zinc or tin; for zinc, the process is called *zincating* in which a thin layer of zinc is deposited on the aluminum surface by chemical replacement by aluminum of zinc ions in an aqueous solution of zinc salts. Zincating is not a durable enough coating to be used alone for parts subjected to outdoor exposure.

2.10   Glossary

age softening   a spontaneous decrease in strength that takes place at room temperature in certain strain hardened alloys containing magnesium that are not stabilized

alclad   an adjective applied to aluminum alloys given a thin aluminum or aluminum alloy coating metallurgically bonded to the surface

alloy   a material with metallic properties and composed of two or more elements, at least one of which is a metal

aluminum   a silvery, lightweight, easily worked metallic element with atomic number 13

annealing   a thermal treatment that reduces the yield strength and softens a metal by relieving stresses induced by cold working or by coalescing precipitates from solid solution

anodizing   forming an oxide coating on a metal by electrochemical treatment

artificial aging   a rapid precipitation from solid solution at elevated temperatures to produce a change in mechanical properties, also called *precipitation heat treating* or *precipitation hardening*

bar   a solid wrought product whose cross section is square, rectangular, or a regular hexagon or octagon, and that may have rounded corners, with at least one perpendicular distance between parallel faces 0.375 in. or more (or over 10 mm)

bar, cold finished   bar made by cold working to obtain better surface finish and dimensional tolerances

billet   an unfinished product produced by hot working that may subsequently be worked by forging, extrusion, or other methods

brazing   joining metals by fusion using filler metals with a melting point above 840°F (450°C), but lower than the base metals being joined

buffing   mechanical finishing done with rotating wheels with abrasives

casting   a product made by pouring molten metal into a mold

casting, permanent mold   a casting made in a mold that may be used multiple times

casting, sand   a casting made in a mold made of sand, typically discarded after one use

coiled sheet   sheet that has been rolled into a coil

cold working   permanent deformation of a metal that produces strain hardening, which results in an increase in strength and a loss in ductility

corrosion, exfoliation   a delamination parallel to the metal surface caused by the formation of corrosion product

corrosion, galvanic   corrosion that occurs when two conductors with different electric potential are electrically connected by an electrolyte

corrosion, pitting   corrosion resulting in small pits in a metal surface

draft   taper on the sides of a die or mold to allow removal of forgings, castings, or patterns from the die or mold

drawing   pulling material through a die to change the cross section or harden the material

ductility   the ability of a material to withstand plastic strain before rupture

elastic   pertaining to behavior of material under load at stresses below the proportional limit, where deformations under load are not permanent

elongation   the percentage increase in the distance between two gage marks of a specimen tensile tested to rupture

extrusion   a product made by pushing material through an opening called a *die*

fatigue   fracture caused by repeated application of stresses

forging   a product worked to a predetermined shape using one or more processes such as hammering, upsetting, pressing, rolling, etc.

fracture toughness   the ability to resist cracking at a notch or crack

gage (or gauge)   the number designating the thickness of sheet, plate, or wire

heat-treating   obtaining desired material properties by heating or cooling under controlled conditions

hot working   plastic deformation at a temperature high enough that strain hardening does not occur

mechanical properties   properties related to the behavior of material when subjected to force

mill finish   the finish on material as produced by the mill without any additional treatment

modulus of elasticity   a measure of the resistance to deflection of a material prior to yielding

natural aging   precipitation from solid solution at room temperature, slowly producing a change in mechanical properties

plate   a rolled product with a rectangular cross section and thickness of at least 0.25 in. (over 6.3 mm) with sheared or sawed edges

profile   a wrought product much longer than its width other than sheet, plate, rod, bar, tube, or wire, synonymous with *shape*

quenching   controlled rapid cooling of a metal from an elevated temperature by contact with a liquid, gas, or solid

reaming   fabricating a hole to final size by enlarging a smaller hole

rod   a solid wrought product circular in cross section with a diameter not less than 0.375 in. (over 10 mm)

roll forming   a fabrication method of forming parts with a constant cross section and longitudinal bends using a series of cylindrical dies in male-female pairs to progressively form a sheet or plate to a final shape in a continuous operation

sheet   a rolled product with a rectangular cross section and a thickness less than 0.25 in. and greater than 0.006 in. (over 0.15 through 6.3 mm) with slit, sheared, or sawed edges

soldering   joining metals by fusion with filler metals with a melting point below 840°F (450°C)

solution heat treating   heating an alloy for a sufficient time to allow soluble constituents to enter into solid solution where they are retained in a supersaturated state after quenching

spinning   a fabrication method of shaping material into a piece with an axis of revolution in a spinning lathe with a mandrel

stabilizing   a low temperature thermal treatment designed to prevent age-softening in certain strain hardened alloys containing magnesium

strain   the deformation of a member under load, referred to its original dimensions

strain hardening   the increase in strength and the loss of ductility that results from cold working

stress corrosion cracking (SCC)   localized directional cracking caused by a combination of tensile stress and corrosive environment

temper   a condition produced by mechanical or thermal treatment

thermal expansion, coefficient of   the measure of the change in strain in a material caused by a change in temperature

tolerance   an allowable deviation from a nominal or specified dimension or property

tube   a hollow wrought product that is longer than its width, that is symmetrical and round, hexagonal, octagonal, elliptical, square, or rectangular, and that has uniform wall thickness

water stains   stains caused by water that is drawn between adjacent pieces of aluminum that are tightly packed together, such as stacks of sheet or plate or rolled coils, and varying in color from white to gray to nearly black

wire   a solid wrought product whose diameter or greatest perpendicular distance between parallel faces is less than 0.375 in. (up through 10 mm), with a round, square, rectangular, or regular octagonal or hexagonal cross section

wrought products   products that result from mechanically working by a process such as rolling, extruding, forging, drawing, etc.

yield strength   the stress at and above which loading causes permanent deformation

References

1. Aluminum Association, *Aluminum Brazing Handbook*, Washington, DC, 1990.
2. Aluminum Association, *Aluminum Design Manual*, Washington, DC, 2000.
3. Aluminum Association, *Aluminum Forging Design Manual*, Washington, DC, 1995.
4. Aluminum Association, *Aluminum Soldering Handbook*, Washington, DC, 1996.
5. Aluminum Association, *Aluminum Standards and Data 2000*, Washington, DC, 2000.
6. Aluminum Association, *Aluminum Standards and Data 1998 Metric SI*, Washington, DC, 1998.
7. Aluminum Association, *Designation System for Aluminum Finishes*, Washington, DC, 1997.
8. Aluminum Association, *Forming and Machining Aluminum*, Washington, DC, 1988.
9. Aluminum Association, *Guidelines for Minimizing Water Staining of Aluminum*, Washington, DC, 1990.
10. Aluminum Association, *Guidelines for the Use of Aluminum With Food and Chemicals*, Washington, DC, 1994.
11. Aluminum Association, *Standards for Aluminum Sand and Permanent Mold Castings*, Washington, DC, 1992.
12. Aluminum Association, *Welding Aluminum: Theory and Practice*, Washington, DC, 1997.
13. Aluminum Association and Aluminum Extrusion Council, *The Aluminum Extrusion Manual*, Washington, DC, 1998.
14. Aluminum Association, Secretariat, *ANSI H35.1 Alloy and Temper Designation Systems for Aluminum*, 1997.
15. Aluminum Association, Secretariat, *ANSI H35.2 Dimensional Tolerances for Aluminum Mill Products*, 1997.
16. American Society for Testing and Materials, Vol. 02.02 *Aluminum and Magnesium Alloys*, Conshohocken, PA, 1999.
17. American Welding Society, *A5.3 Specification for Aluminum and Aluminum Alloy Electrodes for Shielded Metal Arc Welding*, Miami, FL, 1999.
18. American Welding Society, *A5.10 Specification for Bare Aluminum and Aluminum Alloy Welding Electrodes and Rods*, Miami, FL, 1992.
19. American Welding Society, *C3.7 Specification for Aluminum Brazing*, Miami, FL, 1993.

20. American Welding Society, *D1.2 Structural Welding Code—Aluminum*, Miami, FL, 1997.
21. ASM International, *Aluminum and Aluminum Alloys*, Materials Park, OH, 1993.
22. Department of Defense, MIL-HDBK-5 *Military Handbook Metallic Materials and Elements for Aerospace Vehicle Structures*, 1994.
23. Godard, H. P., et al, *The Corrosion of Light Metals*, John Wiley & Sons, New York, 1967.
24. Kissell, J. R. and Ferry, R. L., *Aluminum Structures*, John Wiley & Sons, New York, 1995.
25. Non-Ferrous Founders Society, *The NFFS Guide to Aluminum Casting Design: Sand and Permanent Mold*, Des Plaines, IL, 1994.
26. Sharp, M. L., *Behavior and Design of Aluminum Structures*, McGraw-Hill, New York, 1993.
27. Sharp, M. L., Nordmark, G. E., and Menzemer, C. C., *Fatigue Design of Aluminum Components and Structures*, McGraw-Hill, New York, 1996.

3

Titanium

Steven Yue
Dept. of Metallurgical Engineering
McGill University
Montreal, Quebec

Simon Durham
Materials Engineering
Pratt & Whitney Canada
Longueuil, Quebec

3.1 Introduction

Titanium was discovered as the mineral rutile (TiO_2) at the end of the eighteenth century by Gregor and Klaproth. It is the fourth most abundant metal on the planet, in comparison to, for example, iron, which is the second most abundant metal. There is a large mineral supply, with most of the oxide going into pigmentation applications. Despite the relative profusion of titanium oxide, however, commercial production of the metal only commenced in about 1950. There are many reasons why the commercialization of titanium began so late. One reason is the perceived difficulty in separating the metal from oxygen. It is certainly true that the Kroll extraction process is somewhat daunting compared to that of other "commodity" metals. The Kroll process begins by converting the titanium dioxide to the tetrachloride, commonly called *tickle,* by chlorination in the presence of carbon. Magnesium metal is then added to reduce the chloride in a closed vessel under an inert atmosphere, producing pure sponge Ti, which is consolidated, refined, and alloyed with other elements by vacuum arc melting (VAR). The magnesium chloride is recycled electrolytically. Obviously, this is a technologically complex procedure, which needed the development of electricity, vacuum systems, and magnesium extraction before titanium could be isolated. This is unlike the history of iron, which required only heat and carbon to produce the element.

Since this rather slow beginning, the production growth of Ti has also proceeded at a very slow rate, partly because of the cost of extracting the material. Currently, Ti extraction is relatively energy intensive, requiring 16 times as much energy per ton as steel, but only twice as much as Al. In fact, the heats of formation of the oxides of these three metals are quite similar, in the region of 250 kcal/mol. Therefore, the energy issue stems, at least partly, from the fact that Ti is still regarded as an exotic material rather than a commodity. If a high-tonnage use is ever found for Ti, then economies of scale will come into play, probably immediately bringing the energy requirement for Ti production in line with that of Al production. There have been many attempts at developing a controlled nucleation Kroll process and electrolytic processes, but none to date has been a commercial success.

A recent exciting development[1] has been a novel approach to the electrolytic process that can reduce titanium sponge directly from TiO_2 but, as of this writing, the process has not been scaled up to the point where its advantages over the Kroll process can be determined. If the promise holds out it, may be possible to dramatically reduce the cost of producing titanium sponge by this route, but it is unlikely that this will have the dramatic effect on the cost of producing titanium components, since raw material costs make up a relatively small percentage of the total. The forming of titanium alloys is significantly more complex than for steel or aluminum, which maintains the cost premium.

The conventional routes of forming involve one or more of the following processes:

- casting
- cold forming
- hot forming
- machining
- welding

Any elevated-temperature process has to take into consideration the affinity ot titanium for oxygen, which leads to a deleteriously brittle surface layer, or *alpha case,* which must be removed to preserve the mechanical properties. Titanium alloys are, as a general rule, more difficult to cold form than Al or iron, partly because of the high strength but also because of the unpredictable *springback.* Machining can be a problem, again, because of the reactivity of Ti leading to high tool wear rates. Casting is also difficult, because the molten metal reacts aggressively with most traditional mold-making materials. Having noted all of these problems, it would take only one high-tonnage application of titanium for many of these issues to be decisively (i.e., economically) dealt with, as has happened with all other high-tonnage materials.

It is almost impossible to anticipate a high-tonnage use for titanium, but it has a low density and has superior corrosion resistance to aluminum, and it is also strong enough to replace steel. Perhaps the sheer breadth of the possibilities of titanium has made it difficult to focus on a high-tonnage application. The combination of high strength and low density at moderately high temperatures is, of course, highly desirable in aerospace applications where every pound of weight saved is a significant benefit. However, it is quite possible that the close connection of titanium with the aerospace industry may have hindered the development of titanium as a high-tonnage material. Nevertheless, apart from its high-profile applications in aerospace (in frames, skins, and engine components), titanium has made some inroads in other areas. The uses mainly capitalize on the corrosion resistance and good elevated temperature mechanical properties, a combination that positions titanium uniquely above steel and Al. These uses include steam turbine blades, condenser tubing for power generators (nuclear and fossil fuels), pulp and paper industry applications, and food preparation and chemical engineering applications in general. Corrosion resistance is again to the forefront in applications such as biomaterials, marine submersibles, desalination, and waste treatment plants. Finally, titanium alloys also display more exotic properties, such as interstitial hydride formation, that enable them to be considered for hydrogen storage media, and other properties that are applicable to superconductors and applications involving the shape memory effect. As can be appreciated from this breadth of applications, titanium could become the steel of the twenty-first century.

Having said this, however, the most popular alloys remain the ones that have been developed for the aerospace industry, namely commercial purity, Ti-6Al-4V (which accounts for at least 50% of the usage of titanium), Ti-5Al 2.5 Sn, Ti-8Al 1Mo 1V, Ti-6Al 6V 2Sn, and Ti-13V 11Cr 3Al (see Table 3.1).

TABLE 3.1   Examples of Compositions that Fall into Various TI Alloy Classifications

Alpha	Near alpha	Alpha/beta	Near beta	Beta
Unalloyed Ti	Ti-5Al-6Sn-2Zr-1Mo-.2Si	Ti-6Al-4V	Ti-6Al-2Sn-4Zr-6Mo	Ti-8Mo-8V
Ti-5Al-2.5Sn	Ti-6Al-2Sn-4Zr-2Mo	Ti-6Al-6V-2Sn		Ti-11.5Mo-6Zr-4.5Sn
	Ti-8Al-1Mo-1V			

3.2 Basic Metallurgy

3.2.1 Phases

3.2.1.1 Equilibrium phases.   Pure Ti is HCP α up to 882.5°C, at which point it transforms to β (BCC). Therefore, alloying elements are classified according to how they influence the stability of these two phases:

- α *stabilizers* either increase the β transus temperature (i.e., stabilize the α to higher temperatures) or impart no change to the phase stability. Such alloying elements are aluminum, oxygen, and nitrogen.
- β *stabilizers* are generally transition and noble metals that are similar to titanium with regard to electron shells (i.e., they have unfilled or just filled d-shell), such as molybdenum, iron, vanadium, chromium and manganese.

Why elements are beta and alpha stabilizers is not exactly known, but it has been correlated to size (larger atoms tend to stabilize beta) and the electronic nature of atoms. Depending on the alloying additions, therefore, three types of titanium alloy emerge:

- α or near α
- $\alpha + \beta$
- β or near beta

In general, however, any β alloys tend to be metastable β alloys, since stable β would require considerable concentrations of alloying elements, which would significantly reduce the weight advantage of titanium. Other important phases in Ti alloys are precipitates such as aluminides, chromides, and silicides will be discussed in later sections.

3.2.1.2 Nonequilibrium phases.   Quenching of β to below the martensite start (M_s) temperature leads to HCP martensite (α') in pure and dilute beta stabilized alloys. With increasing beta stabilizer (e.g., 11 V, >4Mo, 14 Nb), quenching from the β region gives orthorhombic martensite (α''). Martensite can also form by plastically deforming β that is metastable at room temperature (i.e., strain induced martensite). Since retaining metastable β to room temperature would require considerable alloying additions, strain-induced martensite is invariably orthorhombic.

The effect of β stabilizer concentration on the β transus and the martensite start temperature is shown schematically in Fig. 3.1. The martensite start temperature (M_s) decreases with increasing solute,

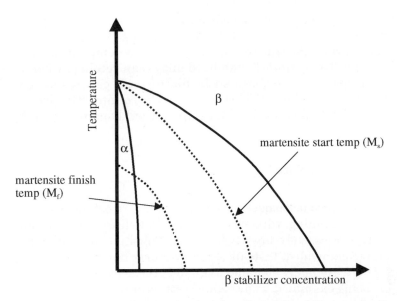

Figure 3.1 Schematic diagram showing the effect of β stabilizer on the β transus temperature and the martensite start and finish temperatures.

and increasing e/a ratio (the electron/atom ratio, which is a function of the periodic table group number; for example, per atm%, vanadium, a Gp V element, is less effective than Mn, a Gp VII element, in depressing the M_s). The M_s decreases with increasing alloy content and can also decrease with increasing quench rate although, above a critical quench rate, the M_s is constant.

As mentioned above, martensite can form by the strain-induced transformation of beta-stabilized alloys if deformation is close to the M_s temperature. If the temperature of deformation is significantly above the M_s, twinning can occur. The resulting twinned structure can be regarded as a *variant* of martensite and, under the optical microscope, it has many morphological similarities to martensite (although, of course, crystallographically, it is BCC).

Note that high M_s temperatures can lead to retained beta due to segregation of beta stabilizers by diffusion during the martensite transformation. In other words, if the martensite transformation were quick enough, there would be no retained β; it would all transform to martensite. However, martensite actually undergoes a nucleation and growth transformation. As the martensite forms, there is a driving force for the β alloying elements to segregate to the remaining β. If the M_s temperature is low, diffusion rates are low, β elements do not move, and martensite will engulf the structure. If the M_s is high, as the martensite is forming, there is time for some β to segregate to the remaining β. Thus, as the martensite formation progresses, the remaining β becomes richer in β stabilizer until the remaining β has reached a level of β stabilizer that will retain the β to room temperature in a metastable form. Thus, *dilute* β stabilized alloys can retain β, whereas an increase in β stabilizer above such dilute levels can reduce and even eliminate the retained β. This is somewhat counterintuitive, although adding still higher levels of β stabilizer would again eventually lead to increases in retained β.

3.2.2   Metallography and Microstructures

3.2.2.1   Metallographic techniques. Metallographic techniques are fairly routine, beginning with conventional Bakelite mounting and grinding with successively finer SiC media. Polishing with diamond paste is not recommended. Instead, alumina slurry down to 0.05 μm is preferred. There are many etchants available, the selection depending on the alloy being examined, and the microstructural features that are of interest.[2] The general all-purpose etch is Kroll's reagent (1–3 mL HF, 2–6 mL HNO_3, water added to 1 L), which usually takes only a matter of seconds at room temperature to reveal the microstructure.

Most of the common Ti alloys and microstructures are very amenable to thin foil preparation for TEM observation by jet polishing 3 mm discs. A good electrolyte to use is 1–6% (by volume) perchloric acid, about 35–37% n-butyl alcohol (to increase the viscosity of the solution so as to obtain uniform coverage of the electrolyte on the disc), and the balance being methanol. Temperatures not greater than 30°C should be employed to minimize the formation of hydrides during polishing.

3.2.2.2   Microstructures. For α or near α alloys, there are two extreme morphologies of α: equiaxed and lamellar. When the alloy is heated into the single-phase β region, equiaxed β forms. As the β cools through the transus, α begins to form at the β boundaries, quickly forming a continuous film of α, the thickness of which, for a given alloy, depends largely on the cooling rate. As the β transformation proceeds, a lamellar α morphology is produced. It occurs in packets or colonies of aligned α laths, the size and shape of the laths depending on the cooling rate during the transformation. The faster the cooling rate, the finer the structure, with a tendency for the plates to adopt more of a *basket weave* configuration as opposed to aligned colonies. The equiaxed structure is formed by thermomechanical processing (hot rolling below the β transus or cold rolling plus *annealing* below the β transus) by a process of recrystallizing the α plates that have formed during cooling from the β region.

For α + β alloys, the microstructure can also exist in the two extremes noted above and, additionally, there will be a small volume fraction of stable and/or metastable β present. In the equiaxed morphology (often referred to as *mill annealed* structure) the β will exist as very small β particles at α triple points (Fig. 3.2). In the 100% lamellar structure (the so-called "β annealed" condition), the β exists as interlamellar layers. Much more frequently, α + β alloys are processed to exist as a mixture of these two morphologies, with the lamellar regions being commonly referred to as "transformed β" since these mixed structures are usually formed by heat treating the mill-annealed ("100%" equiaxed morphology) in the two-phase α + β region to transform the desired volume fraction of α to equiaxed β. On cooling to room temperature, the β will transform to the lamellar α morphology, as noted above (Fig. 3.3).

Above a critical cooling rate, β transforms to one of the martensites described previously (Sec. 3.2.1.2). Hexagonal martensite is more frequently observed, primarily because many alloys are on the lean side with respect to β alloying additions. It has been observed to exist in at least two distinctive morphologies. For high M_s temperatures, i.e., in alloys that are very lean in β stabilizer, or with slower cooling rates, the β phase transforms to grain boundary and Widmanstatten martensite. The latter exists as large colonies of α plates, the so-called *aligned structure* (sometimes known as *massive martensite*). These are colonies of parallel sided platelets, each platelet in the colony having the same variant of Burgers orientation to the β matrix, and the plate-

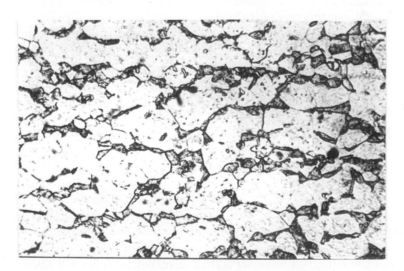

Figure 3.2 Equiaxed α (matrix) plus dark etching equiaxed β grains (Ti-6-4).

Figure 3.3 Equiaxed α plus lamellar transformed β grains (Ti-6-4, 960°C, air cool).

lets within each colony are separated by low angle boundaries. Such colonies are tens of micrometers in dimension, with platelets of the order of 1 μm thick or less.

With decreasing M_s temperature (i.e., increasing β solute levels and/or increasing cooling rates), martensite platelets form individually, each with a different variant of the Burgers orientation relationship. The internal structure changes from dislocations to twins. FCC or FCT martensites have been observed,[3] but these are believed to be an artifact due to spontaneous transformation on the release of bulk constraint by chemical thinning. In all cases, β is retained as thin layers surrounding the α plates; α and near α alloys tend to have discontinuous β layers of about 0.1 μm thick.

3.2.3 Thermomechanical Processing

As mentioned above, thermomechanical processing can lead to an equiaxed structure or a mixture of the equiaxed and lamellar morphologies. As well, although deformation is not required to obtain the lamellar microstructure, thermomechanical processing variables can be used to control or eliminate grain boundary α in coarse lamellar structures. A strain of 0.7 or greater applied in the β region has been found to eliminate grain boundary α and to refine the lamellar structure in Ti-6-4.[4]

To obtain the equiaxed microstructure in α, near α, or α + β alloys, the alloy must be sufficiently worked to break up the lamellar structure. The structure is then annealed to allow for recrystallization of the deformed lamellar structure into equiaxed grains, and for any β to penetrate along the newly created (i.e., nonlamellar) α grain boundaries, in this way transforming the interlamellar β to the particles of β that are found in mill annealed α + β alloys.[5] The equiaxed structure characteristics are very sensitive to temperature of deformation, extent and rate of deformation, mode of deformation, prior microstructure, annealing time, and temperature. Improper choice of these variables can lead to a *spaghetti-like* microstructure with a morphology somewhere between lamellar and equiaxed.

A mixture of equiaxed and lamellar morphologies can be produced by heating and holding the sufficiently deformed titanium alloy into the α + β region, thus creating α and β grains. The β grains transform into the lamellar structure (transformed β) on cooling. Obviously, the volume fractions of transformed β increases with increasing hold temperature. This type of structure is referred to as *bimodal* or *duplex*.

Much work has been done on Ti-6-4 with regard to defining critical thermomechanical processing parameters. The most important parameter is that of the critical amount of deformation required to obtain recrystallized grains. As noted above, this depends on the prior microstructure. For example, starting with a β annealed microstructure, which would essentially be a coarse Widmanstatten morphology, the critical strain was found to be 0.7 if hot deformation is performed in the α + β range (900–1000°C).[4] On the other hand, starting with a martensitic structure, Peters and Lutjering[6] found the critical strains shown in Table 3.2. From these data, it is clear that the finer the initial microstructure and the lower the deformation temperature, the more efficiently the deformation is used to recrystallize the initial α microstructure. The effect of the finer microstructure can possibly be attributed to more nucleation sites for recrystallization leading to a higher rate of recrystallization. The lower the temperature, the lower the amount of recovery during deformation, increasing the driving force for recrystallization in this way.

TABLE 3.2   Effect of Deformation
Temperature on Critical Strains Required
to Recrystallize Martensite (from Ref. 10)

Temperature, °C	Critical strain
Room	0.15
800	0.4–0.5
950	0.5–0.6
1050	1.2

It should be noted that deformation in the β region will never lead to an equiaxed α structure. In addition, it has been found that very coarse lamellar structures cannot be recrystallized to the equiaxed structure. This appears to be due to the α and β phases recrystallizing within the prior lamellae, thus preserving the lamellar morphology.

3.2.4   Tempering of Martensites and Aging

3.2.4.1   α phase aging and tempering of martensites.   Increasing   levels of α stabilizers can lead to formation of coherent intermetallic precipitates on aging the supersaturated α phase. In alloys stabilized with aluminum or tin, the precipitate is usually Ti_3(Al or Sn) and is referred to as $α_2$. Although such precipitates are strengtheners, they are to be avoided because of embrittlement issues. Hammond and Nutting[7] have suggested that the $α_2$ occurs as coherent ellipsoids when the aluminum equivalent (Al_{eq}) concentration is approximately 5 wt% or greater. An Al_{eq} >9% results in a detrimental volume fraction of $α_2$. The aluminum equivalent equation that they use is:

$$Al_{eq} = [Al] + \frac{[Zr]}{6} + \frac{[Sn]}{3} + 10[O + C + 2N] \qquad (3.1)$$

Copper and silicon both have high solid solubilities in α, and therefore aging the quenched structures will lead to intermetallic precipitation. Aging ternary Ti-Si alloys gives Ti_5Si_3, which precipitates as rods. For alloys that contain Cu, Ti_2Cu is the intermetallic that can form.

Aging of both martensites and retained supersaturated α phase occurs by similar mechanisms. Hexagonal martensites temper very quickly to HCP α and then, subsequently, by precipitation of β on the substructure and along α plate boundaries. Where the precipitate is not associated with a boundary, it grows as elongated rods.[3] Orthorhombic martensites can be tempered in two ways. For martensites formed at high M_s temperatures (i.e., near α alloys), aging leads to the precipitation of fine α particles, followed by initial coarsening of these particles and nucleation of α + β lamellar regions at prior β boundaries, which eventually engulf the microstructure.[8] Low M_s alloys age by precipitating β initially on the substructure. The resulting reduction in β stabilizer in the orthorhombic structure (due to partitioning to the nucleated β) converts the orthorhombic structure to hexagonal martensite,[9] which presumably then tempers in the manner described previously.

3.2.4.2 β phase aging. Table 3.3 gives the minimum wt% of various β stabilizers required to retain β on quenching. Aging metastable β alloys that are slightly more solute rich in the range of 300 to 400°C leads to the formation of the ω phase. The ω phase, which is hexagonal, precipitates as particles with a mean size and interparticle spacing of 40–120 Å. A low lattice mismatch results in ellipsoidal ω in Mo and Nb stabilized alloys, for example; a high lattice mismatch (e.g., V) results in a cuboidal ω morphology. Formation of ω is closely associated with the formation of β′, a BCC phase that occurs either when ω formation is sluggish, or when ω cannot form because of α stabilizers and α formation itself is sluggish.

TABLE 3.3   Minimum Concentration of Various β Stabilizers Required to Retain β on Quenching

Element	V	Cr	Mn	Fe	Co	Ni	Cu	Mo
Min. wt% concentration	14.9	6.3	6.4	3.5	7.0	9.0	13.0	10.0

Further aging of β (higher temperatures or longer times) leads to the nucleation and growth of α. prolonged aging nucleates the α at ω/β interfaces in alloys of high lattice misfit. The α then grows to consume the ω phase and can subsequently link up to form a needle-like morphology. Alloys with low lattice misfits will nucleate α at β boundaries, independent of the ω phase. Aging at temperatures above the stability of ω results in α precipitating as Widmanstatten plates, where the alloying levels are low, or rafts of α in highly β stabilized alloys containing little or no α stabilizer. Further aging produces growth and α agglomeration.

It should be noted that the presence of any of the precipitates mentioned above (α_2, ω, and β′) is generally undesirable because of the tendency to embrittlement. Thus, heat treatments are generally designed to at least prevent these precipitates from forming. As noted above, the α and near α and α + β alloys have α stabilizer levels that have a low likelihood of precipitating out intermetallics, although there have been some indications that Ti_3Al may be forming after tempering a fully martensitic structure. With regard to β alloys, some aging treatments are designed to preempt ω formation during elevated temperature service. Thus, for example, for the Ti-11.5Mo-6Zr-4.5Sn alloy, since aging below 400°C leads to ω formation, the structure is stabilized by heat treating at 550–600 (8 hr) to form 40% α and to segregate the Mo to stabilize the β rather than it transforming to ω.

3.3   Alloy Classification and Overview

As already stated, titanium alloys fall under three broad types: α or near α, α+ β, and β. Although the type of alloy can be defined by its microstructure (e.g., the duplex or bimodal structure is often regarded as being characteristic of an α + β alloy), most of the existing alloys can be classified by their composition. The aluminum equivalent, which is measure of the level of α stabilization, has already been defined in Eq. (3.1). The corresponding value for β stabilization is the Mo equivalent, which is defined by the following equation:

$$Mo_{eq} = [Mo] + \frac{[Ta]}{5} + \frac{[Nb]}{3.6} + \frac{[W]}{2.5} + \frac{[V]}{1.5} + 1.25[Cr] + 1.7[Mn] + 1.7[Co] + 2.5[Fe] \quad (3.2)$$

Applying these equations to the alloys listed in Table 3.4, the compositional ranges of the three alloy classifications can be approximately defined in terms of the Al_{eq}/Mo_{eq} ratio as follows:

α/near α, $Al_{eq}/Mo_{eq} > 3$

α/β alloys Al_{eq}/Mo_{eq} between 3 and 1.5

β alloys $Al_{eq}/Mo_{eq} < 1.5$

3.3.1   Alpha Alloys

These are unalloyed (pure or commercially pure) and α stabilized with non transition metals, the most popular being Al (which, of course, also has a low density), Sn, and, to a lesser extent, Ga. These elements also confer significant solid solution strengthening at room temperatures. The extent to which α stabilizers can be added is determined by the solubility of the intermetallics that can form. As noted above, the formation of such precipitates greatly increases the strength but drastically embrittles the structure. Note that all of the alloys listed in Table 3.4 (regardless of the classification) have Al_{eq} values < 9 wt%, as per the guideline of Hammond and Nutting[7] for avoiding embrittling levels of intermetallic precipitates.

It can also be seen that most α or near α alloys contain Al (about 5% on average) as the alloying element, with Sn and Zr being about equal, although α alloys may have as much as 11% Sn. Small amounts of Si (for creep strength) are not uncommon, and microalloying additions (<001%) of B are also seen in couple of alloys.

α alloys have acceptable strength, toughness, and creep properties and are weldable. Hence, they can be used in any application that is not mechanically arduous but requires corrosion resistance, preferably in an oxidizing atmosphere. In particular, because these alloys do not exhibit a ductile-to-brittle transition temperature, they are the most useful titanium alloy for cryogenic applications.

TABLE 3.4 Various Ti Alloys with Corresponding Values of Al_{eq} and Mo_{eq}, and the Ratio of These Two Values (Alloy Types as Classified in Ref. 2)

Desig.	Alloy type	Al	Zr	Sn	Si	B	Al_{eq}	Mo	Ta	Nb	V	Cr	Ni	Mn	Co	Fe	Re	Mo_{eq}	Al_{eq}/Mo_{eq}
	α/near α						0.00	0.3					0.8			0.1		1.30	0.00
48-OT3		4			0.1	0.005	4.00											0.25	16.00
IMI 685		6	5		0.3		6.83	0.5										0.50	13.67
IMI 834		5.5	4	4	0.5		7.50	0.3		1								0.58	12.98
IMI 829		5.5	3	3.5	0.3		7.17	0.3		1								0.58	12.40
VT 5.1	α	5		2.5			5.83											0.00	10.00
VT 5		4.5					4.50											0.00	10.00
		5		2.5			5.83											0.00	10.00
IMI 679	α/near α	2.25	5	11	0.25		6.75	1										1.00	6.75
AT 8		2.25	5	11	0.3	0.01	6.75	1										1.00	6.75
	near α	7					7.00	1				0.6				0.2		1.25	5.60
811		8					8.00	1			1							1.67	4.80
	α/near α	6					6.00	0.8	1	2								1.56	3.86
6242	near α	6	4	2			7.33	2										2.00	3.67
IRM 2		4					4.00			4								1.11	3.60
IRM 1		4					4.00			4							0.1	1.11	3.60
	α/near α	5	2	5			7.00	2										2.00	3.50

TABLE 3.4 Various Ti Alloys with Corresponding Values of Al_{eq} and Mo_{eq}, and the Ratio of These Two Values (Alloy Types as Classified in Ref. 2) (Continued)

Desig.	Alloy type	Al	Zr	Sn	Si	B	Al_{eq}	Mo	Ta	Nb	V	Cr	Ni	Mn	Co	Fe	Re	Mo_{eq}	Al_{eq}/Mo_{eq}
AT 6		6			0.3	0.01	6.00					0.6				0.4		1.75	3.43
AT 4		4			0.4	0.01	4.00					0.6				0.23		1.33	3.02
IMI 318	α + β	6					6.00				4							2.67	2.25
VT 8		6					6.00	3										3.00	2.00
VT 4		4.6					4.60							1.5				2.55	1.80
	α + β	3					3.00				2.5							1.67	1.80
	α + β	7					7.00	4										4.00	1.75
	α + β	6		2			6.67				6							4.00	1.67
AT 3		3		2	0.3	0.01	3.00					0.7				0.4		1.88	1.60
	α + β	6	2	2			7.00	2				2						4.50	1.56
IMI 680		2.25		11	0.25		5.92	4										4.00	1.48
VT 3		4.6					4.60					2.5						3.13	1.47
	near β	6	4	2			7.33	6										6.00	1.22
OT 4		3					3.00							1.5				2.55	1.18
IMI 550		4		2	0.5		4.67	4										4.00	1.17
IRM 3		4					4.00	3.5										3.50	1.14
VT 14		4					4.00	3			1							3.67	1.09

Desig.	Alloy type	Al	Zr	Sn	Si	B	Al_{eq}	Mo	Ta	Nb	V	Cr	Ni	Mn	Co	Fe	Re	Mo_{eq}	Al_{eq}/Mo_{eq}
IRM 4		3.5					3.50	3.5									0.1	3.50	1.00
VT 3.1		4.6					4.60	1.7				2				0.5		5.45	0.84
OT 4.1		1.7	2	2			1.70							1.4				2.38	0.71
	near β	5	2				6.00	4				4						9.00	0.67
	α + β	4.5					4.50	5				1.5						6.88	0.65
VT 2		1.6					1.60					2.5						3.13	0.51
Ti 15-3	β	3		3			4.00				15	3						13.75	0.29
VT 16		2					2.00	7										7.00	0.29
	near β	3					3.00				10					2		11.67	0.26
beta-C	β	3	4				3.67	4			8	6						16.83	0.22
	β		6	4.5			2.50	11.5										11.50	0.22
	β	3		3			4.00				15	3				2		18.75	0.21
	β	3	4				3.67	4			8	8						19.33	0.19
	β	3					3.00	8			8					2		18.33	0.16
VT 15	β	3					3.00	6.5				11						20.25	0.15
	β	3					3.00				13	11						22.42	0.13
	β						0.00							8				13.60	0.00

3.3.2   Beta Alloys

β stabilizing elements come from Gp. V and VI of the periodic table. The most common are Nb (Gp. V) and, as noted earlier, Mo (Gp. VI). The structure of β alloys at room temperature is usually FCC (stable or metastable) at room temperature. These alloys are highly formable and retain strength levels to high temperatures. Hence, β alloys tend to have been developed for elevated temperature applications. These alloys exhibit a ductile-to-brittle transition and hence are usually not considered for cryogenic purposes.

3.3.3   Alpha-Beta Alloys

Although α/β alloys contain α and β stabilizers, the equilibrium phase is usually mainly α, even though the volume fraction of β does vary between 5 and 50%. The simplest and most popular of these alloys is Ti-6Al-4V. These are processed to produce a range of structures encompassing a fully equiaxed structure or a fully lamellar structure and any mixtures of these two phases to form duplex or bimodal structures. They can also be processed to produce tempered martensites. Hence, this class of alloys has a superficial resemblance to plain carbon steels, but they do not possess as large a range of properties as the corresponding steel microstructures. Nevertheless, the simplicity of the alloy composition 6-4, coupled with the considerable range of microstructures and properties, means that it has been used in many different applications. α + β alloys generally are reasonably formable (with the exception being 6-4) and have good elevated temperature properties. It should be noted that alloys with >20% β have very poor weldability.

3.4   Properties

The properties of titanium alloys are quite sensitive to the alloy composition and microstructure, hence the resulting range in property values can be quite considerable, as can be seen from the following summary of properties, adapted from data found in the Cambridge Materials Selector software.

General

Atomic volume (average)	0.01–0.011 m³/kmol
Density	4.38–4.82 Mg/m³
Energy content	750–1250 MJ/kg

Mechanical

Bulk Modulus	102–112 GPa

Compressive strength	130–1400 MPa
Ductility	0.02–0.3 (strain)
Elastic limit	172–1050 MPa
Endurance limit	*176–623 MPa
Fracture toughness	55–123 MPa × √m
Hardness	1030–4700 MPa
Loss coefficient	8.00×10^{-5}–0.002
Modulus of rupture	250–1300 MPa
Poisson's ratio	0.358–0.364
Shear modulus	*35–50 GPa
Tensile strength	241–1280 MPa
Young's modulus	95–125 GPa

Thermal

Latent heat of fusion	360–370 kJ/kg
Max. service temperature	600–970 K
Melting point	1770–1940 K
Min. service temperature	1–2 K
Specific heat	510–650 J/kg × K
Thermal conductivity	4–21.9 W/m × K

3.4.1 Physical Properties

The thermal, magnetic, and electric properties are usually not the primary reason for selection of titanium for an application, although presently there is interest in the superconducting properties of titanium alloys. Hence, these properties are not covered in this chapter. However, such properties are often of secondary consideration and are also useful in indicating the condition of the microstructure. Thus, there is a considerable body of knowledge related to physical properties of titanium alloys. With regard to technical alloys, the reader is directed to the work of Salmon, who has compiled a selection of these properties in his work at NPL.[10]

3.4.2 Mechanical Properties

3.4.2.1 Elastic properties. As a general rule, the elastic modulus increases with increasing solute concentration and second phase, particularly ω, but any intermetallic or nonintermetallic precipitate will also increase the modulus.

3.4.2.2   Plastic properties

Solid solution strengthening of the α phase. α alloys rely significantly on the solid solution strengthening effect of α stabilizers, since the contribution to strength of these elements is high and even compares favorably with precipitation hardening mechanisms. Examples of the magnitude of α stabilizer solid solution strengthening are shown in Table 3.5. The strengthening mechanism is a combination of atomic size misfit (which impedes dislocation motion) and *chemical interaction,* i.e., strong, local atomic bonding (which increases the resistance to bond stretching and breaking).

TABLE 3.5   Solid Solution Strengthening of α Due to α Stabilizers According to the Effect on Vickers Hardness, Hv = a + b (concentration of α stabilizer) (from Ref. 11)

Alloy addn.	Concentration range (at%)	Condition	a (kg/mm^2)	b kg mm^{-2} × at%$^{-1}$	Hardening rate at 1 atm % (kg mm^{-2} × at%$^{-1}$)
Al	0–10	100 hr at 850°C ice brine quench	102	15	15
Ga	0–5	As cast	108	24	24
Sn	0–7	As cast	112	24	24

α can also be strengthened by interstitial atoms in solid solution by the same mechanisms. As can be seen in Table 3.6, the effect of interstitials is much stronger than that of the substitutionals, probably because of increased misfit differences, as well as stronger atomic bonds formed between the interstitials and Ti. In fact, it has been observed

TABLE 3.6   Solid Solution Strengthening of a Due to Interstitials According to the Effect on Vickers Hardness, Hv = a + b (concentration of a stabilizer)1/2 (from Ref. 11)

Alloy addn.	Concentration range (at%)	Condition	a (kg/mm^2)	b kg mm^{-2} × at%$^{-1}$	Hardening rate at 1 atm % (kg mm^{-2} × at%$^{-1}$)
B	0–0.2	120 h/800 C/IBQ	110	15	344
C	0–0.5	120 h/800 C/IBQ	104	24	269
N	0–7	120 h/800 C/IBQ	98	24	378
O	0–7	120 h/800 C/IBQ	100	24	307

that, in very dilute Ti-Al alloys, the strength of the alloy decreases with increasing Al, probably because of a scavenging effect that Al has on oxygen. Titanium has a stronger affinity for C and N compared to Al, but Al has a stronger affinity for O.

As mentioned earlier, the strengthening of the α phase by these elements is restricted by precipitate formation. The latter leads to considerable strengthening, but this is accompanied by embrittlement. In the case of Al, further limitation in the use of Al as a strengthener is that the increase in critical resolved shear stress with increasing Al concentration is accompanied by a decrease in the critical stress for stress corrosion failure that accompanies it.[12]

Solid solution strengthening of α alloys is not effective at elevated temperatures, especially when interstitial strengthening is employed, because the increased diffusion rates of these elements means that dislocations are freer to move. The effect is quite drastic; for example, the UTS of Ti-6-4 drops by about 60% as the temperature is increased from room temperature to 500°C.

Solid solution strengthening of the β phase. The solid solution strengthening effect of the β phase by β stabilizers is relatively weak, as can be seen in Table 3.7, and is soon overshadowed by the formation of the ω phase. As may be expected, solid solution strengthening in the β phase becomes ineffective at elevated temperatures, and the strength must be derived from the formation of precipitates or other microstructural effects.

TABLE 3.7   Solid Solution Strengthening of β due to β Stabilizers

	V	Cr	Mn	Fe	Co	Ni	Cu	Mo
N m^{-2}/wt%	19	21	34	46	48	35	14	27
N m^{-2}/atm%	20	23	39	54	59	43	18	54

Microstructural strengthening. Precipitation strengthening due to formation of α_2 in α stabilized alloys and ω in β stabilized alloys, and the loss of ductility that accompanies the high flow stresses, have been mentioned several times. On further aging of an $\omega + \beta$ structure, a second hardness peak occurs, coinciding with the formation of α precipitates. Continued overaging of either β or α stabilized alloys that have precipitates results in precipitate coarsening, a decrease in strength, and an increase in ductility.

The effect of microstructural morphologies on strengths and ductilities is fairly complex because of the wide range of microstructural variations possible. In alloys treated in the $\alpha + \beta$ two phase region, it is clear that transformed β and primary α grain sizes, phase distribution, volume fraction, and morphology will each contribute to the strength of the structure. Evidence of an inverse relationship between yield stress (YS) and the distance between equiaxed α in a structure of matrix transformed β with second phase equiaxed α has been put forward, but the effect is not pronounced, and the evidence is inconclusive.

The onset of transformed β in an α matrix theoretically should signal a strength increase through grain refinement. However, the start of martensite formation does not coincide with any increase in hardness, and much higher volume fractions of martensite are required to register any significant hardness increase. This observation has been attributed to autotempering of the martensite during quenching, the strength increment being due to β precipitation in the martensite. Certainly, unalloyed (massive) martensite has little effect on the strength. Alloyed martensites contribute more to strength, presumably through a finer structure and a denser substructure.

Another method of strengthening martensites is the retention of β as martensite interplate layers, assuming a heavy segregation of β stabilizer to these layers, since the solid solution strengthening effect of β stabilizers is not very high. As noted just previously, tempering the martensite leads to a strength increment through β precipitation. There is some evidence that α_2 might also form when aging a martensite lean in α stabilizer. Coarse martensitic Widmanstatten (or acicular) α formed by slower cooling rates in alloyed titanium seem to have no significant effect on strength, for reasons that are probably similar to the massive martensite effect.

Given the results of the equiaxed α + martensitic α structures, it is not then surprising that the slow cooled transformed β does not necessarily lead to strength increases and, in some instances, has led to strength decreases. It seems that the interplate boundaries offer no added resistance to dislocation motion, and only refinement of the overall grain size will have a significant effect on strength.

Although strength does not appear to be strongly affected by the microstructure (other than precipitates), ductility does seem to be affected by the presence of transformed β. Margolin et al.[13] developed an empirical equation to define the effect of microstructural characteristics on the stress for tensile fracture (σ_f in MPa) of a duplex structure.

$$\sigma_f = \sigma_0 + 2.3d^{-1/2} + 2.3\lambda^{-1/2} + 3.2D^{-1/2} \tag{3.3}$$

where d is the grain size of equiaxed α, λ is the distance between equiaxed α grains, and D is the grain size of the transformed β matrix.

This equation was explained in terms of the crack sizes generated affecting the stress concentration. For example, a large distance between equiaxed α grains (λ) would lead to a large crack formed when voids, nucleated at α/transformed β boundaries, coalesced, which then would lead to the development of a high stress concentration factor. Note that there is no term for the effect of the Widmanstatten structure size. It was argued that, for a finer Widmanstatten structure, the stress required to nucleate voids would be higher but would be offset by easier crack propagation due to a decreased tortuosity, assuming crack propagation occurs along interplate boundaries.

To summarize this section on strengths, the room temperature tensile strengths of some of the common technical alloys are given in Table 3.8.

Fracture toughness. Many workers have shown that voids nucleate preferentially at α/β boundaries, and cracks tend to propagate along these boundaries. Many of the explanations for the effect of microstructure on fracture toughness (k_{Ic}) hinge on these observations.

Margolin,[15] investigating the effect of grain boundary α on the k_{Ic}, showed that increasing the GB α thickness results in an increasing k_{Ic}, reaching a maximum at a critical GB α thickness. Part of this effect is due to the fact that GB α is softer than transformed β and therefore absorbs more energy prior to void initiation. In equiaxed α/aged β structures, there appears to be no relationship between k_{Ic} and α particle size or distribution. However, increasing the β grain size results in increasing k_{Ic}, which may be interpreted as due to an increasing crack propagation path length, since fracture around β grain boundaries was observed in this structure.

Much work has been focused on the role of the acicular α morphology. It is generally accepted that toughness dramatically increases as the structure is changed from a duplex structure to a β annealed (i.e., fully lamellar) structure. However, there seems to be little change in the k_{Ic} with increasing transformed β in a duplex structure. Many workers have since attributed this superiority of the fully lamellar structure to an increase in crack tortuosity, since an increase in aspect ratio of the lamellae increases the fracture toughness. There is some uncertainty as to whether there is an effect of the absolute size of the α plates, with some workers favoring fine plates and others observing the opposite.

There has been some analysis of the possible interrelationship between yield strength and k_{Ic}, resulting in the conclusion that, for similar microstructures, the fracture toughness is inversely dependent on the yield strength, as may be expected. However, there are exceptions to this rule, mainly because yielding does not involve any fracture

TABLE 3.8 The Room Temperature Tensile Strengths of Some of the Common Technical Alloys (Data from Ref. 14)

Alloy name	Composition	Condition	UTS (Mpa)	YS (MPa)	Elongation (%)
5-2.5	5Al-2.5Sn	Annealed 0.25–4h/710–878°C	830–900	790–830	13–18
3-2.5	6Al-2Nb-1Ta-1Mo	Annealed 1–3h/654–766°C	950	900	22
6-2-1-1	3Al-2.5Sn	Annealed 1–3h/654–766°C	860	760	14
8-1-1	8Al-1Mo-1V	Annealed 8h/794°C	1000	930	12
Corona 5	4.5Al-5Mo-1.5Cr	α/β annealed after β processing	970–1100	930–1030	12–15
Ti-17	5Al-2Sn-2Zr-4Mo-4Cr	α/β annealed or β processed plus aging	1140	1070	8
Ti-6-4	6Al-4V	Annealed 2h/710–878	960	900	17
		Aged	1170	1100	12
6-6-2	6Al-6V-2Sn	Annealed 3h/710–822°C	1070	1000	14
		Aged	1280	1210	10
6-2-4-2	6Al-2Sn-4Zr-2Mo	Annealed 4h/710–850°C	1000	930	15
6-2-4-6	6Al-2Sn-4Zr-6Mo	Annealed 2h/822–878°C	1030	970	11
6-22-22	6Al-2Sn-2Zr-2Mo-2Cr-0.25Si	α/β processed plus aging	1120	1010	14
		Aged	1210	1140	8

10-2-3	10V-2Fe-3Al	Annealed 1h/766°C	970	900	9
		Aged	1240–1340	1140–1240	7
15-3-3-3	15V-3Cr-3Sn-3Al	Annealed 0.25h/794°C	790	770	20–25
		Aged	1140	1070	8
13-11-3	13V-11Cr-3Al	Annealed 0.5h/766–822°C	930–970	860	18
38-6-44	3Al-8V-6Cr-4Mo-4Zr	Aged	1210	1140	7
		Annealed 0.5h/822–934°C	830–900	780–830	10–15
		Aged	1240	1170	7
β-III	4.5Sn-6Zr-11.5Mo	Annealed 0.5h/710–878°C	690–760	650	23

mechanisms, in particular, the crack propagation path. Thus, the general conclusion is that the relationship of yield strength to fracture toughness, if any, is specific to the alloy and the family of microstructures of interest, and is heuristic at best

Finally, it has been suggested that the strain induced transformation of metastable β to martensite has been responsible for toughness increments in certain titanium alloy microstructures. For example, Hall[16] suggested that a k_{Ic} peak observed after in Ti-6-4 after heat treating at 850°C, followed by a water quench, was due to a strain induced transformation, following observations made by in situ straining in a transmission electron microscope. He also suggested that this might be occurring in the coarse, fully lamellar, air-cooled structures of this alloy, since these possess very good k_{Ic} values.

Fatigue. Fatigue failure occurs through three mechanisms: (1) void initiation (2) crack propagation, and (3) final failure. However, loading conditions can vary such that fatigue failure can occur after many cycles (i.e., high cycle fatigue, or HCF, which is taken to mean fracture after 10^8 cycles or more) or after only a few cycles (low cycle fatigue, LCF, which refers to fracture after less than 10^6 cycles). It is generally accepted that failure after a very large number of fatigue cycles (i.e., HCF) is controlled mainly by void initiation, since it generally does not take many fatigue cycles to propagate an existing crack to the reach the critical crack size (the latter being defined by the fracture toughness of the material). The outcome of this conclusion is low cycle fatigue characteristics depend largely on crack propagation mechanisms. As may be expected, the mechanisms of fatigue void initiation and crack propagation are quite distinct, even though they both involve breaking of atomic bonds.

Having said that, HCF and LCF are affected in a similar fashion by the microstructure. This is because, in general, fatigue failure begins with dislocation motion. Thus, anything that stops dislocation motion (i.e., increases the yield strength) tends to increase the HCF. Thus, finer equiaxed α grain sizes improve the fatigue properties, because these not only present dislocation barriers but also reduce the dislocation pile-up length, thus reducing the stress concentration at the head of the pile-up.

The effect of α and β morphology on HCF has been studied extensively. It has been found that the bimodal structure has the best HCF resistance, followed by fine equiaxed and fine lamellar (β quenched) structures, as seen in Fig. 3.3.[17] The optimal amount of equiaxed α is in the range of 35–50%. The size of the equiaxed α should be kept around 15–20 μm. Below this, there is no influence on the HCF. This is due to the reduced effective dislocation glide length of the lamellar part of the structure, which, in this study, was the finest of the structure investigated. It seems that the transformed β of the bimodal structure is always finer than the lamellar structure of a β annealed structure, cooled at the same rate. Thus, the finer the transformed β plates in the bimodal structure, the better the HCF properties.

In the fully lamellar microstructures, preferential crack propagation occurs along α/β interfaces. Therefore, the length of these interfaces should be minimized. Prior β grain boundaries and sub-grain boundaries in the α plates have been found to be particularly favorable for crack initiation. Thus, thinning any grain boundary α and refining the microstructure within prior β grains have been found to increase the HCF.

Referring back to the relationship between the yield strength on the HCF resistance, it is not surprising that both age hardening and higher oxygen levels increase the HCF life, although care must be taken to maintain a balance between increased fatigue properties and increased embrittlement. However, these modes of strengthening are detrimental to LCF, because the nominal stresses are considerably higher than for HCF conditions, which leads to significant intensification at the dislocation pile-ups and accelerated crack propagation.

Fatigue crack propagation.   It has been found that fatigue crack propagation (FCP) resistance is affected by microstructural variables up to a critical stress intensity factor, beyond which there is no effect of microstructure. This is when the reverse plastic zone size becomes equal to or greater than the critical microstructural characteristic (e.g., the α grain size or the average colony size in lamellar microstructures). Moreover, at high load ratios, the microstructural effects become less pronounced.[18]

As may be expected, the morphology of the α and β phases have the largest affect on FCP resistance, with the coarse (β annealed) lamellar morphology having the optimum resistance. The larger the average colony size of the lamellae, the lower the FCP, in part due to a loss in tortuosity moving from colony to colony. The fractured surfaces also are less rough, thus leading to a lowering of the *crack closure* effect. Here, the roughness of the fractured surfaces prevents the cracks from closing entirely when the stress is lowered. This effectively decreases the stress intensity amplitude by increasing the lower limit of the applied stress while the upper limit remains unchanged. The bimodal structure has a slower FCP rate than equiaxed structures with no transformed β, which again is due to increased tortuosity.

Quenched microstructures can retain metastable austenite, which, if transformed to martensite during crack propagation, can retard fatigue crack propagation by inducing crack closure. Aging can negatively affect this by stabilizing the β or transforming it to α.

As for any metal, the surface condition strongly affects the fatigue strength, especially if the surface treatment generates compressive surface stresses, e.g., shot peening or even machining. In most cases, the effect of the surface is more influential than the microstructural effects. In Fig. 3.4, it is shown that shot peening greatly improves the fatigue strength. However, note that stress relieving a shot-peened surface leads to a fatigue strength that is much inferior to the smooth electropolished specimen (Fig. 3.5). This illustrates that shot peening, while creating a surface layer of beneficial compressive stresses, can actually damage the surface. Note also that the surface compression must be balanced by subsurface tension, which may lead to premature subsurface crack initiation.[19]

Anomalous strength properties. The most significant of the anomalous strength properties are thermoelasticity and pseudoelasticity. The thermoelastic martensitic transformation is one in which continuous martensitic transformation occurs as the temperature is raised or lowered (unlike Fe-martensite transformation, which occurs by "bursts"). Pseudoelasticity is the strain-induced martensite analog of thermoelasticity in that a martensite transformation occurs continuously during loading and reverts back to β continuously on unloading.

Both effects combine to generate the shape memory effect, where a stress-induced martensite transformation is completely removed by

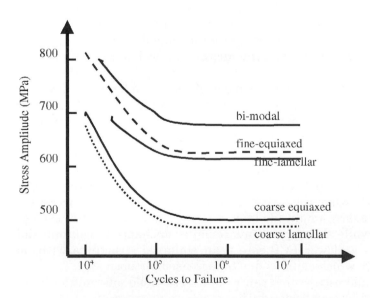

Figure 3.4 Effect of microstructure on fatigue life of Ti-6-4 for rotating bending fatigue tests (R = −1). All specimens heat treated at 500°C for 24 hr after producing the desired microstructure.[17]

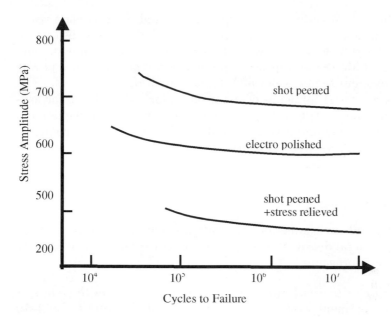

Figure 3.5 Effect of microstructure on fatigue life of Ti-6-4 for rotating bending fatigue tests (R = −1). All specimens heat treated at 500°C for 24 hr after producing the desired microstructure.[17]

raising the temperature of the sample. Nitinol, an alloy based on the Ti-Ni binary) is the best known version of this alloy, but other titanium alloys display the same effect. In fact, the transformation does not have to be martensitic. For example, a shape memory effect was found in Ti-6-4 as a result of $\alpha + \beta$ transforming to β.[20]

3.5 Corrosion and Corrosion Resistance

The corrosion resistance of titanium chiefly stems from its high affinity for oxygen, which enables the formation of a stable, continuous adherent oxide film. It is therefore recommended for mildly reducing to quite highly oxidizing atmospheres, and it is excellent in marine and general industrial environments. However, if the oxidizing atmosphere is too strong (or in the absence of moisture), the oxide film is not protective, and titanium and oxygen can react violently. Titanium alloys are also adversely affected by hot, concentrated chloride salts and all acidic solutions that are reducing in nature.

Small alloying additions have no effect in environments in which titanium is effective. In adverse conditions, some elements can further decrease the corrosion resistance; e.g., Fe and S are detrimental when corrosion rates exceed 0.13 mm/yr.[21]

Minor Ni and Pd additions significantly expand the corrosion resistance of titanium under reducing conditions; Pd also improves crevice corrosion resistance in hot aqueous chlorides. Major alloying elements, V and Mo (>4%) improve the corrosion resistance in aqueous reducing acid, but the presence of Al is detrimental.

There is no strong influence of the product form or welding for low-alloyed titanium or Ti-6-4. In more active environments, weldments may be more susceptible to corrosion, depending on the alloy composition; FeS and O, as well as coarse transformed β microstructures, seem to be detrimental.

3.5.1   General Corrosion

In reducing acid environments, the oxide film can deteriorate, and the unprotected metal dissolves. Corrosion resistance can be increased by a number of approaches.

1. *Anodizing or thermal oxidation to increase oxide film thickness.* A typical treatment would be to heat in air at 600–800°C for 2–10 min.

2. *Anodic polarization.* Sustained impressed potentials of +1 to +4 V (vs. standard H electrode) give full protection in sulfuric, HCl, phosphoric, formic, oxalic, and sulfamic acids. However, complex geometries and the possibility of stray currents can complicate the application of this technique. Also, in an alternating wet and dry atmosphere, anodic polarization is not always effective.

3. *Applying a surface coating (precious metal or metal oxide).*

4. *Alloying.* Additions (wt%) of >0.05 Pd, >4 Mo, and >0.5 Ni all improve corrosion resistance, thus leading to the following commercial alloys: Ti-0.3Mo-0.8Ni (grade 12), Ti-3Al-8V-6Cr-4Zr-4Mo, Ti-15Mo-5Zr, and Ti-6Al-2Sn-4Zr-6Mo. These grades can withstand hotter/stronger HCl, H_2SO_4, and H_3PO_4 acids (especially pitting resistance).

5. *Adding oxidizing species to the environment to promote oxide film formation.*

3.5.1.1   General corrosion in water/seawater.   Titanium alloys are fully resistant to water and seawater up to 315°C (i.e., steam). There is, however a slight weight gain and discoloration due to thickening of the oxide film. Natural contaminants in water (e.g., Fe, carbonates, sulfides, etc.) do not pose a problem; however, chloride concentrations in excess of 1000 ppm at temperatures approaching the boiling point of water 75°C could lead to crevice corrosion.

3.5.1.2 General corrosion in oxidizing environments.

As may be expected, titanium and the alloys are highly resistant to oxidizing acids such as chromic, nitric, and perchloric acids, etc., and oxidizing reagents such as permanganates and thiosulphates, up to boiling points. It is also unique in resisting oxidizing chloride environments. In fact, halide salts (e.g., metal chlorides and bromides) actually enhance passivity. Tests in boiling HNO_3 indicate that increases in Al and β alloying elements decrease corrosion resistance, although these are limited results. Note also that high-purity/low-alloy titanium weldments do not exhibit the accelerated attack that has been observed in less pure, unalloyed titanium and near β grades.

Dry oxidants (e.g., fuming HNO_3) can lead to a violent reaction. Increasing the water content will effectively inhibit the reaction (e.g., for fuming HNO_3, the addition of >2% water is sufficient).

3.5.1.3 General corrosion in reducing acids.

Corrosion behavior is very sensitive to alloy composition and all characteristics of these acids (i.e., acid concentration, temperature, and background chemistry). If the acid characteristics exceed certain critical levels, the protective oxide breaks down, and severe general corrosion occurs. Corrosive acids include

- HCl
- H_2SO_4
- hydrobromic
- hydroiodic
- hydrofluoric
- phosphoric
- sulfamic
- oxalic
- trichloroacetic

As noted before, alloys such as the Pd series, Ti-0.3 Mo-0.8 Ni, Ti-3Al-8V-6Cr-4Zr-4Mo, Ti-15Mo-5Zr, β 21S, and Ti-6Al-2Sn-4Zr-6Mo offer much better resistance to these environments (Mo seems to be a key element in all these alloys). A list of inhibitors against corrosion of reducing acids is given in Table 3.9. Many of these are effective in concentrations as low as 20 to 100 ppm.

TABLE 3.9   Inhibitors against Corrosion of Reducing Acids (from Ref. 22)

Category	Species
Oxidizing metal cations	Ti^{4+}, Fe^{3+}, Cu^{2+}, Hg^{4+}, Ce^{4+}, Sn^{4+}, VP^{2+}(??), $Te^{4+,}$ Te^{6+}, Se^{4+}, Se^{6+}, Ni^{2+}
Oxidizing anions	$(ClO_4)^{2-}$, $(Cr_2O_7)^{2-}$, $(MoO_4)^{2-}$, $(MnO_4)^{2-}$, $(WO_4)^-$, $(IO_3)^-$, $(VO_4)^{3-}$, $(VO_3)^-$, $(NO_3)^-$, $(NO_2)^-$, $(S_2O_3)^{2-}$,
Precious metal ions	Pt^{2+}, Pt^{4+}, Pd^{2+}, Ru^{3+}, Ir^{3+}, Rh^{3+}, Au^{3+},
Oxidizing organics	Picric acid, o-dinitrobenzene, 8-nitroquinoline, m-nitroacetanilide, trinitrobenzoic acid, plus certain nitro, nitroso and quinone organics
Others	O^2, H_2O_2, $(ClO_3)^-$, OCl^-

3.5.1.4 General corrosion in salt solutions.

Titanium is highly resistant to almost all salt solutions from 3–11 pH and up to the boiling point. In fact, titanium is selected in many current applications because of its extremely good resistance to chlorides in processing, marine environments, steam, etc. However, some problems could arise in contact with nonoxidizing acidic or hydrolyzable salt solutions. Some guidelines for acceptable concentration/temperatures of these environments are available in the literature for $AlCl_3$,[23,24] $ZnCl_2$,[25] and $MgCl_2$.[26]

3.5.1.5 General corrosion in alkaline environments.

Titanium is very resistant to solutions of the usual hydroxides (Na, K, etc.). In boiling solutions of NaOH and KOH, corrosion rates do increase, especially in the case of the latter, but titanium still exhibits a high resistance. In these solutions, H pick-up and embrittlement could occur for α and near α alloys (>80°C, pH >12). Dissolved oxidizing species (e.g., nitrates) can act to inhibit H uptake in such environments.

3.5.1.6 General corrosion in organics.

With only small amounts (ppm) of H_2O, the oxide film is maintained, and titanium becomes impervious to organics. Consequently, completely anhydrous organics could be a problem, and H_2O should be added if possible (e.g., absolute methanol requires 1.5% H_2O). Higher molecular weight alcohols and chlorinated hydrocarbons are not a problem.

3.5.2 Crevice Corrosion

Crevice corrosion is due to oxygen depletion in a crevice, leading to the formation of reducing acids in the crevice. Tight crevices (e.g., joints, seals, oxide films) exposed to hot chloride, bromide, iodide, fluoride, or sulfate containing solutions may lead to crevice corrosion. Anionic oxidizing species [e.g., $(NO_3)^-$] can inhibit crevice corrosion caused by halides. Note that oxidizing species (e.g., Fe^{3+}) accelerate crevice corrosion, even though they inhibit general corrosion.

3.5.3 Hydrogen Damage

Embrittlement of α and α/β alloys by hydrogen occurs largely because of interstitial hydride formation. Pure titanium is not strongly affected by H < 200 ppm, but decreasing purity increases the hydrogen embrittlement susceptibility. Also detrimental are conditions that encourage high H diffusion rates (elevated temperatures) and H segregation (high residual stress or stress concentration). Under such a combination of circumstances, embrittlement could occur with H concentrations as low as 30–40 ppm. Note, however, that uniaxial tensile properties are not as affected by H embrittlement, compared to biaxial states (e.g., forming operations such as drawing and cupping) or triaxial states (stress raisers), since these latter conditions will exacerbate embrittlement by enhancing segregation.

Diminished impact resistance as a result of hydrogen embrittlement is due to hydrides that formed *statically* at high temperatures before the impact is delivered. Low strain rate embrittlement is due to *dynamic* (during straining) formation of hydrides. H solubility increases with temperature, so H embrittlement is more likely at higher temperatures.

3.5.3.1 Effect of microstructure and oxide layer on H embrittlement. At slow strain rates, embrittlement in α/β alloys occurs along grain boundaries; in α alloys, it is manifested as transgranular fracture. In α/β microstructures with continuous grain boundary α, embrittlement is not as severe, although it is a function of H pressure, and this effect could be reversed at lower pressures.

The titanium surface oxide is very effective in reducing H penetration. Therefore, traces of moisture or oxygen in the environment are very effective in maintaining a barrier. Conversely, an anhydrous H atmosphere will be a problem.

β alloys have a very high H solubility, therefore thousands of parts per million of H can be sustained before embrittlement occurs. Higher strengths (e.g., aged alloys) will, of course, exhibit greater susceptibility to H embrittlement.

3.5.3.2 Prevention of H embrittlement in aqueous media. As mentioned before, environments involving H, high temperatures, and stress con-

centrations must be avoided, or efforts must be made to minimize the severity of these conditions. In addition, galvanic couples and high temperature alkaline conditions should also be avoided.

3.5.4   Other Forms of Corrosion

3.5.4.1 Stress corrosion cracking. This type of damage begins as a pit or a crevice that corrodes into a crack under tensile stress. Water, seawater, and any neutral aqueous environment can cause SCC in many titanium alloys in the presence of flaws. See Table 3.10 for other environments that can cause stress corrosion cracking.

TABLE 3.10   Stress Corrosion Cracking Environments (from B. Craig, Ref. 2, p. 1075)

Environment	Temperature, °C
Dry chloride salts	260–480
Seawater, distilled water, neutral aqueous solutions	Ambient
Nitric acid, red fuming	Ambient
Nitrogen tetroxide	Ambient to 75
Methanol/ethanol (anhydrous)	Ambient
Chlorine	Elevated
Hydrogen chloride	Elevated
HCl, 10%	Ambient to 40
Trichloroethylene	Elevated
Trichlorofluoroethane	Elevated
Chlorinated diphenyl	Elevated

3.5.4.2 Galvanic corrosion. This is only a problem in strongly reducing environments where titanium severely corrodes. Under these conditions, acceleration of corrosion is possible when titanium is coupled to more noble metals (e.g., C, Pt, Au). If titanium is connected to an anodic metal, accelerated attack of the latter can occur, with the danger of H absorbing in the titanium. Depending on the environment, more active metals could be low-alloy steels—Al, Zn, Cu alloys, or stainless steel in a depassivated or pitting (i.e., corroding) condition.

3.5.4.3 Erosion/corrosion. In general, surface titanium oxide resists erosion quite well, and it is immediately self-healing in neutral aqueous environments. Harder alloys perform better but, if there is a significant erosion loss, coatings should be considered.

3.6   Alloys for Cryogenic Applications

As already noted, for low-temperature or cryogenic applications, α alloys are recommended, since these do not exhibit a ductile-to-brittle transition at low temperatures. Extra low interstitial (<0.1 wt%) versions of any of the recommended titanium alloys are also better, although the strength is reduced. To compensate for this, strength increases can be achieved by alloying additions, and microstructure manipulations (particularly α/β alloys) can additionally be used to increase cryo-toughness. The cryogenic properties of the three classes of alloys are briefly overviewed in the next few paragraphs.

3.6.1   Unalloyed Ti and Dilute Alloys

As noted above, strength increases, if necessary, can be achieved with additions of alloys. However, small levels of some β stabilizers are detrimental to extreme low-temperature properties. Moreover, some residual/contaminant elements must be removed for cryogenic applications. In particular, Fe and Mn should be minimized. Extra low interstitial grades are low in such elements (e.g., <0.25 wt%).

3.6.2 β alloys

All β alloys display a ductile-to-brittle transition, but it is particularly pronounced in alloys containing Fe and Cr. In commercial alloys, the β phase is usually metastable; hence, cleavage fracture is not always observed, but ductilities are generally much lower, even in the absence of such a fracture surface.

3.6.3 α/β alloys

Considerable improvement is possible through microstructural control. For example, a β anneal improves the ductility of 6-4 compared to an anneal in the two phase region (e.g., for the ELI version of this alloy, ductility is 3% at 20 K after heat treat in the two-phase region, but is 4% at 4 K after a β anneal and air cool). In fact, an increasingly slow cooling rate from a β anneal (e.g., in 6-4) radically improves the fracture toughness at 4 K from 40 to 80 MPa$\sqrt{m}$, although the strength decreases from about 2000 MPa to 1600 MPa.[27] This is most likely due to an increase in tortuosity.

The tensile properties of selected alloys at low and cryogenic temperatures are listed in Table 3.11.

TABLE Inhibitors Against Corrosion of Reducing Acids (from Ref. 22)

Alloy	Condition	Test temp. (K)	UTS (10^8 N mm^{-2})	YS (10^8 N mm^{-2})	Elong. (%)
5Al-2,5 Sn (5-2.5)	Annealed 15 min–4 hr @ 707–867°C	295	8.8	8.6	16
		200	11.0	10.6	14
		77	14.0	13.7	12
		20	16.9	16.8	5.1
	Annealed ELI	295	7.6	7.1	17
		200	9.2	8.6	16
		77	12.6	11.9	17
		20	15.4	14.4	15
8Al-1Mo-1V (8-1-1)	Annealed 8 hr at 787°C, fce cool	295	10.3	9.7	16
		200	11.9	11.3	14
		77	15.6	14.4	13
		20	17.5	16.2	2.4
	Duplex annealed 8 hr at 787°C, fce cool + 15 min at 787°C, air cool	295	10.2	9.5	15
		200	11.2	10.3	15
		77	14.9	13.4	22
		20	16.9	16.1	1.2
6-4	Annealed 30 min–4 hr, 707–817°C, air or fce cool	295	9.9	8.9	12
		200	11.6	10.7	11
		77	15.3	14.3	11
		20	17.9	17.3	2.4
	Solution treated 847–957°C and aged 1–10 h at 482–597°C	295	12.2	11.3	8
		200	13.2	12.8	6
		77	17.6	17.0	5
		20	20.4	19.9	0.7
	Annealed, ELI	295	9.9	9.3	12
		200	11.5	10.9	12
		77	15.1	14.6	10
		20	18.2	17.9	2.9
	Solution treated and aged ELI	295	11.2	10.6	9
		200	13.2	13.2	7
		77	17.2	16.7	5
		20	19.6	19.6	1.0
13V-11Cr-3Al (13-11-3)	Annealed or solution treated 10–30 min at 757–787°C	295	9.7	9.4	19
		200	12.5	12.2	12
		77	19.5	18.9	2.1
		20	22.6	—	0.5
	Solution treated and aged 20–100 hr @ 427–507°C	295	13.6	12.4	7
		200	15.7	14.7	2.1
		77	16.5	—	0.2
		20	—	—	—

3.7 Titanium Alloys for High-Temperature Applications

α alloys perform well at low temperatures, but solid solution strengthening is rapidly reduced at high temperatures due to increased diffusion rates of solute atoms. Microstructural strengthening is therefore required, and this has led to the development of α + β alloys that could be used to 550°C. These alloys are based on the athermal strengthening of β by Mo and α by Al Sn or Ga (and perhaps Zr). Ti-8Al-1Mo-1V was a result of this alloying philosophy, as was Ti-6Al-2Sn-4Zr-2Mo (6242), the latter being much superior. Variants of 6242 were developed but were not much of an improvement, with the exception of Si and Bi additions, which improved creep resistance and were designed for aircraft gas turbines.

Apart from corrosion/oxidation damage, a clear temperature limitation is the β transus temperature. To increase this, α stabilizers were added, but the levels reached a maximum due to the formation of embrittling α_2. This detrimental effect caused by the presentation of Al_{eq} from exceeding a level of 9 defined by Hammond and Nutting,[7] as mentioned earlier.

An alternative strengthening approach was to use the (metastable) β phase, because the effect of temperature on UTS is much less as compared to α alloys. For example, β III (based on Ti-Mo system) loses only 20% of its UTS from room temperature to 400°C, compared to Ti-5Al-2.5Sn, which loses about 40% of its strength.

α+β alloys for high-temperature applications were developed for thermomechanical, metallurgical stability and processing reasons. IMI 834 and TIMETAL-1100 have been developed that operate to nearly 600°C; both contain 0.35 wt% Si. In Ti-6Al-2Sn-4Zr-2Mo, the creep resistance is maximized at about 0.1% Si (510°C at 240 MPa)[28] but, for other alloys, the creep resistance can be further increased up to the solubility limit of Si. The effect of Si appears to be due to Si atoms segregating to and the pinning of dislocations, impeding their motion. Figure 3.6 shows the year of introduction and the relative service temperatures of the more commonly used high-temperature alloys. The heat resistance, as determined by the Larson-Miller parameter, is shown in Fig. 3.7.

Currently, the operating temperature limit of conventional alloys is about 500°C. Above this temperature, oxidation is the limiting factor, but this possibly could be overcome by coatings. To date, the oxidation-resistant coatings based on chrome carbide that can be applied are themselves brittle and reduce the surface fatigue initiation life. Even with protection, α-β alloys undergo phase change above 600°C, and the accompanying volumetric cycling becomes a problem. Metastable β alloys are unstable at working temperatures. Increased alloying additions stabilizes the β but also leads to density increases. The use of V additions, which are not as dense as other β stabilizers, is not a good option, because the oxidation resistance diminishes with increasing V content.

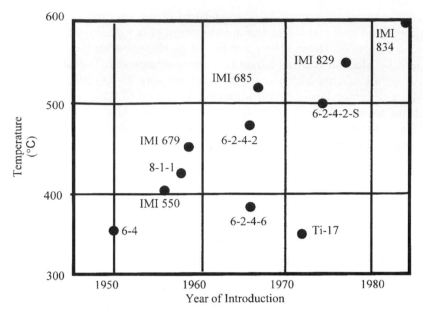

Figure 3.6 Service temperatures of selected alloys and year of introduction (from Ref. 28).

3.7.1   Intermetallics for High-Temperature Applications

The temperature capability of titanium alloys progressed steadily from the time Ti-6-4 was first introduced in 1954 until 1984, when IMI 834 was developed (Fig. 3.6), but further advancements in temperature capability effectively came to a halt at that point, the challenge being the oxidation resistance of the alloy and the β transus temperature. Oxidation resistant coatings have been investigated but, so far, there are none available that do not add a significant fatigue debit to the alloy. The approaches taken from that point have been as follows:

1. Improve the creep strength of alpha alloys by controlling the precipitation of α_2 phase in an alpha alloy.

2. Develop alloys that were based on the intermetallic α_2 (Ti_3Al) composition.

3. Develop alloys using the gamma intermetallic composition (TiAl).

In the late 1980s, the U.S. government introduced several research and development programs for the development of gamma alloys,

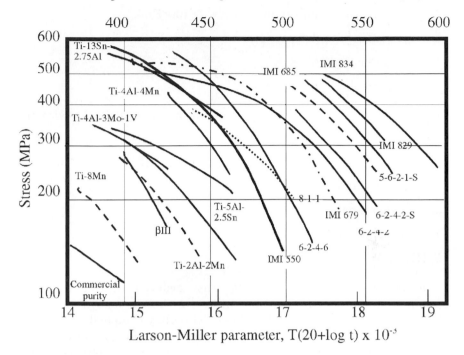

Temperature for total plastic strain = 0.2% in 100h (°C)

Figure 3.7 Creep properties of various alloys, showing the range of elevated temperature properties of Ti alloys (T = temperature, t = time (from Ref. 28).

since this classification of materials offers the greatest potential for replacing nickel-based alloys up to temperatures of 654°C (1200°F). The benefit was dramatic reductions in the weight of turbine engines by almost halving the density turbine blades. α_2 alloy development has also progressed, but problems have been encountered with the stability of the alloys, hence the concentration on gamma alloys.

The U.S. government-led research programs were focused on the development wrought alloy, powder, and casting processes for structural and rotating engine components, and sheet products for the National Aerospace Plane (NASP). The U.S. military also had programs to develop advanced engine components on the Integrated High Performance Turbine Engine Technology (IHPTET) Program to develop compressor blades and exhaust nozzle structures. Of these programs, the investment casting process has had the greatest commercial success, most notably by Howmet Corporation, although other companies such as GE and ABB have developed other casting alloys.

The principal advantages of gamma titanium aluminides are as follows:

1. Fifty percent lower density than nickel alloys

2. Temperature capability up to 766°C (1400°F)

3. High stiffness-to-density ratio

4. Moderate creep properties

5. Not flammable

These advantages are countered by some major drawbacks with the alloys.

1. Low ductility

2. Fracture toughness

3. Poor machinability

4. Poor formability

5. High cost

Gamma titanium alloys are long range ordered intermetallic compounds with a face centered cubic unit cell, whereas α_2 by comparison has a hexagonal close packed unit cell. The alloy composition range is between 46 to 48 atomic percent aluminum with additions of Cr, Mn V, W, and Ta for property enhancement. To generalize, the effects of various additions are as follows:

Ductility	Cr, Mn
Tensile strength	Cr, Si, C, B
Creep strength	W, Ta, Mo, Si, C
Oxidation resistance	Nb, W, Ta, Si
Grain size control	TiB_2

Table 3.11 lists typical properties of gamma aluminides.

Alloys are made as per the route described in Sec. 3.10 on titanium ingot production using double- or triple-melted VAR ingots stock. The ingot is then melted and investment cast in a similar manner to that for investment casting any other titanium alloy, except for changes to the gating system that are designed to prevent the development of residual stresses that, if not controlled, lead to cracking. A selection of alloys that have been produced with the best balance of properties to date is shown in Table 3.12.

Originally, alloy development focused on increasing the room temperature ductility with additions of Cr and Mn, which was followed by enhancing the creep properties with additions such as W, Ta, and Mo, which had the counterproductive effect of reducing the room-tempera-

TABLE 3.11   Typical Properties of Gamma TiAl Alloys

Density	4.0 g/cm^3
Temperature limit	760°C
Oxidation resistance	760°C
Ductility	<3% ambient, 5–16% at 650°C
CTE	1.0 × 10–5 at 20°C 1.7 × 10–5 at 800°C
Thermal conductivity	0.3 W-cm-1 K-1 at 700°C

TABLE 3.12   Common Casting Alloys

Howmet 45XD alloy	Ti-45Al-2Nb-2Mn + 0.8 v% TiB2	Castability, tensile and fatigue strength
Howmet 47XD alloy	Ti-47Al-2Nb-2Mn + 0.8 v% TiB2	Castability RT and high-temperature strength
ABB alloy 2	Ti-47Al-2W-0.5Si	Creep and oxidation resistance
GE 48-2-2	Ti-47Al-2Cr-2Nb	Ductility and toughness
W-Mo-Si	Ti-47Al-2Nb-1Mn-0.5W-0.5Mo-0.2Si	Creep resistance and balance of properties

ture ductility. Second-generation alloys were produced that had additions of Nb to improve the oxidation resistance, but balancing out the properties of the alloys in this way has had the effect of lowering the continuous operating temperature to around 766°C (1400°F). Refractory element additions in the form of W and Ta have were made to further enhance the creep properties but had the overall affect of increasing the density and cost of the alloy. In efforts to find a cheaper method, Howmet has found other methods of improving the creep strength by additions of carbon and/or oxygen.

Adjusting the microstructure of the cast alloy can also alter the creep properties. An as-cast equiaxed microstructure has good ductility properties but poor creep properties, whereas a lamella structure has good creep properties and poor ductility. Thus, a good balance of creep and ductility is achieved by developing a mixed morphology microstructure. This can be accomplished either by altering the casting mold preheat temperature per methods developed by Howmet Corporation or by heat treatment as reported by Zhao et al. (U.S. patent No. 5,653,828, August 1997).[33] Figure 3.8 shows the typical microstruc-

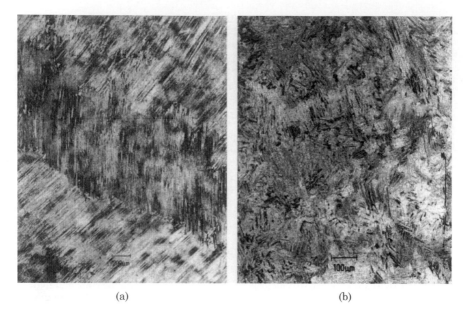

(a)   (b)

Figure 3.8   (a) As-cast Ti-47Al-2Nb-2Cr (at%) exhibiting coarse α_2 + β lamellar structure, and (b) heat-treated Ti-47Al-2Nb-2Cr (at%) showing refined α_2 + β lamellar colonies (HT: 1420°C/20 min/WQ + cyclic heat treatment described in U.S. patent)

tures of a cast Ti-47Al-2Nb-2Cr (at%) alloy before and after heat treatment. The HT applied to this material involves 1420°C/20 min/WQ + cyclic heat treating. The refined lamellar structure will yield a better balance between room-temperature tensile properties and high-temperature creep resistance in comparison with coarse cast lamellar structure or cast duplex structure derived from conventional heat treatment (at 1010°C/50 hr).

Examples of cast γ-TiAl components have been produced ranging from turbocharger rotors and engine valves for the automobile engines, compressor and turbine blades, casings, and other structural components for turbine engines, but widespread applications have been slow to take hold. GE has built and tested a turbine engine with a low-pressure turbine rotor that was fitted with γ-TiAl power turbine blades, but there are still no reports of these alloys being used in commercial engine applications. Table 3.13 shows some data comparing the cast γ-TiAl with two commonly used cast nickel based alloys. From these data it is clear that, although the creep and fatigue properties of γ-TiAl have been improved significantly, engine manufacturers are still looking for an increase in these properties that would allow them to have the same service life as the nickel alloys that are the current alloys of choice.

TABLE 3.13   Comparative Property Data for Nickel Alloys and Cast γ-TiAl

Property/Alloy	IN100	IN713LC	XD alloy
Density (ib/cu. in.)	0.28	0.289	0.144
UTS (ksi)	140	129	69 to 96
Specific UTS (ksi/ib/cu. in.)	500	446	480 to 668
Yield strength (ksi)	112	102	36 to 69
Specific yield strength (ksi/lb/cu. in.)	400	353	250 to 480
Young's modulus	2.4×10^7	2.5×10^7	2.2×10^7
Specific creep properties Time to 1% creep strain	43,000 hr @40 ksi	3000 hr @40 ksi	50 to 350 hr @20 ksi

3.8 Casting Alloys

Casting is an important fabrication method, and there are titanium alloys that are readily castable. This includes the ubiquitous Ti-6-4, but some unalloyed and Ti-6-2-4-2 are also used in the as cast condition. β alloys Ti-3Al-8V-6Cr-4Zr-4Mo (β-C) and 15V (Al_{eq}:Mo_{eq} = 0.22), 3Cr-3Sn-3Al (Ti-15-3) (Al_{eq}:Mo_{eq} = 0.29) are highly castable and heat treatable to strengths as high as 1170 MPa. Titanium aluminides are also being considered in the as-cast condition because of difficulties in hot working and machining. Some mechanical properties of selected alloys in the cast condition are listed in Table 3.14.

3.8.1 Casting Techniques

Melting is usually performed using a consumable Ti electrode in a vacuum furnace. The electrode can be made from forged billet, revert wrought material, selected foundry returns, or a combination of all the above. Titanium is so reactive with many of the commonly used refractory materials that steps have to be taken to prevent the molten metal from coming into contact with the surfaces of the melt crucible. To prepare for casting, the molten Ti is poured into a Cu water cooled crucible so that a Ti lining or *skull* forms on the surface of the crucible and thereby prevents attack by the molten Ti. When ready, molten Ti is poured out of crucible into the mold, and the skull is retained for next batch. Once again, specialized mold facecoat materials such as yttria are required that do not react with molten titanium.

Since the crucible is water cooled, not much superheating is possible, which lead to problems in filling molds. To compensate for this,

TABLE 3.14   Typical Properties of Some Alloys in the As-Cast State

Alloy	Heat treatments	Yield stress (MPa)	UTS (MPa)	Elong. (%)	Reduc. in area (%)
Commercially pure		448	552	18	32
Ti-6Al-4V	Annealed	855	930	12	20
Ti-6Al-2Sn-4Zr-2Mo	Annealed	910	1006	10	21
Ti-6Al-4V(ELI)		758	827	13	22
Ti-1100	β soln. treat; aged	848	938	11	20
IMI 834	β soln. treat; aged	952	1069	5	8
Ti-6Al-2Sn-4Zr-6Mo	β soln. treat; aged	1170	1240	1	1
Ti-3Al-8V-6Cr-4Zr-4Mo	β soln. treat; aged	1241	1330	7	12
Ti-15V-3Al-3Cr-3Sn	β soln. treat; aged	1200	1275	6	12

centrifugal casting and preheated molds can be used to ensure complete filling. The casting is then cooled in a vacuum or an inert gas atmosphere until the temperature drops below the temperature of heavy oxidation.

3.8.2   Molding Methods

Instead of traditional sand, which would react with molten titanium, rammed graphite (powdered graphite mixed with organic binders) is used. However, more traditional sand molds with organic binders have been developed that increase dimensional stability and reduce cost. The lost wax or investment casting process also required development of a ceramic slurry that was inert to molten titanium. Facecoats are made with special refractory oxides and then backed with more traditional (and less expensive) refractory slurries. Despite these developments, some reaction always occurs, generating an α case layer that has to be removed. This problem must be taken into account in the mold design. As a general rule, the depth of the case layer depends on time at temperature. Consequently, in thin sections, the layer is virtually negligible but, in heavy sections, the layer can be as thick as 1 or 2 mm.

With regard to casting design, titanium alloys require no major departures from conventional casting design principles. Shrink porosity can be removed by hot isostatic pressing, but there is a limiting size of the defect beyond which other problems begin, e.g., surface/structural deformation.

3.9   Precipitate and Dispersoid Strengthened Alloys

These alloys have second phase particles that are closely spaced and are physically and chemically stable at high temperatures. Therefore, strength is good at low temperatures and is maintained at elevated temperatures. Alloy manufacture is difficult, since elements that participate in the formation of these particles will tend to segregate during ingot manufacturing. Rapid solidification processing (RSP) can reduce this problem, but the section thickness over which this can be applied is limited. RSP has also been used for other systems, with varying degrees of success, as can be seen in Table 3.15. These microstructures tend to be unstable in the β region, thus RSP is carried out in the α region and used in the α region. Some tensile properties are shown in Table 3.16 for dispersoid/precipitate alloys.

TABLE 3.15   Comparison of Ingot Metallurgy and Rapid Solidification Processing (from Ref. 29)

Alloy type	Alloy system	Result of ingot metallurgy	Result of RSP
Dispersion strengthened alloys	Binary systems with Er, Y, Gd, Nd, Sc, La, Dy	Coarse particles	Fine particles, extended solid solutions
Compound formers	Ti-B, Ti-C	Limited SS; coarse particles	Grain refine, fine dispersoids
Eutectoid formers	Ti-Ni/Ti-Si/Ti-Fe	Seg; coarse grains and precipitates	Controlled eutectoid formation
Combined precipitates and dispersoids	Ti-Al + either Er, Y, Gd, Nd, Sc, La, Dy Ti-Al-Ni, Ti-Al-B, C	Coarse dispersoids	Coherent precipitates and incoherent dispersoids
Conventional Ti alloys	6-4, 6242, 811	Coarse elongated grains	Fine martensites giving fine α+β grains on annealing in two-phase region
Amorphous alloys	Ti- B or Ti-Si with either Mn, Nb, V or Cr	Cannot be made	Amorphous/microcrystalline structures
Intermetallics	Ti_3Al, TiAl	Coarse grains	Grain refine, fine dispersions, decrease in LRO (possibly)

Note the considerable use of rare earth elements. These can also be useful as oxygen scavengers, particularly if the alloy is rapid solidification processed into a powder, and the powder is subsequently made into a product by P/M techniques. Note also that, for the latter, processing must be done at relatively low temperatures (and therefore high pressures) to minimize particle coarsening.

TABLE 3.16   Some Tensile Properties of Dispersoid/Precipitate Alloys (from Ref. 29)

Alloy	Heat treatments	Yield stress (MPa)		UTS (MPa)		Total elongation (%)	
		RSP	I/M	RSP	I/M	RSP	I/M
Ti-5Al-2Er	3 hr 860°C/WQ	670	469	735	536	27.0	—
Ti-7.5Al-2Er	3 hr 860°C/WQ	850	680	920	756	—	7.0
Ti-9Al-2Er	3 hr 860°C/WQ	880	750	928	790	11.0	0.1
Ti-5Al-2Er	3 hr 860°C/WQ + 25 hr at 625°C	700	510	763	564	13.8	10.0
Ti-7.5Al-2Er	3 hr 860°C/WQ + 25 hr at 625°C	952	815	973	843	7.7	6.0
Ti-9Al-2Er	3 hr 860°C/WQ + 25 hr at 625°C	931	802	952	824	1.6	0.2
Ti-5Al-2Er	3 hr 860°C/WQ + 500 hr at 550°C	714	515	780	590	54.0	18.0
Ti-7.5Al-2Er	3 hr 860°C/WQ + 500 hr at 550°C	973	830	990	865	12.0	9.0
Ti-9Al-2Er	3 hr 860°C/WQ + 500 hr at 550°C	—	810	—	835	—	0.3

3.10   Wrought Alloy Processing

The Kroll process, the melting process, and mechanical working of titanium alloys are, in principle, relatively simple processes. However, the properties of titanium alloys are very sensitive to the processing conditions, and great care has to be taken to control each processing step to minimize property variation. The principal steps in the refining process are:

1. Production of pure titanium sponge

2. Alloying

3. Remelting for higher quality grades

4. Mechanical working

3.10.1   The Kroll Process

Titanium sponge is produced by the Kroll process in two stages: first, rutile ore is chlorinated to form $TiCl_2$, which is then reduced using either sodium or magnesium metal to produce a porous form of titanium known as *sponge*. The reduction step is a very rapid and uncontrollable reaction, which leads to the entrapment of $MgCl_2$. Attempts to find more effective ways of controlling the reduction process have met with little success, and thus the sponge still has to be treated to remove residual NaCl or $MgCl_2$.

Sodium-reduced sponge is leached with acid to remove the NaCl by-product, whereas the magnesium-reduced sponge is either leached or vacuum distilled to remove the excess $MgCl_2$. Vacuum distillation is the more costly method, but it results in lower residual levels of hydrogen, magnesium, and chlorine. Most manufacturers of titanium sponge today are using the vacuum distillation method, although this can lead to slightly elevated levels of nickel from the walls of the vacuum retort; this has been reported to reduce the creep properties of alloys such as Ti-6-2-4-2.

Residual elements, such as carbon, iron, nitrogen, and oxygen, contained in the sponge, have a profound affect on titanium alloys by raising the tensile strength and lowering the ductility. Therefore, they must be controlled to the specified levels to maintain the optimum balance between strength and ductility. Figure 3.9 shows the relationship between these elements and the typical strength and ductility of pure titanium.

3.10.2 Alloying and Ingot Production— Vacuum Arc Consumable Electrode Melting

The minimization of residual elements in the alloying process is equally important as it is in sponge production. Care is taken to prevent the addition of elements that would otherwise form refractory or high-density inclusions.

Ingots are prepared by forming compacts of a mixture of crushed titanium sponge and alloying elements. Scrap titanium alloy of suitable composition may also be added as revert, but the proportion of revert allowed depends on the specification to which the melt is being pre-

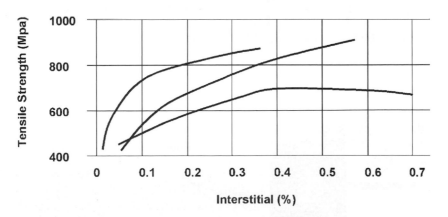

Figure 3.9 Effect of carbon nitrogen and oxygen interstitial content on the strength of CP titanium.

pared. The compacts may be of the form of discs or segments that are stacked and welded into an electrode that may be as large as 14 ft long and 2 ft in diameter. Care is taken to select a welding process that does not introduce refractory elements, such as tungsten for TIG welding. Alternatively, electrodes may be extruded out in a single step, thereby eliminating the need for welding, but this requires very large compaction presses.

The compact is melted in a vacuum arc furnace (Fig. 3.10) principally to remove hydrogen and other volatiles and to reduce compositional segregation. Depending on the grade of the alloy to be made, there may be a number of repeated melting steps or remelts in which

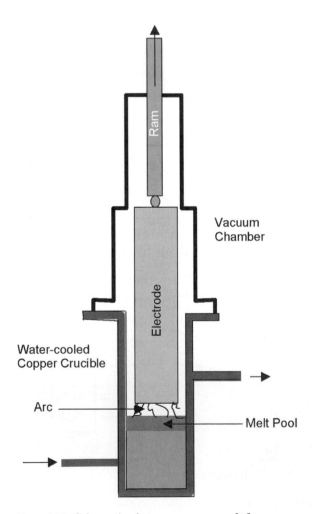

Figure 3.10 Schematic of a vacuum arc remelt furnace.

the ingot is used as the electrode in the following remelt. High-speed critical rotating components of gas turbine engines (compressor discs) typically use triple-melted material, but other applications, such as shafts or structural components, may only need one remelt.

As was shown earlier, the properties of titanium alloys are strongly linked to minor variations in residual elemental composition, but this does not account for all property anomalies. Indeed, there are significant property variations that emerge as a result of melting due to changes in vacuum, the condition of the arc between the electrode and melt pool, arc-outs, power fluctuations, and water leaks from the water cooled mold. The objective therefore is to produce an ingot with homogeneous mechanical properties as a result of uniform melt chemistry with uniform melting parameters.

Ingots may be produced in sizes anywhere from 26 to 36 in. dia. and weigh from 8000 to 15,000 lb. The larger ingots are more economical to produce and may also add in other economies of scale for parts manufacturers that, for reasons of quality assessment, can produce parts only from a single ingot or heat. However, ingots in excess of 36 in. dia. are prone to increasing levels of segregation. The length, too, may be limited by segregation from top to bottom of the ingot and is dependent on the type alloy to be produced. For example, IMI 834 is prone to such behavior.

Electron beam and plasma arc melting are similar refining methods that are becoming credible and lower-cost alternatives to vacuum arc remelting process for aerospace applications. Both processes can greatly reduce the level of both gaseous elements plus high and low density inclusions.

3.10.3 Ingot Defects

The defects that give cause for concern in titanium alloy production are all attributable to segregation of alloying elements. For aerospace grades, these defects are all detrimental to the mechanical properties, and constitute the reason why the ingots are remelted in vacuum arc remelt (VAR) furnaces up to two more times to eliminate any segregation. After the ingot is converted into billet, the bars are inspected ultrasonically to detect internal defects. Any zones containing ultrasonic indications are cut out of the bar and examined metallographically to determine the nature of the defect.

3.10.3.1 Type I defects. Type I defects, also called *high interstitial defects,* result from very high local concentration of nitrogen or oxygen in the original sponge material. These defects are very much harder and have lower ductility than the surrounding matrix and are usually associated with cracks. The higher concentration of nitrogen or oxygen has the effect of locally raising the β transus temperature, and these defects will not deform as readily as the surrounding matrix during thermomechanical processing.

3.10.3.2 Type II defects. Type II defects are regions of high concentrations of metallic α stabilizers such as aluminum. This is caused by the diffusion of alpha stabilizers to shrinkage voids near the top of the ingot that become fully incorporated into the alloy during conversion of the ingot into billet. Since the matrix surrounding the voids is depleted of alpha stabilizers, type II defects are often associated with beta segregation. The inclusions themselves are primary alpha that is only slightly harder than the surrounding matrix, and they may extend over a number of grains. Blue etch anodizing can normally reveal these defects, but care must be taken to ensure that such indications are not etching artifacts.

Type I and II defects are not permitted in aerospace grades of titanium, because they greatly reduce the fatigue properties of the alloy. In 1988, an aircraft crash at Sioux City, Iowa, was directly attributable to the presence of a type II defect in one of the engine compressor discs. The disc fractured, and the pieces thrown from the engine destroyed the plane's hydraulic systems, rendering the aircraft almost entirely uncontrollable. This incident led to a concerted effort among engine manufacturers and titanium alloy producers to improve the overall quality of the rotor grade material, and it resulted in the requirement for triple-melt refining of rotor grade alloys.

3.10.3.3 Beta flecks. Beta flecks, regions of beta stabilized material that has been worked and heat treated in the alpha-beta range, are most often found in large-diameter ingots. They may either be regions that are devoid of alpha grains or that have a significantly lower concentration of alpha relative to the surrounding matrix. Beta flecks are typically seen in alloys that contain strong beta stabilizing alloy additions and therefore form as a result of microsegregation. As a general rule, beta flecks are not considered detrimental in alloys that are lean in beta stabilizers when used in the annealed condition. However, these defects, being regions of local composition variation, do not respond to heat treatment to the same degree as the rest of the matrix and therefore become areas of weakness.

3.10.4 Hot Working

Final hot working must be carried out in the α + β region of the phase diagram. By carrying out the conversion in the α + β range, the microstructure developed has significantly better ductility and fatigue properties than if the hot-working temperature was maintained in the fully beta range. Usually, this temperature is approximately 30°C below the β transus temperature to allow for the effects of adiabatic heating, the furnace temperature variation, and errors in the determination of the β transus. Care must be taken to avoid slow cooling of α/β alloys through the β transus, as α will begin to precipitate at the prior β grain boundaries, which in turn leads to a decrease in strength and ductility.

3.10.5 Heat Treatment

Heat treatment is the means by which the properties of an alloy can be tailored for a specific application, and this is as true for titanium as it is for steel and nickel-based alloys. The heat treatment used is a balance between the target property or properties and practicality. The objective must be to develop a commercially viable and robust process, which means that each step of the heat treatment must be clearly understood. Section 3.2.2, on the microstructure of titanium, has shown that Ti-6-4, the most commonly used alloy, can be varied greatly, depending on the heat treatment used. Heat treating above the β solvus temperature and quenching gives a very different microstructure from that of a slow cool through the β solvus, and the mechanical properties are quite different. Thus, it is possible to enhance the fatigue properties by controlling the grain size and advancing it even further by allowing α_2 to precipitate—that is, if the embrittlement that would accompany this were of no importance.

The reason why Ti-6-4 has been such a popular alloy is that it is so *forgiving*; the alloy has very useful strength up to moderate temperatures, is stable, has good fatigue properties, and does not show any significant interaction between creep and fatigue conditions. All of these properties can be produced on a repeatable basis by a very simple heat treatment process.

By way of illustration, the commonly used aerospace standard for Ti-6-4 alloy (AMS 4928) has a heat treatment as follows:

1. *Solution heat treatment.* When solution heat treatment is used, heat to a range of 50–150°F (28–83°C) below the _-transus, hold at the selected temperature to ±25°F (±14°C) for a time commensurate with the section thickness and the heating equipment and procedure used, and cool at a rate equivalent to air cool or faster.

2. *Annealing.* Heat to within the range of 1300–1450°F (704–788°C), hold at the selected temperature within ±25°F (±14°C) for not less than 1 hr, and cool as required.

By β heat treating at about 1200°C, it is possible to increase the fracture toughness by up to a factor of 2, but the penalty for this is the accompanying reduction in the fatigue and ductility of the alloy to what would normally be considered unacceptably low levels, but with only a minor reduction in the tensile strength (see Figure 3.11).[30]

The terminology used for heat treatment can be confusing, with terms such as *annealing* and *solution heat treatment,* or *stress relieving* and *aging* being used almost interchangeably. The confusion stems from the use of heat treatment for dual purposes and the generic use of the word *anneal* for processes such as solution anneal and recrystallization annealing.

3.10.5.1   Stress relief.   Stress relief is used to remove residual stresses that have been generated by mechanical working, welding, machining processes, and heat treatments. The presence of residual stresses is rather like a two-edged sword; in general, compressive residual stresses are considered beneficial and enhance the fatigue properties of the alloy by increasing the effective crack propagation stress; tensile residual stresses are additive to the applied stress and will decrease the fatigue properties. Thus, it is very important to take into account the required stress state when the component is complete and ready to

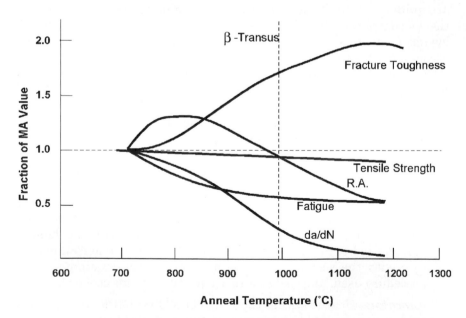

Figure 3.11 Properties of Ti-6-4 as a function of annealing temperature. The properties are ratioed to the properties of the standard mill anneal.

be put into service. The fatigue properties are purposefully enhanced using processes such as shot peening, or other methods of creating compressive residual stresses like sleeve cold working of bolt holes. Residual stresses produced during machining operations can also cause distortion, either during subsequent heat treatment operations or when operating the component at high operating temperatures. This condition can be avoided by a stress-relieving heat treatment prior to final machining where the part has approximately 2 mm of material over the final dimensions.

The stress-relieving heat treatment temperatures for α and α/β alloys is in the range of 460 to 820°C but may be higher for alloys with greater creep strength properties. Ti-6-4 is stress relieved at 675 to 790°C for a period of 30 min. Air cooling is usually sufficient to prevent the formation of deleterious ordered phases such as α_2, although slow cooling of the order of 150°C/hr should be avoided to guarantee this. If it should be necessary to use such slow cooling rates, then it is recommended that the extra low interstitial (ELI) grades of alloy should be used, although it is not certain that this is truly preventive.

The stress relieving process is carried out in the aging temperature range for all titanium alloys, where the α is strengthened by the precipitation of α_2, and if there were metastable β present, α precipitation would occur, strengthening the alloy.

β alloys must be stress relieved at temperatures appropriate to the heat treatment state. Alloys in the solution heat-treated condition are heat treated sufficiently below the β transus to prevent the formation of grain boundary α. If material were in the final aged condition, the stress relief would be carried out at or below the aging heat treatment temperature, as higher temperatures would lead to a reduction in the strength of the alloy.

3.10.5.2 Anneal. *Annealing,* also referred to as a *solution heat treatment,* is similar to the stress relief heat treatment except that it is conducted at higher temperatures, which not only removes residual stresses but also removes the last vestiges of the prior worked microstructures. The mill anneal is a general-purpose treatment given to all mill products. It is not a full anneal and may leave traces of prior worked microstructures of heavily worked products. For α and α/β alloys, this is the final condition of the alloy, since this provides a good property balance for the alloy and is one of the reasons why Ti-6-4 has been so popular. The heat treatment is achieved over a broad temperature range, as can be seen from the Society for Automotive Engineers specification AMS 4928, where the annealing temperature is defined within a range of 700 to 780°C and held at temperature for not less than an hour. β alloys are not maintained in the annealed condition, because they are not stable at higher temperatures, so that and remaining deformed structure. Temperatures up to 400°C can result in the formation of ω or fine α_2 precipitation within the metastable β phase, which can lead to embrittlement after prolonged exposure, whereas temperatures in between 400 and 600°C will generally result in the strengthening of the alloy and an accompanying loss of ductility.

3.10.5.3 Duplex anneal. This heat treatment is used to provide improved creep resistance for alloys such as Ti-6-2-4-2. This two-stage process starts with a high-temperature anneal in the α/β field, then air-cooling, followed by a mill anneal at a lower temperature to provide thermal stability, and again air cooling. The higher the temperature, the more transformed β is present in the final microstructure, which increases the creep strength and fracture toughness but also decreases the ductility of the alloy. The cooling rate from the annealing temperature has a very strong affect on the final microstructure; faster cooling rates result in a finer transformed structure, which increases the yield and ultimate tensile strength and the LCF properties. Also, rapid cooling increases the proportion of retained β in the structure, giving a stronger strengthening potential on aging.

3.10.5.4 Recrystallizing anneal. Recrystallizing anneal processes are used to improve the toughness of an alloy. The alloy is heated into the upper range of the α/β phase field, held for a period, and slowly cooled. The result is a volume fraction of equiaxed α with islands of retained β at the triple points and interfacial β at the α grain boundaries.

3.10.5.5 Beta anneal. The β anneal is carried out above the β transus temperature followed by a mill anneal. This maximizes the damage tolerance of α and α/β alloys but, as seen in Fig. 3.11, this is accompanied by a substantial decrease in the fatigue properties.

3.10.5.6 Solution heat treatment. The purpose of the solution heat treatment is to transform a proportion of the α phase into β and cooling rapidly enough to retain the β such that α will precipitate during the aging or form martensite during heat treatment. The selection of the solution temperature comes down to the practical requirements of the balance in the strength of the alloy with the ductility. The solution treatments for β/α alloys are conducted slightly below the β transus, and it is essential that this temperature be controlled precisely so that the β transus is not exceeded. If the β transus is exceeded, there is a large decrease in the tensile properties that cannot be recovered on subsequent heat treatments.

The solution heat treatment temperature for β alloys is carried out above the β transus. Beta alloy sheet is supplied in solution treated condition that has minimum strength with a 100% β structure to provide the maximum cold formability. Subsequent solution anneals undertaken to restore the fully β structure must be carried out for short durations, because grain growth can occur rapidly, since there is no secondary phase present to restrict grain boundary movement

3.10.5.7   Aging. Aging is carried out to strengthen the solution heat-treated structure through either the precipitation of α phase at temperatures between 800 and 1200°F. To do this, the alloy must be capable of these transformations, thus α alloys are not possible to age. Increasing the β alloy stabilizer content steadily increases the age hardening response. As an example, Ti-6-4 can be hardened in relatively thin sections to strengths of 200 Mpa, whereas the strength of α/β alloy such as Ti-3Al-8V-6Cr-4Mo-4Zr can be increased by 400 Mpa.

3.10.5.8   Special heat treatments. Many other heat treatment processes have been developed that are aimed at improving the strength–ductility relationships for individual alloys such as Ti-15V-3Cr-3Al-3Sn, but these will not be discussed here because of the limited application of these materials. However, there is one interesting thermomechanical treatment of note, developed by Gray et al.[31] for Ti-6-2-4-2, that was able to achieve a good combination of creep properties with high fatigue crack initiation resistance. The lamella microstructure is ideal for developing good creep properties but has poor fatigue initiation life. Gray et al. found that, by shot peening the surface and then annealing the material, the deformed microstructure on the surface would recrystallize as a fine-grained equiaxed structure. The combination of the two microstructures balanced the good creep and fracture toughness of the lamella microstructure with that of the fine-grained fatigue resistant surface and had the effect of enhancing the fatigue properties by about 50 Mpa at 500°C.

3.10.6   Contamination during Heat Treatment

Prior to heat treatment, it is important to ensure that the parts are clean and free of contaminants. Ordinary tap water must not be used, because this leads to the contamination of the parts with chlorides and fluorides. Therefore, deionized water must be used instead. Oil, grease, and other hydrocarbons are the leading causes of embrittlement during heat treatment. Even fingerprints must be removed, since they can create initiation sites for stress corrosion cracking. Titanium alloys become highly reactive with hydrogen and halides at heat treating temperatures, so these contaminants must be removed.

Nitric acid-based or alkaline cleaning solutions can be used effectively, but it is still recommended that sample parts cleaned in the acid bath be checked for hydrogen pickup. Organic solvents may also be used for cleaning, but methanol and chlorinated compounds should be prohibited. In fact anhydrous methanol causes stress corrosion cracking in titanium alloys. This effect can be prevented by using methanol diluted 50:50 with deionized water; however, many manufacturers of titanium components avoid the use of methanol altogether to prevent confusion. A good alternative organic cleaning solvent is isopropanol, which does not have any stress corrosion effect.

3.10.7   Treatment Furnaces

To some extent, all furnace atmospheres contain hydrogen, whether they be oil-fired, gas-fired, or electric furnaces. At temperatures used to heat treat titanium, the incomplete combustion of hydrocarbons produces residual hydrogen; in electric furnaces, moisture in the atmosphere will dissociate to form oxygen and hydrogen. Therefore, it is important for the heating furnace to maintain an oxidizing atmosphere; a reducing atmosphere promotes the formation of hydrogen, which embrittles titanium alloys by the formation of interstitial hydrides. Specifications generally limit the allowable hydrogen content to between 125 and 250 ppm.

To minimize contamination, soak time at temperature must be kept to a minimum to prevent or minimize the formation of an oxide surface layer, or *alpha case,* hydrogen pick-up, and grain growth. However, the formation of the alpha case is to some degree beneficial in that it retards the pick-up of hydrogen. The alpha case must be removed from the part, as this is very brittle in nature and reduces the fatigue properties of the part. However, the alpha case is very abrasive to most machine tools and is costly and time consuming to remove.

Table 13.17 summarizes various types of heat treatments.

3.10.8   Secondary Fabrication

Secondary fabrication is the stage that can have a very profound effect on the mechanical properties of the alloy, and in some cases this is

TABLE 3.17   Summary of the Types of Heat Treatments Carried Out on Wrought Ti Alloys

Heat treat designation	Heat treat cycle	Resulting microstructure
Duplex anneal	Solution treat 50–75°C below β trans; air cool and age for 2–8 hr at 540–675°C	Primary (equiaxed) alpha plus equiaxed grains of transformed β (Widmanstatten α plus β)
Solution treated and aged	Solution treat 40°C below β trans; water quench and age for 2–8 hr at 535–675°C	Primary (equiaxed) alpha plus tempered α′
β anneal	Solution treat 15°C above β trans; air cool and stabilize for 2 hr at 650–760°C	Widmanstatten α + β colonies
β quench (β STOA)	Solution treat 15°C above β trans; WQ and temper for 2 hr at 650–760°C	Tempered α′
Recrystallization anneal	925 ° C for 4 hr, cool at 50°C/s to 760°C, air cool	Equiaxed α with β at GB triple points
Mill anneal	Hot work in α + β region, plus anneal at 705°C for 30 min to several hr; air cool	Incompletely recrystallized α + small fraction of small β particles

where the full benefit of the mechanical properties of titanium alloys can be realized. Secondary fabrication is considered to consist of closed die forging, extrusion, hot and cold forming, machining, chemical milling, and joining

3.10.8.1 Forging.   This is one method for forming parts that are close to the finished dimensions, often called *near net shape*. Generally, it is desirable to be as close to the finished shape as possible so as to minimize the material wasted during machining and maintain minimal production lead times. However, the forging outline or envelope may be substantially different from the final machined part, because the final shape is invariably a compromise between the shape and the development of the mechanical properties.

The complexity of the forging shape can also have an important effect on the mechanical properties. Areas of the forging with a large cross section will receive less mechanical work, and the microstructure may contain features that have been carried over from the billet.

To produce a fine-grained microstructure in thick section areas of a forging, it may be necessary for the forger to order billet from the mill with a highly refined microstructure.

3.10.8.2 Isothermal forging.

Isothermal forging takes advantage of a greatly reduced flow stress at high temperatures in fine-grained materials. This means that the material flow stress is lower and consequently reduces the required forging pressure, and it also ensures that the die is completely filled. This process is far from ideal, because high-temperature die materials have to be used that are based on refractory metals such as molybdenum alloys, which readily oxidize at forging temperatures, and so the forging process has to be carried out in a vacuum or an inert atmosphere. (Forging has to be carried out in inert atmosphere in any event, to prevent oxidation of the forging.) This process is therefore significantly more costly than hammer or press forging because of the need for molybdenum die alloys and a controlled atmosphere.

3.10.8.3 Superplastic forming.

The advantages of superplastic forming all revolve around the very high workability engendered by this technique. The benefits therefore include the following:

1. It renders low flow stresses, leading to reduced die wear and lower loads required for deformation.

2. The as-formed shape is very close to the final dimensions.

3. Metal flow is uniform.

4. Springback is minimal.

5. There are no residual stresses.

The main disadvantage is that it is a slow process and is very sensitive to changes in processing variables.

Not all titanium alloys are superplastic. The main requirement for superplasticity, with respect to materials properties, is high strain rate sensitivity, that is

$$m = \frac{d(\ln\sigma)}{d(\ln\dot{\varepsilon})} \text{ should be high}$$

where m = strain rate sensitivity
σ = flow stress
$\dot{\varepsilon}$ = strain rate

A material that exhibits this property can develop large uniform deformation without necking. Referring to the schematic diagram of Fig. 3.12, the reason why a high m value leads to superplasticity can be explained in the following manner. If m is high, any increase in stress leads to a rapid decrease in the strain rate and a consequent decrease in ductility, and vice versa. Even in alloys that display this behavior, superplasticity exists only over certain temperature and strain rate ranges, because the behavior depends on the microstructure, and the microstructure evolves dynamically. (As a consequence of this, the temperature and strain rate ranges of superplasticity depend on the microstructure as well.) The essential microstructural mechanism is grain rotation, which is accompanied by grain boundary sliding.

Increasing grain size leads to *increases* in flow stress in a superplastic material, because of difficulties in grain boundary rotation. Large grain sizes also reduce the maximum m value and the strain rate at which maximum m occurs. The problem with most metals is that, at elevated temperatures and creep strains, grain coarsening will occur, thus adversely affecting any potential superplastic behavior. Two phase alloys have an inherent grain size stability, especially if the two phases are isolated from each other, since it is much more difficult to move α/β boundaries than α/α or β/β ones. Thus, two-phase alloys have the best superplastic properties. In these alloys, the equiaxed α morphology is the better one, because lamellar α impedes GB sliding. The

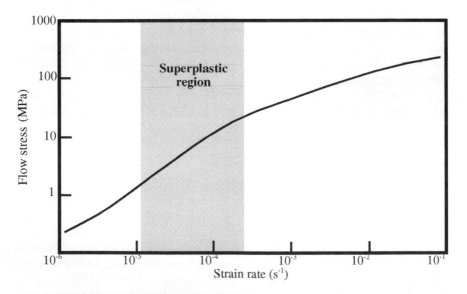

Figure 3.12   Log stress vs. log strain curve for Ti-6-4 at 870°C showing the superplastic region, which is the region of the highest strain rate sensitivity.[32]

optimum proportions of the two phases depend on the alloy and hot deformation conditions. For Ti-6-4, the optimum volume fraction of β is about 20%.

With regard to tooling, the industry standard alloy for dies for superplastic forming is ESCO 49C (Fe-22Cr-4Ni-9Mn-5Co). For higher temperatures, as in forming titanium aluminides, ceramic dies required. The dies are lubricated with boron nitride or yttria sprayed on with isopropyl alcohol as a carrier.

3.10.8.4 Forming. Titanium is not as workable as steel or aluminum, primarily because of springback variability, which is due to wide strength variations coupled with a relatively low modulus of elasticity. Additional problems include notch sensitivity, galling, and a high strain rate sensitivity. Some titanium alloys have been developed for cold formability, e.g., Ti-15V-3Sn-3Cr-3Al, β 21S, and β-C. Other alloys that exhibit acceptable room temperature formability include commercially pure and *ductile* titanium alloys such as Ti-3Al-8V-6Cr-4Zr-4Mo.

Titanium alloys other than those mentioned above possess more springback, requiring stress relieving operations between forming stages and more power to form the alloy. Thus, the slower the speed, the better the formability but the lower the productivity. Because of these ambient temperature problems, forming is done at elevated temperatures either wholly or partially (by cold preforming followed by hot sizing). Care must be exercised to minimize oxygen embrittlement.

For any metal, it is difficult to determine the formability from basic mechanical properties, since specific forming operations require specific property combinations. Nevertheless, as a general guideline, to minimize the possibility of splitting (e.g., in processes such as sheet stretching, brake forming, dimpling, etc.) the ductility is obviously the first indicator to consider. Another indicator is the product of the ductility (tensile elongation to fracture) and UTS. With regard to minimizing the probability of buckling (e.g., in deep drawing, spinning, roll forming, etc.), the ratio of the modulus of elasticity to yield strength is a good initial guide to formability.

Conditioning of sheet for forming. Because of notch sensitivity, scratches deeper than a finish produced by 180-grit particles must be eliminated. Tool or grinding marks can be reduced chemically with an aqueous solution of 30% (by vol.) HNO_3, plus not more than 3% HF. The ratio must be 10:1 or greater to minimize the possibility of H pickup. Edges should be smoothed and scale/heavy oxidation (due to temperatures >500°C) should be removed as well. Mechanical removal should be performed with fine abrasives, or by coarse abrasives and acid "softening" of abrasions, although this latter approach is less desirable. Acid pickling is also an option and, if the oxides are really tenacious, grit blasting followed by pickling could be employed.

Tool materials and lubricants. For cold forming, epoxy-faced aluminum or zirconium tools are used. Lubricants are similar for aluminum alloys. Other techniques include the use of a steel or plastic sheet with an auxiliary lubricant.

Blank preparation. Sheet up to 3.56 mm can be sheared with a cutter rated for low-carbon steel. Up to 4.75 mm, extra care and sharp cutters are required. With increasing thickness, there is an increasing tendency for deviations in straightness and/or cracking to occur. At such thicknesses, greater than 3.2 mm band sawing can be considered using abrasive blades. Cracking is avoided, but it generates large burrs.

Any nicks and burrs must be removed to prevent scratching of dies. This can be done by machining, sanding, or filing followed by a final deburr and polish. For sheared parts, at least 0.25 mm must be removed from edge; for punched holes, 0.35mm; for sawing, 6.35 mm or one thickness of sheet, whichever is smaller.

Forming temperatures. For cold forming, the workpiece is not heated, but the die can be heated to 150°C to facilitate forming. Pure titanium and alloys behave like cold rolled stainless steel. To produce severe contours, hot sizing may be required. Hot sizing or warm forming (260–315°C) can also reduce effect of variability of springback.

Hot sizing and creep forming. Both of these techniques employ creep mechanisms, with temperatures similar to those used for hot forming. Hot sizing, as implied by the name of the technique, can correctly size and shape the component, and it can smooth buckles and wrinkles. In this technique, pressure is applied only to ensure that the part and die surfaces are in contact. No further pressure is required and may even distort the die.

Creep forming differs from hot sizing in that this technique is used to form the component from a blank, as opposed to "finishing" a component formed at a previous stage. Here, pressure is applied to generate plastic flow at creep strain rates. The component is generally formed within a few minutes although, in certain cases, one or two hours may be necessary.

3.10.8.5 Machining. Titanium is one of the more difficult alloys to machine, because it is highly reactive, and this can lead to tool failures via the workpiece "welding" to the tool. Tool deterioration prior to failure will also generate surface defects on the workpiece. An additional problem is the relatively low stiffness of titanium, which requires that the workpiece be held in a stiffer configuration to prevent it from moving. Finally, titanium tends to generate a fairly deep, significantly work hardened layer during machining, which can greatly increase tool loading.

In view of these problems, the following guidelines should be followed:

- Low cutting speeds should be used to minimize temperature and maximize tool life. Titanium alloys should be machined at slower speeds than pure titanium.

- High feed rates should be used, since these enable the tool to get below the work-hardened layer. In addition, high feed rates do not increase the temperature as much as high cutting speeds.

- While tool and workpiece are in contact, feeding should never cease, since this will lead to work hardening, galling, smearing, and seizing.

- Rigid setups must be used to ensure accurate control of the cut depth.

- Use plenty of cutting fluid to cool the workpiece and wash away chips.

- Finally, using sharp tools will alleviate many of the problems encountered with titanium and its alloys.

Tool materials. Since abrasion resistance and hot hardness are required, general-purpose tool steels are not recommended. Instead, carbide tools, submicron sized cemented carbides, and relatively highly alloyed tool steels are appropriate.

Cutting fluids. Water-base fluids are more efficient than oils, since they also provide significant levels of cooling as well as lubrication. Chlorinated fluids can facilitate slow and difficult machining operations, but care must be taken to thoroughly clean the component after machining to avoid stress corrosion cracking problems during heat treatment or in service.

References

1. Chen, G.Z., Fray, D.J. and Farthing, T.W., *Nature,* September 21, 407, 2000, p. 361.
2. *Material Properties Handbook: Ti Alloys,* Boyer, R., Welsch, G. and Collings, E.W., eds., ASM International, Materials Park, Ohio, 1994.
3. Blackburn, M.J., *Trans. ASM,* 59, 1966, p. 876; Hammond, C. and Kelly, P.M., *Acta Met.* 1969, 17, 1969, p. 869.
4. deGélas, B. Séraphin, L., Tricot, R. and Castro, R., *Revue de Métallurgie,* 1, 1974, p. 51.
5. Weiss, I., Froes, F.H., Eylon, D. and Welsch. G.E., *Metall. Trans. A.,* 17A, 1986, p. 1935; Weiss, I., Welsch, G.E., Froes, F.H. and Eylon, D., *Proc. 5th International Conf. on Ti,* Munich, Germany, 1984, p. 1503.

6. Peters, M., and Lutjering, G., *Proc. Titanium 80,* 4th International Conf. on Ti, Kyoto, Japan, 1980, p. 925.
7. Hammond, C. and Nutting, J., Conference on Forging and Properties of Aerospace Materials, Metals Society, London, 1977, paper 5.
8. Williams, J.C. and Hickman, B.S., *Metall. Trans.,* 1, 1970, p. 2648.
9. Young, M., Levine, E. and Margolin, H., *Metall. Trans.,* 5, 1974, p. 1891.
10. Salmon, D.R., *Low Temperature Data Handbook, Ti And Ti Alloys,* National Physical Laboratory, NPL Report QU53 (N 80 23448), May 1979.
11. Collings, E.W., *The Physical Metallurgy of Titanium Alloys,* ASM, 1984.
12. Williams, J.C., *Metall. Trans.,* 4, 1973, p. 675.
13. Margolin, H., Farrar and Greenfield, *The Science, Technology and Application of Ti,* Jaffee and Promisel, eds., Pergamon, 1970, p. 795.
14. *Structural Alloys Handbook: 1982 Supplement,* Battelle's Columbus Labs, Columbus, OH.
15. Margolin, H., *Titanium Science and Technology,* 3/e, Jaffee and Burke, eds., Plenum Press, New York-London, 1973, p. 1709.
16. Hall, I.W., Ph.D. Thesis, Leeds University, Dept. of Metallurgy, UK, 1974.
17. Lutjering, G. and Gysler, A., *Proc. 5th International Conf. on Ti,* Munich, Germany, 1984, p. 2065.
18. Chesnutt, J.C., *Proc. 5th International Conf. on Ti,* Munich, Germany, 1984, p. 2227.
19. Broichaussen, J. and Calles, W., *Proc. 5th International Conf. on Ti,* Munich, Germany, 1984, p. 2195.
20. Nishihara, T. and Iguchi, N., *Nippon Kinzoku Gakkaishi,* 40, 1976, p. 51.
21. Covington, L.C. and Schutz, R.W., "Industrial Applications of Ti and Zr," *STP 728,* ASTM, 1981, p. 163.
22. Petit, J.A. et al., *Corrosion Sci.* 21, no. 4, 1981, p. 113; Gupta, V. P., Process for decreasing the rate of Ti corrosion, U.S. Patent 4,321,231, 1982.
23. Covington, L.C. and Schutz, R.W., *Corrosion Resistance of Ti,* TIMET Corp., 1982.
24. LaQue, R.L. and Copson, H.R., *Corrosion Resistance of Metals and Alloys,* 2/e, ACS Monograph, Reinhold, 1963, p. 646.
25. Smallwood, R.E., Industrial Applications of Ti and Zr, *STP 728,* ASTM, 1981, p. 147.
26. Schutz, R.W. and Grauman, J.S., "Ti 1986—Products and Applications," *Conference Proceedings,* San Francisco, CA, Ti Development Association, 1986.
27. Nagai, K., Hiraga, K., Ogata, T. and Ishikawa, K., *Adv. Cryo. Eng. (Materials),* 30, 1984, p. 375; Nagai, K, Hiraga, K., Ogata, T. and Ishikawa, K., *Trans. Jpn. Inst. Met.,* 26, 1985, p. 405.
28. Blenkinsop, J.C., *Proc. 5th International Conf. on Ti,* Munich, Germany, 1984, p. 2323.
29. Sastry, S.M.L., Peng, T.C., Meschter, P.J., and O'Neal, J.E., *J. Met.,* 35, 1983, p. 21.
30. R.R. Boyer, G. Luttering, *Advances in the Science and Technology of Titanium Alloy Processing,* p. 349–367.
31. Gray, H., Wagner, L., Luttering, G., *Shot Peening* (Oberfufsel, Germany: DGM, 1987) p. 467–475.
32. Hamilton, C.H. and Ghosh, A.K., *Metall. Trans. A.,* 11A, 1980, p. 1494.
33. Zhao, Linruo, Au, Peter, Beddoes, Jonathan C., and Wallace, William. Method to produce fine-grained lamellar microstructures in gamma titanium aluminides. U.S. patent No. 5,653,828, August 5, 1997.

4

Plastics

Jordan I. Rotheiser
Rotheiser Design Inc.
Highland Park, Illinois

4.1 Introduction

The term *plastics* is an extremely broad term, covering approximately three dozen basic materials and nearly 38,000 individual compounds.[1] That total changes daily as new compounds are added and old ones are dropped. However, the bulk of these compounds were specially formulated to compete for specific large-volume applications. Consequently, the number of plastics that most design engineers actually use is a small fraction of the total.

It is necessary for the design engineers to familiarize themselves with this family of materials, because the growth in their usage has been phenomenal. The volume of plastics used surpassed that of steels in 1979, and the gap has been increasing ever since. It is estimated to be two and one-half times that of steels at the time of this writing (2001). The advantages and disadvantages of plastics are as follows.

4.1.1 Advantages of Plastics

1. *Low product cost.* The principal reason engineers look to plastics as the material for their applications is to achieve a lower product cost. Note that the term *product cost* is used. That is because, with the possible exception of a few of the lowest-cost resins, most plastic materials are more expensive than their competitors. However, the processes by which they are converted to usable parts are extremely efficient. Extrusion, thermoforming, blow molding and, in some cases, injection molding can create extremely thin wall sections of less than 1 mm (0.040 in). Most of the remaining processes can produce wall thicknesses less than 2 mm (0.080 in). The ther-

moplastics processes can achieve nearly 100% utilization (there is some small loss due to contamination) through recycling of the in-process scrap. There are even some applications for thermoset scrap, although at a lower value.

2. *Weight reduction.*   Plastics generally weigh far less than natural materials (except for a few woods, such as balsa). This has been the driving force for many applications. The increased use of plastics has enabled automobile manufacturers to reduce the weight of each vehicle by approximately 25% in recent years, resulting in a corresponding improvement in fuel economy.

3. *Wide range of available properties.*   The range of properties available with the various plastics ranges from values too low to test to those that approach low-grade steels. Furthermore, there are plastics materials that offer properties that are unique to this class of materials.

4. *Highly complex integral shapes.*   The processing methods associated with traditional materials cannot create the kind of complex integral shapes that the plastics processing methods can accomplish. From an assembly standpoint, that permits the combination of many parts into one. In this way, many assembly operations are completely eliminated. Changes in wall thickness can create some variation in part stiffness, thereby providing additional opportunities for part consolidation.

5. *Corrosion resistance.*   Plastics do not corrode. Indeed, most of the paints used to protect other materials are essentially plastics themselves. Aside from the obvious benefits for outdoor applications, there are assembly advantages as well, as there need be no concern for galvanic action between metal components.

6. *Integral coloring.*   Painting is an expensive process and one in which consistency is difficult to maintain. Nearly all of the plastics materials and processes permit a vast spectrum of integral coloring. Not only is there a substantial cost savings, but surface abrasions do not remove the paint to reveal a different colored substrate. Thus, over time, a product improvement is provided. There are occasions, however, such as those where an identical color match is required or a special effect is desired, when painting is used on plastics.

   Integral coloring has some impact on assembly considerations as well. Color matching between parts is readily accomplished when they are of the same material, and with varying degrees of precision when they are of dissimilar materials. Multi-part injection molding permits parts of different colors or materials to be molded together, thus eliminating an assembly operation. Co-extrusion permits two-color extrusions that can be used alone, as in striped soda straws, or in combination with another process such as blow molding or thermoforming. This could create a product with clear and opaque or different colored segments.

7. *Transparency with flexibility.*   Glass, the traditional transparent material, is very brittle and breaks readily upon impact. It is also quite stiff and cannot be readily bent without cracking it. Very high temperatures are required to form it. There are a significant number of plastic materials in the acetate, polycarbonate, ABS, cellulosic, polyester, vinyl, polyethylene and polypropylene families that offer varying degrees of transparency with flexibility and/or ductility. Acrylic, while not flexible, is capable of offering even greater transparency than the finest lead glass.

These materials can be formed at relatively low temperatures and by the group of processes unique to the plastics industry. They can permit shapes that are impossible to duplicate with glass processes. These attributes provide product design opportunities and vastly increase the number of assembly techniques available to the design engineer.

8. *Compound customization.*   The chemical composition of plastics often allow them to be modified for specific characteristics for applications whose volume can warrant the cost. The most economical combination of properties can be combined to optimize the material utilization.

9. *New assembly techniques.*   Parts made of thermoplastics can be snap fitted, ultrasonically welded, induction welded, hot die welded, hot gas welded, spin welded, vibration welded, laser welded, and solvent welded. The traditional methods of assembly such as press fits, adhesive joining, staking, swaging and the use of fasteners can also be used for plastics. Threads and threaded inserts can be molded right into the part or can be added as a secondary operation.

10. *Insulation qualities.*   Plastics can provide both thermal and electrical insulating properties. Many plastics provide both, permitting the elimination of parts and assembly operations.

These advantages of plastics have been the driving forces in the success of the plastics industry. They have played a large part in the enhancement of mankind's standard of living. While it is still possible to purchase an automobile manufactured by traditional hand methods, the cost of such vehicles is 10 to 15 times that of the average car. Few could afford the price of products made without plastics. Nonetheless, they do have disadvantages.

4.1.2   Disadvantages of Plastics

1. *Variable properties.*   Plastics' physical and thermal properties can vary considerably with changes in wall thickness, temperature, humidity, processing parameters, gating locations, environmental conditions and chemical exposures. Testing procedures cannot begin to cope with the number of variables existing in product applications. Thus, the data is suspect and cannot be used empirically in engineering computations with high levels of reliability. Experienced plastics engineers rely on significant safety factors and testing of actual assemblies for high-risk applications.

2. *Thermal characteristics.*   Plastics' properties are affected by much smaller changes in temperature than are those of metals. This characteristic can affect product applications at temperatures in the range of normal human living conditions. Data sheets usually list physical properties at 72°F. Properties such as tensile strength and stiffness can drop off noticeably by the time the temperature has reached 100°F. As temperatures drop, ductility declines such that plastics that are normally thought of as relatively ductile at room temperature may actually be quite brittle when the temperature drops below freezing. Experienced engineers learn to take these characteristics into account when designing with plastics.

   The coefficient of linear thermal expansion (CLTE) for most plastics is much higher than it is for metals. The difference in CLTE between plastics and between most plastics and metals can be quite pronounced. This can lead to significant changes in the amount of stress on a given joint over the functional temperature range. The result may be stress cracks or disassembly of the product.

   Thermal degradation can cause plastics to lose their physical properties with prolonged exposure to heat. While this can take years at 23°C (72°F), it can occur in minutes at elevated temperatures. Each time the temperature of a thermoplastic is elevated, some degradation occurs. Therefore, some plastics can be reground and reused only a limited number of times. For many critical applications, only fresh resin, known as *virgin,* must be used. This characteristic must be taken into account when designing with plastics. In certain cases, a form of heat shielding is required.

3. *Chemical characteristics.*   Plastics are essentially chemical in nature. Consequently, most of them are significantly effected by

chemical exposures and environments. This can lead to disaster when the potential for chemical exposure is not taken into account. Cleaning solutions, greases, oils, acids, bases, gases, and other chemicals have led to the downfall of many plastics applications. Many of these exposures occur in the field, far from the eyes of the manufacturer's engineers. The potential chemical exposures must be thoroughly investigated for any product application. This can, however, be a desirable quality, as many plastics can be joined by solvent welding.

4. *Design complexity.*   Designing in plastic is more complex than designing in other materials. Not only are there a staggering number of resin options and a variety of processes to choose from, there are also tooling considerations to take into account. While everyone is eager to enjoy the benefits of cost reductions achievable through the use of plastics, many are unwilling to devote the resources to the project necessary to ensure success. It is important to recognize that one cannot take a design made of metal, for example, and simply convert it to plastic. Most engineers schooled in traditional engineering institutions find it difficult to fully understand the peculiarities of the plastics medium. Their projects encounter development problems that could be avoided, or they fail to take full advantage of the capabilities of the medium. Pressure to speed a new product to market often results in failure to fully investigate all the aspects of a given project. Each step skipped incurs additional risk. In short, it is more work to design in plastic.

5. *Flammability.*   Flammability is defined as relative ease of ignition and the ability to withstand combustion. Materials behave quite differently when exposed to open flame, and standards for flammability are difficult to establish. While a few polymers do not support their own combustion, it is fair to state that, in general, plastics burn—most of them quite readily. Flame retardants can be used, but they are expensive, and they have an effect on the other properties as well. For applications where there is the possible presence of an open flame, or where fire is a major risk, the flammability of the selected resin must be carefully examined.

6. *Ultraviolet light sensitivity.*   Ultraviolet light causes or catalyzes chemical degradation in many plastics. The result is photo-oxidation, which leads to the loss of color, transparency, and physical properties over a period of time. Some polymers exhibit a natural resistance to ultraviolet rays. Others require barrier coatings or additives, such as ultraviolet stabilizers and antioxidants, for pro-

longed life in environments where high concentrations of ultraviolet light are present (i.e., outdoors).

4.2   The Nature of Plastics

Practically stated, a plastic is an organic polymer, available in some resin form or some form derived from the basic polymerized resin. These forms can be liquid or paste-like resins for embedding, coating, and adhesive bonding; or they can be molded, laminated, or formed shapes, including sheet, film, or larger mass bulk shapes. The primary raw material sources of the major plastics are gas, coal, and crude oil.[4,5]

The number of basic plastic materials is large, and the number of variations and modifications to these basic plastic materials is also quite large. Taken together, the resultant quantity of materials available is just too great to be completely understood and correctly applied by anyone other than those whose day-to-day work puts them in direct contact with a diverse selection of materials. The practice of mixing brand names, trade names, and chemical names of various plastics only makes the problem of understanding these materials more troublesome.

Another variable that makes it difficult for those not versed in plastics to understand and properly design with plastics is the large number of processes by which these materials can be fabricated. Fortunately, there is an organized pattern on which an orderly presentation of these variables can be based.

While there are numerous minor classifications for polymers, depending on how one wishes to categorize them, nearly all can be placed into one of two major classifications—thermosetting materials (or thermosets) and thermoplastic materials.[3] Likewise, foams, adhesives, embedding resins, elastomers, and so on can be subdivided into the thermoplastic and thermosetting classifications.

4.2.1   Thermosetting Plastics

As the name implies, thermosetting plastics, or *thermosets,* are cured, set, or hardened into a permanent shape. Curing is an irreversible chemical reaction known as *cross-linking,* which usually occurs under heat. For some thermosetting materials, curing is initiated or completed at room temperature. Even here, however, it is often the heat of the reaction, or the *exotherm*, that actually cures the plastic material. Such is the case, for instance, with a room-temperature-curing epoxy or polyester compound.

The cross-linking that occurs in the curing reaction is brought about by the linking of atoms between or across two linear polymers, result-

ing in a three-dimensional rigidized chemical structure. One such reaction is shown in Fig. 4.1.[4] Although the cured part can be softened
by heat, it cannot be remelted or restored to the flowable state that existed before curing. Continued heating for long times leads to degradation or decomposition. None of the welding techniques used to join
thermoplastics can be used for thermosets.

4.2.2 Thermoplastics

Thermoplastics differ from thermosets in that they do not cure or set
under heat as do thermosets. Thermoplastics merely soften or melt to
a flowable state when heated, and under pressure they can be forced
or transferred from a heated cavity into a cool mold. Upon cooling in a

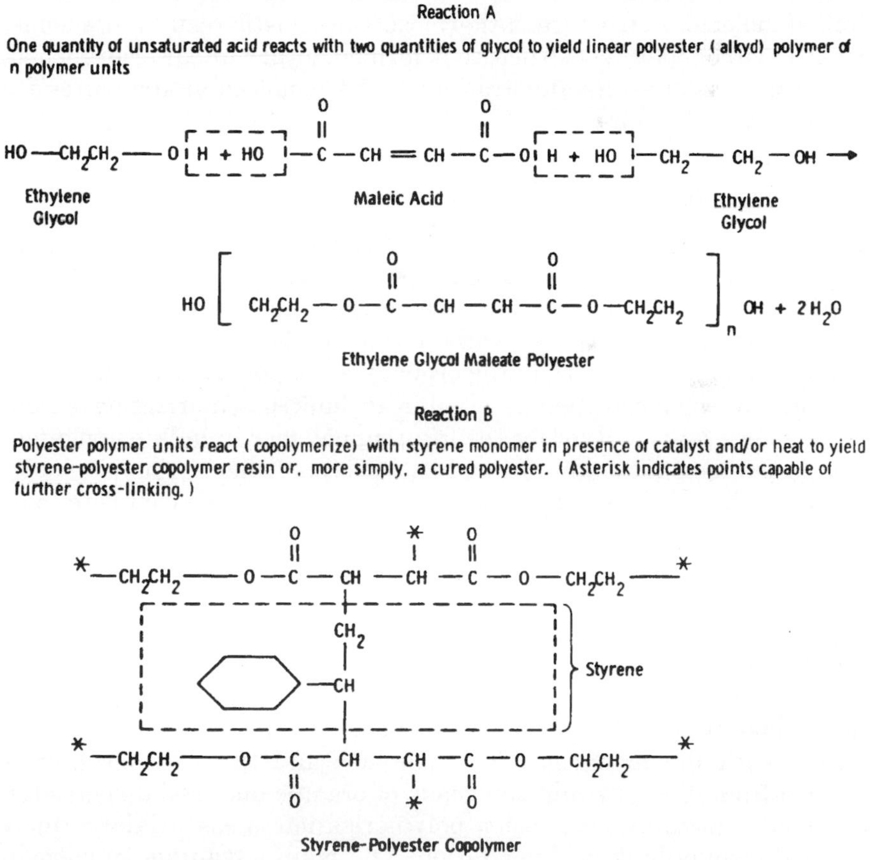

Reaction A

One quantity of unsaturated acid reacts with two quantities of glycol to yield linear polyester (alkyd) polymer of
n polymer units

Ethylene
Glycol

Maleic Acid

Ethylene
Glycol

Ethylene Glycol Maleate Polyester

Reaction B

Polyester polymer units react (copolymerize) with styrene monomer in presence of catalyst and/or heat to yield
styrene-polyester copolymer resin or, more simply, a cured polyester. (Asterisk indicates points capable of
further cross-linking.)

Styrene-Polyester Copolymer

Figure 4.1 Simplified diagrams showing how cross-linking reactions produce polyester
resin (styrene-polyester copolymer resin) from basic chemicals.

mold, thermoplastics harden and take the shape of the mold. Since thermoplastics do not cure or set, they can be remelted and then rehardened by cooling. Thermal aging, brought about by repeated exposure to the high temperatures required for melting, causes eventual degradation of the material and so limits the number of reheat cycles. Most of the common thermoplastic materials are discussed in detail in a following section of this chapter.

4.3   Polymer Structures and Polymerization Reactions

4.3.1   Polymer Structures

All polymers are formed by the creation of chemical linkages between relatively small molecules, or *monomers,* to form very large molecules, or *polymers.* As mentioned, if the chemical linkages form a rigid, cross-linked molecular structure, a thermosetting plastic results. If a somewhat flexible molecular structure is formed, either linear or branched, a thermoplastic results. Illustrations of these molecular structures are presented in Fig. 4.2.[2]

4.3.2   Polymerization Reactions

Polymerization reactions may occur in a number of ways; four common techniques are bulk, solution, suspension, and emulsion polymerization.[2,3] Bulk polymerization involves the reaction of monomers or reactants among themselves, without placing them in some form of extraneous media as is done in the other types of polymerization.

Solution polymerization is similar to bulk polymerization except that, whereas the solvent for the forming polymer in bulk polymerization is the monomer, the solvent in solution polymerization is usually a chemically inert medium. The solvents used may be complete, partial, or nonsolvents for the growing polymer chains.

Suspension polymerization normally is used only for catalyst-initiated or free-radical addition polymerizations. The monomer is dispersed mechanically in a liquid, usually water, which is a nonsolvent for the monomer as well as for all sizes of polymer molecules that form during the reaction. The catalyst initiator is dissolved in the monomer, and it is preferable that it does not dissolve in the water so that it remains with the monomer. The monomer and the polymer that is formed from it stay within the beads of organic material dispersed in the phase. Actually, suspension polymerization is essentially a finely divided form of bulk polymerization. The main advantage of suspension polymerization over bulk is that it allows cooling of the exothermic polymerization reaction and maintains closer control over the

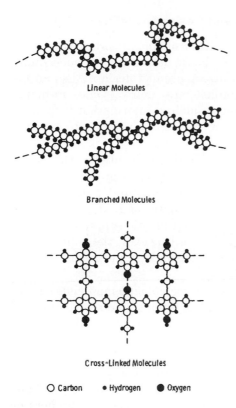

Linear Molecules

Branched Molecules

Cross-Linked Molecules

○ Carbon   ● Hydrogen   ● Oxygen

Figure 4.2 Some possible molecular structures in polymers.

chain-building process. By controlling the degree of agitation, monomer-to-water ratios, and other variables, it is also possible to control the particle size of the finished polymer, thus eliminating the need to reform the material into pellets from a melt, as is usually necessary with bulk polymerization.

Emulsion polymerization is a technique in which addition polymerizations are carried out in a water medium containing an emulsifier (a soap) and a water-soluble initiator. Emulsion polymerization is much more rapid than bulk or solution polymerization at the same temperatures and produces polymers with molecular weights much greater than those obtained at the same rate in bulk polymerizations.

In emulsion polymerization, the monomer diffuses into *micelles*, which are small spheres of soap film. Polymerization occurs within the micelles. Soap concentration, overall reaction-mass recipe, and reaction conditions can be varied to provide control of the reaction rate and yield.

4.4 Plastic-Processing Methods and Design Guidelines

Although most users of plastics buy parts from plastic processors, they should still have some knowledge of plastic processing, as such information can often be helpful in optimizing product design. Also, an increasing number of user companies are doing some in-house processing. For these reasons, some guideline information in plastic processing and some guidelines on the design of plastic parts are presented here.

It should be mentioned that the information presented at this point applies broadly to all classes of plastics and types of processing. Most plastic suppliers will provide very specific data and guidelines for their individual products. This invaluable source of guidance is too often unused. It is strongly recommended that plastic suppliers be more fully utilized for product-design guidance. However, the information presented at this point will be valuable for making initial design and process decisions.

Table 4.1[6] explains the major ways in which plastic materials can be formed into parts, and the advantages, limitations, and relative cost of each processing method. In general, a plastic part is produced by a combination of cooling, heating, flowing, deformation, and chemical reaction. As noted previously, the processes differ, depending on whether the material is a thermoplastic or a thermoset.

The usual sequence of processing a thermoplastic is to heat the material so that it softens and flows, force the material in the desired shape through a die or in a mold, and chill the melt into its final shape. By comparison, a thermoset is typically processed by starting out with partially polymerized material, which is softened and activated by heating (either in or out of the mold), forcing it into the desired shape by pressure, and holding it at the curing temperature until final polymerization reaches the point where the part hardens and stiffens sufficiently to keep its shape when demolded.

The cost of the finished part depends on the material and the process used. A very rough estimate of the finished cost of a part can be obtained by multiplying the material cost by a factor ranging from 1.5 to 10. The cost factors shown in Table 4.1 are based on general industry experience.

Table 4.2 gives guidelines on part design for the various plastic-processing methods listed in Table 4.1. The design of a part frequently depends on the processing method selected to make the part. Also, of course, selection of the best processing method frequently is a function of the part design. Major plastic-processing methods and their respective design capabilities, such as minimum section thicknesses and radii and overall dimensional tolerances, are listed in Table 4.2. The basic purpose of this guide is to show the fundamental design limits of the many plastic-processing methods.

4.4.1   Plastic-Fabrication Processes and Forms

There are many plastic-fabrication processes, and a wide variety of plastics can be processed by each of these processes or techniques. Fabrication processes can be broadly divided into pressure processes and pressureless or low-pressure processes. Pressure processes, which constitute the bulk of plastics production, are either thermoplastic-materials processes (such as injection molding, extrusion, and thermoforming) or thermosetting processes (such as compression molding, transfer molding, and laminating).

Compression molding and transfer molding. Compression molding and transfer molding are the two major processes used for forming molded parts from thermosetting raw materials. The two can be carried out in the same type of molding press, but different types of molds are used. The thermosetting materials are normally molded by the compression or transfer process, but it is also possible to mold thermoplastics by these processes, since the heated thermoplastics will flow to conform to the mold-cavity shape under suitable pressure. These processes are usually impractical for thermoplastic molding, however, since, after the mold cavity is filled to its final shape, the heated mold would have to be cooled to solidify the thermoplastic part. Since repeated heating and cooling of this large mass of metal and the resultant long cycle time per part produced are both objectionable, injection molding is commonly used to process thermoplastics.

Compression molding. In compression molding, the open mold is placed between the heated platens of the molding press, filled with a given quantity of molding material, and closed under pressure, causing the material to flow into the shape of the mold cavity. The actual pressure required depends on the molding material being used and the geometry of the mold. The mold is kept closed until the plastic material is suitably cured. Then the mold is opened, the part ejected, and the cycle repeated. The mold is usually made of steel with a polished or plated cavity. However, special grades of aluminum are sometimes used with low-pressure molding compounds.

Transfer molding. In transfer-molding sequence, the molding material is first placed in a heated pot, separate from the mold cavity. The hot

TABLE 4.1   Descriptions and Guidelines for Plastic-Processing Methods[5]

Process	Description	Key advantages	Notable limitations	Cost factor*
Blow molding	An extruded tube (parison) of heated thermoplastic is placed between two halves of an open split mold and expanded against the sides of the closed mold by air pressure. The mold is open, and the part is ejected	Low tool and die costs; rapid production rates; ability to mold relatively complex hollow shapes in one piece	Limited to hollow or tubular parts; wall thickness and tolerances often hard to control	1.5–5, 2–3
Calendering	Doughlike thermoplastic mass is worked into a sheet of uniform thickness by passing it through and over a series of heated or cooled rolls. Calenders also are used to apply plastic covering to the back of other materials	Low cost; sheet materials are virtually free of molded-in stresses; i.e., they are isotropic	Limited to sheet materials; very thin films not possible	1.5–3, 2–5.5
Casting	Liquid plastic (usually thermoset except for acrylics) is poured into a mold (without pressure), cured, and removed from the mold. Cast thermoplastic films are made by depositing the material, either in solution or in hot-melt form, against a highly polished supporting surface	Low mold cost; ability to produce large parts with thick cross sections; good surface finish; suitable to low-volume production	Limited to relatively simple shapes; except for cast films, becomes uneconomical at high-volume production levels; most thermoplastics not suitable	1.5–3, 2–2.5
Compression molding	A thermoplastic or partially polymerized thermosetting resin compound, usually preformed, is placed in a heated mold cavity; the mold is closed, heat and pressure are applied, and the material flows and fills the mold cavity. Heat completes polymerization, and the mold is opened to remove the part. The process is sometimes used for thermoplastics, e.g., vinyl phonograph records	Little waste of material and low finishing costs; large, bulky parts are possible	Extremely intricate parts involving undercuts, side draws, small holes, delicate inserts, etc., not practical; very close tolerances difficult to produce	2–10, 1.5–3
Cold forming	Similar to compression molding in that material is charged into split mold; it differs in that it uses no heat—only pressure. Part is cured in an oven in a separate operation. Some thermoplastic sheet material and billets are cold-formed in process similar to drop hammer–die forming of metals. Shotgun shells are made in this manner from polyethylene billets	Ability to form heavy or tough-to-mold materials; simple; inexpensive; often has rapid production rate	Limited to relatively simple shapes; few materials can be processed in this manner	

Process	Description	Characteristics	Limitations		
Extrusion	Thermoplastic or thermoset molding compound is fed from a hopper to a screw and barrel where it is heated to plasticity and then forwarded, usually by a rotating screw, through a nozzle having the desired cross-section configuration	Low tool cost; great many complex profile shapes possible; very rapid production rates; can apply coatings or jacketing to core materials, such as wire	Limited to sections of uniform cross section	2–5,	3–4
Filament winding	Continuous filaments, usually glass, in form of rovings are saturated with resin and machine-wound onto mandrels having shape of desired finished part. Once winding is completed, part and mandrel are placed in oven for curing. Mandrel is then removed through porthole at end of wound part	High-strength fiber reinforcements are oriented precisely in direction where strength is needed; exceptional strength/weight ratio; good uniformity of resin distribution in finished part	Limited to shapes of positive curvature; openings and holes reduce strength	5–10,	6–8
Injection molding	Thermoplastic or thermoset molding compound is heated to plasticity in cylinder at controlled temperature; then forced under pressure through a nozzle into sprues, runners, gates, and cavities of mold. The resin solidifies rapidly, the mold is opened and the part(s) ejected. In modified version of process—runnerless molding—the runners are part of mold cavity	Extremely rapid production rates, hence low cost per part; little finishing required; good dimensional accuracy; ability to produce relatively large, complex shapes; very good surface finish	High initial tool and die costs; not practical for small runs	1.5–5,	2–3
Laminating, high pressure	Material, usually in form of reinforcing cloth, paper, foil, etc., preimpregnated or coated with thermoset resin (sometimes a thermoplastic), is molded under pressure greater than 1,000 lb/in.^{2} into sheet, rod, tube, or other simple shape	Excellent dimensional stability of finished product; very economical in large production of parts	High tool and die costs; limited to simple shapes and cross-section profiles	2–5,	3–4
Matched-die molding	A variation of conventional compression molding, this process uses two metal molds having a close-fitting, telescoping area to seal in the plastic compound being molded and to trim the reinforcement. The reinforcement, usually mat or preform, is positioned in the mold, and the mold is closed and heated (pressures generally vary between 150 and 400 lb/in.^{2}). Mold is then opened and part lifted out	Rapid production rates; good quality and reproducibility of parts	High mold and equipment costs; parts often require extensive surface finishing, e.g., sanding	2–5,	3–4

TABLE 4.1   Descriptions and Guidelines for Plastic-Processing Methods (Continued)

Process	Description	Key advantages	Notable limitations	Cost factor*
Rotational molding	A predetermined amount of powdered or liquid thermoplastic or thermoset material is poured into mold. Mold is closed, heated, and rotated in the axis of two planes until contents have fused to inner walls of mold. The mold is opened and part removed	Low mold cost; large hollow parts in one piece can be produced; molded parts are essentially isotropic in nature	Limited to hollow parts; in general, production rates are slow	1.5–5, 2–3
Slush molding	Powdered or liquid thermoplastic material is poured into a mold to capacity. Mold is closed and heated for a predetermined time to achieve a specified buildup of partially cured material on mold walls. Mold is opened, and unpolymerized material is poured out. Semifused part is removed from mold and fully polymerized in oven	Very low mold costs; very economical for small-production runs	Limited to hollow parts; production rates are very slow; limited choice of materials that can be processed	1.5–4, 2–3
Thermoforming	Heat-softened thermoplastic sheet is placed over male or female mold. Air is evacuated from between sheet and mold, causing sheet to conform to contour of mold. There are many variations, including vacuum snapback, plug assist, drape forming, etc.	Tooling costs generally are low; produces large parts with thin sections; often economical for limited production of parts	In general, limited to parts of simple configuration; limited number of materials to choose from; high scrap	2–10, 3–5
Transfer molding	Thermoset molding compound is fed from hopper into a transfer chamber where it is heated to plasticity. It is then fed by means of a plunger through sprues, runners, and gates of closed mold into mold cavity. Mold is opened and the part ejected	Good dimensional accuracy; rapid production rate; very intricate parts can be produced	Molds are expensive; high material loss in sprues and runners; size of parts is somewhat limited	1.5–5, 2–3
Wet lay-up or contact molding	Number of layers, consisting of a mixture of reinforcement (usually glass cloth) and resin (thermosetting), are placed in mold and contoured by roller to mold's shape. Assembly is allowed to cure (usually in an oven) without application of pressure. In modification of process, called spray molding, resin systems and chopped fibers are sprayed simultaneously from spray gun against mold surface; roller assist also is used. Wet lay-up parts sometimes are cured under pressure, using vacuum bag, pressure bag, or autoclave	Very low cost; large parts can be produced; suitable for low-volume production of parts	Not economical for large-volume production; uniformity of resin distribution very difficult to control; mainly limited to simple shapes	1.5–4, 2–3

* Material cost × factor = purchase price of a part: top figure is overall range, bottom is probable average cost.

TABLE 4.2 Guidelines on Part Design for Plastic-Processing Methods[3]

Design rules	Blow molding	Casting	Compression molding	Extrusion	Injection molding	Reinforced plastic molding			Rotational molding	Thermo-forming	Transfer molding
						Wet lay-up (contact molding)	Matched-die molding	Filament winding			
Major shape characteristics	Hollow bodies	Simple configurations	Moldable in one plane	Constant cross-section profile	Few limitations	Moldable in one plane	Moldable in one plane	Structure with surfaces of revolution	Hollow bodies	Moldable in one plane	Simple configurations
Limiting size factor	M	M	ME	M	ME	MS	ME	WE	M	M	ME
Min inside radius, in.	0.125	0.01–0.125	0.125	0.01–0.125	0.01–0.125	0.25	0.06	0.125	0.01–0.125	0.125	0.01–0.125
Undercuts	Yes	Yes[a]	NR[b]	Yes	Yes[a]	Yes	NR	NR	Yes[e]	Yes[a]	NR
Min draft, degrees	0	0–1	>1	NA[b]	<1	0	1	2–3	1		1
Min thickness, in.	0.01	0.01–0.125	0.01–0.125	0.001	0.015	0.06	0.03	0.015	0.02	0.002	0.01–0.125
Max thickness, in.	>0.25	None	0.5	6	1	0.5	1	3	0.5	3	1
Max thickness buildup, in.	NA	2–1	2–1	NA	2–1	2–1	2–1[d]	NR	NA	NA	2–1
Inserts	Yes	Yes	Yes	Yes	Yes	Yes	Yes	Yes	Yes	NR	Yes
Built-up cores	Yes	Yes	No	Yes	Yes	Yes	Yes	Yes	Yes	Yes	Yes
Molded-in holes	Yes	Yes	Yes	Yes[e]	Yes	Yes	Yes	No	Yes	No	Yes
Bosses	Yes	Yes	Yes	Yes	Yes	Yes	No[f]	No[d]	Yes	Yes	Yes
Fins or ribs	Yes	Yes	Yes	Yes	Yes	Yes	No[f]	No	Yes	Yes	Yes
Molded-in designs and nos.	Yes	Yes	Yes	No	Yes	Yes	Yes	No	Yes	Yes	Yes
Overall dimensional tolerance, in./in.	±0.01	±0.001	±0.001	±0.005	±0.001	±0.02	±0.005	±0.005	±0.01	±0.01	±0.001
Surface finish[h]	1–2	2	1–2	1–2	1	4–5	4–5	5	2–3	1–3	1–2
Threads	Yes	Yes	Yes	No	Yes	No	No	No	Yes	No	Yes

M = material. ME = molding equipment. MS = mold size. WE = winding equipment.
[a] Special molds required.
[b] NR—not recommended; NA—not applicable.
[c] Only with flexible materials.
[d] Using premix: as desired.
[e] Only in direction of extrusion.
[f] Using premix: yes.
[g] Possible using special techniques.
[h] Rated 1 to 5:1 = very smooth, 5 = rough.

plastic material is then transferred under pressure from the pot through the runners into the closed cavity of the mold.

The advantage of transfer molding lies in the fact that the mold proper is closed at the time the material enters. Parting lines that might give trouble in finishing are held to a minimum. Inserts are positioned, and delicate steel parts of the mold are not subject to movement. Vertical dimensions are more stable than in straight compression. Also, delicate inserts can often be molded by transfer molding, especially with the low-pressure molding compounds.

Injection molding. Injection molding is the most practical process for molding thermoplastic materials. The operating principle is simple, but the equipment is not.

A material with thermoplastic qualities—one that is viscous at some elevated temperature and stable at room temperature without appreciable deterioration during the cycle—is maintained in a heated reservoir. This hot, soft material is forced from the reservoir into a cool mold. The mold is opened as soon as the material has cooled enough to hold its shape on demolding. The cycle speed is determined by the rapidity with which the temperature of the material used can be reduced, which in turn depends on the thermal conductivity of that material. Acrylics are slow performers, and styrenes are among the fastest.

The machine itself is usually a horizontal cylinder, the bore of which determines the capacity. Within the bore is a screw. Plastic resin enters the cylinder through a hole in the top of the cylinder to replace the charge shot into the mold. The cylinder is heated by electric bands that permit temperature variation along its length. Inside the cylinder, there is a screw similar to that of an extruder. The screw turns, forcing the resin down the cylinder and melting it as it goes. For a shot to enter the mold, the entire screw moves forward, forcing the plastic through the nozzle into the mold. After a delay to permit the part to cool enough to withstand the forces of ejection, the finished part is ejected from the mold. The mold opens and closes automatically, and the whole cycle is controlled by timers.

Thermoset injection molding. Because of the chemical nature of the plastic materials, injection molding has traditionally been the primary molding method for thermoplastics, and compression and transfer molding have been the primary molding methods for thermosetting plastics. Because of the greater molding cycle speeds and lower molding costs in injection molding, thermoplastics have had a substantial molding cost advantage over thermosets. As a result, advances in equipment and in thermosetting molding compounds have resulted in a rapid transition to screw-injection, in-line molding. This has been especially prominent with phenolics, but other thermosets are also included to varying degrees. The growth in screw-injection molding of phenolics has been extremely rapid. The development of this technique allows the molder to automate further, reduce labor costs, improve quality, reduce rejects, and gain substantial overall molding cycle efficiency.

Extrusion and pultrusion. The process of extrusion consists basically of forcing heated, melted plastic continuously through a die that has an opening shaped to produce a desired finished cross section. Normally, it is used for processing thermoplastic materials, but it can also be used for processing thermosetting materials. The main application of extrusion is the production of continuous lengths of film, sheeting, pipe, filaments, wire jacketing, and other useful forms and cross sections. After the plastic melt has been extruded through the die, the extruded material is hardened by cooling, usually by air or water.

Extruded thermosetting materials are used increasingly in wire and cable coverings. The main object here is the production of shapes, parts, and tolerances not obtainable in compression or transfer molding. Pultrusion is a special, increasingly used technique for pulling resin-soaked fibers through an orifice, at it offers significant strength improvements. Any thermoset, B-stage, granular molding compound can be extruded, and almost any type of filler may be added to the compound. In fiber-filled compounds, the length of fiber is limited only by the cross-sectional thickness of the extruded piece.

A metered volume of molding compound is fed into the die feed zone, where it is slightly warmed. As the ram forces the compound through the die, the compound is heated gradually until it becomes semifluid. Before leaving the die, the extruded part is cured by controlling the time it takes to travel through a zone of increasing temperature. The cured material exits from the die at temperatures of 300 to 350°F and at variable rates.

Thermoforming. Thermoforming is a relatively simple basic process that consists of heating a plastic sheet and forming it to conform to the shape of the mold, either by differential air pressure or by some mechanical means. By this processing technique, thermoplastic sheets can be converted rapidly and efficiently to a myriad of shapes, the thicknesses of which depend on the thickness of the film being used and the processing details of the individual operations. Although there are many variations of this process, they generally involve heating the plastic sheet and making it conform to the contour of a male or female form, either by air pressure or a matching set of male and female molds.

4.5   Thermosetting Plastics

Plastic materials included in the thermosetting plastic category and discussed separately in this section are alkyds, diallyl phthalates, epoxies, melamines, phenolics, polyesters, silicones, and ureas. In general, unfilled thermosetting plastics tend to be harder, more brittle, and not as tough as thermoplastics. Thus, it is common practice to add fillers to thermosetting materials. A wide variety of fillers can be used for varying product properties. For molded products, usually compression or transfer molding, mineral or cellulose fillers are often used as lower-cost, general-purpose fillers, and glass fiber fillers are often used for optimum strength or dimensional stability. There are always product and processing trade-offs, but a general guide to the application of fillers is given in Seymour.[7] It should be added that filler form and filler surface treatment can also be major variables. Thus, it is important to consider fillers along with the thermosetting material, especially for molded products. Other product forms may be filled or unfilled, depending on requirements.

4.5.1   Alkyds—Thermosets

Alkyds are available in granular, rope, and putty form, some suitable for molding at relatively low pressures and at temperatures in the range of 300 to 400°F. They are formulated from polyester-type resins, in general reactions as shown in Fig. 4.1. Other possible monomers, aside from styrene, are diallyl phthalate and methyl methacrylate. Alkyd compounds are chemically similar to the polyester compounds but make use of higher-viscosity, or dry, monomers. Alkyd compounds often contain glass fiber filler but may include clay, calcium carbonate, or alumina, for example.

These unsaturated resins are produced through the reaction of an organic alcohol with an organic acid. The selection of suitable polyfunctional alcohols and acids permits selection of a large variation of repeating units. Formulating can provide resins that demonstrate a wide range of characteristics involving flexibility, heat resistance, chemical resistance, and electrical properties. Typical properties of alkyds are shown in Harper[2] and Ref. 8.

Alkyds are easy to mold and economical to use. Molding dimensional tolerances can be held to within ±0.001 in/in. Postmolding shrinkage is small, as shown in Fig. 4.3.[8] Their greatest limitation is in extremes of temperature (above 350°F) and humidity. Silicones and diallyl phthalates are superior here, silicones especially with respect to temperature and diallyl phthalates with respect to humidity.

4.5.2   Diallyl Phthalate (Allyl) (DAP, DAIP)—Thermosets

Diallyl phthalates, or allyls, are among the best of the thermosetting plastics with respect to high insulation resistance and low electrical losses, which are maintained up to 400°F or higher, and in the presence of high-humidity environments. Also, diallyl phthalate resins offer high heat and chemical resistance and are easily molded and fabricated. The coefficient of linear thermal expansion for some DAP compounds is low enough to be comparable to steel.

There are several chemical variations of diallyl phthalate resins, but the two most commonly used are diallyl phthalate (DAP) and diallyl isophthalate (DAIP). The primary application difference is that DAIP will withstand somewhat higher temperatures than will DAP. Typical properties of DAP and DAIP molding compounds are shown in Harper[2] and Ref. 9. The excellent dimensional stability of DAPs is

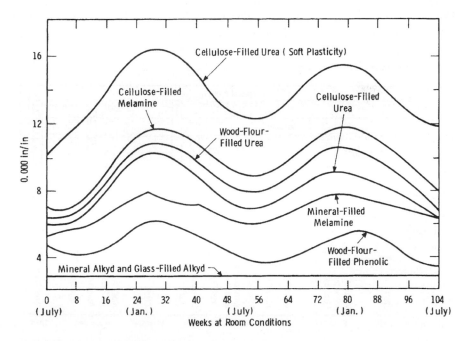

Figure 4.3   Stability of alkyds. Postmolding shrinkage variation of several molding compounds over a period of weeks.[8]

demonstrated in Fig. 4.4,[10] which compares them to other plastic materials at various temperatures.

DAPs are extremely stable, (most stable of all thermosets) having very low after-shrinkage, on the order of 0.1 percent, based on the MIL-M-14F dimensional stability test. The ultimate in electrical properties is obtained by the use of the synthetic fiber fillers. However, these materials are expensive and have high mold shrinkage and a strong, flexible flash that is extremely difficult to remove from the parts. Consequently, the largest commercial volume is in the short-glass-fiber-filled materials, which combine moldability in thin sections with extremely high tensile and flexural strengths. Molded parts are typically joined with adhesives and fasteners.

4.5.3   Epoxies (EP)—Thermosets

Types of epoxies.   Epoxy resins are characterized by the epoxide group (oxirane rings).[11,12] The most widely used resins are diglycidyl ethers of bisphenol A (Fig. 4.5). These are made by reacting epichlorohydrin with bisphenol A in the presence of an alkaline catalyst. By controlling operating conditions and varying the ratio of epichlorohydrin to bisphenol A, products of different molecular weights can be made. For liquid resins, the n of Fig. 4.5 is generally less than 1; for solid resins, n is 2 or greater. Solids with very high melting points have n values as high as 20.

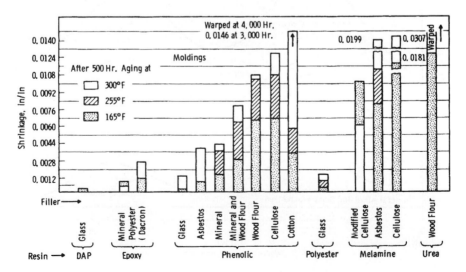

Figure 4.4 Shrinkage of various thermosetting molding materials as a result of heat aging.[10]

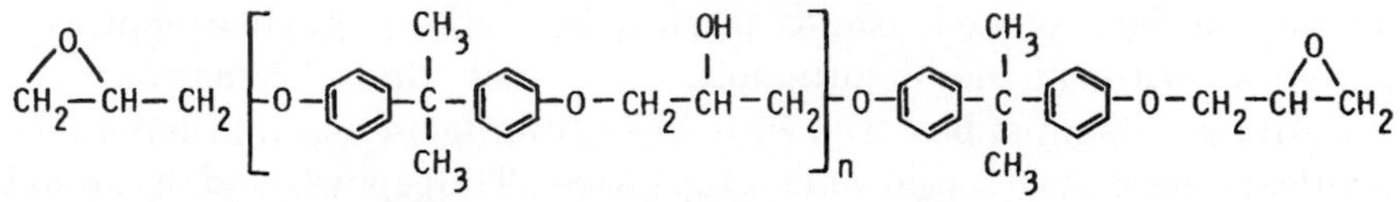

Figure 4.5 General structure of epoxy (diglycidyl ether of bisphenol A).

Another class of epoxy resins is the novolacs, particularly the epoxy cresols (Fig. 4.6) and the epoxy phenol novolacs. These are produced by reacting a novolac resin, usually formed by the reaction of o-cresol or phenol and formaldehyde with epichlorohydrin. These highly functional materials are particularly recommended for transfer-molding powders, electrical laminates, and parts where superior thermal properties, high resistance to solvents and chemicals, and high reactivity with hardeners are needed.

Another group of epoxy resins, the cycloaliphatics (Fig. 4.6), are particularly important when superior arc-track and weathering resistance are necessary requirements.

A distinguishing feature of cycloaliphatic resins is the location of the epoxy group(s) on a ring structure rather than on the aliphatic chain. Cycloaliphatics can be produced by the peracetic epoxidation of cyclic

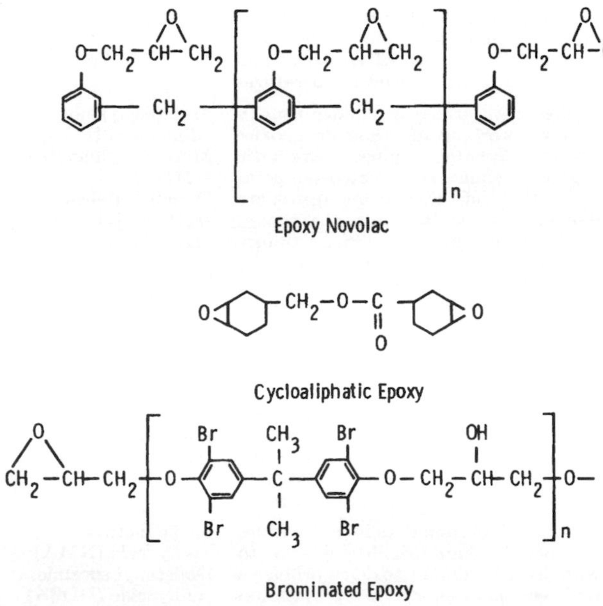

Epoxy Novolac

Cycloaliphatic Epoxy

Brominated Epoxy

Figure 4.6 General structures of novolac and cycloaliphatic and brominated epoxies.

olefins and by the condensation of an acid such as tetrahydrophthalic anhydrite with epichlorohydrin, followed by dehydrohalogenation.

Epoxy resins must be cured with cross-linking agents (hardeners) or catalysts to develop desirable properties. The epoxy and hydroxyl groups are the reaction sites through which cross-linking occurs. Useful agents include amines, anhydrides, aldehyde condensation products, and Lewis acid catalysts. Careful selection of the proper curing agent is required to achieve a balance of application properties and initial handling characteristics.[11,12] Major types of curing agents are aliphatic amines, aromatic amines, catalytic curing agents, and acid anhydrides, as shown in Table 4.3.

Aliphatic amine curing agents produce a resin-curing agent mixture that has a relatively short working life but that cures at room temper-

TABLE 4.3   Curing Agents for Epoxy Resins

Curing-agent type	Characteristics	Typical materials
Aliphatic amines	Aliphatic amines allow curing of epoxy resins at room temperature, and thus are widely used. Resins cured with aliphatic amines, however, usually develop the highest exothermic temperatures during the curing reaction, and therefore the mass of material which can be cured is limited. Epoxy resins cured with aliphatic amines have the greatest tendency toward degradation of electrical and physical properties at elevated temperatures	Diethylene triamine (DETA) Triethylene tetramine (TETA)
Aromatic amines	Epoxies cured with aromatic amines usually have a longer working life than do epoxies cured with aliphatic amines. Aromatic amines usually require an elevated-temperature cure. Many of these curing agents are solid and must be melted into the epoxy, which makes them relatively difficult to use. The cured resin systems, however, can be used at temperatures considerably above those which are safe for resin systems cured with aliphatic amines	Metaphenylene diamine (MPDA) Methylene dianiline (MDA) Diamino diphenyl sulfone (DDS or DADS)
Catalytic curing agents	Catalytic curing agents also have a working life better than that of aliphatic amine curing agents and, like the aromatic amines, normally require curing of the resin system at a temperature of 200°F or above. In some cases, the exothermic reaction is critically affected by the mass of the resin mixture	Piperidine Boron trifluoride ethylamine complex Benzyl dimethyl-amine (BDMA)
Acid anhydrides	The development of liquid acid anhydrides provides curing agents which are easy to work with, have minimum toxicity problems compared with amines, and offer optimum high-temperature properties of the cured resins. These curing agents are becoming more and more widely used	Nadic methyl anhydride (NMA) Dodecenyl succinic anhydride (DDSA) Hexahydrophthalic anhydride (HHPA) Alkendic anhydride

ature or at low baking temperatures in relatively short time. Resins cured with aliphatic amines usually develop the highest exothermic temperatures during the curing reaction; thus, the amount of material that can be cured at one time is limited because of possible cracking, crazing, or even charring of the resin system if too large a mass is mixed and cured. Also, physical and electrical properties of epoxy resins cured with aliphatic amines tend to degrade as the operating temperature increases. Epoxies cured with aliphatic amines find their greatest usefulness where small masses can be used, where room-temperature curing is desirable, and where the operating temperature required is below 100°C.

Epoxies cured with aromatic amines have a considerably longer working life than do those cured with aliphatic amines, but they require curing at 100°C or higher. Resins cured with aromatic amines can operate at a temperature considerably above the temperature necessary for those cured with aliphatic amines. However, aromatic amines are not so easy to work with as aliphatic amines, because of the solid nature of the curing agents and the fact that some (such as metaphenylene diamine) sublime when heated, causing stains and residue deposition.

Catalytic curing agents also have longer working lives than the aliphatic amine materials and, like the aromatic amines, catalytic curing agents normally require curing of the epoxy system at 100°C or above. Resins cured with these systems have good high-temperature properties as compared with epoxies cured with aliphatic amines. With some of the catalytic curing agents, the exothermic reaction becomes high as the mass of the resin mixture increases, as shown in Fig. 4.7.

Acid anhydride curing agents are particularly important for epoxy resins, especially the liquid anhydrides. The high-temperature proper-

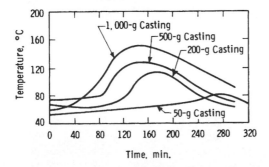

Figure 4.7 Exothermic curves as a function of resin mass for bisphenol epoxy and 5 percent piperidene curing agent (catalytic type) cured at 60°C.

ties of resin systems cured with these materials are better than those of resin systems cured with aromatic amines. Some anhydride-cured epoxy-resin systems retain most electrical properties to 150°C and higher and are little affected physically, even after prolonged heat aging at 200°C. In addition, the liquid anhydrides are extremely easy to work with; they blend easily with the resins and reduce the viscosity of the resin system. Also, the working life of the liquid acid anhydride systems is long compared with that of mixtures of aliphatic amine and resin, and odors are slight. Amine promoters such as benzyl dimethylamine (BDMA) or DMP-30 are used to promote the curing of mixtures of acid anhydride and epoxy resin. The thermal stability of epoxies is improved by anhydride curing agents, as shown in Fig. 4.8.

Applications of Epoxies.   Epoxies are among the most versatile and widely used plastics in the electronics field. This is primarily because of the wide variety of formulations possible and the ease with which these formulations can be made and utilized with minimal equipment requirements. Formulations range from flexible to rigid in the cured state and from thin liquids to thick pastes and molding powders in the uncured state. Conversion from uncured to cured state is made by use of hardeners or heat, or both. The largest application of epoxies is in embedding applications (potting, casting, encapsulating, and impregnating), in molded parts, and in laminated constructions such as metal-clad laminates for printed circuits and unclad laminates for various types of insulating and terminal boards.

Molded parts have excellent dimensional stability, as shown in Fig. 4.4, and some compounds have a coefficient of linear thermal ex-

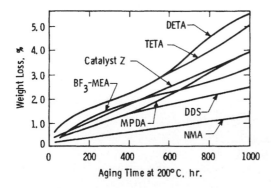

Figure 4.8 Weight-loss data at 200°C for Epon 828 cured with various curing agents. DETA, TETA = aliphatic amines; MPDA, DDS = aromatic amines; NMA = liquid acid anyhdride.

pansion as low as steel. Applications include ignition coils, high-voltage insulators, bushings, switch gear, and semiconductors. Molded epoxy parts are typically joined with adhesives and fasteners.

Epoxies are used in prototype and short-run molds for several plastics processes.

4.5.4 Melamine (MF, MPF) and Urea (UF) (Aminos)—Thermosets

As compared with alkyds, diallyl phthalates, and epoxies, which are polymers created by addition reactions and hence have no reaction by-products, melamines and ureas (also commonly referred to as *aminos*) are polymers that are formed by condensation reactions and do give off by-products. Another example of this type of reaction is the polymerization reaction, which produces phenolics. A typical condensation reaction is shown in Sec. 4.5.5 on phenolics. Melamines and ureas are a reaction product of formaldehyde with amino compounds containing NH_2 groups. Hence, they are often also referred to as melamine formaldehydes and urea formaldehydes. Their general chemical structure is shown in Fig. 4.9.

Amino resins have found applications in the fields of industrial and decorative laminating, adhesives, protective coatings, textile treatment, paper manufacture, and molding compounds. Their clarity permits products to be fabricated in virtually any color. Finished products having an amino-resin surface exhibit excellent resistance to moisture, greases, oils, and solvents; are tasteless and odorless; are self-extinguishing; offer excellent electrical properties; and resist scratching and marring. The melamine resins offer better chemical, heat, and moisture resistance than do the ureas.[13]

Amino molding compounds can be fabricated by economical molding methods. They are hard, rigid, and abrasion-resistant, and they have high resistance to deformation under load. These materials can be exposed to subzero temperatures without embrittlement. Under tropical conditions, the melamines do not support fungus growth.

Amino materials are self-extinguishing and have excellent electrical insulation characteristics. They are unaffected by common organic solvents, greases and oils, and weak acids and alkalis. Melamines are superior to ureas in resistance to acids, alkalis, heat, and boiling water, and are preferred for applications involving cycling between wet and dry conditions or rough handling. Aminos do not impart taste or odor to foods.

Addition of alpha cellulose filler, the most commonly used filler for aminos, produces an unlimited range of light-stable colors and high degrees of translucency. Colors are obtained without sacrifice of basic material properties. Shrinkage characteristics with cellulose filler are shown in Fig. 4.3.

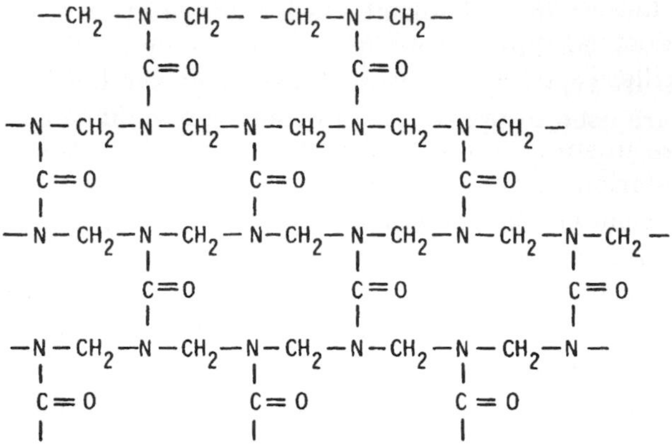

$$-CH_2-N-CH_2- -CH_2-N-CH_2-$$

Urea Formaldehyde

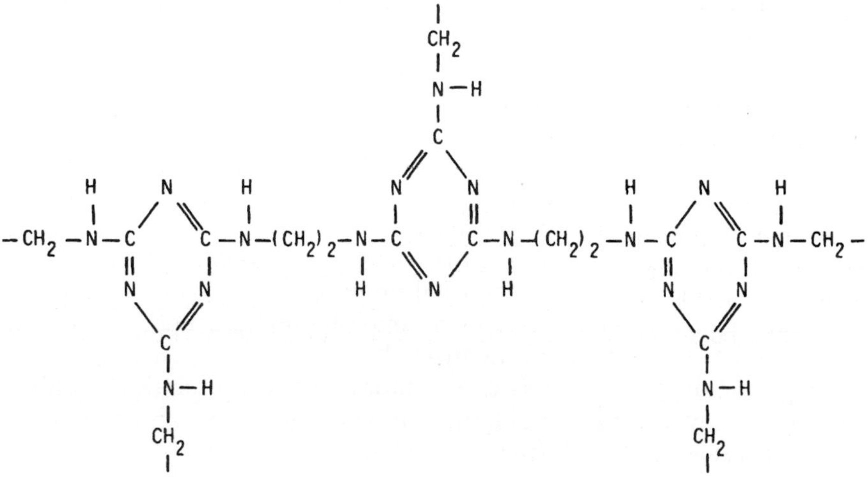

Melamine Formaldehyde

Figure 4.9 General characteristics of urea formaldehyde and malamine formaldehyde.[13]

Melamines and ureas provide excellent heat insulation and resistance; temperatures up to the destruction point will not cause parts to lose their shape. Amino resins exhibit relatively high mold shrinkage and also shrink on aging. Cracks develop in urea moldings subjected to severe cycling between dry and wet conditions. Prolonged exposure to high temperature affects the color of both urea and melamine products.

A loss of certain strength characteristics also occurs when amino moldings are subjected to prolonged elevated temperatures. Some electrical characteristics are also adversely affected; the arc resistance of some industrial types, however, remains unaffected after exposure at 500°F. Fillers are used to improve strength and hardness of amino resins.

Ureas are unsuitable for outdoor exposure. Melamines experience little degradation in electrical or physical properties after outdoor exposure, but color changes may occur.

Typical physical properties of amino plastics are shown in Table 4.4[14] and typical mechanical and electrical properties in Table 4.5.[14]

Product applications for amino resins include dinnerware, buttons, toilet seats, knobs, handles, ashtrays, food utensils, mixing bowls, and military equipment. Parts are typically joined with adhesives and fasteners.[1]

4.5.5 Phenolic (Phenol-Formaldehydes) (PF)—Thermosets

Like melamines and ureas, phenolic resin precursors are formed by a condensation reaction, the general nature of which is shown in Fig. 4.10. The general chemical structure of a cross-linked phenol-formaldehyde resin is given in Fig. 4.11.

Phenolics are among the oldest and best known general-purpose molding materials. They provide an excellent combination of high physical strength, high temperature resistance, scratch resistance, and good dimensional stability, electrical properties, and chemical resistance. Phenolics are also among the lowest plastics in cost and the easiest to mold. An extremely large number of phenolic materials are available, based on the many resin and filler combinations, and they

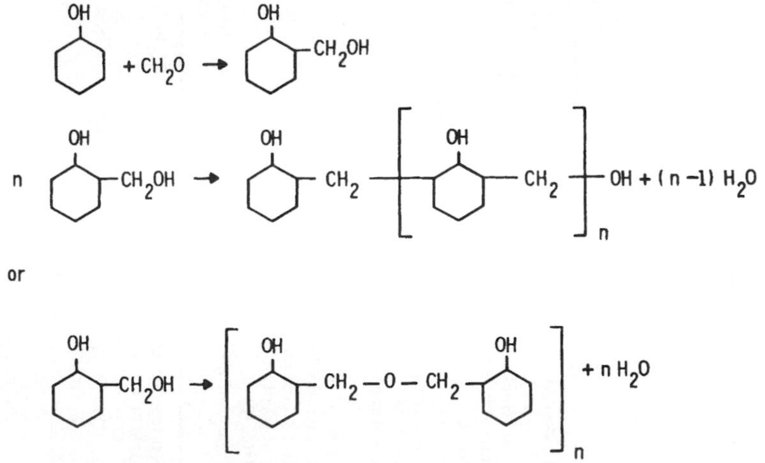

Figure 4.10 General nature of condensation reaction.

TABLE 4.4 Typical Physical Properties of Amino Compounds[14]

	Urea		Melamine					
	Alpha cellulose	Wood flour	Alpha cellulose	Wood flour	Alpha cellulose, modified	Rag	Asbestos	Glass fiber
Specific gravity	1.5	1.5	1.5	1.42	1.43	1.5	1.78	1.94–2.0
Density, g/in.³	24.6	24.6	24.6	23.8	23.5	24.6	29.2	31.8–32.8
Hardness, Rockwell E	94–97	95	110	94	......	100	90	......
Shrinkage,* in./in.:								
Molding	0.006–0.009	0.006–0.014	0.008–0.009	0.007–0.008	0.006–0.008	0.003–0.004	0.005–0.007	0.002–0.004
Postmold	0.006–0.012	0.006–0.012	0.009–0.011	0.004–0.007	0.001–0.002	0.004–0.008	0.002–0.003	0.002–0.005
Deflection temp, °F, at 264 lb/in.²	266	270	361	266	266	310	266	400
Heat resistance, continuous, °F	170†	170	210†	250	250	250	300	300
Coefficient of thermal expansion, per °C × 10⁻⁶	22–36	30	20–57	32–50	34–36	25–30	21–43	12–25
Thermal conductivity, (cal)(cm)/(s²)(cm)(°C) × 10⁻⁴	10.1	10.1	10.1	8.4	......	10.6	13.1	......
Water absorption, 24 h, at 23°C, %	0.4–0.8	0.7	0.3–0.5	0.34–0.6	0.3–0.6	0.3–0.6	0.13–0.15	0.09–0.3
Color possibilities	Unlimited	Brown, black	Unlimited	Brown	Brown	Limited	Brown	Natural gray

* Test specimen: 4-in. diam × ⅛-in disc.
† Based on no color change.

TABLE 4.5 Typical Mechanical and Electrical Properties of Amino Compounds[14]

	Urea		Melamine					
	Alpha cellulose	Wood flour	Alpha cellulose	Wood flour	Alpha cellulose modified	Rag	Asbestos	Glass fiber
Mechanical:								
Tensile strength, 1,000 lb/in.^{2}	5.5–7	5.5–10	7–8	5.7–6.5	5.5–6.5	8–10	5.5–6.5	5.9
Compressive strength, 1,000 lb/in.^{2}	30–38	25–35	40–45	30–35	24.5–26	30–35	25–30	20–29
Flexural strength, 1,000 lb/in.^{2}	11–18	8–16	12–15	6.5–9	11.5–12	12–15	7.4–10	13.2–24
Shear strength, 1,000 lb/in.^{2}	11–12		11–12	10–10.5	11.4–12.2	12–14	7–8	13.0–15.6
Impact strength, Izod ft-lb/in. of notch	0.24–0.28	0.25–0.35	0.30–0.35	0.25–0.35	0.30–0.42	0.55–0.90	0.30–0.40	0.5–6.0
Tensile modulus, 10^5 lb/in.^{2}	1.3–1.4		1.35	1.0	1.0	1.4	1.95	2.4
Flexural modulus, 10^6 lb/in.^{2}	1.4–1.5	1.3–1.6	1.1	1.0	1.1	1.4	1.8	
Electrical:								
Arc resistance, s	80–100	80–100	125–136	70–106	90–120	122–128	120–180	180–186
Dielectric strength, V/mil								
Short time:								
At 23°C	330–370	300–400	270–300	350–370	350–390	250–340	410–430	170–370
At 100°C	200–270		170–210	290–330	140–190	110–130	280–310	90–350*
Step by step:								
At 23°C	220–250	250–300	240–270	200–240	200–250	220–240	280–300	170–270
At 100°C	110–150		90–130	190–210	90–100	60–90	190–210	60–250*
Slow rate of rise:								
At 23°C	250–260		210–240	240–260	280–290	210–240	270–290	170–210
At 100°C	120–170		90–120	170–200	90	70–80	170–190	70–90
Dielectric constant:								
At 60 Hz	7.7–7.9	7.0–9.5	7.9–8.2	6.4–6.6	7.0–7.7	8.1–12.6	10.0–10.2	7.0–11.1
At 10^6 Hz	6.7–6.9	6.4–6.9	7.6–8.0	5.6–5.8	5.2–6.0	6.7–6.9	5.3–6.1	6.6–7.9
At 3×10^{10} Hz							4.9	5.5
Dissipation factor:								
At 60 Hz	0.034–0.043	0.035–0.040	0.052–0.083	0.026–0.033	0.192	0.100–0.340	0.100	0.14–0.23
At 10^6 Hz	0.029–0.031	0.028–0.032	0.026–0.030	0.034–0.035	0.044–0.12	0.036–0.041	0.039–0.048	0.013–0.016
At 3×10^9 Hz							0.032	0.040
Dielectric loss factor:								
At 60 Hz	0.28–0.34	0.24–0.38	0.44–0.78	0.17–0.22	0.90–2.4	2.0–5.0	0.5–1.0	1.5–2.5
At 10^6 Hz	0.19–0.21	0.18–0.22	0.20–0.33	0.20–0.21	0.19–0.28	0.24–0.26	0.21–0.31	0.09–0.19
Volume resistivity, Ω-cm	0.5–5.0 × 10^{11}		0.8–2.0 × 10^{12}	6–10 × 10^{12}	6 × 10^{10}	1.0–3.0 × 10^{11}	1.2 × 10^{12}	0.9–20 × 10^{11}
Surface resistivity, Ω	0.4–3.0 × 10^{11}		0.8–4.0 × 10^{11}	0.3–5.0 × 10^{11}	1.7 × 10^{12}	0.7–7.0 × 10^{11}	1.9 × 10^{12}	3.0–4.6 × 10^{12}
Insulation resistance, Ω	0.2–5.0 × 10^{11}		1.0–4.0 × 10^{10}	1.0–3.0 × 10^{10}	2.0–5.0 × 10^9	0.1–3.0 × 10^{10}	1.0–4.0 × 10^{10}	0.2–6.0 × 10^{10}

* At 50°C.

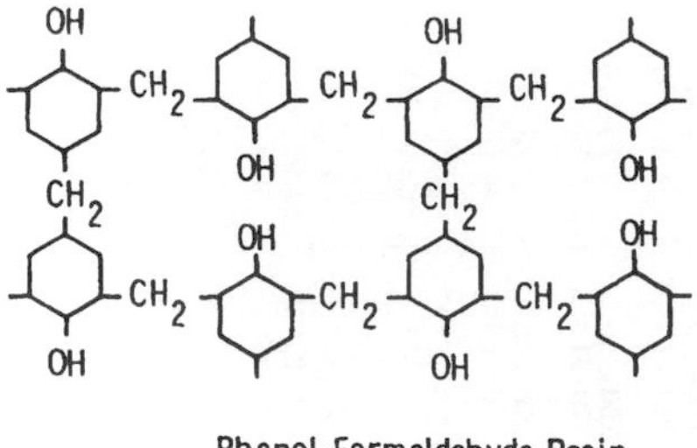

Figure 4.11   General chemical structure of phenol formaldehyde.

Phenol-Formaldehyde Resin

can be classified in many ways. One common way of classifying them is by type of application or grade. Typical properties for some of these common molding-material classifications are shown in Harper[15] and Grafton.[16] Molded phenolics are used for electrical devices, pulleys, commutators, pumps, closures, and automotive ignition components.[1]

In addition to molding materials, phenolics are used to bond friction materials for automotive brake linings, clutch parts, and transmission bands. They serve as binders for wood particle board used in building panels and core material for furniture, as the water-resistant adhesive for exterior-grade plywood, and as the bonding agent for converting both organic and inorganic fibers into acoustical and thermal insulation pads, batts, or cushioning for home, industrial, and automotive applications. They are used to impregnate paper for electrical or decorative laminates and as special additives to tackify, plasticize, reinforce, or harden a variety of elastomers.

Although it is possible to obtain different molding grades of phenolics for various applications, as discussed, phenolics, generally speaking, are not equivalent to diallyl phthalates and epoxies in resistance to humidity and retention of electrical properties in extreme environments. Phenolics are, however, quite adequate for a large percentage of electrical applications. Grades have been developed that yield considerable improvements in humid environments and at higher temperatures. The glass-filled, heat-resistant grades are outstanding in thermal stability up to 400°F and higher, with some being useful up to 500°F. Shrinkage in heat aging varies over a fairly wide range, depending on the filler used. Glass-filled phenolics are the more stable, as shown in Fig. 4.4. Martin[17] and Carswell[18] give considerable background on phenolic resin chemistry and applications.

4.5.6 Polybutadienes (PBS, PBAN)—Thermosets

Polybutadiene polymers that vary in 1,2 microstructure from 60 to 90 percent offer potential as moldings, laminating resins, coatings, and cast-liquid and formed-sheet products. These materials, which are essentially pure hydrocarbon, have outstanding electrical and thermal stability properties. The chemical structure is as shown in Fig. 4.12.

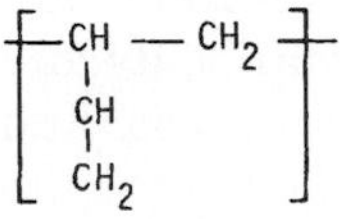

$$\left[\begin{array}{c} CH - CH_2 \\ | \\ CH \\ | \\ CH_2 \end{array} \right]$$

1, 2 - polybutadiene microstructure

Figure 4.12 Polybutadiene

Polybutadienes are cured by peroxide catalysts, which produce carbon-to-carbon bonds at the double bonds in the vinyl groups.[19] The final product is 100 percent hydrocarbon except where the starting polymer is the }OH or }COOH terminated variety. The nature of the resultant product may be more readily understood if the structure is regarded as polyethylene with a cross-link at every other carbon in the main chain.

Use of the high-temperature peroxides maximizes the opportunity for thermoplastic-like processing, because even the higher-molecular-weight forms become quite fluid at temperatures well below the cure temperature. Compounds can be injection-molded in an in-line machine with a thermoplastic screw.

4.5.7 Polyester (Unsaturated) (UP)—Thermoset

Unsaturated thermosetting polyesters are produced by addition polymerization reactions, as shown in simplified form in Fig. 4.1. Polyester resins can be formulated to have a range of physical properties from brittle and hard to tough and resistant to soft and flexible. Viscosities at room temperature may range from 50 to more than 25,000 cP. Polyesters can be used to fabricate a myriad of products by many techniques—open-mold casting, hand lay-up, spray-up, vacuum-bag molding, matched-metal-die molding, filament winding, pultrusion, encapsulation, centrifugal casting, and injection molding.[20]

By the appropriate choice of ingredients, particularly to form the linear polyester resin, special properties can be imparted. Fire retardance can be achieved through the use of one or more of the following: chlorendic anhydride, tetrabromophthalic anhydride, tetrachlorophthalic anhydride, dibromoneopentyl glycol, and chlorostyrene. Chemical resistance is obtained by using neopentyl glycol, isophthalic acid, hydrogenated bisphenol A, and trimethyl pentanediol. Weathering resistance can be enhanced by the use of neopentyl glycol and methyl methacrylate. Appropriate thermoplastic polymers can be added to reduce or eliminate shrinkage during curing and thereby minimize one of the disadvantages historically inherent in polyester systems.[20]

Thermosetting polyesters are the workhorse of the reinforced plastics industry. They are widely used for moldings, laminated or reinforced structures, surface gel coatings, liquid castings, furniture products, and structures. Cast products include furniture, bowling balls, simulated marble, gaskets for vitrified-clay sewer pipe, pistol grips, pearlescent shirt buttons, and implosion barriers for television tubes.

By lay-up and spray-up techniques, large- and short-run items are fabricated. Examples include boats of all kinds (pleasure sailboats and powered yachts, commercial fishing boats and shrimp trawlers, and small military vessels), dune buggies, all-terrain vehicles, custom auto bodies, truck cabs, horse trailers, motor homes, housing modules, concrete forms, and playground equipment.[20]

Molding is also performed with premix compounds, which are dough-like materials generally prepared by the molder shortly before they are to be molded by combining the premix constituents in a sigma-blade mixer or similar equipment. Premix, using conventional polyester resins, is used to mold automotive-heater housings and air-conditioner components. Low-shrinkage resin systems permit the fabrication of exterior automotive components such as fender extensions, lamp housings, hood scoops, and trim rails.

Wet molding of glass mats or preforms is used to fabricate such items as snack-table tops, food trays, tote boxes, and stackable chairs. Corrugated and flat paneling for room dividers, roofing and siding, awnings, skylights, fences, and the like is a very important outlet for polyesters. Pultrusion techniques are used to make fishing-rod stock and profiles from which slatted benches and ladders can be fabricated. Chemical storage tanks are made by filament winding.[20]

Polyesters are typically joined with adhesives or fasteners. Thread cutting self-tapping screws are feasible. Threaded inserts can be molded in or emplaced with adhesives. Expansion inserts are also used.[1]

4.5.8 Silicone (SI)—Thermosets

Silicones are a family of unique synthetic polymers that are partly organic and partly inorganic.[21] They have a quartz-like polymer structure, being made up of alternating silicon and oxygen atoms rather than the carbon-to-carbon backbone, which is a characteristic of the organic polymers. Silicones have outstanding thermal stability.

Typically, the silicon atoms will have one or more organic side groups attached to them, generally phenyl ($C6H5$}), methyl ($CH3$}), or vinyl ($CH2 = CH$}) units. Other alkyl, aryl, and reactive organic groups on the silicon atom are also possible. These groups impart characteristics such as solvent resistance, lubricity and compatibility, and reactivity with organic chemicals and polymers.[21]

Silicone polymers may be filled or unfilled, depending on properties desired and application. They can be cured by several mechanisms, either at room temperature (by room-temperature vulcanization, or

RTV) or at elevated temperatures. Their final form may be fluid, gel, elastomeric, or rigid.[21]

Some of the properties that distinguish silicone polymers from their organic counterparts are (1) relatively uniform properties over a wide temperature range, (2) low surface tension, (3) high degree of slip or lubricity, (4) excellent release properties, (5) extreme water repellency, (6) excellent electrical properties over a wide range of temperatures and frequencies, (7) inertness and compatibility, both physiologically and in electronic applications, (8) chemical inertness, and (9) weather resistance.

Flexible silicone resins. Flexible two-part, solvent-free silicone resins are available in filled and unfilled forms. Their viscosities range from 3000 cP to viscous thixotropic fluids of greater than 50,000 cP.[21] The polymer base for these resins is primarily dimethylpolysiloxane. Some vinyl and hydrogen groups attached to silicon are also present as part of the polymer.

These products are cured at room or slightly elevated temperatures. During cure, there is little if any exotherm, and there are no by-products from the cure. The flexible resins have Shore A hardnesses of 0 to 60 and Bashore resiliencies of 0 to 80. Flexibility can be retained from −55°C or lower to 250°C or higher.

Flexible resins find extensive use in electrical and electronic applications where stable dielectric properties and resistance to harsh environments are important. They are also used in many industries to make rubber molds and patterns.

Rigid silicone resins. Rigid silicone resins exist as solvent solutions or as solvent-free solids. The most significant uses of these resins are as paint intermediates to upgrade thermal and weathering characteristics of organic coatings, as electrical varnishes, glass tape, and circuit-board coatings.

Glass cloth, asbestos, and mica laminates are prepared with silicone resins for a variety of electrical applications. Laminated parts can be molded under high or low pressures, vacuum-bag-molded, or filament-wound.

Thermosetting molding compounds made with silicone resins as the binder are finding wide application in the electronic industry as encapsulants for semiconductor devices. Inertness toward devices, stable electrical and thermal properties, and self-extinguishing characteristics are important reasons for their use. Typical properties are given in Ref. 22, and thermal stability is shown in Fig. 4.13.

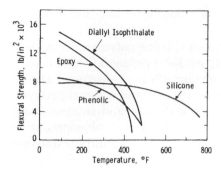

Figure 4.13 Retention of flexural strength with heat aging for silicone and organic molding compounds.[22] All samples aged 250 h and tested at room temperature.

Similar molding compounds containing refractory fillers can be molded on conventional thermoset equipment. Molded parts are then fired to yield a ceramic article. High-impact, long-glass-fiber-filled molding compounds are also available for use in high-temperature structural applications.

In general, silicone resins and composites made with silicone resins exhibit outstanding long-term thermal stabilities at temperatures approaching 300°C, and excellent moisture resistance and electrical properties.

4.6 Thermoplastics

The general nature of thermoplastics has been discussed earlier in this chapter, and specific data for a variety of thermoplastics can be found in Ref. 23. In general, thermoplastic materials tend to be tougher and less brittle than thermosets so that they can be applied without the use of fillers. This permits certain thermoplastic formulations to be used in food, packaging, and medical applications that are closed to thermosets. Also, several thermoplastics are available in transparent or translucent formulations (styrene, polycarbonate, acrylic, cellulosic, polyester, vinyl, polyethylene, and polypropylene).

Traditionally, by virtue of their basic polymer structure, thermoplastics have been much less dimensionally and thermally stable than thermosetting plastics. Hence, thermosets have offered an advantage for high-performance applications, although the lower processing costs for thermoplastics have given the latter a cost advantage and opened many new markets to plastics. In addition, there are a variety of welding techniques (hot plate, vibration, ultrasonic, spin, resistance, induction, laser, and hot gas) that can be used only with thermoplastics. Finally, thermoplastics can be overmolded with a second, and even a third, resin in some cases, thereby eliminating a joining operation.

Four major trends tend to put both thermoplastics and thermosets on a performance-consideration basis. First, much has been done in the development of reinforced, fiber-filled thermoplastics, greatly increasing stability in many areas. Second, much has been achieved in the development of so-called engineering thermoplastics, or high-stability, higher-performance plastics, which can also be reinforced with fiber fillers to increase their stability further. Third, advances in thermoplastic molding and joining technologies have increased the overall capabilities of these materials and their processes. Fourth, and countering these gains in thermoplastics, has been the development of lower-cost processing of thermosetting plastics, especially the screw injection molding and low-pressure compression molding technologies. All these options should be considered in optimizing the design, fabrication, and performance of plastic parts.

Specific data for a variety of formulations is given in Ref. 23. Some of the more common classes of thermoplastics are discussed in this section.

4.6.1 Acrylonitrile-Butadiene-Styrene (ABS)—Amorphous Thermoplastic

ABS plastics are derived from acrylonitrile, butadiene, and styrene. They have the general chemical structure shown in Fig. 4.14.

This class possesses hardness and rigidity without brittleness, at moderate costs. ABS materials have a good balance of tensile strength, impact resistance, surface hardness, rigidity, heat resistance, low-temperature properties, and electrical characteristics[24] and can be tailored to suit an application. They can be joined with snap fits, press fits, fasteners, adhesives, solvents, staking, and virtually all the thermoplastic welding techniques.[1] There are many ABS modifications and many blends of ABS with other thermoplastics. A list of trade names and suppliers is given in Ref. 23 with general data on a variety of ABS-based plastics.

ABS resins are used for drain and pipe fittings, appliance housings, automobile kick panels and headlight housings, refrigerator door liners, luggage, camper tops, sporting goods, and power tool housings. They are the most widely used engineering (as opposed to commodity) thermoplastics in the world.[1]

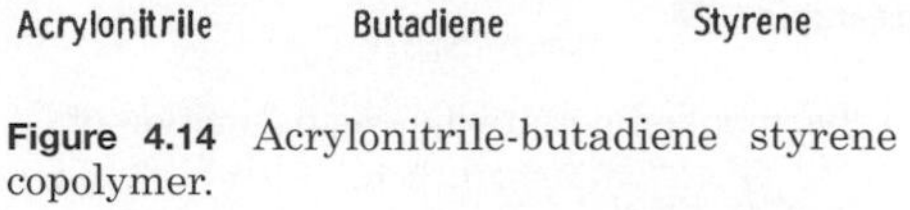

Acrylonitrile   Butadiene   Styrene

Figure 4.14 Acrylonitrile-butadiene styrene copolymer.

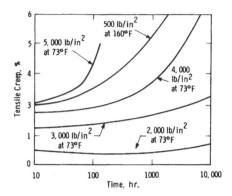

Figure 4.15 Tensile creep strength of ABS under various loads.[25]

Polymer properties. The most outstanding mechanical properties of ABS plastics are impact resistance and toughness. A wide variety of modifications are available for improved impact resistance, toughness, and heat resistance. Impact resistance does not fall off rapidly at lower temperatures. Stability under limited load is excellent, as shown in Fig. 4.15.[25] Heat-resistant ABS is equivalent to or better than acetals, polycarbonates, and polysulfones in room-temperature creep at 3000 lb/in^2. The Izod impact strength at 75°F is in the range of 3 to 5 ft-lb/in notch. This value is gradually reduced to 1 ft-lb/in notch at −40°F. When impact failure does occur, the failure is ductile rather than brittle. The modulus of elasticity versus temperature is shown in Fig. 4.16.[26] Physical properties are little affected by moisture, which contributes greatly to the dimensional stability of ABS materials.

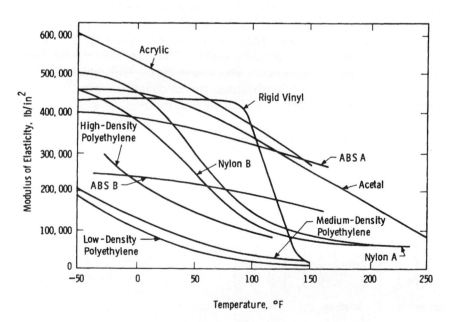

Figure 4.16 Modulus of elasticity of several thermoplastic materials as a function of temperature.[26]

ABS alloys and electroplating grades. Much work has been done to modify ABS plastics by alloying to improve certain properties, and by modifying to enhance adhesion of electroplated coatings.[23] ABS alloyed or blended with polycarbonate combines some of the best qualities of both materials, resulting in a thermoplastic that is easier to process, has high heat and impact resistance, and sells for considerably less than polycarbonate.

The impact strength of the ABS-polycarbonate alloy, 10.7 ft-lb/in notch, is well above average for the high-impact engineering thermoplastics but still not as high as that of polycarbonate. However, unlike polycarbonate, the alloy does not have critical thicknesses with respect to notched impact strength. Its notched Izod value drops by only 2 to 4 ft-lb/in in the 1/8- to 1/4-in range. The notched Izod value for a polycarbonate 1/8 in thick is about 16 ft-lb/in but only 3 to 4 ft-lb/in at thicknesses greater than 1/4 in. The flexural modulus of the ABS-polycarbonate alloy is about 15 percent greater than that of polycarbonate alone. The alloy remains more rigid than polycarbonate up to about 200°F, as shown in Fig. 4.17.[27]

The 264-lb/in^2 heat-deflection temperature of the alloy is 245°F, and the 66-lb/in^2 value is 260°F. These values are 35 and 30°F, respectively, lower than those of polycarbonate. However, the maximum recommended continuous (no-load) temperature of the alloy is only 10°F lower than that of polycarbonate. The good creep resistance of polycarbonate, one of its biggest advantages and shown in Fig. 4.18,[27] is maintained after alloying.

In addition to the polycarbonate alloy, ABS can be alloyed with other plastics to obtain special properties. Furthermore, ABS has been modified to gain improved adhesion of electroplated metals. Advances have been so great in this area that ABS is perhaps the most widely

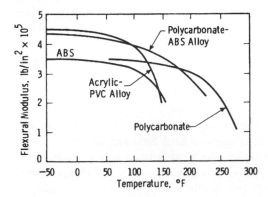

Figure 4.17 Flexural modulus of ABS-polycarbonate alloy and base materials.[27]

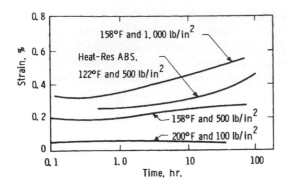

Figure 4.18 Creep resistance of ABS-polycarbonate alloy and heat-resistant ABS.[27]

used material for producing electroplated plastic parts. Electroplated ABS is used extensively for many electrical and mechanical products in many forms and shapes. Adhesion of the electroplated metal is excellent.

4.6.2 Acetal (Polyoxymethylene) (POM)—Semicrystalline Thermoplastic

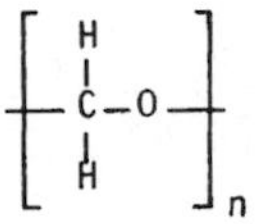

$$\left[\begin{array}{c} H \\ | \\ C-O \\ | \\ H \end{array}\right]_n$$

Figure 4.19 Polyacetal resin.

Acetals are among the group of high-performance engineering thermoplastics that resemble nylon somewhat in appearance but not in properties. The general repeating chemical structural unit is shown in Fig. 4.19.

Acetals are strong and rigid (but not brittle) and have good moisture, heat, and chemical resistance. They have a high melt point and excellent resistance to creep under load. Glass-filled acetals provide higher stiffness, lower creep, higher heat deflection temperature, better arc resistance, and greater dimensional stability. They can be joined with snap fits, press fits, fasteners, adhesives, staking and virtually all the thermoplastic welding techniques.[1] There are two basic types of acetals: the homopolymers by DuPont and the copolymers by Ticona. Reference 23 further identifies these materials and gives typical properties.

The homopolymers are harder, have higher resistance to fatigue, are more rigid, and have higher tensile and flexural strengths with lower elongation. In addition, they are the strongest and stiffest of unreinforced thermoplastics. References made to acetals without identification of polymer type usually imply the homopolymer materials.

The copolymers are more stable in long-term, high-temperature service and more resistant to hot water. Neither type of acetal is resistant to strong mineral acids, but the copolymers are resistant to strong bases. They can handle pH values from 4 to 14 and can be immersed in common solvents, lubricants, and gas.[1] They are among the most creep resistant semicrystalline thermoplastics.

Polymer properties. The most outstanding properties of acetals are high tensile strength and stiffness, resilience, good recovery from deformation under load, and toughness under repeated impact. They exhibit excellent long-term load-carrying properties and dimensional stability and can be used for precision parts. Acetals have low static and dynamic coefficients of friction and are usable over a wide range of environmental conditions. The plastic surface is hard, smooth, and glossy. A fluorocarbon fiber-filled acetal, Delrin AF, is available and offers even better low-friction and resistance properties.

The modulus of elasticity as a function of temperature for acetals and several other thermoplastic materials is shown in Fig. 4.16.[26] The tensile yield strength of acetals is compared in Fig. 4.20[28] with that of some other thermoplastics. The deflection under load for Delrin acetal is compared in Fig. 4.21[28] with that of other thermoplastics.

Effect of moisture. Because acetals absorb a small amount of water, the dimensions of molded parts are affected by their water content. Figure 4.22[29] presents the relationship of dimensional changes with changes in moisture content. The dimensional change resulting from an environmental change is found by subtracting the "percent in length" found at the first humidity-temperature condition from the "percent in length" at the final condition. For example, to go from 77°F (25°C) and 0 percent water (as-molded condition) to 100°F (38°C) and

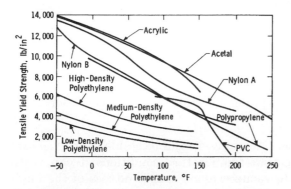

Figure 4.20 Tensile yield strength of several thermoplastic materials as a function of temperature.[28]

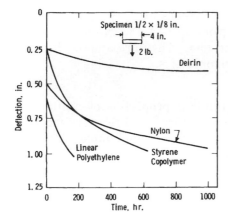

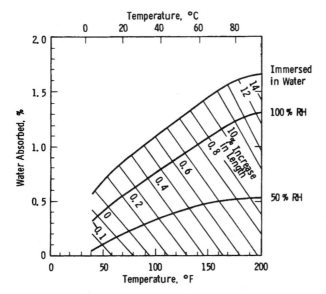

Figure 4.21 Deflection of several thermo-plastic materials as a function of time at 90% RH and 150°F.[28]

Figure 4.22 Dimensional changes of Delrin acetal with variations of temperature and moisture.[29]

100 percent humidity, a change of 0.45 percent will occur. Figure 4.23[29] shows the rate of water absorption of acetals at various conditions.

Effect of space and radiation.   Acetal resin will be stable in the vacuum of space under the same time-temperature conditions it can withstand in air.[29] Exposure to vacuum alone causes no loss of the engineering properties. The principal result is slight outgassing of small amounts of moisture and free formaldehyde. In a vacuum, as in air,

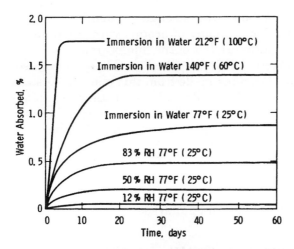

Figure 4.23 Rate of water absorption of Delrin acetal under various conditions.[29]

prolonged exposure to elevated temperatures results in the liberation of increasing amounts of formaldehyde due to thermal degradation of the polymer.

Particulate radiation, such as the protons and electrons of the Van Allen radiation belts, is damaging to acetal resins and will cause a loss of engineering properties.[28] For example, acetals should not be used in a radiation environment where the total electron dose is likely to exceed 1 Mrad. When irradiated with 2-MeV electrons, a 1-Mrad dose causes only slight discoloration, while 2.3 Mrad causes considerable embrittlement. At 0.6 Mrad, however, acetals are still mechanically sound except for a moderate decrease in impact strength.

The regions of the electromagnetic spectrum that are most damaging to acetal resins are ultraviolet light and gamma rays.[29] In space, the deleterious effects of ultraviolet light are of prime consideration. This is due to the absence of the protective air atmosphere, which normally filters out much of the sun's ultraviolet energy. Therefore, the amount of ultraviolet light may be 10 to 100 times as intense in space as on Earth.

Fluorocarbon fiber-filled acetal. This is a modified acetal homopolymer developed to meet the need for a thermoplastic injection-molding material to be used in moving parts in which low friction and exceptional wear resistance are the principal requirements. This resin consists of oriented tetrafluoroethylene (TFE) fluorocarbon fibers uniformly dispersed in a matrix of acetal resin. The result is an injection-molding and extrusion resin that combines the strength, toughness, dimensional stability, and fabrication economy of acetals with the unusual surface and low frictional characteristics of the fluorocarbons.

The outstanding properties of TFE fiber-filled acetal are those associated with sliding friction. Bearings made from this material sustain high loads when operating at high speeds and show little wear. In addition, such bearings are essentially free of slipstick behavior because their static and dynamic coefficients are almost equal. Comparative properties of the filled and unfilled acetal are given in Refs. 23 and 30.

Automotive applications for acetals include fuel system components, seat belts and steering columns. Other uses are gears, bearings, plumbing fixture components, hardware items, butane lighter housings, zippers, mechanical couplings and pulleys.

4.6.3 Acrylic (Polymethylmethacrylate) (PMMA)—Amorphous Thermoplastics

The general properties of acrylics are presented in Ref. 31. Acrylics are based on polymethyl methacrylate, and their chemical structure is as shown in Fig. 4.24.

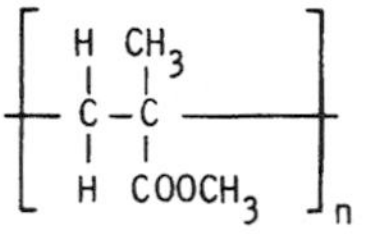

Figure 4.24 Polymethyl methacrylate.

Acrylics have exceptional optical clarity and basically good weather resistance, strength, electrical properties, and chemical resistance. Acrylics are known for their hardness (Rockwell M85-105) and good electrical insulation properties. They do not discolor or shrink after fabrication; they have low water-absorption characteristics, a slow burning rate, and will not flash-ignite. Acrylics are attacked by strong solvents, gasoline, acetone, and other similar fluids.

Acrylics can be injection molded, extruded, cast, vacuum and pressure formed, and machined, although molded parts for load bearing should be carefully analyzed, especially for long-term loading. They can be joined with press fits, fasteners, adhesives, solvents, staking, and virtually all the thermoplastic welding techniques.[1]

Optical properties. Parts molded from acrylic powders in their natural state may be crystal clear and nearly optically perfect. The index of refraction ranges from 1.486 to 1.496. The total light transmittance is as high as 92 percent, and haze measurements average only 1 percent. Transmittance at various wavelengths is shown in Fig. 4.25.[32] Light transmittance and clarity can be modified by the addition of a wide range of transparent and opaque colors, most of these being formulated for long outdoor service.

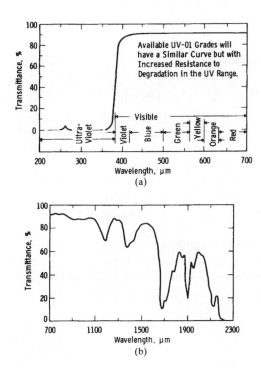

Figure 4.25 Percent transmittance as a function of wavelength for acrylic resin. Crystal, 1/8-in. (3.2-mm) thickness, grades 129, 130, 140, 147, 148, molded only.[32] (a) <700 μm, (b) >700 μm.

Dimensional stability and aging. The amount of dimensional change for a given application is determined from the coefficient of linear thermal expansion of the material. In the case of an exterior glazing unit, which is 1×1 ft (30.4×30.4 cm) and is subjected to a temperature variation of 32 to 100°F (0 to 38°C), the space to be provided in the frame for expansion is calculated as follows (assuming no thermal expansion of the frame material):

$$\Delta L = \alpha L T$$

where ΔL = change in length, in or cm

α = coefficient of thermal expansion [= 4×10^{-5} in/in°F (7.2×10^{-5} cm/cm°C)]

L = initial length [= 12 in (30.4 cm)]

T = temperature variation [= 68°F (38°C)]

Therefore,

$$\Delta L = 4 \times 10^{-5} \times 12 \times 68 = 0.033 \text{ in.}$$

$$= 7.2 \times 10^{-5} \times 30.4 \times 38 = 0.083 \text{ cm}$$

This indicates the need for an expansion space of approximately 0.0165 in (0.0419 cm) at each end for the direction in which this calculation was made (in this case the major length).

The moisture level in acrylics will depend on the relative humidity of the environment. Figure 4.26[32] shows the equilibrium moisture content of acrylics versus the relative humidity in the surrounding air at room temperature.

When the relative humidity of the air changes from a lower value to a higher one, acrylics will absorb moisture, which in turn causes a slight dimensional expansion. Depending on the thickness of the part, the equilibrium moisture level may be reached in a couple of weeks or in months.

Figure 4.27[32] shows the dimensional changes taking place in a 1/8-in. (3.2-mm) thick sheet as a function of time and the environmental relative humidity at room temperature. The flat horizontal part of the curves corresponds to the equilibrium conditions at the specific environmental relative humidity.

For example, assume that a 1/8-in (3.2-mm) thick lighting panel as molded from dry resin is initially 24 in (61 cm) long. This panel, installed in a 50 percent RH atmosphere, will expand approximately 0.1 percent, or 0.024 in (0.61 mm).

Because of their chemical structure, acrylic resins are inherently resistant to discoloration and loss of light transmission. They are unsur-

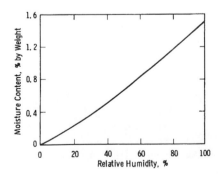

Figure 4.26 Equilibrium moisture content in acrylic resins versus percent RH at 73°F.[32]

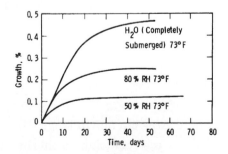

Figure 4.27 Dimensional changes versus time for acrylics under various environmental moisture conditions. Test samples 1/8 in (3.2 mm) thick.[32]

passed in this respect by any other transparent plastic. Their outstanding weatherability has been proved by their long-time performance in such products as automotive lenses, fluorescent street lights, outdoor signs, and boat windshields. They are also used for auto lenses, dials, instrument panels, nameplates, windows, appliance panels, knobs and housings.

4.6.4 Cellulosics: Cellulose Acetate (CA), Cellulose Acetate Butyrate (CAB), Cellulose Acetate Proprionate (CAP), Ethyl Cellulose (EC)—Semicrystalline Thermoplastics

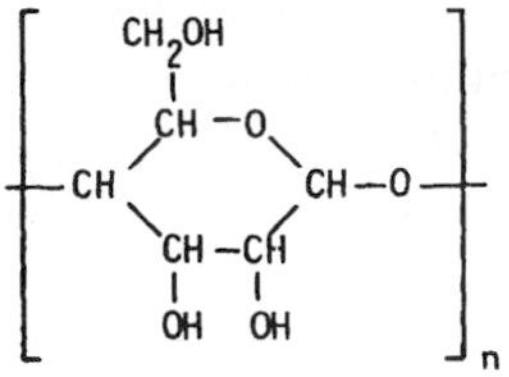

Figure 4.28 Cellulose (natural polymer)

Cellulosics are among the toughest of plastics. They are generally economical and basically good insulating materials. However, they are temperature limited and are not as resistant to extreme environments as many other thermoplastics. Their general structure is shown in Fig. 4.28.[2]

Cellulosics have similar properties that vary in degree. The four most prominent industrial cellulosics are cellulose acetate, cellulose acetate butyrate, cellulose propionate, and ethyl cellulose. A fifth member of this group is cellulose nitrate. Cellulose materials are available in a great number of formulas and flows and are manufactured to offer a wide range of properties. They are formulated with a wide range of plasticizers for specific plasticized properties.

Cellulose butyrate, proprionate, and acetate provide a range of toughness and rigidity that is useful for many applications, especially where clarity, outdoor weatherability, and aging characteristics are needed. The materials are fast-molding plastics and can be provided with hard, glossy surfaces over the full range of color and texture.

Butyrate, propionate, and acetate are rated in that order in dimensional stability in relation to the effects of water absorption and plasticizers. The materials are slow burning, although self-extinguishing forms of acetate are available. Special formulations of butyrate and propionate are serviceable outdoors for long periods. Acetate is generally considered unsuitable for outdoor uses. From an application standpoint, the acetates generally are used where tight dimensional stability, under anticipated humidity and temperature, is not required. Hardness, stiffness, and cost are lower than for butyrate or propionate. Butyrate is generally selected over propionate where weatherability, low-temperature impact strength, and dimensional stability are required. Propionate is often chosen for hardness, tensile strength, and stiffness, combined with good weather resistance.

Ethyl cellulose, best known for its toughness and resiliency at sub-zero temperatures, also has excellent dimensional stability over a wide range of temperature and humidity conditions. Alkalis or weak acids do not affect this material, but cleaning fluids, oils, and solvents are very harmful.

Cellulose acetate is mainly an extrusion (film and sheet) material, but injection applications include premium toys, tool handles, appliance housings, shields, lenses, and eyeglass frames. Cellulose acetate butyrate is used for pen barrels, steering wheels, tool handles, machine guards and skylights. Cellulose acetate propionate applications include lighting fixtures, safety goggles, motor covers, brush handles, face shields, and steering wheels. Ethyl cellulose is used for flashlight cases, fire extinguisher parts, and electrical appliance parts.[1]

4.6.5   Fluorocarbons

Tetrafluoroethylene (TFE) and fluorinated ethylene propylene (FEP).   For practical purposes, there are eight types of fluorocarbons, as summarized in Table 4.6[2] and discussed hereafter. Suppliers and trade names are also given in Table 4.6. Like other plastics, each type is available in several grades. The original, basic fluorocarbon, and perhaps still the most widely known one, is TFE fluorocarbon. It has the optimum of electrical and thermal properties and almost complete moisture resistance and chemical inertness, but it does have the disadvantage of cold flow or creep under mechanical loading. TFE hardness as a function of temperature is shown in Fig. 4.29.[33]

Stronger, filled modifications exist, as do newer, more cold-flow-resistant grades. FEP fluorocarbon is quite similar to TFE in most properties, except that its useful temperature is limited to about 400°F. FEP is much more easily processed, and molded parts are possible

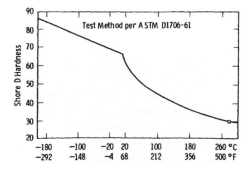

Figure 4.29   Hardness of TFE fluorocarbon as a function of temperature.[33]

TABLE 4.6 Structures, Trade Names, and Suppliers of Fluorcarbons

Fluorocarbon	Structure	Typical trade names and suppliers
TFE (tetrafluoroethylene)....	$\left[\begin{array}{c} F \quad F \\ -C-C- \\ F \quad F \end{array}\right]_n$	Teflon TFE (E. I. du Pont) Halon TFE (Ausimont USA, Inc.)
FEP (fluorinated ethylenepropylene).........	$\left[\begin{array}{c} F \quad F \quad F \quad F \\ C-C-C-C \\ F \quad F \quad F-C-F \\ \qquad\qquad F \end{array}\right]_n$	Teflon FEP (E. I. du Pont)
ETFE (ethylene-tetra-fluoroethylene copolymer)...	Copolymer of ethylene and TFE	Tefzel (E. I. du Pont)
PFA (perfluoroalkoxy).......	$\left[\begin{array}{c} F \quad F \quad F \quad F \quad F \\ C-C-C-C-C \\ F \quad F \quad O \quad F \quad F \\ \qquad\quad R_f* \end{array}\right]_n$	Teflon PFA (E. I. du Pont)
CTFE (chlorotrifluoro-ethylene).................	$\left[\begin{array}{c} Cl \quad F \\ -C-C- \\ F \quad F \end{array}\right]_n$	Kel-F (3M)
E-CTFE (ethylene-chlorotri-fluoroethylene copolymer)...	Copolymer of ethylene and CTFE	Halar E-CTFE (Ausimont USA, Inc.)
PVF₂ (vinylidene fluoride)....	$\left[\begin{array}{c} H \quad F \\ -C-C- \\ H \quad F \end{array}\right]_n$	Kynar (Atochem)
PVF (polyvinyl fluoride)......	$\left[\begin{array}{c} H \quad H \\ -C-C- \\ H \quad F \end{array}\right]_n$	Tedlar (E. I. du Pont)

* $R_f = C_n F_{2n-1}$.

with FEP, which might not be possible with TFE. Linear thermal-expansion curves for both are shown in Fig. 4.30.[34]

'Ethylene-tetrafluoroethylene (ETFE) copolymer. ETFE, a high-temperature thermoplastic fluoropolymer, is readily processed by conventional methods, including extrusion and injection molding. As a copolymer of ethylene and tetrafluoroethylene, it is closely related to TFE. It has good thermal properties and abrasion resistance, and excellent im-

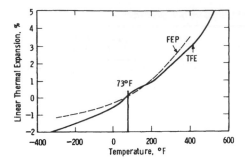

Figure 4.30 Linear thermal expansion of TFE and FEP fluorocarbons as a function of temperature.[34]

pact strength, resistance to chemicals, and electrical-insulation characteristics.

ETFE can be described best as a rugged thermoplastic with an outstanding balance of properties. Mechanically, it is tough and has medium stiffness (200,000 lb/in^2) and excellent flex life, impact, cut-through, and abrasion resistance. The glass-fiber-reinforced compound has even higher tensile strength (12,000 lb/in^2), stiffness (950,000 lb/in^2), and creep resistance, but it is still tough and impact-resistant.

Thermally, ETFE has a continuous temperature rating of 300°F (150°C). It can be used intermittently up to 200°C, depending on exposure time, load, and environment. Glass-fiber-reinforced ETFE appears capable of useful service at 392°F (200°C).

ETFE is weather-resistant, inert to most solvents and chemicals, and hydrolytically stable. It has excellent resistance to high-energy radiation and is ultraviolet-resistant.

The dielectric constant of the fluoropolymer is 2.6, and the dissipation factor is 0.0006, making it an excellent low-loss dielectric. These values are not affected by changes in temperature and other environmental conditions. Dielectric strength is high; resistivity is excellent.

The molding characteristics of the resin are excellent. It exhibits good melt flow, allowing the filling of thin sections (10 mil for small parts). Cycle times are equivalent to those of other thermoplastics.

EFTE is the first fluoroplastic that can be reinforced—not merely filled—with glass fiber. Because the resin will bond to the fibers, strength, stiffness, creep resistance, heat-distortion temperature, and dimensional stability are enhanced. Electrical and chemical properties approach those of the unreinforced resin, while the coefficient of friction is actually lower. Tests indicate that the service temperature limit is approximately 400°F. The fluoropolymer, reinforced with glass, can be molded by conventional methods with rapid molding cycles.

Perfluoroalkoxy (PFA) resins. PFA resins, a class of melt-processible fluoroplastics, combine the ease and economics of thermoplastic processing with high-temperature performance in the range of TFE fluorocarbon resins. PFA resins resemble FEP fluorocarbon resins in having a branched polymer chain that provides good mechanical properties at melt viscosities much lower than those of TFE. However, the unique branch in PFA is longer and more flexible, leading to improvements in high-temperature properties, higher melting point, and greater thermal stability.

In use, PFA resins have the desirable properties typical of fluorocarbons, including resistance to virtually all chemicals, anti-stick characteristics, low coefficient of friction, excellent electrical characteristics, low smoke, excellent flammability resistance, ability to perform in temperature extremes, and excellent weatherability. Their strength and stiffness at high operating temperatures are at least equivalent to those of TFE, while their creep resistance appears to be better over a wide temperature range. They should perform successfully in the 500°F area.

Film and sheet are expected to find use as electrical insulations in flat cables and circuitry and in laminates used in electrical and mechanical applications.

Chlorotrifluoroethylene (CTFE). The CTFE resins, like the FEP materials, are melt processible and can be injection, transfer, and compression molded or screw extruded. Compared with TFE, CTFE has greater tensile and compressive strength within its service-temperature range. However, at the temperature extremes, CTFE does not perform as well as TFE for parts such as seals. At the low end of the range, TFE has somewhat better physical properties; at the higher temperatures, CTFE is more prone to stress cracking and other difficulties.

The electrical properties of CTFE are generally excellent, but dielectric losses are higher than those of TFE. Chemical resistance is poorer than that of TFE, but radiation resistance is better. CTFE does not have the low-friction and bearing properties of TFE.

Ethylene-chlorotrifluoroethylene (E-CTFE) copolymer. E-CTFE copolymer, a copolymer of ethylene and CTFE, is a strong, highly impact-resistant material that retains useful properties over a broad temperature range. E-CTFE is available in pellet and powder form and can be readily extruded, injection, transfer, and compression molded, rotocast, and powder coated. Its tensile strength, impact resistance, hardness, creep properties, and abrasion resistance at ambient temperature are comparable with those of nylon 6. E-CTFE retains its strength and impact resistance down to cryogenic temperatures.

E-CTFE also exhibits excellent ac loss properties. Its dielectric constant measures 2.5 to 2.6 over the frequency range of 10^2 to 10^6 Hz and is unaffected by temperature over the range of -80 to $300°F$. Its dissipation factor is low, ranging from 0.0008 to 0.015, depending on temperature and frequency. It has a dielectric strength of 2000 V/mil in 10-mil wall thickness and a volume resistivity of 10^{16} Ω-cm.

Vinylidene fluoride (PVF2).   PVF2 is another melt processible fluorocarbon capable of being injection and compression molded and screw extruded. Its 20 percent lower specific gravity compared with that of TFE and CTFE and its good processing characteristics permit economy and provide excellent chemical and physical characteristics. The useful temperature range is from -80 to $+300°F$. Although it is stiffer and has higher resistance to cold flow than TFE, its chemical resistance, useful temperature range, antistick properties, lubricity, and electrical properties are lower. Polyvinylidene polymers have seen some use in piezoelectric devices as sensors.[35]

Polyvinyl fluoride (PVF).   PVF is manufactured as a film, the combination of excellent weathering and fabrication properties of which has made it widely accepted for surfacing industrial, architectural, and decorative building materials. It has outstanding weatherability, solvent resistance, chemical resistance, abrasion resistance, and color retention. It is supplied with a variety of surfaces for different bonding and antistick objectives.

4.6.6   Ionomers (IOM)

Outstanding advantages of this polymer class are combinations of toughness, transparency, low-temperature impact, and solvent resistance.[36] Ionomers have high melt strength for thermoforming and extrusion-coating processes, and a broad processing-temperature range. There are resin grades for extrusion, films, injection molding, blow molding, and other thermoplastic processes.

Limitations of ionomers include low stiffness, susceptibility to creep, low heat-distortion temperature, and poor ultraviolet resistance unless stabilizers are added where these properties are important. Most ionomers are very transparent. In 60-mil sections, internal haze ranges from 5 to 25 percent. Light transmission ranges from 80 to 92 percent over the visible region and, in specific compositions, high transmittance extends into the ultraviolet region.

Basically, commercial ionomers are nonrigid, unplasticized plastics. Outstanding low-temperature flexibility, resilience, high elongation, and excellent impact strength typify the ionomer resins.

Deterioration of mechanical and optical properties can occur when ionomers are exposed to ultraviolet light and weather. Some grades are available with ultraviolet stabilizers that provide up to one year of outdoor exposure with no loss in mechanical properties. Formulations containing carbon black provide ultraviolet resistance equal to that of black polyethylene.

Most ionomers have good dielectric characteristics over a broad frequency range. The combination of these electrical properties, high melt strength, and abrasion resistance qualifies these materials for insulation and jacketing of wire and cable. Their excellent mechanical and optical properties make them specially useful for thermoformed packaging and abrasion- and impact-resistant shoe parts. Typical material properties are given in Ref. 31.

The most important applications for ionomers are films and sheets, primarily for food packaging. Injection molded products include golf ball covers, ski-boot shells, bowling pin covers and bumper guards for automobiles.

4.6.7 Nylon (Polyamide) (PA)—Semicrystalline Thermoplastics

Also known as polyamides, nylons are strong, tough thermoplastics with good impact, tensile, and flexural strengths from freezing temperatures up to 300°F; excellent low-friction properties; and good electrical resistivities. They also have excellent chemical and wear resistance and a low coefficient of friction. Nylons are known for the fact that they are extremely hygroscopic and their mechanical property values can drop substantially with increasing moisture content. Nylons have very low viscosities at elevated temperatures, allowing them to mold mechanical parts with very thin sections. They can be joined with snap fits, press fits, fasteners, staking, and virtually all the thermoplastic welding techniques.[1]

The structures of four common nylons are shown in Fig. 4.31.[2] Since all nylons absorb some moisture from environmental humidity, moisture-absorption characteristics must be considered in designing with these materials. They will absorb from 0.5 to nearly 2 percent moisture after a 24-h water immersion. There are low-moisture-absorption

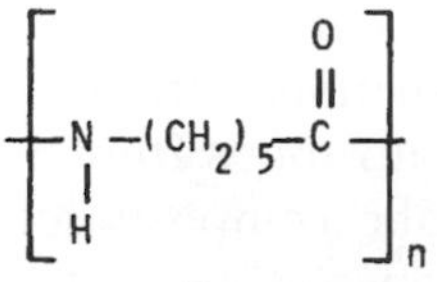

Figure 4.31 Chemical structures for four nylons.

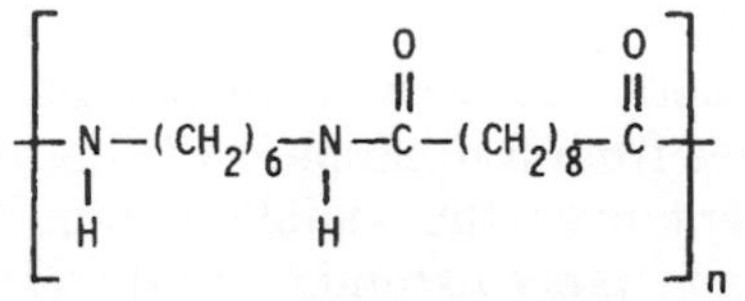

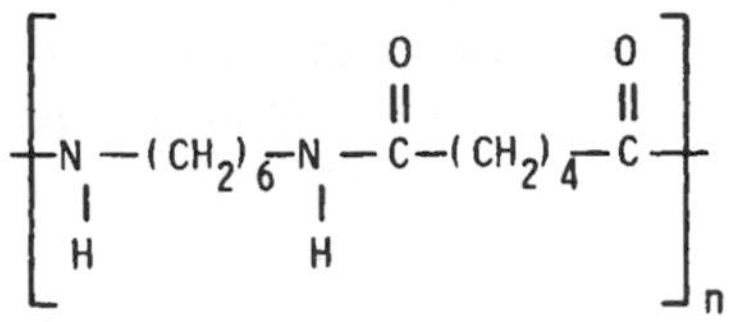

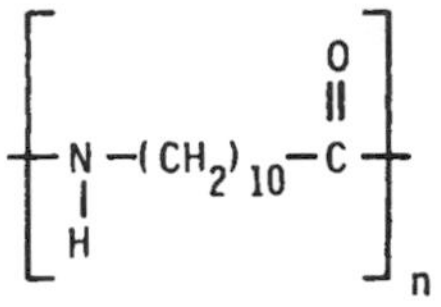

grades, however; hence moisture-absorption properties do not have to limit the use of nylons, especially for the lower moisture-absorption grades. Typical materials and properties are shown in Ref. 23.

Regarding the identification of the various grades of nylon, certain nylons are identified by the number of carbon atoms in the diamine and dibasic acid used to produce that particular grade. For instance, nylon 6/6 is the reaction product of hexamethylenediamine and adipic acid, both of which are materials containing six carbon atoms in their chemical structure. Some common commercially available nylons are 6/6, 6, 6/10, 8, 11, and 12. Grades 6 and 6/6 are the strongest structurally; grades 6/10 and 11 have the lowest moisture absorption, best electrical properties, and best dimensional stability; and grades 6, 6/6, and 6/10 are the most flexible. Grades 6, 6/6, and 8 are heat-sealable, with nylon 8 being capable of cross-linking. Another grade, nylon 12, offers advantages similar to those of grades 6/10 and 11 but lower cost possibilities due to being more easily and economically processed. Also a high-temperature type of nylon exists. It is discussed separately hereafter.

In situ polymerization of nylon permits massive castings. Cast nylons are readily polymerized directly from the monomer material in the mold at atmospheric pressure. The method finds application where the size of the part required or the need for low tooling cost precludes injection molding. Cast nylon displays excellent bearing and fatigue properties as well as the other properties characteristic of other basic nylon formulations, with the addition of size and short-run flexibility advantages of the low-pressure casting process.

One special process exists in which nylon parts are made by compressing and sintering, thereby creating parts having exceptional wear characteristics and dimensional stability. Various fillers, such as molybdenum disulfide and graphite, can be incorporated into nylon to give special low-friction properties. Also, nylon can be reinforced with glass fibers, thus giving it considerable additional strength. These variations are further discussed in a later part of this chapter dealing with glass-fiber-reinforced thermoplastic materials.

Effect of temperature on nylon. One of the major advantages of nylon resins is that they retain useful mechanical properties over a range of temperatures from −60 to +400°F. Both long-term and short-term effects of temperature must, however, be considered. In the short term, there are effects on such properties as stiffness and toughness. There is also the possibility of stress relief and its effect on dimensions. Of most concern in long-term applications at high temperature is gradual oxidative embrittlement. For such cases, the use of heat-stabilized resins is recommended.

The important consideration for design work with nylon resins is that exposure to high temperatures in air for a period of time will result in a permanent change in properties due to oxidation. The degree of change in properties depends on the temperature level, the time exposed, and the composition of the nylon used. The effect on the tensile strength, as measured at room temperature and 2.5 percent moisture content, is that a 25 percent strength reduction can occur in 3 months at 185°F, and 50 percent reduction can occur in 3 months at 250°F. Stabilized nylon does not change appreciably at 250°F aging.

High-temperature oxidation reduces the impact strength even more than the static-strength properties. For instance, the impact strength and elongation of nylon are reduced considerably after several days at 250°F.

A similar change in properties is encountered on exposure to high-temperature water for long periods of time. In this case a reaction with water takes place. There is no significant reaction up to 120°F. This has been confirmed by molecular-weight measurements on 15-year-old samples.

In boiling water, the tensile strength is slowly reduced until, after 2500 h, it levels off at 6000 lb/in^2 (tested at room temperature and 2.5 percent moisture content). The elongation drops rapidly after 1500 h; hence, this time has been taken as the limit for the use of basic nylon. Some compositions are especially resistant to hot-water exposure, however.

Effect of moisture on dimensional control. Freshly molded objects normally contain less than 0.3 percent of water, since only dry molding powder can be molded successfully. These objects will then absorb moisture when they are exposed to air or water. The amount of absorbed water will increase in any environment until an equilibrium condition based on relative humidity (RH) is reached. Equilibrium moisture contents for two humidity levels are approximately as shown in Table 4.7.[37]

TABLE 4.7   Equilibrium Moisture Contents for Nylon

	Zytel 101	Zytel 31
50% RH air	2.5%	1.4%
100 RH air (or water)	8.5%	3.5%

These equilibrium moisture contents are not affected by temperature to any significant extent. Thus, the final water content at equilibrium will be the same whether objects of nylon are exposed to water at room temperature or at boiling temperature.

The time required to reach equilibrium, however, is dependent on the temperature, the thickness of the specimen, and the amount of moisture present in the surroundings. Nylon exposed in boiling water will reach the equilibrium level, 8.5 percent, much sooner than nylon in cool water.

When nylon that contains some moisture is exposed to a dry atmosphere, the loss of water will be the reverse of the changes described, and it will take about the same length of time for a corresponding change to occur.

In the most common exposure, an environment of constantly varying humidity, no true equilibrium moisture content can be established. However, moldings of nylon will gradually gain in moisture content in such an environment until a balance is obtained with the midrange humidities. A slow cycling of moisture content near this value will then occur. In all but very thin moldings, the day-to-day or week-to-week variations in relative humidity will have little effect on the total moisture content. The long period changes, such as between summer and winter, will have some effect, depending on the thickness and the relative-humidity range. The highest average humidity for a month will not generally be above 70 percent. In cold weather, heated air may average as low as 20 percent relative humidity. Even at these extremes, the change in the moisture content of nylon is small in most cases, because of the very low rate of both absorption and desorption.

There are two significant dimensional effects that occur after molding. In some cases, these oppose each other so that critical dimensions may change very little in a typical air environment. The first is a shrinkage in the direction of flow due to the relief of molded-in stresses. The second is an increase due to moisture absorption. In applications where dimensions are critical to performance, and one of these effects predominates, it may be necessary to anneal or moisture-condition the parts to obtain the best performance. For example, a part that will be exposed to high temperatures might require annealing, and an object in water service might be moisture-conditioned to effect dimensional changes before use.

The magnitude of the dimensional change due to molded-in stresses depends on molding conditions and part geometry. This change in size can be determined in a given case only by measuring and annealing a few pieces. In general, long dimensions (in the direction of flow) will shrink, but the short dimensions will increase in an amount that is often too small to measure.

The effect of the moisture content on the dimensions of a molding can be predicted more accurately than the effect of annealing. The changes in dimension at various water contents for Zytel 101 and Zytel 31 are shown in Fig. 4.32.[37] These data are for the annealed or stress-free condition. For most applications in air, the dimensions corresponding to those obtained in equilibrium with 50 percent relative-humidity air are usually chosen as the average size expected when a moisture balance is established. As shown, Zytel 101 will increase 0.006 in/in from the dry condition to equilibrium with 50 percent relative humidity. Zytel 31 under similar conditions will increase 0.0025 in/in. In many applications, these changes are small enough and occur so slowly that they do not affect the operation of the finished parts.

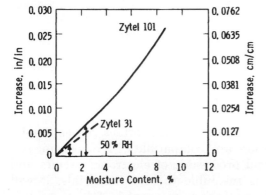

Figure 4.32 Changes in dimensions with moisture content for two nylon materials in stress-free, annealed conditions.[37]

Stress-free objects of Zytel 101 and Zytel 31 will increase approximately 0.026 and 0.007 in/in, respectively, during the change from dry conditions to completely saturated with water. In water service, if dimensions are critical, the molding can be moisture conditioned before assembly to allow for this change. Once a part is at or near saturation, little dimensional change can occur unless it is allowed to dry for long periods.

Applications.   Gears, bearings, and other anti-friction parts are often made of a nylon. Automotive uses include exhaust canisters, transmission covers, headlamp housings, fender extensions, cowl vents, exterior parts, and under-hood mechanical components. Electrical applications include plugs, connectors, computer parts, and bobbins. Other applications include springs, drapery hardware brush bristles, bike wheels, mallet heads, and hot comb and brush handles.[1]

Aramid (high-temperature nylon).   In addition to conventional nylon molding resins, there is a high-temperature nylon now called *aramid*. This high-temperature nylon retains about 60 percent of its strength at temperatures of 475 to 500°F, which would melt conventional nylons. It has good dielectric strength (constant to 400°F and 95 percent relative humidity) and volume resistivity and a low dissipation factor. Fabricated primarily in sheet, fiber, and paper form, this material is being used in wrapped electrical-insulation constructions such as transformer coils and motor stators. It retains high tensile strength, resistance to wear, and electrical properties after prolonged exposure at temperatures up to 500°F.

4.6.8   Parylene

Parylene is the generic name for members of a thermoplastic-polymer series developed by Union Carbide. They are unique among plastics in that they are produced as thin films by vapor-phase polymerization.[38] The polymers are highly crystalline, high-molecular-weight, straight-chain compounds that are known as *tough* materials with excellent dielectric characteristics.

Parylene is extremely resistant to chemical attack, exceptionally low in trace-metal contamination, and compatible with all organic solvents used in the cleaning and processing of electronic circuits and systems. Although parylene is insoluble in most materials, it will soften in solvents having boiling points in excess of 150°C.

The basic member of this thermoplastic polymer family, parylene N or poly-para-xylylene, exhibits superior dielectric strength, exceptionally high surface and volume resistivities, and electrical properties that vary remarkably little with changes in temperature. Its chemical structure is shown in Fig. 4.33.

Since its melting point far exceeds that of many other thermoplastics, parylene N can even be used at temperatures exceeding 220°C in the absence of oxygen. Because parylene can be deposited in very thin coatings, heat generated by coated components is easier to control, and differences in thermal expansions are less of a problem than with conventional coatings.

The other commercially available member of the group, parylene C, is poly-mono-chloro-para-xylylene. Its chemical structure is shown in Fig. 4.34. Parylene C offers significantly lower permeability to moisture and gases, such as nitrogen or oxygen, while retaining excellent electrical properties.

The parylene process.   Unlike most plastics, parylene is not produced and sold as a polymer. It is not practical to melt, extrude, mold, or calender it as with other thermoplastics. Nor can it be applied from solvent systems, since it is insoluble in conventional solvents.

The advantages that characterize this rugged, intractable material are made possible by a fast, relatively simple vacuum-application system. While it is a unique system, it is often less cumbersome, less complex, and easier to use and maintain than many techniques devised for more conventional plastic-coating systems.

The parylene processor starts with a dimer rather than a polymer and polymerizes it on the surface of an object. To achieve this, the dimer must first go through a two-step heating process. The solid dimer is converted to a reactive vapor of the monomer. When passed over room-temperature objects, the vapor will rapidly coat them with polymer.

This process of polymerization on the object surface has much to recommend it, particularly when the goal is coating uniformity. Unlike dip or spray coating, condensation coating does not run off or sag. It is not "line of sight" as in vacuum metallizing; the vapor coats evenly—over edges, points, and internal areas. The vapor is pervasive, but it coats without bridging so that holes can be jacketed evenly. Masking tape can be used if it is desired that some areas not be coated. Moreover, with parylene, the object to be coated remains at or near

Figure 4.33 Parylene N.   **Figure 4.34** Parylene C.

room temperature, eliminating all risk of thermal damage. The coating thickness is controlled easily and very accurately simply by regulating the amount of dimer to be vaporized.

Deposition chambers of virtually any size can be constructed. Those currently in use range from 500 to 28,000 in^3. Large parts (up to 5 ft long and 18 in high) can be processed in this equipment. The versatility of the process also enables the simultaneous coating of many small parts of varying configurations. Because of this, a time savings in labor requirements can readily be achieved.

Parylenes can also be deposited onto a cold condenser and then stripped off as a free film, or they can be deposited onto the surface of objects as a continuous, adherent coating in thicknesses ranging from 1000 Å (about 0.004 mil) to 3 mil or more. The deposition rate is normally about 0.5 mm/min (about 0.02 mil). On cooled substrates, the deposition rate can be as high as 1.0 mil/min.

Parylene properties. The material can be used at both elevated and cryogenic temperatures. The 1000-h service life for the N and the C members is 200 to 240°F. Corresponding 10-year service in air is limited to 140 to 175°F. Parylenes are excellent in having low gas permeability, low moisture-vapor transmission, and low-temperature ductility. Dimensional stability is reported as better than that of polycarbonate, and overall barrier properties to most gases are reported superior to those of many other barrier films. Important properties of this unique material are listed in Tables 4.8 and 4.9.[38]

4.6.9 Polyaryl Sulfone (PASU)—Amorphous Thermoplastic

Polyaryl sulfone offers the unique combination of thermoplasticity and retention of structurally useful properties at 500°F.[39] It offers high long-term stability, creep resistance, stress crack resistance and electrical properties. It can withstand the heat of soldering and can be joined with snap fits, press fits, fasteners, adhesives, staking, and virtually all the thermoplastic welding techniques, although with some difficulty due to the high melting point of the resin.[1] This material is supplied in both pellet and powder form.

Polyaryl sulfone consists mainly of phenyl and biphenyl groups linked by thermally stable ether and sulfone groups. It is distinguished from polysulfone polymers by the absence of aliphatic groups, which are subject to oxidative attack. This aromatic structure gives it excellent resistance to oxidative degradation and accounts for its retention of mechanical properties at high temperatures. The presence of ether-oxygen linkages gives the polymer chain flexibility to permit

TABLE 4.8   Thermal, Physical, and Mechanical Properties of Parylenes[38]

	Parylene N	Parylene C
Typical thermal properties		
Melting temperature, °C	405	280
Linear coefficient of expansion, mm/mm/°C	6.9	3.5
Thermal conductivity, 10^{-4} cal/s/(cm^2) (°C/cm)	~3	
Typical physical and mechanical properties		
Tensile strength, lb/in.^{2}	6,500	10,000
Yield strength, lb/in.^{2}	6,100	8,000
Elongation to break, %	30	200
Yield elongation, %	2.5	2.9
Density, g/cm^3	1.11	1.289
Coefficient of friction:		
Static	0.29	0.25
Dynamic	0.29	0.25
Water absorption, 24 h	0.06 (0.029 in)	0.01 (0.019 in)
Index of refraction, n_D 23°C	1.661	1.639

Data recorded following appropriate ASTM method.

TABLE 4.9   Film-Barrier Properties of Parylenes[38]

Polymer	Gas permeability, cm^3-mil/100 in^2, 24 h-atm (23°C)						Moisture-vapor transmission, g-mil/100 in^2, 24 h, 37°C, 90% RH
	N$_2$	O$_2$	CO$_2$	H$_2$S	SO$_2$	Cl$_2$	
Parylene N	7.7	39.2	214	795	1,890	74	1.6
Parylene C	1.0	7.2	7.7	13	11	0.35	0.5
Epoxies	4	5–10	8	...	...	...	1.8–2.4
Silicones	...	50,000	300,000	...	...	...	4.4–7.9
Urethanes	80	200	3,000	...	...	...	2.4–8.7

Data recorded following appropriate ASTM method.

fabrication by conventional melt-processing techniques. Typical properties are given in Ref. 40.

Thermal properties.   Polyaryl sulfone is characterized by a very high heat-deflection temperature, 535°F at 264 lb/in^2, which is approximately 150°F higher than many other commercially available thermoplastics, as shown in Fig. 4.35. This is a consequence of its high glass-

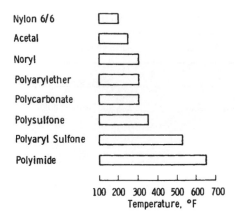

Nylon 6/6
Acetal
Noryl
Polyarylether
Polycarbonate
Polysulfone
Polyaryl Sulfone
Polyimide

100 200 300 400 500 600 700
Temperature, °F

Figure 4.35 Approximate heat-deflection temperatures for some engineering thermoplastics at 264 lb/in^2.

transition temperature, 550°F, rather than the effect of filler reinforcement or a crystalline melting point. At 500°F, it maintains a tensile strength in excess of 4000 lb/in^2 and a flexural modulus of 250,000 lb/in^2. The resistance to oxidative degradation is indicated by the ability of polyaryl sulfone to retain its tensile strength after 2000-h exposure to 500°F air-oven aging.

Chemical resistance. Polyaryl sulfone has good resistance to a wide variety of chemicals, including acids, bases, and common solvents. It is unaffected by practically all fuels, lubricants, hydraulic fluids, and cleaning agents used on or around electrical components. Highly polar solvents such as N,N-dimethylformamide, N,N-dimethylacetamide, and N-methylpyrrolidone are solvents for the material.

Applications. PASU is used in electrical components and printed circuit boards. It has extreme service environment applications.[1]

4.6.10   Polycarbonate (PC)—Amorphous Thermoplastic

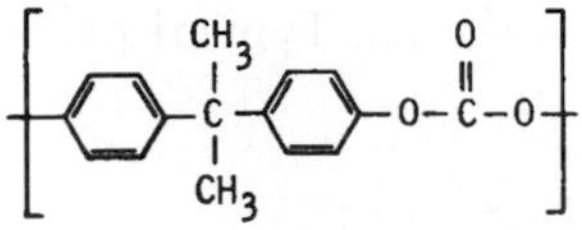

Figure 4.36 Polycarbonate.

This group of plastics is also among those classified as *engineering thermoplastics* because of their high-performance characteristics in engineering designs. The generalized chemical structure is shown in Fig. 4.36.

Polycarbonates are especially outstanding in impact strength, having strengths several times higher than other engineering thermoplastics. Polycarbonates are tough, rigid, and dimensionally stable and are available as transparent or colored parts. They have excellent outdoor dimensional stability but are vulnerable to grease and oils. Polycarbonates are easily fabricated with reproducible results, using molding or machining techniques. An important molding characteristic is the low and predictable mold shrinkage (0.005 to 0.007 in/in), which sometimes gives polycarbonates an advantage over nylons and acetals for close-tolerance parts. They can be joined with snap fits, press fits, fasteners, adhesives, solvents, staking, and virtually all the thermoplastic welding techniques.[1] As with most other plastics containing aromatic groups, radiation stability is high.

The most commonly useful properties of polycarbonates are creep resistance, high heat resistance, dimensional stability, good electrical properties, self-extinguishing properties, product transparency, and exceptional impact strength, which compares favorably with that of some metals and exceeds that of many competitive plastics. In fact, polycarbonate is sometimes considered to be competitive with zinc and aluminum castings. Although such comparisons have limits, the fact that the comparisons are sometimes made in material selection for product design indicates the strong performance characteristics possible in polycarbonates.

In addition to their performance as engineering materials, polycarbonates are also alloyed with other plastics in order to increase the strength and rigidity of these plastics. Notable among the plastics with which polycarbonates have been alloyed are the ABS plastics. In addition to standard grades of polycarbonates, a special film grade exists for high-performance capacitors.[41]

Moisture-resistance properties.   Oxidation stability on heating in air is good, and immersion in water and exposure to high humidity at temperatures up to 212°F have little effect on dimensions. Steam sterilization is another advantage that is attributable to the resin's high heat stability. However, if the application requires continuous exposure in water, the temperature should be limited to 140°F. Polycarbonates are among the most stable plastics in a wet environment, as shown in Figs. 4.37 and 4.38.[42,43]

Applications.   Automotive uses include tail and side marker lights, headlamp support fixtures, instrument panels, trim strips, and exterior body components. It is also used in traffic light housings, optical lenses, glazing, and signal lenses. Food uses include returnable milk containers and microwave ovenware, mugs, ice cream dishes, food storage containers, microwave oven applications, and water cooler bot-

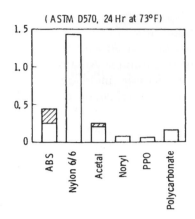

(ASTM D570, 24 Hr at 73°F)

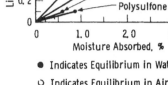

Moisture Absorbed, %

● Indicates Equilibrium in Water at 212°F

○ Indicates Equilibrium in Air at 50 % RH, 72°F

⊢ Indicates Equilibrium in Water at 72°F

Figure 4.37 Water absorption of several thermoplastics.[42,43]

Figure 4.38 Dimensional changes of several thermoplastics due to absorbed moisture.[42,43]

tles. Other applications are intravenous and blood processing equipment, appliance and tool housings, telephone, televisions, and boat and conveyor components.[1]

4.6.11 Polyesters—Polybutylene Terephthalate (PBT), Polyethylene Terephthalate (PET)—Semicrystalline Thermoplastics

Thermoplastic polyesters have been and are currently used extensively in the production of film and fibers. These materials are denoted chemically as polyethylene terephthalate. During the past few years, a new class of high-performance molding and extrusion grades of thermoplastic polyesters has been made available and is becoming increasingly competitive among plastics. These polymers are denoted chemically as poly(1,4-butylene terephthalate) and poly(tetramethylene terephthalate). These thermoplastic polyesters are highly crystalline, with a melting point of about 430°F. They are fairly translucent in thin molded sections and opaque in thick sections, but they can be extruded into transparent thin film. Both unreinforced and reinforced formulations are extremely easy to process and can be molded in very fast cycles. Typical properties are shown in Ref. 44.

The unreinforced resin offers the following characteristics: (1) good tensile strength, toughness and impact resistance; (2) high abrasion resistance, low coefficient of friction; (3) good chemical resistance, very low moisture absorption and resistance to cold flow; (4) good stress crack and fatigue resistance; (5) good electrical properties; and (6) good surface appearance. Electrical properties are stable up to the rated temperature limits. The material can be joined with snap fits, press fits, fasteners, adhesives, staking, and virtually all the thermoplastic welding techniques (with limitations) except hot gas welding.[1]

The glass-reinforced polyester resins are unusual in that they can compare with, or are better than, thermosets in electrical, mechanical, dimensional, and creep properties at elevated temperatures (approximately 300°F), while having superior impact properties.

The glass-fiber concentration usually ranges from 10 to 30 percent in commercially available grades. In molded parts, the glass fibers remain slightly below the surface so that finished items have a very smooth surface finish as well as an excellent appearance.

Unreinforced resins are primarily used in housings requiring excellent impact and in moving parts such as gears, bearings, and pulleys, in packaging applications, and in writing instruments. The flame-retardant grades are primarily aimed at television, radio, and electrical and electronics parts as well as business-machine and pump components. Reinforced resins are being used in automotive (hardware, under-hood components), electrical (switches, relays, coil bobbins, light sockets) electronic (sensors), and general industrial (conveyors) area, where they are replacing thermosets, other thermoplastics, and metals. Electrical and mechanical properties coupled with low finished-part cost are enabling reinforced thermoplastic polyesters to replace phenolics, alkyds, DAP, and glass-reinforced thermoplastics in many applications.

4.6.12 Polyethersulfone (PES)—Amorphous Thermoplastic

Polyethersulfone is a high-temperature engineering thermoplastic with excellent tensile strength, electrical properties, and chemical resistance. It has outstanding long-term resistance to creep at temperatures up to 150°C,[45] and it is capable of being used continuously under load at temperatures of up to about 180°C (and, in some low-stress applications, up to 200°C). Other grades are capable of operating at temperatures above 200°C and for specialized adhesive and lacquer applications. Polyethersulfone is a premium material usually used for high-heat aerospace, automotive, chemical, and electrical components. It can be joined with snap fits, press fits, fasteners, adhesives, solvents, staking, and virtually all the thermoplastic welding techniques.[1]

The polyethersulfone chemical structure shown in Fig. 4.39 gives an amorphous polymer, which possesses only bonds of high thermal and oxidative stability. While the sulfone group confers high-temperature performance, the ether linkage contributes toward practical pro-

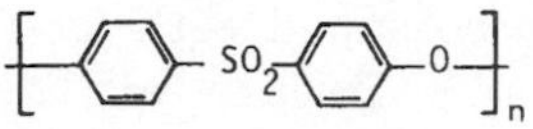

Figure 4.39 Polyethersulfone.

cessing by allowing mobility of the polymer chain when in the melt phase.

Polyethersulfone exhibits low creep. A constant stress of 3000 lb/in^2 at 20°C for 3 years produces a strain of 1 percent, while a stress of 6,500 lb/in^2 results in a strain of only 2.6 percent over the same period of time. Higher modulus values are obtained with polyethersulfone at 150°C than with polysulfone, phenylene oxide-based resins, or polycarbonate at considerably lower temperatures.

Although its load-bearing properties are reduced above 150°C, polyethersulfone can still be considered for applications at temperatures up to 180°C. It remains form-stable to above 200°C and has a heat-deflection temperature of 203°C at 264 lb/in^2.

Polyethersulfone is especially resistant to acids, alkalis, oils, greases, and aliphatic hydrocarbons and alcohols. It is attacked by ketones, esters, and some halogenated and aromatic hydrocarbons.

4.6.13 Polyethylene (PE), Polypropylene (PP), and Polyallomer (PAL)—Semicrystalline Thermoplastics

This large group of polymers is basically divided into the three separate polymer groups listed under this heading; all belong to the broad chemical classification known as *polyolefins*. Polyethylene and polypropylene can be considered as the first two members of a large group of polymers based on the ethylene structure. Their structures are shown in Fig. 4.40.

Molecular changes beyond these two structures give quite different polymers and properties and are covered separately in other parts of this chapter. The chemical changes result from the replacement of the methyl group (}CH3) in polypropylene with substituents such as chlorine (polyvinyl chloride), }OH (polyvinyl alcohol), F (polyvinyl fluoride), and }CN (polyacrylonitrile). There are many categories or types even within each of the three polymer groups discussed in this section. Although property variations exist among these three polymer groups and among the subcategories within these groups, there are also many similarities. The differences or unique features of each are discussed in this section.

The similarities are, broadly speaking, appearance, general chemical characteristics, and electrical properties. The differences are more

(a)   (b)

Figure 4.40 (a) Polyethylene and (b) polypropylene.

notably in physical and thermal-stability properties. Basically, poly-olefins are all wax-like in appearance and extremely inert chemically, and they exhibit decreases in physical strength at somewhat lower temperatures than the higher-performance engineering thermoplas-tics. Polyethylenes were the first of these materials developed and, hence, for some of the original types, have the weakest mechanical properties. The later-developed polyethylenes, polypropylenes, and polyallomers offer improvements. They can be joined with snap fits, press fits, fasteners, hot-melt adhesives, staking, and virtually all the thermoplastic welding techniques, although ultrasonic welding poses some challenges.[1] The unique features of each of these three polymer groups are outlined in the following paragraphs. Typical properties are given in Ref. 23.

Polyethylenes. Polyethylenes are among the most widely used plastics and are regarded as low-cost, commodity plastics. They are available in three main classifications based on density: low, medium, and high. These density ranges are 0.910 to 0.925, 0.925 to 0.940, and 0.940 to 0.965, respectively. These three density grades are also sometimes known as types I, II, and III. All polyethylenes are relatively soft, and hardness increases as density increases. Generally, the higher the density, the better are the dimensional stability and physical proper-ties, particularly as a function of temperature. The thermal stability of polyethylenes ranges from 190°F for the low-density material up to 250°F for the high-density material. Toughness is maintained to low negative temperatures.

Polyethylenes are used for toys, lids, closures, packaging, rotation-ally molded tanks, and medical apparatus. Other applications are pipe, gas tanks, large containers, institutional seating, luggage, out-door furniture, pails, containers and housewares. Polyethylene is the work horse of the rotational molding industry.[1]

Polypropylenes. Polypropylenes are also among the most widely used plastics and regarded as low-cost, commodity plastics. They are chem-ically similar to polyethylenes but have somewhat better physical strength at a lower density. The density of polypropylenes is among the lowest of all plastic materials, ranging from 0.900 to 0.915. Polypropylenes offer more of a balance of properties than a single unique property, with the exception of flex-fatigue resistance. These materials have an almost infinite life under flexing, and hinges made of polypropylenes are often referred to as "living hinges." Use of this characteristic is widespread in the form of plastic hinges. Polypropylenes are perhaps the only thermoplastics surpassing all others in combined electrical properties, heat resistance, rigidity, toughness, chemical resistance, dimensional stability, surface gloss, and melt flow, at a lower cost than that of competing resins.

Because of their exceptional quality and versatility, polypropylenes offer outstanding potential in the manufacture of products through injection molding. Mold shrinkage is significantly less than that of other polyolefins; uniformity in and across the direction of flow is appreciably greater. Shrinkage is therefore more predictable, and there is less susceptibility to warpage in flat sections.

Polypropylenes are among the fastest-growing resins. They are used for tubs, agitators, dispensers, pump housings, and filters in appliances, and in automotive applications (fan shrouds, fan blades, ducts, housings, batteries, door panels, trim glove boxes, seat frames, louvers, and seat belt retractor covers). They are also used in medical, luggage, toy, packaging and housewares applications.

Polyallomers. Polyallomers are also polyolefin-type thermoplastic polymers produced from two or more different monomers, such as propylene and ethylene, which would produce a propylene-ethylene polyallomer. The monomers, or base chemical materials, are similar to those of polypropylene or polyethylene. Hence, as was mentioned, and as would be expected, many properties of polyallomers are similar to those of polyethylenes and polypropylenes. Having a density of about 0.9, they, like polypropylenes, are among the lightest plastics.

Polyallomers have a brittleness temperature as low as $-40°F$ and a heat-distortion temperature as high as $210°F$ at 66 lb/in^2. The excellent impact strength plus exceptional flow properties of polyallomer provide wide latitude in product design. Notched Izod impact strengths run as high as 12 ft-lb/in notch.

Although the surface hardness of polyallomers is slightly less than that of polypropylenes, resistance to abrasion is greater. Polyallomers are superior to linear polyethylene in flow characteristics, moldability, softening point, hardness, stress-crack resistance, and mold shrinkage. The flexural-fatigue-resistance properties of polyallomers are as good as or better than those of polypropylenes.

Polyallomer applications include shoe lasts, automotive body components, closures, and a variety of cases such as tackle boxes, office machine cases, and bowling ball bags.

Cross-linked polyolefins. While polyolefins have many outstanding characteristics, they, like all thermoplastics to some degree, tend to creep or cold-flow under the influence of temperature, load, and time. To improve this and some other properties, considerable work has been done on developing cross-linked polyolefins, especially polyethylenes. The cross-linked polyethylenes offer thermal performance improvements of up to 25°C or more.

Cross-linking has been achieved primarily by chemical means and by ionizing radiation. Products of both types are available. Radiation-cross-linked polyolefins have gained particular prominence in a heat-shrinkable form. This is achieved by cross-linking the extruded or molded polyolefin using high-energy electron-beam radiation, heating the irradiated material above its crystalline melting point to a rubbery state, mechanically stretching to an expanded form (up to four or five times the original size), and cooling the stretched material. Upon further heating, the material will return to its original size, tightly shrinking onto the object around which it has been placed. Heat-shrinkable boots, jackets, and tubing are widely used. Also, irradiated polyolefins, sometimes known as *irradiated polyalkenes,* are important materials for certain wire and cable jacketing applications.

4.6.14   Polyimide (PI) and Poly(amide-Imide) (PA-I)—Amorphous Thermoplastics

Among the commercially available plastics generally considered as having high heat resistance, polyimides can be used at the highest temperatures, and they are the strongest and most rigid. Polyimides have a useful operating range to about 900°F (482°C) for short durations and 500 to 600°F (260 to 315°C) for continuous service in air. Prolonged exposure at 500°F (260°C) results in moderate (25 to 30 percent) loss of original strength and rigidity.

These materials, which can be used in various forms including moldings, laminates, films, coatings, and adhesives, have high mechanical properties, wear resistance, chemical and radiation inertness, and excellent dielectric properties over a broad temperature range. They can be joined with snap fits, press fits, fasteners, adhesives, solvents, staking, and virtually all the thermoplastic welding techniques (some, with difficulty).[1] Material properties are given in Ref. 23. The thermal stability is compared with that of other engineering plastics in Fig. 4.35.

Chemical structures.   Polyimides are heterocyclic polymers, having a noncarbon atom of nitrogen in one of the rings in the molecular chains.[23] The atom is nitrogen and it is in the inside ring as shown in Fig. 4.41.

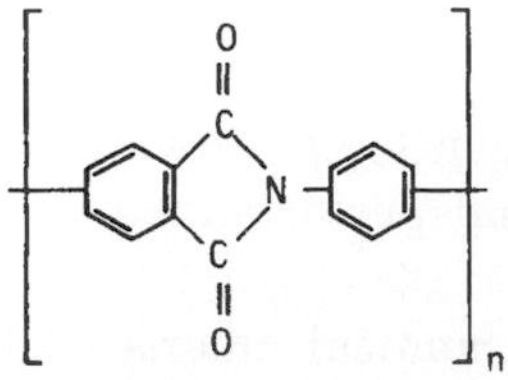

Figure 4.41   Polyimides.

The fused rings provide chain stiffness essential to high-temperature strength retention. The low concentration of hydrogen provides oxidative resistance by preventing thermal degradative fracture of the chain.

The other resins considered as members of this family of polymers are the poly(amide-imide)s. These compositions contain aromatic rings and the characteristic nitrogen linkages, as shown in Fig. 4.42.

There are two basic types of polyimides: (1) condensation and (2) addition resins. The condensation polyimides are based on a reaction of an aromatic diamine with an aromatic dianhydride. A tractable (fusible) polyamic acid intermediate produced by this reaction is converted by heat to an insoluble and infusible polyimide, with water being given off during the cure. Generally, the condensation polyimides result in products having high void contents that detract from inherent mechanical properties and result in some loss of long-term heat-aging resistance.

The addition polyimides are based on short, preimidized polymer-chain segments similar to those comprising condensation polyimides. These prepolymer chains, which have unsaturated aliphatic end groups, are capped by termini that polymerize thermally without the loss of volatiles. The addition polyimides yield products that have slightly lower heat resistance than the condensation polyimides.

The condensation polyimides are available as either thermosets or thermoplastics, and the addition polyimides are available only as thermosets. Although some of the condensation polyimides technically are thermoplastics, which would indicate that they can be melted, this is not the case, since they have melting temperatures that are above the temperature at which the materials begin to decompose thermally.

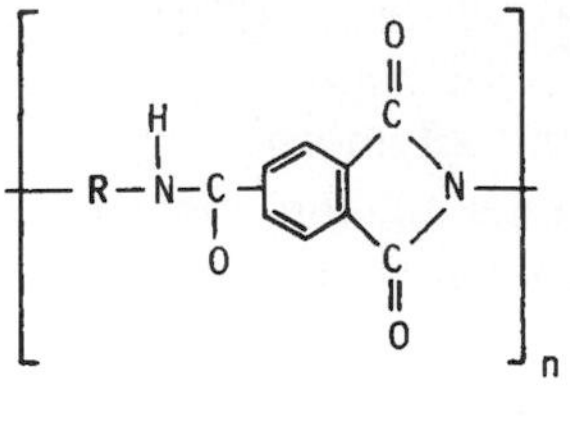

Figure 4.42   Poly(amide-imide).

Properties. Polyimides and polyamide-imides exhibit some outstanding properties due to their combination of high-temperature stability (up to 500 to 600°F continuously and to 900°F for intermediate use), excellent electrical and mechanical properties that are also relatively stable from low negative temperatures to high positive temperatures, dimensional stability (low cold flow) in most environments, excellent resistance to ionizing radiation, and very low outgassing in high vacuum. They have very low coefficients of friction, which can be further improved by use of graphite or other fillers. Materials and properties are shown in Ref. 23.

Polyamide-imides and polyimides have very good electrical properties, although not as good as those of TFE fluorocarbons, but they are much better than TFE fluorocarbons in mechanical and dimensional-stability properties. This provides advantages in many high-temperature electronic applications. All these properties also make polyamide-imides and polyimides excellent material choices in extreme environments of space and temperature. These materials are available as solid (molded and machined) parts, films, laminates, and liquid varnishes and adhesives. Since the data are relatively similar, except for the form factor, the data presented are for solid polyimides unless indicated otherwise. Films are quite similar to Mylar except for improved high-temperature capabilities.

Applications. These materials have been used in extreme service applications in aerospace, automotive under-hood, and transmission elements and in electrical, nuclear, business machine, and military components. They are also used for industrial hydraulic equipment, jet engines, automobiles, recreation vehicles, machinery, pumps, valves, and turbines.

4.6.15   Polymethylpentene (PMP)—Semicrystalline Thermoplastic

Another thermoplastic based on the ethylene structure, polymethylpentene, has special properties due to its combination of transparency and relatively high melting point. This polymer has four combined properties of (1) a high crystalline melting point of 464°F, coupled with useful mechanical properties at 400°F, and retention of form stability to near melting; (2) transparency with a light-transmission value of 90 percent in comparison with 88 to 92 percent for polystyrene and 92 percent for acrylics; (3) a density of 0.83, which is close to the theoretical minimum for thermoplastics materials; and (4) excellent electrical properties with power factor, dielectric constant (2.12) and volume resistivity of the same order as PTFE fluorocarbon. It can be joined with snap fits, press fits, fasteners, staking, and virtually all the thermoplastic welding techniques (some, with difficulty).[1]

Polymethylpentene properties are given in Ref. 23. Applications for polymethylpentene have been developed in the field of lighting and in the automotive, appliance, and electrical industries. It is used for laboratory and medical ware (syringes, connectors, hollowware, disposable curettes), lenses, food (freezing to cooking range), and liquid level and flow indicators.

4.6.16   Polyphenylene Oxide (PPO)—Amorphous Thermoplastics

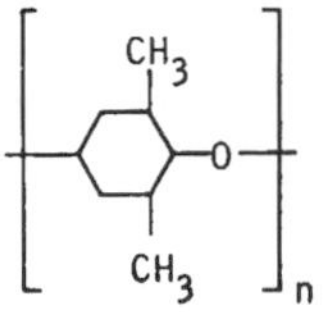

Figure 4.43   Phenylene oxide.

A patented process for oxidative coupling of phenolic monomers is used in formulating Noryl phenylene oxide-based thermoplastic resins.[46] The basic phenylene oxide structure is shown in Fig. 4.43.

This family of engineering materials is characterized by outstanding dimensional stability at elevated temperatures, toughness, broad temperature-use range, outstanding hydrolytic stability, and excellent dielectric properties over a wide range of frequencies and temperatures. They can be joined with snap fits, press fits, fasteners, adhesives, solvents, staking, and virtually all the thermoplastic welding techniques.[1] Several grades are available that have been developed to provide a choice of performance characteristics to meet a wide range of engineering-application requirements.

Among their principal design advantages are (1) excellent mechanical properties over temperatures from below −40°F to above 300°F; (2) self-extinguishing, nondripping characteristics; (3) excellent dimensional stability with low creep, high modulus, and low water absorption; (4) good electrical properties; (5) excellent resistance to aqueous chemical environments; (6) ease of processing with injection-molding and extrusion equipment; and (7) excellent impact strength. Properties are shown in Ref. 23. Thermal stability and moisture absorption are compared with those of other engineering thermoplastics in Figs. 4.35 and 4.37, respectively.

These materials are used for automobile dashboards, electrical connectors, grilles, and wheel covers. They are also used for hot water pumps, underwater components, shower heads, appliances, and electrical and appliance housings.[1]

4.6.17   Polyphenylene Sulfide (PPS)—Semicrystalline Thermoplastic

Polyphenylene sulfide (PPS), has a symmetrical, rigid backbone chain consisting of recurring para-substituted benzene rings and sulfur atoms. Its chemical structure is shown in Fig. 4.44.

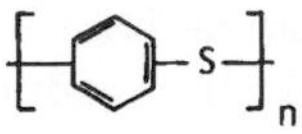

Figure 4.44 Polyphenylene sulfide

This chemical structure is responsible for the high melting point (550°F), outstanding chemical resistance, thermal stability, and nonflammability of the polymer. The polymer is characterized by high stiffness, impact resistance, and good retention of mechanical properties at elevated temperatures, which provide utility in coatings as well as in molding compounds. Polyphenylene sulfide is available in a variety of grades suitable for slurry coating, fluidized-bed coating, flocking, electrostatic spraying, and injection and compression molding.[47] The properties of unfilled and glass-filled varieties of this material are detailed in Ref. 23.

At normal temperatures, the unfilled polymer is a hard material with high tensile and flexural strengths. Substantial increases in these properties are realized by the addition of fillers, especially glass. Tensile strength and flexural modulus decrease with increasing temperature, leveling off at about 250°F, with good tensile strength and rigidity retained up to 500°F. With increasing temperature, there is a marked increase in elongation and a corresponding increase in toughness.

The mechanical properties of PPS are unaffected by long-term exposure in air at 450°F. For injection-molding applications, a 40 percent glass-filled grade is recommended. Coatings of PPS require a baking operation. Nonstick formulations can be prepared when a combination of hardness, chemical inertness, and release behavior is required.

Polyphenylene sulfide can be joined with snap fits, press fits, fasteners, adhesives, staking, and virtually all the thermoplastic welding techniques.[1] There are no known solvents below 375 to 400°F. Good adhesion to aluminum requires grit blasting and degreasing treatment. Good adhesion to steel is obtained by grit blasting and degreasing, followed by treatment at 700°F in air. Polyphenylene sulfide adheres well to titanium and to bronze after the metal surface has been degreased.

Molded items have applications where chemical resistance and high-temperature properties are of prime importance. Polyphenylene sulfide is used for electrical (connectors, coil forms, bobbins), mechanical (chemical processing equipment and pumps, including submersibles), and automotive (under hood) applications.

4.6.18 Polystyrene (PS)—Amorphous Thermoplastics

Polystyrenes are commodity plastics that are very easy to process and low in cost with good rigidity and dimensional stability. They have low moisture absorption, glossy surface, good clarity, and are easy to decorate. General-purpose styrene is brittle without a modifier; butadiene is added to improve impact resistance. Clear is not ultraviolet light resistant. Weather exposure discolors the material and reduces its strength. Improvement can be gained with pigments (finely disbursed carbon black). Limit outdoor use to applications where parts can be replaced or exposure is intermittent. Polystyrene is available in heat resistant, ultraviolet-light-resistant and flame-retardant grades.[1]

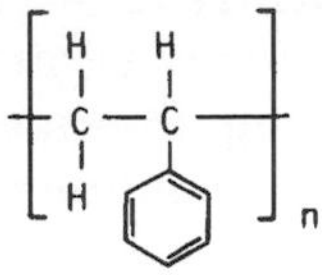

Commercial polystyrene is produced by continuous bulk, suspension, and solution polymerization techniques or by combining various aspects of these techniques.[2,48] Its structure is shown in Fig. 4.45

Figure 4.45   Polystyrene.

The polymerization is a highly exothermic, free-radical reaction. The homopolymer is characterized by its rigidity, sparkling clarity, and ease of processibility; however, it tends to be brittle. Impact properties are improved by copolymerization or grafting polystyrene chains to unsaturated rubbers such as polybutadiene. Rubber levels typically range from 3 to 12 percent. Commercially available impact-modified polystyrene is not as transparent as the homopolymers, but it has a marked increase in toughness.

The versatility of the styrene polymerization processes allows manufacturers to produce products with a wide variety of properties by varying the molecular-weight characteristics, additives, plasticizer content, and rubber levels. Heat resistance ranges from 170 to 200°F. Polystyrenes with tensile elongations from near zero to over 50 percent are produced. Various melt viscosities are also available.

Since properties can be varied so extensively, polystyrene is used in sheet and profile extrusion, thermoforming, injection and extrusion blow molding, heavy and thin-wall injection molding, direct-injection foam-sheet extrusion, biaxially oriented sheet extrusion, and extrusion of structural foam, and rotational molding. Polystyrenes can be printed; painted; vacuum-metallized and hot-stamped; sonic, solvent, adhesive, and spin welded; and screwed, nailed, and stapled. Polystyrenes are most attractive when considered on a cost-performance comparison with other thermoplastics. Limitations of polystyrene include poor weatherability, loss of clarity with impact modification, limited heat resistance, and flammability. The properties of these polymers are shown in Ref. 36.

Polystyrenes represent an important class of thermoplastic materials in the electronics industry because of very low electrical losses. Mechanical properties are adequate within operating-temperature limits, but polystyrenes are temperature-limited with normal temperature capabilities below 200°F. Polystyrenes can, however, be cross-linked to produce a higher-temperature material.

Cross-linked polystyrenes are actually thermosetting materials and hence do not remelt, even though they may soften. The improved thermal properties, coupled with the outstanding electrical properties, hardness, and associated dimensional stability, make cross-linked polystyrenes the leading choice of dielectric for many high-frequency-radar-band applications.

Conventional polystyrenes are essentially polymerized styrene monomer alone. By varying manufacturing conditions or by adding small amounts of internal and external lubrication, it is possible to vary such properties as ease of flow, speed of setup, physical strength, and heat resistance. Conventional polystyrenes are frequently referred to as normal, regular, or standard polystyrenes.

Since conventional polystyrenes are somewhat hard and brittle and have low impact strength, many modified polystyrenes are available. Modified polystyrenes are materials in which the properties of elongation and resistance to shock have been increased by incorporating into their composition varying percentages of elastomers, as was described. Hence, these types are frequently referred to as high-impact (HIPS), high-elongation, or rubber-modified polystyrenes. The so-called *superhigh-impact* types can be quite rubbery. Electrical properties are usually degraded by these rubber modifications.

Polystyrenes are subject to stresses in fabrication and forming operations and often require annealing to minimize such stresses for optimized final-product properties. Parts can usually be annealed by exposing them to an elevated temperature approximately 5 to 10°F lower than the temperature at which the greatest tolerable distortion occurs.

Polystyrenes generally have good dimensional stability and low mold shrinkage and are easily processed at low costs. They have poor weatherability and are chemically attacked by oils and organic solvents. Resistance is good, however, to water, inorganic chemicals, and alcohols.

General-purpose polystyrene is used for home furnishings (mirror and picture frames and moldings), housewares (personal care, flower pots, toys, cutlery, bottles, combs, disposables such as tumblers, dishes and trays), consumer electronics (cassettes, reels, and housings), and medical uses (sample collectors, petri dishes, test tubes). Impact styrene (with flame retardants) is used for televisions, smoke detectors, and small appliance housings.

4.6.19 Polysulfone (PSU)—Amorphous Thermoplastics

Polysulfones offer good transparency and high mechanical strengths, heat resistance, and electrical strengths. They have unusual resistance to strong mineral acids and alkalis and retention of properties on heat aging; however, weatherability is poor without coating.[1]

In the natural and unmodified form, polysulfone is a rigid, strong thermoplastic[23] that can be molded, extruded, or thermoformed (in sheets) into a wide variety of shapes. Characteristics of special significance to the design engineer are their heat-deflection temperature of 345°F at 264 lb/in^2 and long-term use temperature of 300 to 340°F. This is compared with some other engineering thermoplastics in Fig. 4.35. The properties of these polymers are shown in Ref. 23.

Thermal gravimetric analyses show polysulfone to be stable in air up to 500°C. This excellent thermal resistance of polysulfones, along with outstanding oxidation resistance, provides a high degree of melt stability for molding and extrusion.

Some flexibility in the polymer chain is derived from the ether linkage, thus providing inherent toughness. Polysulfone has a second, low-temperature glass transition at −150°F, similar to other tough, rigid thermoplastic polymers. This minor glass transition is attributable to the ether linkages. The linkages connecting the benzene rings are hydrolytically stable in polysulfones. These polymers therefore resist hydrolysis and aqueous acid and alkaline environments.

Polysulfone is produced by the reaction between the sodium salt of 2,2-bis(4-hydroxyphenol) propane and 4,4'-dichlorodiphenyl sulfone.[49] The sodium phenoxide end groups react with methyl chloride to terminate the polymerization. The molecular weight of the polymer is thereby controlled and thermal stability is assisted. Polysulfone has found markets in high-temperature automotive, office machine, consumer electronics, appliance, and medical applications.

4.6.20   Vinyls—Polyvinyl Acetal, Polyvinyl Acetate (PVAC), Polyvinyl Alcohol (PVOH), Polyvinyl Carbazole (PVK), Polyvinyl Chloride (PVC), Polyvinyl Chloride-Acetate (PVAC), and Polyvinylidene Chloride (PVDC)—Semicrystalline Thermoplastics

Vinyls are structurally based on the ethylene molecule through substitution of a hydrogen atom with a halogen or other group. The material's properties are outlined in Ref. 23. Basically, the vinyl family comprises the seven major types listed above.

Polyvinyl acetals consist of three groups, namely polyvinyl formal, polyvinyl acetal, and polyvinyl butyral. These materials are available as molding powders, sheet, rod, and tube. Fabrication methods include molding, extruding, casting, and calendering. Polyvinyl chloride (PVC) is perhaps the most widely used and highest-volume type of the vinyl family. PVC and polyvinyl chloride-acetate are the most commonly used vinyls for electronic and electrical applications.

Vinyls are basically tough and strong. They resist water and abrasion and are excellent electrical insulators. Special tougher types provide high wear resistance. Excluding some nonrigid types, vinyls are not degraded by prolonged contact with water, oils, foods, common chemicals, or cleaning fluids such as gasoline or naphtha. Vinyls are affected by chlorinated solvents.

Generally, vinyls will withstand continuous exposure to temperatures ranging up to 130°F; flexible types, filaments, and some rigids are unaffected by even higher temperatures. Some of these materials, in some operations, may be health hazards. These materials also are slow-burning, and certain types are self-extinguishing—but direct contact with an open flame or extreme heat must be avoided.

PVC is a material with a wide range of rigidity or flexibility. One of its basic advantages is the way it accepts compounding ingredients. For instance, PVC can be plasticized with a variety of plasticizers to produce soft, yielding materials to almost any desired degree of flexibility. Without plasticizers, it is a strong, rigid material that can be machined, heat formed, or welded by solvents or heat. It is tough, with high resistance to acids, alcohol, alkalis, oils, and many other hydrocarbons. It is available in a wide range of colors. Molded rigid vinyl is used for pipe fittings, toys, dinnerware, sporting goods, toys, shoe heels, credit cards, gate ball valves, and electrical applications in appliances, television sets, and electrical boxes.

Flexible PVC is easier to process but offers lower heat resistance and lesser physical and weathering properties. It provides the unusual combination of transparency with flexibility. Typical uses include profile extrusions, film, and wire insulation.

PVC raw materials are available as resins, latexes, organosols, plastisols, and compounds. Fabrication methods include injection, compression, blow or slush molding, extruding, calendering, coating, laminating, rotational and solution casting, and thermoforming.

4.7   Glass-Fiber-Reinforced Thermoplastics

Basically, thermoplastic molding materials are developed and can be used without fillers, as opposed to thermosetting molding materials, which are more commonly used with fillers incorporated into the compound. This is primarily because shrinkage, hardness, brittleness, and other important processing and use properties require the use of fillers in thermosets.

Thermoplastics, on the other hand, do not suffer from the same shortcomings as thermosets and hence can be used as molded products without fillers. However, thermoplastics do suffer from creep and dimensional stability problems, especially under elevated temperature and load conditions. Because of this weakness, most designers find difficulty in matching the techniques of classical stress-strain analysis with the nonlinear, time-dependent strength-modulus properties of thermoplastics. Glass-fiber-reinforced thermoplastics (FRTPs) help to simplify these problems. For instance, 40 percent glass-fiber-reinforced nylon outperforms its unreinforced version by exhibiting two and one-half times greater tensile and Izod impact strengths, four times greater flexural modulus, and only one-fifth of the tensile creep. There is, however, a drop in impact resistance and the cost of the glass reinforced material is greater than that of the neat resin.

Thus, FRTPs fill a major materials gap in providing plastic materials that can be used reliably for strength purposes, and which in fact can compete with metal die castings. Strength is increased with glass-fiber reinforcement, as are stiffness and dimensional stability. The thermal expansion of the FRTPs is reduced, creep is substantially reduced, and molding precision is much greater.

The dimensional stability of glass-reinforced polymers is invariably better than that of the nonreinforced materials. Mold shrinkages of only a few mils per inch are characteristic of these products; however, part distortion may be increased, because the glass cools at different rate from the polymer. Low moisture absorption of reinforced plastics ensures that parts will not suffer dimensional increases under high-humidity conditions. Also, the characteristic low coefficient of thermal expansion is close enough to that of such metals as zinc, aluminum, and magnesium that it is possible to design composite assemblies without fear that they will warp or buckle when cycled over temperature extremes. In applications where part geometry limits maximum wall thickness, reinforced plastics almost always afford economies for similar strength or stiffness over their unreinforced equivalents. A comparison of some important properties for unfilled and glass-filled (20 and 30 percent) thermoplastics is given in Ref. 23.

Chemical resistance is essentially unchanged, except that environmental stress-crack resistance of such polymers as polycarbonate and polyethylene is markedly increased by glass reinforcement.

4.8   Plastic Films and Tapes

4.8.1   Films

Films are thin sections of the same polymers described previously in this chapter. Most films are thermoplastic in nature because of the great flexibility of this class of resins. Films can be made from most thermoplastics.

Films are made by extrusion, casting, calendering, and skiving. Certain of the materials are also available in foam form. The films are sold in thicknesses from 0.5 to 10 mil. (0.0005 to 0.010 in). Thicknesses in excess of 10 mil are more properly called *sheets*.

4.8.2 Tapes

Tapes are films slit to some acceptable width and are frequently coated with adhesives. The adhesives are either thermosetting or thermoplastic. The thermoset adhesives consist of rubber, acrylic, silicones, and epoxies, whereas the thermoplastic adhesives are generally acrylic or rubber. Tackifying resins are generally added to increase the adhesion. The adhesives all deteriorate with storage. The deterioration is marked by loss of tack or bond strength and can be inhibited by storage at low temperature.

4.8.3 Film Properties

Films differ from similar polymers in other forms in several key properties but are identical in all others. Since an earlier section of this chapter described in detail most of the thermoplastic resins, this section will be limited to film properties. The properties of common films are presented in Ref. 50. To aid in the selection of the proper films, the most important features are summarized in Table 4.10.

TABLE 4.10   Film Selection Chart

Film	Cost	Thermal stability	Dielectric constant	Dissipation factor	Strength	Electric strength	Water absorption	Folding endurance
Cellulose	L	L	M	M	H	M	H	L
FEP fluorocarbon	H	H	L	L	L	H	VL	M
Polyamide	M	M	M	M	H	L	H	VH
PTFE polytetra-fluoroethylene	H	H	L	L	L	L	VL	M
Acrylic	M	L	M	M	M	L	M	M
Polyethylene	L	L	L	L	L	L	L	H
Polypropylene	L	M	L	L	L	M	L	H
Polyvinyl fluoride	H	H	H	H	H	M	L	H
Polyester	M	M	M	L	H	H	L	VH
Polytrifluoro-chloroethylene	H	H	L	L	M	M	VL	M
Polycarbonate	M	M	M	M	M	L	M	L
Polyimide	VH	H	M	L	H	H	H	M

VL = low, L = low, M = medium, H = high, VH = very high.

Films differ from other polymers chiefly in improved electric strength and flexibility. Both of these properties vary inversely with the film thickness. Electric strength is also related to the method of manufacture. Cast and extruded films have higher electric strength than skived films. This is caused by the greater incidence of holes in the latter films. Some films can be oriented, which improves their physical properties substantially. Orientation is a process of selectively stretching the films, thereby reducing the thickness and causing changes in the crystallinity of the polymer. This process is usually accomplished under conditions of elevated temperature, and the benefits are lost if the processing temperatures are exceeded during service.

Most films can be bonded to other substrates with a variety of adhesives. Films that do not readily accept adhesives can be surface-treated for bonding by chemical and electrical etching. Films can also be combined to obtain bondable surfaces. Examples of these combined films are polyolefins laminated to polyester films and fluorocarbons laminated to polyimide films.

4.9   Plastic Surface Finishing

While the greatest majority of plastic parts can be, and often are, used either with their as-molded natural-colored surface or with colors obtained by use of precolored resins, color concentrate, or dry powder molded into the resin, competitive design factors may require surface finishing of plastics after molding to provide color or metallization. Some important points related to painting and plating are presented in the following sections.

4.9.1   Painting of Plastics

Plastics are often difficult to paint, and proper consideration must be given to all the important factors involved. In Harper[2] (Tables 37 and 38), a selection guide to paints for plastics is presented, and application ratings are given for various paints. Some important considerations related to painting plastics are given in the following.

- *Heat-distortion point and heat resistance.* This determines whether a bake-type paint can be used and, if so, the maximum baking temperature the plastic can tolerate.

- *Solvent resistance.* The susceptibility of the plastic to solvent attack dictates the choice of paint system. Some softening of the substrate is desirable to improve adhesion, but a solvent that attacks the surface aggressively and results in cracking or crazing obviously must be avoided.

- *Residual stress.*   Molding operations often produce parts with localized areas of stress. Application of coating to these areas may swell the plastic and cause crazing. Annealing of the part before coating will minimize or eliminate the problem. Often, it can be avoided entirely by careful design of the molded part to prevent locked-in stress.

- *Mold-release residues.*   Excessive amounts of mold-release agents often cause surface-finishing adhesion problems. To ensure satisfactory adhesion, the plastic surface must be rinsed or otherwise cleaned to remove the release agents.

- *Plasticizers and other additives.*   Most plastics are compounded with plasticizers and chemical additives. These materials usually migrate to the surface and may eventually soften the coating, destroying adhesion. A coating should be checked for short- and long-term softening or adhesion problems for the specific plastic formulation on which it will be used.

- *Other factors.*   Stiffness or rigidity, dimensional stability, and coefficient of expansion of the plastic are factors that affect the long-term adhesion of the coating. The physical properties of the paint film must accommodate those of the plastic substrate.

4.9.2   Plating on Plastics

The advantages of metallized plastics in many industries, coupled with major advances in both platable plastic materials and plating technology, have resulted in a continuing and rapid growth of metallized plastic parts. Some of the major problems have been adhesion of plating to plastic, differential expansion between plastics and metals, failure of plated part in thermal cycling, heat distortion and warpage of plastic parts during plating and in system use, and improper design for plating. The major plastics that are plated, and their characteristics for plating, are identified in Table 4.11.[51] Improvements are being made continuously, especially in ABS and polypropylene, that yield generally lower product costs. Thus, the guidelines of Table 4.11 should be reviewed at any given time and for any given application. Aside from the commercial plastics described in Table 4.11, excellent plated plastics can be obtained with other resins. Notable is the plating of TFE fluorocarbon, where otherwise unachievable electrical products of high quality are reproducibly made. Examples are corona-free capacitors and low-loss high-frequency electronic components.[52]

Design considerations.   Proper design is extremely important in producing a quality plated-plastic part, and some important design considerations are presented in Ref. 53.

TABLE 4.11   Characteristics of Major Plated Plastics[65]

	ABS	Polypropylene	Polysulfone	Polyarylether	Modified PPO
Flow	AA	AA	BA	BA	A
Heat distortion under load	A	BA	AA	AA	A
Platability	AA	A	BA	AA	BA
Thermal cycling	BA	A	AA	AA	AA
Warpage	A	BA	A	A	A
Mold definition	AA	BA	A	A	A
Coefficient of expansion	A	A	A	A	A
Water absorption	BA	AA	A	BA	AA
Material cost	A	AA	BA	BA	BA
Finishing cost	AA	BA	BA	AA	BA
Peel strength	A	AA	AA	BA	BA

Polymers are rated according to relative desirability of various characteristics: AA = above average, A = average, BA = below average.

Appearance.   Because most plated-plastic parts now being produced are decorative (such as washer end caps, escutcheons) rather than functional (such as copper-plated conductive plastic automotive distributor parts), appearance is extremely critical. For a smooth, even finish, one-piece or integral parts should be designed. Mechanical welds are difficult to plate. If they are necessary, they should be hidden on a noncritical surface. Gates should be hidden on noncritical surfaces or should be disguised in a prominent feature. Gate design should minimize flow and stress lines, which may impair adhesion.

4.10   Material Selection

This section looks at material selection from the design engineer's perspective. The vast number of plastics compounds on the market is enough to stagger the mind of the designer trying to make a material selection. Fortunately, only a small percentage of these are actually serious contenders for any given application. Some of them were developed specifically for a single product, particularly in the packaging industry. Others became the material of choice for certain applications because of special properties they offer that are required for that product or process. For example, the vast majority of rotomolded parts are made of polyethylene, while glass-fiber-reinforced polyester is the workhorse of the thermoset industry. A bit of research should reveal if there is a material of choice for any given product application.

First, a bit of a review of the basic categories of plastics materials. In general, they fall into one of two categories: thermosets and thermoplastics. Thermosets undergo a chemical reaction when heated and cannot return to their original state. Consequently, they are chemical resistant and do not burn. Cross-linked plastics are thermosets. Thermoplastics constitute the bulk of the polymers available. Although some degradation does occur, they can be remelted. Most are readily attacked by chemicals, and they burn readily.

Thermoplastics can also be broken down into two basic categories: amorphous and semicrystalline (hereafter referred to as *crystalline*). The names refer to their structures; amorphous having molecular chains in random fashion, and crystalline having molecular chains in a regular structure. Polymers are referred to as *semicrystalline* because they are not completely crystalline in nature. Amorphous resins soften over a range of temperatures, whereas crystallines have a definite point at which they melt. Amorphous polymers can have greater transparency and lower, more uniform post-molding shrinkage. Chemical resistance is, in general, much greater for crystalline resins than for amorphous resins, which are sufficiently affected to be solvent welded. The triangle illustrated in Figure 4.46[55] provides an easy way to categorize the thermoplastics.

The cost of plastics generally increases with a corresponding improvement in thermal properties. (Other properties typically go up as well.) The lowest-cost plastics are the most widely used. The triangle is organized with the least temperature-resistant plastics at the base and those with the highest temperature resistance at the top. Therefore, the plastics designated *Standard* at the base of the triangle, often referred to as *commodity* plastics, are the lowest in cost and most widely used. They can be used in applications with temperatures up to 150°F. (Note: These are very loose groupings, and the precise properties of a specific resin must be evaluated before specifying it.)

The next level shows the *Engineering* plastics, which can be used for applications ranging up to 250°F. ABS is often considered an engineering plastic for its other properties, although it cannot withstand this temperature level. For applications requiring temperature resistance up to 450°F, there is the next step, the *Advanced Engineering* level. The amorphous plastics at this level are often used in steam environments, and the crystalline plastics have improved chemical resistance. The top level, the *Imidized* plastics, can withstand temperatures up to 800°F and have excellent stress and wear properties as well.

There has been considerable development work done on high-temperature plastics in recent years. Table 4.12[54] lists the properties of these materials.

Table 4.13[55] lists many of the other principal properties and some of the polymers that are noted for those properties. While incomplete, this table should at least provide a beginning. They are listed in their natural state without reinforcements, such as glass or carbon fibers.

TABLE 4.12 Properties of Representative High-Temperature Thermoplastics[54]

Polymer	Common designation	Morphology*	Glass transition, °F	Tensile strength, ksi	Tensile modulus, ksi	Elongation, %	Fracture toughness, G_{JC}, in·lb/in²	Notched Izod, ft·lb/in
Polyimide	N†	A	700	16.0	580	6	–	–
Polyimide	LARC-TPI	A	507	17.3	540	4.8	–	1.0
Polyimide	K-III†	A	484	14.8	546	14	11	–
Polyetherimide	PEI	A	423–518	15.2	430	60	19	1.0
Polyamide-imide	PAI	A	527	9.2–13.0	400–66	1.4–30	19.4	2.7
Polyarylimide	J-2†	A	320	15.0	7	25	–	–
Polyimidesulfone	PISO₂	A	523	9.1	460	1.3	8	–
Polysulfone	PSF	A	374	10.2	719	>50	14	1.2
Polyarylsulfone	PASF	A	428	10.4	360	60	20	1.2
Polyarylene sulfide	PAS	SC	419	14.5	310	7.3	–	0.8
Polyphenylene sulfide	PPS	SC	194	12.0	470 630	5	–	3.0
Polyether sulfone	PES HTA‡	A	446–500	12.2	380	>40	11	1.6
Polyetherketone	PEK	SC	329	16.0	580	–	–	1.52
Polyetherketoneketone	PEKK	SC	311	–	580	–	–	–
Polyetheretherketone	PEEK	SC	289	14.5	–	>40	>23	1.6
Poly(EKEKK)	PEKEKK	SC	343	–	450	–	–	–
Polyarylene ketone	PAK, HTK‡	SC	509	12.7	–	13	–	–
Liquid crystal polymer	LCP, SRP‡	C	662	20.0	360 2400	4.9	6.9	2.4

*A = amorphous, SC = semicrystalline, C = crystalline.
†Trade name of E.I. du Pont.
‡Trade name of ICI.

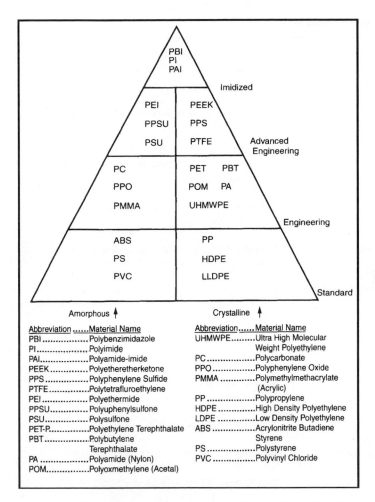

Figure 4.46 Classification of thermoplastics. *(Source: Laura Pugliese, Defining Engineering Plastics, Plastics Machining and Fabrication, Jan.–Feb. 1999, courtesy of ESM Engineering Plastic Products.)*

These reinforcements can be used to increase mechanical strength, maximum use temperature, impact resistance, stiffness, mold shrinkage, and dimensional stability.

Generally, the resin prices increase with improved mechanical and thermal properties. When there is no clear-cut material of choice, plastics designers generally follow the practice of looking for the lowest-cost material that will meet the product's requirements. If there is a reason that polymer is not acceptable, designers start working up the cost ladder until they find one that will fulfill their needs. In thermoplastics, there are the so-called *commodity resins*. These are the low-cost resins

TABLE 4.13   Recommended Materials

Property	Recommendation
Abrasion, resistance to (high)	Nylon
Cost:weight (low)	Urea, phenolics, polystyrene, polyethylene, polypropylene, PVC
Compressive strength	Polyphthalamide, phenolic (glass), epoxy, melamine, nylon, thermoplastic polyester (glass), polyimide
Cost:volume (low)	Polystyrene, polyethylene, urea, phenolics, polypropylene, PVC
Dielectric constant (high)	Phenolic, PVC, fluorocarbon, malamine, alkyd, nylon, polyphthalamide, epoxy
Dielectric strength (high)	PVC, fluorocarbon, polypropylene, polyphenylene ether, phenolic, TP polyester, nylon (glass), polyolefin, polyethylene
Dissipation factor (high)	PVC, fluorocarbon, phenolic, TP polyester, nylon, epoxy, diallyl phthalate, polyurethane
Distortion, resistance to under load (high)	Thermosetting laminates
Elastic modulus (high)	Melamine, urea, phenolics
Elastic modulus (low)	Polyethylene, polycarbonate, fluorocarbons
Electrical resistivity (high)	Polystyrene, fluorocarbons, polypropylene
Elongation at break (high)	Polyethylene, polypropylene, silicone, ethylene vinyl acetate
Elongation at break (low)	Polyether sulfone, polycarbonate (glass), nylon (glass), polypropylene (glass), thermoplastic polyester, polyetherimide, vinyl ester, polyetheretherketone, epoxy, polyimide
Flexural modulus (stiffness)	Polyphenylene sulfide, epoxy, phenolic (glass), nylon (glass) polyimide, diallyl phthalate, polyphthalamide, TP polyester
Flexural strength (yield)	Polyurethane (glass), epoxy, nylon (carbon fiber) (glass), polyphenylene sulfide, polyphthalamide, polyetherimide, polyetheretherketone, polycarbonate (carbon fiber)

TABLE 4.13   Recommended Materials *(Continued)*

Property	Recommendation
Friction, coefficient of (low)	Fluorocarbons, nylon, acetal
Hardness (high)	Melamine, phenolic (glass) (cellulose), polyimide, epoxy
Impact strength (high)	Phenolics, epoxies, polycarbonate, ABS
Moisture resistance (high)	Polyethylene, polypropylene, fluorocarbon, polyphenylene sulfide, polyolefin, thermoplastic polyester, polyphenylene ether, polystyrene, polycarbonate (glass or carbon fiber)
Softness	Polyethylene, silicone, PVC, thermoplastic elastomer, polyurethane, ethylene vinyl acetate
Tensile strength, break (high)	Epoxy, nylon (glass or carbon fiber), polyurethane, thermoplastic polyester (glass), polyphthalamide, polyetheretherketone, polycarbonate (carbon fiber), polyetherimide, polyether-sulfone
Tensile strength, yield (high)	Nylon (glass or carbon fiber), polyurethane, thermoplastic polyester (glass), polyetheretherketone, polyetherimide, polyphthalamide, polyphenylene sulfide (glass or carbon fiber)
Temperature (maximum use)	Ref. Table 4.12
Thermal conductivity (low)	Polypropylene, PVC, ABS, polyphenylene oxide, polybutylene, acrylic, polycarbonate, thermoplastic polyester, nylon
Thermal expansion, coefficient of (low)	Polycarbonate (carbon fiber or glass), phenolic (glass), nylon (carbon fiber or glass), thermoplastic polyester (glass), polyphenylene sulfide (glass or carbon fiber), polyetherimide, polyetheretherketone, polyphthalamide, alkyd, melamine
Transparency, permanent (high)	Acrylic, polycarbonate
Weight (low)	Polypropylene, polyethylene, polybutylene, ethylene vinyl acetate, ethylene methyl acrylate
Whiteness retention (high)	Melamine, urea

used in great volume for housewares, packaging, toys and so on. This group is made up of polyethylene, polypropylene, polystyrene and PVC. Reinforcements can improve the properties of these resins at moderate additional cost. A lower-priced resin with reinforcement will often provide properties comparable to a more expensive resin.

Table 4.14[55] is a list of the approximate cost of a number of plastics in increasing order of cost per cubic inch. This is regarded as a more useful figure than cost per pound in selecting a plastic material.

Thermosets usually provide higher mechanical and thermal properties at a lower material cost than do thermoplastics—"more bang for the buck," so to speak. However, most of the processes used to fabricate thermoset parts are slower and more limited in design freedom than the thermoplastic processes. Furthermore, the opportunity to utilize 100% of the material that thermoplastics provide is simply not available with thermosets, because the regrind cannot be reused. Recycling possibilities are far more limited for thermosets for the same reason. Nonetheless, glass-fiber-reinforced thermoset polyester is the material of choice for many severe environment outdoor applications such as boats and truck housings.

4.10.1   About the Data

Comparison of resins is usually done with data sheets supplied by the resin manufacturers. It is extremely important that the plastics design engineer understand the limitations of this data. Since the properties of polymers change with temperature, the data sheet does not provide the total picture of a given compound. Instead, think of it as a "snapshot" of the material taken at 72°F. As the temperature goes down from this point, the material becomes harder and more brittle. Increasing the temperature makes the polymer softer and more ductile. These are general statements and the effect of temperature will vary widely between resins. For one material, tensile strength at 140°F may be only half that at 72°F. For another polymer, it may change only slightly.

The graph depicted in Fig. 4.47,[55] "Effect of temperature on tensile yield strength," illustrates this phenomenon. The upper curve indicates that the value at 0°F is 14,000 psi. At 72°F, it has dropped to around 12,000 psi. By the time it reaches 140°F, the tensile yield strength is approximately 7,000 psi. This data is for nylon, a polymer particularly effected by moisture. The lower curve illustrates the effect of 2.5% moisture. In the range of temperatures between 30°F and 100°F, the tensile yield strength appears to be about 20% lower for the moist material. Note that the curves begin to run together beyond 150°F as most of the water has been driven off by that point.

TABLE 4.14   Approximate* Cost of Plastics in Dollars per Cubic Inch

Plastic	$/in³
Polypropylene	0.010
Polyethylene (HD)	0.010
Polyvinylchloride	0.013
Polyethylene (LD)	0.014
Polystyrene	0.015
ABS	0.032
Acrylic	0.034
Phenolic	0.038
Thermoplastic elastomer	0.039
Styrene acrylonitrile	0.043
Polyester (TS)	0.043
Melamine	0.049
Polyphenylene ether	0.050
Polyphenylene oxide	0.050
Urea	0.053
Polyester (PET)	0.055
Nylon	0.058
Styrene maleic anhydride	0.059
Vinyl ester	0.060
Polyurethane (TS)	0.062
Alkyd	0.068
Polycarbonate	0.073
Polyurethane (TP)	0.074
Acetal	0.074
Cellulosics	0.075
Polycarbonate/ABS	0.078
Epoxy	0.081
Polyphenylene sulfide	0.119
Dially phthalate	0.154
Polysulfone	0.201
Polyetherimide	0.253
Liquid crystal polymer	0.434
Fluorpolymer	0.481
Polyamideimide	0.949
Polyetheretherketone	1.720

*These values are very approximate. They were arrived at by multiplying the average density by the average price at the time this was being written. In many cases, the range from which the average was taken was quite wide.

Time is also a significant factor. Figure 4.48,[55] "Long-term behavior of delrin under load at 23°C (73°F) air," illustrates the effect of time on the stress-strain relationship of Delrin, an acetal polymer, at room temperature. Note that the strain rate increases with time.

Figures 4.49,[55] "Long-term behavior of delrin under load at 45°C (115°F) air," and 4.50,[55] "Long-term behavior of delrin under load at

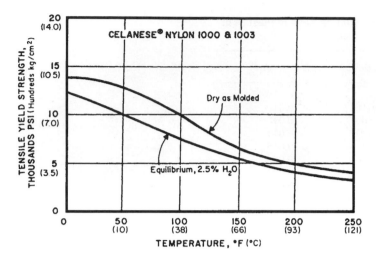

Figure 4.47 Effect of temperature on tensile yield strength. (Courtesy of Ticona.)[55]

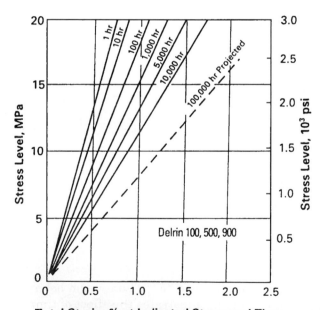

Figure 4.48 Long-term behavior of delrin under load at 23°C (73°F) air. (Courtesy of DuPont.)[55]

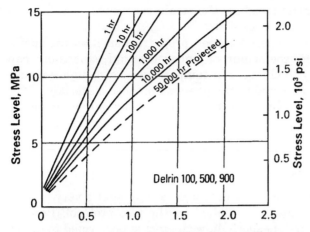

Figure 4.49 Long-term behavior of delrin under load at 45°C (115°F) air. (Courtesy of DuPont.)[55]

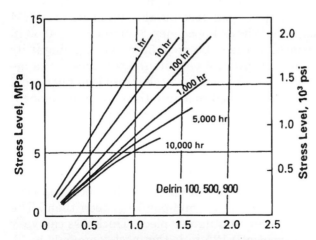

Figure 4.50 Long-term behavior of delrin under load at 85°C (185°F) air. (Courtesy of DuPont.)[55]

85°C (185°F) air," demonstrate how the strain rate of this material increases as the temperature rises.

Most of the physical properties, even properties such as the coefficient of friction or the coefficient of linear thermal expansion, can change significantly with changes in temperature—although not necessarily to the degree depicted in the graphs for these materials.

Temperature and humidity differences are not the only phenomena that can affect the data. Chemical and ultraviolet light exposure will also affect most resins, in some cases to a very high degree.

4.10.2 Interpreting the Test Data

It is important to recognize that the test data represent a very precise set of circumstances, those established by the test protocol. Product designers must be aware of exactly how the test is performed to determine how well the results relate to the conditions experienced by the product under development. The following sections discuss the test procedures for the mechanical and thermal properties most commonly required. Table 4.15[55] represents a typical property sheet as supplied by the resin manufacturer.

4.10.3 Tensile Test—ASTM D638

The first mechanical property most product designers look for in evaluating a potential material is its strength—and by this they mean its tensile strength at yield or break. Therefore, it is often found at the top of the data sheet. The principal test for this property is ASTM D638; it calls for a "dog bone" shaped specimen 8.50 in long by 0.50 in wide. The gripping surfaces at the ends are 0.55 in wide, giving it its characteristic shape. The test protocol permits the thickness to range from 0.12 in to 0.55 in and the rate at which the stress is applied from 0.5 to 20 in/min. This test is also used to obtain the percentage elongation at break and produce the stress strain curve, from which the modulus of elasticity is derived.

4.10.4 Flexural Properties of Plastics ASTM D790

This test is performed by suspending a specimen between supports and applying a downward load at the mid-point between them. The specimen is a 0.50 in by 5.00 in rectangular piece. Thickness can vary form 0.06 in to 0.25 in, however 0.125 in is the most commonly used. The distance between the supports is 16 times the specimen thickness. The load is applied at rates defined by the specimen size until fracture occurs or until the strain in the outer fibers reaches 5%. The flexural modulus is the flexural stress at 5% strain. In the event of failure be-

TABLE 4.15   Celcon™ Acetal Copolymer—Typical Properties[55*]

Property	ASTM test method	Units	M90™
Physical:			
Specific gravity	D792	—	1.41
Water absorption: 24 h			
@73°F	D570	%	0.22
@equilibrium	D570	%	0.8
Mold shrinkage:			
Flow direction	D955	mils/in	22
Transverse direction	D955	mils/in	18
Mechanical and thermal:			
Tensile strength @ yield			
−40°F	D638	lb/in^2	13,700
73°F	D638	lb/in^2	8,800
160°F	D638	lb/in^2	5,000
Elongation @ break:			
−40°F	D638	%	20
73°F	D638	%	60.0
160°F	D638	%	>250
Flexural stress @ 5% deformation	D790	lb/in$^2 \times 10^3$	13.0
Flexural modulus:			
73°F	D790	lb/in$^2 \times 10^4$	37.5
160°F	D790	lb/in$^2 \times 10^4$	18.0
220°F	D790	lb/in$^2 \times 10^4$	10.0
Fatigue endurance (limit @ 10^7 cycles)	D671	lb/in^2	4,100
Compressive strength:			
@1% deflection	D695	lb/in^2	4,500
@10% deflection	D695	lb/in^2	16,000
Rockwell hardness	D785	—	M80
Izod impact strength:			
−40°F (notched)	D256	ft-lb/in	1.0
73°F (notched)	D256	ft-lb/in	1.3
Tensile impact strength	D1822	ft-lb/in^2	70
Heat deflection temperature:			
@66 lb/in^2	D648	°F	316
@264 lb/in^2	D648	°F	230
Shear strength: 73°F	D732	lb/in^2	7,700

*Source: Courtesy Ticona. These data are based on testing of laboratory test specimens and represent data that fall within the standard range of properties for natural material. Colorant and other additives may cause significant variations in data values.

fore that point, the flexural strength is the tensile stress in the outer-most fibers at the break point.

4.10.5   Fatigue Endurance ASTM D671

The test covers determination of the effect of repeated flexural stress of the same magnitude with a fixed-cantilever apparatus designed to produce a constant amplitude of force on the plastic test specimen. The results are suitable for application in design only when all of the application parameters are directly comparable to the those of the test.

4.10.6 Compressive Strength ASTM D695

The apparatus for this test resembles a C-clamp with the specimen compressed between the jaws of the apparatus, which close at the rate of 0.05 in per minute until failure occurs. A wide range of specimen sizes is permitted for this test.

4.10.7 Rockwell Hardness ASTM D785

An indenter is placed on the surface of the test specimen, and the depth of the impression is measured as the load on the indenter is increased from a fixed minimum value to a higher value and then returned to the previous value. A number of different diameter steel balls and a diamond cone penetrator are used. The Rockwell scale refers to a given combination of indenter and load; M70 for example.

A number of scales are used within the plastics industry. Figure 4.51[55] illustrates the relationship between them.

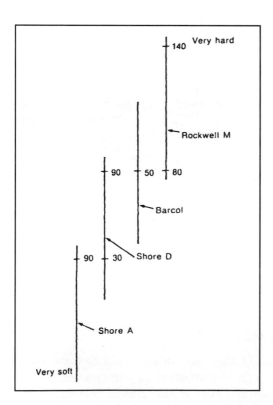

Figure 4.51 Range of hardness common to plastics. *(Source: Dominick V. Rosato,* Rosato's Plastics Encyclopedia and Dictionary, *Carl Hanser Verlag, Munich, 1993.)*[55]

4.10.8   The IZOD Impact Test—ASTM D256

D256, the IZOD impact test, is the most common variety of impact test. It is a pendulum test with the pendulum dropping from the 12 o'clock position to hit a sample held in a clamp at the 6 o'clock position. The pendulum breaks the sample, and the distance it travels beyond the specimen is a measure of the energy absorbed in breaking the sample. The value calculated from this test is usually expressed in ft-lb/in. of sample width.

The plastics design engineer must be wary of the fact that there are five different methods of performing this test, and the results will vary with each method. The four used by design engineers are as follows:

Method A.   The specimen for this method is 2.50 in long by 0.50 in thick. There is a 45° included angle notch at mid-point that is 0.10 in deep and has a 0.01 in radius at the V. The notch faces the pendulum. The impact point is just above the notch.

Method B.   This procedure is also known as the "Charpy" test. It is similar to the previous method except that the bar is laid horizontally, and the impact is directly behind the notch. The length of the specimen is increased to 5.0 in for this method.

Method D.   The principal difference between this method and Method A is that it permits a larger radius at the V of the notch, which substantially affects the results. It is used for highly notch-sensitive polymers.

Method E.   In this case, the same size specimen and procedure applied in Method A is used, except that the notch faces the pendulum.

The difference between the results of the notched IZOD test and the unnotched IZOD test (when available) can be used as a rough measure of notch sensitivity for a given material.

4.10.9   The Falling Dart (Tup) Impact Test—ASTM D3029

The falling dart test, ASTM D3029, is not on this particular data sheet. However, it may be more appropriate to reveal the behavior of materials on impact for many product applications such as appliance housings and the like. Unfortunately, this test is usually performed only for extrusion grades of resins that are to be made into sheet. Therefore, it may be necessary to request that this test be performed on a material under consideration.

For testing, a flat specimen is suspended over a circular opening below a graduated column with a cantilever arm attached. A weight, also known as a *dart* or *tup,* is attached to the arm, from which it is released to strike the sample. The arm can be raised or lowered and the weight of the tup varied until 50% of the sample quantity fails the test. Method A of this test calls for the opening below the sample to be 5.00 inches in diameter. For method B, it is 1.50 inches in diameter.

4.10.10   The Tensile Impact Strength ASTM D1822

This test uses an apparatus very similar to that used for the IZOD tests except that, in this case, the specimen is attached to the pendulum on one end and has a T-bar attached to the other end. When the pendulum drops, the T-bar catches on the apparatus at its base, causing the specimen to undergo tensile impact. For this test, the specimen is 2.50 in long and necks to 0.125 in at the center. The thickness can vary. The gripping surfaces at the ends are 0.50 in wide. This test is typically performed on materials which are too elastic to fail in the IZOD test and is normally found on data sheets. It can be performed on request if it best represents the product's performance requirements.

4.10.11   Heat Deflection Temperature ASTM D648

The apparatus for this test somewhat resembles that of the flexural test in that the specimen is suspended between two supports four inches apart with a downward load at the mid-point. However, in this case, the entire structure is immersed in a liquid whose temperature is increased at the rate of 2°C. per minute. Two loadings are used, 66 psi and 264 psi. Consequently, the plastics engineer must be careful to compare values for the same loading. The heat deflection temperature is the temperature at which the specimen deflects 0.010 in. The specimen for this test is 0.50 in wide by 5.00 in long. Thicknesses vary from 0.125 in to 0.50 in.

Different sample thicknesses and processes can produce significant differences in values. The author recalls a project where the field had been narrowed to two competing materials. One of them had a 15% heat deflection temperature advantage at a slightly higher cost. However, the other was produced by a long-standing supplier and, before taking the business away from him, it seemed only fair to call and ask if he had a comparable resin that was not in the current brochure. The discussion revealed that the competing material was, in fact, the very same resin that the competitor bought from our supplier and resold under his own brand. Why then the difference in test values? Further research revealed that one supplier had used an injection-molded sample 0.125 in thick, and the other had tested an extruded sample 0.50 in thick, which resulted in higher values.

4.10.12 Vicat Softening Point ASTM D1525

The Vicat softening point is not provided on this particular data sheet. However, it is a method of determining the softening point of plastics that have no definite melting point. A 1000 g load is placed on a needle with a 0.0015 in^2 circular or square cross section. The softening point is taken as the point where the needle penetrates the specimen to a depth of 1 mm.

4.10.13 Glass Transition Temperature ASTM D3418

Amorphous thermoplastics exhibit a characteristic whereby they change from a material that behaves like glass (strong, rigid, but brittle) to one with generally reduced physical properties (weaker and more ductile). This is known as the glass transition temperature (T_g) and is actually a range of temperatures, as the value is different for each property and is significantly affected by variations in the test protocol. Usually, a single value is provided; therefore, it should be treated as an approximation.

4.10.14 Relative Temperature Index UL746B (Maximum Continuous Use Temperature)

Underwriter's Laboratories Inc. has devised a thermal aging test protocol whereby a subject material is tested in comparison to a material with an acceptable service experience and correlates numerically with the temperatures above which the material is likely to degrade prematurely. The end of life of a material is regarded as the point where the value of the critical properties have dropped to half their original values. The resin manufacturer must submit his material to Underwriters Laboratories to have it tested. When this has not been done, the designer can use the generic value for the polymer, which is usually regarded as conservative.

4.10.15 Shear Strength ASTM D732

Shear strength is determined using a 2-in. diameter or 2-in. square test specimen, ranging in thickness from 0.005 to 0.500 in., placed in a punch-type shear fixture. Pressure is applied to the punch at the rate of 0.005 in/min until the moving part of the sample clears the stationary portion. The force divided by the area sheared determines the shear strength.

4.10.16 Flammability and Flame Retardancy of Plastics

Flammability is not among the properties listed in Table 4.15, however the issue of flammability is becoming a key requirement for continued growth in many areas, such as home furnishings, clothing, wire and cable, automobiles, and aircraft. Some major fires have been partially attributed to plastics,[56] and numerous deaths have been attributed to toxic smoke generated by burning plastics.

Strong feelings have been expressed that small-scale laboratory tests do not accurately predict results for large-scale fire tests.[57] Likewise, some investigators feel that many descriptive terms for flammability rating, such as self-extinguishing (SE), tend to give the user an unjustified sense of security.[58,59]

To get products approved, a number of companies have been created that will measure flame test results to specific standards. While large-scale tests have been and are being developed, small-scale tests continue to be used widely. The available test methods are manifold and have been created by numerous sources for many applications.

A summary of some of the major flammability tests is given in Table 4.16.[60] Several updated tests aimed specifically at plastics in aircraft passenger compartments are described in ASTM E906[61] and the *Federal Register*.[62]

TABLE 4.16   Summary of Some of the Major Flammability Tests[60]

Test	Material
Ignition	
ASTM D2863, oxygen index test	All
UL hot-wire ignition test	Plastics
UL high-current arc ignition test	Plastics
UL high-voltage arc ignition test	Plastics
ASTM D2859, methenamine pill test	Carpets and floor coverings
Flame propagation	
ASTM D635, FTM 2021	Plastics, sheet
ASTM E84, UL 723, NFPA 255, 25-ft tunnel test	All
UL 94, test for self-extinguishing polymers	Plastics, sheet
MVSS302, horizontal burn test	All
FAA vertical test	All
FAA horizontal test	All
Fire endurance	
ASTM E119, UL 263, MFPA 251, fire endurance test	All
Bureau of Mines, flame penetration	Plastics, foams
Heat contribution factory mutual calorimeter	All
Smoke generation	
ASTM D2843, Rohm & Haas XP2 smoke density chamber	Plastics
NBS chamber	All

Flame Retardants.   When plastics burn, heat from an external source pyrolyzes the solid plastic to produce gases and liquids that act as fuel for the fire. Fire-retardant chemicals affect the burning rate of a solid in one or more ways:[63]

1. By interfering with the combustion reactions

2. By making the products of pyrolysis less flammable

3. By reducing the transfer of heat from the flame to the solid

4. By reducing the rate of diffusion of pyrolysis products to the flame front

Since various plastics burn differently, and often at different temperatures and rates, there is no single universal fire retardant. Almost every application requires either a different agent or different amounts of agent to obtain the desired flame retardancy with minimum effect on other properties.

A number of elements can act as flame retardants, but they must be incorporated in a structure that enables them to become active at the proper temperature. They include nitrogen, phosphorus, arsenic, antimony, bismuth, fluorine, bromine, chlorine, iodine, and boron. Of these, phosphorus, bromine, chlorine, and antimony are currently considered to be the most efficient.[63]

Table 4.17[55] shows the levels at which these elements are used as flame retardants in common polymers. Notice that, in many cases, several of these elements are used in combination to achieve a synergistic effect; that is, the combination is more effective than any one element used at the same level of loading.[63] A current summary on flame retardants and fire testing is given in Ref. 64. Often, plastic parts, construction beams, and so on are protected by use of intumescent materials, which foam and form a heat-resistant char.[65,66]

4.10.17   Other Properties

There are a number of other properties commonly used to evaluate plastic materials. Space limitations prevent a detailed description of the tests used to establish values for these properties. However, a listing of them is provided for the reader to research independently in Table 4.18.[55] More information can be found in Ref. 67 and can be obtained from the American Society for Testing and Materials (ASTM, 100 Barr Harbor Dr., West Consohocken, PA 19428, www.astm.org).

4.10.18   The Material Selection Process

To avoid unpleasant surprises that can cause a design to fail, it is necessary to know everything possible about the conditions to which the

TABLE 4.17 Flame Retardants for Plastics and Typical Composition Ranges[63]

Plastic	Typical flame retardants	Typical percentages of flame-retardant elements for equivalent retardance							
		BR	Cl	P	P + Br	P + Cl	Sb_2O_3 + Br	Sb_2O_3 + Cl	Sb_2O_3
ABS	Bromine and chlorine-containing organic additive compounds (such as chlorinated paraffins, brominated biphenyl, phosphate-containing aliphatic and aromatic compounds). Antimony trioxides, hydrated aluminum oxide, and zinc borate with halogenated and phosphate-containing organic compounds. Terpolymerization with halogen-containing monomers such as bis (2,3 dibromo propyl) fumerate.	3	23					5+7	
Acrylics	Halogen- and phosphorus-containing organic compound. PVC to form alloy.	16	20	5	1+3	2+4	7+5		
Cellulose acetate, cellulose butyrate, cellulose propionate	Halogenated and/or phosphate-containing organic plasticizers and halogenated compounds (such as brominated biphenyl).		24	2.5–2.5	1+9			12–15+9–12	
Epoxies	Halogenated and/or phosphate-containing compounds. Chlorinated brominated bisphenol A. Halogenated anhydrides (like chlorendic anhydride). Antimony trioxide, hydrated aluminum oxide with halogenated and phosphate-containing organic compounds.	13–15	26–20	5–6	2+5	2+6		10+6	
Phenolic compounds	Halogens and/or phosphorus organic compounds (like chlorinated paraffins). Tris (2,3-dibromo propyl) phosphate. Hydrated aluminum oxide, zinc borate.		16	6					
Polyamides	Chlorinated and brominated biphenyls. Antimony trioxide with halogenated compounds.		3.5–7	3.5				10+6	
Polycarbonate	Brominated biphenyl, chlorinated paraffin. Tetrabromo bisphenol A. Antimony trioxide with halogenated compounds.	4–5	10–15		7+7–8				

Polymer	Description							
Polyesters	Halogenated phosphate-containing plasticizers. Halogenated compounds (like chlorinated paraffin and brominated biphenyl). Halogen-containing intermediates (such as tetrachlorophthalic anhydride, tetrabromophthalic anhydride, and chlorendic anhydride). Antimony trioxide, hydrated aluminum oxide, hydrated zinc borate with halogenated and phosphate-containing organic compounds.	12–15	26	5	2+6	1+15–20	2+8–9	1+16–18
Polyolefins	Halogen-containing compounds (such as chlorinated paraffins and brominated biphenyl). Halogenation of polymer as chlorinated polyethylene. Antimony trioxide, hydrated aluminum oxide, zinc borate, or barium metaborate in conjunction with halogenated compounds.	20	40	5	2.5+7	2.5+9	3+6	5+8
Polystyrene	Halogen- and phosphate-containing compounds like tris (2,3-dibromo propyl) phosphate and brominated biphenyls. Halogen efficiency as flame retardant is increased by incorporating small amount of free radical initiators such as peroxides. Chemical bonding of halogenated compounds into the polystyrene chain. Antimony trioxide and hydrated aluminum oxide.	4–5	10–15		0.2+3	0.5+5	7+7–8	7+7–8
Polyurethanes	Halogenated phosphorus-containing plasticizers where flexibility is not a problem. Phosphorus-containing intermediates (such as phosphate, phosphite, phosphonates, and amino phosphonate polyols). Brominated isocyanates and halogen-containing prepolymers/polyols. Chlorine- and fluorine-containing compounds as blowing agents in foam. Antimony trioxide hydrated aluminum oxide, and zinc borate in conjunction with halogenated compound.	12–14	18–20	1.5	0.5+4–7	1+10–15	2.5+2.5	4+4
Polyvinyl chloride	Replace other plasticizer such as DOP with halogen- and/or phosphorus-containing plasticizer. Antimony trioxides, hydrated aluminum oxide, and zinc borate.		40	2–4				5–15

Additional references: C.J. Hilado, *Flammability Handbook for Plastics*, Technomic Publishers; J.W. Lyons, *The Chemistry and Uses of Fire Retardants*, Wiley-Interscience.

TABLE 4.18   Other Tests of Interest[55]

Test	Standard
Coefficient of linear thermal expansion	ASTM D696, E228
Creep	
Stress relaxation	ASTM D2991
Tensile, compressive, and flexural	
creep and creep-rupture	ASTM D2990
Crystallization, heat of	ASTM D3417
Cure kinetics of thermosets	ASTM D4473
Deformation under load	ASTM D621
Dimensional stability	ASTM D756
Electrical properties	
Arc resistance	ASTM D495
Dielectric constant	ASTM D150
Insulation resistance	ASTM D257
Friction, coefficient of	ASTM D1894
Fatigue	
Flexural fatigue	ASTM D671
Tension-tension fatigue	ASTM D3479
Flammability	
Flooring radiant panel test	ASTM E684
Smoke emission	ASTM E662
Vertical test for cellular plastics	ASTM D3014
Flowability of thermosets	ASTM D3123
Fracture toughness	ASTM D5045
Moisture vapor transmission	ASTM D675
Optical properties	
Color	ASTM E308
Haze	ASTM D1003
Specular gloss	ASTM D253
Shear modulus	ASTM D5279
Viscosity-shear rate	ASTM D3835
Wear	
Abrasion resistance	ASTM D1242
Mar resistance	ASTM D673
Weathering	
Accelerated	ASTM G23, G26, G53
Light and water exposure	ASTM D1499, D2565
Outdoor	ASTM D1435, E838

product will be exposed in its lifetime. Armed with that information, the plastics designer can determine the appropriate material (as well as the design, process, and tooling) for the application.

That is, the designer can determine the appropriate material at least to the limits of the available information. A certain degree of risk is inherent in plastics design, because the cost in time and resources is too great to permit the accumulation of enough information to eliminate that risk. Higher levels of risk are acceptable where tooling investment is low and where product failure results only in very low levels of property loss. As the cost of failure increases, more resources are devoted to risk reduction, and greater safety factors are used. When product failure could result in serious injury or loss of life, exhaustive testing and greater safety factors are employed.

Determination of material requirements. Most product structural failures result from conditions the designer did not anticipate. Thus, the first order of business is to establish the design parameters. This is achieved by the rather tedious process of considering all the conditions to which the product will be exposed. A checklist such as the one shown in Table 4.19[55] is a useful means of reminding the designer of all these conditions. This is a general checklist, and the individual designer will no doubt find it necessary to make additions appropriate to the specific product area.

Note 2 of the checklist serves to remind the designer that not all of the conditions encountered by the product occur in its use. The highest temperature the product may be called upon to endure may actually occur in a boxcar in the desert in mid-summer, in cleaning, or in a mandated test procedure. Decorating, cleaning, or assembly may result in unforeseen chemical exposures. The author recalls one project where he was "blindsided" by the discovery, long after production had commenced, that the product was washed in an acid solution, which was followed by a base solution, out in the field. Fortune smiled on that occasion, and the material used happened to be one that could withstand those exposures.

Impacts may also occur under odd circumstances. One time, the author found that the sales department of a company had brought a baseball bat to the product's introduction and was merrily inviting customers to take a swing at the housing to demonstrate its superiority over the competitors' metal housing. The engineers joined in a silent nondenominational prayer that the housing would be able to withstand three days of that kind of treatment.

Apparently, their prayer was answered, and the product did survive the onslaught. However, there is a well known story in the automobile industry about a bumper mold that had to be rebuilt because the engineers did not account for the heat of the oven that dried the paint. The original mold had been built for a material that would not withstand the oven temperatures, and the resin with a higher temperature resistance had a significantly different mold shrinkage. The mold could not be salvaged, and a new one had to be made.

Fear often strikes the heart of those who are called upon to fill out a checklist, and they find themselves specifying higher levels of performance than are really required. Performance has a price, and that practice leads to unnecessary cost, particularly when the designer in-

TABLE 4.19   Product Design Checklist[55]

Notes:
1. Items left blank presume "does not apply."
2. Include conditions encountered in use, cleaning, shipping, assembly, testing, and decorating.
3. Remember: Overspecification results in unnecessary cost.

Part name and description of application _____

A. Physical Limitations

Length: _____   Width: _____   Height: _____   Weight: _____   Density: _____

Mating parts: _____   Mating fitments: _____

B. Mechanical Requirements

Functional (dynamic) life: _____   Nonfunctional (static life): _____   Shelf life: _____

Tensile strength: −40°F _____   72°F _____   140°F _____

Flexural strength: −40°F _____   72°F _____   140°F _____

Compressive strength: −40°F _____   72°F _____   140°F _____

Flexural modulus (stiffness): _____   Hardness: _____

Maximum allowable creep: _____   Deflection: _____

Impact strength: −40°F _____   72°F _____   140°F _____

Shear strength: _____   Abrasion resistance: _____

C. Environmental Limitations

Chemical resistance: Continuous _____   Intermittent: _____   Occasional: _____

Temperature: Max. + duration: _____   Min. + duration: _____   Operating + duration: _____

Ultraviolet light: _____   Water immersion: _____   Moisture vapor transmission: _____

Radiation: _____   Flammability: _____

D. Electrical Requirements

Volume resistivity: _____   Surface resistivity: _____   Dielectric constant: _____

Dissipation factor: _____   Dielectric loss: _____   Arc resistance: _____

Electrical conductance: _____   Microwave transparency: _____   EMI/RFI shielding: _____

E. Appearance Requirements

Surface finish:

Inside (SPI#): _____ Outside (SPI#): _____ Depth: _____

Texture no.: _____

Match to: _____ Color: _____

Color maintenance: _____ Transparency: _____ Translucency: _____

Metallizing: _____ Decoration: _____

Identification: Model: _____ Production date: _____ Recycling: _____

Warnings: _____ Nameplates: _____ Instructions: _____

Manufacturing limits: Flash: _____ Mismatch: _____

Gate location: _____ Ejector location: _____

F. Assembly Requirements

Parts to be assembled to, method prepared for (screws, solvents, etc.), and type (permanent, serviceable, occasionally reopenable, water-tight, or hermetic).

1. _____ 2. _____

3. _____ 4. _____

G. Other Design Parameters

Anticipated volume: Annual: _____ Order: _____

Anticipated production date: Start: _____ Volume: _____

Unusual legal exposure: Liability: _____ Patents: _____

Regulatory approvals required: 1. _____ 2. _____ 3. _____ 4. _____ 5. _____ 6. _____

Specifications: Military: _____ Building codes: _____

Foreign production: _____ Foreign sales areas: _____

H. List All Tests To Be Performed on This Product

1. _____ 2. _____ 3. _____

4. _____ 5. _____ 6. _____

7. _____ 8. _____ 9. _____

I. Additional Conditions

_____

_____

SOURCE: Adapted from Jordan I. Rotheiser, *Joining of Plastics Handbook for Designers and Engineers*, Hanser Publications, Munich-Hanser/ Gardner Publications, Inc., Cincinnati, 1999.

corporates a safety factor to ensure that the required standard is met. (Safety factors are added in the design stage.) In some cases, requirements are raised to a level that causes a more expensive material or process to be used. Occasionally, they are lifted to a level that cannot be reached using plastics.

In the checklist, Section A, Physical Limitations, set the basic parameters of the design. The sizes alone are enough to eliminate some of the processes. Since none of the plastics can be used with all of the processes, a change in process can affect the material selected. Section B, Mechanical Requirements, will reveal some interesting, and often not considered, aspects of the design. For example, there is the question whether it is more important to have a long functional life, static life, or shelf life.

Sometimes one type of performance must be sacrificed to another. An example of this type of phenomenon might be a shaft seal in place on a product. Its shelf life would be the length of time the product could remain unused and still have the seal function properly when used. Its static life would be the length of time it would be able to function after it had been first used. Finally, its functional life would be the period of time or number of rotations the seal could withstand while the shaft was rotating. Bear in mind that the temperatures would rise when the shaft was turning, and there would also be abrasion taking place. It is, therefore, quite likely that the ideal polymer for one circumstance might be less favorable for another. Compromises often must be made.

Which of the product requirements in this section is most likely to result in a structural failure? Of course, any of them can, or they would not be listed. However, in the author's experience, the most common failures (short of gross design errors) occur due to weakening of the material at elevated temperature, impact failure at low temperature, or creep failure over time.

Designers often forget that plastics are, themselves, chemical compounds. Hence, they are particularly vulnerable to environmental exposures. Section C, Environmental Limitations, attempts to reveal the exposures that could result in failure. The chemical exposures are the most difficult to discover, because there are so many of them, and they are often hidden from view. Many common cleaning, cosmetic, and food preparations contain unusual and secret ingredients. Nonetheless, we identify what we can and list the commercial names for the others. In some cases, resin manufacturers have actually tested their polymers in these commercial preparations. However, care must be taken to examine the test used, as the exposure tested (diluted sample, limited time, etc.) may have no relationship to the application in question.

Often, these exposures take place in combination as in products to be used outdoors. In fact, design for products to be used outdoors is practically a category in itself, as ultraviolet light significantly affects most plastics over time. In a few cases, even indoor applications have been severely affected.

Section D, Electrical Requirements, is largely limited to electrical applications. However, Section E, Appearance Requirements, affects nearly every application to some degree. Designers often overlook these details in the early stages of the project, because they are preoccupied with the structural aspects of the design. Nonetheless, appearance requirements can lead to expensive material and/or mold changes. For example, gate blush may be nearly impossible to eliminate for some resins, and a change in material can result in a change in shrinkage requiring mold alterations. Section F deals with Assembly Requirements.

There are some other design parameters as indicated in Section G. For example, the anticipated volume will often dictate the process, and therefore the polymer, to be used. Section H, List All Tests To Be Performed on This Product, may reveal test protocols that are more stringent than the demands on the product in the field, particularly if they are archaic in nature.

Selection of resin candidates. With the application's property requirements established, the design engineer then proceeds to list those polymers that can best meet them. The material provided in this chapter can be effectively used for this purpose. However, for most applications, cost is a prime consideration. Therefore, the design engineer establishes a list of resin candidates and arranges them in order of cost. Table 4.14, Approximate Dollars per Cubic Inch of Plastics,[55] has been provided for this purpose. (Due to the effect of market price fluctuations, the engineer is encouraged to recalculate these figures at the time of application.)

With a pecking order so established, the fine points of the lowest-cost candidate must be examined in detail with regard to properties not usually listed. Such matters as chemical exposures, weathering, and appropriateness to manufacturing and joining processes need to be considered. If the lowest-cost candidate cannot meet the requirements, the engineer moves on to the next one, and so on.

The final selection. Compromises are usually necessary, since it is rare that a given resin will meet every product requirement. Furthermore, there is always some risk involved, since most compounds do not come with enough information available to cover all the product requirements. For this reason, it is wise to establish alternate material selections to be used in the event that the first choice is not successful.

The molding shrinkage for any alternative must be very close to that of the original choice, since that property determines the usability of the mold that has been built. An alternative with a lower shrinkage rate will result in parts that are larger than those of the original material. While small adjustments can often be made with processing conditions, any significant difference will call for a new mold.

A higher shrinkage rate for an alternate resin will produce a smaller part. In addition to adjustments in processing conditions, the addition of fillers, such as mineral or glass, can result in a lower shrinkage rate for the same resin, because they replace polymer that shrinks with material that does not. However, the addition of such fillers will affect the properties of the resin.

References

1. J. I. Rotheiser, *Joining of Plastics, Handbook for Designers and Engineers,* Hanser Gardner Publications, Cincinnati, OH, 1999.
2. C. A. Harper, *Handbook of Plastics, Elastomers, and Composites, 3rd ed.,* McGraw-Hill, New York, 1996.
3. P. J. Flory, *Principles of Polymer Chemistry,* Cornell Univ. Press, Ithaca, NY, 1953.
4. "The Plastic Industry and the Energy Shortage," *Celanese Plastics Co. Bull.,* 1974.
5. H. F. Mark, *Giant Molecules,* Time Inc., New York, 1966.
6. J. E. Hauck, "Engineers' Guide to Plastics," *Mater. Eng.,* Feb. 1967.
7. R. B. Seymour, "Fillers for Molding Compounds," *Mod. Plast.,* MPE Suppl., 1966–1967.
8. "Plaskon Plastics and Resins," *Tech. Bull., Allied Chemical Corp.,* Plastics Div. (regularly updated).
9. "Dapon Diallyl Phthalate Resin," *Tech. Bull., FMC Corp.* (regularly updated).
10. J. Chottiner, "Dimensional Stability of Thermosetting Plastics," *Mater. Eng.,* Feb. 1962.
11. S. Sherman, "Epoxy Resins," *Mod. Plast.,* MPE Suppl., 1976.
12. C. A. May and Y. Tanaka, *Epoxy Resins,* Marcel Dekker, New York, 1973.
13. F. Petruccelli, "Aminos," *Mod. Plast.,* MPE Suppl., 1974.
14. G. B. Sunderland and A. Nufer, "Aminos," *Mach. Des.,* Plastic Reference Issue, June 1966.
15. C. A. Harper, *Handbook of Materials and Processes for Electronics,* McGraw-Hill, New York, 1972.
16. P. Grafton, "Commonly Used Plastics," Boonton Molding Co., Boonton, NJ, 1973.
17. R. W. Martin, *The Chemistry of Phenolic Resins,* Wiley, New York, 1956.
18. T. S. Carswell, *Phenoplastics—Their Structure, Properties and Chemical Technology,* Interscience, New York, 1947.
19. "Ricon High Vinyl 1,2 Liquid Polybutadiene," Product Bull., CCS-110, Advanced Resins, Inc., Grand Junction, CO, Feb. 1980.
20. T. J. Czarnomski, "Unsaturated Polyesters," *Mod. Plast.,* MPE Suppl., 1974.
21. G. J. Kookootsedes, "Silicones," *Mod. Plast.,* MPE Suppl., 1974.
22. "Dow Corning Molding Compounds," Tech. Bull., Dow Corning Corp. (regularly updated).
23. *Mod. Plast.,* MPE2000 Suppl., 1999
24. "Marbon ABS Plastics," Tech. Data Bull., Borg-Warner Corp.
25. J. E. Hauck, "Long-Term Performance of Plastics," *Mater. Eng.,* Nov. 1965.

26. G. A. Patten, "Heat Resistance of Thermoplastics," *Mater. Eng.,* July 1962.
27. J. E. Hauck, "Alloy Plastic to Improve Properties," *Mater. Eng.,* July 1967.
28. H. E. Barkén and A. E. Javitz, "Plastics Molding Materials for Structural and Mechanical Applications," *Elec. Manuf.,* May 1962.
29. "Delrin Acetal," *Design Handbook,* E. I. du Pont de Nemours and Co. (regularly updated).
30. Tech. Data Bull. on AF Delrin, E. I. du Pont de Nemours and Co. (regularly updated).
31. Staff Report, "Contemporary Thermoplastic Materials Properties," *Plast. World,* Sept. 17, 1973.
32. "Lucite Acrylic Resins," Design Handbook, Tech. Bull., E. I. du Pont de Nemours and Co. (regularly updated).
33. G. P. Koo et al., "Engineering Properties of a New Polytetrafluoroethylene," *SPE Journal,* Sept. 1965.
34. "Mechanical Design Data for Teflon Resins," Tech. Booklet, E. I. du Pont de Nemours and Co. (regularly updated).
35. P. West, "Wide Space Ranging Applications Developing for Piezoelectric Polymers," *Adv. Mater.,* vol. 8, May 26, 1986.
36. "Surlyn Ionomers," Tech. Bull., E. I. du Pont de Nemours and Co. (regularly updated).
37. "Zytel Nylon Resins," Tech. Bull., E. I. du Pont de Nemours and Co. (regularly updated).
38. "Parylene Conformal Coatings," Tech. Bull., Union Carbide Corp. (regularly updated).
39. W. B. Isaacson, "Polyarylsulfone," *Mod. Plast.,* MPE Suppl., 1974.
40. "Astrel 360 Plastic," Tech. Bull., 3M Corp.
41. A. G. Bager, "Bayer Polycarbonate Films," Tech. Bull., Mobay Chemical Co. (regularly updated).
42. "Lexan Polycarbonate Resin," Tech. Booklet, General Electric Co.
43. "Noryl and PPO Resins," Tech. Bull., General Electric Co.
44. "Celanex," Tech. Data Bull., Celanese Plastic Co.
45. V. J. Leslie, "Polyethersulfone," *Mod. Plast.,* MPE Suppl., 1974.
46. J. S. Eickert, "Phenylene Oxide-Based Resin," *Mod. Plast.,* MPE Suppl., 1974.
47. R. V. Jones, "Polyphenylene Sulfide," *Mod. Plast.,* MPE Suppl., 1974.
48. L. R. Guenin, "Polystyrene," *Mod. Plast.,* MPE Suppl., 1974.
49. R. K. Walton, "Polysulfone," *Mod. Plast.,* MPE Suppl., 1974.
50. "Film Properties Charts," *Mod. Plast.,* 1974.
51. F. A. Leary, "Characteristics of Platable Plastics," *Plast. Des. Process.,* June 1971.
52. "Polyflon Plated TFE," Tech. Bull., Polyflon Corp.
53. Staff Rep., "Proper Part Design for Profitable Plating," *Plast. World,* Feb. 1971.
54. L. K. English, "Raising the Ceiling of High-Temperature Thermoplastics," *Mater. Eng.,* pp. 67, 69, 70, May 1988.
55. C. A. Harper, *Modern Plastics Handbook,* McGraw-Hill, New York,1999
56. Staff Rep., "What Are Fire Officials Squawking About?," *Mod. Plast.,* Aug. 1973.
57. R. H. Wehrenberg, "Flammability of Plastics: The Heat Is on for Large-Scale Tests," *Mater. Eng.,* pp. 35–40, Feb. 1979.
58. R. C. Masek, "Flammability Ratings. What They Mean and Don't Mean," *Insul. Circuits,* pp. 25–28, Nov. 1978.
59. A. Tewarson, "Fire Behavior of Polymeric Materials," *IEEE Elec. Insul. Mag.,* vol. 6, pp. 20–23, May/June 1990.
60. Staff Rep., "Flammability Report," Mod. Plast., Nov. 1972.
61. ASTM E906, "Standard Test Method for Heat and Visible Smoke Release Rate for Materials and Products," Am. Soc. for Testing and Materials, Philadelphia, Pa., 1983.
62. "Improved Flammability Standards for Materials Used in the Interiors of Transport Category Airplane Cabins," *Federal Resister,* pt. II, vol. 50, no. 73, Apr. 16, 1985; "Final Rule; Findings Concerning Comments," ibid., vol. 53, no. 165, Aug. 25, 1988.

63. J. T. Howarth et al., "Flame-Retardant Additives," *Plast. World,* Mar. 1973.
64. British Plastics Federation, *Flame Retardants '90,* Elsevier Appl. Sci., London, 1990.
65. L. K. English, "Intumescent Coatings: First Line of Defense against Fines," *Mater. Eng.,* pp. 39–43, Feb. 1986.
66. "Fire Protection Coatings, Work by Forming Foam Like Chars," *Prod. Eng.,* pp. 36–37, Feb. 1976.
67. *Plastics Engineering Handbook of the Society of the Plastics Industry,* Chapman & Hall, New York, 1991
68. D. V. Rosato, *Rosato's Plastics Encyclopedia and Dictionary,* Carl Hanser Verlag, Munich, 1993.

5

Composite Materials and Processes

S. T. Peters
Process Research
Mountain View, California

5.1 Introduction

There are two general types of composites, distinguished by the type of materials that are used in construction and by the general market in which they can be found. The more prevalent composites, such as used in printed circuit boards, shower enclosures, and pleasure boats, are generally reinforced with fiberglass fabric, use a type of polyester resin as the matrix, and can be referred to as *commodity* composites. Large overlaps exist for the two types; for instance, there is a significant weight percent of fiberglass-reinforced plastic in most commercial airliners, and carbon/graphite or aramid have been used in reinforcing laminated truss beams for home building. Modern structural composites, frequently referred to as *advanced composites,* can be distinguished from *commodity* composites because of their frequent use of more exotic or expensive matrix materials and higher-priced reinforcements such as carbon/graphite, and they can be found in more structurally demanding locations that have a greater need for weight savings. They are a blend of two or more components. One component is made up of stiff, long fibers, and the other, for polymeric composites, is a resinous binder or *matrix* that holds the fibers in place. The fibers are strong and stiff relative to the matrix and are generally orthotropic (having different properties in two different directions). These properties are most evident when the components are shown in a breakdown view as in Fig. 5.1.

The fiber for advanced structural composites is long, with length-to-diameter ratios >100. Predominately, for advanced structural compos-

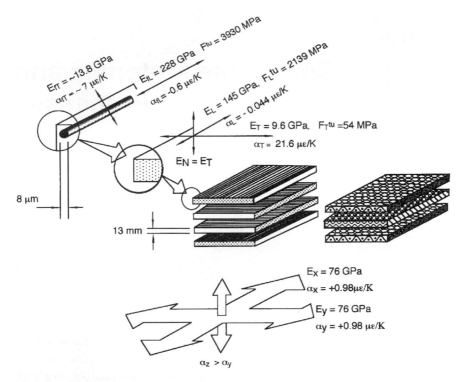

Figure 5.1 The anatomy of a composite laminate.

ites, the fiber has been continuous, but there is an increased awareness that discontinuous fibers allow potentially huge savings in
manufacturing costs, so there are now efforts to incorporate them in
areas previously reserved for continuous fiber. The fiber's strength
and stiffness are much greater—many times more than the matrix
material. For instance, the tensile strength quoted for 3502 resin is
4.8 ksi, and the longitudinal elastic modulus (Young's modulus) is
526 ksi. The "B" basis unidirectional tensile strength of the AS-4/3502
lamina is 205 ksi, and the fiber (tow) tensile strength and modulus are
580 ksi and 36 msi, respectively. Thus, for this combination of "typical"
advanced composite materials, the fiber is >100 times the strength of
the resin and >50 times the modulus.[1–3] (These data will change over
time due to manufacturing improvements, and more detailed data, including design allowables, will be shown later in this chapter.) When
the fiber and matrix are joined to form a composite, they both retain
their individual identities, and both directly influence the composite's
final properties. The resulting composite is composed of layers (laminae) of the fibers and matrix stacked to achieve the desired properties
in one or more directions.

Designers of aircraft structures have been quick to realize that the high strength-to-weight or modulus-to-weight ratios of composites could result in lighter structural components with lower operating costs and better maintenance histories. The first high-performance aircraft to use advanced composites (boron/epoxy horizontal stabilizers, 1500 built) in a production contract was the F-14A. The use of composites has continued and has resulted in a preponderance of the structure being fabricated from composite materials in one aircraft, the Bell-Boeing V-22, of which approximately 60% of the weight of the craft is composite (Fig. 5.2). Some of the advantages in the use of composites are shown in Table 5.1. These advantages translate not only into aircraft production but also into everyday activities. For example, a carbon/graphite-shafted golf club produces longer drives, because more of the mass is concentrated at the club head. Tennis players also experience less fatigue and pain, because a carbon/graphite composite tennis racquet is lighter and has some inherent vibration damping. Generally, the advantages shown in Table 5.1 can be realized for most fiber/composite combinations, and the disadvantages are more obvious

Figure 5.2 The Bell-Boeing V-22. *(Courtesy of the Boeing Company)*

TABLE 5.1 Reasons for Using Composites

Reason for use	Material selected	Application/driver
Lighter, stiffer, stronger	Boron, all carbon/graphites, some aramid	Military aircraft, better performance; commercial aircraft, operating costs
Controlled or zero thermal expansion	Very high modulus carbon/graphite	Spacecraft with high positional accuracy requirements for optical sensors
Environmental resistance	Fiberglass, vinyl esters, bisphenol-a fumarates, chlorendic resins	Tanks and piping, corrosion resistance to industrial chemicals, crude oil, gasoline at elevated temperatures
Lower inertia, faster startups, less deflection	High-strength carbon/graphite, epoxy	Industrial rolls, for paper, films
Light weight, damage tolerance	High-strength carbon/graphite, fiberglass, (hybrids), epoxy	CNG tanks for "green" cars, trucks and busses to reduce environmental pollution
More reproducible complex surfaces	High-strength or high-modulus carbon graphite/epoxy	High-speed aircraft; metal skins cannot be formed accurately
Less pain and fatigue	Carbon/graphite/epoxy	Tennis, squash and racquetball racquets; metallic racquets no longer available
Reduces logging in "old growth" forests	Aramid, carbon/graphite	Laminated "new" growth wooden support beams with high-modulus fibers incorporated
Reduces need for intermediate support and resists constant 100% humidity atmosphere	High-strength carbon/graphite-epoxy	Cooling tower driveshafts
Tailorability of bending and twisting response	Carbon/graphite-epoxy	Golf club shafts, fishing rods
Transparency to radiation	Carbon/graphite-epoxy	X-ray tables
Crashworthiness	Carbon/graphite-epoxy	Racing cars
Higher natural frequency, lighter	Carbon/graphite-epoxy	Automotive and industrial driveshafts
Water resistance	Fiberglass (woven fabric), polyester, or isopolyester	Commercial boats
Ease of field application	Carbon/graphite, fiberglass- epoxy, tape, and fabric	Freeway support structure repair after earthquake

with some. These advantages have now resulted in composite applications far outside the aircraft industry, with many more reasons for use as shown in Table 5.2. Proper design and material selection can circumvent many of the disadvantages.

5.2 Material Systems

An advanced or commodity composite laminate can be tailored so that the directional dependence of strength and stiffness matches that of the loading environment. To do that, layers of unidirectional material called lamina, or woven fabric with fibers predominately in the expected loading directions, are oriented to satisfy the strength or stiffness requirements. These laminae and fabrics contain both fibers and a matrix. Because of the use of directional laminae, the tensile, flexural, and torsional shear properties of a structure can be disassociated from each other to some extent, and a golf shaft, for example, can be changed in torsional stiffness without changing the flexural or tensile stiffness. This allows for almost infinite variations in the shafts to accommodate individual needs. It also allows for altering the stiffness of a forward-swept aircraft wing to respond to the incoming loads (Fig. 5.3). This is not an option with isotropic metal.

TABLE 5.2   Advantages and Disadvantages of Advanced Composites

Advantages	Disadvantages
Weight reduction	Cost of raw materials and fabrication
High strength- or stiffness-to-weight ratio)	Transverse properties may be weak
	Matrix weakness, low toughness
Tailorable properties: can tailor strength or stiffness to be in the load direction	Matrix subject to environmental degradation
Redundant load paths (fiber to fiber)	Difficult to attach
Longer life (no corrosion)	Analysis for physical properties and mechanical properties difficult, analysis for damping efficiency has not reached a consensus
Lower manufacturing costs because of lower part count	
Inherent damping	
Increased (or decreased) thermal or electrical conductivity	Nondestructive testing tedious
Better fatigue life	Acceptable methods for evaluation of residual properties have not reached a consensus

5.2.1 Fibers

Fibers can be of the same material within a lamina or several fibers mixed (hybrid). The common commercially available fiber classes are as follows:

Figure 5.3 NASA-Grumman forward-swept wing, X-29 aircraft. *(Courtesy of Dryden Flight Research Center)*

- Carbon/graphite
- Fiberglass
- Organic

  Aramid

  Polyethylene

  PBO
- Boron
- Silicon carbide
- Silicon nitride, silica, alumina, alumina silica

5.2.1.1 Fiberglass.   The most widely used fiber for commodity composites, and the fiber that has had the longest period for development, is fiberglass, which has been marketed in several grades in the United States for more than 40 years. During that time, many different glasses were developed, including, leaded glass, and beryllium high-modulus glass, all of which are no longer produced or are in limited supply. The various types of glass that continue to be useful for composite structures are shown in Table 5.3.[4]

TABLE 5.3   Glass Fibers*

Type	Nominal tensile modulus, GPa (psi × 10⁶)	Nominal tensile strength, MPa (psi × 10³)	Ultimate strain,%	Fiber density, Mg/m³ (lb/in³)	Suppliers
E	72.5 (10.5)	3447 (500)	4.8	2600 (0.093)	PPG Manville Co. Owens Corning Fiberglass
R	85.2 (12.5)	2068 (300)	5.1	2491 (0.089)	Vetrotex Certainteed
Te	84.3 (12.2)	4660 (675)	5.5	2491 (0.089)	Nittobo
S-2	86.9 (12.6)	4585 (665)	5.4	2550 (0.092)	Owens Corning
Zentron high silica	94 (13.5)	3970 (575)		2460 (0.089)	Owens Corning

*In order of ascending modulus normalized to 100% fiber volume (vendor data) (see Ref. 4, p. 2-3).

The table shows the common description for the glass and the nominal tensile strength and tensile modulus of strands and composite. The composite data are average values generated from a series of tests by the manufacturer of the fiber, and they reflect the "ideal" or maximum strength and stiffness of a single fiber or tow or thin unidirectional laminated composite. A tensile test by the user will generally not reflect these values for strength. The maximum number of fibers per strand is usually important information for applications (e.g., filament winding or pultrusion) that use dry fibers mixed with the matrix resin at the point of fiber laydown, since it influences handling ease and per ply thickness. The fiber density is included so that the rule of mixtures equations involving fiber volume and resin volume can be used to evaluate void volume and theoretical mechanical properties. Compressive strength of glass-reinforced composites is relatively high and has led to their selection for use in underwater deep diving applications. The electrical properties of glass-reinforced composites have allowed their use as radomes and printed circuit boards, and in many other areas that require high dielectric strength.

Fiberglass is a product of silica sand, limestone, boric acid, and other ingredients that are dry-mixed, melted (at approximately 1260°C), and then drawn into fibers. Fiberglass, for structural use, is marketed in the form of fiber strand, unidirectional fabric, woven and knitted fabrics, and as preform shapes. It is by far the most widely used fiber, primarily because of its low cost, but its mechanical properties and specific mechanical properties (i.e., its strength or modulus to weight ratio) are not comparable with other structural fibers. Fiber properties for any of the materials, by themselves, are of little use to most designers, given that they prefer to work with laminate or lamina data. However, with the equations shown later, the designer can use these fiber values to compute the preliminary laminate properties, without the benefit of a computer program or extensive laminate data, which may not be available for newer fibers.

Glass fibers, like all other continuous composite reinforcements, are coated with a thin coating, somewhat like a paint and called a *finish* or *sizing*, that forms a bond between the fiber and the matrix material, improves the handling characteristics, and protects the fiber in the composite from some environmental effects. Each fiber type has a series of unique finishes that are formulated specifically for that type of fiber. Finishes are of little interest to the composite manufacturer, but they are somewhat critical to some processes that use dry strands or tow. For instance, if the fiber finish on the strand or tow for wet filament winding is changed, the spread of the fiber will be changed. If there are no other changes in the process, the resulting width of the applied band will be wider or narrower, and there will be overlaps or gaps in the laminae. Usually, the operator of the machine will adjust to remove laps and gaps (they are usually covered in a specification). Thus, the final laminate thickness will be less than specified and could result in a problem such as premature burst of a pressure vessel.

5.2.1.2 Carbon/graphite. Carbon/graphite fibers have demonstrated the widest variety of strengths and moduli and have the greatest number of suppliers (Table 5.4).[4] The carbon/graphite fiber manufacturers, as a result of the turmoil in the composite industry induced by the severe decline in the U.S. defense demand for advanced composites, have gone through drastic price declines, inducing the departure of several manufacturers during the past 10 years. The fibers begin as an organic fiber, such as rayon, polyacrylonitrile (PAN), or pitch (a derivative of crude oil) that is called the *precursor*. The precursor is then stretched, oxidized, carbonized, and graphitized. There are many ways to produce these fibers,[5] but the relative amount of exposure at temperatures from 2500–3000°C results in greater or less graphitization of the fiber. The degree of graphitization in most high-strength or intermediate fibers is low (less than 10%), so the appellation *graphite* fiber is a misnomer and has been promulgated as a marketing tool for sporting goods. Higher degrees of graphitization usually result in a stiffer fiber (higher modulus) with greater electrical and thermal conductivities. Pitch fibers above 689 GPa (100 msi) tensile modulus have thermal conductiv-

TABLE 5.4a   Carbon/Graphite Fibers (Pan) (from Ref. 4, p. 2-7)

Class of fiber	Nominal tensile modulus, GPa (psi $\times 10^6$)	Nominal tensile strength, MPa (psi $\times 10^3$)	Ultimate strain,%	Fiber density, Mg/m^3 (lb/in^3)	Suppliers/typical products
High tensile strength	227 (33)	3996 (580)	1.60	1.750 (0.063)	Amoco T-300 Hexcel AS-4
High strain	234 (34) 248 (36)	4500 (650) 4550 (660)	1.9 1.7	1.800 (0.064) 1.77	Mitsubishi Grafil 34-700 Amoco T-650-35
Intermediate modulus	275 (40) 290	5133 (745) 5650	1.75 1.8	1740 (0.062) 1.81	Hexcel IM-6 Amoco T-40
Very high strength	289 (42.7)	6370 (924)	2.1	1800 (0.06)	Toray T-1000G
High modulus	390 (57) 436	2900 (420) 4.210	0.70 1.0	1810 (0.065) 1.84	Amoco T-50 Toray M-46J
Very high modulus	540	3920	0.7	1.750	Toray M55-J

ity greater than copper and have been used in spacecraft for thermal control as well as for structural applications. Figures 5.4 and 5.5[5,6] show the effect of processing temperature on tensile strength, tensile modulus, and thermal and electrical conductivity.

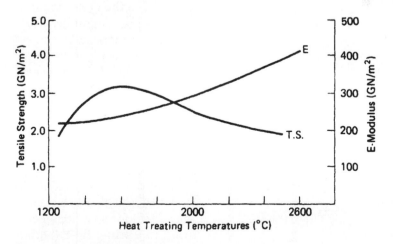

Figure 5.4 Graphitization temperature vs. modulus and tensile strength for carbon graphite fibers.[6]

TABLE 5.4b Carbon/Graphite Fibers (Pitch) (from Ref. 4, p. 2–7)

Class of fiber	Nominal tensile modulus, GPa (psi × 10^6)	Nominal tensile strength, MPa (psi × 10^3)	Ultimate strain, %	Fiber density, Mg/m^3 (lb/in^3)	Thermal conductivity,*,† K W/m-K (Btu/h/°F)	CTE,† ppm/K (10^{-6}/°F)	Suppliers/typical products
High modulus (55–65 msi)	379 (55)	2068 (300)	0.50	2000 (0.072)	120	−1.3	Amoco P-55 Nippon Granoc XN-40
	441		0.76	2080	n/a‡	n/a	Mitsubishi-Kasei Dialead K-1334U
Very high modulus (75–85 msi)	517 (75)	2068 (300)	0.40	2000 (0.072)	185	−1.4	Mitsubishi K-9354U Amoco P-75
Ultra high modulus (90–100 msi)	689 (100)	2240 (325)	0.31	2150 (0.077)	520	−1.6	Granoc XN-70 Mitsubishi K1374U Amoco P-100
Extreme high modulus (>100 msi)	895–999 (130–145)	2411–3789 (350–550)	n/a	2150–2250 (0.077–0.081)	1000	−1.6	Amoco Thornel K-1100X
	780	3530	0.5	n/a	n/a		Granoc XN 80
	826	2239	0.3	2180	640	−1.45	Amoco P-120K
	900 (130)	3800 (550)	0.42	2200	620	−1.2	Mitsubishi Dialead K13C2U

*K and CTE for copper are 400 and 17, respectively.
†Unidirectional composite property, not fiber.
‡Not available.

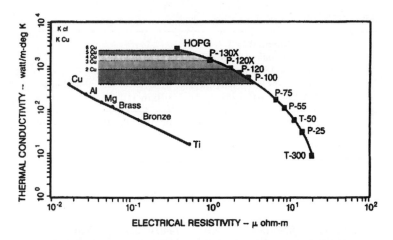

Figure 5.5 Electrical and thermal conductivity of carbon fibers and metals.[7]

5.2.1.3 Aramid fibers. Table 5.5[4] shows the properties of several organic reinforcing fibers. Aramid fibers were introduced by DuPont in 1972. Kevlar 49, one aramid, essentially revolutionized pressure vessel technology because of its great tensile strength and consistency coupled with low density, resulting in much more weight-effective designs for rocket motors. The specific tensile strength of Kevlar was, at its introduction, the highest of any fiber. Carbon/graphite fibers, because of advances in processing, have the highest values now. (Specific strength and modulus based on fiber values are simply tensile strength or tensile modulus/density and are a good measure of structural efficiency of a fiber for airborne applications.) The values for tensile modulii may be near to those developed in a composite structure, but the tensile strength values for fibers may be quite different because of factors such as translation efficiency, possibility of flaws, processing damage, or incorrect fiber orientations. Aramid composites are still widely used for pressure vessels but have been largely supplanted by the very high-strength carbon/graphite fibers. Aramids have outstanding toughness and creep resistance, and their failure mode in compression, shear, or flexure is not in a brittle manner and requires a relatively great deal of work. Aramid composites have relatively poor shear and compression properties; careful design is required for their use in structural applications that involve bending.

5.2.1.4 PBO fibers. A new fiber, poly(p-phenylene-2-6-benzobisoxazole) (PBO) was introduced in the 1990s. It was developed by Dow Chemical

TABLE 5.5   Organic Fibers (from Ref. 4, p. 2-5)

Type	Nominal tensile modulus, GPa (psi × 10⁶)	Nominal tensile strength, MPa (psi × 10³)	Ultimate strain,%	Fiber density, Mg/m³ (lb/in³)	Suppliers/ products
Aramid (medium modulus)	62 (9.0)	3617 (525)	4.0	1440 (0.052)	DuPont Kevlar 29
	80 (11.6)	3150 (457)	3.3	1440	Enka Twaron
	70 (10.1)	3000 (440)	4.4	1390	Teijin Technora
Oriented polyethylene	117 (17)	2585 (375)	3.5	968 (0.035)	Allied Fibers Spectra 900
Aramid (intermediate modulus)	121 (18)	3792 (550)	2.9	1440 (0.052)	DuPont Kevlar 49
	121(18)	3150 (457)	2.0	1450	Enka Twaron HM
Oriented polyethylene	172 (25)	3274 (471)	2.7	968 (0.035)	Allied Fibers Spectra 1000
Aramid (high modulus)	186(27)	3445 (500)	1.8	1440 (0.052)	DuPont Kevlar 149
PBO	180 (26)	5800 (840)	2.5	1540 (0.056)	Toyobo Zylon AS*

*High-modulus grade also available.

Co. under U.S. Air Force funding and was tested in pressure vessels by Lincoln Composites. It has shown great promise due to its high strength (5.5 GPa, 798 ksi) and low density (1560 kg/m³, 0.056 lb/in³). Pressure vessels (Fig. 5.6) fabricated with the fiber and a proprietary Brunswick Composites resin system, LRF-0092, demonstrated a performance factor 30% better than test vessels fabricated with the highest-performing carbon/graphite fibers.[7] There are two grades of the fiber, with tensile strengths and moduli almost double those of p-aramid fibers, along with attractive strain to failure and low moisture regain, making them competitive with carbon high-strength fibers for pressure vessels. The performance factor of the fiber, using strength and density values quoted by the vendor, shows that the PBO fiber is 3% less efficient than T-1000 carbon/graphite fiber. Pressure vessel testing may result in a higher efficiency factor because of the higher strain-to-failure and other considerations. Development ended in the U.S., and now the fiber is manufactured in Japan and marketed by Toyobo under the trade name *Zylon*.

5.2.1.5 Polyethylene fibers. The polyethylene fibers have the same shear and compression property drawbacks as the aramids, but they

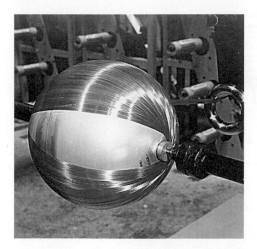

Figure 5.6 Completed PBO spherical test vessel. *(Courtesy of Lincoln Composites)*

also suffer from a low melting temperature that limits their use to composites that cure or operate below 149°C (300°F) and a susceptibility to degradation by ultraviolet light exposure. Both the aramids and the polyethylene fibers have wide use in personal protective armor, and the polyethylene fibers have found wide use as ropes and lines for boating and sailing due to their high strength and low density. They float on water and have a pleasant feel or *hand* as a rope or line. In spite of the drawbacks, both of these fibers are enjoying strong worldwide growth.

5.2.1.6 Boron fibers. Boron fibers, the first fibers to be used on production aircraft (rudders for USAF F-14A fighter, and horizontal stabilizers for the F-111 in approximately 1964–1970), are produced as individual monofilaments on a tungsten or carbon substrate by pyrolylic reduction of boron trichloride (BCl_3) in a sealed glass chamber. (Fig. 5.7). Because the fiber is made as a single filament rather than as a group or tow, the manufacturing process is slower, and the prices are, and will continue to be, higher than for most carbon/graphite fibers. The relatively large-cross-section fiber is used today primarily in polymeric composites that undergo significant compressive stresses (combat aircraft control surfaces) or in composites that are processed at temperatures that would attack carbon/graphite fibers (i.e., metal matrix composites). The carbon/graphite core is protected by the unreactive boron (Table 5.6).[4]

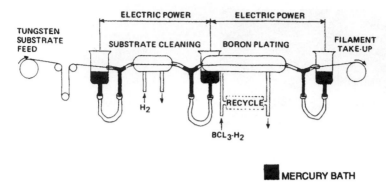

Figure 5.7   Production of boron fiber (from Ref. 8).

5.2.1.7 Ceramic fibers. The other fibers shown in Table 5.6[4] have varying uses, and several are still in development. Silicon carbide continuous fiber is produced in a chemical vapor deposition (CVD) process similar to that for boron, and it has many mechanical properties identical to those of boron. The other fibers show promise in metal matrix composites, as high-temperature polymeric ablative reinforcements, in ceramic-ceramic composites, and in microwave transparent structures (radomes or microwave printed wiring boards).

5.2.2 Matrix Materials

If parallel and continuous fibers are combined with a suitable matrix and cured properly, unidirectional composite properties such as those shown in Table 5.7 are the result. The functions for and requirements of the matrix are to:

- Help to distribute or transfer loads
- Protect the filaments, both in the structure and before and during structure fabrication
- Control the electrical and chemical properties of the composite
- Carry interlaminar shear

The requirements of and for the matrix, which will vary somewhat with the purpose of the structure, are as follows. It must:

- Minimize moisture absorption
- Have low shrinkage
- Wet and bond to fiber

TABLE 5.6 Boron and Ceramic Fibers (from Ref. 4, p. 2-4)

Class of fiber	Nominal tensile modulus, GPa (psi × 10^6)	Nominal tensile strength, MPa (psi × 10^3)	Ultimate strain,%	Fiber density, Mg/m^3 (lb/in^3)	Thermal conductivity,* K W/m-K (Btu/h/°F)	CTE, ppm/K (10^{-6}/°F)	Suppliers/typical products
Alumina	206 (30)	1760 (255)	n/a	3200	1.32	n/a	Sumitomo Altex
	150 (22)	1700 (250)	1.2	2700	0.06	3	3M Nextel 312
SiC	167 (24.3)	2962 (430)	1.4–1.5	2300–2400	n/a	3.1	UBE Tyranno
	186 (27)	2962 (430)	1.6	2360	n/a	n/a	Nippon (DC) HVR Nicolon
SiO$_2$	69 (10)	3600 (530)	n/a	2200	n/a	n/a	J.P. Stevens Astroquartz II
	72 (10)	3600 (530)	n/a	2200	n/a	n/a	Quartz Products Quartzel

*Unidirectional composite property, not fiber.

TABLE 5.7   Properties of Typical Unidirectional Graphite/Epoxy Composites (Fiber Volume Fraction, V_f = 0.60) (from Ref. 10)

	High strength	High modulus
Elastic constants, GPa (psi $\times 10^6$)		
Longitudinal modulus, E_L	145 (21)	220 (32)
Transverse modulus, E_T	10 (1.5)	6.9 (1.0)
Shear modulus, G_{LT}	4.8 (0.7)	4.8 (0.7)
Poisson's ratio (dimensionless) υ_{LT}	0.25	0.25
Strength properties, MPa (10^3 psi)		
Longitudinal tension, F^{tu}_L	1240 (180)	760 (110)
Transverse tension, F^{tu}_T	41 (6)	28 (4)
Longitudinal compression, F^{cu}_L	1240 (180)	690 (100)
Transverse compression, F^{cu}_T	170 (25)	170 (25)
In-plane shear, F^{su}_{LT}	80 (12)	70 (10)
Interlaminar shear, F^{Lsu}	90 (13)	70 (10)
Ultimate strains, %		
Longitudinal tension, $\in^{tu}_L$	0.9	0.3
Transverse tension, $\in^{tu}_T$	0.4	0.4
Longitudinal compression, $\in^{cu}_L$	0.9	0.3
Transverse compression, $\in^{cu}_T$	1.6	2.8
In-plane shear	2.0	—
Physical properties		
Specific gravity	1.6	1.7
Density (lb/in^3)	0.056	0.058
Longitudinal CTE, 10^{-6} in/in/°F (10^{-6} m/m/°C)	−0.2	−0.3
Transverse CTE, 10^{-6} m/m/°C (10^{-6} in/in/°F)	32 (18)	32 (18)

- Have a low coefficient of thermal expansion
- Flow to penetrate the fiber bundles completely and eliminate voids during the compacting/curing process

- Have reasonable strength, modulus, and elongation (elongation should be greater than fiber)
- Be elastic to transfer load to fibers
- Have strength at elevated temperature (depending on application)
- Have low-temperature capability (depending on application)
- Have excellent chemical resistance (depending on application)
- Be easily processable into the final composite shape
- Have dimensional stability (maintain its shape)

There are two alternates in matrix selection, thermoplastic and thermoset, and there are many matrix choices available within the two main divisions. The basic difference between the two is that thermoplastic materials can be repeatedly softened by heat, and thermosetting resins cannot be changed after the chemical reaction to cause their cure has been completed. The two alternatives differ profoundly in terms of manufacture, processing, physical and mechanical properties of the final product, and the environmental resistance of the resultant composite.

5.2.2.1 Thermoplastic matrices. Several thermoplastic matrices were developed to increase hot-wet use temperature and the fracture toughness of aerospace, continuous-fiber composites. There are also many thermoplastic matrices, such as polyethylene, ABS, and nylon, that are common to the *commodity* plastics arena. Although continuous-fiber, high-performance "aerospace" thermoplastic composites are still not in general usage, their properties are well documented because of sponsorship of development programs by the U.S. Air Force. Table 5.8 shows the relative advantages and disadvantages of both thermoplastics and thermoset matrices. Thermoplastic matrix choices range from nylon and polypropylene in the commodity arena to those matrices selected for extreme resistance to high temperature and aggressive solvents encountered in the commercial aircraft daily environment, such as the polyether-ether-ketone (PEEK) resins. There is a decided difference in the costs of the commodity resins and the resins that would be used for aerospace use—in a similar order as the differences in fiber prices, for instance, (~U.S.$1.00/lb for polypropylene to >U.S.$100.00/lb for PEEK). Some manufacturers have elected to propose the use of a commodity approach to manufacturing aerospace structures such as small aircraft with polypropylene/glass.[8] The aerospace, high-performance thermoplastic composites have a relatively high potential advantage, because their large-scale use is still in the future. Some special considerations must be made for thermoplastics, as follows:

TABLE 5.8   Composite Matrix Trade-Offs

Property	Thermoset	Thermoplastic	Notes
Resin cost	Low to medium-high, based on resin requirements	Low to high premium thermoplastic prepregs are more than thermoset prepregs	Will decrease for thermoplastics as volume increases
Formulation	Complex	Simple	
Melt viscosity	Very low	High	High melt viscosity interferes with fiber impregnation
Fiber impregnation	Easy	Difficult	
Prepreg tack/drape	Good	None	Simplified by co-mingled fibers
Prepreg stability	Poor	Good	
Composite voids	Good (low)	Good to excellent	
Processing cycles	Long	Short to long (long processing degrades polymer)	
Fabrication costs	High for aerospace, low for pipes and tanks with glass fibers	Low (potentially); some shapes still cannot be processed economically	
Composite mechanical properties	Fair to good	Good	
Interlaminar fracture toughness	Low	High	
Resistance to fluids/solvents	Good	Poor to excellent; choose matrix well	Thermoplastics stress craze
Damage tolerance	Poor to excellent	Fair to good	
Resistance to creep	Good	Not known	
Data base	Very large	Small	
Crystallinity problems	None	Possible	Crystallinity affects solvent resistance
Other		Thermoplastics can be reformed to make an interference joint	

- Because high temperatures (up to 300°C) are required for processing the higher-performance matrices, special autoclaves, processes, ovens, and bagging materials may be needed.

- The fiber finishes used for thermosetting resins may not be compatible with thermoplastic matrices, requiring alternative treatment.

- Thermoplastic composites can have greater or much less solvent resistance than a thermoset material. If the stressed matrix of the composite is not resistant to the solvent, the attack and destruction of the composite may be nearly instantaneous. (This is due to *stress corrosion cracking,* a common concern for commodity thermoplastics. Thermoplastic liquid detergent bottle materials must undergo rigorous testing to verify their resistance to stress cracking with the contained material, and the addition of fibers into the matrix aggravates the propensity to crack).

5.2.2.2 Thermoset matrices. Thermoset matrices do not necessarily have the same stress corrosion problems but have a completely different and just as extensive set of environmental and physical-mechanical concerns. To provide solutions for these potential problems, a great number of matrices have been under development for over 50 years.

The common thermoset matrices for composites include the following:

- Polyester and vinylesters

- Epoxy

- Bismaleimide

- Polyimide

- Cyanate ester and phenolic triazine

Each of the resin systems has some drawbacks that must be accounted for in design and manufacturing plans. Polyester matrices have been in use for the longest period, and they are used in the widest variety and greatest number of structures. These structures have included storage tanks with fiberglass and many types of watercraft, ranging from small fishing or speed boats to large minesweepers. The usable polymers can contain up to 50% by weight of unsaturated monomers and solvents such as styrene. These can cause a significant shrinkage on matrix cure. Polyesters cure via a catalyst (usually a peroxide), which results in an exothermic reaction. This reaction can be initiated at room temperature. Because of the large shrinkage with the polyester-type matrices, they are generally not used with the high-modulus fibers.

The most widely used matrices for advanced composites have been the epoxy resins. These resins cost more than polyesters and do not have the high-temperature capability of the bisimalimides or polyimides; but, because of the advantages shown in Table 5.9, they are widely used.

TABLE 5.9   Epoxy Resin Selection Factors

Advantages	Disadvantages
Adhesion to fibers and resin	Resins and curatives somewhat toxic in uncured form
No by-products formed during cure	Moisture absorption:
Low shrinkage during cure	Heat distortion point lowered by moisture absorption
Solvent and chemical resistance	Change in dimensions and physical properties due to moisture absorption
High or low strength and flexibility	Limited to about 200°C upper temperature use (dry)
Resistance to creep and fatigue	Difficult to combine toughness and high temperature resistance
Good electrical properties	High thermal coefficient of expansion
Solid or liquid resins in uncured state	High degree of smoke liberation in a fire
Wide range of curative options	May be sensitive to UV light degradation
	Slow curing

There are two resin systems in common use for higher temperatures, bismaleimides and polyimides. New designs for aircraft demand a 177°C (350°F) operating temperature that is not met by the other common structural resin systems. The primary bismaleimide (BMI) in use is based on the reaction product from methylene dianiline (MDA) and maleic anhydride: bis (4 maleimidophenyl) methane (MDA BMI).

Two newer resin systems have been developed and have found applications in widely diverse areas. The cyanate ester resins, marketed by Ciba-Geigy, have shown superior dielectric properties and much lower moisture absorption than any other structural resin for composites. The dielectric properties have enabled their use as adhesives in multilayer microwave printed circuit boards and the low moisture absorbance have caused them to be the resin of universal choice for structurally stable spacecraft components.

The PT resins also have superior elevated temperature properties, along with excellent properties at cryogenic temperatures. Their resistance to proton radiation under cryogenic conditions was a prime cause for their choice for use in the superconducting supercollider, subsequently canceled by the U.S. Congress. They are still available from the Lonza Company.

Polyimides are the highest-temperature polymer in general advanced composite use, with a long-term upper temperature limit of

232°C (450°F) or 316°C (600°F). Two general types are *condensation* polyimides, which release water during the curing reaction, and *addition* type polyimides, with somewhat easier process requirements.

Several problems consistently arise with thermoset matrices and prepregs that do not apply to thermoplastic composite starting materials. Because of the problems shown below, if raw material and processing costs were comparable for the two matrices, the choice would probably always be thermoplastic composites, without regard to the other advantages resulting in the composite. These problems lead to a great increase in quality control efforts that may result in the bulk of final composite structure costs. They are as follows:

Problems Associated with Thermoset Matrices

1. Frequent variations from batch to batch
   - Effects of small amounts of impurities
   - Effects of small changes in chemistry
   - Change in matrix component vendor or manufacturing location
2. Void generation, caused by
   - Premature gelation
   - Premature pressure application
   - Effects on interlaminar shear and flexural modulus because of water absorption
3. Change in processing characteristics
   - Absorbed water in prepreg
   - Length of time under refrigeration
   - Length of time out before cure
   - Loss of solvent in wet systems

Some other resins that are in general commercial and aerospace use are not treated here, because they are not in wide use with the modern fibers.

The following general notes are more or less applicable to all thermoset matrices:

- The higher the service temperature limitation the less strain to failure.
- The greater the service temperature, the more difficult the processing that may be due to:
   1. Volatiles in matrix

2. Higher melt viscosity
3. Longer heating curing cycles

- The greater the service temperature or the greater the curing temperature, the greater the chance for development of color in the matrix.

- Higher service temperatures and higher curing temperatures may sometimes result in better flame resistance (although this is not evident for epoxies with curing temperatures between 250°F and 350°F).

5.2.3   Fiber Matrix Systems

The end-user sees a composite structure. Someone else, probably a prepregger, combined the fiber and the resin system, and someone else caused the cure and compaction to result in a laminated structure. A schematic of the steps is shown in Fig. 5.8. In many cases, the end-user of the structure has fabricated the composite from prepreg. The three types of continuous fibers, roving or tow, tape, and woven fabric available as prepregs give the end user many options in terms of design and manufacture of a composite structure. Although the use of dry fibers and impregnation at the work (i.e., filament winding pultrusion or hand layup) is very advantageous in terms of raw material costs, there are many advantages to the use of prepregs, as shown in Table 5.10, particularly for the manufacture of modern composites. In general, fabricators skilled in manufacturing from prepreg will not care to use wet processes.

TABLE 5.10   Advantages of Prepregs over Wet Impregnation

Prepregs reduce the handling damage to dry fibers.
They improve laminate properties via better dispersion of short fibers.
Prepregs allow the use of hard-to-mix or proprietary resin systems.
They allow more consistency, because there is a chance for inspection before use.
Heat curing provides more time for the proper laydown of fibers and for the resin to move and degas before cure.
Increasing curing pressure reduces voids and improves fiber wetting.
Most prepregs have been optimized as individual systems to improve processing.

The prepreg process for thermoset matrices is accomplished by feeding the fiber continuous tape, woven fabric, strands, or roving through a resin-rich solvent solution and then removing the solvent by hot tower drying. The excess resin is removed via a doctor blade or metering rolls, and then the product is staged to the cold-stable prepreg form (B stage). The newer technique, the hot-melt procedure for prepregs, has substantially replaced the solvent method because of en-

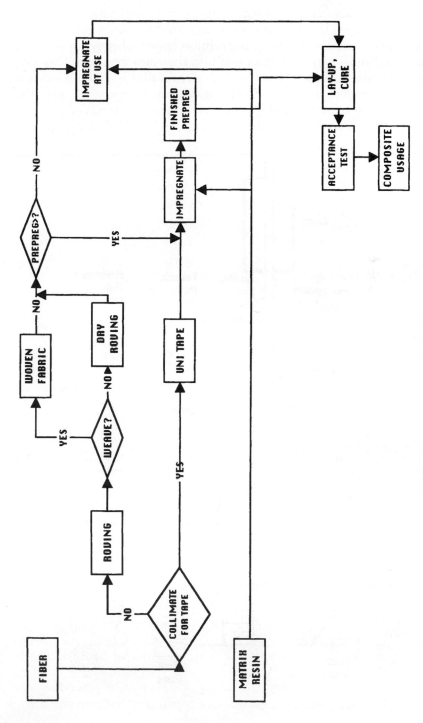

Figure 5.8 The manufacturing steps in composite structure fabrication.

vironmental concerns and the need to exert better control over the amount of resin on the fiber. A film of resin that has been cast hot onto release paper is fed, along with the reinforcement, through a series of heaters and rollers to force the resin into the reinforcement. Two layers of resin are commonly used so that a resin film is on both sides of the reinforcement; one of the release papers is removed, and the prepreg is then trimmed, rolled, and frozen. The two types of prepregging techniques, solvent and film are shown in Figs. 5.9 and 5.10.[9]

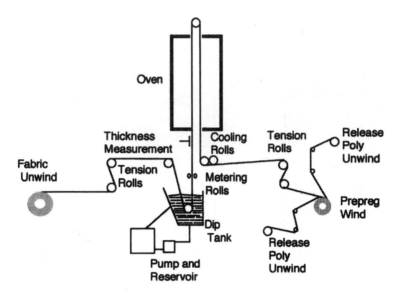

Figure 5.9 Schematic of the typical solution prepregging process.

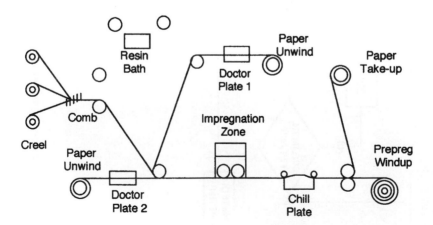

Figure 5.10 Schematic of the typical film prepregging process.

5.2.4 Unidirectional Ply Properties

The manufacturer of the prepreg reports an areal weight for the prepreg and a resin percentage, by weight. Since fiber volume is used to relate the properties of the manufactured composites, the following equations can be used to convert between weight fraction and fiber volume.

$$W_f = \frac{w_f}{w_c} = \frac{\rho_f V_f}{\rho_c V_f} = \frac{\rho_f}{\rho_c} V_f \qquad (5.1)$$

$$V_f = \frac{\rho_c}{\rho_f} W_f = 1 - V_m \qquad (5.2)$$

where W_f = weight fraction of fiber
  w_f = weight of fiber
  w_c = weight of composite
  ρ_f = density of fiber
  ρ_c = density of composite
  V_f = volume fraction of fiber
  V_m = volume fraction of matrix
  ρ_m = density of matrix

A percentage fiber that is easily achievable and repeatable in a composite and convenient for reporting mechanical and physical properties for several fibers is 60%. The properties of unidirectional fiber laminates are shown in Tables 5.7, 5.11, 5.12, and 5.13.[10] These values are for individual lamina or for a unidirectional composite, and they represent the theoretical maximum (for that fiber volume) for longitudinal in-plane properties. Transverse, shear, and compression properties will show maximums at different fiber volumes and for different fibers, depending on how the matrix and fiber interact. These properties are not reflected in strand data. These values may also be used to calculate the properties of a laminate that has fibers oriented in several directions. Using the techniques shown in Sec. 5.5.1, the methods of description for ply orientation must be introduced.

5.3 Ply Orientations, Symmetry, and Balance

5.3.1 Ply Orientations

One of the advantages of using a modern composite is the potential to orient the fibers to respond to the load requirements. This means that the composite designer must show the material, the fiber orientations

TABLE 5.11 Properties of Typical Unidirectional Glass/Epoxy Composites (Fiber Volume Fraction, V_f = 0.60); Elastic Constants, Strengths, Strains, and Physical Properties (from Ref. 10)

	E-glass	S-glass
Elastic constants, GPa (psi $\times 10^6$)		
Longitudinal modulus, E_L	45 (6.5)	55 (8.0)
Transverse modulus, E_T	12 (1.8)	16 (2.3)
Shear modulus, G_{LT}	5.5 (0.8)	7.6 (1.1)
Poisson's ratio (dimensionless) υ_{LT}	0.19	0.28
Strength properties, MPa (10^3 psi)		
Longitudinal tension, $F^{tu}{}_L$	1020 (150)	1620 (230)
Transverse tension, $F^{tu}{}_T$	40 (7)	40 (7)
Longitudinal compression, $F^{cu}{}_L$	620 (90)	690 (100)
Transverse compression, $F^{cu}{}_T$	140 (20)	140 (20)
In-plane shear, $F^{su}{}_{LT}$	60 (9)	60 (9)
Interlaminar shear, F^{Lsu}	60 (9)	80 (12)
Ultimate strains,%		
Longitudinal tension, $\epsilon^{tu}{}_L$	2.30.9	2.9
Transverse tension, $\epsilon^{tu}{}_T$	0.4	0.3
Longitudinal compression, $\epsilon^{cu}{}_L$	1.4	1.3
Transverse compression, $\epsilon^{cu}{}_T$	1.1	1.9
In-plane shear	—	3.2
Physical properties		
Specific gravity	2.1	2.0
Density (lb/in^3)	0.075	0.72
Longitudinal CTE, 10^{-6} in/in/°F (10^{-6} m/m/°C)	3.7 (6.6)	3.5 (6.3)
Transverse CTE, 10^{-6} m/m/°C (10^{-6} in/in/°F)	30 (17)	32 (18)

in each ply, and how the plies are arranged (ply stackup). A shorthand "code" (Fig. 5.11b) for ply fiber orientations has been adapted for use in layouts and studies.

Each ply (lamina) is shown by a number representing the direction of the fibers in degrees, with respect to a reference (x) axis. 0° fibers of both tape and fabric are normally aligned with the largest axial load (axis) (Fig. 5.11a).

TABLE 5.12   Properties of Unidirectional Aramid/Epoxy Composites
(Fiber Volume Fraction, V_f = 0.60) (from Ref. 10)

	Kevlar 49
Elastic constants, GPa (psi $\times 10^6$)	
Longitudinal modulus, E_L	76 (11)
Transverse modulus, E_T	5.5 (0.8)
Shear modulus, G_{LT}	2.1 (0.3)
Poisson's ratio (dimensionless) υ_{LT}	0.34
Strength properties, MPa (10^3 psi)	
Longitudinal tension, $F^{tu}{}_L$	1380 (200)
Transverse tension, $F^{tu}{}_T$	30 (4.3)
Longitudinal compression, $F^{cu}{}_L$	280 (40)
Transverse compression, $F^{cu}{}_T$	140 (20)
In-plane shear, $F^{su}{}_{LT}$	60 (9)
Interlaminar shear, F^{Lsu}	60 (9)
Ultimate strains,%	
Longitudinal tension, $\epsilon^{tu}{}_L$	1.8
Transverse tension, $\epsilon^{tu}{}_T$	0.5
Longitudinal compression, $\epsilon^{cu}{}_L$	2.0
Transverse compression, $\epsilon^{cu}{}_T$	2.5
In-plane shear	—
Physical properties	
Specific gravity	1.4
Density (lb/in^3)	0.050
Longitudinal CTE, 10^{-6} in/in/°F (10^{-6} m/m/°C)	−4 (−2.2)
Transverse CTE, 10^{-6} m/m/°C (10^{-6} in/in/°F)	70 (40)

Individual adjacent plies are separated by a slash in the code if their angles are different (Fig. 5.11b).

The plies are listed in sequence, from one laminate face to the other, starting with the ply first on the tool and indicated by the code arrow with brackets indicating the beginning and end of the code. Adjacent plies of the same angle of orientation are shown by a numerical subscript (Fig. 5.11c).

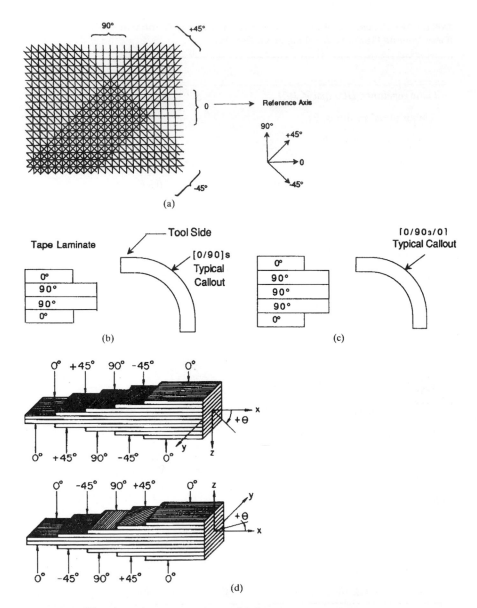

Figure 5.11   Ply orientations, symmetry, and balance.

When tape plies are oriented at angles equal in magnitude but opposite in sign, (+) and (−) are used. Each (+) or (−) sign represents one
ply. A numerical subscript is used only when there are repeating angles of the same sign. Positive and negative angles should be consistent with the coordinate system chosen. An orientation shown as

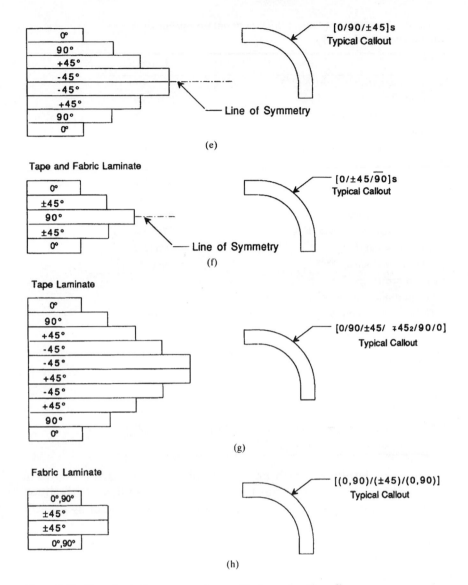

Figure 5.11 Ply orientations, symmetry, and balance *(continued)*.

positive in one right-handed coordinate system may be negative in another. If the Y and Z axis directions are reversed, the ±45 plies are reversed (Fig. 5.11d).

Symmetric laminates with an even number of plies are listed in sequence, starting at one face and stopping at the midpoint. A subscript "S" following the bracket indicates only one half of the code is shown (Fig. 5.11e).

TABLE 5.13   Properties of Typical Unidirectional Boron/Epoxy Composites (from Ref. 10)

	Boron
Elastic constants, GPa (psi $\times 10^6$)	
Longitudinal modulus, E_L	207 (30)
Transverse modulus, E_T	19 (2.7)
Shear modulus, G_{LT}	4.8 (0.7)
Poisson's ratio (dimensionless) υ_{LT}	0.21
Strength properties, MPa (10^3 psi)	
Longitudinal tension, $F^{tu}{}_L$	1320 (192)
Transverse tension, $F^{tu}{}_T$	72 (10.4)
Longitudinal compression, $F^{cu}{}_L$	2430 (350)
Transverse compression, $F^{cu}{}_T$	276 (40)
In-plane shear, $F^{su}{}_{LT}$	105 (15)
Interlaminar shear, F^{Lsu}	90 (13)
Ultimate strains,%	
Longitudinal tension, $\in^{tu}{}_L$	0.6
Transverse tension, $\in^{tu}{}_T$	0.4
Longitudinal compression, $\in^{cu}{}_L$	—
Transverse compression, $\in^{cu}{}_T$	—
In-plane shear	—
Physical properties	
Specific gravity	2.0
Density (lb/in^3)	0.072
Longitudinal CTE, 10^{-6} in/in/°F (10^{-6} m/m/°C)	4.1 (2.3)
Transverse CTE, 10^{-6} m/m/°C (10^{-6} in/in/°F)	19 (11)

Symmetric laminates with an odd number of plies are coded as a symmetric laminate except that the center ply, listed last, is overlined to indicate that half of it lies on either side of the plane of symmetry (Fig. 5.11f–h).

5.3.2 Symmetry

The geometric midplane is the reference surface for determining if a laminate is symmetrical. In general, to reduce out-of-plane strains, coupled bending and stretching of the laminate, and complexity of analysis, symmetric laminates should be used. However, some composite structures (e.g., filament wound pressure vessels) are geometrically symmetric, so symmetry through a single laminate wall is not necessary if it constrains manufacture. To construct a midplane symmetric laminate, for each layer above the midplane there must exist an identical layer (same thickness, material properties, and angular orientation) below the midplane (Fig. 5.11e).

5.3.3 Balance

All laminates should be balanced to achieve in-plane orthotropic behavior. To achieve balance, for every layer centered at some positive angle $+\theta$, there must exist an identical layer oriented at $-\theta$ with the same thickness and material properties. If the laminate contains only 0° and/or 90° layers, it satisfies the requirements for balance. Laminates may be midplane symmetric but not balanced, and vice versa. Figure 5.11e is symmetric and balanced, whereas Fig. 5.11g is balanced but unsymmetric.

5.4 Quasi-isotropic Laminate.   The goal of composite design is to achieve the lightest, most efficient structure by aligning most of the fibers in the direction of the load. Many times, there is a need, however, to produce a composite that has some isotropic properties, similar to metal, because of multiple or undefined load paths or for a more conservative design. A *quasi-isotropic* laminate layup accomplishes this for the x and y planes only; the z, or through-the-laminate thickness plane, is quite different and lower. Most laminates produced for aircraft applications have been, with few exceptions, quasi-isotropic. One exception was the X-29 (Fig. 5.3). As designers become more confident and have access to a greater database with fiber-based structures, more applications will evolve. For a quasi-isotropic (QI) laminate, the following are requirements:

- It must have three layers or more.
- Individual layers must have identical stiffness matrices and thicknesses
- The layers must be oriented at equal angles. For example, if the total number of layers is n, the angle between two adjacent layers should be $360°/n$. If a laminate is constructed from identical sets of three or more layers each, the condition on orientation must be satisfied by the layers in each set, for example: $[0°/\pm60°]_S$ or $[0°/\pm45°/90°]_S$ [Ref. 11, p. 199].

Table 5.14[12] shows mechanical values for several composite laminates with the high-strength fiber of Table 5.4 and a typical resin system. The first and second entries are for simple 0/90 laminates and show the effect of changing the position of the plies. The effect of increasing the number of 0° plies is shown next, and the final two laminates demonstrate the effect of ±45° plies on mechanical properties, particularly the shear modulus. The last entry is a quasi-isotropic laminate. These laminates are then compared to a typical aluminum alloy. This effectively shows that there is a strength and modulus penalty that goes with the conservatism of the use a QI laminate.

TABLE 5.14   High-Strength Carbon Graphite Laminate Properties

Laminate	Longitudinal modulus, E_{11}, GPa	Bending modulus, E_B, GPa	Shear modulus, G_{xy}, GPa
$[0/90_2/0]$	76.5	126.8	5.24
$[90/0_2/90]$	76.5	26.3	5.24
$[0_2/90_2/0]$	98.5	137.8	5.24
$[0_2/\pm45_2/0]$	81.5	127.5	21.0
$[0/\pm45/90]_s$	55.0	89.6	21.0
Aluminum	41.34	41.34	27.56

When employing the data extracted from tables, there are some cautions that should be observed by the reader. The values seen in many tables of data may not always be consistent for the same materials or the same group of materials from several sources for the following reasons:

1. Manufacturers have been refining their production processes so that newer fibers may have greater strength or stiffness. These new data may not be reflected in the compiled data.

2. The manufacturer may not be able to change the value quoted for the fiber because of government or commercial restrictions imposed by the specification process of his customers.

3. Many different high-strength fibers are commercially available. Each manufacturer has optimized its process to maximize the mechanical properties, and each of the processes may by different from that of the competitor, so all vendor values in a generic class may differ widely.

4. Most tables of values are presented as "typical values." Those values and the values that are part of the menu of many computer analysis programs should be used with care. Each user must find the most appropriate set of values for design, develop useful design allowables, and apply appropriate "knock down" factors, based on the operating environments expected in service.

5.5   Analysis

5.5.1   Micromechanical Analysis

A number of methods are in common use for the analysis of composite laminates. The use of micromechanics, i.e., the application of the properties of the constituents to arrive at the properties of the composite ply, can be used to:

1. Arrive at "back of the envelope" values to determine if a composite is feasible

2. Arrive at values for insertion into computer programs for laminate analysis or finite element analysis

3. Check on the results of computer analysis

The rule of mixtures holds for composites. The micromechanics formula to arrive at the Young's modulus for a given composite is

$$E_C = V_f E_f + V_m E_m$$

and

$$V_f + V_m = 1$$

$$E_c = V_f E_f + E_m(1 - V_f) \tag{5.3}$$

where E_c = composite or ply Young's modulus in tension for fibers oriented in direction of applied load

V = volume fraction of fiber (f) or matrix (m)

E = Young's modulus of fiber (f) or matrix (m)

But, since the fiber has much higher Young's modulus than the matrix, (Table 5.7 vs. the value for the 3502 matrix on p. 5.1), the second part of the equation can be ignored.

$$E_f \gg E_m$$

$$E_c = E_f V_f \tag{5.4}$$

This is the basic rule of mixture and represents the highest Young's modulus composite, where all fibers are aligned in the direction of load. The minimum Young's modulus for a reasonable design (other than a preponderance of fibers being orientated transverse to the load direction) is the quasi-isotropic composite and can be approximated by

$$E_c \cong \frac{3}{8} E_f V_f \tag{5.5}$$

Note: the quasi-isotropic modulus, E, of a composite laminate is

$$\frac{3}{8} E_{11} + \frac{5}{8} E_{22} \text{ (see Ref. 13)}$$

where E_{11} is the modulus of the lamina in the fiber direction and E_{22} is the transverse modulus of the lamina. The transverse modulus for polymeric-based composites is a small fraction of the longitudinal modulus (see E_t in Table 5.7) and can be ignored for preliminary estimates, resulting in a slightly lower-than-theoretical value for E_c for a quasi-isotropic laminate. This approximate value for quasi-isotropic modulus represents the lower bound of composite modulus. It is useful for comparisons of composite properties to those of metals and to establish if a composite is appropriate for a particular application.

The following formulas also can be used to obtain important data for unidirectional composites:

Density,
$$\rho_c = V_f \rho_f + V_m \rho_m \tag{5.6}$$

Poisson's ratio,
$$\nu_{12} = \nu_f V_f + \nu_m V_m \tag{5.7}$$

Transverse Young's modulus,
$$E_2 = \frac{E_{2m}(1 + \xi \eta_2 V_f)}{1 - \eta_2 V_f} \tag{5.8}$$

and values for η_2 and ξ can be seen in Ref. 14 and Ref. 11, pp. 76–78. The matrix is isotropic.

5.5.2 Carpet Plots

The analysis of a multilayered composite, if attempted by hand calculations, is not trivial. Fortunately, there are a significant number of computer programs to perform the matrix multiplications and the transformations.[14–16] However, the use of carpet plots is still in practice in U.S. industry, and these plots are useful for preliminary analysis. The carpet plot shows graphically the range of properties available with a specific laminate configuration. For example, if the design options include [±0/90]$_S$ laminates, a separate carpet plot for each value of θ would show properties attainable by varying percentage of ±θ plies versus 90° plies. A sequence of these charts would display attainable properties over a range of θ values. The computer programs described above can be programmed to produce such charts for arbitrary laminates.

Figure 5.12 shows a sample carpet plot[17] of extensional modulus of elasticity E_x for Kevlar 49/epoxy with [0/±45/90]$_s$ construction. As expected, the chart shows $E_x = 76$ GPa (11×10^6 psi) with all 0° plies, and $E_x = 5.5$ GPa (0.8×10^6 psi) with all 90s. With all 45s, an axial modulus is only slightly higher, 8 GPa (1.1×10^6 psi), than the all 90s value predicted for this material. A quasi-isotropic laminate (Sec.

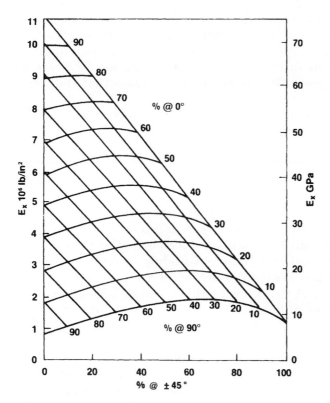

Figure 5.12 Predicted axial modulii for [0/±45/90] kevlar epoxy laminates.

5.5.2) with 25% 0s, 50% ±45s, and 25% 90s, produces an intermediate value of $E_x = 29$ GPa (4.2×10^6 psi).

5.6   Composite Failure and Design Allowables

5.6.1   Failure[18–20]

Composite failure modes are different from those of isotropic materials such as metals. Because of the fibers, they do not tend to fail in only one area, they do not have the strain-bearing capacity of most metals, and they are prone to premature failure if stressed in a direction that was not anticipated in the design. Useful structures nearly always have been constructed from ductile materials such as steel or aluminum, with fairly well defined strengths. This allows designers to accurately comprehend and specify safety factors that provide some assurance that the structures will not fail in service.

It has became necessary, in the practical design of structures for demanding environments, to use brittle materials such as glass and ceramics to take advantage of special properties such as high-temperature strength When brittle materials are employed in practical structures, the designer still has the need to ensure that the structure will not fail prematurely.

The data that provide the background for the design confidence can be obtained from various sources. They can be derived from previous designs that have proven reliable and resulted in data being published in a reference work such as Mil-Handbook-5 for Aerospace Metals (Ref. 21) or industry journals. Or the data can be obtained through testing conducted by the designer's own organization. Typically, on the basis of laboratory experiments on a statistically determined number of small specimens tested in simple tension or bending, the probability of failure can be calculated for structural members of other sizes and shapes, often under completely different loading conditions. The tool for accomplishing this is *statistical fracture theory.*

To predict strength of the ply with the laminate, it is usually assumed that knowledge of failure of a ply by itself under simple tension, compression, or shear will allow prediction of failure of that ply under combined loading in the laminate.

The matrix plays a special role in the failure of the composite. The matrix is extremely weak compared to the fibers (particularly if they are the *advanced composite* fibers) and cannot carry primary loads, but it efficiently allows the transfer of the loads in the composite. This is demonstrated by the experimental observation that the strength of matrix-impregnated fiber bundles can be on the order of a factor of 2 higher than the measured tensile strength of dry fiber bundles without matrix impregnation. The key to this apparently contradictory evidence lies in a synergistic effect between fiber and matrix. The first and primary design rule for composites of this type is that the fibers must be oriented to carry the primary loads. A comparison of the tensile strengths illustrates this point. High-strength carbon fibers have tensile strengths that approach 1×10^6 psi (6900 MPa), while the tensile strength of typical polymer matrices may be on the order of 3×10^4 psi (200 MPa) or less. Clearly, the tensile strength of the matrix is insignificant in comparison.

A number of investigators have provided an explanation for the above observation. It can be explained by noting that the strength of individual brittle fibers varies widely because of a statistical distribution of flaws. The fibers can be considered to be brittle and sensitive to surface imperfections randomly distributed over the fiber length. The strength of individual fibers varies widely and will decrease with increasing length. These are characteristics that are typical of brittle materials failing at random defects, and they are changed dramatically through the addition of the matrix. The matrix acts to almost double the apparent strength of a fiber bundle, and it significantly reduces the variability.

In a dry fiber bundle, when a fiber breaks, it loses all of its load-carrying ability over its entire length, and the load is shifted to the remaining fibers. When enough of the weaker fibers fail, the strength of the remaining fibers is exceeded, and the bundle fails. In matrix-impregnated fiber bundles, the matrix acts to bridge around individual fiber breaks so that adjacent fibers quickly pick up the load. Thus, the adjacent fibers have to carry an increased load over only a small axial distance. Statistical distribution of fiber defects makes it unlikely that each fiber would be weakest at the same axial location, so failure will occur at a higher load value after enough fibers have failed in adjacent locations. Because of the small diameter of individual fibers (5.7 mm for some typical carbon fibers), there are many millions of fibers in a structure. This makes statistical effects important.

5.6.2 Failure Theories

For over three decades, there has been a continuous effort to develop a more universal failure criterion for unidirectional fiber composites and their laminates. A recent FAA publication lists 21 of these theories.[22] The simplest choices for failure criteria are maximum stress or maximum strain. With the maximum stress theory, the ply stresses, in-plane tensile, out-of-plane tensile, and shear are calculated for each individual ply using lamination theory and compared with the allowables. When one of these stresses equals the allowable stress, the ply is considered to have failed. Other theories use more complicated (e.g., quadratic) parameters, which allow for interaction of these stresses in the failure process.

Although long-fiber composites typically fail at low tensile strains, they are generally not considered to be brittle, i.e., in the realm of glass or ceramics. The fibers do have strain to failure, and the failures can be predicted. A bundle of fibers bound together by a matrix does not usually fail when the first fiber ruptures. Instead, the final failure is preceded by a period of progressive damage.

The basic assumption of statistical fracture theory is that the reason for the variations in strength of nominally identical specimens is their varying content of randomly distributed (and generally invisible) flaws. The strength of a specimen thus becomes the strength of its weakest flaw, just as the strength of a chain is that of its weakest link.

Since it is not possible to obtain strengths in all possible lamina orientations or for all combinations of lamina, a means must be established by which these characteristics can be determined from basic layer data. Theories of failure are hypotheses concerning the limit of load-carrying ability under different load combinations. Using expressions derived from these theories, it is possible to construct failure envelopes or, if in three dimensions, failure surfaces that represent the limit of usefulness of the material as a load-bearing component, i.e., if a given loading condition is within the envelope, the material will not fail. The suitability of any proposed criterion is determined by a number of factors, the most important of which has to do with the nature of the failure mode. As a result, it is important that proposed failure criteria be accompanied by a definition of material behavior.

5.6.3 Design Allowables

The design of composites involves knowledge of a significantly greater number of material properties than those needed for conventional isotropic metals. As mentioned previously, these data are not always conveniently available from a single source of data such as a handbook. The data at the maximum, for the design of aerospace structures, takes the form of those shown in Table 5.15.[23]

Data for design use requires statistical significance with a known confidence level. The old MIL-Hdbk-17B[24] provides a guide concerning the number and type of tests sufficient to establish statistically based material properties along with some limited data that is now somewhat out of date. The (new) Composites Materials Handbook, Mil-17,[25] in preparation, has somewhat enhanced statistical treatment approaches. Three classes of allowables, pertinent to current usage of composites for many applications, are:

- *A-Basis Allowable.* The value above which 99% of the population of values is expected to fall, with a confidence of 95%.

- *B-Basis Allowable.* The value above which 90% of the population of values is expected to fall, with a confidence of 95%.

- *S-Basis Allowable.* The value that is usually the specified minimum value of the appropriate government specification.

For most flightworthy composites, material properties are usually required to be either A-Basis or B-Basis allowables.

The effort is still in progress to provide a new family of design allowables including the most advanced fiber composites and the background guidance for their use. This is a reprise of the Mil-Hdbk-17 effort initiated in 1972, which had property values primarily for fiberglass fibers. The new Mil-Hdbk-17 committee has published the first interim report on the effort. The original Mil-Hdbk-17 treated

TABLE 5.15 Lamina Properties and Equations Used To Calculate Material Properties[23]

Lamina material properties	Definition	Equation used to calculate material property
Elastic		
E_1	Elastic modulus in the fiber direction	Property based on test data
E_2	Elastic modulus transverse to the fiber direction	Property based on test data
E_3	Elastic modulus through-the-thickness	Transverse isotropy: $E_3 = E_2$
G_{12}	Shear modulus in the 1-2 plane	Property based on test data
G_{23}	Shear modulus in the 2-3 plane	$G_{23} = \dfrac{E_3}{2(1 + v_{23})}$
G_{13}	Shear modulus in the 1-3 plane	Transverse isotropy: $G_{13} = G_{12}$
v_{12}	Poisson's ratio in 1-3 plane	Property based on test data
v_{23}	Poisson's ratio in 2-3 plane	$v_{23} = v_f V_f + v_m (1 - V_f) \left[\dfrac{1 + v_m - v_{12} \dfrac{E_m}{E_1}}{1 - v_m^2 + v_m v_{12} \dfrac{E_m}{E_1}} \right]$
v_{13}	Poisson's ratio in 1-3 plane	Transverse isotropy: $v_{13} = v_{12}$
Strength		
σ_1	Tensile strength in the fiber direction	Property based on test data
$-\sigma_1$	Compressive strength in the fiber direction	Property based on test data
σ_2	Tensile strength transverse to the fiber	Property based on test data
$-\sigma_2$	Compressive strength transverse to the fiber	Property based on test data
σ_3	Tensile strength through-the-thickness	$\sigma_3 = \sigma_2$
$-\sigma_3$	Compressive strength through-the-thickness	$-\sigma_3 = -\sigma_2$
τ_{12}	Shear strength in 1-2 plane (in-plane)	Property based on test data
τ_{13}	Shear strength in the 1-3 plane (interlaminar)	Property based on test data
τ_{23}	Shear strength in the 2-3 plane (interlaminar)	$\tau_{23} = \tau_{13}$

unidirectional and woven fabric laminates, and two typical entries are shown in Tables 5.16 and 5.17. The first table shows the properties of 3M XP 2515 fiberglass epoxy with 100% unidirectional fibers. Although the resin portion of this laminate is no longer produced, and the data has not been upgraded, this is one of the few areas in which one can see statistically significant data for unidirectional fiberglass that is of a quality suitable for inclusion in computer analysis programs for laminate analysis. The second table is for materials that still exist in the marketplace and was continued in the 1999 edition of the handbook. This table shows the data for a woven fabric with a heat-curing epoxy resin. The 7781 fabric is a reasonably balanced fabric with good drape qualities in wide use. Tables 5.18 and 5.19, presented for carbon/graphite/epoxy materials, are extracted from the new MIL-Hdbk. They show the "B" basis allowables for strength and modulus of a unidirectional and a woven fabric laminate for three temperatures of interest. If the values of the fiber and the resin as shown by the vendor are used for contrast, a full picture of the laminate materials emerges and further simplifies analysis of laminates made with these materials or similar materials for extrapolation

5.7   Composite Fabrication Techniques

5.7.1   Choosing the Manufacturing Method

There is a history of choosing the composite manufacturing technique for the wrong reasons. Sometimes the choice is good regardless of the method, but often the end product or the schedule suffers and, in turn, the customer is unhappy. The rationales for choice have historically been as outlined below.

Design needs.   This is the best reason for choosing a manufacturing method. The key to attaining a good composite design with a manufacturing process that can operate with minimum dysfunction is the choice of the method based on the design of the composite component. Thus, the manufacturing process must be kept in mind during the component design phase and must also be a consideration in laminate design.

Part configuration.   This must have a great influence on manufacturing technique. In no other manufacturing endeavor does the finished configuration of the structure play such an important role. Some component configurations, such as pressure vessels, drive the process deci-

TABLE 5.16 Unidirectional Fiberglass Laminate Data

Fabrication	Layup: unidirectional	Vacuum	Pressure: 50 psi	Bleedout: vertical	Cure: 2 hr/300°F 4 hr & 350°F	Postcure:	Plies: 6
Physical properties	Weight percent resin: 21.4	Avg. specific gravity: 1.96		Avg. percent voids:		Avg. thickness: 0.045 inch	
Test methods	Tension: see text	Compression: see text	Shear: rail	Flexure: ASTM – D790	Bearing: ASTM – D953	Interlaminar shear: short beam	

Temperature		–65°F				75°F				160°F				300°F	
Condition		Dry		Wet		Dry		Wet		Dry		Wet		Dry	
		Avg	SD	Avg	SD	Avg	SD	Avg	SD	Avg	SD	Avg	SD	Avg	SD
Tension															
ultimate stress, ksi	0°	288.6	15.5	304.3	11.9	276.0	11.2	274.0	8.2	247.8	7.2	231.2	16.6	248.2	11.3
	90%	12.5	1.4	11.6	1.4	9.0	1.6	9.2	0.8	7.0	1.55	5.2	1.1	2.2	0.3
ultimate stress,%	0°	3.85	0.29	4.38	0.18	3.57	0.22	3.83	0.12	3.82	0.31	3.27	0.36	3.13	0.26
	90°	0.49	0.07	0.46	0.05	0.35	0.07	0.37	0.04	0.41	0.10	0.29	0.12	0.61	0.13
proportional limit, ksi	0°	247.6	16.9	237.0	19.8	201.8	15.8	213.4	16.0	195.2	11.4	167.9	18.8	187.2	22.6
	90°	10.6	0.8	10.6	1.2	5.7	0.70	6.2	0.36	4.7	0.73	3.7	0.46	0.8	0.1
initial modulus, 10^6 psi	0°	7.64	0.29	7.64	0.21	7.99	0.28	7.69	0.26	7.62	0.18	7.60	0.18	8.30	0.44
	90°	2.59	0.16	2.57	0.09	2.69	0.14	2.65	0.13	1.91	0.10	2.03	0.28	0.69	0.15
secondary modulus, 10^6 psi	0°														
	90°														
Compression															
ultimate stress, ksi	0°	115.3	5.8	119.9	10.8	100.0	7.7	92.4	10.0	76.0	5.5	70.0	4.1	48.6	4.1
	90°	41.4	2.2	37.5	2.8	29.3	0.6	28.0	0.7	22.0	0.8	19.8	0.7	8.4	0.8
ultimate stress,%	0°	1.66	0.22	1.89	0.21	1.36	0.06	1.41	0.11	1.06	0.09	1.08	0.07	0.76	0.06
	90°	2.36	0.27	2.03	0.29	1.95	0.20	1.94	0.19	1.99	0.32	2.04	0.34	4.08	0.33
proportional limit, ksi	0°	75.6	5.6	87.0	10.2	84.2	5.5	66.3	6.1	68.5	4.2	54.2	3.4	31.7	5.0
	90°	14.8	1.6	14.8	1.5	10.4	0.8	11.1	1.1	8.4	0.8	8.0	0.4	2.8	0.3
initial modulus, 10^6 psi	0°	7.73	0.20	7.07	0.24	7.52	0.18	6.78	0.22	7.24	0.30	6.68	0.26	7.50	0.47
	90°	2.60	0.10	2.53	0.09	2.69	0.09	2.49	0.07	2.21	0.11	2.08	0.08	0.57	0.16
secondary modulus, 10^6 psi	0°														
	90°														

TABLE 5.16 Unidirectional Fiberglass Laminate Data (Continued)

Fabrication	Layup: unidirectional	Vacuum	Pressure: 50 psi	Bleedout: vertical	Cure: 2 hr/300°F 4 hr & 350°F	Postcure:	Plies: 6
Physical properties	Weight percent resin: 21.4	Avg. specific gravity: 1.96	Avg. percent voids: Figure 4.32.5	Avg. thickness: 0.045 inch			
Test methods	Tension: see text	Compression: see text	Shear: rail	Flexure: ASTM – D790	Bearing: ASTM – D953	Interlaminar shear: short beam	

Temperature		−65°F				75°F				160°F				300°F			
Condition		Dry		Wet		Dry		Wet		Dry		Wet		Dry		Wet	
		Avg	SD	Avg	SD	Avg	SD	Avg	SD	Avg	SD	Avg	SD	Avg	SD	Avg	SD
Shear ultimate stress, ksi	0–90°					9.9				8.2							
	90–0°					13.4				12.0							

		−65°F dry			75°F dry			160°F dry		
		Avg	Max	Min	Avg	Max	Min	Avg	Max	Min
Flexure										
ultimate stress, ksi	0°	262.2	271.5	251.9	219.6	224.2	216.1	193.2	196.7	198.2
proportional limit, ksi	0°	177.6	181.3	173.8	181.8	187.9	170.6	165.0	173.8	154.7
initial modulus, 10⁶ psi	0°	7.00	7.11	6.85	6.78	6.82	6.75	6.95	7.18	6.80
Bearing										
ultimate stress, ksi	0°	77.3	80.9	74.8	68.8	71.4	65.3	53.6	54.6	52.1
stress at 4% elong. ksi	0°	20.5	21.5	19.1	17.7	19.7	17.2	17.0	18.5	15.8
Interlaminar shear										
ultimate stress, ksi	0°	14.6	15.2	13.7	11.9	12.6	11.0	14.2	14.5	13.9

TABLE 5.17 Woven Fiberglass (Style 7781) Laminate Data

Summary of Mechanical Properties of U.S. Polymeric E-720E/7781 (ECDE–1/0–550) Fiberglass Epoxy

Fabrication	Layup: parallel	Pressure: 55–65 psi	Vacuum: none	Bleedout: edge & vertical	Cure: 2 hr/350°F	Postcure: 4 hr/400°F	Postcure: 4 hr/400°F	Plies: 8
Physical properties	Weight percent resin: 34.9	Avg. specific gravity: 1.78	Avg. specific gravity: 1.78	Cure:	Avg. specific gravity: 1.78	Avg. percent voids: 2.0	Avg. thickness: 0.082 inches	
Test methods	Tension: ASTM D 638 TYPE-1	Compression: MIL–HDBK–17		Shear: Rail	Flexure: ASTM D 790	Bearing: ASTM D 953	Interlaminar shear: short beam	

Temperature		−65°F				75°F				160°F				400°F	
Condition		Dry		Wet		Dry		Wet		Dry		Wet		Dry	
		Avg	SD	Avg	SD	Avg	SD	Avg	SD	Avg	SD	Avg	SD	Avg	SD
Tension															
ultimate stress, ksi	0°	69.2	1.6	69.1	1.7	60.4	1.7	55.7	1.5	52.5	1.0	42.9	0.8	44.8	2.0
	90°	56.0	2.0	56.5	2.0	49.0	1.8	45.9	1.4	42.3	1.2	36.9	1.1	34.9	1.6
ultimate strain,%	0°	2.93	0.08	2.70	0.11	2.43	0.14	2.12	0.08	2.05	0.08	1.61	0.06	1.80	0.20
	90°	2.92	0.22	2.54	0.19	2.33	0.09	2.04	0.09	1.98	0.08	1.70	0.13	1.72	0.22
proportional limit, ksi	0°														
	90°														
initial modulus, 10⁶ psi	0°	3.30		3.38		3.12		3.12		2.95		2.76		2.60	
	90°	2.90		3.02		2.82		2.78		2.50		2.65		2.30	
secondary modulus, 10⁶ psi	0°	2.30		2.85		2.45		2.50		2.46		2.37			
	90°	1.90		1.74		2.05		2.19		2.01		1.97			
Compression															
ultimate stress, ksi	0°	77.1	4.0	75.0	3.7	64.8	2.9	57.3	3.8	54.0	1.4	46.2	1.4	23.8	2.2
	90°	57.2	2.7	53.9	2.7	50.2	2.9	45.2	2.4	40.8	2.9	36.2	3.1	14.7	1.6
ultimate strain,%	0°	2.48	0.16	2.44	0.15	2.14	0.11	1.99	0.09	1.86	0.08	1.62	0.06	1.12	0.22
	90°	1.93	0.16	1.81	0.19	1.70	0.14	1.58	0.14	1.46	0.17	1.37	0.15	0.91	0.08
proportional limit, ksi	0°														
	90°														
initial modulus, 10⁶ ksi	0°	3.50		3.45		3.25		3.10		3.15		3.03		2.45	
	90°	3.20		3.26		3.21		3.03		2.99		2.85		1.85	

TABLE 5.17   Woven Fiberglass (Style 7781) Laminate Data (Continued)

Summary of Mechanical Properties of U.S. Polymeric E-720E/7781 (ECDE–1/0–550) Fiberglass Epoxy

Fabrication	Layup: parallel	Vacuum: none	Pressure: 55–65 psi	Bleedout: edge & vertical	Cure: 2 hr/350°F	Postcure: 4 hr/400°F	Postcure: 4 hr/400°F	Postcure: 4 hr/400°F	Plies: 8
Physical properties	Weight percent resin: 34.9		Avg. specific gravity: 1.78	Avg. specific gravity: 1.78	Avg. specific gravity: 1.78	Avg. percent voids: 2.0		Avg. thickness: 0.082 inches	
Test methods	Tension: ASTM D 638 TYPE-1		Compression: MIL–HDBK–17		Shear: Rail	Flexure: ASTM D 790	Bearing: ASTM D 953	Interlaminar shear: short beam	

Shear, Flexure block

	0°	−65°F Avg	−65°F Max	−65°F Min	75°F Avg	75°F Max	75°F Min	160°F Avg	160°F Max	160°F Min
Shear ultimate stress, ksi	0–90°	17.5			14.3			11.2		
Flexure ultimate stress, ksi	0°	115.6	119.4	111.5	91.7	93.4	90.3	69.4	71.1	67.2
proportional limit, ksi	0°	88.1	100.7	77.5	32.5	36.2	30.8	56.2	62.8	49.4
initial modulus, 10^6 ksi	0°	2.87	2.91	2.74	3.21	3.36	3.03	2.81	2.87	2.76

Bearing, Interlaminar shear block

	0°	−65°F Avg	−65°F Max	−65°F Min	75°F Avg	75°F Max	75°F Min	160° dry Avg	160° dry Max	160° dry Min
Bearing ultimate stress, ksi	0°	74.1	78.4	70.7	64.4	70.7	60.8	50.0	53.0	47.9
stress at 4% elong., ksi	0°	32.1	34.8	29.1	29.1	34.2	23.9	18.1	21.5	15.9
Interlaminar shear ultimate stress, ksi	0°	7.09	7.36	6.80	5.90	6.07	5.72	6.05	6.16	5.91

TABLE 5.18 Unidirectional Carbon/Graphite Laminate Data (AS4/3502)

Material:	AS4 12k/3502 unidirectional tape	Comp. density:	1.56–1.59 g/cm³
Resin content:	30–33 wt%	Void content:	0.0–1.0%
Fiber volume:	59–61%		
Ply thickness:	0.0049–0.0071 in		

Test method: ASTM D 3039-76
Normalized by: Ply thickness to 0.0055 in. (58%)

Modulus calculation: Linear portion of curve

Table 4.2.8(a)
C/Ep 147–UT
AS4/3502
Tension, 1-axis
$[0]_s$
75/A, −65/A, 180/W
Fully approved

		75 ambient		−65 ambient		180 1.1–1.3	
Temperature, °F		75		−65		180	
Moisture content, %		ambient		ambient		1.1–1.3	
Equilibrium at T, RH						(1)	
Source code		49		49		49	
		Normalized	Measured	Normalized	Measured	Normalized	Measured
F_1^{tu} ksi	Mean	258	253	231	227	261	255
	Minimum	191	196	162	151	140	135
	Maximum	317	302	285	280	317	315
	C.V.,%	9.83	9.13	13.4	13.6	14.8	15.2
	B-value	204		173		200	
	Distribution	Weibull		Weibull		Weibull	
	C_1	269		244		276	
	C_2	11.2		8.82		9.39	
	No. specimens	36		38		40	
	No. batches	5		5		5	
	Approval class	Fully approved		Fully approved		Fully approved	
E_1^t msi	Mean	19.3		19.2		19.7	
	Minimum	15.6		16.8		15.1	
	Maximum	21.0		23.2		23.3	
	C.V.,%	5.74		6.31		6.87	
	No. specimens	35		38		40	
	No. batches	5		5		5	
	Approval class	Fully approved		Fully approved		Fully approved	

TABLE 5.19 Woven Fabric Carbon/Graphite Laminate Data (AS4/3502)

Material:	AS4 6k/3502 5-harness satin weave	Comp. density:	1.55–1.56 g/cm³
Resin content:	36–37 wt%	Void content:	0.0–1.0 0.2
Fiber volume:	56–57%		
Ply thickness:	0.0146–0.0157 in		

Test method:	Modulus calculation:
BMS 8-168D	Linear portion of curve
Normalized by:	Ply thickness to 0.0145 in. (57%)

> Table 4.2.16(a)
> C/Ep 365-5HS
> AS4/3502
> Tension, 1-axis
> [/90/0/90/90/0]f
> 75/A, −65/A, 180/W
> Fully approved

Temperature, °F	75		−65		180	
Moisture content, %	ambient		ambient		1.1–1.3	
Equilibrium at T, RH					(1)	
Source code	49		49		49	
	Normalized	Measured	Normalized	Measured	Normalized	Measured
F_1^{tu} ksi						
Mean	114	109	105	101	117	111
Minimum	126	91.9	87.9	84.0	102	96.1
Maximum	6.87	121	116	112	128	123
C.V.,%		7.01	5.33	5.53	5.29	5.36
B-value	91.9		95.0		102	
Distribution	ANOVA		Normal		ANOVA	
C_1	8.15		104.9		6.31	
C_2	2.70		5.59		2.33	
No. specimens	30		30		30	
No. batches	5		5		5	
Approval class	Fully approved		Fully approved		Fully approved	
E_1^t msi						
Mean	9.61		9.67		10.5	
Minimum	9.29		9.09		9.74	
Maximum	10.4		10.1		10.9	
C.V.,%	3.08		2.35		2.75	
No. specimens	30		30		30	
No. batches	5		5		5	
Approval class	Fully approved		Fully approved		Fully approved	

sion to only one method (e.g., filament winding), while others, such as driveshafts, have more processing options (e.g., filament winding, pultrusion, roll wrapping, or RTM).

Prior experience.   Experience provides a good guide to manufacturing method selection, but it is not the best. We tend to stay in the same track because of inertia; however, it imperative that each new design be thoroughly analyzed to select the optimum and not the wrong manufacturing method.

Facilities available or underutilized.   This is considered the least effective reason for manufacturing choice, but it regrettably seems to be the most prevalent. The history of the composite manufacturing business has been that an independent composite manufacturer would never be expected to reject a contract because of the lack of the optimum facilities or experienced personnel. The independent contractor would find a way to build it or would, in turn, subcontract. The same comment can be extended to most larger companies, such as the large aerospace manufacturers with adequate, but not infinite, facilities in house. Usually, the capital cost and overhead rate of a particular machine will be a strong consideration in the choosing process; manufacturers do not want to see valuable machinery idle.

This approach to composite manufacturing was demonstrated in the last decade when a commercial aircraft supplier selected two companies to fabricate composite fuselage prototypes with different approaches. One manufacturer chose a filament-wound isogrid, and the other chose a filament-wound honeycomb-stiffened structure with carbon/graphite-epoxy skins. Both of these approaches were viable and very cost effective, and the stiffened honeycomb structure approach was chosen for advanced development. Seven fuselages were then built, and one was flown on a subscale test aircraft. However, the autoclave that had been a long-lead acquisition for the aircraft manufacturer, and that was the reason for considering the subcontracts, became available earlier than anticipated. The subcontracts were cancelled, and the manufacturing was taken in house. Subsequently, a few years ago, the manufacturer acquired a fiber placement machine. It is difficult to say whether the fuselage by either method is lighter, stiffer, or stronger, but we know that the filament-wound structure did fly, and we also know that the fiber-placed structure has to cost considerably more than the filament-wound structure, because there is up to a 10:1 acquisition cost differential for the fiber placement machine vs. a suitable filament winding machine. The cost of most fiber placement machines is in the realm of $3 million to $5 million, and the cost of

most CNC filament winding machines with 5 to 7 axes of motion is
$300 thousand to $700 thousand. The machine, in this case, used to
wind the fuselages was a two-axis winder that had been modified with
a tracer-lathe-type mechanism to add a third axis of movement. The
cost of the two-axis filament winding machine is unknown but would
be expected to be less than $100 thousand. We also know that the dif-
ferential in cost of the materials was greater than 3:1. The fuselage
was wet wound; the fiber placed fuselage uses prepreg tow or slit
prepreg tape.

This is the reason why all participants in the design or acquisition
process for composites should have some familiarity with the composite manufacturing processes. All participants in the design process, at
a minimum, should include the program manager, the materials engineer, the manufacturing engineer, and the purchasing agent. In
smaller companies, many of these tasks are combined. This knowledge
can be used for the following objectives:

- For help in vendor selection
- To have a knowledge of the limitations of the different manufactur-
  ing processes
- To influence the design so as to simplify manufacture
- To exert the proper control over vendors

The goals any composite manufacturing process are to

- Achieve a consistent product by controlling:
     composite thickness
     fiber volume
     fiber directions
- Minimize voids
- Reduce internal residual stresses
- Process in the least costly manner

The procedures to reach these goals involve iterative processes to select the three key components:

- Composite material and its configuration
- Tooling
- Process

Composite material selection.   Composite material selection will be dictated by the requirements of the structure and may be constrained by

the manufacturing process. It involves both the selection of the material and the form of material. The forms of material can be described as dry fiber (with wet resin), prepreg, and fiber preform (prepreg or dry). Often, the form-of-material choice is also constrained by the manufacturing process selected; i.e., the wet process hand-layup technique is constrained to use woven fabric instead of unidirectional tape, and filament winding uses tow or strand rather than fabric (fabric can be used for local reinforcement). Because these choices for any significant structure are not trivial, this is another reason for concurrent engineering. The reinforcement configurations and their relation to the several manufacturing processes are shown in Table 5.20.

TABLE 5.20 Common Reinforcement Configuration for Manufacturing Process

Reinforcement configuration	Prepreg tape	Prepreg or (dry) tow	Prepreg or (dry) woven or nonwoven fabric	Other woven preforms, chopped fibers
Hand layup	X		X, (X)	X
Automatic tape lay-down	X			
Filament winding		X, (X)	X, (X)	X, (X)
Resin transfer molding		(X)	(X)	X
Pultrusion		(X)		
Fiber placement		X		

5.7.2 Tooling

Once material selection has been completed, the first step leading to the acceptable composite structure is the selection of tooling, which is intimately tied to process and material. For all curing techniques, the tool must be:

- Strong and stiff enough to resist the pressure exerted during cure
- Dimensionally stable through repeated heating and cooling cycles
- Light enough to respond reasonably quickly to the changes in cure cycle temperature and to be moved in the shop (thermal conductivity may also be a concern)
- Leakproof so that the vacuum and pressure cycles are consistent

Table 5.21 shows some thermal, physical, and mechanical properties for different tool materials.

TABLE 5.21   Typical Tooling Materials for Advanced Composites

Material	Density, kg/m^3	Young's modulus in tension E, GPa	Temp. limitation, °C	Coefficient of thermal expansion, m/m°C × 10^{-6}
Silicone rubber	1605		260	81–360
Aluminum alloy	2714	68.9	204	22.5
Steel	7833		Note[*]	12.5
Electroless nickel	–	–	Note[*]	12.6
Cast iron	7474	165	Note[*]	10.8
Fiberglass	1950	20	177[†]	11.7–13.1
Carbon fiber epoxy (T-300)	1577	66	177[†]	2.8–3.6
Cast ceramic	3266	10	Note[*]	0.81
Monolithic graphite	1522	13	Note[*]	2.7–3
Low-expansion nickel alloys	8137	144	Note[*]	1.4–1.7
Carbon fiber epoxy pitch 55	1720	227	177[†]	<1

[*]Above composite
[†]Limited by resin matrix

The tool face is commonly the surface imparted to the outer surface of the composite and must be smooth, particularly for aerodynamic surfaces. The other surface frequently may be of lower finish quality and is imparted by the disposable or reusable vacuum bag. This surface can be improved by the use of a supplemental metal tool known as a *caul plate*. (Press curing, resin transfer molding, injection molding, and pultrusion require a fully closed or two-sided mold.) Figure 5.13 shows the basic components of the tooling for vacuum bag or autoclave processed components. Table 5.22 shows the function of each part of the system. Tooling options have been augmented by the introduction of elastomeric tooling wherein the thermal expansion of an elastomer provides some or all of the pressure curing cure, or a rubber blanket is used as a reusable vacuum bag. The volumetric expansion of an elastomer can be used to fill a cavity between the uncured composite and an outer mold. The use of elastomeric tooling can provide the means for fabricating complex box-like structures such as integrally stiffened skins with a co-cured substructure in a single curing operation.[26]

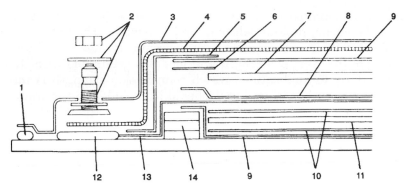

Figure 5.13 Typical vacuum bag layup components.[25]

TABLE 5.22   Functions of Vacuum Bag Components

Component of process	Functions
1. Bag sealant[*]	Temporarily bonds vacuum bag to tool
2. Vacuum fitting and hardware	Exhausts air, provides convenient connection to vacuum pump
3. Bagging film	Encloses part, allows for vacuum and pressure
4. Open-weave breather mat	Allows air or vacuum transfer to all of part
5. Polyester tape (wide)	Holds other components of bag in place
6. Polyester tape (narrow)	Holds components in place
7. Caul sheet	Imparts desired contour and surface finish to composite
8. Perforated release film	Allows flow of resin or air without adhesion
9. Nonperforated release film	Prevents adhesion of laminate resin to tool surface
10. Peel ply	Imparts a bondable surface to cured laminate
11. Laminate	
12. 1581-style glass breather manifold	Allows transfer of air or vacuum
13. 1581-style glass bleeder ply	Soaks up excess resin
14. Stacked silicone rubber edge dam	Forces excess resin to flow vertically, increasing fluid pressure

[*]Numbers refer to Fig. 5.13.

Tooling and the configuration of the reinforcement have a great influence on the curing process selected, and vice-versa. The choice between unidirectional tape and woven fabric frequently has been made on the basis of the greater strength and modulus attainable with the tape, particularly in applications which compression strength is important. There are other factors that should be included in the trade, as shown on Table 5.23.

TABLE 5.23   Fabric vs. Tape Reinforcement

Tape advantages	*Tape disadvantages*
Best modulus and strength efficiency	Poor drape on complex shapes
High fiber volume achievable	Cured composite more difficult to
Low scrap rate	machine
No discontinuities	Lower impact resistance
Automated layup possible	Multiple plies required for balance
Available in thin plies	and symmetry
Lowest cost prepreg form	Higher costs for hand layup
Less tendency to trap volatiles	
Fabric advantages	*Fabric disadvantages*
Better drape for complex shapes	Fiber discontinuities (splices)
Single ply is balanced and may be	Less strength and modulus efficient
essentially symmetric	Lower fiber volume than tape
Can be laid up without resin	Greater scrap rates
Plys stay in line better during cure	Warp and fill properties differ
Cured parts easier to machine	Fabric distortion can cause part
Better impact resistance	warpage
Many forms available	

5.7.3   Layup Technique

5.7.3.1   Automation considerations.   The layup of a composite, the actual placement of plies in their expected final position in the laminate, cannot be separated from the total manufacturing process in some procedures, e.g., filament winding or pultrusion. In other techniques, e.g., automatic tape laydown or RTM, the layup process is a separate *batch* process and is completely separate from the compaction and cure phase. All composite manufacturing processes can be automated to some extent, and the amount of automation depends on how amenable the optimum manufacturing technique is to total or partial automation, the capital costs of the automation machinery, eventual total number of piece parts to be manufactured, the expected cost of each, the time frame available, and a host of other factors.

Layup techniques, along with composite cure control, have received the greatest attention in terms of cost control. In efforts to reduce labor costs of composite fabrication, to which layup has traditionally been cast as the largest contributor, mechanically assisted, controlled tape laying and automated integrated manufacturing systems have been developed. Table 5.24 shows some of the considerations for choosing a layup technique. In addition to any cost savings by the use of automated layup technique for long production runs, there are two key quality assurance factors that favor the use of them. These are the greatly reduced chance that release paper or film could be retained, which would destroy shear and compressive strength if undetected, and the reduced probability of the addition or loss of an angle ply that would cause warping due to the laminate's lack of symmetry and balance. New ATL machines also have a laser ply mapping accessory that can verify the position of the surface ply.

The U.S. Air Force (USAF) was quick to realize that the composite materials that added so much performance were also very costly. The USAF supported, and still supports, many development programs to identify the excessive cost contributors and to find techniques to reduce them. A recent presentation showed the present accretion of the cost elements. These costs, referred to as the *composite cost chain,* have been fairly consistent over the past 30 years for aircraft and aerospace applications and are broken down as follows:[28]

Material or operation	Est. cost, $U.S.
Polyacrylonitrile (PAN) fiber precursor	$2/lb
Conversion PAN to carbon fiber	$18/lb
Woven into fabric	$22/lb
Preimpregnation	$200/lb
Assembly	$300/lb

As the largest user and the principal supporter of programs to quantify and lower composite costs, the USAF has tried to lower costs through automation and a number of other ways. One of these other ways was to lower the costs of the raw materials. They attempted to do this by ensuring that there were always multiple suppliers for carbon fibers. However, the "buy America" programs, the loss of U.S. suppliers because of the turndown in aerospace demand, and the extensive quality assurance and specification requirements have ensured that the fibers supplied to Air Force programs have always been at the high end of the fiber cost spectrum. This was in spite of the fact that some commercial fibers are sharply decreasing in cost. (It is expected that there will be a high-strength, high-tow-count carbon fiber on the market at $5.00/lb within three years.) A more successful approach that the Air Force has been taking to reduce these costs has been to stress the use of automation in the composite manufacturing

TABLE 5.24   Considerations in Composite Layup Technique

	Manual	Flat tape	Contoured tape
Orientation accuracy	Least accurate	Automatic Dependent on tape	Somewhat dependent on tape accuracy and computer program
Ply count	Dependent on operator, count mylars, and separator films	Dependent on operator	Program records
Release film retention	Up to operator	Automatic	Automatic removal
Labor costs	High	86% improvement	Slight additional improvement
Machine costs	N/A	Some costs	Approximately $2–3 million or greater
Production rate	Low (1.5 lb/h)	10 lb/h	Up to 30 lb/h quoted
Machine uptime	N/A	Not a consideration	Complex program and machine make this a consideration
Varying tape widths	Not a concern	Easily changed	Difficulty in changing
Tape lengths	Longer tapes more difficult	Longer is more economical	Longer tape is more economical
Cutting waste	Scrap on cutting	Less scrap	Least scrap due to back-and-forth laydown
Compaction pressure	No pressure	Fewer voids because of compaction at laydown	Fewer voids because of compaction at laydown
Programming	N/A	N/A	Necessary

process, and to include the life cycle costs in the ultimate cost considerations. The Air Force, along with NASA, has funded many programs to quantify and/or lower the costs of composite structures. These programs usually had a common goal of reducing composite acquisition costs by 25% minimum. They have all had some success, but the acquisition costs for high-performance military aircraft remain high and close to the estimate above.

These automation efforts have traditionally been directed toward the prepreg tape-laydown—autoclave cure processes that were neces-

sary for complex aircraft shapes. This effort has resulted in the emergence of fiber placement systems available from two vendors
(Cincinnati Machines and Ingersoll); contour tape layers from Cincinnati Machines; and flat tape layers from Cincinnati Machines, Ingersoll, and Brisard Machine (Fig. 5.14).[27] To adequately take advantage
of these expensive machines (flat tape layers have been estimated to
cost U.S.$2–4 million, and contour tape layers U.S.$3–5 million),[28] a
series of support machines are needed. These can include cutting machines for prepreg and core materials; robots and conveyor material
handling systems and automated methods of edge trimming and drilling; such as water jet or abrasive jet; and better autoclave controls.
Also, in conjunction with some of the support machines mentioned
above, there are automated methods that can assist hand layup, the
principal method being laser projection. Finally, there are faster techniques that can come into play when the starting material forms are
not predisposed. Thus, if the form of fiber to be considered can be expanded to discontinuous fibers, other automation techniques become

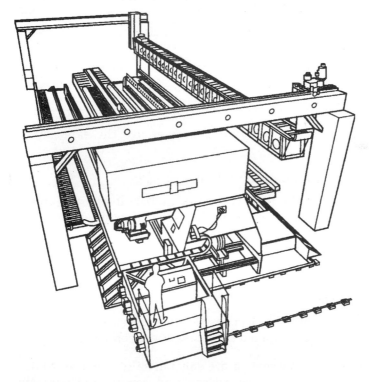

Figure 5.14 Typical contour tape layout machine layout *(from Sarl,
B., Moore, B., and Riedell, J.,* Proc. 40th International SAMPE Symposium, *May 1995, p. 384).*

apparent, as discussed later and as shown in Table 5.25. Obviously, these same machines can be used in support of the other process techniques that have inherent automation but are not widely used in the aircraft industry, such as filament winding and pultrusion.

TABLE 5.25   Types of Automation Based on Starting Fiber Form

Reinforcement	Fiber type	Proximate use	Technique for use	Application (potential)
Continuous	Glass	Preforms	Matched metal dies, VARTM, SCRIMP™	Commercial, aerospace, automotive
		Direct	Prepregs, wet layup (fabrics), pultrusion, filament winding	Aerospace, commercial, (tanks and pipes), automotive
	High-performance fiber	Preforms	SCRIMP™, autoclave	Aerospace (Navy ship structure), commercial
		Direct	Prepregs, filament winding, pultrusion	Aerospace, commercial, sporting goods
Discontinuous	Glass	Preforms	P-4 process	Automotive, commercial
		Direct	Glass fabrics, dry and prepreg sprayup	Commercial, automotive
	High-performance fiber	Preforms	P-4 process	Aerospace
		Direct	Metal matrix composites, thermoplastics	Aerospace, commercial, (automotive)

The goals in the curing-compacting process are the same for all techniques, namely,

- Composite with no or minimal voids and porosity

    This is to assure the preservation of matrix-dominated properties and matrix-influenced properties such as compression.

- Full cure of the composite

    This results in the most environmentally stable composite as well as maximizing matrix-dominated properties.

- Reproducible, consistent fiber volumes

    This assures consistent composite mechanical response along with constant or anticipated thickness of the composite.

When these goals are not met, or are only partially met, the result is production slowdowns, excessive rework and repair, and extensive nondestructive testing, in addition to the production of expensive scrap composite. Obviously, there have been many studies to determine the critical parameters in the process and for the autoclave curing of thin, composite aircraft components. There are now recommendations dealing with prepreg properties such as tack and drape, layup environment (moisture), laminate thickness, ply orientation and drop-offs, and bleeder amount and placement.[29]

The cure/consolidation techniques are somewhat different for thermosets and thermoplastics. The process with thermosets is generally to squeeze the excess resin out of the composite. An exception is resin transfer molding (RTM) processes, where the procedure is to infiltrate the proper resin content into the dry fiber preform. Thermoplastics are completely "cured" in the prepreg, and the process to form them to the final shape involves melting or softening and forming to the final desired shape. Because of the much higher melt viscosity, thermoplastic matrices must have close to the final composite resin content within the layup, with very little resin bleed. There are a number of unique routes to the formation of a thermoplastic composite; these are shown in Table 5.26.[30] All curing techniques use heat and pressure to cause the matrix to flow and wet out all the fibers before the matrix solidifies or cools.

Generally, for thermosets, the percent matrix weight is higher before cure initiation; the matrix flows out of the laminate and takes the excess resin with the potential voids. The matrix exhibits a low-viscosity phase during the cure of a thermoset and then advances in viscosity rapidly after a gel period at an intermediate temperature (Fig. 5.15). Because of the many changes going on in the matrix, determining the actual point at which cure occurs is very difficult, and cure techniques (by almost all methods) have evolved into a stepped cure with gradual application of heat to avoid formation of voids. An arbitrary 1% void limit has been adopted for most autoclaved composites; filament-wound and pultruded composites may have higher void volumes, which may be acceptable in several product applications (i.e., 3–4% is acceptable in many filament-wound pressure vessel applications) depending on the situation.

5.7.3.2 Curing.   Each technique for compaction and cure is unique in terms of types of structures for which it can be used, and also in terms of the forms of starting materials. For instance, a unidirectional prepreg is optimum for automated laydown methods with autoclave cure but, because of potential fiber washout by low viscosity resin movement, which can move fibers, and for better handling, a prepreg fabric may be more appropriate for a press-cured laminate. The advantages and disadvantages of fabric vs. unidirectional tape fiber forms for composite manufacture were shown earlier in Table 5.23.

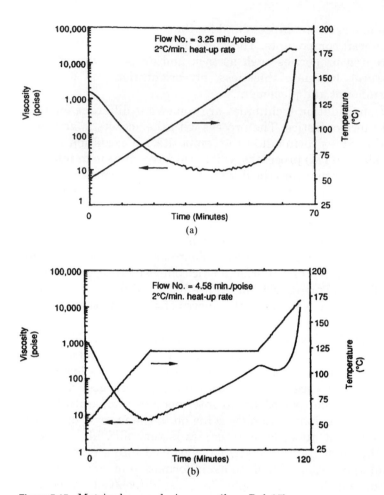

Figure 5.15   Matrix changes during cure *(from Ref. 27)*.

An autoclave is essentially a closed, pressurized oven; the most common epoxy laminates are cured at an upper temperature of 177°C (350°F) and 6 MPa (100 psig). Figure 5.16 shows a typical cure schedule. Autoclaves are still the primary tool in advanced composite curing and have been built up to 16m (55 ft) long at 6.1 m (20 ft) diameter. Since autoclaves are so expensive to build and operate, other methods of curing and compacting composites have been promoted for large aircraft components.

Each resin fiber combination has one or more optimum cure techniques, depending on the proposed environment. The prepreg manufacturer supplies a time-temperature cycle that may have been used to manufacture the test specimens employed in generating preliminary

TABLE 5.26   Manufacturing Routes for Thermoplastic Composites Based on Processing Methods

Manufacturing route	Outline of fabrication and processing methods
Open-mold processes	
1. Autoclave	Unidirectional or woven fibers pre-impregnated by the resin (prepreg) are used. Other forms of prepreg have reinforcing fibers in combination with the resin as fibers or as powder. The prepreg layers are stacked on the mold surface and covered with a flexible bag. Consolidation is obtained by external pressure applied in an autoclave at elevated temperature.
2. Filament winding	Prepreg tape or tape with the resin as fibers or powder are wound onto a mandrel at pre-determined angles. Heat and pressure are applied to the tape in order to continuously weld it onto the underlying material.
3. Folding	Preconsolidated sheets are heated. Simple fixtures are then used to shape the sheets into the desired geometry.
Closed-mold processes	
4. Injection molding (short fibers, 0.1–10 mm)	A mixture of molten thermoplastic and short fibers is injected into a colder metal mold at very high pressure. The component is allowed to solidify and is automatically ejected.
5. Compression molding (short fibers, 5–50 mm)	Semi-finished sheets of glass mat thermoplastics are heated and placed in the lower part of the mold in a fast press. The press is quickly closed and pressure is applied so that the material can flow to fill the mold. Technology is also available where the hot molding compound reaches the mold from an extruder.
6. Compression molding (continuous fibers)	The same principle as for short fiber materials. Continuous fibers require special clamping fixtures for the sheets and can primarily be used for simple geometries.
7. Diaphragm forming	A stack of prepreg is placed in between two diaphragms (superplastic aluminium or polymer film). The diaphragms are fixed whereas the prepreg can move freely. The material is slowly deformed by external pressure and the mold.
8. Pultrusion	Prepreg tape or tape with the resin as fibers or powder is pulled through a heated die to form beams or similar continuous structures with constant cross-section geometry. The material is allowed to cool and solidify.
9. Resin injection	Dry reinforcing fibers are placed in the mold. Monomers and/or low-molecular-weight polymer with low viscosity are injected, the reinforcement is impregnated. Polymerization to a high molecular weight thermoplastic occurs by mixing of reactive components and/or thermal activation.

SOURCE: Bergland, Lars, in S.T. Peters, ed., *Handbook of Composites,* 2nd ed., p. 116, Chapman & Hall, London, 1998.

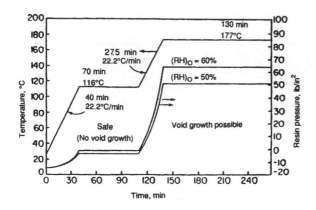

Figure 5.16 Time-temperature cure cycle for autoclave cure to prevent moisture-induced void growth.[28]

mechanical and physical properties. Frequently, it has been necessary to modify the preliminary cure cycle because of part configuration (thickness) or because of production economics.[31] There are several developments that help to inject science into the otherwise hit-or-miss, time-consuming procedure of cure cycle optimization. They are

- Development of dielectric sensors and signal processing to gain a knowledge of the resin condition (viscosity) at any point during the cure process

- Use of thermocouple data, transducer outputs, and dielectric information to interactively control the curing process within an autoclave

- Development of computer programs supported by test results to predict the physical changes that will occur during cure

- Analytical techniques, such as rheometric dynamic scanning and differential scanning calorimetry, that give information on changes in viscosity and the amount of heat absorbed or liberated during cure

5.7.4   Resin Transfer Molding

Previous discussions have centered on moving resin out of the laminate to reduce voids. Resin transfer involves the placement of dry fiber reinforcement into a closed mold and then injecting a catalyzed resin into the mold to encapsulate the reinforcement and form a composite. The impetus for the use of this process comes from the large cost reductions that can be realized in raw materials and layup. The process can utilize low injection pressures, i.e., 55 MPa (80 psig); therefore, the tooling can be lower-cost plastic rather than metal. The process is most appropriately used for non-aerospace composites but has been extended to many advanced applications. RTM manufacturing considerations are shown in Table 5.27. The advantages of RTM include the possibility of producing very large (Fig. 5.17) and complex shapes efficiently and inexpensively, and reducing production times with the ability to include inserts in the composite. Effective for large structures such as boats, the SCRIMP™ process uses a vacuum bag on one side of the laminate instead of the two plates of a typical mold. Advantages quoted for this technique are one sided mold, better control of and higher fiber volume, and lower porosity in the composite. It has been used for structures that need higher quality than can be obtained by other RTM processes. Table 5.28 shows the range of applications for the RTM technique. RTM is also a way of preparing a composite structure from a knitted preform. Knitting and braiding and sewn tridimensionally reinforced preforms offer complex shapes that are not attainable by other techniques. The techniques can possibly lower costs due to reduction of labor. The product may also gain increased impact resistance due to the multiple, interlocked directions of fiber.

TABLE 5.27   RTM Manufacturing Considerations

Materials	Tooling	Technology
Fiber type	Mold material	Resin viscosity
Preform complexity	Mold surface finish	Flow modeling
Preform cost	Resin pump type	Gating and vent design
Inserts	Integral or oven heating	Composite strength, stiffness,
	Clamping method	fiber volume, transverse properties
	Tool durability	ties
		Vacuum assist or not

5.7.5   Fiber Placement

Fiber placement was invented by Hercules Aerospace Co. Now, machines are marketed by Cincinnati Machines and Ingersoll in the U.S.A. The process is a cross between filament winding and automatic tape laydown, retaining many of the advantages of both processes. Fiber placement, the natural outgrowth of adding multiple axes of control to filament winding machines, results in control of the fiber laydown so that nonaxisymmetric surfaces can be wound. This involves the addition of a modified tape laydown head to the filament winding machine and other sub-machine additions that include in-process compaction, individual tow cut/start capabilities, a resin tack control system, differential tow payout, low tension on fiber, and enhanced off-line programming. Several manufacturers

Figure 5.17 RTM (SCRIMP™) process for injecting a boat hull. *(Copyright Billy Black, courtesy of Seemann Composites)*

TABLE 5.28   RTM Composite End Uses

Composite use	Part
Industrial	Solar collectors Electrostatic precipitator plates Fan blades Business machine cabinetry Water tanks
Recreational	Canoe paddles Large yachts Television antennas Snowmobile bodies
Construction	Seating Baths and showers Roofing
Aerospace	Airplane wing ribs Cockpit hatch covers Speed brakes Escape doors
Automotive	Crash members Leaf springs Car bodies Bus shelters

now use slit prepreg tape rather than tow. The machines are capable of winding the shapes shown in Fig. 5.18 and can change fiber paths such as shown on Fig. 5.19. Table 5.29[32] shows the advantages quoted for the technique. The present disadvantages are the cost of the machines (very high when compared to some filament winding machines), the dependence on computers and electronics rather than mechanical means of directing fiber laydown, and the cost and complexity of mandrels.

5.7.6 Filament Winding

Filament winding is a process by which continuous reinforcements in the form of rovings or tows (gathered strands of fiber) are wound over a rotating mandrel. The mandrel can be cylindrical, round, or any other shape as long as it does not have re-entrant curvature. Special machines (Fig. 5.20) traversing a wind eye at speeds synchronized with the mandrel rotation, control winding angle of the reinforcement and the fiber lay-down rate. The reinforcement may be wrapped in adjacent bands or in repeating bands that are stepped the width of the band and that eventually cover the mandrel surface. Local reinforcement can be added to the structure using circumferential windings or

Figure 5.18 Versatility of shapes fabricated by fiber placement. *(Courtesy of Hercules Aerospace Co., now Alliant Techsystems.)*

Radius:

Fiber steering with differential tow payout

Figure 5.19 Versatility of shapes fabricated by fiber placement. *(Courtesy of Hercules Aerospace Co., now Alliant Techsystems.)*

TABLE 5.29   Fiber Placement Processing Advantages (adapted from Ref. 32)

Flexibility	Compaction	Material usage
Full range of fiber orientations Non-geodesic path generation Constant ply thickness over complex shapes Localized reinforcements Continuous fibers over three-dimensional shapes (without joints) Large structures	Continuous in-process debulking Complex surface fabrications: ■ Concave/convex ■ Nonaxisymmetric and axisymmetric shapes	Wide use of advanced thermoset material systems Near prepreg tape equivalent

local helical bands, or by the use of woven or unidirectional cloth. The wrap angle can be varied from low-angle helical to high-angle circumferential or *hoop,* which allows winding from about 4–90° relative to the mandrel axis for older mechanical machines; newer machines can *place* fiber at 0°.

There are advantages and disadvantages to filament winding as compared to other methods. The most obvious advantages, summarized in Table 5.30, are cost savings (both capital and recurring labor) and the ability to build a structure that is larger than autoclave capac-

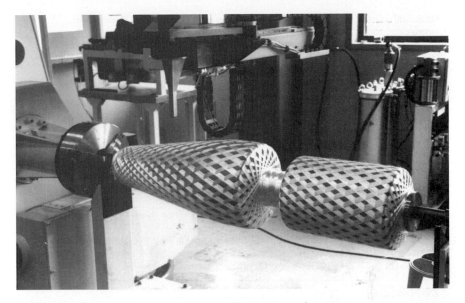

Figure 5.20 Filament winding. *(Courtesy of Plastrex)*

ity. The disadvantages of filament winding, in most cases, can be worked around by innovative engineering and manufacture. Fabricators of large rocket motors have used plaster mandrels that can be stripped, reduced in size, and passed out through the relatively small port. Reverse curvature can be formed into a positive curvature by the addition of oriented fibers or mats or, if the curvature is necessary to the design, such as on an airfoil, it can be accomplished by removing the uncured structure from the mandrel and using alternate means of compaction to form the composite. This is the fabrication of a continuous multisided preform.

Newer filament winding machines have the capacity to change the wind angle at any point over the part surface. This gives the option of actually winding the fiber into a reverse curvature by selecting the wind angle that will follow a hyperboliodal path into a smooth recess, without bridging or slipping. This technique has been used to wind in an *in-situ* metallic end ring for a composite-to-metal joint, eliminating the need for further bolting or pinning and providing a measure of failsafe operation The fiber path can be altered by pins or sawtooth to avoid slipping or bridging. Mandrels are less expensive than two-sided molds, and their reusability contributes to the cost-effectiveness of filament winding. The expansion of the mandrel during the cure cycle provides the compaction necessary to result in a fully compacted, dense laminate, without external compacting measures such as autoclave. Filament wound laminates, without special curing conditions,

TABLE 5.30   Comparison of Filament Winding with Other Fiber Deposition, Compacting, and Curing Processes

Advantages

Filament winding is highly repetitive and accurate in fiber placement (from part to part and from layer to layer).

It can use continuous fibers over the whole component area (without joints); can orient fibers easily in load direction. This simplifies the fabrication of aircraft fuselages and reduces the joints.

It avoids capital expense and the recurring expense for inert gas of autoclave.

Large and thick-walled structures can be built—larger than any autoclave.

Mandrel costs can be lower than other tooling costs. There is only one tool, the male mandrel, which sets the inside diameter and the inner surface finish. The outer surface is uncontrolled and may be rough.

The cost is lower for large numbers of components, since there is less labor than in many other processes.

Material costs are relatively low, since fiber and resin can be used in their lowest-cost form rather than as prepreg, and no preforming is necessary. (Preforming is necessary for RTM and may be a significant recurring expense.)

Disadvantages

The shape of the component must permit mandrel removal. Long, tubular shapes will generally have a taper. Different mandrel materials, because of differing thermal expansion and differing laminate layup percentages of hoops versus helical plies, will demonstrate varying amounts of difficulty in removal of the part from the mandrel.

One generally cannot wind reverse curvature. To wind a reverse curvature, wind the exact shape on a positive dummy mandrel insert and then remove the insert and place the fiber.

One cannot change fiber path easily (in one lamina). It can be done via the use of pins or slip of the tow. Fiber placement is the only fabrication method capable of "steering" the fiber.

The process requires a mandrel, which sometimes can be complex or expensive. Usually, the mandrel is less costly than the dies or molds for forming methods other than pultrusion or RTM.

Generally poor external surfaces are produced, which may hamper aerodynamics. A better outside surface can be obtained by
- Use of outer clamshell molds
- External hoop plies or thinner tows on last ply
- Shrink tape or porous TFE-glass tape overwrap

can have void contents on the order of 3 to 7%. Because there is no external mold, there is a poor external or bumpy surface that can be improved by mild compaction as exerted by shrink tape. The poor external surface can be smoothed somewhat by proper selection of resin and fiber, use of surfacing mat or filled smoothing compounds at some weight penalty, or by compaction and cure in a female die mold using vacuum bag or autoclave pressure. Figure 5.21 shows the Beech Starship carbon/graphite epoxy fuselage that was filament wound then expanded into tooling during cure to form a smooth outer skin.

Figure 5.21 Filament wound aircraft fuselage with smooth skin. (*Courtesy of Fibertec Div. of Alcoa*)

Thermoset resins generally have been used as the binders for the reinforcements. These resins can be applied to the dry roving at the time of winding (wet winding) or applied previously and gelled to a "B" stage as prepreg. The fiber can be impregnated and rerolled without B staging and used promptly or refrigerated. Prepreg and wet rerolled materials are useful because of the opportunity to perform quality control checks early.

The cure of the filament wound composite is generally conducted at elevated temperatures without the addition of any process for composite compaction. The filament winding process, like pultrusion, can employ wet resin systems to result in potentially lower cost composite structures.

Handling guidelines that are unique to wet filament winding for a wet resin system are:

- Viscosity should be 2 PaS or lower.
- Pot life should be as long as possible (preferably >6 hr).
- Toxicity should be low.

The other approach is to use a wet-rerolled system, essentially a wet resin that is applied to the fiber beforehand and then kept in a freezer until use.

Fibers are used in tow (carbon/graphite) or (roving) glass and Kevlar. The two terms define a gathered parallel bunch of fibers with essentially no twist. The fibers that are made in individual processes (e.g., boron) have not been used extensively in filament winding applications due to their stiffness.

5.7.7   Pultrusion Braiding and Weaving

5.7.7.1 Pultrusion. Pultrusion is an automated process for the manufacture of constant-volume/shape profiles from composite materials. The composite reinforcement is continuously pulled through a heated die and shaped (Fig. 5.22) and cured simultaneously. If the cross-sectional shape is conducive to the process, it is the fastest and most economical method of composite production. Straight and cured configurations can be fabricated with square, round, hat-shaped, angled *I*, or *T*-shaped cross sections from vinylester, polyester, or epoxy matrices with E and S-glass, Kevlar and carbon/graphite reinforcements. Some of the available cross-sectional shapes and limitations are shown in Table 5.31.[33] The curing is effected by combinations of dielectric preheating and microwave or induction (with conductive reinforcements like carbon graphite) while the shape traverses the die. The resin systems, predominantly polyester, can be wet or prepreg, but the cure

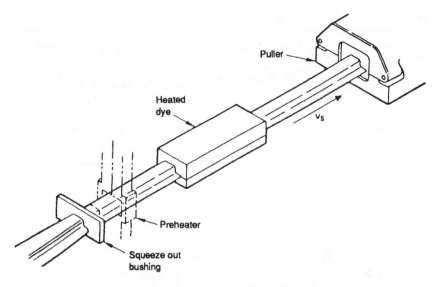

Puller

Heated
dye

v_s

Preheater

Squeeze out
bushing

Figure 5.22 Schematic of pultrusion elements. *(From Ref. 34)*

rates will be much more rapid than for processes that use thermal conduction for heat transfer into the laminate. The process lends itself to long component lengths with the need for reinforcement in the 0° direction. Many uses have been found for the process; pultrusion is the primary fabrication technique for reinforced plastic booms, ladder components, light poles, boundary stakes for snow roads, and conduits. The primary reasons for the use of the technique are design considerations driven by commercial uses, such as cost, weight, electrical properties, and environmental resistance, but the process has also been used to produce some components from advanced composite materials. An application that uses the cost-effective technique for high-volume production is the pultruded carbon/graphite-reinforced drive shaft that replaces two metal shafts, universal joints, and hangers and results in a quieter product because of inherent composite damping (Fig. 5.23). The process has some limitations:

- The part must be of constant cross section over its length; it cannot be tapered.
- Transverse strength will be somewhat lower than for other manufacturing methods.
- Curved shapes require special machines.
- Thick sections are difficult because of exotherm with rapidly curing resins.
- Cure shrinkage may cause dimensional and mechanical problems.

TABLE 5.31   Pultrusion Design Guidelines

Minimum inside radius		Roving, 0.79 mm mat, 1.6 mm	Corrugated sections		Yes, longitudinal
Molded-in holes		No	Metal inserts		No
Trimmed in mold		Yes	Bosses		No
Core pull and slides		Yes	Ribs		Yes, longitudinal
Undercuts		Yes	Molded-in labels		Yes, but not recessed
Minimum recommended draft		No limitation	Raised numbers		No
Minimum practical thickness		Roving, 1.0 mm Mat, 1.5 mm	Finished surfaces (reproduces mold surface)		Two
Maximum practical thickness		Roving, 75 mm Mat, 25 mm	Hollow sections		Yes, longitudinal
Normal thickness variation		See ASTM D 3917[22]	Wire inserts		Yes, longitudinal
Maximum thickness buildup		As required	Embossed surface		No

SOURCE: From J. D. Martin and J. E. Sumerak in Reinhart,[33] p. 542.

- There generally will be a cut edge, and this must be cut carefully to reduce delamination and then sealed to prevent moisture or other attack to fiber or interface.
- Joints are more difficult because of fiber orientation and lack of section changes that would allow geometric locking.

Pultrusion has also been combined with filament winding to achieve high transverse properties with advanced composite starting materials as shown in Fig. 5.24.

5.7.7.2 Braiding, weaving, and other preform techniques. Braiding, weaving, knitting, and stitching (with low- or high-modulus fibers)

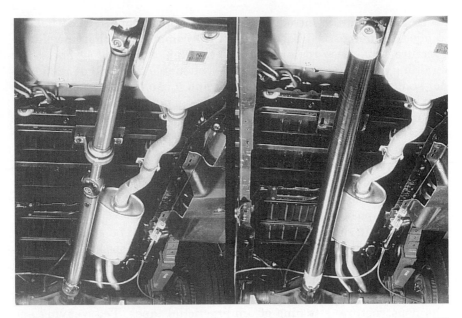

Figure 5.23 Carbon/graphite reinforced pultruded driveshaft (right) for GMT-400 trucks. *(Courtesy of Spicer Div. of Dana Corp.)*

Figure 5.24 Filament-wound/protruded graphite epoxy bridge beam. *(Courtesy of Sci. Div. of Harsco.)*

represent methods of forming a shape, generally referred to as a *preform* (a complete fiber layup representing the total laminate thickness with a small amount of resin or transverse fiber to hold it in place before resin infusion). The shape may be the final product or some intermediate form such as a woven fabric. New techniques allow prepregs to be used, and the introduction of three-dimensional braids has extended braiding to airborne structural components that meet high fracture toughness requirements with high damage tolerance. The braiding process is continuous and is amenable to round or rectangular shapes or smooth curved surfaces, and it can transition easily from one shape to another. Resin systems are generally epoxies or polyester, and the fiber options are similar to those for filament winding; the single stiff fibers such as boron and ceramics cannot endure the tight bend radii.

In the process, the mandrel is fed through the center of the machine at a controlled rate, and fibers from a large number of rolls are deposited at multiple angles, usually 45° and 90° (Fig. 5.25). 0° plies can be laid from tape or placed by a triaxial braider.

Braiding, in some applications, has turned out to be almost half the cost of filament winding because of labor savings in assembly and simplification of design.[34] The fiber volume of a braided composite will generally be lower than for other methods.

The other fabric preforming techniques are weaving, knitting, and the nonstructural stitching of unidirectional tapes. The weaving and knitting are compared to braiding in Table 5.32.[34] Stitching simply

TABLE 5.32   Comparison of Fabric Formation Techniques (from Ref. 34)

	Braiding	Weaving	Knitting
Basic direction of yarn introduction	One (machine direction)	Two (0°/90°) (warp and fill)	One (0° or 90°) (warp or fill)
Basic formation technique	Intertwining (position displacement)	Interlacing (by selective insertion of 90° yarns into 0° yarn system)	Interlooping (by drawing loops of yarns over previous loops)

Figure 5.25   Continuous braiding machine. *(Courtesy of B.C.M.E. Ltd.)*

uses a nonstructural thread such as nylon or Dacron or a stiff fiber such as Kevlar to hold dry tapes or fibers at selected fiber angles. Preforming in this manner results in a higher-cost raw material but saves labor costs. The stitched preform has known, stable fiber orientations similar to woven fabric, without the crossovers.

5.7.7.3 Short fiber composites.

Short fiber composites have only recently arrived in the advanced composite field. There is an increased interest in short fiber composites because of the advantages in laydown speed inherent in SCRIMP™ and other RTM resin infusion techniques. The P-4 process *(programmable powdered preforming process)* has great promise for automotive uses with the use of a chopper gun to lay down the fiber preform of a thermoplastic composite. Short fiber composites have been dismissed for advanced composites because of the weight, stiffness, and strength penalties inherent with the use of short fibers.

The speed of fabricating a composite short fiber preform is substantially increased when the P-4 technology is applied. This technology was jointly developed by Owens Corning (Battice, Belgium) and Applicator System AB (Mölnlycke, Sweden) and is now under further development at the National Composites Center (Kettering, OH). The process uses chopped fiberglass with a thermoplastic matrix for applications such as a structural automotive components at high production rates of up to 50,000 units per year for a pickup truck box. This high production rate has been the driving force in the process development. A parallel interest is for the quick and lower-cost manufacture of structural advanced composites with discontinuous carbon/graphite fibers. Besides the control of physical and mechanical properties exerted by fiber strength, stiffness, direction, and position in the laminate, a new consideration has to be added for advanced structural applications: fiber length. There are equations for predicting the properties the result from varying the fiber lengths,[34] and studies have determined some relationships empirically. Figure 5.26[35] shows the relation between fiber length and ultimate tensile strength for a discontinuous, unidirectional, glass fiber composite. Note that the composite has a low fiber volume. Another penalty is evident when the fibers are not aligned. Thus, one of the main development tasks for aerospace applications is the need to align the fibers.

5.7.8 Machining Techniques

The machining of composite materials also poses some challenges. The unique machining requirements for composites starts with the need for harder rather than high-speed drills. PCD-coated (polycrystalline

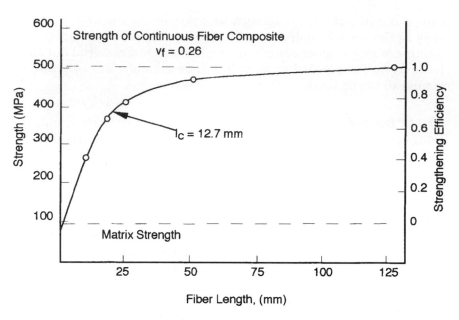

Figure 5.26 Relationship between short fiber length and composite strength. *(After Hancock, P., and Cuthbertson, R.C., J. Mat. Sci., 5, 76–768, 1970)*

diamond), carbide, or some other hard coating is necessary because of tool wear and the high temperatures generated by dull drills in the relatively low thermal conductivity of composite materials. Of practical interest, it is recommended that high-speed steel not be used. Drilling polymeric composites can result in delamination from too high or low feed-to-speed ratios (Figs. 5.27[36] and 5.28[37]), from heat buildup (composites generally have lower thermal conductivities than metallic materials), and because of inadequate backup at the exit point. The hole quality can also suffer, resulting in fiber pullout and fuzzing, matrix cratering, thermal alterations, and delaminations. The recommendations are even more relevant in the case of aramid composites. These composites require a different drill technique, such as the *spade* drill, also hard coated, because the fiber is so hard to shear—it tends to recede back into the composite matrix. In general, aramid composites can be expected to generate copious quantities of fluff under most machining inputs—this is natural and in most cases does not represent any significant degradation of the remaining composite. A summary of some relevant parameters for drilling of several composite materials with different drilling conditions is shown in Table 5.33.[37]

The precautions described for drilling merit repeating for machining with the additional proviso that, when machining unidirectional com-

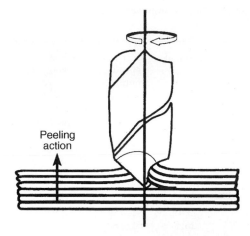

Figure 5.27 Typical composite drilling problem: peelup delamination at entrance. *(Kohkonen, K.E., and Potdar, N., in S.T. Peters, ed.,* Handbook of Composites, *2nd ed., Chapman & Hall, p. 598, London, 1998)*

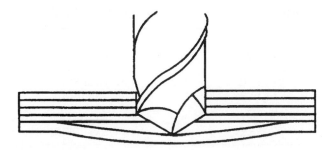

Figure 5.28 Typical composite drilling problem: pushout delamination at exit. *(Abrate, S., in P.K. Mallick, ed.,* Composites Engineering Handbook, *Marcel Dekker, New York, 1997, p. 783)*

posites, the operator must attend to the fiber direction to avoid excessive fuzzing and potential splitting of the laminate. Table 5.34[37] shows some machining parameters for two typical composites. Grinding is generally recommended as an approach to avoid many of the foregoing problems. Grinding does generate fine airborne dust that poses an environmental hazard and should be controlled or contained.

There are many other approaches to composite machining that involve some significant capital outlays and should be justified by a reasonably long production run. These include water jet or abrasive jet, and laser cutting.

TABLE 5.33 Typical Machining Parameters for Drilling Polymeric Composite Materials[37]

Workpiece material	Tool material	Hole dia., mm	Material thickness, mm	Cutting speed, m/min	Feed rate, mm/rev
Unidirectional	Carbide	4.85–7.92	0–12.7	42.7	0.0254–0.0508
graphite-epoxy	PCD	4.85–7.92	0–12.7	61.0	0.0508–0.0889
Multidirectional	PCD	4.85–7.924	0–12.7	61.0	0.0254–0.0508
graphite-epoxy	Carbide	85–7.92	0–12.7	68.6	0.0508–0.0889
Glass-epoxy	HSS	3	10	33.0	0.05
Boron-epoxy	PCD	6.35	2.0	91–182	25.4*
		6.35	25.4	91–182	
Boron-epoxy	PCD	6.35	10.4	79	41.91*
Kevlar-epoxy	Carbide	5.6	—	158	0.05

*Units = mm/min

TABLE 5.34 Typical Machining Parameters for Sawing Polymeric Composites[37]

Material	Thickness, mm	Cutting speed, m/s	Feed rate, mm/min	Type of saw
Graphite-epoxy	4	2–12	50.8	Circular (HSS)
	0–25.4	15.24	50.8	Band saw (PCD)
Boron-epoxy	2.0	30.48	254	Circular* (PCD)
	25.4	15.24	50.8	Band saw* (PCD)

*With coolant.

5.8 Analysis

5.8.1 Overview of Mechanics of Composite Materials

The 1, 2, and 3 axes in Fig. 5.29 are special and are called the *ply axes,* or *material axes.* The 1 axis is in the direction of the fibers and is called the *longitudinal axis* or the *fiber axis.* The longitudinal axis typically has the highest stiffness and strength of any direction. Any direction perpendicular to the fibers (in the 2, 3 plane) is called a *transverse direction.* Sometimes, to simplify analysis and test requirements, ply properties are assumed to be the same in any transverse direction. This is the transverse isotropy assumption; it is approximately satisfied for most unidirectional composite plies.

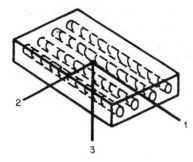

Figure 5.29 The unidirectional ply.

These properties are typically modified by transformation relative to the laminate axis, where these may not be the same as the ply axes. These calculations involve matrix manipulations. For the simplest ply, stresses along the ply axes, the strains produced by the stresses are

$$
\begin{bmatrix} \varepsilon_1 \\ \varepsilon_2 \\ \gamma_{12} \end{bmatrix} = \begin{bmatrix} \dfrac{1}{E} & -\dfrac{\upsilon_{21}}{E_2} & 0 \\ -\dfrac{\gamma_{12}}{E_1} & \dfrac{1}{E_2} & 0 \\ 0 & 0 & \dfrac{1}{G_{12}} \end{bmatrix} \begin{bmatrix} \sigma_1 \\ \sigma_2 \\ \tau_{12} \end{bmatrix}
\tag{5.9}
$$

where ε = strain
υ = Poisson's ratio
E = Young's modulus
G = shear modulus
σ = tensile stress
τ = shear stress
γ = shear strain

In a multidirectional laminate, there can be as many as 21 stiffness constants. Strength predictions are equally complicated (a) because of directional differences, i.e., compression is not always equal to tension, and (b) because of the complexity of the several failure theories. As the complexity of the matrix calculations increase, it becomes evident that errorless mathematical manipulations are impossible without the aid of computers. Fortunately, there are several software packages that can accomplish these manipulations. The output of the computer, however, is only as good as the input information. Users must view the data in handbooks and computer material property files as "preliminary" and should verify the necessary constants and failure properties of the composite materials and processes by subscale and full-scale tests. Since there are no standard, accepted "design allowables" for composite materials, most organizations that extensively use the materials have had to develop their own data to supply "A" or "B" basis allowables. Because of the possible numbers of permutations of resins, fibers, and curing techniques, it will be some time before standardized strength and modulii values can be published as is done now for most metallic structural materials.

5.9   Design of Composite Structures

5.9.1   Composite Laminate Design

The design process for composites involves both laminate design and component design and must also include considerations of manufacturing process and eventual environmental exposure. The following are some steps that can simplify the process.

1. Take advantage of the orthotropic nature of the fiber composite ply.

   - To carry in-plane tensile or compressive loads, align the fibers in the directions of these loads.

   - For in-plane shear loads, align most fibers at ±45° to these shear loads.

   - For combined normal and shear in-plane loading, provide multiple or intermediate ply angles for a combined load capability.

2. Intersperse the ply orientations.

   - If a design requires a laminate with 16 plies at ±45°, 16 plies at 0°, and 16 plies at 90°, use the interspersed design $[90_2/\pm45_2/0_2]_{4S}$ rather than $[90_8/\pm45_8/0_8]_S$. Concentrating plies at nearly the same angle (0° and 90° in the above example) provides the opportunity for large matrix cracks to form. These produce lower laminate allowables, probably because large cracks are more injurious to the fibers and more readily form delaminations than the finer cracks occurring in interspersed laminates.

   - If a design requires all 0° plies, some 90° plies (and perhaps some off-angle plies) should be interspersed in the laminate to provide some biaxial strength and stability and to accommodate unplanned loads. This improves handling characteristics and also serves to prevent large matrix cracks from forming.

   - Locally reinforce with fabric or mat in areas of concentrated loading. (This technique is used to locally reinforce pressure vessel domes.)

- Use fabric, particularly fiberglass or Kevlar, as a surface ply to restrict surface (handling) damage.

- Ensure that the laminate has sufficient fiber orientations to avoid dependence on the matrix for stability. A minimum coverage of 6 to 10% of total thickness in 0, ±45, 90 directions is recommended.

3. Select the layup to avoid a mismatch of laminate properties with those of the adjoining structures, or provide a shear/separator ply.

- Poisson's ratio—if the transverse strain of a laminate greatly differs from that of adjoining structure, large interlaminar stresses are produced under load. Large Poisson's ratios can be achieved with some carbon/graphite-reinforced laminates. This effect is graphically portrayed in Fig. 5.30. The solution to the problem is to analyze the effects of induced loads on the metal/composite couple and to consider using 90° plies to reduce the Poisson's ratio.

- Coefficient of thermal expansion—temperature change can produce large interlaminar stresses if the coefficient of thermal expansion of the laminate differs greatly from that of adjoining structure. If the adjacent structure is metallic, try to adjust the

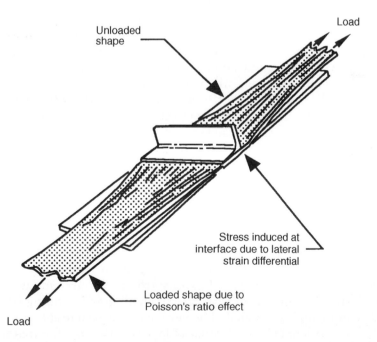

Figure 5.30 Poisson's ratio problem.

laminate CTE to more closely approximate the CTE of the metal. If 90° plies are not effective, hybridizing may solve the problem. Fiberglass woven fabric has been successfully used to reduce the CTE of a carbon/graphite-epoxy joint to titanium. If the composite and the metal's CTE cannot be reasonably matched, allow for the differences in the joint. Note that the composite will generally have the lower CTE.

- Composite curing stresses, called *residual stresses* are significant and, in the case of high-modulus composites, can cause part failure on cool-down from cure. These stresses will also cause Poisson's ratio strains.

- The ply layer adjacent to most bonded joints should not be perpendicular to the direction of loading. Reduce composite section thickness in the joint area, soften the composite by adding fiberglass or angle plies, and select the highest-strain-capability adhesive.

- Hygrothermal stresses, induced in the laminate when it absorbs moisture, should also be considered. These are generally factored in by most computer analysis programs.

4. Use multiple ply angles.

- Typical composite laminates are constructed from multiple unidirectional or fabric layers that are positioned at angular orientations in a specified stacking sequence. From many choices, experience suggests a rather narrow range of practical construction from which the final laminate configuration is usually selected. The multiple layers are usually oriented in at least two different angles, and possibly three or four (±θ, 0°/±θ, or 0°/±θ/90° cover most applications, with θ between 30 and 60°). Unidirectional laminates are rarely used except when the basic composite material is only mildly orthotropic (e.g., certain metal matrix applications) or when the load path is absolutely known or carefully oriented parallel to the reinforcement (e.g., stiffener caps or cryogenic vessel support straps).

- An important observation concerning practical laminate construction is the concept of *midplane symmetry* with respect to stacking sequence, i.e., uncoupled response, and *balanced construction* with regard to the in-plane angular orientations, i.e., orthotropic behavior as opposed to anisotropic. If a laminate does not exhibit midplane symmetry, the stretching and bending behavior can be highly coupled. The severity of this coupling is inversely proportional to the number of layers within the laminate. The fewer the layers, the worse the problem. This heterogeneous nature should be avoided. In general, seek a symmetric laminate to satisfy the design. When symmetric and balanced laminates are used, bending, twisting, and warping effects are reduced.

- When unsymmetrical laminates are used, such as for the faces of a sandwich structure or the sides of a cylinder, the entire laminate is generally made symmetric because of geometry. If an unsymmetric laminate must be used alone, it is very prudent practice to avoid those of less than eight layers. Problems inherent with unsymmetric laminates may be more severe with higher-modulus fibers

5. Use midplane symmetry.

- The usual reference surface for determining if a laminate is symmetrical is the geometric midplane. A midplane symmetric laminate is highly desirable and to be preferred. Extremely few cases exist where an unsymmetrical stacking sequence should even be considered. To construct a midplane symmetric laminate, for each layer above the midplane there must exist an identical layer the same distance below the midplane, with a repetition of thickness, material properties, and angular orientation. Thus, the lamination stacking sequence will possess a mirror image about the geometric midsurface. As Figs. 5.11e and 5.11f show, both even- and odd-number-ply laminates can satisfy these criteria. Note that the use of an identical woven fabric at identical distances from the midplane may not confer symmetry. One of the layers will have to "flipped," because the fabric may not identical on both surfaces.

6. Observe the effects of stacking sequence.

- Once the orientations of a laminate are specified, the stacking sequence will control the flexural rigidities of a laminated plate. Thus, the stability, vibration, and static bending behavior are controlled by the relative dispersion and thickness distribution of these oriented layers.

- Interlaminar stresses (normal and shear) are influenced by the laminate stacking sequence. For example, a $(\pm 45°_2/0°_2)_S$ laminate generates compressive normal stresses at the laminate boundaries that are 10 times as high as the compressive stresses produced by a $(45°/0°/\pm 45°/0°/-45°)_S$ laminate, when each is loaded in compression. Tensile free-edge stresses would result if these laminates were subjected to axial tensile loads. However, if the 0° layers were stacked on the outside, tensile (delamination) stresses would be induced at the edges as a result of compressive loads. The interlaminar normal stresses can be minimized, particularly for fatigue applications, by optimizing the stacking sequence in reference to the direction of load. The interlaminar stress problem is normally referred to as the free-edge effect. Such stresses are usually dominant at the free-edges of laminates (Fig. 5.31). Their magnitude can be significant, especially for fatigue conditions such as thermal cycling. The major free-edge-effect width is over a very narrow region (approximately equal to the laminate thickness). Furthermore, the magnitudes of such stresses are proportional to the thicknesses of nondispersed oriented layers. Hence, avoid stacks of layers at the same orientation; alternate or disperse the layers.

- If plies must be terminated (in the laminate), provide for equal steps of 0.10 or greater and cover the step area with an additional ply (preferably fabric) to avoid interlaminar shear failure, to provide some load redistribution, and to avoid ply edge peeling.

- Adjacent layers, if possible, should be positioned with a maximum orientation differential of 60°. Potential macrocracking of layers, from induced thermal stress during cool-down from curing, is sufficient reason to avoid this problem. As the adjacent-orientation differential increases, so do thermal stresses (e.g., ±45° or 0°/90° laminates are representative of the worst-case situations).

- Macrocracking of layers within a laminate is an irreversible process. Although it may not affect static strengths, it can lead to re-

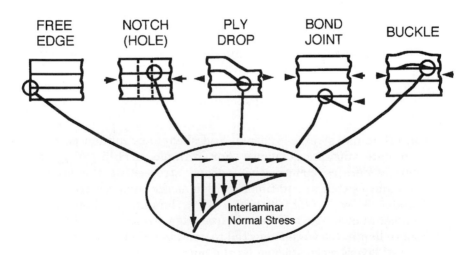

Figure 5.31 Sources of out-of-plane loads.

duced fatigue life. Generally, the frequency of macrocracking increases with thermal cycling and can quickly lead to reduced laminate stiffnesses and changes in Poisson's ratio and coefficients of thermal expansion. In addition, layer cracking may interact with free-edge stresses that can trigger delamination. Fabric layers (as opposed to unidirectional) are very resistant to cracking.

5.9.2   Component Design Recommendations

These recommendations relate to material and processes interactions with environmental concerns and cost-effective fabrication.

- Consider electrode potential; corrosive galvanic cells can be produced whenever two materials of different electrode potential are in electrical contact in the presence of an electrolyte. This is an important consideration for carbon/graphite/epoxy laminates, which are electrically conductive. Many metals are anodic with respect to carbon/graphite/epoxy and are corroded when they are part of a galvanic cell with carbon/graphite/epoxy. An electrolyte, a fluid capable of transferring electrons, is very difficult to avoid in practice. In designs with carbon/graphite/epoxy laminates, observe the following:

   1. Avoid magnesium and magnesium alloys because of their high potential relative to carbon/graphite/epoxy. Aluminum has been used in conjunction with carbon/graphite/epoxy composites with a designed separation to ensure that there is no conductive path from the composite to the aluminum. Titanium has been generally recommended as a replacement for aluminum alloys in areas where galvanic corrosion is expected.
   2. Carefully investigate the potential of other metals and provide special inter-surface protection, such as a fiberglass ply or other nonconductor, as required.
   3. Consider the use of a nonconductive structural fiber composite such as aramid or glass rather than carbon/graphite/epoxy in areas where the presence of electrolytes is expected.

5.9.3   Component Fabrication Recommendations

- Consider fabrication requirements during preliminary design.
- Avoid selecting a design that might be technically elegant but could not be built consistently or reliably. The design of a composite structure requires early concurrent engineering to ensure that the composite can be manufactured.

- Review the design carefully to avoid out-of-plane loads that can cause high levels of interlaminar shear. These, in turn, can cause delaminations.

- Select the simplest design to manufacture.

- If there is a hard choice between ease of analysis and ease of manufacture, choose the latter.

  Design for a composite structure rather than using composite as a substitute for a metal.

  The designer/composites technologist should perform preliminary analysis.

  Many tools are available for preliminary composite sizing and layout, including excellent computer programs for use on microcomputers. Use these computer programs and closed form analyses early, and defer finite element analysis.

- Reduce the part count.

  Consolidate as many substructures as possible, co-cure adhesive bonded assemblies, and build in stiffeners. This is the primary method of reducing costs. (Co-curing is not universally accepted in the industry due to the inability to perform an inspection on the co-cured joint and the possible cost ramifications if the joint is found to be defective.)

- Use the greatest tow size and thickest tape or fiber as practical.

  Larger tow sizes will generally be much less expensive on a weight basis and will tend to decrease fabrication time, but they may have a negative effect on physical properties and surface finish.

- Use the most appropriate fabrication method for the part.

  Note the constraints associated with each technique, e.g., filament winding results in non-smooth outer surface; pultrusion does not normally result in a quasi-isotropic laminate. For large structures, the relative cost of the filament winding is less than one-fourth that of hand layup and less than one-half that of the best tape laying machine.[38]

- Use the least expensive form of composite raw materials.

  Use a wet resin and dry fiber, if appropriate to the process. This method reduces the cost of filament winding. In Ref. 38, the cost for the large MX launch canister wet filament wound by Hercules in the 1990–95 time period was $45.00/lb—less than the cost of just the prepreg tape material for many composite parts.

- Consider the use of fabric, if appropriate to the structure.

  Although raw material costs are higher (some estimates are 18 to 20%), and some physical properties are reduced, woven fabric shortens layup times, avoids tape spreadout problems at abrupt contour changes, remains in place during cure (no distortion or fiber wash-out), and is frequently easier to drill or machine.

- Use composite and metals together to obtain a synergistic part.

  Avoid making the component of either metal or composite only when the judicious use of both will make a better, cheaper part.

- Make in-situ joints if possible.

  A metal-to-composite in-situ joint, made concurrently with the composite structure, saves costs in terms of machining, surface preparation quality assurance, and adhesive application. Generally, the technique will require the use of a film adhesive. This is similar to co-curing, but one side of the joint can be metal or composite, and the joint will generally have a mechanical locking feature.

- Select verification methods carefully.

  For coupon testing, avoid extensive tests for matrix-dominated properties when loads are almost totally reacted by fiber, and vice-versa. Make witness panels as a part of the component if possible. Coupon testing does not remove the need for full-scale component testing, and vice versa.

- Water absorption will change the mechanical and physical proper-ties of a composite laminate. Matrix-dominated shear, compression and bending modulii, and strength may be reduced, and coefficient of expansion may be increased. Account for these changes in design.

- Carbon/graphite composites have poor impact resistance. If impact is a consideration, add glass plies to the exterior surfaces.

5.10   Damage Tolerance

One of the primary hurdles for the widespread use of composites, in addition to cost, has been concern about the damage tolerance. Com-posite materials show no yield behavior, defects are hard to find, and their effects are not always predictable.

Almost all applications for advanced composites, except one-time-use structures like rocket motors, involve repeated cyclic application of stresses. Defects can be introduced by the manufacturing processes or by unplanned occurrences such as impacts. The effects of these de-fects must be evaluated, because their growth during cyclic stresses could cause delaminations or other failures of the structure. Thus, the structure must be damage tolerant to ensure that it will not fail cata-strophically during its operating life. Three suggestions for safety of advanced composite structures are:

1. Undetectable defects or damage should not affect the life of the structure.

2. Detectable damage should be sustainable by the composite structure for a period of time until it is found.

3. Damage during a mission or flight should be sustainable to as great an extent as possible to complete the mission.

The operational stresses in the composite structure can be kept low by increasing wall thicknesses, but that approach would negate the obvious weight advantage of composites and would also make the product more expensive. Also, some stresses, such as interlaminar or transverse, will not be reduced but may be actually increased by additional plies. Increased attention to processing controls can reduce or eliminate many defects (e.g., voids, delaminations, and ply buckles) but not all (e.g., ply drops and free edges). Costs escalate when nonstandard controls are exercised in production. The advantage of low-maintenance costs of composites over product lifetime is enhanced because of good fracture-toughness properties of the composite.

5.10.1 Reducing Damage Concerns

Probably the easiest way to reduce concerns over damage to composites is to impose a ply of an impact absorbing material on the exposed surfaces. Thus, carbon/graphite laminates have received a measure of protection from impacts and other environmental insults by the addition of a fiberglass or aramid ply on the exterior surfaces. This can also have the effect of increasing the laminate allowables by reducing the *knockdown* values for many environmental insults. Another advantage of a thin fiberglass external ply is the telltale stress whitening that happens with external impact when the ply is added to the exterior a carbon/graphite composite.[39] Other ways are as follows:[40]

- Use higher strain fibers and tougher matrices.

    This has been the approach used by the U.S. Air Force to increase the fracture toughness and impact resistance of carbon/graphite laminates for new high-performance aircraft. The bulk of effort has centered on higher-strain carbon/graphite fibers and advanced thermoplastic matrices with their higher levels of toughness.

- Modify the stacking sequence.

    Modify the stacking sequence of the laminate plies to reduce interlaminar stresses or damage openings. One of the layup arrangements studied has been $(30/-30_2/30/90)_S$. This layup places the 90° ply at the midplane of the laminate, which may be a better position than others in terms of longitudinal tensile strength and modulus. However, it must be considered in light of the following potential problems:

Ply thickness.   Since the failure may originate more predominately at a 90° ply, reduce the thickness of the 90° plies and do not stack them together. Possibly move them away from the mid-plane of the laminate.

Stitching.   The benefit of through-the-laminate-thickness stitching has already been demonstrated to lock the laminate plies together and to reduce the propensity for transverse tensile failure

Braiding.   The elimination of a weak ply interface can be accomplished by braiding. It also has some drawbacks, namely a complicated manufacturing process and a loss of the ability to tailor other properties of the laminate.

Edge cap reinforcement.   A layer of aramid or fiberglass can act to reduce interlaminar normal stresses and increase both static and fatigue strengths at a cut edge of the laminate. Capping will also increase bending rigidity, manufacturing, and NDT costs.

Notched edges.   Some investigators have found that narrow notches along the edges of a thin laminate seem to disrupt the edge load path. It can result in stress concentrations, also.

Critical ply termination.   When a laminate with a 90° ply at mid-plane is subjected to tension, peak interlaminar stresses are generated at the interface between the ply or within it. This is because the Poisson's ratio is so different from the surrounding plies. One solution has been to terminate the 90° ply away from the free edge.

Discrete critical ply.   If there is a material mismatch at the mid-plane, such as a 90° ply, it can cause delamination, but if the discrete ply is cut, the delamination will be contained

Hybridization.   Replace the 90° ply with a softer material, such as fiberglass. Do not use an aramid, due to the poorer shear strength.

Adhesive layer interleave.   The adhesive layer softens the center of the laminate and also increases the impact resistance but, in turn, it decreases the bending stiffness dramatically.

The fracture toughness of the laminate material can be measured by a number of methods, some of which are shown in Fig. 5.32. The results will vary with different layup orientations. Other methods may be similar to the test specimens shown in ASTM-E-399 and may involve impact. Presently, the most widely accepted technique for evaluating the fracture toughness of a laminate or structure is a compression test after impact. Compression is a valid indicator; with-

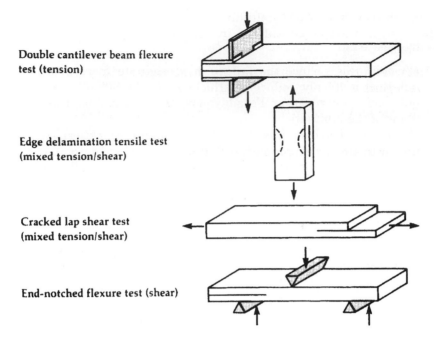

Double cantilever beam flexure
test (tension)

Edge delamination tensile test
(mixed tension/shear)

Cracked lap shear test
(mixed tension/shear)

End-notched flexure test (shear)

Figure 5.32 Typical methods of measuring fracture toughness of composites.
(From Ref. 12)

out matrix support, the fibers will buckle, while the fibers do not need
as much matrix support for tensile loading.

Fracture toughness depends on matrix and fiber properties, fiber
orientations, and stacking sequence. The emphasis on thermoplastic
composites by the U.S. Air Force has been directed toward increasing
the fracture toughness of composites.

5.11   Composite Repairs

There will be damage to composite structures, particularly in aircraft,
that may not be visible but are of concern. The damage to a composite
structure may be more than that incurred by a comparable metal
structure under an identical impact, due to the lack of strain capability in the composite. The metal may yield to show the impact site, but
the composite may delaminate within and not reveal the damaged
area. In-plane tensile strength may not always be compromised by impact damage to the matrix, but damaged matrix cannot stabilize fibers
under compressive bending or shear stresses, and compression is usually the critical loading mode in aircraft structures. A modest impact
to a composite can result in severe undetected internal damage in the
form of delamination.

The objectives of most repair/maintenance programs are as follows:

1. Investigate and quantify the extent of damage.

2. Determine the comprehensiveness of the repair effort and where the repairs will be performed.

   ▪ Repair to full properties (with all flaws detected and corrected), or

   ▪ Repair to acceptable properties that will allow the structure to be usable to some high, predetermined percentage of full-scale operation, or

   ▪ Repair to some emergency level that will allow use at a low percentage of operational effectiveness (i.e., fly the damaged aircraft to depot maintenance base).

3. Select the repair configuration.

4. Define the materials and processes to be used, i.e., adhesives, cure cycle limitations.

The repair options after detection and quantification are shown schematically in Fig. 5.33.[41] Most repairs to aircraft composites involve damaged and possibly wet honeycomb in addition to the composite. A nomograph for determining acceptable conditions for these repairs has been prepared (Fig. 5.34).[42]

The first priority is preparing the composite for repair and may involve drilling holes for a bolt-on patch or reducing the moisture to a level that will allow heating of the composite and the adhesive for cure.

Repair can involve the simple injection of a wet resin through drilled holes into the delaminations then adding blind fasteners to defer the onset of local buckling, or the use of wet or dry prepregs in conjunction with adhesives to form a plug or flush on bonded-on patch. A summary of the available techniques for repair of solid and honeycomb laminates is shown in Table 5.35.[43] Several of the most important points for consideration, summarized in Ref. 43, are:

1. Match the modulus of the original material and the fiber direction as closely as possible.

2. Repair techniques are sufficiently complex that a high degree of technician skill and training are necessary.

3. For aircraft, nonmetallic repairs must incorporate lightning strike protection.

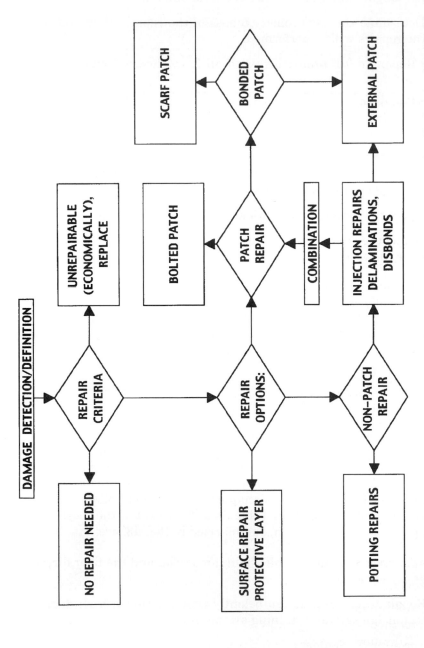

Figure 5.33 Flow diagram for repair options. (From Baker, A., in Mallik, P.K., ed., Composites Engineering Handbook, Marcel Dekker, New York, 1997, p. 747.

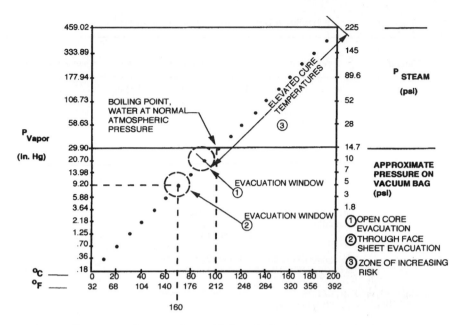

Figure 5.34 Pressure and evacuation guidelines for honeycomb core repair.

5.12 Adhesive Bonding and Mechanical Fastening

There are overwhelming reasons to adhesive bond composites to themselves and to metals, and conversely there are substantial reasons for using mechanical methods such as bolts or rivets (Table 5.36).

The studies on adhesives, surface preparation, test specimen preparation, and design of bonded joints reported for the PABST Program[44] gave much more credibility to the concept of a bonded aircraft and provided reliable methods of transferring loads between composites and metals or other composites.

A number of decisions must be made before bonding to a composite is attempted. These are as follows:

1. Surface preparation technique

   ▪ *Composite.*   Peel ply or manual abrasion or, if possible, co-cure.

   ▪ *Metal:*

   Aluminum—Phosphoric anodize is accepted, preferred airframe method.
   Titanium—Several types of etches are available.
   Steel—Time between surface preparation and bonding is concern.

TABLE 5.35   Summary of Available Composite Repair Techniques[42]

Method	Advantages	Disadvantages	Ease of repair	Structural integrity
Bolted patch	No surface treatment; no refrigeration, heating blankets, or vacuum bags required	Bolt holes weaken structure; bolts can pull out	Fast	Low
Bonded patch	Flat or curved surface; field repair	Not suitable for high temperatures or critical parts	Fast, but depends on cure cycle of adhesive	Low to medium
Flush aerodynamic	Restores full design strength; high-temp capability	Time consuming; usually limited to depot; requires refrigeration	Time consuming	High
Resin injection	Quick; may be combined with an external patch	May cause plies to separate further	Fast	Low
Honeycomb, fill in with body filler	Fast; restores aerodynamic shape	Limited to minimal damage	Fast	Low
Honeycomb, remove damage, replace with synthetic foam	Restores aerodynamic shape and full compressive strength	Some loss of impact strength, gain in weight	Relatively quick	High
Honeycomb, remove damage, replace with another piece of honeycomb	Restores full strength with nominal weight gain	Time consuming; requires spare honeycomb	More difficult	High

TABLE 5.36   Reasons for and against Adhesive Bonding

For	Against
Higher strength-to-weight ratio	Sometimes difficult surface preparation techniques cannot be verified as 100% effective
Lower manufacturing cost	
Better distribution of stresses	Formulation may change
Electrical isolation of components	May require heat and pressure
Minimal strength reduction of composite	Must track shelf life and out-time
	Adhesives change values with temperature
Reduced maintenance costs	May be attacked by solvents or cleaners
Reduced corrosion of metal adherend (no drilled holes)	Common attitude: "I won't ride in a glued-together airplane"
Better sonic fatigue resistance	

2. Type of adhesive

- *Film adhesive.* Reproducible chemistry; early, frequent quality assurance (because resin is premixed by the manufacturer); requires special storage, and it generally is more difficult to accommodate varying bond line thicknesses.

- *Paste adhesive.*   Longer shelf life, no or less changes in storage, accommodates varying bond line thicknesses, but mixing errors are possible, and quality assurance of the adhesive (mix ratio) is generally performed after the structure is bonded.

3. Cure temperature of adhesive

The maximum cure temperature of adhesives should be below the composite cure temperature unless they are co-cured. Cure temperature or upper use temperature of the adhesive may dictate the maximum environmental exposure temperature of the component. There are several cure temperature ranges:

- Room temperature to 225°F. Generally for paste adhesives for noncritical structures.

- 225°F to 285°F. For nonaircraft critical structures. Cannot be used for aircraft generally, because moisture absorption may lower HDT (heat deflection temperature) below environmental operating temperature.

- 350°F and above for aircraft structural bonding. Higher cure temperature may mean more strain discontinuity between adhesive and both mating surfaces (residual stresses in bond).

4. Joint design

The primary desired method of load transfer through an adhesive bonded joint is by shear. This means that the design must avoid peel, cleavage, and normal tensile stresses. One practical application of a method avoiding other than shear stresses is the use of a rivet bonded construction (Fig. 5.35). The direction of the fibers in the outer ply of the composite (against the adhesive) should not be 90° to the expected load path.

5.12.1   Design Guidelines for a Bolted or Bonded Joint

Observe the unique design requirements of a bolted or bonded joint from a composite to a metal or other composite: composites are different from metals. Metals can be relied on to reduce high stress concentrations because of the metal's plasticity. In a brittle material (composite), bolt tolerances and hole drilling are very important. Inadequate attention can result in a few bolts being forced to carry the load. Brittle failure at a low strain can happen at the bolts before the load can be redistributed. Some general design rules for a bolted joint are as follows:

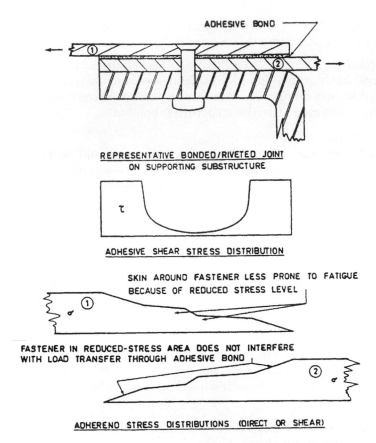

Figure 5.35 Load transfer in rivet-bonded construction. *(From Ref. 45)*

- Design the joint to be critical in bearing with a non-catastrophic failure (Fig. 5.36).[45]

- Fasteners must have sufficient diameter and strength. Locally reinforce as required. Maintain fastener-to-edge distance (e/d) and spacing distance (s/d) as follows: e/d = 2.5 to 3 and s/d = 6 vs. 1.7 to 2 and 4, respectively, for aluminum alloy. Figure 5.37 shows these physical dimension suggestions.

- Keep nearest row of pins at least three fastener diameters from the free edge. Note that multiple rows of holes do not share the load equally. This suggests that two rows of holes would be more efficient than, and preferred to, multiple rows. Keep bearing and tensile values well within the elastic range of expected laminate properties.

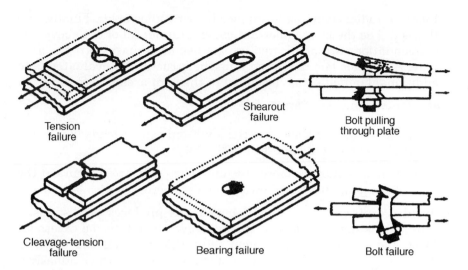

Figure 5.36 Failure modes for bolted joints. *(From Nelson, W.D., Bunin, B.L., and Hart-Smith, L.J., in* Proc. 4th Conf. Fibrous Composites in Structural Design, *Army Materials and Mechanics Research Center, Manuscript Report AMMRC MS 83-2, 1983, pp. U-2 through II-38)*

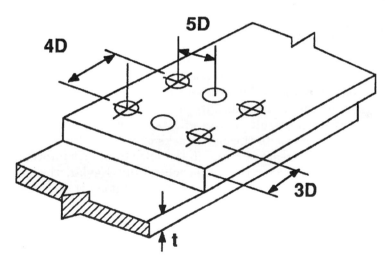

Figure 5.37 Recommended fastener spacings and edge margins for bolted joints.

- Eliminate, as much as possible, the joint eccentricity. Do not introduce bending into the joint. Design for tight pin-to-hole tolerance. Ensure a smooth fastener shank loading against composite. Fastener-to-composite thickness ratio should be greater than ~0.70.

- Increase radial clamping for higher bearing allowables. Finally, and this could be the most important consideration: All of the above recommendations are predicated on the use of a quasi-isotropic laminate. Any other laminate type must be structurally evaluated, and additional laminate must be added to the joint area if necessary.

Rules for a bonded joint are as follows:

- Do not allow tensile loads into the joint, only shear loads. For a tensile load, use a bolted joint.

- Short bonded joints are more efficient. The peak stress occurs at the ends of the joint, and the allowable stress decreases with length.

- Avoid eccentricity in the joint, because it produces peel stresses. This means the thickness of the adhesive should not be excessive. Most manufacturers recommend joint thickness on the order of 0.005–0.010 in. Besides changing of the stress type, thick adhesives generally are weaker. Thick laminates are more affected than thin. If a thick laminate has to be bonded, chamfer. There are concerns for both bolting and bonding; they are summarized in Table 5.37.[46] Further details are provided in Ref. 44.

5.13   Environmental Effects

The primary environmental concern for all polymeric matrix composites is the effect of moisture intrusion both into the raw materials and into the finished components. Moisture in the raw materials causes voids and delaminations of the composite, and moisture in the finished composite structure can result in the effects shown in Table 5.38. Moisture enters the composite through cracks or voids in the matrix, and diffuses through the resin or through the fiber-to-matrix interface (if open to atmosphere). Except for fiberglass and Kevlar, the advanced composite fibers do not transport water, and none of the fibers except for fiberglass are substantially affected by water intrusion. Figure 5.38 shows typical degradation techniques for fiberglass and Kevlar.[47] Most of the effects of moisture are reversible, and the affected property will be restored when the composite is dried. It may take a long time to dry, since the moisture must diffuse out.

Many users of composites require a systematic evaluation of the reaction of each composite (fiber and resin) to a broad list of environments. Many of these environments are listed in test documents such as MIL-STD-810. The evaluation can result in the approval of a composite structure, because it is the same or very close to one already tested (similarity), because it is theoretically unresponsive (analysis), or because the actual structure or a series of coupons has been tested

TABLE 5.37   Comparison of Bolting and Bonding (from Ref. 46)

Advantages	Disadvantages
Bonded joints	
Small stress concentration in adherends	Limits to thickness that can be joined with simple joint configuration
Stiff connection	Inspection difficult other than for gross flaws
Excellent fatigue properties	Prone to environmental degradation
No fretting problems	Sensitive to peel and through-the-thickness stresses
Sealed against corrosion	Residual stress problems when joining to metals
Smooth surface contour	Cannot be disassembled
Relatively lightweight	May require costly tooling and facilities
Damage tolerant	Requires high degree of quality control
	May be environmental concerns
Bolted joints	
Positive connection, low initial risk	Considerable stress concentration
Can be disassembled	Prone to fatigue cracking in metallic component
No thickness limitations	Hole formation can damage composite
Simple joint configuration	In composites, relatively poor bearing properties
Simple manufacturing process	Prone to fretting in metal
Simple inspection procedure	Prone to corrosion in metal
Not environmentally sensitive	May require extensive shimming with thick composites
Provides through-the-thickness reinforcement, not sensitive to peel stresses	
No major residual stress problem	

TABLE 5.38   Possible Effects of Absorbed Moisture on Polymeric Composites

- Plasticization of epoxy matrix
  Reduction of glass transition temperature reduction of usable range
- Change in dimensions due to matrix swelling
- Enhanced creep and stress relaxation
  Increased ductility
- Change in coefficient of expansion
- Reduction in ultimate strength and stiffness properties in matrix-dominated properties
  Transverse tension
  In-plane shear
  Interlaminar shear
  Longitudinal compression
  Fatigue properties (change may be beneficial)
- Fiber-dominated properties are generally not affected
- Change in microwave transmissibility properties

by the composite manufacturer (test). It is in the best interest of both the fabricator and the user whenever the resolution of the composites' response to an environment can be obtained at minimum cost, i.e.,

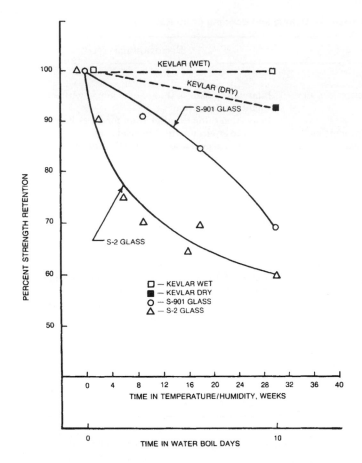

Figure 5.38   Composite response to humidity. *(From Ref. 46)*

verification by similarity. Some of these environments, the predicted effects, and references are shown in Table 5.39.[47]

5.14   Composite Testing

To ensure consistent, reproducible components, three levels of testing are employed: incoming materials testing, in-process testing, and control and final structure verification.

5.14.1   Incoming Materials Testing

Incoming materials testing seeks to verify the conformance of the raw materials to specifications and to ensure processability. The levels of knowledge of composite raw materials do not approach those for met-

TABLE 5.39   Environmental Effects on Composite Structures (adapted from Ref. 47, pp. 9-2 through 9-44)

Environment	Effects (matrix of fiber)	Comments
Moisture related		
Rain	Matrix, softens and swells	Rain may cause water intrusion into composite and joints
Humidity	Matrix, softens and swells	Effect can be aggravated by high, >1%, void content.
Salt fog	Matrix, softens and swells	Corrosion of metal attached to graphite epoxy composite can be increased.
Deep submergence	Matrix	Interface may be affected, low void content is requirement for compressive inputs.
Rain and sand erosion	Both	Protect surfaces with paint or elastomer, composite response is significantly different from metals.
Galvanic corrosion	Neither	Graphite composites can cause corrosion of metals; composite is unaffected.
Radiation		
Solar radiation (Earth)	Matrix	Unprotected aramid or polyethylene is affected.
Ultraviolet	Matrix	Unprotected aramid or polyethylene is affected.
Nonionizing space	Both	Composites can be destroyed, most metals are unaffected.
Solar (LEO)	Both	Composites can be destroyed, most metals are unaffected.
Temperature		
High	Matrix	Fibers (except polyethylene) are generally more resistant than matrix.
Low	Matrix	
Extreme (ablation)	Both	
Thermal cycling (shock)	Matrix	
Miscellaneous environments		
Solvents, fuels	Matrix	Determine for each matrix, stressed and unstressed; effects may be aggravated by voids.
Vacuum	Matrix	
Fungus	Matrix	Each material must be evaluated. May affect organic fibers.
Fatigue	Both	Effects may be diminished by voids.

als, which can be bought to several consensus specifications and will appear to be generally identical, even if purchased from many manufacturers. Although there are fewer suppliers for composite raw materials, the numbers of permutations of resins, fibers, and manufacturers prevents the kind of standardization necessary to be able to buy composite raw materials as if they were alloys.

The fabricators of composites will rely on specifications for control of fiber, resin and/or the prepreg as shown in Table 5.40. Many prepreg resin and fiber vendors will certify only to their own specifications, which may differ from those shown; users should consult the vendors to determine what certification limits exist before committing to specification control.

The purpose of incoming testing is to achieve a consistently reliable product that can be verified to meet the user's requirements and processing techniques; thus, the testing should reflect, if possible, the individuality of the processing method.

As part of raw materials verification, composite design effort, and final product verification, mechanical testing of composite test specimens will be performed. The testing of composite materials offers unique challenges because of the special characteristics of composites. Factors not considered important in metals testing are very important in testing composites. For example, composites are anisotropic, with properties that depend on the direction in which they are measured. Strain and load rate must be carefully monitored, specimen conditioning (drying, storage, etc.) can substantially affect results, and even humidity conditions at the time of specimen fabrication or test can significantly affect the material. Fiber content and void content, which can also vary with manufacturing conditions, have important effects on material properties. All personnel involved in the generation and use of test results, from fabrication of the material to final data interpretation, should be familiar with these factors and their influence on test results. References 47 (pp. 7-1 through 7-31) and 48 provide reference monographs on composites testing; Refs. 49 and 50 provide excellent descriptions of the factors that influence test results.

5.14.2 In-Process Testing

In-process testing seeks to ensure the repeatability of the process so as to reach a consistent fiber volume, resin content and fiber angular, and correct layer placement. Since the processing from start of the cure is a continuous part of the batch processing, there can be no test sample testing. The control is essentially interrogation of the curing process, modification of the process to accommodate the batch to-batch inconsistencies of the input materials, and correlation of the effects of

TABLE 5.40   Test Methods for Composite Raw Materials

Component	Test	Test method	Reference
Resin	Moisture content	Karl Fisher Titration[*]	ARP[†] 1610
	Comparison identity	Chromatography (several types)[*]	ARP 1610
	Chlorine (epoxies)	Hydrolyzable chlorine[*]	ARP1610
	Comparison	Color[*]	ASTM D-1544
	Comparison	Density	ASTM D-1475
	Comparison	Refractometric	User defined
	Comparison	Infrared identity[*]	ARP 1610
	Comparison	Viscosity	ASTM D-445, 1545
	Reactivity	Weight per epoxide	ASTM D-1652
Fiber	Strand mechanicals	Strand tensile strength and modulus	AMS 3892
	Yield	Weight/length	AMS 3892
	Sizing	Extraction	ARP 1610, AMS 3901, AMS 3892
	Fuzz		User defined
	Ribbonization		User defined
	Packaging		AMS 3901, AMS 3894
	Density		ASTM D-3800
Prepreg	Fiber mechanicals	Tow test or unidirectional tensile	ASTM D-2343, D-3379, D-4018, D-3039
	Resin content	Extraction	ASTM C-613, D-1652, D-3529
	Resin identity	Infrared identity	ARP 1610
	Curative identity,%	Extraction, infrared	ARP 1610
	Nonvolatiles		ASTM D-3530
	Resin flow		ASTM D-3531
	Gel time		ASTM D-3532
	Tack		AMS 3894
	Visual imperfection	Optical	AMS 3894
	Areal weight	Weight	ASTM D-3529, D-3539

[*]Also applicable to prepreg.
[†]ARP = Aerospace Recommended Practice, Society of Automotice Engineers (SAE), 400 Commercial Dr., Warrendale, PA 15096. AMS = Aerospace Materials Specification, SAE. ASTM = Americal Society for Tesging and Materials, 1916 Race St., Philadelphia, PA 19193.

changes in input materials and processes to arrive at limits on the process variables and inputs. There have been frequent problems of resin exotherm and cure temperature overshoot with large or thick structures that could ruin a component. New autoclaves have been provided with interactive controls and "smart" operating software to accommodate changes in cure dynamics. An interactive "smart" curing machine can respond to the curing temperature overshoot automatically and rapidly. Table 5.41[48] shows techniques for in-process monitoring.

TABLE 5.41   In-Process Monitoring of Advanced Composites (see Ref. 48 for further details)

Parameter controlled (measured)	Desired location control[*]	Method	Notes
Temperature	In laminate, top bottom center on tool, air temp.	Thermocouple	Includes rate of heating and cooling[†]
Vacuum	Composite surface	Gage (transducer)	Note[†]
Pressure	Composite surface in composite vessel	Gage (transducer) Springback go/no-go gage	Note[†]
Cure rate	In composite	Dielectric analysis Optical (fluorescence)[‡] FTIR spectra[‡]	Note[†]
Viscosity	In composite	Dielectric analysis Ultrasonic[‡] Acoustic Waveguides[‡]	Note[†]

[*]Frequency and number of monitored factors will decrease as product confidence is gained.
[†]Information is part of adaptive control system.
[‡]Not in general use yet.

Since composite structures can be large and complex to cost-effectively replace metal assemblies, the labor input value of these structures can be considerable. The benefit of testing the incoming materials and of verifying the processing is obvious. Ideally, with feed-back control and complete control over the incoming materials, there would be reduced need for final product inspection.

5.14.3   Final Product Verification

Rarely does composite testing include actual mechanical verification of the final product. The exceptions are filament-wound pressure vessels and some small composite structures, such as golf shafts, that are tested in bending. Many pressure vessels experience some deterioration due to the proof pressure test, which causes transverse stress in the hoop plies that ends up fracturing the matrix in the outer plies and results in a *first-ply* failure. This is one reason why another failure criterion is invoked for pressure vessels, namely the *last-ply* failure criterion. However, the failure in the outermost plies reduces the composite's resistance to water intrusion, and consequently the product may have to be recoated with a moisture-resistant paint to satisfy some environmental fears.

Nearly all final composite structure verification includes mechanical tests of tag end or coprocessed coupons and nondestructive tests on the structural laminate. Mechanical testing is also used to verify the materials and processes prior to committing to the layup and cure of a larger composite structure. Table 5.42, adapted from Ref. 50, shows the key data and test methods that are needed to support design or production of a composite structure.

It can be seen that there are several tests in current use to determine some properties. Many of the tests are still evolving and will eventually become standardized through ASTM or other consensus organizations. Also, this list does not exhaust the available test methods, and nonstandard tests abound for composites fabricated by alternative techniques (e.g., filament winding, Ref. 47).

Nondestructive test techniques are preferred to verify the final structure, because coprocessed or tag end test specimens suffer from edge effects and may not mirror the fiber angles, resin, or fiber content. The component may also have discrepancies such as fiber kinking and washout that may not be reflected in the test specimens. Confidence is gained in a structure because of careful control of incoming materials, controlled processing, adherence of tag end or co-processed test specimens to required test values, and the meeting of NDI standards. The simplest NDI techniques, which involved coin tapping and visual observance, have evolved into those shown in Table 5.43 (adapted from Ref. 51). They can be used individually or, in some cases, concurrently.

5.15   Safety Issues with Composite Materials

Composite materials, particularly in the raw material form, present toxic hazards to users and should be handled carefully. Essentially, there is a need for management inputs and controls on the following aspects of composite fabrication:

- Material handling
- Training

TABLE 5.42   Advanced Composite Test Methods in Current Use (adapted from Ref. 50)

Test	Test method(s)
Tensile strength and modulus axial and transverse	ASTM D-3039
Compression	ASTM D-695 (modified for high-modulus composites); Celanese,[*] ASTM D-3410; IITRI, ASTM D-3410; sandwich beam, ASTM D3410
Shear	Iosipescu, ASTM D-3518 Rail shear, ASTM D-4255 Short beam, ASTM D-2344
Flexure	ASTM D-790
Fracture toughness	ASTM E-399 NASA 1092[*] End notched flexure mode II Cracked lap shear (mixed modes I and II) Edge delamination NASA 1092
Impact	Instrumented drop-weight impact Tensile impact Compression after impact (Boeing BSS-7260)
Fatigue	ASTM D-3479, ASTM D-671
Coefficient of thermal expansion	User defined
Coefficient of moisture expansion	User defined
Single fiber tension	ASTM D-3379
Single fiber tensile creep	Note[†]
Open-hole tension, compression	Boeing BMS 8-276
Thermal conductivity	User defined
Bolt bearing	ASTM D-953

[*]Refers to the original design agency or the agency that prepared the reference specification or to the commercial company whose internal specification was used throughout the industry.
[†]No general consensus specification exists.

- Gaining awareness of hazards and proper use of toxic materials
- Isolation of some operations
- Use of personal protective equipment (if necessary)
- Personal hygiene
- Significance of warnings and labels

TABLE 5.43   NDE Test Methods for Composites (adapted from Ref. 51)

	Principal characteristic detected	Advantages	Limitations
Radiography	Differential absorption of penetrating radiation	Film provides record of inspection, extensive data base	Expensive, depth of defect not indicated, rad. safety
Computer tomography	Conventional x-ray technology with computer digital processing	Pinpoint defect location; image display is computer controlled	Very expensive, thin wall structure might give problems
Ultrasonics	Changes in acoustic impedance caused by defects	Can penetrate thick materials, can be automated	Water immersion or couplant needed
Acoustic emission	Defects in part stressed generate stress waves	Remote and continuous surveillance	Requires application of stress for defect detection
Acoustic ultrasonics	Uses pulsed ultrasound stress wave stimulation	Portable, quantitative, automated, graphic imaging	Surface, contact surface geometry critical
Thermography	Mapping of temperature distribution over the test area	Rapid, remote measurement, need not contact part, quantitative	Poor resolution for thick specimens
Optical holography	3-D imaging of a diffusely reflecting object	No special surface preparation or coating required	Vibration-free environment required, heavy base needed

- Housekeeping
- Dispensing and storage
- Emergency instructions

The typical materials that are encountered are shown on Table 5.44. This table is not exhaustive, and the user should consult the Material Safety Data Sheets (MSDS) and more in-depth references (e.g., Ref. 52). The liquid resins and catalysts are the highest noted hazards, because there are so many avenues of attack to our bodies, and because there are many different effects. Once the composite structure is fabricated, the fibers and resins are rendered essentially innocuous, and machining, drilling, etc. pose minimal hazards. Machining dusts should be contained and properly disposed of to reduce the nuisance.

References

1. *Mil-Hdbk-17/2D.*
2. Hercules Data Sheet, H050-377/GF.
3. *Carbon and High Performance Fibres*, 6th ed., Chapman and Hall, London, 1995.
4. Peters, S.T., Humphrey, W.D. and Foral, R.F., *Filament Winding, Composite Structure Fabrication,* 2nd ed., SAMPE Publishers, Covina, CA, 1999.

TABLE 5.44 Commonly Encountered Hazards with Advanced Composite Materials

Hazard	Form	Reported hazards
Epoxy resins	Liquid or as prepreg	Dermatitis, some may be potential skin carcinogens
Epoxy curing agents		
Aromatic amines	Liquids, but primarily in prepreg	Liver, kidney damage, jaundice, some are carcinogens, anemia
Aliphatic amines	Liquids	Severe bases and irritants and visual disturbances
Polyaminoamides	Liquids	Somewhat less irritating, may cause sensitization
Amide		Slight irritant
Anhydride	Liquid, solid, prepreg	Severe eye irritants, strong skin irritants
Phenolic and amino resins	Usually as dry prepreg	Free Phenol, also released on cure. Formaldehyde is strong eye, skin and respiratory irritant
Bismaleimide resins	Liquid or paste, generally as prepreg	Skin irritant and sensitization, not fully characterized yet
Thermoplastic resins	Solids or prepreg	Exercise care with molten materials & provide good ventilation. Consult MSDS
Reinforcing materials		
Graphite fibers	Airborne dusts	Skin irritation, contain dusts, prevent exposure and protect skin
Aramid fibers		Contain airborne dusts
Fiberglass		Mechanical irritation to skin, eyes, nose, and throat

5. Delmonte, John, *Technology of Carbon and Graphite Fiber Composites*, Van Nostrand Reinhold, New York, 1981, p. 59.
6. Bacon, R., Towne, M., Amoco Performance Products, personal communication.
7. Humphrey, W. Donald and Vedula, Murali, *SAMPE Proceedings*, Vol. 40, May 1995, pp. 14785–7.
8. Goldsworthy, W.B. and Hiel, C., *Proc. 44th SAMPE Symposium*, May 1999, pp. 931–942.
9. Mayorga, G.D., in Lee, S.M., ed. *International Encyclopedia of Composites*, Vol. 4, VCH Publishers, New York, NY.
10. Foral, Ralph F., and Peters, Stanley T., Composite Structures and Technology Seminar Notes, 1989.

11. Agarwal, Bhagwan D., and Broutman, Lawrence J., *Analysis and Performance of Fiber Composites*, 2nd ed., Wiley Interscience, New York, 1990, p. 199.
12. Peters, S.T., in S.T. Peters, ed. *Handbook of Composites*, 2nd ed., p. 16, Chapman & Hall, London, 1998.
13. Tsai, Stephen W. and Pagano, Nicholas J., in *Composite Materials Workshop*, S. W. Tsai, J.C. Halpin and Nicholas J. Pagano, eds., Technomic Publishing Co., Lancaster, PA, 1978 p. 249.
14. Tsai, Steven W., *Theory of Composites Design*, 1st ed., Think Composites, Dayton, Ohio, 1992.
15. Brown, Richard T., in *Engineered Materials Handbook,* Vol. 1, Composites, Theodore Reinhart, Tech. Chairman, ASM International, 1987, p. 268–274.
16. *Genlam*, General Purpose Laminate Program, Think Composites, P.O. Box 581, Dayton, Ohio 45419.
17. DuPont, *Data Manual for Kevlar 49 Aramid*, May 1986, p. 95.
18. G. Eckold, *Design and Manufacture of Composite Structures*, pp. 76–96, McGraw-Hill, New York, NY, 1994.
19. S. Swanson, in P.K. Mallick, ed. *Composites Engineering Handbook*, Marcel Dekker, pp. 1183 et seq, New York, NY, 1997.
20. S.B. Batdorf, in S.M. Lee, ed., *International Encyclopedia of Composites*, VCH Publishers, New York, NY, 1989.
21. *Military Handbook 5 for Aerospace Metals*.
22. Sun, C.T, Quinn, B.J., Tao, J., and Oplinger, D.W., *Comparative Evaluation of Failure Analysis Methods for Composite Laminates,* DOT/FAA/AR-95/109, May 1996.
23. Kirchner–Lapp in *Handbook of Composites*, 2nd ed., S. T. Peters, ed., Chapman and Hall, London, 1998, p. 760.
24. Military Handbook 17 B, *Plastics for Aerospace Vehicles Part 1, Notice-1*, 1973.
25. *The Composites Materials Handbook-Mil 17*, Technomic Publishing Co. Inc., Lancaster, PA, 1999.
26. Foston, Marvin and Adams, R.C., in *Engineered Materials Handbook,* Vol. 1, Composites, Theodore Reinhart, Tech. Chairman, ASM International, 1987, pp. 591–601.
27. B. Sarh, B. Moore, and J. Riedell, 40th International SAMPE Symposium, May 8–11, 1995, p. 381.
28. Habermeier, J., Verbal Presentation at SAMPE 44th ISSS&E, May 1999.
29. Campbell, Flake C., et al, *J. Adv. Materials*, July 1995, pp.18–33.
30. Bergland, Lars in S.T. Peters, ed. *Handbook of Composites*, 2nd ed., p. 116, Chapman & Hall, London, 1998.
31. Loos, Alfred C. and Springer, George, S.J., *Comp. Mater.*, March 17, 1983, pp. 135–169.
32. Enders, Mark L. and Hopkins, Paul C., *Proc. SAMPE International Symposium and Exhibition,* Vol. 36, April 1981, pp. 778–790.
33. Martin, Jeffrey D. and Sumerak, Joseph E., Composite Structures and Technology Seminar Notes, 1989, p. 542.
34. Ko, Frank K., in *Engineered Materials Handbook,* Vol. 1, Composites, Theodore Reinhart, Tech. Chairman, ASM International, 1987, p. 519.
35. Hancock, P. and Cuthbertson, R.C., *J. Mat Sci.*, 5, 762–768, 1970.
36. Kohkonen, K.E. and Potdar, N., in S.T. Peters, ed. *Handbook of Composites*, 2nd ed., Chapman and Hall, p. 598, London, 1998.
37. Abrate, S., in P.K. Mallick, ed., *Composites Engineering Handbook*, Marcel Dekker, New York, NY, 1997, p. 783.
38. Freeman, W.T. and Stein, B. A., *Aerospace America*, Oct. 1985, pp. 44–49.
39. Heil, C., Dittman, D., and Ishai, O., *Composites* (24) no. 5, 1993, pp. 447–450.
40. Chan, W.S., in P.K. Mallick, ed., *Composites Engineering Handbook*, Marcel Dekker, New York, NY, 1997, pp. 357–364.
41. Baker, A., P.K. Mallick, ed., *Composites Engineering Handbook,* Marcel Dekker, p. 747, New York, NY 1997.
42. Seidl, A.L., in *Handbook of Composites*, 2nd ed., S. T. Peters, ed., Chapman and Hall, London, 1999, p. 864.

43. Seidl, A.L., Repair of Composite Structures on Commercial Aircraft, 15th Annual Advanced Composites Workshop, Northern California Chapter of SAMPE, 27 Jan. 1989.
44. Potter, D.L., *Primary Adhesively Bonded Structure Technology (PABST) Design Handbook for Adhesive Bonding*, Douglas Aircraft Co., McDonnell Douglas Corporation, Long Beach, CA, Jan. 1979.
45. Nelson, W.D., Bunin, B.L., and Hart-Smith, L.J., Critical Joints in Large Composite Aircraft Structure, in *Proc. 6th Conf. Fibrous Composites in Structural Design,* Army Materials and Mechanics Research Center Manuscript Report AMMRC MS 83-2 (1983), pp. U-2 through II-38.
46. Baker, A., P.K. Mallick, ed., *Composites Engineering Handbook,* Marcel Dekker, p. 674, New York, NY, 1997.
47. Peters, S.T., Humphrey, W. D., and Foral, R., *Filament Winding, Composite Structure Fabrication,* 2nd ed., SAMPE Publishers, 1999, Covina, CA, pp. 9–13.
48. Kranbuehl, David E., in *International Encyclopedia of Composites*, Vol. 1, pp. 531–543, Stuart M. Lee, ed., VCH Publishers, New York.
49. Whitney, J.M., Daniel, I.M., and Pipes, R.B., *Experimental Mechanics of Fiber Reinforced Composite Materials*, SESA Monograph No. 4, The Society for Experimental Stress Analysis, Brookfield Center, Connnecticut, 1982.
50. Carlsson, L.A. and Pipes, R.B., *Experimental Characterization of Advanced Composite Materials*, Prentice-Hall, Englewood Cliffs, N.J., 1987.
51. Munjal, A., *SAMPE Quarterly*, Jan. 1986.
52. Safe Handling of Advanced Composite Materials Components, Health Association, Arlington, VA, April 1989.

6

Part 1
Natural and Synthetic Rubbers

Robert Ohm
Uniroyal Chemical
Naugatuck, CT

R. J. Del Vecchio
Technical Consulting Services
Fuquay-Varina, North Carolina

Rubber is a material that is capable of recovering from large deformations quickly and forcibly. To achieve this characteristic, the rubber polymer must be of high molecular weight, controlled crystallinity, and low glass transition temperature, and it must be crosslinked. The importance of each property will be shown after a historical perspective of the uses of rubber.

6.1 Historical

In his second voyage, Christopher Columbus became the first western person to see natives playing with bouncing balls of "cau-uchu" or "weeping wood." In the mid-eighteenth century, François Fresneau discovered a tree in the Amazon valley that provides a sap whose solids are rubbery. Cultivation of the Hevea Brasilienis is still the source of most of the world's natural rubber. Joseph Priestly coined the term *rubber* in 1770 after its ability to remove pencil marks.

The nineteenth century saw several developments that were important in the initial development of the rubber industry. Charles Macintosh developed the first rubber raincoat in 1823. Thomas Hancock developed a process for masticating the tough raw rubber to a smooth processible state. Charles Goodyear is credited with the process of crosslinking or *vulcanization* of rubber to prevent its becoming sticky in summer and brittle in winter. Henry Wickam brought samples of Hevea nuts to Malaysia in 1876. This became important when a leaf blight wiped out all of the plantations in South America. John Boyd Dunlop made the first pneumatic tire in 1888.

The historical consumption of rubber is shown in Figure 6.1. It basically parallels the growth of motor vehicles. Today, the majority of new rubber is used in tires, and a comparable amount of non-tire rubber is used in the modern automobile.

Technical developments as well as the widespread adoption of the motor vehicle have fueled growth of the rubber industry in the twentieth century. Some early milestones are S. C. Mote's finding of reinforcement by carbon black in 1904, George Onslager's discovery of organic accelerators in 1906, plantation rubber overtaking the harvesting of wild rubber in 1914, and Germany's production of 5 million pounds of *methyl rubber* in 1918. Antidegradants to protect rubber during service were discovered by Herbert Winkelmann, Harold Gray, and Sidney Cadwell in 1924. The first oil-resistant rubber was introduced in 1931, based on work by Rev. Julius A. Nieuwland at Notre Dame and Drs. Wallace Carothers and Arnold Collins at DuPont. In 1948, the United States produced 1.5 billion pounds of synthetic rubber for the war effort. Stereospecific catalysts developed by Karl Zeigler and Giulio Natta in the 1950s gave the polymer scientist greater control over the polymer structure and ushered in solution polymers. The last decade of the twentieth century saw the introduction of even greater polymerization control by the use of constrained geometry catalysts.

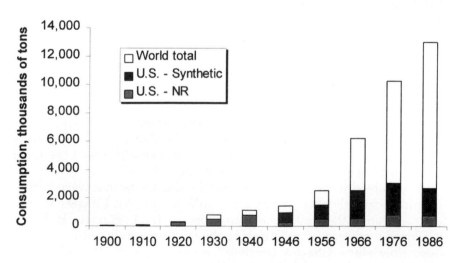

Figure 6.1 Historical rubber consumption.[1]

6.2   Properties of Polymers

Both rubbers and plastics are in the family of *polymers,* a term from the Greek meaning *many units.* Table 6.1 shows how properties change as an increasing number (n) of repeating units (CH_2-CH_2) are joined to form a high-molecular-weight (MW) polymer.

TABLE 6.1   Properties of Hydrocarbons -(CH_2-Ch_2)$_n$-

Chemical name	n	MW	Appearance	Melting point, °C	Boiling point, °C
Ethane	1	30	gas	−183	−89
Butane	2	58	gas	−138	0
Hexane	3	86	liquid	−95	68
Decane	5	142	liquid	−30	174
Eicosane	10	282	grease	38	343
Low-MW PE	25	700	grease	92	dec.
Low-MW PE	75	2,100	wax	106	dec.
High-MW PE	7,500	210,000	solid	120	dec.

As the number of units joined together and the molecular weight increase, the melting and boiling points increase, and the products go from gases to liquids to waxes. Only at sufficiently high molecular weight is the polymer capable of high strength to make useful load-bearing parts.

High-molecular-weight polyethylene (PE) is used to make milk jugs. The regularity of its structure allows adjacent chain segments to align in perfect order to form crystals, which are the source of opacity or cloudiness in articles made from PE. To become rubbery and recover from large deformation, the amount of crystallinity must be controlled.

One approach to controlling crystallinity is to add a differently shaped co-monomer such as propylene. Ethylene propylene co- and ter-polymers (EPM and EPDM, respectively) are rubber polymers used in weatherstrip door seals and the white sidewalls of tires. The optional third monomer in EPDM is a diene that allows crosslinking by sulfur cure systems. EPM copolymers must be crosslinked with peroxides (see Fig. 6.2).

Many rubbers are based on diene monomers in which only one of the two double bonds polymerizes. The double bond remaining in the poly-

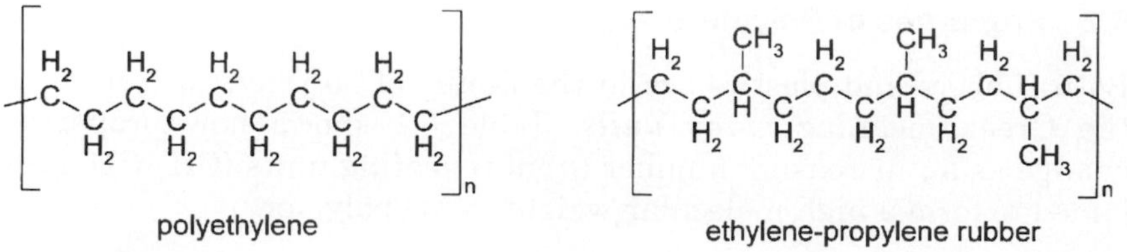

polyethylene   ethylene-propylene rubber

Figure 6.2 Structure of polyethylene plastic and ethylene propylene rubber.

mer prevents free rotation of the polymer chain and minimizes the possibility of crystallization. There are two possible isomers, cis and trans, depending on whether the polymer continues on the same side (cis) or the opposite side (trans) of the double bond. Normal emulsion polymerization gives a mixture of cis and trans structures, and no crystallization occurs in these rubbers.

If the cis or trans content is very high (>90%), then some crystallization can occur on stretching, which provides high strength in gum (unfilled) compounds. Two strain-crystallizing rubbers are shown in Fig. 6.3.

The glass transition is the temperature at which a polymer becomes stiff and brittle. As such, it determines the low temperature service limit of rubbers. The effect on glass transition (Tg) as the polymer composition changes from pure polybutadiene (BR, a rubber) to pure polystyrene (a plastic) is shown in Table 6.2. Copolymers of 23% styrene and 77% butadiene (SBR) are used in tires.

Crosslinking or joining adjacent polymer chains is necessary to prevent flow. Adhesives and chewing gum are applications of uncrosslinked rubber. Charles Goodyear used molten sulfur to cure

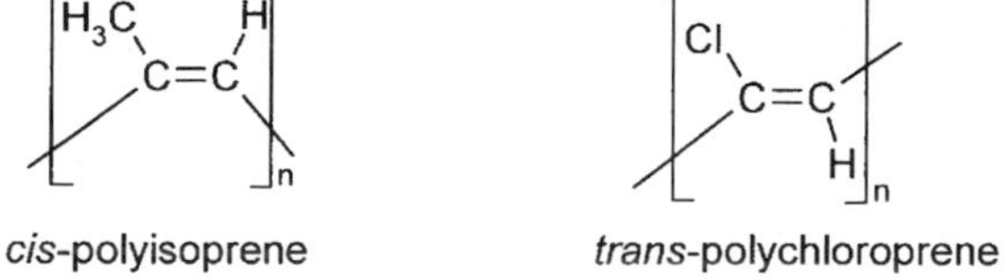

cis-polyisoprene   *trans*-polychloroprene

Figure 6.3 Structure of natural rubber and polychloroprene rubber.

TABLE 6.2   Glass Transition of Styrene and Butadiene (Co)polymers

% styrene	0	23	36	53	75	100
% butadiene	100	77	64	47	25	0
Tg, °C	−79	−52	−38	−14	+13	+100

rubber of its tendency to soften and flow. Elemental sulfur is still the most widely used means to crosslink or vulcanize rubber. Other cure systems have been developed over the years to improve certain properties or to crosslink fully saturated polymers that cannot be crosslinked with sulfur.

6.3 General-Purpose Rubbers

General-purpose rubbers are low-cost hydrocarbon polymers that find use in tires as well as other large-volume applications. The 1994 world consumption of general purpose rubbers is shown in Table 6.3.

TABLE 6.3   Consumption of General-Purpose Rubbers[1]

Rubber	Abbrev.	Commercialized	Consumption, thousands of metric tons
Natural rubber	NR	—	5,403
Styrene butadiene	SBR	1941	4,220
Polybutadiene	BR	1960	1,473
Polyisoprene	IR	1960	982
Ethylene propylene	EPDM, EPM	1962	630
Butyl	IIR, CIIR, BIIR	1943	558
Total			13,266

Natural rubber (NR) was the only available rubber for many years. It is produced primarily in the Far East (Malaysia, Indonesia, and Thailand), either as a concentrated liquid latex or coagulated, dried, and baled. Latex is used to make thin-walled articles such as gloves and balloons. Rubber bales are usually mixed with fillers for tires and mechanical goods. But NR can also be used unfilled to make translucent articles such as rubber bands and baby bottle nipples.

Articles made from natural rubber possess high strength and abrasion resistance and are very resilient with low heat buildup in dynamic applications. Their heat resistance is limited, and the rubber parts are susceptible to attack by oxygen, ozone, and sunlight.

Polyisoprene (IR) is the synthetic equivalent of natural rubber and possesses many of the same characteristics and limitations. IR is free of the nonrubber components contained in NR, including tree proteins that cause allergic reactions in some individuals. IR is also more consistent, whereas NR can vary seasonally, and different Hevea clones may provide slightly different properties. The nonrubber components in NR also provide some acceleration, and antioxidant properties that must be taken into account when compounding IR.

Styrene butadiene rubber (SBR) was an outgrowth of the war effort when supplies of NR were cut off. Private companies later purchased the government rubber production facilities, many of which are still in operation today.

SBR is offered as a latex or in baled form. The baled rubber can be pure, clear polymer as well as having carbon black and/or processing oil incorporated. These low-cost polymers are extensively used in tires and general mechanical goods. The use of a reinforcing filler is necessary to develop good tensile and tear strength. It may be blended with NR, IR, or other polymers for cost or performance purposes.

Polybutadiene (BR) is a polymer of 1,3-butadiene, which can have varying amounts of cis, trans, and vinyl 1,2 structures incorporated in the polymer. The pendant vinyl structure can also be incorporated in different ways, leading to an array of polymers with varying physical properties and processing characteristics.

Polybutadiene is mainly used in polymer blends, with the major consumption in tires. High-cis polybutadiene is used in tire components because of its high resilience, abrasion resistance, and good flex fatigue. Polybutadiene with high vinyl content is used in tire treads for low rolling resistance and good fuel economy. Non-tire applications include high-impact polystyrene and solid-core golf ball centers.

Butyl rubber is a copolymer of isobutylene with a few percent of a cure site monomer. The cure site is typically isoprene (IIR), which may be halogenated to produce bromobutyl (BIIR) and chlorobutyl (CIIR) rubbers. Halobutyl rubbers have faster cure rates and so may be blended and co-cured with high-diene polymers such as NR, SBR, and BR. Polyisobutylene with a brominated para-methylstyrene cure site monomer (Exxpro® BIMS) has recently been introduced.

The polyisobutylene polymers have improved heat resistance compared to the foregoing high-diene rubbers with double bonds in the repeating structural unit. The polymers have low air permeability, leading to their use in inner tubes (butyl) and tire liners (halobutyl). Polyisobutylene rubbers also are very energy absorbent, which provides ideal characteristics for articles in dynamic service.

Ethylene propylene rubbers may be either a fully saturated copolymer (EPM) or a terpolymer containing <10% of a diene (EPDM), typically ethylidene norbornene, to enable vulcanization with sulfur curing systems.

EP rubbers are the largest-volume rubber used in non-tire applications. They combine the heat resistance of a fully saturated polymer backbone with the ability to use high levels of low-cost fillers and plasticizers. Examples of EP uses are hose, automotive weatherstrip, single-ply roofing membranes, and high-temperature-service wire and cable insulations.

6.4 Specialty Rubbers

Specialty rubbers have chlorine, fluorine, nitrogen, oxygen, or sulfur incorporated into the repeating structure. These polar atoms provide resistance to swelling in hydrocarbon fluids such as gasoline and motor oil. The 1994 world consumption of specialty rubbers is shown in Table 6.4.

TABLE 6.4   Consumption of Specialty Rubbers[1]

Rubber	Abbrev.	Commercialized	Consumption, thousands of metric tons
Polychloroprene	CR	1931	306
Nitrile-butadiene	NBR	1941	252
Polyurethane	AU, EU	1945	129
Acrylates/acrylics	ACM, EAM	1947	63
Chlorinated/chlorosulfonated (alkylated) polyethylene	CPE, CSM, ACSM	1951	54
Silicone	MQ, VMQ, FVMQ	1944	48
Fluorocarbon	FKM	1957	24
Others	ECO, T,	—	36
Total			912

Polychloroprene (CR) was the first oil-resistant rubber. It may be likened to natural rubber in which the pendant methyl group is replaced with a polar chlorine atom. Like NR, CR has high strength in unfilled (gum) compounds. Copolymerization with sulfur leads to high resistance to flex fatigue, whereas using a thiuram polymerization modifier improves heat resistance. Some grades use 2,3-dichlorobutadiene as a co-monomer to obtain resistance to crystallization and hardening at low temperature.

Polychloroprene is used in adhesives, v-belts, molded goods, and jackets for electrical wire and cables. Latex grades are available for dipped goods manufacture or foaming into mattress applications.

Nitrile rubber (NBR) is a copolymer of butadiene with 20 to 40% acrylonitrile, typically 33%. Oil resistance increases in proportion to the amount of acrylonitrile in the copolymer; low-temperature resistance improves in proportion to the amount of butadiene. Nitrile rubber containing carboxyl functionality has exceptionally good toughness and abrasion resistance. Built-in antioxidants can improve heat resistance, and hydrogenation of the double bonds can maximize high-temperature performance.

Nitrile rubber is used in the tube and cover of fuel hose, curb pump hose, hydraulic hose, and oil-resistant molded parts. Hydrogenated nitrile rubber (HNBR) is used in automotive power transmission belts.

Polyurethane has exceptional toughness and abrasion resistance. There are two main types, produced by the reaction of an isocyanate with a diol, either an ether (EU) or an ester (AU). Ether-based polyurethanes have higher resilience and somewhat better low-temperature and water resistance.

Solid tire applications are a mainstay of polyurethane uses, including fork lift tires, caster wheels, and skate wheels. Polyurethanes are also used to cover rubber rolls and line pumps and pipes in abrasive service.

Polyacrylates (ACM) and acrylic elastomers (EAM) have carboxyl ester groups in the repeating structural unit. A small percentage of a cure site monomer is also incorporated during polymerization. The polar ester group provides oil resistance with the usual sacrifice in low-temperature resistance.

These polymers are widely used for high-temperature oil seals such as transmission lip seals and shaft seals. They are energy absorbent for dynamic applications and are used in wire and cable.

Chlorinated polyethylene (CPE) has a fully saturated polymer backbone for improved heat resistance as compared to the first two oil-resistant polymer families discussed. For crosslinking flexibility, chlorosulfonated grades (Hypalon® CSM and an analog containing branching, Acsium® ACSM) are available.

CPE, CSM, and ACSM are used for improved heat resistance in hose and belt applications. Colorable compounds can be provided that are resistant to outdoor exposure.

Silicone rubber (MQ) has a repeating polymer backbone of alternating silicon and oxygen atoms. Each silicon atom has two methyl groups attached. For improved low-temperature properties, some methyl groups are replaced with phenyl groups (PMQ). For crosslinking with peroxides, a vinyl silicone monomer is incorporated (VMQ). Silicone rubber has the broadest temperature range of any rubber. It is as good as polybutadiene on the low-temperature side and is superior to most all hydrocarbon based rubbers on the high-temperature side.

The uses of silicone include high-temperature seals and gaskets, electrical insulation for spark plug and appliance wires, and aerospace (both aircraft and spacecraft). These take advantage of the broad service temperature range.

Fluorocarbon rubber (FKM) replaces the oxidizable carbon-hydrogen bond with a thermally stable carbon-fluorine bond. The polar fluorine atom provides exceptionally good resistance to oils and solvents that would attack most all other rubbers.

Many fluorocarbon applications involve parts that are small but provide a critical function. And they are used in applications where no other material will work, such as flue duct expansion joints. The modern automobile uses fluoroelastomer-lined hose in fuel-injected engines.

Other rubbers include epichlorohydrin (CO), which is usually a copolymer with ethylene oxide (ECO) or a terpolymer containing a sulfur or peroxide crosslinking site (GECO); polysulfide copolymers with ethylene dichloride (T); polynorbornene (PNR); tetrafluouroethylene-propylene copolymers (Aflas®); and fluorosilicone (FVMQ).

6.5 Thermoplastic Elastomers

Thermoplastic elastomers have two phases that are intimately intermixed. One phase is a rubbery phase that provides elastic recovery from deformation. The other phase is a hard phase that softens and flows at elevated temperature. Above the melting point of the hard phase, the polymers will flow and can be shaped. Below the melting point of the hard phase, the material behaves like a conventional rubber.

Unlike conventional rubbers, the hard phase can be melted many times, and the scrap can be recycled. The melting of the hard phase limits high-temperature service and detracts from compression set. The 1994 world consumption of thermoplastic elastomers is shown in Table 6.5.

Styrene block copolymers have a polystyrene hard phase at each end of the polymer with a midblock of butadiene (SBS), isoprene (SIS), or hydrogenated butadiene (SEBS). They are used in footwear and adhesives.

Thermoplastic polyolefins (TEO or TPO) have a polyolefin hard phase, typically polypropylene, physically mixed with a rubbery phase such as EPDM. The rubber phase has little or no crosslinking. TEOs are used in automotive exterior panels and in lower-temperature wire and cable applications.

Thermoplastic vulcanizates (TPV) also have a polyolefin hard phase with a crosslinked elastomer phase. The crosslinking provides

TABLE 6.5   Consumption of Thermoplastic Elastomers[1]

Rubber	Abbrev.	Consumption, thousands of metric tons
Styrene block copolymers	SBS, SIS, SEBS	294
Thermoplastic polyolefins	TEO, TPO, TPV	192
Polyurethane	EU	84
Polyester	Hytrel®	30
Total		600

improved resistance to compression set and creep. The improved temperature resistance permits use in under-the-hood automotive applications.

Thermoplastic polyurethanes combine the toughness and abrasion resistance of urethanes with the ability to be recycled.

Thermoplastic polyesters have a terephthalate ester hard phase and soft phase, the difference being the length of the alkylene diol joining terephthalate groups. The polymers are very stiff relative to conventional rubbers, which allows less material to be used to realize weight and cost savings. Applications that take advantage of the polymer's high strength and flexibility include fuel tanks, gear wheels, and ski boots.

6.6   Characterizing Heat and Oil Resistance

The heat and oil resistance of natural and synthetic rubbers may be characterized for automotive applications by a specification system that has been jointly developed by the American Society for Testing and Materials (ASTM) and the Society of Automotive Engineers (SAE). ASTM Test Method D2000, or the corresponding SAE Method J200, characterizes the heat and oil resistance by the retention of properties after exposure to a standard time and temperature. The composition of the oil is well characterized and supplied by ASTM. In addition to property retention minimums, the volume change upon oil immersion is a key requirement.

The relative heat and oil resistance for rubbers is shown in Fig. 6.4 according to the ASTM/SAE scheme. Both the heat resistance and oil resistance of the polymers shown are not absolute, immutable properties.

During exposure to high temperature, the properties of the rubber vulcanizate will continue to change with time. And, within a particu-

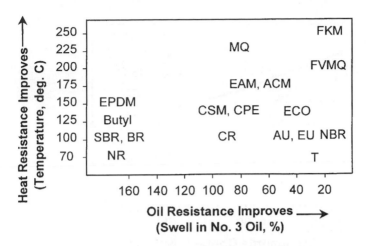

Figure 6.4 Heat and oil resistance per ASTM D200/SAE J200 scheme.

lar rubber type, the amount of change may vary to some extent depending on the rubber formulation—particularly the heat resistance of the cure system and the use of antidegradants. The typical range and variation with time for three rubbers of different recipes is shown in Figure 6.5.

The composition of polymer and the immersion fluid affect the volume swell and change in properties of a rubber compound. This is illustrated in Figs. 6.6 and 6.7 for compounds based on nitrile-butadiene rubber (NBR) and fluoroelastomer (FKM), respectively.

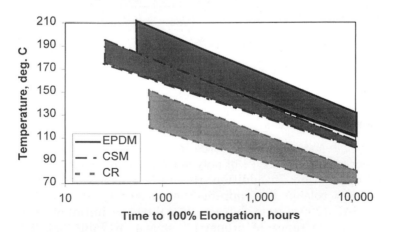

Figure 6.5 Hours to 100% elongation for three rubbers.

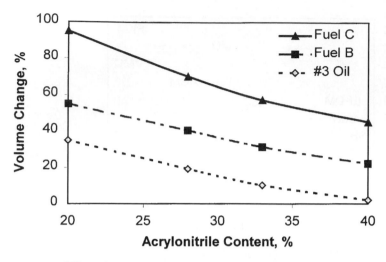

Figure 6.6   Effect of acrylonitrile content on volume change of NBR.

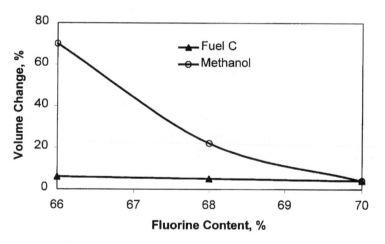

Figure 6.7   Effect of fluorine content on volume change of FKM.

6.6.1   Initial Physical Properties

The heat and oil resistance encountered in the application helps the design engineer to select the type of polymer most likely to perform in the intended application. In addition, the initial physical properties play a significant role in determining the suitability for use. The ASTM D2000/SAE J200 system characterizes the basic initial physical properties across the range of properties shown in Table 6.6. The durometer A hardness is measured by an indentor of specific radius.

TABLE 6.6   Basic Initial Physical Properties of Rubbers

Polymer	Hardness durometer A	Min. tensile strength, MPa	Min. elongation at break, %
NR, IR	30–90	3–24	75–500
SBR, BR	30–90	3–17	75–400
IIR, CIIR, BIIR	30–90	3–17	75–400
EPM, EPDM	30–90	7–17	100–500
T	40	3–7	100–400
CR	30–90	3–24	50–500
AU, EU	40–90	7–28	50–400
NBR	60–90	3–17	50–300
CSM	50–80	7–17	200–400
EAM	50–90	6–14	100–500
ACM	40–80	6–9	100–250
MQ	30–80	3–8	50–400
FVMQ	60	6	150
FKM	60–90	7–14	100–200

The durometer measurement generally correlates to the load-bearing capability of the rubber article. A tolerance of ±5 durometer A points from the specified target is typically allowed.

Accuracy declines toward either end of the 0 to 100 durometer A scale. Because rubber includes soft sponge and extremely hard, plastic-like urethane, other hardness measurements are typically used for these materials. The OO durometer has a blunter indentor for sponge; the more pointed D durometer is used for hard rubber. Some applications use hardness measurements specific to their industry, such as the Pusey and Jones (P&J) scale in rubber-covered rolls for paper mills.

Minimum values for the ultimate tensile strength and the elongation at break are two other basic properties in the ASTM D2000/SAE J200 system. These two properties are useful to control the uniformity of a particular compound but generally do not correlate with end-use performance. At a particular hardness, not all levels of tensile and elongation may be obtainable.

To measure tensile and elongation, the test specimen is secured in the jaws of a tensile test machine and stretched at a specified rate until it breaks. The final force required for the break is recorded, along with the amount of stretch that was achieved at the break point. The forces in effect at various degrees of elongation of the specimen usually are also recorded. These forces are used to calculate the stresses per unit area at those elongations, which are reported as tensile moduli. These are typically written as the M-100 (or L-100), which is the stress at 100% strain, the M-200, M-300, etc.

It is very important to understand that none of these numbers is a classic modulus, that is, a basic ratio of stress to strain that applies across a wide range of strains. With the possible exception of a narrow region of moderately low strain, the stress-strain plot for elastomers is always nonlinear, and these are secant moduli drawn to various points of the particular curve that applies at that temperature and rate of strain. In Figure 6.8, a typical stress-strain plot is displayed, with the lines drawn on it that illustrate what the secant and tangent moduli are. (The tangent modulus, which estimates the force required to deform the rubber in the strain region of interest, is more meaningful for many engineering applications but is not commonly used.)

For a few materials, such as steel, a comparatively pure modulus can be measured that is not a function of strain, temperature, or rate of strain. It is a single point of information, so to speak, that applies very broadly. In contrast, the response of rubber (and many other polymeric materials) to a deforming force is not a point; it is a three dimensional surface, the axes of which are temperature, amount of strain (deformation), and rate of strain.

What can be said simply about responses of rubber to different conditions is that the material will become stiffer as the temperature drops or as the rate of strain is increased. There is also the well known Mullins effect, which says that, for many rubber compounds, the force

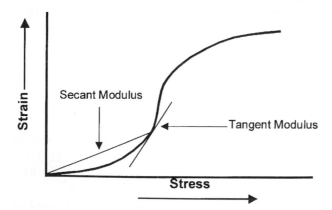

Figure 6.8 Stress/strain plot for a rubber specimen.

necessary to deform them the very first time will be significantly more than will be required on subsequent deformations. Also, for moderate deformations (10–50%), rubber undergoes what engineers refer to as *strain softening;* this means that, as the rubber is forced to deform, it takes less force per unit of deformation to achieve a high strain than a lower one. As an example, the dynamic modulus of a compound in shear at a level of ±10% might be 150 psi, but under the greater strain of ±20%, that modulus will be less than 300 psi. It will still take more force to make the rubber deform to the greater strain, but not twice as much force.

These special characteristics of rubber add up to the important concept. The single-cycle stress-strain curve to rupture of a rubber specimen at room temperature cannot generate data with the kind of meaning that exists in tensile testing other materials—definitely not the kind of meaning that exists in tensile testing of metals.

6.6.2 Specifying Heat and Oil Resistance

The ASTM D2000/SAE J200 system specifies a maximum change in properties after heat aging. The allowable change is measured after 70 hours at the maximum service temperature (see Table 6.9 for the test temperature). The basic change in properties that is permitted is identical for all rubbers. For particularly severe applications, the allowable change in properties after heat aging may be tightened by incorporating suffix requirements. The basic and most stringent suffix property changes are summarized in Table 6.7.

For oil-resistant rubbers, the ASTM D2000/SAE J200 system specifies a maximum change in volume after immersion an oil of standard composition, called ASTM No. 3 oil. (Since the supply of No. 3 oil has been depleted, testing is currently being performed with its replacement, 903 oil.) The allowable change is generally is measured after 70 hours immersion at the normal maximum service temperature. Above 150°C, the oil tends to degrade, and immersion tests on higher temperature rubbers are not run above this limit. The basic property is a maximum volume swell that depends on the specific rubber. Incorporating suffix requirements that reduce the allowable change on volume swell and limit the change in physical properties may help performance in severe oil immersion applications. The basic and most stringent suffix property changes for immersion in No. 3 Oil are summarized in Table 6.8.

6.7 Other Properties

The heat and oil resistance encountered in the application may not define all the requirements necessary for successful use. Depending on

TABLE 6.7 **Basic and Suffix Heat Resistance of Rubbers**

Polymer	Hardness Δ, pts.		Tensile Δ, %		Elongation Δ, % max.	
	Basic	Suffix	Basic	Suffix	Basic	Suffix
NR, IR	±15	+10 max.	±30	−25 max.	−50	−25
SBR, BR	±15	+10 max.	±30	−25 max.	−50	−25
IIR, CIIR, BIIR	±15	+10 max.	±30	−25 max.	−50	−25
EPM, EPDM	±15	+10 max.	±30	−20 max.	−50	−40
T	±15	+15 max.	±30	−15 max.	−50	−40
CR	±15	+15 max.	±30	−15 max.	−50	−40
AU, EU	±15	±5	±30	±15	−50	−15
NBR	±15	±10	±30	−20 max.	−50	−30
CSM*	±15	±20	±30	±30	−50	−60
EAM	±15	+10 max.	±30	−30 max.	−50	−50
ACM	±15	+10 max.	±30	−30 max.	−50	−40
MQ	±15	+10 max.	±30	−25 max.	−50	−25
FVMQ*	±15	+15 max.	±30	−45 max.	−50	−45
FKM	±15	+10 max.	±30	−25 max.	−50	−25

*Suffix heat aging determined at 25°C above basic requirements.

the particular application, certain additional properties may be required.

Power transmission belts often require good tear strength and resistance to crack growth during flexing. O-rings and seals generally specify a maximum compression set, or the allowable unrecovered deformation after aging while compressed between two steel plates (see Table 6.9). Burst strength is an important property in hose. As with many applications, both part design and polymer selection are important for best performance.

Frequently, the attainment of one property involves a trade-off or a sacrifice in other areas. An example is the balance that must be struck in tire treads among abrasion resistance, fuel economy, and wet traction.

Some of the comparative properties of rubbers are shown in Table 6.10. The ratings are only a general guideline, because the specific

TABLE 6.8   Basic and Suffix No. 3 Oil Resistance of Rubbers

Polymer	Vol. Δ, % max.		Hardness Δ, pts. (suffix)	Tensile Δ, % max. (suffix)	Elong. Δ, % max. (suffix)
	Basic	Suffix			
CR	+120	+80	−10 to +15	−45	−30
CSM	+80				
EAM	+80	+50		−50	−50
MQ	+80	+60	−30 max.		
AU, EU	+40	0 to +6	−10 to +5	−35	−40
ECO	+30	0 to +15	−5 to +10	−10	−50
ACM	±30	+25	−20 max.	−40	−30
T	+10	—	−5 to +10	−30	−50
NBR	+10	0 to +5	−10 to +5	−20	−30
FVMQ	+10	0 to +10	0 to −10	−35	−30
FKM	+10				

compounding, processing, and part design can affect actual performance.

For example, electrical properties are highly dependent on the type of filler used. The addition of nonconductive mineral fillers is employed in wire and cable insulations for high resistivity and good dielectric strength. Conversely, the use of high-structure carbon blacks achieves antistatic or electrical conductive properties to drain static charges from walk-off mats and mouse pads. Electrical conductivity versus carbon black concentration in EPDM is shown in Figure 6.9 for four high-structure blacks.

Another example of the general nature of the ratings shown in Table 6.10 is the resistance to hydrocarbons and oils. Crankcase lubricants are based on paraffinic hydrocarbon base oils. The fully formulated lubricant may contain about 20% functional additives that improve specific performance properties: lower friction, increased viscosity index, dispersancy and/or detergency provided, etc. These additives can attack certain rubbers that one might expect to be impervious to the base oil itself. In any contemplated use, the rubber part should be experimentally tested under actual use conditions before adoption in production.

TABLE 6.9   Basic and Suffix Compression Set of Rubbers

Polymer	Test temp., °C	Compression set, % max.	
		Basic	Suffix
NR, IR	75	50	25
T	75	50	50
SBR, BR	100	50	25
IIR, CIIR, BIIR	100	50	25
CR	100	80	35
AU, EU	100	50	25
NBR	100	50	25
CSM	125	50*	60
ECO	125	50	25
EPM, EPDM	150	60	60
EAM	175	75	50
ACM	175	75	60
FVMQ*	200	50	—
MQ	225	50	25
FKM	250	35	15

*Compression set determined at 70°C.

6.7.1   Rubber in Motion

When rubber is deformed and then allowed to recover, not all the energy input is recovered. That is, rubbers are not purely elastic but exhibit a significant viscous component that can be used for energy management purposes. The ratio of elastic to viscous response depends on several factors: temperature, frequency, strain amplitude (both static and dynamic), the particular polymer, and how it is compounded.

To characterize rubber's dynamic response, it is helpful to examine both purely elastic and purely viscous responses as shown in Figure 6.10. In the deformation of a purely elastic Hookean spring, there is a linear response of stress to strain. The ratio of stress to strain does not change, no matter how rapidly the strain is applied. Metal springs typically exhibit Hookean elasticity.

TABLE 6.10   Comparative Properties of Rubbers (from Ref. 2)

	1	2	3	4	5	6	7	8
ASTM classifications D1418 D2000	NR IR AA	SBR AA	EPR EPDM DA	CR BC	IIR AA	BIIR CIIR BA	NBR BK	CSM CE
Density, mg/m^3	0.93	0.94	0.86	1.23	0.92	0.92	1.00	1.10
Hardness, Shore A	20–90	40–90	40–90	20–95	40–75	40–75	20–95	45–95
Typical tensile strength								
Pure gum, MPa	21	7	3	21	10	10	7	14
Pure gum, psi	3000	1000	400	3000	1500	1500	1000	2100
Reinforced, MPa	21	14	21	21	14	14	14	19
Reinforced, psi	3000	2000	3000	3000	2000	2000	2000	2800
Resilience								
Room temp.	E	G	VG	VG	L	L	G	G
Hot	E	G	VG	VG	VG	G	G	G
Resistance to								
Tear	E	F	G	G	G	G	F	F
Abrasion	E	G	G	E	G	F-G	G	E
Compression set	G	G	VG	F-G	F	G	G	F
Weathering	E	VG	E	E	VG	VG	F-G	E
Oxidation	G	G	E	VG	E	VG	G	E
Ozone	P	P	E	VG	G	VG	P	O
Temperature range								
High temp.	G	G	E	G	G	VG	VG	VG
Low temp.	E	G	E	G	G	F	G	F
Aqueous fluid resistance								
Dilute acid	E	F-G	E	VG	E	E	G	E
Conc. acid	F-G	F-G	G	G	E	G	G	E
Water	VG	VG	E	G	VG	VG	F-G	G
Organic fluid resistance								
Aliphatic (A)	P	P	P	G	P	P	E	G
Oxygenated (B)	F	F	G	P	G	G	P	P
Chlorinated (C)	P	P	P	P-F	P	P	P-F	P-F
Aromatic (D)	P	P	P	F	P	P	G	F-G
Fuels (E)	P	P	P	G	P	P	E	E
Fats and oils (F)	P-G	P-G	G	G	VG	VG	E	G
Permeability	F	F	F-P	L	VL	VL	L	L
Flame resistance	P	P	P	G	P	P	P	F-G
Dielectric properties	E	G	E	VG	G-E	VG	P	E

A = hexane, isooctane, etc., B = acetone, methyl-ethyl ketone, etc., C = chloroform, etc., D = toluene, xylene, etc., E = kerosone, gasoline, etc., F = animal and vegetable oils.

TABLE 6.10   Comparative Properties of Rubbers *(Continued)*

	9	10	11	12	13	14	15	16
ASTM classifications D1418 D2000	CO ECO CH	CM BC	ACM EH	AU EU BG	T AK	MQ GE	FKM HK	FVMQ FK
Density, mg/m^3	1.27–1.36	1.16–1.32	1.09	1.02	1.20	1.1–1.6	1.85	1.47
Hardness, Shore A	40–90	40–95	40–90	60–95	20–80	10–85	60–95	40–70
Typical tensile strength								
Pure gum, MPa	—	10	3	42	1	1	14	—
Pure gum, psi	—	1500	400	6000	200	200	2000	—
Reinforced, MPa	14	14	12	42	9	8	14	10
Reinforced, psi	2000	2000	1800	6000	1300	1100	2000	1500
Resilience								
Room temp.	P-F	F	L	L-G	F	VG	L	G
Hot	P-F	F	VG	G	F-G	VG	VG	VG
Resistance to								
Tear	F	G	P	O	P	P	F	P
Abrasion	G	VG	F	O	P-F	P	G	P
Compression set	F	VG	G	F	F	VG	VG	VG
Weathering	E	E	E	E	E	E	E	E
Oxidation	VG	E	E	G	E	E	O	E
Ozone	VG	E	G	E	E	E	O	E
Temperature range								
High temp.	E	VG	E	G	F-G	O	O	O
Low temp.	F-G	G	P	P-G	E	O	P-G	O
Aqueous fluid resistance								
Dilute acid	G	VG	F	P	G	F	E	E
Conc. acid	F	VG	F	P	P	F	E	G
Water	G	G	P	P	F	G	VG	VG
Organic fluid resistance								
Aliphatic (A)	G	G	E	E	E	P	E	E
Oxygenated (B)	P	P	P	P-F	F	F	P	P
Chlorinated (C)	P-F	P	P	P-F	P-F	P-F	E	G
Aromatic (D)	G	F	F	F-G	VG	P	E	E
Fuels (E)	E	G	G	G	E	F	E	E
Fats and oils (F)	E	G	VG	E	G	F	E	E
Permeability	L	L	L	L	L	F	L	F
Flame resistance	F-P	G	P	F	P	F	G	G
Dielectric properties	G	G	F	F	F-G	E	G	G

A = hexane, isooctane, etc., B = acetone, methyl-ethyl ketone, etc., C = chloroform, etc., D = toluene, xylene, etc., E = kerosone, gasoline, etc., F = animal and vegetable oils.

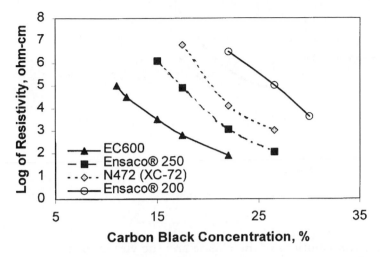

Figure 6.9   Effect of carbon blacks on electrical conductivity in EPDM.[3]

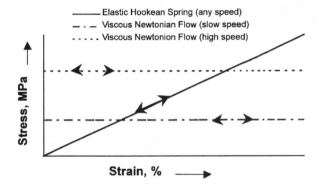

Figure 6.10   Elastic and viscous stress/strain responses.

Viscous behavior is typified by a shock absorber, i.e., a cylinder filled with a fluid through which a piston is moved. For pure viscous behavior, the fluid must be Newtonian. That is, the fluid will exhibit a linear response to strain rate and show no dependency on displacement.

The viscoelastic stress/strain response of a typical rubber is shown in Figure 6.11. Initially, stress increases in response to strain. The stress will then almost plateau at a level that depends on strain rate. At higher deformation, the finite extensibility of the polymer chains is reached, and the curve bends upward toward the break point. This latter part of the curve can not be readily modeled theoretically, and rubbers are generally not used at strains of this magnitude—at least for long periods of time.

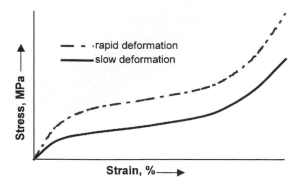

Figure 6.11 Viscoelastic stress/strain response of rubber.

The specific contributions of the elastic and the viscous components can be separated by repeatedly cycling the rubber through a deformation range as indicated in Figure 6.12. The observed stress response will lead the strain deformation by a certain amount, the phase angle δ. With this angle and the force measured at maximum deformation, the respective elastic and viscous components can be calculated. The elastic component is in phase with the applied deformation; the viscous component is 90° out of phase. Often, the ratio of the viscous to elastic components (loss factor or tangent δ) is computed, because it is related to the amount of kinetic energy that is converted to heat energy.

When the deformation of rubber occurs to a constant energy, the hysteresis (heat generated) is equal to tangent δ. When the deformation is to constant strain, the hysteresis is determined by the product

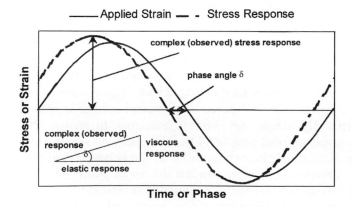

Figure 6.12 Determining the viscous and elastic components.

of the elastic response (E' if in tension or G' if in shear) times tangent δ. Finally, for a constant dynamic stress or load input, hysteresis is equal to the viscous response (E'' or G'') divided by the square of the complex (observed) response (E^* or G^*). Because the elastic response normally is much larger than the viscous response, the constant load hysteresis is approximated by the ratio of tangent δ to E' (or G').

As the temperature is lowered, rubbers become stiff and leather-like, where the tangent δ goes through a maximum. The maximum in tangent δ occurs slightly before the rubber transitions to a rigid, glassy state.

Increasing the test frequency has the same effect as lowering the temperature. For ideal viscoelasticity, a tenfold increase in frequency is approximately equal to a 10°C decrease in temperature. Some gum (unfilled) compounds exhibit nearly ideal viscoelasticity in which time (frequency) and temperature can be superimposed on a single master curve of dynamic behavior.

However, the incorporation of fillers complicates the situation, and ideal viscoelasticity is not observed in these compounds, which are representative of most rubber articles. The addition of fillers significantly increases the low strain dynamic modulus. However, at higher (1 to 10%) dynamic strain amplitude, the filler network breaks down, resulting in a rapid decrease in dynamic modulus and a maximum in tangent δ. At still higher strain amplitudes, the filled compound approximates the dynamic response of an unfilled gum compound.

The geometric design of the part also determines the dynamic response. A shape factor is calculated as the ratio of the loaded area of the part (A) to the area that is free to deform (L). The shape factor can be used to estimate various moduli (shear, compression, etc.), spring rates, and damping coefficients for simple shapes. However, complex shapes are less readily modeled. In this case, finite element analysis is applied for static deformation estimations, and dynamic simulations may be possible in the future.

For a given rubber part, the spring rate (K) and the damping coefficient (C) characterize the dynamic response. For simple geometries, K is equal to E' times A divided by L, where A and L are the loaded area and the area free to deform. The damping coefficient is the viscous response, calculated from E'' times A and divided by the product of L times ω, the test frequency. The "C to K ratio" then becomes the E'' divided by E' times ω (or tangent δ/ω).

In many applications, the rubber part supports a vibrating machine. As the speed of the machine changes, it affects the frequency of the vibrations that the rubber article partly absorbs and partly transmits. Transmissibility is the ratio of the transmitted force to the applied force ($T = F_t/F_a$).

The transmitted force is greater than the applied force in the low frequency attenuation region. The natural or resonant frequency (f_n) is determined by the spring rate of the rubber part (K') and the mass of the system (M) by the equation

$$f_n = 15.76\left(\frac{K'}{W}\right)^{1/2}$$

At resonance, transmissibility goes through a maximum that is proportional to the reciprocal of the damping coefficient or tangent δ. (see Figure 6.13). Normally, it is desirable to design the system so that resonance is experienced only occasionally, such as during startup.

In a plot of transmissibility versus frequency (Fig. 6.14), the high-frequency region is called the *isolation region,* and transmissibility is less than 1.0. The spring rate increases with frequency, and the amount of increase is generally greater with higher damping rubber compounds. In a log-log plot, the rolloff rate is therefore greater for lower damping compounds, and they transmit less vibration in the isolation region where the machines are typically designed to operate.

Over time, the flexing of a rubber part can cause fatigue, as evidenced by the development and growth of cracks. The fatigue life is strongly dependent on the dynamic strain (or stress, if deformed to constant load). In the extreme, a total failure can occur in one cycle. Alternatively, the part may last indefinitely at very low dynamic deformations.

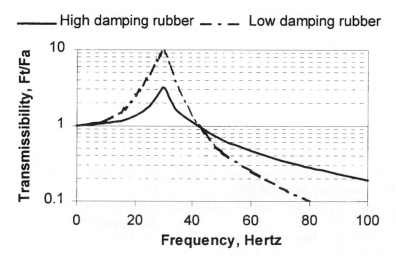

Figure 6.13   Transmissibility versus frequency.

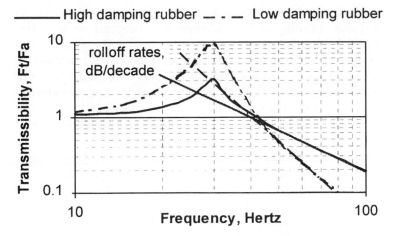

Figure 6.14   Transmissibility in the high-frequency isolation region.

Generally, higher tearing energy gives longer flex fatigue life, but the relationship is not always linear. Higher temperatures may shorten flex life due to lower tear strength of the rubber at elevated temperature. However, as modulus also declines with an increase in temperature, if the deformation is to constant strain, less energy is input per cycle, and longer flex life may ensue.

Strain-crystallizing rubbers (such as NR and CR) exhibit longer flex life if the minimum strain does not go to zero. In this case, it is believed that crystallites form at the tip of the growing crack, where maximum strain is encountered. The crystals blunt the crack tip and force the tear to travel around the crystals.

Both the vulcanization system and the antidegradant package can affect flex life. In general, flex life improves in going from peroxide to sulfur-donor to elemental sulfur cures. The use of antioxidants and antiozonants can improve fatigue life.

A final complication occurs in the measurement of flex fatigue life. Laboratory tests can have poor correlation with actual application results. Reproducibility can also be a problem, since crack initiation is thought to occur at microscopic inhomogenities in the rubber sample, which depend on how well each test specimen is prepared. To achieve more reproducible results, the laboratory specimen is often cut or nicked to measure crack growth rather than crack initiation.

When rubber moves in relation to a contacting surface, wear of the rubber can occur. Again, laboratory tests generally do not correlate well with application results, because the loss of rubber from the surface depends on the service conditions, which typically vary with time, temperature, frequency, strain, etc.

The mechanism by which wear or loss of rubber occurs depends on service conditions. Sliding abrasion, such as observed with a tire tread, is caused by hard surface projections cutting the rubber. Impingement abrasion, such as encountered in a sandblasting hose, is due to high-speed particles impacting the rubber. Less frequently encountered is adhesive wear in which rubber particles are transferred to another surface because of high adhesion to the surface.

Sliding abrasion can be improved by adding reinforcing carbon blacks of small particle size to the rubber compound. Generally, there is an optimal level of carbon black that depends on the specific black(s), polymer(s), and operating conditions. On the other hand, the resistance to impingement abrasion is often best in unfilled gum compounds.

6.7.2 Friction

Dry rubber surfaces are generally accepted as having high coefficients of friction, but measurement of COF can be done in many ways (ASTM D1894 is one method), which will generate very different numbers. For instance, static COF, the force needed to start movement across a rubber surface, can be quite high, with levels ranging easily as high as 1 and often appreciably greater, such as 2–4. Dynamic COF, the force required to maintain movement, is always less and can range from 0.2 up toward 1. It should be noted that actual movement across a rubber surface is almost always in the mode of a stick-slip process, and usually the COF is calculated from average force recorded.

Furthermore, many factors affect the frictional force measured, which include the rubber hardness, the load and speed used in the test, and the particular material and surface morphology of the surface against which the rubber is pressed. Compounding differences of polymer type and especially additives (which can bloom to the rubber surface and act to lubricate or tackify) have major effects. Lubrication of rubber surfaces by light oils or soapy water can render them very slippery, with the COF dropping to levels below 0.1 and the disappearance of the stick-slip phenomenon. With all these variables in play, the determination of frictional properties of any rubber compound should be carefully considered in light of the actual application and its environment.

6.7.3 Permeability

The ability to retain air is a key property of the modern tire and other articles. Relative to other materials of construction, however, rubber is relatively permeable to the migration of small molecules.

Permeability is mainly determined by polymer, and specifically its Tg. In general, the higher is the temperature (above Tg), the greater is the permeability to gases. However, use of plasticizer and the selection of filler also modify permeability. Plasticizers tend to depress Tg and therefore increase permeability. Fillers, particularly platy ones such as talc and clay, can decrease permeability by creating a longer, more tortuous diffusion path. The permeability of selected rubbers to various penetrant molecules is shown in Table 6.11.

TABLE 6.11   Permeability, $(10^{-9})(m^2)/(sec)(Pa)$, of Some Rubbers to Various Gases (from Ref. 4)

	Helium		Oxygen		Nitrogen		Carbon dioxide	
Gas temp., °C	25	50	25	50	25	50	25	50
Butyl rubber (IIR)	6.3	17.1	0.98	3.98	0.25	1.25	3.89	14.1
Nitrile rubber (39% ACN)	5.1	14.0	0.72	3.45	0.18	1.07	5.60	22.1
Polychloroprene (CR-G)			2.96	9.97	0.88	3.50	19.2	55.8
SBR (23% styrene)	17.3	41.5	12.8	34.0	4.74	14.3	92.8	192
Natural rubber (NR)	23.4	51.6	17.5	46.4	6.04	19.1	98.3	218
Silicone (VMQ)			395	493	197	276	1580	1530

6.7.4   Flame Resistance

Being composed of oxidizable carbon and hydrogen, most polymers will burn. Incorporating antimony oxide (Sb_2O_3) and a halogen (typically at a 1:3 ratio) is a cost-effective method to achieve flame retardance. Polymers such as CR and CPE, as well as halogenated plasticizers, can supply the halogen source. However, this technique of Sb_2O_3 and halogen generates smoke (which can obscure exit signs) and acid that corrodes electrical equipment (and can be toxic).

Other methods for low-smoke (halogen-free) rubber compounds include materials that form a glassy char on the surface, such as zinc borate or phosphate plasticizers, and/or the use of magnesium salts, such as carbonates or hydroxides, that release carbon dioxide and water, respectively. The use of polymers with a high oxygen content, as well as inorganic fillers such as clay to dilute the oxidizable components, is also beneficial. Due to the large quantity of additives required, the nonhalogen systems usually have inferior properties.

Alumina trihydrate (ATH) is incorporated in rug backing compounds to release water on heating and help the material pass a *pill* test, when a hot metal disk is placed on the carpet, which should not ignite. The decomposition temperature of ATH is too low for many applications, and it detracts from physical properties.

6.8 Compounding Rubber

A typical rubber compound may contain a dozen different ingredients, each with a specific function.

Fillers represent the largest volume category of materials added to polymers and may be subdivided into *reinforcing* or *extending* types. Reinforcing fillers increase hardness and modulus, improve tensile and tear strength, and provide abrasion resistance. The purpose of extending fillers is to lower costs without sacrificing key performance properties. Many compounds use a mixture of both reinforcing and extending fillers.

Reinforcing fillers are typically carbon blacks, because they provide the best properties. Carbon blacks are characterized by two parameters: particle size and *structure,* the way in which the particles are fused together. Particle size is typically measured by surface area, either nitrogen absorption or iodine number, with larger surface area corresponding to smaller particle size. Structure is measured by dibutyl phthalate (DBP) absorption. The more adsorption of the relatively large DBP molecule into the chain-like structure of the aggregates it takes to wet the black, the greater is the black's structure.

Smaller particle, larger surface area blacks provide higher hardness and tensile strength, with poorer resilience and more difficulty in obtaining a uniform dispersion in the rubber. Higher structure blacks increase hardness and modulus, decrease ultimate elongation, disperse more rapidly in the rubber, and give higher dimensional stability or "green" strength to the uncured compound.

There are also non-black reinforcing fillers, such as precipitated silica and hard clay, that can approximate the properties of carbon black. These or the extending fillers can be used to give bright white or colorable compounds for ready identification. Treating some non-black fillers with silanes, molecules that increase polymer-to-filler bonding, can bring the performance of non-black fillers even closer to that of carbon black or, for some properties, even surpass carbon blacks.

Plasticizers represent a second large volume category of compounding ingredients. These are typically either hydrocarbon oils or ester plasticizers.

Oils are used in hydrocarbon polymers to provide lower hardness, better low-temperature performance, and improved processing. Oils for rubber are available in three general types (aromatic, naphthenic, and paraffinic), although each type contains a mixture of the three structures. Each type is available in various viscosities, with the higher-viscosity oils giving low volatility and good permanence in the finished vulcanizate. Lower-viscosity oils are more volatile and affect processing more than end-use properties.

The ester plasticizers are more polar and provide similar properties in the oil-resistant specialty rubbers. A wide variety of esters are available to serve a range of cost and performance goals. In some critical seal applications, plasticizers should be avoided, because they are not permanently bound into the polymer and can either be extracted by the fluid being contained or volatilized (lost) during high temperature exposure.

Vulcanizing agents provide a critical function, that of crosslinking adjacent polymer chains together, which prevents flow and permits the recovery from deformation. The vulcanizing system can be the simple addition of one additive such as a peroxide or metal oxide, or it can be a complex mixture of several ingredients, the components of which may diffuse into different phases of a polymer blend before reacting. Literally hundreds of materials are available for curing rubbers.

Vulcanizing agents may be broadly grouped into five categories: peroxides, elemental sulfur, sulfur donors, metal oxides, and multifunctional additives—typically amines, phenols, or thiophenols. In addition to the crosslinking agent itself, various accelerators, activators, and retarders are used in the vulcanizing system.

Peroxides generate the most thermally stable carbon-to-carbon crosslink bonds. Peroxide-cured vulcanizates exhibit outstanding heat resistance and compression set. They are deficient in some failure properties, particularly flex fatigue, and can cause problems in processing, such as tacky surfaces if exposed to air during cure.

Sulfur donor cure systems have good thermal stability, approaching that of peroxides. Table 6.12 compares sulfur and peroxide cure systems in nitrile rubber (NBR). The high-temperature performance of these systems can be improved by adding a good antioxidant package, as indicated in compounds B and D.

The sulfur donor cure depends on the ability of certain molecules to donate an atom of sulfur for crosslinking purposes. This monosulfide crosslink has better heat resistance that the polysulfide crosslinks formed when elemental sulfur (an eight-membered sulfur ring) is used.

Elemental sulfur is the most widely used cure system for general-purpose rubbers. It provides good failure properties and high flex fa-

TABLE 6.12   Comparison of Peroxide and Sulfur Donor Cures in NBR (from Ref. 5)

Compound	A	B	C	D
Hycar® 1042	100	100	100	100
Zinc oxide	5	5	5	5
CORAX® N774 carbon black	65	65	65	65
Paraplex® G-50 plasticizer	15	15	15	15
VAROX® DCP-40KE	4	4	—	—
Sulfur	—	—	0.3	0.3
ETHYL CADMATE®	—	—	2.5	2.5
MORFAX®	—	—	2.5	2.5
AGERITE® SUPERFLEX® SOLID	—	1	—	1
VANOX® ZMTI	—	2	—	2
Press cured physical properties				
Cured 10 min at 171°C (340°F)				
100% modulus, MPa (psi)	6.9 (1000)	5.7 (830)	4.2 (610)	6.7 (530)
Tensile strength, MPa (psi)	18.9 (2740)	19.2 (2780)	15.8 (2300)	15.6 (2270)
Elongation, %	210	250	530	340
Hardness, Shore A	70	67	66	65
Aged 1,000 hr at 100°C (212°F)				
Tensile change, %	−19	+4	+8	+5
Elongation change, %	−62	−20	−66	−39
Hardness change, points	+11	+8	+10	+7
Aged 70 hr at 150°C (302°F)				
Compression set, %	54	42	77	58

tigue resistance, and it is very economical. Its deficiencies are in poor high-temperature aging and compression set. See Table 6.13 for a comparison of a normal sulfur and a sulfurless efficient vulcanizing (EV) cure system in an SBR/IR blend.

Metal oxides can be used to crosslink certain polymers, notably some grades of polychloroprene, chlorosulfonated polyethylene, polysulfides, and carboxylated polymers. The properties provided by metal oxides are generally quite good. However, heat resistance is often limited.

Multifunctional additives are mainly used to crosslink specialty oil-resistant rubbers. Generally, the crosslinking additive is specific to one or two polymers. In addition, an accelerator is often used to speed up the rate of crosslinking. In many cases, an oven post-cure after a typical press cure is required to optimize compression set and other performance properties with these cures.

Antidegradants are used to prolong the useful life of a rubber article. They may protect against oxidative attack (a bulk effect) or ozone attack (a surface effect). Mixtures of several antioxidants can be used; one example is shown in Table 6.12. Similarly, more than one antiozonant is frequently found in rubber recipes. The addition of antiozonants and, to a lesser extent, antioxidants may involve the use of discoloring additives.

Many other additives are used in typical rubber compounds to provide improved mold flow, release from processing equipment, bonding to metal inserts, color, tack, resistance to antimicrobial growth, or other desired properties.

References

1. James V. Fusco, "The World of Elastomers, An Industry Overview," Paper No. G presented at the Rubber Division Meeting of the A.C.S., May 1996.
2. R.F. Ohm, ed., *The Vanderbilt Rubber Handbook,* Vol. 13, pp. 429–430, 1990.
3. Anon., Ensaco Carbon Blacks in Rubber Compounds, Erachem Europe, s.a. technical literature.
4. G. J. van Amerongen, Diffusion in Elastomers, *Rubber Chemistry and Technology,* Vol. 37, pp. 1065–1152, 1964.
5. Anon., *Vanderbilt News,* Vol. 39, no. 1, p.15, 1983.
6. R.F. Ohm, ed., *The Vanderbilt Rubber Handbook,* Vol. 13, p. 459, 1990.

Suggested Readings

Alan N. Gent, ed., *Engineering with Rubber,* A.C.S. Rubber Division, 1992. A practical introduction and theoretical reference.

P. B. Lindley, Engineering design with natural rubber, NR Technical Bulletin, The Malaysian Rubber Producers' Research Association, 1984.

Maurice Morton, ed., *Rubber Technology,* Van Nostrand Reinhold, 3rd ed., 1987. A basic introduction to rubber technology.

Harry Long, ed., *Basic Compounding and Processing of Rubber,* A.C.S. Rubber Division, 1985.

Frederick R. Eirich, ed., *Science and Technology of Rubber,* Academic Press, 2nd ed., 1994. An advanced text.

Robert F. Ohm, ed., *The Vanderbilt Rubber Handbook,* 13th ed., 1990, also available on CD-ROM, 14th ed., 1998. An introduction for neophytes and a reference for technologists.

Robert F. Ohm, "Introduction to Rubber Technology," presented at the Conference on Engineering Design with Elastomers, organized by the New York Rubber Group, Aug. 1990.

TABLE 6.13   Comparison of Sulfur and EV Cure Systems in SBR/IR Blend (from Ref. 6)

	Black filled		Mineral filled	
	Sulfur	EV	Sulfur	EV
SBR 1500C	70	70	—	—
SBR 1502	—	—	70	70
NATSYN® 2200	30	30	30	30
VANPLAST® R	2	2	2	2
VANFRE® AP-2 SPECIAL	2	2	2	2
Stearic acid	2	2	3	3
Zinc oxide	5	5	5	5
AGERITE® STALITE® S	2	2	—	—
AGERITE SUPERLITE® SOLID	—	—	2	2
Process oil (high aromatic)	25	25	—	—
Process oil (naphthenic)	—	—	5	5
Cumar® resin	—	—	5	5
VANWAX® H SPECIAL paraffin wax	—	—	1	1
CORAX® N-330 carbon black	60	60	—	—
N-990 carbon black	20	20	—	—
DIXIE CLAY®	—	—	50	50
Whiting (water ground)	—	—	50	50
HiSil® 233 precipitated silica	—	—	30	30
Sulfur	2.5	—	2.1	—
ALTAX®	1.25	—	—	—
AMAX®	—	—	—	1
ETHYL CADMATE®	—	1	—	1
METHYL TUADS®	0.15	0.6	—	—
ETHYL TUADS	—	0.4	—	—
MORFAX®	—	—	1.3	1
VANAX® A RODFORM®	—	2	—	—
Cured 20 min at 153 °C (307 °F)				
300% modulus, MPa (psi)	11.3 (1640)	11.0 (1600)	2.0 (290)	1.8 (260)
Tensile, MPa (psi)	16.9 (2450)	16.5 (2400)	8.4 (1220)	9.4(1440)
Elongation, %	450	440	960	830
Hardness	61	60	51	54
Aged 96 hr at 100 °C (212 °F)				
Tensile, % change	−24	−8	−26	+6
Elongation, % change	−60	−20	−62	−13
Hardness, points change	+11	+6	+11	+1
Aged 70 hr at 100 °C (212 °F)				
Compression set, %	42	12	46	29
Rheometer at 143°C (290°F)				
t´c90, minutes	16	11	9	11

Robert F. Ohm, "Compounding Rubber for Dynamic Applications," presented at the Educational Symposium on Designing, Compounding and Testing for Dynamic Properties, organized by the Akron and Northeast Ohio Rubber Groups, April 1995.

Robert F. Ohm, "Review of Antioxidants," presented to the Northeast Regional Rubber & Plastics Exposition, Sept. 1994.

Robert F. Ohm, "Review of Antiozonants," *Rubber World,* Vol. 208, no. 8, pp. 18–22, Aug. 1993, based on Paper No. D at the Rubber Division Meeting of the A.C.S., May, 1993.

R. P. Brown, *Physical Testing of Rubbers,* Applied Science, 1979.

TABLE 6.14   Some Common Trade Names for Elastomers and Compound Ingredients

Trade Name	Composition	Abbr.
ACSIUM[1]	Alkylated chlorosulfonated polyethylene	ACSM
Aflas[2]	Tetrafluouroehtylene-propylene copolymer	—
AGERITE[3] STALITE[3] S	Octylated diphenylamine	ODPA
AGERITE SUPERFLEX[3]	Acetone-diphenylamine reaction product	ADPA
AGERITE SUPERLITE[3]	Polybutylated Bisphenol A mixture	—
ALTAX[4]	Mercaptobenzothiazole disulfide	MBTS
AMAX[4]	N-oxydiethylenebenzothiazole-2-sulfenamide	OBTS
CORAX[5]	carbon black	—
Cumar[6]	tackifying resin	—
DIXIE CLAY[4]	Air floated hard clay	—
ENSACO[7]	Conductive carbon black	—
ETHYL CADMATE[4]	Cadmium diethyl dithiocarbamate	CdEC
ETHYL TUADS[4]	Tetraethyl thiuram disulfide	TETD
EXXPRO[8]	Polyisobutylene/brominated p-methylstyrene	BIMS
HiSil[9] 233	Precipitated silica	—
Hycar[10]	Nitrile-butadiene rubber	NBR
HYPALON[1]	Chlorosulfonated polyethylene	CSM
Hytrel[1]	Polyester thermoplastic elastomer	—
METHYL TUADS[4]	Tetramethyl thiuram disulfide	TMTD
MORFAX[4]	2-Morpholino benzothiazole disulfide	MBS
NATSYN[11] 2200	Synthetic polyisoprene	IR
Paraplex[12] G-50	Polymeric ester plasticizer	—
VANAX[4] A RODFORM[4]	Dithiodimorpholine, pelleted form	DTDM

TABLE 6.14 Some Common Trade Names for Elastomers and Compound Ingredients

Trade Name	Composition	Abbr.
VANFRE[4] AP-2 SPECIAL	Processing aid	—
VANOX[4] ZMTI	Zinc 2-mercaptotoluimidazole	ZMTI
VANPLAST[4] R	Plasticizer	—
VANWAX[4] H SPECIAL	Paraffin wax	—
VAROX[4] DCP-40KE	Dicumyl peroxide, 40% active	—

[1] ACSIUM, HYPALON, and Hytrel are registered trademarks of DuPont Dow Elastomers, L.L.C., Wilmington, DE.

[2] Aflas is a registered trademark of Dyneon, Minneapolis, MN.

[3] AGERITE, STALITE, SUPERFLEX and SUPERLITE are registered trademarks of The B.F. Goodrich Company, Akron, OH.

[4] ALTAX, AMAX, DIXIE CLAY, CADMATE, MORFAX, RODFORM, TUADS, VANAX, VANFRE, VANOX, VANPLAST, VANWAX and VAROX are registered trademarks of the R. T. Vanderbilt Company, Inc., Norwalk, CT.

[5] CORAX is a registered trademark of Degussa-Hüls Corp., Ridgefield Park, NJ.

[6] Cumar is a registered trademark of Neville Chemical Co., Pittsburgh, PA.

[7] ENSACO is a registered trademark of Erachem Europe s.a., Brussels, Belgium.

[8] EXXPRO is a registered trademark of Exxon Corp., Houston, TX .

[9] HiSil is a registered trademark of PPG Industries, Inc., Pittsburgh, PA.

[10] Hycar is a registered trademark of Zeon Chemicals, Inc., Louisville, KY.

[11] NATSYN is a registered trademark of The Goodyear Tire and Rubber Company, Akron, OH.

[12] Paraplex is a registered trademark of C.P. Hall Co., Chicago, IL.

Note: The R. T. Vanderbilt Company, Inc. sells the capitalized products.

6

Part 2
Elastomeric Materials and Processes*

James M. Margolis
Montreal, Province of Quebec, Canada

6.9 Introduction

Another important group of polymers is the group that is elastic or rubberlike, known as *elastomers*. This chapter discusses this group of materials, including TPEs, MPRs, TPVs, synthetic rubbers, and natural rubber.

6.10 Thermoplastic Elastomers (TPEs)

Worldwide consumption of TPE for the year 2000 is estimated to be about 2.5 billion pounds, primarily due to new polymer and processing technologies, with an annual average growth rate of about 6% between 1996 and 2000. About 40% of this total is consumed in North America.[4]

TPE grades are often characterized by their hardness, resistance to abrasion, cutting, scratching, local strain, and wear. A conventional measure of hardness is Shore A and Shore D, shown in Fig. 6.15. Shore A is a softer, and Shore D is a harder TPE, with ranges from as soft as Shore A 28 to as hard as Shore D 82. Durometer hardness (ASTM D 2240) is an industry standard test method for rubbery

---

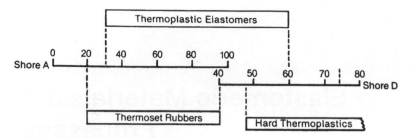

Figure 6.15   TPEs bridge the hardness ranges of rubbers and plastics. *(Source: Ref. 10, p. 5.2.)*

materials, covering two types of durometers, A and D. The *durometer* is the hardness measuring apparatus, and the term *durometer hardness* is often used with Shore hardness values. There are other hardness test methods such as Rockwell hardness for plastics and electrical insulating materials (ASTM D785 and ISO 2039), and Barcol hardness (ASTM D2583) for rigid plastics. While hardness is often a quantifying distinction between grades, it does not indicate comparisons between physical/mechanical, chemical, and electrical properties.

Drying times depend on moisture absorption of a given resin. TPE producers suggest typical drying times and processing parameters. Actual processing temperature and pressure settings are determined by resin melt temperatures and rheological properties, mold cavity design, and equipment design such as screw configuration.

Performance property tables provided by suppliers usually refer to compounded grades containing property enhancers (additives) such as stabilizers, modifiers, and flame retardants. Sometimes, the suppliers' property tables refer to a polymer rather than a formulated compound.

6.10.1 Styrenics

Styrene block copolymers are the most widely used TPEs, accounting for close to 45% of total TPE consumption worldwide at the close of the twentieth century.[1] They are characterized by their molecular architecture, which has a "hard" thermoplastic segment (block) and a "soft" elastomeric segment (block) (see Fig. 6.16). Styrenic TPEs are usually styrene butadiene styrene (SBS), styrene ethylene/butylene styrene (SEBS), and styrene isoprene styrene (SIS). Styrenic TPEs usually have about 30 to 40% (wt) bound styrene; certain grades have a higher bound styrene content. The polystyrene endblocks create a network of reversible physical cross links that allow thermoplasticity for melt

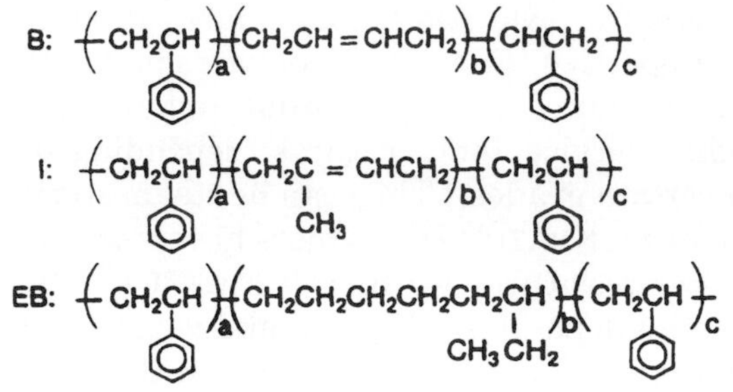

B: $\left(\text{CH}_2\text{CH}\right)_a\left(\text{CH}_2\text{CH} = \text{CHCH}_2\right)_b\left(\text{CHCH}_2\right)_c$

I: $\left(\text{CH}_2\text{CH}\right)_a\left(\text{CH}_2\text{C} = \text{CHCH}_2\right)_b\left(\text{CH}_2\text{CH}\right)_c$
$\qquad\qquad\qquad\quad\ \ \text{CH}_3$

EB: $\left(\text{CH}_2\text{CH}\right)_a\left(\text{CH}_2\text{CH}_2\text{CH}_2\text{CH}_2\text{CH}_2\text{CH}\right)_b\left(\text{CH}_2\text{CH}\right)_c$
$\qquad\qquad\qquad\qquad\qquad\qquad\quad \text{CH}_3\ \text{CH}_2$

Figure 6.16 Structures of three common styrenic block copolymer TPEs: a and c = 50 to 80; b = 20 to 100. *(Source: Ref. 10, p. 5.12.)*

processing or solvation. With cooling or solvent evaporation, the polystyrene domains reform and harden, and the rubber network is fixed in position.[2]

Principal styrenic TPE markets are: molded shoe soles and other footwear; extruded film/sheet and wire/cable covering; pressure-sensitive adhesives (PSA) and hot-melt adhesives; and viscosity index (VI) improver additives in lube oils, resin modifiers, and asphalt modifiers. They are also popular as grips (bike handles), kitchen utensils, clear medical products, and personal care products.[1,4] Adhesives and sealants are the largest single market.[1] Styrenic TPEs are useful in adhesive compositions in web coatings.[1]

Styrenic block copolymer (SBC) thermoplastic elastomers were produced by Shell Chemical (KRATON®) and are available from Firestone Synthetic Rubber and Latex, Division of Bridgestone/Firestone (Stereon®*), Dexco Polymers (Vector®†), and EniChem Elastomers (Europrene®‡). SBC properties and processes are described for these four SBCs.

KRATON§ TPEs are usually SBS, SEBS, and SIS, as are SEP (styrene ethylene/propylene) and SEB (styrene ethylene/butylene).[2] The polymers can be precisely controlled during polymerization to meet property requirements for a given application.[2]

Two KRATON types are chemically distinguished: KRATON G and (raton D, as described below. A third type, KRATON Liquid®, poly(ethyl-

---

* Stereon is a registered trademark of Firestone Synthetic Rubber and Latex Company, Division of Bridgestone/Firestone.

† Vector is a registered trademark of Dexco, A Dow/Exxon Partnership.

‡ Europrene is a registered trademark of EniChem Elastomers.

§ KRATON styrenics were developed by Shell Chemical. In March 2001, the company formally announced finalization of sale to Ripplewood Holdings. For information, contact Kate Hill, Shell International Media Relations, telephone 44(0)2079342914.

ene/butylene), is described thereafter. KRATON G and D have different performance and processing properties. KRATON G polymers have saturated midblocks with better resistance to oxygen, ozone and ultraviolet (UV) radiation, and higher service temperatures, depending on load, up to 350°F (177°C) for certain grades.[2] They can be steam sterilized for reusable hospital products. KRATON D polymers have unsaturated midblocks with service temperatures up to 150°F (66°C).[2] SBC upper service temperature limits depend on the type and weight percent (wt%) thermoplastic and type and wt% elastomer, and the addition of heat stabilizers. A number of KRATON G polymers are linear SEBS, while several KRATON D polymers are linear SIS.[2] KRATON G polymer compounds' melt process is similar to polypropylene; KRATON D polymer compounds' process is comparable to polystyrene (PS).[2]

Styrenic TPEs have strength properties equal to those of vulcanized rubber, but they do not require vulcanization.[2] Properties are determined by polymer type and formulation. There is wide latitude in compounding to meet a wide variety of application properties.[2] According to application-driven formulations, KRATONs are compounded with a hardness range from Shore A 28 to 95 (Shore A 95 is approximately equal to Shore D 40), sp gr from 0.90 to 1.18, tensile strengths from 150 to 5000 lb/in^2 (1.03 to 34.4 MPa), and flexibility down to −112°F (80°C) (see Table 6.15).[2]

TABLE 6.15   Typical Properties of a KRATON D and KRATON G Polymer for Use as Formulation Ingredients and as Additive (U.S. FDA Compliance)

Property [74°F (23°C)]	KRATON D D1101 (linear SBS)	KRATON G G1650 (linear SEBS)
Specific gravity, g/cm^3	0.94	0.91
Hardness, Shore A	71	75
Tensile strength, lb/in^2 (MPa)	4600 (32)	5000 (34)
300% modulus, lb/in2 (MPa)	400 (2.7)	800 (5.5)
Elongation, %	880	550
Set @ break, %	10	—
Melt flow index, g/10 min	<1	—
Brookfield viscosity, Hz @ 77°F (25°C), toluene solution	4000	8000
Styrene/rubber ratio	31/69	29/71

KRATONs are resistant to acids, alkalis, and water, but long soaking in hydrocarbon solvents and oils deteriorates the polymers.[2]

Automotive applications range from window seals and gasketing to enhanced noise/vibration attenuation.[1] The polymers are candidates for automotive seating, interior padded trim and insulation, hospital padding, and topper pads.[1] SEBS is extruded/blown into 1-mil films for disposable gloves for surgical/hospital/dental, food/pharmaceutical, and household markets.[1]

KRATONs are used in PSAs, hot-melt adhesives, sealants, solution-applied coatings, flexible oil gels, modifiers in asphalt, thermoplastics, and thermosetting resins.[2] When KRATONs are used as an impact modifier in nylon 66, notched Izod impact strength can be increased from 0.8 ft-lb/in for unmodified nylon 66 to 19 ft-lb/in. Flexural modulus may decrease from 44,000 lb/in^2 (302 MPa) for unmodified nylon 66 to about 27,000 lb/in^2 (186 MPa) for impact-modified nylon 66.

SBCs are injection molded, extruded, blow molded, and compression molded.[2]

KRATON Liquid polymers are polymeric diols with an aliphatic, primary OH$^-$ group on each terminal end of the poly(ethylene/butylene) elastomer. They are used in formulations for adhesives, sealants, coatings, inks, foams, fibers, surfactants, and polymer modifiers.[13]

Two large markets for Firestone's styrenic block copolymer SBS Stereon TPEs are (1) impact modifiers (enhancers) for flame-retardant polystyrene and polyolefin resins and (2) PSA and hot-melt adhesives. Moldable SBS block copolymers possess high clarity and gloss, have good flex cycle stability for "living hinge" applications, and come in FDA-compliant grades for food containers and medical/hospital products.[1] Typical mechanical properties are 4600-lb/in^2 (31.7-MPa) tensile strength, 6000-lb/in^2 (41.4-MPa) flexural strength, and 200,000-lb/in^2 (1.4-GPa) flexural modulus.[1]

Stereon stereospecific butadiene styrene block copolymer is used as an impact modifier in PS, high-impact polystyrene (HIPS), polyolefin sheet and films, such as blown film grade linear low-density polyethylene (LLDPE), to achieve downgauging and improve tear resistance and heat sealing.[1] Blown LLDPE film modified with 7.5% stereospecific styrene block copolymers has a Dart impact strength of 185°F per 50 g, compared with 135°F per 50 g for unmodified LLDPE film. These copolymers also improve environmental stress crack resistance (ESCR) (especially to fats and oils for meat/poultry packaging trays), increase melt flow rates, increase gloss, and meet U.S. FDA 21 CFR 177.1640 (PS and rubber modified PS) with at least 60% PS for food contact packaging.[1] When used with thermoformable foam PS, flexibility is improved without sacrificing stiffness, allowing deeper draws.[1] The stereospecific butadiene block copolymer TPEs are easily dispersed and improve blendability of primary polymer with scrap for recycling.

Vector SBS, SIS, and SB styrenic block copolymers are produced as diblock-free and diblock copolymers.[29] The company's process to make linear SBCs yields virtually no diblock residuals. Residual styrene butadiene and styrene isoprene require endblocks at both ends of the polymer to have a load-bearing segment in the elastomeric network.[29] However, diblocks are blended into the copolymer for certain applications.[29] Vector SBCs are injection molded, extruded, and formulated into pressure-sensitive adhesives for tapes and labels, hot-melt product-assembly adhesives, construction adhesives, mastics, sealants, and asphalt modifiers.[29] The asphalts are used to make membranes for single-ply roofing and waterproofing systems, binders for pavement construction and repair, and sealants for joints and cracks.[29] Vector SBCs are used as property enhancers (additives) to improve the toughness and impact strength at ambient and low temperatures of engineering thermoplastics, olefinic and styrenic thermoplastics, and thermosetting resins.[29] The copolymers meet applicable U.S. FDA food additive 21 CFR 177.1810 regulations and United States Pharmacopoeia (USP) (Class VI medical devices) standards for health-care applications.[29]

The company's patented hydrogenation techniques are developed to improve SBC heat resistance as well as ultraviolet resistance.[29]

EniChem Europrene SOL T products are styrene butadiene and styrene isoprene linear and radial block copolymers.[1] They are solution-polymerized using anionic type catalysts.[33] The molecules have polystyrene endblocks with central elastomeric polydiene (butadiene or isoprene) blocks.[33] The copolymers are *(S-B)nX* type where S = polystyrene, B = polybutadiene and polyisoprene, and X = a coupling agent. Both configurations have polystyrene (PS) endblocks, with bound styrene content ranging from 25 to 70% (wt).[1] Polystyrene contributes styrene hardness, tensile strength, and modulus; polybutadiene and polyisoprene contribute high resilience and flexibility, even at low temperatures.[1] Higher molecular weight (MW) contributes a little to mechanical properties but decreases melt flow characteristics and processibility.

The polystyrene and polydiene blocks are mutually insoluble, and this shows with two T_g peaks on a cartesian graph with tan δ (y axis) versus temperature (x axis): one T_g for the polydiene phase and a second T_g, for the polystyrene phase. A synthetic rubber, such as SBR, shows one T_g.[33] The two phases of a styrenic TPE are chemically bound, forming a network with the PS domains dispersed in the polydiene phase. This structure accounts for mechanical/elastic properties and thermoplastic processing properties.[33] At temperatures up to about 167°F (80°C), which is below PS T_g of 203 to 212°F (95 to 100°C), the PS phase is rigid.[33] Consequently, the PS domains behave as cross-linking sites in the polydiene phase, similar to sulfur links in vulcanized rubber.[33] The rigid PS phase also acts as a reinforcement, as noted here.[33] Crystal PS, HIPS, poly-alpha-methylstyrene, ethylene vinyl acetate (EVA) copolymers, low-density polyethylene (LDPE), and high-density polyethylene (HDPE) can be used as organic reinforcements. $CaCO_3$, clay, silica, and silicates act as inorganic fillers, with little reinforcement, and they can adversely affect melt flow if used in excessive amounts.[33]

The type of PS, as well as its percent content, affect properties. Crystal PS, which is the most commonly used, and HIPS increase hardness, stiffness, and tear resistance without reducing melt rheology.[1] High-styrene copolymers, especially Europrene SOL S types produced by solution polymerization, significantly improve tensile strength, hardness, and plasticity, and they enhance adhesive properties.[33] High styrene content does not decrease the translucency of the compounds.[33] Poly-alpha-methylstyrene provides higher hardness and modulus, but abrasion resistance decreases.[33] EVA improves resistance to weather, ozone, aging, and solvents, retaining melt rheology and finished product elasticity. The highest Shore hardness is 90 A, the highest melt flow is 16 g/10 min, and specific gravity is 0.92–0.96.[1]

Europrene compounds can be extended with plasticizers that are basically a paraflinic oil containing specified amounts of naphthenic and aromatic fractions.[33] Europrenes are produced in both oil-extended and dry forms.[1] Oils were specially developed for optimum mechanical, aging, processing, and color properties.[33] Increasing oil content significantly increases melt flow properties, but it reduces mechanical properties. Oil extenders must be incompatible with PS so as to avoid PS swelling, which would decrease mechanical properties even more.[33]

The elastomers are compounded with antioxidants to prevent thermal and photo-oxidation, which can be initiated through the unsaturated zones in the copolymers.[33] Oxidation can take place during melt processing and during the life of the fabricated product.[33] Phenolic, or phosphitic antioxidants, and dilauryldithiopropionate as a stabilizer during melt processing, are recommended.[33] Conventional UV stabilizers are used such as benzophenone and benzotriazine.[33] Depending on the application, the elastomer is compounded with flow enhancers such as low MW polyethylene (PE), microcrystalline waxes or zinc stearate, pigments, and blowing agents.[33]

Europrene compounds, especially oil-extended grades, are used in shoe soles and other footwear.[1] Principal applications are impact modifiers in PS, HDPE, LDPE, polypropylene (PP), other thermoplastic resins and asphalt; extruded hose, tubing, O-rings, gaskets, mats, swimming equipment (eye masks, snorkels, fins, "rubberized" suits) and rafts; and pressure-sensitive adhesives (PSA) and hot melts.[1] SIS types are used in PSA and hot melts; SBS types are used in footwear.[1]

The copolymer is supplied in crumb form, and mixing is done by conventional industry practices, with an internal mixer or low-speed room temperature premixing and compounding with either a single- or twin-screw extruder.[33] Low-speed premixing/extrusion compounding is the process of choice.

Europrenes have thermoplastic polymer melt processing properties and characteristics of TPEs. At melt processing temperatures, they behave as thermoplastics, and below the PS T_g of 203 to 212°F (95 to 100°C) the copolymers act as cross-linked elastomers, as noted earlier. Injection-molding barrel temperature settings are from 284 to 374°F (140 to 190°C). Extrusion temperature at the head of the extruder is maintained between 212 and 356°F (100 and 180°C).

6.10.2 Olefenics and TPO Elastomers

Thermoplastic polyolefin (TPO) elastomers are typically composed of ethylene propylene rubber (EPR) or ethylene propylene diene "M" (EPDM) as the elastomeric segment and polypropylene thermoplastic segment.[18] LDPE, HDPE, and LLDPE; copolymers ethylene vinyl acetate (EVA), ethylene ethylacrylate (EEA), ethylene, methyl-acrylate (EMA); and polybutene-1 can be used in TPOs.[18] Hydrogenation of polyisoprene can yield ethylene propylene copolymers, and hydrogenation of 1,4- and 1,2-stereoisomers of S-B-S yields ethylene butylene copolymers.[1]

TPO elastomers are the second most widely used TPEs on a tonnage basis, accounting for about 25% of total world consumption at the close of the twentieth century (according to what TPOs are included as thermoplastic elastomers).

EPR and polypropylene can be polymerized in a single reactor or in two reactors. With two reactors, one polymerizes propylene monomer to polypropylene, and the second copolymerizes polypropylene with ethylene propylene rubber (EPR) or EPDM. Reactor grades are (co)polymerized in a single reactor. Compounding can be done in the single reactor.

Montell's in-reactor Catalloy®* ("catalytic alloy") polymerization process alloys propylene with comonomers, such as EPR and EPDM, yielding very soft, very hard, and rigid plastics, impact grades, or elastomeric TPOs, depending on the EPR or EPDM percent content. The term *olefinic* for thermoplastic olefinic elastomers is arguable because of the generic definition of olefinic. TPVs are composed of a continuous

---

* Catalloy is a registered trademark of Montell North America Inc., wholly owned by the Royal Dutch/Shell Group.

thermoplastic polypropylene phase and a discontinuous vulcanized rubber phase, usually EPDM, EPR, nitrile rubber, or butyl rubber.

Montell describes TPOs as flexible plastics, stating "TPOs are not TPEs."[32a] The company's Catalloy catalytic polymerization is a cost-effective process, used with propylene monomers that are alloyed with comonomers, including the same comonomers with different molecular architecture. Catalloy technology uses multiple gas-phase reactors that allow the separate polymerization of a variety of monomer streams.[32] Alloyed or blended polymers are produced directly from a series of reactors that can be operated independently from each other to a degree.[32]

Typical applications are flexible products such as boots, bellows, drive belts, conveyor belts, diaphragms, and keypads; connectors, gaskets, grommets, lip seals, O-rings, and plugs; bumper components, bushings, dunnage, motor mounts, sound deadening; and casters, handle grips, rollers, and step pads.[9]

Insite[®]* technology is used to produce Affinity[®]† polyolefin plastomers (POPs), which contain up to 20 wt% octene co-monomer.[14] Dow Chemical's 8-carbon octene polyethylene technology produces the company's ULDPE Attane[®]‡ ethylene-octene-1 copolymer for cast and blown films. Alternative copolymers are 6-carbon hexene and 4-carbon butene, for heat-sealing packaging films. Octene copolymer POP has lower heat-sealing temperatures for high-speed form-fill-seal lines and high hot-tack strength over a wide temperature range.[14] Other benefits cited by Dow Chemical are toughness, clarity, and low taste/odor transmission.[14]

Insite technology is used for homogeneous single-site catalysts that produce virtually identical molecular structure, such as branching, co-monomer distribution, and narrow molecular weight distribution (MWD).[14] Solution polymerization yields Affinity polymers with uniform, consistent structures, resulting in controllable, predictable performance properties.[14]

Improved performance properties are obtained without diminishing processibility, because the Insite process adds long-chain branching onto a linear short-chain, branched polymer.[14] The addition of long-chain branches improves melt strength and flow.[14] Long-chain branching results in polyolefin plastomers processing at least as smooth as LLDPE and ultra-low-density polyethylene (ULDPE) film extrusion.[14] Polymer design contributes to extrusion advantages such as enhanced shear flow, drawdown, and thermoformability.

---

* Insite is a registered trademark of DuPont Dow Elastomers LLC.
† Affinity is a registered trademark of The Dow Chemical Company.
‡ Attane is a registered trademark of The Dow Chemical Company.

For extrusion temperature and machine design, the melt temperature is 450 to 550°F (232 to 288°C), the feed zone temperature setting
is 300 to 325°F (149 to 163°C), 24/1 to 32/1 *L/D;* for the sizing gear
box, use 5 lb/h/hp (1.38 kg/h/kW) to estimate power required to extrude POP at a given rate; for single-flight screws, line draw over the
length of the line, 10 to 15 ft/min (3 to 4.5 m/min) maximum. Processing conditions and equipment design vary according to the resin selection and finished product. For example, a melt temperature of 450 to
550°F (232 to 288°C) applies to cast, nip-roll fabrication using an ethylene alpha-olefin POP[14a]), while 350 to 450°F (177 to 232°C) is recommended for extrusion/blown film for packaging, using an ethylene
aipha-olefin POP.[14b]

POP applications are sealants for multilayer bags and pouches to
package cake mixes, coffee, processed meats and cheese, and liquids;
overwraps; shrink films; skin packaging; heavy-duty bags and sacks;
and molded storage containers and lids.[14]

Engage®* polyolefin elastomers (POEs), ethylene octene copolymer
elastomers, produced by DuPont Dow Elastomers, use Insite catalytic
technology.[5] Table 6.16 shows their low density and wide range of
physical/mechanical properties (using ASTM test methods).[5]

**TABLE 6.16   Typical Property Profile for Engage Polyolefin Elastomers
(Unfilled Polymer, Room Temperature Except where Indicated)**

Property	Values
Specific gravity, g/cm^3	0.857–0.91
Flexural modulus, lb/in^2 (MPa), 2% secant	435–27,55 (3–190)
100% modulus, lb/in^2(MPa)	145–>725 (1–>5)
Elongation, %	700+
Hardness, Shore A	50–95
Haze, %, 0.070 in (1.8 mm) injection-molded plaque	<10–20
Low-temperature brittleness, °F (°C)	<–104 (<–76)
Melt flow index, g/10 min	0.5–30
Melting point, °F (°C)	91–225 (33–107)

The copolymer retains toughness and flexibility down to 40°F
(–40°C).[5] When cross-linked with peroxide or silane, or by radiation,
heat resistance and thermal aging increase to >302°F (>150°C).[5]
Cross-linked copolymer is extruded into covering for low- and medium-
voltage cables. POE elastomers have a saturated chain, providing inherent UV stability.[5] Ethylene octene copolymers are used as impact
modifiers, for example, in polypropylene. Typical products are foams

---

* Engage is a registered trademark of DuPont Dow Elastomers LLC.

and cushioning components, sandal and slipper bottoms, sockliners and midsoles, swim fins, and winter and work boots; TPO bumpers, interior trim and rub strips, automotive interior air ducts, mats and liners, extruded hose and tube, interior trim, NVH applications, primary covering for wire and cable voltage insulation (low and medium voltage), appliance wire, semiconductive shields, nonhalogen flame-retardant and low smoke emission jackets, and bedding compounds.[5]

Union Carbide elastomeric polyolefin flexomers combine flexibility, toughness, and weatherability, with properties midrange between polyethylene and EPR.

6.11 Polyurethane Thermoplastic Elastomers (TPUs)

TPUs are the third most widely used TPEs, accounting for about 15% of TPE consumption worldwide.

Linear polyurethane thermoplastic elastomers can be produced by reacting a diisocyanate [methane diisocyanate (MDI) or toluene diisocyanate (TDI)] with long-chain diols such as liquid polyester or polyether polyols and adding a chain extender such as 1,4-butanediol.[17,18c] The diisocyanate and chain extender form the hard segment, and the long-chain diol forms the soft segment.[18c] For sulfur curing, unsaturation is introduced, usually with an allyl ether group.[17] Peroxide curing agents can be used for cross linking.

The two principal types of TPUs are polyether and polyester. Polyethers have good low-temperature properties and resistance to fungi; polyesters have good resistance to fuel, oil, and hydrocarbon solvents.

BASF Elastollan®* TPU elastomer property profiles show typical properties of polyurethane thermoplastic elastomers (see Table 6.17).

Shore hardness can be as soft as 70 A and as hard as 74 D, depending on the hard/soft segment ratio. Specific gravity, modulus, compressive stress, load-bearing strength, and tear strength are also hard/soft ratio dependent.[18c] TPU thermoplastic elastomers are tough, tear resistant, abrasion resistant, and exhibit low-temperature properties.[4]

Dow Plastics Pellethane®† TPU elastomers are based on both polyester and polyether soft segments.[3]

Five series indicate typical applications:

1. Polyester polycaprolactones for injection-molded automotive panels, painted (without primer) with urethane and acrylic enamels, or water-based elastomeric coating

---

* Elastopllan is a registered trademark of BASF Corporation.
† Pellethane is a registered trademark of The Dow Chemical Company.

TABLE 6.17   Mechanical Property Profile of Elastollan

Property	Value
Specific gravity, g/cm^3	1.11–1.21
Hardness, Shore	70 A–74 D
Tensile strength, lb/in^2 (MPa)	4600–5800 (31.7–40.0)
Tensile strength, lb/in^2 (MPa):	
100% elongation	770–1450 (5.3–10.0)
300% elongation	1300–1750 (9.0–12.0)
Elongation @ break, %	550–700
Tensile set @ break, %	45–50
Tear strength, Die C pli	515–770
Abrasion-resistance, mg loss (Tabor)	25

2. Polyester polycaprolactones for seals, gaskets, and belting

3. Polyester polyadipates extruded into film, sheet, and tubing

4. Polytetramethylene glycol ethers with excellent dielectrics for extruded wire and cable covering, and also for films, tubing, belting, and caster wheels

5. Polytetramethylene glycol ethers for healthcare applications[3]

Polyether-polyester hybrid specialty compounds are the softest non-plasticized TPU (Shore hardness 70 A), and they are used as impact modifiers.

Polycaprolactones possess good low-temperature impact strength for paintable body panels, good fuel and oil resistance, and hydrolytic stability for seals, gaskets, and belting.[3] Polycaprolactones have fast crystallization rates and high crystallinity, and they are generally easily processed into complex parts.

Polyester polyadipates have improved oil and chemical resistance but slightly lower hydrolytic stability than polycaprolactones, which are used for seals, gaskets, and beltings.

Polytetramethylene glycol ether resins for wire/cable covering have excellent resistance to hydrolysis and microorganisms, compared with polyester polyurethanes. Healthcare grade polyether TPUs are resistant to fungi, have low levels of extractable ingredients, offer excellent hydrolysis resistance, and can be sterilized for reuse by gamma irradiation, ethylene oxide, and dry heat—but not with pressurized steam (autoclave).[3] Polyether TPUs are an option for sneakers and athletic footwear components such as outer soles.

Bayer Bayflex®* elastomeric polyurethane reaction injection molding (RIM) is a two-component diphenylmethane diisocyanate- (MDI)-

---

* Bayflex is a registered trademark of Bayer Corporation.

based liquid system produced in unreinforced, glass-reinforced, and mineral-/microsphere-reinforced grades.[15] They possess a wide stiffness range, relatively high impact strength, and quality molded product surface, and they can be in-mold coated. Room temperature properties are:

- Specific gravity, 0.95 to 1.18
- Ultimate tensile strength, 2300 to 4200 lb/in^2 (16 to 29 MPa)
- Flexural modulus, 5000 to 210,000 lb/in^2 (34 to 1443 MPa)
- Tear strength, Die C, 230 to 700 lb/in (40 to 123 kN/m)[15]

Related Bayer U.S. patents are TPU-Urea Elastomers, U.S. Patent 5,739,250 assigned to Bayer AG, April 14, 1998; and RIM Elastomers Based on Prepolymers of Cyoloaliphatic Diisocyanates, U.S. Patent 5,738,253 assigned to Bayer Corp., April 14, 1998.

Representative applications are tractor body panels and doors, automotive fascia, body panels, window encapsulation, heavy-duty truck bumpers, and recreation vehicle (RV) panels.

Bayer's Texin$^{®*}$ polyester and polyether TPU and TPU/polycarbonate (PC) elastomers were pioneer TPEs in the early development of passenger car fronts and rear bumpers. PC imparts Izod impact strength toughness required for automotive exterior body panels. Extrusion applications include film/sheet, hose, tubing, profiles, and wire/cable covering. Hardness ranges from Shore A 70 to Shore D 75. Texin can be painted without a primer.

Morton International's Morthane$^{®†}$ TPU elastomers are classified into four groups: polyesters, polyethers, polycaprolactones, and polyblends. Polyester polyurethanes have good tear and abrasion resistance, toughness, and low-temperature flexibility, and Shore hardness ranges from 75 A to 65 D. They are extruded into clear film and tubing and fuel line hose. Certain grades are blended with acrylonitrile buta-diene styrene (ABS), styrene acrylonitrile (SAN), nylon, PC, polyvinyl chloride (PVC), and other thermoplastic resins. Polyethers possess hydrolytic stability, resilience, toughness, good low-temperature flexibility, easy processibility and fast cycles. They also have tensile strength up to 7500 lb/in^2 (52 MPa) and melt flow ranges from about 5 to 60 g/10 min. Hardnesses are in the Shore A range up to 90. Certain grades can be used in medical applications. Aliphatic polyester and polyether grades provide UV resistance for pipes, tubing, films, and liners. They can be formulated for high clarity.

---

* Texin is a registered trademark of Bayer Corporation.
† Morthane is a registered trademark of Morton International Inc.

The polyblends are polyester TPUs blended with ABS, SAN, PC, nylon, PVC, and other thermoplastics for injection molding and extrusion. A 10 to 20% loading into PVC compositions can increase mechanical properties 30 to 40%.[1] Although the elastomeric TPUs are inherently flexible, plasticizers may be recommended, for example, in films.

TPU elastomers are processed on rubber equipment, injection molded, extruded, compression molded, transfer molded, and calendered. To be fabricated into products, such as athletic shoe outer soles, the elastomer and ingredients are mixed in conventional rubber equipment (two-roll mills, internal mixers) and compounded.[17] Subsequently, the compound is processed; for example, injection molded.[17]

Typical melt processing practices are described with Pellethane TPU (see Table 6.18). The moisture content is brought to <0.02% before molding or extruding.[3] Desiccant, dehumidifying hopper dryers that can produce a −40°F (40°C) dew point at the air inlet are suggested. A dew point of −20°F (−29°C) or lower is suggested for TPU elastomers.[3] The suggested air-inlet temperature range is 180 to 230°F (82 to 110°C)[3]: 180 to 200°F (82 to 93°C) for the softer Shore A elastomers, and 210 to 230°F (99 to 110°C) for harder Shore D elastomers.

TABLE 6.18 Typical Injection-Molding Settings* for Pellethane TPU

Temperature, °F (°C)	Shore A 80	Shore D 55
Melt temperature max	415 (213)	435 (224)
Cylinder zone		
Rear (feed)	350–370 (177–188)	360–380 (183–193)
Middle (transition)	360–380 (183–193)	370–390 (188–198)
Front (metering)	370–390 (188–199)	390–410 (204–210)
Nozzle	390–410 (199–210)	400–410 (204–210)
Mold temperature	80–140	(27–60)

*Typical temperature and pressure settings are based on Ref. 3. Settings are based on studies using a reciprocating screw, general-purpose crew, clamp capacity of 175 tons, and rated shot capacity of 10oz (280 g). Molded specimen thicknesses ranged from 0.065 to 0.125 in (1.7 to 3.2 mm).

Drying time to achieve a given moisture content for resin used directly from sealed bags is shown in Fig. 6.17: about 4 h to achieve <0.02% moisture content @ 210°F (99°C) air-inlet temperature and −20°F (−29°C) dew point.[3] When TPU elastomers are exposed to air just prior to processing, the pellets are maintained at 150 to 200°F (65 to 93°C), a warmer temperature than the ambient air.[3] A polymer temperature that is warmer than the ambient air reduces the ambient moisture absorption.[3]

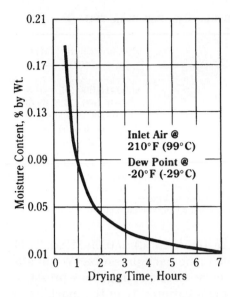

Figure 6.17 Typical drying curve for Pelle-thane elastomers. *(From Ref. 3, p. 6.)*

Melt temperature is determined by T_m of resin and processing and equipment specifications, including machine capacity, rated shot size, screw configuration *(LID,* flight number, and design), part design, mold design (gate type and runner geometry), and cycle time.[3] Shear energy created by the reciprocating screw contributes heat to the melt, causing the actual melt temperature to be 10 to 20°F (6 to 10°C) higher than the barrel temperature settings.[3] Temperature settings should take shear energy into account. To ensure maximum product quality, the processor should discuss processing parameters, specific machinery, equipment, and tool and product data with the resin supplier. For example, if it is suspected that an improper screw design will be used, melt temperature gradient may be reversed. Instead of increasing temperature from rear to front, it may be reduced from rear to front.[3]

A higher mold temperature favors a uniform melt cooling rate, minimizing residual stresses, and improves the surface finish, mold release, and product quality. The mold cooling rate affects finished product quality. Polyether type TPU can set up better and release better.

High pressures and temperatures fill a high surface-to-volume ratio mold cavity more easily, but TPU melts can flash fairly easily at high pressures (Table 6.19). Pressure can be carefully controlled to achieve a quality product by using higher pressure during quick-fill, followed by lower pressure.[3] The initial higher pressure may reduce mold shrinkage by compressing the elastomeric TPU.[3]

TABLE 6.19   Typical Temperature Pressure Settings* for Pellethane TPU

	Pressure, lb/in^2 † (MPa)
Injection pressure	
First stage	8000–15,000 (55.0–103)
Second stage	5000–10,000 (34.5–69.60)
Back pressure	0–100 (0–0.69)
Cushion, in/mm	0.25 (6.4)
Screw speed, rpm	50–75
Cycle time, s (injection, relatively slow to avoid flash, etc.)	3–10

*Typical temperature and pressure settings are based on Ref. 3. Settings are based on studies using a reciprocating screw, general-purpose screw, clamp capcity of 175 tons, and rated shot capacity of 10 oz (280 g). Molded speciment thickness ranged from 0.065 to 0.125 in (1.7– to 3.2–mm).
†U.S. units refer to line pressures; metric units are based on the pressure on the (average) cross-sectional area of the screw.

The back pressure ranges from 0 to 100 lb/in^2 (0 to 0.69 MPa). TPU elastomers usually require very little or no back pressure.[3] When additives are introduced by the processor prior to molding, back pressure will enhance mixing, and when the plastication rate of the machine is insufficient for shot size or cycle time, a back pressure up to 200 lb/in^2 (1.4 MPa) can be used.[3]

Product quality is not as sensitive to screw speed as it is to process temperatures and pressures. The rotating speed of the screw, along with flight design, affects mixing (when additives have been introduced) and shear energy. Higher speeds generate more shear energy (heat). A speed above 90 rpm can generate excessive shear energy, creating voids and bubbles in the melt, which remain in the molded part.[3]

Cycle times are related to TPU hardness, part design, temperatures, and wall thickness. Higher temperature melt and a hot mold require longer cycles, when the cooling gradient is not too steep. The cycle time for thin-wall parts, <0.125 in (<3.2 mm), is typically about 20 s.[3] The wall thickness for most parts is less than 0.125 in (3.2 mm), and a wall thickness as small as 0.062 in (1.6 mm) is not uncommon. When the wall thickness is 0.250 in (6.4 mm), the cycle time can increase to about 90 s.[3]

Mold shrinkage is related to TPU hardness and wall thickness, part and mold designs, and processing parameters (temperatures and pressures). For a wall thickness of 0.062 in (1.6 mm) for durometer hardness Shore A 70, the mold shrinkage is 0.35%. Using the same wall thickness for durometer hardness Shore A 90, the mold shrinkage is 0.83%.[3]

Purging when advisable is accomplished with conventional purging materials, polyethylene, or polystyrene. Good machine maintenance includes removing and cleaning the screw and barrel mechanically with a salt bath or with a high-temperature fluidizedsand bath.[3]

Reciprocating screw injection machines are usually used to injection mold TPU, and these are the preferred machines, but ram types can be successfully used. Ram machines are slightly oversized to avoid (1) incomplete melting and (2) steep temperature gradients during resin melting and freezing. Oversizing applies especially to TPU durometers harder than Shore D 55.[3]

Molded and extruded TPU have a wide range of applications, including:

Automotive: body panels (tractors) and RVs, doors, bumpers (heavy-duty trucks), fascia, and window encapsulations

Belting

Caster wheels

Covering for wire and cable

Film/sheet

Footwear and outer soles

Seals and gaskets

Tubing

6.11.1 Copolyesters

Thermoplastic copolyester elastomers are segmented block copolymers with a polyester hard crystalline segment and a flexible soft amorphous segment with a very low Tg.[35] Typically, the hard segments are composed of short-chain ester blocks such as tetramethylene terephthalate, and the soft segments are composed of aliphatic polyether or aliphatic polyester glycols, their derivatives, or polyetherester glycols. The copolymers are also called *thermoplastic etheresterelastomers (TEEEs)*.[35] The terms *COPE* and *TEEE* are used interchangeably (see Fig. 6.18).

TEEEs are typically produced by condensation polymerization of an aromatic dicarboxylic acid or ester with a low MW aliphatic diol and a polyakylene ether glycol.[35] Reaction of the first two components leads

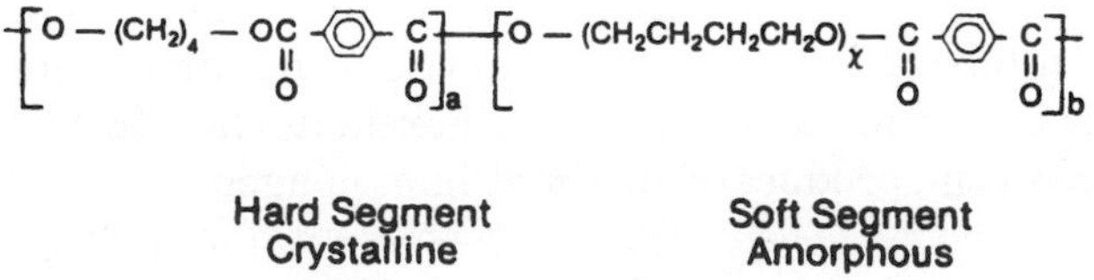

Hard Segment
Crystalline

Soft Segment
Amorphous

Figure 6.18 Structure of a commercial COPE TPE: a = 16 to 40, x = 10 to 50, and b = 16 to 40. *(Source: Ref. 10, p. 5.14.)*

to the hard segment, and the soft segment is the product of the diacid or diester with a long-chain glycol.[35] This can be described as a melt transesterification of an aromatic dicarboxylic acid, or preferably its dimethyl ester, with a low MW poly(alklylene glycol ether) plus a short-chain diol.[35]

An example is melt phase polycondensation of a mixture of dimethyl terephthaate (DMT) + poly(tetramethylene oxide) glycol + an excess of tetramethylene glycol. A wide range of properties can be built into the TEEE by using different mixtures of isomeric phthalate esters, different polymeric glycols, and varying MW and MWD.[36] Antioxidants, such as hindered phenols or secondary aromatic amines, are added during polymerization, and the process is carried out under nitrogen, because the polyethers are subject to oxidative and thermal degradation.[35]

Hytrel®* TEEElastomer block copolymers' property profile is given in Table 6.20.

TABLE 6.20   Typical Hytrel Property Profile[1]—ASTM Test Methods

Property	Value
Specific gravity, g/cm^3	1.01–1.43
Tensile strength @ break, lb/in^2 (MPa)	1,400–7,000 (10–48)
Tensile elongation @ break, %	200–700
Hardness Shore D	30–82
Flexural modulus, lb/in ^{2} (MPa)	
−40°F (−40°C)	9,000–440,000 (62–3,030)
73°F (22.8°C)	4,700–175,000 (32–1,203)
212°F (100°C)	1,010–37,000 (7.0–255)
Izod impact strength, ft-lb/in (J/m), notched	
−40°F (−40°C)	No break–0.4 (No break–20)
−73°F (22.8°C)	No break–0.8 (No break–40)
Tabor abrasion, mg/1000 rev	
CS–17 wheel	0–85
H–18 wheel	20–310
Tear resistance, lb/in, initial Die C	210–1,440
Vicat softening temperature, °F (°C)	169–414 (7601–212)
Melt point, °F (°C)	302–433 (150–223)

The mechanical properties are between rigid thermoplastics and thermosetting hard rubber.[35] Mechanical properties and processing parameters for Hytrel, and for a number of other materials in this chapter, can be found on the producers' Internet home pages.

---

* Hytrel is a registered trademark of DuPont for its brand of thermoplastic polyester elastomer.

Copolymer properties are largely determined by the soft/hard segment ratio; as with any commercial resin, properties are determined with compound formulations.

TEEEs combine flexural fatigue strength, low-temperature flexibility, good apparent modulus (creep resistance), DTUL and heat resistance, resistance to hydrolysis, and good chemical resistance to nonpolar solvents at elevated temperatures. A tensile stress/percent elongation curve reveals an initial narrow linear region.[19] COPEs are attacked by polar solvents at elevated temperatures. The copolymers can be completely soluble in meta-cresol, which can be used for dilute solution polymer analysis.[19]

TEEEs are processed by conventional thermoplastic melt-processing methods, injection molding, and extrusion, requiring no vulcanization.[35] They have sharp melting transitions and rapid crystallization (except for softer grades with higher amount of amorphous segment), and apparently melt viscosity decreases slightly with shear rate (at low shear rates).[35] The melt behaves like a Newtonian fluid.[35] In a true Newtonian fluid, the coefficient of viscosity is independent of the rate of deformation. In a non-Newtonian fluid, the apparent viscosity is dependent on shear rate and temperature.

TPE melts are typically highly non-Newtonian fluids, and their apparent viscosity is a function of shear rate.[10] TPE's apparent viscosity is much less sensitive to temperature than it is to shear rate.[10] The apparent viscosity of TPEs as a function of apparent shear rate and as a function of temperature are shown in Figs. 6.19 and 6.20.

TEEEs can be processed successfully by low-shear methods such as laminating, rotational molding, and casting.[35] Standard TEEElas-

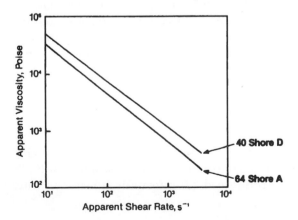

Figure 6.19 Viscosity as a function of shear rate for hard and soft TPEs. (*Source: Ref. 10, p. 5.31.*)

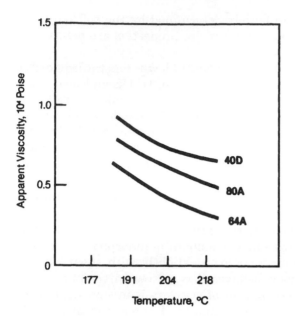

Figure 6.20   TPE (with different hardnesses) viscosity as a function of temperature. *(Source: Ref. 10, p. 5.31.)*

tomers are usually modified with viscosity enhancers for improved melt viscosity for blow molding.[35]

Riteflex®* copolyester elastomers have high fatigue resistance, chemical resistance, good low-temperature [−40°F (−40°C)] impact strength, and service temperatures up to 250°F (121°C). Riteflex grades are classified according to hardness and thermal stability. The typical hardness range is Shore D 35 to 77. They are injection molded, extruded, and blow molded. The copolyester can be used as a modifier in other polymer formulations. Applications for Riteflex copolyester and other compounds that use it as a modifier include bellows, hydraulic tubing, seals, wire coating, and jacketing; molded air dams, automotive exterior panel components (fender extensions, spoilers), fascia and fascia coverings, radiator panels; extruded hose, belting, and cable covering; and spark plug and ignition boots.

Arnitel®† TPEs are based on polyether ester or polyester ester, including specialty compounds as well as standard grades.[34] Specialty grades are classified as (1) flame-retardant UL 94 V/0 @ 0.031 in (0.79 mm), (2) high modulus glass-reinforced, (3) internally lubricated with

---

* Riteflex is a registered trademark of Ticona.
† Arnitel is a registered trademark of DSM.

polytetrafluoro ethylene (PTFE) or silicone for improved wear resistance, and (4) conductive, compounded with carbon black, carbon fibers, nickel-coated fibers, stainless-steel fibers, for ESD applications.

Standard grades have a hardness range of about Shore D 38 to 74 for injection molding, extrusion, and powder rotational molding.[1] Arnitels have high impact strength, even at subzero temperatures, near-constant stiffness over a wide temperature range, and good abrasion.[34] They have excellent chemical resistance to mineral acids, organic solvents, oils, and hydraulic fluids.[34] They can be compounded with property enhancers (additives) for resistance to oxygen, light, and hydrolysis.[34] Glass fiber-reinforced grades, like other thermoplastic composites, have improved DTUL, modulus, and coefficient of linear thermal expansion (CLTE).[34]

Typical products are automotive exterior trim, fascia components, spoilers, window track tapes, boots, bellows, underhood wire covering, connectors, hose, and belts; appliance seals, power tool components, ski boots, and camping equipment.[1]

Like other thermoplastics, processing temperatures and pressures and machinery/tool designs are adjusted to the compound and application.

The following conditions apply to Arnitel COPE compounds, for optimum product quality: melt temperature range, 428 to 500°F (220 to 260°C); cylinder (barrel) temperature setting range, 392 to 482°F (200 to 250°C); mold temperature range for thin-wall products, 122°F (50°C) and for thick-wall products, 68°F (20°C).

Injection pressure is a function of flow length, wall thickness, and melt rheology, and it is calculated to achieve uniform mold filling. The Arnitel injection pressure range is <5000 to >20,000 lb/in^2 (<34 to >137 MPa). Thermoplastic elastomers may not require back pressure, and when back pressure is applied, it is much lower than for thermoplastics that are not elastomeric. Back pressure for Arnitel is about 44 to 87 lb/in^2 (0.3 to 0.6 MPa). Back pressure is used to ensure a homogeneous melt with no bubbles.

The screw configuration is as follows: thread depth ratio, approximately 1:2, and L/D ratio, 17/1 to 23/1 (standard three-zone screws: feed, transition or middle, and metering or feed zones).[34] Screws are equipped with a nonreturn valve to prevent backflow.[34] Decompression-controlled injection-molding machines have an open nozzle.[34] A short nozzle with a wide bore (3-mm minimum) is recommended to minimize pressure loss and heat due to friction.[34] Residence time should be as short as possible, and this is accomplished with barrel temperatures at the lower limits of recommended settings.[34]

Tool design generally follows conventional requirements for gates and runners. DSM recommends trapezoidal gates or, for wall thickness more than 3 to 5 mm, full sprue gates.[34] Vents approximately 1.5 × 0.02 mm are located in the mold at the end of the flow patterns, either in the mold faces or through existing channels around the ejector pins and cores.[34] Ejector pins and plates for thermoplastic elastomers must take into account the molded product's flexibility. Knock-out pins/plates for flexible products should have a large enough face to distribute evenly the minimum possible load. Prior to ejection, the part is cooled, carefully following the resin supplier's recommendation. The cooling system configuration in the mold base, and the cooling rate, are critical to optimum cycle time and product quality. The product is cooled as fast as possible without causing warpage. Cycle times vary from about 6 s for a wall thickness of 0.8 to 1.5 mm to 40 s for a wall thickness of about 5 to 6 mm. Drying temperatures and times range from 3 to 10 h at 194 to 248°F (90 to 120°C).

In general, COPEs can require drying for 4 h @ 225°F (107°C) in a dehumidifying oven to bring the pellet moisture content to 0.02% max.[1] The melt processing range is typically about 428 to 448°F (220 to 231°C); however, melt processing temperatures can be as high as 450 to 500°F (232 to 260°C). A typical injection-molding grade has a T_m of 385°F (196°C).[1] The mold temperature is usually between 75 and 125°F (24 and 52°C).

Injection-molding screws have a gradual transition (center) zone to avoid excess shearing of the melt and high metering (front) zone flight depths [0.10 to 0.12 in (2.5 to 3.0 mm)], a compression ratio of 3.0:1 to 3.5:1, and an L/D of 18/1 min (24/1 for extrusion).[18a] Barrier screws can provide more efficient melting and uniform melt temperatures for molding very large parts and for high-speed extrusions.[18a] When Hytrel is injection molded, molding pressures range from 6000 to 14,000 lb/in^2 (41.2 to 96.2 MPa). When pressures are too high, overpacking and sticking to the mold cavity wall can occur.[18a] Certain mold designs are recommended: large knock-out pins and stripper plates, and generous draft angles for parts with cores.[18a]

6.12   Polyamides

Polyamide TPEs are usually either polyester-amides, polyetheresteramide block copolymers, or polyether block amides (PEBA) (see Fig. 6.21). PEBA block copolymer molecular architecture is similar to typical block copolymers.[10] The polyamide is the hard (thermoplastic) segment, whereas the polyester, polyetherester, and polyether segments are the soft (elastomeric) segment.[10]

Polyamide TPEs can be produced by reacting a polyamide with a polyol such as polyoxyethylene glycol or polyoxypropylene glycol, a polyesterification reaction.[1] Relatively high aromaticity is achieved by

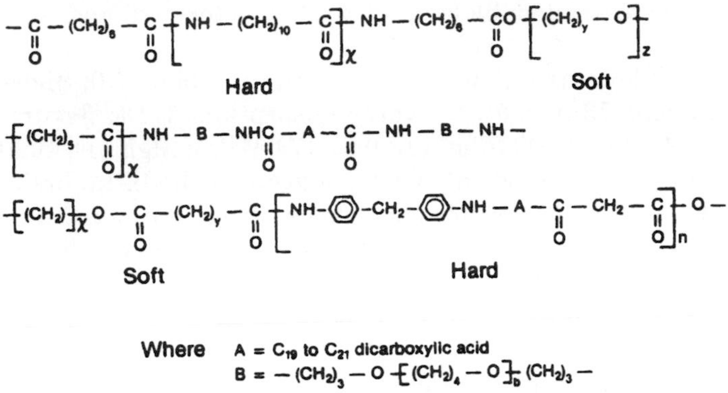

Figure 6.21 Structure of three PEBA TPEs. *(Source: Ref. 10, p. 5.17.)*

esterification of a glycol to form an acid-terminated soft segment, which is reacted with a diisocyanate to produce a polyesteramide. The polyamide segment is formed by adding diacid and diisocyanate.[1] The chain extender can be a dicarboxylic acid.[1] Polyamide TPEs can be composed of lauryl lactam and ethylene-propylene rubber (EPR).

Polyamide thermoplastic elastomers are characterized by their high service temperature under load, good heat aging, and solvent resistance.[1] They retain serviceable properties >120 h @ 302°F (150°C) without adding heat stabilizers.[1] Addition of a heat stabilizer increases service temperature. Polyesteramides retain tensile strength, elongation, and modulus to 347°F (175°C).[1] Oxidative instability of the ether linkage develops at 347°F (175°C). The advantages of polyether block amide copolymers are their elastic memory, which allows repeated strain (deformation) without significant loss of properties, lower hysteresis, good cold-weather properties, hydrocarbon solvent resistance, UV stabilization without discoloration, and lot-to-lot consistency.[1]

The copolymers are used for waterproof/breathable outerwear; air-conditioning hose; underhood wire covering; automotive bellows; flexible keypads; decorative watch faces; rotationally molded basketballs, soccer balls, and volleyballs; and athletic footwear soles.[1] They are insert-molded over metal cores for nonslip handle covers (for video cameras) and coinjected with polycarbonate core for radio/TV control knobs.[1]

Pebax®* polyether block amide copolymers consist of regular linear chains of rigid polyamide blocks and flexible polyether blocks. They

---

* Pebax is a registered trademark of Elf Atochem.

are injection molded, extruded, blow molded, thermoformed, and rotational molded.

The property profile is as follows: specific gravity about 1.0; Shore hardness range about 73 A to 72 D; water absorption, 1.2%; flexural modulus range, 2600 to 69,000 lb/in^2 (18.0 to 474 MPa); high torsional modulus from −40° to 0°C; Izod impact strength (notched), no break from −40to 68°F (−40 to 20°C); abrasion resistance; long wear life; elastic memory, allowing repeated strain under severe conditions without permanent deformation; lower hysteresis values than many thermoplastics and thermosets with equivalent hardness; flexibility temperature range, −40to 178°F (−40 to 81°C), and flexibility temperature range is achieved without plasticizer (it is accomplished by engineering the polymer configuration); lower temperature increase with dynamic applications; chemical resistance similar to polyurethane (PUR); good adhesion to metals; small variation in electrical properties over service temperature range and frequency (Hz) range; printability and colorability; tactile properties, such as good "hand," feel; and nonallergenic.[1]

The T_m for polyetheresteramides is about 248 to 401°F (120 to 205°C) and about 464°F (240°C) for aromatic polyesteramides.[18b]

Typical Pebax applications are one-piece, thin-wall soft keyboard pads; rotationally molded, high-resiliency, elastic memory soccer balls, basketballs, and volleyballs; flexible, tough mouthpieces for respiratory devices, scuba equipment, frames for goggles, and ski and swimming breakers; and decorative watch faces. Pebax offers good nonslip adhesion to metal and can be used for coverings over metal housings for hand-held devices such as remote controls, electric shavers, camera handle covers; coinjected over polycarbonate for control knobs; and employed as films for waterproof, breathable outerwear.[1]

Polyamide/ethylene-propylene, with higher crystallinity than other elastomeric polyamides, has improved fatigue resistance and improved oil and weather resistance.[1] T_m and service temperature usually increase with higher polyamide crystallinity.[1]

Polyamide/acrylate graft copolymers have a Shore D hardness range from 50 to 65, and continuous service temperature range from −40 to 329°F (−40 to 165°C). The markets are underhood hose and tubing, seals and gaskets, and connectors and optic fiber sheathing, snap-fit fasteners.[1] Nylon 12/nitrile rubber blends were commercialized by Denki Kagaku Kogyo as part of the company's overall nitrile blend development.[1]

6.13 Melt Processable Rubber (MPR)

MPRs are amorphous polymers, with no sharp melt point,[1] which can be processed in both resin melt and rubber processing machines, injection molded, extruded, blow molded, calendered, and compression molded.[*][1] Flow properties are more similar to rubber than to thermoplastics.[1] The polymer does not melt by externally applied heat alone but becomes a high-viscosity, intractable semifluid. It must be subjected to shear to achieve flowable melt viscosities, and shear force applied by the plasticating screw is necessary. Without applied shear, melt viscosity and melt strength increase too rapidly in the mold. Even with shear and a hot mold, as soon as the mold is filled and the plasticating screw stops or retracts, melt viscosity and melt strength increase rapidly.

Melt rheology is illustrated with Aclryn®.[†] The combination of applied heat and shear-generated heat brings the melt to 320 to 330°F (160 to 166°C). The melt temperature should not be higher than 360°F (182°C). New grades have been introduced with improved melt processing.

Proponents of MPR view its rheology as a processing cost benefit by allowing faster demolding and lower processing temperature settings, significantly reducing cycle time.[1] High melt strength can minimize or virtually eliminate distortion and sticking, and cleanup is easier.[1] MPR is usually composed of halogenated (chlorinated) polyolefins, with reactive intermediate-stage ethylene interpolymers that promote H^+ bonding.

Alcryn is an example of single-phase MPR with overall midrange performance properties, supplementing the higher-price COPE thermoplastic elastomers. Polymers in single-phase blends are miscible, but polymers in multiple-phase blends are immiscible, requiring a compatibilizer for blending. Alcryns are partially cross-linked halogenated polyolefin MPR blends.[1] The specific gravity ranges from 1.08 to 1.35.[1] MPRs are compounded with various property enhancers (additives), especially stabilizers, plasticizers, and flame retardants.[1]

The applications are automotive window seals and fuel filler gaskets, industrial door and window seals and weatherstripping, wire/cable covering, and hand-held power tool housing/handles. Nonslip soft-touch hand-held tool handles provide weather and chemical resistance and vibration absorption.[16] Translucent grade is extruded into films for face masks and tube/hosing and injection-molded into flexible keypads for computers and telephones.[1] Certain grades are paintable without a primer. Typical durometer hardnesses are Shore A 60, 76, and 80.

The halogen content of MPRs requires corrosion-resistant equipment and tool cavity steels along with adequate venting. Viscosity and

---

* MPR is a trademark of Advanced Polymer Alloys, Division of Ferro Corporation.

† Alcryn is a registered trademark of Advanced Polymer Alloys Division of Ferro Corporation.

melt strength buildup are taken into account with product design, equipment, and tooling design: wall thickness gradients and radii, screw configuration (flights, L/D, length), gate type and size, and runner dimensions.[1] The processing temperature and pressure setting are calculated according to rheology.[1]

To convert solid pellet feed into uniform melt, moderate screws with some shallow flights are recommended. Melt flow is kept uniform in the mold with small gates (which maximize shear), large vents, and large sprues for smooth mold filling.[1] Runners should be balanced and radiused for smooth, uniform melt flow.[1] Recommendations, such as balanced, radiused runners, are conventional practice for any mold design, but they are more critical for certain melts such as MPRs. Molds have large knock-out pins or plates to facilitate stripping the rubbery parts during demolding. Molds may be chilled to 75°F (24°C). Mold temperatures depend on grades and applications; hot molds are used for smooth surfaces and to minimize orientation.[1]

Similar objectives of the injection-molding process apply to extrusion and blow molding, namely, creating and maintaining uniform, homogeneous, and properly fluxed melt. Shallow-screw flights increase shear and mixing. Screws that are 4.5 in (11.4 cm) in diameter with L/D 20/1 to 30/1 are recommended for extrusion. Longer barrels and screws produce more uniform melt flux, but L/D ratios can be as low as 15/1. The temperature gradient is reversed. Instead of the temperature setting being increased from the rear (feed) zone to the front (metering) zone, a higher temperature is set in the rear zone, and a lower temperature is set at the front zone and at the adapter (head).[1] Extruder dies are tapered, with short land lengths, and die dimensions are close to the finished part dimension.[1] Alcryns have low to minimum die swell.

The polymer's melt rheology is an advantage in blow molding during parison formation, because the parison is not under shear, and it begins to solidify at about 330°F (166°C). High melt viscosity allows blow ratios up to 3:1 and significantly reduces demolding time.

MPRs are thermoformed and calendered with similar considerations described for molding and extrusion. Film and sheet can be calendered with thicknesses from 0.005 to 0.035 in (0.13 to 0.89 mm).

6.14   Thermoplastic Vulcanizate (TPV)

TPVs are composed of a vulcanized rubber component, such as EPDM, nitrile rubber, and butyl rubber, in a thermoplastic olefinic matrix. TPVs have a continuous thermoplastic phase and a discontinuous vulcanized rubber phase. TPVs are dynamically vulcanized during a melt-mixing process in which vulcanization of the rubber polymer takes place under conditions of high temperatures and high shear. Static vulcanization of thermoset rubber involves heating a compounded rubber stock under zero shear (no mixing), with subsequent cross-linking of the polymer chains.

Advanced Elastomer Systems' Santoprene®* thermoplastic vulcanizate is composed of polypropylene and finely dispersed, highly vulcanized EPDM rubber. Geolast®† TPV is composed of polypropylene and nitrile rubber, and the company's Trefsin®‡ is a dynamically vulcanized composition of polypropylene plus butyl rubber.

EPDM particle size is a significant parameter for Santoprene's mechanical properties, with smaller particles providing higher strength and elongation.[1] Higher cross-link density increases tensile strength and reduces tension set (plastic deformation under tension).[1] Santoprene grades can be characterized by EPDM particle size and cross-link density.[1]

These copolymers are rated as midrange with overall performance generally between the Tower cost styrenics and the higher-cost TPUs and copolyesters.[1] The properties of Santoprene, according to its developer (Monsanto), are generally equivalent to the properties of general purpose EPDM, and oil resistance is comparable to that of neoprene.[1] Geolast has higher fuel/oil resistance and better hot oil aging than Santoprene (see Tables 6.21, 6.22, and 6.23).

TABLE 6.21   Santoprene Mechanical Property Profile—ASTM Test Methods—Durometer Hardness Range, Shore 55A to 50 D

	Shore hardness		
Property	55A	80A	50D
Specific gravity, g/cm^3	0.97	0.97	0.94
Tensile strength, lb/in^2 (MPa)	640 (4.4)	600 (11)	4000 (27.5)
Ultimate elongation, %	330	450	600
Compression set, %, 168 h	23	29	41
Tension set, %	6	20	61
Tear strength, pli			
77°F (25°C)	108 (42)	194 (90)	594 (312)
212°F (100°C)	42 (5.6)	75 (24)	364 (184)
Flex fatigue megacycles to failure	>3.4	—	—
Brittle point, °F (°C)	<−76 (<−60)	−81 (−63)	−29 (−34)

---

* Santoprene is a registered trademark of Advanced Elastomer Systems LP.
† Geolast is a registered trademark of Advanced Elastomer Inc. Systems LP.
‡ Trefsin is a registered trademark of Advanced Elastomer Systems LP.

TABLE 6.22   Santoprene Mechanical Property Profile—Hot Oil Aging[*]/ Hot Air Aging—Durometer Hardness Range Shore 55 A to 55 D *(from Ref. 1)*

	Shore hardness		
Property	55A	80A	50D
Tensile strength, ultimate			
lb/in^2 (MPa)	470 (3.2)	980 (6.8)	2620 (18.10)
Percent retention	77	73	70
Ultimate elongation, %	320	270	450
Percent retention	101	54	69
100% modulus, lb/in^2 (MPa)	250 (1.7)	610 (4.2)	1500 (10.3)
Percent retention	87	84	91

[*]Hot oil aging (IRM 903), 70 h @ 257°F (125°C).

TABLE 6.23   Santoprene Mechanical Property Profile—Hot Oil Aging/ Hot Air Aging[*]—Durometer Hardness Range Shore 55 A to 55 D *(from Ref. 1)*

	Shore hardness		
Property	55A	80A	50D
Tensile strength, ultimate			
lb/in^2 (MPa)	680 (4.7)	1530 (10.6)	3800 (26.2)
Percent retention	104	109	97
Ultimate elongation, %	370	400	560
Percent retention	101	93	90
100% modulus, lb/in^2 (MPa)	277 (1.9)	710 (4.9)	1830 (12.6)
Percent retention	105	111	117

[*]Hot air aging, 168 h @ 257°F (125°C).

Tensile stress-strain curves for Santoprene at several temperatures for Shore 55 A and 50 D hardnesses are shown in Fig. 6.22.[8]

Generally, tensile stress decreases with temperature increase, while elongation at break increases with temperature. Tensile stress at a given strain increases with hardness from the softer Shore A grades to the harder Shore D grades. For a given hardness, the tensile stress-strain curve becomes progressively more rubber-like with increasing temperature. For a given temperature, the curve is progressively more rubberlike with decreasing hardness. Figure 6.23 shows dynamic mechanical properties for Shore 55 A and 50 D hardness grades over a wide range of temperatures.[8]

TPVs composed of polypropylene and EPDM have a service temperature range from −75 to 275°F (−60 to 135°C) for more than 30 days

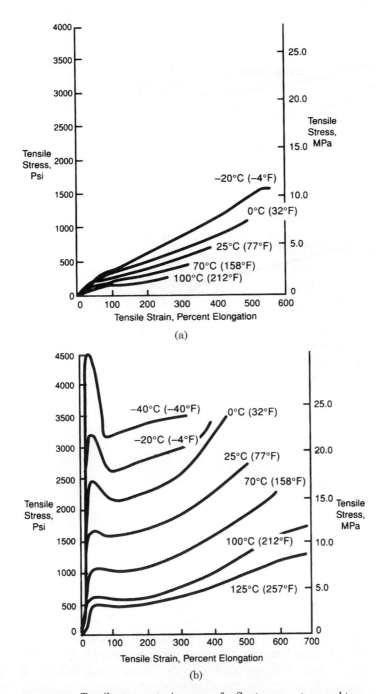

Figure 6.22 Tensile stress-strain curves for Santoprene at several temperatures for different hardness grades. *(a)* 55 Shore A grades (ASTM D 412), *(b)* 50 Shore D grades (ASTM D 412). *(Source: Ref. 8, pp. 3–4.)*

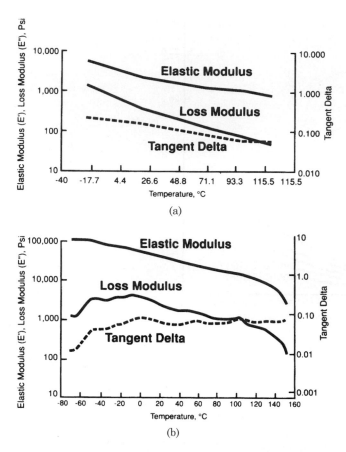

Figure 6.23 Dynamic mechanical properties for different hardness grades over a range of temperatures. *(a)* 55 Shore A grades, *(b)* 50 Shore D grades. *(Source: Ref. 8, pp. 12–13.)*

and 302°F (150°C) for short times (up to 1 week). Reference 8 reports further properties, including tensile and compression set, fatigue resistance, and resilience and tear strength. Polypropylene/nitrile rubber high/low service temperature limits are 257°F (125°C)/–40°F (–40°C).

Santoprene automotive applications include air ducts, body seals, boots (covers), bumper components, cable/wire covering, weatherstripping, underhood and other automotive hose/tubing, and gaskets. Appliance uses include diaphragms, handles, motor mounts, vibration dampers, seals, gaskets, wheels, and rollers. Santoprene rubber is used in building/construction for expansion joints, sewer pipe seals, valves for irrigation, weatherstripping, and welding line connectors. Prominent electrical uses are in cable jackets, motor shaft mounts, switch boots, and terminal plugs. Business machines, power tools, and plumbing/hardware provide TPVs with numerous applications. In healthcare applications, it is used in disposable bed covers, drainage bags, pharmaceutical packaging, wound dressings (U.S. Pharmacopoeia Class VI rating for biocompatibility). Special-purpose Santoprene grades meet flame retardance, outdoor weathering, and heat aging requirements.

Santoprene applications of note are a nylon-bondable grade for the General Motors GMT 800 truck air-induction system; driveshaft boot in Ford-F Series trucks, giving easier assembly, lighter weight, and higher temperature resistance than the material it replaced; and Santoprene cover and intermediate layers of tubing assembly for hydraulic oil hose. Nylon-bondable Santoprene TPV is coextruded with an impact modified (or pure) nylon 6 inner layer.

Polypropylene/EPDM TPVs are hygroscopic, requiring drying at least 3 h at 160°F (71°C) and avoiding exposure to humidity.[1] They are not susceptible to hydrolysis.[1] Moisture in the resin can create voids, disturbing processing and finished product performance properties. Moisture precautions are similar to those for polyethylene or polypropylene.[1]

Typical of melts with a relatively low melt flow index (0.5 to 30 g/ 10 min for Santoprene), gates should be small, and runners and sprues should be short; long plasticating screws are used with an L/D ratio typically 24/1 or higher.[1] The high viscosity at low shear rates (see Fig. 6.24) provides good melt integrity and retention of design dimensions during cooling.[1]

Similar injection-molding equipment design considerations apply to extrusion equipment such as long plasticating screws with 24/1 or higher L/D ratios and approximately 3:1 compression ratios.[1]

Equipment/tool design, construction, and processing of TPVs differ from that of other thermoplastics. EPDM/polypropylene is thermally stable up to 500°F (260°C), and it should not be processed above this temperature.[1] It has a flash ignition temperature above 650°F (343°C).

TPV's high shear sensitivity allows easy mold removal; thus, sprays and dry powder mold release agents are not recommended.

Geolast TPVs are composed of polypropylene and nitrile rubber. Table 6.24 profiles the mechanical properties for these TPVs with Shore hardness range of 70 A to 45 D.

Geolast (polypropylene plus nitrile rubber) has a higher resistance than Santoprene (polypropylene plus EPDM) to oils (such as IRM 903) and fuels, plus good hot-oil/hot-air aging.[1] Geolast applications include molded fuel filler gasket (Cadillac Seville), carburetor components, hydraulic lines, and engine parts such as mounts and tank liners.

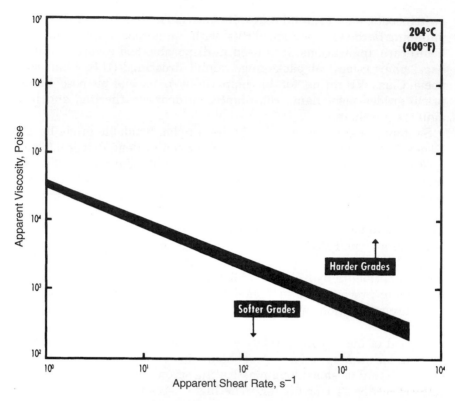

Figure 6.24 Apparent viscosity vs. apparent shear rate @ 400°F (04°C). *(Source: Ref. 7, pp. 36.)*

TABLE 6.24   Geolast Mechanical Property Profile—ASTM Test Methods—Room Temperature—Durometer Hardness Range, Shore 70 A to 45 D

Property	Shore hardness		
	70 A	87 A	45 D
Specific gravity, g/cm^3	1.00	0.98	0.97
Tensile strength, lb/in^2 (MPa)	900 (6.2)	1750 (12)	2150 (15)
Ultimate elongation, %	265	380	350
Compression set @ 22 h, %			
212°F (100°C)	28	39	52
257°F (125°C)	37	48	78
Tension set (%)	10	24	40
Tear strength, pli			
73°F (23°C)	175 (79)	350 (177)	440 (227)
212°/F (100°C)	52 (11)	150 (66)	220 (104)
Brittle point, °F (°C)	−40 (−40)	−33 (−28)	−31 (−36)

Three property distinctions among Trefsin grades are (1) heat aging; (2) high energy attenuation for vibration damping applications such as automotive mounts, energy absorbing fascia and bumper parts, and sound deadening; and (3) moisture and O_2 barrier. Other applications are soft bellows; basketballs, soccer balls, and footballs; calendered textile coatings; and packaging seals. Since Trefsin is hygroscopic, it requires drying before processing. Melt has low viscosity at high shear rates, providing fast mold filling. High viscosity at low shear during cooling provides a short cooling time. Overall, cycle times are reduced.

Advanced Elastomer Systems L.P. (AES) is the beneficiary of Monsanto Polymers' TPE technology and business, which included Monsanto's earlier acquisition of BP Performance Polymers' partially vulcanized EPDM/polypropylene (TPR), and Bayer's partially vulcanized EPDM/polyolefin TPEs in Europe.

6.15   Synthetic Rubbers (SRs)

A second major group of elastomers is that group known as *synthetic rubbers*. Elastomers in this group, discussed in detail in this section, are

Acrylonitrile butadiene copolymers (NBR)

Butadiene rubber (BR)

Butyl rubber (IIR)

Chiorosulfonated polyethylene (CSM)

Epichiorohydrin (ECH, ECO)

Ethylene propylene diene monomer (EPDM)

Ethylene propylene monomer (EPM)

Fluoroelastomers (FKM)

Polyacrylate (ACM)

Polybutadiene (PB)

Polychloroprene (CR)

Polyisoprene (IR)

Polysulfide rubber (PSR)

Silicone rubber (SiR)

Styrene butadiene rubber (SBR)

Worldwide consumption of synthetic rubber can be expected to be about 11 million metric tons in 2000 and about 12 million metric tons in 2003, based on earlier reporting (1999) by the International Institute of Synthetic Rubber Producers.[26] About 24% is consumed in North America.[1] Estimates depend on which synthetic rubbers are included and reporting sources from world regions.

New synthetic rubber polymerization technologies replacing older plants and increasing world consumption are two reasons new production facilities are being built around the world. Goodyear Tire & Rubber's 110,000-metric tons/y butadiene-based solution polymers went onstream in 2000 in Beaumont Texas.[25] Goodyear's 18,200-metric tons/y polyisoprene unit went onstream in 1999 in Beaumont.[25] Sumitomo Sumika AL built a 15,000-metric tons/y SBR plant in Chiba, Japan, adding to the company's 40,000-metric tons/y SBR capacity at Ehime.[25] Haldia Petrochemical Ltd. of India is constructing a 50,000-metric tons/y SBR unit and a 50,000-metric tons/y PB unit using BASF technology.[25]

Bayer Corporation added a 75,000-metric tons/y SBR and PB capacity at Orange, Texas, in 1999, converting a lithium PB unit to produce solution SBR and neodymium PB.[25] Bayer AG increased SBR and PB-capacity from 85,000 to 120,000 metric tons/y at Port Jerome, France, in 1999.[25] Bayer AG will complete a worldwide butadiene rubber capacity increase from 345,000 metric tons/y in 1998 to more than 600,000 metric tons/y by 2001.[25] Bayer AG increased EPDM capacity at Orange, Texas, using slurry polymerization and at Marl, Germany, using solvent polymerization in 1999.[25] Bayer Inc. added 20,000–metric tons/y butyl rubber capacity at Sarnia, Ontario, to the company's 70,000-metric tons/y butyl rubber capacity and 50,000-metric tons/y halo-butyl capacity at Sarnia. Bayer's 90,000-metric tons/y halo- or regular butyl capacity was scheduled to be restarted in 2000.[25]

Mitsui Chemicals goes onstream with a 40,000-metric tons/y metallocene EPDM in Singapore in 2001.[25] The joint venture Nitrilo SA between Uniroyal Chemical subsidiary of Crompton and Girsa subsidiary of Desc SA (Mexico) went onstream with a 28,000-metric tons/y NBR at Altamira, Mexico, in 1999.[25] Uniroyal NBR technology and Girsa process technology were joined.[25] Chevron Chemical went onstream in 1999 with a 60,000-metric tons/y capacity polyisobutylene (PIB) at Belle Chase, Louisiana, licensing technology from BASF.[25] BASF is adding a 20,000–metric tons/y medium MW PIB at its Lufwigshafen complex, which will double the unit's capacity to 40,000 metric tons/y. This addition will be completed in 2001. BASF has 70,000–metric tons/y low MW PIB capacity. The company is using its own selective polymerization technology, which allows MW to be controlled.[25]

BST Elastomers, a joint venture of Bangkok Synthetics, Japan Synthetic Rubber (JSR), Nippon Zeon, Mitsui, and Itochu, went onstream in 1998–1999 with 40,000–metric tons/y PB capacity and 60,000–metric tons/y SBR at Map Ta Phut, Rayong, Thailand.[25] Nippon Zeon completed adding 25,000–metric tons/y SBR capacity to its existing 30,000-metric tons/y at Yamaguchi, Japan, in 1999, and the company licensed its solution polymerization technology to Buna Sow Leuna Olefinverbund (BSLO).[25] BSLO will start 60,000-metric tons/y SBR capacity at Schkopau, Germany, in 2000.[25]

Sinopec, China's state-owned petrochemical and polymer company, is increasing synthetic rubber capacity across the board, including butyls, SBRs, nitrile, and chloroprene. Sinopec is starting polyisoprene and EPR production, although the company did not produce polyisoprene or EPR prior to 1999.[25] Total synthetic rubber capacity will be 1.15 million metric tons/y by 2000. China's synthetic rubber consumption is forecast by the company to be almost 7 million metric tons/y in 2000.[25] Dow chemical purchased Shell's E-SBR and BR. Dow is the fastest growing SR producer with the broadest portfolio of SBR and BR.

Synthetic rubber is milled and cured prior to processing such as injection molding. Processing machinery is designed specifically for synthetic rubber.

Engel (Guelph, Ontario) ELAST®* technology includes injection-molding machines designed specifically for molding cross-linked rubbers.[31] Typical process temperature settings, depending on the polymer and finished product, are 380 to 425°F (193 to 218°C). Pressures, which also depend on polymer and product, are typically 20,000 to 30,000 lb/in^2 (137 to 206 MPa). A vertical machine's typical clamping force is 100 to 600 U.S. tons, while a horizontal machine's typical clamping force is 60 to 400 U.S. tons. They have short flow paths, allowing injection of rubber very close to the cross-linking temperature.[31] The screw L/D can be as small as 10/1.31 ELAST technology includes tiebarless machines for small and medium capacities and proprietary state-of-the-art computer controls.[31]

6.15.1 Acrylonitrile Butadiene Copolymers

Nitrile butadiene rubbers (NBRs) are poly(acrylonitrile-co-1,3-butadiene) copolymers of butadiene and acrylonitrile.[23] Resistance to swelling caused by oils, greases, solvents, and fuels is related to percent bound acrylonitrile (ACN) content, which usually ranges from 20 to 46%.[6] Higher ACN provides higher resistance to swelling but diminishes compression set and low temperature flexibility.[6] ACN properties are related to percent acrylo and percent nitrile content. Nitrile increases compression set, flex properties, and processing properties.[1]

---

* ELAST is a registered trademark of Engel Canada.

The rubber has good barrier properties due to the polar nitrile groups.[23] Continuous-use temperature for vulcanized NBR is up to 248°F (120°C) in air and up to 302°F (150°C) immersed in oil.[23]

NBR curing, compounding, and processing are similar to those for other synthetic rubbers.[6]

Fine-powder NBR grades are ingredients in PVC/nitrile TPEs and in other polar thermoplastics to improve melt processibility; reduce plasticizer blooming (migration of plasticizer to the surface of a finished product); and improve oil resistance, compression set, flex properties, feel, and finish of the plastic product.[1] Chemigum®* fine powder is blended with PVC/ABS and other polar thermoplastics.[1] The powders are typically less than 1 mm in diameter (0.5 nominal diameter particle size), containing 9% partitioning agent. Partitioning agents may be SiO_2, $CaCO_3$, or PVC. Their structures may be linear, linear/cross linked, and branched/cross linked (see Table 6.25).

TABLE 6.25   Typical Chemigum/PVC Formulations and Properties for General-Purpose and Oil/Fuel-Resistant Hose *(from Ref. 1)*

	General–purpose	Oil/fuel resistant
Ingredient, parts by weight		
PVC	100	100
Chemigum	25	100
DOP plasticizer	78	40
$CaCO_3$	—	20
Epoxidized soya oil	3	5
Stabilizer, lubricant	2	3
Properties		
Specific gravity, g/cm^3	1.17	1.20
Tensile strength, lb/in^2 (MPa)	2566 (17.6)	1929 (13.3)
Elongation @ break, %	390	340
Hardness Shore A	57	73

Nitrile rubber applications are belting, sheeting, cable jacketing, hose for fuel lines and air conditioners, sponge, gaskets, arctic/aviation O-rings and seals, precision dynamic abrasion seals, and shoe soles. Nitrile rubbers are coextruded as the inner tube with chlorinated polyethylene outer tube for automotive applications.[10d] Nitrile provides resistance to hydrocarbon fluids, and chlorinated polyethyl-

---

* Chemigum is a registered trademark of Goodyear Tire and Rubber Company.

ene provides ozone resistance.[10d] Other automotive applications are engine gaskets, fluid- and vapor-resistant tubing, fuel filler neck inner hose, fuel system vent inner hose, oil, and grease seals. Nitrile powder grades are used in window seals, appliance gasketing, footwear, cable covering, hose, friction material composites such as brake linings, and food contact applications.[1,6]

Blends based on nitrile rubbers are used in underground wire/cable covering; automotive weatherstripping, spoiler extensions, foam-integral skin core-cover armrests, and window frames; footwear; and flexible, lay-flat, reinforced, rigid, and spiral hose for oils, water, food, and compressed air.

6.15.2 Butadiene Rubber (BR) and Polybutadiene (PB)

Budene[®*] solution polybutadiene (solution polymerized) is cis-1,4-poly(butadiene) produced with stereospecific catalysts which yield a controlled MWD, which is essentially a linear polymer.[6] Butadiene rubber, polybutadiene, is solution-polymerized to stereospecific polymer configurations[10a] by the additional polymerization of butadiene monomer. The following cis- and trans-1,4-polybutadiene isomers can be produced: cis-1,4-polybutadiene with good dynamic properties, low hysteresis, and good abrasion resistance; trans isomers are tougher, harder, and show more thermoplasticity.[10,23] Grades are oil and nonoil extended and vary according to their cis-content.[6]

The applications are primarily tire tread and carcass stock, conveyor belt coverings, V-belts, hose covers, tubing, golf balls, shoe soles and heels, sponges, and mechanical goods.[6] They are blended with SBR for tire treads to improve abrasion and wear resistance.[10a] The tread is the part of the tire that contacts the road, requiring low rolling resistance, abrasion and wear resistance, and good traction and durability.[22]

Replacement passenger car shipments in the United States are expected to increase from 185.5 million units in 1998 to 199 million units in 2004, according to the Rubber Manufacturers Association. Synthetic rubber choices for tires and tire treads are related to tire design. Composition and design for passenger cars, sport utility vehicles (SUV), pickup trucks, tractor trailers, and snow tires are continually under development, as illustrated by the following sampling of U.S. Patents.[16] Tire Having Silica Reinforced Rubber Tread Containing Carbon Fibers to Goodyear Tire & Rubber, U.S. Patent 5,718,781, Feb-

---

* Budene is a registered trademark of Goodyear Tire and Rubber Company.

ruary 17, 1998; Silica Reinforced Rubber Composition to Goodyear Tire & Rubber, U.S. Patent 5,719,208, February 17, 1998; and Silica Reinforced Rubber Composition and Tire With Tread to Goodyear Tire & Rubber, U.S. Patent 5,719,207, February 17, 1998.

Other patents include "Ternary Blend of Polyisoprene, Epoxidized Natural Rubber and Chlorosulfonated Polyethylene" to Goodyear Tire & Rubber, U.S. Patent 5,736,593, April 7, 1998, and "Truck Tire With Cap/Base Construction Tread" to Goodyear Tire & Rubber, U.S. Patent 5,718,782, February 17, 1998; "Tread Of Heavy Duty Pneumatic Radial Tire" to Bridgestone Corporation, U.S. Patent 5,720,831, February 24, 1998; "Pneumatic Tire With Asymmetric Tread Profile" to Dunlop Tire, U.S. Patent 5,735,979, April 7, 1998; and "Tire Having Specified Crown Reinforcement" to Michelin, U.S. Patent 5,738,740, April 14, 1998. Silica improves a passenger car's tread rolling resistance and traction when used with carbon black.[28] High dispersible silica (HDS) in high vinyl solution polymerized SER compounds show improved processing and passenger car tread abrasion resistance.[28] Precipitated silica with carbon black has been used in truck tire tread compounds which are commonly made with natural rubber (NR).[28]

Modeling is the method of choice for analyzing passenger car cord-reinforced rubber composite behavior. Large-scale three-dimensional finite element analysis (FEA) improves understanding of tire performance, including tire and tread behavior when "the rubber meets the road."

BR is extruded and calendered. Processing properties and performance properties are related to polymer configuration: cis- or trans-stereoisomerism, MW and MWD, degree of crystallization (DC), degree of branching, and Mooney viscosity.[23] Broad MWD and branched BR tend to mill and process more easily than narrow MWD and more linear polymer.[23] Lower Mooney viscosity enhances processing.[23] BR is blended with other synthetic rubbers such as SBR to combine BR properties with millability and extrudability.

6.15.3 Butyl Rubber

Butyl rubber (IIR) is an isobutylene-based rubber, which includes copolymers of isobutylene and isoprene, halogenated butyl rubbers, and isobutylene/p-methylstyrene/bromo-p-methylstyrene terpolymers.[22] IIR can be slurry polymerized from isobutylene copolymerized with small amounts of isoprene in methyl chloride diluent at -130 to $-148°F$ (-90 to $-100°C$). Halogenated butyl is produced by dissolving butyl rubber in a hydrocarbon solvent and introducing elemental halogen in gas or liquid state.[23] Cross-linked terpolymers are formed with isobutylene + isoprene + divinylbenzene.

Most butyl rubber is used in the tire industry. Isobutylene-based rubbers are used in underhood hose for the polymer's low permeability and temperature resistance, and high damping, resilient butyl rubbers are used for NVH (noise, vibration, harshness) applications such as automotive mounts for engine and vehicle/road NVH attenuation.[22]

Butyl rubber is ideal for automotive body mounts that connect the chassis to the body, damping road vibration.[10d] Road vibration generates low vibration frequencies. Butyl rubber can absorb and dissipate large amounts of energy due to its high mechanical hysteresis over a useful temperature range.[10d]

Low-MW "liquid" butyls are used for sealants, caulking compounds, potting compounds, and coatings.[23] Depolymerized virgin butyl rubber is high viscosity and is used for reservoir liners, roofing coatings, and aquarium sealants.[10b] It has property values similar to conventional butyl rubber: extremely low VTR (vapor transmission rate); resistance to degradation in hot, humid environments; excellent electrical properties; and resistance to chemicals, oxidation, and soil bacteria.[10b] To make high-viscosity depolymerized butyl rubber pourable, solvents or oil is added.[10b]

Chlorobutyl provides flex resistance in the blend chlorobutyl rubber/ EPDM rubber/NR for white sidewall tires and white sidewall coverstrips.[22] An important application of chlorobutyl rubber in automotive hose is extruded air conditioning hose to provide barrier properties to reduce moisture gain and minimize refrigerant loss.[22] The polymer is used in compounds for fuel line and brake line hoses.[22] Brominated isobutylene-p-methylstyrene (BIMS) was shown to have better aging properties than halobutyl rubber for underhood hose and comparable aging properties to peroxide-cured EPDM, depending on compound formulations.[22] Bromobutyls demonstrate good resistance to brake fluids for hydraulic brake lines and to methanol and methanol/gasoline blends.[22]

6.15.4   Chlorosulfonated Polyethylene (CSM)

Chlorosulfonated polyethylene is a saturated chlorohydrocarbon rubber produced from Cl_2, SO_2, and a number of polyethylenes, and contains about 20 to 40% chlorine and 1 to 2% sulfur as sulfonyl chloride.[23] Sulfonyl chloride groups are the curing or cross-linking sites.[23] CSM properties are largely based on initial polyethylene (PE) and percent chlorine. A free-radical-based PE with 28% chlorine and 1.24% S has a dynamic shear modulus range from 1000 to 300,000 lb/in² (7 MPa to 2.1 GPa).[23] Stiffness differs for free-radical-based PE and linear PE, with chlorine content: at about 30%, Cl_2 free-radical-based PE stiffness decreases to minimum value, and at about 35%, Cl_2 content linear PE stiffness decreases to minimum value.[23] When the Cl_2 content is increased more than 30 and 35%, respectively, the stiffness (modulus) increases.[23]

Hypalon[®*] CSMs are specified by their Cl_2, S contents, and Mooney viscosity.[23] CSM has an excellent combination of heat and oil resistance and oxygen and ozone resistance. CSM, like other polymers, is compounded to meet specific application requirements. Hypalon is used for underhood wiring and fuel hose resistance.

6.15.5 Epichlorohydrin (ECH, ECO)

ECH and ECO polyethers are homo- and copolymers, respectively: chloromethyloxirane homopolymer and chloromethyloxirane copolymer with oxirane.[23] Chloromethyl side chains provide sites for cross-linking (curing and vulcanizing). These chlorohydrins are chemically 1-chloro-2,3-epoxypropane. They have excellent resistance to swelling when exposed to oils and fuels; good resistance to acids, alkalis, water, and ozone; and good aging properties.[10a] Aging can be ascribed to environments such as weathering (UV radiation, oxygen, ozone, heat, and stress).[10a] High chlorine content provides inherent flame retardance,[10a] and, like other halogenated polymers, flame-retardant enhancers (additives) may be added to increase UL 94 flammability rating.

ECH and ECO can be blended with other polymers to increase high- and low-temperature properties and oil resistance.[23] Modified polyethers have potential use for new, improved synthetic rubbers. ECH and ECO derivatives, formed by nucleophilic substitution on the chloromethyl side chains, may provide better processing.

6.15.6 Ethylene Propylene Copolymer (EPDM)

EPM [poly(ethylene-co-propylene)] and EPDM [poly(ethylene-copropylene-co-5-ethylidene-2-norbornene)][23] can be metallocene catalyst polymerized. Metallocene catalyst technologies include (1) Insite, a constrained geometry group of catalysts used to produce Affinity polyolefin plastomers (POP), Elite[®†] PE, Nordel[®‡] EPDM, and Engage polyolefin elastomers (POP) and (2) Exxpol[®§] ionic metallocene catalyst compositions used to produce Exact[¶] plastomer octene copolymers.[24] Insite technology produces EPDM-based Nordel IP with property consistency and predictability[16] (see Sec. 6.10.2).

---

* Hypalon is a registered trademark of DuPont Dow Elastomers LLC.
† Elite is a registered trademark of Dow Chemical Company.
‡ Nordel is a registered trademark of DuPont Dow Elastomers LLC.
§ Exxpol is a registered trademark of Exxon Mobil Corporation.
¶ Exact is a registered trademark of Exxon Mobil Corporation.

Mitsui Chemical reportedly has developed "FI" catalyst technology, called a *phenoxycyimine complex,* with 10 times the ethylene polymerization activity of metallocene catalysts, according to *Japan Chemical Weekly* (summer, 1999).[25]

EPM and EPDM can be produced by solution polymerization, while suspension and slurry polymerization are viable options. EPDM can be gas-phase 1,-4 hexadiene polymerized using Ziegler-Natta catalysts. Union Carbide produces ethylene propylene rubber (EPR) using modified Unipol® low-pressure gas-phase technology.

The letter "M" designates that the ethylene propylene has a saturated polymer chain of the polymethylene type, according to the ASTM.[12] EPM (copolymer of ethylene and propylene) rubber and EPDM (terpolymer of ethylene, propylene, and a nonconjugated diene) with residual side chain unsaturation, are subclassified under the ASTM "M" designation.[12]

The diene ethylidiene norbornene in Vistalon®* EPDM allows sulfur vulcanization (see Table 6.26).[12] 1,4-Hexadiene and dicyclopentadiene (DCPD) are also used as curing agents.[18] The completely saturated polymer "backbone" precludes the need for antioxidants that can bleed to the surface (bloom) of the finished product and cause staining.[12] Saturation provides inherent ozone and weather resistance, good thermal properties, and a low compression set.[12] Saturation also allows a relatively high-volume addition of low-cost fillers and oils in compounds while retaining a high level of mechanical properties.[12] The ethylene/propylene monomer ratio also affects the properties.

EPM and EPDM compounds, in general, have excellent chemical resistance to water, ozone, radiation, weather, brake fluid (nonpetroleum based), and glycol.[12]

EPM is preferred for dynamic applications, because its age resistance retains initial product design over time and environmental exposure.[12] EPDM is preferred for its high resilience.[12] EPM is resistant to acids, bases (alkalis), and hot detergent solution. EPM and EPDM are resistant to salt solutions, oxygenated solvents, and synthetic hydraulic fluids.[12] Properties are determined by the composition of the base compound. A typical formulation includes Vistalon EP(D)M, carbon black, process oil, zinc oxide, stearic acid, and sulfur. [12]

EPDM formulations are increasingly popular for medium-voltage, up to 221°F (105°C) continuous-use temperature wire and cable covering.[20] Thinner walls, yet lower (power) losses and better production rates, are sought by cable manufacturers.[20] Low-MW (Mooney viscos-

---

* Vistalon is a registered trademark of Exxon Chemical Company, Division of Exxon Corporation.

TABLE 6.26 Typical Properties of EPM/EPDM Compounds[12] Based on Vistalon

Property	Value
Hardness Shore A	35–90
Tensile strength, lb/in^2 (MPa)	580–3200 (4–22)
Compression set (%), 70 h @ 302°F (150°C)	15–35
Elongation, %	150–180
Tear strength, lb/in (kN/m)	86–286 (15–50)
Continuous service temperature, °F (°C)	302 (150) max
Intermediate service temperature, °F (°C)	347 (175) max
Resilience (Yerzley), %	75
Loss tangent (15 Hz), % (dynamic)	0.14
Elastic spring rate, lb/in (kN/m) (15 Hz)	3143 (550)
Dielectric constant, %	2.8
Dielectric strength, kV/mm	26
Power factor, %	0.25
Volume resistivity, Ω–cm	1×10^{16}

ity, ML), high-ethylene-content copolymers and terpolymers are used In medium-voltage cable formulations.[20] With ethylidiene norbornene or hexadiene, EPDMs are good vulcanizates, providing improved wet electrical properties.[20] When the diene vinyl norbornene is incorporated on the EPDM backbone by a gel-free process, a significantly improved EPDM terpolymer is obtained for wire/cable applications.[20] Other applications are automotive body seals, mounts, weatherstripping, roofing, hose, tubing, ducts, and tires. Molded EPDM rubber is used for bumpers and fillers to dampen vibrations around the vehicle, such as deck-lid over-slam bumpers, for its ozone and heat resistance.[10d] EPDM can be bonded to steels, aluminum, and brass, with modified poly(acrylic acid) and polyvinylamine water-soluble coupling agents.[27]

EPDM is a favorable selection for passenger car washer-fluid tubes and automotive body seals, and it is used for automotive vacuum tubing.[10d] EPDM has good water-alcohol resistance for delivering fluid from the reservoir to the spray nozzle and good oxygen and UV resistance.[10d] Random polymerization yields a liquid with a viscosity of 100,000 centipoise (cP) @ 203°F (95°C), room temperature-cured with para-quinone dioxime systems or two-component peroxide systems, or cured at an elevated temperature with sulfur.[10b] They can be used as automotive and construction sealants, waterproof roofing membranes, and for encapsulating electrical components.[10b]

6.15.7 Fluoroelastomers (FKM)

Fluoroelastomers can be polymerized with copolymers and terpolymers of tetrafluoroethylene, hexafluoroethylene, and vinylidene fluoride. The fluorine content largely determines chemical resistance and T_g, which increases with increasing fluorine content. Low-temperature flexibility decreases with increasing fluorine content.[1] The fluorine content is typically 57 wt%.[11]

TFE/propylene copolymers can be represented by Aflas®* TFE, produced by Asahi Glass. They are copolymers of tetrafluoroethylene (TFE) + propylene, and terpolymers of TFE + propylene + vinylidene fluoride. Fluoroelastomer dipolymer and terpolymer gums are amine- or bisphenol-cured and peroxide-cured for covulcanizable blends with other peroxide curable elastomers. They can contain cure accelerators for faster cures, and they are divided into three categories: (1) gums with incorporated cures, (2) gums without incorporated cures, and (3) specialty master batches used with other fluoroelastomers.

Aflas products are marketed in five categories according to their MW and viscosities.[11] The five categories possess similar thermal, chemical, and electrical resistance properties but different mechanical properties.[11] The lowest viscosity is used for chemical process industry tank and valve linings, gaskets for heat exchangers and pipe/flanges, flue duct expansion joints, flexible and spool joints, and viscosity improver additives in other Aflas grades.[11]

The second-lowest viscosity grade is high-speed extruded into wire/cable coverings, sheet, and calendered stock.[11] Wire and cable covering are a principal application, especially in Japan. The third grade, general-purpose, is molded, extruded, and calendered into pipe connector gaskets, seals, and diaphragms in pumps and valves.[11] The fourth grade, with higher MW, is compression molded into O-rings and other seal applications.

The fifth grade, with the highest MW, is compression molded into oil field applications requiring resistance to high-pressure gas blistering.[11] It is used for down-hole packers and seals in oil exploration and production. Oilfield equipment seals are exposed to short-term temperatures from 302 to 482°F (150 to 250°C) and pressures above 10,000 lb/in^2 (68.7 MPa) in the presence of aggressive hydrocarbons H_2S, CH_4, CO_2, amine-containing corrosion inhibitors, and steam and water.[11]

Synthetic rubbers, EPM/EPDM, nitrile, polychloroprene (neoprene), epichlorohydrin, and polyacrylate have good oil resistance, heat stability, and chemical resistance. Fluoropolymers are used in oil and gas

---

* Aflas is a registered trademark of Asahi Glass Company.

wells 20,000 ft (6096 m) deep. These depths can have pressures of 20,000 lb/in^2 (137.5 MPa), which cause "extrusion" failures of down-hole seals by forcing the rubber part out of its retaining gland. TFE/propylene jackets protect down-hole assemblies, which consist of stainless steel tubes that deliver corrosion-resistant fluid into the well.

Aircraft jet engine O-rings require fluoropolymer grades for engine cover gaskets that are resistant to jet fuel, turbine lube oils, and hydraulic fluids.

Dyneon®* BREs (base-resistant elastomers) are used in applications exposed to automotive fluids such as ATF, gear lubricants, engine oils shaft seals, O-rings, and gaskets.

DuPont Dow Elastomers fuel-resistant Viton®† fluoroelastomers are an important source for the applications described previously. The company's Kalrez®‡ perfluoroelastomers with reduced contamination are widely used with semiconductors and other contamination-sensitive applications. Contamination caused by high alcohol content in gasoline can cause fuel pump malfunction. The choice of polymer can determine whether an engine functions properly.

The three principal Viton categories are (1) Viton A dipolymers composed of vinylidene fluoride (VF$_2$) and hexafluoropropylene (HFP) to produce a polymer with 66% (wt%) fluorine content, (2) Viton B terpolymers of VF$_2$ + HFP + tetrafluoroethylene (TFE) to produce a polymer with 68% fluorine, and (3) Viton F terpolymers composed of VF$_2$ + HFP + TFE to produce a polymer with 70% fluorine.[36] The three categories are based on their resistance to fluids and chemicals.[36] Fluid resistance generally increases but low-temperature flexibility decreases with higher fluorine content.[36] Specialty Viton grades are made with additional or different principal monomers in order to achieve specialty performance properties.[36] An example of a specialty property is low-temperature flexibility.

Compounding further yields properties to meet a given application.[36] Curing systems are an important variable affecting properties. DuPont Dow Elastomers developed curing systems during the 1990s, and the company should be consulted for the appropriate system for a given Viton grade.

FKMs are coextruded with lower-cost (co)polymers such as ethylene acrylic copolymer.[10d] They can be modified by blending and vulcanizing with other synthetic rubbers such as silicones, EPR and EPDM,

---

epichlorohydrin, and nitriles. Fluoroelastomers are blended with modified NBR to obtain an intermediate performance/cost balance. These blends are useful for underhood applications in environments outside the engine temperature zone such as timing chain tensioner seals.

Fluoroelastomers are blended with fluorosilicones and other high-temperature polymers to meet engine compartment environments and cost/performance balance. Fine-particle silica increases hardness, red iron oxide improves heat resistance, and zinc oxide improves thermal conductivity. Hardness ranges from about Shore 35 A to 70 A. Fluorosilicones are resistant to nonpolar and nominally polar solvents, diesel and jet fuel, and gasoline, but not to solvents such as ketones and esters.

Typical applications are exhaust gas recirculating and seals for engine valve stems and cylinders, crankshaft, speedometers, and O-rings for fuel injector systems.

FKMs are compounded in either water-cooled internal mixers or two-roll mills. A two-pass mixing is recommended for internally mixed compounds with the peroxide curing agent added in the second pass.[11] Compounds press-cured 10 min @ 350°F (177°C) can be formulated to possess more than 2100-lb/in^2 (14.4-MPa) tensile strength, 380% elongation, 525% @ 100% modulus, and higher values when postcured 16 h @ 392°F (200°C).[11] Processing temperatures are >392°F (200°C).[30]

6.15.8   Polyacrylate Acrylic Rubber (ACM)

Acrylic rubber can be emulsion- and suspension-polymerized from acrylic esters such as ethyl, butyl, and/or methoxyethyl acetate to produce polymers of ethyl acetate and copolymers of ethyl, butyl, and methoxyl acetate. Polyacrylate rubber, such as Acron®* from Cancarb Ltd., Alberta, Canada, possesses heat resistance and oil resistance between nitrile and silicone rubbers[23] Acrylic rubbers retain properties in the presence of hot oils and other automotive fluids, and they resist softening or cracking when exposed to air up to 392°F (200°C). The copolymers retain flexibility down to –40°F (–40°C). Automotive seals and gaskets constitute a major market.[23] These properties and inherent ozone resistance are largely due to the polymer's saturated "backbone" (see Table 6.27). Polyacrylates are vulcanized with sulfur or metal carboxylate, with a reactive chlorine-containing monomer to create a cross-linking site.[23]

Copolymers of ethylene and methyl acrylate, and ethylene acrylics, have a fully saturated "backbone," providing heat-aging resistance

---

* Acron is a registered trademark of Cancarb Ltd.

TABLE 6.27   Property Profile of Polyacrylic Rubbers[23]

Property at room temperature[*]	Value
Tensile strength, lb/in^2 (MPa)	2212 (15.2)
100% modulus	1500 (10.3)
Compression set, % [70 h @ 302°F (150°C)]	28
Hardness Shore A	80

[*]Unless indicated otherwise.

and inherent ozone resistance.[23] They are compounded in a Banbury mixer and fabricated by injection molding, compression molding, resin transfer molding, extrusion, and calendering.

6.15.9   Polychloroprene (Neoprene) (CR)

Polychloroprene is produced by free-radical emulsion polymerization of primarily trans-2-chloro-2-butenylene moieties.[23] Chloroprene rubber possesses moderate oil resistance, very good weather and oil resistance, and good resistance to oxidative chemicals.[10a] Performance properties depend on compound formulation, with the polymer providing fundamental properties. This is typical of any polymer and its compounds. Chloride imparts inherent self-extinguishing flame retardance.

Crystallization contributes to high tensile strength, elongation, and wear resistance in its pure gum state before CR is extended or hardened.[10a]

6.15.10   Polyisoprene (IR)

Polymerization of isoprene can yield high-purity cis-1,4-polyisoprene and trans-1,4-polyisoprene. Isoprene is 2-methyl-1,3-butadiene, 2-methyldivinyl, or 2-methylerythrene.[23] Isoprene is polymerized by 1,4 or vinyl addition, the former producing cis-1,4 or trans-1,4 isomer.[23]

Synthetic polyisoprene, isoprene rubber (IR), was introduced in the 1950s as odorless rubber with virtually the same properties as natural rubber. Isoprene rubber product and processing properties are better than natural rubber in a number of characteristics. MW and MWD can be controlled for consistent performance and processing properties.

Polyisoprene rubber products are illustrated by Natsyn®,[*] which is used to make tires and tire tread (cis isomer). Tires are the major ci-

---

* Natsyn is registered trademark of Goodyear Tire and Rubber Company.

spolyisoprene product. Trans-polyisoprene can be used to make golf ball covers, hot-melt adhesives, and automotive and industrial products.

Depolymerized polyisoprene liquid is used as a reactive plasticizer for adhesive tapes, hot melts, brake linings, grinding wheels, and wire and cable sealants.[10b]

6.15.11 Polysulfide Rubber (PSR)

PSR is highly resistant to hydrocarbon solvents, aliphatic fluids, and aliphatic-aromatic blends.[10a] It is also resistant to conventional alcohols, ketones, and esters used in coatings and inks and to certain chlorinated solvents.[10d] With these attributes, PSR is extruded into hose to carry solvents and printing rolls, and, due to its good weather resistance, it is useful in exterior caulking compounds.[10a] Its limitations, compared with nitrile, are relatively poor tensile strength, rebound, abrasion resistance, high creep under strain, and odor.[10a]

Liquid PSR is oxidized to rubbers with service temperatures from −67 to 302°F (−55 to 150°C), excellent resistance to most solvents, and good resistance to water and ozone.[10b] It has very low selective permeability rates to a number of highly volatile solvents and gases and odors. Compounds formulated with liquid PSR can be used as a flexibilizer in epoxy resins, and epoxy-terminated polysulfides have better underwater lap shear strength than toughened epoxies.[10b] Other PSR applications are aircraft fuel tank sealant, seals for flexible electrical connections, printing rollers, protective coatings on metals, binders in gaskets, caulking compound ingredient, adhesives, and to provide water and solvent resistance to leather.

6.15.12 Silicone Rubber (SiR)

Silicone rubber polymers have the more stable Si atom compared with carbon. Silicone's property signature is its combined (1) high-temperature resistance [>500°F (260°C)], (2) good flexibility at <−100°F (−73°C), (3) good electrical properties, (4) good compression set, and (5) tear resistance and stability over a wide temperature range.[10a] When exposed to decomposition level temperature, the polymer forms SiO_2, which can continue to serve as an electrical insulator.[10a] Silicone rubber is used for high-purity coatings for semiconductor junctions, high-temperature wire, and cable coverings.[1]

RTV (room temperature vulcanizing) silicones cure in about 24 h.[10b] They can be graded according to their room-temperature viscosities, which range from as low as 1500 cP (general-purpose soft) up to 700,000 cP (high-temperature paste). Most, however, are between 12,000 and 40,000 cP.[10b] RTV silicone has a low modulus over a wide temperature range from −85 to 392°F (−65 to 200°C), making them suitable for encapsulating electrical components during thermal cycling and shock.[10b] Low modulus minimizes stress on the encapsulated electrical components.[10b]

High-consistency rubber (HCR) from Dow Corning is injection-molded into high-voltage insulators, surge arrestors, weather sheds, and railway insulators. The key properties are wet electrical performance and high tracking resistance.

Liquid silicone rubbers (LSRs) are two-part grades that can be coinjection-molded with thermoplastics to make door locks and flaps for vents.[10b] LSRs can be biocompatible and have low compression set, low durometer hardness, and excellent adhesion.[10b] One-part silicones are cured by ambient moisture. They are used for adhesives and sealants with plastic, metal, glass, ceramic, and silicone rubber substrates.[10b] A solventless, clear silicone/PC has been developed that requires no mixing and can be applied without a primer.[1]

6.15.13 Styrene Butadiene Rubber (SBR)

SBR is emulsion- and solution-polymerized from styrene and butadiene, plus small volumes of emulsifiers, catalysts and initiators, endcapping agents, and other chemicals. It can be sulfur-cured SBR types are illustrated with Plioflex®* emulsion SBR (emulsion polymerized) and Solflex®† solution SBR.[6] Emulsion SBR is produced by hot polymerization for adhesives and by cold polymerization for tires and other molded automotive and industrial products.[6] Solution SBR is used for tires.

SBR is a low-cost rubber with slightly better heat aging and wear resistance than NR for tires.[10a] SBR grades are largely established by the bound styrene/butadiene ratio, polymerization conditions such as reaction temperature, and auxiliary chemicals added during polymerization.

SBR/PVC blends that employ nitrile rubber (NBR) as a compatibilizer show improved mechanical properties at lower cost than NBI/PVC.[21] This was the conclusion of studies using a divinylbenzene cross-linked, hot-polymerized emulsion polymer with 30% bound styrene and a cold-polymerized emulsion polymer with 23% bound styrene; PVC with inherent viscosity from 0.86 to 1.4; NBR with Mooney viscosity from 30 to 86; acrylonitrile content of 23.5, 32.6, and 39.7%; and ZnO, stabilizers, sulfur, and accelerators.[21]

---

* Plioflex is a registered trademark of Goodyear Tire and Rubber Company.

† Solflex is a registered trademark of Goodyear Tire and Rubber Company.

6.16 Natural Rubber (NR)

Natural rubber, the original elastomer, still plays an important role among elastomers. Worldwide consumption of NR in 2000 is expected to be about 7 million metric tons/y, based on earlier reporting by the International Rubber Study Group. Chemically, natural rubber is cis1,4-polyisoprene and occurs in Hevea rubber trees. NR tapped from other rubber trees (gutta-percha and balata) is the trans isomer of polyisoprene.[23] NR's principal uses are automotive tires, tire tread, and mechanical goods. Automotive applications are always compounded with carbon black to impart UV resistance and to increase mechanical properties.[10d] Latex concentrate is used for dipped goods, adhesives, and latex thread.[23] Latex concentrate is produced by centrifuge-concentrating field latex tapped from rubber trees. The dry rubber content is subsequently increased from 30 to 40 to 60% minimum.[23]

Vulcanization is the most important NR chemical reaction.[23] Most applications require cross-linking via vulcanization to increase resiliency and strength. Exceptions are crepe rubber shoe soles and rubber cements.[23] There are a number of methods for sulfur vulcanization, with certain methods producing polysulfidic cross-linking and other methods producing more monosulfidic cross-links.[10d]

NR is imported from areas such as Southeast Asia to the world's most industrial regions, North America, Europe, and Japan, since it is not indigenous to these regions. The huge rubber trees require about 80 to 100 in/y (200 to 250 cm/y) rainfall, and they flourish at an altitude of about 1000 ft (300 m).[23] As long as NH is needed for tires, industrial regions will be import dependent.

NR has good resilience; high tensile strength; low compression set; resistance to wear and tear, cut-through and cold flow; and good electricalproperties.[10a] Resilience is the principal property advantage compared with synthetic rubbers.[10a] For this reason, NH is usually used for engine mounts, because NR isolates vibrations caused when an engine is running. NH is an effective decoupler, isolating vibrations such as engine vibration from being transmitted to another location such as the passenger compartment.[10d] With decoupling, vibration is returned to its source instead of being transmitted through the rubber.[10d] Polychloroprene is used for higher under-hood temperatures above NR service limits; butyl rubber is used for body mounts and for road vibration frequencies, which occur less frequently than engine vibrations or have low energy; EPDM is often used for molded rubber bumpers and fillers throughout the vehicle, such as deck-lid over-slam bumpers.

Degree of crystallinity (DC) can affect NH properties, and milling reduces MW. MW is reduced by mastication, typically with a Banbury mill, adding a peptizing agent during milling to further reduce MW, which improves NR solubility after milling.[23] NR latex grades are provided to customers in low (0.20 wt %) and high (0.75 wt %), with ammonia added as a preservative.[23] Low NH_4 has reduced odor and eliminates the need for deammoniation.[23]

Properties of polymers are improved by compounding with enhancing agents (additives), and NR is not an exception. Compounding NR with property enhancers improves resistance to UV oxygen, and ozone, but formulated TPEs and synthetic rubbers overall have better resistance than compounded NR to UV, oxygen, and ozone.[10a] NR does not have satisfactory resistance to fuels, vegetable, and animal oils, while TPEs and synthetic rubbers can possess good resistance to them.[10a] NR has good resistance to acids and alkalis.[10a] It is soluble in aliphatic, aromatic, and chlorinated solvents, but it does not dissolve easily because of its high MW. Synthetic rubbers have better aging properties; they harden over time, while NR softens over time (see Table 6.28).[10a]

TABLE 6.28   Typical Thermal and Electrical Property Profile of NR[23]

Property	Value
Specific gravity	
@ 32°F (0°C)	0.950
@ 68°F (20°C)	0.934
T_g, °F (°C)	−98 (−72)
Specific heat	0.502
Heat of combustion, cal/g (J/g)	10,547 (44,129)
Thermal conductivity, (BTU-in) (h-ft^2-°F)	0.90
W/(m · K)	0.13
Coefficient of cubical expansion, in^3/°C	0.00062
Dielectric strength, V/mm	3.937
Dielectric constant	2.37
Power factor @ 1000 cycles	0.15–0.20
Volume resistivity, Ω · cm	1015
Cohesive energy density, cal/cm^3 (J/cm^3)	64 (266.5)
Refractive index	
68°F (20°C) RSS*	1.5192
68°F (20°C) pale crepe	1.5218

*RSS = ribbed smoked sheet.

There are several visually graded latex NRs, including ribbed smoked sheets (RSS) and crepes such as white and pale, thin and thick brown latex, etc.[23] Two types of raw NR are field latex and raw coagulum, and these two types comprise all NR ("downstream") grades.[23]

Depolymerized NR is used as a base for asphalt modifiers, potting compound, and cold-molding compounds for arts and crafts.[10b]

6.17 Conclusion

Producers can engineer polymers and copolymers, and compounders can formulate recipes for a range of products that challenges the designers' imaginations. Computer variable-controlled machinery, tools, and dies can meet the designers' demands. Processing elastomeric materials is not as established as the more traditional thermoplastic and thermosetting polymers. Melt rheology, more than just viscosity, is the central differentiating characteristic for processing elastomeric materials. Processing temperature and pressure settings are not fixed ranges; they are dynamic, changing values from the hopper to the demolded product. Operators and management of future elastomeric materials processing plants will be educated to the finesse of melt processing these materials. Elastomeric materials industries, welcome to the twenty-first century.

References

1. James M. Margolis, "Elastomeric Polymers 2000 to 2010: Properties, Processes and Products" Report, 2000.
2. *KRATON Polymers and Compounds, Typical Properties Guide,* Shell Chemical Company, Houston, Texas, 1997.
3. *Products, Properties and Processing for PELLETHANE Thermoplastic Polyurethane Elastomers,* Dow Plastics, The Dow Chemical Company Midland, Michigan, ca. 1997.
4. *Modern Plastics Encyclopedia '99,* McGraw-Hill, New York, 1999, pp. B-51, B-52.
5. Engage, A Product of DuPont Dow Elastomers, Wilmington, Delaware, December 1998.
6. *Product Guide,* Goodyear Chemical, Goodyear Tire & Rubber Company, Akron, Ohio, October 1996.
7. *Injection Molding Guide for Thermoplastic Rubber–Processing, Mold Design, Equipment,* Advanced Elastomer Systems LP, Akron, Ohio, 1997.
8. *Santoprene Rubber Physical Properties Guide,* Advanced Elastomer Systems LP, Akron, Ohio, ca. 1998.
9. Hifax MXL 55A01 (1998), FXL 75A01 (1997) and MXL 42D01 Developmental Data Sheets secured during product development and subject to change before final commercialization. Montell Polyolefins Montell North America Inc., Wilmington, Delaware.
10. Charles B. Rader, "Thermoplastic Elastomers," in *Handbook of Plastics, Elastomers, and Composites,* 3d ed., Charles A. Harper, ed., McGraw-Hill, New York, 1996.
10a. Joseph F. Meier, "Fundamentals of Plastics and Elastomers," in *Handbook of Plastics, Elastomers, and Composites,* 3d ed., Charles A. Harper, ed., McGraw-Hill, New York, 1996.
10b. Leonard S. Buchoff, "Liquid and Low-Pressure Resin Systems," in *Handbook of Plastics, Elastomers, and Composites,* 3d ed., Charles A. Harper, ed., McGraw-Hill, New York, 1996.

10c. Edward M. Petrie. "Joining of Plastics, Elastomers, and Composites," in *Handbook of Plastics, Elastomers, and Composites,* 3d ed., Charles A. Harper, ed., McGraw-Hill, New York, 1996.

10d. Ronald Toth, "Elastomers and Engineering Thermoplastics for Automotive Applications," in *Handbook of Plastics, Elastomers, and Composites,* 3d ed., Charles Harper, ed., McGraw-Hill, New York, 1996.

11. Aflas TFE Elastomers Technical Information and Performance Profile Data Sheets, Dyneon LLC, A 3M-Hoechst Enterprise, Oakdale, Minnesota, 1997.

12. Vistalon User's Guide, Properties of Ethylene-Propylene Rubber, Exxon Chemical Company, Houston, Texas, Division of Exxon Corporation, ca. 1996.

13. KRATON Liquid L-2203 Polymer, Shell Chemical Company, Houston, Texas, 1997.

14. Affinity Polyolefin Plastomers, Dow Plastics, The Dow Chemical Company, Midland, Michigan, 1997.

14a. Affinity HF-1030 Data Sheet, Dow Plastics, The Dow Chemical Company Midland, Michigan, 1997.

14b. Affinity PF 1140 Data Sheet, Dow Plastics, The Dow Chemical Company Midland, Michigan, 1997.

15. Bayer Engineering Polymers Properties Guide, Thermoplastics and Polyurethanes, Bayer Corporation, Pittsburgh, Pennsylvania, 1998.

16. *Rubber World Magazine,* monthly, 1999.

17. Jim Ahnemiller, "PU Rubber Outsoles for Athletic Footwear," *Rubber World,* December 1998.

18. Charles D. Shedd, "Thermoplastic Polyolefin Elastomers," in *Handbook of Thermoplastic Elastomers,* 2d ed., Benjamin M. Walker and Charles P. Rader, eds., Van Nostrand Reinhold, New York, 1988.

18a. Thomas W. Sheridan, "Copolyester Thermoplastic Elastomers," *Handbook* of *Thermoplastic Elastomers,* 2d ed., Benjamin M. Walker and Charles P. Rader, eds., Van Nostrand Reinhold, New York, 1988.

18b. William J. Farrisey "Polyamide Thermoplastic Elastomers," in *Handbook of Thermoplastic Elastomers,* 2d ed., Benjamin M. Walker and Charles P. Rader, eds., Van Nostrand Reinhold, New York, 1988.

18c. Eric C. Ma, "Thermoplastic Polyurethane Elastomers, in *Handbook of Thermoplastic Elastomers,* 2d ed., Benjamin M. Walker and Charles P. Rader, eds., Van Nostrand Reinhold, New York, 1988.

19. N. R. Legge, G. Holden, and H. E. Schroeder, eds., *Thermoplastic Elastomers, A Comprehensive Review,* Hanser Publishers, Munich, Germany, 1987.

20. P S. Ravisbanker, "Advanced EPDM for W & C Applications," *Rubber World,* December 1998.

21. Junling Zbao, G. N. Chebremeskel, and J. Peasley "SBR/PVC Blends With NBR As Compatibilizer," *Rubber World,* December 1998.

22. John E. Rogers and Walter H. Waddell, "A Review of Isobutylene-Based Elastomers Used in Automotive Applications," *Rubber World,* February 1999.

23. *Kirk-Othmer Concise Encyclopedia of Chemical Technology,* John Wiley & Sons, New York, 1999.

24. *PetroChemical News (PCN),* weekly, William F. Bland Company Chapel Hill, North Carolina, September 14,1998.

25. *PeroChemical News (PCN),* weekly, William F. Bland Company, Chapel Hill, North Carolina, 1998 and 1999.

26. *PetroChemical News (PCN),* weekly William F. Bland Company, Chapel Hill, North Carolina, February 22, 1999.

27. C. P. J. van der Aar, et al., "Adhesion of EPDMs and Fluorocarbons to Metals by Using Water-Soluble Polymers," *Rubber World,* November 1998.

28. Larry R. Evans and William C. Fultz, "Tread Compounds with Highly Dispersible Silica," *Rubber World,* December 1998.

29. Vector Styrene Block Copolymers, Dexco Polymers, A Dow/Exxon Partnership, Houston, Texas, 1997.

30. *Fluoroelastomers Product Information Manual* (1997), *Product Comparison Guide* (1999), Dyneon LLC, A 3M-Hoechst Enterprise, Oakdale, Minnesota, 1997.

31. Engel data sheets and brochures, Guelph, Ontario, 1998.
32. Catalloy Process Resins, Montell Polyolefins, Wilmington, Delaware.
32a. Catalloy Process Resins, Montell Polyolefins, Wilmington, Delaware, p.7.
33. EniChem Europrene SOL T Thermoplastic Rubber, styrene butadiene types, styrene isoprene types, EniChem Elastomers Americas Inc., Technical Assistance Laboratory, Baytown, Texas.
34. "Arnitel Guidelines for the Injection Molding of Thermoplastic Elastomer TPE-E," DSM Engineering Plastics, Evansville, Ind., ca, 1998.
35. Correspondence from DuPont Engineering Polymers, July 1999.
36. Correspondence from DuPont Dow Elastomers, Wilmington, Delaware, August 1999.

7

Ceramics and Ceramic Composites

Dr. Jerry E. Sergent
TCA, Inc.
Corbin, Kentucky

7.1 Introduction

Ceramics are crystalline in nature, with a dearth of free electrons. They have a high electrical resistivity, are very stable (chemically and thermally), and have a high melting point. They are formed by the bonding of a metal and a nonmetal and may exist as oxides, nitrides, carbides, or silicides. An exception is diamond, which consists of pure carbon subjected to high temperature and pressure. Diamond substrates meet the criteria for ceramics and may be considered as such in this context.

The primary bonding mechanism in ceramics is ionic bonding. An ionic bond is formed by the electrostatic attraction between positive and negative ions. Atoms are most stable when they have eight electrons in the outer shell. Metals have a surplus of electrons in the outer shell, which are loosely bound to the nucleus and readily become free, creating positive ions. Similarly, nonmetals have a deficit of electrons in the outer shell and readily accept free electrons, creating negative ions. Figure 7.1 illustrates an ionic bond between a magnesium ion with a charge of +2 and an oxygen ion with a charge of –2, forming magnesium oxide (MgO). Ionically bonded materials are crystalline in nature and have both a high electrical resistance and a high relative dielectric constant. Due to the strong nature of the bond, they have a high melting point and do not readily break down at elevated temperatures. By the same token, they are very stable chemically and are not attacked by ordinary solvents and most acids.

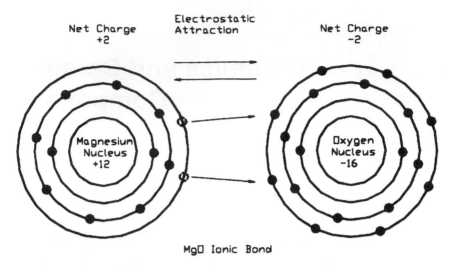

Figure 7.1 Magnesium oxide ionic bond.

A degree of covalent bonding may also be present, particularly in some of the silicon and carbon-based ceramics. The sharing of electrons in the outer shell forms a covalent bond. A covalent bond is depicted in Fig. 7.2, illustrating the bond between oxygen and hydrogen to form water. A covalent bond is also a very strong bond and may be present in liquids, solids, or gases.

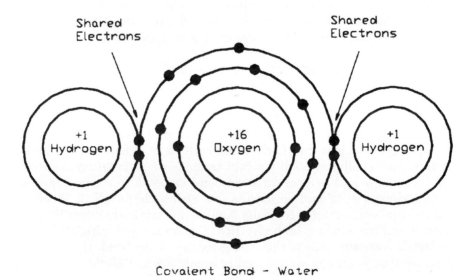

Figure 7.2 Covalent bond between oxygen and hydrogen to form water.

A composite is a mixture of two or more materials that retain their original properties but, in concert, offer parameters that are superior to either. Composites in various forms have been used for centuries. Ancient peoples, for example, used straw and rocks in bricks to increase their strength. Modern day structures use steel rods to reinforce concrete. The resulting composite structure combines the strength of steel with the lower cost and weight of concrete.

Ceramics are commonly used in conjunction with metals to form composites for electronic applications, especially thermal management. Ceramic-metal (cermet) composites typically have a lower TCE than metals, possess a higher thermal conductivity than ceramics, and are more ductile and more resistant to stress than ceramics. These properties combine to make cermet composites ideal for use in high-power applications.

This chapter considers the properties of ceramics used in microelectronic applications, including aluminum oxide (alumina, Al_2O_3), beryllium oxide (beryllia, BeO), aluminum nitride (AlN), boron nitride (BN), diamond (C), and silicon carbide (SiC). Several composite materials, aluminum silicon carbide (AlSiC) and Dymalloy, a diamond/copper structure, are also described. Although the conductive nature of these materials prevents them from being used as a conventional substrate, they have a high thermal conductivity and may be used in applications where the relatively low electrical resistance is not a consideration.

7.2   Ceramic Fabrication

It is difficult to manufacture ceramic substrates in the pure form. The melting point of most ceramics is very high, as shown in Table 7.1, and most are also very hard, limiting the ability to machine the ceramics. For these reasons, ceramic substrates are typically mixed with fluxing and binding glasses, which melt at a lower temperature and make the finished product easier to machine.

The manufacturing process for Al_2O_3, BeO, and AlN substrates is very similar. The base material is ground into a fine powder, several microns in diameter, and mixed with various fluxing and binding glasses, including magnesia and calcia, also in the form of powders. An organic binder, along with various plasticizers, is added to the mixture, and the resultant slurry is ball-milled to remove agglomerates and to make the composition uniform.

The slurry is formed into a sheet, the so-called *green state,* by one of several processes as shown in Fig. 7.3[1] and sintered at an elevated temperature to remove the organics and to form a solid structure.

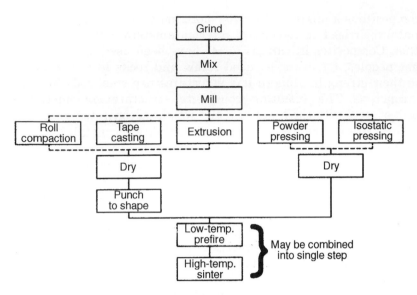

Figure 7.3   Flow chart for ceramic substrate processing.

TABLE 7.1   Melting Points of Selected Ceramics

Material	Melting point, °C
SiC	2700
BN	2732
AlN	2232
BeO	2570
Al_2O_3	2000

Roll compaction.   The slurry is sprayed onto a flat surface and partially dried to form a sheet with the consistency of putty. The sheet is fed through a pair of large parallel rollers to form a sheet of uniform thickness.

Tape casting.   The slurry is dispensed onto a moving belt that flows under a knife-edge to form the sheet. This is a relatively low-pressure process compared to the others.

Powder pressing.   The powder is forced into a hard die cavity and subjected to very high pressure (up to 20,000 psi) throughout the sintering process. This produces a very dense part with tighter as-fired tolerances than other methods, although pressure variations may produce excessive warpage.

Isostatic powder pressing. This process utilizes a flexible die surrounded with water or glycerin and compressed with up to 10,000 psi. The pressure is more uniform and produces a part with less warpage.

Extrusion. The slurry, less viscous than for other processes, is forced through a die. Tight tolerances are hard to obtain, but the process is very economical and produces a thinner part than is attainable by other methods.

In the green state, the substrate is approximately the consistency of putty and may be punched to the desired size. Holes and other geometries may also be punched at this time.

Once the part is formed and punched, it is sintered at a temperature above the glass melting point to produce a continuous structure. The temperature profile is very critical, and the process may actually be performed in two stages: one stage to remove the volatile organic materials and a second stage to remove the remaining organics and to sinter the glass/ceramic structure. The peak temperature may be as high as several thousand degrees celsius and may be held for several hours, depending on the material and the type and amount of binding glasses. For example, pure alumina substrates formed by powder processing with no glasses are sintered at 1930°C.

It is essential that all the organic material be removed prior to sintering. Otherwise, the gases formed by the organic decomposition may leave serious voids in the ceramic structure and cause serious weakening. The oxide ceramics may be sintered in air. In fact, it is desirable to have an oxidizing atmosphere to aid in removing the organic materials by allowing them to react with the oxygen to form CO_2. The nitride ceramics must be sintered in the presence of nitrogen to prevent oxides of the metal from being formed. In this case, no reaction of the organics takes place; they are evaporated and carried away by the nitrogen flow.

During sintering, a degree of shrinkage takes place as the organic is removed and the fluxing glasses activate. Shrinkage may range from as low as 10% for powder processing to as high as 22% for sheet casting. The degree of shrinkage is highly predictable and may be considered during design.

Powder pressing generally forms boron nitride substrates. Various silica and/or calcium compounds may be added to lower the processing temperature and improve machinability. Diamond substrates are typically formed by chemical vapor deposition (CVD). Composite substrates, such as AlSiC, are fabricated by creating a spongy structure of SiC and forcing molten aluminum into the crevices.

7.3   Surface Properties of Ceramics

The surface properties of interest, surface roughness and camber, are highly dependent on the particle size and method of processing. Surface roughness is a measure of the surface microstructure, and camber is a measure of the deviation from flatness. In general, the smaller the particle size, the smoother will be the surface.

Surface roughness may be measured by electrical or optical means. Electrically, surface roughness is measured by moving a fine-tipped stylus across the surface. The stylus may be attached to a piezoelectric crystal or to a small magnet that moves inside a coil, inducing a voltage proportional to the magnitude of the substrate variations. The stylus must have a resolution of 25.4 nm (1 μin) to read accurately in the most common ranges. Optically, a coherent light beam from a laser diode or other source is directed onto the surface. The deviations in the substrate surface create interference patterns that are used to calculate the roughness. Optical profilometers have a higher resolution than the electrical versions and are used primarily for very smooth surfaces. For ordinary use, the electrical profilometer is adequate and is widely used to characterize substrates in both manufacturing and laboratory environments.

The output of an electrical profilometer is plotted as shown in schematic form in Fig. 7.4 and in actual form in Fig. 7.5. A quantitative interpretation of surface roughness can be obtained from this plot in one of two ways: by the rms value and by the arithmetic average.

The rms value is obtained by dividing the plot into n small, even increments of distance and measuring the height, m, at each point, as shown in Fig. 7.4. The rms value is calculated by

$$\text{rms} = \sqrt{\frac{m_1^2 + m_2^2 + \ldots + m_n^2}{n}} \tag{7.1}$$

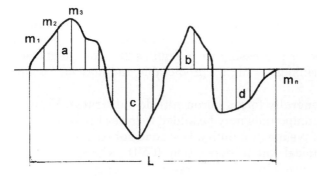

Figure 7.4   Schematic of surface trace.

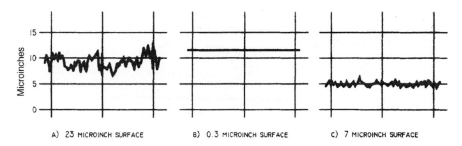

Figure 7.5 Surface trace of three substrate surfaces.

and the average value (usually referred to as the *center line average, CLA*) is calculated by

$$\text{CLA} = \frac{a_1 + a_2 + a_3 + \dots + a_n}{L} \qquad (7.2)$$

where $a_1, a_2, a_3, \dots$ = areas under the trace segments (Fig 7.4)
$\qquad L$ = length of travel

For systems where the trace is magnified by a factor M, Eq. (7.2) must be divided by the same factor.

For a sine wave, the average value is $0.636 \times$ peak and the rms value is $0.707 \times$ peak, which is 11.2% larger than the average. The profilometer trace is not quite sinusoidal in nature. The rms value may be greater than the CLA value from 10 to 30%.

Of the two methods, the CLA is the preferred method of use, because the calculation is more directly related to the surface roughness. However, it also has several shortcomings.

- The method does not consider surface waviness or camber as shown in Fig. 7.6.[2]

- Surface profiles with different periodicities and the same amplitudes yield the same results, although the effect in use may be somewhat different.

- The value obtained is a function of the tip radius.

Surface roughness has a significant effect on the adhesion and performance of thick and thin film depositions. For adhesion purposes, it is desirable to have a high surface roughness to increase the effective interface area between the film and the substrate. For stability and repeatability, the thickness of the deposited film should be much greater than the variations in the surface. For thick films, which have

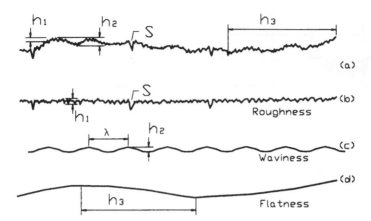

Figure 7.6 Surface characteristics.

a typical thickness of 10–12 μ, surface roughness is not a consideration, and a value of 25 μin (625 nm) is desirable. For thin films, however, which may have a thickness measured in angstroms, a much smoother surface is required. Figure 7.7 illustrates the difference in a thin film of tantalum nitride (TaN) deposited on both a 1 μin surface and a 5 μin surface. Tantalum nitride is commonly used to fabricate resistors in thin film circuits and is stabilized by growing a layer of tantalum oxide, which is nonconductive, over the surface by baking the resistors in air. Note that the oxide layer in the rougher surface represents a more significant percentage of the overall thickness of the film in areas where the surface deviation is the greatest. The result is a wider variation in both the initial and post-stabilization resistor values and a larger drift in value with time.

Camber and waviness are similar in form in that they are variations in flatness over the substrate surface. Referring to Fig. 7.6, camber can be considered as an overall warpage of the substrate, while wavi-

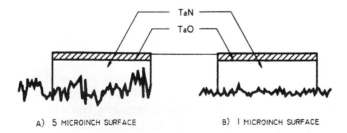

Figure 7.7 TaN resistor with TaO passivation on substrates with different surface roughness.

ness is more periodic in nature. Both of these factors may occur as a result of uneven shrinkage during the organic removal/sintering process or as a result of nonuniform composition. Waviness may also occur as a result of a "flat spot" in the rollers used to form the green sheets.

Camber is measured in units of length/length, interpreted as the deviation from flatness per unit length, and is measured with reference to the longest dimension by placing the substrate through parallel plates set a specific distance apart. Thus, a rectangular substrate would be measured along the diagonal. A typical value of camber is 0.003 in/in (also 0.003 mm/mm), which for a 2×2 inch substrate represents a total deviation of $0.003 \times 2 \times 1.414 = 0.0085$ in. For a substrate that is 0.025 thick, a common value, the total deviation represents a third of the overall thickness!

The nonplanar surface created by camber adversely affects subsequent metallization and assembly processes. In particular, screen printing is made more difficult due to the variable snap-off distance. Torsion bar printing heads on modern screen printers can compensate to a certain extent, but not entirely. A vacuum hold-down on the screen printer platen also helps but only flattens the substrate temporarily during the actual printing process. Camber can also create excessive stresses and a nonuniform temperature coefficient of expansion. At temperature extremes, these factors can cause cracking, breaking, or even shattering of the substrate.

Camber is measured by first measuring the thickness of the substrate and then placing the substrate between a series of pairs of parallel plates set specific distances apart. Camber is calculated by subtracting the substrate thickness from the smallest distance that the substrate will pass through and dividing by the longest substrate dimension. A few generalizations can be made about camber.

- Thicker substrates will have less camber than thinner ones.

- Square shapes will have less camber than rectangular ones.

- The pressed methods of forming will produce substrates with less camber than the sheet methods.

7.4 Thermal Properties of Ceramic Materials

7.4.1 Thermal Conductivity

The thermal conductivity of a material is a measure of its ability to carry heat and is defined as

$$q = -k\frac{dT}{dx} \tag{7.3}$$

where k = thermal conductivity in W/m-°C
q = heat flux in w/cm^2

$\dfrac{dT}{dx}$ = temperature gradient in °C/m in steady state

The negative sign denotes that heat flows from areas of higher temperature to areas of lower temperature.

Two mechanisms contribute to thermal conductivity, (1) the movement of free electrons and (2) lattice vibrations, or phonons. When a material is locally heated, the kinetic energy of the free electrons in the vicinity of the heat source increases, causing the electrons to migrate to cooler areas. These electrons undergo collisions with other atoms, losing their kinetic energy in the process. The net result is that heat is drawn away from the source toward cooler areas. In a similar fashion, an increase in temperature increases the magnitude of the lattice vibrations, which, in turn, generate and transmit phonons, carrying energy away from the source. The thermal conductivity of a material is the sum of the contributions of these two parameters.

$$k = k_p + k_e \tag{7.4}$$

where k_p = contribution due to phonons
k_e = contribution due to electrons

In ceramics, the heat flow is primarily due to phonon generation, and the thermal conductivity is generally lower than that of metals. Crystalline structures, such as alumina and beryllia, are more efficient heat conductors than amorphous structures such as glass. Organic materials used to fabricate printed circuit boards or epoxy attachment materials are electrical insulators and highly amorphous, and they tend to be very poor thermal conductors.

Impurities and other structural defects in ceramics tend to lower the thermal conductivity by causing the phonons to undergo more collisions, lowering the mobility and lessening their ability to transport heat away from the source. This is illustrated by Table 7.2, which lists the thermal conductivity of alumina as a function of the percentage of glass. Although the thermal conductivity of the glass binder is lower than that of the alumina, the drop in thermal conductivity is greater than expected from the addition of the glass alone. If the thermal conductivity is a function of the ratio of the materials alone, it follows the rule of mixtures.

TABLE 7.2 Thermal Conductivity of Alumina Substrates with Different Concentrations of Alumina

Volume percentage of alumina	Thermal conductivity, W/m-°C
85	16.0
90	16.7
94	22.4
96	24.7
99.5	28.1
100	31.0

$$k_T = P_1 k_1 + P_2 k_2 \qquad (7.5)$$

where k_T = net thermal conductivity
P_1 = volume percentage of material one in decimal form
k_1 = thermal conductivity of material one
P_2 = volume percentage of material two in decimal form
k_2 = thermal conductivity of material two

In pure form, alumina has a thermal conductivity of about 31 W/m-°C, and the binding glass has a thermal conductivity of about 1 W/m-°C. Equation (7.5) and the parameters from Table 7.2 are plotted in Fig. 7.8.

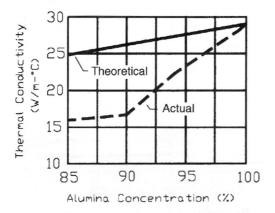

Figure 7.8 Thermal conductivity of alumina vs. concentration, theoretical and actual.

By the same token, as the ambient temperature increases, the number of collisions increases, and the thermal conductivity of most materials decreases. A plot of the thermal conductivity vs. temperature for several materials is shown in Fig. 7.9.[3] One material not plotted in this graph is diamond. The thermal conductivity of diamond varies widely with composition and the method of preparation, and it is much higher than those materials listed. Diamond will be discussed in detail in a later section. Selected data from Fig. 7.9 was analyzed and extrapolated into binomial equations that quantitatively describe the thermal conductivity vs. temperature relationship. This data is summarized in Table 7.3.

7.4.2 Specific Heat

The specific heat of a material is defined as

$$c = \frac{dQ}{dT} \tag{7.6}$$

where c = specific heat in W-s/g-°C
Q = energy in watt-s
T = temperature in Kelvins (K)

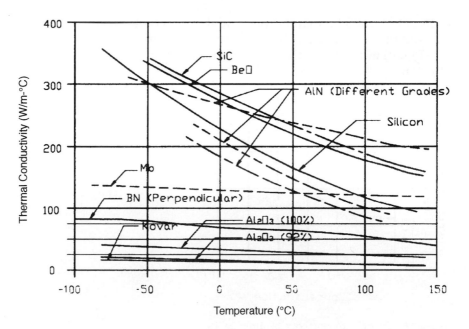

Figure 7.9 Thermal conductivity vs. temperature for selected materials.

TABLE 7.3 Approximate Thermal Conductivity vs. Temperature for
Selected Ceramic Materials (Binomial Relationship)

Material	Constant	T Coefficient	T^2 Coefficient
SiC	285	−1.11	1.55×10^{-3}
BeO	275	−1.10	1.06×10^{-3}
AlN (Pure)	271	−0.60	3.81×10^{-4}
AlN (Grade 1)	210	−1.39	3.45×10^{-3}
AlN (Grade 2)	185	−1.37	4.40×10^{-3}
BN (Perpendicular)	73	−0.06	2.17×10^{-4}
Al_2O_3 (99%)	34	−0.12	2.00×10^{-4}
Al_2O_3 (96%)	17	−0.07	5.70×10^{-5}

The specific heat, c, is defined in a similar manner and is the amount of heat required to raise the temperature of one gram of material by one degree, with units of watt-s/gm-°C. The quantity "specific heat" in this context refers to the quantity, c_V, which is the specific heat measured with the volume constant, as opposed to c_P, which is measured with the pressure constant. At the temperatures of interest, these numbers are nearly the same for most solid materials. The specific heat is primarily the result of an increase in the vibrational energy of the atoms when heated, and the specific heat of most materials increases with temperature up to a temperature called the Debye temperature, at which point it becomes essentially independent of temperature. The specific heat of several common ceramic materials as a function of temperature is shown in Fig. 7.10.

The heat capacity, C, is similar in form, except that it is defined in terms of the amount of heat required to raise the temperature of a mole of material by one degree and has the units of watt-s/mol-°C.

7.4.3 Temperature Coefficient of Expansion

The temperature coefficient of expansion (TCE) arises from the asymmetrical increase in the interatomic spacing of atoms as a result of increased heat. Most metals and ceramics exhibit a linear, isotropic relationship in the temperature range of interest, while certain plastics may be anisotropic in nature. The TCE is defined as

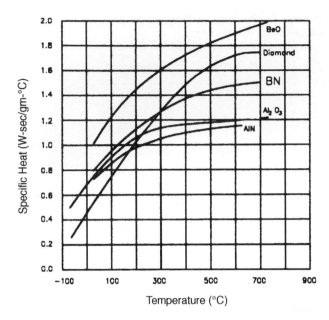

Figure 7.10   Specific heat vs. temperature for selected materials.

$$\alpha = \frac{\ell(T_2) - \ell(T_1)}{\ell(T_1)(T_2 - T_1)} \tag{7.7}$$

where α = temperature coefficient of expansion in ppm/°C^{-1}
   T_1 = initial temperature
   T_2 = final temperature
   $\ell(T_1)$ = length at initial temperature
   $\ell(T_2)$ = length at final temperature

The TCE of most ceramics is isotropic. For certain crystalline or single-crystal ceramics, the TCE may be anisotropic, and some may even contract in one direction and expand in the other. Ceramics used for substrates do not generally fall into this category, as most are mixed with glasses in the preparation stage and do not exhibit anisotropic properties as a result. The temperature coefficient of expansion of several ceramic materials is shown in Table 7.4.

7.5 Mechanical Properties of Ceramic Substrates

The mechanical properties of ceramic materials are strongly influenced by the strong interatomic bonds that prevail. Dislocation mech-

TABLE 7.4   Temperature Coefficient of Expansion of Selected Ceramic Substrate Materials

Material	TCE, ppm/°C
Alumina (96%)	6.5
Alumina (99%)	6.8
BeO (99.5%)	7.5
BN	
Parallel	0.57
Perpendicular	–0.46
Silicon carbide	3.7
Aluminum nitride	4.4
Diamond, type IIA	1.02
AlSiC (70% SiC loading)	6.3

anisms, which create slip mechanisms in softer metals, are relatively scarce in ceramics, and failure may occur with very little plastic deformation. Ceramics also tend to fracture with little resistance.

7.5.1   Modulus of Elasticity

The temperature coefficient of expansion (TCE) phenomenon has serious implications in the applications of ceramic substrates. When a sample of material has one end fixed, which may be considered to be a result of bonding to another material that has a much smaller TCE, the net elongation of the hotter end per unit length, or *strain (E),* of the material is calculated by

$$E = \text{TCE} \times \Delta T \tag{7.8}$$

where E = strain in length/length
ΔT = temperature differential across the sample

Elongation develops a stress *(S)* per unit length in the sample as given by Hooke's Law.

$$S = E\,Y \tag{7.9}$$

where S = stress in psi/in $(N/m^2/m)$
Y = modulus of elasticity in lb/in^2 (N/m^2)

When the total stress, as calculated by multiplying the stress/unit length by the maximum dimension of the sample, exceeds the strength of the material, mechanical cracks will form in the sample that may even propagate to the point of separation. The small elongation that occurs before failure is referred to as *plastic deformation*. This analysis is somewhat simplistic in nature but serves as a basic understanding of the mechanical considerations. The modulus of elasticity of selected ceramics is summarized in Table 7.5, along with other mechanical properties.

TABLE 7.5   Mechanical Properties of Selected Ceramics

Material	Modulus of elasticity, GPa	Tensile strength, MPa	Compressive strength, MPa	Modulus of rupture, MPa	Flexural strength, MPa	Density, g/cm^3
Alumina (99%)	370	500	2600	386	352	3.98
Alumina (96%)	344	172	2260	341	331	3.92
Beryllia (99.5%)	345	138	1550	233	235	2.87
Boron nitride (normal)	43	2410	6525	800	53.1	1.92
Aluminum nitride	300	310	2000	300	269	3.27
Silicon carbide	407	197	4400	470	518	3.10
Diamond (type IIA)	1000	1200	11000	940	1000	3.52

7.5.2   Modulus of Rupture

Ordinary stress-strain testing is not generally used to test ceramic substrates, since they do not exhibit elastic behavior to a great degree. An alternative test, the modulus of rupture (bend strength) test, as described in Fig. 7.11, is preferred. A sample of ceramic, either circular or rectangular, is suspended between two points, a force is applied in the center, and the elongation of the sample is measured. The stress is calculated by

$$\sigma = \frac{Mx}{I} \tag{7.10}$$

where σ = stress in Mpa
   M = maximum bending moment in N-m
   x = distance from center to outer surface in m
   I = moment of inertia in N-m^2

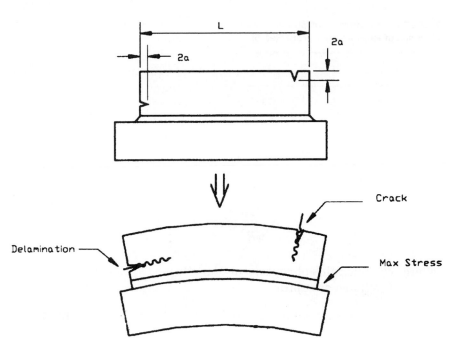

Figure 7.11   Modulus of rupture test setup.

The expressions for σ, M, x, and I are summarized in Table 7.6. When these are inserted into Equation (7.10), the result is

$$\sigma = \frac{3FL}{2xy^2} \text{ (rectangular cross section)} \tag{7.11}$$

$$\sigma = \frac{FL}{\pi R^3} \text{ (circular cross section)} \tag{7.12}$$

where F = applied force in newtons
   x = long dimension of rectangular cross section in m
   y = short dimension of rectangular cross section in m
   L = length of sample in m
   R = radius of circular cross-section in m

The modulus of rupture is the stress required to produce fracture and is given by

$$\sigma_r = \frac{3F_rL}{2xy^2} \text{ (rectangular)} \tag{7.13}$$

TABLE 7.6   Parameters of Stress in
Modulus of Rupture Test (from Ref. 5)

Cross section	M	c	I
Rectangular	$\dfrac{FL}{4}$	$\dfrac{y}{2}$	$\dfrac{xy^3}{12}$
Circular	$\dfrac{FL}{4}$	R	$\dfrac{\pi R^2}{4}$

$$\sigma_r = \frac{F_r L}{\pi R^3} \text{ (circular)} \tag{7.14}$$

where σ_r = modulus of rupture in n/m^2
     F_r = force at rupture

The modulus of rupture for selected ceramics is shown in Table 7.5.

7.5.3   Tensile and Compressive Strength

A force applied to a ceramic substrate in a tangential direction may product tensile or compressive forces. If the force is tensile, in a direction such that the material is pulled apart, the stress produces plastic deformation as defined in Equation (7.9). As the force increases past a value referred to as the *tensile strength,* breakage occurs. Conversely, a force applied in the opposite direction creates compressive forces until a value referred to as the *compressive strength* is reached, at which point breakage also occurs. The compressive strength of ceramics is, in general, much larger than the tensile strength. The tensile and compressive strength of selected ceramic materials is shown in Table 7.5.

In practice, the force required to fracture a ceramic substrate is much lower than predicted by theory. The discrepancy is due to small flaws or cracks residing within these materials as a result of processing. For example, when a substrate is sawed, small edge cracks may be created. Similarly, when a substrate is fired, trapped organic material may outgas during firing, leaving a microscopic void in the bulk. The result is an amplification of the applied stress in the vicinity of the void that may exceed the tensile strength of the material and create a fracture. If the microcrack is assumed to be elliptical with the major axis perpendicular to the applied stress, the maximum stress at the tip of the crack may be approximated by[4]

$$S_M = 2S_o \leq \left(\frac{a}{\rho_t}\right)^{\frac{1}{2}} \tag{7.15}$$

where S_M = maximum stress at the tip of the crack
   S_O = nominal applied stress
   a = length of the crack as defined in Fig. 7.11
   ρ_t = radius of the crack tip

The ratio of the maximum stress to the applied stress may be defined as

$$K_t = \frac{S_M}{S_o} = 2\left(\frac{a}{\rho_t}\right)^{\frac{1}{2}} \tag{7.16}$$

where K_t = stress concentration factor

For certain geometries, such as a long crack with a small tip radius, K_t may be much larger than 1, and the force at the tip may be substantially larger than the applied force.

Based on this analysis, a material parameter called the *plain strain fracture toughness,* a measure of the ability of the material to resist fracture, can be defined as

$$K_{IC} = ZS_c\sqrt{\pi a} \tag{7.17}$$

where K_{IC} = plain strain fracture toughness in psi-in$^{1/2}$ or Mpa-m$^{1/2}$
   Z = dimensionless constant, typically 1.2 (Ref. 4)
   S_c = critical force required to cause breakage

From Eq. (7.17), the expression for the critical force can be defined as

$$S_c = Z\frac{K_{IC}}{\sqrt{\pi a}} \tag{7.18}$$

When the applied force on the die due to TCE or thermal differences exceeds this figure, fracture is likely. The plain strain fracture toughness for selected materials is presented in Table 7.7. It should be noted that Eq. (7.13) is a function of thickness up to a point but is approximately constant for the area to thickness ratio normally found in substrates.

TABLE 7.7   Fracture Toughness for Selected Materials

Material	Fracture toughness, MPA-m$^{1/2}$
Silicon	0.8
Alumina (96%)	3.7
Alumina (99%)	4.6
Silicon carbide	7.0
Molding compound	2.0

7.5.4   Hardness

Ceramics are among the hardest substances known, and the hardness is correspondingly difficult to measure. Most methods rely on the ability of one material to scratch another, and the measurement is presented on a relative scale. Of the available methods, the Knoop method, is the most frequently used. In this approach, the surface is highly polished, and a pointed diamond stylus under a light load is allowed to impact on the material. The depth of the indentation formed by the stylus is measured and converted to a qualitative scale called the *Knoop* or *HK* scale. The Knoop hardness of selected ceramics is given in Table 7.8.

TABLE 7.8   Knoop Hardness for Selected Ceramics

Material	Knoop hardness, 100 g
Diamond	7000
Aluminum oxide	2100
Aluminum nitride	1200
Beryllium oxide	1200
Boron nitride	5000
Silicon carbide	2500

7.5.5   Thermal Shock

Thermal shock occurs when a substrate is exposed to temperature extremes in a short period of time. Under these conditions, the substrate is not in thermal equilibrium, and internal stresses may be sufficient to cause fracture. Thermal shock can be liquid to liquid or air to air, with the most extreme exposure occurring when the substrate is transferred directly from one liquid bath to another. The heat is more rapidly absorbed or transmitted, depending on the relative temperature of the bath, due to the higher specific of the liquid as opposed to air.

The ability of a substrate to withstand thermal shock is a function of several variables, including the thermal conductivity, the coefficient of thermal expansion, and the specific heat. Winkleman and Schott[5] developed a parameter called the *coefficient of thermal endurance* that qualitatively measures the ability of a substrate to withstand thermal stress.

$$F = \frac{P}{\alpha Y}\sqrt{\frac{k}{\rho c}} \qquad (7.19)$$

where F = coefficient of thermal endurance
   P = tensile strength in MPa
   α = thermal coefficient of expansion in 1/K
   Y = modulus of elasticity in MPa
   k = thermal conductivity in W/m-K
   ρ = density in kg/m^3
   c = specific heat in W-s/kg-K

The coefficient of thermal endurance for selected materials is shown in Table 7.9. The phenomenally high coefficient of thermal endurance for BN is primarily a result of the high tensile strength to modulus of elasticity ratio as compared to other materials. Diamond is also high, primarily due to the high tensile strength, the high thermal conductivity, and the low TCE.

TABLE 7.9   Thermal Endurance Factor for Selected Materials at 25°C

Material	Thermal endurance factor
Alumina (99%)	0.640
Alumina (96%)	0.234
Beryllia (99.5%)	0.225
Boron nitride ("a" axis)	648
Aluminum nitride	2.325
Silicon carbide	1.40
Diamond (Type IIA)	30.29

The thermal endurance factor is a function of temperature in that several of the variables, particularly the thermal conductivity and the specific heat, are functions of temperature. From Table 7.9, it is also noted that the thermal endurance factor may drop rapidly as the alumina to glass ratio drops. This is due to the difference in the thermal conductivity and TCE of the alumina and glass constituents that increase the internal stresses. This is true of other materials as well.

7.6   Electrical Properties of Ceramics

The electrical properties of ceramic substrates perform an important task in the operation of electronic circuits. Depending on the applications, the electrical parameters may be advantageous or detrimental to circuit function. Of most interest are the resistivity, the breakdown voltage or dielectric strength, and the dielectric properties, including the dielectric constant and the loss tangent.

7.6.1   Resistivity

The electrical resistivity of a material is a measure of the ability of that material to transport charge under the influence of an applied electric field. More often, this ability is presented in the form of the electrical conductivity, which is the reciprocal of the resistivity as defined in Eq. (7.20).

$$\sigma = \frac{1}{\rho} \qquad (7.20)$$

where σ = conductivity in siemens/unit length
        ρ = resistivity in ohm-unit length

The conductivity is a function primarily of two variables: the concentration of charge and the mobility—the ability of that charge to be transported through the material. The current density and the applied field are related by the expression defined in Eq. (7.21).

$$J = \sigma E \qquad (7.21)$$

where J = current density in amperes/unit area
        E = electric field in volts/unit length

It should be noted that both the current density and the electric field are vectors, since the current is in the direction of the electric field.
    The current density may also be defined as

$$J = nv_d \tag{7.22}$$

where n = free carrier concentration in coulombs/unit volume
v_d = drift velocity of electrons in unit length/second

The drift velocity is related to the electric field by

$$V_d = \mu E \tag{7.23}$$

where μ = mobility in length^2/volt-second

In terms of the free carrier concentration and the mobility, the current density is

$$J = n\mu E \tag{7.24}$$

Comparing Eq. (7.19) with Eq. (7.23), the conductivity can be defined as

$$\sigma = n\mu \tag{7.25}$$

The free carrier concentration may be expressed as

$$n = n_t + n_i \tag{7.26}$$

where n_t = free carrier concentration due to thermal activity
n_i = free carrier concentration due to field injection

The thermal charge density, n_t, in insulators is a result of free electrons obtaining sufficient thermal energy to break the interatomic bonds, allowing them to move freely within the atomic lattice. Ceramic materials characteristically have few thermal electrons as a result of the strong ionic bonds between atoms. The injected charge density, n_I, occurs when a potential is applied and is a result of the inherent capacity of the material. The injected charge density is given by

$$n_i = \varepsilon \tag{7.27}$$

where ε = dielectric constant of the material in farads/unit length

Inserting Eq. (7.27) and Eq. (7.26) into Eq. (7.24), the result is

$$J = \mu n_t E + \mu \varepsilon E^2 \tag{7.28}$$

For conductors, $n_t \gg n_i$ and Ohm's law applies. For insulators, $n_i \gg n_t$, and the result is a square law relationship between the voltage and the current.[6]

$$J = \mu \varepsilon E^2 \tag{7.29}$$

The conductivity of ceramic substrates is extremely low. In practice, it is primarily due to impurities and lattice defects and may vary widely from batch to batch. The conductivity is also a strong function of temperature. As the temperature increases, the ratio of thermal to injected carriers increases. As a result, the conductivity increases and the V-I relationship follows Ohm's law more closely. Typical values of the resistivity of selected ceramic materials are presented in Table 7.10.

TABLE 7.10   Electrical Properties of Selected Ceramic Substrates

	Property			
	Electrical resistivity (Ω-cm)	Breakdown voltage (ac kV/mm)	Dielectric constant	Loss tangent (@ 1 MHz)
Alumina (96%)				
25°C	$>10^{14}$		9.0	
500°C	4×10^9	8.3	10.8	0.0002
1000°C	1×10^6			
Alumina (99.5%)				
25°C	$>10^{14}$			
500°C	2×10^{10}	8.7	9.4	0.0001
1000°C	2×10^6		10.1	
Beryllia				
25°C	$>10^{14}$	6.6	6.4	0.0001
500°C	2×10^{10}		6.9	0.0004
Aluminum nitride	$>10^{13}$	14	8.9	0.0004
Boron nitride	$>10^{14}$	61	4.1	0.0003
Silicon carbide*	$>10^{13}$	0.7	40	0.05
Diamond (Type II)	$>10^{14}$	1000	5.7	0.0006

*Depends on method of preparation; may be substantially lower.

7.6.2   Breakdown Voltage

The term *breakdown voltage* is very descriptive. While ceramics are normally very good insulators, the application of excessively high potentials can dislodge electrons from orbit with sufficient energy to allow them to dislodge other electrons from orbit, creating an *avalanche effect.* The result is a breakdown of the insulation properties of the material, allowing current to flow. This phenomenon is accelerated by elevated temperature, particularly when mobile ionic impurities are present.

The breakdown voltage is a function of numerous variables, including the concentration of mobile ionic impurities, grain boundaries, and the degree of stoichiometry. In most applications, the breakdown voltage is sufficiently high as to not be an issue. However, there are two cases in which it must be a consideration:

1. At elevated temperatures created by localized power dissipation or high ambient temperature, the breakdown voltage may drop by orders of magnitude. Combined with a high potential gradient, this condition may be susceptible to breakdown.

2. The surface of most ceramics is highly *wettable,* in that moisture tends to spread rapidly. Under conditions of high humidity, coupled with surface contamination, the effective breakdown voltage is much lower than the intrinsic value.

7.6.3 Dielectric Properties

Two conductors in proximity with a difference in potential have the ability to attract and store electric charge. Placing a material with dielectric properties between them enhances this effect. A dielectric material has the capability of forming electric dipoles (displacements of electric charge) internally. At the surface of the dielectric, the dipoles attract more electric charge, thus enhancing the charge storage capability, or capacitance, of the system. The relative ability of a material to attract electric charge in this manner is called the *relative dielectric constant,* or *relative permittivity,* and is usually given the symbol K. The relative permittivity of free space is 1.0 by definition, and the absolute permittivity is

$$\varepsilon_o = \frac{1}{36\pi} \times 10^{-9} \text{ farads/meter} \tag{7.30}$$

where ε_o = permittivity of free space

The relationship between the polarization and the electric field is

$$\bar{P} = \varepsilon_o(K-1)\bar{E}\frac{Q}{m^2} \tag{7.31}$$

where P = polarization, coulombs/m^2
E = electric field, V/m

Four basic mechanisms contribute to polarization.

1. *Electronic polarization.* In the presence of an applied field, the cloud of electrons is displaced relative to the positive nucleus of the atom or molecule, creating an induced dipole moment. Electronic polarization is essentially independent of temperature and may occur very rapidly. The dielectric constant may therefore exist at very high frequencies, up to 10^{17} Hz.

2. *Molecular polarization.* Certain molecule structures create permanent dipoles that exist even in the absence of an electric field. These may be rotated by an applied electric field, generating a degree of polarization by orientation. Molecular polarization is inversely proportional to temperature and occurs only at low to moderate frequencies. Molecular polarization does not occur to a great extent in ceramics and is more prevalent in organic materials and liquids such as water.

3. *Ionic polarization.* Ionic polarization occurs in ionically bonded materials when the positive and negative ions undergo a relative displacement to each other in the presence of an applied electric field. Ionic polarization is somewhat insensitive to temperature and occurs at high frequencies, up to 10^{13} Hz.

4. *Space charge polarization.* Space charge polarization exists as a result of charges derived from contaminants or irregularities that exist within the dielectric. These charges exist to a greater or lesser degree in all crystal lattices and are partly mobile. Consequently, they will migrate in the presence of an applied electric field. Space charge polarization occurs only at very low frequencies.

In a given material, more than one type of polarization can exist, and the net polarization is given by

$$\bar{P}_t = \bar{P}_e + \bar{P}_m + \bar{P}_i + \bar{P}_s \qquad (7.32)$$

where P_t = total polarization
P_e = electronic polarization
P_m = molecular polarization
P_i = ionic polarization
P_s = space charge polarization

Normally, the dipoles are randomly oriented in the material, and the resulting internal electric field is zero. In the presence of an external applied electric field, the dipoles become oriented as shown in Fig. 7.12.

There are two common ways to categorize dielectric materials: as polar or nonpolar, and as paraelectric or ferroelectric. Polar materials include those that are primarily molecular in nature, such as water, and nonpolar materials include both electronically and ionically polarized materials. Paraelectric materials are polarized only in the presence of an applied electric field and lose their polarization when the field is removed. Ferroelectric materials retain a degree of polarization after the field is removed. Materials used as ceramic substrates are usually nonpolar and paraelectric in nature. An exception is silicon carbide, which has a degree of molecular polarization.

In the presence of an electric field that is changing at a high frequency, the polarity of the dipoles must change at the same rate as the polarity of the signal to maintain the dielectric constant at the same level. Some materials are excellent dielectrics at low frequencies, but the dielectric qualities drop off rapidly as the frequency increases. Electronic polarization, which involves only displacement of free charge and not ions, responds more rapidly to the changes in the direction of the electric field and remains viable up to about 10^{17} Hz. The polarization effect of ionic displacement begins to fall off at about 10^{13} Hz, and molecular and space charge polarizations fall off at still lower frequencies. The frequency response of the different types is shown in Fig. 7.13, which also illustrates that the dielectric constant decreases with frequency.

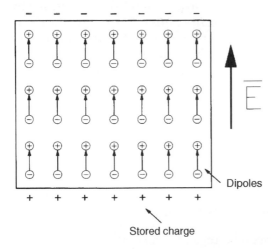

Figure 7.12 Orientation of dipoles in an electric field.[1]

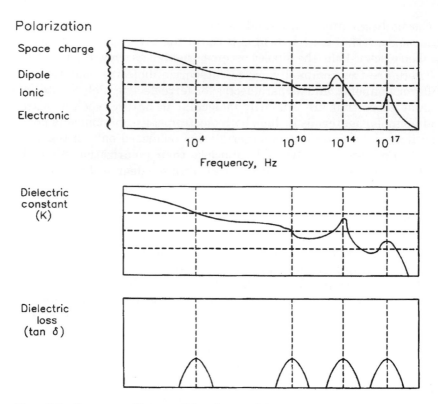

Figure 7.13 Frequency effects on dielectric materials.

Changing the polarity of the dipoles requires a finite amount of energy and time. The energy is dissipated as internal heat, quantified by a parameter called the *loss tangent* or *dissipation factor*. Furthermore, dielectric materials are not perfect insulators. These phenomena may be modeled as a resistor in parallel with a capacitor. The loss tangent, as expected, is a strong function of the applied frequency, increasing as the frequency increases.

In alternating current applications, the current and voltage across an ideal capacitor are exactly 90° out of phase, with the current leading the voltage. In actuality, the resistive component causes the current to lead the voltage by an angle less than 90°. The loss tangent is a measure of the real or resistive component of the capacitor and is the tangent of the difference between 90° and the actual phase angle.

$$\text{Loss tangent} = \tan(90° - \delta) \tag{7.33}$$

where δ = phase angle between voltage and current

The loss tangent is also referred to as the *dissipation factor (DF)*.

The loss tangent may also be considered as a measure of the time required for polarization. It requires a finite amount of time to change the polarity of the dipole after an alternating field is applied. The resulting phase retardation is equivalent to the time indicated by the difference in phase angles.

7.7 Metallization of Ceramic Substrates

There are three fundamental methods of metallizing ceramic substrates; thick film, thin film, and copper, which includes direct bond copper (DBC), plated copper, and active metal braze (AMB). Not all of these processes are compatible with all substrates. The selection of a metallization system depends on both the application and the compatibility with the substrate material.

7.7.1 Thick Film

The thick film process is an additive procedure by which conductive, resistive, and dielectric (insulating) patterns in the form of a viscous paste are screen printed, dried, and fired onto a ceramic substrate at an elevated temperature to promote the adhesion of the film. In this manner, by depositing successive layers as shown in Fig. 7.14, multilayer interconnection structures can be formed that may contain integrated resistors, capacitors, or inductors.

The initial step is to generate 1:1 artworks corresponding to each layer of the circuit. The screen is a stainless steel mesh with a mesh count of 80–400 wires/inch. The mesh is stretched to the proper tension and mounted to a cast aluminum frame with epoxy. It is coated with a photosensitive material and exposed to light through one of the artworks. The unexposed portion is rinsed away, leaving openings in the screen mesh corresponding to the pattern to be printed.

Thick film materials in the fired state are a combination of glass ceramic and metal, referred to as *cermet* thick films, and are designed to be fired in the range 850–1000°C. A standard cermet thick film paste has four major ingredients.

1. An active element, which establishes the function of the film

2. An adhesion element, which provides the adhesion to the substrate and a matrix that holds the active particles in suspension

3. An organic binder, which provides the proper fluid properties for screen printing

4. A solvent or thinner, which establishes the viscosity of the vehicle phase

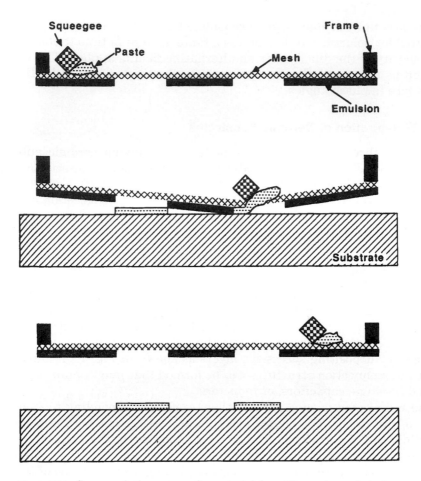

Figure 7.14 Screen printing process for material deposition onto a substrate.

7.7.1.1 The active element. The active element within the paste dictates the electrical properties of the fired film. If the active element is a metal, the fired film will be a conductor; if it is a conductive metal oxide, a resistor; and, if it is an insulator, a dielectric. The active element is most commonly found in powder form ranging from 1 to 10 μ in size, with a mean diameter of about 5 μ.

7.7.1.2 The adhesion element. There are two primary constituents used to bond the film to the substrate: glass and metal oxides, which may be used singly or in combination. Films that use a glass, or *frit*, are referred to as *fritted* materials and have a relatively low melting point (500–600°C). There are two adhesion mechanisms associated with the fritted materials: a chemical reaction and a physical reaction. In the chemical reaction, the molten glass chemically reacts with the glass in the substrate to a degree. In the physical reaction, the glass flows into and around the irregularities in the substrate surface. The total adhesion is the sum of the two factors. The physical bonds are more susceptible to degradation by thermal cycling or thermal storage than the chemical bonds and are generally the first to fracture under stress. The glass also creates a matrix for the active particles, holding them in contact with each other to promote sintering and to provide a series of three-dimensional continuous paths from one end of the film to the other. Principal thick film glasses are based on B_2O_3-SiO_2 network formers with modifiers such as PbO, Al_2O_3, Bi_2O_3, ZnO, BaO, and CdO added to change the physical characteristics of the film, such as melting point, viscosity, and coefficient of thermal expansion. Bi_2O_3 also has excellent wetting properties, both to the active element and to the substrate, and is frequently used as a flux. The glass phase may be introduced as a pre-reacted particle or formed *in-situ* by using glass precursors such as boric oxide, lead oxide, and silicon. Fritted conductor materials tend to have glass on the surface, making subsequent component assembly processes more difficult.

A second class of materials utilizes metal oxides to provide the adhesion to the substrate. In this case, a pure metal, such as copper or cadmium, is mixed with the paste and reacts with oxygen atoms on the surface of the substrate to form an oxide. The conductor adheres to the oxide and to itself by sintering, which takes place during firing. During firing, the oxides react with broken oxygen bonds on the surface of the substrate, forming a Cu or Cd spinel structure, such as $CuAl_2O_4$. Pastes of this type offer improved adhesion over fritted materials and are referred to as *fritless, oxide-bonded,* or *molecular-bonded* materials. Fritless materials typically fire at 900–1000°C, which is undesirable from a manufacturing aspect. Ovens used for thick film firing degrade more rapidly and need more maintenance when operated at these temperatures for long periods of time.

A third class of materials utilizes both reactive oxides and glasses. The oxides in these materials react at lower temperatures but are not as strong as copper. A lesser concentration of glass than found in fritted materials is added to supplement the adhesion. These materials, referred to as *mixed bonded* systems, incorporate the advantages of both technologies and fire at a lower temperature.

The selection of a binding material is strongly dependent on the substrate material. For example, the most common glass composition used with alumina is a lead/bismuth borosilicate composition. When this glass is used in conjunction with aluminum nitride, however, it is rapidly reduced at firing temperatures.[1] Alkaline earth borosilicates must be used with AlN to promote adhesion

7.7.1.3 Organic binder. The organic binder is generally a thixotropic fluid and serves two purposes: it holds the active and adhesion elements in suspension until the film is fired, and it gives the paste the proper fluid characteristics for screen printing. The organic binder is usually referred to as the *nonvolatile* organic, since it does not evaporate but begins to burn off at about 350°C. The binder must oxidize cleanly during firing, with no residual carbon that could contaminate the film. Typical materials used in this application are ethyl cellulose and various acrylics.

For nitrogen-fireable films, where the firing atmosphere can contain only a few ppm of oxygen, the organic vehicle must decompose and thermally depolymerize, departing as a highly volatile organic vapor in the nitrogen blanket provided as the firing atmosphere, since oxidation into CO_2 or H_2O is not feasible due to the oxidation of the copper film.

7.7.1.4 Solvent or thinner. The organic binder in the natural form is too thick to permit screen printing, which requires the use of a solvent or thinner. The thinner is somewhat more volatile than the binder, evaporating rapidly above about 100°C. Typical materials used for this application are terpineol, butyl carbitol, or certain of the complex alcohols into which the nonvolatile phase can dissolve. The low vapor pressure at room temperature is desirable to minimize drying of the pastes and to maintain a constant viscosity during printing. Additionally, plasticizers, surfactants, and agents that modify the thixotropic nature of the paste are added to the solvent to improve paste characteristics and printing performance.

To complete the formulation process, the ingredients of the thick film paste are mixed together in proper proportions and milled on a three-roller mill for a sufficient period of time to ensure that they are thoroughly mixed and that no agglomeration exists.

Thick film conductor materials may be divided into two broad classes; air fireable and nitrogen fireable. Air fireable materials are made up of noble metals that do not readily form oxides, gold and silver in the pure form, or alloyed with palladium and/or platinum. Nitrogen fireable materials include copper, nickel, and aluminum, with copper being the most common.

Thick film resistors are formed by adding metal oxide particles to glass particles and firing the mixture at a temperature/time combination sufficient to melt the glass and to sinter the oxide particles together. The resulting structure consists of a series of three-dimensional chains of metal oxide particles embedded in a glass matrix. The higher the metal oxide-to-glass ratio, the lower the resistivity, and vice versa. The most common materials used are ruthenium based, such as ruthenium dioxide, RuO_2, and bismuth ruthenate, $BiRu_2O_7$.

Thick film dielectric materials are used primarily as insulators between conductors, either as simple crossovers or in complex multilayer structures. Small openings, or *vias,* may be left in the dielectric layers so that adjacent conductor layers may interconnect. In complex structures, as many as several hundred vias per layer may be required. In this manner, complex interconnection structures may be created. Although the majority of thick film circuits can be fabricated with only three layers of metallization, others may require several more. If more than three layers are required, the yield begins dropping dramatically, with a corresponding increase in cost.

Dielectric materials used in this application must be of the *devitrifying* or *recrystallizable* type. Dielectrics in the paste form are a mixture of glasses that melt at a relatively low temperature. During firing, when they are in the liquid state, they blend together to form a uniform composition with a higher melting point than the firing temperature. Consequently, on subsequent firings they remain in the solid state, which maintains a stable foundation for firing sequential layers. By contrast, vitreous glasses always melt at the same temperature and would be unacceptable for layers to either "sink" and short to conductor layers underneath, or "swim" and form an open circuit. Additionally, secondary loading of ceramic particles is used to enhance devitrification and to modify the temperature coefficient of expansion (TCE).

Dielectric materials have two conflicting requirements in that they must form a continuous film to eliminate short circuits between layers and, at the same time, they must maintain openings as small as 0.010 in. In general, dielectric materials must be printed and fired twice per layer to eliminate pinholes and prevent short circuits between layers.

The TCE of thick film dielectric materials must be as close as possible to that of the substrate to avoid excessive *bowing,* or warpage, of the substrate after several layers. Excessive bowing can cause severe problems with subsequent processing, especially where the substrate must be held down with a vacuum or where it must be mounted on a heated stage. In addition, the stresses created by the bowing can cause the dielectric material to crack, especially when it is sealed within a package. Thick film material manufacturers have addressed this problem by developing dielectric materials that have an almost exact TCE match with alumina substrates. Where a serious mismatch exists, matching layers of dielectric must be printed on the bottom of the substrate to minimize bowing, which obviously increases the cost.

7.7.2   Thin Film

The thin film technology is a subtractive technology in that the entire substrate is coated with several layers of metallization, and the unwanted material is etched away in a succession of selective photoetching processes. The use of photolithographic processes to form the patterns enables much finer and better defined lines than can be formed by the thick film process. This feature promotes the use the thin film technology for high-density and high-frequency applications.

Thin film circuits typically consist of three layers of material deposited on a substrate. The bottom layer serves two purposes: it is the resistor material, and it also provides the adhesion to the substrate. The adhesion mechanism of the film to the substrate is an oxide layer that forms at the interface between the two. The bottom layer must therefore be a material that oxidizes readily. The most common types of resistor material are nichrome (NiCr) and tantalum nitride (TaN). Gold and silver, for example, are noble metals and do not adhere well to ceramic surfaces.

The middle layer acts as an interface between the resistor layer and the conductor layer, either by improving the adhesion of the conductor or by preventing diffusion of the resistor material into the conductor. The interface layer is for TaN is usually tungsten (W); for NiCr, a thin layer of pure Ni is used.

Gold is the most common conductor material used in thin film circuits because of the ease of wire and die bonding and the high resistance of the gold to tarnish and corrosion. Aluminum and copper are also frequently used in certain applications. It should be noted that copper and aluminum will adhere directly to ceramic substrates, but gold requires one or more intermediate layers, since it does not form the necessary oxides for adhesion.

The term *thin film* refers more to the manner in which the film is deposited onto the substrate as opposed to the actual thickness of the film. Thin films are typically deposited by one of the vacuum deposition techniques, sputtering or evaporation, or by electroplating.

Sputtering is the prime method by which thin films are applied to substrates. In ordinary dc triode sputtering, as shown in Fig. 7.15, a current is established in a conducting plasma formed by striking an arc in an inert gas, such as argon, with a partial vacuum of approximately 10 μ pressure. A substrate at ground potential and a target material at high potential are placed in the plasma. The potential may be ac or dc. The high potential attracts the gas ions in the plasma to

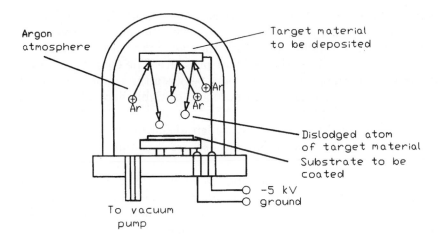

Figure 7.15 A dc sputtering chamber.

the point at which they collide with the target with sufficient kinetic energy to dislodge microscopically sized particles with enough residual kinetic energy to travel the distance to the substrate and adhere.

The adhesion of the film is enhanced by presputtering the substrate surface by random bombardment of argon ions prior to applying the potential to the target. This process removes several atomic layers of the substrate surface, creating a large number of broken oxygen bonds and promoting the formation of the oxide interface layer. The oxide formation is further enhanced by the residual heating of the substrate as a result of the transfer of the kinetic energy of the sputtered particles to the substrate when they collide.

Direct current triode sputtering is a very slow process, requiring hours to produce films with a usable thickness. By utilizing magnets at strategic points, the plasma can be concentrated in the vicinity of the target, greatly speeding up the deposition process. In most applications, an RF potential at a frequency of 13.56 MHz is applied to the target. The RF energy may be generated by a conventional electronic oscillator or by a magnetron. The magnetron is capable of generating considerably more power with a correspondingly higher deposition rate.

By adding small amounts of other gases, such as oxygen and nitrogen to the argon, it is possible to form oxides and nitrides of certain target materials on the substrate. It is this technique, called *reactive sputtering,* that is used to form tantalum nitride, a common resistor material.

Evaporation of a material into the surrounding area occurs when the vapor pressure of the material exceeds the ambient pressure and can take place from either the solid state of the liquid state. In the thin film process, the material to be evaporated is placed in the vicinity of the substrate and heated until the vapor pressure of the material is considerably above the ambient pressure. The evaporation rate is directly proportional to the difference between the vapor pressure of the material and the ambient pressure and is highly dependent on the temperature of the material.

There are several techniques by which evaporation can be accomplished. The two most common of these are resistance heating and electron-beam (E-beam) heating.

Evaporation by resistance heating, as depicted in Fig. 7.16, usually takes place from a boat made with a refractory metal, a ceramic crucible wrapped with a wire heater, or a wire filament coated with the evaporant. A current is passed through the element, and the heat generated heats the evaporant. It is somewhat difficult to monitor the temperature of the melt by optical means due to the propensity of the evaporant to coat the inside of the chamber, and control must be effected by empirical means. There exist closed-loop systems that can control the deposition rate and the thickness, but these are quite expensive. In general, adequate results can be obtained from the empirical process if proper controls are used.

The E-beam evaporation method takes advantage of the fact that a stream of electrons accelerated by an electric field tends to travel in a circle when entering a magnetic field. This phenomenon is utilized to direct a high-energy stream of electrons onto an evaporant source. The kinetic energy of the electrons is converted into heat when they strike the evaporant. E-beam evaporation is somewhat more controllable,

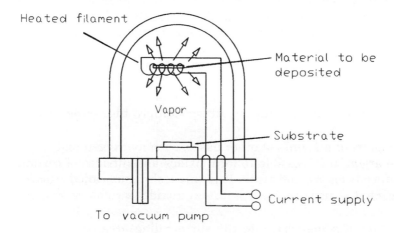

Figure 7.16 A thermal vacuum evaporation system.

since the resistance of the boat is not a factor, and the variables controlling the energy of the electrons are easier to measure and control. In addition, the heat is more localized and intense, making it possible to evaporate metals with higher 10^{-2} torr temperatures and lessening the reaction between the evaporant and the boat.

While evaporation provides a more rapid deposition rate, there are certain disadvantages when compared with sputtering.

1. It is difficult to evaporate alloys such as nichrome (NiCr) due to the difference between the 10^{-2} torr temperatures. The element with the lower temperature tends to evaporate somewhat faster, causing the composition of the evaporated film to be different from the composition of the alloy. To achieve a particular film composition, the composition of the melt must contain a higher portion of the material with the higher 10^{-2} torr temperature, and the temperature of the melt must be tightly controlled. By contrast, the composition of a sputtered film is identical to that of the target.

2. Evaporation is limited to the metals with lower melting points. Refractory metals, ceramics, and other insulators are virtually impossible to deposit by evaporation.

3. Reactive deposition of nitrides and oxides is very difficult to control.

Electroplating is accomplished by applying a potential between the substrate and the anode, which are suspended in a conductive solution of the material to be plated. The plating rate is a function of the potential and the concentration of the solution. In this manner, most metals can be plated to a metal surface.

In the thin film technology, it is a common practice to sputter a film of gold a few Å thick and to build up the thickness of the gold film by electroplating. This is considerably more economical and results in much less target usage. For added savings, photoresist can be applied to the substrate and gold electroplated only where actually required by the pattern.

The interconnection and resistor patterns are formed by selective photoetching.

The substrate is coated with a photosensitive material exposed to ultraviolet light through a pattern formed on a glass plate. The photoresist may be of the positive or negative type, with the positive type being prevalent due to its inherently higher resistance to the etchant materials. The unwanted material that is not protected by the photoresist may be removed by "wet" (chemical) etching or by "dry" (sputter) etching.

In general, two masks are required, one corresponding to the conductor pattern and one corresponding to a combination of both the conductor and resistor patterns, generally referred to as the *composite* pattern. As an alternative to the composite mask, a mask that contains only the resistor pattern plus a slight overlap onto the conductor to allow for misalignment may be used. The composite mask is preferred, since it allows a second gold etch process to be performed to remove any bridges or extraneous gold to be removed that might have been left from the first etch.

Sputtering may also be used to etch thin films. In this technique, the substrate is coated with photoresist and the pattern exposed in exactly the same manner as with chemical etching. The substrate is then placed in a plasma and connected to a potential. In effect, the substrate acts as the target during the sputter etching process, with the unwanted material being removed by the impingement of the gas ions on the exposed film. The photoresistive film, being considerably thicker than the sputtered film, is not affected.

7.7.3   Copper Metallization Technologies

The thick film and thin film technologies are limited in their ability to deposit films with a thickness greater than 1 mil (25 µ). This factor directly affects the ohmic resistance of the circuit traces and affects their ability to handle large currents or high frequencies. The copper metallization technologies provide conductors with greatly increased conductor thickness, which offers improved circuit performance in many applications. There are three basic technologies available to the hybrid designer: direct bond copper (DBC), active metal braze (AMB), and the various methods of plating copper directly to ceramic.

7.7.3.1   Direct bond copper. Copper may be directly bonded to alumina ceramic by placing a film of copper in contact with the alumina and heating to about 1065°C, just below the melting point of copper, 1083°C. At this temperature, a combination of 0.39% O_2 and 99.61% Cu form a liquid that can melt, wet, and bond tightly to the surfaces in contact with it when cooled to room temperature. In this process, the copper remains in the solid state during the bonding process, and a strong bond is formed between the copper and the alumina with no intermediate material required. The metallized substrate is slowly cooled to room temperature at a controlled rate to avoid quenching. To prevent excessive bowing of the substrate, copper must be bonded to both sides of the substrate to minimize stresses due to the difference in TCE between copper and alumina.

In this manner, a film of copper from 5 to 25 mils thick can be bonded to a substrate and a metallization pattern formed by photolithographic etching. For subsequent processing, the copper is usually plated with several hundred microinches of nickel to prevent oxidation. The nickel-plated surface is readily solderable, and aluminum wire bonds to nickel is one of the most reliable combinations.[7] Aluminum wire bonded directly to copper is not as reliable and may result in failure on exposure to heat and/or moisture.[8]

Multilayer structures of up to four layers have been formed by etching patterns on both sides of two substrates and bonding them to a common alumina substrate. Interconnections between layers are made by inserting oxidized copper pellets into holes drilled or formed in the substrates prior to firing. Vias may also be created by using one of the copper plating processes.

The line and space resolution of DBC is limited due to the difficulty of etching thick layers of metal without substantial undercutting. Special design guidelines must be followed to allow for this factor.[1] While the DBC technology does not have a resistor system, the thick film technology can be used in conjunction with DBC to produce integrated resistors and areas of high-density interconnections.

Aluminum nitride can also be used with copper, although the consistency of such factors as grain size and shape is not as good as aluminum oxide at this time. Additional preparation of the AlN surface is required to produce the requisite layer of oxide necessary to produce the bond. This can be accomplished by heating the substrate to about 1250°C in the presence of oxygen.

7.7.3.2 Plated copper technology. The various methods of plating copper to a ceramic all begin with the formation of a conductive film on the surface. This film may be vacuum deposited by thin film methods, screen printed by thick film processes, or deposited with the aid of a catalyst. A layer of electroless copper may be plated over the conductive surface, followed by a layer of electrolytic copper to increase the thickness.

A pattern may be generated in the plated surface by one of two methods. Conventional photolithographic methods may be used to etch the pattern, but this may result in undercutting and loss of resolution when used with thicker films. To produce more precise lines, a dry film photoresist may be utilized to generate a pattern on the electroless copper film that is the negative of the one required for etching. The traces may then be electroplated to the desired thickness using the photoresist pattern as a mold. Once the photoresist pattern is removed, the entire substrate may be immersed in an appropriate etchant to remove the unwanted material between the traces. Plated copper films created in this manner may be fired at an elevated temperature in a nitrogen atmosphere to improve the adhesion.

7.7.3.3 Active metal brazing copper technology. The active metal brazing (AMB) process utilizes one or more of the metals in the IV-B column of the periodic table, such as titanium, hafnium, or zirconium, to act as an activation agent with ceramic. These metals are typically alloyed with other metals to form a braze that can be used to bond copper to ceramic. One such example is an alloy of 70Ti/15Cu/15Ni that melts at 960 to 1000°C. Numerous other alloys can also be used.[2]

The braze may be applied in the form of a paste, a powder, or a film. The combination is heated to the melting point of the selected braze in a vacuum to minimize oxidation of the copper. The active metal forms a liquidus with the oxygen in the system that acts to bond the metal to the ceramic. After brazing, the copper film may be processed in much the same manner as DBC.

7.8   Ceramic Materials

The characteristics of various substrate materials have been summarized in previous sections. However, there may be substantial variations in the parameters due to processing, composition, stoichiometry, or other factors. This section covers the materials in more detail and also describes some common applications.

7.8.1   Aluminum Oxide

Aluminum oxide, Al_2O_3, commonly referred to as *alumina*, is by far the most common substrate material used in the microelectronics industry, because it is superior to most other oxide ceramics in mechanical, thermal, and electrical properties. The raw materials are plentiful, low in cost, and are amenable to fabrication by a wide variety of techniques into a wide variety of shapes.

Alumina is hexagonal close-packed with a corundum structure. Several metastable structures exist, but they all ultimately irreversibly transform to the hexagonal alpha phase. Alumina is stable in both oxidizing and reducing atmospheres up to 1925°C.

Weight loss in vacuum over the temperature range of 1700°C to 2000°C ranges from 10^{-7} to 10^{-6} g/cm^2-s. It is resistant to attack by all gases except wet fluorine to at least 1700°C. Alumina is attacked at elevated temperatures by alkali metal vapors and halogen acids, especially the lower-purity alumina compositions that may contain a percentage of glasses.

Alumina is used extensively in the microelectronics industry as a substrate material for thick and thin film circuits, for circuit packages, and as multilayer structures for multichip modules. Compositions exist for both high- and low-temperature processing. High-temperature cofired ceramics (HTCC) use a refractory metal, such as tungsten or molybdenum/manganese, as a conductor and fire at about 1800°C. The circuits are formed as separate layers, laminated together, and fired as a unit. Low-temperature cofired ceramics (LTCCs) use conventional gold or palladium silver as conductors and fire as low as 850°C. Certain power MOSFETs and bipolar transistors are mounted on alumina substrates to act as electrical insulators and thermal conductors. The parameters of alumina are summarized in Table 7.11.

TABLE 7.11   Typical Parameters of Aluminum Oxide

Parameter	Units	Test	Percentage			
			85	90	96	99.5
Density	g/cm^3	ASTM C20	3.40	3.60	3.92	3.98
Elastic modulus	GPa	ASTM C848	220	275	344	370
Poisson's ratio		ASTM C848	0.22	0.22	0.22	0.22
Compressive strength	MPa	ASTM C773	1930	2150	2260	2600
Fracture toughness	Mpa-m$^{0.5}$	Notched beam	3.1	3.3	3.7	4.6
Thermal conductivity	W/m-°C	ASTM C408	16	16.7	24.7	31.0
TCE	10^{-6}/°C	ASTM C372	5.9	6.2	6.5	6.8
Specific heat	W-s/g-°C	ASTM E1269	920	920	880	880
Dielectric strength	ac kV/mm	ASTM D116	8.3	8.3	8.3	8.7
Loss tangent (1 MHz)		ASTM D2520	0.0009	0.0004	0.0002	0.0001
Volume resistivity	Ω-cm,	ASTM D1829				
25°C			>10^{14}	>10^{14}	>10^{14}	>10^{14}
500°C				4 × 10^8	4 × 10^9	2 × 10^{10}
1000°C			4 × 10^8	5 × 10^5	1 × 10^6	2 × 10^6

7.8.2   Beryllium Oxide

Beryllium oxide (BeO, beryllia) is cubic close-packed and has a zinc blende structure. The alpha form of BeO is stable to above 2050°C. BeO is stable in dry atmospheres and is inert to most materials. It hydrolyzes at temperatures greater than 1100°C with the formation and volatilization of beryllium hydroxide. BeO reacts with graphite at high temperature, forming beryllium carbide.

Beryllia has an extremely high thermal conductivity, higher than aluminum metal, and is widely used in applications where this parameter is critical. The thermal conductivity drops rapidly above 300°C but is suitable for most practical applications.

Beryllia is available in a wide variety of geometries formed using a variety of fabrication techniques. While beryllia in the pure form is perfectly safe, care must be taken when machining BeO, however, as the dust is toxic if inhaled.

Beryllia may be metallized with thick film, thin film, or by one of the copper processes. However, thick film pastes must be specially formulated to be compatible. Laser or abrasive trimming of BeO must be performed in the presence of a vacuum to remove the dust. The properties of 99.5% beryllia are summarized in Table 7.12.

TABLE 7.12   Typical Parameters for 99.5% Beryllium Oxide

Parameter	Units	Value
Density	g/cm^3	2.87
Hardness	Knoop 100g	1200
Melting point	°C	2570
Modulus of elasticity	GPa	345
Compressive strength	MPa	1550
Poisson's ratio		0.26
Thermal conductivity 25°C 500°C	W/m-K	 250 55
Specific heat 25°C 500°C	W-s/gm-K	 1.05 1.85
TCE	10^{-6}/K	7.5
Dielectric constant 1 MHz 10 GHz		 6.5 6.6
Loss tangent 1 MHz 10 GHz		 0.0004 0.0004
Volume resistivity 25°C 500°C	Ω-cm	 $>10^{14}$ 2×10^{10}

7.8.3   Aluminum Nitride

Aluminum nitride is covalently bonded with a wurtzite structure and decomposes at 2300°C under one atmosphere of argon. In a nitrogen atmosphere of 1500 psi, melting may occur in excess of 2700°C. Oxidation of AlN in even a low concentration of oxygen (<0.1%) occurs at temperatures above 700°C. A layer of aluminum oxide protects the nitride to a temperature of 1370°C, above which the protective layer cracks, allowing oxidation to continue. Aluminum nitride is not appreciably affected by hydrogen, steam, or oxides of carbon to 980°C. It dissolves slowly in mineral acids and decomposes slowly in water. It is compatible with aluminum to 1980°C, gallium to 1300°C, iron or nickel to 1400°C, and molybdenum to 1200°C.

Aluminum nitride substrates are fabricated by mixing AlN powder with compatible glass powders containing additives such as CaO and Y_2O_3, along with organic binders, and casting the mixture into the desired shape. Densification of the AlN requires very tight control of both atmosphere and temperature. The solvents used in the preparation of substrates must be anhydrous to minimize oxidation of the AlN powder and prevent the generation of ammonia during firing.[9] For maximum densification and maximum thermal conductivity, the substrates must be sintered in a dry reducing atmosphere to minimize oxidation.

Aluminum nitride is primarily noted for two very important properties: a high thermal conductivity and a TCE closely matching that of silicon. There are several grades of aluminum nitride with different thermal conductivities available. The prime reason is the oxygen content of the material. It is important to note that even a thin surface layer of oxidation on a fraction of the particles can adversely affect the thermal conductivity. Only with a high degree of material and process control can AlN substrates be made consistent.

The thermal conductivity of AlN does not vary as widely with temperature as that of BeO. Considering the highest grade of AlN, the crossover temperature is about 20°C. Above this temperature, the thermal conductivity of AlN is higher; below, 20°C, BeO is higher.

The TCE of AlN closely matches that of silicon, an important consideration when mounting large power devices. The second level of packaging is also critical. If an aluminum nitride substrate is mounted directly to a package with a much higher TCE, such as copper, the result can be worse than if a substrate with an intermediate, although higher, TCE were used. The large difference in TCE builds up stresses during the mounting operation that can be sufficient to fracture the die and/or the substrate.

Thick film, thin film, and copper metallization processes are available for aluminum nitride. Certain of these processes, such as direct bond copper (DBC), require oxidation of the surface to promote adhesion. For maximum thermal conductivity, a metallization process should be selected that bonds directly to AlN to eliminate the relatively high thermal resistance of the oxide layer.

Thick film materials must be formulated to adhere to AlN. The lead oxides prevalent in thick film pastes that are designed for alumina and beryllia oxidize AlN rapidly, causing blistering and a loss of adhesion. Thick film resistor materials are primarily based on RuO_2 and MnO_2

Thin film processes available for AlN include NiCr/Ni/Au, Ti/Pt/Au, and Ti/Ni/Au.[10] Titanium in particular provides excellent adhesion by diffusing into the surface of the AlN. Platinum and nickel are transition layers to promote gold adhesion. Solders such as Sn60/Pb40 and Au80/Sn20 can also be evaporated onto the substrate to facilitate soldering.

Multilayer circuits can be fabricated with W or MbMn conductors. The top layer is plated with nickel and gold to promote solderability and bondability. Ultrasonic milling may be used for cavities, blind vias, and through vias. Laser machining is suitable for through vias as well.

Direct bond copper may be attached to AlN by forming a layer of oxide over the substrate surface, which may require several hours at temperatures above 900°C. The DBC forms a eutectic with aluminum oxide at about 963°C. The layer of oxide, however, increases the thermal resistance by a significant amount, partially negating the high thermal conductivity of the aluminum nitride. Copper foil may also be brazed to AlN with one of the compatible braze compounds. Active metal brazing (AMB) does not generate an oxide layer. The copper may also be plated with nickel and gold. The properties of aluminum nitride are summarized in Table 7.13.

7.8.4 Diamond

Diamond substrates are primarily grown by chemical vapor deposition (CVD). In this process, a carbon-based gas is passed over a solid surface and activated by a plasma, a heated filament, or by a combustion flame. The surface must be maintained at a high temperature, above 700°C, to sustain the reaction. The gas is typically a mixture of methane (CH_4) and hydrogen (H_2) in a ratio of 1–2% CH_4 by volume.[11] The consistency of the film in terms of the ratio of diamond to graphite is inversely proportional to the growth rate of the film. Films produced by plasma have a growth rate of 0.1 to 10 µ/hr and are very high quality, while films produced by combustion methods have a growth rate of 100 to 1000 µ/hr and are of lesser quality.

TABLE 7.13 Typical Parameters for Aluminum Nitride (Highest Grade)

Parameter	Units	Value
Density	g/cm^3	3.27
Hardness	Knoop 100g	1200
Melting point	°C	2232
Modulus of elasticity	GPa	300
Compressive strength	MPa	2000
Poisson's ratio		0.23
Thermal conductivity	W/m-K	
25°C		290
150°C		195
Specific heat	W-s/gm-K	
25°C		0.76
150°C		0.94
TCE	10^{-6}/K	4.4
Dielectric constant		
1 MHz		8.9
10 GHz		9.0
Loss tangent		
1 MHz		0.0004
10 GHz		0.0004
Volume resistivity	Ω-cm	
25°C		>10^{12}
500°C		2 × 10^8

The growth begins at nucleation sites and is columnar in nature, growing faster in the normal direction than in the lateral direction. Eventually, the columns grow together to form a polycrystalline structure with microcavities spread throughout the film. The resulting substrate is somewhat rough, with a 2 to 5 μ surface. This feature is detrimental to the effective thermal conductivity, and the surface must be polished for optimum results. An alternative method is to use an organic filler[12] on the surface for planarization. This process has been shown to have a negligible effect on the overall thermal conductivity from the bulk, and it dramatically improves heat transfer. Substrates as large as 10 cm^2 and as thick as 1000 μ have been fabricated.

Diamond can be deposited as a coating on refractory metals, oxides, nitrides, and carbides. For maximum adhesion, the surface should be a carbide-forming material with a low TCE.[13]

Diamond has an extremely high thermal conductivity, several times that of the next highest material. The primary application is, obviously, in packaging power devices. Diamond has a low specific heat, however, and works best as a heat spreader in conjunction with a heat sink. For maximum effectiveness,[13]

$$t_D = 0.5 \rightarrow 1 \times r_h$$

$$r_D = 3 \times r_h$$

where t_D = thickness of diamond substrate
      r_h = radius of heat source
      r_D = radius of diamond substrate

Applications of diamond substrates include heat sinks for laser diodes and laser diode arrays. The low dielectric constant of diamond coupled with the high thermal conductivity makes it attractive for microwave circuits as well. As improved methods of fabrication lower the cost, the use of diamond substrates is expected to expand rapidly. The properties of diamond are summarized in Table 7.14.

7.8.5   Boron Nitride

There are two basic types of boron nitride (BN). Hexagonal (alpha) BN is soft and is structurally similar to graphite. It is white in color and is sometimes called *white graphite*. Cubic (beta) BN is formed by subjecting hexagonal BN to extreme heat and pressure, similar to the process used to fabricate synthetic industrial diamonds. Melting of either phase is possible only under nitrogen at high pressure.

Hot pressed BN is very pure (>99%), with the major impurity being boric oxide (BO). Boric oxide tends to hydrolyze in water, degrading the dielectric and thermal shock properties. Calcium oxide (CaO) is frequently added to tie up the BO to minimize the water absorption. When exposed to temperatures above 1100°C, BO forms a thin coating on the surface, slowing further oxide growth.

Boron nitride in the hot pressed state is easily machinable and may be formed into various shapes. The properties are highly anisotropic and vary considerable in the normal and tangential directions of the pressing force. The thermal conductivity in the normal direction is very high and the TCE is very low, making BN an attractive possibility for a substrate material. However, it has not yet been proven possible to metallize BN,[14] thereby limiting the range of applications. It can be used in contact with various metals, including, copper, tin, and aluminum, and it may be used as a thermally conductive electrical in-

TABLE 7.14 Typical Parameters for CVD Diamond

Parameter	Units	Value
Density	g/cm^3	3.52
Hardness	Knoop 100g	7000
Modulus of elasticity	GPa	1000
Compressive strength	MPa	11000
Poisson's ratio		0.148
Thermal conductivity normal tangential	W/m-K	 2200 1610
Specific heat 25°C 150°C	W-s/gm-K	 0.55 0.90
TCE	10^{-6}/K	1.02
Dielectric constant 1 MHz 10 GHz		 5.6 5.6
Loss tangent 1 MHz 10 GHz		 0.001 0.001
Volume resistivity 25°C 500°C	Ω-cm	 >10^{13} 2 × 10^{11}

sulator. Applications of BN include microwave tubes and crucibles. The properties of boron nitride are summarized in Table 7.15.

7.8.6 Silicon Carbide

Silicon carbide (SiC) has a tetrahedral structure and is the only known alloy of silicon and carbon. Both elements have four electrons in the outer shell, with an atom of one bonded to four atoms of the other. The result is a very stable structure that is not affected by hydrogen or nitrogen up to 1600°C. In air, SiC begins decomposing above 1000°C. As with other compounds, a protective oxide layer forms over the silicon, reducing the rate of decomposition. Silicon carbide is highly resistant to both acids and bases. Even the so-called *white etch* (hydrofluoric acid mixed with nitric and sulfuric acids) has no effect.

Silicon carbide structures are formed by hot pressing, dry and isostatic pressing (preferred), by CVD, or by slip casting. Isostatic pressing using gas as the fluid provides optimum mechanical properties.

TABLE 7.15   Typical Parameters for Boron Nitride

Parameter	Units	Value
Density	g/cm^3	1.92
Hardness	Knoop 100g	5000
Modulus of elasticity	GPa	
normal		43
tangential		768
Compressive strength	MPa	
normal		110
tangential		793
Poisson's ratio		0.05
Thermal conductivity	W/m-K	
normal		73
tangential		161
Specific heat	W-s/gm-K	
25°C		0.84
150°C		1.08
TCE	10^{-6}/K	
normal		0.57
tangential		−0.46
Dielectric constant, 1 MHz		
		4.1
Loss tangent, 1 MHz		
		0.0003
Volume resistivity	Ω-cm	
25°C		1.6×10^{12}
500°C		2×10^{10}

Silicon carbide in pure form is a semiconductor, and the resistivity depends on the impurity concentration. In the intrinsic form, the resistivity is less than 1000 Ω-cm, which is unsuitable for ordinary use. The addition of a small percentage (<1%) of BeO during the fabrication process[15] increases the resistivity to as high as 10^{13} Ω-cm by creating carrier-depleted layers around the grain boundaries.

Both thick and thin films can be used to metallize SiC, although some machining of the surface to attain a higher degree of smoothness is necessary for optimum results. The two parameters that make SiC attractive as a substrate are the exceptionally high thermal conductivity, second only to diamond, and the low TCE, which matches that of silicon to a higher degree than any other ceramic. SiC is also less expensive than either BeO or AlN. A possible disadvantage is the high dielectric constant, 4 to 5 times higher than that of other substrate materials. This parameter can result in cross-coupling of electronic signals or in excessive transmission delay. The parameters of SiC are summarized in Table 7.16.

TABLE 7.16   Typical Parameters for Silicon Carbide

Parameter	Units	Value
Density	g/cm^3	3.10
Hardness	Knoop 100g	500
Modulus of elasticity	GPa	407
Compressive strength	MPa	4400
Fracture toughness	$MPa\text{-}m^{1/2}$	7.0
Poisson's ratio		0.14
Thermal conductivity	W/m-K	
25°C		290
150°C		160
Specific heat	W-s/gm-K	
25°C		0.64
150°C		0.92
TCE	$10^{-6}/K$	3.70
Dielectric constant		40
Loss tangent		0.05
Volume resistivity	$\Omega\text{-cm}$	
25°C		$>10^{13}$
500°C		2×10^9

7.9   Composite Materials

Composite materials have been in use for thousands of years. Ancient peoples used straw and rocks to strengthen adobe bricks for building shelters, and concrete reinforced with steel rods is used extensively in the construction industry. Only recently have composites been introduced into electronics applications. In particular, the demands for improved thermal management and higher packaging density have driven the development of composite materials. Today, several composite materials are available to meet the needs of electronic packaging engineers.

A two-component composite consists of a matrix and a filler, which may be in the form of long or short fibers, large or small particles, or as laminates. The theory of composite materials is well documented. For this discussion it is sufficient to simply describe composites by the so-called "rule of mixtures," stated as

$$P_C = P_M V_{fM} + P_F V_{fF} \tag{7.34}$$

where P_C = parameter of the composite
P_M = parameter of the matrix
V_{fM} = volume fraction of the matrix
P_F = parameter of the filler
V_{fF} = volume fraction of the filler

The degree to which the composite will obey the rule of mixtures depends on a number of factors, including the size and orientation of the filler and the extent of the interaction between the matrix and the filler. Consider the structure in Fig. 7.17, consisting of continuous fibers uniformly dispersed in a matrix. Assuming that no slippage between the two occurs, the strain, E, as defined in Eq. (7.8) is the same for both materials when a longitudinal force is applied. For this case,

$$E_C = E_M = E_F = \frac{\Delta L}{L_0} \tag{7.35}$$

where E_C = strain in composite
E_M = strain in matrix
E_F = strain in filler
ΔL = change in length
L_0 = initial length

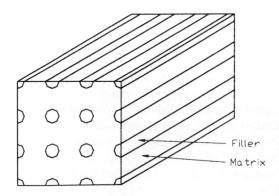

Figure 7.17 Matrix with continuous filler.

A structure of this type will closely approximate the rule of mixtures. Where the filler is not uniformly distributed as in Fig. 7.17, a critical length may be defined as

$$L_C = \frac{D_F S_{max}}{2\sigma_M}$$

(7.36)

where L_C = critical length

D_F = diameter of filler

S_{max} = maximum shear strength of filler

σ_M = shear strength of matrix

To be effective, the filler must be greater than the critical length.

The rule of mixtures may also be used to approximate other parameters, as illustrated in Eq. (7.5) to calculate the thermal conductivity of a glass-alumina matrix.

It is possible to tailor composite materials to obtain desirable properties superior to either component for specific applications. Composite materials may be divided into several categories, based on the nature of the matrix and the filler.

7.9.1 Organic-Organic Composites

Composites composed of organic materials typically use an epoxy as the matrix and a variety of organic fibers as the filler. The epoxy may be in the liquid or powder form. For example, nylon reinforced plastics are commonly used to mold plastic parts for a number of applications.

7.9.2 Organic-Ceramic/Glass Composites

Organic-ceramic composites may use an epoxy as the matrix and glass or ceramic powder as the filler. A common example is the fiber-glass-reinforced epoxy used as a printed circuit laminate. An epoxy substrate filled with alumina and carbon black has also been developed.[14] By weight, the composition is 10.8% epoxy resin, 89% alumina, and 0.2% carbon black. This material has a thermal conductivity of 3.0–4.0 W/m-K as compared to both glass epoxy printed circuit material (0.2 W/m-K) and glass-alumina low temperature cofired substrates (2.5 W/m-K). The TCE (17 ppm) is substantially below that of PCB material (25–50 ppm). This composite can be utilized to make multilayer interconnection structures.

Organic-ceramic composites are also commonly used as materials for housings for electronic systems. Glass-reinforced plastics, for example, are very strong and are capable of withstanding considerable mechanical shock without breaking.

7.9.3 Ceramic-Ceramic Composites

Ceramic-based composites are somewhat difficult to fabricate due to their high melting point. Lawrence Berkeley National Laboratory[15] has developed a method of coating platelets of silicon carbide with alumina and forming them into a composite matrix with exceptional fracture strength and fracture toughness. This approach was demonstrated by coating alpha SiC platelets with about 2 μ of alumina and combining them with a matrix of beta SiC. The coating prevents the alpha SiC from being assimilated into the beta SiC during processing and diverts cracks to the perimeter of the particles rather than through the body. The result is a SiC-SiC composite ceramic that is 2–3 times stronger than conventional SiC.

7.9.4 Ceramic-Glass Composites

In all but the simplest electronic circuits, it is necessary to have a method for fabricating multilayer interconnection structures to enable all the necessary points to be connected. The thick film technology is limited to three layers for all practical purposes due to yield and planarity considerations, and thin film multilayer circuits are quite expensive to fabricate. The copper technologies are limited to a single layer due to processing limitations.

Ceramic-glass composite materials may be used to economically fabricate very complex multilayer interconnection structures. The materials in powder form are mixed with an organic binder, a plasticizer, and a solvent and formed into a slurry by ball or roll milling. The slurry is forced under a doctor blade and dried to form a thin sheet, referred to as *green tape* or *greensheet*. Further processing is dependent on the type of material. There are three basic classes of materials: high temperature cofired ceramic (HTCC), low temperature cofired ceramic (LTCC), and aluminum nitride.

High temperature cofired ceramic (HTCC). HTCC multilayer circuits are primarily alumina based. The green tape is blanked into sheets of uniform size, and holes are punched where vias and alignment holes are required. The metal patterns are printed and dried next. Despite their relatively high electrical resistance, refractory metals such as tungsten and molybdenum are used as conductors, due to the high firing temperature. Via fills may be accomplished during conductor printing or during a separate printing operation. The process is repeated for each layer.

The individual layers are aligned and laminated under heat and pressure to form a monolithic structure in preparation for firing. The structure is heated to approximately 600°C to remove the organic materials. Carbon residue is removed by heating to approximately 1200°C in a wet hydrogen atmosphere. Sintering and densification take place at approximately 1600°C.

During firing, HTCC circuits shrink anywhere from 14 to 17%, depending on the organic content. With careful control of the material properties and processing parameters, the shrinkage can be controlled to within 0.1%. Shrinkage must be taken into consideration during the design, punching, and printing processes. The artwork enlargement must exactly match the shrinkage factor associated with a particular lot of green tape.

Processing of the substrate is completed by plating the outer layers with nickel and gold for component mounting and wire bonding. The gold is plated to a thickness of 25 μin for gold wire and 5 μin for aluminum wire. Gold wire bonds to the gold plating, while aluminum wire bonds to the nickel underneath. The gold plating in this instance is simply to protect the nickel surface from oxidation or corrosion. The properties of HTCC materials are summarized in Table 7.17.

Low temperature cofired ceramic (LTCC).   LTCC circuits consist of alumina mixed with glasses with the capability to simultaneously sinter and crystallize.[17] These structures are often referred to as *glass-ceramics*. Typical glasses are listed in Table 7.17. The processing steps are similar to those used to fabricate HTCC circuits, with two exceptions:[16] the firing temperature is much lower, 850–1050°C, and the metallization is gold-based or silver-based thick film formulated to be compatible with the LTCC material. Frequently, silver-based materials are used in the inner layers with gold on the outside for economic reasons. Special via fill materials are used between the gold and the silver layers to prevent electrolytic reaction.

The shrinkage of LTCC circuits during firing is in the range of 12–18%. If the edges are restrained, the lateral shrinkage can be held to 0.1% with a corresponding increase in the vertical shrinkage.

One advantage that LTCC has over the other multilayer technologies is the ability to print and fire resistors. Where trimming is not required, the resistors can be buried in intermediate layers with a corresponding saving of space. It is also possible to bury printed capacitors of small value.

Aluminum nitride.   Aluminum nitride multilayer circuits are formed by combining AlN powder with yttria or calcium oxide.[17] Glass may also be added. Sintering may be accomplished in three ways:

TABLE 7.17   Properties of Multilayer Ceramic Materials

	Low temperature cofired ceramic (LTCC)	High temperature cofired ceramic (HTCC)	Aluminum nitride

Material	Cordierite
	  MgO, SiO_2, Al_2O_3
	Glass filled composites
	  SiO_2, B_2O_3, Al_2O_3
	  PbO, SiO_2, CaO, Al_2O_3
	Crystalline phase ceramics
	  Al_2O_3, CaO, SiO_2, MgO, B_2O_3
		88–92% alumina	AlN, yttria, CaO

Firing temperature	850–1050°C	1500–1600°C	1600–1800°C
Conductors	Au, Ag, Cu, PdAg	W, MoMn	W, MoMn
Conductor resistance	3–20 mΩ/	8–12 mΩ/	8–12 mΩ/
Dissipation factor	15–30×10^{-4}	5–15×10^{-4}	20–30×10^{-3}
Relative dielectric constant	5–8	9–10	8–9
Resistor values	0.1 Ω–1 MΩ	N/A	N/A
Firing shrinkage X,Y Z	12.0 ± 0.1% 17.0 ± 0.5%	12–18% 12–18%	15–20% 15–20%
Repeatability	0.3–1%	0.3–1%	0.3–1%
Line width	100 μm	100 μm	100 μm
Via diameter	125 μm	125 μm	125 μm
Number of metal layers	33	63	8
CTE	3–8 ppm/°C	6.5 ppm/°C	4.4 ppm/°C
Thermal conductivity	2–6 W/m-°C	15–20 W/m-°C	180–200 W/m-°C

1. Hot pressing during sintering.

2. High temperature (>1800°C) sintering without pressure

3. Low temperature (<1650°C) sintering without pressure

Tungsten or molybdenum pastes are used to withstand the high firing temperatures. Control of the processing parameters during sintering is critical if optimum properties are to be attained. A carbon atmosphere helps in attaining a high thermal conductivity by preventing oxidation of the AlN particles. Shrinkage is in the range of 15–20%. The properties of aluminum nitride multilayer materials are summarized in Table 7.17.

7.9.5 Metal-Ceramic Composites

Metal-ceramic composites are primarily formed by one of two methods. One is to form *in situ* whereby the filler and the matrix are mixed and fused together by a combination of heat and pressure. A more common approach is to form the filler into a porous shell and subsequently filling the shell with the matrix material in the molten state.

Ceramics typically have a low thermal conductivity and a low TCE, while metals have a high thermal conductivity and a high TCE. It is a logical step to combine these properties to obtain a material with a high thermal conductivity and a low TCE. The ceramic in the form of particles or continuous fibers is mixed with the metal to form a structure with the desirable properties of both. The resultant material is referred to as a *metal matrix composite (MMC)*. MMCs are primarily used in applications where thermal management is critical.

The most common metals used in this application are aluminum and copper, with aluminum being more common due to lower cost. Fillers include SiC, AlN, BeO, graphite, and diamond. Compatibility of the materials is a prime consideration. Graphite, for example, has an electrolytic reaction with aluminum but not with copper.[16] Two examples will be described here: AlSiC, a composite made up of aluminum and silicon carbide, and Dymalloy®, a combination of copper and diamond.

Aluminum silicon carbide (AlSiC) is produced by forcing liquid aluminum into a porous SiC preform. The preform is made by any of the common ceramic processing technologies, including dry pressing, slip molding, and tape casting. The size and shape of the preform is selected to provide the desired volume fraction of SiC. The resulting combination has a thermal conductivity almost as high as pure aluminum, with a TCE as low as 6.1 ppm/°C. AlSiC is also electrically conductive, prohibiting its use as a conventional substrate.

The mechanical properties of the composite are determined by the ratio of SiC to aluminum as shown in Fig. 7.18.[18] A ratio of 70–73% of SiC by volume provides the most optimum properties for electronic packaging.[16] This ratio gives a TCE of about 6.5 ppm/°C, which closely matches that of alumina and beryllia. This allows AlSiC to be used as a baseplate for ceramic substrate, using its high thermal conductivity to maximum advantage.

AlSiC, being electrically conductive, may be readily plated with aluminum to provide a surface for further processing. The aluminum coating may be plated with nickel and gold to permit soldering or may be anodized where an insulating surface is required.[21] An alternative approach is to flame-spray the AlSiC with various silver alloys for solderability.

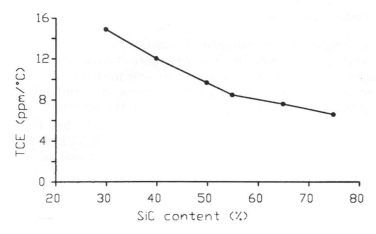

Figure 7.18 TCE vs. SiC content for AlSiC.

Two other advantages of AlSiC are strength and weight. The aluminum is somewhat softer than SiC and reduces the propagation of cracks. The density is only about 1/3 that of Kovar®, and the thermal conductivity is over 12 times greater.

AlSiC has been used to advantage in the fabrication of hermetic single-chip and multichip packages and as heat sinks for power devices and circuits. While difficult to drill and machine, AlSiC can be formed into a variety of shapes in the powder state. It has been successfully integrated with patterned AlN to form a power module package.[21]

The TCE linearly increases with temperature up to about 350°C and then begins to decrease. At this temperature, the aluminum matrix softens, and the SiC matrix dominates. This factor is an important feature for power packaging.

The parameters of AlSiC are summarized in Table 7.18, along with those of aluminum and silicon carbide. Table 7.18 provides an interesting comparison of how the properties of a composite compare to those of the constituents.

Dymalloy is a matrix of Type I diamond and Cu20/Ag80 alloy.[22] The diamond is ground into a powder in the 6 to 50 μm range. The powder is coated with W74/Rh26 to form a carbide layer approximately 100 Å thick followed by a 1000 Å coating of copper. The copper is plated to a thickness of several μm to permit brazing.

The powder is packed into a form and filled in a vacuum with Cu20/Ag80 alloy that melts at approximately 800°C. This material is selected over pure copper, which melts at a much higher temperature, to minimize graphization of the diamond. The diamond loading is approximately 55% by volume. The parameters of Dymalloy are summarized in Table 7.19.

TABLE 7.18   Typical Parameters for AlSiC (70% SiC by Volume), SiC, and Aluminum

Parameter	Units	AlSiC	SiC	Al
Density	g/cm^3	3.02	3.10	2.70
Modulus of elasticity	GPa	224	407	69
Tensile strength	MPa	192	Note[*]	55
Thermal conductivity, 25°C	W/m-K	218	290	237
TCE	10^{-6}/K	7.0	3.70	23
Volume resistivity, 25°C	μΩ-cm	34	>10^{13}	2.8

[*]Depends on method of preparation and number/size of defects.

TABLE 7.19   Typical Parameters for Dymalloy (55% Diamond by Volume)

Parameter	Units	Value
Density	g/cm^3	6.4
Tensile strength	MPa	400
Specific heat[*]	W-s/g-°C	$0.316 + 8.372 \times 10^{-4}$ T
Thermal conductivity	W/m-K	360
TCE[†]	10^{-6}/K	$5.48 + 6.5 + 10^{-3}$ T

[*]Temperature in °C from 25 to 75°C.
[†]Temperature in °C from 25 to 200°C.

7.10   Forming Ceramics and Composites to Shape

Ceramic substrate materials are typically hard and brittle, making them difficult to machine by conventional means. Sawing and drilling are virtually impossible to accomplish, while grinding often results in surface voids. Certain composites, on the other hand, are relatively easy to machine. The high metal content of these structures permits sawing and drilling to shape.

The most common methods of forming ceramics to shape are punching in the green state and laser scribing. The ceramic in the green state is formed to the proper thickness by one of the described methods and punched to shape with a carbide die in the desired shape or drilled with a carbide-tipped drill. During the firing process, the ceramic may shrink anywhere from 10–22%, depending primarily on the organic content, the ratio of glass to ceramic, the particle size distribution, and the particle shape distribution. The amount of shrinkage during firing is highly predictable and consistent, and it may be compensated for during the design process. Typical tolerances after firing are typically 1–5%.

Ceramic substrates may also be formed to shape after firing with a CO_2 laser. Holes and patterns may be created to a high degree of tolerance, typically 1%. It is common to partially scribe a groove in the substrate so that a single process may create several patterns. After processing, the circuits may readily be separated by a gentle force applied along the scribe line. The surface finish of laser scribe lines may vary greatly from the bulk of the ceramic, tending to have a higher glass content due to the nature of the laser machining process. Annealing at an elevated temperature will return the scribed surface of the substrate to the original state.[1] A diagram of a scribed substrate and typical examples are found in Figs. 7.19 and 7.20, respectively.

The surface finish of ceramic substrates may be made smoother by grinding, honing, ultrasonic machining, lapping, or polishing.[23] Grinding is accomplished by exposing the substrate surface to abrasive particles bonded to a wheel that is rotating at high speed. The orientation of the wheel to the substrate is obviously critical, as is the particle size of the abrasive particles. The surface finish of the substrate is limited to a large extent by the size of the particles in the substrate itself. When a particle is dislodged from the substrate, as frequently happens, a pit equal to the size of the particle will be present on the substrate surface.

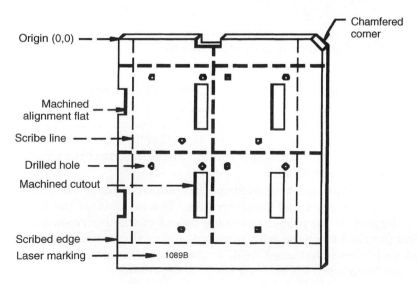

Figure 7.19 Laser scribing.

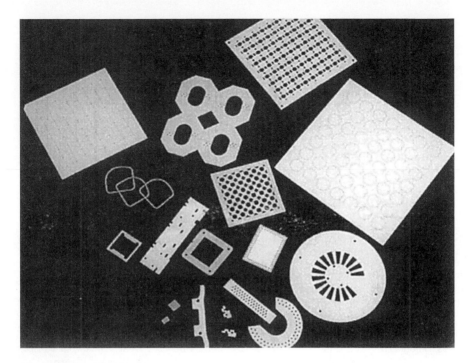

Figure 7.20   Examples of laser machined substrates.

Honing utilizes a cylinder or other surface formed from the abrasive particles. The cylinder is moved rapidly back and forth across the surface of the substrate to form the finish. Honing provides a smoother, more accurate surface than grinding.

In ultrasonic machining, a tool in the desired shape is formed from a soft, malleable metal. The abrasive in the form of a slurry is applied to the substrate and ultrasonic energy is applied to the tool. This process is somewhat slow but permits round holes and other shapes to be formed in fired ceramic.

Lapping utilizes a fine abrasive powder, such as aluminum oxide or silicon carbide, suspended in a vehicle. The substrate surface is placed in contact with the powder, and a weight is placed on the back of the substrate. The apparatus is placed on a rotating wheel made of cast iron that is rotated to produce a planetary rotation of the substrate/weight.

Polishing is similar to lapping in principle, except that the abrasive powder is placed in a soft material, such as felt, and the substrate is rotated against the powder. Typical powders used in this application are cerium oxide and ferrous oxide. These materials can be made into very fine powders, producing a very smooth surface.

References

1. Jerry Sergent and Charles Harper, *Hybrid Microelectronics Handbook,* 2nd ed., McGraw-Hill, 1995.
2. Richard Brown, "Thin Film Substrates," in *Handbook of Thin Film Technology,* Leon Maissel and Reinhard Glang, eds., McGraw-Hill, 1971.
3. Philip Garrou and Arne Knudsen, "Aluminum Nitride for Microelectronic Packaging," *Advancing Microelectronics,* Vol. 21, No. 1, Jan-Feb 1994.
4. C.G.M. Van Kessel, S.A. Gee, and J.J. Murphy, "The Quality of Die Attachment and Its Relationship to Stresses and Vertical Die-cracking," *Proc. IEEE Components Conf.,* 1983.
5. A. Winkleman and O. Schott, *Ann. Phys. Chem.,* Vol. 51, 1984.
6. Jerry Sergent and H. Thurman Henderson, "Double Injection in Semi-Insulators," *Proc. Solid State Materials Conference,* 1973.
7. George Harman, *Wire Bond Reliability and Yield,* ISHM Monograph, 1989.
8. Craig Johnston, Robin A. Susko, John V. Siciliano, and Robert J. Murcko, "Temperature Dependent Wear-Out Mechanism for Aluminum/Copper Wire Bonds," *Proc. ISHM Symposium,* 1991.
9. Ellice Y. Yuh, John W. Lau, Debra S. Horn, and William T. Minehan, "Current Processing Capabilities for Multilayer Aluminum Nitride," *International Journal of Microelectronics and Electronic Packaging,* Vol. 16, No. 2, 2nd qtr., 1993.
10. Nobuyiki Karamoto, "Thin Film and Co-Fired Metallization on Shapal Aluminum Nitride," *Advancing Microelectronics,* Vol. 21, No. 1, January/February, 1994.
11. Paul W. May, "CVD Diamond—A New Technology for the Future?" *Endeavor Magazine,* Vol. 19, No. 3, 1995.
12. Ajay P. Malshe, S. Jamil, M. H. Gordon, H. A. Naseem, W. D. Brown, and L. W. Schaper, "Diamond for MCMs," *Advanced Packaging,* September/October, 1995.
13. Thomas Moravec and Arjun Partha, "Diamond Takes the Heat," *Advanced Packaging,* Special Issue, October, 1993.
14. Koichi Hirano, Seiichi Nakatani, and Jun'ichi Kato, "A Novel Composite Substrate with High Thermal Conductivity for CSP, MCM, and Power Modules," *Proc. International Microelectronic and Packaging Society,* 1998.
15. C. De Johghe, T. Mitchell, W.J. Moberly-Chan and R.O. Ritchie, "Silicon Carbide Platelet/Silicon Carbide Composites," *Journal American Ceramics Society,* 1994.
16. S. Fuchs and P. Barnwell, "A Review of Substrate Materials for Power Hybrid Circuits," *The IMAPS Journal of Microcircuits and Electronic Packaging,* Vol. 20, No. 1, 1st qtr., 1997.
17. Mitsuru Ura and Osami Asai, internal report, Hitachi Research Laboratory, Hitachi Industries, Ltd.
18. Jerry E. Sergent, "Materials for Multichip Modules," in *Electronic Packaging and Production,* December, 1996.
19. Philip E. Garrou and Iwona Turlik, *Multichip Module Technology Handbook,* McGraw-Hill, 1998.
20. M. K. Premkumar and R. R. Sawtell, "Alcoa's AlSiC Cermet Technology for Microelectronics Packaging," *Advancing Microelectronics,* July/August, 1995.
21. M. K. Premkumar and R. R. Sawtell, "Aluminum-Silicon Carbide," *Advanced Packaging,* September/October 1996.
22. J. A. Kerns, N. J. Colella, D. Makowiecki, and H. L. Davidson, "Dymalloy: A Composite Substrate for High Power Density Electronic Components," *International Journal of Microcircuits and Electronic Packaging,* Vol. 19, No. 3, 3rd qtr., 1996.
23. Ioan D. Marinescu, Hans K. Tonshoff, and Ichiro Inasaki, *Handbook of Ceramic Grinding and Polishing,* Noyes Publications, 1998.

8

Inorganic Glasses: Commercial Glass Families, Applications, and Manufacturing Methods

Thomas P. Seward III and Arun K. Varshneya

New York State College of Ceramics
Alfred University
Alfred, New York

8.1 Commercial Glass Families

8.1.1 Introduction

Historically, glasses of many different compositions and properties were developed based on product (end use) needs and manufacturing capabilities. Composition inventions and process inventions went hand in hand. This is still true today, but now employee safety and environmental concerns have considerable impact on development as well. Among other things, this chapter will show how glass composition and manufacturing processes are interrelated, through chemistry and physics, and how both must be considered when selecting or developing a glass for a particular application.

Viscosity reigns supreme. We say this because almost every process step in glass manufacturing is performed most effectively within a certain range of melt viscosity. For example, melting of batch raw materials is generally performed best at a temperature where the melt has a viscosity of about 100 Poise (P). On the other hand, *pressing* of molten glass into metal molds is most effective at viscosities in the range from 1,000 to 10,000 Poises. Figure 8.1 shows the useful viscosity ranges for a number of process steps that will be described later in this chapter.

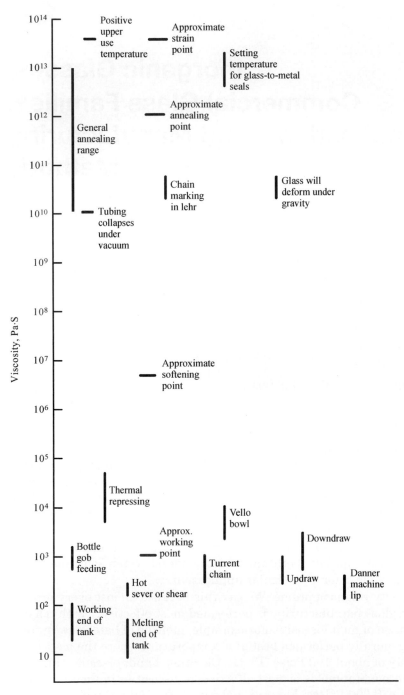

Figure 8.1 Approximate viscosity values for forming and processing methods (Pa · s × 10 = P). *(Courtesy of Corning Inc.)*

Another key factor is the viscosity of the melt at the *liquidus* temperature. The liquidus temperature of a melt is defined thermodynamically as the temperature above which the melt is stable as a liquid. Whether this liquid is very fluid or very viscous is not of thermodynamic importance. At temperatures below the liquidus temperature, the melt begins to *devitrify* (i.e., develop crystals); the extent of devitrification depends on the glass chemical composition, the temperature, and the time allowed.

If, to operate a certain manufacturing process at the required viscosity, the glass temperature has to be decreased below the liquidus temperature, then there is the risk that, given enough time, the molten glass will crystallize. So, for any melting and forming process, the glass composition must be designed so that the liquidus temperature is less than any processing temperature at which some or all of the molten glass will be held for long periods of time.

It is customary in glass science and technology to refer to glasses as being *soft* or *hard*, not in terms of physical hardness, but rather in terms of the temperatures required to soften the glass and make it flow. Soft and hard are relative terms; one glass is harder than another if we must heat it to higher temperatures to make it flow. Nevertheless, on a technology scale, some glasses are considered "soft" and others "hard." Additionally, glasses are called *long* or *short*, depending on the slope of the viscosity curve (vs. temperature) within the *working range*, the temperature interval between the softening and working points. The terms *short* and *long* thus refer to the relative amount of time available to work the glass as it cools. The wide variety of viscosity-temperature relationships exhibited by commercial glasses is illustrated in Fig. 8.2. The glass code numbers shown in the figure correspond to those of glass compositions listed in Tables 8.1 and 8.2.

In the following sections, we present several families of glasses, generally progressing from the softer to the harder. We discuss each family, giving examples and ranges of useful compositions. We also list key properties of importance, current commercial applications, and some manufacturing processes used to form articles from those glasses. Applications and processes are discussed more thoroughly in later sections of the chapter.

8.1.2 Soft Glasses

8.1.2.1 Soda-lime-silica glass

Brief history. Soda-lime-silica glass, sometimes called simply *soda-lime* glass, is the generic name for a family of glass compositions based

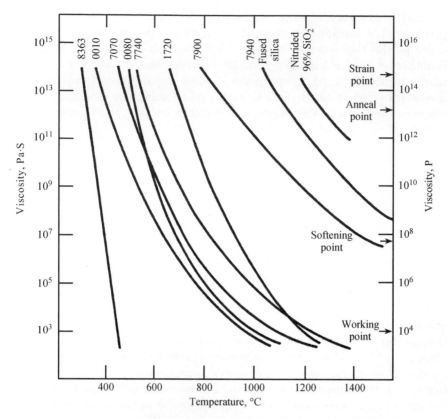

Figure 8.2 Viscosity vs. temperature for some commercial glasses designated by glass codes (see Tables 8.1 and 8.2). *(Courtesy of Corning Inc.)*

on the network former silica (SiO_2) and modified by *soda* (Na_2O) and *lime* (CaO). It is the oldest commercial type of glass and has roots with the earliest commercial glasses developed in the Middle East thousands of years B.C. Silica is too refractory a material to be melted and formed as a glass by wood or fossil fuel fires. However, it was discovered that additions of soda ash (Na_2CO_3) and limestone ($CaCO_3$) to silica sand (crystalline quartz) acted as fluxes to help melt the sand over simply fueled fires. The resulting melt was sufficiently fluid that it could be *worked* to form useful objects.

The relative ease of discovery and manufacture are key to the almost universal use of soda-lime-silica glass products. For example, sand is plentiful and at the outset contained sufficient calcium oxide as impurities, often as calcium carbonate from shell fragments, that when combined with soda and potash from wood and plant fires allowed the discovery of good stable glasses that could be melted over

TABLE 8.1 Glass Compositions Listed by Glass Type, wt%

Glass identification	Description/use	SiO$_2$	Al$_2$O$_3$	B$_2$O$_3$	Li$_2$O	Na$_2$O	K$_2$O	MgO	CaO	SrO	BaO	PbO	ZnO	Other	References
Soda-lime glasses															
Typical generic	Multipurpose	73	2			15			10						Boyd et al.[9]
Corning 0070	Elect./lamp tubing	71	3			13	1	5	7						Boyd et al.[9]
Corning 0080	Lamp bulbs	73	1			17		4	5						Boyd et al.[9]
Corning 0091	Elect./lamp bulbs	73.5	1			15	0.5	4	6						Boyd et al.[9]
Typical container	Green tinted	73	0.5			16			9.3					Fe$_2$O$_3$ = 0.5; Cr$_2$O$_3$ = 0.2; SO$_2$ = 0.5	
Typical flat glass	Clear float	73.1	0.1			13.7	0.1	3.8	8.9					Fe$_2$O$_3$ = 0.5; TiO$_2$ = 0.5; F = 0.2	
Lead glasses															
Corning 0010	Lamp tubing	63	1			8	6					22			Boyd et al.[9]
Corning 0120	Lamp tubing	56	2			4	9					29			Boyd et al.[9]
Corning 8160	Electron tubes (discontinued)	56	2			3	10		1		5	23		Some F	Hutchins et al.[9a]
Corning 8161	Electron tubes (discontinued)	40					5				2	51		Rb$_2$O = 2	Hutchins et al.[9a]
Corning 8363	Radiation shielding	5	3	10	1							82			Hutchins et al.[9a]
Corning 8871	Capacitors	42				2	6					49			Hutchins et al.[9a]
Lead sealing															
Corning 1990	To iron	41			2	5	12					40			Boyd et al.[9]
Corning 7570	Solder sealing	3	11	12								74			Boyd et al.[9]

TABLE 8.1 Glass Compositions Listed by Glass Type, wt% (Continued)

Glass identification	Description/use	SiO$_2$	Al$_2$O$_3$	B$_2$O$_3$	Li$_2$O	Na$_2$O	K$_2$O	MgO	CaO	SrO	BaO	PbO	ZnO	Other	References
Lead TV glasses															
Corning 0120	Stem tubing	56	2			4	9					29		Sb$_2$O$_3$ = 0.5	Connelly et al. in Schneider[7]
Corning 0137	Neck tubing	52.5	1			0.8	12			5		28		Sb$_2$O$_3$ = 0.5	Connelly et al.[7]
Corning 0138	Funnel	54	2			6	8	2.5	3.5			23		Sb$_2$O$_3$ = 0.1	Connelly et al.[7]
Corning 7580	Sealing frit	2	0.1	8.4							1.9	75.4	12.2		Connelly et al.[7]
Borosilicate glasses															
Corning 0211	Microsheet	65	2	9		7	7						7	TiO$_2$ = 3	Boyd et al.[9]
Corning 7070	Low-loss electrical	72	1	25	0.5	0.5	1								Boyd et al.[9]
Corning 7251	Sealed beam lamps	82	2	12		4									Boyd et al.[9]
Corning 7740	General purpose	81	2	13		4									Boyd et al.[9]
Kimble KG33	Laboratory ware	81	2	13		4									Boyd et al.[9]
Corning 7720	Tungsten sealing	74	1	15		4						6			Boyd et al.[9]
Corning 7760	General purpose	78	2	15		3	1							As$_2$O$_3$ = 1	Boyd et al.[9]
Corning 9741	UV transmitting	65	5	27	1	2								Some F	Boyd et al.[9]
Aluminoborosilicates															
Corning 7059	Electronic display substrates	49	10	15							25			As$_2$O$_3$ = 1	Boyd et al.[9]

Glass identification	Description/use	SiO2	Al2O3	B2O3	Li2O	Na2O	K2O	MgO	CaO	SrO	BaO	PbO	ZnO	Other	References
Aluminosilicates															
Corning 1720	Ignition tubes	62	17	5		1		7	8						Boyd et al.[9]
Corning 1723	Electron tubes, spacecraft windows	57	16	4				7	10	6					Boyd et al.[9]
Corning 6720	Tableware (opal)	60	10	1		8	2		5				10	F=4	Boyd et al.[9]
Corning 6750	Lighting ware	61	11			15								F=3	Boyd et al.[9]
Corning 0317	Ion exchange strengthened	61.4	16.8			12.7	3.6	3.7	0.2		9			TiO2=0.8	Dumbaugh in Boyd[5]
E-type fiberglass	Continuous textile fiber—typical	54	14	10				4.5	17.5						Boyd et al.[9]
S-type fiberglass	High-strength continuous textile fiber	65	25					10							Aubourg and Wolf in Boyd[5]
High-silica glasses															
Typical fused quartz	Fused quartz	>99.9													
Typical synthetic	Synthetic fused silica	>99.9													
Corning 7940	Synthetic fused silica	>99.9												H2O=0.08	Corning prod. lit.
Corning 7980	Synthetic fused silica	>99.9												H2O=0.08	Corning prod. lit.
Corning 7913	Vycor brand 96% silica	96.5	0.5	3											Corning prod. lit.
Corning 7971	ULE brand (ultralow expansion)	93												TiO2=7	Corning prod. lit.

TABLE 8.2 Glass Physical Properties Listed by Glass Type

Glass identification	Description/use	Density, g/cm³	CTE, 0 to 300°C 10⁻⁷/°C	Strain point, °C	Annealing point, °C	Softening point, °C	Working point, °C	Young's modulus, GPa	Poisson's ratio	Refractive index	Resistivity at 350°C log(Ω·cm)	References
Soda-lime glasses												
Typical generic												
Corning 0070	Multipurpose Elect./lamp tubing	2.5	91	487	527	715				1.513		Boyd et al.[9]
Corning 0080	Lamp bulbs	2.47	93.5	473	514	696	1005	70	0.22	1.512	5.1	Boyd et al.[9]
Corning 0091	Elect./lamp tubing	2.48	91	485	523	705						Boyd et al.[9]
Lead glasses												
Corning 0010	Lamp tubing	2.79	93.5	395	435	628	985	62	0.21	1.54	7	Boyd et al.[9]
Corning 0120	Lamp tubing	3.05	89.5	395	435	630	985	59	0.22	1.56	8.1	Boyd et al.[9]
Corning 8160	Electron tubes (discontinued)	2.98	91	395	435	630	975			1.553	8.4	Hutchins et al.[9a]
Corning 8161	Electron tubes (discontinued)	4.00	90	400	435	600	860	54	0.24	1.659	9.9	Hutchins et al.[9a]
Corning 8363	Radiation shielding	6.22	104	300	315	380	460	51	0.27	1.97	7.5	Hutchins et al.[9a]
Corning 8871	Capacitors	3.84	102	350	385	525	785	58	0.26		8.8	Hutchins et al.[9a]
Lead sealing												
Corning 1990	To iron	3.5	124	340	370	500	756	58	0.25		7.7	Boyd et al.[9]
Corning 7570	Sealing frit	5.42	84	342	363	440	558	55	0.28	1.86	8.7	Boyd et al.[9]
Lead TV glasses												
Corning 0120	Stem tubing	3.05	89.5	395	435	630	986	59	0.22	1.56	8	Connelly et al. in in Schneider[7]
Corning 0137	Neck tubing	3.18	97	436	478	661	978	65	0.24	1.55	8.3	Connelly et al.[7]
Corning 0138	Funnel	2.98	97	435	474	654	965	69	0.23	1.565	7.6	Connelly et al.[7]
Corning 7580	Sealing frit	6.47	98	293	311	374	890		0.25	1.65	8.2	Connelly et al.[7]

Glass identification	Description/use	Density, g/cm³	CTE, 0 to 300°C 10^{-7}/°C	Strain point, °C	Annealing point, °C	Softening point, °C	Working point, °C	Young's modulus, GPa	Poisson's ratio	Refractive index	Resistivity at 350°C log(Ω·cm)	References
Borosilicate glasses												
Corning 0211	Microsheet	2.57	74	508	550	720	1008	74	0.21	1.52		Boyd et al.[9]
Corning 7070	Low-loss dielectric	2.13	32	456	496		1068	51	0.22	1.469	9.1	Boyd et al.[9]
Corning 7251	Sealed beam lamps	2.26	36.5	521	565	808	1192	64	0.19	1.476	6.5	Boyd et al.[9]
Corning 7740	General purpose	2.23	32.5	510	560	821	1252	63	0.2	1.474	6.6	Boyd et al.[9]
Kimble KG33	Laboratory ware	2.23	32	513	565	827	1240			1.47	6.6	Boyd et al.[9]
Corning 7720	Tungsten sealing	2.35	36	484	523	755	1146	63	0.2	1.487	7.3	Boyd et al.[9]
Corning 7760	General purpose	2.24	34	478	523	780	1198	62	0.2	1.473	7.7	Boyd et al.[9]
Corning 9741	UV transmitting	2.16	39.5	408	450	705	1161	49	0.23	1.468	7.6	Boyd et al.[9]
Aluminoborosilicates												
Corning 7059	Electronic display substrates	2.76	46	593	639	844	1160	68	0.28	1.53	11	Boyd et al.[9]
Aluminosilicates												
Corning 1720	Ignition tubes	2.52	42	667	712	915	1202	87	0.24	1.53	9.5	Boyd et al.[9]
Corning 1723	Electron tubes, spacecraft window	2.64	46	665	710	908	1168	86	0.24	1.547	11.3	Boyd et al.[9]
Corning 6720	Tableware (opal)	2.58	78.5	505	540	780	1023	70	0.21	1.507		Boyd et al.[9]
Corning 6750	Lighting ware	2.59	88	447	485	676	1040			1.513		Boyd et al.[9]
Corning 0317	Ion-exchange strengthenable	2.46	88	576	622	870		72	0.22	1.506	5.5 @ 250°C	Corning prod. lit.
E-type fiberglass	Typical	2.62	54	615	657	846		72		1.562		Auborg et al. in Boyd and MacDowell[5]
S-type fiberglass	Typical	2.50	16	760	810	970		89		1.525		Auborg et al.[5]
High-silica glasses												
Typical fused quartz	Ranges			1040–1108		1582–1813						
Typical synthetic	Ranges			987–1020	1075–1120	1585–1625						
Corning 7940	Synth. fused silica	2.202	5.6	990	1075	1585		73	0.17	1.459	10.5	Boyd et al.[9]
Corning 7980	Synth. fused silica	2.202	5.6	990	1075	1585		73	0.17	1.459	10.5	Corning prod. lit.
Corning 7913	Vycor® brand	2.18	7.5	890	1020	1530		68	0.19	1.458	8.1	Boyd et al.[9]
Corning 7971	ULE™ brand	2.21	0.5	890	1000	1500		68	0.17	1.484	10.1	Boyd et al.[9]

wood fires. The fluxing agent magnesia (MgO) was also present as a minor component in the ash. The relatively low cost (compared to most other commercial glass compositions) and the relatively low required melting temperatures (about 1150°C) combined with physical and chemical properties acceptable for a wide variety of applications continue to make it the major volume of glass manufactured today.

Most commercial soda-lime glass compositions are based on a eutectic composition in the Na_2O-CaO-SiO_2 phase diagram located at about 22% Na_2O, 5% CaO, and 73% SiO_2 by weight. The more useful compositions are higher in CaO and lower in Na_2O for several reasons, one of which is improved chemical durability. A typical composition is $15Na_2O \cdot 10CaO \cdot 2Al_2O_3 \cdot 73SiO_2$ (wt.%); the alumina (Al_2O_3) is present in part to help improve chemical durability and in part to decrease the tendency for crystallization. This composition lies in the thermodynamic stability field of the mineral *devitrite* ($Na_2O \cdot 3CaO \cdot 6SiO_2$). While the term *devitrification* is applied to all families of glasses to describe the process whereby the molten glass gradually crystallizes if held at temperatures somewhat below the liquidus, the name devitrite is reserved for devitrification in this particular soda-lime-silica composition field.

In commercial glass compositions, one often finds other alkali oxides substituted for some of the Na_2O and magnesia (MgO) substituted for some of the lime. This is in part to provide more desirable properties, including higher viscosities at the liquidus temperature, and sometimes, especially in the case of the magnesia, because of more economically available raw materials. Typical composition ranges for soda-lime-silica glasses and example commercial glass compositions are given in Table 8.1. Properties for those glasses are given in Table 8.2.

Chemistry and properties. This section, and similar sections in the discussions of other glass families, contains a synopsis of the types of chemical components used in typical glass compositions and the properties or performance characteristics that depend on them.

Glass former. SiO_2

Modifiers. Alkali and alkaline earth oxides

Intermediates. Alumina

Alkalis. They act as fluxes, making the glass easier to melt. The melt viscosity is greatly decreased relative to that of silica at all temperatures. This makes melting and forming easier. Hence, alkali-containing glass is more economical as a product. Among adverse effects are lower use temperatures (deformation temperatures), increased thermal expansion coefficient [about 5 points ($5 \times 10^{-7}/°C$) per 1% soda], which is frequently undesirable, and greatly decreased chemical durability (attack by aqueous solutions). Furthermore, since the alkali ions are somewhat mobile within the glass network, the glass becomes more electrically (ionically) conductive, especially as temperature increases, thus decreasing its effectiveness as an insulator. (This, however, can be an advantage in electrically boosted melting).

- *Soda, Na_2O (from soda ash and/or nitre).*   This is the most important alkali for glass making.

- *Lithia, Li_2O (from lithium carbonate).*   This is a better flux than soda, and provides slightly better durability, but it is much more expensive.

- *Potassia, K_2O (from potash and/or potassium nitrate).*   This has a larger ionic size, so it is less mobile and therefore better for electrical insulation properties, but not quite as good a flux as soda.

- *Alkaline earths.*   They also provide fluxing action (but not as good as the alkalis). They are significantly better than alkalis for chemical durability, and for dielectric properties (i.e., the electrical conductivity of alkaline earth ions is less than alkalis; they have a double ionic charge, thus are less mobile).

- *Lime, CaO (from limestone).*   This is the most important alkaline earth for glass making.

- *Magnesia, MgO.*   MgO is often present with CaO in natural batch materials, for example in *dolomite* [$MgCa(CO_3)_2$] or dolomitic limestone.

- *Alumina, Al_2O_3 (generally derived from feldspar or nepheline syenite).*   Alumina can be considered a network modifier, but often a network former, especially if present in amounts less than the alkalis. It improves chemical durability and generally decreases the tendency for crystallization (devitrification). Rarely is alumina included in the batch as the oxide, but rather as a compound containing two or more of the major glass components.

Key characteristics.   Soda-lime-silicates are generally useful glasses. Density is about 2.3 g/cm^3. Typically, they are hard (5–7 on the Mohs scale), stiff (Young's modulus about 70 GPa, 10^7 lb/in^2), and strong (design strength > 7 MPa, 1,000 lb/in^2). They are transparent, with refractive index about 1.51 to 1.53. They are electrical insulators with good dielectric strength. They have a dielectric constant of about 7 and a dielectric loss of about 7% at 1 MHz. They are chemically durable against water and acid and have fairly good weathering resistance.

However, they have high thermal expansion (~90 × 10^{-7}/°C) and only fair upper use temperature (softening point ~700°C; strain point ~475°C). They provide benchmarks against which all property improvements may be judged.

Advantages. It is a general, all-around glass; easy to melt; perhaps the least expensive and most widely used family of glasses

Disadvantages. Because of the high thermal expansion soda-lime-silica glasses are prone to thermal shock failure and because of their relative softness, have limited high temperature usefulness relative to the harder families of glasses to be discussed later. They are somewhat inferior electrically (have high dielectric loss).

Applications

- Architectural glazing (windows and spandrels)
- Automotive glazing (flat glass and reshaped glass)
- Lighting (household lamp and fluorescent tubing envelopes; tubing for neon lighting)
- Electrical (radio tube envelopes—a reemerging application)
- Containers (bottles, drinkware, tableware)
- Decorative and art
- Variations—tempered (thermal and chemical), coated, laminated

For each of these applications, one could list all the required glass properties to see why soda-lime glass is suited to the application. We will do this just for the lighting application. Some of the key glass requirements are that it:

- Transmit light and other wavelength radiation
- Be impervious to gas diffusion
- Be readily degassed in a vacuum
- Be hermetically sealable to other glass and metal parts
- Be relatively chemically durable
- Be able to withstand the high temperatures generated by the light source
- Have good dielectric properties
- Be easily fabricated into shapes suitable for the product application

Soda-lime glass satisfactorily meets all these requirements for incandescent and fluorescent lamp applications (but not for halogen cycle and arc lamp applications) and is inexpensive compared to higher temperature (harder) glasses. This explains its almost universal use in these applications. Glass compositions and properties of soda-lime-silica glasses used for some of the above applications are given in Tables 8.3 through 8.6.

Forming processes. Forming processes include pressing, blowing, casting, rolling, tube drawing, sheet drawing, floating on a tin bath. (These and other processes are described in Section 8.3.)

8.1.2.2 Lead silicate glasses

Brief history. The first commercial use of lead oxide as a major glass component was by George Ravenscroft in seventeenth century England. For many reasons, the supply of glass from continental Europe had become unreliable, as had the importing of plant ashes for soda. Ravenscroft developed high-quality potassia-lead-silica glasses. The glass was exceptionally brilliant in appearance. Ravenscroft called it *crystal* and kept secret the fact that lead was the key ingredient; so did his successor. The glass was also called *flint* glass, because the silica came from flint nodules found in the chalk deposits of southeast England. This source of silica was highly pure, a key factor in the clarity and brilliance of Ravenscroft's products. The term *flint glass* is still used today in two contexts: (1) for high-dispersion optical glasses, often containing lead oxide, and (2) clear (uncolored) beverage bottles, generally made of soda-lime silica glass containing no lead at all.

Chemistry and properties

Glass former. SiO_2

Modifiers. Alkali, alkaline earth, and lead oxide

Intermediates. Alumina

Lead oxide, PbO. At low molar compositions, PbO acts as a network modifier. It is a good flux. Like alkaline earth oxides, it has less adverse effect on thermal expansion or chemical durability than alkalis. It significantly increases the refractive index and elastic modulus. It increases the working range ("*longer*" glass, see Sec. 8.1.1). A typical composition (alkali lead silicate with intermediate lead concentration) is $30PbO. 9K_2O. 4Na_2O. 2Al_2O_3. 55SiO_2$ (wt.%). On a mole percent basis, this composition would look similar to a mixed alkali-lime-silica glass, with lead oxide substituted for calcium oxide.

TABLE 8.3 Glass Compositions for Laboratory, Electric Lighting, and Container Applications, wt%

Glass identification	Description	SiO$_2$	Al$_2$O$_3$	B$_2$O$_3$	Li$_2$O	Na$_2$O	K$_2$O	MgO	CaO	SrO	BaO	PbO	ZnO	Other	References
Laboratory glassware															
Borosilicate	Typical, low TCE	81	2	13		4									Bardhan et al. in Schneider[7]
	Corning 7740	81	2	13		4									Boyd et al.[9]
Aluminoborosilicate	Jena G20	75.7	5.1	6.9		6.2	1.2		1.3		3.6				Hutchins et al.[9a]
	O-I N-51a	74.7	5.6	9.6		6.4	0.5		0.9		2.2				Hutchins et al.[9a]
Soda-lime	Typical	72	6	11		7	1		1						Bardhan et al.[7]
	Typical	73	2			14		4	7						Bardhan et al.[7]
Zinc-titania cover glass	Typical	65	2	9		7	7						7	TiO$_2$ = 3	Bardhan et al.[7]
High silica	Corning 7913	96.5	0.5	3											Boyd et al.[9]
Electric lighting															
Soda-lime	Typical	72	2			16	1	4	3						Van Reine et al. in Boyd[5]
Lead	Typical	63	2			7	7					21			Van Reine et al.[7]
Borosilicate	Typical	78	2	15		5									Van Reine et al.[7]
Aluminosilicate	Typical	61	16	1		1		7	10		12				Van Reine et al.[7]
	Corning 1720	62	17	5				7	8						Schneider[7]
	Osram 742a	51.3	25.3	1				4.2	8.3		5.3			P$_2$O$_5$ = 4.6	Dumbaugh et al.[5]
	AEI C37	55.8	23	5.1					13		3.1				Dumbaugh et al.[5]
	GEC H.26	54	21	8					14		3				Dumbaugh et al.[5]
High silica	Corning 7913	96.5	0.5	3											Boyd et al.[9]
Fused silica/fused quartz	Typical	>99.9													Van Reine et al.[7]
Container															
Soda-lime	Typical	74.1	1.3			13.5	0.3		10.5	Tr.	Tr.				Vergano[7]
Borosilicate	Typical	80.5	2.2	12.9		3.8	0.4			Tr.					
Pharmaceutical	Corning 7800	72	6	11		7	1		1		2				Boyd et al.[9]

Tr = Trace, ≪1.0%

TABLE 8.4 Physical Properties of Glasses for Laboratory, Electric Lighting, and Container Applications

Glass identification	Description	Density, g/cm³	CTE, 0 to 300°C, $10^{-7}/°C$	Strain point, °C	Annealing point, °C	Softening point, °C	Working point, °C	Young's modulus, GPa	Refractive index, n	Resistivity 350°C, log (Ω·cm)	References
Laboratory glassware											
Borosilicate	Typical, low TCE	2.23	32.5	510	570	821	1252	63	1.474		Bardhan et al in Schneider[7]
	Corning 7740	2.23	32.5	510	560	821	1252	63	1.474	6.6	Boyd et al.[9]
Aluminoborosilicate	Jena G20	2.39		524	569	794	1190	71		6	Hutchins et al.[9a]
	O-I N51a	2.36	50	540	580	795	1175		1.49	5.6	Hutchins et al.[9a]
	Typical	2.36		533	576	795	1189		1.491		Bardhan et al.[7]
Soda-lime	Typical	2.4	89	511	545	724			1.515		Bardhan et al.[7]
Zinc-titania cover glass	Typical	2.57	74	508	550	720		74	1.52		Bardhan et al.[7]
High silica	Corning 7913	2.18	7.5	890	1020	1530	1015	68	1.458	8.1	Boyd et al.[9]
Electric lighting											
Soda-lime	Typical	2.5	94	490	520	700	1015	72		5	van Reine et al. in Schneider[7]
Lead	Typical	2.8	93	410	445	635	1000	61		7	van Reine et al.[7]
Borosilicate	Typical	2.3	40	520	570	800	1200	64		7	van Reine et al.[7]
Aluminosilicate	Typical	2.6	45	770	810	1025	1250	88		11	van Reine et al.[7]
	Corning 1720	2.52	42	667	712	915	1202	87	1.53	9.5	Dumbaugh et al. in Boyd[5]
	Osram 742a	2.62	31		760						Dumbaugh et al.[5]
	AEI C37	2.55	42.5		760						Dumbaugh et al.[5]
	GEC H.26		43								Dumbaugh et al.[5]
High silica	Corning 7913	2.18	7.5	890	1020	1530		68	1.458	8.1	Boyd et al.[9]
Fused silica/fused quartz	Typical	2.2	5.5	1070	1140	1670		73	1.459	10	van Reine et al.[7]
Container											
Soda-lime	Typical	(see laboratory glassware, above)									
Borosilicate	Typical, low TCE	(see laboratory glassware, above)									
Pharmaceutical	Corning 7800	2.36	50	533	576	795	1189		1.491	5.7	Boyd et al.[9]

TABLE 8.5 Glass Compositions for Houseware Applications, wt%

Glass identification	Description	SiO_2	Al_2O_3	B_2O_3	Li_2O	Na_2O	K_2O	MgO	CaO	SrO	BaO	PbO	ZnO	As_2O_3	Sb_2O_3	F^-	Other	References
Tableware—glass																		
Durand table	NaF opal	71.5	5.6			13					6.3					3.7		Fine in Schneider[7]
Durand Arcopal	Glass-glass opal	75.3	3.2	13.5		6.6			0.7							0.8		Fine[7]
Corning 6720	CaF_2 opal	58.9	10.5	1.4		8.4			5.8				8.7			4.2		Fine[7]
Durand soda-lime	Soda-lime	71.1	1.6			14.2	0.5	1.1	9.8									Fine[7]
Corning corelle—	Borosilicate (skin)	58	15	6		3	3	6	15							3		Fine[7]
a glass-glass laminate	CaF_2 opal (core)	64	6	5		3	3	1	15									
Tableware—glass-ceramic																		
Corning 9609 (Pyroceram)	Nephelene	43.3	29.8	5.5		14								0.9			$TiO_2 = 6.5$	Fine[7]
Corning 0308 (Pyroceram)	Potassium-richterite	67.1	1.8	0.3	0.8	3	4.8	14.3	4.7						0.2	3.5	$P_2O_5 = 1$	Fine[7]
Drinkware																		
Durand	Lead crystal	59.9				3.3	10.3		1.3			24.4						Fine[7]
Libbey	Soda-lime	72.6	1.3			13.5		4.3	6.8									Fine[7]
Schott	Alkaline earth silicate	68.4	0.2	0.8		10.6	5.5		10		4.5							Fine[7]
Rosenthal	Alkaline earth silicate	69.4	0.6			10.7	7.2		4.9		6							Fine[7]
Bormioli	Alkaline earth silicate	69.4	1.4	0.7		9.8	5.6	1.2	7.6		5.7							Fine[7]
Anchor/Libbey/ Durand/Pasabache	Soda-lime range	72–73.1	1.1–1.5			13.5–14.4	Tr.–0.28	1.0–3.4	8.5–11.3	0–0.2							$Fe_2O_3 < 0.03$; $SO_3 = 0.2$–0.3	McGaughey[12]
Ovenware—glass																		
Anchor Hocking	Tempered borosilicate	76.4	1.8	14.7		5.4			0.7					0.4				Fine[7]
Corning 0281	Tempered soda-lime	74	2	0.5		13.5			10						0.2			Fine[7]
Corning 7251	Tempered borosilicate	82	2	12		4												Fine[7]

Glass identification	Description	SiO₂	Al₂O₃	B₂O₃	Li₂O	Na₂O	K₂O	MgO	CaO	SrO	BaO	PbO	ZnO	As₂O₃	Sb₂O₃	F⁻	Other	References
Ovenware—glass-ceramic																		
Corning Ware	β-spodumene s.s.	69.7	17.8		2.8			2.6					1	0.6			$TiO_2 = 4.7$	Fine[7]
Corning Visions	β-quartz s.s.	68.8	19.2		2.7	0.2	0.1	1.8			0.8		1	0.8			$TiO_2 = 2.7$; $ZrO_2 = 1.8$	Fine[7]
Durand Arcoflam—Clear Line	β-quartz s.s.	66.1	20.2		3.9	0.4		1			1.8		1	1			$TiO_2 = 3.0$; $ZrO_2 = 1.7$	Fine[7]
Tableware (dinnerware) glazes																		
Semivitreous lead-free		59.09	13.53	4.3	0.51	1.81	3.92	0.96	11.76	4.12								Fine[7], Eppler in Boyd[5]
Vitreous lead-free		55.79	7.37	5.47		1.81	2.71	0.62	9.16	3.07	2.5		10.94				$ZrO_2 = 0.57$	Fine[7], Eppler in Boyd[5]
Leaded		55.88	9.57	6.04		3.06	1.72		7.65			16.08						Fine[7], Eppler[5]
Leaded, low-temperature (Cone 06)		42.45	7.04	8.93		2.46			3.09			35.3					$ZrO_2 = 0.72$	Fine[7], Eppler[5]

s.s. = Solid solution.

TABLE 8.6   Physical Properties of Glasses for Houseware Applications

Glass identification	Description	CTE 25–300°C, 10^{-7}/°C	Softening point, °C	References
Tableware—glass				
Durand table	NaF opal	75	715	Fine in Schneider[7]
Durand Arcopal	Glass-glass opal	44	764	Fine[7]
Corning 6720	CaF_2 opal	80	792	Fine[7]
Durand soda-lime	Soda-lime			Fine[7]
Corning Corelle—	Borosilicate (skin)	48	890	Fine[7]
glass-glass laminate	CaF_2 opal (core)	71		
Tableware—glass-ceramic				
Corning 9609 (Pyroceram)	Nephelene	95	NA (crystallized)	Fine[7]
Corning 0308 (Pyroceram)	Potassium-richterite	118	NA (crystallized)	Fine[7]
Drinkware				
Durand	Lead crystal	89	647	Fine[7]
Libbey	Soda-lime	86	736	Fine[7]
Schott	Alkaline earth silicate	98	695	Fine[7]
Rosenthal	Alkaline earth silicate	99	682	Fine[7]
Bormioli	Alkaline earth silicate	89	704	Fine[7]
Anchor/Libbey/ Durand/Pasabache	Soda-lime range			McGaughey[12]
Ovenware—glass				
Anchor Hocking	Tempered borosilicate			Fine[7]
Corning 0281	Tempered soda-lime	86		Fine[7]
Corning 7251	Tempered borosilicate	37		Fine[7]
Ovenware—glass-ceramic				
Corning Corning Ware	β-spodumene s.s.	4–20	NA (crystallized)	Fine[7]
Corning Visions	β-quartz s.s.	5–7		Fine[7]
Durand Arcoflam— Clear Line	β-quartz s.s.	1		Fine[7]

At high lead concentrations, say greater than 50% PbO by weight, lead can be a network former. This is especially true in binary lead silicates, borate glasses, and phosphate glasses, which will be discussed in later sections. Up to 70 wt.% lead oxide is used in high refractive index optical glasses and in radiation shielding windows. Typical composition ranges for lead silicate glasses and example commercial glass compositions are given in Table 8.1. Properties for those glasses are given in Table 8.2.

Key characteristics. The main characteristics of lead silicate glasses are high expansion, long working range, good dielectric properties, high refractive index in combination with high dispersion, and fairly good chemical durability.

Advantages. Glasses containing lead are generally easy to melt. Compared to soda-lime glasses of similar softening points, they have a longer working range, higher refractive index, better electrical resistivity, and lower dielectric loss, but they are more expensive than soda lime. Lead provides X-ray absorption in television cathode ray tube (CRT) bulbs.

Disadvantages. Lead oxide is considered to be a health and environmental concern. It is especially so in the glass-manufacturing environment where particles of lead oxide may be airborne. Once incorporated into a commercial glass product, the concerns are less unless the glass itself is ground into a fine powder such as during sawing, grinding and engraving operations.

Applications. Major uses include

- Glass tubing (e.g., neon sign tubing).
- Electric filament mount structures (incandescent and fluorescent lamps)
- Cathode ray tube (television bulb neck and funnel, radar screens)
- Optical/ophthalmic (refractive index adjustment)
- Solder glass and frits (for relatively low temperature joining or sealing of glass/metal, glass/ceramic and glass/glass; conductive, resistive, and dielectric pastes in electronic circuits; decorative enamels)
- Drinkware ("hollow ware")
- Decorative and art glass
- Radiation shielding (nuclear "hot" labs)

Glass compositions and properties of lead-silicate glasses used for some of the above applications are given in Tables 8.3 through 8.10. Two example applications are discussed below.

Cathode ray tube application. The construction of a color television bulb is illustrated in Fig. 8.3. Several different glass components are involved: panel, funnel, neck tubing, electron gun mount, and evacuation tube. Some key design requirements of the glasses are as follows:

- *Strength.* The bulbs are designed to withstand three atmospheres of external pressure.
- *X-ray absorption.* Assembled tubes must have sufficiently high X-ray absorption to meet federal standards. Lead is generally used in neck and funnel glass for good electrical properties as well as radia-

TABLE 8.7 TV Glass Compositions, wt%

Glass ID	Description	SiO_2	Al_2O_3	B_2O_3	Li_2O	Na_2O	K_2O	MgO	CaO	SrO	BaO	PbO	ZnO	As_2O_3	Sb_2O_3	TiO_2	ZrO_2	CeO_2	Other	Refs.
Lead TV glasses																				
Corning 0120	Stem tubing	56	2			4	9					29			0.5					Ref. 7
Corning 0137	Neck tubing	52.2	1			0.8	12					28			0.5					Ref. 7
Corning 0138	Funnel	54	2			6	8	2.5	3.5	5		23			0.1					Ref. 7
Corning 7580	Sealing frit	2	0.1	8.4							1.9	75.4	12.2							Ref. 7
TV panel glasses																				
Corning 9008	Panel, B&W	55.1	3.6		0.5	6.8	6.4				12	1.7		.02	0.4					Ref. 7
Corning 9039	Panel, proj.	57.5	1.7		1	6.3	5.8			8.7	15				0.4		3	0.7		Ref. 7
Corning 9061*	Panel, color	62	2			7	9	0.7	1.7	10.3	2.4	2.2		0.2	0.7	0.5		0.2	†	Ref. 7
Corning 9068*	"	62.9	2.1			6.99	8.9	0.85	1.79	10.31	2.37	2.31		0.16	0.38	0.52		0.21	F=0.29,†	Ref. 7
Ranges (1990)	"	59–65	1.3–2.3			7.0–8.5	6.6–8.9	0–1.5	0.1–2.8	8.0–10.3	2.0–8.7	0.05–2.8	0–1.0	0–0.3	0–0.6	0.3–0.5	1–2.6	0.2–0.5	F=0–0.3,† Nd_2O_5=0–1.00	
Typical	"	61.3	2		0.01	7.6	7.6		0.05	9.2	9.2		0.5		0.3	0.4	1.4	0.3	*	

*Obsolete

†Typical colorant levels for contrast enhancement are $Fe_2O_3 = 0.04$, $NiO = 0.012$, $CO_3O_4 = 0.002$, and $Cr_2O_3 = 0.0006$.

TABLE 8.8 TV Glass Properties, wt%

Glass ID	Description	Density g/cm³	CTE 0 to 300°C 10⁻⁷/°C	Strain point °C	Annealing point °C	Softening point °C	Working point °C	Young's modulus GPa	Poisson's ratio	Refractive index	Resistivity 350°C log(Ω·cm)	Absorption @0.06 nm	Refs.
Lead TV glasses													
Corning 0120	Stem tubing	3.05	89.5	395	435	630	986	59	0.22	1.56	8	75	Ref. 7
Corning 0137	Neck tubing	3.18	97	436	478	661	978		0.23	1.55	8.3	90	Ref. 7
Corning 0138	Funnel	2.98	97	435	474	654	965		0.24	1.565	7.6	62	Ref. 7
Corning 7580	Sealing frit	6.47	98	293	311	374	890		0.25	1.65	8.2	40	Ref. 7
TV panel glasses													
Corning 9008	Panel, B&W	2.64	89	406	444	646	1004		0.24	1.506	7.4	20	Ref. 7
Corning 9039	Panel, proj.	2.9	96.4	458	500	680	983		0.24	1.553	7.7	28	Ref. 7
Corning 9061*	Panel, color	2.7	99	460	501	689	1000		0.23	1.518	7.5	35	Ref. 7
Corning 9068*	"	2.696	98.5–99.5	455–470	510–525	687–707					>7		Ref. 7
Ranges (1990)	"	2.7–2.8	"	"	"	"					"	29.1	Ref. 7
Typical	"	2.8	102	"	"	"					"	~29	~29

*Obsolete

8.21

TABLE 8.9   Sealing and Solder Glass Compositions, wt%

Glass identification	Application	SiO$_2$	Al$_2$O$_3$	B$_2$O$_3$	Li$_2$O	Na$_2$O	K$_2$O	BaO	PbO	ZnO	Other	References
Sealing glasses												
Lead glasses												
Corning 1990	To iron	41			2	5	12		40			Boyd et al.[9]
Corning 7570	Solder sealing	3	11	12					74			Boyd et al.[9]
Corning 7575		X		X					X	X		Corning prod. lit.
Corning 7580	Sealing frit	2	0.1	8.4				1.9	75.4	12.2		Connelly et al. in Schneider[7]
Corning 7583		X		X					X	X		Corning prod. lit.
Corning 7593				X					X		TiO$_2$	Corning prod. lit.
Borosilicate glasses												
Corning 3320	To tungsten	76	2	15		4	2				U$_3$O$_8$ = 1	Boyd et al.[9]
Corning 7040	To Kovar	67	3	23		4	3					Boyd et al.[9]
Corning 7050	Miscellaneous	68	2	24		6						Boyd et al.[9]
Corning 7052	To Kovar	64	8	19	1	2	3	3			Some F	Boyd et al.[9]
Corning 7056	To Kovar	68	3	18	1	1	9	3				Boyd et al.[9]
Corning 7070	Low-loss (elect.)	72	1	25	0.5	0.5	1					Boyd et al.[9]
Corning 7720	To tungsten	74	1	15		4			6			Boyd et al.[9]
Corning 7723		X		X					X			Corning prod. lit.
Corning 8830	To Kovar	65	5	23		7						Boyd et al.[9]
Corning 7574		X		X						X		Corning prod. lit.

Note: X = components present but amounts not indicated by manufacturer.

TABLE 8.10 Physical Properties of Sealing and Solder Glasses

Glass identification	Application	Density, g/cm³	CTE, 0–300°C, 10⁻⁷/°C	CTE, 25°C to set point, 10⁻⁷/°C	Strain point, °C	Annealing point, °C	Softening point, °C	Working point, °C	Young's modulus, Gpa	Poisson's ratio	Refractive index	Resistivity, 350C log (Ω·cm)	References
Sealing glasses													
Lead glasses													
Corning 1990	To iron	3.5	124	136	340	370	500	756	58	0.25		7.7	Boyd et al.[9]
Corning 7570	Solder sealing	5.42	84	92	342	363	440	558	55	0.28	1.86	8.7	Boyd et al.[9]
Corning 7575		3.80	89				380		51	0.25		8.6*	Corning prod. lit.
Corning 7580	Sealing frit	6.47	98		293	311	374	890		0.25	1.65	8.2	Connelly et al. in Schneider[7]
Corning 7583		6.00	84				370					8.5*	Corning prod. lit.
Corning 7593													Corning prod. lit.
Borosilicate glasses													
Corning 3320	To tungsten	2.27	40	43	493	540	780	1171	65	0.19	1.481	7.1	Boyd et al.[9]
Corning 7040	To Kovar	2.24	47.5	54	449	490	702	1080	59	0.23	1.48	7.8	Boyd et al.[9]
Corning 7050	Miscellaneous	2.24	46	51	461	501	703	1027	60	0.22	1.479	6.8	Boyd et al.[9]
Corning 7052	To Kovar	2.27	46	53	436	480	712	1128	57	0.22	1.484	7.4	Boyd et al.[9]
Corning 7056	To Kovar	2.29	51.5	56	472	512	718	1058	64	0.21	1.487	8.3	Boyd et al.[9]
Corning 7070	Low-loss (elect.)	2.13	32	39	456	496		1068	51	0.22	1.469	9.1	Boyd et al.[9]
Corning 7720	To tungsten	2.35	36	43	484	523	755	1146	63	0.20	1.487	7.3	Boyd et al.[9]
Corning 7723		2.90	35		480	522	770				1.548	12.1*	Corning prod. lit.
Corning 8830	To Kovar	2.24	49.5		460	501	708	1042	56	0.22		6.3	Boyd et al.[9]
Corning 7574		3.78	42.2		537	560	644		63		1.668	13.7*	Corning prod. lit.

* at 250°C.

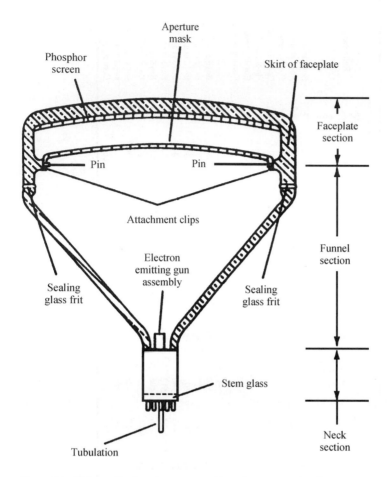

Figure 8.3 Schematic showing cross section of components of a conventional color television tube. The glass envelope consists of a funnel section, a faceplate section, and a neck section. The tubulation is used to evacuate the tube and is removed after vacuum processing. *(From Ref. 7, Fig. 1, p. 1039, reproduced with permission of publisher.)*

tion absorption (see Sec. 8.2.3.5); it is not used as much in the panel glass (screen), because it "browns" under irradiation.

- *Radiation damage.* The panel glass must not change color or "brown" under X-ray or electron irradiation. Since the presence of lead promotes electron browning, only small amounts of lead can be used in a panel glass. Cerium oxide is often added to the composition to help minimize the browning. Cerium ions act as electron traps that do not absorb light in the visible portion of the spectrum.

- *Electrical properties.* Properties include high bulk electrical resistivity (excellent insulator) and high dielectric strength to resist electrical puncture or discharge through the glass under high potential gradients (anode voltage can exceed 25 kV for color television).

- *Optical properties.*   The luminous transmittance and chromaticity of the screen is specified and varies among TV set designers and manufacturers (absorption of light is needed to improve viewing contrast); doping elements used for tinting include Ni, Co, Cr, and Fe.

- *Viscosity.*   It has a working range that allows forming into large sizes.

- *Thermal expansion.*   All parts (panel, funnel, neck, solder glass, and shadow mask metal alignment pins) must match fairly closely, 90 to $100 \times 10^{-7}/°C$.

Compositions and properties of lead-containing glasses useful for television bulb manufacture are given in Tables 8.7 and 8.8.

Drinkware ("hollow ware") and art glass applications of lead glass. Historically, the term *crystal* has been used for any clear glass of high transparency, especially when cut so that it has many light-reflecting surfaces, as in crystal chandeliers. We should note here that this use of the term *crystal* indicates nothing about the atomic level structure of the material. As is true for all the glasses in this section, the structure is amorphous, not crystalline.

ASTM defines crystal as "(1) colorless, highly transparent glass, which is frequently used for art or tableware. (2) colorless, highly transparent glass historically containing lead oxide." In Europe, *lead crystal* must contain at least 24% (wt.) PbO and *full lead crystal* at least 30%. Steuben™ glass has traditionally contained about 30% PbO.

Attributes.   Attributes of lead crystal ware include brilliance, clarity, sonority (acoustic resonance), density, meltability, and workability. Brilliance is related to the refractive index, transmittance, and the degree that surface polish is maintained. Sonority is related to the high elastic modulus and low internal friction (acoustic damping) of the glass, a function of the mixed alkalis present. Lead is not an absolute requirement for any of these attributes. A negative: high-lead glasses are more easily scratched or abraded. Most lead crystal manufacturers are developing non-lead compositions for at least some of their products.

Forming processes.   Forming processes include pressing, blowing, tube drawing, casting, centrifugal casting, and fritting but are not limited to these.

8.1.3   Hard Glasses

8.1.3.1   Borosilicate glasses

Brief history.   Boric oxide was used as a fluxing agent in glass compositions in Europe in the eighteenth century. Michael Faraday used it as a component in his pioneering optical glass developments in the mid-nineteenth century. Beginning in the 1880s, Otto Schott and Ernst Abbe developed boron-containing optical glasses including borate crowns, borosilicate crowns, and borate flints. (See Sec. 8.2.4 for definitions of the terms *crown* and *flint*.) Schott was perhaps the first to recognize the usefulness of boric oxide additions to silicate glasses for the purpose of reducing their thermal expansion, which led to Schott's development of thermal shock resistant borosilicate laboratory ware. However, it was Corning Glass Works' development of chemically durable, low-thermal-expansion borosilicate glasses during the first two decades of the twentieth century that opened the way for wide commercialization of borosilicate glasses. Between 1908 and 1912, Corning developed a soft, lead-containing, low-expansion borosilicate glass in response to the practical problem of thermal shock breakage of railroad lantern globes and lenses. The uncolored version, Nonex^{TM} (for *nonexpanding* glass), was also used for battery jars and even baking dishes. A harder, lead-free version, first introduced in 1915 as a heat-resistant glass under the trade name Pyrex^{TM}, soon became the standard for laboratory ware and oven-safe bakeware throughout the world.

Typical composition ranges for borosilicate glasses and example commercial glass compositions are given in Table 8.1. Properties for those glasses are given in Table 8.2.

Chemistry and properties

Glass formers. SiO_2 and B_2O_3 [B_2O_3 from hydrated sodium tetraborate (borax), boric acid, or other boron oxide containing compounds; rarely from anhydrous B_2O_3].

Modifiers.   Alkali, alkaline earth, and lead oxides.

Intermediates.   Alumina.

Boron oxide, B_2O_3.   This is a network former. Alone, it readily forms a soft glass, having low deformation temperature, high thermal expansion, and good electrical properties. Pure B_2O_3 glass is very hygroscopic, even *soluble* in water; it is usually used in combination with silica, SiO_2.

In borosilicate glasses, B_2O_3 acts as a flux for silica, and when combined with alkali oxides, the thermal expansion can be controlled to match various materials. Alkali borosilicates tend to phase-separate. Alumina helps to reduce this tendency. A typical composition (alkali borosilicate) is $81SiO_2$. $13B_2O_3$. $4Na_2O$. $2Al_2O_3$ (wt.%). This is close to that of Corning Code 7740 - used for Pyrex^{TM} brand products. It's approximate properties are listed in Table 8.2.

Key characteristics. Borosilicate glasses provide low expansion (can be tailored over a range by varying the ratio of triangularly coordinated boron to the tetrahedrally coordinated silicon), moderate hardness (higher deformation temperatures than soda-lime-silica glasses, several hundred degrees higher melting temperature), good electrical properties, and excellent chemical durability when the composition is not phase separated.

Advantages. Thermal expansion can be "tailored" over a wide range by suitable composition changes. This allows excellent thermal shock resistance and excellent chemical durability to be combined for a product application. It also allows for designing glasses that can be sealed directly to certain metals such as tungsten, molybdenum, and Kovar^{TM}. Many borosilicate glasses have higher use temperatures than do soda-lime-silica glasses. Borosilicates are perhaps the second most widely used family of glasses.

Disadvantages. They are more expensive than soda-lime glasses. This is primarily due to costs associated with melting at higher temperatures and the price of boron-containing raw materials.

Low-alkali, low-alumina borosilicate glasses are subject to phase separation when held for prolonged periods of time above their glass transformation range (say, between the annealing and softening points). This can lead to decreased chemical durability and sometimes to haziness or opacity of the glass. However, the manufacture of some commercial products (e.g., Vycor^{TM} brand 96% silica glass and controlled-pore-size porous glass), is based on the phase separation phenomenon. This will be discussed in Sec. 8.2.9.2.

Applications. Applications include chemical ware, pharmaceutical ware, cosmetic containers, housewares, optical and ophthalmic lenses, photochromic glass, lighting and electrical applications, telescope and other mirror substrates, solar energy systems, and flat-panel display substrates. Glass compositions and properties of borosilicate glasses used for some of the above applications are given in Tables 8.3 through 8.6, 8.9, and 8.10. Borosilicate glasses for fiberglass applications are discussed in Sec. 8.6.

Forming processes. Forming processes include casting, pressing, blowing, tube-drawing, rolling, and sheet-drawing but are not limited to these. Borosilicates are a very versatile family of glass.

8.1.3.2 Aluminosilicate glasses

Chemistry and properties

Glass formers. SiO_2, B_2O_3, and sometimes Al_2O_3.

Modifiers. Alkali, alkaline earth, and lead oxides.

Intermediates. Al_2O_3.

Alumina (aluminum oxide, Al_2O_3). This is a very refractory oxide, even more so than silica. It has a very high Al-O bond strength. When incorporated in the glass network, it makes the glass more refractory (*harder* in glassmakers' terms).

Aluminosilicates typically (not always) contain boron oxide but are termed *aluminosilicate* if they contain more alumina than boron oxide on a molar basis. Generally, they are the most refractory (high-temperature) of glasses containing alkali and alkaline earth modifiers. They have a relatively steep *(short)* viscosity-temperature curve giving high annealing and strain points but are still meltable in conventional furnaces. A typical composition (lime aluminosilicate): $58SiO_2 \cdot 20Al_2O_3 \cdot 16CaO \cdot 5B_2O_3 \cdot 1Na_2O$ (wt.%).

The *molar ratio of alumina to alkali modifiers* is key to the glass properties. At ratios less than 1, alumina tends to enter the glass network, replacing the NBOs (nonbridging oxygens) caused by the presence of the alkalis. Many glass properties are sensitive to this ratio, including density, viscosity, thermal expansion, internal friction, Knoop hardness, electrical conductivity, and ionic diffusion. For example, adding aluminum oxide to alkali silicate glasses increases the viscosity.

Composition ranges for aluminosilicate glasses and examples of commercial glass compositions are given in Table 8.1. Properties for those glasses are given in Table 8.2.

Key characteristics. They are very hard (high deformation temperature) glasses but still meltable at reasonable temperatures; have good dielectric properties, high elastic moduli, and good resistance to chemical attack (particularly to caustic alkalis); and are useful for high-temperature applications. The alkali aluminosilicates can be readily strengthened chemically by ion exchange (see Sec. 8.5.8).

Advantages. All characteristics listed above are advantages for some applications. Ready availability of alkali-free compositions is important for electronic applications where the presence of alkali would degrade silicon semiconductor device performance.

Disadvantages. The required higher melting temperatures and more costly batch materials make these glasses more expensive to produce relative to soda-lime-silica and some borosilicates.

Applications. Applications include electrical and electronic insulators and dielectrics, aircraft and space craft windows, stovetop cookware, high-temperature lamp envelopes, chemically strengthened *frangible* components, chemical apparatus, glass electrodes, and load-bearing members in fiberglass reinforced plastics (E-glass and S-glass). An important emerging area of application is as substrate glass for flat-panel/active matrix liquid crystal displays (AM-LCDs). Compositions and properties of aluminosilicate glasses used for some of the above applications are given in Tables 8.3 through 8.6, 8.9, and 8.10.

Alkali aluminosilicates are used for ion-selective electrodes, chemically strengthened (ion-exchanged) glass and nuclear waste fixation.

Chemical (ion exchange) strengthening of glass is discussed in Sec. 8.5.8. Examples of chemically strengthenable aluminosilicate glasses are Corning codes 0317, 0331, 0417, and PPG 947M. The composition of 0317 is listed in Table 8.11.

Calcium-based *alkaline earth aluminosilicates* are the most important commercially. Eutectic compositions in the calcia-alumina-silica system have been the basis for a number of commercial glasses. The eutectic at 1170°C ($62.1SiO_2 \cdot 14.6Al_2O_3 \cdot 23.3CaO$ wt.%) was the basis for the original OCF (Owens Corning Fiberglas) E-glass fiber, tungsten-halogen lamp envelopes, space shuttle window inner panes, and the cladding glass for Corning's Corelle^{TM} tableware. Some of the lower-silica eutectics ($31.7SiO_2 \cdot 29.1Al_2O_3 \cdot 27.1CaO$ wt.% and $7SiO_2 \cdot 43.4Al_2O_3 \cdot 49.6CaO$ wt.%) provided the basis for infrared transmitting glasses at 4 to 5 μm wavelengths.

Eutectic compositions are further modified by additions of other network modifiers and boron oxide. For high-temperature lamp applications, such as the tungsten-halogen lamp ampoule in automobile headlamps, the thermal expansion is matched to that of molybdenum, the electric lead wire metal.

Typical types are 1710 (once used for top-of-stove ware), 1720 (ignition furnace tubes), 1723 (spacecraft windows), and 1724 (halogen headlight inner bulb). Compositions and properties of these glasses are given in Tables 8.1 and 8.2. Aluminosilicate glasses for fiberglass applications (for example, E-glass and S-glass) are discussed in Sec. 8.6.

TABLE 8.11 Ion-Exchange-Strengthenable Glass Compositions wt%

Glass identification	Description	SiO2	Al2O3	B2O3	Li2O	Na2O	K2O	MgO	CaO	BaO	PbO	ZnO	Other	Typical exchange bath	References
Aluminosilicate	Corning 0317	61.4	16.8			12.7	3.6	3.7	0.2				$TiO_2 = 0.8$	KNO_3	Dumbaugh and Danielson in Boyd[5]
Borosilicate															
Photochromic base	Ophthalmic	56.5	6.2	18.2	1.8	4.1	5.7						$ZrO_2 = 5.0$ $TiO_2 = 2.1$	60% KNO_3/40% $NaNO_3$ (wt)	Borrelli and Seward in Schneider[7]
Photochromic base	Ophthalmic	55.9	9	16.2	2.6	1.9				6.7	5.1		$ZrO_2 = 2.3$ Other = 0.3	60% KNO_3/40% $NaNO_3$ (wt)	Kozlowski and Chase[13]
Photochromic base	Sheet glass	60.4	11.8	17.7	2.1	5.9	1.6				0.3		Other 0.2	60% KNO_3/40% $NaNO_3$ (wt)	Borrelli and Seward[7]
Pharmaceutical	Corning 7800	72	6	11		7	1	1	1	2				KNO_3	Boyd et al.[9]
Ophthalmic crown	Corning 8361	68.3	2			8	9.4		8.4			3.5	Other = 0.4	KNO_3	Kozlowski and Chase[13]

Liquid crystal display applications. Some key requirements for AM-LCD applications are as follows:

- Thin (< 1.0 mm)
- Flat (low warp and bow)
- Smooth surface (low roughness)
- Excellent dimensional tolerances
- High strain point (600–800°C)
- Low thermal shrinkage (often called *compaction*) during customer's processing
- "Zero" alkali (avoids contamination of silicon electronics; no need for barrier layers)
- Defect-free (interior and at the surface)
- Excellent chemical durability (resistant to etchants)
- Preferred thermal expansion match to silicon (<40 × 10^{-7}/°C)

Corning codes 7059 (alumino-borosilicate), 1729, 1733, 1737, 2000, and other manufacturers' glasses are applicable (See Table 8.12.)

Glass-ceramics. Lithium aluminosilicate and magnesium aluminosilicate glass compositions form the bases for several families of commercially important glass-ceramic products, as described in Sec. 8.2.7.

Forming processes. Forming processes include pressing, blowing, downdraw from a slot, "fusion" (Corning's overflow sheet drawing process, see Sec. 8.4.), mini-float, redraw, and others.

8.1.4 Fused Silica and High-Silica Glasses

Fused silica (fused quartz) is the glassy form of the chemical compound SiO_2. Its natural raw materials include quartz sand and the mineral quartzite. However, because of its very high viscosity and high volatility at the temperatures required to melt the raw materials (1723°C is the melting point of cristobalite, the high-temperature crystalline form of silica), silica cannot be melted to form good quality products by conventional, large-scale glass melting techniques. (Most commercial glasses are melted at temperatures corresponding to a viscosity of about 100 P; the viscosity of silica at 1725°C is about 10^7 P, five orders of magnitude greater.) Consequently, specialized small-scale melting techniques have been devised. Fused silica can also be prepared by high-temperature synthesis from silicon-containing halide or metal-organic precursors. These specialized manufacturing

TABLE 8.12 LCD Glasses, Designations and Properties

Glass identification	Description	Density, g/cm³	CTE 0 to 300°C, 10^{-7}/°C	Strain point, °C	Annealing point, °C	Softening point, °C	Working point, °C	Young's modulus, GPa	Poisson's ratio	Refractive index	Resistivity at 350°C log(Ω · cm)	References
Display manufacturing temperature <350C												
Corning 0211	Alkali-zinc borosilicate	2.57	74	508	550	720		74.5	0.22	1.523	6.7	Bocko & Whitney in Schneider[7]
Corning 7740	Soda borosilicate	2.23	32.5	510	560	821		62.7	0.20	1.473	6.6	Bocko & Whitney[7]
Asahi AS	Soda-lime silicate	2.49	81*	511	554	740		71.7	0.21	1.52	5.9	Bocko & Whitney[7]
Asahi AX	Borosilicate	2.42	49*	522	568	789		68.9	0.18	1.5		Bocko & Whitney[7]
NEG BLC	Alkali borosilicate	3.36	51†	535	575	775				1.493		Bocko & Whitney[7]
Display manufacturing temperature <450C												
Corning Code 7059	Barium aluminosilicate	2.76	46‡	593	639	844	1160	68	0.28	1.53	11	Boyd et al[9]
Hoya NA 45	Barium aluminosilicate	2.78	46	610	658	859		68.9	0.24	1.533		Bocko & Whitney[7]
Schott AF 45	Barium borosilicate	2.72	45	627	663	876		66	0.235	1.5276	11.5	Bocko & Whitney[7]
Asahi AN635	Alkaline-earth aluminosilicate	2.77	48	635				73				Nakao[14]
Corning Code 1733	Alkaline-earth boroaluminosilicate	2.49	37	640	689	928		66.9	0.235	1.516	11.4	Bocko & Whitney[7]
Display manufacturing temperature <550C												
NEG OA2	Alkaline-earth aluminosilicate	2.76	47†	650	700					1.54	12.5	NEG Product into sheet 11/96
NH Techno NA35	Alkaline-earth boroaluminosilicate	2.49	37‡	650	705			70.2	0.24	1.516		Hoya product info sheet, 1993
Hoya NA 40	Alkaline-earth zinc lead aluminosilicate	2.87	43‡	656				92.4	0.26	1.574		Bocko & Whitney[7]
Corning Code 1737	Alkaline-earth boroaluminosilicate	2.54	37.6	666	721	975	1312	69.9	0.22	1.5186	11.4	Corning info sheet 1/99
Corning Code 2000	Alkaline-earth aluminosilicate	2.37	31.8	666	722	985	1321	69.2	0.23	1.5068	10.5	Corning info sheet 8/00
Asahi AN100	Alkaline-earth boroaluminosilicate	2.51	38	670				77.4				Nakao[14]
Corning Code 1724	Alkaline-earth boroaluminosilicate	2.64	44	674	726	926		82.7		1.54	11.6	Bocko & Whitney[7]
Display manufacturing temperature <750C												
Corning Code 1729	Alkaline-earth aluminosilicate	2.56	35	799	855	1107		80.6	0.216	1.52	11	Bocko & Whitney[7]
Display manufacturing temperature <950C												
Corning Code 7940	Synthetic fused silica	2.2	5.6	990	1075	1585		72.4	0.16	1.458	10.7	Bocko & Whitney[7]

*50–200°C.
†30–380°C.
‡100–300°C.
§at 300°C.

techniques will be discussed in Sec. 8.2.9, as will techniques for manufacturing special high silica (>90 wt.%) glasses.

Key characteristics. It is the most refractory glass (annealing point >1000°C, depending on purity and thermal history), with very high thermal shock resistance (due to the low thermal expansion, 5.5×10^{-7}/°C), high chemical durability (water, acid, base resistance), high optical transparency, very broad spectral transmittance range (deep UV to near IR, approximately 170 to 3,500 nm wavelength), good radiation damage resistance, low dielectric constant, and low electrical loss tangent.

Transport properties of fused silica (for example, viscosity and electrical conductivity) are controlled by the purity of the silica, in part as a result of nonbridging oxygens (NBOs) introduced into the glass network by the impurities. Values for these properties in absolutely pure silica have probably never been measured. The reported values depend on the method of manufacture (as described in Sec. 8.2.9). Table 8.2 shows the range of viscosity-related properties reported for commercial fused silica glasses.

Advantages. Each of the key characteristics listed above give silica glass different advantages over other types of glass, depending on the application.

Disadvantages. One of its main advantages is also its main disadvantage. As described above, the refractoriness (high softening point) makes melting of the raw material, crystalline quartz, extremely difficult, specialized, and hence expensive. Furthermore, silica glass is not easily worked in a flame to change its shape. These difficulties cause silica products to be rather expensive compared to other glasses. Silica is used only when the applications warrant the cost.

Applications. Applications include lighting (high-intensity-discharge lamp envelopes), the semiconductor industry (crucibles, substrates, coatings and annealing furnace tubing, and "furniture"), optical (lenses, prisms, windows, mirrors, fiber optic, and photonic devices for communications), high-energy laser optics, spacecraft windows, and others (labware, furnace, and other tubing; specialty fiber and wool).

Some properties. See Sec. 8.2.9 for a further discussion of silica properties.

8.1.5   Borate, Phosphate, Aluminate, and Germanate Glasses

8.1.5.1   Borate glasses. As noted in Sec. 8.1.3.1, pure B_2O_3 glass is very hygroscopic, even *soluble* in water. The same is true for many binary alkali and alkaline earth borate glasses. The solubility of sodium borate glasses has been utilized to prevent rot in telephone poles and fence poles. When inserted in the wood below ground level, the glass slowly dissolves, producing a fungicidal solution. An older, perhaps archaic, application is the Lindeman glass (B_2O_3 + Li_2O + BeO), developed in the early days of radiography as an X-ray transmitting window.

Rare earth borate glasses of high refractive index (e.g., lanthanum borates) have unusually low dispersions, making them very useful optical glasses. These glasses are much more chemically durable. There are also useful chemically stable aluminum borate optical glasses.

Other commercial applications of boric oxide in glass include the lead and zinc borosilicate solder glasses discussed in Secs. 8.1.2.2 and 8.2.2, which are quite chemically durable. Boric oxide is also used in many fiberglass compositions, discussed in Sec. 8.6.

8.1.5.2 Phosphate glasses. P_2O_5 is *deliquescent* (dissolves in absorbed atmospheric moisture), so phosphate glasses are noted for their lack of durability. The structure of P_2O_5 is characterized by rings, chains, and sheets of PO_4 tetrahedra.

Some commercial applications. "Slow-release" glasses for treating mineral deficiencies in ruminant animals and as fertilizer. A general composition range for animal treatment is 28-50Na_2O · 0-28CaO · 0-28MgO · 28-50P_2O_5 (mol%) and 0.1–20 of required nutrient elements, e.g., cobalt, copper, selenium, and iodine.

Bioactive and *bioresorbable* glasses are a recent development. Since the middle of this century, biologically inert materials such as surgical stainless steel, certain organic polymers, and alumina ceramics have been used to repair or replace damaged body parts such as bones, bone joints, and teeth. Such materials, while stable and nontoxic in the body environment, are not truly inert. They are gradually encapsulated by a thin, nonadherent fibrous layer, which progressively loosens the implant and limits its useful lifetime. In the late 1960s, Hench and coworkers at the University of Florida discovered that glasses within a certain composition range of the soda-lime-phosphate-silica system developed mechanically strong chemical bonding to bone surface. Such glasses are now called *bioactive glasses,* some of which have been manufactured under the trademark Bioglass^{TM}. Phosphate glass compositions have also been discovered that bond to soft tissue. Applications include prostheses to replace bones in the middle ear and roots of teeth and materials to repair damaged or diseased jawbone. More recently, compositions have been developed that are gradually absorbed by the body as they catalyze the regeneration of the bone they replaced. These discoveries have opened up major new areas for biomedical materials science and engineering.

Possible emerging applications. Emerging applications include the following:

1. Low-temperature zinc-alkali-phosphate glasses for polymer glass melt-blends have been developed by Corning. Key characteristics are glass transitions near 325°C and water durability comparable to or exceeding common soda-lime glass. Most durable compositions lie in the orthophosphate and pyrophosphate regions.

2. Non-lead solder glasses for use in color TV tube assembly, based on SnO-ZnO-P_2O_5, have been patented, tested, and found suitable for the application. The cost is higher, so commercialization may be difficult unless absolutely required by environmental legislation.

Aluminophosphates. Al_2O_3 and P_2O_5 can combine to give an $AlPO_4$ structural unit isomorphous to $SiSiO_4 = 2(SiO_2)$. $AlPO_4$ alone does not produce a glass but, when combined with suitable modifiers, it yields good glasses (higher temperature capability and improved chemical durability compared to the straight phosphates).

Heat-absorbing glasses are made from iron-doped aluminophosphates. (In phosphate glasses, the iron ion absorption bands, located in the UV and IR regions, are much sharper than they are in silicate glasses. Hence, almost clear glasses containing several percent of iron are possible.) HF acid resistant glasses have been made from zinc aluminophosphates.

Laser host glasses are typically neodymium-doped aluminum phosphates. An example is $12Na_2O \cdot 10Al_2O_3 \cdot 6La_2O_3 \cdot 2Nd_2O_3 \cdot 70P_2O_5$ (mol%). An advantage over silicate-based glass is a high solubility for platinum. Glasses that are melted in platinum for purity dissolve some platinum during melting. Some of the dissolved platinum precipitates in silicate glasses upon cooling, causing the glass to fracture due to heat generated during the lasing action, rendering the silicate glasses less desirable.

Fluorophosphates. These are used as specialized optical glasses. For example, fluorophosphate glasses, such as those designated FK-5 or FK-50 by Schott, have very low optical dispersion.

Silicophosphates. These also are used as optical glasses. [Example: ophthalmic crowns, $20P_2O_5 \cdot 21SiO_2 \cdot 22Al_2O_3 \cdot 12B_2O_3$ (wt.%) plus alkali and alkaline earth oxides, n = 1.523.]

8.1.5.3 Aluminate glasses. Al_2O_3 does not form a glass (other than by vapor deposition processes). However, it can make a major contribution to the glass network structure as in the aluminosilicates (Sec. 8.1.3.2) and the aluminophosphate glasses discussed above. It can even be the primary network former and as such has found several commercial applications. Calcium aluminate glasses are used as IR (infrared) transmitting glasses, boroaluminates have been used as optical glasses and as non-discoloring envelopes for sodium vapor lamps, and calcium boroaluminates *(Cabal glasses)* have electrical resistivities greater than those of silica.

8.1.5.4 Germanate glasses.

Germanate glasses are useful as IR transmitting glasses. As such, they are of considerable interest in heat-seeking missile applications.

8.1.6 Nonoxide Glasses

Other commercial glass systems include fluoride based glasses; chalcogenide and chalcohalide glasses; the amorphous semiconductors, silicon and germanium; and glassy metals.

In fluoride glasses, fluorine rather than oxygen is the primary glass network anion. BeF_2 (beryllium fluoride), alone and in combination with alkali fluorides, has sometimes been considered a low-temperature model for silica. Those glasses transmit even farther into the ultraviolet region and have lower refractive index and lower dispersion than silica but, because of health hazards associated with handling beryllium compounds and the difficulty of producing high-purity melts, these glasses have seen little commercialization. Heavy metal fluoride (HMF) glasses also transmit farther into the infrared region than silica and can be produced with sufficiently high purity that they have found optical applications, including optical fiber (mostly for short haul and sensors).

The chalcogen elements, sulphur (S), selenium (Se), and tellurium (Te), are elements from group 16 (previously called VIA) of the chemical periodic table. Sulphur and selenium themselves form glasses. All three, in combination with certain group 14 (IV) and group 15 (V) elements, such as arsenic (As) and antimony (Sb), form glasses over considerably broad composition ranges. When modified by adding halogens, the materials are known as *chalcohalides*. The major interest in these glasses is for their semiconducting, photoconducting, and IR-transmitting properties. The photoconductivity of amorphous selenium was the historical basis for the xerographic approach to photocopying. Although generally opaque to visible light, the IR transmitting capabilities, in some cases extending to 18 μm wavelength or more, are unique among glasses and have made them candidates for optic fiber transmission of high-intensity CO_2 laser light (10.6 μm wavelength) for laser-assisted surgery. (See Sec. 8.6.4, traditional fiber optics.)

Amorphous silicon, in thin-film form, is a widely used electronic semiconductor. Its excellent photovoltaic properties, grain-boundary-free structure, and relative ease of fabrication have made it widely used for solar energy conversion (solar cells) and as the thin-film transistor (TFT) switching elements in AM-LCD television screens and computer monitors, particularly portable units. However, for some AM-LCD applications, electronic properties of crystalline rather than amorphous silicon are needed, the requirement being achieved either during film deposition or by subsequent heat treatment steps. Chalcogenides and amorphous semiconductors have been considered for other electronic and electro-optic applications such as computer memories.

Glassy metals, which are essentially metals, metal alloys, or metals in combination with metalloid elements, having a glass-like atomic structural arrangement, are generally prepared by extremely rapid quenching (10^5 to 10^8 °C/s) from the molten state. While they are of commercial value for their significantly enhanced electrical, magnetic, and structural strength aspects, a full discussion of these materials is beyond the scope of this chapter.

Oxyhalide, oxynitride, and oxycarbide glasses have also been made and studied. The high anionic electrical conductivity observed for some oxyhalides has raised interest in them as possible solid electrolytes.

For the glasses described in this section (8.1.6), the glass manufacturing processes described in Sec. 8.3 (glass melting), and many described in Sec. 8.4 (glass forming), will not be applicable.

8.2   Special Glasses

8.2.1   Introduction

In this section, we group together some glasses that are somewhat special, either because of their rather unique chemical, physical, or optical properties, or because they have been developed for specific applications requiring special combinations of properties.

8.2.2   Sealing Glasses and Solder Glasses

Sealing glasses, as the name implies, are used to form a seal with (or to) another material, such as electrical wires entering a glass light bulb envelope. Typically, the seal must be mechanically strong and hermetic, requiring that the thermal expansions of the glass and the material being sealed to match over the temperature range between the *set-point* of the glass and the use temperature of the seal, which is generally room temperature. Because some mismatch in thermal contraction between the sealed components from the set-point to room temperature and/or the use temperature is inevitable, the process of sealing leaves both components in a stressed condition. A risk of glass fracture exists when tensile stresses so produced in one of the glass components exceed the engineering strength of the glass. An often-used design criterion is to ensure that the mismatch is no greater than 200×10^{-6} cm/cm ($\Delta L/L$). Modification of these stresses by a suitable annealing schedule may be helpful (see Sec. 8.5.4). Be it glass-to-metal or glass-to-glass, if a sufficiently close match in expansion is not directly possible, a graded seal utilizing a zone containing one or more glass layers of intermediate thermal expansion is sometimes used.

When two dissimilar glasses or other materials are joined using a layer of glass, that sealing glass is sometimes called a *solder glass.* However, the term is generally reserved for sealing glasses used at relatively low temperatures when parts that are being sealed or encapsulated would be damaged if subjected to higher temperatures. Solder glasses are used for IC (integrated circuit) packaging, color TV picture tube assembly and other mechanical seals, coatings, and wire feed-through seals.

Sealing glasses are available in two forms: vitreous and devitrifying. Vitreous sealing glasses are thermoplastic materials (glasses) that soften and flow at the same temperatures each time they are processed. A seal may sometimes be undone (unsoldered) by heating it to temperatures at or slightly above those used to make the seal. Devitrifying sealing glasses are *thermosetting* materials that crystallize according to a designed time-temperature relationship. Due to the crystalline nature, the devitrified sealing glass has a thermal and often chemical stability greater than that of the parent glass. Seals using devitrifying glasses cannot be undone by reheating to their sealing temperature. Devitrifying sealing glasses are a type of the glass-ceramic materials that are discussed more fully in Sec. 8.2.7.

Vitreous sealing glasses are available with softening points between 330 and 770°C; coefficients of thermal expansion between 35 and 125 $\times 10^{-7}$/°C. For devitrifying sealing glasses, softening points range from 310 to 645°C; expansion from 42 to 100×10^{-7} /°C. Thermal expansion of sealing glasses is often modified by use of low-expansion fillers (fine particles of low-expansion crystals or glasses). Compositions and properties of some commercial sealing glasses are given in Tables 8.9 and 8.10.

Typically, sealing glasses are supplied as 100-mesh powder (pass through screen of 100 mesh/inch), called *frit,* which is mixed with an organic vehicle and applied like paint, paste, or slurry before firing.

Generally, lead-borate based glasses are used for low-sealing-temperature vitreous sealing glasses and lead-zinc-borates for the devitrifying type. Because of environmental and workplace health concerns, non-lead-based sealing glasses are being developed.

8.2.3 Colored and Opal Glasses

8.2.3.1 Overview and applications.   Glasses based on the glass formers SiO_2, B_2O_3, and P_2O_5 generally transmit light well across a broad spectral region extending from within the ultraviolet region to well within the near-infrared region. Thus, they are inherently clear, *water-white* materials. Certain impurities and intentionally added network modifying components can alter this situation, giving color, grayness, and sometimes dullness to the glass as an unintended and undesired consequence. Other constituents can be added to the glass to color or opacify it to produce specific effects for intended applications.

Generally, a glass is considered to be *colored* if it absorbs some portions of the visible spectrum more than it does others. For example, a glass that preferentially absorbs light in the blue and green regions of the spectrum is colored red.

For many applications, particularly the more scientific or technical, detailed knowledge of the full transmittance and reflectance spectra of the glass may be important. Such information can be obtained using UV, visible, and infrared spectrophotometers. For other applications, consistent color matching of different lots of product, often to a well defined target, is sufficient. Here, the science of colorimetry is important. The observed color of an object depends on the emission spectrum of the illuminating source, the transmittance and/or reflectance spectra of the material composing the object, and the spectral sensitivity of the eyes of the observer.

Applications include:

- Colored lenses and filters (e.g., traffic lights, sunglasses, photography, scientific applications)
- Color television panel glass (gray color for screen contrast enhancement)
- Decorative/art glass
- Bottles and other containers
- Drinkware, tableware, and cookware
- Automotive and architectural glazing

8.2.3.2 Glasses colored by transition metal and rare earth ions.   Probably the most common method of imparting color to glass is to include various multivalent ions, such as transition metal and lanthanide series metal ions, in the glass composition. The coloration results from light absorption that occurs when electrons are excited from one energy state to another, either within a single ion or between pairs of ions. The color depends on the wavelength dependence of these absorption processes. Colors resulting from some common coloring ions are shown in Table 8.13.

It must be emphasized that the electronic energy levels of a given ion, and consequently the color it imparts, are dependent on both the valence state of the ion and the static electric field at the ion produced by surrounding ions, generally oxygen ions. This electric field is often referred to as the *ligand field,* where *ligand* is another name for *near-neighbor.* For any coloring ion, the ligand fields depend on the internal atomic structure of the glass and consequently on its chemical composition and thermal history. Thus, colors produced by a given ion will vary from glass system to glass system. The colors will also vary somewhat depending on the size and valence of the network modifying ions used to make the glass. For example, progressively interchanging the alkali ions from lithium to sodium to potassium in a series of alkali borosilicate glasses containing nickel as a colorant will change the color from a straw yellow color due predominantly to Ni^{++} ions to a purple color due predominantly to Ni^+ ions. The Ni^{++}/Ni^+ equilibrium is shifted by the change in alkali ion size.

Coloration due to chromium ions in certain borosilicate glasses illustrates the effect of thermal history on color. Chromium enters the glass structure in a Cr^{+3}/Cr^{+6} equilibrium. Cr^{+6} is tetrahedrally coordinated with oxygen, giving the glass a yellow color; Cr^{3+} is octahedrally coordinated, producing an emerald green color. Heat treatment of certain rapidly cooled borosilicate glasses at temperatures somewhat above their annealing point will shift the chromium ion equilibrium toward Cr^{3+}, giving the glass a more greenish color. Whether this is related to a glass-glass phase separation or merely to an overall change in oxygen coordination of the boron within the glass has been debated.

Zinc oxide is often used as a major component in colored glasses. Structurally, ZnO is an RO-type modifier oxide similar in behavior to PbO, whereby at high concentrations it acts more as a network former (like silica) than a modifier (like lime). Zinc-containing batch materials are generally more expensive than their calcium or lead counterparts; hence, it tends to be little used commercially. One exception is in the manufacture of colored glass—especially color filter glasses for scientific/technical use. For example, to get a good purple color using NiO as the coloring agent, the nickel must occupy a network-forming position rather than act as a network modifier. (As a network modifier,

TABLE 8.13   G ass Co orants

Color effect	Colorant	Typical additive
UV absorbing	Ce^{3+} and Ce^{4+} ions	CeO_2
	Ti^{4+} ion	TiO_2
	Fe^{2+} ion	Fe_2O_3 in P_2O_5 glasses
	Cu^{2+} ion	CuO
Blue/turquoise	Co^{2+} ion	Co_3O_4
	Cu^{2+} ion	$Cu^2 + CuO$
Purple/violet	Mn^{3+} ion	Mn_2O_3
	Ni^{2+} (tetrahedral)	NiO in potassium borosilicates
	Nd^{3+} ion	Nd_2O_3
	Ti^{3+} ion	
Green	Cr^{3+} ion	Cr_2O_3
	Fe^{3+} ion	$Fe_2O_3 + Cr_2O_3 + CuO$
	V^{5+} ion	V_2O_3
		CuO
	Pr^{3+} ion	
	Sm^{2+} ion	
Brown	Mn^{2+} ion	MnO (reducing conditions)
	Ion complex	$MnO + Fe_2O_3$
	Ion complex	$TiO_2 + Fe_2O_3$
	Ni^{2+} ion	NiO
		$MnO + CeO_2$
	Eu^{2+} ion	
Amber	FeS_x	Na_2S with Fe_2O_3
	Ionic complex	$Fe_2O_3 + TiO_2$
Pale yellow	V^{5+} ion	
	Cr^{6+} ion	
Yellow	Colloidal CdS	CdS
	$Ce^{4+} + Ti^{4+}$	$CeO_2 + TiO_2$
	Colloidal Ag	$AgNO_3 + SnO_2$
	U^{6+} ion	UO_3
Orange	Colloidal CdS:Se	
Red	Colloidal CdS:Se	
	Colloidal Au	$AuCl + SnO_2$
	Colloidal Cu	CuO
Pink	Er^{3+} ions	
	Se^{2-} ions	
Gray and black	Co^{2+} + other ions	$Co_3O_4 + Mn$, Ni, Fe, Cu and Cr oxides
	PbS, FeS, $CoSe_x$	Various sulfides and selenides
Near IR (heat) absorbing	Fe^{2+} ion	FeO

Source: Taken in part from "Glass," D. C. Boyd, P. S. Danielson, and D. A. Thompson, in *Kirk-Othmer Encyclopedia of Chemical Technology*, 4th ed., vol. 12, John Wiley & Sons, New York, 1994, p. 591.

Ni^{++} gives a yellow color, as discussed above.) When used in combination with relatively high concentrations of ZnO instead of CaO as the chemically stabilizing flux oxide, the nickel seems to enter the network positions along with the zinc. Similar ZnO-containing glasses are also used to for cobalt-containing blue filters and cadmium sulfoselenide containing sharp cut-off red and orange filter glasses. The latter will be discussed in Sec. 8.2.3.3 below.

Due to the presence of impurities in commercial glassmaking batch materials, iron is commonly present at sufficient concentrations to color the resulting glass. When the iron impurity is small (up to about 0.075%), a judicious control of the Fe2+/Fe3+ (redox control, see Sec. 8.3.2.2) can bring about decoloration. The two oxidation states of iron provide complementary colors. Selenium (1 to 5 ppm), which provides a pink color in the metallic form, often mixed with cobalt oxide (Se:CoO = about 1:3), nickel oxide, or cerium oxide, is added to bring about decolorization when larger amounts of the iron impurity are present in the batch. (Older use of MnO_2 has now been abandoned because of Mn solarization.) Decolorization produces a more neutral color but somewhat less brilliant glass because of overall transmission decrease in the glass. The more brilliant, higher-transmission glasses are made using iron-free raw materials.

8.2.3.3 Glasses colored by precipitated colloidal particles.

Glasses may also be colored by creating a dispersion of colloidal-sized particles of some light-absorbing pigment within the glass. Such colloidal colorants include semiconducting particles such as CuCl/CuBr and CdS/CdSe solid solutions and the noble metals copper, silver, and gold. The colors produced by these colorants are listed in Table 8.13. They tend to be relatively independent of the base glass structure and composition.

Colloidal-sized particles cannot simply be added to a glass batch with the expectation that they will survive the melting process. Rather, the chemical components are added to the batch and the particles generated within the glass melt by nucleation and growth processes. Sometimes the precipitation is controlled by rapidly cooling the melt to a rigid state without particle precipitation occurring, followed by a reheating to temperatures above the annealing point to nucleate and grow the particles. Occasionally, the precipitation is allowed to occur during the initial cooling of the melt. Such types of processing can result in very uniform dispersions of particles within the glass with consequent uniform coloration. The color often appears very rapidly during the cooling or reheating process step, leading to the terms *striking* or *striking-in* of the color. Sharp cut-off orange and red color filter glasses are produced in this manner by precipitating cadmium sulfo-selenide solid-solution particles; ruby red colors are often produced using gold or copper metal particles. Alkali-zinc borosilicate base glass composition have been found particularly useful for such filters, e.g., Corning code 2405 [$1Al_2O_3 \cdot 12\ B_2O_3 \cdot 5Na_2O \cdot 11ZnO \cdot 70SiO_2$ (wt.%)] with small amounts of CdS and Se added. An important application for colloidal-particle colored glasses is red and yellow traffic light lenses.

Colloidal metal coloration generally requires careful control of the overall oxidation state of the glass throughout the process. The melt must be maintained sufficiently oxidized at the melting temperatures to keep the metal dissolved in its oxidized state until the melt is cooled, then it must be made sufficiently reduced to allow precipitation of the metal particles. [The metal must transform from the +1 oxidation state to the neutral (metallic) oxidation state as the glass cools.] The oxidation state is generally controlled by incorporating multivalent noncoloring ions such as tin or antimony in the batch. As the glass cools, these ions become more oxidized at the expense of the noble metals which become reduced. (Such oxidation-state changes of antimony and tin are key to their roles as glass fining agents, as discussed in Sec. 8.3.2.)

8.2.3.4 Light-polarizing, dichroic glasses.

If a glass containing non-spherical, non-equiaxed light-absorbing particles of refractive index different from the glass matrix is illuminated by polarized white light, each of the particles will absorb light to a different degree, and generally with a different wavelength dependence, depending on the polarization orientation of the light. If most of the particles within the glass are aligned along a common axis, the glass itself will be a light polarizer. The glass will also be *dichroic*; it will show different colors, depending upon the polarization of the light with which it is viewed. The effectiveness of the polarizer will depend upon the effectiveness of the individual particles to absorb light of one polarization orientation in comparison to the other, upon the degree of their common alignment, and upon the concentration of the particles in the glass. Submicron-sized elongated particles of silver metal have been found most effective for this purpose. Elongated particles of copper-doped silver halide have provided glasses which are both photochromic and polarizing (see Sec. 8.2.5).

8.2.3.5 Radiation absorbing glasses

X-ray absorbing: In color television (CTV) picture tubes, electrons hitting the aperture mask or screen have about 25 keV of kinetic energy. Some of this energy is converted to X-rays, 0.5 Å or greater wavelength, as the electrons are scattered or absorbed. The X-rays can be effectively absorbed by the neck, funnel, and panel glasses of the picture tube by incorporating heavy metal (high atomic number) elements such as strontium, barium, zirconium, and lead into the compositions. While X-ray absorption generally increases with the atomic number of the element, the wavelength locations of the K and L absorption edges vary from element to element and must be taken into consideration when seeking the most effective elements for absorption of specific X-ray wavelength ranges. In Fig. 8.4, we show the mass absorption coefficient as a function of wavelength for these elements.

We note here that CTV neck and funnel glasses often contain between 20 and 30 wt.% PbO. Because of the tendency for electron browning mentioned above in Sec. 8.1.2.2, panel glasses should contain less than 5% PbO. Barium oxide (BaO), strontium oxide (SrO) and zirconium oxide (ZrO_2) provide the required magnitude of the X-ray absorption. Almost all American, European, and Asian manufacturers now use a similar lead-free base-glass composition (with sometimes differing levels of the coloring ions, depending on the CTV set manufacturers' specifications for luminous transmission). This greatly facilitates the use of recycled glass (post-consumer and in-house cullet) in manufacturing. Tables 8.7 and 8.8 show compositions and properties for commercial TV panel glasses.

Ultraviolet absorbing. UV-absorbing species such as cerium oxide (CeO_2) are used as glass composition additives to reduce the amount of UV radiation transmitted by the glass. Mixed semiconductor precipitates of CuCl/CuBr provide a sharp cut-off of short-wavelength

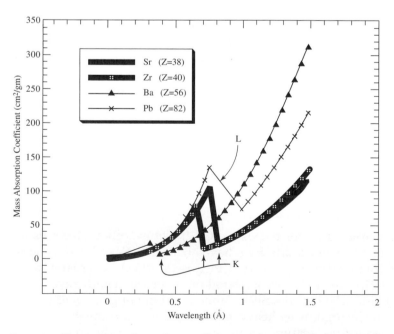

Figure 8.4 X-ray mass absorption coefficient vs. wavelength for several chemical elements. *(From data in International Critical Tables)*

radiation. The spectral cut-off wavelength region can be tailored by adjusting the chlorine/bromine ratio to provide good UV absorption while maintaining a clear, "white" appearance of the glass.

Heat (infrared) absorbing. Most silicate glasses are naturally opaque at wavelengths longer than ~4.5 microns, so they are inherently good absorbers for mid- and long-wavelength infrared heat radiation. Infrared-absorbing species, such as ferrous oxide (FeO), that absorb in the near-infrared spectral region are used to make so-called *heat absorbing glass.* Phosphate glasses are particularly effective, because the absorption of the ferrous ion in phosphate glasses is less strong in the visible region than it is in silicates, thus providing good near-IR absorption without adding significant coloration to the glass.

Hot cell windows. Large concentrations (high weight fractions) of PbO have been used to make radiation shielding glass for use in nuclear *hot cells,* primarily as viewing windows behind which mechanical and chemical operations are conducted with radioactive components. An example is Corning code 8363 glass ($3Al_2O_3 \cdot 10B_2O_3 \cdot 82PbO \cdot 5SiO_2$ wt.%) shown in Tables 8.1 and 8.2.

8.2.3.6 Opal glasses. Opal glasses are characterized by their milky appearance. They range from translucent (light is transmitted, but visual images are not) to fully opaque. The opal nature arises from light scattering due to the presence of inhomogeneities or inclusions within the glass. Generally, the inclusions themselves transmit light but are of different refractive index from the matrix glass. Thus, the inclusions scatter the light but absorb relatively little of it. If there is a high concentration of scatterers, or if the glass is very thick, it will be opaque due to most of the light being scattered back toward the source. Examples of such opalizing or opacifying agents are TiO_2, NaF, or CaF_2 crystallites, or glassy phase-separated particles such as sodium-borate glass particles in a sodium borosilicate glass matrix. If either the particulate phase or the matrix phase is colored, the opal product appears colored, although generally only with low color saturation that is pastel-like, because most of the incoming light is scattered from the glass before much wavelength-selective absorption can occur. (Examples of or CaF_2-based white opal glass compositions are the Corning code 6720 and Corelle body glass dinnerware opals shown in Tables 8.5 and 8.6.)

8.2.4 Optical Glass

8.2.4.1 Introduction. Broadly defined, an optical glass could be any glass used in an optical device, instrument, or system. However, as used by the optical glass industry (and in this section), the terminology refers to glasses designed for use in optical imaging systems such as microscopes, telescopes, and a wide variety of camera types. Properties important for optical glasses—in addition to light transmittance, refractive index, dispersion, and birefringence—are temperature dependence of the refractive index, chemical durability of the glass, and glass quality, especially homogeneity of refractive index and birefringence throughout the body of the glass. All these factors affect image quality. Often, the refractive index is specified to four or more decimal places and, for most applications, birefringence must be extremely low. These quality requirements require careful melting and annealing of the glass during manufacture.

To obtain the required ranges of refractive indices and dispersions, optical glass compositions sometimes contain heavy batch ingredients that tend to settle toward the bottom of the melt and volatile species that tend to evaporate from the surface of the melt. The melts often are corrosive, gradually dissolving (into the melt) the refractory container used for melting them. While these phenomena occur to some extent during the melting of all glasses (see Sec. 8.3), the strict homogeneity requirements of optical glass make its manufacture a greater challenge.

For a given glass composition, the refractive index of the glass depends on its density, which in turn depends on its thermal history, particularly how rapidly it has been cooled through its glass transformation range. Careful annealing of the glass is a must, especially when the refractive index must be held within tolerances to the fourth or higher decimal place. Much of what we understand today about the fine annealing and volume-temperature relationships of glass was learned because of the striving for higher-quality optical glass for sight telescopes and other military optics during the World Wars.

The design and manufacture of optical glass and the techniques for generating and polishing precision lens and mirror surfaces constitute some of the more technically sophisticated areas of glass science and technology. Many treatises and books have been written on those subjects, so the coverage in this chapter can only be superficial.

8.2.4.2 Brief history. Glass has been an important element for optical devices beginning with its first use in spectacle lenses to aid aging eyes in Italy, around the year 1280. Magnifying glasses came into use about the same time. It has been said that these two inventions or discoveries contributed to the widespread use of the printed word that followed the invention of the printing press in 1450. Galileo Galilei invented, or reinvented, the two-lens-element telescope in the summer of 1609. Advances in optical instruments depended on developments in optical glass light refracting properties and manufacturing quality, and on advances in the techniques for grinding and polishing the glass to the required lens shapes. Pierre-Louis Guinand (1803) pioneered mechanical stirring of melts for improved homogeneity. Michael Faraday (1820s) is credited with the development of glasses with different refractive indices and dispersions, and the use of platinum crucibles and stirrers for increased homogeneity. William Vernon Harcourt (1870s) explored the effects of many elements including phosphorus, boron, tin, and zinc on the properties of glasses. He melted in platinum and worked with combustion-gas-free atmospheres above the melts. Ernst Abbe and Otto Schott (1880s) are credited with the development of a wide variety of glass compositions and first attempts to scientifically define composition factors that affect glass properties. Corning Glass Works (1940s) developed continuous melting of optical glass in platinum lined melters incorporating platinum finers and stirrers. Annealing schedules were developed at the U.S. National Bureau of Standards and Corning Glass Works from the 1920s through the 1950s.

8.2.4.3 Applications of optical glass

Applications.   Lenses, prisms, windows (for instruments), and mirrors are typical major applications.

While glass remains the material of choice for most imaging applications, many other applications for glass in the fields of optics and optoelectronics have emerged during the past half century. Some have matured, others are growing, and yet others are just now emerging. Some of the currently most active developmental areas include UV-transmitting glasses, IR-transmitting glasses, laser glasses, ophthalmic glasses, and special glasses for atomic and nuclear technology applications.

UV-transmitting glasses.   The trend of the semiconductor industry (electronics) to more and therefore smaller features on a semiconductor computer chip (memory chips and microprocessors) has driven the designs to submicron feature sizes. The optical lithography techniques involved require imaging wavelengths in the near and deep UV regions. While pure borosilicate crown glasses and special fluorophosphate glasses have been applied to lens systems operating at wavelengths longer than 300 nm, only high-purity fused silica is capable of meeting all lens design needs at shorter wavelengths, specifically 248 and 193 nm, which are the emission wavelengths of the KrF and ArF pulsed excimer laser, respectively. Some fluoride glasses transmit at wavelengths shorter than 250 nm, but the percent transmittance of these glasses is not yet sufficient for these applications because of difficulties in manufacturing them at sufficient purity. To move to wavelengths shorter than about 185 nm will be a serious challenge for transmissive optical materials, glass or otherwise.

IR-transmitting glasses.   Glasses that transmit well in various spectral regions throughout the infrared are needed for thermography, pyrometry, IR spectroscopy, sensing, and a variety of specialized military applications. The wavelength limits for transmission depend, of course, on the thickness of the glass, since there are no sharp cut-off edges to the absorption spectra. However, one can say that, in general, pure silica has a long wavelength limit of about 4.5 μm. HMO (heavy metal oxide) glasses extend this into the 6 to 9 μm range, with refractive indices reaching 2.4. Heavy metal fluoride glasses are good to about 8 μm. Some chalcogenide glasses can transmit to about 25 μm, and have refractive indices as high as about 3.

Laser glass.   For some applications, glass has an advantage over crystals as a host for the *lasing* ions: It can be produced in large volumes and large sizes with high homogeneity and free of absorbing particles or other absorbing defects. The concentration of active ions in glass can often be greater than in crystals. Also, adverse nonlinear refractive index effects can be kept low in glass. Glass lasers are used in industrial applications, mostly for materials processing. Neodymium-doped glasses have been chosen for the large, multilaser systems being developed for inertial confinement nuclear fusion energy studies in the U.S. and Europe. Erbium-doped silica core glasses are used in optical fiber amplifiers (OFAs) operating in the 1.55 μm communications band, and other lasing core glasses are being developed for the 1.31 μm band.

Ophthalmic glasses.   The term *ophthalmic* here refers to glasses made for spectacles (eyeglasses) intended for vision correction, as opposed to nonprescription sunglasses. Trends in recent decades have been to "smart" photochromic glasses (as described in Sec. 8.2.5) and to lighter weight. The lighter weight has been achieved by using higher-refractive-index glass, which allows thinner lenses, a particular advantage for "strong" corrections, and by using glass compositions with lower mass densities. To further lower the mass density will be a continuing challenge because of the conflicting need to maintain high index, which itself requires the presence of heavy, highly electrically polarizable atoms. Ophthalmic glasses are discussed in more detail in Sec. 8.2.4.5.

Atomic and nuclear technology glasses.   In the atomic and nuclear technology area, special glasses are used as particle detectors, dosimeters, X-ray imaging screens, and radiation-absorbing windows that shield against nuclear and X-radiation.

8.2.4.4 Characteristics, properties and qualities. The key property of optical glass that relates to its refractive (light bending) ability is the refractive index. The higher the index, the more light is refracted for a given lens geometry. Complex lens designs call for glasses with a variety of refractive indices. With any optical material, the refractive index varies with the wavelength of light, the phenomenon known as *dispersion*. This means light of different wavelengths focuses differently through the same lens. To develop achromatic lens systems (i.e., ones that equally focus all wavelengths, or at least several widely spaced wavelengths), combinations of lenses of different refractive indices and complementary dispersive ability are needed. Much effort has been expended in developing a wide range of suitable optical glass compositions. An equally great effort went into developing methods for melting the glasses from raw materials that would ensure the homogeneity of properties over a volume large enough for the required lens. Since refractive index is a sensitive measure of density of the glass, careful annealing of optical glasses is needed to assure uniform density and refractive index.

Glass compositions and nomenclature. Glass composition is key to generating the variety of required refractive indices and dispersions needed by lens designers. Figure 8.5 shows the historical evolution of the range of refractive properties. Glasses at the upper right are alkali-lead-silicates and alkali-barium-lead silicates; glasses at the lower left tend to be fluorosilicates with high fluorine or P_2O_5 content. Fluoride-based glasses tend to have low index and low dispersion (high *Abbe number*, v_d). Heavy metal and rare earth lanthanum-borate glasses tend to have high index and medium to high dispersion (low to medium Abbe number).

Historically, optical glasses have been classified as either *flint* or *crown* according to the following criteria: All glasses with a refractive index n_d less than 1.60 and a v value of 55 or greater (low index, low dispersion) are called *crown*. Glasses with an index greater than 1.60 are also considered as crown, provided the v value is at least 50. All glasses with a v value less than 50 (high dispersion) are called *flint*. The line of demarcation between crown and flint glasses is shown in Fig. 8.5. Alphanumeric labels such as K3, or BK7 have for a long time been used to identify optical glasses; K stood for crown (die Krone or das Kronglas in German) and B for boron. Hence, BK7, a widely used glass, is a particular boron crown glass whose properties can be found in manufacturers' tables; these include $n_d = 1.517$ and $v_d = 64.2$. Another system of nomenclature more recently developed is based on the actual refractive index and dispersion of the glass. The scheme uses the first three digits after the decimal point in refractive index fol-

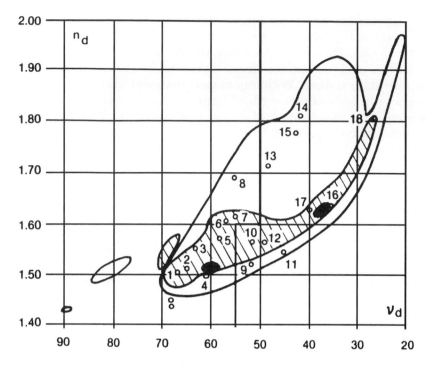

Figure 8.5   Representation of optical glasses in the $\{n_d, \nu_d\}$ plane. Glasses listed by numbers correspond approx. to nominal compositions given in Tables 8.5 and 8.6. 0 = fused silica. Solid dividing line separates crowns (left) and flints (right). Diagram also contains historically significant boundaries: solid areas = 1870, hatched area 1920, and solid boundaries only = 1984. *(From E.W. Deeg in Ref. 5)*

lowed by the first three digits of the Abbe value, ignoring the decimal points. So BK7 becomes 517642. Actually, the latter nomenclature provides a broader definition, since it specifies only index and dispersion; the former often specifies chemical composition and properties such as physical density and transmittance ranges. There are more than 750 different optical glass types, from five major manufacturers, listed in a recent compilation by Schott Glas of Germany (see Ref. 11). Some typical optical glass compositions are shown in Table 8.14.

The refractive index of glass changes with temperature, a factor that must be considered when designing lenses that operate over a range of temperature. Stress and strain also affect optical properties. Generally any nonhydrostatic stress develops *birefringence* in the glass, the refractive index for light of one polarization orientation being different from that of another. The proportionality coefficients for stress birefringence differ, depending on glass composition. Stress birefringence provides an optical method for determining stress within a body of glass, but the analysis can sometimes be quite complicated.

TABLE 8.14 Optical and Ophthalmic Glass Compositions, wt%

Generic type	Description	Fig. 6.5 designation	SiO_2	Al_2O_3	B_2O_3	Li_2O	Na_2O	K_2O	MgO	CaO	SrO	BaO	PbO	ZnO	TiO_2	ZrO_2	La_2O_3	Other	References
Optical crown glasses																			
PK	Phosphorous crown	1	68		14		8	8				1						•	Deg[5]
BK	Boron crown	2	72		12		4	8				1							Deg[5]
PSK	Dense phosphorus crown	3	55	2	12		5	4				22		2					Deg[5]
K	Crown	4	75	1			9	11		4									Deg[5]
BaK	Barium crown	5	48	1	4		1	7				29		9					Deg[5]
SK	Dense crown	6	41	2	5							42		9					Deg[5]
SSK	Extra-dense crown	7	37	2	6							40	4	8					Deg[5]
LaK	Lanthanum crown	8	6		40					17						6	30	†	Deg[5]
Optical flint glasses																			
KF	Crown flint	9	67				16						13	3					Deg[5]
BaLF	Barium light flint	10	54				2	9				14	11	10					Deg[5]
LLF	Extra-light flint	11	61				5	8					26						Deg[5]
BaF	Barium flint	12	56				2	13				12	17						Deg[5]
LaF	Lanthanum flint	13	4		32							8	6	12		6	19		Deg[5]
LaSF	Dense lanthanum flint	14	6		17					14		11	6	4		6	24	$WO_3 = 7$ $Ta_2O_5 = 8$ $CdO = 14$	Deg[5]
LF	Light flint	15	53				4	7					34						Deg[5]
F	Flint	16	44				2	7					46						Deg[5]
BaSF	Dense barium flint	17	42				1	7				11	34	5					Deg[5]
SF	Dense flint	18	27				1	1					71						Deg[5]

*Additional oxides found in crowns include P_2O_5, CeO_2, TiO_2, MgO, SrO, and Li_2O.
†Additional oxides found in optical flints include P_2O_5, Nb_2O_5, TiO_2, Al_2O_3, Sb_2O_3, and Li_2O.
‡Sb_2O_3 and As_2O_3 are often used in small quantities as fining agents and the halides CaF_2, KHF_2, NaCl, KBr, and NH_4Cl in small amounts as fluxes.

TABLE 8.14 Optical and Ophthalmic Glass Compositions wt% *(Continued)*

Generic type	Description	Fig. 6.5 designation	SiO_2	Al_2O_3	B_2O_3	Li_2O	Na_2O	K_2O	MgO	CaO	SrO	BaO	PbO	ZnO	TiO_2	ZrO_2	La_2O_3	Other	References
Ophthalmic crowns																			
Refractive index = 1.523 for all			67	2			7	11	<1	9				2	<1				Deeg[5]
			68	2			8	9		9				3	<1				Deeg[5]
			70				9	8		6				5	1				Deeg[5]
			56	9	16	3	2					7							Deeg[5]
			21	22	12	<1	2	7	<1	3		9	6		<1	2	2	$P_2O_5 = 20$	Deeg[5]
	Corning code 8361		68.3	2			8	9.4		8.4				3.5		2		Others total 0.4	Kozlowski & Chase[b]
	photochromic		56.46	6.19	18.15	1.81	4.08	5.72							2.07	4.99		Ag = 0.21; Cl = 0.17; Br = 0.14; CuO = 0.006	Borrelli & Seward[7]
Ophthalmic flints																			
Refractive index = 1.59			47	1	5		6	7		4		22		2	3	2			Deeg[5]
= 1.60			48		4		7	2		6	<1	13	14		4				Deeg[5]
= 1.62			45	1			3	8					40	1	1				Deeg[5]
= 1.66			39		2		6	1		6		19	16	4	6				Deeg[5]
= 1.68			34		6		8	<1		8	<1	14	18	7	5				Deeg[5]

Minimization of birefringence is another reason for careful annealing during manufacture of optical glasses (see Sec. 8.5).

Chemical durability of optical glasses is an important design factor. Some optical glasses, especially those containing large amounts of alkali oxides, P_2O_5, B_2O_3, or fluorides, have rather poor durability and can only be used in protected environments. Others, while generally considered chemically durable, will suffer gradual degradation of their optical surface properties in acidic or alkaline environments.

The optical materials discussed in this section are all passive or static materials; their properties normally remain constant in use. As mentioned in Sec. 8.2.4.1, modern optics has evolved to use many non-static, or active, properties developed in certain glasses. Important effects of this type include the ability to amplify light (laser glasses), the ability to change refractive index as a function of light intensity (non-linear optics), and the ability to change optical properties as a function of externally applied electric or magnetic fields. These properties and effects, important for optical fiber-based communication, are beyond the scope of this chapter and are not covered here.

8.2.4.5 Ophthalmic glass. *Ophthalmic glass* refers to glass used for prescription eyeglasses or *spectacles*. While often considered optical glass, ophthalmic glasses are not manufactured to tolerances as tight as those discussed above. Corrective lens prescriptions are not determined by the physician to less than 1/8 *diopter*, and often ±1/4 diopter differences cannot be perceived by the wearer.

Diopter, D, is a measure of the magnification power of a lens. It is the inverse of the focal length measured in meters. For a thin lens, it can be calculated as

$$D = \frac{1}{f} = \frac{n-1}{\dfrac{1}{R_1} - \dfrac{1}{R_2}}$$

where f = the focal length
$\quad\quad\quad n$ = the refractive index
$\quad\quad\quad R_i$ = radii of curvature of the lens surfaces

Homogeneity is important, but considerable gradual index variations across a lens can be tolerated, as evidenced by the popularity of the "progressive" type lenses that many wearers prefer to *multifocal* (bifocal and trifocal) lenses.

Refractive indices of ophthalmic lenses have been standardized at about six levels, with 1.523 being the lowest and most popular, and ranging to 1.9. The higher indices allow "stronger" prescriptions without requiring great differences between front and back lens surface curvatures and the consequent thick (and heavy) lenses. The glasses are generally manufactured to a three-decimal-place tolerance, but tighter tolerances are specified for fused multifocal *segment* glasses described below. Dispersion is generally not a very important consideration for ophthalmic lenses, but it ranges between 50 and 60 for the 1.523 index "white crown" glass. The composition of a typical chemically (ion exchange) strengthenable white crown ophthalmic glass is given in Table 8.14.

Multifocal eyeglasses may be made by generating (grinding) different curvatures into the upper and lower portions of the lenses. Alternatively, a glass of different composition and refractive index (the segment glass) can be fused into the lens blank (major glass) onto which a single lens curvature may be generated, the different index regions producing different magnifications at the same curvature. Design of such "fused" multifocal lenses requires matching of thermal expansion to prevent residual stress after sealing that could lead to debonding of the seal or fracture of the lens.

8.2.5 Photochromic and Polarizing Glasses

Photochromic glasses are glasses that darken, or decrease their transmittance, when exposed to light, particularly ultraviolet light or light in the violet and blue spectral regions. These glasses generally recover their high transmittance when removed from the darkening radiation, an effect often called *clearing* or *fading*. The speed of recovery depends on the particular glass design and on temperature; higher temperatures yield faster clearing. The degree of darkening is not linear with light intensity, which partly explains the limited commercial applications for this material other than as eyeglasses, including prescription and nonprescription sunglasses. For prescription eyewear, the photochromic glasses must meet all the requirements described above for ophthalmic glass (Sec. 8.2.4.5).

Commercial photochromic glasses depend for their behavior on many very small crystallites (~10 nm) of copper-doped silver chloride/bromide, uniformly dispersed throughout their volume. The components for the crystals are dissolved in the glass melt during manufacture and are precipitated as molten silver halide droplets by heat treatment of the resulting glass. The droplets crystallize as the glass is cooled to room temperature. The darkening results from a process similar to latent image formation in silver halide-based photographic emulsions. The light in a sense decomposes some of the silver halide to produce metallic silver, which absorbs visible light. Because each silver halide crystal is trapped in a small cavity within the glass, all the reaction products are available to recombine when the light source is removed. (This photochromic process is similar to that of latent image formation in silver-based photography but, in photography, some of the reaction products are lost in the organic emulsion, rendering the process irreversible. Photochromic plastics, on the other hand, rely on reversible photochemical reactions of organic dye molecules captured within their structure.)

The performance of photochromic glass strongly depends on the size, concentration, and composition of the silver halide particles dispersed within the glass. These factors in turn depend on the overall composition of the glass and the heat treatment used to precipitate the particles. The glass is often formed as a homogeneous glass, followed by a special heat treatment, which sometimes includes separate particle nucleation and growth steps at temperatures somewhat above the glass transition temperature (usually between the softening and annealing point temperatures). Borosilicate glasses are especially useful for preparing photochromic glasses, because they show a large difference between the high- and low-temperature solubility of silver chloride; this difference results from a change of boron-oxygen coordination with temperature. Borosilicate glasses with the required refractive indices for ophthalmic (prescription) eyeglasses have been developed, as have ion-exchange strengthenable glasses for both prescription eyewear and for sunglasses. Two such compositions are shown in Table 8.14.

Light-polarizing materials, in the context we use here and as introduced in Sec. 8.2.3.4, are materials that preferentially transmit light of one linear polarization compared to light of a different polarization. A sheet of such material, when rotated in a beam of linear polarized light will have an orientation of maximum transmittance and, at 90° from that, an orientation of minimum transmittance. The effectiveness of the polarizer can be characterized by the difference in these two transmittances; the greater the difference, the better the polarizer. (Or, if one considers optical absorption, the greater the ratio of the two absorbances, the more effective the polarizer.)

Commercially produced polarizing glass, developed by Corning Inc. and sold as Polarcor^{TM}, is made by stretching photochromic-type glass at a very high viscosity, using redraw techniques described in Sec. 8.4.5 so as to elongate and align the silver halide particles within the glass. The stretched glass is then treated in a hydrogen or forming gas (H_2-N_2) atmosphere at temperatures below the melting point of the silver halide crystals to chemically reduce the silver halide to silver metal. The polarizing efficiencies of these glasses tend to be greater in the near-infrared than in the visible portions of the spectrum and have found applications in photonic devices. In particular, they have become key components in optical isolators used in optical fiber communications.

Glasses containing precipitated particles of silver or other metals can also be deformed to produce polarizing glass; silver gives the best performance. In all cases, since the size of the precipitated particles are small, surface tension forces work to keep the particles spherical throughout the manufacturing process. Thus, it is necessary to redraw (or extrude) the glasses under conditions that provide viscous elongation forces sufficiently great to overcome the surface tension effects.

8.2.6 Photosensitive Glass

Photosensitive is a more general term than is *photochromic*. Photochromic glasses are certainly sensitive to light, but the term *photosensitive* is more commonly used to describe glasses in which a latent image can be produced in the glass by selective exposure to ultraviolet light, such as through a suitable photomask or other imaging system. Once the latent image is recorded, subsequent heat treatment can generate colored patterns or opal patterns. Commercial products made this way are the Corning Inc. Polychromatic^{TM} and Fotalite^{TM} glasses. Another photosensitive glass, in which the opal regions can be dissolved in mineral acids to make intricate mechanical parts, was developed by Corning and sold under the trade name Fotoform^{TM} but is now produced by other manufacturers. Some photosensitive glass compositions are also shown in Table 8.15.

Commercial photosensitive glasses typically have compositions near $17Na \cdot 7Al_2O_3 \cdot 5ZnO \cdot 71SiO_2$ (wt.%) with $0.2\ Sb_2O_3$, 0.01 to 0.1 Ag, and $0.05\ CeO_2$ added in excess of 100%. The cerium ions absorb the sensitizing ultraviolet radiation and release electrons to reduce the silver to the metallic state needed to grow the silver colloidal particles. The antimony oxide helps establish the appropriate oxidizing-reducing conditions in the glass. Exposure is generally done at room temperature, but heating at temperatures above about 500°C is required to grow (develop) the metallic colloids that color the glass. Gold, or gold in combination with palladium, can be substituted for the silver. Such photosensitive glasses produce monochrome photographic images.

Overall glass composition is important, because it controls the glass structure and the photosensitivity. Generally, a significant number of nonbridging oxygens (NBOs), such as found in the above soda-zinc-aluminosilicate glasses, must be present for effective photosensitivity. It has been suggested that these NBOs act as deep hole traps to prevent premature recombination of the photoelectrons with the cerium ions from which they were released. No commercially successful pho-

TABLE 8.15   Photosensitive Glass Compositions

Type	Description	SiO_2	Al_2O_3	Li_2O	Na_2O	K_2O	ZnO	SnO	Sb_2O_3	CeO_2	Other (as excess over 100%)	References
Typical photosensitive glasses												
Colored	Monochrome	71	7		17		5	0.05	0.2	0.05	Ag = 0.01 to 0.1	Borrelli & Seward in Schneider[7]
Opal	(White)	71	7		17		5	0.05	0.2	0.05	F = 2.5; Ag = 0.1	
Colored	Polychrome	71.6	7.1		16.3		5	0.05	0.2	0.05	F = 2.3; Ag = 0.1; Br = 1	
Photosensitive glass-ceramics												
Fotoform	Corning code 8603	79.6	4	9.3	1.6	4.1		0.003	0.4	0.014	Ag = 0.11; Au = 0.001	Beall in Boyd and MacDowell[5]

tosensitive glasses have been made in the borosilicate glass system, presumably because they lack sufficient concentrations of NBOs.

If fluorine is added to the above composition at about the 2 to 3% level, sodium fluoride crystals can nucleate and grow on the metal colloids to give the glass opacity in a pattern controlled by the initial exposure. Such glasses are referred to as *photo-opals*. One version was sold by Corning Glass Works under the trade name Fotalite[TM]. If, in addition, bromine is added, a second exposure and heat treatment can impart a wide variety of low-saturation colors to the glass, again in patterns controlled by the exposure. Such a glass, developed by Stookey, was briefly marketed by Corning Glass Works under the trade name Polychromatic[TM] glass.

Crystallization in certain glass-ceramics (see Sec. 8.2.7) can be controlled photosensitively to create patterns top close dimensional tolerance. In the Fotoform[TM] brand products briefly described above, lithium disilicate crystals grow on photosensitively generated colloidal silver nuclei. A typical Fotoform-type glass composition is discussed in Sec. 8.7.2.

8.2.7 Glass-Ceramics

Glass-ceramics can be described as polycrystalline materials formed by the controlled crystallization of glass. Internally nucleated glass-ceramics were discovered at Corning Glass Works in the late 1950s by Stookey. For the purposes of this section, we will define a glass-ceramic as a material or product that has been manufactured and/or formed as a glass, using typical, often highly automated, glass-forming techniques, and subsequently using a suitable heat treatment, caused to crystallize in a controlled manner. The resulting product consists of a fully dense (no pores or voids) ceramic body, often of a shape that cannot be easily obtained by normal ceramic processing techniques. Some definitions of *glass-ceramic* require that the final product be at least 50% crystalline; often, the percentage exceeds 90%. For our definition, we require only that the properties of the product be significantly determined or controlled by the crystals that are present. This being said, however, the overall glass composition is important for glass formation, workability, and control of nucleation, and the composition of the residual glassy phase after processing is important for chemical durability.

Because of the requirements of various glass-forming processes that the molten glass be stable against crystallization during the forming steps, not every ceramic composition can be formed as a glass. Perhaps the most notable example is alumina (Al_2O_3), which cannot be formed from the melt as a glass, even under conditions of extremely rapid quenching.

Glass-ceramics can also be formed by powder processing methods in which glass frits are sintered and crystallized. This procedure somewhat extends the range of possible glass-ceramic compositions. It also allows for surface as well as internal nucleation. The devitrifying solder glasses of Sec. 8.2.2 are examples of powder-processed glass-ceramics.

In general, efficient bulk (internal) nucleation is necessary for fine, uniform grain size in the final product. Nucleation can occur heterogeneously (on a nucleation catalyzing surface, such as another previously precipitated crystalline phase), or homogeneously (spontaneously throughout the volume of the glass). Homogeneous nucleation is sometimes preceded by a fine-scale glass-in-glass phase separation in which one of the separated phases is more unstable with respect to nucleation and growth of the desired crystalline phase than was the parent glass.

8.2.7.1 Thermal, Mechanical and Optical Properties. The unusual and, in some cases, unique thermal, mechanical, and optical properties of glass-ceramics have contributed much to their commercial success. Almost all of these properties relate to the characteristics of the crystalline phases and to their microstructural arrangement.

Low coefficient of thermal expansion and the consequent resistance to thermal shock arises when the thermal expansion of the crystalline phase is very low. In the lithium-alumina-silicate system described in Table 8.16, the crystals are either β-quartz or β-spodumene solid solutions that have very low or even negative volume thermal expansion. [Expansion is positive along one crystal axis (c) and negative along the other two (a and b).] These properties led to applications as varied as cookwear, stovetops and stove windows, and giant (~8 m dia.) Earth-based telescope mirror substrates.

The strength and fracture toughness of a glass-ceramic depends very much on the size and shape of the crystallites within the manufactured body. Crack propagation generally follows the intercrystalline grain boundaries, regions that often contain a glassy phase. If the crystallite size is small, say less than 100 nm, the cracks do not deviate much from a plane surface and propagate much as they would within a homogeneous glass. For much larger grain sizes, the cracks deviate around the crystals, giving the crack significantly greater surface area and requiring a higher energy for fracture. In glass-ceramics composed of three-dimensional network-type crystal structures (sometimes referred to as *framework-type* structures), such as the lithium-alumino-silicate spodumene and quartz solid solutions, the crystallites grow in an approximately equiaxed manner (all crystal growth directions having similar dimensions). However, growth of

TABLE 8.16   Glass-Ceramics by Primary Crystal Type

Network crystals (framework silicates)	Description
1. Lithium aluminosilicates (Li_2O-Al_2O_3-SiO_2)	- Based on stuffed β-quartz or β-spodumine (keatite) solid solutions - Three-dimensional crystalline phases - Nucleated by TiO_2 and ZrO_2 - Low thermal expansion (< 0 to $12 \times 10^{-7}/°C$) - Somewhat expensive because of lithium content
a. β-quartz isomorphs (solid solutions), sometimes referred to as "stuffed" β-quartz	- $(Li_2,R)O \cdot Al_2O_3 \cdot nSiO_2$, with n between 2 and 10 (commercially between 6 and 8) and R = divalent cations such as Mg^{2+} or Zn^{2+} - 3-dimensional, hexagonal crystal structure - $ZrTiO_4$ common nucleating agent - Unstable above 900°C, therefore heat-treated <900°C - Transparent material when crystal sizes <100 nm, sometimes <50 nm - Tends to be brownish in color because of nucleating agents - Can have very low thermal expansions, near zero, or even negative - High chemical durability - Applications include: transparent cookware (Visions™, thermal expansion = $7 \times 10^{-7}/°C$), optical materials (e.g., ring-laser gyroscopes), telescope mirror blanks (Zerodur™), infrared-transmitting electric range tops, wood-stove windows, fire door glazing
b. β-spodumene (keatite) solid solutions	- 3-dimensional, tetragonal crystal structure - $Li_2O \cdot Al_2O_3 \cdot nSiO_2$, with n between 4 and 10 - Generally nucleated by TiO_2 - Crystallized at temperatures between 1000°C and 1200°C - Opaque white (or tinted with colorants); opacity aided by crystalline anatase or rutile coming from the TiO_2 nucleating agent - Low in thermal expansion (about $12 \times 10^{-7}/°C$) - High chemical durability - Strong and tough; toughness aided by the larger crystal size (1–2 μm) and resulting tortuous paths for crack propagation - Applications: cookware (e.g., Corning Ware™, Neoceram), opaque cooking range tops, building wall cladding, laboratory bench tops, ceramic regenerators in turbine engines
2. Magnesium aluminosilicates (MgO-Al_2O_3-SiO_2)	- Based on hexagonal crystal, cordierite ($Mg_2Al_4Si_5O_{18}$) - Nucleated by TiO_2 - Other crystals present: spinel ($MgO \cdot Al_2O_3$), stuffed β-quartz, quartz, cristobalite - Excellent dielectric properties, transparent to radar - Good thermal stability and shock resistance, moderate thermal expansion ($45 \times 10^{-7}/°C$) - Applications: cooking ware, radomes (e.g., Corning code 9606)
3. Lithium silicates *a.* Photosensitivity nucleated	- Basis for Fotoform/Fotoceram™ "chemically machined" (etched) products

TABLE 8.16 Glass-Ceramics by Primary Crystal Type *(Continued)*

Network crystals (framework silicates)	Description
	- Nucleated by gold and/or silver nanoparticles, themselves photonucleated with the aid of UV-absorbing cerium (3⁺) ions Crystallized (at about 600°C) to dendritic form of lithium metasilicate (Li_2SiO_3), which is soluble in HF acid
	- Photonucleation plus heat treatment at higher temperatures (850°C) produces stable lithium disilicate ($Li_2Si_2O_5$) and α-quartz
	- Applications include: fluidics devices (fluid amplifiers), electronic components, magnetic recording head and inkjet printer head components, SMILE™ microlens arrays
b. Nucleated via glass-in-glass phase separation	- Additional components present to improve the chemical durability of the glass-ceramic via the glass matrix phase
	- Applications: dental prosthesis (inlays, onlays, crowns, bridges); example—IPS Empress 2 (Ivoclar)
4. Calcium silicates	- Based on blast furnace slag, chiefly in eastern Europe (referred to as slag-sitall in the former Soviet Union)
	- Crystal phases: diopside ($CaMgSi_2O_6$) and wollastonite ($CaSiO_3$), nucleated by sulfides of Zn, Mn, and Fe present in the slag
	- Inexpensive, high hardness (abrasion resistance), good chemical durability
	- Applications: interior and exterior wall cladding and tile
5. Sodium aluminosilicates	- Based on soda nephelene ($NaAlSiO_4$)
	- High in thermal expansion (about $95 \times 10^{-7}/°C$)
	- Can be glazed with lower expansion (e.g., $65 \times 10^{-7}/°C$) glaze to provide additional strength
	- Applications: institutional (hotels, restaurants, etc.) tableware (the original Pyroceram™ and Centura™ brand products)

Sheet silicates*	Description
1. Fluorophlogopite solid solutions	- Based on the trisilicic fluorophlogopite ($KMg_3AlSi_3O_{10}F_2$)
	- Other components (B_2O_3 and excess SiO_2) added to form the glass
	- Other minor crystals present: mullite ($3Al_2O_3 \cdot 2SiO_2$)
	- Tough, machinable (with conventional metalworking tools)
	- House of cards structure, only local mechanical damage
	- Applications: precision dielectric components, electrical insulators, high-vacuum components, other electronic and mechanical parts
	- Example: Macor™ (Corning code 9658)
2. Tetrasilicic fluormicas	- Based on the tetrasilicic fluormica $KMg_{2.5}Si_4O_{10}F_2$
	- Tough, translucent, very chemically durable
	- Application: dental restorations (inlays, crowns, etc.)
	- Cerium oxide added to simulate fluorescent character of natural teeth
	- Example: Dicor™ (Dentsply)

TABLE 8.16   Glass-Ceramics by Primary Crystal Type *(Continued)*

Chain silicates[†]	Description
1. Potassium fluorrichterite	- Principal phase: potassium fluorrichterite ($KNaCaMg_5Si_8O_{22}F_2$), an amphibole - Minor additions such as Al_2O_3, P_2O_5, Li_2O, and BaO in some combination with excess SiO_2 are needed to form the glass - Random acicular (rod-like) structure - Strong and tough (cracks must follow very tortuous path) - Commercial compositions also contain minor amounts of cristobalite - Thermal expansion $\approx 115 \times 10^{-7}/°C$ - Articles can be strengthened with a compressive glaze - Applications include high-performance institutional tableware (latest version of Pyroceram™) and cups for the Corelle™ line of tableware
2. Fluorcanasite	- Principal phase: $Ca_5K_{2-3}Na_{3-4}Si_{12}O_{30}F_2$ - Can be synthesized from glasses close to this stoichiometry - Structure characterized by four silicate chains running parallel to the *b* axis to form a tubular unit - CaF_2 crystallites act as nuclei - Very fine grained, strong, and tough - Thermal expansion about $125 \times 10^{-7}/°C$ - Potential applications: architectural building cladding, thin tableware, magnetic memory substrates

[*]Fluormica crystals composed of 2-dimensional tetrahedral-silica layers. The OH in the natural mica structure is replaced by F in these synthetic materials.

[†]Crystals in which single or higher order chains of silica tetrahedra form the structural backbone, like natural nephrite jade. The crystals tend to grow in acicular (needle-like) form in the glass.

sheet-silicate crystal structures, such as fluorine-substituted micas and clays, tends to be nonequiaxed, almost two-dimensional. Here, the internal microstructure of a glass-ceramic often resembles a "house of cards" structure. Fracture toughness for these materials is generally much greater than for homogeneous glass, as illustrated in Table 8.18. This is due in part to the extremely tortuous path the fracture surface must follow. This effect can be even more pronounced in glass-ceramics containing crystals of chain-silicate-type structure. The structure of these glass-ceramics resemble tightly interwoven arrays of acicular or fiber-like crystals. Glass-ceramics of the chain-silicate types have shown the highest body strengths and fracture toughness of any glass-ceramic. A limit to the advantage of the tortuous fracture surface along the crystallite boundaries is reached in each case at some crystallite size where it becomes easier (requires less energy) for the crack to propagate through the crystallites, rather than around them.

As discussed in Sec. 8.2.3.6, glasses tend to be cloudy or even opaque because of light scattering effects when they contain many small particles of refractive index different from that of the matrix glass. The smaller the size and concentration (number per unit volume) of the particles, the less the effect. Crystallites within glass-ceramics behave as such light-scattering particles. The difference between a very transparent manufactured body and an extremely opaque one of the same composition is often only a matter of crystallite size, as discussed below for the lithium-alumino-silicate glass-ceramic composition systems. Extremely bright "white" opaque bodies often result when crystalline phases of very high refractive index, such as zirconia and titania phases, are present. So the presence of titania and zirconia is important not only for their role as nucleating agents but also as opacifiers.

It has recently been found that, if the crystallite sizes are small, even with considerable difference in refractive indices between the crystals and the residual glass, transparency is often preserved even at very high concentrations of the crystalline phase. This would be surprising based on the mathematical descriptions of light scattering most often appearing in the literature, but those descriptions strictly apply only to materials in which the particles scatter independently of each other. For highly concentrated systems, different mathematics applies. In these cases, it can be shown that, for sufficiently small particle sizes, there is a range of concentrations for which the material behaves as a transparent body with a refractive index equal to an appropriately weighted average of the indices of the two phases. Materials of this type that contain microcrystalline host phases for lasing ions are finding interest for photonic devices.

Commercial compositions and applications. Many glass-ceramics have been produced in university and industrial laboratories, but only a few have found commercial applications so far. Those that have been applied can be grouped according to their type of crystal structure as shown in Table 8.16. Commercial applications for each type are listed in the table.

Applications. The wide variety of applications include electric range tops, woodstove windows, telescope mirrors, cooking utensils, dinnerware, building facing materials, radomes, precision electronic parts, fluid amplifiers, ink-jet printer heads, dental prostheses, and many others.

Manufacturing considerations. To obtain uniformity of the microstructure throughout the glass, objects are generally formed and cooled sufficiently rapidly to produce a homogeneous glass body. This body is then subjected to a controlled heat treatment at temperatures between the annealing and softening points. This heat treatment generally consists of two or more sequential steps at progressively higher temperatures, the first to generate a uniform dispersion of crystal nuclei throughout the glass, then a second at a higher temperature during which the final crystalline assemblage grows on these nuclei to achieve the desired crystal sizes. The composition of the nuclei may bear little resemblance to the major glass-ceramic phase. Sometimes the preferred nuclei are noble metal particles, sometimes very refractory materials like titania and zirconia, and sometimes very small particles of the final crystal phase itself. The latter is referred to as *homogeneous nucleation,* and the first two examples as *heterogeneous nucleation.* For some compositions and products, the growing crystals pass through a series of phases (differing in both crystal structure and composition) before reaching the final goal. Often, the temperature or sequence of temperatures chosen for the crystal growth stage must be high enough to maintain sufficient fluidity in the composite structure to relax any strains generated by volume changes during crystallization, but low enough that the product is sufficiently viscous to maintain its shape [viscosity = 10^{10} to 10^{12} P].

Forming methods.   Common forming methods include rolling, pressing, blowing, and casting. Some glass-ceramics are not suitable for forming processes such as sheet and tube drawing, because the required forming viscosities occur at temperatures below the liquidus temperature of the melt, leading to uncontrolled crystallization during the forming operations.

Dental and bioactive applications.   In addition to the fluormica and lithium disilicate materials shown in Tables 8.16 through 8.18, glass-ceramic dental restoratives based on the systems SiO_2-Al_2O_3-Na_2O-K_2O-CaO-P_2O_5-F (leucite and apatite phases), SiO_2-Li_2O-ZrO_2-P_2O_5, and SiO_2-Li_2O-ZnO-K_2O-P_2O_5 have been developed by Ivoclar in Liechtenstein. Others based on CaO-P_2O_5-Al_2O_3 have been developed in Japan (Kyushu Refractories Co.). See Table 8.18.

As discussed in Sec. 8.1.5.2, CaO and P_2O_5 are key components in bioactive glasses. Correspondingly, bioactive glass-ceramics have been developed. Ceravital^{TM} is a bioactive glass-ceramic containing crystalline apatite [$Ca_5(PO_4)_3(OH,F)$], Cerabone^{TM} A-W contains the crystals apatite and wollastonite ($CaSiO_3$), and Biovert^{TM} contains apatite and fluorphlogopite. All have been shown to bond with human bone; some are in clinical use as bone-repairing materials. Note that hydroxyapatite ($Ca_5(PO_4)_3OH$ is the major mineral constituent of bone.

Photonic applications.   Several areas of application of glass-ceramic materials are emerging in the photonic and optical communications fields. These include lasers and frequency up-conversion devices in which the optical advantages of fluorescent rare-earth ions in crystalline host materials are combined with manufacturing and structural

TABLE 8.17 Glass-Ceramic Composition, wt%

Glass identification	Description	SiO_2	Al_2O_3	B_2O_3	P_2O_5	Li_2O	Na_2O	K_2O	MgO	CaO	BaO	ZnO	TiO_2	ZrO_2	As_2O_3	F	Other	References
Aluminosilicates: Transparent	*β*-quartz solid solutions																	
Corning Visions		68.8	19.2			2.7	0.2	0.1	1.8		0.8	1.0	2.7	1.8	0.8		$Fe_2O_3 = 0.1$	Pinckney[7], Beall[5]
Schott Zerodur		55.5	25.3		7.9	3.7	0.5		1.0			1.4	2.3	1.9	0.5		$Fe_2O_3 = 0.03$	Pinckney[7], Beall[5]
Shott Ceran		63.4	22.7		N.A	3.3	0.7	N.A	N.A		2.2	1.3	2.7	1.5	N.A	N.A	N.A	Pinckney[7]
Nippon Electric Narumi		65.1	22.6		1.2	4.2	0.6	0.3	0.5				2.0	2.3	1.1	0.1	$Fe_2O_3 = 0.03$	Pinckney[7], Beall[5]
Aluminosilicates: Opaque	*β*-spodumene solid solutions																	
Corning 9608 (Corning Ware)		69.7	17.8			2.8	0.4	0.2	2.6			1.0	4.7	0.1	0.6		$Fe_2O_3 = 0.1$	Pinckney[7], Beall[5]
Corning 9606	Cordierite	56.1	19.8						14.7	0.1			8.9		0.3		$Fe_2O_3 = 0.1$	Pinckney[7], Beall[5]
Corning 9609 (Pyroceram)	Nephelene	43.3	29.8				14.0				5.5		6.5		0.9			Beall[5]
Lithium disilicates: Fotoform/ Fotoceram																		
Corning 8603		79.6	4		0–11	9.3	1.6	4.1										$Ag = 0.11$, $Au = 0.001$, $CeO_2 = 0.014$, $SnO_2 = 0.003$, $Sb_2O_3 = 0.4$ $La_2O_3 = 0.1$ to 6, + others	Beall[5]
Ivoclar Empress 2		57–80	0–5		0–11	11–19		0–13	0–5			0–8						Holand[15]
Sheet silicates: fluoromicas																		
Corning Macor		47.2	16.7	8.5				9.5	14.5							6.3		Pinckney[7], Beall[5]
Dentsply Dicor		56–64	0–2					12–18	15–20					0–5		4–9	$CeO_2 = 0.05$	Pinckney[7], Beall[5]
Chain silicates: Slagsial	White (Russia) Wollastonte	55.5	8.3				5.4	0.6	2.2	24.8		1.4					$MnO = 0.9$, $Fe_2O_3 = 0.3$, $S = 0.4$	Pinckney[7], Beall[5]
Minelbite	Gray (Hungary) diopside	60.9	14.2				3.2	1.9	5.7	9.0							$MnO = 2.0$, $Fe_2O_3 = 2.5$, $S = 0.6$	Pinckney[7], Beall[5]
K-richterite	Corning	67.3	1.8		1.0	0.8	3.0	4.8	14.3	4.7	0.3					3.5		Pinckney[7]
Canasite	Corning	54–62	1–4				6–10	6–12	0–2	17–25		0–2				4–8		Pinckney[7]
Other Calcium aluminophosphate	Kyushu Refractories		10		70					20								Abe[16]

NA = Data not available.

TABLE 8.18 Glass-Ceramic Properties

Glass identification	Description	Modulus of rupture, MPa	Young's Modulus, GPa	Hardness, HK_{100}	Fracture toughness MPa·m$^{1/2}$	Fracture energy, J·m^2	CTE, 10^{-7}/°C	References
Aluminosilicates:								
Transparent	β-quartz solid solutions							
Corning Visions								
Schott Zerodur								
Shott Ceran								
Nippon Electric Narumi								
Aluminosilicates:								
Opaque								
Corning 9608 (Corning Ware)	β-spodumene solid solutions	100	81	660			12 (0–500°C)	Pinckney[7]
Corning 9606 (Cercor)	Cordierite	250	120	700	2.2		45 (0–700°C)	Pinckney[7]
Corning 9609 (Pyroceram)	Nephelene							
Lithium disilicates:								
Fotoform/Fotoceram								
Empress 2—Ivoclar	fluoromicas	350 ± 50			3.2 ± 0.3		106 ± 5	Holand[d]
Sheet silicates:								
Corning Macor	Corning 9658	100	65	250		8.2	129 (25–600°C)	Pinckney[7]
Corning/Dentsply Dicor		152	70.3	362	2.5		72 (25–600°C)	Pinckney[7], Abe[16]
Chain silicates:								
Slagsital	White (Russia)	65–100	75–88	590–700			91–95 (0–500°C)	Pinckney[7]
	Gray (Hungary)	80–120	88–108	640–740			72–76 (0–500°C)	Pinckney[7]
K-richterite	Corning	150–200	87–95		3.2	60	115 (0–300°C)	Pinckney[7], Beall[23]
Canasite	Corning	250–300	80–82	500	4–5	150	125 (0–300°C)	Pinckney[7], Beall[23]
Other								
Calcium aluminophosphate	Kyushu Refractories	160–230	80		2.7		118	Abe[e]
For comparison								
Ion-exchanged	Soda-lime-silica	170	70		0.5	2		Beall[23]
Corning code 0313*	Aluminosilicate	340	70		0.5	2		Beall[23]

*Code 0313 is an ion exchanged product made from code 0317 glass.

advantages of glass. Their technical importance and commercial impact is not yet clear.

The ability to make a structurally stable material with a designed thermal expansion is being applied to temperature compensating substrates for coupling and decoupling (mixing) devices in WDM (wavelength division multiplexed) communications systems. For this application, a stuffed β-eucryptite ($Li_2O:Al_2O_3:SiO_2$ = 1:1:2.5) with a thermal expansion of about $-7 \times 10^{-6}/°C$ (note negative expansion coefficient) has been proposed.

8.2.8 Strengthened Glasses

8.2.8.1 Introduction. The strength of glass tends to be limited by the presence of surface flaws, which when stressed in tension, propagate as cracks that lead to fracture of the glass. The generally accepted approaches to improving the use strength of glass include fire polishing to remove mechanical surface flaws; acid polishing and/or etching to remove surface flaws or make the crack tips more blunt; coating the newly formed pristine glass surface with a protective coating, such as the multilayer polymer coatings applied to optical communications fiber; and providing a compressively stressed layer at and beneath the glass surface whose effects must be overcome when the glass is stressed in tension during use.

Strengthening by removal of surface flaws can only be temporary; it is generally limited to an in-process step.

Many methods have been demonstrated for effectively generating surface compressive layers, some of which have been applied commercially. These include thermal tempering, ion exchange at temperatures either above or below the glass transition, surface crystallization, lamination, and glazing.

Thermal tempering and ion exchange strengthening are discussed along with annealing in Sec. 8.5.

8.2.8.2 Glazed and cased glasses. Crystalline ceramic articles may be strengthened considerably by *glazing* them with a glassy coating of lower thermal expansion than the bulk ceramic body. This process has also been used to strengthen glass and glass-ceramic commercial products. The glaze is applied as a glass frit in a suitable binder that is then fired to give a smooth glassy coating. Glass objects can also be clad with a lower expansion layer during their manufacturing process. This is a relatively old technique. In its earliest hand shop versions, the gob of glass on the end of the blowing iron would be dipped into a pot containing the cladding glass, which would leave a layer covering the entire surface of the glass when it was removed from the pot. After blowing, the cladding glass formed a continuous layer over the surface of the object. If this cladding glass was colored, it was often referred to as *cased* glass; if thin and colored, it was sometimes referred to as *flashed.* If the thermal expansion of the glass closely matched that of the body glass, little strengthening would result, but the colored surface glass could be cut away to give decorative patterns of colored and clear regions.

8.2.8.3 Laminated and wired glasses. A variation on cased glass has been commercialized by Corning Inc. in the form of products made from laminated glass sheet. In the Corning process, illustrated in Fig. 8.6, molten core (body) and skin glasses, prepared in separate melters, are brought together in a special laminating delivery orifice from which the core glass flows as a wide ribbon, simultaneously clad on both sides by the skin glass. This ribbon is fed between counter-rotating rollers to generate a sheet of controlled thickness, which is then fed onto a special machine consisting of a hub-like device that serves to vacuum-form articles of various shapes. For its Corelle™ line of tableware products, the Corning Consumer Products Company uses a special combination of glasses for laminating: an opal core glass, containing sodium fluoride microcrystals as the opacifying agent, and a clear, lower-expansion skin glass. Examples of these compositions are shown in Table 8.6. Although the difference in thermal expansion between the core and skin glass provides improved strength, Corelle products are additionally strengthened by thermal tempering the laminated glass.

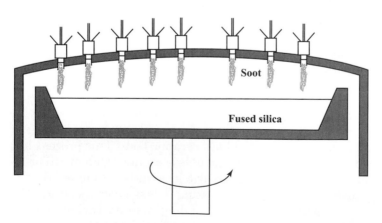

Figure 8.6 Corning Inc. code 7940 synthetic fused silica process. *(Courtesy of John Rowe, Corning Inc., ca. 1993)*

Sheets of glass are sometimes sealed together using a polymer adhesive interlayer to form "safety" and other products. A PVB (polyvinyl butyral) elastomeric sheet is often used to seal together sheets of annealed, partially tempered, or fully tempered glass to produce versions of safety glass. While this approach does not directly reduce the probability of surface flaws leading to fracture, the PVB layer does add several safety features: (1) it adds toughness to the composite, (2) upon fracture of the product, the PVB layer keeps the fracture-produced shards from flying about, or from having their sharp edges readily accessible to cause injury to humans, and (3) it serves as a restraint to prevent objects from being propelled through the glass from inside or outside a room or vehicle. This is one form of "safety" glass used for automobile windshields and, in earlier times, for side lights as well. It is both "anti-lacerative" and passenger restraining.

Such laminated glass is gaining acceptance as safety and security glazing for buildings, the intent being to prevent shards of glass from being propelled into the interior of buildings during hurricanes or following nearby bomb blasts. A simpler laminate, consisting of a single sheet of annealed glass coated on one side (the inside) by retrofitted polymer films, is also gaining popularity. These films are sometimes known as *shatter resistant window film (SRWF)* and *fragment retention film (FRF)*.

More complex versions of laminated glass, consisting of multiple, thick, alternating layers of glass and polymer (plastic) are used as a type of transparent armor, most commonly called *bullet-proof* glass. Such armor is inherently thick and heavy, thus providing an opportunity for future weight-reduction methods research. Aircraft windshields, which are designed to withstand a 400-knot bird impact, are typically composed of two or three layers of chemically strengthened glass laminated with PVB interlayers.

Wired glass (or *wire* glass) consists of a metallic "chicken wire-type" mesh imbedded along its center plane (see Sec. 8.4.7.) It is often seen in skylights and fire doors. The purpose of the wire is not to increase breaking strength but rather to prevent the fractured glass from falling out of the frame, an especially important function in preventing the spread of a building fire.

8.2.9 High-Silica Glasses

8.2.9.1 Fused Silica

Manufacturing history. As pointed out in Sec. 8.1.4, because of the high viscosity and high volatility of silica (SiO_2) at the temperatures needed to melt its crystalline phases, specialized techniques are required to prepare silica glass products. Some key highlights in process evolution include the following:

- *Nineteenth century.* The introduction of oxygen-injected or oxygen/fuel (as opposed to air/fuel) burners and torches allowed sufficient energy concentration to melt crystalline quartz.

- *1899–1910.* Commercial development began in England, France, and Germany. Key issues were purity and continuous fabrication.

- *Early twentieth century.* Carbon arc and carbon resistance electric heating methods were developed.

- *1930s.* Hyde (Corning Glass Works) first prepared silica glass by flame hydrolysis of $SiCl_4$. His U.S. patent was issued in 1942 and became the basis for all synthetic fused silica processes used commercially today.

Classifications of fused silica and methods of manufacture. Methods of manufacture that involve gas-oxygen combustion as a heat source (as compared to electric melting methods) allow considerably more water vapor (a product of combustion) to be incorporated within the glass structure as hydroxyl ions. Synthetic precursors, as opposed to naturally occurring minerals, allow greater chemical purity. These two factors, heat source and raw material source (natural vs. synthetic), have led to a commonly accepted classification of fused silica types (after Heatherington et al., 1964, and Bruckner, 1970):

Type I

- Produced from natural quartz
- Electric fusion (arc or resistance heating) under vacuum or inert gas
- Sometimes using tungsten or molybdenum containers
- "Dry"—contains <5 ppm of OH
- Contains the metallic impurities present in the raw materials and the furnace refractories, e.g., 100 ppm Al, 5 ppm Na, etc.
- Examples: Infrasil (Heraeus), IR-Vitreosil (TSL), GE codes 105, 124, 201, and 214

Type II

- Produced from crystalline quartz powder by oxy-hydrogen *flame fusion* (similar to Verneuille crystal growth process)
- "Wet"—contains 150 to 400 ppm OH from combustion products

- Lower in metallic impurities than Type I .
- Examples: Herasil, Homosil, Optosil (Heraeus), OG Vitreosil (TSL), GE 104
- Special oxygen treatment can improve UV transmittance; example, Ultrasil (Heraeus)

Type III

- "Synthetic" fused silica produced by flame hydrolysis of $SiCl_4$ in oxy-hydrogen flame [variations: other chemical precursors; natural gas (methane)-oxygen flame]
- "Wet"—contains OH up to 1200 ppm
- Very few metallic impurities, a few ppm or less
- With $SiCl_4$ precursor, about 100 ppm Cl retained
- Examples: Suprasil (Heraeus), Spectrosil (TSL), Corning 7940 and 7980

Type IV

- "Synthetic" fused silica produced from $SiCl_4$ (or other precursors) in water-free plasma oxidation torch
- Like Type III, but "dry," containing <5 ppm OH
- Examples: Suprasil W (Heraeus) and Spectrosil WF (TSL)

Type IVa

- Not a traditional classification; coined here by Seward
- Synthetic, like Types III and IV
- Made by a two-step synthetic process:
  - Step 1. Deposition of a porous synthetic preform
  - Step 2. Drying and consolidation under flow of gases, often chlorine (for drying) and helium (for completeness of densification)
- "Very dry," OH measured in ppb
- "Very pure," impurities measured in ppb

Type V

- Sol-gel derived silica
- Classification V introduced by Sempolinski and Schermerhorn
- Many processing variations exist
- Glass properties depend on the process techniques

Types I and II are often referred to as *fused quartz,* and Types III and IV as *synthetic fused silica,* although this nomenclature is not always observed. Table 8.19 shows some of the current commercially available fused silicas on a process method/purity matrix.

Process methods

Electric arc melting (type I).   In this process, grains of crystalline quartz are fed into an electric arc; they soften and fall onto a heated rotating surface where they form a dense glassy body by fluid flow (continuous viscous sintering).

The portion of the process in which the molten glass particles drop and are collected and consolidated onto/into a layer of previously deposited material to eventually build up a large mass of glass is called the *boule* process; the mass of solid glass produced being called a boule. After the initial layers of glass are deposited, the glass is no longer in contact with any supporting refractory materials, so the boule process is in a sense a "containerless" process, without the inherent contamination that comes from melting in a container.

Often the raw materials are beneficiated to improve purity. *Beneficiation* steps include:

- Quartz crystals extracted from the earth
- Hand sorted for defects (contaminants)
- Rinsed in dilute HF
- Calcined at 800°C
- Water quenched
- HF acid washed

Resistance electric heating (type I).   In one version, crushed quartz crystals are continuously fed into the top of an electrically heated refractory metal crucible (Ta, W, or Mo), enclosed by an inert gas atmosphere. The molten glass is drawn from the bottom through a refractory metal die to continuously form solid rod. (If tubing is needed, a suitable core with flowing inert gas can be inserted within the orifice. See Sec. 8.4.5 for a discussion of tube drawing techniques.) These techniques are used to make silica rods and tubing for a variety of applications. In more recent versions, radio frequency (RF) induction heating of the crucible has been used.

Type I processes are used by GE, Nippon Silica Glass, Shin-Etsu, Heraeus, Heraeus Amersil, TSL, Toshiba, and Quartz et Silice.

Flame fusion (type II).   This also is a boule process. The fine grains of crystalline quartz are fed into the flame of a hydrogen-oxygen

TABLE 8.19 Vitreous Silica, Purity and Manufacturing Methods

Raw material, increasing purity →

Process	Sand (Alk = 50, M = 400, Al = 200)	Crystal (Alk = 5, M < 10, Al = 10–50)	Beneficiated sand (Alk = 5, M < 10, Al = 20)	Synthetic (Alk = 1, M = 1, Al = 1)
Flame Fusion: Decreasing OH⁻ — $CH_4, H_2/O_2$		$OH^- = 200$ H/A—TO8, Ultrasil, Homosil TSL/TAFQ—Vitreosil 055 TOS–T1030, 1070, 1130(d), 1170(d) Q/S—981		$OH^- = 1000$ $Cl^- = 100$ CGW–7940, 7980, ULE (d) Dynasil 1000 GE(WQS)—Synsil H/A—Suprasil TSL/TAFQ—Spectrosil TOS–T–4040 Q/S—Tetrasil A,B
Plasma				$OH^- < 10$ $Cl^- < 200$ H/A—Suprasil W TSL/TAFQ—Spectrosil WF TOS–T–4042 Q/S—Tetrasil SE
Electric Fusion: Arc atmosphere			$OH^- = 30$ GE–510 GTE crucibles QSI crucibles	
Decreasing OH⁻ — Resistance/induction atmosphere	$OH^- = 200$ GE–318 H/A—Rotosil TSL/TAFQ—Vitreosil TOS–T100, 200, 800 Q/S—opaque		$OH^- = 30$ GE–511 Pyro crucibles	
Vacuum Rebake		$OH^- = 30$ TOS–T–2030 Q/S–453, Purposil, Pursil, Germiosil(d), 676	$OH^- = 30$ GE–124,214 $OH^- < 5$ GE–214, 982, 219(d) GTE–SG 255C TOS–T–7082	

Alk = alkali content, ppmw total; M = transition metal content, ppmw; A = aluminum metal content, ppmw; GTE = Sylvania; QSI = Quartz Scientific Inc.; Q/S = Quartz et Silice; H/A = Heraeus, Amersil; TSL = Thermal Syndicate Ltd.; TAFQ = Thermal American Fused Quartz; TOS = Toshiba; GE = General Electric; WQS (West Deutsche Quarzschmelze); CGW = Corning Inc.; and d = doped. *Source*: After Bihuniak in Ref. 5.

burner to soften and drop onto the surface of a rotating rod or boule, where they flow and merge. This approach was developed by Heraeus and is presently used by that company.

Flame hydrolysis (types III and IVa). In this process, in its simplest form, a liquid silica precursor such as $SiCl_4$ is volatilized and entrained in a suitable carrier gas, such as oxygen or helium, and fed into a specially designed high-temperature gas-oxygen burner. In the resulting burner flame, the precursor reacts to form SiO_2 molecules as described by the following chemical equations:

$$SiCl_4 + H_2O \Rightarrow SiO_2 + 4HCl$$

$$SiCl_4 + O_2 \Rightarrow SiO_2 + 2Cl_2$$

The first process is called *flame hydrolysis,* and the second *flame oxidation.* Because of the high concentrations of water vapor generated by the combustion process, the first reaction, flame hydrolysis, dominates.

The gas phase within the flame becomes supersaturated with respect to solid SiO_2, which precipitates as molten droplets of transparent soot approximately 10 nm in diameter. These droplets are subsequently expelled from the flame. (If this expelled soot is given sufficient room to cool to the solid state, it can be collected as a fine white powder, called *fumed silica,* such as is sold under the trade name Cab-O-Sil^{TM}.) In the Type III silica processes, the silica soot is deposited on a hot (>1600°C) surface where it continuously consolidates to produce a homogeneous boule.

Figure 8.6 shows schematically the deposition process whereby Corning Inc. deposits boules of its codes 7940 and 7980 fused silica exceeding 5 ft in diameter and approximately 1 ft thick.

Heraeus, Shin-Etsu, and Nippon Silica Glass employ variations of this process called *vapor axial deposition (VAD),* whereby the silica soot is deposited on the end of a continuously withdrawing rotating mandrel and is consolidated to full density in a separately heated zone of the processing apparatus. This process is illustrated in Fig. 8.7.

In the Type IVa process, the soot is deposited on a relatively cool substrate (near the annealing point) so that the arriving soot particles bond to the layer of particles beneath them but do not sinter to full density. This intermediate product has about 20 to 30% porosity, which allows it to be reacted at high temperatures with suitable gaseous drying agents such as Cl_2, HCl, thionyl chloride ($SOCl_2$), or ammonia diluted with nitrogen. The dry porous body is subsequently consolidated to full density at temperatures >1400°C in vacuum or in a flowing helium atmosphere. Such drying and consolidation leads to incorporation of either chlorine or ammonia ions within the glass

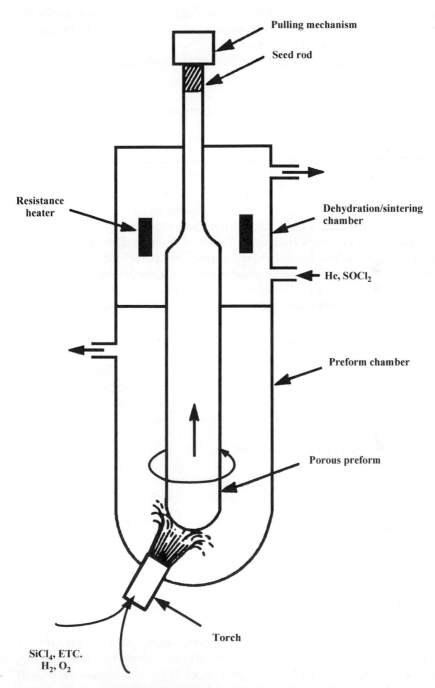

Figure 8.7 Vapor axial deposition (VAD) process for synthetic fused silica. *(From Ref. 24, reproduced by permission of the publisher)*

structure. (It is a variation of this type IVa process that is used in the OVD (outside vapor deposition) process for optical communications fiber manufacture described in Sec. 8.7.

In summary, these processes consist of these steps:

- Vapor generation
- Chemical reaction using specialized burners
- Generation of soot particles
- Deposition of particles
- Consolidation (sintering/densification)

Plasma torch (type IV).   In this synthetic process, the precursor chemical vapor stream is fed into the high-temperature flame of an oxygen gas plasma, generated by microwave energy. The chemical precursors react via the flame oxidation reaction described above. (There is no water vapor present to promote flame hydrolysis.) A dry silica soot is generated and collected on a hot boule rotating below. Heraeus Amersil, TSL, Toshiba, and Quartz et Silice use these techniques.

Sol-gel techniques (type V).   Sol-gel methods for glassmaking consist of chemical and thermal processes conducted at temperatures considerably less than those required for melting glass from crystalline raw materials. They involve solution chemistry processes, sometimes augmented by traditional ceramic processing techniques. The steps for most methods include, in the following sequence:

1. Preparation of a solution or sol
2. Casting in a mold
3. *Gelation* (formation of a colloidal gel network)
4. Drying (formation of a three-dimensional porous solid)
5. Consolidating (*sintering*) to fully dense glass

Gelation can be produced by destabilization of colloidal silica sols wherein the colloidal size can range between 1 and 1,000 nm. Alternatively, polymeric gels can be produced by the hydrolysis of metal-organic compounds in solution.

The sol-gel concepts have been known since the late nineteenth century but have been intensively investigated since the 1970s, a motivator having been global concerns over high quantities of energy consumed in traditional glass manufacture. Although multicomponent glasses can be produced by sol-gel techniques, the greatest commercial successes have been with silica and high-silica glasses; maintaining chemical homogeneity in multicomponent glasses adds to the complexity and expense of the processing. While energy savings are still of interest, the ability to generate highly pure silica glass in "near net shape" form, albeit with high degrees of dimensional shrinkage, have been strong drivers for sol-gel processing of silica. While purity of the precursors is key to the purity of the product, the fact that the gels can be dried and consolidated at relatively low temperatures lessens the likelihood of contamination from the furnace refractories and crucibles. The various forms of sol-gel derived silica have been collectively termed Type V silica.

Drying of the gel to produce a porous solid is perhaps the most challenging part of any sol-gel process. The small pore sizes of the gels can generate great internal capillary forces that in turn can lead to cracking and destruction of the dried porous structure unless very tedious or expensive techniques, such as supercritical drying or special DCCA (drying control chemical additives), are employed. In general, the larger the gel pore sizes the easier the drying and sintering. We can correspondingly classify silica sol-gel preparation techniques in the order of increasing pore size, depending on the solution or sol precursor used:

- *Polymerizing gel routes.*   Precursor: metal organic compounds, e.g., alkoxides such as TEOS (tetraethylorthosilicate). Process: acid catalyzed hydrolysis and condensation reactions to form gel. Pore sizes: 1 to 5 nm.

- *Colloid particle routes.*   Precursor: colloidal silica sol. Process: flocculation of silica particles, sometimes with chemically induced bonding. Pore sizes: 100 to 250 nm.

- *Dispersion processing (aqueous or nonaqueous).*   Precursor: fumed silica/ silica soot dispersed in polymerizing gels or potassium silicate aqueous solutions. Colloidal silica for dispersion can be prepared from aerosols as soot or "fumed silica" by flame hydrolysis, or other suitable methods. Examples of fumed silica are Cab-O-Sil^{TM} (Cabot Corp.) and Aerosil OX-50 (Degussa Corp.). Process Pore sizes: 100 to 300 nm.

- *Alkoxide-derived gel.*   The gel is dried rapidly to form high-surface-area particles, reheated and milled in water to form slurries that are cast, dried, and consolidated.

Consolidation (sintering) can be performed at temperatures less than 1200°C for pore sizes less than 10 nm; 1400°C is generally required for pore sizes > 100 nm. Prior to consolidation, the porous body can be treated in Cl_2 gas to remove residual OH, dechlorinated by heating in O_2, and consolidated to full density in He or under vacuum. Details of the various sol-gel methods are beyond the scope of this handbook.

Applications of sol-gel-derived fused silica include thin films (thickness <1 μm), coatings, controlled-size spherical particles, fibers, and bulk silica glass. Among product examples for bulk silica glasses are lightweight space-based telescope mirror blanks (Corning), optical components (Corning and GelTech, Alachua, Florida), integrated circuit *photomask* substrates (Seiko-Epson, Japan) and outer cladding layers of optical telecommunications fiber preforms (AT&T/Lucent Technologies).

8.2.9.2 Porous and "reconstructed" high-silica glasses.

In the 1930s, a technique was developed by Corning Glass Works for making high-silica glass that did not require the high melting temperatures described above. It is based on the principles of liquid-liquid or glass-glass phase separation. An alkali borosilicate glass of appropriate composition is melted, formed into a desired shape, and heat treated to generate the two intertwining matrix phases, one very high in silica content and the other high in boron oxide. This two-phase product is then leached in a hot dilute acid (such as nitric acid), rinsed, and carefully dried, yielding a porous material having very high silica content. Several products have resulted including the following:

Porous glass. By careful control of the initial glass composition, heat treatment, and chemical leaching, glass products having uniform interconnected (open) porosity can be made with a wide selection of pore sizes, generally less than 5 nm diameter (common range 4 to 6 nm) although, by special processing, pore sizes of 20 nm have been achieved. When dry, these materials display a strong tendency to absorb water from the atmosphere due to their very high internal surface area (up to ~300 m^2/g), giving them the name *thirsty glass*. Applications include drying agents, salt bridges for electrochemistry, filtration media, catalyst supports (including immobilization of enzymes), and adsorption chromatography substrates. For these latter two applications, the porous glass has been generated in the form of granules with pore sizes available in the range of about 7 to 300 nm and is often referred to as *controlled pore glass (CPG)*. Potential applications include time-release capsules used to enclose drugs, vaccines, and fertilizer.

96% silica (Vycor™ brand products)

- These are made by phase separating alkali borosilicate glass, chemically leaching out the nonsilica phase and consolidating to a fully dense body, approximately 96 wt.% SiO_2, 3% B_2O_3, and < 0.5% alkali.

- The product is consolidated by firing in vacuum at temperatures ~1200°C.

- Unlike fused silica, it can be produced using standard glass-forming methods like tube draw, pressing, blowing, and rolling.

- An example is Corning code 7900 glass.

Primarily because of the slow leaching and rinsing times required (hours to days), the process is generally applicable only to relatively thin-walled bodies, usually less that 10 mm thick, although 25 mm has been achieved when the cost of the slow chemical processing was justified.

Near-IR transmitting properties can be improved by drying the porous glass with special treatments, such as flowing chlorine gas at high temperatures to remove the hydroxyl ions just prior to consolidation.

Colored glass can be made by impregnating the porous glass with aqueous or organic solutions containing suitable coloring ions (e.g., nitrate or chloride salt in water), then heating to drive off the carrier media (evaporate the water, evaporate or oxidize the organic materials), leaving the coloring ions to be incorporated into the glass structure during consolidation. A variation is to impregnate only a surface layer so as to produce a product with a clear body and colored surface layer. These techniques are used for certain high temperature lamp envelopes.

Properties.   It is similar to fused silica but not quite as refractory. It is suitable for continuous use at 900°C, intermittent at up to about 1200°C. The coefficient of thermal expansion is $7.5 \times 10^{-7}/°C$. See Tables 8.1 and 8.2.

Applications.   The material has applications in defroster tubes (refrigerators), radiant electric heater substrates, lab ware, high-temperature furnace/stove windows, UV transmitting windows, and UV lamps. (These are examples only; many specialized applications exist.)

8.2.9.3   Ultra-low-expansion (ULE™) glass.

When TiO_2 (titania) is substituted into the silica structure in amounts less than about 15 wt.%, it enters into the glass network. An unusual property, not yet fully explained on a theoretical basis, is that the glass has a region of negative thermal expansion extending from cryogenic temperatures to above room temperature, the upper limit of the range depending on the concentration of the titania. At about 7% TiO_2, the expansion changes from negative to positive somewhere between about 0°C and 30°C, being near zero over most of that range. Large pieces of such titania-doped silica can be made by the Type III synthetic process, describe above, by feeding suitable concentrations of titania precursors (e.g., $TiCl_4$) into the burner along with the silica precursors. Corning Inc. sells such glass products under the ULE^{TM} trade name, glass code 7971.

Properties.   Thermal expansion is near zero from 5°C to 35°C, thus it can be fusion welded at room temperature, enabling fabrication of complex structures having extreme dimensional stability. (See Tables 8.1 and 8.2.)

Applications.   Applications include telescope mirrors (currently up to 8 m diameter), lightweight mirrors (for space deployment), precision instrument stages, and athermal (temperature inert) mountings.

8.2.9.4   Doped silica for optical communications (fiber and planar waveguides).   This is a very specialized, highly technical topic that is further developed in Sec. 8.7. The desired optical properties of the various optical system components are often produced by a controlled geometrical distribution of dopant ion concentrations in the glass. Here, we briefly discuss the needs for such doping and methods for achieving it.

Communications at optical wavelengths.   Transmission of data at higher rates is possible using the higher frequencies (shorter wavelengths) of light radiation than is possible with radio or television frequencies. What is needed is a transmission medium (other than air or vacuum) having low loss. An early threshold target for long distance communication was 20 dB/km loss (equivalent to 1% transmission through 1 km of fiber), consistent with an economical number of electronic repeaters (amplifiers) along the path. This was achieved using glass waveguides made from high-purity synthetic silica, now called *optical communications fiber,* or just *optical fiber.*

Lightguiding over optical fiber requires that the fiber have a higher refractive index at its core and a lower refractive index at its outer surface, often referred to as the *cladding.* This can be achieved by doping the silica core regions to increase the index, by doping the silica cladding regions to decrease the index, or by doing both. For success, one must dope with components that (a) will not produce optical absorption in the glass at the transmitted wavelengths (specifically near 1.31 and 1.55 µm), (b) can be incorporated synthetically by the vapor phase route along with the silica, and (c) are available with sufficiently high purity. This limits the choices to glass-forming oxides (boron, phosphorous, and germanium), fluorine (which can substitute for oxygen in the network), and possibly titania and alumina. It is advantageous from a strength point of view to use dopants that give the core a slightly higher thermal expansion than the cladding. Figure 8.8 shows the effects various concentrations of these components have upon refractive index of silica.

Synthetic silicas can be doped during manufacture by incorporating appropriate precursors along with the silica precursor in the vapor stream fed to the deposition burners. Methods for doing so are described in Sec. 8.7.4. Example precursors include the chlorides $GeCl_4$, BCl_3, and $POCl_3$, the fluoride SiF_4, and the solid $AlCl_3$, which can be delivered by subliming at high temperatures.

Fiber amplifiers and all-optical systems. The future directions of optical communication are toward all-optical systems. That is, the trend is to utilize optical signal processing (coupling/splitting, switching, amplifying) rather than converting light signals to electronic signals, processing electronically, then converting back to light signals. Such signal processing can be done with fiber devices (constructed from optical fiber) or planar devices (constructed by building optical paths on flat substrates using patterned layers of different refractive index glass, generally doped silica, often in combination with vapor-deposited conducting and semiconducting films).

Light amplifiers are essential to all-optical systems (no electronic repeaters, no light-to-electronic conversion), since they compensate for optical losses within the fibers and the associated connectors, couplers, and switches. They are achieved by doping the fiber core with

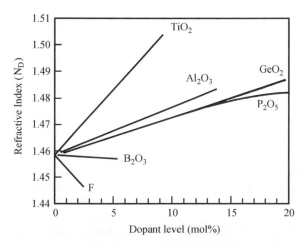

Figure 8.8 Refractive indexes (n_D) as a function of dopant level in bulk SiO_2. F levels are in atomic percent.[25]

fluorescent *lasing* ions such as erbium or neodymium, and "pumping" using laser diodes. The operation of the erbium optical fiber amplifier (OFA) is based on the lasing action at 1.55 µm of the Er^{3+} (erbium) ion. The fiber is optically pumped at either 980 or 1480 nm by a suitable solid state laser diode. Light emission by electronic transitions between the $^4I_{13/2}$ and $^4I_{15/2}$ levels is stimulated by the light signal at 1.55 µm and transfers power to that signal. To construct the erbium OFA, the fiber core is co-doped with Er^{3+} at levels up to about 300 ppm, in addition to the index adjusting elements (for example, germania). A major challenge is to obtain uniform gain (amplification) across a wide band of wavelengths, such as is required for WDM (wavelength division multiplexing) at 1.5 and 1.6 µm. Some methods for achieving this use separate gain-leveling devices.

Nd (neodymium) doped fibers have been investigated for operation within the 1.31 µm communications band but have not yet been commercially successful. For amplifier operation at those wavelengths, non-silica-based fibers may be of advantage, leading to hybrid (mixed) glass fiber systems.

8.3   Glass Making I—Glass Melting

8.3.1   Introduction and General Nature

The term *glass melting* as generally used applies collectively to all the steps used to convert raw materials into a molten mass that can be subsequently formed as an object. The process of preparing a quality melt from which glass products are to be made consists of several stages, as follows:

- *Batch preparation* (raw material selection, weighing and mixing)
- *Batch melting* (conversion of the batch raw materials into a viscous liquid that is essentially free of any crystalline material)
- *Fining* (the removal of bubbles)
- *Homogenizing* (the removal of chemical and thermal variations within the melt, often occurring simultaneously with fining, the two processes together sometimes being called *refining*)
- *Conditioning* (bringing the melt to the uniform temperature required for whatever forming process will be used to make the product)

Each of these stages is performed more or less in sequence, either at one location, as in a crucible, or at a progression of locations, as in a continuous glass melting tank consisting of several or many zones.

The various types of melters are described in Sec. 8.3.3. Here, we will describe glass-melting steps in more detail.

8.3.2   Steps in Glass Melting

8.3.2.1   Glass batch considerations.   Rarely does a commercial process start with an all-oxide batch. This is in part because the expense of the materials would be too great, but mostly because oxides melt at much higher temperatures than the salts of the corresponding metallic elements, thus requiring longer times at higher temperatures to complete the chemical reactions. Many different combinations of raw materials can yield the same final glass composition. Final choice of raw materials is based on factors such as chemical composition, the level of impurities tolerated, particle size, particle size distribution, the precision to which these characteristics are controlled (maintained) by the vendor, and price.

Batch materials.   Because of space restrictions, we cannot discuss all of the batch materials used in glass manufacturing, but we will discuss several of the most relevant to give a feeling for some of the key considerations.

Silica (SiO_2)

- Silica consists of quartz sand, crushed quartzite, or beneficiated sandstone.
- For sandstone, a combination of quartz sand and clay, beneficiation consists of crushing, washing, and froth flotation to remove refractory heavy metal oxides.
- In the U.S.A., it is shipped and handled dry; in Europe, wet.
- Generally, it is > 99% SiO2; often > 99.5% with Fe_2O_3 < 0.03% (by weight).
- Generally, it is 40 to 140 mesh (coarser is harder to melt; finer leads to "dusting" effects). Also, finer particles may melt too fast and raise the melt viscosity before air trapped in the batch can escape.
- Cost trade-offs include iron content and particle size distribution.
- Chemical analysis of raw materials, either in-house or by supplier, is important.

Limestone/calcite ($CaCO_3$)

- This is either high calcium ($\sim CaCO_3$) or dolomitic (< 46% $MgCO_3$).
- Dolomite is $CaMg(CO_3)_2$.

- Sometimes the oxide (CaO) is used (lime, quicklime, burned lime, or dolomitic quicklime).
- Sometimes aragonite, a different mineral form of $CaCO_3$ is used.
- Limestone generally > 96% $CaCO_3$, with SiO_2 and $MgCO_3$ as major impurities.

Soda ash (Na_2CO_3)

- Sources for the U.S.A. and most of world are trona deposits in Wyoming. (Trona is a hydrated sodium carbonate sodium bicarbonate ore; $Na_2CO_3 \cdot NaHCO_3 \cdot 2H_2O$.)
- The ore is calcined and treated to remove insoluble impurities, generally achieving > 99.5% Na_2CO_3.
- Na_2CO_3 is also made by the Solvay process (react salt with limestone to get $CaCl_2$ as a by-product). Solvay is still used as a source in Europe and other parts of the world.
- Hydration of the raw materials at temperatures < 110°C is a problem, so materials are often supplied as the monohydrate ($Na_2CO_3 \cdot H_2O$).

Alumina (Al_2O_3)

- Alumina is a very refractory material and slow to dissolve, so it is generally batched as a mixed-oxide mineral.
- The source is generally feldspar or nepheline syenite (minerals containing 60 to 70% silica, 18 to 23% alumina, calcia, magnesia, soda, potassia, and iron oxide (0.1%).
- Since alumina is generally a minor batch component, the accompanying oxides can usually be easily assimilated in the composition.
- Accurate chemical analysis is important.
- It is relatively expensive but, when refined from the ore bauxite, it is even more expensive—often too expensive.

Borates

- Borates are obtained mostly from California or Turkey.
- Borax ($Na_2O \cdot 2B_2O_3 \cdot 5H_2O$), sometimes called *5 mol borax.*
- Boric acid (H_3BO_3 or $B_2O_3 \cdot 3H_2O$).
- Colemanite ($Ca_2B_6O_{11} \cdot 5H_2O$ or $2CaO \cdot 3B_2O_3 \cdot 5H_2O$), from Turkey.
- The mixed sodium-calcium borate minerals, ulexite and probertite, are used in the fiberglass industry.

- Anhydrous borax ($Na_2O \cdot 2B_2O_3$), also known as anhydrous sodium tetraborate ($Na_2B_4O_7$), is generally not used because it absorbs water. The fully hydrated form of borax ($Na_2O \cdot 2B_2O_3 \cdot 10H_2O$), on the other hand, is generally not used because it loses water, decomposing to 5 mol borax at temperatures above 60°C. Such changes in water content make batch calculations and weighing somewhat uncertain.

- Anhydrous B_2O_3 is too hygroscopic and too expensive to be a good commercial batch material.

Others

- Litharge (PbO) and red lead (Pb_3O_4) are being replaced as batch materials by various lead silicates, which are safer to handle.

- Salt cake (Na_2SO_4) is used as a melting accelerator (forms eutectic with Na_2CO_3) and fining agent.

- Gypsum ($CaSO_4 \cdot 2H_2O$), melting accelerator, fining agent.

- Sodium nitrate (Na_2NO_3), oxidizing agent, stabilizes color.

Cullet. Waste or broken glass softens readily in glass melters to bring batch particles together, thereby increasing batch reaction rates, often immensely. This batch material is referred to as *cullet*. There are many economic and environmental factors favoring use of cullet in batch melting operations. Sources include the following:

- *In-house cullet*, generally of known composition. This is usually the same as the glass being produced, so few special batch adjustments are required (exceptions being laminated/clad glass).

- *Foreign cullet*, from sources other than the manufacturing plant (other plants, customers, other manufacturers, post-consumer).

- *Post-consumer cullet,* which is increasing in importance as recycling expands. It must be carefully processed to remove metal, ceramic, and other detrimental contaminants. Transparent glass-ceramics mixed in with the glass cullet are a concern for container manufactures, especially in Europe.

Cullet can be used in any proportion, but 30 to 60% is generally considered most effective. Sizes coarser than the raw materials seem to work best; sometimes they are added as chunks. Cullet can be added in the mixer, at the fill (doghouse), under, over, or layered with the batch. It must be added in a continuous enough manner to maintain melting stability.

Batch formulation. A combination of batch materials is selected based on the desired glass composition and oxidation state in combination with manufacturing process requirements (including those of mixing, melting, fining, etc.) and product cost. The quantities of each component are calculated to give the required weight of oxide after the volatile components and reaction products, such as water, carbon dioxide, sulfur dioxide, and nitrogen oxides, are lost to the atmosphere.

Raw materials handling

Transportation and storage. Raw materials that are used in large quantities are often delivered by railroad car and stored in large vertical silos or moderate-sized storage bins. A conveyor or elevating mechanism is used for loading the silos. Lesser batch components are delivered and stored in bags or other containers.

Collecting and weighing. The trend in large manufacturing operations is to automate these steps as much as possible. Batch components can be weighed in hoppers at individual silos or storage bins and conveyed to the mixing location. More often, the batch components are fed into a single scale for weighing. The scales themselves must be accurate and regularly calibrated. The weight sensors are often electronic, with digital readouts and modern computer interfacing and control. Such electronic weighing systems can be sensitive to 1 part in 4,000 (250 ppm).

Mixing. A wide variety of mixers are used. Most are of rotary (like a cement mixer) or pan-type construction. A variety of paddle and mixer blade designs are used. The trend is to mix the batch at a location very close to the melting furnace to minimize opportunity for batch segregation (see below) to occur.

Filling or batch feeding. A variety of filling devices or machines are used. They often involve a hopper and a mechanism such as an auger or mechanical vibrator to feed the batch into the furnace via a chute or tube. Multiple feeders are generally used to supply batch to large melters. Reciprocating feeders, whereby a pile of batch is deposited on a shelf extending the full width of the melter entrance and then pushed into the melter by a bar, are used for large container and float-glass furnaces. This reciprocating action results in a sequence of batch "logs" being fed into the tank. The resulting undulating batch surface improves the melting process by allowing rapidly melted batch to run downhill, exposing fresh batch to the heat.

Many variations are possible for all the operations described above. In smaller factories and hand shops, the automation tends to be less, but care and accuracy remain important.

Batch segregation. The term *batch segregation* refers to some batch components becoming spatially separated from the rest on some relatively large scale. (It can also refer to a separation of different particle sizes of the same batch component.) It is a type of demixing, which is problematic, because different regions of segregated batch may then melt to produce glasses of different compositions. Some regions may not melt at all. The resulting inhomogeneities must be removed later in the process. Segregation can occur inside the mixing machine, as the mixed batch is being transported to the melter, or as it is being fed into the melter. Forces causing demixing operate whenever moving streams of particles are present. Often, they rapidly create steady-state patterns of segregation under constant flow conditions. These patterns are reproducible.

Factors affecting batch segregation include the following:

- Size difference is the main contributing factor. (A size difference ratio of as little as 5% can lead to measurable demixing.)
- Nominal grain size and density differences are lesser factors.
- Other much lesser contributing factors include particle shape, lubricity, and surface charge.

Methods for controlling batch segregation include the following:

- Use multiple discharge silos.
- Use raw materials having a narrow size distribution.
- Match their particle size and size distribution.
- Do not mix too long.
- Minimize the movement of mixed batch (locate the mixer at the furnace, not in the batch house).
- Wet the batch.

Wetting of the batch helps reduce batch segregation by suppressing the free-flowing characteristics of batch. (It provides a thin liquid coating on each particle.) Wetting also helps decrease dusting and batch carry-over into the exhaust system. Generally, water or a 50% NaOH solution is used. Of course, wetting complicates the batch handling system and adds expense. The technique must be used with care when hydratable batch components are present, since they can absorb water, which may lead to "setting up" of the batch.

Batch preheating.   Several manufacturing groups are currently investigating the advantages of preheating of batch using heat from the furnace exhaust gases. The goals are at least twofold: improve the efficiency of the batch melting step, and recover waste heat. This will, of course, complicate the process and add some expense.

8.3.2.2 Processes occurring within the melter.   *Batch melting* involves heat transfer from a source of heat energy; for example, fuel combustion in a burner or electric resistance (Joule) heating, and complex chemical reactions among the batch components. Since a formulated batch often contains water (moisture), and many batch components such as boric acid contain chemically combined water, the first step of melting consists of drying and dehydrating the powdered raw materials. Many chemical reactions among the dry batch materials occur, some simultaneously. Solid state reactions between batch components occur, sometimes leading to molten phases and other times to new crystalline phases. This is followed or accompanied by melting the salts and other low-melting-temperature ingredients (*fluxes*); decomposing some of the fluxes, producing gases; reacting silica sand and other refractory materials with the melted fluxes (for example, reaction of sodium carbonate with silica sand); and dissolving the remaining refractory batch components (e.g., silica, zirconia) into the melt.

Fining involves the removal of the bubbles generated during batch melting. Such bubbles consist of air that was trapped between grains of batch and gases such as CO_2, SO_2, and H_2O generated by the reactions of the batch materials; e.g., decomposition of sodium carbonate releases CO_2 gas. Gases can also be generated by reaction of the melt with materials comprising the melter, i.e., the container and electrodes. The bubbles tend to rise to the surface of the melt, driven by buoyant forces according to Stokes' law. This is often referred to as *Stokes fining*. If the bubbles are small, say 0.2 mm diameter or less, this can be a very slow process. To aid the fining, the temperature of the melt is often increased so as to decrease the viscosity of the melt and accordingly to decrease the viscous drag on the rising bubbles. Occasionally, pressure changes above the melt may also be utilized to aid in fining.

During fining, gases produced by further chemical reactions in the melt enlarge the bubbles. Generally, the glass composition is designed to contain chemical compounds, known as *fining agents*, which release these additional gases during fining. Examples of such fining agents are As_2O_5, Sb_2O_5, and Na_2SO_4. The elements As, Sb, and S can exist in the glass in different oxidation states (see the discussion of oxidation state below), depending on the temperature and oxygen activity of the glass. If the glass temperature is raised above that at which the glass was melted, these elements become more reduced, giving off oxygen (O_2) or sulfur dioxide (SO_2) gas according to the following equations:

$$As_2O_5 \Rightarrow As_2O_3 + O_2 \text{ (gas)}$$

$$Sb_2O_5 \Rightarrow Sb_2O_3 + O_2 \text{ (gas)}$$

$$2Na_2SO_4 \Rightarrow 2Na_2O + 2SO_2 \text{ (gas)} + O_2 \text{ (gas)}$$

The gases generated by the above equations remain dissolved within the glass in their molecular state until supersaturation becomes sufficient to nucleate bubbles of gas. However, if bubbles are already present, the dissolved gas molecules diffuse to those bubbles, thereby enlarging them. The enlarged bubbles then rise to the surface of the melt and break. Bubbles not reaching the surface dissolve when the melt is allowed to cool from the fining temperature. This dissolution process is aided by the reverse action of the fining agents.

Sometimes, during batch melting and fining, the gas bubbles do not burst when they reach the melt surface but rather create a foam layer. Such foaming adversely affects heat transfer into the melt and can lead to glass defects, so it should be avoided by any effective means.

Homogenization. Sources of inhomogeneity in the melt include the physical segregation of chemical components during batch mixing or batch melting, preferential volatilization of certain chemical components from the melt surfaces, and reaction of the melt with the furnace wall refractories. Mechanisms operating to improve homogeneity are composition-gradient-driven mass diffusion in combination with stirring (forced or thermally driven convection). Stirring serves to reduce the diffusion distances required for homogenization.

Thermally driven convective motion in melters serves several functions. It increases the melting rate by moving hot molten glass under the floating batch layer (called the *batch blanket*), supplying it with a new source of heat. Upward convection part way along the tank can serve to help separate the batch melting and fining sections of the melter. Simple patterns of convective motion in a large tank-type melter are illustrated in Fig. 8.9. Sometimes a row of bubblers is built into the tank bottom to release large air bubbles that forcefully rise to the top aiding the convective motion. And, as described above, convection helps to mix and homogenize the melt.

Conditioning and delivery. For a molten glass to be processed into a product by any of the many commercial forming processes, it first must be adjusted to the temperature required for that process. In gen-

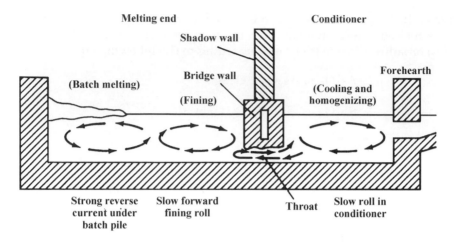

Figure 8.9 Major convection flows in a fuel-fired tank furnace (longitudinal vertical section on tank centerline). *(From F. E. Woolley, Ref. 7, Fig. 3, p. 388, reproduced with permission of publisher)*

eral, all the glass being delivered from the tank must be at that same temperature. This is referred to as *conditioning* (or *thermal conditioning*) the glass. Conditioning is usually done either in a final zone of the glass-melting tank or in a separate unit, called a *forehearth*, attached to the exit of the melter. Heat input and heat losses are adjusted to yield glass of uniform temperature. To achieve the needed balance, the forehearth is often heated electrically or by fossil-fuel burners. Mechanical stirrers are often placed in the forehearth to improve the chemical and thermal homogeneity.

The *forehearth* often has more than one exit from which molten glass is *delivered*; thereby several different forming machines can be *fed* from one melter. Forehearth exit configurations vary depending on the forming process that follows. They can be simple single or multiple circular orifices of refractory material, an orifice containing a flow controlling *needle*, a tube or pipe, or a horizontal "lip" over which the glass flows. When the forming process requires delivery of discrete gobs of molten glass, automated metallic shears are generally used to cut the stream of glass into sections as it exits the melter. Often, the motion of a needle or plunger within the delivery orifice is synchronized with the shearing action to better control the gob weight and shape. One common type of gobbing feeder is illustrated in Fig. 8.10.

Oxidation state (*redox*). When a glass is melted, a chemical equilibrium is approached between the glass composition and that of the surrounding atmosphere. If there are polyvalent ions present within the glass, their average oxidation state is influenced by the partial pressure of

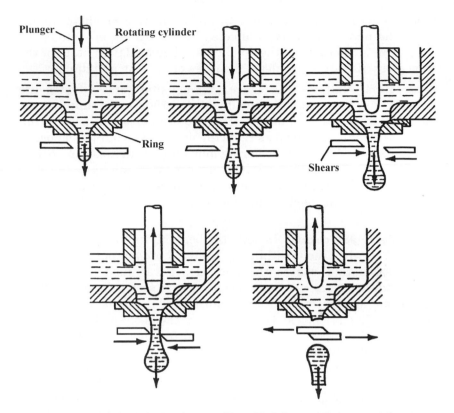

Figure 8.10   A typical gob feeder process. *(From H. J. Stevens, Ref. 7, p. 396)*

oxygen in the atmosphere above the melt. Thus, a glass melted in a fuel-fired furnace, where the atmosphere has been depleted of oxygen by the combustion process, tends to be more reduced than one that has been electrically melted in a standard atmosphere containing about 20% oxygen. The degree to which a glass is oxidized is often referred to as its *oxidation state*. The fining equations listed above describe reversible oxidation-reduction reactions, called *redox reactions*. Thus, the oxidation state of a glass is sometimes referred to as its *redox state*.

When the temperature of a glass melt is decreased below its melting temperature during glass manufacture, it generally has insufficient time to equilibrate with the atmospheres it encounters at the various steps. Consequently, it tends to retain the overall oxidation state that it established during melting. However, as the glass cools, different multivalent elements (for example, iron and manganese) compete with each other for the available oxygen. The relative oxidation states (percent oxidized) of those elements can change with respect to each other during cooling of the glass. Thus, properties such as glass color can change as a glass melt cools. In photosensitive or photochromic glasses, the photosensitivity is affected by the oxidation state of the melt, and in some cases of the solid glass. The oxidation state of the molten glass also strongly affects the rates of melting and fining.

Redox control. Sulfates (*salt cake* and gypsum) and nitrates are sometimes used as oxidizing agents, and carbon or calumite (a reduced calcium-aluminum silicate slag from steel manufacture containing carbon and sulfides) as reducing agents. Of course, the use of cullet containing polyvalent coloring ions and impurities can also alter the oxidation state of a melt. Oxygen sensors, usually in the form of galvanic cell probes inserted into the melt, are being used with increasing success as devices for monitoring and controlling redox state.

Melter-created glass defects. Despite the best efforts to generate a perfectly uniform glass melt, within any melter there are naturally occurring processes that oppose those efforts. These include refractory corrosion (dissolution), electrode corrosion, and preferential volatilization of some species from the melt surface. These produce localized and sometimes more global deviations from the desired glass composition. Localized composition deviations lead to inhomogeneities in the product, called *cord* and *striae*.

Other glass defects originating in the melter are solid inclusions called *stones*. These may be particles of nonmelted batch or pieces of eroded furnace refractory. Devitrification (crystallization of the melt resulting from it having been cooled to temperatures below its thermodynamic liquidus temperature and held there for extended periods of time) is another source of solid inclusions, generally called *devit*. Bubbles and blisters may result from incomplete fining or from *reboil* (the exsolution of bubbles in an otherwise bubble-free melt, generally the result of thermally driven chemical or electrochemical reactions). Such reboil occurs either homogeneously within the melt or heterogeneously at a melt-refractory interface.

The selection rate (percentage of "good" ware) and the profitability and competitiveness of a glass manufacturing operation often depend on how successfully these melting-related defects are avoided. The design of the melter and its method of operation are key factors for success.

8.3.3 Types of Melters

8.3.3.1 General. Glass can be melted on greatly different scales, ranging from a few grams in a laboratory experiment to hundreds of tons per day in a large-scale manufacturing operation. A great deal of glass is melted on a scale somewhere in between. It is important to distinguish between *continuous* and *discontinuous* melting of glass. In a discontinuous process, the batch is placed in a container, such as a crucible, pot, or a tub-like vessel, and heated according to a prescribed time-temperature cycle to carry the contents through the melting stages described above, namely batch melting, fining, and conditioning. Homogenization may be enhanced by mechanical stirring. The vessels may be constructed of metal, refractory ceramic blocks, or cast fireclay; the stirrers generally of metal, fireclay, and high-density ceramic refractories, occasionally with platinum cladding.

In a continuous process, batch materials are continually fed into a melter at one location, and molten glass is continually withdrawn from the melter at another location. All the steps required for melting take place approximately sequentially as the molten mass progresses through the melter, either vertically or horizontally. Large glass melters, or *tanks*, are generally constructed so that the melt passes through them horizontally, with melting, fining, homogenizing, and conditioning occurring progressively as the melt passes through. In some tank designs, attempts are made to confine the process steps to zones separated from each other by physical walls or partial walls, or by thermal convective patterns within the melt itself.

Examples of simple discontinuous melters include crucibles of various sizes and configurations and pot melters. Crucibles are generally made of refractory ceramic materials such as alumina, or refractory metals such as platinum alloys. If small, the crucibles can be inserted and removed from a heated furnace and the glass delivered to a suitable forming operation by the simple means of tipping the crucible and pouring the glass. Larger crucibles are generally fixed in position and molten glass delivered, after it has been suitably homogenized, by the means of an orifice, or tap, at the bottom of the crucible. Large crucible and tub-like containers operated discontinuously are sometimes referred to as *day tanks*.

Pot melters generally consist of single or multiple refractory ceramic pots located in a single furnace. The pots are shaped something like tall bee hives with an opening at the top front through which batch or cullet (broken glass) can be inserted and molten glass withdrawn using a ladle or blow pipe. Such pots have been used for hand glassmaking operations, for both artistic and utilitarian products, for centuries and are still used for such purposes today.

Fuel-fired continuous melters, often called *tank melters*, or simply *glass tanks*, generally consist of large refractory swimming-pool shaped tanks, heated from above the melt by gas or oil flames from multiple burners. The top of the melter is covered by a refractory superstructure consisting of walls and a self-supporting roof called a *crown*. In the simplest version, the so-called *direct-fired furnace,* the gas flames pass across the top of the melt, and the melt is heated by radiation from the flames and from the hot crown of the furnace. The combustion gases are then exhausted from the furnace and dispersed, generally via a tall *stack* and often after passing through electrostatic bag precipitators to reduce particulate emissions and chemical scrubbers to remove acidic vapors.

Such fuel-fired continuous melters can be very large. For example, a soda-lime-silica float glass melter may contain as much as 2,000 tons of molten glass and deliver it at a rate of 800T/day. The top surface area of the glass in the melter could be more than 5,000 ft^2. Large-container glass melters can deliver 100 to 300 tons of glass per day with melt surface areas between 600 and 2,000 ft^2. Because of the rather slow forward motion of the glass, in combination with the convection-driven mixing effects, it is not unusual to expect an average of 3 to 7 days residence time for the glass in such large melters.

8.3.3.2 Heat recovery—regenerative and recuperative furnaces.

Historically, the heat energy required for melting glass was produced by wood or coal fires. Today, the heat is generally provided by burning natural gas or oil, or by immersed electrodes utilizing electric resistance (Joule) heating.

Obviously, not all of the heat generated by a flame is transferred to the glass. Some of it is used to heat the nitrogen gas contained in the combustion air fed to the burners (which is about 80% nitrogen by volume), and much is carried away by the nitrogen and by the gaseous products of combustion (H_2O and CO_2) as they go up the stack. For economic reasons, especially when operating large melting tanks, it is important to recover some of the exhausted heat. This is generally done by using the exhaust gases to preheat the incoming combustion air, either regeneratively or recuperatively, in rather massive refractory heat exchangers.

In a cross-fired *regenerative* glass furnace, of the type first developed by Siemens toward the end of the nineteenth century (see Fig. 8.11), the furnace is fired for a period of time by a row of burners along one sidewall, while the hot gasses are exhausted from the opposite wall and passed through large chambers called *regenerators.* The regenerators consist of stacks of firebrick in open three-dimensional checkerboard patterns, called *checkers*, that absorb heat. When the checkers have become suitably heated, the burners are shut down. A set of burners on the opposite side of the furnace is then fired and fed with combustion air drawn in through the hot regenerators. The combustion gases are now exhausted from the opposite side of the furnace

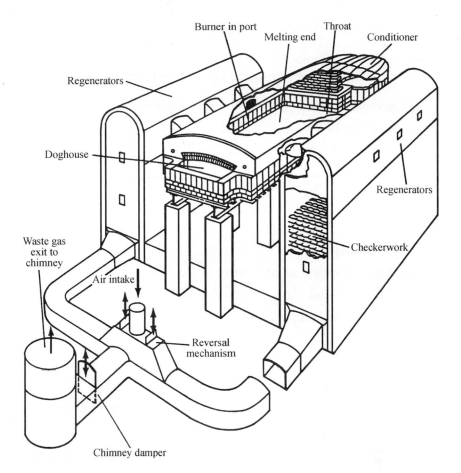

Figure 8.11 Cross-fired regenerative-type furnace construction typical of glass melters.[26]

through another set of regenerators. This process is reversed about every 15 to 20 min.

In a *recuperative* furnace, the burners are fired continuously, and heat is exchanged continuously in the exhaust system where the exhaust gases flow outward through a central tube or tubes with the incoming air channeled along the outside of the tube(s).

8.3.3.3 Construction materials/refractories. The refractory materials serve three purposes: they contain the glass melt, provide thermal insulation and heat transfer as needed, and act as key structural elements of the tank. The refractories are generally held in place by an open steel framework, but much of the mechanical load is born by the refractory materials themselves. Crowns are often self-supported refractory structures. The refractories must withstand or resist high temperatures, heavy loads, abrasion, and corrosion.

Types of oxide refractories used in glass melters are as follows:

1. Clay refractories (generally used for insulation)

   - Fireclay: kaolinite ($Al_2O_3 \cdot 2SiO_2 \cdot 2H_2O$) plus minor components; classified as low-, medium-, high-, or super-duty; (25 to 45% Al_2O_3)

   - High alumina (50 to 87.5% Al_2O_3)

2. Nonclay refractories (generally used for glass contact or furnace superstructure)

   - Magnesia (MgO)

   - Silica (cristobalite and trydimite)

   - Stabilized zirconia

   - Extra-high alumina (>87.5%)

   - Mullite ($3Al_2O_3 \cdot 2SiO_2$)

   - AZS (alumina-zirconia-silica) containing 30 to 42% zirconia

   - Zircon ($ZrO_2 \cdot SiO_2$)

   - Chrome-magnesite and magnesite-chrome (combinations of Cr_2O_3 and MgO)

Classifications based on methods of of refractory manufacture (in order of increasing density) are as follows:

- Bonded

- Sintered (sometimes densified by cold isostatic pressing before firing)

- Fusion cast (cast as blocks from arc-melted raw materials); also referred to as *fused* or *fused cast*

Many factors affect the choice of refractory for a particular application. These include melting temperature, thermal conductivity, mechanical strength, creep resistance, and resistance to corrosion and spalling, to name a few. Generally, different refractories are used in different regions of a glass-melting tank, because the requirements are different. However, *the ultimate design consideration* is resistance to corrosion by molten glass and by the hot gas atmosphere within the melter. Corrosion determines tank lifetimes and affects the rates at which certain glass defects (such as stones) are generated. In selecting a refractory the considerations are, in order,

1. glass quality

2. tank lifetime

3. initial cost

Many different corrosion mechanisms operate within a melter. The types and severity depend on the glass composition, the composition and microstructure of the refractory, and the temperature. Corrosion rates tend to increase dramatically with temperature. This is an important reason why different refractories are often selected for different portions of the melter. It is also an important reason why the harder (higher melting temperature) glasses are generally more expensive to manufacture. Corrosion types include:

- *Front surface attack* (frontal attack).   This is direct attack at the glass/refractory interface. Its mechanisms include alkali diffusion into the refractory with consequent fluxing and dissolution of the refractory crystals. Porous refractories are generally more susceptible.

- *Melt line corrosion* (attack where glass surface, air and refractory meet).   Corrosion at this location is enhanced by localized convection currents and fluctuations in glass level within the melter.

- *Upward drilling.*   This form of corrosion occurs where bubbles form under horizontal refractory surfaces, such as throat cover blocks or submerged horizontal refractory joints. Corrosive vapor species concentrate in the bubbles. As with melt line corrosion, the greatest corrosive activity is believed to take place where the vapor, refractory, and glass touch.

- *Downward drilling.*   This type of corrosion results when droplets of molten metal settle on the bottom of the tank. Sources of metal can be contaminants in batch raw materials or cullet, chemical reduction of certain glass components (such as lead oxide), and even tools or metal parts accidentally dropped into the tank.

Glass contact refractories.   The most common glass contact refractories include the following:

- Fused AZS (alumina-zirconia-silica). This is the most common today (41% ZrO_2 in high-wear areas and electrode blocks, as opposed to less expensive 34% variety). The oxidation state is critical (the refractory contains a residual glassy phase which, if produced in reduced condition, will oxidize in use, swell, and exude from the brick).

- Dense sintered zircon ($ZrO_2 \cdot SiO_2$). This is used in some low expansion borosilicate melters.

- Clay, fused alumina, bonded AZS, and dense sintered alumina. These are used for lower-melting specialty glasses.

Typical glass contact refractories used for melting various glass types are:

- *Container glass.*   Fused AZS, life 8 to 10 years; also, fused α-β alumina; finer bottoms are sometimes bonded AZS, zircon, and clay.

- *Float glass.*   Fused AZS in melting zones; fused alumina in conditioner zones, life 10 to 12 years.

- *Hard borosilicates.*   Fused AZS and zirconia, sometimes dense sintered zircon.

- *Fiberglass wool.*   Highly corrosive, melted electrically in fused chrome-AZS or fused alumina-chrome refractories (coloration due to chromium is of little consequence in this application).

- *E-glass (textile fiber).*   Less corrosive, melted in dense sintered chrome oxide.

- *Lead crystal.*   Tendency to electric melting in AZS.

Several refractory sidewall design considerations are based on corrosion concerns. First, the thicker the wall, the lesser the heat lost and the lesser the energy consumed. But thicker walls create a smaller temperature gradient, allowing chemical attack to penetrate more deeply into the refractory. Thus, the design thickness of a wall must be a compromise between heat loss and wear. To give all sections of the wall approximately equal lifetimes, thickness, and type of refractory are often varied from location to location. These techniques are sometimes referred to as *zoning by thickness* and *zoning by type*.

To avoid melt infiltration of horizontal refractory seams and the consequent increased opportunity for upward drilling of corrosion, glass contact wall refractories are often large, full-height blocks arranged adjacent to each other in a "soldier" course fashion. However, if multiple courses are required, close-fitted diamond ground horizontal joints can help minimize melt infiltration. Similarly, to avoid horizontal joints, the *paver* blocks composing the top layer of the tank bottom are butted up against the side blocks, not placed under them.

Superstructure and crown refractories. The superstructure, which includes all furnace walls above the melt line and the crown, are subject to corrosion by aggressive vapor species such as NaOH, KOH, PbO and HBO_2, batch dust particles, liquid condensates and liquid reaction products running down from refractories higher up. Superstructure temperatures in fuel fired furnaces are often 60-100°C hotter than the glass. Typically, walls and ports made of fused AZS from the *back wall* (where the batch enters) to the *hot spot*; fused β-alumina is used downstream. But these are not hard and fast rules. Crowns typically consist of sintered silica block. Although silica crowns are attacked by alkali vapors, the drips are homogenized into melt. Crown life is more of a concern. This is especially true with gas-oxy firing (more aggressive vapors, see Sec. 8.3.3.5), in which case more costly refractories may be justified. Silica and alumina blocks should never be in direct contact, for example at the joint where the crown and superstructure walls meet (they will react); zircon is used as a buffer.

Regenerative heat exchanger refractories. Special refractory considerations are needed because of the large temperature gradient (top–bottom) within the checker chambers and the corrosive nature of the exhaust gases. As the gases cool, a temperature is reached at which the corrosive vapors condense on the refractory surfaces, enhancing the corrosion. Fine batch particles carried over with the exhaust gases also tend to react with and corrode the regenerator refractories. High thermal conductivity and heat capacity are also important characteristics. Checker construction for a typical soda-lime-silica melting tank is: top third, bonded 95 to 98% MgO bricks; middle third, lower magnesia content bricks; bottom third (where alkali vapors condense), sintered chrome (chromic oxide) or magnesium-chrome bricks.

A relatively new approach to checker construction, especially in Europe, is with special interlocking shapes (cruciforms) of AZS or high-alumina fused cast refractories.

8.3.3.4 Electric boosting and all-electric melting.
Crucibles, pots, and day tanks for glass melting can be heated in electric furnaces where the heat is generated by resistance heating in windings or bars. In come cases, heat is produced by the flow of electricity through the metal crucible itself, and in others by the flow of electricity through the molten glass, which is a moderately good ionic conductor at high temperatures, between submerged electrodes.

These principles are applied to varying degrees in large continuous melters as well. In some fuel-fired furnaces, electrodes are installed in the walls below the glass line so as to provide a source of heat below the batch layer, thus the batch is actively melted from below as well as above. By such means, the melting rate for the tank can be increased, or boosted, leading to the term *electric boosting*. Resistance heating by external windings or bars is often used to control temperatures at the orifice or delivery tube by which the molten glass leaves the melter.

Sometimes, all the heat required for melting is supplied electrically within the molten glass. In this case, electrodes are positioned as plates at the walls of the furnace, or as water-cooled metal rods extending upward through the bottom of the furnace. The electric current passes from electrode to electrode, through the glass, with the amount of heat generated depending on the applied voltages, the shape and spacing of the electrodes, and, very importantly, the electrical resistivity of the molten glass. In this case of all-electric melting, the batch components are melted solely by heat flow up from below. It is possible and desirable to maintain a continuous layer of batch across the top of the melt, eliminating the need for a refractory roof or crown to contain and reflect the heat. However, in many cases, the crown is there for other reasons, but it is cold in temperature. Hence, all-electric melting in this manner is sometimes referred to as *cold-top* or *cold-crown* melting.

It should be noted that the electrical resistivity of a pile of non-melted batch, or even a glass melt at temperatures well below 1,000°C, is too great to allow for efficient heat generation. Consequently, all-electric melters are started in a more or less traditional way using fossil-fuel burners and a hot crown. Once the molten glass has reached sufficient temperature, the burners (and sometimes the crown) are removed.

Cold-top melting is valuable for two reasons. First, the batch layer acts as a thermal insulating blanket (the *batch blanket*), which helps reduce heat loss out the top of the melter (thus enabling it to operate cold-top). Second, the top layers of the batch blanket, being much cooler than the molten glass below, act to condense volatile vapor species that might otherwise escape into the atmosphere. This is especially valuable for fiberglass and other specialty glass melting where the compositions contain fluorides and other very volatile, and sometimes unhealthy, components.

Electrodes are made of materials such as carbon, tin oxide, molybdenum, and platinum, with the choice depending on the temperature of operation and the composition of the glass being melted. Melt temperatures, temperature gradients, and convection currents are greater at and near the electrodes; therefore, better refractories (higher temperature, more corrosion resistant) are required at these locations.

8.3.3.5 Oxygen for combustion. Over the past decade, there has been a trend toward the use of oxygen instead of air, in combination with natural gas, to heat fuel-fired furnaces. This is called *oxy-fuel firing*. It has several advantages. First, since one is not using air with its 80% nitrogen content, much less polluting NO_x gases are produced. In the face of increasingly stringent air quality legislation, this factor alone is often sufficient to justify conversion to oxy-fuel firing. Second, since there is not the large volume of nitrogen to heat and expel from the furnace, much less waste heat is generated than with gas-air. Some operations have been able to eliminate the massive, costly regenerators as a consequence. Third, higher flame temperatures are possible. Fourth, as claimed by some manufacturers, more stable furnace operation, with an associated improvement in glass quality, is achieved. This is especially the case when regeneration, with its inherent periodic reversals of gas and heat flow, is eliminated.

There are some disadvantages to oxy-fuel firing. One is the need for liquid oxygen storage or oxygen generation on site. A second is that, without the large volumes of air moving through the furnace, the concentrations of water vapor (a product of combustion) and corrosive volatile species from the melt (e.g., NaOH) are much higher, in some cases leading to increased deterioration of the refractory superstructure.

8.3.3.6 Furnaces for specific applications. Furnaces designed for specific applications include the following:

- *Container glass.* Typically cross-fired regenerative; maximum melt temperatures ~1600°C; large, up to 500T/day.

- *Float glass.* Typically cross-fired regenerative; no bridge wall, but rather an open surfaced narrow region called a *waist* to keep inhomogeneities running parallel to the surface of the glass sheet; maximum melt temperatures ~1600°C; larger, up to 800T/day.

- *Fiberglass.* Smaller gas-fired recuperative or all-electric.

- *Lead crystal.* Small electric boosted or all-electric.

- *Hard borosilicates.* Tending to all-electric or heavily boosted regenerative; melt temperatures > 1600°C.

- *Aluminosilicate glass-ceramics.* Regenerative gas-fired; temperatures near 1700°C required for efficient fining.

- *Optical glass.* Small fuel-fired or electric heated; fining and conditioning often done in platinum tubes to avoid refractory contact and resulting inclusions and inhomogeneity.

With the advent of oxy-fuel, or more specifically gas-oxygen, firing, many of the above listed regenerative and recuperative furnaces have been converted to use this new technology. However, as of this writing, float glass manufacturing is just beginning to convert to gas-oxygen firing.

Trends. As is typical in most industries, new designs are aimed at lower overall cost of operation (the calculation of which includes initial cost, melter lifetime, and costs of repairs as well as the daily operating costs) less energy consumption and less overall environmental impact. Lower cost almost always must be achieved in combination with improved glass quality and less adverse environmental impact.

8.4   Glass Making II—Glass Forming

The term *forming* collectively refers to all the processes of glass making used to form a solid object or product from the molten glass. Historically, all glass objects were formed by hand using relatively simple implements. Over time, the techniques were modified, automated, and scaled up. While several glass forming methods in use today have no precedent in early glass history, most still bear important resemblance to their forbearers. Due to space limitations here, this section will describe only processes used in today's manufacturing plants and, when relevant, the early hand-forming operations. Little attention will be paid to the many processes that have intervened. We will first discuss processes involving molds.

8.4.1   Blowing

By far, containers (bottles and like products) account for the largest volume of glass production. Almost all these products are manufactured using some form of a *blowing* process.

Historically, glass containers have been blown to shape by gathering a gob of molten glass on the end of a hollow iron pipe, the blowpipe or blowing iron, and blowing a puff of air into the soft glass to form a bubble, which is gradually expanded and *worked* into shape by the combined effects of gravity and the forces of tools pressed against it. Generally, the blowing iron, with the soft glass attached, is rotated to balance the effects of gravity and provide an axial symmetry to the product. While useful containers of remarkably repeatable shapes and dimensions can be created in this manner, for rapid and precise production, it is preferable to use a two-step process. First, a hollow preform, called a *parison*, is prepared using a simple blowing process. Second, the parison is blown to the final shape in a mold.

This process has been automated to a very high degree in modern times, to the point where more than a dozen containers per minute can be generated from each mold. Generally, rotating *split* molds are used for shapes involving bodies of revolution whenever visible seam lines from the molds are undesirable, such as for light bulbs or high-quality drinkware. Stationary split molds must be used for containers having handles, flutes, or other nonrotationally symmetric shapes. The rotating split molds are generally *paste* molds, called that because their molding surface is coated with a thin layer of cork or similarly permeable substance, which is saturated with water after each molding cycle. When the mold surface is contacted by hot, molten glass, a steam layer results, which provides a low-friction layer between glass and mold, giving the product a highly polished appearance without seam lines.

The stationary molds are generally *hot iron* molds. These metal molds are operated at a temperature hot enough to keep the molten glass from being chilled so quickly that surface cracks or *checks* result, but cold enough to quickly extract heat from the glass and allow it to become rigid before removal. Any metal mold surface defects, as well as the mold seam lines, are transferred to the ware, but production rates can be much faster than with paste molds. Also, on the plus side, intentional designs such as logos can be molded or embossed into the glass surface.

When blowing by a hand-type operation, the final product must be separated from the blowing iron, usually by cracking it off. This leaves a rough surface that must be properly finished by grinding or *fire polishing*, a process step that involves locally reheating the glass to a point at which it will flow to a smooth surface under the influence of surface tension. In modern automated container production, free gobs of glass are handled in the molds, so separation from a blowing iron is not required. Two common processes are called *blow-and-blow* and *press-and-blow*, depending on the method used to form the parison. Blow-and-blow is generally used for narrow-neck containers such as beverage bottles. The parison is blown in one mold in a way that forms the neck and then, held by the relatively cold newly formed neck, it is transferred into a second mold to blow the body of the container. One of the more common machines featuring these operations is Hartford Empire's (now Emhart Corporation's) *Individual Section (IS)* machine. This mechanism may have as many as 12 sections driven in tandem by a cam with overlapped timing or, more recently, by electronically synchronized operation. Each section operates on as many as four gobs. Processing speeds are about 10 s per section. In addition to speed, an advantage of the IS machine (as opposed to a rotating turret machine) is that the machine can be programmed to run the remaining sections while one is being repaired. The operation of a single two-mold IS section is shown in Fig. 8.12.

Press forming of the parison before blowing to final shape is used for wide-mouthed containers such as food jars. Press forming will be described in the next section. For container manufacture, while pressing of the parison is complicated by the need for an additional tool (the

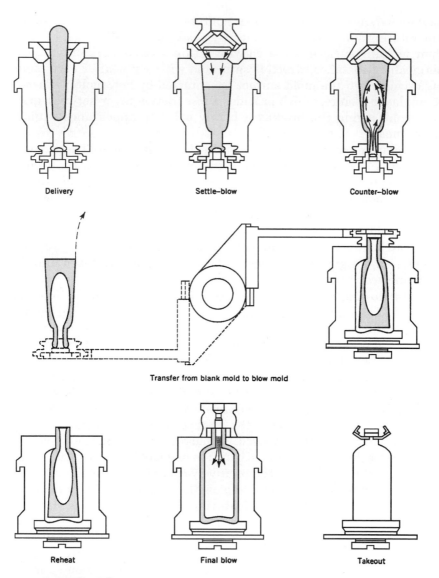

Figure 8.12 The H.E. IS (individual section) blow-and-blow machine. The gob is delivered into a blank mold, settled with compressed air, and then preformed with a counter-blow. The parison or preform is then inverted and transferred into the blow mold where it is finished by blowing.[27]

plunger), this disadvantage is offset by yielding a product of more uniform wall thickness, hence a more efficient utilization of glass and a lighter-weight product than produced by blow-and-blow.

A very high-speed process for blowing light bulb envelopes and the like, known as the *ribbon machine*, was developed in the 1920s by

Corning Glass Works (now Corning Inc.) and is still in use worldwide. In this machine, a stream of molten glass is continuously fed between a set of rollers, one flat and the other with pocket-like indentations. These rollers form a ribbon of glass several inches wide, containing regularly spaced circular mounds of glass down the centerline. The parison for each light bulb is formed by inserting a synchronously moving blow head (analogous to a *blowpipe*) into each mound of glass and blowing it through a synchronously moving orifice plate. As the ribbon travels horizontally along the machine, the parison is enclosed in an also synchronously moving rotating paste mold, and the blowing process is completed. The moving molds open and swing away to allow the finished glass envelope to be cracked off the ribbon at the machine exit. The operation of the ribbon machine is illustrated in Fig. 8.13. Incandescent lamp envelopes (for example A-19, 60-W bulbs) can be made at speeds in excess of 1,200 per minute on a single machine using this technique. Small automotive and other specialty lighting bulbs can be made at rates exceeding 2,000 per minute.

8.4.2 Pressing

In simplest terms, *pressing* or *press forming* of glass involves placing a gob of molten glass in a hot metal mold and pressing it into final shape with a plunger. Sometimes a ring is used, as illustrated in Fig. 8.14, to limit the flow of glass up the side of the mold and produce a rim of well

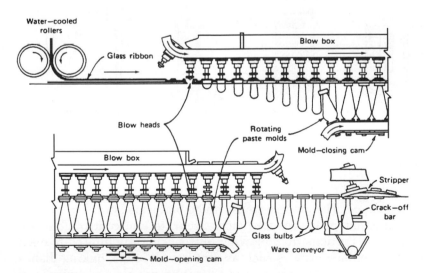

Figure 8.13 The "ribbon machine" used for light bulb envelope manufacture. U.S. patent 1,790,397 (Jan. 27, 1931), W. J. Woods and D. E. Gray (to Corning Inc.). *(Courtesy of Corning Inc.)*

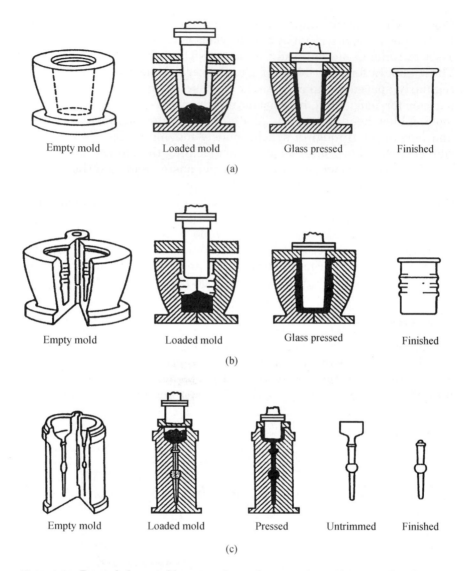

Empty mold   Loaded mold   Glass pressed   Finished

(a)

Empty mold   Loaded mold   Glass pressed   Finished

(b)

Empty mold   Loaded mold   Pressed   Untrimmed   Finished

(c)

Figure 8.14 Pressed glass, mold types and pressing operations. *(Courtesy of McGraw-Hill)*[28]

controlled shape. The process steps can be performed entirely by hand or fully automated. It produces more accurate and controllable wall thickness distributions than blowing but is generally limited to open, moderately shallow articles such as dinnerware, cups, baking dishes, sealed-beam headlamp lenses, and television panels and funnels, or for solid objects. Pressing is capable of generating intricate and accurate patterns in the glass surface, such as found in sealed beam spotlight, floodlight, and automotive headlamp lenses and in street and traffic light refractors and lenses. Large objects, such as 27- and 35-in. (diagonal) color television bulb panels weighing more than 25 pounds, can be made using automatic pressing equipment.

Glass is generally pressed at a viscosity between 2000 and 3000 P with an applied pressure of about 100 lb/in^2 of article surface area. For large television panels, the total force on the plunger can exceed 20 T. Temperature control of the molds and plunger is crucial; too cold leads to brittle fracture of the glass under the pressing forces, and too hot leads to sticking of the glass to the mold surface, requiring it to be physically broken free. Vents within the mold body, through which cooling air or water may flow, are often used to maintain uniform temperature distribution across the mold surface.

8.4.3 Casting

Casting is a relatively little used process, found mostly in hand shops and for the production of very large pieces of glass such as glass sculptures and astronomical telescope mirrors. For the large pieces, glass is poured into hot ceramic refractory molds (often sand with a small amount of binder) that are slowly cooled after the mold is completely filled. Alternatively, chunks of rigid glass may be placed in a cold mold and raised in temperature until the glass is sufficiently fluid to flow and fill the mold. This latter method is more susceptible to entrapment of bubbles. Generally, slow cooling and long annealing times are required. The mold can be used only once. The glass surfaces in contact with the mold are generally rough.

8.4.4 Centrifugal Forming

Centrifugal forces have often been utilized by the glassmaker. A glass bubble on the end of a blowing iron can be elongated by swinging the iron back and forth to aid gravity in elongating the bubble to generate the parison. A thick-walled bubble on the end of a rod can be cut open at the point opposite to the rod, and the rod rotated to generate sufficient centrifugal force to open the bubble and spin it into a relatively flat, circular sheet of glass. This is one of the earliest flat glass manufacturing methods, the *crown* process. Glass made this way is often found in old European churches. A droplet of very fluid glass placed at the center of a rotating turntable will also spread under centrifugal force, a process utilized in *spin coating* or *spin casting*. The latter is sometimes simply called *spinning*.

If molten glass partially fills a rotating container such as a mold or a crucible, the molten glass will tend to climb the walls, propelled by the centrifugal forces, giving the glass surface the shape of a paraboloid of revolution. This method, called *centrifugal casting,* is used to form the parabolic shapes for thin astronomical telescope mirrors. It has also been used to spin, rather than press, large, deep television tube funnels and glass-ceramic missile radomes. Six- to 8-meter diameter mirror blanks are spun at about 10 rpm, a television bulb funnel at about 200 rpm. In the 1960s, Corning Glass Works used this spinning process to make large, 56-inch diameter glass hemispheres of 1.5-inch wall thickness for use in undersea exploration.

A continuous centrifugal process of forming tubing from very short (steep viscosity) glasses or easily devitrifiable glasses has been devised. It is somewhat analogous to the Danner tubing process (see below) except that the stream of glass is fed into the open end of an inclined rapidly rotating pipe. The very fluid entering glass is flattened against the wall and the adjacent layer of previously deposited glass and is maintained in position by centrifugal forces until it is cooled to sufficient rigidity to be withdrawn from the end of the pipe.

8.4.5   Rod and Tube Drawing

Drawing is the term for a process in which a preshaped blank, or glass flowing from an orifice, is elongated (stretched) in one dimension while diminishing in orthogonal dimensions without losing its cross-sectional characteristics.

The above statement is exactly true for the drawing of *cane* (rods) or fiber. It is not so for tubing or sheet, where the ratios of inside to outside diameter or width to thickness are not the same as they were at the *root.* (The solid section of the blank or the glass at the orifice is often referred to as the "root.")

Redrawing is the specific case of drawing from a solid *preform* (or *blank*) rather than from a melt. This involves reheating the end of the blank to provide glass sufficiently fluid to be stretched and attenuated. In a continuous process, the blank is replaced at the volume rate at which it is used up by gradually feeding it into the hot zone of the redraw furnace.

In a steady-state process, the volume flow per unit time, the quantity Q, is constant and equals A (area) $\times v$ (velocity) at any point in the process. As we will show, A is a very important parameter in the tube drawing process.

Tube drawing processes. In a hand process, a gob of glass is gathered on the end of a blow pipe, a bubble is blown within the glass, and an assistant attaches a rod to the side of the gob opposite the blow pipe (or grabs the gob with a pair of tongs) and walks across the room to stretch out the glass and the bubble within it. The final diameter of the resulting tubing, and its wall thickness, depend on several factors, including how fast the assistant walks (compared to how rapidly the glass cools) and how much pressure the blower maintains in the bubble. If faster cooling is needed, a second assistant may fan the tubing as it is drawn out. The air pressure resists tubing collapse from the draw forces and surface tension. After the drawing step is completed, the hollow glass tubing is cut away from the bulky pieces at each end. There is only about 10 to 20% glass utilization. The rest of the glass remains on the blowpipes or is of unusable dimensions and is generally recycled as cullet. The process is highly labor intensive.

The *Danner* process, named after its inventor, Edward Danner, was developed by the Libbey Glass Company. It is one of the oldest continuous tubing drawing processes still in use today and is the common method for forming fluorescent lamp tubing. The process is somewhat unique in the manner of preparing the root of glass from which the tubing is drawn. See Fig. 8.15. A ribbon-shaped stream of glass is fed onto a slowly rotating (~10 rpm) hollow clay (or metal) mandrel, inclined downwards perhaps 15° from horizontal. The glass stream flows onto the mandrel at a viscosity about 1,000 P, wraps around the mandrel, and overlaps itself to form a cylinder, which is smoothed by the forces of gravity and surface tension. The glass cools (to a viscosity of

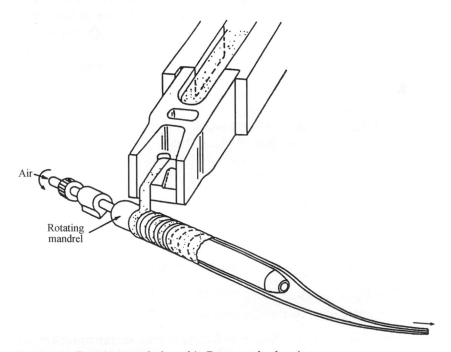

Figure 8.15 Rotating mandrel used in Danner tube drawing.

about 50,000 P) as it moves along the mandrel and is drawn off the mandrel end in a horizontal direction forming a catenary. Air is fed through the mandrel, so the latter acts somewhat like a hand blow pipe with a continually replenished supply of glass. The tubing is drawn (stretched to smaller dimensions) by a tractor device located many feet from the mandrel and is cut into lengths after it passes through the tractor.

The Danner process is capable of drawing 1/16- to 2.5-inch diameter tubes. Because of temperature nonuniformity, coupled with gravity effects, Danner tubes often exhibit some ovalness and wall thickness (called *siding*) variations. Solid rods of similar diameters can be drawn by stopping the airflow through the mandrel or even drawing a slight vacuum. Composition variations can produce hairpin-shaped cord defects.

The *Vello* process, after inventor Sanchez Vello, was developed by Corning Glass Works and dates back to the early part of the 20th century. Here, the glass is delivered from the glass-melting furnace, at about 100,000 P viscosity, through an annular orifice created by the spacing between a conical *bowl* and a bell-shaped blowpipe, called the *bell*, centered within the bowl. See Fig. 8.16. The drawn tubing first extends vertically downward then turns horizontally, following a catenary curve as it stretches under its own weight, then is transported on a runway of "V" rollers often several hundred feet long as the glass cools. As with Danner, the tubing is cut into lengths after it passes through the tractor at the end of the runway.

The Vello process allows drawing of precision-bore tubing, such as for thermometers and burettes. It is fast (for example, 800 52-inch sticks/min at a 2000 lb/hr flow rate) and can draw tubing of diameters up to 3 in. without significant oval.

Control of diameter and wall thickness is based on a mathematical equation relating volume flow of glass, tubing velocity, and tubing cross-sectional area, which determines how fast one must run the tractor pulling the tubing to give the requires cross-sectional area of the glass in the tubing. This area is given by

$$\frac{Q}{v} = A = \pi w (D - w)$$

where Q = volume flow of glass from the melter
v = speed of the tractor
w = wall thickness
D = outer diameter of the tubing

Only in the case of a rod, where $w = D/2$, is the diameter uniquely determined by the pulling speed. In all other cases, the diameter and

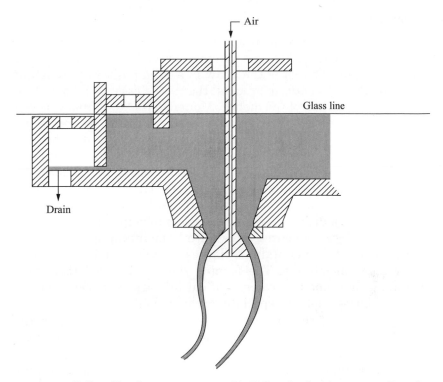

Figure 8.16 Bell and bowl arrangement used in Vello tube drawing process. From lecture notes, E. H. Wellech, Corning Glass Works, 1963. *(Courtesy of Corning Inc.)*

wall thickness interact, and their ratio is maintained within specification by the pressure of the air flowing through the tubing, just as with hand drawing.

The *downdraw* tubing process is essentially a vertical Vello. The glass tubing is cut off one or more floors below the draw. It is useful for tubing that is too large in diameter or wall thickness to be successfully turned horizontally (i.e., without breaking or deforming from cylindrical shape). For example, Corning has drawn 6-in. dia., 3/8-in. wall borosilicate tubing for Pyrex™ brand pipe by this process. Molds and dies can be added to give controlled cross-sectional shapes.

The *updraw* tubing process is analogous to the updraw processes, flat or cylinder, used to manufacture sheet glass, which will be described in the next section. The air pressure needed to keep the tube bore open is supplied from below, through a refractory cone positioned in the melt just beneath and on axis with the drawn tubing. The height of the cone helps control the tubing wall thickness. Continuous updraw of thermometer tubing with enclosed colored glass ribbons is only slightly more complicated.

8.4.6 Sheet Drawing

Sheets of glass can be drawn either upward or downward from a bath of molten glass by a tractive mechanism, provided a method can be found to maintain the root of the draw in a fixed position at constant dimensions. For a *downdraw* process, the root is essentially a rectangular slot in the bottom of the melter. Molten glass flows from this slot under the combined effects of hydrostatic pressure from above and tension from below. Below the melter, the sheet is cooled to become rigid, after which its weight is supported by pairs of horizontal rollers. As the glass is cooled by radiation beneath the slot-shaped orifice, the rollers stretch it to the desired final thickness. The sheet also tends to become narrower in width during stretching, but this effect can be minimized by the judicious use of edge coolers to help hold the edges of the sheet out. These edgewise forces, while maintaining the desired sheet width, also help maintain its flatness. One is essentially pulling on a stretched membrane. The formed sheet is lowered through a heated annealing zone (*annealer*) and cut into separate sheets below, often in a subbasement or a specially excavated pit.

A disadvantage of the process is that any nonuniformities in the slot, such as might be caused by erosion or corrosion, lead to vertical streaks in the glass surface. In part to overcome this difficulty, and in part to have better thickness control over the resulting sheet, Corning Glass Works (now Corning Inco.) developed their *fusion* downdraw process. In this process, molten glass is fed into one end of a slightly inclined refractory trough at a viscosity of about 40,000 P and allowed to overflow both sides, as shown in Fig. 8.17. (Sometimes this trough is called the *overflow pipe* for obvious reasons.) The outside of the trough tapers to a line at the bottom where the two layers of overflowing liquid meet and fuse together, forming the root of the draw, hence the name *fusion*. The outside surfaces of the glass are generated from within the interior of the melt and are therefore never subsequently in contact with other materials; thus, they are pristine and defect free. A key element of Corning's initial patent for this process is the mathematical design of the tapering cross-sectional profile of the trough which, in combination with the incline of the top of the trough, assures that the volume flow of glass over the pipe is uniform along its length. Along with a method to precisely control the temperature along the root of the draw, this assures uniform sheet thickness across its width.

Below the root, the process is somewhat similar to a downdraw from a slot. One notable difference is that the glass at the root is far more viscous. Having been cooled to a viscosity of 500,000 P or more as it descended the tapered lower refractory of the trough, it must be pulled downward with greater tensile stress. The pulling forces are provided

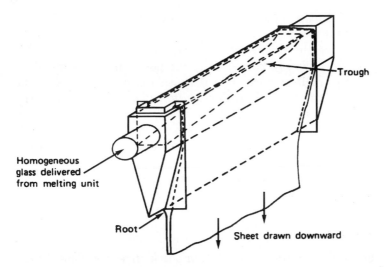

Figure 8.17 Illustration of Corning's "fusion" overflow sheet drawing process. *(Courtesy of Corning Inc.)*[29]

by the weight of the sheet and edge rollers only, so the glass surface is not subject to roller marking. A wide range of thicknesses may be made, ranging from less than 1 mm to greater than 1 cm. This process is currently used to manufacture thin, flat glass of exceptional quality for active matrix liquid crystal displays (AM-LCDs) for flat-panel computer screens and televisions.

A variety of updraw processes have also evolved. The *Fourcault* process, invented in 1910 by the Belgian Emile Fourcault, uses a partially submerged refractory block containing a long machined slot to form the root of the draw. This block is called the *debiteuse*. The molten glass that forms the root of the draw is forced up through the slot by the buoyant force of gravity. The draw is started by lowering a metal mesh, called the *bait,* to the slot and then, once it is wetted by the glass, drawing it upward to form the sheet. Once the process has been started, the sheet of glass is pulled upward by pairs of horizontal rollers, drawn through an annealing zone, and cut into separate sheets above. The bath of molten glass is held at temperatures providing a viscosity of about 100,000 P. Water-cooled edge rollers, located somewhat above the melt surface, help prevent the sheet from narrowing in width as it is pulled upward. This process is still in use throughout the world for the manufacture of window glass. Process disadvantages include vertical streaks, called *peignage* or *music* lines, caused by erosion of the slot in the debiteuse.

The Pennvernon process, developed by PPG, uses a fully submerged draw bar to control the location and straightness of the root. Viscous forces confine the drawn sheet to a position over the center of the bar. In the Colburn-LOF process, the soft glass is bent 90° over a roller a short distance above the melt surface. This roller, along with water-cooled edge rollers, serves to keep steady the position where the glass is drawn from the melt surface. In all these updraw processes, the pulling rollers and the bending rollers, when used, tend to mar the glass. Another disadvantage of updraw is that whenever the sheet breaks in the rolls, the broken glass falls back into the machine and into the melt. The updraw processes are more difficult to restart than are the downdraw processes. With the possible exception of Fourcault, these updraw processes are in relatively little use today, primarily because they have been superseded by float (see Sec. 8.4.8).

8.4.7   Rolling

Rolling can be used to manufacture thin or thick sheets of glass. In the simplest form, a puddle of glass can be poured onto a metal table, and a metal roller is used to spread it to a constant thickness. Generally, parallel spacer bars are used to limit the thinness to which the roller can spread the glass. In a continuous process, a stream of glass is fed between a pair of counter-rotating water-cooled metal rollers. The spacing between the rollers determines the thickness of the resulting sheet; the width is controlled by the amount of glass fed to the rollers. The sheet is generally rolled vertically downward, or at a sufficient incline that a puddle of molten glass can be maintained at the entrance to the rollers. The speed of the operation and the diameter of the rolls must be sized to the thickness of the sheet being rolled. The glass enters the rolls relatively fluid, but it must be considerably less fluid when it leaves so as to maintain its shape. The thicker the glass, the more heat that must be removed by the rollers. Generally, thick sheet requires large rollers, maybe 4 to 10 ft in diameter, and forming rates of only a few feet per minute. Very thin ribbon can be made with small rollers, a few inches in diameter, at rates of several feet per second. After rolling, the continuous sheet is transported horizontally through a heated annealer.

The surface finish quality and thickness uniformity is generally insufficient for mirror, automotive, and architectural applications. Prior to the development of the float process, described below, rolled glass sheet was ground and polished on both sides, sometimes simultaneously, to meet these requirements. These products were known as *plate* glass. The process was inherently wasteful and expensive.

Patterned glass is made by applying texture to one or both glass surfaces using suitably embossed rollers. Applications include shower doors, furniture tops, room dividers, and windows. There is even an application for roller-applied Fourcault-type sheet texture for use in restorations of nineteenth century homes.

As a variation, *wired* glass, such as used in fire doors and building skylights, can be manufactured by continuously feeding wire mesh between the rollers along with the molten glass.

8.4.8   The Float Process

Grinding and polishing of rolled glass was very expensive, labor intensive, and wasteful of materials. In the 1950s and 1960s, the Pilkington company in England developed a much more economical process based on floating a continuous ribbon of molten glass on a bath of molten tin as the glass cooled and solidified. This process is illustrated in Fig. 8.18. A detailed description of the process and the difficulties that were overcome in its development lie outside the intent of this handbook, but some key points should be made. The glass product, known as *float* glass, has excellent surface properties, the upper surface having flowed freely without contact with rollers or any other forming de-

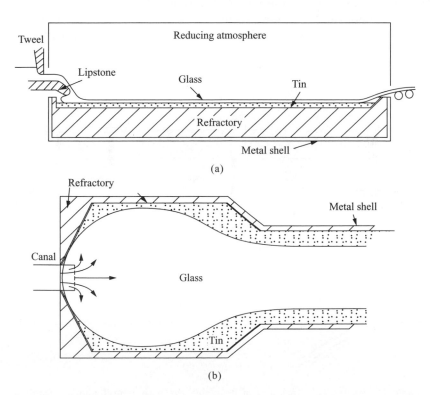

Figure 8.18   Float process: Pilkington-type tin bath: (a) side view and (b) top view.[30]

vices before its solidification, and the lower surface similarly having been in contact only with a flat, smooth liquid metal surface that was incapable of marring it. The product also has exceptionally uniform thickness.

Regarding thickness, a freely spreading puddle of glass suspended on molten tin (by buoyant forces in a gravity environment) will reach an equilibrium thickness determined by the tin and glass densities and the various surface and interfacial tensions. For soda-lime-silica glass on tin in the Earth's gravity, this thickness is between 6 and 7 mm, approximately that of traditional plate glass. Several techniques have evolved to make thinner and thicker float glass. These essentially involve pulling the glass off the bath at a rate faster or slower than would maintain the above-defined equilibrium thickness, and doing so in a manner that preserves the thickness uniformity. Several techniques for "stretching" the glass in this manner have evolved. All employ gripping the edge of the spreading glass puddle on its top surface with knurled rollers to assist, restrict, or redirect the glass flow. One method is illustrated in Fig. 8.19.

It should be noted that many early references to the float process describe the glass sheet as being formed by rolling between two rollers before it is fed onto the tin bath. This approach was tried initially by

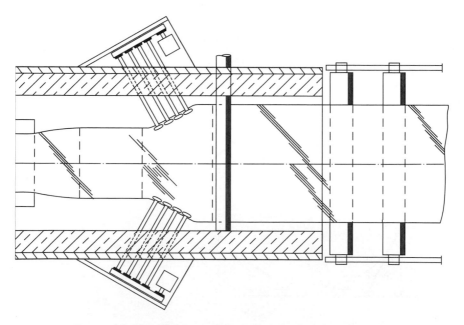

Figure 8.19 Decreasing the thickness of float glass by both lateral and longitudinal stretching, with knurled wheels pressing on the edges of the ribbon.[31]

Pilkington, but it proved unsuccessful. (It led to surface defects in the glass.) In the commercialized processes, the molten glass is fed onto the tin bath by flowing it over a refractory block. The Pilkington and PPG designs differ in how this is done.

The Pilkington and PPG float processes have proven so effective and so economical that they have virtually replaced all plate glass manufacture (ground and polished rolled sheet) throughout the world, and most of the drawn sheet products as well. Most of the flat glass producers in the world have been licensed to use the float process. Today's largest float glass plants can produce about 1,000 T of finished glass per day at widths up to about 12 ft and thickness between about 2 and 25 mm. The overall length of the production line, including melter, tin bath, and annealing *lehr* can exceed 700 ft, with the tin bath itself occupying between 100 and 200 of those feet.

Specially designed float lines can produce glass less than 1 mm thick. While initially developed for the manufacture of soda-lime-silica glass, several manufacturers have successfully applied the techniques to borosilicate glasses. However, because of temperature limitations of the tin bath, and its required chemically reducing atmosphere, not all commercially useful glass compositions can be manufactured by the float process.

8.4.9 Fritting

Techniques used for making glass *frit* (granules) include *dry gauging* or *dry gaging* (drizzling or pouring a stream of molten glass into cold water) and rolling as very thin ribbon, followed by particle attrition or comminution (size reduction) processes. These techniques involve a rapid quenching of the melt and can be used to vitrify (make glass from) compositions that tend to crystallize readily. Cooling the glass quickly, directly from the melter, creates high thermal stresses, which shock and often break the glass into small pieces suitable for charging into a ball mill. Dry gauging sometimes forms clinker-like pieces that are difficult to mill and may require an intermediate process step. Thin rolled ribbon often provides the better, more uniform mill feed.

8.4.10 Spheres, Marbles, and Microspheres

Glass spheres can range in size from a few nanometers in diameter to a meter or more. They can be solid, porous, or hollow, in all but the smallest sizes. The application range includes

1. The extremely small solid precursor particles used in Types III, IV, and V fused silica manufacture

2. The small, hollow spheres used to contain fuel in inertial confinement nuclear fusion research

3. The small, <0.2 mm *microspheres* used in reflective signs and projection screens or as fillers in plastics and elastomeric composites

4. The ~1/2-in. marbles used in games

5. Fish net floats

6. Deep ocean submersible vessels of military interest

Because of the wide range of size and the need for solid, porous, and hollow variations, manufacturing techniques necessarily vary. Directing a high-temperature, high-velocity flame across a vertically descending stream of molten glass can generate small solid spheres. If the glass is sufficiently fluid, strands of glass are formed that quickly break apart into droplets. These droplets spheroidize under the forces of surface tension and cool as they leave the flame. The resulting particle size distribution is not easily controlled. Similarly, molten droplets can form and detach from an orifice at the bottom of a crucible or glass-melting tank. If the droplets have sufficiently low viscosity and are released from a great enough height, they will spheroidize and cool before reaching the ground, where they are collected. Particle size control is better.

Microspheres (<0.2 mm dia.) of controlled size and composition can also be prepared by fritting, sieving, and injecting into a heated region to remelt, spheroidize, and cool, somewhat as described above. The precursor particles can be fed into the top of a tall column having a thermal gradient decreasing downward and collected at the bottom, or they can be injected into an upward-directed flame, whereby the molten droplets are drafted to cooler higher altitudes and collected.

To mass produce *marbles* (solid spheres about 1/2 in. in diameter), a more viscous stream of glass is delivered from the melter and mechanically cut (sheared) into mini-gobs having the required volume. The soft gobs fall into the space between two counter-rotating cylinders in which there are machined opposing spiral grooves. The gobs of glass are simultaneously rolled into spheres and cooled to temperatures near their annealing point as they are transported down the length of the rolls. Streams of different colored glass can be partially mixed together before gobbing to give variegated appearance. Less spherically perfect marbles can be generated by dropping the mini-gobs into cylindrical holes in vibrating molds, sometimes mounted on a conveyor belt. Steady vibration, maintained until the gobs have been well cooled, generates near-spherical shapes. Of course, decorative marbles of varying sizes can be created as novelties or works of art by studio or hand-shop techniques.

Perfectly spherical spheres of a wide variety of sizes with optical finish can be produced from near-spherical starting blocks by grinding and lapping on optical finishing machines.

Hollow spheres of moderate size can be hand blown freely or in molds, but the location where the blowpipe is separated after forming is seldom perfect. Hemispheres can be pressed and then pairs fused together. Very large, thick-walled spheres can be made from pairs of centrifugally cast hemispheres.

At the other end of the size spectrum, very small, hollow glass spheres such as used in inertial confinement nuclear fusion (ICF) research are also prepared by a variety of techniques, all relying on generating small volumes of precursor materials that are injected into a hot zone. The surface of the precursor particle melts or otherwise reacts to generate a viscous liquid layer. As the interior material heats, it evolves gasses that serve to blow the hollow sphere. Solid precursors can be formed as small aggregates of batch material, sometimes by spray drying or sol-gel techniques; alternatively, the required chemical species can be dissolved in an aqueous or organic solution with precursor droplets of the required size being generated by an appropriate means such as an ultrasonic *nebulizer*.

Porous spheres can be prepared by leaching one glassy phase from a two-phase spherical product or by processes similar to those used for hollow microspheres whereby the surface layer never forms in a fully continuous manner; i.e., the blowing bubbles are exposed at the surface.

8.5 Annealing and Tempering

8.5.1 Development of Permanent Stresses in Glass

Stresses in an unconstrained elastic solid develop only if there is a nonlinear temperature gradient across the body. Such stresses are temporary, or transient; they exist as long as the temperature gradient exists. Liquids, on the other hand, cannot sustain shear stresses for any finite length of time; such stresses relax by viscous flow. Glasses behave like a liquid when heated into the liquid state; i.e., all stresses relax due to viscous flow. However, upon cooling through the glass transition range into the solid state, stresses are likely to develop within the body, and such stresses no longer relax in the absence of a viscous flow. The various mechanisms for such permanent stress development are as follows:

1. Cooling from the outside results in a "frozen" temperature gradient with a higher temperature in the interior (Fig. 8.20). Inner lay-

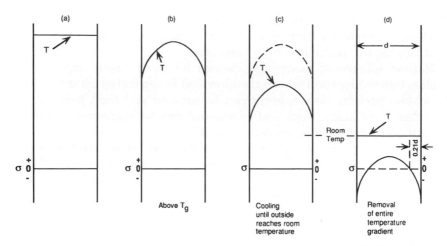

Figure 8.20 Simplified concept of permanent stress production in glass due to a "frozen temperature gradient." (a) Glass with no temperature gradient well above the transition region has no stress. (b) Temperature gradient develops on cooling; however, no stress develops because of rapid relaxation while the glass remains well above T_g. (c) Cooling to below T_g while maintaining the same "frozen temperature gradient" produces no stress yet. (d) Final removal of the temperature gradient produces stresses that are now permanent.[32]

ers continue to relax from fluid flow while the outer layers gradually freeze. In effect, the stress-free state is a solid with a temperature gradient. The need to shrink the inside more relative to the outside at room temperature and the enforcement of the elastic compatibility criteria between the layers cause the appearance of compression on the outside and tension on the inside. *The magnitude of the stresses so developed is related to the linear expansion coefficient of the solid.* This mechanism is also called the *viscoelastic mechanism.*

2. The outside layers cool faster than the inside layers during normal cooling. Hence, the outside layers tend to possess a "faster-cooled" structure having a higher volume in the free state (see Ref. 8, p. 15). *This, in principle, is a permanent structural heterogeneity.* Again, the enforcement of the elastic compatibility criteria causes the outside layers to develop compression and the inside layers to develop tension. *The magnitude of the stresses so developed is related to the difference between the volumes of the fast-cooled and the slow-cooled solids.*

3. The fact that the various layers travel through the glass transition range at different instants in time causes the development of a "frozen fictive temperature gradient." *This is a transient structural heterogeneity.* As in (1), the removal of the transient fictive temper-

ature gradient causes the appearance of compression on the out-side and tension on the inside. *The magnitude of the stresses so developed is related to the configurational (or the structural) contri-bution to the linear expansion coefficient of the supercooled liquid* (= linear expansion coefficient of the liquid – the linear expansion coefficient of the solid).

During glass forming, various regions of glass indeed go through varying cooling rates due to the nonuniform applied heating/cooling and the usually nonuniform thickness; hence, stress generation is likely. There is little chance of fracture while the glass remains some-what viscous fluid but, once the material starts solidifying, stresses begin to accumulate. For a common soda lime silicate glass product cooled through the glass transition region at "normal" cooling rates, nearly 60% or so of the total stress is due to the mechanisms (1) and (2). Warping may result to relieve uneven bending moments across the body of a glass product. In turn, the warping generates bending stresses, some of which end up being tensile on an outer surface. This then poses a potential risk for glass fracture during the manufacture and, worse yet, during service. There is additionally a longer-term risk for property and dimensional changes due to the continuing stabiliza-tion of glass. (Optical glass components, particularly those that are space-based, need to have minimized refractive index and dimensional changes.) Glass products are therefore customarily reheated in the proximity of the glass transition range to allow the release of any in-ternal stresses that might have developed, and cooled at a rate slow enough that prevents their rebuilding. This procedure is termed *an-nealing.* An additional objective of annealing is to homogenize the thermal history across the body.

Tempering (also called *toughening,* which is a misnomer in this case), on the other hand, implies strengthening and is accomplished by introducing surface compression into the glass. Since flaws that cause strength degradation usually occur on the surface, the introduc-tion of surface compression strengthens a glass product. Unless the volume-temperature diagram for the glass at hand is somewhat un-usual (like that of fused silica), one expects to have some degree of temper obtained by normal cooling through the glass transition, which is clearly beneficial for ordinary handling of glass products.

8.5.2 Stress Profiles in a Symmetrically Cooled Glass Plate during Annealing and Tempering

If a glass plate is cooled symmetrically from both sides, then the tem-perature distribution across the section is a parabola (for low cooling rates); see Fig. 8.21. The magnitude of the temperature difference ΔT between the outside and the inside layers of glass is given by

$$\Delta T = \frac{d^2 R}{2\kappa}$$

where d = half-thickness of the section
          κ = thermal diffusivity[*]
          R = cooling rate

The outside surface compression σ_c is proportional to $2\,\Delta T/3$ and is approximately given by[†]

$$\sigma_c = \frac{E\alpha d^2 R}{\kappa(1-v)} \tag{8.1}$$

where E = Young's modulus
          v = Poisson's ratio of the glass

The center tension is proportional to $\Delta T/3$ (= half of the surface compression magnitude) and is therefore given by one-half of the above quantity. The above equation may also be used to calculate a constant cooling rate that would give rise to a specified center tension at room temperature. For instance, for a soda lime silicate glass with α = $90 \times 10^{-7}/°C$, E = 70 GPa, κ = 0.0084 cm²/s, v = 0.2, and d = 0.3 cm, a 2.8°C/min cooling rate will give rise to a center tension of about 0.2 MPa.

For glass tempering, a high cooling rate is deliberately employed to introduce a level of surface compression in the glass. Because of the high cooling rates, the temper stress profile is often not a rectangular parabola of the form $y = ax^2$ but may be approximated by $y = ax^n$. It can be shown that the ratio of the magnitude of the surface compression to that of the midplane tension is simply n. Depending on the Biot number (= hd/k; where h = the heat transfer coefficient and k = thermal conductivity), n values as high as 4.5 have been obtained (see Fig. 8.22). Most commonly, n = 2.2 to 2.4. For a rectangular parabolic distribution of stress, the zero stress ("neutral") line is about one-fifth of the glass thickness below the surface from each side (see Fig. 8.21). The thickness of the compression layer from the surface up to the

---

   [*] A typical value of κ for soda lime silica glass is 0.0084 cm²/s.
   [†] Note that the calculations assume that the room temperature stress is the inverse of that established during fluid state when the thermal expansion coefficient was roughly three times that in the solid state.

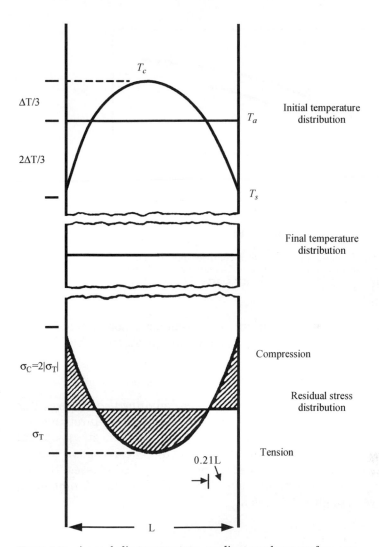

Figure 8.21 A parabolic temperature gradient produces surface compression that is twice the interior tension in magnitude; d = glass thickness. *(From Ref. 32, p. 304)*

nearest neutral line is called *case depth*. Thus, a 6 mm plate may have surface compression layers roughly 1.2 mm thick on each side, which represents a significant depth of protection.

One of the fascinating examples of tempered glass is the *Prince Rupert Drop*. Teardrops, generally 4 to 10 mm in diameter, of molten soda lime silicate glass with a tail are allowed to fall into water. The high degree of surface compression that develops into the teardrop allows hammering on it. The drop explodes into flying fine powder of glass

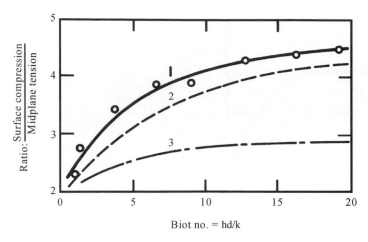

Figure 8.22 Ratio of "plateau level" surface compression to midplane tension as a function of Biot number. Curve 1 = experimental results. Curve 2 = predictions by Indenbom's theory. Curve 3 = predictions by Bartenev's theory.[33]

when the tail is snapped. (Caution: such demonstration should be carried out under protective wrapping or behind a protective shield.)

8.5.3 Standards of Annealing

Commercial glass products are generally called *annealed* if the center tension is no greater than about 500 psi. In a stress-birefringence setup (see Sec. 8.5.9), and with the assumption that the stress-optical coefficient is 3 brewsters, this stress would roughly correspond to a retardation of 100 nm/cm of light path. For most optical glasses, the optical birefringence in the middle should be no more than 5 nm/cm or, even better yet, roughly 2.5 nm/cm.

8.5.4 Annealing Practices

Coarse annealing may be accomplished, particularly in a laboratory setting, using the following guidelines:

1. Hold the glass product isothermally in a box furnace at a temperature within the annealing range. The holding time t may be estimated by the KWW relaxation formula:

$$\sigma_t = \sigma_o \exp\left[-\left(G\frac{t}{\eta}\right)^{\frac{1}{2}}\right] \tag{8.2}$$

where σ_t is the stress at time t, and η and G are the viscosity and shear modulus of the glass at the holding temperature, respectively. The stress σ_t is normally taken as 5% of σ_o, the initial stress, and the high-temperature value of G is taken as 20 to 25% lower than that at room temperature. Thus, at the annealing temperature, where viscosity is 10^{13} P, one needs to hold for 10 to 15 min. However, as much as 6 to 8 hr are needed for a comparable relaxation at the strain point ($\eta = 10^{14.5}$ P).

2. Cool the furnace usually at 5°C/min, which can be accomplished by simply turning the box furnace off to cool overnight.

On an industrial scale, a continuous belt lehr is often employed for annealing. One of the several suggested heat treatment schedules for the annealing of glass products is shown in Fig. 8.23. In region A, the glass is rapidly heated or cooled to about 5°C above the rated annealing point, depending on whether the product is hot coming out of the forming machine or is cold coming out of storage. It is held at that temperature for a time period t (region B), following which it is slowly cooled to a temperature $a°$ below the strain point (region C). Once the glass is well below the strain point, any risk of rebuilding permanent stresses is slight, and the glass may be cooled faster (regions D and E). The limiting consideration for the cooling rate in the two regions is the avoidance of thermal shock. The recommended schedule for the various regions is as follows:

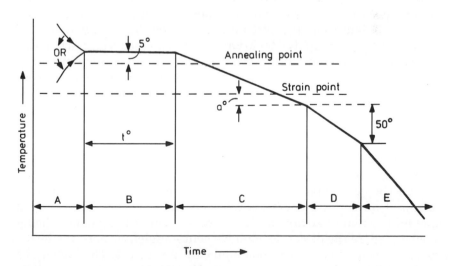

Figure 8.23 Suggested schedules for commercial annealing of soda-lime-silica glassware. *(Courtesy of Corning, Inc., from Ref. 8, p. 307)*

A (heating rate) $= 500/[\alpha d^2]$ °C/min

B (holding time) $= 15z$ min for cooling from one side

                                    $= 30d$ min for cooling from both sides

C (slow cool) $= 42.6/[\alpha d^2]$ °C/min

$a°$ (°C below
  the strain point) $= 5$°C for 0.3 cm thick plate

                                    $= 10$°C for 0.6 cm thick plate

                                    $= 20$°C for 1.3 cm thick plate

D (fast cool) $= 2$ times the slow cool

E (final cool) $=$ no more than 10 times the slow cool

where α is the thermal expansion coefficient in $\times 10^{-7}$/°C units, z is the glass thickness (cm), and $d = z$ when heated/cooled from one side only and $= z/2$ when heated/cooled symmetrically from both sides. Computations for a few glasses with different plate thickness and cooling conditions are tabulated in Table 8.20. The total time for the completion of a commercial quality annealing is expected to be about 20 min.

8.5.5 Standards of Temper

According to ASTM-C1048, *Kind FT* (fully tempered) glass shall have a surface compression no less than 69 MPa (10,000 psi) or an edge compression of not less than 67 MPa (9700 psi). Such a glass is up to five times as strong as a normally annealed product of the same thickness and configuration. It is often recommended that a fully tempered glass should have roughly 100 MPa (14,500 psi) surface compression. This corresponds to roughly 45 MPa (~6,400 psi) midplane tension, or about 1350 nm/cm birefringence in the midplane. Glass should meet American National Standards Institute (ANSI) standard Z97.1 or Consumer Product Safety Commission (CPSC) standard 16CFR1201. The ANSI standard specifies that a tempered automotive glass must withstand the impact of a 1/2-lb steel ball and an 11-lb shot-filled bag, both dropped from specified heights. In the ball drop test, 10 out of 12 specimens must remain unbroken. The drop height is then increased until all the specimens have broken. Upon initiation of the fracture, the stored mechanical energy is released virtually instantly, causing rapid bifurcation of advancing crack fronts. This then yields a dicing behavior, where the pieces do not have acute-angled corners and hence are not expected to cause serious injury by sharp cuts. The higher the midplane stress, the smaller the pieces. For the glass to be acceptable, none of the diced pieces may weigh more than 0.15 oz (4.3 g) for a

TABLE 8.20 Suggested Annealing Schedules for Glass Products[*][†]

Expansion coefficient of glass, per °C	Thickness of glass, mm	Cooling on one side						Cooling on two sides					
		Heat rate, °C/min	Time t, min	Temp. a,°C	Cool rate, °C/min			Heat rate, °C/min	Time t, min	Temp. a,°C	Cool rate, °C/min		
		A	B	C	C	D	E	A	B	C	C	D	E
33×10^{-7}	3	130	5	5	12	24	130	400	5	5	39	78	400
	6	30	15	10	3	6	30	130	15	10	12	24	130
	13	8	30	20	0.8	1.6	8	30	30	20	3	6	30
50×10^{-7}	3	85	5	5	8	16	85	260	5	5	26	52	260
	6	21	15	10	2	4	21	85	15	10	8	16	85
	13	5	30	20	0.5	1.0	5	21	30	20	2	4	21
90×10^{-7}	3	50	5	5	4	8	50	140	5	5	14	28	140
	6	11	15	10	1	2	11	50	15	10	4	8	50
	13	3	30	20	0.3	0.6	3	11	30	20	1	2	11

[*]After G. M. McLellan and E. B. Shand, *Glass Engineering Handbook.* McGraw-Hill, New York, 1984.
[†]Annealing period: A, heating to 5°C above annealing point. B, hold temperature for time t. C, initial cooling to $a°$ above strain point. D, cooling next 50°C. E, final cooling.
Source: Reference 8.

6-mm thick glass. The dependence of the break pattern on midplane stress is shown in Fig. 8.24.

Kind HS (heat-strengthened) glass has a surface compression between 24 and 52 MPa (3500 to 7500 psi) for a thickness of 6 mm. Such

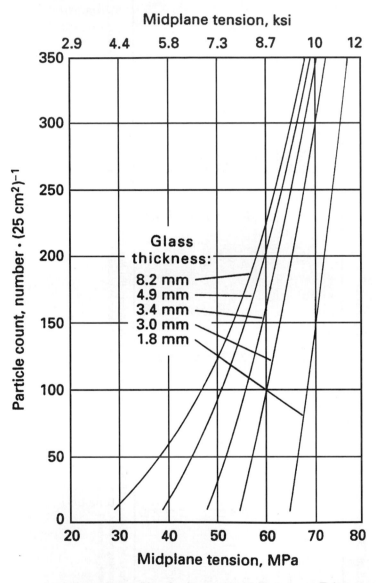

Figure 8.24 Dependence of break pattern on midplane stress. Exit temperature, 660°C (1220°F). *(From R. A. McMaster et al., Ref. 7, Fig. 10, p. 458, reproduced with permission of publisher)*

a glass may be up to three times as strong as a normally annealed product of the same thickness and configuration.

The U.S. Food and Drug Administration (FDA) requires that all eyewear (prescription or otherwise) sold in the United States pass safety ball drop tests for impact resistance described by ANSI Z80.1-1995 and ANSI Z80.3-1996. This testing requires a fully finished and edged glass lens to survive impact testing with a 16-mm dia. steel ball, dropped free fall from a height not less than 1.27 m, onto the central 16 mm diameter area of the outer surface of the horizontally placed lens. Certain specialty lenses are exempt from this test, as are plastic lenses. Industrial safety glasses must pass a more severe test. Both thermal tempering and ion-exchange strengthening are used to meet these impact requirements.

8.5.6 Commercial Tempering Practices

Tempering is usually accomplished by heating the glass product, generally to a viscosity of about $10^{9.5}$ to 10^{10} P for a short while, and then quenching symmetrically from both sides using forced air jets. For a typical soda lime silicate glass plate, these viscosity values correspond to about 640 to 620°C. The normal practice is to run a thinner glass hotter; however, it must be remembered that higher temperatures bring about noticeable optical distortion. Air power requirement is roughly 40 kw · m^{-2} for a 3-mm plate and is only about 12 kw · m^{-2} for a 6-mm plate. The quenching to a viscosity below $10^{14.5}$ P takes roughly 5 s for a 3-mm plate and ~12 s for a 6-mm thick plate (see Fig. 8.25). Cooling to room temperature occurs over longer periods. A typical tempering furnace is about 40 m long in which the heating zone may be ~20 m long, the quenching zone ~5 m long, and the slower cooling zone ~15 m long. With a width of ~1.5 m, the furnace can deliver about 15 T of fully tempered glass per day.

8.5.7 Limitations of Thermal Tempering

The most important drawbacks of the thermal tempering process are:

1. Surface compression magnitudes much higher than about 140 MPa (~20,000 psi) are accompanied by a center tension of about 65 MPa. At this magnitude of tension, a defect such as a stone in the interior could cause the glass product to fail spontaneously.

2. It is difficult to exceed cooling rates of about 100°C/s through the glass transition by forced air jets. Use of Eq. (8.1) suggests that plates thinner than about 1 mm develop no more than about 24 MPa of surface compression, which is not much of a protection.

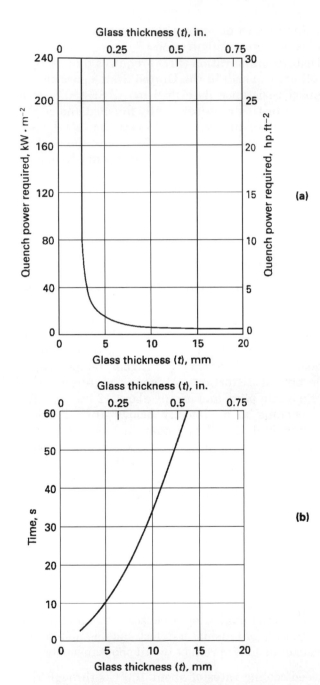

Figure 8.25 Effect of glass thickness on (a) air power and (b) quenching time required for full temper. Exit temperature, 620°C (1150°F). *(From R. A. McMaster et al., Ref. 7, Fig. 5, p. 456, reproduced with permission of publisher)*

3. Because of the lack of symmetry in achievable cooling, glass tubes, containers, and other complex-shaped products cannot be meaningfully tempered using thermal means.

8.5.8 Chemical Strengthening of Glass

The exchange of large alkali ions from an external source such as a molten salt bath with comparatively smaller host alkali ions in a glass at low temperatures leaves the glass in a state of surface compression that is an effective means of strengthening the glass. *The level of strengthening achieved could be 2 to 10 times that of an unstrengthened product.* Large ions are essentially "stuffed" in glass network interstitial spaces occupied previously by the small ions (Fig. 8.26). Stress generation is closely linked to the kinetics of the interdiffusion and the difference in the size of the exchanging ions. Although the diffusion rates increase exponentially with temperature, the rate of stress buildup as the exchange temperature approaches the glass transition temperature is actually reduced by stress relaxation arising from glass fluidity and perhaps some network plasticity. Hence, *chemical strengthening of glass must be carried out at temperatures well below the glass transition range.* A typical ion penetration and stress profile in a glass is shown in

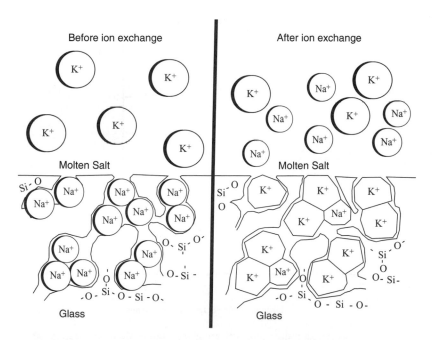

Figure 8.26 Crowding from low-temperature exchange of K^+ for Na^+ ions. *(After W. H. Dumbaugh and P. S. Danielson, Ref. 5, p. 120)*

Fig. 8.27. Surface compression on the order of 100 to 1000 MPa (= 14,500 to 145,000 psi) can be generated using this technique in many glasses having wall thickness as small as about 1 mm. For consumer glass products, a minimum case depth of about 30 μm is generally recommended to provide an effective protection from the surface flaws. This, unfortunately, means that *a successful ion exchange strengthening process would usually involve 2 to 100 hr of diffusion, which does not encourage a large-scale continuous process.* Nonetheless, some commercial

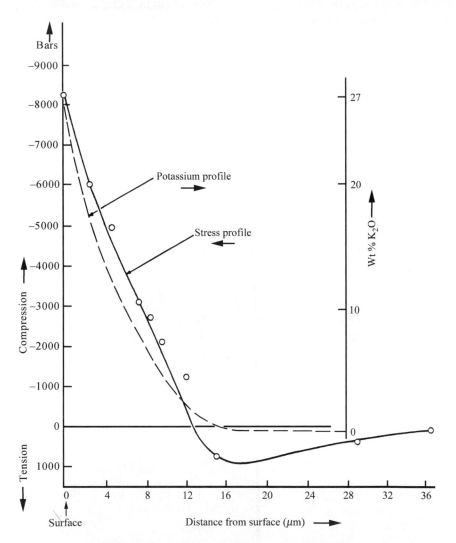

Figure 8.27   Stress profile in a sodium borosilicate glass sample after ion exchange at 400°C for 12 h. *(After A.K. Varshneya,* J. Non-cryst. Sol. *19, 355, 1975. Reproduced with permission of Elsevier Science, from Ref. 8)*

glass products are ion exchange strengthened: windshields for aircraft, borosilicate glass syringes, ophthalmic products, and large-capacity returnable beverage containers are some examples.

Ion exchange strengthening of glass is typically carried out by immersion of the glass in a bath of molten salt. For instance, a common commercial soda lime silicate glass would be strengthened by immersion in a bath of molten KNO_3 at temperatures higher than 328°C (melting point of KNO_3) but lower than about 480°C (the onset of glass transition). Lithium aluminosilicate glass is strengthened by immersion in $NaNO_3$ similarly. The alkali aluminosilicate glasses are the best candidates for ion exchange; the soda lime silica glasses do not strengthen well because of rapid relaxation processes, despite the rather large alkali content in the host glass. Low-alkali-content glasses, such as the alkali borosilicates of the Pyrex type, make even poorer candidates for ion exchange strengthening.

As the glass ions are continuously rejected into the bath, the concentration of this contaminant builds up. This usually has deleterious consequences on the percent surface exchange even at very low level of contamination. Hence, the salt bath requires careful control for the "contamination" level.

The strength is commonly measured in four-point bending mode or a ring-on-ring method using a mechanical tester on suitably sized test coupons, with or without surface abrasion. (It is strongly recommended to employ edge-independent techniques.) A steel ball drop test for ophthalmic lenses and a bird impact test for aircraft windshields are also carried out.

To measure stresses in a chemically strengthened glass, one needs to obtain a thin slice across the wall thickness, which has been ceramographically prepared to have sharp square edges at the surface. The observed stresses in such a thin slice must be divided by $(1 - v)$ to obtain stresses in the unsliced plate. It may also be possible to employ techniques such as the differential surface refractometer (DSR) or grazing angle surface polarimeter (GASP) to measure surface compression and the laser scatter to measure internal tension without the need of slicing (see Sec. 8.5.9).

Although immersion in molten salt is the more popular technique of ion exchange, other techniques have been tried. Examples of these are pastes using mixtures of inert clays, vapor phase, electrical field-assisted strengthening, multistep (such as thermal and chemical or two-step chemical), sonic-assisted, and plasma-assisted ion exchange.

8.5.8.1 Standards of chemical strengthening. Chemically strengthened glass is sold on the consumer's expectation for the improvement of strength; however, it is believed that the varying conditions for the application, such as the level of abrasion to be tolerated and the edge conditions, make it difficult to classify the products on the basis of strength. For this reason, the ASTM's recent standard for chemically strengthened flat glass products is based on the measurement of surface compression as well as case depth. These are:

Surface compression σ_c:

7 MPa (1,000 psi) < σ_c < 172 MPa (25,000 psi)	Level 1
172 MPa (25,000 psi) < σ_c < 345 MPa (50,000 psi)	Level 2
345 MPa (50,000 psi) < σ_c < 517 MPa (75,000 psi)	Level 3
517 MPa (75,000 psi) < σ_c < 690 MPa (100,000 psi)	Level 4
690 MPa (100,000 psi) < σ_c	Level 5

Case depth ε:

<50 microns (0.002 in.)	Level A
50 microns (0.002 in.) < ε < 150 μm (0.006 in.)	Level B
150 microns (0.006 in.) < ε < 250 μm (0.010 in.)	Level C
250 microns (0.010 in.) < ε < 350 μm (0.014 in.)	Level D
350 microns (0.014 in.) < ε < 500 μm (0.020 in.)	Level E
500 microns (0.020 in.) < ε	Level F

Thus, a glass plate that has 60,000 psi surface compression and a case depth of 70 μm will be termed Level 3B chemically strengthened glass.

8.5.9 Examination of Stresses in Glass

Stresses in glass are best examined using photoelastic techniques. In the absence of external forces, a well annealed and homogeneous specimen of glass is isotropic. However, when a nonhydrostatic stress is applied to the glass or in the presence of unannealed internal stresses, perpendicular vibrations of an unpolarized wave of light travel at different velocities along planes of principal stresses. Glass develops two refractive indices, also termed *double refraction* or *birefringence*. The magnitude of the stress-birefringence is measured using one of several methods and related to the stress using the stress-optic equation:

$$\sigma_{33} = \frac{\delta}{B_\lambda d}$$

where d = thickness of the specimen
         B = the stress-optical coefficient or the Brewster's constant
         δ = the optical path difference

The optical path difference is the product of the physical path length times the difference in effective refractive index for each polarization of light, and it describes the retardation on polarized beam will experience relative to the other. The subscript λ indicates that B may vary with the wavelength.

In a simple polariscope setup, light rays from a lamp are first allowed to pass through a polarizer whose sole function is to allow vibrations along only one plane. When an analyzer is oriented such that its plane of allowed vibrations is at 90° to that of the polarizer, all of the light intensity is cut off, leading to a field of view that is dark (i.e., the analyzer is *crossed* with respect to the polarizer to achieve extinction). A stressed glass specimen is now inserted between the polarizer and the analyzer. The principal stress axes of the specimen are to be at 45° to that of the polarizer. The light rays entering the specimen may be considered to be composed of two equal-intensity, in-phase, and mutually perpendicular vibrations. The vibrations travel with different velocities through the specimen, and recombine upon exiting. Depending on the path difference δ (or the phase difference $\phi = 360° \, \delta/\lambda$) that is introduced between the two vibrations while traveling through glass, the recombination yields some intensity along the 90° plane of the analyzer. Consequently, the field of view is no longer dark. For instance, a phase difference of full 360°, i.e., a path difference of λ, brings the two vibrations in the same manner as they were upon entering, with the result that the analyzer cuts off that particular wavelength λ (i.e., extinction is achieved for λ). When a white light source is used, the particular λ will be missing from the field of view. The balance color of light, therefore, may be judged to estimate δ from a *Michelle-Levy chart*, and the use of the stress-optic relation gives the particular stress responsible for it. Sensitivity is greatly increased if a "full wave" retardation plate (*sensitive tint* plate for $\lambda = 565$ nm) is introduced between the specimen and the analyzer. With a zero-stress specimen, the analyzer cuts off only the 565 nm wavelength, which is between the green and yellow color of the spectrum. The absence of this light causes the balance reds and blues to form a magenta color. Introduction of stress that adds to 565 nm will cause wavelengths at the red end to be cut off, yielding a bluish field of view. On the other hand, a stress that subtracts from 565 nm will cause bluish light to be cut off, thus changing the field of view to orange-red colors. Since the human eye is most sensitive to changes of about 565 nm, even small magnitudes of stresses can give noticeable color change. The identification of tensile or compressive stresses by the shift of the color from the magenta tint to either side of the green-yellow line in the white spectrum is the essence of *strain viewing* in glass. The appearance of colors with and without the sensitive tint plate is shown in Table. 8.21.

TABLE 8.21   Co ors in Strained G ass when Observed in a Po ariscope

Without Tint Plate		With Tint Plate (565 nm retardation)		
Phase difference or retardation, nm	Color	Phase difference or retardation, nm	"Blue" position, Color rising	"Orange" position, Color falling
50	Iron-gray	0	Violet-red	Violet-red
200	Grayish-white	23	Violet-blue	Reddish-violet
300	Yellow	46	Blue	Reddish-orange
425	Orange	69	Greenish-blue	Orange
530	Red	92	Bluish-green	Yellowish-orange
565†	Violet-red	115	Yellowish-green	Gold-yellow
640	Blue	150	Deep green	Gold-yellow
675	Blue-green	180	Green	Yellow
740	Green	220	Pale green	Pale yellow
840	Yellowish-green	290	Greenish-yellow	White
880	Yellow	330	Pale yellow	

†Color with tint plate of 565 mμ retardation.

Source: J. H. Partredge and W. E. S. Turner, *Glass-to-Metal Seals*, Society of Glass Technology, Sheffield, England, 1949, p. 123.

More precise measurements of stress in glass are possible with the use of stress (actually, birefringence) compensators. When a compensator is introduced between the specimen and the analyzer, its function is to introduce a retardation that is exactly the same in magnitude but opposite in sign to that introduced by the specimen. A precalibration of the compensator readily yields the unknown value of δ. Popular compensators are quartz wedges (or a stretched polymer wedge), Babinet, Babinet-Soleil, Berek, and Sénarmont.

8.5.9.1 Wedge compensator. Quartz crystal is a naturally birefringent material. It is cut into a slightly tapered wedge such that the directions of the "fast" and the "slow" rays are along and perpendicular to the grade of the wedge. (Alternatively, a transparent epoxy resin is stretched while warm and "frozen" into solid form.) The compensator is introduced such that its fast and slow axes are lined up with that of the specimen (and at 45° to the polarizer or the analyzer). Light rays of a particular wavelength are compensated by the thickness of the quartz wedge at a particular location, giving a field that comprises the balance of the spectrum. In the presence of a properly oriented stressed specimen, the colored bands are shifted. The shift is precalibrated in terms of δ using a monochromatic light without the stressed specimen. The field of view in such a case would be a series of bands that represent exactly one wavelength path difference from each other.

A more useful version of the quartz wedge is shown in Fig. 8.28, which comprises a graduated wedge and a flat mineral plate having

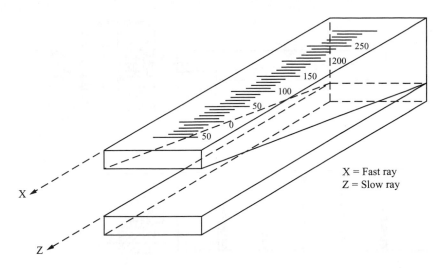

Figure 8.28 Combination wedge with wedge-shaped quartz section above and mineral plate below. *(From Ref. 34, p. 246)*

the opposite sign of the optical birefringence. The path difference created by one is exactly compensated by that created by the other at the *zero* mark.

8.5.9.2 The Babinet compensator: The Babinet compensator (Fig. 8.29) comprises two quartz wedges with opposite grades; one of the wedges is fixed, the other is movable. At the location where the two wedges have equal thickness, the retardation introduced by one is cancelled by the other, giving rise to a black line (*zero* mark on the movable wedge dial). On either side of the black line, the net retardation increases. The black line will therefore be displaced either to the left or to the right, depending on the sign of birefringence of the specimen, i.e., whether the stress is tensile or compressive. The movable wedge is then moved to bring the black line back to its original location, and the dial is read to obtain the corresponding retardation. Again, the compensator is calibrated using a monochromatic source that yields equispaced black lines corresponding exactly to one wavelength path difference.

8.5.9.3 The Babinet-Soleil compensator. The Babinet-Soleil compensator comprises two quartz wedges with parallel optic axes, one of the wedges being fixed and the other movable, plus a fixed quartz plate with perpendicular optic axes (Fig. 8.30). The double wedge combination adds a constant retardation all across the field of view. Dial rota-

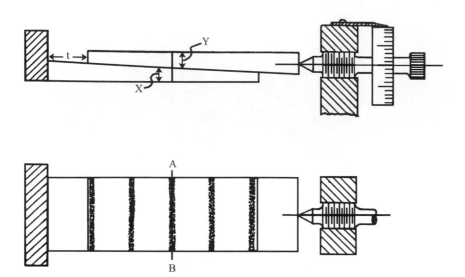

Figure 8.29 Arrangement of wedges in the Babinet compensator. *(From Ref. 34, p. 247)*

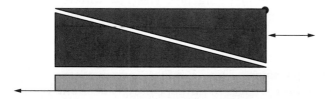

Figure 8.30 Arrangement of the optical elements of a Babinet-Soleil compensator.[35]

tion causes the upper wedge to move, producing a change in the uniform total thickness of the double wedge. Thus, instead of seeing black and colored bands as in the Babinet compensator, one views a uniform colored field corresponding to the net retardation. In essence, the Babinet-Soleil compensator acts as a variable retardation plate.

8.5.9.4 The Berek compensator. The Berek compensator (shown in Fig. 8.31) is a calcite (naturally birefringent crystal) plate, cut perpendicular to the optic axis, about 0.1 mm thick, which can be tilted about ±20° using a graduated dial motion. In the "null" position (generally 30° reading on the dial), the plate is horizontal and introduces no net retardation. A tilt, however, introduces a net retardation between the perpendicular vibrations of light. When viewing a stressed specimen, the dial is rotated on either side of the 30° null position to bring the

Figure 8.31 Berek compensator.

black line back to the center of the crosshairs. It should be noted that the Berek can compensate only one sign of birefringence; the specimen may have to be rotated by 90° for compensation to occur. Nonetheless, Berek is a very widely used instrument, primarily because of its compactness and also because the compensator can be designed (procured) to measure a range of birefringence magnitudes such as 0 to $\lambda/10$, 0 to λ, and 0 to 10λ.

8.5.9.5 The Sénarmont compensator. A Sénarmont compensator utilizes a quarter-wave plate introduced between the specimen and the analyzer (Fig. 8.32). Its fast axis is lined up with that of the polarizer. When a stressed glass specimen is introduced (with its axes at 45°), the field of view appears bright in the presence of stresses. The analyzer is rotated by an angle $\theta°$ to achieve extinction. The phase difference introduced due to the specimen is $2\theta°$, which implies that the path difference introduced is $2\lambda\theta°/360°$. If $\lambda = 540$ nm, then the path difference introduced by the specimen is 3θ nm. The direction of rotation gives the sign of the birefringence. Thus, tension and compression can both be measured without changing the orientation of the specimen. However, the maximum permissible rotation is 180°, hence the Sénarmont compensator is limited to measuring a maximum of λ nm retardation. The usefulness of the Sénarmont compensator lies in the high level of accuracy with which small retardations can be measured.

8.5.9.6 Measurement of surface stresses. An instrument that allows surface stresses to be measured *individually* is the differential surface refractometer (DSR) shown in Fig. 8.33. It is quite useful for surface compression measurements in strengthened glass windows. A specially designed prism rests on the flat glass specimen. The refractive index of the prism glass must be higher than that of the glass specimen to allow a total internal reflection at the glass surface. Optical

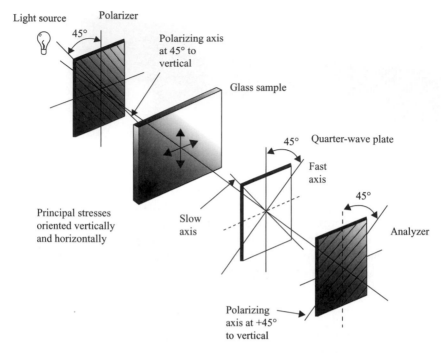

Figure 8.32   Polariscope arrangement for Sénarmont compensation. *(From Ref. 35, p. 41)*

contact between the prism and the glass is assured by using a few drops of an intermediate refractive index oil. Some of the incoming light vibrations punch through the glass surface and are internally reflected back (with mirage-like bending), exiting out of the prism at critical angles corresponding to the two refractive indices of the stressed glass. Since the critical angles for the two polarized transverse vibration components (one perpendicular, the other parallel to the test glass surface) are different in general, the exiting rays come out in two slightly shifted bundles, which are then focused using a telescope. The shift between the two (see Fig. 8.33b) is measured and calibrated in terms of the optical retardation. For the surface, velocity of the vertical vibration corresponds to refractive index of the glass in a stress-free condition (since the normal stress component must be zero). Thus, the entire measured optical retardation corresponds to the in-plane stress (normal to the plane of incidence) in the test glass. Compression or tension may be identified by the direction of the shift relative to the neutral beam. The DSR requires specific prisms for various glass compositions. It is also a common experience that measurements on a float glass are possible only on the tinned side—the refractive index increases with depth on the air side, causing the light

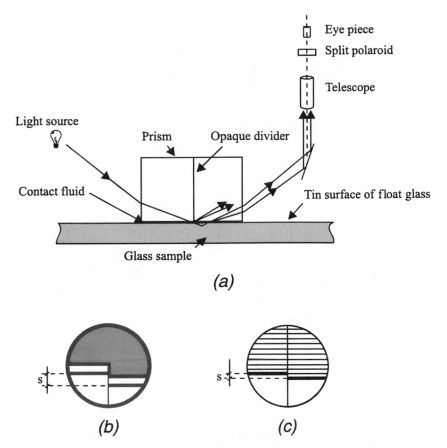

Figure 8.33 Optical arrangement for differential surface refractometer and diagrammatic representation of fringes for thermally and chemically toughened glass. *(From Ref. 35, p. 64)*

rays to bend more into the depth rather than be bent mirage-like to exit. Nonetheless, the DSR requires little sample preparation and is essentially nondestructive.

Another instrument that enables the measurement of surface stresses is the grazing angle surface polarimeter (GASP), see Fig. 8.34. The GASP instrument is much like the DSR except that a polarized light beam is incident on the glass at critical angle. Thus, the beam of light travels in the surface and is continuously attenuated and refracted back into the prism; the longer the travel, the larger its optical retardation. The optical retardation in these rays is compared with the main beam of light, which has been specularly reflected by the top surface using a Babinet compensator. The no-stress fringe pattern of the Babinet compensator rotates. The rotation is related to the

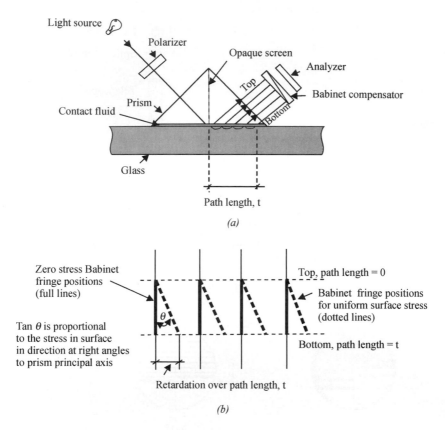

Figure 8.34 Optical arrangements and fringe pattern for surface polarimetry. *(From Ref. 35, p. 67)*

optical retardation due to the in-plane stress normal to the plane of incidence. The sign of stress may be deduced from the direction of the rotation. Again, the instrument can be used only on the tin-side of a float glass; however, it requires no specimen preparation. The GASP instrument is more sensitive than a DSR, since traveling on the surface over a path length t allows a greater amount of optical retardation to be accumulated by multiple refractions.

8.5.9.7 Measurement of internal stresses. In tempered glass products, the measurement of the magnitude of center tension is often desired. This can be achieved using a laser light scattering method, as shown in Fig. 8.35. A laser beam that has been polarized to have its intensity vector at 45° to the midplane of the glass is allowed to enter glass from the edge, making sure that propagation of the beam occurs through the central plane. Two transverse vibrations propagate (one

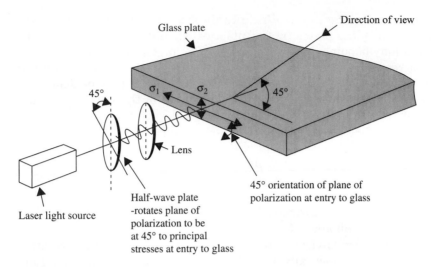

Direction of view

Glass plate

45° σ_1 σ_2 45°

Lens

45° orientation of plane of
polarization at entry to glass

Half-wave plate
-rotates plane of
polarization to be
at 45° to principal
stresses at entry to glass

Laser light source

Figure 8.35 Optical arrangement for scattered light measurements in thin glass plates. *(From Ref. 35, p. 60)*

parallel to the midplane and the other perpendicular to it) generally at different velocities due to the difference in the stress magnitudes along these planes. Because of the gradual development of phase difference between the two with continuing propagation, there will be a location when the vibrations develop exactly one wavelength retardation. This means that the original vector of vibrations will be reconstituted (which is oriented at 45° to the midplane). When the scattered light is viewed from a point outside along this vector, no intensity will be observed (since light rays may not have longitudinal vibrations). At locations corresponding to a half-wavelength retardation, the reconstituting vector makes 90° angle to the original vector, and one will observe maximum scattered intensity (allowed transverse vibration) when viewing along the original vector. Thus, one observes an interference pattern consisting of 10 to 20 bright bands of light separated by dark bands. The center-to-center distance between two adjacent fringes is measured (by averaging over several fringes) and related to stress using the stress-optic equation where $\delta = \lambda$ nm. Laser light scattering method is useful where the stress field does not change greatly, for instance, in the midplane region of a chemically strengthened glass. Small, 10 to 20 mW output, portable polarized light lasers are well suited for the purpose.

8.5.9.8 Practice and limitations of the through-transmission photoelastic techniques. *Light transmission through a glass product to view birefringence is greatly aided (and often necessary) by immersion in a* *matching refractive index oil.* Suitably sized rectangular immersion cells of clear glass plates (not of plastics) should be constructed. (The corners may be joined using an epoxy.) The matching refractive index allows rectilinear propagation of light without suffering bending at curved surfaces of the glass product. Occasionally, it may be desired to view birefringence through the wall thickness of the glass. In such a case, slices or ring sections having parallel surfaces containing the wall should be sawed out of the glass using a low-speed diamond wheel saw. (*A considerable amount of experimental skill is needed to obtain wall sections with well-preserved edges*). The flat surfaces may be smoothed using fine abrasive slurry on a ceramographic grinding/polishing wheel. Again, use of a drop of a matching refractive index oil on both surfaces sandwiched between a microscope slide or a glass cover slip helps photoelastic measurements.

In a laboratory environment, photoelastic measurements are often made using a commercially available transmission-type polarizing microscope. Care should be taken in procuring strain-free optics. For most measurements, it will be necessary to remove the condenser mechanism to allow incidence of light rays on the specimen in a collimated manner.

It should be readily realized that only transparent glasses can be subjected to photoelastic techniques. Darkly colored glasses or sintered glasses are not amenable to these techniques.

A serious problem in the measurement of stresses in glass is the fact that the stress field is rarely uniaxial. One usually needs to consider the effect of a biaxial stress field perpendicular to the direction of light propagation *(stress component along the light propagation has no effect)*. The stress-optic relation has to be modified to read

$$\sigma_{33} - \sigma_{11} = \frac{\delta}{B_\lambda d} \qquad (8.3)$$

This means that only the difference of principal stresses integrated along the light path d can be estimated using the through-transmission techniques. Since fracture in glass occurs mostly perpendicular to the maximum principal tension, no information about stress magnitudes or sign is obtained in a random stress field. Individual stress components can be separated only at a surface, where normal stress component must be zero, or the second component adds to zero over the light path d. The integration over d also implies that a biaxial stress field, which varies along the light path, would be impossible to resolve. The sign of the observed birefringence changes if the specimen is rotated by 90°. Hence, even if one of the stress components is zero, the type of the stress (tension or compression) can be mixed up due to an "unknown" orientation. It is strongly advised to use a stressed-glass standard to calibrate a polarized light microscope setup.

Having extinction at some location does not always imply the absence of stresses. It may also mean that the principal stress axes are lined up with the polarizer axes.

Finally, most polarization equipment rarely provide a spatial resolution better than ~100 μm. Glass-to-metal seals often require stress information on a scale of 10 μm or better. Nonetheless, photoelastic methods of measuring stresses in glass remain the most useful means of glass product quality control and assurance.

8.6   Glass Fiber

This and the following section (8.7) discuss a variety of types of glass fiber, including discontinuous filament glass "wool" used for thermal insulation, continuous filament fiber used for textiles, and the very technically sophisticated optical fiber used for modern high-data-rate long distance communications. The compositions and manufacturing processes used for each application differ from each other and from the more common glass manufacturing processes described in Sec. 8.4. For these reasons, we are discussing fiber applications, compositions, and manufacturing together in two separate main chapter sections.

In this section (8.6), we will describe the manufacture of the continuous and discontinuous fiber used for textiles, reinforcement, insulation, and filter media. All of these materials are often referred to by the one word, *fiberglass*. This section also discusses traditional (noncommunication) fiber optics. Optical communications fiber is discussed in Sec. 8.7

8.6.1   Discontinuous Fiberglass (Wool and Textile)

Terminology.   *Wool* is a fluffy mass of discontinuous fibers, generally held together by an organic binder.

Brief history.   Glass fibers have been manufactured for insulation and other purposes since the nineteenth century. In the earliest forms, natural mineral compositions like *basalt* and slag from iron-smelting blast furnaces were melted and *fiberized* (fibrillated) by directing a jet of steam or air perpendicularly across a falling stream of molten glass (Fig. 8.36). The shear forces from the vapor jet would disrupt the glass stream into droplets and attenuate some droplets into fiber. Whether the molten material remained as droplets or was attenuated into fiber, and whether the attenuated fiber remained as such (or disintegrated into arrays of smaller droplets), depended on the size of the droplets,

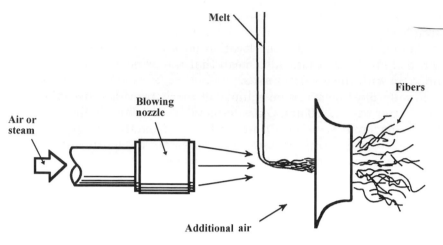

Figure 8.36 Fiber-blowing process to make glass and refractory ceramic fibers. *(From Ref. 36, p. 16)*

the shear forces exerted by the jet, and the rates of cooling. Many small, unusable droplets of glass, called *shot,* often as much as 50-vol.% of the melt, were produced, resulting in rather poor process efficiency. The range of fiber diameter produced in any given operation was quite broad.*

Dramatic improvements to the process were made in the 1930s. In modern steam-blown fiberglass manufacture, the molten glass is delivered through multiple holes in the bottom of a heated platinum trough. Each hole is a millimeter or so in diameter and delivers glass at the rate of a few pounds per hour. Pairs of downward-directed blowers, one on each side of the trough, impinge on the streams of glass, attenuating them into thin fiber (3 to 6 µm diameter, 3 to 10 cm length). The percentage of shot or nonfiberized glass is still significant. This process, illustrated in Fig. 8.37, continues to be used to varying extent throughout the world. In the United States, it is used primarily for specialized fiber applications but has generally been replaced for building insulation by the rotary process described in the next paragraph.

Centrifugal fiber forming processes date back to the 1920s, when the Hager process was invented, whereby a stream of molten glass is fed onto a horizontal rotating disk from which it is flung off as strands or fibers by centrifugal force. Hager fibers were rather coarse and not as well controlled as steam-blown fiber. In the 1930s, St.-Gobain in

---

* This same type of process, adjusted differently, was described above for the manufacture of glass microspheres.

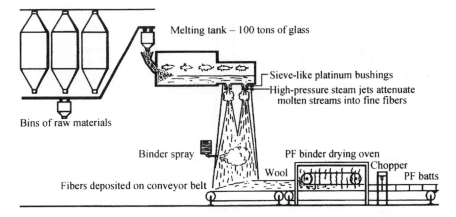

Figure 8.37   Steam blowing (Owens) process for making fiberglass. *(From Ref. 37, p. 722)*

Europe and Owens-Corning Fiberglas in the United States developed the rotary (or modern centrifugal) process whereby molten streams of glass are delivered by centrifugal force through multiple fine orifices in the wall of a shallow rotating cylinder, called a *spinner*. An annular downward-directed burner, concentric with and exterior to the rotating cylinder, attenuated the glass streams into fine fibers. A later version developed by Owens-Corning uses an annular burner to heat the cylinder and an annular steam blower to attenuate the fiber. Variations on these two processes, with spinners up to 2 ft in diameter and containing many thousands of holes (orifices), are now used throughout the world to produce hundreds of thousands of tons of product annually. The rotary process is illustrated in Fig. 8.38.

In a specialized two-step process, called the flame attenuation process, primary fibers approximately 1 mm in dia. are drawn from multiple orifices (in a manner similar to that described in Sec. 8.6.2) and then further attenuated in a high-velocity burner flame to produce a myriad of very fine-diameter discountinuous fibers in a well controlled manner. This more expensive process is used when fibers much finer than those produced by the rotary process are needed.

Applications.   Applications include thermal and acoustical insulation, filtration media, and staple fiber for textiles.

Common compositions

- *A-glass.*   This soda-lime-silica composition, with little or no B_2O_3, is related to container and window glass and was used in early glass wool manufacture. Because of its relatively low cost, it is still in use in some parts of the world. However, for many applications, its rela-

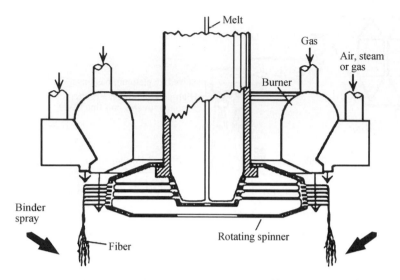

Figure 8.38   Rotary fiberglass wood forming process.[36]

tively poor chemical durability leads to mechanical failure due to stress corrosion. For example, when it was used as thermal insulation in refrigerated railroad cars, condensed moisture and constant mechanical vibration often led to fiber fracture, crumbling of the wool, and settling of the insulation at the bottom of the insulating panels. (Because of the small fiber diameter and consequent high surface-to-volume ratio, fibers are much less resistant to water attack than are bulk glasses.)

- *T-glass.* Thermal-glass was developed in the 1930s to supersede A-glass. It contained less silica, less alkali, and more lime than A-glass; B_2O_3 was added to improve acid and water durability and to lower viscosity. It was more expensive because of the boron content. It was manufactured by the Owens steam-blowing process; however, because of a relatively low liquidus viscosity, it was unsuitable for manufacture using the high-throughput rotary process developed in the 1950s.

- *Modern wool fiberglass compositions.* These stem from Welsh's patent (U.S. Patent no. 2,877,124, September 25,1955). They tend to contain even more B_2O_3 than T-glass. (It should be noted that some authors refer to these modern compositions as T2-glass, calling the earlier compositions T1-glass.)

- *Rock wool and slag wool.* These materials are still manufactured to varying degrees throughout the world. The raw materials are basalt, basalt in combination with limestone, and blast furnace slag.

Composition ranges and typical physical properties for wool-type fiberglass compositions (including rock wool and slag wool) are shown in Tables 8.22. and 8.23.

Manufacturing processes can be summarized as follows:

1. Steam or air blowing (Owens process, Fig. 8.37)

   - Blows high-pressure steam or air downward to entrain a stream of glass

   - Produces fibers and lots of unwanted glass droplets, called *shot*

   - Advantage: high-viscosity and high-liquidus temperature glasses such as rock wool and slag wool can be *fiberized* this way

2. Rotary process (Fig. 8.38)

   - Most commonly used process

   - Free of shot

   - Molten glass fed into rotating cylinder (spinner cup) with many holes in the sidewall

   - Centrifugal force extrudes glass laterally

   - High-velocity gas stream (air, steam, combustion gases) entrain and attenuate fiber

   - Disadvantage: viscosity and liquidus temperature must be low enough for spinning (rock wool and slag wool liquidus temperatures too high to be made by this process)

3. Mechanical attenuation (see Fig. 8.39)

   - Also called the rock wool, wheel centrifuge, cascade, or mechanical spinning processes

   - Stream of glass falls on rotating disks or series of rotating wheels/drums

   - Produces unwanted shot

4. Flame attenuation (not illustrated)

   - 1 mm (approx. dia.) primary fibers drawn from bushings

   - Aligned and passed through burner jet to elongate

   - Passed through binder spray and collected

   - Very fine fibers possible (<0.1 to 8 μm diameter)

TABLE 8.22 Fiberglass Compositions

Glass identification		SiO_2	Al_2O_3	B_2O_3	Li_2O	Na_2O	K_2O	MgO	CaO	BaO	ZnO	F	Other	References
Continuous fiberglass														
A glass	Ranges	63–72	0–6	0–6		14–16	~Na+K	0–4	6–10			0–0.4	$TiO_2 = 0{-}0.6$; $Fe_2O_3 = 0{-}0.5$	Hofmann[17]
C glass	Ranges	64–68	3–5	4–6		7–10	~Na+K	2–4	11–15	0–1			$Fe_2O_3 = 0.8$	Hofmann[17]
	Typical	65	4	5.5		8.5	~Na+K	3	14					Aubourg & Wolf[5]
D glass	Ranges	72–75	0–1	21–24		0–4	~Na+K		0–1				$Fe_2O_3 = 0{-}0.3$	Hofmann[17]
E glass	Ranges	52–62	12–16	0–10		0–2	~Na+K	0–5	16–25			0–1	$TiO_2 = 0{-}1.5$; $Fe_2O_3 = 0{-}0.8$	Aubourg & Wolf[5], Aubourg et al.[7] Boyd et al.[9]
	Typical Advantex	54 X	14 X	10 None				4.5 X	17.5 X				X	Owens Corning
E CR glass	Ranges	54–62	9–15	0–8		0–1	~Na+K	0–4	17–25		0–5		$TiO_2 = 0{-}4$; $Fe_2O_3 = 0{-}0.8$	Hofmann[17], Aubourg et al.[7]
AR glass	Typical Ranges	55–75	0–5		0–1.5	11–21	~Na+K		1–10			0–5	$TiO_2 = 0{-}12$; $ZrO_2 = 0{-}18$; $Fe_2O_3 = 0{-}5$	Hofmann[17]
R glass	Ranges	55–65	15–30			0–1	~Na+K	3–8	9–25			0–0.3	$Fe_2O_3 = 0{-}0.2$	Hofmann[17]
S glass	Ranges	64–65	24–25			0–0.3	~Na+K	10–11	0–0.1					Aubourg & Wolf[5], Aubourg et al.[7]
	Typical	65	25					10						

Glass identification		SiO$_2$	Al$_2$O$_3$	B$_2$O$_3$	Li$_2$O	Na$_2$O	K$_2$O	MgO	CaO	BaO	ZnO	F	Other	References
Wool fiberglass	Ranges	55–70	0–7	3–12		13–18	0–2.5	0–5	5–13	0–3		0–1.5	TiO$_2$ = 0–0.5; Fe$_2$O$_3$ = 0.1–0.5; S = 0–0.5	Aubourg et al.[7]; TIMA/NAIMA
Welsh's patent (1995)	Ranges	50–65	0–8	5–15		10–20	~Na+K	0–10	3–14	0–8	0–2		TiO$_2$/ZrO$_2$ = 0–8; Fe$_2$O$_3$ = 0–12; MnO = 0–12	Aubourg & Wolf[5]
	Typical	58.6	3.2	10.1		15.1	~Na+K	4.2	8					Aubourg & Wolf[5]
T$_2$ glass	Typical	59	4.5	3.5		11	0.5	5.5	16			N.A.	ZrO$_2$ = 4; TiO$_2$ = 8	Boyd et al.[9]
SF glass	Typical	59.5	5	7		14.5								Boyd et al.[9]
Rock wool														
Basalt (glass furnace)	Ranges	45–48	12–13.5			2.5–3.3	0.8–2	8–10	10–12				TiO$_2$ = 2.5–3; FeO = 5–6; S = 0–0.2	Aubourg et al.[7]; TIMA/NAIMA
Basalt + limestone (cupola)	Ranges	41–43	6–14			1.1–3.5	0.5–2	6–16	10–25				TiO$_2$ = 0.9–3.5; FeO = 3–8; S = 0–0.2	Aubourg et al.[7]; TIMA/NAIMA
Slag wool (cupola)	Ranges	38–52	5–15			0–1	0.3–2	4–14	20–43				TiO$_2$ = 0.3–1; FeO = 0–2; P$_2$O$_5$ = 0–0.5; S = 0–2	TIMA/NAIMA

Note: X = Specific composition unavailable from manufacturer.

TABLE 8.23 Fiberglass properties

Glass identification		Density, g/cm³	CTE, 0–300°C, 10^{-7}/°C	Strain point, °C	Annealing point, °C	Softening point, °C	At 10^3 P, °C	Liquidus, °C	Young's modulus, GPa	Tensile strength, GPa (23°C)	Refractive index	References
Textile fiberglass												
A glass	Typical	2.44				705			68.9	3.3	1.538	Hartman[18]
C glass	Typical	2.56	63	552	588	750			70	3.3	1.537	Aubourg et al.[7]
D glass	Typical	2.11		477	521	771			51.7	2.4	1.465	Hartman[18]
E glass	Ranges	2.52–2.62	54	600–630	640–675	830–860			76–78	3.1–3.8	1.547–1.562	Owens Corning
	Typical	2.62	54	615	657	846			72.3	3.4	1.562	Aubourg et al.[7]
	Advantex	2.62	60	691	736	916			80–81	3.1–3.8	1.560–1.562	Owens Corning
E CR glass	Ranges	2.66–2.68	59			880			80–81	3.1–3.8	1.576	Owens Corning
AR glass	Typical	2.76				882			81.3	3.4	1.583	Aubourg et al.[7]
	Typical	2.7				773			73.1	3.2	1.562	Hartman[18]
R glass	Typical	2.54				952			85.5	4.1	1.546	Hartman[18]
S glass	Ranges	2.46–2.49	29	766	816	1056			88–91	4.6–4.8	1.523–1.525	Owens Corning
	Typical	2.50	16	760	810	970			88.9	4.6	1.525	Aubourg et al.[7], TIMA/NAIMA
Wool fiberglass	Ranges	2.40–2.55				650–700	915–1085	880–955	55–62	2.4	1.51–1.54	Aubourg & Wolf[5]
Welsh's patent	Ranges							870–980				
	Typical						1015	925				Aubourg & Wolf[5]
T₂ glass	Typical	2.54	80			715					1.541	Boyd et al.[9]
SF glass	Typical	2.57	75			675					1.537	Boyd et al.[9]
Rock wool												
Basalt (electric furnace)												
Basalt + limestone (cupola)												
Slag Wool (cupola)												

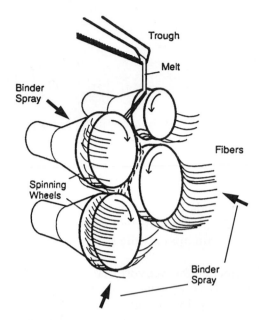

Figure 8.39 Mechanical attenuation method for making rock wool and slag wool. *(From Ref. 36, p. 17).*

Thermal insulation and filter media are manufactured continuously following fiberization. The glass may be fiberized by any of the processes described above. As the fiber falls after fiberization, it is sprayed with a binder, which increases its volume by 5 to 15%. The sprayed fiber is then collected as a wool mat on a conveyor belt, compressed by rollers, cured in an oven and cut into lengths. This is illustrated for the steam-blowing process in Fig. 8.37. The velocity of the conveyor and the height of the roller control mat density. It ranges about 0.5 to 5 lb/ft^3.

Fiber diameters for insulating materials range between 0.2 and 30 μm, with specific surface areas of 0.1 to 0.2 m^2/g. Upper use temperature for glass wool is about 550 to 600°C.

Staple textile yarn (a yarn composed of overlapping discontinuous fibers) can be produced on small scale by collecting discontinuous fiber on a rotating, perforated vacuum drum, then cutting it loose and carding it (somewhat like is done with animal wool and cotton), and finally twisting (spinning) it into *strands* and *yarns*. The product is generally of a lower quality and strength than those made by continuous fiber processes (to be described next) because of diameter variations and short fiber length.

Nonwoven *mats* are formed by collecting the fiber on a continuous suction belt.

8.6.2   Continuous Fiber (Textile and Reinforcement)

Textile terminology

- Filament (single fiber)
- Strand (made up of many filaments)
- Size or sizing (protective lubricant; may also be a coupling agent)
- Yarn (twisted and/or plied from strands)
- Roving (strands bundled together, but not twisted or plied)
- Textured (strands with fiber spacing opened/varied)

Brief history. Glass fiber in long strands has been produced by hand since ancient times, the length of fiber being determined pretty much by how far and fast two people could run apart while stretching a rapidly cooling gob of glass.

Continuous yarns of glass fiber needed for weaving of textiles can be produced from discontinuous fiber, as described in the previous section. Such yarn has significant strength limitations and the disadvantage of many abrasive fiber ends at the yarn's surface. The goal of yarn made from continuous fiber of arbitrary length was realized in the 1930s through a joint effort by Owens-Illinois Glass Company and Corning Glass Works, which lead to the formation of Owens-Corning Fiberglas Corporation. The process consists of simultaneously drawing continuous filaments (fiber) from multiple (up to about 4000) electrically heated precious metal orifices (called *bushings)*, spraying these filaments with a lubricating sizing as they cool, combining into a single multifilament strand, and gathering by winding onto a spool. The wound product is referred to as spun *cake.* This is sometimes referred to as the bushing fiberizing process. It is shown schematically in Fig. 8.40. The melter supplying the molten glass either melts batch, as described in Sec. 8.4 (direct melt), or remelts glass marbles (15 to 20 mm diameter) produced at a different location (marble-melt).

Applications

1. Woven: electrical insulation, fabrics (drapes, clothing), screening, battery separator mats, fiber reinforced composites

2. Nonwoven: mats, tire cord, gaskets, fiber reinforced composites

3. Chopped fiber-reinforced composites

Useful physical properties of textile fiberglass include a linear stress-strain relationship up to the yield stress (3,400 MPa for E-glass and 4,500 MPa for S-glass; approximately 5% deformation), heat resistance, and lack of flammability.

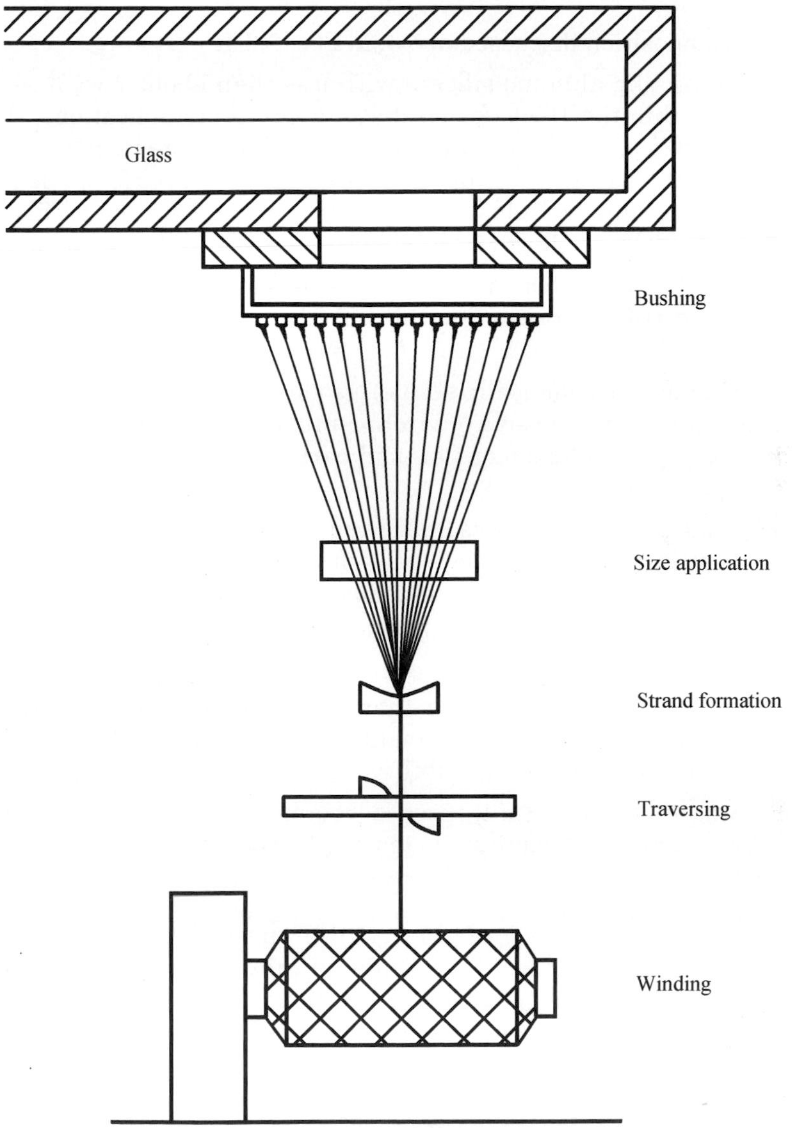

Forehearth

Glass

Bushing

Size application

Strand formation

Traversing

Winding

Figure 8.40 Mechanical attenuation method for making rock wool and slag wool. *(From P.F. Aubourg and W.W. Wolf, Ref. 5, p. 56)*

Common compositions. Several different glass composition types are currently manufactured as continuous filament fiber. The choice of type depends primarily on the intended application. Traditionally, each type has been given letter name indicative of its original intended application. The names and some key characteristics follow.

- *A-glass*—composition discussed above in Sec. 8.6.1.

- *E-glass*—a calcium alumino-silicate with less than about 2 wt.% alkali; developed in the 1930s for electrical insulation applications. It has good electrical resistivity due to its low alkali content. Because of its excellent mechanical properties, its use has spread, particularly for glass-reinforced plastics. (Over 90% of all continuous filament fiber is E-glass.) However, it has poor acid durability.

- *E-CR glass* (CR indicates corrosion resistant)—similar to E-glass, but alkali and B_2O_3 are each less than 1%. It has improved acid durability.

- *C-glass* (C indicates chemically stable)—soda-lime borosilicate; good chemical stability in corrosive environments; used in highly acid environments (e.g., reinforcement of acid carrying vessels) and other special applications.

- *S-glass* (S indicates strong)—essentially pure magnesium alumino-silicate; high strength/weight ratio and high use temperature; used especially as reinforcement of containers having high internal pressure.

- *R-glass*—compositions showing high tensile strength in combination with resistance to acid corrosion

- *Advantex*™—a boron-free E-glass that, according to manufacturer (Owens-Corning), combines the advantages of E- and E-CR glasses and has higher temperature capability.

- Other—A.R. (alkali-resistant, used to reinforce cement); L-glass (lead, for radiation shielding); D-glass (dielectric); M-glass (high modulus).

Typical compositions and glass properties for most of these types are given in Tables 8.22 and 8.23. The tabulated ranges have been taken from several sources. The precise compositions used by any single manufacturing plant often depend on its specific manufacturing process and the available raw materials. The fiberglass manufacturers publish few compositions and, with the exception of E-glass, there are no published standards for composition.

Bushing fiberizing manufacturing process summarized

- Melt and deliver glass.
- Draw from bushings (Pt-Rh alloy, electrically heated) containing 200 to 4000 orifices (nozzles), each 0.8 to 3.0 mm dia., at speeds to 180 mph (80 m/s).

- A single filament is formed from each bushing.
- Cool by air or water to form fiber (solid, continuous, 3 to 25 μm dia.)
- Apply sizing (for abrasion resistance, may be designed for final application).
- Gather into strand (bundle).
- Wind on reel as spun cake. (Alternatively, rather than collecting it on a spool, the strand can be directly processed into rovings, continuous strand mats, or chopped strands.)

During the fiber draw, the melt is highly deformed by attenuation under relatively high stress, the attenuation (area reduction) ratios being about 40,000:1. The elongated filaments are cooled under load; consequently, internal structural changes induced by the deformation are quenched-in. Hence, the internal structure and physical properties of the fiberglass can differ from those of annealed glass of the same composition.

Glass processing requirements

1. A high quality melt:
   - Few inclusions—could block bushing orifices or act as stress concentrators in fibers, leading to fiber breakage during draw and consequent process interruption. (Breakage of a single filament between the bushing nozzles and the winder generally causes breakage of others.)
   - Homogeneous (uniform viscosity)—same reasons.

2. As low a temperature as possible to limit deformation and wear of bushings (typically fiberize at about 500 to 1500 P; for E-glass, that requires temperatures of about 1250°C.)

3. A low liquidus temperature (must operate the process at temperatures above the liquidus temperature to avoid devitrification in the bushing).

Other manufacturing methods. Continuous single filaments of glass fiber can, of course, be drawn upward from the surface of a melt, downward from an orifice at the bottom of a glass melting tank or a crucible, or in almost any direction off the heated end of a solid glass rod. For some applications, such single fibers are useful, and the production methods are efficient. But for most applications, the simultaneous downdrawing of multiple filaments, as described above, is preferred.

For composite reinforcement applications such as fiber reinforced polymers (FRP), also known as glass-fiber reinforced plastic (GRP), the fiber can be supplied as chopped strands (at several standard lengths from 3 to 12 mm) or continuous roving, which can be chopped at the point of composite manufacture. It can also be supplied as milled fiber, with lengths of about 0.2 mm. Roving can be woven into a wide variety of two-dimensional and three-dimensional shapes for polymer impregnation. The relationships between the various process steps and the resulting types of reinforcement fiber products are shown in Fig. 8.41.

Fiber dimensions. For textile products, such as textile yarns and plied yarns, filament diameters range from 5 to 13 µm. For plastics reinforcement, such as mats, rovings and chopped strands, the diameters range from 9 to 25 mm.

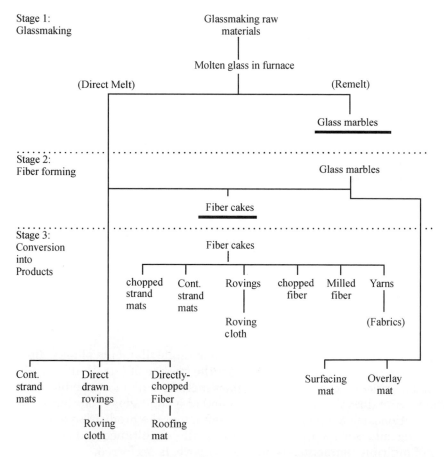

Figure 8.41 Flowchart of glass fiber products manufacture. Some products appear in more than one sequence; this is because they are made by both the remelt and the direct-melt processes.[38]

8.6.3 Traditional Fiber Optics

Principles of operation. With optical fibers, light is transmitted within the fiber, along its length, without losing a significant fraction of its energy through the walls of the fiber. This behavior is due to a principle called *total internal reflection,* which can be explained with reference to Fig. 8.42, recalling that refraction and reflection at an interface between two transparent media behave according to Snell's law and Fresnel's equations.

For light traveling through a fiber surrounded by air (refractive index = 1) or clad by a polymer or glass medium of refractive index less than that of the fiber core, there will be at the wall a maximum internal angle of incidence, α_c, called the *critical angle,* for which the angle of refraction, α', becomes 90°. For angles less than critical, some light is transmitted through the wall; at all greater angles, the light is totally internally reflected. α_c can be calculated using Snell's law. By that same law, the entrance angle into the fiber (relative to the axis of the fiber), θ_c (corresponding to the angle of total internal reflection,

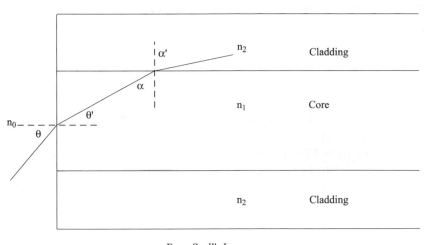

From Snell's Law:
$$n_1 \sin \alpha = n_2 \sin \alpha'$$
$$n_0 \sin \theta = n_1 \sin \theta'$$
$$\text{Sin } \theta = \frac{n_1}{n_0} \sin \theta' = \frac{n_1}{n_0} \cos \alpha = \frac{n_1}{n_0} \sqrt{1 - \sin^2\alpha}$$
$$\text{Sin } \alpha = \frac{n_2}{n_1} \sin \alpha'$$

For critical α and θ :
$$\alpha' = 90°$$
$$\sin \alpha_c = \frac{n_2}{n_1}$$

$$\sin \theta_c = \frac{n_1}{n_0} \sqrt{1 - \left(\frac{n_2}{n_1}\right)^2} = \frac{1}{n_0} \sqrt{n_1^2 - n_2^2} \equiv NA$$

Figure 8.42 Reflection relationships in clad fiber optics.

α_c), is determined. All light entering the fiber at angles less than this critical entrance angle will be transmitted through the fiber. Correspondingly, light entering the fiber at angles greater than this critical angle will eventually, after multiple reflections, be lost through the walls of the fiber. The sine of this angle is called the *numerical aperture (NA)* of the fiber. The higher the numerical aperture, the greater the light-gathering power of the fiber. An NA = 1 means that all light incident on the face of the fiber will be conducted or *piped* through the fiber. These same definitions and terminology apply to the optical communications fiber discussed in Sec. 8.7.

It should be pointed out that a simple unclad glass rod or fiber would channel light as described above. The maximum possible NA for a given glass composition results when it is surrounded by air or vacuum. However, for that case, any damage to the fiber surface, such as abraded scratches, disrupts the reflection conditions and allows light to leak (scatter) out of the fiber. Hence, the surface of the fiber generally requires protection by a lower-index glass or polymer cladding. Also, if unclad fibers are packed close to each other, some of the light traveling in one fiber may cross over into the adjacent fiber. This is undesirable in applications where images are to be transmitted coherently by the fiber bundle.

Traditional fiber optics such as those described in this section can be designed to transmit images, illuminating light, or both. They consist either of flexible bundles of fibers or arrays of fibers stacked and fused together along their length. The latter are appropriately called *fused fiber optics*. Either type may be used to transmit light of various wavelengths or intensity.

Either type (fused fiber optics or flexible bundles fused at each end, called *coherent bundles*) may also be used to transmit images. Since each fiber carries information (light intensity) related to one geometric element of an image (a single *pixel*), the fiber size must be made smaller than the finest detail to be resolved; core diameters are often as small as 10 µm; cladding sometimes is less than 0.5 µm thick.

Special capabilities. Flexible fiber optics can transmit light and images around bends and in controlled paths. They can also and can scramble and unscramble images. Fused fiber optics can magnify, demagnify, rotate and invert images.

Applications

1. Illumination
   - Medical (diagnostics and surgery)
   - Scientific instrumentation

- Automobile and aircraft instrument panels

- Flat-panel display backlighting

- Future automotive lighting (interior and driving lights)

2. Image transmission, magnification, and inversion

- Medical (diagnostics and surgery; endoscopes and the like)

- Technical (e.g., chimney, pipeline, and chemical system inspection; remote sensing)

- Optoelectronic devices (including vacuum-tight electron tube faceplates for TV cameras and CRTs; advantages include flat fields and greater image intensity)

Process. Single filaments of clad fiber may be made by redraw of rod in tubing or, alternatively, by a double-crucible method, which is essentially a Vello-type delivery system where a second glass rather than air is delivered through the center of the bell. A version of the latter method is illustrated in Fig. 8.43.

A flexible fiber optic is formed by bundling many of these clad filaments or *monofibers* together. Often, the entrance and exit to the bundle are made more mechanically strong by locally fusing the fibers together, either by heating under pressure, with or without a glass sealing frit as a binder, or with a suitable organic polymer. The fused ends can then be polished to a smooth optical surface. Scrambling and unscrambling devices may be made by disarraying the fibers at some location along the length of the bundle, fusing them together at that location, then cutting the bundle at the fused point.

Fused fiber optic devices are formed by stacking together in parallel alignment many (hundreds of) individual monofibers, then fusing them together under heat and pressure. These fused multifiber stacks are redrawn (to smaller cross-sectional dimensions), cut, stacked, fused, and redrawn again. This process is repeated until the target core diameter (or number of fibers per unit area) is achieved. The final multifiber elements are then cut, stacked, and fused (but not redrawn) to give whatever size device is required. Sometimes an EMA (extramural absorbing) cladding is used to improve image contrast. This prevents leakage of light from one fiber core to another and attenuates light that enters the fiber stack through the cladding; a light-absorbing second cladding can be made part of the original (first-draw) fibers. More often, absorbing (black) glass rods are substituted for some of the fibers in the first stacking, in either a random or designed pattern. This is often sufficient to provide the extramural absorption needed for contrast enhancement.

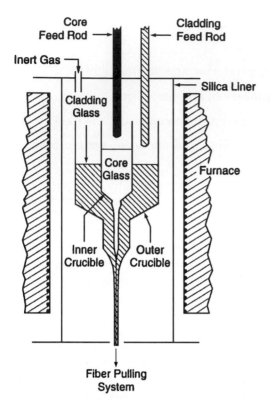

Figure 8.43 The double-crucible method for making clad optical fiber.[39]

Fused fiber optic tapers can be used to magnify (or demagnify) images. Tapers are made by locally heating the central region of a length of fused fiber optic and stretching it to form an hourglass shape. From this hourglass, two tapered sections can be cut. Each section is capable of transferring an image from one surface to the other, the magnification being determined by the ratios of the two diameters (assuming a circular cross section). Image rotators or inverters are made by rotating one end of the hourglass with respect to the other as it is being drawn; not much necking is required.

Near-optical-quality glass is generally required for both core and cladding glasses. Examples of refractive index combinations and the corresponding fiber characteristics are:

$$n_1 = 1.700; \quad n_2 = 1.512; \quad NA = 0.78; \quad \theta_1 = 51°$$

$$n_1 = 1.650; \quad n_2 = 1.560; \quad NA = 0.54; \quad \theta_1 = 32°$$

The thermal expansions of core and cladding glasses must be designed to match from set point of softer glass to room temperature.

8.7   Optical Communications Fiber

8.7.1   Introduction

The subjects of optical communication of information and all the associated applications of glass are beyond the scope of this chapter. But because the invention, manufacture, and application of low-loss optical communications fiber have been so important to the development of worldwide telecommunications and computer networking for the past three decades, and promise to remain so well into the new century, we devote the final section of this chapter to the materials and techniques used to manufacture low-loss optical fiber.

The use of glass as the medium for long distance communication of information by optical signals was pioneered by Corning Glass Works (now Corning Inc.) and AT&T Bell Telephone Laboratories (now Lucent Technologies). It was Corning that first demonstrated in the early 1970s that optical fiber could be made with an optical signal loss rate less than 20 dB/km (approximately 1% of the input light transmitted through 1 km of fiber), the upper limit imposed by the practicalities of telecommunications system design. This was done using fiber made from titania-doped synthetic silica glass of otherwise extremely high purity. Each company, Corning and AT&T, proceeded to develop different approaches to manufacturing the fiber, and soon others entered the business, sometimes with new techniques as well. In this chapter section, we describe the various key methods of optical fiber manufacture.

8.7.2   Materials

Silica glass is the material of choice for long distance optical communications applications. A significant reason is that silica is one of the few optically transmitting materials that can be made with sufficient purity that light absorption in the near-IR spectral region can be kept at an acceptably low level. The silica glass for optical communications fiber is not manufactured by the traditional melting approaches used for most other oxide glasses but rather by methods based on those described in Sec. 8.2.9 above for Types III, IV, and V fused silica.

Other glass composition types, such as fluorides, oxynitrides, and chalcogenides, have been considered, mainly because of their ability to transmit even longer infrared wavelengths, but they are not now manufactured in great volume for long-distance communication. This is partly because of difficulties met in obtaining the extremely high purity needed for low optical loss, and partly because of manufacturing obstacles yet to be overcome.

8.7.3 Types of Optical Fiber Design

There are basically two types of optical communications fiber design, each having many variations. The first type, called *step-index* fiber, is similar in construction to that described in Sec. 8.6 for traditional fiber optics, with several key differences.

1. Communications fiber dimensions are generally much smaller. (The outside diameter is now a standard 125 µm; the core diameter can be as small as about 8 µm.)

2. The refractive index difference (Δn) between *core* and *cladding* is much smaller, on the order of 1% or less.

3. Much greater precision is required for the critical dimensions, including core concentricity within the fiber.

The second type is called *graded index* (sometimes *GRIN*) fiber. Here, the refractive index is varied from the central axis location gradually out into the cladding. Often, the index gradient is parabolic in shape, highest at the axis and decreasing outwardly toward the cladding, but sometimes it is more complex in profile. The core region occupies about 50 to 70 µm of the fiber diameter; the outside diameter is generally, again, 125 µm.

Light propagates through an optical communications fiber as electromagnetic waves in a manner similar to how microwave energy propagates through a hollow microwave *waveguide*. Consequently, optical fibers are sometimes referred to as *optical waveguides,* and fiber types are correspondingly classified according to whether they are *single-mode* or *multimode*. This waveguide terminology refers to the manner in which the light energy is transmitted down the fiber, in particular how the energy is distributed in space, not to how many different signals the fiber can carry. In fact, a single-mode fiber can carry information at much higher rates over a given distance than can a multimode fiber.

8.7.4 Manufacturing Processes

The manufacturing techniques for all the above types of fiber are generally similar. We list the key steps here as follows:

1. Vapor generation

2. Preform preparation (silica deposition, including refractive index adjusting elements)

3. Blank preparation (glass rod with required index gradient)

4. Fiber drawing

5. Coating

6. Testing

Step 1. As discussed previously, extremely high-purity chemical precursors are required and used—even better than semiconductor manufacturing quality. The precursor for silica is often liquid $SiCl_4$. The refractive index controlling *dopants*, germania and fluorine, are provided by $GeCl_4$ and SiF_4, respectively. (Other precursor chemicals can be used; there is a trend away from the chlorides, partly for environmental reasons.) These liquids are converted to the vapor phase by bubbling a carrier gas through the liquids at controlled temperatures, as illustrated in Fig. 8.44. Alternatively, the needed chemicals may be sublimed from the surface of a solid in the presence of a car-

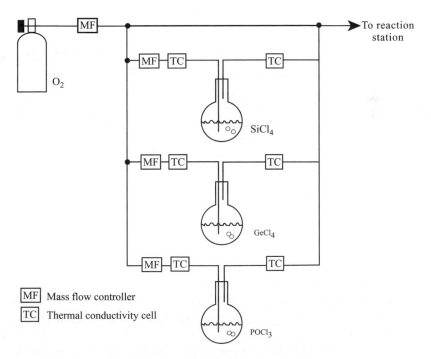

Figure 8.44 Reactant delivery system for vapor-phase techniques.[40]

rier gas. In either case, the materials must have a high vapor pressure at reasonable process temperatures. This requirement actually limits the dopants that can be incorporated easily into the silica. During preform preparation, the ratio of the mix of chemical precursors is varied over time to control the eventual refractive index gradient within the fiber.

Steps 2 and 3. In these steps, the glass blank, the material that eventually will be drawn into fiber, is generated. It is here that the two major approaches to optical fiber manufacture first differ in a very significant way.

The first approach is called the *outside* process. In step 2, preform preparation, the chemical precursors are reacted in the flame of a specially designed gas-oxygen burner, the combustion gas being either natural gas or hydrogen, just as for all the Type III synthetic silicas described in Sec. 8.2.9.1. The doped silica is generated as a fine white amorphous (molten glass) soot that is progressively deposited on a rotating bait rod, as shown in Fig. 8.45a. The deposition begins with the material that will eventually lie at the axis of the fiber core and ends with the material that will form the outside of the cladding. In other words, in the outside process, the preform is created from the inside outward. The resulting preform, inherently containing the desired chemical gradient, is porous and moderately fragile.

To consolidate this porous preform into a solid glass, in Step 3, the bait rod is removed and the remaining preform heated to temperatures near the softening point of fused silica to allow sintering of the soot via viscous flow. To ensure that air does not become trapped in interstices and thereby create bubbles in the glass, the preform is consolidated in a helium gas atmosphere. The rapid diffusion of helium through the silica glass assures that all voids shrink to zero dimensions under the forces of surface tension. As mentioned also in Sec. 8.2.9.1, Type III silicas contain water in their structure in the form of OH bonds. Light absorption by such OH is extremely detrimental to communications optical fiber performance, so it is removed from the glass by flowing chlorine (Cl_2) gas through the preform at high temperatures just prior to consolidation.

This "outside" process is sometimes referred to as outside vapor deposition (OVD). This is somewhat a misnomer, because the glass is not deposited directly from the vapor phase but rather by the progressive aggregation of submicroscopic particles that were themselves created from the vapor phase. The process has also been called radial flame hydrolysis (RFH).

In a variation of the outside process, called the *axial* process, the soot is deposited at the end of the rotating preform as it is gradually pulled

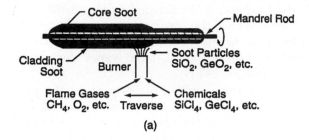

(a)

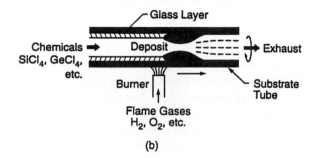

(b)

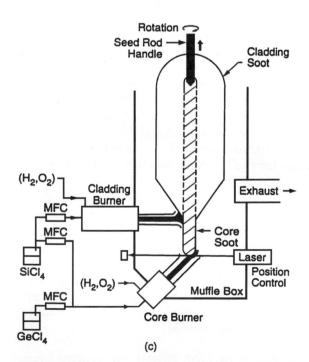

(c)

Figure 8.45 Schematic representation of (a) OVD, (b) MCVD, and (c) VAD preform fabrication routes. *(From A.J. Bruce, Ref. 10, p. 44)*

away (withdrawn) from the burners, as illustrated in Fig. 8.45c. The preform is thus created from the end, the core and cladding materials being deposited essentially simultaneously. This process is sometimes referred to as vapor axial deposition (VAD), axial vapor deposition (AVD), or even axial flame hydrolysis (AFH).

In the second major approach, the so-called *inside* process, the preform is built up on the inside of a silica glass tube, beginning with the materials that will compose the bulk of the fiber cladding, and ending with what will become the innermost region of the core. (So, for the inside process, the preform is generated from the outside inward.) In this process, the heat-providing burner traverses the outside of the rotating silica tube and the reactive gases travel down the inside, as shown in Fig. 8.45b. The gases inside the tube react to form silica soot in the space within the tube. The soot migrates to the walls under thermophoretic forces (diffusion down a temperature gradient). The inner surface of the glass tube is sufficiently hot that the soot particles immediately flow and consolidate into a void-free molten layer (just as in Type III and Type IV synthetic fused silica manufacture). The water vapor generated by the gas-oxygen burner never reaches the inside of the tube, so the deposited glass is very dry, requiring no further drying steps. After the deposition is complete, in process Step 3, the resulting thick-walled tubing is collapsed under heat and vacuum to form a solid rod. The outside silica layer (the starting tube) is left in place and becomes part of the cladding.

This *inside* process is sometimes called *inside vapor phase oxidation (IVPO)* or *inside vapor deposition (IVD)*. However, it is perhaps most commonly referred to as *modified chemical vapor deposition (MCVD)*. In one variation, perhaps not currently in commercial use, an argon/oxygen plasma is generated within the tube by microwave radiation to provide the thermal energy for the chemical reaction rather than relying on heat from a burner outside the tube.

Another technique proposed for the inside process is plasma chemical vapor deposition (PCVD). In this version of the process, a low-pressure plasma is generated inside the tube that does not lead to soot generation but rather allows a heterogeneously nucleated chemical reaction to occur at the inner surface of the tube so that the glass is built up in molecular-scale layers. This is a true chemical vapor deposition (CVD) process. More layers are required than with soot, but the process can be controlled more precisely. Unfortunately, the process has not yet proven to be commercially economical.

Step 4. Fiber drawing from the preform follows the general procedures described in Sec. 8.4.5. Notable differences may be higher temperatures, greater cleanliness, greater speed, and more precise dimensional control. Large commercial draw towers can be 20 m high, with process preforms of dimensions exceeding 10 cm diameter and 2 m length. They draw at speeds exceeding 10 m/s to produce more than 50 km of fiber from a single preform. A fiber draw tower is illustrated schematically in Fig. 8.46.

Step 5—Coating. To protect the fiber's freshly created (pristine) glass surface and thus preserve its inherent strength, the fiber is coated on the draw tower with a polymeric layer as soon as it is cool enough to do so. Often, dual UV-curable coatings are used: a soft inner coating and a hard outer coating. Sometimes, thin hermetic coatings (metal, ceramic, or amorphous carbon) are applied by special equipment just prior to applying the polymeric coatings.

Step 6—Testing. Often, the coated fiber is continuously proof-tested for strength on the draw tower or as it is rewound onto other spools. Optical testing is also critically important. For multimode fibers, the tests include attenuation, bandwidth, numerical aperture, and core diameter; for single-mode fibers, they include such additional characteristics as chromatic dispersion, cutoff wavelength, and mode field diameter.

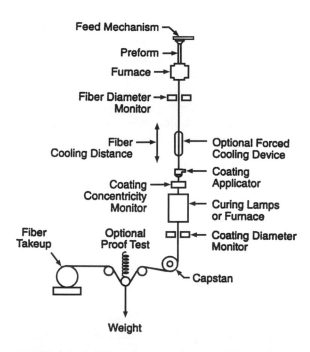

Figure 8.46 Schematic representation of a draw tower for high-silica preforms. *(From A.J. Bruce, Ref. 10, p. 44)*

8.8   Notes and Acknowledgments

8.8.1   Notes

1. Glass manufacturers generally do not sell bulk quantities of glass, that is, chunks of glass by the pound, but rather they sell glass products, e.g., beverage bottles, electric lamp envelopes, television bulbs, optical lens blanks, and telescope mirror blanks. However, some specialty glass manufacturers sell glass in limited forms such as sheet, tubing, rod, frit, and bars of optical glass. Often, these items are not stocked in inventory and must be special ordered. There are several specialty glassmakers that will melt glasses and fabricate shapes to customer's specifications, within the range of their capabilities.

2. Glass manufacturers generally specify their glass types and products by their physical and chemical properties; it is the properties that they control and sell, not the compositions (although careful composition control is needed to assure the desired properties). Often, the same properties can be produced by compensated variations in composition, i.e., increase of one alkaline earth ion at the expense of another. Such choices are often made on the basis of batch component cost or availability. For these reasons, many manufacturers do not disclose the compositions of their products. In other cases, the compositions are considered proprietary. Thus, in handbooks such as this one, compositions for some glass types are not known, not given, or only approximately given; others are only known because of independent chemical analysis, sometimes by competitors and sometimes by scientific researchers.

8.8.2   Acknowledgments

One of the authors (TPS) wishes to acknowledge that much of his perspective on glass forming processes came through working with several Corning Inc. process engineers: G. Clinton Shay, Stuart Dockerty, Richmond Wilson, Leonard Anderson, William Lentz, William Pardue, and Frank Coppola. Similarly, his thinking about glass composition has been strongly influenced by close working relationships with Corning researchers Roger Araujo, Nicholas Borrelli, George Beall, and George Hares.

References

1. R. W. Douglas and Susan Frank, *A History of Glassmaking,* G. T. Foulis & Co. Ltd., Henley-on-Thames, England, 1972.

2. D. R. Uhlmann and N. J. Kreidl, eds., *Glass: Science and Technology, vol. 2, Processing I,* Academic Press, Orlando, 1983.

2a. D. R. Uhlmann and N. J. Kreidl, eds., *Glass: Science and Technology, vol. 5, Elasticity and Strength of Glasses,* Academic Press, Orlando, 1980.

3. E. B. Shand and G. W. McLellan, *Glass Engineering Handbook,* 3d ed., McGraw Hill, New York, 1984.

4. F. V. Tooley, ed., *The Handbook of Glass Manufacture,* vols. I and II, Books for the Glass Industry Division, Ashlee Publishing Co., New York, 1984.

5. D. C. Boyd and J. F. MacDowell, eds., *Advances in Ceramics,* vol. 18, Commercial Glasses, The American Ceramic Society, Columbus, Ohio, 1986.

6. N. P. Bansal and R. H. Doremus, *Handbook of Glass Properties,* Academic Press, Orlando, 1986.

7. S. J. Schneider, Jr., vol. chairman, *Engineered Materials Handbook,* vol. 4, Ceramics and Glasses, ASM International, Materials Park, Ohio, 1991.

8. A. K. Varshneya, *Fundamentals of Inorganic Glasses,* Academic Press, San Diego, 1994.

9. D. C. Boyd, P. S. Danielson, and D. A. Thompson, "Glass," in *Kirk-Othmer Encyclopedia of Chemical Technology,* 4th ed, vol. 12, John Wiley and Sons, New York, 1994.

9a. J. R. Hutchins, III and R. V. Harrington, "Glass," in *Kirk-Othmer: Encyclopedia of Chemical Technology,* 2d ed., vol. 10, John Wiley and Sons, New York, pp. 533–604, 1966.

10. J. S. Sanghera and I. D. Aggarwal, eds., *Infrared Fiber Optics,* CRC Press, Boca Raton, Fla., 1998.

11. H. Bach and N. Neuroth (eds.), *The Properties of Optical Glass,* Springer-Verlag, Berlin and New York, 1998.

12. J. McGaughey, Libbey Inc., personal communication, 1999.

13. T. R. Kozlowski and G. A. Chase, "Parameters of Chemical Strengthening and Impact Performance of Corning Code 8361 (White Crown) and Corning Code 8097 (Photogray®) Lenses," *Am J. of Optometry* and *Archives of Am. Academy of Optometry,* vol. 50, 1973, pp. 273–282.

14. Y. Nakao, "Next-Generation Glass Substrate," presented at Electronic Display Forum 98, Yokohama, Japan, April 29, 1998.

15. W. Holand, "Materials Science Fundamentals of the IPS Empress 2 Glass Ceramic," *Ivoclar-Vivident Report,* no. 12, December 1998, Ivoclar Aktiengesellschaft, Schaan, Liechtenstein.

16. Y. Abe, "Glass-ceramics based on calcium phosphates—artificial dental crown and microporous glass-ceramics," *Glastech. Ber. Glass Sci. Technol.,* vol. 73, no. CI, 2000.

17. D. Hoffman, Owens Corning Science and Technology Center, private communication, April 1999.

18. D. R. Hartman, "Evolution and Application of High Strength Fibers," *The Glass Research: Bulletin of Glass Science and Engineering,* vol. 4, no. 2, Winter 1995, NSF Industry-University Center for Glass Research, Alfred University, Alfred, N.Y.

19. G. Hetherington, K. Jack, and M. W. Ramsay, "The high-temperature electrolysis of vitreous silica: Part I. Oxidation, ultra-violet induced fluorescence, and irradiation color," *Phys, Chem Glasses,* vol. 6, 1965, pp. 6–15.

20. R. Bruckner, "Properties and Structure of Vitreous Silica, I," *J. Non-Cryst. Solids,* vol. 5, 1970, pp. 123–175.

21. H. W. McKenzie and R. J. Hand, *Basic Optical Stress Measurements in Glass,* Society of Glass Technology, Sheffield, England, 1999.

22. A. G. Pincus and T. R. Holmes, *Annealing and Strengthening in the Glass Industry,* 2d ed., Ashlee Publishing Co., New York, 1987.

23. G. H. Beall, "Chain silicate glass-ceramics," *J. Non-Cryst. Solids,* vol. 129, 1991, pp. 163–173.

24. R. M. Klein, "Optical fiber waveguides," in *Glass: Science and Technology, vol. 2, Processing I.,* R. R. Uhlmann and N. J. Kreidl, eds., © 1983, Academic Press.

25. A. J. Bruce, in *Infrared Fiber Optics,* J. S. Sanghera and I. D. Aggarwal, eds., CRC Press, Boca Raton, FL, 1998, p. 40. His references 21 and 22 are P. C. Shultz, Wiss. Z. Friederich-Schiller-Univ. Jena, *Math. Naturwiss. Reihe* 32, 215 (1983), and H. Takahashi, A. Oyobe, M. Kusuge, and R. Setaka, *Tech. Dig.,* ECOC '86, 3 (1986).

26. F. E. Woolley, p. 390, in S. J. Schneider, Jr., vol. chairman, *Engineered Materials Handbook,* vol. 4, Ceramics and Glasses, ASM International, Materials Park, Ohio, 1991. His reference is W. Trier, *Glass Furnaces (Design, Construction and Operation),* trans. K. L. Lowenstein, Society of Glass Technology, Sheffield, England, 1987.

27. D. C. Boyd, P. S. Danielson, and D. A. Thompson, "Glass," in *Kirk-Othmer Encyclopedia of Chemical Technology,* 4th ed, vol. 12, John Wiley and Sons, New York, 1994, p. 602. Their reference is W. Giegerich and W. Trier, *Glass Machine Construction and Operation of Machines for the Forming of Hot Glass,* trans. N. J. Kreidl, Springer-Verlag, Berlin, 1969.

28. J. R. Hutchins, III and R. V. Harrington, "Glass," in *Kirk-Othmer: Encyclopedia of Chemical Technology,* 2d ed., vol. 10, John Wiley and Sons, New York, pp. 533–604, 1966, p. 558. Their reference is E. B. Shand, *Glass Engineering Handbook,* McGraw-Hill, New York, 1958, p. 164.

29. U.S. patent 3,338,696 (Aug. 29, 1967), S. M. Dockerty (to Corning Inc.).

30. R. A. McCauley, "Float glass production: Pilkington vs PPG," *The Glass Industry,* April 1980, 18–22.

31. M. Cable, "Mechanism of Glass Manufacture," *J. Am. Ceram. Soc.* 82, 1093–1112. His reference is Ford Motor Co., Float Method of Manufacturing Glass, Br. patent no. 1,085,010, 1967).

32. A. K. Varshneya, "Glass Transition Range Behavior," in *Fundamentals of Inorganic Glasses,* p. 298, ©1994 by Academic Press, reproduced by permission of the publisher.

33. R. Gardon, "Thermal Tempering of Glass," in *Glass: Science and Technology, vol. 5, Elasticity and Strength of Glasses,* D. R. Uhlmann and N. J. Kreidl, eds., ©1980 by Academic Press, reproduced by permission of the publisher.

34. A. G. Pincus and T. R. Holmes, *Annealing and Strengthening in the Glass Industry,* 2nd ed., Ashlee Publishing Co., New York, 1987.

35. H. W. McKenzie and R. J. Hand, *Basic Optical Stress Measurement in Glass,* Society of Glass Technology, Sheffield, England, 1999, p. 33.

36. *Man-Made Vitreous Fibers: Nomenclature, Chemical and Physical Properties,* Nomenclature Committee of TIMA Inc. (courtesy of North American Insulation Manufacturers Assoc. [NAIMA]).

37. G. R. Machlan and F. V. Tooley, *Handbook of Glass Manufacture,* vol. II, F. V. Tooley, ed., 1974.

38. Loewenstein, K. L., *The Manufacturing Technology of Continuous Glass Fibers,* 3d ed., Elsevier, New York, 1993.

39. A. J. Bruce, in *Infrared Fiber Optics,* J. S. Sanghera na dI. D. Aggarwal, eds., CRC Press, Boca Raton, FL, 1998, p. 35. His reference is L. G. Cohen, *J. Lightwave Technology* LT3, 1985, p. 958.

40. R. M. Klein, "Optical Fiber Waveguides," in *Glass: Science and Technology, vol. 2, Processing I,* D. R. Uhlmann and N. J. Kreidl, eds., ©1983 by Academic Press, reproduced by permission of the publisher.

Bibliography

Bamford, C. R., *Color Generation and Control in Glasses,* Elsevier Scientific, New York, 1977.

Cable, M., "A Century of Developments in Glassmelting Research," *J. Am. Ceram. Soc.* vol. 81, no. 5, pp.1083–1094 (1998).

Cable, M., "Mechanization of Glass Manufacture," *J. Am. Ceram. Soc.* vol. 82, no. 5, pp. 1093–1112 (1999).

De Jong, B. H. W. S., J. W. Adams, B. G. Aitken, J. E. Dickinson, and G. J. Fine, "Glass-Ceramics," in *Ullmann's Encyclopedia of Industrial Technology,* vol. A12, VCH Verlagsgesellschaft mbH, Weinheim, Germany, 1989, pp. 433–448.

De Jong, B. H. W. S., "Glass," in *Ullmann's Encyclopedia of Industrial Technology,* vol. A12, VCH Verlagsgesellschaft mbH, Weinheim, Germany, 1989, pp. 365–432.

Doremus, R. H., *Glass Science,* John Wiley & Sons., New York, 1973.

Giegerich, G. and W. Trier, *Glass Machines* (English translation), Springer, Berlin-Heidelberg, Germany, 1969.

Kleinholz, R. and H. Tiesler, "Glass Fibers," in *Ullmann's Encyclopedia of Industrial Technology,* vol. A11, VCH Verlagsgesellschaft mbH, Weinheim, Germany, 1989, pp. 11–27.

Kurkjian, C. R. and W. R. Prindle, "Perspectives on the History of Glass Composition," *J. Am. Ceram. Soc.* vol. 81, no. 4, pp. 795–813 (1998).

Morey, G. W., *The Properties of Glass,* 2d ed., Reinhold, New York, 1954.

McMillan, P. W., Glass-Ceramics, 2d ed., *Academic Press,* London, 1979.

Nakahara, T., M. Hoshikawa, T. Sugawa, and M. Watanabe, "Fiber Optics," in *Ullmann's Encyclopedia of Industrial Technology,* vol. A10, VCH Verlagsgesellschaft mbH, Weinheim, Germany, 1989, pp. 433–450.

Perkins, W. W., ed., *Ceramic Glossary,* 2d ed., American Ceramic Society, Columbus, Ohio (1984).

Pilkington, L. A. B., "The float glass process," *Proc. Roy. Soc. Lond. A.,* vol. 314, pp. 1–25, 1969.

Pilkington, Alastair, "Flat glass: evolution and revolution over 60 years," *Glass Technology,* vol. 17, no. 5, pp. 182–193, 1976.

Rawson, H., *Inorganic Glass-Forming Systems*, Academic Press, London, 1967.

Standard Terminology of Glass and Glass Products, ASTM Standards 1995, ASTM C162-94c, vol. 15.02, pp. 27–41.

Weyl, W. A., *Colored Glasses,* Society of Glass Technology, Sheffield (Reprinted by Dawson's of Pall Mall, London) 1951.

Wachtman, J. B., Jr., ed., *Ceramic Innovations in the 20th Century,* American Ceramic Society, Westerville, Ohio, 1999.

9

Coatings and Finishes

Carl P. Izzo
Industrial Paint Consultant
Export, Pennsylvania

9.1 Introduction

Coatings and finishes today are engineering materials. From cave dwellers decorating their walls with earth pigments ground in egg whites to factory workers protecting products with E-coat primers and urethane acrylic enamels, these coatings are still composed of film-forming vehicles, pigments, solvents, and additives.[1] Unlike Noah, who received divine instructions to paint his ark with pitch; the Egyptians, who mixed amber with oils to make decorative varnishes; or the Romans, who mixed white lead with amber and with pitch, coatings scientists and engineers using these materials design coatings today for specific applications on products. Because of concern for environmental regulations, safety in the workplace, performance requirements and costs, the design of these coatings must be optimized.

The first 50 years of the twentieth century were the decades of discovery. Significant changes were made in the vehicles, which are the liquid portions of the coatings composed of binder and thinner.[2] Since the 1900s and the introduction of phenolic synthetic resin vehicles, coatings have been designed to increase production and meet performance requirements at lower costs. These developments were highlighted by the introduction of nitrocellulose lacquers for the automotive and furniture industries; followed by the alkyds, epoxies, vinyls, polyesters, acrylics, and a host of other resins; and finally the polyurethanes.

In the 1950s, the decade of expansion, manufacturers built new plants to supply the coatings demand for industrial and consumer products. Coatings were applied at low solids using inefficient conventional air-atomized spray guns. The atmosphere was polluted with volatile organic compounds (VOCs), but no one cared as long as finished products were shipped out of the factories. Coatings suppliers were fine-tuning formulations to provide faster curing and improved performance properties. In 1956, powder coatings were invented. By 1959, there were several commercial conveyorized lines applying powder coatings by the fluidized bed process.

In the 1960s, the decade of technology, just as coatings were becoming highly developed, another variable—environmental impact—was added to the equation. Someone finally noticed that the solvents, which were used in coatings for viscosity and flow control, and which evaporate during application and cure, were emitted to the atmosphere. Los Angeles County officials, who found that VOC emissions were a major source of air pollution, enacted Rule 66 to control the emission of solvents that cause photochemical smog. To comply with Rule 66, the paint industry reformulated its coatings using exempt solvents, which presumably did not produce smog. California's Rule 66 was followed by other local air quality standards and finally by the establishment of the U.S. Environmental Protection Agency (EPA), whose charter, under the law, was to improve air quality by reducing solvent emissions. During this decade, there were three notable developments which would eventually reduce VOC emissions in coating operations, electrocoating, electrostatically sprayed powder coatings and radiation curable coatings.

In the 1970s, the decade of conservation, the energy crisis resulted in shortages and price increases for solvents and coatings materials. Also affected was the distribution of natural gas, the primary fuel for curing ovens, which caused shortages and price increases. In response to those pressures, the coatings industry developed low-temperature curing coatings in an effort to reduce energy consumption. Of greater importance to suppliers and end users of coatings and finishes was the establishment of the Clean Air Act of 1970. An important development during this period was radiation curable coatings, mostly clears, for flat line applications using electron beam (EB) and ultraviolet (UV) radiation sources.

In the 1980s, the decade of restriction, the energy crisis was ending, and the more restrictive air quality standards were beginning. However, energy costs remained high. The importance of transfer efficiency, the percentage of an applied coating that actually coats the product, was recognized by industry and the EPA. This led to the development and use of coating application equipment and coating methods that have higher transfer efficiencies. The benefits of using higher transfer efficiency coating methods are threefold: lower coating material usage, lower solvent emissions, and lower costs. The automotive industry switched almost exclusively to electrocoating for the application of primers. Powder coating material and equipment suppliers worked feverishly to solve the problems associated with automotive topcoats. Other powder coating applications saw rapid growth. Radiation curable pigmented coatings for three-dimensional products were developed.

In the 1990s, the decade of compliance, resin and coating suppliers developed compliance coatings—electrocoating, high-solids, powder, radiation curable, and waterborne. Equipment suppliers developed devices to apply and cure these new coatings. These developments were in response to the 1990 amendments to the Clean Air Act of 1970. The amendments established a national permit program that made the law more enforceable, ensuring better compliance and calling for nationwide regulation of VOC emissions from all organic finishing operations. The amendments also established *Control Technique Guidelines* to allow state and local governments to develop *Attainment Rules*. Electrocoating was the process of choice for priming many industrial and consumer products. Powder coatings were used in a host of applications where durability was essential. UV curable coatings were applied to three-dimensional objects. Equipment suppliers developed more efficient application equipment.

In the 2000s, the beginning of the green millennium, coatings and equipment suppliers' investments in research and development will pay dividends. Improvements in coating materials and application equipment have enabled end users to comply with air quality regulations. Primers are applied by electrocoating. One-coat finishes are replacing two coats in many cases. High-solids and waterborne liquid coatings are replacing conventional solvent-thinned coatings. Powder coatings usage has increased dramatically. Radiation-cured coatings are finding more applications. Coatings and solvent usage, as well as application costs, are being reduced. Air quality standards are being met. Coatings and equipment suppliers, as well as end users, are recognizing the cost savings bonus associated with attainment. Compliance coatings applied by more efficient painting methods will reduce coatings and solvent usage, thereby effecting cost savings.

Since coatings today are considered to be engineering materials, their performance characteristics not only must match service requirements, they must also meet governmental regulations and production cost considerations. In the past, the selection of a coating depended mainly on the service requirements and application method. Now, more than ever before, worker safety, environmental impact, and economics must be considered. For this reason, compliance coatings-electrocoating, high-solids, powders, radiation-cured, and waterborne coatings are the most sensible choices.

Coatings are applied to most industrial products by spraying. Figure 9.1 shows a typical industrial spray booth. In 1890, Joseph Binks

Figure 9.1 A typical industrial spray booth used for applying industrial coatings. *(Courtesy of George Koch Sons, LLC.)*

invented the cold-water paint-spraying machine, the first airless sprayer, which was used to apply whitewash to barns and other building interiors. In 1924, Thomas DeVilbiss used a modified medical atomizer, the first air-atomizing sprayer, to apply a nitrocellulose lacquer on the Oakland automobile. Since then, these tools have remained virtually unchanged and, until the enactment of the air quality standards, they were used to apply coatings at 25 to 50% volume solids at transfer efficiencies of 30 to 50%. Using this equipment, the remainder of the nonvolatile material, the overspray, coated the floor and walls of spray booths and became hazardous or nonhazardous waste, while the solvents—the VOCs—evaporated from the coating during application and cure to become air pollutants. Today, finishes are applied by highly transfer-efficient application equipment. The choice of application equipment must be optimized.

Even the best coatings will not perform their function if they are not applied on properly prepared substrates. For this reason surfaces must first be cleaned to remove oily soils, corrosion products, and particulates, and then pretreated before applying any coatings and finishes.

After coatings are applied, they form films and cure. Curing mechanisms can be as simple as solvent evaporation or as complicated as free-radical polymerization. Basically, coatings can be classified as *baking* or *air drying,* which usually means room-temperature curing. The curing method and times are important in coating selection, because they must be considered for optimizing equipment and production schedules choices.

The purposes of this chapter are threefold: (1) to aid the designer in optimizing the selection of coating materials and application equipment; (2) to acquaint the reader with surface preparation, coating materials, application equipment, and curing methods; and (3) to stress the importance of environmental compliance in coating operations.

9.2   Environment and Safety

In the past, changes in coating materials and coating application lines were discussed only when lower prices, novel products, new coating lines, or new plants were considered. Today, with rising material costs, rising energy costs, and more restrictive governmental regulations, they are the subject of frequent discussions. During these discussions, both in-house and with suppliers, choices of coating materials and processing equipment are optimized. Coating material and solvent costs, which are tied to the price of crude oil, have risen since the 1970s, as has the cost of natural gas, which is the most frequently used fuel for coating bake ovens. The EPA has imposed restrictive air quality standards. The Occupational Safety and Health Act (OSHA) and the Toxic Substances Control Act (TSCA) regulate the environment in the workplace and limit workers' contact with hazardous materials. These factors have increased coating costs and the awareness of product finishers. To meet the challenge, they must investigate and use alternative coating materials and processes for compliance and cost effectiveness.

Initial attempts to control air pollution in the late 1940s resulted in smoke-control laws to reduce airborne particulates. The increased use of the automobile and industrial expansion during that period caused a condition called *photochemical smog* (smog created by the reaction of chemicals exposed to sunlight in the atmosphere) in major cities throughout the United States. Los Angeles County officials recognized that automobile exhaust and VOC emissions were major sources of smog, and they enacted an air pollution regulation called Rule 66. Rule 66 forbade the use of specific solvents that produced photochemical smog and published a list of exempt solvents for use in coatings. Further study by the EPA has shown that, if given enough time, even the Rule 66 exempt solvents will produce photochemical smog in the atmosphere.

The Clean Air Act of 1970, and its 1990 amendments, formulated by the EPA, established national air quality standards that regulate the amount of solvents emitted. The EPA divided the 50 states into 250 air quality regions, each of which is responsible for the implementation of the national air quality standards. It is important to recognize that many of the local standards are more stringent than the national ones. For this reason, specific coatings that comply with the air quality standards of one district may not comply with another's. Waterborne, high-solids, powder, electrophoretic, and radiation-cured coatings will comply. The use of precoated metal can eliminate all the compliance problems.

Not only because the EPA mandates the reduction of VOC emissions, but also because of economic advantages, spray painting, which is the most used application method, must be done more efficiently. The increased efficiency will reduce the amount of expensive coatings and solvents used, thereby reducing production costs.

9.3   Surface Preparation

The most important step in any coating operation is surface preparation, which includes cleaning and pretreatment. For coatings to adhere, surfaces must be free from oily soils, corrosion products, and loose particulates. New wood surfaces are often coated without cleaning. Old wood and coated wood must be cleaned to remove oily soils and loose, flaky coatings. Plastics are cleaned by using solvents and chemicals to remove mold release. Metals are cleaned by media blasting, sanding, brushing, and by solvents or aqueous chemicals. The choice of a cleaning method depends on the substrate and the size and shape of the object.

After cleaning, pretreatments are applied to enhance coating adhesion and, in the case of metals, corrosion resistance. Some wood surfaces require no pretreatment, while others require priming of knots and filling of nail holes. Cementitious and masonry substrates are pretreated using acids to remove loosely adhering contaminants and to passivate the surfaces. Metals, still the most common industrial substrates, are generally pretreated using phosphates, chromates, and oxides to passivate their surfaces and provide corrosion resistance. Plastics, second only to steel, are gaining rapidly in use as industrial substrates. Some are paintable after cleaning to remove mold release and other contaminants, while others require priming, physical treatments, or chemical etching to ensure coating adhesion. Since most of the industrial substrates coated are metals and plastics, their cleaning and pretreatment are described in the next sections. Because of their complexities, detailed descriptions of cleaning and pretreatment processes are beyond the scope of this chapter. Enough detail will be given to allow the reader to make a choice. As with the choice of a cleaning method, the choice of a pretreatment method depends on the composition, size, and shape of the product.

9.3.1 Metal Surface Cleaning

Oily soils must be removed before any other surface preparation is attempted. Otherwise these soils may be spread over the surface. These soils can also contaminate abrasive cleaning media and tools. Oily soils can be removed faster using liquid cleaners that impinge on the surface or in agitated immersion baths. It is often necessary to heat liquid cleaners to facilitate soil removal.

9.3.1.1 Abrasive cleaning. After removal of the oily soils, surfaces are abrasive cleaned to remove rust and corrosion by media blasting, hand or power sanding, and hand or power brushing. Media blasting consists of propelling materials, such as sand, metallic shot, nut shells, plastic pellets, and dry ice crystals, by gases under pressure, so that they impinge on the surfaces to be cleaned. High-pressure water-jet cleaning is similar to media blasting.

9.3.1.2 Alkaline cleaning. To remove oily soils, aqueous solutions of alkaline phosphates, borates, and hydroxides are applied to metals by immersion or spray. After cleaning, the surfaces are rinsed with clear water to remove the alkali. These materials are not effective for removing rust and corrosion.

9.3.1.3 Detergent cleaning. Aqueous solutions of detergents are used to remove oily soils in much the same way as alkaline cleaners. Then they are rinsed with cold water to flush away the soils.

9.3.1.4 Emulsion cleaning. Heavy oily soils and greases are removed by aqueous emulsions of organic solvents such as mineral spirits and kerosene. After the emulsified solvent has dissolved the oily soils, they are flushed away using a hot-water rinse. Any remaining oily residue must be removed using clean solvent, alkaline, or detergent cleaners.

9.3.1.5 Solvent cleaning. Immersion, hand wiping, and spraying using organic solvents are effective methods for removing oily soils. Since these soils will contaminate solvents and wipers, it is important to change them frequently. Otherwise, oily residues will remain on substrates. Safe handling practices must be followed because of the hazardous nature of most organic solvents.

9.3.1.6   Manual spray cleaning.   For large products, detergent and alkaline cleaners applied using steam cleaners are a well-known degreasing method. In addition to oily soils, the impingement of the steam and the action of the chemicals will dissolve and flush away heavy greases and waxes. Hot-water spray cleaning using chemicals is nearly as effective as steam cleaning.

9.3.1.7   Vapor degreasing.   Vapor degreasing has been a very popular cleaning method for removing oily soils. Boiling solvent condenses on the cool surface of the product and flushes away oily soils, but does not remove particulates. Since this process uses chlorinated solvents, which are under regulatory scrutiny by government agencies, its popularity is declining. However closed-loop systems are still available.

9.3.2   Metal Surface Pretreatment

Cleaning metals will remove oily soils but generally will not remove rust and corrosion from substrates to be coated. Abrasive cleaning will remove corrosion products, and for this reason it is also considered a pretreatment, because the impingement of blasting media and the action of abrasive pads and brushes roughen the substrate and therefore enhance adhesion. The other pretreatments use aqueous chemical solutions, which are applied by immersion or spray techniques. Pretreatments for metallic substrates used on industrial products are discussed in this section. Because they provide corrosion protection to ferrous and nonferrous metals, chromates are used in pretreatment stages and as conversion coatings. They are being replaced by non-chromate chemicals.

9.3.2.1   Aluminum.   Aluminum is cleaned by solvents and chemical solutions to remove oily soils and corrosion products. Cleaned aluminum is pretreated using chromate conversion coating and anodizing. Phosphoric acid-activated vinyl wash primers, which are also considered pretreatments, must be applied directly to metal and not over other pretreatments.

9.3.2.2   Copper.   Copper is cleaned by solvents and chemicals and then abraded to remove corrosion. Bright dipping in acids will also remove corrosion. Cleaned surfaces are often pretreated using chromates and vinyl wash primers.

9.3.2.3   Galvanized steel.   Galvanized steel must be cleaned to remove the oil or wax that is applied at the mill to prevent white corrosion. After cleaning, the surfaces are pretreated using chromates and phosphates. Vinyl wash primer pretreatments can also be applied on galvanized steel surfaces having no other pretreatments.

9.3.2.4 Steel. Steel surfaces are cleaned to remove oily soils and, if necessary, pickled in acid to remove rust. Clean steel is generally pretreated with phosphates to provide corrosion resistance. Other pretreatments for steel are chromates and wash primers.

9.3.2.5 Stainless steel. Owing to its corrosion resistance, stainless steel usually is not coated. Otherwise, the substrate must be cleaned to remove oily soils and then abraded to roughen the surface. Wash primers will enhance adhesion.

9.3.2.6 Titanium. Cleaned titanium is pretreated like stainless steel.

9.3.2.7 Zinc and cadmium. Zinc and cadmium substrates are pretreated like galvanized steel.

9.3.3 Plastic Surface Cleaning

9.3.3.1 Alkaline cleaning. Aqueous solutions of alkaline phosphates, borates, and hydroxides are applied to plastics by immersion or spray to remove oily soils and mold release agents. After cleaning, the surfaces are rinsed with clear water to remove the alkali.

9.3.3.2 Detergent cleaning. Aqueous solutions of detergents are used to remove oily soils and mold release agents in much the same way as with alkaline cleaners. Then they are rinsed with cold water to flush away the soils.

9.3.3.3 Emulsion cleaning. Heavy, oily soils, greases, and mold release agents are removed by aqueous emulsions of organic solvents such as mineral spirits and kerosene. After the emulsified solvent has dissolved the oily soils, they are flushed away using a hot-water rinse. The remaining oily residue must be removed using clean solvent, alkaline, or detergent cleaners.

9.3.3.4 Solvent cleaning. Immersion, hand wiping, and spraying, using organic solvents, are effective methods for removing oily soils and mold release agents. Since these soils will contaminate solvents and wipers, it is important to change them frequently. Otherwise, oily residues will remain on substrates. Compatibility of cleaning solvents with the plastic substrates is extremely important. Solvents that affect plastics are shown in Table 9.1. Suppliers of mold release agents are the best source for information on solvents that will remove their materials. Safe handling practices must be followed because of the hazardous nature of most organic solvents.

9.3.3.5 Manual spray cleaning. Detergent and alkaline cleaners applied using steam and hot-water spray cleaners are a well known degreasing method. The method can also be used for removing mold release agents. The impingement of the steam and hot water and the action of the chemicals will dissolve and flush away the contaminants. Manual spray cleaning is used for large products.

9.3.4 Plastic Surface Pretreatment

Cleaning will remove oily soils and mold release agents, but additional pretreatment may be needed on certain plastic surfaces to ensure adhesion. Many of the plastic substrates are chemically inert and will not accept coatings because of their poor wettability. Depending on their

TABLE 9.1   Solvents That Affect Plastics

Resin	Heat-distortion point, °F	Solvents that affect surface
Acetal	338	None
Methyl methacrylate	160–195	Ketones, esters, aromatics
Modified acrylic	170–190	Ketones, esters, aromatics
Cellulose acetate	110–209	Ketones, some esters
Cellulose propionate	110–250	Ketones, esters, aromatics, alcohols
Cellulose acetate butyrate	115–227	Alcohols, ketones, esters, aromatics
Nylon	260–360	None
Polyethylene:		
High density	140–180	
Medium density	120–150	None
Low density	105–121	
Polypropylene	210–230	None
Polycarbonate	210–290	Ketones, esters, aromatics
Polystyrene (GP high heat)	150–195	Some aliphatics, ketones, esters, aromatics
Polystyrene (impact, heat-resistant)	148–200	Ketones, esters, aromatics, some aliphatics
Acrylonitrile butadiene styrene (ABS)	165–225	Ketones, esters, aromatics, alcohol

chemical composition, they will require mechanical, chemical, and physical pretreatment or priming to enhance coating adhesion. Since mechanical pretreatment consists of abrasion, its effect on the substrate must be considered. Chemical pretreatments involve corrosive materials that etch the substrates and can be hazardous. Therefore, handling and disposal must be considered. Physical pretreatments consist of plasma, corona discharge, and flame impingement. Process control must be considered.

9.3.4.1 Abrasive cleaning. After removal of the oily soils, surfaces are abrasive pretreated to roughen the substrate by media blasting, hand or power sanding, and hand or power brushing. Media blasting consists of propelling materials such as sand, metallic shot, nut shells, plastic pellets, and dry ice crystals by gases under pressure so that they impinge on the surfaces to be pretreated.

9.3.4.2 Chemical etching. Chemical pretreatments use solutions of corrosive chemicals, which are applied by immersion or spray techniques, to etch the substrate.

9.3.4.3 Corona discharge. During corona discharge pretreatment, the plastic is bombarded by gases directed toward its surface.

9.3.4.4 Flame treating. During the flame pretreatment, an open flame impinging on the surface of the plastic product causes alterations in the surface chemistry.

9.3.4.5 Plasma pretreatment. Low-pressure plasma pretreatment is conducted in a chamber, while atmospheric plasma pretreatment is done in the open. In both cases, ablation alters the surface chemistry and causes changes in surface roughness.

9.3.4.6 Laser pretreatment. Laser pretreatment ablates the plastic substrate causing increased surface roughness and changes in the surface chemistry.

9.3.5 Priming

Priming involves the application of a coating on the surface of the plastic product to promote adhesion or to prevent attack by the solvents in a subsequent protective or decorative coating. In some cases, priming can be done after cleaning. In others, it must be done after pretreatment.

9.4   Coating Selection

To aid in their selection, coatings will be classified by their use in finish systems, physical state, and resin type. Coatings are also classified by their use as electrical insulation. It is not the intent of this chapter to instruct the reader in the chemistry of organic coating but rather to aid in selection of coatings for specific applications. Therefore, the coating resin's raw material feed stock and polymerization reactions will not be discussed. On the other hand, generic resin types, curing, physical states, and application methods are discussed.

9.4.1   Selection by Finish Systems

Finish systems can be one-coat or multicoat schemes that use primers, intermediate coats, and topcoats. Primers provide adhesion, corrosion protection, passivation, and solvent resistance to substrates. Topcoats provide weather, chemical, and physical resistance and generally determine the performance characteristics of finish systems. Performance properties for coatings, formulated with the most commonly used resins, are shown in Table 9.2.

In coating selection, intended service conditions must be considered. To illustrate this point, consider the differences between service conditions for toy boats and for battleships. Table 9.3 shows the use of industrial finish systems in various service conditions.

9.4.2   Selection by Physical State

A resin's physical state can help determine the application equipment required. Solid materials can be applied by powder coating methods.

TABLE   9.2   Performance Properties* of Common Coating Resins[3]

Resin type	Humidity resistance	Corrosion resistance	Exterior durability	Chemical resistance	Mar resistance
Acrylic	E	E	E	G	E
Alkyd	F	F	P	G	G
Epoxy	E	E	G	E	E
Polyester	E	G	G	G	G
Polyurethane	E	G	E	G	E
Vinyl	E	G	G	G	G

* Note: E—excellent; G—good; F—fair; P—poor.

TABLE 9.3   Typical Industrial Finish Systems[3]

Service conditions	Primer	One-coat enamel	Intermediate coat	Topcoat
Interior				
Light duty		X		
Heavy duty	X			X
Exterior				
Light duty		X		
Heavy duty	X			X
Extreme duty	X		X	X

Table 9.4 lists resins applied as powder coatings. Liquids can be applied by most of the other methods, which are discussed later. Many of the coating resins exist in several physical states. Table 9.5 lists the physical states of common coating resins.

9.4.3   Selection by Resin Type

Since resin type determines the performance properties of a coating, it is used most often. Table 9.6 shows the physical, environmental, and film-forming characteristics of coatings by polymer (resin) type. It

TABLE 9.4   Plastics Used in Powder Coatings

	Fluidizing conditions			Fluidized bed powder		
	Preheat tempera-	Cure or fusion		Maximum operating tempera-		Weather
Resin	ture, °F	Tempera-ture, °F	Time, min	ture, °F	Adhesion	resistance
Epoxy	250–450	250–450	1–60	200–400	Excellent	Good
Vinyl	450–550	400–600	1–3	225	Poor	Good
Cellulose acetate butyrate	500–600	400–550	1–3	225	Poor	Good
Nylon	550–800	650–700	1	300	Poor	Fair
Polyethylene	500–600	400–600	1–5	225	Fair	Good
Polypropylene	500–700	400–600	1–3	260	Poor	Good
Penton	500–650	450–600	1–10	350	Poor	Good
Teflon	800–1000	800–900	1–3	500	Poor	Good

TABLE 9.5   Physical States of Common Coating Resins[3]

Resin type	Conventional solvent	Waterborne	High-solids	Powder coating	100% solids liquid	Two-component liquid
Acrylic	X	X	X	X		
Alkyd	X	X	X			
Epoxy	X	X	X	X	X	X
Polyester	X		X	X	X	X
Polyurethane	X	X	X	X	X	X
Vinyl	X	X	X	X	X	

TABLE 9.6 Properties of Coatings by Polymer Type

Coating type	Electrical properties				Maximum continuous service temperature, °F	Physical characteristics			
	Volume resistivity, Ω-cm (ASTM D 257)	Dielectric strength, V/mil	Dielectric constant	Dissipation factor		Adhesion to metals	Flexibility	Approximate Sward hardness (higher number is harder)	Abrasion resistance
Acrylic	10^{14}–10^{15}	450–550	2.7–3.5	0.02–0.06	180	Good	Good	12–24	Fair
Alkyd	10^{14}	300–350	4.5–5.0	0.003–0.06	200 / 250 TS	Excellent	Fair to good / Low temperature—poor	3–13 (air dry) / 10–24 (bake)	Fair
Cellulosic (nitrate butyrate)		250–400	3.2–6.2		180	Good	Good / Low temperature—poor	10–15	
Chlorinated polyether (Penton*)	10^{15}	400	3.0	0.01	250	Excellent	Good		
Epoxy-amine cure	10^{14} at 30°C / 10^{10} at 105°C	400–550	3.5–5.0	0.02–0.03 at 30°C	350	Excellent	Fair to good / Low temperature—poor	26–36	Good to excellent
Epoxy-anhydride, dicy		650–730	3.4–3.8	0.01–0.03	400	Excellent	Good to excellent / Low temperature—poor	20	Good to excellent
Epoxy-polyamide	10^{14} at 30°C / 10^{10} at 105°C	400–500	2.5–3.0	0.008–0.02	350	Excellent	Good to excellent / Low temperature—poor	20	Fair to good
Epoxy-phenolic	10^{12}–10^{13}	300–450			400	Excellent	Good / Low temperature—fair		Good to excellent
Fluorocarbon TFE	10^{18}	430	2.0–2.1	0.0002	500	Can be excellent; primers required	Excellent		
FEP	10^{18}	480	2.1	0.0003–0.0007	400				
CTFE	10^{18}	500–600	2.3–2.8	0.003–0.004	400	Can be excellent; primers required	Excellent		
Parylene (polyxylylenes)	10^{16}–10^{17}	700	2.6–3.1	0.0002–0.02	240°F (air) / 510°F (inert atmosphere)	Good	Excellent		
Phenolics	10^{9}–10^{12}	100–300	4–8	0.005–0.5	350	Excellent	Poor to good / Low temperature—poor	30–38	Fair

Material	Volume resistivity	Dielectric strength	Dielectric constant	Dissipation factor	Temp	Adhesion	Flexibility		
Phenolic-oil varnish					250	Excellent	Good; Low temperature—fair		Poor to fair
Phenoxy	10^{13}–10^{14}	500	3.7–4.0	0.001	180	Excellent	Excellent		
Polyamide (nylon)	10^{13}–10^{15}	400–500	2.8–3.6	0.01–0.1	225–250	Excellent	Excellent		
Polyester	10^{12}–10^{14}	500	3.3–8.1	0.008–0.04	200	Good on rough surfaces; poor to polished metals	Fair to excellent	25–30	Good
Chlorosulfonated (polyethylene Hypalon)†		400	6–10	0.03–0.07	250	Good	Elastomeric	Less than 10	
Polyimide	10^{16}–10^{18}	3000 (10 mil)	3.4–3.8	0.003	500	Good	Fair to excellent		
Polystyrene	10^{10}–10^{19}	500–700	2.4–2.6	0.0001–0.0005	140–180		Poor to fair		
Polyurethane	10^{12}–10^{13}	450–500; 3800 (1 mil)	6.8 (1 kHz); 4.4 (1 MHz)	0.02–0.08	250	Often poor to metals (excellent to most nonmetals)	Good to excellent; Low temperature—poor	10–17 (castor oil); 50–60 (polyester)	Good
Silicone	10^{14}–10^{16}	550	3.0–4.2	0.001–0.008	500	Varies, but usually needs primer for good adhesion	Excellent; Low temperature—excellent	12–16	Fair to excellent
Vinyl chloride (poly-)	10^{11}–10^{15}	300–800	3–9	0.04–0.14	150	Excellent, if so formulated	Excellent; Low temperature—fair to good	5–10	
Vinyl chloride (plastisol, organisol)	10^{9}–10^{16}	400	2.3–9	0.10–0.15	150	Requires adhesive primer	Excellent; Low temperature—fair to good	3–6	
Vinyl fluoride	10^{13}–10^{14}	260; 1200 (8 mil)	6.4–8.4	0.05–0.15	300	Excellent, if fused on surface	Excellent; Low temperature—excellent		
Vinyl formal (Formvar‡)	10^{13}–10^{15}	850–1000	3.7	0.007–0.2	200	Excellent			

*Trademark of Hercules Powder Co., Inc., Wilmington, Del.
†Trademark of E. I. du Pont de Nemours & Co., Wilmington, Del.
‡Trademark of Monsanto Co., St. Louis, Mo.

TABLE 10.6 Properties of Coatings by Polymer Type (Continued)

Coating type	Resistance to environmental effects					Repairability	Film formation		Application method	Typical uses
	Chemical and solvent resistance	Moisture and humidity resistance	Weatherability	Resistance to micro-organisms	Flamma-bility		Method of cure	Cure schedule		
Acrylic	Solvents—poor Alkalies—poor Dilute acids—poor to fair	Good	Excellent resistance to UV and weather	Good	Medium	Remove with solvent	Solvent evaporation	Air dry or low-temperature bake	Spray, brush, dip	Coatings for circuit boards. Quick dry protection for markings and color coding.
Alkyd	Solvents—poor Alkalies—poor Dilute acids—poor to fair	Poor	Good to excellent	Poor	Medium	Poor	Oxidation or heat	Air dry or baking types	Most common methods	Painting of metal parts and hardware.
Cellulosic (nitrate butyrate)	Solvents—good Alkalies—good Acids—good	Fair		Poor to good	High	Remove with solvents	Solvent evaporation	Air dry or low-temperature bake	Spray, dip	Lacquers for decoration and protection. Hot-metal coatings.
Chlorinate polyether (Penton*)		Good			Low		Powder or dispersion fuses	High-temperature fusion	Spray, dip, fluid bed	Chemically resistant coatings.
Epoxy-amine cure	Solvents—good to excellent Alkalies—good Dilute acids—fair	Good	Pigmented—fair; clear—poor (chalks)	Good	Medium	No	Cured by catalyst reaction	Air dry to medium bake	Spray, dip, fluid bed	Coatings for circuit boards. Corrosion-protective coatings for metals.
Epoxy-anhydride, Dicy	Solvents—good Alkalies— Dilute acids—	Good		Good	Medium	No	Cured by chemical reaction	High bakes 300 to 400°F	Spray, dip, fluid bed, impregnate	High-bake, high-temperature-resistant dielectric and corrosion coatings.
Epoxy-polyamide	Solvents—good Alkalies—good Dilute acids—poor	Good		Good	Medium	No	Cured by coreactant	Air dry or medium bake	Spray, dip	Coatings for circuit boards. Filleting coating.
Epoxy-phenolic	Solvents—excellent Alkalies—fair Dilute acids—good	Excellent	Pigmented—fair; clear—poor	Good	Medium	No	Cured by coreactant	High bakes 300 to 400°F	Spray, dip	High-bake solvent and chemical resistant coating.
Fluorocarbon TFE	Solvents—excellent Alkalies—good Dilute acids—excellent	Excellent		Good	None	No	Fusion from water or solvent dispersion	Approximately 750°F	Spray, dip	High-temperature resistant insulation for wiring.
FEP CTFE		Excellent		Good	None	No	Fusion from water or solvent dispersion	500–600°F	Spray, dip	High-temperature-resistant insulation. Extrudable.
Parylene (polyxylylenes)		Excellent			None		Vapor phase deposition and polymerization requiring special license from Union Carbide.			Very thin, pinhole-free coatings, possible semiconductable coating.
Phenolics	Solvents—good to excellent Alkalies—poor Dilute acids—good to excellent	Excellent	Fair	Poor to good	Medium	No	Cured by heat	Bake 350–500°F	Spray, dip	High-bake chemical and solvent-resistant coatings.

Material		Chemical resistance	Outdoor weathering	Fungus resistance		Removal/solderability	Curing reaction	Curing schedule	Application	Typical uses
Phenolic-oil varnish	Good	Solvents—poor; Alkalies—poor; Dilute acids—good to excellent	Good	Poor, unless toxic additive	Medium	Poor	Oxidation or heat		Spray, brush, dip-impregnate	Impregnation of electronic modules, quick protective coating.
Phenoxy	Good	Good	Good	Good			Cured by heat			Chemical resistant coating.
Polyamide (nylon)	Fair	Fair				Fairly solderable				Wire coating.
Polyester	Fair	Solvents—poor; Alkalies—poor to fair; Dilute acids—good	Very good	Good	Medium	Poor	Cured by heat or catalyst	Air dry or bake 100–250°F	Spray, brush, dip	
Chlorosulfonated polyethylene (Hypalon†)	Good	Good	Good	Good	Low		Solvent evaporation	Air dry or low-temperature bake	Spray, brush	Moisture and fungus proofing of materials.
Polyimide	Good	Solvents—excellent; Alkalies—poor to fair; Dilute acids—good	Good	Good	Low	Poor	Cured by heat	High bake	Dip, impregnate, wire coater	Very high temperature resistant with insulation.
Polystyrene	Good	Good	Good	Good	High	Dissolve with solvents	Solvent evaporation	Air dry or low bake	Spray, dip	Coil coating, low dielectric constant, low loss in radar uses.
Polyurethane	Good	Solvents—good; Dilute alkalies—fair; Dilute acids—good	Good	Poor to good	Medium	Excellent; melts, solder-through properties	Coreactant or moisture cure	Air dry to medium bake	Spray, brush, dip	Conformal coating of circuitry, solderable wire insulation.
Silicone	Excellent	Solvents—good; Alkalies—good (dilute) poor (concentrated); Dilute acids—good	Excellent	Good	Very low (except in O₂ atmosphere)	Fair to excellent. Cut and peel	Cured by heat or catalyst	Air dry (RTV) to high bakes	Spray, brush, dip	Heat-resistant coating for electronic circuitry. Good moisture resistance.
Vinyl chloride (poly-)	Good	Solvents—alcohol, good; Alkalies—good	Pigmented—fair to good; Clear—poor	Poor to good (depends on plasticizer)	Very low	Dissolve with solvents	Solvent evaporation	Air dry or elevated temperature for speed	Spray, dip, roller coat	Wire insulation. Metal protection (especially magnesium, aluminum).
Vinyl chloride (plastisol, organisol)	Good	Good	Good	Poor to good (depends on plasticizer)	Low	Poor	Fusion of liquid to gel	Bake 250–350°F	Spray, dip, reverse roll	Soft-to-hard thick coatings, electroplating racks, equipment.
Vinyl fluoride	Good	Good	Excellent	Good	Very low	Poor	Fusion from solvent dispersion	Bake 400–500°F	Spray, roller coat	Coatings for circuitry. Long-life exterior finish.
Vinyl formal (Form-var‡)	Good	Good	Good	Good	Medium	Poor	Cured by heat	Bake 300–500°F	Roller coat, wire coater	Wire insulation (thin coatings) coil impregnation.

*Trademark of Hercules Powder Co., Inc., Wilmington, Del.
†Trademark of E. I. du Pont de Nemours & Co., Wilmington, Del.
‡Trademark of Monsanto Co., St. Louis, Mo.
SOURCE: This table has been reprinted from Machine Design, May 25, 1967. Copyright, 1967, by The Penton Publishing Company, Cleveland, Ohio.

is important to realize that in, selecting coatings, tables of performance properties of generic resins must be used only as guides, because coatings of one generic type, such as acrylic, epoxy, or polyurethane, are often modified using one or more of the other generic types. Notable examples are acrylic alkyds, acrylic urethanes, acrylic melamines, epoxy esters, epoxy polyamides, silicone alkyds, silicone epoxies, silicone polyesters, vinyl acrylics, and vinyl alkyds. While predicting specific coating performance properties of unmodified resins is simple, predicting the properties of modified resins is difficult, if not impossible. Parameters causing these difficulties are resin modification percentages and modifying methods such as simple blending or copolymerization. The performance of a 30% copolymerized silicone alkyd is not necessarily the same as one which was modified by blending. These modifications can change the performance properties subtly or dramatically.[3]

There are more than 1200 coating manufacturers in the United States, each having various formulations that could number in the hundreds. Further complicating the coatings selection difficulty is the well known practice of a few coating manufacturers who add small amounts of a more expensive, better performing resin to a less expensive, poorer performing resin and call the product by the name of the former. An unsuspecting person, whose choice of such a coating is based on properties of the generic resin, can be greatly disappointed. Instead, selections must be made on the basis of performance data for specific coatings or finish systems. Performance data are generated by the paint and product manufacturing industries when conducting standard paint evaluation tests. Test methods for coating material evaluation are listed in Table 9.7.

9.4.4   Selection by Electrical Properties

Electrical properties of organic coatings vary by resin (also referred to as *polymer*) type. When selecting insulating varnishes, insulating enamels, and magnet wire enamels, the electrical properties and physical properties determine the choice.

Table 9.8 shows electric strengths, Table 9.9 shows volume resistivities, Table 9.10 shows dielectric constants, and Table 9.11 shows dissipation factors for coatings using most of the available resins. Magnet wire insulation is an important use for organic coatings. National Electrical Manufacturer's Association (NEMA) standards and manufacturers' trade names for various wire enamels are shown in Table 9.12. This information can be used as a guide in the selection of coatings. However, it is important to remember the aforementioned warnings about blends of various resins and the effects on performance properties.

TABLE 9.7 Specific Test Methods for Coatings[*]

Test	ASTM	Federal Std. 141a, method	MIL-Std.-202, method	Federal Std. 406, method	Others
Abrasion	D968	6191 (Falling sand) 6192 (Taber)		1091	Fed. Std. 601, 14111
Adhesion	D2197	6301.1 (Tape test, wet) 6302.1 (Microknife) 6303.1 (Scratch adhesion) 6304.1 (Knife test)		1111	Fed. Std. 601, 8031
Arc resistance	D495		303	4011	
Dielectric constant	D150		301	4021	Fed. Std. 101, 303
Dielectric strength (breakdown voltage)	D149 D115			4031	Fed. Std. 601, 13311
Dissipation factor	D150			4021	
Drying time	D1640 D115	4061.1			
Electrical insulation resistance	D229 D257		302	4041	MIL-W-81044, 4.7.5.2
Exposure (interior)	D1014	6160 (On metals) 6161.1 (Outdoor rack)			
Flash point	D56, D92 D1310 (Tag open cup)	4291 (Tag Closed Cup) 4294 (Cleveland Open Cup)			Fed. Std. 810, 509
Flexibility		6221 (Mandrel) 6222 (Conical Mandrel)		1031	Fed. Std. 601, 11041
Fungus resistance	D1924				MIL-E-5272, 4.8 MIL-STD-810, 508.1 MIL-T-5422, 4.8
Hardness	D1474	6211 (Print Hardness) 6212 (Indentation)			
Heat resistance	D115 D1932	6051			
Humidity	D2247	6071 (100% RH) 6201 (Continuous Condensation)	103 106A		MIL-E-5272, proc. 1 Fed. Std. 810, 507

TABLE 9.7 Specific Test Methods for Coatings[*] (Continued)

Test	ASTM	Federal Std. 141a, method	MIL-Std.- 202, method	Federal Std. 406, method	Others
Impact resistance		6226 (G.E. Impact)	1074		
Moisture-vapor permeability	E 96 D1653	6171	7032		MIL-STD-810, 509.1 MIL-E-5272, 4.6 Fed. Std. 151, 811.1 Fed. Std. 810, 509
Nonvolatile content		4044			
Salt spray (fog)	B117	6061	101C	6071	
Temperature-altitude					MIL-E-5272, 4.14 MIL-T-5422, 4.1 MIL-STD-810, 504.1
Thermal conductivity	D1674 (Cenco Fitch) C177 (Guarded hot plate)				MIL-I-16923, 4.6.9
Thermal shock			107		MIL-E-5272, 4.3 MIL-STD-810,503.1
Thickness (dry film)	D1005 D1186	6181 (Magnetic Gage) 6183 (Mechanical Gage)		2111, 2121, 2131, 2141, 2151	Fed. Std. 151, 520, 521.1
Viscosity	D1545 D562 D1200 D88	4271 (Gardner Tubes) 4281 (Krebs-Stormer) 4282 (Ford Cup) 4285 (Saybolt) 4287 (Brookfield)			
Weathering (accelerated)	D822	6151 (Open Arc) 6152 (Enclosed Arc)		6024	

[*]Note: A more complete compilation of test methods is found in J.J. Licari, *Plastic Coatings for Electronics*, McGraw-Hill, New York, 1970. The major collection of complete test methods for coatings is *Physical and Chemical Examination of Paints, Varnishes, Lacquers, and Colors*, by Gardner and Sward, Gardner Laboratory, Bethesda, MD. This has gone through many editions.

TABLE 9.8   Electric Strengths of Coatings

Material	Dielectric strength, V/mil	Comments*	Source of information
Polymer coatings:			
Acrylics	450–550	Short-time method	[a]
	350–400	Step-by-step method	[a]
	400–530		[b]
	1700–2500	2-mil-thick samples	Columbia Technical Corporation, Humiseal Coatings
Alkyds	300–350		[b]
Chlorinated polyether	400	Short-time method	[a]
Chlorosulfonated polyethylene	500	Short-time method	[a]
Diallyl phthalate	275–450		[b]
	450	Step-by-step method	[a]
Diallyl isophthalate	422	Step-by-step method	[a]
Depolymerized rubber (DPR)	360–380		H. V. Hardman Company, DPR Subsidiary
Epoxy	650–730	Cured with anhydride–castor oil adduct	Autonetics, Division of North American Rockwell
Epoxy	1300	10-mil-thick dip coating	
Epoxies, modified	1200–2000	2-mil-thick sample	Columbia Technical Corporation, Humiseal Coatings
Neoprene	150–600	Short-time method	[a]
Phenolic	300–450		[b]
Polyamide	780	106 mil thick sample	
Polyamide-imide	2700		
Polyesters	250–400	Short-time method	[a]
	170	Step-by-step method	[a]
Polyethylene	480		[b]
	300	60-mil-thick sample	[c]
	500	Short-time method	[a]
Polyimide	3000	Pyre-ML, 10 mils thick	[d]
	4500–5000	Pyre-ML (RC-675)	[e]
	560	Short-time method, 80 mils thick	
Polypropylene	750–800	Short-time method	[a]
Polystyrene	500–700	Short-time method	[a]
	400–600	Step-by-step method	[a]
	450	60-mil-thick sample	[c]
Polysulfide	250–600	Short-time method	[a]
Polyurethane (single component)	3800	1-mil-thick sample	[f]
Polyurethane (two components)/castor oil cured	530–1010		[g]
Polyurethane (two components, 100% solids)	275	125-mil-thick sample	Products Research & Chemical Corporation (PR-1538)
	750	25-mil thick sample	
Polyurethane (single component)	2500	2-mil-thick sample	Columbia Technical Corporation, Humiseal 1A27
Polyvinyl butyral	400		
Polyvinyl chloride	300–1000	Short-time method	
	275–900	Step-by-step method	[b]
Polyvinyl formal	860–1000		
Polyvinylidene fluoride	260	Short-time, 500-V/s, $1/_8$-in sample	[h]
	1280	Short-time, 500-V/s, 8-mil sample	[h]
	950	Step by step (1-kV steps)	[h]
Polyxylylenes:			
Parylene N	6000	Step by step	Union Carbide Corporation
	6500	Short time	Union Carbide Corporation
Parylene C	3700	Short time	Union Carbide Corporation
	1200	Step by step	Union Carbide Corporation
Parylene D	5500	Short time	Union Carbide Corporation
	4500	Step by step	Union Carbide Corporation
Silicone	500	Sylgard 182	Dow Corning Corporation
Silicone	550–650	RTV types	General Electric & Stauffer Chemical Company bulletins
Silicone	800	Flexible dielectric gel	Dow Corning Corporation
Silicone	1500	2-mil-thick sample	Columbia Technical Corporation, Humiseal 1H34
TFE fluorocarbons	400	60-mil-thick sample	[c]
	480	Short-time method	[a]
	430	Step by step	[a]
Teflon TFE dispersion coating	3000–4500	1- to 4-mil-thick sample	E. I. du Pont de Nemours & Company
Teflon FEP dispersion coating	4000	1.5-mil-thick sample	E. I. du Pont de Nemours & Company

TABLE 9.8   Electric Strengths of Coatings *(Continued)*

Material	Dielectric strength, V/mil	Comments*	Source of information
Other materials used in electronic assemblies:			
Alumina ceramics	200–300		b
Boron nitride	900–1400		
Electrical ceramics	55–300		b
Forsterite	250		b
Glass, borosilicate	4500	40-mil sample	c
Steatite	145–280		b

*All samples are standard 125 mils thick unless otherwise specified.
[a]*Insulation*, Directory Encyclopedia Issue, no. 7, June–July 1968.
[b]*Material Engineering*, Materials Selector Issue, vol. 66, no. 5, Chapman-Reinhold Publication, mid-October 1967–1968.
[c]W. H. Kohl, *Handbook of Materials and Techniques for Vacuum Devices*, p. 586, Reinhold Publishing Corporation, New York, 1967, p. 586.
[d]J. R. Learn and M. P. Seegers, "Teflon-Pyre-M. L. Wire Insulation System," *13th Symposium on Technical Progress in Commun. Wire and Cables*, Atlantic City, NJ, December 2–4, 1964.
[e]J. T. Milek: Polyimide Plastics: A State of the Art Report, *Hughes Aircraft Report*, S-8, October, 1965.
[f]*Hughson Chemical Co.* Bulletin 7030A.
[g]*Spencer-Kellogg* (Division of Textron, Inc.) Bulletin TS-6593.
[h]*Pennsalt Chemicals Corp. Prod. Sheet* KI-66a, Kynar Vinylidene Fluoride Resin, 1967.

9.5   Coating Materials

Since it is the resin in the coating's vehicle that determines its performance properties, coatings can be classified by their resin types. The most widely used resins for manufacturing modern coatings are acrylics, alkyds, epoxies, polyesters, polyurethanes, and vinyls. [3] In the following section, the resins used in coatings are described.

9.5.1   Common Coating Resins

9.5.1.1 Acrylics. Acrylics are noted for color and gloss retention in outdoor exposure. Acrylics are supplied as solvent-containing, high-solids, waterborne, and powder coatings. They are formulated as lacquers, enamels, and emulsions. Lacquers and baking enamels are used as automotive and appliance finishes. Both these industries use acrylics as topcoats for multicoat finish systems. Thermosetting acrylics have replaced alkyds in applications requiring greater mar resistance, such as appliance finishes. Acrylic lacquers are brittle and therefore have poor impact resistance, but their outstanding weather resistance allowed them to replace nitrocellulose lacquers in automotive finishes for many years. Acrylic and modified acrylic emulsions have been used as architectural coatings and also on industrial products. These medium-priced resins can be formulated to have excellent hardness, adhesion, abrasion, chemical, and mar resistance. When acrylic resins are used to modify other resins, their properties are often imparted to the resultant resin system.

Uses. Acrylics, both lacquers and enamels, were the topcoats of choice for the automotive industry from the early 1960s to the middle 1980s.

TABLE 9.9   Volume Resistivities of Coatings

Material	Volume resistivity at 25°C, Ω-cm	Source of information
Acrylics	10^{14}–10^{15}	[a]
	$>10^{14}$	[b]
	7.6×10^{14}–1.0×10^{15}	Columbia Technical Corporation, Humiseal
Alkyds	10^{14}	[a]
Chlorinated polyether	10^{15}	[b]
Chlorosulfonated polyethylene	10^{14}	[b]
Depolymerized rubber	1.3×10^{13}	H. V. Hardman, DPR Subsidiary
Diallyl phthalate	10^{8}–2.5×10^{10}	[a]
Epoxy (cured with DETA)	2×10^{16}	[c,d]
Epoxy polyamide	1.1–1.5×10^{14}	
Phenolics	6×10^{12}–10^{13}	[a]
Polyamides	10^{13}	
Polyamide-imide	7.7×10^{16}	[e]
Polyethylene	$>10^{16}$	
Polyimide	10^{16}–10^{18}	[f]
Polypropylene	10^{10}–$>10^{16}$	[a]
Polystyrene	$>10^{16}$	[b]
Polysulfide	2.4×10^{11}	[g]
Polyurethane (single component)	5.5×10^{12}	[h]
Polyurethane (single component)	2.0×10^{12}	[h]
Polyurethane (single component)	4×10^{13}	Columbia Technical Corporation
Polyurethane (two components)	1×10^{13}	Products Research & Chemical Corporation (PR-1538)
	$5 \times 10^{9}(300°\text{F})$	
Polyvinyl chloride	10^{11}–10^{15}	[b]
Polyvinylidene chloride	10^{14}–10^{16}	[a]
Polyvinylidene fluoride	2×10^{14}	[h]
Polyxylylenes (parylenes)	10^{16}–10^{17}	Union Carbide Corporation
Silicone (RTV)	6×10^{14}–3×10^{15}	Stauffer Chemical Company, Si-O-Flex SS 831, 832, & 833
Silicone, flexible dielectric gel	1×10^{15}	Dow Corning Corporation
Silicone, flexible, clear	2×10^{15}	Dow Corning Corporation
Silicone	3.3×10^{14}	Columbia Technical Corporation Humiseal 1H34
Teflon TFE	$>10^{18}$	[b]
Teflon FEP	$>2 \times 10^{18}$	[b]

[a] *Materials Engineering,* Materials Selector Issue, vol. 66, no. 5, Chapman-Reinhold Publication, mid-October 1967–1968.

[b] *Insulation,* Directory Encyclopedia Issue, no. 7, June–July 1968.

[c] H. Lee and K. Neville, *Epoxy Resins,* McGraw-Hill, New York, 1966.

[d] Tucker, Cooperman, and Franklin, Dielectric Properties of Casting Resins, *Electronics Equipment,* July 1956.

[e] J. H. Freeman, "A New Concept in Flat Cable Systems," *5th Annual Symposium on Advanced Technology for Aircraft Electrical Systems,* Washington, D.C., October 1964.

[f] J. T. Milek, "Polyimide Plastics: A State of the Art Report," *Hughes Aircraft Rep.* S-8, October 1965.

[g] L. Hockenberger, *Chem.-Ing. Tech.,* vol. 36, 1964.

[h] *Hughson Chemical Company Technical Bulletin* 7030A; *Pennsalt Chemicals Corporation Product Sheet* KI-66a, Kynar Vinylidene Fluoride Resin, 1967.

TABLE 9.10 Dielectric Constants of Coatings

Coating	60–100 Hz	10^6 Hz	>10^6 Hz	Reference source
Acrylic		2.7–3.2		[a]
Alkyd			3.8 (10^{10} Hz)	[b]
Asphalt and tars			3.5 (10^{10} Hz)	[b]
Cellulose acetate butyrate		3.2–6.2		[a]
Cellulose nitrate		6.4		[a]
Chlorinated polyether	3.1	2.92		[a]
Chlorosulfonated polyethylene (Hypalon)	6.19 7–10 (10^3 Hz)	~5		E. I. du Pont de Nemours & Company
Depolymerized rubber (DPR)	4.1–4.2	3.9–4.0		H. V. Hardman, DPR Subsidiary
Diallyl isophthalate	3.5	3.2	3 (10^8 Hz)	[a,c]
Diallyl phthalate	3–3.6	3.3–4.5		[a,c]
Epoxy-anhydride— castor oil adduct	3.4	3.1	2.9 (10^7 Hz)	Autonetics, Division of North American Rockwell
Epoxy (one component)	3.8	3.7		Conap Inc.
Epoxy (two components)	3.7			Conap Inc.
Epoxy cured with methyl nadic anhydride (100:84 pbw)	3.31			[d]
Epoxy cured with dodecenylsuccinic anhydride (100:132 pbw)	2.82			[d]
Epoxy cured with DETA	4.1	4.2	4.1	[e]
Epoxy cured with m-phenylenediamine	4.6	3.8	3.25 (10^{10} Hz)	[e]
Epoxy dip coating (two components)	3.3	3.1		Conap Inc.
Epoxy (one component)	3.8	3.5		Conap Inc.
Epoxy-polyamide (40% Versamid* 125, 60% epoxy)	3.37	3.08		[e]
Epoxy-polyamide (50% Versamid 125, 50% epoxy)	3.20	3.01		[e]
Fluorocarbon (TFE, Teflon)	2.0–2.08	2.0–2.08		E. I. du Pont de Nemours & Company
Phenolic		4–11		[a]
Phenolic	5–6.5	4.5–5.0		[c]
Polyamide	2.8–3.9	2.7–2.96		
Polyamide-imide	3.09	3.07		
Polyesters	3.3–8.1	3.2–5.9		[c]
Polyethylene		2.3		[a]
Polyethylenes	2.3	2.3		[c]
Polyimide-Pyre-M.L.† enamel	3.8	3.8		[f]
Polyimide-Du Pont RK-692 varnish	3.8			[g]
Polyimide-Du Pont RC-B-24951	3.0 (10^3 Hz)			[g]
Polyimide-Du Pont RC-5060	2.8 (10^3 Hz)			[g]
Polypropylene		2.1		[a]

TABLE 9.10 Dielectric Constants of Coatings *(Continued)*

Coating	60–100 Hz	10^6 Hz	$>10^6$ Hz	Reference source
Polypropylene	2.22–2.28	2.22–2.28		c
Polystyrene	2.45–2.65	2.4–2.65	2.5(10^{10} Hz)	b
Polysulfides	6.9			h
Polyurethane (one component)	4.10	3.8		Conap Inc.
Polyurethane (two components— castor oil cured)	6.8(10^3 Hz)	2.98–3.28		i
Polyurethane		4.4		Products Research & (two components) Chemical Corporation (PR-1538)
Polyvinyl butyral	3.6	3.33		a
Polyvinyl chloride	3.3–6.7	2.3–3.5		a
Polyvinyl chloride— vinyl acetate copolymer	3–10			a
Polyvinyl formal	3.7	3.0		a
Polyvinylidene chloride		3–5		a
Polyvinylidene fluoride	8.1	6.6		i
Polyvinylidene fluoride	8.4	6.43	2.98(10^9 Hz)	k
p-Polyxylylene:				
Parylene N	2.65	2.65		Union Carbide Corporation
Parylene C	3.10	2.90		Union Carbide Corporation
Parylene D	2.84	2.80		Union Carbide Corporation
Shellac (natural, dewaxed)	3.6	3.3	2.75(10^9 Hz)	l
Silicone (RTV types)	3.3–4.2	3.1–4.0		General Electric and Stauffer Chemical Companies
Silicone (Sylgard‡ type)	2.88	2.88		Dow Corning Corporation
Silicone, flexible	3.0			Dow Corning dielectric gel Corporation
FEP dispersion coating	2.1(10^3 Hz)			E. I. du Pont de Nemours & Company
TFE dispersion coating	2.0–2.2(10^3 Hz)			E. I. du Pont de Nemours & Company
Wax (paraffinic)	2.25	2.25	2.22(10^{10} Hz)	l

*Trademark of General Mills, Inc., Kankakee, Ill.
†Trademark of E. I. du Pont de Nemours & Company, Wilmington, Del.
‡Trademark of Dow Corning Corporation, Midland, Mich.
[a] *Materials Engineering,* Materials Selector Issue, vol. 66, no. 5, Chapman-Reinhold Publication, mid-October 1967–1968.
[b] M. C. Volk, J. W. Lefforge, and R. Stetson, *Electrical Encapsulation,* Reinhold Publishing Corporation, New York, 1962.
[c] *Insulation,* Directory Encyclopedia Issue, no. 7, June–July 1968.
[d] C. F. Coombs, ed., *Printed Circuits Handbook,* McGraw-Hill, New York, 1967.
[e] H. Lee and K. Neville, *Handbook of Epoxy Resins,* McGraw-Hill, New York, 1967.
[f] J. R. Learn and M. P. Seeger, Teflon-Pyre-M.L. Wire Insulation System, *13th Symposium of Technical Progress in Communications Wire and Cable,* Atlantic City, NJ, Dec. 2–4, 1964.
[g] *Du Pont Bulletin* H65-4, Experimental Polyimide Insulating Varnishes, RC-B-24951 and RC-5060, January 1965.
[h] L. Hockenberger, *Chem.-Ing. Tech.* vol. 36, 1964.
[i] *Spencer-Kellogg* (division of Textron, Inc.) Bull. TS-6593.
[j] W. S. Barnhart, R. A. Ferren, and H. Iserson, 17th ANTEC of SPE, January 1961.
[k] *Pennsalt Chemicals Corporation Product Sheet* KI-66a, Kynar Vinylidene Fluoride Resin, 1967.
[l] A. R. Von Hippel, ed., *Dielectric Materials and Applications,* Technology Press of MIT and John Wiley & Sons, Inc., New York, 1961.

TABLE 9.11 Dissipation Factors of Coatings

Coating	60–100 Hz	10^6 Hz	$>10^6$ Hz	Reference source
Acrylics	0.04–0.06	0.02–0.03		[a]
Alkyds	0.003–0.06			[a]
Chlorinated polyether	0.01	0.01		[a]
Chlorosulfonated polyethylene	0.03	$0.07(10^3$ Hz)		[b]
Depolymerized rubber (DPR)	0.007–0.013	0.0073–0.016		H. V. Hardman, DPR Subsidiary
Diallyl phthalate	0.010	0.011	0.011	[b]
Diallyl isophthalate	0.008	0.009	$0.014(10^{10}$ Hz)	[b]
Epoxy dip coating (two components)	0.027	0.018		Conap Inc.
Epoxy (one component)	0.011	0.004		Conap Inc.
Epoxy (one component)	0.008	0.006		Conap Inc.
Epoxy polyamide (40% versamid 125, 60% epoxy)	0.0085	0.0213		
Epoxy polyamide (50% Versamid 115, 50% epoxy)	0.009	0.0170		
Epoxy cured with anhydride–castor oil adduct	0.0084	0.0165	0.0240	Autonetics, North American Rockwell
Phenolics	0.005–0.5	0.022		[a]
Polyamide	0.015	0.022–0.097		
Polyesters	0.008–0.041			[a]
Polyethylene (linear)	0.00015	0.00015	$0.0004(10^{10}$ Hz)	[d]
Polymethyl methacrylate	0.06	0.02	$0.009(10^{10}$ Hz)	[d]
Polystyrene	0.0001–0.0005	0.0001–0.0004		
Polyurethane (two component, castor oil cure)		0.016–0.036		
Polyurethane (one component)	0.038–0.039	0.068–0.074		Conap Inc.
Polyurethane (one component)	0.02			Conap Inc.
Polyvinyl butyral	0.007	0.0065		
Polyvinyl chloride	0.08–0.15	0.04–0.14		[a]
Polyvinyl chloride, plasticized	0.10	0.15	$0.01(10^{10}$ Hz)	[d]
Polyvinyl chloride-vinyl acetate copolymer	0.6–0.10			

Material				Source	
Polyvinyl formal	0.007	0.02			f
Polyvinylidene fluoride	0.049	0.17			g
	0.049	0.159	0.110		
Polyxylylenes:					
Parylene N	0.0002	0.0006		Union Carbide Corporation	
Parylene C	0.02	0.0128		Union Carbide Corporation	
Parylene D	0.004	0.0020		Union Carbide Corporation	
Silicone (Sylgard 182)	0.001	0.001		Dow Corning Corporation	
Silicone, flexible dielectric gel	0.0005			Dow Corning Corporation	
Silicone, flexible, clear	0.001			Dow Corning Corporation	
Silicone (RTV types)	0.011–0.02	0.003–0.006		General Electric	
Teflon FEP dispersion coating	0.0002–0.0007			E. I. du Pont de Nemours & Company	
Teflon FEP	<0.0003	<0.0003			a
Teflon TFE	<0.0003				a
Teflon TFE	0.00012	0.00005		Union Carbide Corporation	
Other materials:					
Alumina (99.5%)	0.0001	0.0001			h
Beryllia (99.5%)	0.0003	0.0003			h
Glass silica	0.0001	0.0001	$0.00017(10^{10}$ Hz)		d
Glass, borosilicate	0.013–0.016			Corning Glass Works	
Glass, 96% silica	0.0015–0.0019			Corning Glass Works	

[a] *Machine Design*, Plastics Reference Issue, vol. 38, no. 14, Penton Publishing Co., 1966.

[b] *Insulation*, Directory, Encyclopedia Issue, no. 7, June–July 1968.

[c] H. Lee and K. Neville, *Handbook of Epoxy Resins*, McGraw-Hill, New York, 1967.

[d] K. Mathes, Electrical Insulation Conference, 1967.

[e] *Spencer-Kellogg* (Division of Textron, Inc.) *Technical Bulletin* TS-6593.

[f] W. S. Barnhart, R. A. Ferren, and H. Iserson, 17th ANTEC of SPE, January 1961.

[g] *Pennsalt Chemicals Corporation Product Sheet* KI-66a, Kynar Vinylidene Fluoride Resin, 1967.

[h] *Machine Design*, Design Guide, September 28, 1967.

TABLE 9.12   NEMA Standards and Manufacturers' Trade Names for Magnet Wire Insulation

Manufacturer	Plain enamel	Polyvinyl formal	Polyvinyl formal modified	Polyvinyl formal with nylon overcoat	Polyvinyl formal with butyral overcoat	Polyamide	Acrylic	Epoxy
Thermal class NEMA Standard*	105°C MW 1	105°C MW 15	105°C MW 27	105°C MW 17	105°C MW 19	105°C MW 6	105°C MW 4	130°C MW 9
Anaconda Wire & Cable Company	Plain enamel	Formvar	Hermetic Formvar	Nyform	Cement-coated Formvar	....	....	Epoxy, epoxy-cement-coated
Asco Wire & Cable Company	Enamel	Formvar	....	Nyform	Formbond	Nylon	Acrylic	Epoxy
Belden Manufacturing Company	Beld enamel	Formvar	....	Nyclad	....	....	....	Epoxy
Bridgeport Insulated Wire Company	....	Formvar	....	....	Quickbond	Quick-Sol	....	....
Chicago Magnet Wire Corporation	Plain enamel	Formvar	....	Nyform	Bondable Formvar	Nylon	Acrylic	Epoxy
Essex Wire Corporation	Plain enamel	Formvar	Formetex	Nyform	Bondex	....	Ensolex/ESX	Epoxy
General Cable Corporation	Plain enamel	Formvar	Formetic	Formlon	Formese	....	Solderable acrylic	Epoxy
General Electric Corporation	....	Formex	....	....	....	Nylon	....	....
Haveg-Super Temp Division	....	....	....	....	....	....	....	....
Hitemp Wires Company Division Simplex Wire & Cable Company	....	....	....	....	....	....	....	....
Hudson Wire Company	Plain enamel	Formvar	....	Nyform	Formvar AVC	Ezsol	....	....
New Haven Wire & Cable, Inc.	Plain enamel	....	....	....	....	....	....	....
Phelps Dodge Magnet Wire Corporation	Enamel	Formvar	Hermeteze	Nyform	Bondeze	....	....	....
Rea Magnet Wire Company, Inc.	Plain enamel	Formvar	Hermetic Formvar special	Nyform	Koilset	Nylon	....	Epoxy
Viking Wire Company, Inc.	Enamel	Formvar	....	Nyform	F-Bondall	Nylon	....	....

*National Electrical Manufacturers Association.
SOURCE: Courtesy of Rea Magnet Wire Company, Inc.

Thermosetting acrylics are still used by the major appliance industry. Acrylics are used in electrodeposition and have largely replaced alkyds. The chemistry of acrylic-based resins allows them to be used in radiation-curing applications alone or as monomeric modifiers for other resins. Acrylic-modified polyurethane coatings have excellent exterior durability.

Teflon	Poly-urethane	Polyure-thane with friction surface	Polyure-thane with nylon overcoat	Polyure-thane with butyral overcoat	Polyure-thane with nylon and butyral overcoat	Polyester	Polyester with overcoat	Poly-imide	Polyester polyimide	Ceramic, ceramic-Teflon, ceramic-silicon
200°C MW 10	105°C MW 2	105°C	130°C MW 28	105°C MW 3 (PROP)	130°C MW 29 (PROP)	155°C MW 5	155°C MW 5	220°C MW 16	180°C	180°C+ MW 7
....	Analac	....	Nylac	Cement-coated analac	Cement-coated nylac	....	Anatherm D Anatherm 200	AL 220 ML	Anatherm N	
....	Poly	....	Nypol	Asco bond-P	Asco bond	Ascotherm	Isotherm 200	ML	Anamid M (amide-imide), Ascomid	
....	Beldure	....	Beldsol	....	....	Isonel	Polyther-malese	ML		
....	Polyure-thane	Uniwind	Poly-nylon	Polybond	....	Isonel 200				
....	Soderbrite	....	Nysod	Bondable polyurethene	....	Polyester 155				
....	Soderex	....	Soderon	Soder-bond	Soder-bond N	Thermalex F	Polyther-malex/ PTX 200	Allex		
....	Enamel "G"	....	Genlon	....	....	Gentherm	Polyther-maleze 200			
....	....	....	....	....	....	Alkanex				
Teflon	....	....	....	....	....	Isonel	....			
Temprite										
....	Hudsol	Gripon	Nypoly	Hudsol AVC	Nypoly AVC	Isonel 200	Isonel 200-A	ML	Isomid	
....	Impsol	....	Impsolon	....	....	....	Imp-200	....	....	
....	Sodereze	Gripeze	Nyleze	....	S-Y Bondeze	Polyther-maleze 200 II	ML			
....	Solvar	....	Nylon solvar	Solvar Koilset	....	Isonel 200	Polyther-maleze 200	Pyre ML	Isomid	Ceroc
....	Polyure-thane	....	Polynylon	P-Bondall	....	Isonel 200	Iso-poly	ML	Isomid Isomid-P	

9.5.1.2 Alkyds. Alkyd resin-based coatings were introduced in the 1930s as replacements for nitrocellulose lacquers and oleoresinous coatings. They offer the advantage of good durability at relatively low cost. These low- to medium-priced coatings are still used for finishing a wide variety of products, either alone or modified with oils or other resins. The degree and type of modification determine their performance properties. They were used extensively by the automotive and appliance industries through the 1960s. Although alkyds are used in outdoor applications, they are not as durable in long-term exposure, and their color and gloss retention is inferior to that of acrylics.

Uses. Once the mainstay of organic coatings, alkyds are still used for finishing metal and wood products. Their durability in interior exposures is generally good, but their exterior durability is only fair. Alkyd resins are used in fillers, sealers, and caulks for wood finishing because of their formulating flexibility. Alkyds have also been used in electrodeposition as replacements for the oleoresinous vehicles. They are still used for finishing by the machine tool and other industries. Alkyds have also been widely used in architectural and trade sales coatings. Alkyd-modified acrylic latex paints are excellent architectural finishes.

9.5.1.3 Epoxies. Epoxy resins can be formulated with a wide range of properties. These medium- to high-priced resins are noted for their adhesion, make excellent primers, and are used widely in the appliance and automotive industries. Their heat resistance permits them to be used for electrical insulation. When epoxy topcoats are used outdoors, they tend to chalk and discolor because of inherently poor ultraviolet light resistance. Other resins modified with epoxies are used for outdoor exposure as topcoats, and properties of many other resins can be improved by their addition. Two-component epoxy coatings are used in environments with extreme corrosion and chemical conditions. Flexibility in formulating two-component epoxy resin-based coatings results in a wide range of physical properties.

Uses. Owing to their excellent adhesion, they are used extensively as primers for most coatings over most substrates. Epoxy coatings provide excellent chemical and corrosion resistance. They are used as electrical insulating coatings because of their high electric strength at elevated temperatures. Some of the original work with powder coating was done using epoxy resins, and they are still applied using this method. Many of the primers used for coil coating are epoxy resin-based.

9.5.1.4 Polyesters. Polyesters are used alone or modified with other resins to formulate coatings ranging from clear furniture finishes (replacing lacquers) to industrial finishes (replacing alkyds). These moderately priced finishes permit the same formulating flexibility as alkyds but are tougher and more weather resistant. There are basically two types of polyesters: two-component and single-package. Two-component polyesters are cured using peroxides that initiate free-radical polymerization, while single-package polyesters, sometimes called *oil-free alkyds,* are self-curing, usually at elevated temperatures. It is important to realize that, in both cases, the resin formulator can adjust properties to meet most exposure conditions. Polyesters are also applied as powder coatings.

Uses. Two-component polyesters are well known as gel coats for glass-reinforced plastic bathtubs, lavatories, boats, and automobiles. Figure 9.2 shows tub and shower units using a polyester gel coat. High-quality one-package polyester finishes are used on furniture, appliances, automobiles, magnet wire, and industrial products. Polyester powder coatings are used as high-quality finishes in indoor and outdoor applications for anything from tables to trucks. They are also used as coil coatings.

9.5.1.5 Polyurethanes. Polyurethane resin-based coatings are extremely versatile. They are higher in price than alkyds but lower than epoxies. Polyurethane resins are available as oil-modified, moisture-curing, blocked, two-component, and lacquers. Table 9.13 is a selection

Figure 9.2 Polyester gel coats are used to give a decorative and protective surface to tub shower units that are made out of glass fiber-reinforced plastics. *(Courtesy of Owens-Corning Fiberglas Corporation.)*

TABLE 9.13 Guide to Selecting Polyurethane Coatings

Property	One-component			Two-component	Lacquer
	Urethane oil	Moisture	Blocked		
Abrasion resistance	Fair–good	Excellent	Good–excellent	Excellent	Fair
Hardness	Medium	Medium–hard	Medium–hard	Soft–very hard	Soft–medium
Flexibility	Fair–good	Good–excellent	Good	Good–excellent	Excellent
Impact resistance	Good	Excellent	Good–excellent	Excellent	Excellent
Solvent resistance	Fair	Poor–fair	Good	Excellent	Poor
Chemical resistance	Fair	Fair	Good	Excellent	Fair–good
Corrosion resistance	Fair	Fair	Good	Excellent	Good–excellent
Adhesion	Good	Fair–good	Fair	Excellent	Fair–good
Toughness	Good	Excellent	Good	Excellent	Good–excellent
Elongation	Poor	Poor	Poor	Excellent	Excellent
Tensile	Fair	Good	Fair–good	Good–excellent	Excellent
Weatherability					
Aliphatic	Good	Poor–fair		Good–excellent	Good
Conventional	Poor–fair	Poor–fair	Poor–fair	Poor–fair	Poor
Pigmented glass	High	High	High	High	Medium
Cure rate	Slow	Slow	Fast	Fast	None
Cure temp	Room temperature	Room temperature	300–390°F	212°F	150–225°F
Work life	Infinite	1 y	6 months	1 s–24 h	Infinite

guide for polyurethane coatings. Two-component polyurethanes can be formulated in a wide range of hardnesses. They can be abrasion resistant, flexible, resilient, tough, chemical resistant, and weather resistant. Abrasion resistance of organic coatings is shown in Table 9.14. Polyurethanes can be combined with other resins to reinforce or adopt their properties. Urethane-modified acrylics have excellent outdoor weathering properties. They can also be applied as air-drying, forced-dried, and baking liquid finishes as well as powder coatings.

Uses. Polyurethanes have become very important finishes in the transportation industry, which includes aircraft, automobiles, railroads, trucks, and ships. Owing to their chemical resistance and ease of decontamination from chemical, biological, and radiological warfare agents, they are widely used for painting military land vehicles, ships, and aircraft. They are used on automobiles as coatings for plastic parts and as clear topcoats in the basecoat-clearcoat finish systems. Low-temperature baking polyurethanes are used as mar-resistant finishes for products that must be packaged while still warm. Polyurethanes are used in an increasing number of applications. They are also used in radiation curable coatings.

9.5.1.6 Polyvinyl chloride. Polyvinyl chloride (PVC) coatings, commonly called vinyls, are noted for their toughness, chemical resistance, and durability. They are available as solutions, dispersions, and lattices. Properties of vinyl coatings are listed in Table 9.15. They are applied as lacquers, plastisols, organisols, and lattices. PVC coating powders have essentially the same properties as liquids. PVC organisol, plastisol, and powder coatings have limited adhesion and require primers.

Uses. Vinyls have been used in various applications, including beverage and other can linings, automobile interiors, and office machine exteriors. They are also used as thick film liquids and as powder coatings for electrical insulation. Owing to their excellent chemical resistance, they are used as tank linings and as rack coatings in electroplating shops. Typical applications for vinyl coatings are shown in Fig. 9.3. Vinyl-modified acrylic latex trade sale paints are used as trim enamels for exterior applications and as semigloss wall enamels for interior applications.

9.5.2 Other Coating Resins

In addition to the aforementioned materials, there are a number of other important resins used in formulating coatings. These materials,

TABLE 9.14   **Abrasion Resistance of Coatings**

Coating	Taber ware index, mg/1000 rev
Polyurethane type 1	55–67
Polyurethane type 2 (clear)	8–24
Polyurethane type 2 (pigmented)	31–35
Polyurethane type 5	60
Urethane oil varnish	155
Alkyd	147
Vinyl	85–106
Epoxy-amine–cured varnish	38
Epoxy-polyamide enamel	95
Epoxy-ester enamel	196
Epoxy-polyamide coating (1:1)	50
Phenolic spar varnish	172
Clear nitrocellulose lacquer	96
Chlorinated rubber	200–220
Silicone, white enamel	113
Catalyzed epoxy, air-cured (PT-401)	208
Catalyzed epoxy, Teflon-filled (PT-401)	122
Catalyzed epoxy, bake-Teflon–filled (PT-201)	136
Parylene N	9.7
Parylene C	44
Parylene D	305
Polyamide	290–310
Polyethylene	360
Alkyd TT-E-508 enamel (cured for 45 min at 250°F)	51
Alkyd TT-E-508 (cured for 24 h at room temperature)	70

Figure 9.3 Vinyl plastisols and organisols are used extensively for dip coating of wire products. The coatings can be varied from very hard to very soft. *(Courtesy of M & T Chemicals.)*

TABLE 9.15 Properties of Vinyl Coatings

Coating type	Outstanding characteristics	Mechanical properties[a]					Color and gloss[a]			Weathering[a] properties	
		Hardness	Abrasion resistance	Adhesion	Flexibility	Toughness	Film color	Color retention	Gloss	Weather resistance	Gloss retention
Solution[b]	Excellent color, flexibility, chemical resistance; tasteless, odorless	F	E	F to G	E	E	E[f]	E	G	E[f]	E
Plastisol[c]	Toughness; resilience; abrasion resistance; can be applied without solvents	F	E	E to cloth[e]	E	E	E[f]	E	F	E[f]	F to G
Organosol[d]	High solids content; excellent color, flexibility; tasteless, odorless	F	E	E to cloth[e]	E	E	E[f]	E	P to G	E[f]	G

[a] E = excellent; G = good; F = fair; P = poor.
[b] Vinyl chloride acetate copolymers; resins vary widely in compatibility with other materials.
[c] Vinyl chloride acetate copolymer and vinyl chloride resins.
[d] Vinyl chloride acetate copolymers; require grinding for good dispersions.
[e] Requires primer for use on metal.
[f] Pigmented.

used alone or as modifiers for other resins, provide coating vehicles with diverse properties.

9.5.2.1 Aminos. Resins of this type, such as urea formaldehyde and melamine, are used in modifying other resins to increase their durability. Notable among these modified resins are the superalkyds used in automotive and appliance finishes.

Uses. Melamine and urea formaldehyde resins are used as modifiers for alkyds and other resins to increase hardness and accelerate cure.

9.5.2.2 Cellulosics. Nitrocellulose lacquers are the most important of the cellulosics. They were introduced in the 1920s and used as fast-drying finishes for a number of manufactured products. Applied at low solids using expensive solvents, they will not meet air quality standards. By modifying nitrocellulose with other resins such as alkyds and ureas, the VOC content can be lowered, and performance properties can be increased. Other important cellulosic resins are cellulose acetate butyrate and ethyl cellulose.

Uses. Although no longer used extensively by the automotive industry, nitrocellulose lacquers are still used by the furniture industry because of their fast-drying and hand-rubbing properties. Cellulose acetate butyrate has been used for coating metal in numerous applications. In 1959, one of the first conveyorized powder coating lines in the United States coated distribution transformer lids and hand-hole covers with a cellulose acetate butyrate powder coating.

9.5.2.3 Chlorinated rubber. Chlorinated rubber coatings are used as swimming pool paints and traffic paints.

9.5.2.4 Fluorocarbons. These high-priced coatings require high processing temperatures and therefore are limited in their usage. They are noted for their lubricity or nonstick properties due to low coefficients of friction, and also for weatherability. Table 9.16 gives the coefficients of friction of typical coatings.

Uses. Fluorocarbons are used as chemical-resistant coatings for processing equipment. They are also used as nonstick coatings for cookware, friction-reducing coatings for tools, and as dry lubricated surfaces in many other consumer and industrial products, as shown in Fig. 9.4. Table 9.17 compares the properties of four fluorocarbons.

TABLE 9.16   Coefficients of Friction of Typical Coatings

Coating	Coefficient of friction, μ	Information source
Polyvinyl chloride	0.4–0.5	a
Polystyrene	0.4–0.5	a
Polymethyl methacrylate	0.4–0.5	a
Nylon	0.3	a
Polyethylene	0.6–0.8	a
Polytetrafluoroethylene (Teflon)	0.05–0.1	a
Catalyzed epoxy air-dry coating with Teflon filler	0.15	b
Parylene N	0.25	c
Parylene C	0.29	c
Parylene D	0.31–0.33	c
Polyimide (Pyre-ML)	0.17	d
Graphite	0.18	d
Graphite-molybdenum sulfide:		
Dry-film lubricant	0.02–0.06	e
Steel on steel	0.45–0.60	e
Brass on steel	0.44	e
Babbitt on mild steel	0.33	e
Glass on glass	0.4	e
Steel on steel with SAE no. 20 oil	0.044	e
Polymethyl methacrylate to self	0.8 (static)	e
Polymethyl methacrylate to steel	0.4–0.5 (static)	e

[a]F. P. Bowder, *Endeavor,* vol. 16, no. 61, 1957, p. 5.
[b]Product Techniques Incorporated, Bulletin on PT-401 TE, October 17, 1961.
[c]Union Carbide data.
[d]*DuPont Technical Bulletin* 19, Pyre-ML Wire Enamel, August 1967.
[e]Electrofilm, Inc. data.

Figure 9.4 Nonstick feature of fluorocarbon finishes makes them useful for products such as saws, fan and blower blades, door-lock parts, sliding- and folding-door hardware, skis, and snow shovels. *(Courtesy of E.I. DuPont de Nemours & Company.)*

TABLE 9.17   Properties of Four Fluorocarbons

Property	Polyvinyl fluoride (PVF) $(CH_2-CHF)_n$	Polyvinyl-idene fluoride (PVF-2) $(CH_2-CF_2)_n$	Polytrifluoro-chloroethylene (PTFC1) $(CC1F-CF_2)_n$	Polytetra-fluoroethylene (PTFE) $(CF_2-CF_2)_n$
Physical properties:				
Density	1.4	1.76	2.104	2.17–2.21
Fusing temperature, °F	300	460	500	750
Maximum continuous service and temperature, °F	225	300	400	550
Coefficient of friction	0.16	0.16	0.15	0.1
Flammability	Burns	Nonflammable	Nonflammable	Nonflammable
Mechanical properties:				
Tensile strength, lb/in²	7000	7000	5000	2500–3500
Elongation, %	115–250	300	250	200–400
Izod impact, ft-lb/in		3.8	5	3
Durometer hardness		80	74–78	50–65
Yield strength at 77°F, lb/in²	6000	5500	4500	1300
Heat-distortion temperature at 66 lb/in² °F	NA	300	265	250
Coefficient of linear expansion	2.8×10^5	8.5×10^5	15×10^5	8×10^5
Modulus (tension) $\times 10^5$ lb/in²	2.5–3.7	1.2	1.9	0.6
Electrical properties:				
Dielectric strength, V/mil Short time, V/mil, in	3400 (0.002)	260 (0.125)	500 (0.063)	600 (0.060)
Dielectric constant, 10^3 Hz	8.5	7.72	2.6	2.1
Arc resistance (77°F) ASTM D 495	NA	60	300	300
Volume resistivity Ω-cm at 50% relative humidity, 77°F	10^{12}	10^{14}	10^{16}	10^{18}
Dissipation factor, 100 Hz	1.6	0.05	0.022	0.0003

9.5.2.5 Oleoresinous. Oleoresinous coatings, based on drying oils such as soybean and linseed, are slow curing. For many years prior to the introduction of synthetic resins, they were used as the vehicles in most coatings. They still find application alone or as modifiers to other resins.

Uses. Oleoresinous vehicles are used in low-cost primers and enamels for structural, marine, architectural, and, to a limited extent, industrial product finishing.

9.5.2.6 Phenolics. Introduced in the early 1900s, phenolics were the first commercial synthetic resins. They are available as 100% phenolic baking resins, oil-modified, and phenolic dispersions. Phenolic resins, used as modifiers, will improve the heat and chemical resistance of other resins. Baked phenolic resin-based coatings are well known for their corrosion, chemical, moisture, and heat resistance.

Uses. Phenolic coatings are used on heavy-duty air-handling equipment, chemical equipment, and as insulating varnishes. Phenolic resins are also used as binders for electrical and decorative laminated plastics.

9.5.2.7 Polyamides. One of the more notable polyamide resins is nylon, which is tough and wear resistant and has a relatively low coefficient of friction. It can be applied as a powder coating by fluidized bed, electrostatic spray, or flame spray. Table 9.18 compares the properties of three types of nylon polymers used in coatings. Nylon coatings generally require a primer. Polyamide resins are also used as curing agents for two-component epoxy resin coatings. Film properties can be varied widely by polyamide selection.

Uses. Applied as a powder coating, nylon provides a high degree of toughness and mechanical durability to office furniture. Other polyamide resins are used as curing agents in two-component epoxy resin-based primers and topcoats, adhesives, and sealants.

9.5.2.8 Polyolefins. These coatings, which can be applied by flame spraying, hot melt, or powder coating methods, have limited usage.

Uses. Polyethylene is used for impregnating or coating packaging materials such as paper and aluminum foil. Certain polyethylene-coated composite packaging materials are virtually moisture-proof. Table 9.19 compares the moisture vapor transmission rates of various coat-

TABLE 9.18   Properties of Nylon Coatings

	Nylon 11	Nylon 6/6	Nylon 6
Elongation (73°F), %	120	90	50–200
Tensile strength (73°F), lb/in^2	8,500	10,500	10,500
Modulus of elasticity (73°F), lb/in^2	178,000	400,000	350,000
Rockwell hardness	R 100.5	R 118	R 112–118
Specific gravity	1.04	1.14	1.14
Moisture absorption, %, ASTM D 570	0.4	1.5	1.6–2.3
Thermal conductivity Btu/(ft^2)(h)(°F/in)	1.5	1.7	1.2–1.3
Dielectric strength (short time), V/mil	430	385	440
Dielectric constant (10 Hz)	3.5	4	4.8
Effect of:			
Weak acids	None	None	None
Strong acids	Attack	Attack	Attack
Strong alkalies	None	None	None
Alcholos	None	None	None
Esters	None	None	None
Hydrocarbons	None	None	None

ings and films. Polyethylene powder coatings are used on chemical-processing and food-handling equipment.

9.5.2.9 Polyimides.   Polyimide coatings have excellent long-term thermal stability, wear, mar and moisture resistance, and electrical properties. They are high in price.

Uses.   Polyimide coatings are used in electrical applications as insulating varnishes and magnet wire enamels in high-temperature, high-reliability applications. They are also used as alternatives to fluorocarbon coatings on cookware, as shown in Fig. 9.5.

9.5.2.10 Silicones.   Silicone resins are high in price and are used alone or as modifiers to upgrade other resins. They are noted for their high temperature resistance, moisture resistance, and weatherability. They can be hard or elastomeric, baking or room temperature curing.

Uses.   Silicones are used in high-temperature coatings for exhaust stacks, ovens, and space heaters. Figure 9.6 shows silicone coatings on fireplace equipment. They are also used as conformal coatings for printed wiring boards, moisture repellents for masonry, weather-resistant finishes for outdoors, and thermal control coatings for space vehicles. The thermal conductivities of coatings are listed in Table 9.20.

TABLE 9.19   Moisture-Vapor Transmission Rates per 24-h Period of Coatings and Films in g/(mil)(in^2)

Coating or film	MVTR	Information source
Epoxy-anhydride	2.38	Autonetics data (25°C)
Epoxy-aromatic amine	1.79	Autonetics data (25°C)
Neoprene	15.5	Baer (39°C)
Polyurethane (Magna X-500)	2.4	Autonetics data (25°C)
Polyurethane (isocyanate-polyester)	8.72	Autonetics data (25°C)
Olefane,* polypropylene	0.70	Avisun data
Cellophane (type PVD uncoated film)	134	DuPont
Cellulose acetate (film)	219	DuPont
Polycarbonate	10	FMC data
Mylar†	1.9	Baer (39°C)
	1.8	DuPont data
Polystyrene	8.6	Baer (39°C)
	9.0	Dow data
Polyethylene film	0.97	Dow data (1-mil film)
Saran resin (F120)	0.097–0.45	Baer (39°C)
Polyvinylidene chloride	0.15	Baer (2-mil sample, 40°C)
Polytetrafluoroethylene (PTFE)	0.32	Baer (2-mil sample 40°C)
PTFE, dispersion cast	0.2	DuPont data
Fluorinated ethylene propylene (FEP)	0.46	Baer (40°C)
Polyvinyl fluoride	2.97	Baer (40°C)
Teslar	2.7	DuPont data
Parylene N	14	Union Carbide data (2-mil sample)
Parylene C	1	Union Carbide data (2-mil sample)
Silicone (RTV 521)	120.78	Autonetics data
Methyl phenyl silicone	38.31	Autonetics data
Polyurethane (AB0130-002)	4.33	Autonetics data
Phenoxy	3.5	Lee, Stoffey, and Neville
Alkyd-silicone (DC-1377)	6.47	Autonetics data
Alkyd-silicone (DC-1400)	4.45	Autonetics data
Alkyd-silicone	6.16–7.9	Autonetics data
Polyvinyl fluoride (PT-207)	0.7	Product Techniques Incorp.

*Trademark of Avisun Corporation, Philadelphia, Pa.

†Trademark of E. I. du Pont de Nemours & Company, Wilmington, Del.

9.6   Application Methods

The selection of an application method is as important as the selection of the coating itself. Basically, the application methods for industrial liquid coatings and finishes are dipping, flow coating, and spraying, although some coatings are applied by brushing, rolling, printing, and silk screening. The application methods for powder coatings and finishes are fluidized beds, electrostatic fluidized beds, and electrostatic spray outfits. In these times of environmental awareness, regulation, and compliance, it is mandatory that coatings be applied in the most

Figure 9.5 Polyimide coating is used as a protective finish on the inside of aluminum, stainless steel, and other cookware. *(Courtesy of Mirro Aluminum Company.)*

Figure 9.6 Silicone coatings are used as heat-stable finishes for severe high-temperature applications such as fireplace equipment, exhaust stacks, thermal control coatings for spacecraft, and wall and space heaters. *(Courtesy of Copper Development Assn.)*

TABLE 9.20   **Thermal Conductivities of Coatings**

Material	k value,[*] cal/(s)(cm^2) (°C/cm) $\times$ 10^4	Source of information
Unfilled plastics:		
Acrylic	4–5	[a]
Alkyd	8.3	[a]
Depolymerized rubber	3.2	H. V. Hardman, DPR Subsidiary
Epoxy	3–6	[b]
Epoxy (electrostatic spray coating)	6.6	Hysol Corporation, DK-4
Epoxy (electrostatic spray coating)	2.9	Minnesota Mining & Manufacturing, No. 5133
Epoxy (Epon† 828, 71.4% DEA, 10.7%)	5.2	
Epoxy (cured with diethylenetriamine)	4.8	[c]
Fluorocarbon (Teflon TFE)	7.0	DuPont
Fluorocarbon (Teflon FEP)	5.8	DuPont
Nylon	10	[d]
Polyester	4–5	[a]
Polyethylenes	8	[a]
Polyimide (Pyre-ML enamel)	3.5	[e]
Polyimide (Pyre-ML varnish)	7.2	[f]
Polystyrene	1.73–2.76	[g]
Polystyrene	2.5–3.3	[a]
Polyurethane	4–5	[n]
Polyvinyl chloride	3–4	[a]
Polyvinyl formal	3.7	[a]
Polyvinylidene chloride	2.0	[a]
Polyvinylidene fluoride	3.6	[h]
Polyxylylene (Parylene N)	3	Union Carbide
Silicones (RTV types)	5–7.5	Dow Corning Corporation
Silicones (Sylgard types)	3.5–7.5	Dow Corning Corporation
Silicones (Sylgard varnishes and coatings)	3.5–3.6	Dow Corning Corporation
Silicone (gel coating)	3.7	Dow Corning Corporation
Silicone (gel coating)	7 (150°C)	Dow Corning Corporation
Filled plastics:		
Epon 828/diethylenetriamine = A	4	[b]
A + 50% silica	10	[b]
A + 50% alumina	11	[b]
A + 50% beryllium oxide	12.5	[b]
A + 70% silica	12	[b]
A + 70% alumina	13	[b]
A + 70% beryllium oxide	17.8	[b]
Epoxy, flexibilized = B	5.4	[i]
B + 66% by weight tabular alumina	18.0	[i]
B + 64% by volume tabular alumina	50.0	[i]
Epoxy, filled	20.2	Emerson & Cuming, 2651 ft
Epoxy (highly filled)	15–20	Wakefield Engineering Company
Polyurethane (highly filled)	8–11	International Electronic Research Company

TABLE 9.20   Thermal Conductivities of Coatings *(Continued)*

Material	k value,* cal/(s)(cm^2) (°C/cm) $\times 10^4$	Source of information
Other materials used in electronic assemblies:		
Alumina ceramic	256–442 (20–212°F)	a
Aluminum	2767–5575	a
Aluminum oxide (alumina), 96%	840	i
Beryllium oxide, 99%	5500	i
Copper	8095–9334	a
Glass (Borosill, 7052)	28	k
Glass (pot-soda-lead, 0120)	18	k
Glass (silica, 99.8% SiO$_2$)	40	l
Gold	7104 (20–212°F)	a
Kovar	395	m
Mica	8.3–16.5	a
Nichrome‡	325	m
Silica	40	k
Silicon nitride	359	m
Silver	9995 (20–212°F)	a
Zircon	120–149	a

*All values are at room temperature unless otherwise specified.

†Trademark of Shell Chemical Company, New York, N.Y.

‡Trademark of Driver-Harris Company, Harrison, N.J.

[a]*Materials Engineering,* Materials Selector Issue, vol. 66, no. 5, Chapman-Reinhold Publication, mid-October 1967.

[b]D. C. Wolf, *Proceedings, National Electronics and Packaging Symposium,* New York, June 1964.

[c]H. Lee and K. Neville, *Handbook of Epoxy Resins,* McGraw-Hill, New York, 1966.

[d]R. Davis, *Reinforced Plastics,* October 1962.

[e]*DuPont Technical Bulletin* 19, Pyre-ML Wire Enamel, August 1967.

[f]*DuPont Technical Bulletin* 1, Pyre-ML Varnish RK-692, April 1966.

[g]W. C. Teach and G. C. Kiessling, *Polystyrene,* Reinhold Publishing Corporation, New York, 1960.

[h]W. S. Barnhart, R. A. Ferren, and H. Iserson, 17th ANTEC of SPE, January 1961.

[i]A. J. Gershman and J. R. Andreotti, *Insulation,* September 1967.

[j]*American Lava Corporation* Chart 651.

[k]E. B. Shand, *Glass Engineering Handbook,* McGraw-Hill, 1958.

[l]W. D. Kingery, "Oxides for High Temperature Applications," *Proceedings, International Symposium,* Asilomar, Calif., October 1959, McGraw-Hill, New York, 1960.

[m]W. H. Kohl, *Handbook of Materials and Techniques for Vacuum Devices,* Reinhold Publishing Company, New York, 1967.

[n]*Modern Plastics Encyclopedia,* McGraw-Hill, New York, 1968.

efficient manner[3]. Not only will this help meet the air quality standards, it will also reduce material costs. The advantages and disadvantages of various coating application methods are given in Table 9.21.

Liquid spray coating equipment can be classified by its atomizing method: air, hydraulic, or centrifugal. These can be subclassified into air atomizing, airless, airless electrostatic, air-assisted airless electrostatic, rotating electrostatic disks and bells, and high-volume, low-pressure types. While liquid dip coating equipment is usually simple, electrocoating equipment is fairly complex, using electrophoresis as the driving force. Other liquid coating methods include flow coating, which can be manual or automated, roller coating, curtain coating, and centrifugal coating. Equipment for applying powder coatings is not as diversified as for liquid coatings. It can only be classified as fluidized bed, electrostatic fluidized bed, and electrostatic spray.

It is important to note that environmental and worker safety regulations can be met, hazardous and nonhazardous wastes can be reduced, and money can be saved by using compliance coatings (those that meet the VOC emission standards) in equipment having the highest transfer efficiency (the percentage of the coating used that actually coats the product and is not otherwise wasted). The theoretical transfer efficiencies (TEs) of coating application equipment are indicated in the text and in Table 9.22, where they are listed in descending order.[4] The aforementioned transfer efficiencies are meant to be used only as guidelines. Actual transfer efficiencies depend on a number of factors that are unique to each coating application line.

In the selection of a coating method and equipment, the product's size, configuration, intended market, and appearance must be considered. To aid in the selection of the most efficient application method, each will be discussed in greater detail.

9.6.1   Dip Coating

Dip coating (95 to 100% TE) is a simple coating method wherein products are dipped in a tank of coating material, withdrawn, and allowed to drain in the solvent-rich area above the coating's surface and then allowed to dry. The film thickness is controlled by viscosity, flow, percent solids by volume, and rate of withdrawal. This simple process can also be automated with the addition of a drain-off area, which allows excess coating material to flow back to the dip tank.

Dip coating is a simple, quick method that does not require sophisticated equipment. The disadvantages of dip coating are film thickness differential from top to bottom, resulting in the so-called wedge effect; fatty edges on lower parts of products; and runs and sags. Although

TABLE 9.21 Application Methods for Coatings

Method	Advantages	Limitations	Typical applications
Spray	Fast, adaptable to varied shapes and sizes. Equipment cost is low.	Difficult to completely coat complex parts and to obtain uniform thickness and reproducible coverage.	Motor frames and housings, electronic enclosures, circuit boards, electronic modules.
Dip	Provides thorough coverage, even on complex parts such as tubes and high-density electronic modules.	Viscosity and pot life of dip must be monitored. Speed of withdrawal must be regulated for consistent coating thickness.	Small- and medium-sized parts, castings, moisture and fungus proofing of modules, temporary protection of finished machined parts.
Brush	Brushing action provides good "wetting" of surface, resulting in good adhesion. Cost of equipment is lowest.	Poor thickness control; not for precise applications. High labor cost.	Coating of individual components, spot repairs, or maintenance.
Roller	High-speed continuous process; provides excellent control on thickness.	Large runs of flat sheets or coil stock required to justify equipment cost and setup time. Equipment cost is high.	Metal decorating of sheet to be used to fabricate cans, boxes.
Impregnation	Results in complete coverage of intricate and closely spaced parts. Seals fine leaks or pores.	Requires vacuum or pressure cycling or both. Special equipment usually required.	Coils, transformers, field and armature windings, metal castings, and sealing of porous structures.
Fluidized bed	Thick coatings can be applied in one dip. Uniform coating thickness on exposed surfaces. Dry materials are used, saving cost of solvents.	Requires preheating of part to above fusion temperature of coating. This temperature may be too high for some parts.	Motor stators; heavy-duty electrical insulation on castings, metal substrates for circuit boards, heat sinks.

Screen-on	Deposits coating in selected areas through a mask. Provides good pattern deposition and controlled thickness.	Requires flat or smoothly curved surface. Preparation of screens is time-consuming.	Circuit boards, artwork, labels, masking against etching solution, spot insulation between circuitry layers or under heat sinks or components.
Electrocoating	Provides good control of thickness and uniformity. Parts wet from cleaning need not be dried before coating.	Limited number of coating types can be used; compounds must be specially formulated ionic polymers. Often porous, sometimes nonadherent.	Primers for frames and bodies, complex castings such as open work, motor end bells.
Vacuum deposition	Ultrathin, pinhole-free films possible. Selective deposition can be made through masks.	Thermal instability of most plastics; decomposition occurs on products. Vacuum control needed.	Experimental at present. Potential use is in microelectronics, capacitor dielectrics.
Electrostatic spray	Highly efficient coverage and use of paint on complex parts. Successfully automated.	High equipment cost. Requires specially formulated coatings.	Heat dissipators, electronic enclosures, open-work grills and complex parts.

TABLE 9.22   Theoretical Transfer Efficiencies TE
of Coating Application Methods

Coating method	TE, %
Autodeposition	95–100
Centrifugal coating	95–100
Curtain coating	95–100
Electrocoating	95–100
Fluidized-bed powder	95–100
Electrostatic fluidized-bed powder	95–100
Electrostatic-spray powder	95–100
Flow coating	95–100
Roller coating	95–100
Dip coating	95–100
Rotating electrostatic disks and bells	80–90
Airless electrostatic spray	70–80
Air-assisted airless electrostatic spray	70–80
Air electrostatic spray	60–70
Airless spray	50–60
High-volume, low-pressure spray	40–60
Air-assisted airless spray	40–60
Multicomponent spray	30–70
Air-atomized spray	30–40

this method coats all surface areas, solvent reflux can cause low film build. Light products can float off the hanger and hooks and fall into the dip tank. Solvent-containing coatings in dip tanks and drain tunnels must be protected by fire extinguishers and safety dump tanks. The fire hazard can be eliminated by using waterborne coatings.

9.6.2 Electrocoating

Electrocoating (95 to 100% TE) is a sophisticated dipping method that was commercialized in the 1960s to solve severe corrosion problems in the automotive industry. In principle, it is similar to electroplating, except that organic coatings rather than metals are deposited on products from an electrolytic bath. Electrocoating can be either anodic (deposition of coatings on the anode from an alkaline bath) or cathodic (deposition of coatings on the cathode from an acidic bath). The bath is aqueous and contains very little VOCs. The phenomenon called *throwing power* causes inaccessible areas to be coated with uniform film thicknesses. Electrocoating has gained a significant share of the primer and one-coat enamel coatings market.

Advantages of the electrocoating method include environmental acceptability owing to decreased solvent emissions and increased corrosion protection to inaccessible areas. It is less labor intensive than other methods, and it produces uniform film thickness from top to bottom and inside and outside on products with a complex shape. Disadvantages are high capital equipment costs, high material costs, and more thorough pretreatment. Higher operator skills are required.

9.6.3 Spray Coating

Spray coating (30 to 90% TE), which was introduced to the automobile industry in the 1920s, revolutionized industrial painting. The results of this development were increased production and improved appearance. Electrostatics, which were added in the 1940s, improved transfer efficiency and reduced material consumption. Eight types of spray-painting equipment are discussed in this section. The transfer efficiencies listed are theoretical. The actual transfer efficiency depends on many variables, including the size and configuration of the product and the airflow in the spray booth.

9.6.3.1 Rotating electrostatic disks and bell spray coating. Rotating electrostatic spray coaters (80 to 90% TE) rely on centrifugal force to atomize droplets of liquid as they leave the highly machined, knife-edged rim of an electrically charged rotating applicator. The new higher-rotational-speed applicators will atomize high-viscosity, high-solids coatings (65% volume solids and higher). Disk-shaped applicators are almost always used in the automatic mode, with vertical reciprocators, inside a loop in the conveyor line. Bell-shaped applicators are used in automated systems in the same configurations as spray guns and can also be used manually.

An advantage of rotating disk and bell spray coating is its ability to atomize high-viscosity coating materials. A disadvantage is maintenance of the equipment.

9.6.3.2 High-volume, low-pressure spray coating. High-volume, low-pressure (HVLP) spray coaters (40 to 60% TE) are a development of the early 1960s that has been upgraded. Turbines rather than pumps are now used to supply high volumes of low-pressure heated air to the spray guns. Newer versions use ordinary compressed air. The air is heated to reduce the tendency to condense atmospheric moisture and to stabilize solvent evaporation. Low atomizing pressure results in lower droplet velocity, reduced bounceback, and reduced overspray.

The main advantage of HVLP spray coating is the reduction of overspray and bounce back and the elimination of the vapor cloud usually associated with spray painting. A disadvantage is the poor appearance of the cured film.

9.6.3.3 Airless electrostatic spray coating. The airless electrostatic spray coating method (70 to 80% TE) uses airless spray guns with the addition of a dc power source that electrostatically charges the coating droplets. Its advantage over airless spray is the increase in transfer efficiency due to the electrostatic attraction of charged droplets to the product.

9.6.3.4 Air-assisted airless electrostatic spray coating. The air-assisted airless electrostatic spray coating method (70 to 80% TE) is a hybrid of technologies. The addition of atomizing air to the airless spray gun allows the use of high-viscosity, high-solids coatings. Although the theoretical transfer efficiency is in a high range, it is lower than that of airless electrostatic spray coating because of the higher droplet velocity.

The advantage of using the air-assisted airless electrostatic spray method is its ability to handle high-viscosity materials. An additional advantage is better spray pattern control.

9.6.3.5 Air electrostatic spray coating. The air electrostatic spray coating method (60 to 70% TE) uses conventional equipment with the addition of electrostatic charging capability. The atomizing air permits the use of most high-solids coatings.

Air electrostatic spray equipment has the advantage of being able to handle high-solids materials. This is overshadowed by the fact that it has the lowest transfer efficiency of the electrostatic spray coating methods.

9.6.3.6 Airless spray coating. When it was introduced, airless spray coating (50 to 60% TE) was an important paint-saving development. The coating material is forced by hydraulic pressure through a small orifice in the spray gun nozzle. As the liquid leaves the orifice, it expands and atomizes. The droplets have low velocities, because they are not propelled by air pressure as in conventional spray guns. To reduce the coating's viscosity without adding solvents, in-line heaters were added.

Advantages of airless spray coating are reduced solvent use, less overspray, less bounceback, and compensation for seasonal ambient air temperature and humidity changes. A disadvantage is its slower coating rate.

9.6.3.7 Multicomponent spray coating. Multicomponent spray coating equipment (30 to 70% TE) is used for applying fast-curing coating system components simultaneously. Since they can be either hydraulic or air-atomizing, their transfer efficiencies vary from low to medium. They have two or more sets of supply and metering pumps to transport components to a common spray head.

Their main advantage, the ability to apply fast-curing multicomponent coatings, can be overshadowed by disadvantages in equipment cleanup, maintenance, and low transfer efficiency.

9.6.3.8 Air-atomized spray coating. Air-atomized spray coating equipment (30 to 40% TE) has been used to apply coatings and finishes to products since the 1920s. A stream of compressed air mixes with a stream of liquid coating material, causing it to atomize or break up into small droplets. The liquid and air streams are adjustable, as is the spray pattern, to meet the finishing requirements of most products. This equipment is still being used.

The advantage of the air-atomized spray gun is that a skilled operator can adjust fluid flow, air pressure, and the coating's viscosity to apply a high-quality finish on most products. The disadvantages are its low transfer efficiency and ability to spray only low-viscosity coatings, which emit great quantities of VOCs to the atmosphere.

9.6.4 Powder Coating

Powder coating (95 to 100% TE), developed in the 1950s, is a method for applying finely divided, dry, solid resinous coatings by dipping products in a fluidized bed or by spraying them electrostatically. The fluidized bed is essentially a modified dip tank. When charged powder particles are applied during the electrostatic spraying method, they adhere to grounded parts until fused and cured. In all cases, the powder coating must be heated to its melt temperature, where a phase change occurs, causing it to adhere to the product and fuse to form a continuous coating film. Elaborate reclaiming systems to collect and reuse oversprayed material in electrostatic spray powder systems boost transfer efficiency. Since the enactment of the air quality standards, this method has grown markedly.

9.6.4.1 Fluidized bed powder coating. Fluidized bed powder coating (95 to 100% TE) is simply a dipping process that uses dry, finely divided plastic materials. A fluidized bed is a tank with a porous bottom plate, as illustrated in Fig. 9.7. The plenum below the porous plate supplies low-pressure air uniformly across the plate. The rising air surrounds and suspends the finely divided plastic powder particles, so the powder-air mixture resembles a boiling liquid. Products that are

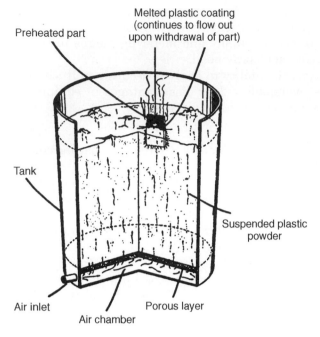

Figure 9.7 Illustration of fluidized-bed process principle.

preheated above the melt temperature of the material are dipped in the fluidized bed, where the powder melts and fuses into a continuous coating. Thermosetting powders often require additional heat to cure the film on the product. The high transfer efficiency results from little dragout and consequently no dripping. This method is used to apply heavy coats, 3 to 10 mil, in one dip, uniformly, to complex-shaped products. The film thickness is dependent on the powder chemistry, preheat temperature, and dwell time. It is possible to build a film thickness of 100 mils using higher preheat temperatures and multiple dips. An illustration of film build is presented in Fig. 9.8.

Advantages of fluidized bed powder coating are uniform and reproducible film thicknesses on all complex-shaped product surfaces. Another advantage is a heavy coating in one dip. A disadvantage of this method is the 3-mil minimum thickness required to form a continuous film.

9.6.4.2 Electrostatic fluidized bed powder coating. An electrostatic fluidized bed (95 to 100% TE) is essentially a fluidized bed with a high-voltage dc grid installed above the porous plate to charge the finely divided particles. Once charged, the particles are repelled by the grid

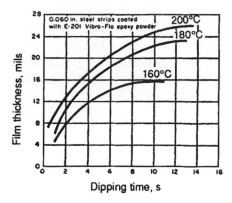

Figure 9.8 Effect of preheat temperature and dipping time on film build in coating a steel bar with epoxy resin.

and they repel each other, forming a cloud of powder above the grid. These electrostatically charged particles are attracted to and coat products that are at ground potential. Film thicknesses of 1.5 to 5 mil are possible on cold parts, and 20 to 25 mil are possible on heated parts.

The advantage of the electrostatic fluidized bed is that small products, such as electrical components, can be coated uniformly and quickly. The disadvantages are that the product size is limited and inside corners have low film thicknesses owing to the well known Faraday cage effect.

9.6.4.3 Electrostatic spray powder coating. Electrostatic spray powder coating (95 to 100% TE) is a method for applying finely divided, electrostatically charged plastic particles to products that are at ground potential. A powder-air mixture from a small fluidized bed in the powder reservoir is supplied by a hose to the spray gun, which has a charged electrode in the nozzle fed by a high-voltage dc power pack. In some cases, the powder is electrostatically charged by friction. The spray guns can be manual or automatic, fixed or reciprocating, and mounted on one or both sides of a conveyorized spray booth. Electrostatic spray powder coating operations use collectors to reclaim overspray. Film thicknesses of 1.5 to 5 mil can be applied on cold products. If the products are heated slightly, 20- to 25-mil-thick coatings can be applied on these products. As with other coating methods, electrostatic spray powder coating has limitations. Despite these limitations, powder coatings are replacing liquid coatings in a growing number of cases. A variation of the electrostatic spray powder coater is the electrostatic disk.

The advantage of this method is that coatings, using many of the resin types, can be applied in low (1.5 to 3 mil) film thicknesses with no VOC emission at extremely high transfer efficiency. Disadvantages include the difficulty in obtaining less than a 1-mil-thick continuous coating and, owing to the complex powder reclaiming systems, color changes are more difficult than with liquid spray systems.

9.6.5   Other Application Methods

9.6.5.1   Autodeposition coating.   Autodeposition (95 to 100% TE) is a dipping method whereby coatings are applied on the product from an aqueous solution. Unlike electrocoating, there is no electric current applied. Instead, the driving force is chemical, because the coating reacts with the metallic substrate.

Advantages of autophoretic coating are no VOC emissions, no metal pretreatment other than cleaning, and uniform coating thickness. This technique requires 30% less floor space than electrocoating, and capital equipment costs are 25% lower than for electrocoating. Disadvantages of autophoretic coatings are that they are only available in dark colors, and corrosion resistance is lower than for electrocoated products.

9.6.5.2   Centrifugal coating.   A centrifugal coater (95 to 100% TE) is a self-contained unit. It consists of an inner basket, a dip coating tank, and exterior housing. Products are placed in the inner basket, which is dipped into the coating tank. The basket is withdrawn and spun at a high enough speed to remove the excess coating material by centrifugal force. This causes the coating to be flung onto the inside of the exterior housing, from which it drains back into the dip coating tank.

The advantage of centrifugal coating is that large numbers of small parts can be coated at the same time. The disadvantage is that the appearance of the finish is a problem, because the parts touch each other.

9.6.5.3   Flow coating.   In a flow coater (95 to 100% TE), the coating material is pumped through hoses and nozzles onto the surfaces of the product, from which the excess drains into a reservoir to be recycled. Flow coaters can be either automatic or manual. Film thickness is controlled by the viscosity and solvent balance of the coating material. A continuous coater is an advanced flow coater using airless spray nozzles mounted on a rotating arm in an enclosure.

Advantages of flow coating are high transfer efficiency and low volume of paint in the system. Products will not float off hangers, and extremely large products can be painted. As with dip coating, the disadvantages of flow coating are coating thickness control and solvent refluxing.

9.6.5.4 Curtain coating. Curtain coating (95 to 100% TE), which is similar to flow coating, is used to coat flat products on conveyorized lines. The coating falls from a slotted pipe or flows over a weir in a steady stream or curtain while the product is conveyed through it. Excess material is collected and recycled through the system. Film thickness is controlled by coating composition, flow rates, and line speed.

The advantage of curtain coating is uniform coating thickness on flat products with high transfer efficiency. The disadvantage is the inability to uniformly coat three-dimensional objects.

9.6.5.5 Roller coating. Roller coating (95 to 100% TE), which is used mainly by the coil coating industry for prefinishing metal coils that will later be formed into products, has seen steady growth. It is also used for finishing flat sheets of material. There are two types of roller coaters, *direct* and *reverse,* depending on the direction of the applicator roller relative to the direction of the substrate movement. Roller coating can apply multiple coats to the front and back of coil stock with great uniformity.

The advantages of roller coating are consistent film thickness and elimination of painting operations at a fabricating plant. The disadvantages are limited metal thickness, limited bend radius, and corrosion of unpainted cut edges.

9.7 Curing

No dissertation on organic coatings and finishes is complete without mentioning film formation and cure. It is not the intent of this chapter to fully discuss the mechanisms, which are more important to researchers and formulators than to end users, but rather to show that differences exist and to aid the reader in making selections. Most of the organic coating resins are liquid, which cure or dry to form solid films. They are classified as *thermoplastic* or *thermosetting*. Thermoplastic resins dry by solvent evaporation and will soften when heated and harden when cooled. Thermosetting resins will not soften when heated after they are cured. Another classification of coatings is by their various film-forming mechanisms, such as solvent evaporation, coalescing, phase change, and conversion. Coatings are also classified as room temperature curing, sometimes called *air drying;* or heat curing, generally called *baking* or *force drying,* which uses elevated temperatures to accelerate air drying. Thermoplastic and thermosetting coatings can be both air drying and baking.

9.7.1   Air Drying

Air drying coatings will form films and cure at room or ambient temperatures (20°C) by the mechanisms described in this section.

9.7.1.1   Solvent evaporation.

Thermoplastic coating resins that form films by solvent evaporation are shellac and lacquers such as nitrocellulose, acrylic, styrene-butadiene, and cellulose acetate butyrate.

9.7.1.2   Conversion.

In these coatings, films formed as solvents evaporate, and they cure by oxidation, catalyzation, or cross-linking. Thermosetting coatings cross-link to form films at room temperature by oxidation or catalyzation. Oxidative curing of drying oils and oil-modified resins can be accelerated by using catalysts. Monomeric materials can form films and cure by cross-linking with polymers in the presence of catalysts, as in the case of styrene monomers and polyester resins. Epoxy resins will cross-link with polyamide resins to form films and cure. In the moisture-curing polyurethane resin coating systems, airborne moisture starts a reaction in the vehicle, resulting in film formation and cure.

9.7.1.3   Coalescing.

Emulsion or latex coatings, such as styrene-butadiene, acrylic ester, and vinyl acetate acrylic, form films by coalescing and dry by solvent evaporation.

9.7.2   Baking

Baking coatings will form films at room temperature, but they require elevated temperatures (150 to 200°C) to cure. Most coatings are baked in gas-fired ovens, although oil-fired ovens are also used. Steam-heated and electric ovens are used on a limited basis. Both electric and gas-fired infrared elements are used as heat sources in paint bake ovens.

9.7.2.1   Conversion.

The cure of many oxidative thermosetting coatings is accelerated by heating. In other resins systems, such as thermosetting acrylics and alkyd melamines, the reactions do not occur below temperature thresholds of 135°C or higher. Baking coatings (those that require heat to cure) are generally tougher than air-drying coatings. In some cases, the cured films are so hard and brittle that they must be modified with other resins.

9.7.2.2 Phase change. Thermoplastic coatings that form films by phase changes, generally from solid to liquid then back to solid, are polyolefins, waxes, and polyamides. Plastisols and organisols undergo phase changes from liquid to solid during film formation. Fluidized bed powder coatings, both thermoplastic and thermosetting, also undergo phase changes from solid to liquid to solid during film formation and cure.

9.7.3 Radiation Curing

Films are formed and cured by bombardment with ultraviolet and electron beam radiation with little increase in surface temperature. Infrared radiation, on the other hand, increases the surface temperature of films and is therefore a baking process. The most notable radiation curing is UV curing. This process requires the use of specially formulated coatings. They incorporate photoinitiators and photosensitizers which respond to specific wavelengths of the spectrum to cause a conversion reaction. Curing is practically instantaneous with little or no surface heating. It is, therefore, useful in coating temperature-sensitive substrates. Since the coatings are 100% solids, there are no VOCs. Although most UV coating are clears (unpigmented), paints can also be cured. Figure 9.9 shows radiation curing equipment.

9.7.4 Force Drying

In many cases, the cure rate of thermoplastic and thermosetting coatings can be accelerated by exposure to elevated temperatures that are below those considered to be baking temperatures.

9.7.5 Vapor Curing

Vapor curing is essentially a catalyzation or cross-linking conversion method for two-component coatings. The product is coated with one component of the coating in a conventional manner. It is then placed in an enclosure filled with the other component, the curing agent, in vapor form. It is in this enclosure that the reaction occurs.

9.7.6 Reflowing

Although not actually a curing process, certain thermoplastic coating films will soften and flow to become smooth and glossy at elevated

Figure 9.9 Radiation curing is fast, allowing production-line speeds of 2000 ft/min. This technique takes place at room temperature, and heat-sensitive wooden and plastic products and electronic components and assemblies can be given the equivalent of a baked finish at speeds never before possible. *(Courtesy of Radiation Dynamics.)*

temperatures. This technique is used on acrylic lacquers by the automotive industry to eliminate buffing.

9.8 Summary

The purpose of this chapter is to aid product designers in selecting surface preparation methods, coating materials, and application methods. It also acquaints them with curing methods and helps them to comply with environmental regulations.

Coating selection is not easy, owing to the formulating versatility of modern coating materials. This versatility also contributes to one of their faults, which is the possible decline in one performance property when another is enhanced. Because of this, the choice of a coating must be based on specific performance properties and not on generalizations. This choice is further complicated by the need to comply with governmental regulations. Obviously, the choice of compliance coatings, electrocoating, waterborne, high solids, powder, radiation curable, and vapor cure is well advised.

To apply coatings in the most effective manner, the product's size, shape, ultimate appearance, and end use must be addressed. This chapter emphasizes the importance of transfer efficiency in choosing a coating method.

To meet these requirements, product designers and finishers have all the tools at their disposal. They can choose coatings that apply easily, coat uniformly, cure rapidly and efficiently, and comply with governmental regulations at lower costs. By applying compliance coatings using methods that have high transfer efficiencies, they will not only comply with air and water quality standards, but they will also provide a safe workplace and decrease the generation of hazardous wastes.[4] Reducing hazardous wastes, using less material, and emitting fewer VOCs can also effect significant cost savings.

References

1. C. P. Izzo, "Today's Paint Finishes: Better than Ever," *Products Finishing,* Vol. 51, No. 1, October. 1986, pp. 48–52.
2. *Coatings Encyclopedic Dictionary,* Federation of Societies for Coatings Technology, 1995.
3. C. Izzo, "How Are Coatings Applied," in *Products Finishing Directory,* Gardner Publications, Cincinnati, OH, 1991.
4. C. Izzo, "Overview of Industrial Coating Materials," in *Products Finishing Directory,* Gardner Publications, Cincinnati, OH, 1991.

10

Metallic Finishes and Processes

Thomas A. Andersen
Northrop Grumman Corporation
Baltimore, Maryland

10.1 Fundamentals

10.1.1 Introduction

Commercial and military demands on the manufacturing industry continue to force product miniaturization, increased reliability, and productivity. The advent of products such as home computers, cell phones, and numerous other electronic devices have inspired us to refer to the "high-tech" era or "computer age." However, the selection of coatings that guarantee customer satisfaction at a reasonable manufacturing cost challenges the design engineer to discover improved resources and coatings that perhaps are worthy of proclaiming this a "materials age." New concerns confront the design engineer's decisions for coating types and materials, including environmental effects, shelf life, product and process certification, manufacturing facility location and workforce, and the anticipation of future requirements.

The need to change the outside surface of an object after it has been formed is a factor of the part's intended use. Seldom are we fortunate enough to cast, stamp, extrude, mold, or fabricate by other methods the exact object for the required purpose. For example, the rapidly stamped and etched copper conductors used in the early designs of commercial and military communication equipment were easily mass-produced. The unprotected copper hardware functioned normally until its conductivity was compromised by environmental and assembly factors such as corrosion and poor solderability. Protecting the copper proved essential. Similarly, using cheap, lightweight plastics as electronic housings failed to prevent interference from electromagnetic radiation. Addition of a metal coating solved the problem.

10.1.2   Design Approach

To determine the proper coating for an object one must carefully consider the following elements, generally in the following order:

- Intended use or specifications for the part
- Expected final cost (part quantity affects cost) for the part
- Composition or material
- Part condition or finish
- Part geometry
- Process needed for coating (including coating options)
- Part environmental and safety impact

10.1.3   Chassis Example

The design approach (D/A) based on an example part, such as a chassis, begins with the review of the part specifications for its intended use. Economics have a major influence on the choice of coating, coating method, and the planned volume of parts. Often, the cost-per-part value becomes the starting point, thus limiting the coating to a select few processes or sending the designer on an extensive search for a high-volume coating vendor who is willing to deal with limited part volumes. Once the coating selection process is initiated based on a known number of parts and cost, substrate geometry and surface finish may pose additional concerns about the part's successful fabrication. Both the substrate geometry and surface finish must complement the coating process.

Environmental and safety concerns constitute the final portion of the coating selection D/A. We must consider factors that range from personnel contact to final disposal in a landfill, in the atmosphere, in space, or at sea. In finalizing our selection of the coating process, we must also consider any possible rework of the particular part and changes in its intended use, specifications, substrate material, or geometry. When all of the D/A factors are considered simultaneously, the overall success level is increased in terms of part performance and customer satisfaction.

10.2   Metallic Finishes and Processes

During the late 1990s, emphasis on *customer satisfaction* became a dominant manufacturing industry trend. *Outsourcing* and *off-shoring* of repetitive operations to achieve lower costs increased company profits and shareholder satisfaction. This increased accountability to customers' needs resulted in an abundance of work for some companies, and an overnight loss of product from the shop floor for others. Coating types and process selection served as cost cornerstones for customer satisfaction and often determined a company's success within a highly competitive business environment. Time considerations also magnified the importance of good coating decision-making, since the coating operation was usually the final operation prior to shipment, and speedy completion of this operation was generally required. Preserving orders and workplace production required enhanced knowledge of alternative coating methods, correct substrate materials, surface preparation, knowledge of the coatings' limitations, and an understanding of what the competition was offering. The 1990s emphasized statistical process control (SPC), ISO standards, and yield improvement as consequential factors for the success and prestige of a business. Manufacturing during the twenty-first century will continue to demand compliance with the highest standards, including "Six Sigma" manufacturing industry excellence, certification to applicable international standards, reliable computer traceability for all parts, and robust manufacturing capabilities that target customer satisfaction (which may be better defined as *COSTomer satisfaction*). Economics and competition continue to reward companies that measure up, and close businesses that do not. To remain in business, the "old reliable," trusted coating operations must meet the growing competitive challenges for lower-cost finishing and coating.

There are numerous reasons to require a metallic finish for commercial or military hardware. The properties of the substrate are changed for a desired end use, because seldom does the base material's surface meet the customers' demands. Compare, for example, cast bronze decorative fittings that undergo cleaning to remove core sand and molding and investment material from exterior and interior surfaces. Compare this to steel plate that is punched, deburred, cleaned, and bronze electroplated. When bubble packaged and displayed on store shelves, both fittings the same customer-desired appearance but at different prices. The solid bronze fittings normally carry a premium price, and bronze-plated steel fittings are less expensive. Both products meet service weight limits and offer bright appearances and easy installation. But the effects of exposure to indoor and outdoor environmental conditions test both fittings. Customers trust the fitting manufacturer as a reliable supplier. They base their selection on the correct size and bright appearance. Rarely do they have any knowledge of the life expectancy values or technical specifications of the fittings. A buyer may comprehend that the higher price for solid bronze reflects a longer life expectancy, but the choice will depend on various considerations in the purchaser's mind, such as

- "Why pay the higher price if the fittings will be installed indoors?"
- "Maybe the plated fittings will be good enough, since we will be moving next month."
- "The fittings are only for show, and who can tell the difference?"
- "Maybe the solid fittings will be less likely to corrode in my ship's cabin.

Customer demands can be difficult to satisfy. "Lifetime" and "money-back" guarantees can help to sell products to a hesitant customer, but if the parts fail within a few months, the customer will still be dissatisfied. Such catastrophic failures should be avoided by careful design engineering, as today's customer expects a reliable product at a competitive price. It is therefore hoped that this chapter will be useful to those who seek either a general or technical understanding of the types of available metallic finishes, common specifications, and their advantages and disadvantages.

10.2.1   Metallic Coating Characteristics

Part designers attempt to minimize the manufacturing steps to produce the ideal part in a manner that holds down manufacturing costs and allows an affordable price. The intended function of the part is the main consideration in the design effort, and so the part should not be designed for any particular coating process. The part designer must consider all of the requirements of the part, such as cosmetic appeal, thickness, hardness, and wear resistance. Ideally, the fabricated substrate or part should be suitable for its intended use, which requires attention to the following, and related, coating characteristics:

1. Corrosion protection
2. Wear protection
3. Adhesion
4. Appearance (smoothness and brightness)
5. Thermal conductivity
6. Electrical and temperature resistance
7. Lubricity
8. Solderability
9. Weldability
10. Bondability
11. Contact and abrasion resistance

12. Thickness uniformity

13. Reflectivity

14. Absorbency

15. Radiation shielding

16. Diffusion barrier

17. Applicability as a base for other finishes

18. Hazardous qualities

19. Toxicity

20. Hardness, ductility, and tensile strength

21. Cost

These represent some of the most obvious surface properties that must be considered when initiating a design approach selection process for coating various types of substrate materials. The purpose of this chapter is to aid the design approach for coating selection by relating the coating requirements for a particular substrate to different metallic finishing processes that meet the selected specification(s).

10.3 Aluminum Coatings

Aluminum finishes continue to gain acceptance as deposits that offer corrosion resistance, wear resistance, and high purity, and they are capable of occluding particles such as aluminum oxide and can be alloyed with zinc for special applications. The benefits of aluminum coatings include alternatives to less environmentally desirable finishes such as cadmium, lead, lead-tin, and chromium. The popularity of aluminum finishes is increasing in aerospace hardware such as enclosures and fasteners, for the protection of steel, and for electroforming and molten salt dip brazing. Additional commercial applications include automobile trim, home appliances such as washers and dryers, and agricultural storage bins, livestock shelters, and fences. Aluminum coatings are versatile enough to be used in sacrificial (cathodic) coatings and for decorative, engineering, and protective finishes. Table 10.1 lists common aluminum coating methods and typical process cost indications for a popular group of substrates.

10.3.1 Aluminum Clad Coatings

Aluminum clad coatings provide nearly the exact aluminum outside coating composition desired for the substrate, including desired coating characteristics such as uniformity of thickness, thermal conductiv-

TABLE 10.1   Aluminum Coating Methods and Costs for Various Substrates

Method	Steel	Copper	Brass	Aluminum	Nickel	Silver	Titanium	Ceramic	Plastic
				Substrate					
Cladding	M	M	M	M	L	L	M	H	M
Diffusion	M	H	H	—	—	H	—	—	—
Electrodeposition	H	H	H	H	H	H	H	—	—
Electrophoretic deposition	M	M	M	—	M	—	H	—	H
Hot dipping	M	—	—	—	—	—	—	—	—
Sputtering PVD	H	—	—	M	—	—	M	—	H
Evaporative PVD	H	—	—	M	—	—	M	—	H
Ion plating PVD	H	—	—	M	—	—	M	—	H
Spraying	L	L	L	—	L	L	H	—	—

Est. cost ratings: H = high, L = low, M = modest, — = unknown or not applicable.

ity, abrasion resistance, corrosion protection, electrical conductivity, and weldability. The aluminum clad material is normally formed to an exact thickness by rolling, extrusion, or pressing methods and then secured to substrate materials using similar methods that may also include hot stamping, adhesives, and detonation bonding.

Aluminum clad coatings on substrates such as steel, copper base alloys, and high-strength aluminum alloys offer good corrosion protection in the atmosphere or neutral aqueous solutions, and they have good electrical and thermal conductivity.[1] Typically, aluminum sheets greater than 10 μm thick are bonded to one or both sides of another alloy. The aluminum clad often serves as a sacrificial coating similar to zinc coatings on carefully selected steel. Intermetallic growth is often avoided by the inclusion of a minimum of 1.2% silicon within the aluminum. Alternative methods of intermetallic avoidance include the introduction of a silver-plated coating onto the steel, followed by roll bonding the aluminum clad at approximately 230°C.[2] Aluminum clad stainless steel serves well for certain types of cookware, household appliances, electronic housings, and heat exchangers. The demands of aerospace applications for light weight, high strength, and corrosion resistance rely on aluminum-to-aluminum clad sandwiching for single-sided clad brazing sheets and double-sided clad aluminum structures. Single-sided clads normally have a hot- or cold-rolled corrosion-resistant aluminum alloy bonded to a tempered, stronger aluminum core. Double-sided cladding permits an additional selected clad bonded to the opposite side. A typical single-sided aluminum clad would include a series 6951 core with a 4045 (10% silicon), or 7075 (7% silicon) cladding. Double-sided clad aluminum might offer a 7075 core and 7072 series cladding, used to protect heat exchangers from corrosion by the internal aqueous coolant and exterior atmospheric exposure (see Tables 10.2 and 10.3).

TABLE 10.2   Example of Typical Single-Sided Aluminum Cladding

Core sheet alloy		Cladding alloy	
6951	Chemical composition, %	4045 (10% Si)	Chemical composition, %
Silicon	0.2–0.5	Silicon	9.0–11.0
Iron	0.8	Iron	0.8
Copper	0.15–0.4	Copper	3.3–4.7
Manganese	0.1	Manganese	0.15
Magnesium	1.0–1.5	Magnesium	0.15
Chromium	—	Chromium	0.15
Zinc	0.2	Zinc	0.2
Other elements	0.05	Other elements	0.05
Aluminum	remainder	Aluminum	remainder

Brazing temperature: 582–604°C (1080–1120°F)

Aluminum cladding of organic materials represents a large segment of manufactured products for commercial and military needs (e.g., laminates, composites, and electronic structures), all of which require the cladding process technology but are beyond the scope of this discussion.

Prerequisites for successful bonding include proper surface preparation, cleaning, and drying. Aqueous abrasive scrub cleaning and mechanical abrasion by blast, or brush, operations are the most popular surface conditioning techniques. Dust and hazardous-chemical-free cleaning operations should be considered for operator safety and environmental impact. Acid and alkaline aluminum etching should be avoided to minimize chances of cladding delamination by corrosive liquids wicking between the core and clad alloys.

10.3.1.1   Advantages of aluminum clad coatings.   The advantages of aluminum clad coatings are as follows:

TABLE 10.3   Typical Double-Sided Aluminum Alloy, Solution Heat Treated and Artificially Aged Aluminum Alloy (7075) Sheet with Cladding Layers of Aluminum Alloy (7072) on Both Surfaces; Intended for Exfoliation Resistance

Core sheet alloy		Cladding alloy	
7075	Chemical composition, %	7072	Chemical composition, %
Zinc	5.1–6.1	Zinc	0.8–1.3
Magnesium	2.1–2.9	Magnesium	0.1 max.
Copper	1.2–2.0	Copper	0.1 max.
Chromium	0.18–0.28	Chromium	—
Manganese, max.	0.30	Manganese, max.	0.10
Iron, max.	0.50	Iron + silicon, max.	0.70
Silicon, max.	0.40		
Titanium, max.	0.20	Titanium	—
Others, each, max.	0.05	Others, each, max.	0.05
Others, total, max.	0.15	Others, total, max.	0.15
Aluminum	remainder	Aluminum	remainder

Brazing temperature: 582–604°C (1080–1120°F)

- Uniformity of thickness, brightness, and exact alloy composition
- Thickness ranging from 25 μm to >2,500 μm
- Relatively inexpensive for large-volume requirements
- Environmentally agreeable process and finish
- Excellent adhesion
- Good ductility
- Does not cause hydrogen embrittlement
- Serves for excellent corrosion protection
- Additional aluminum surface treatments may be added, e.g., anodizing and conversion coating

10.3.1.2   Disadvantages of aluminum clad coatings.   The   disadvantages of aluminum clad coatings are as follows:

- Complex shapes may be too difficult to clad.
- Substantial deformations occur when the surface geometry changes.
- Achieving the desired thickness often requires precise control and is difficult to attain.
- Considerable capital equipment expenses are involved.
- Low volume runs are expensive.
- They offer poor marine corrosion protection.

10.3.2 Aluminum Diffusion

Diffusion methods force the aluminum atoms into the crystalline lattice of the substrate. This penetration of aluminum requires a method that may employ heat, pressure, time, chemical catalysts, and special atmospheres, resulting in an enormous strength of adhesion between the aluminum coating and substrate. Diffusion is an irreversible and spontaneous reaction discussed in a comprehensive manner by Owen,[3] Pinnel,[4] and Lewis.[5]

Special parts demand maximum adhesion, and diffused aluminum coatings can meet severe temperature cycling, corrosion, and oxidation conditions without distortion in, for example, automotive engine parts, venturis, aluminum and stainless steel tubing, honeycomb structures, aircraft fasteners, and exhaust systems.

Aluminizing of steels and cast irons (often termed *calorizing*) serves as a popular example aluminum diffusion coating steel by mixing together certain powders in an enclosed vessel at an elevated temperature. Powdered aluminum, alumina, and ammonium chloride tumble with the steel parts at 850–950°C for up to 6 hr, or up to 24 hr without tumbling, followed by a post-bake at near the coating process temperature.[2] Here, small particles of aluminum join together and migrate into the steel to reduce all pores to nonexistence provided time and temperature are held for the proper duration. The mechanism for the powder-to-metal bonding approaches a surface sintering process[6] that produces a porous and brittle deposit containing 50–60% aluminum at 25–125 μm thick (see Fig. 10.1).

10.3.2.1 Advantages of aluminum diffusion coatings. The advantages of aluminum diffusion coatings are as follows:

- Excellent adhesion

- Good corrosion and high-temperature oxidation resistance

- Additions of chromium, boron and zirconium, and molybdenum offer corrosion protection for nickel and cobalt-based high-temperature alloys *(superalloys)*

- Low risk of hydrogen embrittlement

- Uniform thickness in the 25–100 μm range

- Economical at high part volume

10.3.2.2 Disadvantages of aluminum diffusion coatings. The disadvantages of aluminum diffusion coatings are as follows:

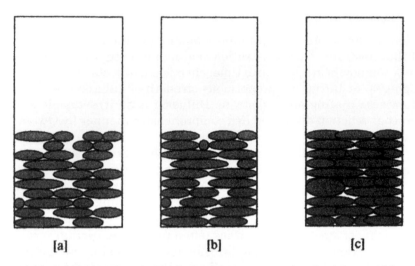

[a]   [b]   [c]

Figure 10.1   During sintering, high temperatures cause the aluminum particles to join together and eventually reduce the voids between the grains. Here, aluminum ions begin to diffuse into grain boundaries where the particles touch to form connections between the individual grains. Pores between the grains become smaller, and the coating density increases. The steps are (a) initial points of contact area decrease following solvent evaporation, and bridging of ions connects grains, (b) grain boundaries diffuse, causing pores to decrease, and (c) near total elimination of pores occurs.

- Unsuitable for fragile parts
- Corrosion resistance limited by volume of sacrificial aluminum contained in the coating
- Expensive to control the various materials-processing techniques
- Possibility of substrate degradation by the high-temperature diffusion process

10.3.3   Electrodeposited Aluminum Coatings

Increased interest in aluminum-plated coatings has been driven by industry's need to find replacements for heavy metals such as cadmium, nickel, chromium, and zinc. Aluminum coatings have been used for coating special parts used in nuclear reactors, for manufacturing aluminum mandrels used in electroforming, and for electroformed aluminum structures such as waveguides and parabolic mirrors.[7] The electrodeposition of aluminum employs several unique and successful plating bath formulations that can avoid water because of the highly electropositive nature of aluminum. Only special plating facilities attempt this aluminum deposition process, because it requires critical bath control and safety guidelines. The nonaqueous solvent requires the use of organic solvents with soluble aluminum salts. Considerable planning before the decision to electroplate aluminum is required to measure the effects of the substrate material, geometry, and expected coating results. The commercial or military need for an electroplated aluminum finish must be very exceptional to justify the manufacturing safety risks, processes, and costs.

Before the processes are set up, or the parts are sent to a capable plating establishment, a careful evaluation of the coating characteristics described in Sec. 10.2.1 should establish the coating requirements. Harper and Pinkerton[8] and Durney[9] discuss typical substrate geometry and material considerations that should be examined before the type of plating process is selected. Safranek[7] and Lowenheim[10] discuss the electroplating processes and deposit characteristics of several aluminum plating and electroforming bath formulations. Four candidate aluminum plating baths include a previously developed bath consisting of aluminum halide salt dissolved in an aromatic solvent such as benzene or toluene.

A second bath, called the *hydride bath,* contains aluminum chloride, lithium aluminum hydride, and lithium aluminum chloride dissolved in tetrahydrofuran. This hydride, or REAL process *(Room-temperature Electrodeposition of ALuminum)* permits the plating of aluminum at room temperature.[11]

The third or *alkyl* bath uses an aluminum organic salt and sodium fluoride dissolved in toluene in what is called the Sigel process.[10] Bath operation requires soluble aluminum anodes, a dry nitrogen atmosphere, and an operating temperature of 100°C. The popularity of the these aluminum organic processes stems from their special corrosion protection for steel, with a deposit thickness between 5 and 25 μm and minimum hydrogen embrittlement. Typical deposits of the organic baths have comparable or higher purity and greater hardness than wrought aluminum. Additional literature describes the electrodeposition of aluminum[12–14] and research on electroforming waveguides, large parabolic mirrors, and aerospace components.[15]

The fourth type of bath consists of a fused salt mixture of aluminum chloride, lithium chloride, and sodium chloride with the option of adding manganese chloride[7] for higher corrosion resistance, hardness, and electrical resistivity. Fused salt baths operate between 150 and 175°C and have been used successfully for coating light-gage steel strip with coatings up to 25 μm in thickness. Other complex part shapes may prove challenging for fused salt aluminum deposition because of the limited anode-to-cathode, 2–3 cm spacing and poor bath throwing power.

Additional information on fused salt research appears in reports from Murphy,[16] Castle,[17] and Couch and Brenner.[18] Recent demands for aluminum coatings with improved cost and environmental benefits center on candidate areas of electroforming, replacements for heavy metal protective coatings, fastener and sheet steel corrosion protection, new alloy development from co-deposition of metals, co-deposition of fibers and alumina, aluminum composites for solar collectors, and numerous other commercial applications.

10.3.3.1 Advantages of electrodeposited and fused salt aluminum coatings. The advantages include the following:

- Deposits from the organic and fused salt processes provide coatings that are pure, smooth, fine-grained, nonporous and ductile.
- Coating metallurgical values can equal or exceed those of wrought aluminum with co-deposition of certain additives and by altering process operations such as the application of periodic reverse-current plating.
- Electrical resistivity values are slightly higher for electrodeposited aluminum than for wrought aluminum.
- Thermal expansion values are nearly the same as for wrought aluminum.
- Excellent corrosion resistance results when the coatings are applied to steel, especially with the addition of a chromate conversion coating or an anodizing post-treatment.[19]
- Satisfactory adhesion of the organic coating to base metals of aluminum, titanium, copper-base alloys, silver, nickel, chromium, and uranium has been reported.[20–22]

10.3.3.2 Disadvantages of electrodeposited and fused salt coatings. The disadvantages include the following:

- Both organic and fused salt processes require extensive financial commitments for the process equipment, safeguards, control of the high-purity chemicals, and normal operation.
- The purity of the deposit can change because of co-deposition of impurities during aluminum plating.
- Water serves as a contaminant for these processes and must be removed by chemical additions and the use of a dry, inert atmosphere and/or an enclosed system.
- Preplating substrate surface preparation, after normal degreasing, cleaning, and surface activation steps, is difficult and may require a proprietary rapid water removal/drying step followed by nonaqueous plating.

- Adhesion loss is difficult to monitor and may occur between the substrate and aluminum.

10.3.4 Electrophoretic Aluminum Coatings

Legend describes a discovery at the turn of the nineteenth century by a Russian physicist, Reuss, who witnessed the movement of suspended particles in water by electricity. This early observation of clay particles moving toward the positive electrode led to experimental depositions of metal oxides in colloidal suspensions that could be deposited onto cathodes.[23] Electrophoretic deposition differs from electroplating because particles of colloidal size (usually less than 1 μm) are deposited with electric current instead of ions. The polar charge of the particles depends on their molecular structure within an applied electric field. The more positively charged particles migrate to the cathode, and negatively charged particles toward the anode. Small particles form thin films at the electrodes, and these compact against each other to squeeze out the liquid media. Following the deposition, the coated part is withdrawn from the bath for further processing, such as compacting curing or baking and sintering to increase the coating-to-substrate bond and to improve the adhesion.

Aluminum particles, or powder, dispersed in an aqueous organic media have been deposited onto steel wire and sheet followed by pressure rolling and sintering (see Fig. 10.1) at high temperature. Aluminum electrophoretic deposited coatings exhibit uniformly smooth and even surfaces that offer good protection from atmosphere and exposure to high temperatures.[24] A process for applying aluminum by reel-to-reel processed steel strip has been developed by the British Steel Corporation (the Elphal process).[25] Automobile and food storage container manufacturers have used electrophoretic deposition of polar organic resins. Additional processes that deposit metal powders include aluminum alloys,[26] copper base alloys, barium-strontium oxide,[27] barium titanate,[28] chromium-nickel,[29] and ceramic frits.[30,31] Numerous electrophoretic references have been published and include studies from Raney,[32] Yeates,[33] and Brewer,[34,35] and much valuable information and updates on recent progress can be obtained from the Electrocoat Association.

10.3.4.1 Advantages of aluminum electrophoretic coatings. The advantages of these coatings are as follows:

- It provides excellent salt spray resistance that can exceed 300 hr for cold rolled steel.

- It produces good bonding to the substrate from post-bake or cure and/or sintering.

- Good corrosion protection results from uniform film thickness via the good *throwing power* that deposits the coating into cavities and corners, along edges, and in blind holes without sags, drips, or runs.

- The coating is uniform for parts with complex shapes.

- It offers a lower cost per part as compared to powder, dip, and spray coating methods.[36]

- It has the ability to deposit various metal powders for special surface requirements.

- It has the capability to deposit alloys that may be diffused for special surface requirements.

- The processes are the most economical, avoiding labor-intensive operations via automation.

- It is environmentally friendly, avoiding heavy and toxic metal processes.

- There are no fire hazards involved.

- Water-borne organic media contain little or no hazardous air pollutants (HAPs) or volatile organic compounds (VOCs).

- This is the process of choice for "bulk" coating fasteners and spring metal stampings for the automotive industry.

- It permits process flexibility for additional coatings, such as the addition of a sealant for increased corrosion resistance and reduced friction.

10.3.4.2 Disadvantages of aluminum electrophoretic coatings. The disadvantages of these coatings are as follows:

- Large capital outlay is required for equipment.

- Considerable control over the bath chemistry may be required because of controlling "binders" and "activators."

- The bath life may require frequent new bath replacement.

- Shrink cracking of the coating may occur if metal oxide is used nstead of a metallic powder.[37,38]

- Bulk processing of parts involves the same problems as "barrel" electrodeposition.

- Parts must be dried and baked separately to avoid sticking.

- Experimentation is required to meet certain coating thickness specifications.

- Rinse waters may require waste treatment costs.

10.3.5 Hot-Dipped Aluminum Coatings

Hot dipping substrate surfaces by exposure to molten aluminum is mostly limited to steel wire and fastener products, because aluminum's high melting point increases the formations of oxide films that tend to contaminate the coating. Commercial applications of hot-dipped aluminum include continuous coating of steel wire and steel strip. Products such as automobile exhaust systems and oven, furnace, and engine parts have found uses for this type of coating. The coating process requires temperatures above 700°C followed by rapid cooling. Special techniques, additions of silicon, halogenated fluxes, and particular substrate surface preparations improve the aluminum-to-steel alloy bond and minimize oxide and formations of aluminum-iron alloys. The expected coating thickness ranges from 25 to 100 µm. The brittle, hard aluminum-iron intermetallic alloy restricts the mechanical properties of the product and is often minimized by the addition of silicon to the molten bath.[39] Beryllium can be added to accomplish the similar intermetallic thickness and hardness decreases, but toxicity hazards outweigh the benefits of its use.

The primary uses for this type of coating include corrosion resistance to industrial and nonalkaline moisture atmospheres. The coatings are coarse and dull and are brightened by mechanical abrasion or surface altering methods. Exposure to outside conditions favors aluminum hot-dipped coatings over similar zinc applied coatings, with corrosion resistance decreasing with coating loss and less diffusion of aluminum into the steel.

10.3.5.1 Advantages of hot-dipped aluminum coatings. The advantages of these coatings include the following:

- These highly adherent aluminum coatings can be used as a replacement for zinc and cadmium.

- This is a good, low-cost method for protecting certain steels that are exposed to outside conditions where thermal cycling may be involved.

- Coating adhesion is improved by diffusion.

- There is a low risk of hydrogen embrittlement.

- Corrosion pits normally stop at the intermetallic alloy layer.
- It is economical for high-volume product coating needs.

10.3.5.2 Disadvantages of hot-dipped aluminum coatings. The disadvantages of these coatings include the following:

- Complex product shapes may be difficult to coat evenly.
- The process requires significant capital equipment investment.
- The process is best suited for automated high-volume coating operations.
- Corrosion resistance is limited by coating thickness.
- Substrate degradation may occur from the high coating process temperature.
- Initial coating appearance is rough.
- Coating thickness is difficult to control.

10.3.6 Vacuum Deposited Aluminum Coatings

Vacuum deposition represents several widely accepted methods for aluminum coating onto both metal and nonmetal substrates. Introduced by electronics and aerospace industries during the 1960s, demands for this form of metal deposition have created a dynamic technological interest in numerous processes that deposit aluminum to meet varied product requirements. The coating needs for thin film decorative, high-temperature superconductor, and corrosion-resistant finishes have advanced vacuum deposition methods to meet high-volume product demands using a more environmentally acceptable process. The introduction of thin film aluminum (several millionths of an inch in thickness) vapor deposited coatings, coated with a protective resin, satisfied the need for large volumes of plastic automotive parts,[40] semiconductor materials, toys, housewares, commercial hardware, jewelry, holiday decorations, compact discs, magnetic films, and trophies. High volumes of compression and injection molded plastic substrates have replaced electroplating processing via rapid racking, primer coating, vacuum sputtering aluminum, and final resin coating.

Significant research funding by industrial nations throughout the world led to the development of numerous vacuum deposition processes for aluminum deposits for the corrosion protection for steel strip[42] and aluminum coatings that visually duplicate bright chrome, gold, and copper surface appearances in multichamber operations.[41] To be respectful of the numerous emerging vacuum deposition processes and techniques for aluminum deposition, we must direct our attention to the difference between methods of chemical vapor deposition (CVD) and physical vapor deposition (PVD). CVD and PVD differ from each other in the way by which the element is transferred in a vaporized state to the intended substrate. CVD depends on reactions such as pyrolysis (thermal decomposition), hydrolysis, oxidation, and reduction of metal compounds to deposit metal coatings. Aluminum compounds such as alumina can be deposited by reacting gaseous aluminum chloride, carbon dioxide, and hydrogen at a given temperature, pressure, and plasma enhanced to yield alumina, carbon monoxide, and hydrogen chloride. Numerous CVD processes are employed for difficult-to-coat materials such as silicon from silicon tetrahydride, titanium from titanium tetrachloride, carbon from methane, and boron from borontrichloride. However, CVD processes for aluminum coatings have not gained wide acceptance, because of factors including significant costs for materials, equipment, and process control.

Many methods of aluminum deposition can be classified as PVD. Aluminum coatings applied by vacuum deposition processes, as described in this chapter, include sputter deposition, vacuum evaporation, ion plating, or ion vapor deposition (IVD).

Sputter deposition of aluminum coatings is performed in a vacuum chamber by vaporizing aluminum particles from the surface of an aluminum source, or *target,* created by a volley of positive gas ions, usually from a plasma or ion gun. The ions are normally generated from an inert gas such as argon, or they can be produced from an ion beam or localized plasma surrounding the target. An electrical field is normally configured to set the target as the cathode coupled with a direct current path to the substrate (conductive material). Groups and individual atoms of aluminum are dislodged from the target by the volleys of positive gas ions into the vapor phase, causing aluminum nucleation sites to form on the substrate surface. Operating pressures can be as high as 30 millitorr and as low as less than 5 millitorr. Figure 10.2 illustrates a typical planar electrode sputter coating process. The introduction of magnetrons has served to increase the coating deposition rate by generating strong magnetic fields that confine the plasma and increase the ionization, causing increased collisions on the target. The bombarding ions now strike the target with energies from 100 to 1000 electron volts (eV), causing the sputtered atoms of aluminum to strike the substrate surface at 10 to 50 eV.

Sputter aluminum deposition has been used to deposit thin aluminum films onto plastics, glass, and transparent electronic films, and for numerous decorative applications.

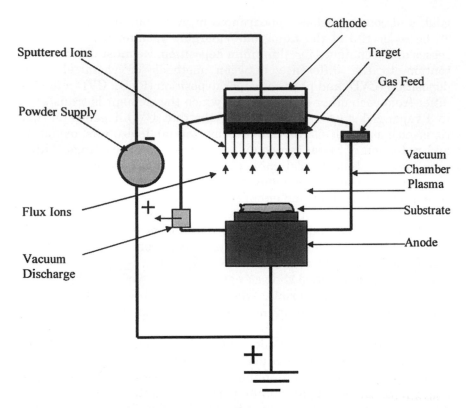

Figure 10.2 Simplified sputtering process showing vaporization of aluminum by bombardment of ions from a plasma. The ions dislodge aluminum particles, causing them to coat the substrate. Typical aluminum coatings and thin and deposit at rates of 0.5 to 5 μm/hr.

10.3.6.1 Advantages of thin film, sputtered aluminum. The advantages are as follows:

- Nonconductors can be metallized, include plastics, glass, ceramics, and all substrates that remain stable in vacuum conditions.
- Lightweight and inexpensive substrate materials can be used.
- Electroplating processes can be avoided.
- Polishing and buffing processes can be avoided.
- Once difficult-to-plate aluminum is easily deposited.
- Adhesion of aluminum to the substrate is good.
- Complex substrate geometries can be coated.
- Additional metals may be deposited.

- Use of magnetrons can increase deposition to 100 µm/hour.[2]

- Very little radiant heat is generated.

- Substrate cleaning may be performed by equipment modifications that reverse the current.

10.3.6.2 Disadvantages of thin film sputtered aluminum. The disadvantages are as follows:

- Sputtering deposition rates are low, typically 0.5 to 5 µm/hr.

- Abrasion and corrosion resistances have been problems because of the thin deposit.

- Cleaning of the substrate surface must be thorough for good aluminum deposit adhesion.

- Decomposition and recombination of the aluminum must be free of contaminants.

- Aluminum film thicknesses are low and depend on line-of-sight deposition.

- Sputtering targets are expensive, and different shapes may be required for complex parts.

- Heat from the metal deposition may change or degrade the substrate properties.

- Gaseous contaminant formations are possible.

- Equipment costs of $500,000 must be justified by high part volume.

- Highly trained personnel are required.

Vacuum evaporation of aluminum coatings has provided one of the most popular PVD processes for industrial applications.[42] Typical finishes include mirror coatings, decorative coatings, barrier coatings for electronics, and corrosion-protective coatings. Aluminum's low melting point makes it an ideal element for a high deposition rate of more than 2000 µin/min (2 mils/min). During vacuum evaporation, one or more thermally heated sources can be used, coupled with magnetron enhanced plasma and electron beam excitation. The aluminum source is normally heated by a tungsten wire. Uniform thickness is often challenged by substrate geometry and may require locating multiple sources in close proximity within the chamber to balance the line-of-sight deposition. Figure 10.3 illustrates evaporation system basics.

The coating properties and rate of aluminum deposition depend on the aluminum source temperature and the contact angles of the aluminum particles bombarding the substrate surface. Good bonding to

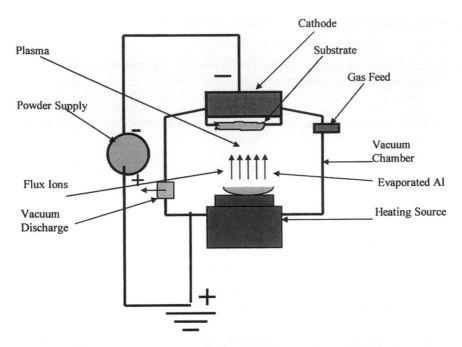

Figure 10.3 General illustration of a vacuum evaporating process showing the evaporation of aluminum by resistance heating of ions from a plasma. The ions dislodge aluminum particles, causing them to coat the substrate. Typical aluminum coatings have high deposition rates of greater than 200 μin/min (2 mils/min).

the substrate and coating growth result from the high-kinetic-energy particles, accelerated from the electric field, that collide on the substrate surface.[43]

The actual deposit structure largely depends on the ratio of the substrate temperature to the melting point of aluminum. As the aluminum deposition temperature ratio increases, small-grained deposit structures are replaced with larger equiaxed or columnar grains. The higher-temperature-ratio depositions are desired for aluminum physical properties that resemble pure aluminum, in contrast to lower-temperature-ratio deposits that contain stresses.

10.3.6.3 Advantages of vacuum evaporated aluminum coatings. The advantages are as follows:

- Vacuum evaporation equipment is expensive, but it has good versatility and is probably the least expensive of the PVD processes.[44]

- Control of the deposition process is not difficult.

- High-purity aluminum deposits are possible.

- Good corrosion protection can be provided for steel as an alternative to cadmium.

- A more environmentally acceptable process is involved.

- The aluminum deposition rate is high.

- Multiple sources permit a uniform coating thickness on complex parts.

10.3.6.4 Disadvantages of vacuum evaporated aluminum coatings. The disadvantages are as follows:

- Aluminum deposit properties depend on the direction of angle of the depositing metal.

- Source material may be expensive and require early replacement.

- Elaborate fixturing may be necessary to achieve uniform thickness.

- Part size is limited by chamber dimensions.

- The porosity of the aluminum coating can be reduced by increased deposit thickness.

- Highly trained personnel are required.

Ion vapor deposition (IVD) of aluminum offers alternative methods to increase the energy of the aluminum deposition, affect the coating distribution and composition and changing the deposit properties.[45] Additionally, certain IVD advantages include inherent substrate cleaning prior to aluminum deposition. IVD uses simultaneous and repeated energetic particle bombardment of the vaporized aluminum from sputtering, evaporation, or other vaporization sources to change, or maintain, the deposit composition, aluminum properties, and adhesion to the substrate.[46] The energetic aluminum atoms and groups of atoms are ejected from the source by inert or reactive gas ions from a plasma vacuum environment and are accelerated by an ion gun or beam. Figure 10.4 illustrates an ion beam deposition system.

IVD aluminum coatings received an eclat status when a major aerospace corporation developed an IVD process for commercial applications. IVD aluminum coating processes, developed during the 1960s, were intended to replace cadmium and aluminum electroplating processes and have been established successfully in more than 70 facilities throughout the world.[47] Production lines for large-scale IVD incorporated multistage operations that included substrate cleaning, abrasive blasting, masking, chamber loading, ion bombardment cleaning, aluminum coating, abrasive blast polishing, and chromate conversion coating.

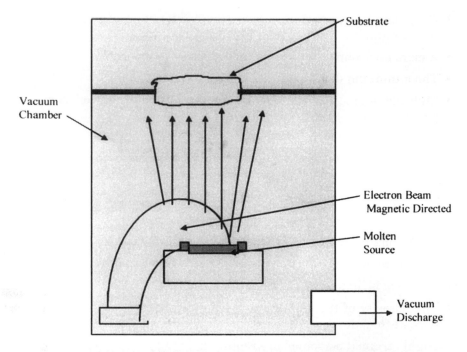

Figure 10.4 Ion beam vapor deposition that uses an external beam of ions of the coating material that bombard the substrate. The beam can be aimed at various locations. Ion implantation processes are similar except that higher-energy ion beams and temperatures are applied, which causes the metal to be driven into the substrate at a depth of 0.1 to 0.3 μm.

IVD aluminum coating properties depend on substrate surface preparation, cleaning, equipment operation, and effective process control. Glow discharge cleaning constitutes an important advantage of IVD substrate cleaning that improves coating adhesion. Adhesion of IVD aluminum coatings on steel is good, as reported by Sebastian.[47] This good adhesion is attributed to highly energetic atom bombardment at magnitudes greater than 50 times the energies of sputtered and evaporated deposits. Corrosion resistance studies[2,48–51] have supported IVD aluminum as an alternative protective coating on ferrous, titanium, and aluminum alloys without causing hydrogen embrittlement. Increased corrosion protection is offered by post-processes such as glass bead peening, chromating, organic coatings, organic chelate coating, cathodic arc deposition (CAP) of hard coatings,[52] anodizing, and IVD aluminum-magnesium/zinc (less than 10%) alloy depositions.[50,51]

10.3.6.5 Advantages of ion vapor deposited aluminum coatings. The advantages are as follows:

- The process is accepted by U.S. Military Specification MIL-C-83488.

- High-energy atom bombardment improves aluminum-to-base adhesion as compared to sputter and evaporative coatings.[45]

- Thick (normally less than 100 μm) and uniform coatings result.

- The deposition rate is rapid, typically similar to evaporative processes.

- The process provides protection for steel that is as good as, or better than, that of cadmium.

- It will not cause hydrogen embrittlement.

- It can be used at higher temperatures (up to 950°F) than cadmium.

- Surface coverage is improved, as compared to sputtering and evaporative coatings.

- Weakly adherent aluminum atom groups are removed along with oxides and other surface gases during the deposition.

- The process combines well with electroplating processes when difficult-to-plate substrates such as titanium receive IVD aluminum followed by electroplating.

- Aluminum deposit properties such as surface morphology, composition, density, and residual stress[53] can be changed by controlling the bombardment.

- Substrate heating temperature is low, permitting coating of polymeric materials.

- The process can be used for porous substrates (powered metal and similar composites) where plating solutions may be absorbed.

10.3.6.6 Disadvantages of ion vapor deposited aluminum coatings. The disadvantages are as follows:

- Substrate heating may affect some materials.

- Careful control is required to minimize deposit stresses.

- Avoidance of entrapped gasses during the deposition minimizes the chances of forming compressive stresses.

- The porosity of the aluminum coating should be tested to ensure adequate corrosion protection.

- Careful control of all process variables is necessary.

- Expect approximately 10% variability from the mean in aluminum deposit thickness.

- Expect similar and greater equipment costs as compared to evaporative aluminum coating processes.

- Adequate coating coverage on complex parts may require equipment modifications.

- Before aluminum coating, certain substrate surfaces may require wet chemical cleaning and thorough drying.

- Highly trained personnel are required.

IVD aluminum coatings have an excellent reputation for adhesion and corrosion protection of steel. Adhesion benefits from the sputter cleaning are realized by contaminant removal, high-energy creation of nucleation sites, enhancing diffusion, and increasing the substrate temperature. Continuing aluminum deposition results in dense and adherent structures.

10.3.7   Sprayed Aluminum Coatings

Thermal sprayed aluminum coatings have furnished manufacturing with methods that add new coating versatility for hard, corrosion-resistant, high-temperature, and abrasion-resistant coatings. Aluminum molten metal spraying is employed by arc metal spray, plasma torch, and detonation gun. All of these processes involve deposition by line-of-sight applications in which aluminum in powder or wire form is heated to melting and propelled by gas pressure or detonation wave onto the substrate. The angle of coating impingement is held as close as possible to 90° (perpendicular to the substrate surface). The molten sprayed aluminum deposits in layers on impact, forming a thick and tenaciously bonded coating (see Fig. 10.5). Plasma thermal spraying has advanced the technology via the use of high-temperature electrodes, which are surrounded by an inert gas that aspirates the metal powder into an ionized electrode arc between the electrodes to form a plasma. The aspirated powdered metal from the flame increases the energy of the deposition by accelerating the droplets onto the substrate surface. Plasma arc temperatures can reach 2200 to 2800°C[54] and may influence the deposit and substrate properties. Effects of the plasma temperatures must be considered for the particular substrate (see Fig. 10.6). Similarly, thermal spraying in a vacuum chamber, called *low pressure plasma spraying (LPPS)*, offers an alternative technique for aluminum deposition. LPPS has gained acceptance for metals such as tantalum and titanium and may be applicable for special aluminum depositions. The d-gun process is a Union Carbide Corp. development, and these coatings can be applied only via Union Carbide products. Oxygen and acetylene gasses pass into a barrel that

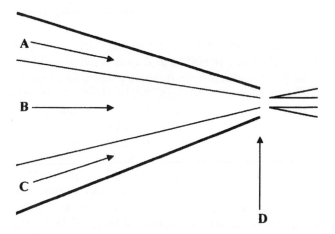

Figure 10.5 Arc spray gun with feed wire introduction at "A" and "C," atomized by electric arc at "D" and gas "B," sprayed to deposit a fine, medium, or coarse pattern.

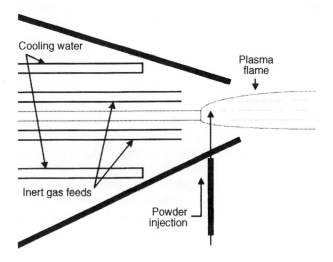

Figure 10.6 Plasma high-temperature gas is passed through an electric arc with powder injected into the flame. Inert gas reduces oxidation of the metal coating.

also contains the metal powder. Here, the ignition causes the gasses to explode, heating the metal to its melting point and expelling the droplets at a velocity greater than 700 m/s. The droplets deposit in a circular pattern averaging 25 mm in diameter from a barrel inside diameter of 2.5 cm. The d-gun explosion is repeated 5 to 10 times per second to produce another coating diameter of 2 to 5 microns thick.

Automation of this process has proven successful for industrial applications on large parts.

The more popular aluminum spray coatings deposit by wire and powder spray technologies that are portable and inexpensive, seldom heat the substrate above 150°C, and offer a deposit thickness ranging between 0.05 and 5 mm. Complex shaped parts with edges, sharp angles, and narrow grooves pose problems. Penetration of the coating inside tubular areas is limited in depth to roughly the tube diameter unless special nozzles can be positioned in the tube interior. Detonation gun (d-gun) applications are restricted to external part surfaces.

Aluminum deposits from thermal spray processes offer good adhesion, provided that the substrate surfaces are clean and roughened from grit blasting or other surface abrasion operations. Mechanical interlocking between coating and substrate governs adhesion. Deposits are rough, porous, and contain oxides and entrapped gas. Some coating stresses may be expected because of the layered deposition and rapid cooling rate. Powder spray coatings are influenced by powder particle size distribution, the carrier gas, and temperature. Finer powder particle size normally results in increased coating density and hardness.

Wire and powder sprayed aluminum coatings are reported to produce nearly the same corrosion protection for steel (in distilled water), provided a 75 to 100 μm thick coating is applied, with less than 50 μm of metal deposited in a single pass. Protection of steel from seawater has been demonstrated with a 200 μm thick aluminum coating that was post-coated with a silicone material. Other applications include high-temperature protection of steel and protection of high-strength alloys, containing zinc.

Emphasis on aluminum thermal spray by wire and powder technologies for aluminum deposits dominates this discussion; moreover, numerous publications dealing with refractory coatings report that aluminum intermetallic coatings have provided unique high-temperature protection of aerospace hardware. Plasma sprayed coatings such as aluminides ($NiAl$, Ni_3Al, $CbAl_3$, and $TaAl_3$) have been used to prevent high-temperature oxidation of space vehicle, rocket engine, and nuclear reactor component surfaces.[55]

10.3.7.1 Advantages of thermal sprayed aluminum coatings. The advantages are as follows:

- The coatings are relatively inexpensive to apply with portable wire and powder equipment.

- Good corrosion protection is provided for steel (normally large parts/ structures).

- Rapid coating results in reduced labor.
- A good wear-resistant coating is achieved.
- Many post-coating treatments are available, such as lacquers and sealants, for inexpensive protection.
- Good erosion resistance is provided for gas turbine blades.
- One may expect the lowest costs with wire and powder gas combustion spraying, followed by plasma and d-gun processes, which involve the highest cost.

10.3.7.2 Disadvantages of thermal sprayed aluminum coatings. The disadvantages are as follows:

- Adhesion of coating depends on mechanical keying or interlocking.
- Special cleaning is required, coupled with special blast medium for a particular surface topography.
- There is a danger of the aluminum coating flaking at greater than 300 μm thicknesses.
- Porosity and contamination of the coating are possible.[56]
- Line-of-sight application limits part complexity.
- Expect variations in coating thickness, especially at edges, angles, and grooves.
- Operator technique may influence results.
- Operator safety is a concern.

10.3.8 Thermal Sprayed Aluminum Oxide Coatings

Powders composed of aluminum oxide (Al_2O_3) can be sprayed by plasma spraying, gas powder, and d-gun methods. Often, other powders are introduced, such as titanium dioxide and silicon dioxide, for required surface properties (hardness, density, chemical and abrasion resistance). Sprayed aluminum oxide coatings offer good electrical, abrasion, and high-temperature resistance and high hardness. These coatings tend to be brittle, lack the adhesion of the sprayed metals, and are deficient in corrosion protection because of 1 to 10% porosity.

10.4 Cadmium Coatings

Cadmium coatings gained popularity from the 1940s through the 1980s by providing sacrificial (cathodic) corrosion protection for mainly iron and steel, with an attractive silver or white metal appearance. Deposits can be mirror bright, from electrodeposited cadmium cyanide or acid baths, to a semibright or dull gray, from mechanical and vacuum technology methods. Major end users of the metal include the United States, Japan, the German Democratic Republic, and the United Kingdom. These countries consume nearly 10,000 metric tons per year.[2] Demand for cadmium finishes for corrosion protection of steel ranks second behind zinc for industrial environmental protection and is superior for marine exposure. Cadmium challenges zinc for corrosion protection of steel and offers the following advantages and disadvantages compared to zinc, regardless of the method of coating application.

10.4.1   Advantages of Cadmium versus Zinc Coatings

The advantages are as follows:

- Cadmium offers better protection for steel in marine environments, given equal coating thickness.[57]

- Cadmium does not form bulky corrosion products that can contaminate or interfere with equipment mechanisms.[58]

- Solderability of cadmium (non-acid fluxes) is better than zinc.[58]

- Cadmium is resistant to alkalis (unlike zinc).

- Cadmium coatings have good lubricity and resist galling between sliding surfaces.[59]

- Fasteners with cadmium coatings offer high tension with low torque, which is preferred by aerospace, military, and automotive manufacturers.

- Cadmium offers good galvanic compatibility with aluminum, provided the aluminum area is sufficiently larger.

- Heat treated cast iron and steel are easier to plate with cadmium than with zinc.

- Cadmium is less sensitive to organic vapors such as formic acid from paints, except with the addition of high humidity.

10.4.2   Disadvantages of Cadmium versus Zinc Coatings

The disadvantages are as follows:

- **It is hazardous!** Cadmium is extremely toxic and presents process, environmental, and ecological concerns.

- Cadmium and its compounds are highly toxic as compared to zinc.
- Cadmium should never be used for finishes that may contact food or beverages.
- Cadmium sacrificial corrosion products are often dusts to which personnel should never be exposed.
- Corrosion products are often cadmium carbonates[60] that become airborne or accidentally contacted by touch.
- Cadmium sublimes at relatively low temperatures (1.0 mm/yr @ 99°C) and cannot be used for space applications.
- Cadmium hardware should not be designed for use at temperatures greater than 225°C.[61]
- Hardware coated with cadmium should not be welded, spot-welded, soldered, or greatly heated without adequate ventilation to remove the toxic fumes.
- Most automobile manufacturers have replaced cadmium coatings.
- Cadmium coatings are more expensive than zinc.

10.4.3 Cadmium Coating Methods

Cadmium coatings normally serve as the outer or finish coating and are rarely used as an undercoating for other metals. There are no advantages to using cadmium as an undercoat for nickel or silver; however, cadmium has provided a thin (approximately 2.5 µm) undercoat for zinc when electroplated onto iron. Besides serving as a corrosion protective coating for steel, cadmium coatings on brass and steel minimize voltaic couple corrosion.[62] Three popular methods to apply cadmium include electrodeposition, mechanical deposition, and vapor deposition. Two of the three methods are capable of depositing cadmium thickness ranges of 2 to 12 µm for threaded fasteners and normal hardware, and 5 to 20 µm for marine exposed hardware. Mechanical deposition is normally limited to a thickness maximum of 12 µm.

10.4.4 Electrodeposition of Cadmium Coatings

Electroplated cadmium coatings dominate the application methods because of deposit thickness distribution on complex parts and overall equipment and material costs that permit the inclusion of a cadmium process into most plating companies. Cadmium deposit thickness ranges from 5 to 25 µm; deposits are soft, smooth, and ductile and offer moderate tensile strength (approximately 70 MPa). Popular plating processes include cyanide (with and without additives), sulfate, and fluoroborate. Cyanide processes are preferred over the other plating methods, since the bright, matte, and nonbrightened cyanide deposits have received strong industry support. Factors such as ease of control, minimal equipment corrosion, room temperature operation, high efficiency, excellent coverage of complex parts, dense fine-grained deposit, and use of a single additive constitute some of the assets of cadmium cyanide plating processes. Wide acceptance of the cyanide process has placed it as the preferred procedure for commercial cadmium plating.

Disadvantages include the hazards of cadmium and cyanide, carbonate increase and removal, and high alkalinity of the plating bath. Of highest importance is the hydrogen embrittlement of certain steels. Hydrogen embrittlement by absorbed hydrogen may be defined as a *latent brittle fracture,* occurring during a loading condition less than the steel ultimate strength. Certain steel alloys with greater than 30 HRC are susceptible to hydrogen embrittlement. Cadmium plating of steel parts must be carefully understood to prevent hydrogen absorption by interstitial diffusion into the metal lattices, or hydrogen embrittlement. Absorption of hydrogen into the steel occurs in seconds, at room temperature, and most often from the aqueous plating solutions. Platers must be careful to prevent their cleaning, plating, and post-plating operations from allowing hydrogen to come in contact with the steel surface. Several mechanisms that explain the brittle fracture theory include studies from Johnson and Birnbaum.[69] Cathodic cleaning, pickling, activating, strike plating, and finish plating can introduce atomic hydrogen into the surface of the steel. Even small areas of corrosion can react to form hydrogen that can enter the steel.[70] Unfortunately, some metal fabrication operations can cause hydrogen embrittlement. These include machining, cold working, use of a susceptible steel microstructure, moisture contact following casting and furnace operations, and contaminated lubricants. ASTM literature, Aerospace Industries Association guidelines, chemical supplier support documents, and numerous published studies offer precautions to prevent hydrogen embrittlement before and after plating.

Cadmium sulfate plating baths replaced the cyanide concerns by early developments.[63–66] Proprietary acid cadmium formulations have found some commercial use, such as the Aldoa acid sulfate process[67] and others.[75] The acid sulfate cadmium bath formulations could lower, but not eliminate, hydrogen embrittlement from certain steel alloys.

Cadmium fluoborate plating baths similarly have removed cyanide in the formulations[71,76] and offered less embrittlement than the cyanide; however, complaints included poor anode corrosion, difficult process control, uneven deposit thickness distribution, inferior appearance, and higher cost than cyanide cadmium.

Acid-based cadmium baths have reduced hydrogen embrittlement in steels, but not to the extent of the nonbrightened cyanide formulations. A typical nonbrightened cyanide formulation is referenced in ANSI/ASTM F519, Std. Method for Mechanical Hydrogen Embrittlement Testing of Plating Processes and Aircraft Maintenance Chemicals. Regardless of the plating bath use, given absolute minimum hydrogen embrittlement on high-strength steel parts following cleaning, activation, and plating, a hydrogen embrittlement bake must be administered within four hours of plating, at a temperature of 200 to 230°C, for a period of from 8 to 24 hr (time depends on tensile strength).[81] In spite of efforts to eliminate hydrogen embrittlement, given a compromising situation when diffused hydrogen has entered the steel at a critical concentration capable of initiating cracks, there is no repair for the initiated crack formation, and strength is permanently lost.[82]

10.4.5 Electrodeposition of Cadmium Alloy Coatings

Cadmium-titanium alloys. Cadmium alloy coatings aimed at the elimination of hydrogen embrittlement, absence of cyanide, and better deposit protection at lower cost have received attention from a limited number of customers. Cadmium-titanium (Cd-Ti) alloy cyanide baths that deposited cadmium with 0.1 to 0.7% titanium gained popularity for preventing hydrogen embrittlement.[72] High-strength steel parts used for supporting aircraft components that perform under stress have been coated with a Cd-Ti cyanide formulated bath.[73] Mil-Std-1500B describes the requirements for a Cd-Ti cyanide bath that deposits 0.07 to 0.5% titanium. A noncyanide, neutral, ammonical, cadmium-titanium bath[74] produces higher corrosion protection, improved deposit coverage, and lower hydrogen embrittlement as compared to the cyanide baths.

Cadmium-tin alloys. Alloy deposits of cadmium and tin can be deposited individually or simultaneously and fused to produce a coating with good salt spray corrosion protection.[77] The ratio of cadmium to tin can vary from 20/80 to 80/20, and various processes include fluoborate,[77] sulfate,[78] fluoride-fluosilicate,[79] and cyanide-stannate.[80] Two specifications for cadmium-tin plating include Mil-P-23408B, which describes requirements for a fused 25 to 50% tin deposit from a separate or alloy plated Cd-Sn bath. In addition, FORD ESA-M1P72-A describes a Cd-Sn specification for automotive hardware needs.

Cadmium-nickel diffused alloys. Electrodeposits from cadmium-nickel (Cd-Ni) diffusion processes were designed to protect carbon, low-alloy, and corrosion-resistant steels such as used for jet engine parts.[83,84] Sulfamate nickel has served as the underplate between the steel and cadmium finish plate. Careful control of the nickel thickness (5 to 10 μm) and cadmium coating (2.5 to 5 μm) achieved the desired thickness ratio before the diffusion bake. Chromate conversion coating completes the plating for the 30 minute (air atmosphere) diffusion bake at 322 ± 6°C. Modifications of this process using electroless nickel improved the metal thickness distribution, which proved beneficial for complex shaped parts.[85]

10.4.5.1 Advantages of electrodeposited cadmium coatings. The advantages include:

- Electroplating presents many reliable plating processes for cadmium.

- Excellent thickness distribution and throw are suitable for complex parts.

- Low-cost operation is provided, suitable for a large array of part shapes and sizes.

- The coatings have received military and commercial specification approvals QQP-416F (1995), ASTM B-766 (1986), MIL-STD-870B (USAF), (1986), ISO 2082-1986, AMS2400 (1999), AMS 2401 (1986), NAS 672, and SAE AMS 2451/4, Brush Plating (1998).

10.4.5.2 Disadvantages of electrodeposited cadmium coatings. The disadvantages include:

- It is difficult to plate high-strength steel because of hydrogen embrittlement.

- The plating process exposes operators to toxic materials.

- Cadmium requires waste treatment equipment for environmental regulation compliance.

10.4.6 Mechanical Deposition of Cadmium Coatings

Mechanical plating bonds powdered cadmium particles to other metal surfaces by a cold-weld union that is formed by mechanical barrel tumbling. This form of barrel finishing uses an aqueous slurry of cadmium metal, impact media such as glass beads, proprietary surface activators and wetting agents, and the weight of the plated hardware to deposit cadmium following an extended tumbling period. Cadmium deposit thickness normally ranges between 5 and 25 µm. Cadmium coatings applied by mechanical plating benefit situations where large volumes of small parts with nonintricate geometry (e.g., fasteners, guide pins, nails, and stampings with insignificant hydrogen embrittlement).[86] High-strength steel parts still must avoid any preplate cleaning processes, such as cathodic cleaning and acid activation, that could introduce hydrogen embrittlement into the steel before mechanical plating. Mechanical plating offers an alternative coating process for electroplating used by vehicle part manufacturers and military part suppliers for corrosion protection of steel.[87,88] Other metal powders such as tin and zinc have been combined with cadmium for special coating characteristics.[89,90] Steel may be the most popular base metal; however, other metals may be mechanically coated if we are mindful of the rule of thumb that most successful mechanical processes require the base metal to be harder than the coating metal.

10.4.6.1 Advantages of mechanical cadmium plating. The advantages are as follows:

- Large volumes of parts can be coated more economically than with electroplating or vapor deposition methods.
- There is less chances of hydrogen embrittlement as compared to electroplating.
- The process avoids several corrosive and toxic chemicals.
- Operator safety is increased.
- Following slurry and media separation from the parts, additional coatings may be applied.

10.4.6.2 Disadvantages of mechanical cadmium plating. The disadvantages are as follows:

- The coating does not have an even thickness distribution.
- The coating has a rough appearance and may be porous.
- Small part volumes are not economical.
- Parts cannot be fragile or complex.
- Capital expenditure for automated equipment is necessary.
- Cadmium metal and powders are toxic and could pose handling and disposal problems.

- Barrel finishing techniques are difficult to control because of small amounts of coating being removed by part-to-part abrasion, exact slurry composition, and equipment wear.

- Some steels require an electroplated *flash* plate of copper (approximately 1 µm) or other metal prior to mechanical plating.

10.4.7   Vapor Deposition (Ion Vapor Deposition) of Cadmium Coatings

Ion vapor deposition (IVD) cadmium coatings require vacuum conditions that preclude the chances of hydrogen embrittlement. This advantage is important for plating high-strength steels. Positive ions from inert gas plasma bombard the various parts as the evaporated cadmium condenses on the part surfaces. Coating thickness is uniform and ranges from 5 to 20 µm.

10.4.7.1   Advantages of IVD cadmium coatings.   The advantages are as follows:

- Hydrogen embrittlement of cadmium-coated high-strength steel is avoided.

- Some cleaning of the substrate is possible before coating using certain IVD techniques.

- IVD coatings offer good adhesion.

- Proven acceptance exists per military specification MIL-C-8837B, Coating, Cadmium (Vacuum Deposited) [refer to SAE AMS C8837 (1999)].

- IVD cadmium can be applied to other, difficult-to-plate substrate materials.

- Substrate heating temperature is low.

10.4.7.2   Disadvantages of IVD coatings.   The disadvantages are as follows:

- Cleaning of the substrate may be a problem, and care must be taken to prevent hydrogen embrittlement before cadmium coating.

- IVD equipment costs are high, and metal targets are expensive.

- IVD coating of complex parts often requires special fixtures.

- Many process variables require control.

- Porosity may be a problem because, under some conditions, the chamber gas may become occluded within the coating.

- Highly trained personnel are required.

10.4.8 Cadmium Coating Alternatives

Cadmium's toxic properties have drawn national and international attention, with the result of cadmium and cadmium compounds being banned or restricted in a wide group of applications. Cadmium carbonates and oxides constitute the sacrificial corrosion products that permit unwelcome opportunities for exposure by intimate contact and water leaching events. These examples account for the majority of cadmium concerns related to military hardware. Military Standard QQ-P-416F warns that cadmium's toxicity should disqualify it as a finish for any part that might be used for food storage, in cooking utensils, and as a part of any other object that may come in contact with food. Cadmium-plated parts should never be heated by soldering, brazing, or welding operations because of the danger of poisonous vapors.

Whenever possible, alternative coatings should be considered as replacements for cadmium finishes. Numerous publications have explored possible replacements for cadmium, but no one coating has been an effective substitute. For example, cadmium's toxicity prevents fungus growth on critical components of military hardware. Here, a cadmium alternate would need to remain toxic and environmentally unfriendly. Several substitutes have been found that partially meet the requirements.[90–101]

10.5 Chromium Coatings

Chromium coatings serve as finishes that may be classified as *decorative* and *engineering* types, often referred to as *hard chromium* and *industrial chromium*. Mirror-bright, tarnish- and corrosion-resistant deposits characterize decorative chromium coatings. A subtle, blue-colored haze or tint that is visible in a reflective mirror distinguishes decorative chromium finishes. Hard chromium finishes lack the brightness yet offer the special characteristics of chromium that include resistance to corrosion, heat, wear, and erosion, plus a low coefficient of friction. Chromium coatings have a special *air passivity* property due to a tenacious thin film of chromic oxide that serves as a self-healing, corrosion-resistant film. This ability to generate protective films is realized by small percentages of chromium added to iron alloys to form heat- and corrosion-resistant steels.

Decorative chromium coatings retain their tarnish resistance and mirror brightness out-of-doors, which explains their growing popularity for automotive and marine metal and plastic trim. Other popular contenders for chromium use include appliances, hardware, plumbing, furniture, and motorcycle parts.

Hard chromium deposits find uses on internal engine part surfaces, cutting tools, hydraulic shafts, aircraft landing gear supports, press and turbine shafts, and for dimension restoration on worn/undersize parts. Surface properties offered by hard chromium include good wear and abrasion resistance, increased corrosion resistance, and prevention of galling and seizing.

Four popular methods for applying hard and decorative chromium coatings exist: electroplating, diffusion, physical vapor deposition (PVD), and flame spray. Whether it is more economical to use an outside chromium plating source or to establish an in-house operation depends on many factors, including government regulations; part material, size, quantity, and complexity; and the type of chromium desired. Electroplating hard, or decorative, chromium by contracting an outside plating source is generally the more economical approach. Purchasing and installing a chromium plating plant may be cost prohibitive under today's standards because of municipal permits and requirements to set up cleaning and underplating capability for copper and nickel plating processes. Additionally, waste treatment equipment, operator safety/medical monitoring, chemical handling liabilities, insurance needs, and government monitoring at federal, state, and local levels all increase the costs of operating a chromium plating facility. OSHA also regulates in-plant worker exposure levels to chromic acid, and the EPA restricts the amount of chromic acid mist f that can leave the building.[118] A regular review of applicable regulations is necessary for all shops. The concept of adding a chromium plating capability to complement an existing plating operation deserves careful study because of the aforementioned controls and safeguards that unique for plating chromium.

Diffusion is another alternative method for chromium coating. It drives chromium into the surface of the substrate material, usually steel, at a high temperature without loss of the substrate properties. Here, chromium fulfills the need for protection from corrosion and high-temperature oxidation. *Chromising*[102] costs are moderate and less than electroplating costs; however, coating appearance, substrate limitations, and the nature of the process restrict this method to special applications.

PVD chromium coatings are limited to thin (about 1000 Å), decorative finishes that usually are coated with a special lacquer. Acceptance for these thin finishes has been realized in automobile trim, bicycles, and appliances. Additionally, ion implantation has found success by driving chromium ions into the surface of existing chromium coatings to form a thin surface alloy that improves corrosion protection and wear resistance.

Unfortunately, chromium, like cadmium, is a toxic poison—a suspected carcinogen that is dangerous to employees operating the processes and is also a dangerous soil contaminant that must not be allowed to contaminate groundwater. Even chromate conversion coatings, chromate primers, and nearly all chromium-containing products and finishes are currently under examination by commercial and military investigators for replacement and total chromium recycling. Developments to improve the safety of chrome plating operations have included fume suppressants to prevent chromic acid droplets from entering the shop atmosphere, exhaust system revisions, new waste treatment equipment advances, and trivalent chrome plating developments that avoid the chromic acid formulations.

10.5.1 Decorative Chromium Electroplated Coatings

Conflicting with efforts to replace chromium and its coating processes by other, more environmentally amiable coatings, vehicle manufacturers want to satisfy a universal customer desire for the elegance, shine, sparkle, distinction, and the beauty of chrome. Only gold finishes can rival the pride that customers derive from the bluish-white dazzle that chrome offers for cosmetic purposes.

Decorative chromium finishes offer thin deposits that range in thickness from 500 Å to 1 µm, often plated on nickel and copper barrier electroplates or underplates and further protected with an ultraviolet protective lacquer. Zinc die castings, aluminum, copper alloys. and low-carbon steel served as some of the first substrates to be decorated with chromium coatings. Following the early years of buffed and polished nickel coatings (late 1800s to mid 1920s), chromium coatings improved the surface of the polished nickel by eliminating tarnishing problems. Further improvements in corrosion resistance included developments such as high-leveling copper and nickel plating, semi-bright and bright nickel processes, crack-controlled and crack-free chromium processes, and multilayer nickel and chromium plating processes.

Automobile industry efforts to combat the corrosion problems of steel led to the understanding that corrosion resistance depends on the underplate thickness and composition. The increased use of salt on roads boosted research to prevent corrosion and led to the development of multilayer nickel coatings coupled with crack-controlled (microcracked) chromium. Considerable published material is available concerning corrosion tests and multilayer copper-nickel-chromium plating processes.[103–106] One three-year, marine atmosphere corrosion study found the best performance with double-layer nickel and micro-cracked chromium.[120] Manufacturers of automobile hardware soon directed their attention to eliminating the corrosive potential of the substrate. Beginning in the 1970s, plastic replacement parts employed new chromium coating processes by electroplating, vacuum deposition, and flame spray. Chromium deposits of less than 1 μm were mirror bright. Unfortunately, the porous character of this thin finish offered little corrosion or abrasion protection until a clear top coat added increased protection.

Bright chromium plate is unique because of the formation of visible cracks observed when the thickness of the deposit exceeds 0.5 μm.[107] The crack patterns often overlap with vestiges of plated-over cracks, and the porosity of the deposit increases. Cross sections of thick, cracked deposits show discontinuities of small separated layers throughout the deposit. Elimination of the cracks has proven to be a challenge, coupled with controlling stress, wear, and corrosion resistance. Introduction of numerous improvements such as *duplex chrome plating,* catalyst formulations,[108–112] and pulsed current[113,114] advanced decorative chromium plating to three types of decorative plating processes. The three processes that improved the traditional chromic acid/sulfate bath include the standard mixed catalyst, self-regulating mixed catalyst, and trivalent chromium baths.

10.5.1.1 Advantages of electroplated decorative chromium coatings. The advantages are as follows:

- The thin chromium electrodeposit on nickel, coated with a protective lacquer, provides a cheap, eye-pleasing, protective coating that normally endures for the intended life of the part.

- Large volumes of small or large, complex parts can be coated economically.

- The hexavalent chromium/sulfate, or trivalent plating processes, offer increased plating capabilities for existing plating businesses.

- The self-regulated, mixed-catalyst chromium plating processes offer less required catalyst control by providing a fluoride catalyst that has limited solubility.

- The self-regulated mixed catalyst processes can provide the type of microcrack-controlled deposit that causes the galvanic corrosion effect to be spread out evenly over the entire part surface[103] instead of at concentrated points.

- Crack-free chromium deposits exhibit increased corrosion and wear resistance.

- Trivalent chromium processes avoid the hazards of hexavalent chromium plating and are one-tenth as concentrated as the hexavalent processes, and trivalent chromium chemical storage avoids the fire hazards of hexavalent chromium.

- Trivalent chromium deposits have a "whitish" decorative appearance as opposed to the "bluish-white" appearance of hexavalent chromium.

- Trivalent will not form "burned" deposits and offers the best throwing power.

- Different heat treatments can be added to change the chromium deposit hardness and stress.

- Trivalent chromium bulk, or barrel, plating of small parts offers good adhesion in the event of electrical current interruptions.

10.5.1.2 Disadvantages of electroplated decorative chromium coatings. The disadvantages are as follows:

- Mist suppressants must be used for hexavalent chromium, coupled with meeting all federal, state, and local chromium emission and reporting guidelines.

- The efficiency of chromium plating is quite low (5 to 20% for sulfate and 20 to 25 percent for fluoride baths) as compared to the other metals.

- Decorative chromium is a thin deposit that offers little or no protection without a underplating such as nickel and an organic top coat.

- Cracking of the deposit is a problem in spite of "crack-free" coatings, because minor flexing or expansion of the substrate will cause the chrome to crack.

- Throwing power is low, and special anodes are often necessary.

- Control of the processes can be difficult, especially for trivalent chrome.

- Trivalent chromium is more sensitive to contamination.

- Tank ventilation with air scrubbing capability is necessary for hexavalent chromium processes.

- Decorative chromium plating on plastics requires careful control of the plastic materials for consistent polymer properties plus the costs on an additional tank line for pre-chromium-plating processes.

10.5.2   Hard Chromium Electrodeposited Coatings

Hard chromium plating is often referred to as *industrial* or *functional* chromium plating, and it is used to apply extremely hard finishes that offer moderate corrosion protection and surfaces with a low coefficient of friction. Good wear properties rank hard chromium as one of the best metal finishes for wear applications to extend the life of service parts. Typical examples of commonly coated hard chromium items and approximate thicknesses include:

- Piston rings, cylinder walls, hydraulic shafts, rollers, and crankshafts (12 to 50 μm)

- Cutting tools and molding dies (12 μm)

- Car engine valves (5 to 8 μm)

Rebuilding worn parts, followed by processes such as remachining, grinding, and polishing constitutes the primary purpose for hard chromium coatings. Normally, 100 to 300 μm of chromium is deposited before grinding to size and polishing. Hard chromium plating serves businesses in areas of overhauling aircraft landing gear equipment, correcting mismachined and worn parts, and salvaging engine exhaust valves.

Characteristics of hard chromium consist of (1) wear and corrosion resistance, (2) hardness, (3) and anti-galling features that change the decorative chromium electroplating processes by using slightly different bath formulations, plating times, thickness, and pollution prevention methods. The plating thickness for hard chromium ranges from 5 to 500 μm, as compared to a decorative thickness of less than 1 μm. The thicker coatings require often >24 hr of plating time. Hard chromium deposits may be applied directly over the base metal, often without copper and nickel underplating. Hardness values of *hard* chromium-plated deposits are comparatively the same as those of the decorative chromium deposits;[115] however, different bath formulations, bath temperatures, current density, and heat treatments have been reported to vary chromium deposit hardness from the 300s to greater than 1200 kg/mm². Most hard chromium deposits aim for 900 to 1200 kg/mm². Numerous helpful studies of electrodeposited hard chromium have been published (see Refs. 7 and 58). Also, Refs. 116 and 117 are helpful guides.

Three popular hard chromium baths require chromic acid with a sulfate catalyst. The conventional bath is the chromic acid and sulfate catalyst (old standby). Fluoride salts, added to the conventional bath, created the *mixed catalyst* bath for improved plating efficiencies. The third bath uses a proprietary nonfluoride catalyst. Trivalent chromium baths are currently in development stages to permit the deposition of chromium coatings greater than 7 μm.

10.5.2.1 Advantages of hard chromium electroplated coatings. The advantages include the following:

- Economical process for salvaging worn, damaged, and mismachined parts.

- Parts of varied size, volume, and complexity can be hard chromium plated.

- Corrosion protection is improved with the >50 μm thick coatings that are ground to desired size and polished.

- Hard chromium coating can be etched, creating pores for lubricant retention purposes as needed for some engine piston rings and cylinder walls.

- Gun bore hard chromium coatings applied with pulsed current to produce crack-free and porous deposits have performed well.[119]

10.5.2.2 Disadvantages of hard chromium electroplated coatings. Disadvantages include the following:

- The toxicity and pollution concerns of hexavalent chromium for hard chromium are analogous to the concerns of decorative chromium electroplating.

- The less-toxic trivalent chromium bath is not useful for the thicker deposits.

- Plating efficiencies are low, similar to decorative plating.

- Trivalent hard chromium baths show promise for the near term.[121]

- Fluoride salts of the mixed-catalyst chromium plating bath will etch iron from steel in the low-current-density areas, and these areas must be masked.

- Iron contamination from steel limits the life of the chromium bath.

- Ductility is very low, with percent elongation values of below 0.1 percent.

- Careful monitoring of all plating parameters is necessary.

- Chromium plating of high-strength steels requires hydrogen embrittlement relief.

10.5.3 Physical Vapor Deposition Chromium Coatings

Thin coatings (300 to 1000 Å) of vacuum deposited chromium find applications for automobile exterior plastic surfaces. Both flexible polyesters and thermoplastic urethanes have been used for automobile grilles, trim molding, and bumper strips that are metallized, followed by a clear vinyl or UV-inhibited top coat.

Employment of physical vapor deposition (PVD) chromium has increased for coating parts that were traditionally hard chromium plated. Advancements in PVD offer ion implantation to form a thin surface alloy[122] that increases the wear, corrosion, and fatigue resistance of the chromium coating. Nitrogen is implanted into the chromium surface to improve wear resistance by forming Cr_2N without changing the chromium thickness dimension.[123,124] Other ions have been implanted into the chromium surface for similar property enhancements. Properties such as wear resistance, hardness, corrosion resistance, low coefficient of friction, and abrasion resistance were improved by PVD thin coatings (usually less than 10 μm) of carbon (diamond and similar carbide coatings), molybdenum disulfide, titanium nitride, and chromium nitride.

Multilayer coatings have been used by combining a hard coating such as chromium nitride with a soft finish coating for lubricity. The hard PVD coating could be 40 μm or more of a nitride of chromium, tungsten, or titanium, and the soft coating could be a low-coefficient-of-friction coating consisting of 2 to 3 μm of a material such as molybdenum disulfide. Automotive engine parts such as turboshafts, fuel injectors, and camshafts have avoided hard chromium electroplating, replacing it with PVD of more than 40 μm thick chromium nitride, overplated with tungsten that is then overplated with 2 to 3 μm of a carbide-containing metal. Greater thickness involves a danger of cracking.

PVD coatings used to enhance or replace chromium require significant investment in equipment that must be able to reproduce run after run of the desired product with a minimum reject rate. Large chambers accommodate large volumes of small parts, or selected large parts, and require additional investments for specialized cleaning processes, monitoring equipment, specialized employee training, and large magnetron sources. Deposits are limited to line-of-sight coating coverage. Replacement of hard chromium electroplating by PVD continues where current costs are justified and by future advances in PVD technology.

CVD (chemical vapor deposition) is a process in which the reactant gas contains a metal such as chromous chloride that condenses on the part, which is enclosed in a special vacuum chamber. Other metals such as nickel, tungsten, and titanium carbide may be deposited similarly with a typical thickness of 1 μm. Advantages include the ability to uniformly coat complex parts (which often contain internal cavities) and bore interiors. Disadvantages include equipment and special training costs. Special cleaning of the parts requires additional equipment and the use of chemical cleaning processes. High process temperatures often reach 1000°C, which may degrade the coated items. Some safety issues are encountered, such as equipment noise level and high temperatures.

10.5.4 Sprayed Chromium Coatings

Sprayed chromium coatings, applied by thermal spraying and plasma arc spraying, use solid and powdered chromium to coat substrates. This "dry" alternative chromium coating process uses a spray gun to deposit the powder or metal by heating it in a flame or plasma and allowing pressurized gas to bombard the substrate. High deposit thicknesses can be obtained with chromium and many other alloys. Normally, excess chromium is applied for a final machining-to-size operation. Deposition temperatures for thermal spray can be lower than 200°C as compared to plasma arc spray temperatures, which can reach greater than 10,000°C.[127] Improved deposit properties, such as corrosion and wear resistance, are offered by sprayed coatings, and wastewater treatment is avoided. Disadvantages include operator training and safety concerns. Capital investment is moderate, coating is restricted to line-of-sight applications, and there are some machining problems with excess or uneven chromium deposits.

10.5.5 Other Chromium Coatings

Black chromium plate meets certain requirements for wear and temperature resistance, high solar energy absorption, low emissivity (low radiation back to the outside), and low reflectivity. Decorative applications include furniture and building, plumbing, and electrical hardware. Solar energy collectors and anti-glare surfaces are some of the more functional uses of black chromium. Proprietary formula modifications of hexavalent and trivalent chromium baths are required to produce the black coating composed of chromium metal and chromium oxide crystals. Normally, a nickel underplate is required to improve the corrosion protection of the somewhat porous coating. Black chromium and several other solar energy-absorbing coatings were tested in outdoor tests, with black chromium producing highly favorable results.[126]

Because of environmental concerns, continuing studies have aimed to eliminate or replace the use of chromium coatings in automobiles. This has led to the development of the trivalent chromium baths, which have gained more acceptance among the environmentally conscious. Other alternate alloys are challenging the chromium coatings, such as tin/nickel, tin/cobalt, nickel/tungsten, cobalt/tungsten,[128] heat-treated electroless nickel, and physical vapor deposited coatings. Hard chromium remains irreplaceable for its unique mechanical properties but it must be carefully and safely deposited, with all pollution concerns in control.

10.6   Copper Coatings

Copper coatings receive most of their popularity because of their high electrical and thermal conductivity and high melting point. The electronics industry requires copper for properties such as solderability, low cost, high ductility, and corrosion resistance (which is often enhanced by corrosion-inhibited lacquers). Copper conductors for electrical and microwave pathways consume significant quantities of copper. Construction industry copper products that combine strength, ductility, and corrosion resistance find such uses as piping, building vias for seawater, and pipes for fresh hot and cold water and soft or aerated waters that are low in carbonic and other acids. Copper is in demand for outdoor protective flashing, cooking utensils, jewelry, furniture, and many other consumer products. The hazards of copper metal are considered not very significant; however, compounds associated with coating applications can be extremely hazardous. Federal, state, and local regulations include restrictions for copper discharge.[129]

Copper coatings applied by cladding electrodeposition, electroless deposition, or flame spray are richly colored and often reflective immediately following deposition. Brightness is rapidly lost in most atmospheres by oxidation, and slow corrosion occurs. Conversion coatings such as benzotriazole and oxides are often necessary to preserve the finish for copper appearance and for functional reasons such as soldering or adhesive bonding. Seldom is copper used as a final finish, but copper's good corrosion resistance, excellent electrical and thermal conductivity, good mechanical workability, and ease of bonding and soldering widened its use as a supportive coating for nickel that is finished with an outer protective coating.

10.6.1   Electrodeposition of Copper Coatings

Copper electroplating by acid, alkaline, and neutral formulations is selected to meet the requirements for coating certain substrates while being mindful of the safety and pollution concerns associated with each plating operation. Electrodeposition of copper serves as one of the prime methods for coating parts of all sizes and complexities economically and within moderate monitoring, safety, and pollution guidelines. Both copper cladding and flame-spray operations serve special applications such as direct copper clad bonding to other materials and rapid line-of-sight flame-spray copper coatings.

Acid copper electroplating processes have received increased attention because of the electronics industry's need for highly ductile (greater than 8 percent), high-throwing, high-speed (deposition rate), highly conductive, and high-purity electrodeposited copper. Electronics uses for acid copper coatings include the sulfate bath, which is the choice of the printed circuit board industry for high-throwing and good leveling formulations, high conductivity, ease-of-use, and effortless waste treatment. Commercial copper coatings also demand acid copper sulfate for cooking vessels that need good heat distribution, steel rolls that are engraved and used for printing, plastic master discs for vinyl records, zinc-based die castings, and to produce an undercoat for nickel and chromium plating. The sulfate bath, operated at ambient temperature and formulated with proprietary addition agents, deposits copper by direct current, pulse, square wave, or periodic reverse to a degree of success that the surface smoothness is normally smoother and brighter than the substrates to which it is applied. Substrates such as copper, copper-base alloys, and previously copper strike (cyanide or neutral strike) plated steels can be plated. Zinc and steels cannot be directly plated without a cyanide or neutral copper strike.

10.6.1.1 Advantages of acid copper sulfate coatings as compared to other acid and alkaline processes. The advantages include the following:

- Lower cost (chemicals and equipment) as compared all others.
- Highly efficient plating that operates at a low voltage.
- It operates at room temperature.
- It is easy to operate and treat waste.
- There are no impurity build-ups from side reactions that limit the bath life, such as orthophosphates (pyrophosphate copper) and carbonates (cyanide copper).
- High brightness is produced.
- It will not attack photoresists, masking, or rack materials (organic polymers that are sulfuric acid resistant).

- Although it is corrosive because of sulfuric acid, acid copper fluoborates can be worse.

- Bright formulations do not require buffing.

- It actually fills in small substrate surface imperfections to produce a level deposit.

- It serves well in electroforming where fine details must be duplicated.

- Additives and some contaminants can be monitored by cyclic voltammetry stripping.[130–132]

- It produces good elongations of 15 to 20 percent for circuit board bath formulations at 80 g/L copper pentahydrate with strength of 40,000 psi.

- The fine-grained and low-stress deposits are good for electroforming.

10.6.1.2 Disadvantages of acid copper sulfate coatings compared to other acid and alkaline processes. The disadvantages are as follows:

- Sulfuric acid solutions are corrosive to equipment.

- Special copper phosphorized anodes must be used for the optimal deposit.

- Special *napped* anode bags must be used to prevent deposition of particles from anodes.

- It cannot be deposited directly onto aluminum, zinc, or steels.

Alkaline cyanide copper plating[133] has a history of decorative and engineering finishing successes because of its ability to plate onto many base substrate materials such as aluminum, steels, and alloys of copper, lead, magnesium, nickel, and zinc, with excellent adhesion. Good adherence to the base materials may be aided by alkaline cyanide activating the base metal and trace removal of soils. Cyanide copper coatings offer good electrical conductivity, solderability, and ductility, and good underplate for copper-nickel-gold, copper-silver, and copper-nickel-chromium electronic and industrial needs. Automotive requirements have depended on copper cyanide for the initial coating on zinc die-castings used for trim. Other products, such as millions of zinc alloy pennies, miles of wire, ammunition, and appliance and plumbing fixtures, receive a copper cyanide plated finish. Improvements such as periodic reverse and pulse plating have increased the leveling property of cyanide copper to permit more uniform thickness distribution. Shop introductions of this equipment for copper have been scarce. Recently, cyanide copper has been on the decline because of replacements largely by acid copper processes for reasons of safety, environment, and overall cost. Other drawbacks for cyanide copper plating include:

- At least 140°F and possibly higher operating temperatures are required.
- Carbonates require removal (depending on part volume).
- Equipment costs are higher (separate rinse, cyanide destruct, and exhaust system).
- The operator training for use and handling cyanide materials presents considerable risks.

The properties of cyanide copper deposits equal and can excel those of acid copper, particularly in the barrel plating processes, in spite of the safety hazards of the cyanide anion. Periodic reverse plating in cyanide copper processes provides good leveling of the deposit by reduction of the excessive copper deposits where direct-current, high-current density areas are located, such as along part edges and projections. Copper cyanide plating circuit board manufacturing at Westinghouse Aerospace (Baltimore) during the 1970s and 1980s successfully applied periodic reverse at 80°C from and mixed salt (sodium and potassium copper cyanide) at 75 g/L copper cyanide concentration using a selenium and betaine additive. This produced fine-grained deposits for *panel* plating that averaged 10 percent elongation at 60,000 psi.

Pyrophosphate copper (pyro copper) plated coatings was popular in the printed circuit industry for several years until acid copper sulfate plating improvements in ductility, plus the overall lower cost, reduced the use of pyrophosphate copper. Pyro copper plating offered the high elongation needed for the thicker printed circuit boards that experienced greater expansion during soldering, and strong deposits that resisted flexing during thermal cycling operations. Later introduction of new materials, designs, and assembly changes for these thick, or *multilayer,* boards reduced the circuit board z-axis expansion, permitting less ductile and lower cost, non-pyro plating baths to become competitive. The slight alkalinity of the pyro copper bath dissolved many of the plating resists required for *pattern* circuit plating, adding to the industry's desire for acid copper processes.

10.6.1.3 Advantages of copper pyrophosphate coatings. Advantages include the following:

- High purity and ductility are produced, with nearly 40 percent elongation.

- High strength of 40,000 ± 2000 psi and low internal stresses result.
- The process produces low electrical resistivity of 1.75 μΩ-cm, which is slightly lower than that of silver deposits.
- Pyro copper plating can be used to deposit copper directly upon zinc, steels, and zincate-coated aluminum.
- It is less toxic than cyanide copper coating processes.
- Good leveling is produced where small surface imperfections need to be concealed.
- There is less corrosive attack on equipment.

10.6.1.4 Disadvantages of copper pyrophosphate coatings. Disadvantages include the following:

- The process is difficult to control, with more than seven variables to control, such as pH, ammonia content, temperature, organic contamination, and pyro-to-copper ratio.
- During plating, pyrophosphate decomposes to form an orthophosphate contaminant that lowers the plated deposit ductility and eventually leads to total replacement of the bath.[134]
- Only high-purity OFHC (oxygen-free, high-purity copper) anodes can be used.
- Good knowledge of the control of all addition agents is necessary for consistent deposit quality.

Electroless copper coatings are deposited by chemical reduction of copper ions to copper metal without the outside use of an applied electric current. A reducing agent such as formaldehyde is normally used in an alkaline medium to reduce chelated copper (usually from a sulfate) to copper metal. Catalyzed surfaces such as palladium initiate the copper reduction to copper metal, and copper continues to deposit because of the understanding that the copper must also be catalytic for the plating to continue (autocatalytic). This autocatalytic copper plating theoretically continues indefinitely until all of the copper is depleted. Electroless copper will adhere to metallic substrates, but the most popular use is for metallizing nonconductors. Many electroless copper formulations, coupled with their pre-electroless plate chemical treatments, are available for deposition onto numerous polymers such as circuit boards and plastics for decorative and EMI shielding. Circuit board requirements for plated-through hole electrical connections increased the demands for electroless copper. In addition, metallization of dielectric materials requiring numerous catalyzation initiators such as palladium, carbon, copper complexes, and polysulfides were formulated for *panel, pattern,* and *fully additive* copper plating processes. Nonconductors such as epoxies, polyimides, and fluorinated polymers received electroless copper metallization by any one of a number of highly competitive chemical processes. While elongation values of 3 to 10 percent and tensile strengths similar to those of electrodeposited copper are possible, careful control of all deposition parameters is necessary to maintain the elongation required to meet thermal stress tests that ensure reliable circuit boards. Circuit boards built to standards such as Mil-C-55110 and ANSI/IPC-CF-150 require qualification testing that applies temperature cycling and thermal shocks to the sample boards for metallographic examination.[135]

10.6.1.5 Advantages of electroless copper coatings. Advantages are as follows:

- The process is effective for depositing onto nonconductors.
- Uniform copper thickness is achieved for complex parts because of autocatalytic copper deposition.
- Good elongation, high strength, low stress deposits are produced.
- Low electrical resistivity is provided.

10.6.1.6 Disadvantages of electroless copper coatings. Disadvantages are as follows:

- They are moderately costly because of copper replenishment chemicals.
- There are added costs for the preplating line, including proprietary surface cleaners, conditioners, catalyst, and accelerator solutions.
- Adhesion to nonconductors relies on mechanical bonding or interlocking of metal to surface.
- Pollution is increased because of the limited bath life.
- Attempts to regenerate the copper depositing solution have not been successful.
- The process requires additional control for chemical adds and monitoring of approximately six critical bath variables such as pH, reducer, hydroxide, copper, and stabilizer concentrations.
- Some employee safety concerns may be raised because of copper formulations that may contain such hazardous materials as copper salts, formaldehyde, thiourea, cyanide, and mercury compounds.

10.6.2 Spraying of Copper Coatings

Sprayed copper coatings do not require complex equipment and elaborate controls, but copper applications are limited because of the line-of-sight application and copper deposit porosity. Steels and iron castings can be sprayed with copper, but tiny openings in the copper coating tend to increase substrate corrosion. Even sprayed copper deposits of up to 300 μm are subject to failure. Sprayed copper at thicknesses of 700 to 1000 μm can be ground or machined and polished to size for excellent substrate protection. Copper-sprayed coatings rely on mechanical interlocking for bonding to the substrate. Therefore, it is important to prepare the substrate surface by a roughening technique, using such methods as mechanical abrasion, grit blasting, or etching. Surface roughening for maximum adhesion requires creation of the optimal roughened finish, immediately followed by copper spraying.

10.7 Copper Alloy Coatings

Numerous copper alloys have been applied as decorative and protective coatings for commercial needs. The most widely recognized alloys include copper-zinc, copper-tin, copper-nickel, and copper-aluminum. The high popularity of brass coatings stemmed from the need for a nickel substitute during World War II. Low-cost methods of application include electrodeposition, spraying, and cladding coupled with a lacquer coating to minimize tarnishing.

10.7.1 Copper-Zinc (Brass) Coatings

Copper-zinc, or brass alloys and copper-tin, or bronze alloys can be economically deposited by electrodeposition or wire and powder spraying. Electrodeposition of brass or bronze provides an abundance of household steel items with a low-cost, protective, adhesive bondable and solderable coating. Steel and zinc alloy parts can be nickel plated to a thickness of 2.5 to 5 μm followed by less than 12.5 μm of brass or bronze. Often, copper plating precedes the nickel and copper alloy when plating parts with porosity.

Brass coatings gained popularity in the automotive industry for brass-plated bumpers as a nickel conservation effort during World War II. Additional uses included plating steel wire reinforcements for rubber bonding, shock absorbers, antivibration spacers, and mounting blocks that required bonding to rubber. Bright brass coatings can be treated to produce a variety of colors and the finishes that are often protected by a thin transparent lacquer. Chromate conversion coatings and other antitarnish treatments can be applied.

Brass alloy coatings can be electrodeposited or sprayed for characteristics such as metallic composition, corrosion protection, hardness, brightness, solderability, electrical and thermal characteristics, contact, heat resistance, ductility, and tensile strength. These and other characteristics become part of the coating specification with respect to the substrate. Guidelines for specifying brass coatings must establish the brass alloy composition, and handbook values of wrought brass agree that as the percentage of copper increases, the density of the alloy increases. Electroplating brass alloy processes can deviate from or match this relationship, depending on the chemical formulations and operating conditions.

Electrodeposited brass alloy demands are greatest for decorative applications where they provide a wide range of final finish colors ranging from yellow to yellow-green, to pink, to white. This is achieved with various alloy concentrations and plating bath formulations.[136] Decorative brass finishes simply offer finishes that make the part appear as solid brass. This is done by depositing copper, then nickel, then brass flash plating, usually followed by an antitarnish coating. Here, the plated deposits for each metal ranges from 0.5 to 25 µm. Thicker brass coatings are normally required for hardware such as fasteners, lighting fixtures, hinges, and bathroom and door fixtures, which need additional brass thickness for such final operations as buffing, polishing, and antique finishing. Protection of the brass from rapid tarnishing is necessary and is usually accomplished by applying a chromate conversion coating, chemical coating, or lacquer.

10.7.1.1 Advantages of copper-zinc brass coatings. The advantages are as follows:

- They function well as low-cost decorative coatings for commercial hardware.

- Brass electroplated coatings can replace other, more costly or toxic metal coatings.

- Brass electroplated coatings offer a wide variety of colors.

- Brass may be formulated to co-deposit other metals such as tin, lead, nickel, and aluminum for certain alloy characteristics; selenium, molybdenum, tellurium, arsenic, bismuth, and lead additives serve as brighteners.

- It can be plated in alternate copper and zinc layers followed by high-temperature diffusion to achieve the exact alloy composition and avoid cyanide plating processes.[137]

10.7.1.2   Disadvantages of copper-zinc brass coatings. The disadvantages are as follows:

- Most successful brass electroplating processes require the cyanide formulations that require care when handling the chemicals.

- Careful control of the deposition process is necessary to form the desired copper-to-zinc concentration.

- Brass corrodes by dezincification in high-copper alloys (>85 percent copper), permitting the zinc to selectively corrode away, leaving a porous copper deposit with low cohesive strength.

- High-zinc alloys (>30 percent zinc) corrode in seawater with increasing zinc content and temperature.

- Most brass alloys are attacked by acid, alkali, and salt solutions.

- Brass finishes can not make contact with food because of possible contamination from the brass corrosion products.

10.7.2   Copper-Tin (Bronze) Alloy Coatings

Commercial decorative coatings often rely on copper-tin coatings that offer colored finishes from reddish-yellow to gold, to silver-white. They normally receive an outer protective lacquer. Functional uses of bronze finishes include stopoff coatings for nitrided steel, gears, and rotor shafts, and spark-resistant coatings for steel in mine hydraulic supports and similar flammable environments. Bronze finishes achieved increased attention during the 1950s, when a nickel scarcity highlighted copper-tin coatings for undercoats of chromium finishes used by automotive and hardware suppliers. Literature supporting the use of bronze electrodeposits has been reported by W. H. Safranek.[141–143] Copper-tin coatings in the range of 8 to 15 percent tin (red bronze) proved most favorable for inexpensive jewelry, doorplates, and trophies. Speculum bronze coatings (40 to 50 percent tin) are used for household fixtures and tableware; they are not recommended for outdoor use, where they can turn from bright to dull and gray. During this short period of increased interest, bronze finishes gained popularity from research-supported efforts that claimed better-than-copper corrosion resistance, abrasion resistance, and hardness. Typical bronze electrodeposited coatings vary in appearance, hardness, and electrical resistivity as shown in Table 10.4. More complete information is provided in Ref. 140.

The corrosion resistance of various compositions of electrodeposited bronze coatings is influenced by such variables as the electroplating process chemistry, equipment, post-treatment baking, deposit characteristics, and the substrate surface. Salt spray testing in accordance

TABLE 10.4   Bronze Electrodeposited Coating Characteristics

Alloy	Composition	Appearance	Hardness	Resistivity, $\mu\Omega$-cm
Copper-tin	8–15% Sn	red bronze[138]	250–300 HV	20–40
	16–19% Sn	gold/yellow bronze	300–400 HV	40–80
	≥20% Sn	white, or "Bell" bronze	300–500 HV	80–30
	42–45% Sn	silver-white, "Speculum"	~500 HV[139]	80–100
	46–60% Sn	silver-white	>500–400 HV	90–60
	60–65% Sn	silver-white	250–130 HV	60–40
Copper (electrodeposited)		red	110–120 HV	3–8
Silver (electrodeposited)		silver-white	50–150 Bhn	1.6
Copper-zinc	1–45% Zn	red to yellow	soft	3–11

with ASTM B117, B287, B368, D609, and D1654 serves as one of the most popular bases for coating endurance. The nonimpinging 5 percent NaCl in the water mist chamber serves to compare the corrosion resistance of the metal coatings. However, the test is not indicative of corrosion resistance in other environments, since other chemical reactions may be involved. The overall corrosion resistance of bronze coatings is low in various atmospheres; however, atmospheric attack is significant in industrial and marine environments. Bronze surfaces have been known to become brown in nonindustrial areas, green in seacoast areas, and black in industrial areas. Higher temperatures cause attack of the bronze surfaces by sulfur, oxygen, halogens, ammonium compounds, and some acid vapors.

Metal-sprayed bronze offers decorative coatings for steel or wrought iron hardware and is usually followed by chemical conversion coloring and lacquering. Often, a sprayed zinc or aluminum coating is applied to the steel before bronze spraying. The sprayed bronze coatings must be applied to clean, grit blasted surfaces for adhesion that depends on *mechanical keying*. The surface blasting cannot be *shot blasted*. Grit blasting, using the correct, sharp, temperature-controlled blast media, is necessary to obtain the exact surface topography. Metallic compositions of the bronze metal wire, ribbon, or powder can be varied for certain bronze coating characteristics, e.g., the addition of zinc for corrosion improvement. Typical coating thicknesses range from 75 to 250 μm.

10.7.3 Copper-Nickel Alloy Coatings

Copper-nickel, or *cupronickel,* alloys may be specified for any copper-to-nickel coating composition by casting, electrodeposition, or metal spraying. Typical uses for the cupronickels include applications for strength and corrosion resistance such as in condenser tubes, heat exchangers, fasteners, holloware, surgical and dental instruments, and seawater-resistant parts. Railways, power plants, desalination plants, refineries, and marine industries find extensive use for the cupronickels for their corrosion resistance to natural waters, condensates, boiler waters, and atmospheres in rural, marine, and industrial locations. Selection of the desired cupronickel alloy coating or cladding composition can be based on the required strength and corrosion resistance. Increasing the percentage of nickel increases the deposit strength and lowers conductivity. Additions of lower percentages of other metals such as iron, cobalt, manganese, tin, and zinc can increase strength and corrosion resistance.

The selection of a cupronickel alloy coating may begin with a consideration of the assets of 70–90% Cu/10–30% Ni alloys, known for their resistance to attack by high-velocity seawater, fresh water, polluted water, steam, phosphoric and sulfuric acids, sulfates, nitrates, and chlorides. Additionally, the 30% Ni/70% Cu alloy is resistant to stress corrosion cracking.[144] The application of cupronickel alloys finds its greatest popularity in the use of metal spraying and cladding. Spraying methods have been used for repairs on iron castings, and claddings have been used to coat nickel-coated steel sheets to protect marine exposed parts from corrosion.

10.8 Tin and Tin Alloy Coatings

Recent emphasis on tin and tin alloy coatings rests largely with efforts to eliminate or minimize the use of toxic properties associated with metals such as cadmium, lead, and to a lesser extent zinc and nickel. The nontoxic characteristics of tin coatings, coupled with corrosion protection and good solderability, have qualified tin as acceptable for certain food industry finishes. Electroplated tin, as a coating for steel "tin cans," combined with a lacquer coating, was considered to be the largest single application of in terms of tonnage of plated product.[145] Other significant users of tin include electronics, printed wiring boards, wire coating, automotive parts, and the hardware and refinery industries. The use of tin and tin alloy coatings continues to increase for reasons of nontoxicity, low cost, corrosion resistance, solderability, lubricity, and ductility. The following application methods for tin and tin alloy coatings are popular procedures, ranked here by cost from lowest (electrochemical displacement) to highest (mechanical):

- Electrochemical displacement (immersion deposition)
- Electrodeposition of bright and matte tin
- Hot dipping
- Autocatalytic deposition
- Spraying
- Mechanical

The following tin alloys represent environmentally popular tin coatings that have gained more attention as replacements for lead, cadmium, and nickel coatings:

- Tin
- Tin-lead
- Tin-zinc
- Tin-nickel
- Tin-cadmium
- Tin-cobalt
- Tin-bismuth

The solderability provided by tin and tin alloy coatings satisfies one of the main requirements of electronics industry manufacturers for product assembly success. The increased attention demanded by soldered connections costs vast sums annually in terms of the extra time required to touch-up, rework, or repair poor solder connections. Experience shows that a major solderability defect is dewetting of the solder surface. This is a condition seen during the molten soldering or reflowing when the solder recedes, leaving irregular mounds of solder separated by areas covered with a thin solder film over the base metal. Nonwetting is similar, with exposure of the bare metal base.

Most pure tin coating specifications list corrosion resistance, deposit thickness, solderability, and appearance (brightness and smoothness) as major considerations. Additional protection may consist of reflowing or a protective lacquer. Typical coating thickness for corrosion protection of steel and other alloys are spelled out in many publications. However, some specific applications offer proven advantages such as pore-free deposits and higher tin purity by the minimization of organic impurities and hydrogen embrittlement. When solderability is required for tin-coated parts that have been in storage for over one year, *tinning* the parts in molten tin prior to assembly has served to greatly improve the assembly soldering operations. Parts that fail to *tin* properly may be rejected from bins that are scheduled for automatic or hand-soldering operations. A second method used to restore solderability to aged parts is to clean them in an acidic (usually fluoboric acid) soap solution, followed by an immersion tin dip. Stripping and retinning the parts is necessary when solderability cannot be restored following the *tinning* or immersion tin efforts. Many tin solderability problems in the electronics industry continue to increase labor costs for touching up soldered connections.

Tin specifications ISO 2093 and ASTM B545 enumerate required thicknesses for "service conditions" such as exposure to a particular atmosphere (see Table 10.5). MIL-HDBK-1250 restricts the use of bright tin and allows acceptance of only the nonbrightened (matte) tin coatings. Agreement between the tin coating supplier and purchaser may require an *average* thickness to be met. Also, other demands may be made, such as porosity testing, solderability, and age testing.

TABLE 10.5   Tin Specifications

Food and beverage contact
  0.5–2.0 µm plus lacquer

Food, water, and beverage constant contact
  30 µm

Automotive aluminum engine pistons
  2.5–5.0 µm[146]

Exceptionally severe service on steel and other metal alloys
  30 µm (International Std. ISO 2093)
  30 µm (ASTM B545)
  45 µm average (ASTM B545)

Severe service on steel and other alloys
  20/15 µm (ISO 2093)
  20/15 µm (ASTM B545)
  30/20 µm steel and other average alloys (ASTM B545)

Moderate service on steel and other alloys
  12/8 µm (ISO 2093) for steel and other alloys
  10/8 µm steel and other alloys (ASTM B545)
  15/12 µm steel and other alloys average (ASTM B545)

Mild service on steel and other alloys
  5/5 µm (ISO 2093) for steel and other alloys
  5/5 µm steel and other alloys (ASTM B545)
  8/8 µm steel and other alloys average (ASTM B545)

Other specifications and tests for tin coatings cover single-reduced and double-reduced electrolytic tin plate for low-carbon, cold-reduced coil and cut steel. ASTM A624, A624M, A626, and A626M provide specifications for tin mill products with different surface conditions of electrolytic deposited tin. Tin deposit thickness is measured by coating weight per unit area (ASTM A630). AMS 2408E covers the properties and engineering requirements for an electrodeposited tin mainly used for preventing galling and seizing, preventing nitriding, and for solderability. Coating thicknesses normally call out multiples of 2.5 μm.

10.8.1   Immersion Tin Coatings

Immersion tin coatings deposit thin coatings by chemical replacement until all surface replaceable ions are essentially tin. Typically, tin deposits onto copper by serving as "less noble" (i.e., higher in the emf series). Copper's potential is rendered significantly lower than tin's by the presence of thiourea or other complexing compounds, enabling tin deposition onto copper. The reaction ceases when all replaceable ions have been removed or, as is the case with aluminum, all evolution of hydrogen has ceased. Thickness can be increased by part contact with additional electronegative metals held in contact with the immersion solution. AMS 2409 specifies a thin (about 2.5 μm thick) immersion coating on aluminum normally applied for anti-galling and abrasion resistance. Relative to a second immersion tin process, MIL-T-81955 covers immersion tin for copper and copper alloys with thin (less than 0.6 to 2.5 μm) coatings. Immersion tin coatings have served successfully when their limitations have received proper consideration. Coatings include copper wire and circuits on printed wiring boards for solderability and adhesive bonding, aluminum alloy engine pistons, tinning the inside of copper tubing, paper clips, steel wire, and small electronic parts. Limitations include the need for good base metal preparation for immersion tin protection. The thin and often porous coatings offer minimal protection when humidity and temperatures increase, and the immersion tin solutions become contaminated, requiring frequent solution replacement.

Immersion tin publications proliferated during the 1980s and 1990s due to circuit board industry requirements for protecting copper circuitry from oxidation and corrosion. Other reports highlight immersion tin as a copper etching resist and promote its copper-to-polymer bonding and use in maintaining the solderability of copper and reflowed electroplated tin and tin-lead plated copper circuitry.[147–154]

Many printed circuit automated assembly operations rely on immersion tin to protect copper circuitry during component wave and hand soldering of components. Successful assembly relies on immersion coating to produce acceptable solder joints where unacceptable ones have occurred due to the following factors:

- Poor cleaning and activation of the copper circuitry before immersion tin application

- Immersion tin coating that is too thin (should be 1.25 μm minimum)

- Contaminated tinning bath, usually from dissolved base metal ions

- Time between immersion tin coating and soldering that has exceeded three months

- Situations in which the immersion coated printed circuit board or hardware experienced heating prior to assembly, or extended heating during assembly

- Problems when protective shipping materials (e.g., sulfur-free paper, dry nitrogen atmosphere bags, and humidity indicator strips) were not used

- Copper-tin intermetallics formed to adversely affect coating adhesion and solderability[159–161]

Overall, immersion tin coatings provide low cost, environmentally acceptable and selective plating when isolated copper circuits on a dielectric (such as epoxy or polyimide laminates) must be coated. Competition for an optimal solderable surface for copper circuits continues with alternatives such as organic conversion coatings,[155] silver immersions with tarnish inhibitors,[156] electroless palladium,[157,164] and electroless nickel-gold immersion for solderability and wire bondability.[158] The *solder mask over bare copper (SMOBC)* processes that use immersion tin coatings continue to compete successfully against the alternative coatings as a more economical method for solderability preservation than tin reflowing or hot air leveling (HAL). More information on surface finishes may be obtained from six publications presented by the Surface Mount Technology Association.[165]

10.8.2 Electrodeposited Tin Coatings

Electroplated tin coating processes normally produce bright or matte deposits via different bath chemistries that serve as alternatives for tin-lead and cadmium coatings. Good solderability and corrosion resistance is offered for the nontoxic coating; however, caution must be employed because of the formation of tin whiskers. Tin coating use in printed wiring board and semiconductor packaging has been limited because of the danger of conductors *bridging* or shorting out. Numerous studies concerning tin whiskering are listed at the following web site: http://nepp.nasa.gov/whisker/. Some of the earliest whisker problems were observed during the 1950s in the communications industry, when tin whiskers shorted out resistors in telephone equipment.

Complete freedom from whisker growth cannot be guaranteed for pure tin deposits, but the following remedies to limit whisker growth:

- Reflowing or fusing the tin deposit before soldering

- Using a tin alloy (tin-bismuth, tin + 3% minimum lead)

- Prohibiting the use of bright tin coatings

- Use of matte tin deposits without co-deposited organic materials (e.g., brighteners and grain refiners)

- Tin thickness >5 μm

Solderability of electroplated tin and tin-lead coatings is concern in the electronics industry during initial assembly stages and especially when resoldering component replacement parts. The United States Department of Defense, its suppliers, and the commercial electronics manufacturing industry adopted standards (Mil-M-38510, IPC test method S-801, Ele. Ind. Spec. RS-319, and Mil-Std-202, method 208) for acceptable solder joints and workmanship that improved product yields and reduced repair costs. Efforts on the part of government and industry initiated numerous discussions on how to improve solderability, and some of the following problems were addressed:

- Tin fusing, or reflowing, in a nitrogen atmosphere for improved solderability removes the co-deposited organic materials; also, wetting and nonwetting conditions are exposed before further assembly.

- Organic materials within the tin, tin-lead, and underplate coatings can produce solder defects such as voids, dewetting, nonwetting, and gas holes (keep carbon to less than 0.05%).[162]

- Cleaning and conditioning of the base metal before coating offers similar problems, along with poor adhesion (good process controls needed).

- Intermetallic formations between metals within the base metal-to-coating contribute to poor adhesion and dewetting.[163]

- Effects of aging, storage environment, and packaging materials must be addressed.

- Proper control of the coating equipment and materials is essential.

- Proper control of the fusing, leveling, and reflow equipment and chemicals such as fusing fluids must be maintained.

- Temperature history of the hardware prior to the final solderability problems should be considered.

- Coating thickness should be at least 5 μm.

- Sequential electrochemical reduction analysis (SERA), nondestructive surface testing equipment has been introduced to determine solderability.[175–176]

10.8.3 Hot-Dipped Tin Coatings

Hot-dipped tin coatings cover limited applications for steel and copper base alloys such as large sheets, buckets, outside steel containers, tanks, housings, and small items that can be held in baskets or racks. Parts are first cleaned then centrifuged to remove the excess metal, quenched in water, and coated with a thin film of paraffin. Meticulous base metal cleaning and fluxing is essential to prevent nonwetting and dewetting. The tin coating is thin, may contain pores, and often is non-uniform in thicknesses that range from 1.5 to 30 μm. Part geometry and hot-tin operating variables influence the thickness. Several specifications for tin coatings combine the hot-dipped and electrodeposited requirements and present an array of tests for both types of coatings for "as-required" needs (MIL-T-10727; ASTM B545; ASTM A624, A624M, A626, and A626M; FORD WSB-M1P9-B).

10.8.4 Sprayed Tin Coatings

Sprayed tin coatings are applied by wire or powder fed into a flame torch, using techniques to minimize occlusion of oxygen in the coating. Applications include large vessels as often used for food manufacturing, tanks, and support brackets. The coating thickness should be greater than 200 μm to minimize porosity and to serve with good corrosion resistance. Following the spray application, the coating is often scratch-brushed for a more uniform appearance.

10.8.5 Electroless Tin Coatings

Electroless tin coatings deposited by autocatalytic reduction processes have been reported that use titanium trichloride as a reducing agent.[166] Limited use of electroless tin processes is noted because of their lack of commercial usefulness. An electroless solder process based on tin and lead fluoborates, thiourea, sodium hypophosphite reducing agent, EDTA, and a surfactant has been reported.[167]

10.8.6 Tin-Lead Coatings

Tin-lead coatings, being mostly lead, offer good atmospheric corrosion protection for steel and serve as overlays for bearings (*terne plate* with tin 12 to 25%) and wire coatings.[168] Tin-lead coatings, with tin as a majority at 50 to 70 percent, present good solderability requirements for wire and act as an etch resist for printed wiring boards. Tin concentrations of 95% tin/5% lead have been used for solderability and to prevent whisker growth on the surfaces of semiconductor devices. Some guidelines offered for tin-lead include specifications MIL-T-10727, ASTM B200, ASTM B605, ASTM B579, and ISO 7587.

Metallic, organic impurities and aging can affect the solderability of tin-lead coatings. Small accumulations of iron, copper, oxides, and especially gold in molten hot-dipping operations must be minimized. Similar metals, and particularly organic co-deposited materials in electrodeposited tin-lead coatings, greatly affect solderability—enough to cause some electronics specifications to only accept non-bright (matte) coatings that contain only trace amounts of organic materials. Time and temperature can increase intermetallic growth between tin and the base metal and oxide formations on the exposed coating surface. Post-plating processes such as fusing, reflowing, and hot-air leveling improve the cosmetics of the matte tin appearance and enhance solderability.[169]

Lead is considered to be very toxic, as it can enter the ground via leaching rain water and similar waters at pH values of less than 7. Lead compounds and coatings are under investigation with the aim of finding nontoxic replacement finishes.

10.8.7 Tin-Zinc Coatings

Tin-zinc coatings continue to attract interest as direct replacements for cadmium coatings because of toxicity problems discovered during the 1980s. Compositions in the 25 ± 5 percent zinc range have proven to be the most popular for the corrosion protection of steel and other metals. Higher concentrations of zinc (up to 80 percent) could be used for special applications. Electrodeposition is the primary application process, and the alkaline, non-cyanide, stannate processes are improvements over the cyanide baths. Deposits in the 25 percent zinc range have good wear resistance, 37 kg/mm^2 hardness,[171] corrosion protection with minimal corrosion product, and a satin, white-colored appearance. Comparisons with cadmium and zinc of equal thickness report that tin-zinc coatings last longer in industrial atmospheres than cadmium, but not as long as zinc. Marine atmospheres prove tin-zinc to be better than zinc and equal to cadmium. Tin-zinc is not as protective in city environment as cadmium and zinc. Static potential measurements in a 5 percent salt solution reported a lower corrosion rate for tin-zinc than cadmium and zinc.[170] Post-chromate conversion coating improves corrosion protection in humidity and salt-spray tests.

Tin-zinc coatings in the 25 percent zinc concentration serve as good candidates to replace cadmium in terms of corrosion protection, lack of whisker formation, lubricity, and solderability. References 172 through 174 present additional tin-zinc information.

10.8.8 Tin-Nickel Coatings

Tin-nickel coatings, for most practical purposes, are electroplated to produce a nickel concentration of 33 to 35 percent nickel, which presents hard, brittle, and often stressed deposits. Applications for the pink-white-tinged colored coating have been in electrical equipment, watches, lamps, light fixtures, jewelry, furniture decorations, and printed wiring boards. Tin-nickel circuit boards, finished with 1 μm or less of gold, provide a tin-nickel, whisker-free, solderable surface[177] that withstands etchants and corrosive substances with a rating of 650 to 700 Vickers hardness (kg/mm^2).[178] Electroplated tin-nickel plating baths have good throwing power for low-current-density recesses and plated-through holes. Tin-nickel's corrosion resistance is not as good as its chemical resistance. Resistance to tarnish in sulfide atmospheres is good; however, moist environments require the tin-nickel coating to be pore-free, to have a copper underplate, or to be sufficiently thick (>25 μm) when protecting steel or nickel. High temperatures above 250°C must be avoided because of transformation changes that cause the plating to crack and peel.[179] ASTM B605 (ASTM Philadelphia, PA, Vol. 2.05, 1992) can be consulted.

10.8.9 Tin-Cobalt Coatings

Tin-cobalt alloy coatings electroplated with approximately 20 percent cobalt have gained popularity as replacements for decorative chromium finishes because of the undesirable features of chromium plating. Other replacements for chromium finishes include blue-tinged passivated zinc, which does not have the abrasion resistance, and tin-nickel alloys that offer protection but do not have the desirable chromium color. Tin-cobalt finishes present good tarnish resistance, high hardness ratings of approximately 300 to 400 Vickers, and have the distinctive blue-white color of chromium. Additional advantages of tin-cobalt coatings are that they provide overall safer operation and lower operating costs, it is easier to control plating and waste treatment, and it is possible to plate parts with complex geometries without burning the part edges.[180] Popular tin-cobalt bath chemistries consist of sulfate,[181,182] fluoride, and pyrophosphate baths. Selection of the proper tin-cobalt process and thickness is important to avoid later darkening of the coating and loss of corrosion resistance in certain environments. Tin-cobalt coating thicknesses of at least 0.25 μm are recommended over nickel underplates.

10.8.10   Tin-Bismuth Coatings

Tin-bismuth coatings recently gained interest as a result of efforts to replace lead in tin-lead coatings and solder for electronics industry products. Tin-bismuth and other "environmentally friendly" alloys are part of the search for non-lead solders and solder coatings. The National Center for Sciences completed a study of more than 75 different alloys without finding an exact alternate.[183] Special applications may warrant the use of tin-bismuth alloy coatings, but this coating has not received acceptance for printed wiring board manufacturing.[184]

10.9   Silver Coatings

Electroplated silver coatings were originally reported circa 1800,[185] but the first patent for silver plating was awarded approximately 40 years later to the Elkington's,[186] which initiated the electroplating industry. The greatest use of silver electroplated coatings includes flatware, holloware, artistic structures, and jewelry applications. Silver has been considered suitable for food contact. Silver coatings also find use on bearing surfaces, electrical contacts, semiconductor lead frames, and waveguides. Silver finishes are usually classified as bright, semibright, and matte, and they are applied by cladding, electroplating, immersion, and (rarely) by electroless methods. Silver has superior thermal and electrical conductivity, excellent solderability, low hardness of 25 Vickers, and they can be polished to high reflectivity.

10.9.1   Silver Tarnish Resistance

Resistance to corrosion, tarnishing, and migration must be considered when specifying a silver coating. Tarnish appearances of black, brown, and yellow discolorations mar silver surfaces as a result of atmospheric hydrogen sulfide and moisture. Sulfide films will increase contact resistance. Tarnishing defects are serious problems for decorative applications, bonding chips to silver-plated lead frames, and reduction of microwave transmissions in waveguides. Moist environments that contain ammonia, halogen gases, and ozone will corrode silver. The following treatments have reduced or prevented silver tarnishing:

- Lacquers can be added to protect decorative items.
- Electrophoretic additions of acrylic and epoxy coatings can protect silver.

- Chromate conversion coatings reduce corrosion as per MIL-S-365, Grade A.

- Organic solderability preservatives (OSPs) can be used for silver coatings.[187]

- Cathodic treatment in an alkali chromate solution can be used to form a chromate conversion coating.

- Consider using a silver alloy coating such as silver-palladium (~40% Pd)[188] or silver-tin.[189]

- Overplate the silver with 0.2 ± 0.1 μm of rhodium. This is expensive, but it prevents tarnish.

10.9.2 Silver Migration

Silver migration is an additional concern in the use of silver for conductors and printed circuits on plastics and ceramics. Evidently, silver ions can migrate along surfaces and through insulated materials to another conductor, given high humidity and a positive direct current potential. The silver ions reach the opposite conductor and are reduced back to silver metal, which leads to electrical shorts. Military and communications industry specifications banned the use of silver coatings where these migrating circuit paths can be formed. Some methods to minimize silver migration have been reported in MIL-STD-1250, Corrosion Prevention and Deterioration Control in Electronic Components and Assemblies. Suggestions include the use of the widest possible conductor spacings, protecting the conductors with a moisture barrier or conformal coating, overplating the conductors with gold or platinum, the use of nonhygroscopic insulation, and adding effective humidity control. Or you can select an alternative conductor coating.

10.9.3 Silver Coating Methods

Silver cladding, electroplating, autocatalytic (electroless), and immersion plating serve as popular applications. Cladding methods have the advantages of nonhydrogen embrittlement, as compared to electroplating; however, cladding methods are not easily adapted for coating complex shapes. Adhesive and soldering processes must be employed to bond silver foil to the base. Spinning, drawing, and hot-rolling techniques can be used to form a variety of clads for bonding to aluminum for lightweight electrical contacts; bonding to copper for jewelry, RF cables and transmission lines; bonding to nickel wire for high-temperature conductors; and bonding to carbon and stainless steels for electrical conductivity and strength. Typical thickness requirements for silver coatings are listed in Table 10.6.

TABLE 10.6   Thickness Requirements for Silver Coatings

Jewelry	20 µm
Cutlery and flatware	10–50 µm for ferrous or aluminum bases with Cu and/or Ni undercoat
Hollowware	5–25 µm for ferrous or aluminum bases with Cu and/or Ni undercoat
Decorative reflectors	<10 µm
Atmospheric corrosion	10–25 µm over 10–20 µm of nickel (electrical industries)
Electronic contacts	1–10 µm
Relay switch contacts	20–40 µm
Heavy relay switches	20–400 µm

10.9.3.1 Electrodeposition of silver coatings. Electrodeposition of silver coatings is usually from an alkaline cyanide-silver electrolyte that contains brighteners that can produce bright, semibright, and matte deposits. Nonbrightened cyanide-silver baths offer matte deposits. Cyanide silver processes are almost always the choice for electrodeposited silver coatings; however, other processes have been used such as nitrate, halide, lactate, citrate, thiosulfate, tartrate, methyl sulfate, sulfamate, thiocyanate, ferrocyanide-pyrophosphate, and sulfasalicylate. Replacements for cyanide silver processes continue to appear that avoid toxic cyanide and detrimental carbonate formations. Most alternative cyanide silver baths have been successful when specified for high volumes of similar base metal parts where positive experience is repeated and diverse base alloys are avoided.

Numerous silver electroplating coating specifications highlight the essential technical and decorative characteristics, including the procedures and tests, for the required coating. These specifications evolved from early commercial and decorative standards into industrial, government, and federal and military standards primarily intended as procurement documents. Basic silver plating specifications agree on deposit thickness values, weight, appearance (bright, semibright, and matte), smoothness, purity (98.0–99.9 percent), corrosion resistance, solderability, conductivity, heat resistance, risk of hydrogen embrittlement, and fitness for food contact. Some specifications demand the avoidance of silver coatings in contact with titanium and constant temperatures exceeding 1200°F. Additional requirements for a silver tarnish-resistant chemical conversation coating has increased the options in most of the standard specifications.

Techniques to change the characteristics of the electrodeposited silver coating (e.g., electrical conductivity, hardness, tarnish resistance, tensile strength, ductility, wear resistance, and solderability) by considering different silver processes include the following:

- Periodic reverse plating the can increase electrical conductivity and deposit density.
- Pulse plating will serve in a similar manner.
- Ultrasonic agitation can be added.
- The bath temperature can be increased.
- The plating current density can be increased or decreased.
- One can add an alloy to the bath such as antimony (0.7 to 10.0 percent), bismuth (1.0 to 2.5 percent), cadmium (2 to 50 percent), cobalt (20 to 50 percent), copper (20 to 85 percent), lead (4 to 10 percent), nickel (8 to 80 percent), palladium (12 to 90 percent), and thallium (9 to 12 percent).[190]
- A silver tarnish resistant chemical conversion coating can be added, such as referenced in U.S. Military Standard QQ-S-365.

10.9.3.2 Electroless deposition of silver coatings. Autocatalytic or electroless silver processes are rare; however, an alkaline dimethylamine borane and silver cyanide formulation can be obtained through some plating chemical suppliers. Immersion silver coatings have increased in importance as an alternative to nickel/gold coatings onto copper alloys for assembly soldering and wire bonding. These immersion silver coatings, including one formulation with an oxide/sulfide tarnish inhibitor, have increased in demand for printed circuits compared to hot air solder leveling (HASL) and other organic solderability preservatives (OSPs).[191]

References

1. H.H.Uhlig, *Corrosion and Corrosion Control,* 2nd ed., pp. 334–335, John Wiley & Sons Inc., New York (1971).
2. J. Edwards, *Coating and Surface Treatment Systems for Metals: A Comprehensive Guide to Selection,* Finishing Publications Ltd., Stevenage, Hertfordshire, England and ASM International, Materials Park, Ohio, U.S.A., 1997.
3. E. L. Owen, "Interdiffusion," *Properties of Electrodeposits: Their Measurement and Significance,* R. Sard, H. Leidheiser, Jr., and F. Ogborn, eds., The Electrochemical Society (1975).
4. M. R. Pinnel, "Diffusion Related Behavior of Gold in Thin Film Systems," *Gold Bulletin,* 12, No. 2, 62 (April 1979).
5. G. Lewis, *Selection of Engineering Materials* (Englewood Cliffs, New Jersey: Prentice Hall Division of Simon & Schuster, 1990) pp. 472–473.

6. D. R. Askeland, *The Science and Engineering of Materials* (Boston, MA: PWS Publishing Co., 1994), pp. 125–128.

7. W. H. Safranek, *The Properties of Electrodeposited Metals and Alloys* (Orlando, Florida: American Electroplaters and Surface Finishers Society, 2nd ed., 1986), p. 21.

8. C. A. Harper and H. L. Pinkerton, *Handbook of Materials and Processes for Electronics* (New York, NY: McGraw-Hill Book Company, 1970), Chapter 10, "Metallic and Chemical Finishes on Metals and Nonconductors," pp. 10.1–10.62.

9. L. J. Durney, *Electroplating Engineering Handbook* (London, UK: Chapman & Hall Publishing Co., 4th ed., 1984).

10. U. Landau, *Proc. Institute of Metal Finishing Conference,* 16 April 1986, 57.

11. T. Daenen, J v d Berg and G v Dijk, *Trans Inst Met Finishing,* 1985, 63(3,4), 104.

12. F. A. Clay, W. B. Harding, and C. J. Stimetz, "Electrodeposition of Aluminum," *Plating,* 56, No. 9, 1027–1037 (September 1969).

13. R. Suchentrunk, "Corrosion Protection by Electrodeposited Aluminum," AGARD Lecture series No. 6, Materials Coating Techniques, March 1980, Defense Technical Information Center Technical Report AD A085603. Suchentrunk, R., "Corrosion Protection by Electrodeposited Aluminum," Z. Werkstofftech, 12, 190–206 (1981) (in English).

14. S. Jayakrishnan, M. Pushparanam, and B. A. Shenoi, "Electrodeposition from Organic Solutions of Metals that are Difficult to Deposit from Aqueous Solutions," *Surface Technology,* 13, 225–240 (1981).

15. W. H. Safranek, W. C. Schickner, and C. L. Faust, "Electroforming Aluminum Wave Guides Using Organic-Aluminum Plating Baths," *Journal of the Electrochemical Society,* 99, (2), 55–59 (February 1952).

16. N. F. Murphy, *Metal Finishing,* 50, 76 (April 1952).

17. A. W. Castle, *Electroplating,* 8, 291 (1954).

18. D. E. Couch and A. Brenner, *J. Electrochem. Soc.,* 99, 234, (1952).

19. C. A. Waine, AB Connectors (Northhampton), R&D Report No 88/06, March 1988.

20. J. H. Connor and A. Brenner, *J. Electrochem. Soc.,* 103, 657 (1956).

21. G. V. Alm and M. H. Binstock, North American Aviation-SR-2704, "Electroclad Aluminum on Uranium."

22. J. G. Beach and C. L. Faust, *J. of Electrochem. Soc.,* 106, 654 (1959).

23. E. Harsanyi, U.S. patent 1,897,902 (1933).

24. A. E. Jackson, *Trans Inst Met Finishing,* 1963, 40(1), 1.

25. J. Edwards, *Coating and Surface Treatment Systems for Metals: A Comprehensive Guide to Selection,* Finishing Publications Ltd., Stevenage, Hertfordshire, England and ASM International, Materials Park, Ohio,U.S.A., 353, 1997.

26. Fahnoe and coworkers, U. S. patent 2,858,256 (1953).

27. J. Benjamin and A. B. Osborn, *Trans. Faraday Soc.,* 36, 287 (1940).

28. S. Senderoff and W. E. Reid, Jr., U. S. patent 2,843,541 (1958).

29. A. C. Werner and R. J. Ableson, "Preparation of Protective Coatings by Electrophoretic Methods," WADC Tech. Rept. 58–11, (Feb. 1958).

30. N. F. Cerulli, *J. Am. Ceram. Soc.,* 33, 12, 373 (1954).

31. E. J. W. Verway et al., U. S. patent 2,321,439 (1943).

32. M. W. Raney, *Electrodeposition and Radiation Curing of Coatings,* Park Ridge, N. J.: Noyes Data Corp., 1970.

33. R. L.Yeates, *Electropainting.* Teddington (England): Robert Draper Ltd., 1970.

34. G. E. F. Brewer, and R. D. Hamilton, *Paint for Electrocoating, ASTM Gardner-Sward Paint Testing Manual,* 13th ed. Philadelphia, PA, 1972.

35. G. E. F. Brewer, (Editor). "Electrodeposition of Coatings," *Advances in Chemistry,* Vol. 119, Washington D.C.: American Chemical Soc., 1973.

36. B. Graves, ed., and S. R. Kline, Jr., assoc. ed., "What Does it Really Cost?" *Products Finishing,* November 1999, pp. 12s–15s.

37. V. L. Lamb, and W. E. Reid, Jr., *Plating,* 47, 291 (1960)

38. J. J. Shyne, et al., "Electrokinetic Processes-Nuclear Aspects," Quarterly Progress Report for Nov. 1, 1955–Jan. 31, 1956, KLX-10021 (decl.).

39. J. Edwards, *Coating and Surface Treatment Systems for Metals: A Comprehensive Guide to Selection,* Finishing Publications Ltd., Stevenage, Hertfordshire, England and ASM International, Materials Park, Ohio,U.S.A., p. 24, 1997.

40. D. A. Swalheim and M. Schwartz, Finishing Highlights, American Electroplaters Society, "Special Report: Plating on Plastics, The Future Brightens for Plating on Plastics," pp. 29 and 31, Nov./Dec. 1977, Report from: Plating for the Electronic Industry Symposium, Los Angles, CA, Jan. 30–Feb. 3, 1978.

41. W. Bialojan and M. Geisler, *Products Finishing,* "Vacuum Metallizing Plastic Parts," pp. 46–52,Oct. 1992.

42. J. A. Thornton, L. J. Durney, ed., *Electroplating Engineering Handbook,* "Vapor Phase Methods," (London, UK: Chapman & Hall Publishing Co., 4th ed., 1984), pp. 426–428.

43. B. Lee, "Decorative Finishing Using PVD," *Products Finishing,* pp. 46–50, October 1995.

44. D. M. Mattox, Education Guide to Vacuum Deposition Technology, Society of Vacuum Coaters publications, 1994 ed.

45. J. W. Dini, "Ion Plating Can Improve Coating Adhesion," Inc. This work was performed under the auspices of the U. S. Department od Energy by Lawrence Livermore National Laboratory under Contact No. W-7405-Eng-48, Metal Finishing, Elsevier Publishing Co., September, pp. 15–20, 1993.

46. D. M. Mattox, "Vacuum Deposition Processes," *Products Finishing,* June, pp. 48–58, 1999.

47. G. Legge, "Ion-Vapor-Deposited Coatings for Improved Corrosion Protection," *Products Finishing,* October, pp. 59–64, 1995.

48. V. L. Holmes, D. E. Muelberger and J. J. Reilly, The Substitution of IVD Aluminum for Cadmium, McDonnell Aircraft Company, P. O. Box 516, St. Louis, MO 63166 and C. J. Carpenter, Department of the Air Force, Headquarters—Air Force Environmental Services Center, Panama City, FL 32403.

49. M. S. White, Cadmium Association report on NBS Government-Industry Workshop on alternatives to cadmium electroplating in metal finishing, October 1977.

50. F. H. Meyer, *Plating and Surface Finishing,* 69(11),39, 1982; D. E. Muelberger, ibid, 40.

51. P. L. Lane and C. J. E. Smith, Engineering the Surface Conference, London, 1986.

52. P. L. Lane and C. J. E. Smith, RAE Technical Report 88026, March 1988.

53. D. M. Mattox, "Residual Film Stress," Management Plus, Inc., donmattox@svc.org, Society of Vacuum Coaters, *News Bulletin,* 7, Spring, 1997.

54. H. Herman, "Plasma-sprayed Coatings," *Scientific American,* 9, 78–83, 1988.

55. E. Proverbio, G. Bonifazi and T. Valente, Dept.ICMMPM, University of Rome "La Sapienza," Italy, "Evaluation of Plasma-Sprayed Coating Quality by SEM and Image Analysis," USA Microscopy and Analysis, 19–21, July 1997.

56. N. A. Tiner, "Refractory Coatings for Aerospace Applications," Aerospace Techniques, Chemical Engineering Progress Symposium Series, 60, 54–60.

57. W. H. Safranek, CEF, "Cadmium Plating," Plating and Surface Finishing, *Journal of the American Electroplaters and Surface Finishers Society, Inc.,* Vol. 86, No. 11, p. 34, Nov. 1999.

58. F. A. Lowenheim, *Electroplating,* Sponsored by the American Electroplaters Soc., McGraw-Hill Book Co., New York, pp. 184–187, 1978.

59. D. Crotty, "Torque and Tension Control for Automotive Fasteners," *Metal Finishing,* pp. 44–46, 48–50, May 1999.

60. S. L. Law, "Corrosion of Cd-Plated Components of Radar Shelter," Research Report, Gascoyne Laboratories, Inc., Baltimore, MD, 21224, Northrop Grumman Proprietary Report. 1989.

61. ASTM B766, Standard Specification for Electrodeposited Coatings of Cadmium, Under the jurisdiction of ASTM Comm. B-8 on Metallic and Inorganic Coatings and is the direct responsibility of Subcommittee B08.04 on Soft Metals, 1998

62. F. A. Lowenheim, *Electroplating,* Sponsored by the American Electroplaters Soc., McGraw-Hill Book Co., New York, p. 64, 1978.

63. N. A. Isgarischev, *Journal Electrodepositors' Tech. Soc.,* 13 (1–6),(1937).

64. F. Forester and K. Klemm, Z. *Electrochem.,*35, 409,(1929).

65. E. Muller and H. Barchmann, Z. *Electrochem.,*39, 341,(1933).

66. S. Wernick, *Trans. Am. Electrochem. Soc.,* 55, 333,(1932).

67. P. G. Humphreys, "New Line Plates Non-Cyanide Cadmium," *Products Finishing,* pp. 80–90, May 1989.

68. J. W. Dini, *Electrodeposition, The Materials Science of Coatings and Substrates,* Noyes Publications, Park Ridge, New Jersey, pp. 11–45, 1993.

69. H. H. Johnson, J. G. Morlet and A. R. Troiana, "Hydrogen Crack Initiation and Delayed Failure in Steel," *Trans. Metallergical Soc. AIME,* 212, 528–536, 1958.

70. R. P. Thierry, "Re-Embrittlement of Cadmium-Plated High Strength Steel," presented at Eleventh Airlines Plating Form, San Francisco, Feb. 1975.

71. H. Narcus, *Metal Finishing,* 43, 188, 199, 242,(1945).

72. K. Takata, Japanese patents SHO-35 18260 (1960) and SHO-38 20703 (1963).

73. AMS-2419A, "Cadmium-Titanium Alloy Plating," Soc. of Automotive Engineers, Warrendale, PA (1979).

74. W. Sheng-Shui, C. Jing-Kun, S. Yuing-Mo and L. Jin-Keul, "Cd-Ti Electrodeposits From a Non-Cyanide Bath," *Plating and Surface Finishing,* 68,62 (December 1981).

75. H. Morrow, "Cadmium Electroplating," internal report from The International Cadmium Association.

76. L. R. Westbrook, U. S. Patent 2,893,934 (1959).

77. B. E. Scott and R. D. Grey, Jr., *Iron Age,* p. 59, Jan. 18, 1951; R. D. Grey, Jr. and W. A. Paecht, U. S. Patent 2,609,338 (1952).

78. P. S. Bennet, *J. Electrodepositors' Tech. Soc.,* 26, 91 (1950).

79. A. E. Davies, *Trans. Inst. Metal Finishing,* 33, 72 (1956).

80. A. E. Davies, *Trans. Inst. Metal Finishing,* 33, 85 (1956).

81. A. W. Grobin, "Other ASTM Committees and ISO Committees Involved in Hydrogen Embrittlement Test Methods," Hydrogen Embrittlement: Prevention and Control, ASTM STP 962, L. Raymond, ed., ASTM, Philadelphia, PA, 46 (1988).

82. A. W. Grobin, Jr., "Hydrogen Embrittlement Problems," *ASTM Standardization News,* 18, 30 (March 1990).

83. R. E. Moellier and W. A. Snell, "Diffused Nickel-Cadmium as a Corrosion Preventive Plate for Jet Engine Parts," *Proc. Am Electroplaters Soc.,* 42, 189–192 (1955).

84. Aerospace Material Specification (AMS) 2416H, Plating, Nickel-Cadmium Diffused (1995).

85. G. Reinhardt, "Electroless Nickel Applications in Aircraft Maintenance," *Proc. 18th Annual Airline Plating and Metal Finishing Forum,* Soc. Automotive Engineers (1982).

86. L. Coch, "Plating Fasteners, Avoiding Embrittlement," *Products Finishing,* 51,56,May,1987.

87. MIL-C-81562B, Coatings, Cadmium, Tin-Cadmium, and Zinc (Mechanically Deposited) 1992.

88. FORD ESF-M1P69-A, Plating, Mechanical—Cadmium (1979).

89. FORD ESF-M1P68-A, Plating, Mechanical—(Tin/Cadmium) (1979).

90. FORD ESV-M1P74-A, Plating, Mechanical—Cadmium/Zinc, (1982)

91. Eric W. Brooman, "Alternatives to Cadmium Coatings For Electrical/Electronic Applications," *Plating and Surface Finishing,* Vol. 80, pp. 29–35, Feb. 1993.

92. Glen H. Graham, CEF, OC-ALC/LPPNP, Tinker Air Force Base, OK, and John Stropki, Battelle Memorial Institute, Columbus, OH, "Cadmium Plating Substitution with Zinc Alloys," 31st Aerospace/Airline Plating & Metal Finishing Forum & Exposition, Denver, CO, April 1995.

93. M. Ingle and J. Ault, "Corrosion Control Performance Evaluation of Environmentally Acceptable Alternatives for Cadmium Plating," AESF Surface Finishing Annual Tech. Conf. and Exhibit, Session U, 1992, e-mail: education@aesf.org.

94. NDCEE (National Defense Center for Environmental Excellence), "Cadmium Alternatives," Update, Vol. I, No. 5, August 1995, e-mail: http://www.ndcee.ctc.com.

95. George Shaw, U. S. Army Tank Automotive and Armaments Command Materials Engineering Team, (AMSTA-TR-E/MEPS), Warren, Michigan 48397-5000, "Long

Term Performance of Cadmium Alternatives," *Products Finishing,* pp. 66–81, February 1999.

96. Gerald G. Kraft, Kraft Chemical Co., Melrose Park, IL, "The Future of Cadmium Electroplating," *Metal Finishing,* pp. 29–31, July 1990.

97. Christopher Helwig, Brooktronics Engineering Corp., Valencia, CA, "Zinc-Tin: A Choice for Replacement of Cadmium For Brush Plating Applications," AESF Technical Paper, 1998.

98. C. H. Ko, C. C. Chang, L. C. Chen, and T. S. Lee,"A Comparison of Cadmium Electroplate and Some Alternatives," *Plating and Surface Finishing,* Vol. 78, p. 46, Oct. 1991.

99. Steve Schachameyer, R. Bauer, T. Halmstead and J. Newman, "The Search for Cadmium Plating Alternatives—A Cooperative Approach," AESF Surface Finishing Tech. Conf. and Exhibit, Session G, 1993.

100. Keith Cramer, Lisa Whiting and Beau Brinckerhoff, "Evaluation of Cadmium Plating Alternatives," 17th AESF/EPA Pollution Prevention and Control Conference, Orlando, FL, Feb. 1996.

101. D. A. Wright, N. Gage, and B. A. Wilson, "Zinc-Nickel Electroplate as a Replacement for Cadmium on High Strength Steels," *Plating and Surface Finishing,* Vol. 81, p. 18, July 1994.

102. J. Edwards, *Coating and Surface Treatment Systems for Metals: A Comprehensive Guide to Selection,* Finishing Publications Ltd., Stevenage, Hertfordshire, England and ASM International, Materials Park, Ohio, U.S.A., 1997.

103. J. Mazia and D. S. Lashmore, "Electroplated Coatings" in *Metals Handbook,* 9th ed., Volume 13, Corrosion, ASM International, Metals Park Ohio, 1987.

104. L. Schlossberg, "Corrosion Theory and Accelerated Testing Procedures," *Metal Finishing,* 62,57, April 1964.

105. F. Altmayer, "Choosing an Accelerated Corrosion Test," *Metal Finishing,* 83, 57, October 1985.

106. Testing Of metallic and Inorganic Coatings, W. B. Harding and G. A. DiBari, eds., ASTM STP 947, American Society for Testing and Materials, 1986.

107. D. L. Snyder, "Decorative Chromium Plating," *Metal Finishing,* 64th Guidebook and Directory Issue, Vol. 94, Jan. 1996.

108. J. E. Stareck, E. J. Seyb, and A. C. Tulumello, *Plating,* 41, 1171, 1954; *Proc. Am. Electroplaters Soc.,* 41, 209, 1954.

109. E. J. Seyb, A. A. Johnson, and A. C. Tulumello, *Proc. Am. Electroplater's Soc.,* 44, 29, 1957. See also J. E. Stareck, E. J. Seyb, A. A. Johnson, and W. H. Rowan, U.S. patents 2,916,424 (1959) and 2,952,590 (1960).

110. H. Brown, M. Weinberg, and R. J. Clauss, *Plating,* 45, 144, 1958.

111. W. H. Safranek and C. L. Faust, *Plating,* 45, 1027, 1958.

112. E. J. Seyb, *Proc. Am. Electroplaters Soc.,* 47, 209, 1960.

113. M. E. Beckmann and F. Maass-Graefe, "Effect of the Wave Form of Rectified Alternating Current on the Growth of Electrodeposits," *Metalloberflaeche,* 5A, 161–169, 1951.

114. B. Sutter, "Pulsed Electrodeposition of Nickel and Chromium," *Theory and Practice of Pulsed Plating,* Jean-Claude Puippe and Frank Leaman, eds., Pub. by Am. Electroplaters and Surface Finishers Soc., Orlando, FL, pp. 109–117, 1986.

115. D. T. Gawne and T. G. P. Gudyanga, "Wear Behaviour of Chromium Electrodeposits," in *Coatings and Surface Treatment for Corrosion and Wear Resistance,* K. N. Stafford, P. K. Datta and C. G. Googan, eds., Inst. of Corrosion Science and Technology, Chapter 2 (1984).

116. R. Weiner and A. Walmsley, *Chromium Plating,* Finishing Publications LTD., Teddington, Middlesex, England, 1980.

117. J. K. Dennis and T. E. Such, *Nickel and Chromium Plating,* 3rd ed., Woodhead Publishing LTD., Cambridge, England, 1993.

118. USEPA A Guidebook on How to Comply with the Chromium Electroplating and Anodizing National Emission Standards for Hazardous Air Pollutants, EPA-453/B-95-001.

119. B. A. Wilson and D. M. Turley, *Trans. Inst. Metal Finishing,* 67(4), 104, 1989.

120. G. A. DiBari, "Decorative Electrodeposited Nickel-Chromium Coatings," *Metal Finishing,* 75, 17 (June 1977) and 75, 17 (July 1977).
121. Dr. J. H. Lindsay, AESF Fellow, "Decorative and Hard Chromium Plating," *Plating and Surface Finishing, The 1999 Shop Guide,* Volume 86, No. 11, pp. 40–42, 1999.
122. R. B. Alexander, "Combined Hard Chrome Plating and Ion Implantation for Improving Tool Life," *Plating and Surface Finishing,* 77, 18, Oct. 1990.
123. W. Lohmann and J. G. P. Van Valkenhoef, "Improvement in Friction and Wear of Hard Chromium Layers by Ion Implantation," *Materials Science and Engineering,* A116, 177, 1989.
124. W. C. Oliver, R. Hutchings and J. B. Pethica, "The wear Behaviour of Nitrogen-Implanted Metals," *Metallurgical Transactions,* 15A, 2221, 1984.
125. J. A. Kubinski, CEF, T. Hurkmans, T. Trinh, Dr. W. Fleischer and Dr. G. J. van der Kolk, "Perspective for Replacement of Hard Chromium by PVD," *Plating and Surface Finishing,* Vol. 86, No. 10, pp. 20–25, Oct. 1999.
126. L. W. Masters, J. F. Seiler, and W. E. Roberts, NBSIR Rep. 82-2583, National Bureau of Standards, U. S. Department of Commerce, October 1982.
127. L. E. Gatzek, "New Developments in Wear-Resistant Finishes and Coatings," 1964.
128. G. E. Shahin, CEF, "Alloys are Promising as Cadmium or Chromium Substitutes," *Plating and Surface Finishing,* pp. 8–14, August 1998.
129. Azita Yazdani, "Pollution Prevention for Copper Plating," A Pollution Prevention Training Course for the "Common Sense" Initiative—Metal Plating and Finishing Sector—Under the Environmental Technology Initiative, Developed through a partnership between the Office of Research and Development of the U. S. Environmental Protection Agency and The American Electroplaters and Surface Finishers Society, Module #10, 1997.
130. W. O. Freitag, C. Ogden, D. Tench and J. White, "Determination of the Individual Additive Components in Acid Copper Plating Baths," *Plating and Surface Finishing,* 70, 55, Oct. 1983.
131. D. Tench and J. White, "Cyclic Pulse Voltammetric Stripping Analysis of Acid Copper Plating Baths," *J. Electrochem. Soc.,* 132, 831,1985.
132. C. Odgen and D. Tench, "Cyclic Voltammetric Stripping Analysis of Copper Plating Baths," *Application of Polarization Measurements in the Control of Metal Deposition,* I. H. Warren, Editor, Elsevier, 1984.
133. Jack Horner, "Cyanide Copper Plating," *Plating and Surface Finishing, The 1999 Shop Guide,* Volume 86, No. 11, pp. 36–38, 1999.
134. B. F. Rothschild, "The Effect of Orthophosphate in Copper Pyrophosphate Plating Solutions and Deposits," *Metal Finishing,* 76, 49, 1978.
135. L. Zakraysek, "Embrittlement in Electrodeposited Copper," *Circuits Manufacturing,* pp. 38–40, Dec. 1983.
136. Arthur J. Kowalski, CEF, "Brass Plating," *Plating and Surface Finishing, The 1999 Shop Guide,* Volume 86, No. 11, pp. 30–32, 1999.
137. J. W. Dini, Electrodeposition, *The Materials Science of Coatings and Substrates,* Noyes Publications, Park Ridge, New Jersey, p. 91, 1993.
138. Battelle Memorial Institute, Instructions for Bright LUSTRALITE Plating.
139. Tin Research Institute: Working Instructions for Speculum Plating, Greenford, England, 1947; J. W. Cuthbertson, *J. Electrodepositors' Tech. Soc.,* 23, 143 (1948); W. H. Sawyer, ibid., 23, 151 (1948); P. S. Bennett, ibid., 26, 107 (1950).
140. W. H. Safranek, *The Properties of Electrodeposited Metals and Alloys* (Orlando, Florida: American Electroplaters and Surface Finishers Society, 2nd ed., pp. 141–145, 1986.
141. W. H. Safranek, W. G. Hespenheide, and C. L. Faust, "Bronze and Speculum Plates Provide Good Corrosion Protection for Steel," *Metal Finishing,* 52, 70–73, 78 (1954).
142. W. A. Safranek, and C. L. Faust, "Copper-Tin Alloy Plating," *Proc. Am. Electroplaters Soc.,* 41, 1159–1170 (1954).
143. W. A. Safranek, W. J. Neill, and D. E. Seelbach, "Bronze Plating Solves Design and Corrosion Problems," *Steel,* 133, (35), 102–109 (1952).

144. H.H.Uhlig, *Corrosion and Corrosion Control,* 2nd ed., p. 333, John Wiley & Sons Inc., New York (1971).

145. F. A. Lowenheim, *Electroplating,* Sponsored by the American Electroplaters Soc., McGraw-Hill Book Co., New York, p. 118, 1978.

146. D. E. Weimer and J. W. Price, *Trans. Inst. Met. Finishing,* 1954, 30, 95.

147. A. C. Vitale "DuraBOND Process Eliminates Pink Ring and Wedge Void Defects," IPC 32nd Annual Meeting (April 1989).

148. Holtzman, "Use of Immersion Tin and Tin Alloys as a Bonding Medium for Multi-layer Circuits," USP 4, 715, 894.

149. Lucia H. Shipley, "Tin Coating of Copper Surfaces by Replacement Plating," USP 3,303,029.

150. Hiroki Uchida et al. "Process for Electroless Plating Tin, Tin-Lead Alloy," USP 5,248,527.

151. OMIKRON Improved Immersion White Tin Process, Designed as a Replacement for Hot Air Leveling, Florida CirTech, Inc. Report.

152. Richard Edgar, "Immersion White Tin," *Printed Circuit Fabrication,* pp. 38–41, Vol. 21, No. 12, Dec. 1998.

153. Jack D. Fellman, "Applications for Selective Plating with a Novel Tin Process," presented at Institute for Interconnecting and Packaging Electronic Circuits, Printed Circuit World Convention III, Washington D.C., May 1984.

154. Craig V. Bishop, G. Bokisa, R. Durante, J. Kochilla, A. Vitale, and G. Keers, "Rapid Deposition, High Build, Non-Electrolytic Tin for Use in the PWB Industry," McGean-Rocho, Inc., UK report.

155. Art Burkhart, "Recent Developments—Taking a Closer Look at Solderable Finishes," *Printed Circuit Fabrication,* Vol. 21, No. 8, pp. 18–21, Aug. 1998.

156. Doug Pauls, edited by Tim Toole, "Tin/Lead Finish," *Printed Circuit Fab.,* Vol. 21, No. 8, pp. 24–30, Aug. 1998.

157. Donald C. Abbot, Raymond A. Frechette, Gardner Haynes and Douglas W. Romm, "Shelf-Life Evaluation for Palladium-Finish ICs," *Circuits Assembly,* pp. 40–45, July 1998.

158. Ian Young, "Nickel/Gold Yield Maximization," *Printed Circuit Fabrication,* Vol. 21, No. 9, Sept. 1998.

159. Paul E. Davis, Malcolm E. Warwick and P. J. Kay, "Intermetallic Compound Growth and Solderability," *Plating and Surface Finishing,* pp. 72–76, Sept. 1982.

160. Paul E. Davis, Malcolm E. Warwick and Stephen J. Muckett, "Intermetallic Compound Growth and Solderability of Reflowed Tin and Tin-Lead Coatings," *Plating and Surface Finishing,* pp. 49–53, August 1983.

161. "Copper-Tin Intermetallics, growth and thermal expansion affect PCB reliability," *Circuits Manufacturing,* pp. 56–64, Sept. 1980, Information was derived from material supplied by the Tin Research Institute, Palo Alto, CA, and from the papers, "Thermal Expansion of the Copper-Tin Intermetallic Compound Cu3Sn in the Temperature Range −195C to +300C, by W. J. Reichenecker, Westinghouse R&D Center, Pittsburgh, PA; and, "Contamination Affecting Integrity of Solder Adhesion of Printed Wiring Boards," by R. B. Keyson, Hughes Aircraft Co., appearing in Proceedings, PC '80.

162. Reginald K. Asher, Sr., "Improving Solderability of Tin and Tin-Lead Electrodeposits," *Products Finishing,* pp. 26–28, August 1984.

163. Dr. Donald H. Daebler, "An Overview of Glod Intermetallics in Solder Joints," *Surface Mount Technology,* pp. 43–46, October 1991.

164. Martin Theriault and Philippe Blostein, Reducing the Cost of Inert Soldering," *Circuits Assembly,* web site: http://www.cassembly.com, pp. 4647, July 1998.

165. Jerry Murray, editor, "Surface Finishes," 1996 Surface Mount International Conference in San Jose, CA, Surface Mount Technology Association, 612/920–1819, a review of five-of-six parers, Printed Circuit Fabrication, pp. 20–24, Vol. 20, No. 4, April 1997.

166. M. E. Warwick and B. J. Shirley, *Trans Inst Met Finishing,* 58(1), 9, 1980.

167. Hidekatsu Koyano, Tokyo Electroplating Co., Ltd., "Electroless Solder Plating," Disclosure No. 211566-1984, Application No. 84248-1983, 1983.

168. Burns and Bradley, "Protective Coatings for Metals," p. 269, Reinhold, New York, 1967.
169. Michael Carano "Tin-Lead Plating," *Plating and Surface Finishing, The 1999 Shop Guide,* Volume 86, No. 11, pp. 68–69, November 1999.
170. George E. Shahin, CEF, "Alloys are Promising as Chromium or Cadmium Substitutes," *Plating and Surface Finishing,* pp. 11–14, August 1998.
171. A. Whittaker, "Production Tin-Zinc Alloy Platings," Machinery (London), 84, (2158), pp. 639–642, 1954.
172. Mark Ingle and James Ault, "Corrosion Control Performance Evaluation of Environmentally Acceptable Alternatives for Cadmium Plating," AESF SUR/FIN, Session U, 1992.
173. Eric W. Brooman, "Alternatives to Cadmium Coatings For Electrical/Electronic Applications," *Plating and Surface Finishing,* Vol. 80, Feb. 1993.
174. "Pollution Prevention for Cadmium Plating/Comparison of Substitutes."
175. Jesse Cheng and Jim Reed, Process Characterization And Troubleshooting Using SERA," Originally presented at the IPC Printed Circuits EXPO '95, San Diego, Calif., May 1995, *Electronic Packaging and Production,* pp. 53–56, July 1995.
176. Peter Bratin, Michael Pavalov, and Gene Chalyt, "Evaluating Finishes Using SERA," *Printed Circuit Fabrication,* Vol. 22, No. 5, pp. 30–37, May 1999.
177. E. Raub and K. Muller, *Fundamentals of Metal Deposition,* Elsevier Publishing Co., Ltd., Barking, Essex, 268 pp., 1967.
178. R. W. Couch and R. G. Bikales, "Plating of Printed Circuits With Pyrophosphate Copper and Tin-Nickel," *Proc. of the American Electroplaters Society,* 48, 176–181, 1961.
179. J. W. Price, *Tin and Tin-Alloy Plating,* Electrochemical Publications Ltd., 1983.
180. K. N. Strafford and A. Reed, "Engineering, Economic and Practical Aspects of the Use of Tin-Cobalt Alloy Electrodeposits as Substitutes for Decorative Chromium," Ch. 5, pp. 74–83, *Coatings and Surface Treatment for Corrosion and Wear Resistance,* K. N. Strafford, P. K. Datta, and C. G. Googan, eds., published for the Institution of Corrosion Science and Technology, Ellis Horwood, Ltd., publishers, Chichester, UK, 1984.
181. US Patent No. 4,035,249,1976.
182. J. D. C. Hemsley and M. E. Roper, "Tin-Cobalt Alloy Plating From a Sulfate Electrolyte," *Trans. Inst. Metal Finish.,* 57, 77, 1979.
183. Jim D. Raby and R. Wayne Johnson, "Is a Lead-Free Future Wishful Thinking?" *Electronic Packaging and Production,* pp. 18–20, August, 1999.
184. George Cushnie, "Pollution Prevention for Tin & Tin-Lead Electroplating," A Pollution Prevention Training Course for the "Common Sense" Initiative—Metal Plating and Finishing Sector—Under the Environmental Technology Initiative, Developed through a partnership between the Office of Research and Development of the U. S. Environmental Protection Agency and The American Electroplaters and Surface Finishers Society, 1997.
185. L. Brugnatelli, *Ann. Chim. (Piva),* 18, 152, 1800.
186. G.Elkington and H. Elkington, British patent 8447, 1840.
187. Fang Jing Li and Cai Zi, "Tarnish Protection for Silver Electrodeposits," *Plating and Surface Finishing,* pp. 58–61, Feb. 1988.
188. I. F. Kushevich, et al, "Electrodeposition of a Silver-Palladium Alloy from a Bath Containing Trilon B and Ammonia," Elektrokhim, Osaz, Prim, Polrytii, Dragotsen, I Redk, Metal., 33–35, 1972.
189. W. Reksc et al, "Silver-Tin Electrodeposition," *Chem. Inz. Chem.,* 13, 157–169, 1979.
190. W. H. Safranek, The Properties of Electrodeposited Metals and Alloys (Orlando, Florida: American Electroplaters and Surface Finishers Society), 2nd ed., p. 430, 1986.
191. Industry News, "Immersion Silver—a Low Cost Alternative to Nickel/Gold Use," *Electronic Packaging & Production,* July 1996.

11

Plastics Joining Materials and Processes

Edward M. Petrie

ABB Power T & D Company, Inc.
Raleigh, North Carolina

11.1 Introduction

Plastics are common substrates found in all industries. Plastic materials range from inexpensive, commodity-type products, such as those used for packaging, to a group of resins that have become known as *engineering plastics* by virtue of their outstanding mechanical and chemical properties. Engineering plastics are often used as an alternative to metals in applications requiring high specific strength, low cost, and fast production methods.

To make large plastic assemblies, the most cost-effective method often involves molding smaller subsections and joining them. In such cases, the manufacturer has a variety of joining options including:

- adhesive bonding
- thermal welding
- solvent cementing
- mechanical fastening

The plastic assembly relies heavily on the characteristics of the joint. In most applications, the joint must be nearly as strong as the substrate throughout the expected life of the product. The method used to join plastics should be carefully evaluated by the designer. In addition to strength and permanence, consideration must be given to tooling cost, labor and energy cost, production time, appearance of the final part, and disassembly requirements. Some plastic materials will be more suited for certain joining processes than others due to their physical and chemical characteristics.

A consideration in plastic assembly that is usually not dominant when joining other substrates is the speed of the joining operation. Plastic products generally require very fast, high-volume assembly processes in industries such as consumer and automotive products, and packaging. Speed, simplicity, and reliability are key concerns in most of these high-volume assembly processes. Speed and simplicity are usually considered of greater value than reliability or durability when bonding commodity plastic substrates. Because of the nature of the polymeric substrate and the type of applications for which such materials are best suited, exceeding high strengths and durability in hash environments are not generally necessary. These properties will be sacrificed gladly for faster, lower-cost production methods. Often, there is not enough time for critical surface preparation or nondestructive testing of every part. In certain industries, such as the automotive industry, plastic materials are often chosen because of their fast joining ability. Thus, thermoplastics are often preferred over thermosets, because they can be joined via thermal welding processes in a few seconds.

Although adhesive bonding often proves to be an effective method for bonding plastics, there are various other ways of joining plastics to themselves or to other materials. Thermal welding, solvent cementing, and mechanical fastening are usually faster than adhesive bonding and, as a result, are often preferred in high-volume assembly operations. These nonadhesive methods of joining will be the subject of this chapter. Adhesive bonding methods for plastics as well as other substrates will be considered in the next chapter.

The sections that follow will describe the various processes that can be used for joining plastic materials. Information will be provided on how to choose the most appropriate process for a specific substrate and application. The plastic materials that are best suited for each process will be identified. Important process parameters and test results will also be reviewed. Recommendations regarding joining processes for specific types of plastic materials (e.g., polyethylene, glass reinforced epoxy, polysulfone) will also be given.

11.2 General Types of Plastic Materials

There are many types of plastic materials with a range of properties that depend on the base polymer and the additives used. Plastics are used routinely in many commodity items such as packaging, pipe, clothing, appliances, and electronics. They are also increasingly being used in structural and engineering applications such as aerospace, building, and automotive. Other chapters in this handbook will cover specific plastic materials, design practices, and applications. This chapter will concentrate on how variations in such materials affect assembly considerations.

All plastics can be classified into two categories: thermoplastics and thermosets. Thermoplastics are not crosslinked, and the polymeric molecules making up the thermoplastic can easily slip by one another. This slip or flow can be caused by thermal energy, by solvents or other chemicals, and by the application of continuous stress. Thermoplastics can be repeatedly softened by heating and hardened on cooling. Hence, they can be welded by the application of heat. Thermoplastics can also be dissolved in solvents so that it is also possible to join thermoplastic parts by solvent cementing. Typical thermoplastics are polyethylene, polyvinyl chloride, polystyrene, polypropylene, nylon, and acrylic.

Engineering thermoplastics are a class of thermoplastic materials that have exceptionally high temperature and chemical resistance. They have properties very similar to those of thermoset plastics and to metals. As a result, they are not as easy to weld thermally or to solvent cement as are the more conventional thermoplastics. Typical engineering thermoplastics are polysulfone, acetal, amide-imide, and thermoplastic polyimide.

Thermoset plastics are crosslinked by chemical reaction so that their molecules cannot slip by one another. They are rigid when cool and cannot be softened by the action of heat. If excessive heat is applied, thermoset plastics will degrade. Consequently, they are not weldable. Because of their chemical resistance, they cannot be solvent cemented. Thermoset plastics are usually joined by either adhesive bonding or mechanical fastening. Typical thermosetting plastics are epoxy, urethane, phenolic, and melamine formaldehyde.

There are but a few typical basic polymer resin suppliers. However, many companies formulate filled plastic systems from these basic resins. These smaller companies are quite often the ultimate supplier to the manufacturer of the end product. Both the basic resin manufacturer and the formulator have considerable influence on the joining characteristics of the final material. Because of this, they should be considered the primary source of information regarding joining processes and expected end results.

Plastics offer several important advantages compared with traditional materials, and their joining process should not detract from these benefits. The greatest advantage is low processing cost. Low weight is also a major advantage. Relative densities of most unfilled plastics materials range from 0.9 to 1.4, compared with 2.7 for aluminum and 7.8 for steel. Relative densities for highly filled plastic compounds can rise to 3.5.

Other advantages of plastics are their low frictional resistance, good corrosion resistance, insulation properties, and ease with which they can be fabricated into various shapes. The chemical resistance of plastics is specific to the type of plastic. Some plastics have chemical resistance comparable to metals, others are attacked by acids, and others are attacked by solvents and oils. Certain plastics are also attacked by moisture, especially at moderately elevated temperatures. The nature of this attack is generally first a swelling of the substrate and finally dissolution. Some plastics are transparent, and others are translucent. These plastics can pass UV radiation so that the adhesive or joint could be affected by UV exposure. In certain cases, the plastic itself can be degraded by UV. Resistance to UV degradation can be improved by the addition of UV absorbers such as carbon black. All plastics can all be colored by the addition of pigments. The addition of pigments generally eliminates the need for painting.

The most serious disadvantage of plastics is that, compared to metals, they have low stiffness and strength and are not suitable for use at high temperatures. Some plastics are unusable at temperatures above 50°C. A very few are capable of resisting short-term exposure to temperatures up to 400°C. Surface hardness of plastics is generally low. This can lead to indentation and compression under local stress. The elastic modulus of plastic materials is relatively high compared to metals. They can be elongated and compressed with applied stress. The thermal expansion coefficient of plastics is generally an order of magnitude greater than that of metals. This results in internal joint stress when joined to a substrate having a much lower coefficient of expansion.

11.3   Types of Plastic Joining Processes

The joining of plastics with adhesives is generally made difficult because of the low surface energy, poor wettability, and presence of weak boundary layers associated with these substrates. Adhesive bonding is a relatively slow process that could be a significant drawback in many industries that produce high-volume plastic assemblies. However, with plastic substrates, the designer has a greater choice of joining techniques than with many other substrate materials. Thermoset plastics usually must be bonded with adhesives or mechanically fastened, but many thermoplastics can be joined by solvent or heat welding as well as by adhesives or mechanical fasteners.

In the thermal or solvent welding processes, the plastic resin that makes up the substrate itself acts as the adhesive. These processes require that the surface region of the substrate be made fluid so that it can wet the mating substrate. If the mating substrate is also a polymer, both substrate surfaces can be made fluid so that the resin can molecularly diffuse into the opposite interfaces. This fluid interface region is usually achieved by thermally heating the surface areas of one or both substrates or by dissolving the surfaces in an appropriate solvent. Once the substrate surface is in a fluid condition, they are then brought together and held in place with moderate pressure. At this point, the molecules of substrate A and substrate B will diffuse into one another and form a very tight bond. The fluid polymer mix then returns back to the solid state, usually by the dissipation of solvent or by cooling from the molten condition.

Thermal or solvent welding can be used effectively when at least one substrate is polymeric. In the case of thermal welding, the molten polymer surface wets the nonpolymeric substrate and acts as a hot-melt type of adhesive. Internal stresses that occur on cooling the interface from the molten condition are the greatest detriment to this method of bonding. Solvent welding can also be used only if the nonpolymeric substrate is porous. If it is not porous, the solvent may become entrapped at the bond line and cause very weak joints.

Table 11.1 indicates the most common joining methods that are suitable for various plastic materials. Descriptions of these joining techniques are summarized both in Table 11.2 and in the section to follow. In general, nonadhesive joining methods for plastics can be classified as one of the following:

1. Welding by direct heating (heated tool, hot gas, resistance wire, infrared, laser, extrusion)

2. Induced heating (induction, electrofusion, dielectric)

3. Frictional (ultrasonic, vibration, spin)

4. Solvent welding

5. Mechanical fastening

Equipment costs for each method vary considerably, as do the amount of labor involved and the speed of the operation. Most techniques have limitations regarding the design of the joint and the types of plastic materials that can be joined.

11.4 Direct Heat Welding

Welding by the direct application of heat provides an advantageous method of joining many thermoplastic materials that do not degrade

TABLE 11.1   Assembly Methods for Plastics (from Ref. 1)

Plastics	Adhesives	Dielectric welding	Induction bonding	Mechanical fastening	Solvent welding	Spin welding	Thermal welding	Ultrasonic welding
Thermoplastics								
ABS	X	—	X	X	X	X	X	X
Acetals	X	—	X	X	X	X	X	X
Acrylics	X	—	X	X	X	X	—	X
Cellulosics	X	—	—	—	X	X	—	—
Chlorinated polyether	X	—	—	X	—	—	—	—
Ethylene copolymers	—	X	—	—	—	—	X	—
Fluoroplastics	X	—	—	—	—	—	—	—
Ionomer	—	X	—	—	—	—	X	—
Methylpentene	—	—	—	—	—	—	—	X
Nylons	X	—	X	X	X	—	—	X
Phenylene oxide–based materials	X	—	—	X	X	X	X	X
Polyesters	X	—	—	X	X	X	—	X
Polyamide-imide	X	—	—	X	—	—	—	—
Polyaryl ether	X	—	—	X	—	—	—	X
Polyaryl sulfone	X	—	—	X	—	—	—	X
Polybutylene	—	—	—	—	—	—	—	X
Polycarbonate	X	—	X	X	X	X	X	X
Polycarbonate/ABS	X	—	—	X	X	X	X	X
Polyethylene	X	X	X	X	—	X	X	X
Polyimide	X	—	—	X	—	—	—	—
Polyphenylene sulfide	X	—	—	X	—	—	—	—
Polypropylenes	X	X	X	X	—	X	X	X
Polystyrene	X	—	X	X	X	X	X	X
Polysulfone	X	—	—	X	X	—	—	X
Polypropylene copolymers	X	X	X	X	—	X	X	—
PVC/acrylic alloy	X	—	X	X	—	—	—	X
PVC/ABS alloys	X	X	—	—	—	—	X	—
Styrene acrylonitrile	X	—	X	X	X	X	X	X
Vinyl	X	X	X	X	X	—	X	—
Thermosets								
Alkyd	X	—	—	—	—	—	—	—
Allyl diglycol carbonate	X	—	—	—	—	—	—	—
Diallyl phthalate	X	—	—	X	X	—	—	—
Epoxies	X	—	—	X	X	—	—	—
Melamines	X	—	—	—	—	—	—	—
Phenolics	X	—	—	X	X	—	—	—
Polybutadienes	X	—	—	—	—	—	—	—
Polyesters	X	—	—	X	X	—	—	—
Silicones	X	—	—	X	X	—	—	—
Ureas	X	—	—	—	—	—	—	—
Urethanes	X	—	—	X	X	—	—	—

TABLE 11.2 Bonding or Joining Plastics: What Techniques Are Available, and What Do They Have To Offer? [2]

Technique	Description	Advantages	Limitations	Processing considerations
Solvent cementing and dopes	Solvent softens the surface of an amorphous thermoplastic; mating takes place when the solvent has completely evaporated. Bodied cement with small percentage of parent material can give more workable cement, fill in voids in bond area. Cannot be used for polyolefins and acetal homopolymers.	Strength, up to 100% of parent materials, easily and economically obtained with minimum equipment requirements.	Long evaporation times required; solvent may be hazardous; may cause crazing in some resins.	Equipment ranges from hypodermic needle or just a wiping medium to tanks for dip and soak. Clamping devices are necessary, and air dryer is usually required. Solvent-recovery apparatus may be required. Processing speeds are relatively slow because of drying times. Equipment costs are low to medium.
Thermal bonding				
Ultrasonics	High-frequency sound vibrations transmitted by a metal horn generate friction at the bond area of a thermoplastic part, melting plastics just enough to permit a bond. Materials most readily weldable are acetal, ABS, acrylic, nylon, PC, polyimide, PS, SAN, phenoxy.	Strong bonds for most thermoplastics; fast, often less than 1 s. Strong bonds obtainable in most thermal techniques if complete fusion is obtained.	Size and shape limited. Limited applications to PVCs, polyolefins.	Converter to change 20 kHz electrical to 20 kHz mechanical energy is required, along with stand and horn to transmit energy to part. Rotary tables and high-speed feeder can be incorporated.
Hot-plate and hot-tool welding	Mating surfaces are heated against a hot surface, allowed to soften sufficiently to produce a good bond, then clamped together while bond sets. Applicable to rigid thermoplastics.	Can be very fast, e.g., 4–10 s in some cases; strong,	Stresses may occur in bond area.	Use simple soldering guns and hot irons, relatively simple hot plates attached to heating elements up to semiautomatic hot-plate equipment. Clamps needed in all cases.
Hot-gas welding	Welding rod of the same material being joined (largest application is vinyl) is softened by hot air or nitrogen as it is fed through a gun that is softening part surface simultaneously. Rod fills in joint area and cools to effect a bond.	Strong bonds, especially for large structural shapes.	Relatively slow; not an "appearance" weld.	Requires a hand gun, special welding tips, an air source, and welding rod. Regular hand-gun speeds run 6 in/min; high-speed hand-held tool boosts this to 48–60 in/min.

TABLE 11.2 Bonding or Joining Plastics: What Techniques Are Available, and What Do They Have To Offer? (Continued)[2]

Technique	Description	Advantages	Limitations	Processing considerations
Spin welding	Parts to be bonded are spun at high speed, developing friction at the bond area; when spinning stops, parts cool in fixture under pressure to set bond. Applicable to most rigid thermoplastics.	Very fast (as low as 1–2 s); strong bonds.	Bond area must be circular.	Basic apparatus is a spinning device, but sophisticated feeding and handling devices are generally incorporated to take advantage of high-speed operation.
Dielectrics	High-frequency voltage applied to film or sheet causes material to melt at bonding surfaces. Material cools rapidly to effect a bond. Most widely used with vinyls.	Fast seal with minimum heat applied.	Only for film and sheet.	Requires rf generator, dies, and press. Operation can range from hand-fed to semiautomatic with speeds depending on thickness and type of product being handled. Units rated 3–25 kW are most common.
Induction	A metal insert or screen is placed between the part to be welded and energized with an electromagnetic field. As the insert heats up, the parts around it melt, and when cooled form a bond. For most thermoplastics.	Provides rapid heating of solid sections to reduce chance of degradation.	Because metal is embedded in plastic, stress may be caused at bond.	High-frequency generator, heating coil, and inserts (generally 0.02–0.04 in thick). Hooked up to automated devices, speeds are high. Work coils, water cooling for electronics, automatic timers, multiple-position stations may also be required.
Adhesives *				
Liquid solvent, water base, anaerobics	Solvent- and water-based liquid adhesives, available in a wide number of base (e.g., polyester, vinyl) in one- or two-part form fill bonding needs ranging from high-speed lamination to one-of-a-kind joining of dissimilar plastics parts. Solvents provide more bite but cost much more than similar base water-type adhesive. Anaerobics are a group of adhesives that cure in the absence of air.	Easy to apply; adhesives available to fit most applications.	Shelf and pot life often limited. Solvents may cause pollution problems; water-base not as strong; anaerobics toxic.	Application techniques range from simply brushing on to spraying and roller coating-lamination for very high production. Adhesive application techniques, often similar to decorating equipment, from hundreds to thousands of dollars with sophisticated laminating equipment costing in the tens of thousands of dollars. Anaerobics are generally applied a drop at a time from a special bottle or dispenser.

Type	Description	Advantages	Disadvantages	Application
Pastes, mastics	Highly viscous single- or two-component materials that cure to a very hard or flexible joint, depending on adhesive type.	Does not run when applied.	Shelf and pot life often limited.	Often applied via a trowel, knife, or gun-type dispenser; one-component systems can be applied directly from a tube. Various types of roller coaters are also used. Metering-type dispensing equipment in the $2,500 range has been used to some extent.
Hot melts	100% solids adhesives that become flowable when heat is applied. Often used to bond continuous flat surfaces.	Fast application; clean operation.	Virtually no structural hot melts for plastics.	Hot melts are applied at high speeds via heating the adhesive then extruding (actually, squirting) it onto a substrate, roller coating, using a special dispenser or roll to apply dots or simply dipping.
Film	Available in several forms including hot melts, these are sheets of solid adhesive. Mostly used to bond film or sheet to a substrate.	Clean, efficient.	High cost.	Film adhesive is reactivated by a heat source; production costs are in the medium–high range, depending on the heat source used.
Pressure-sensitive	Tacky adhesives used in a variety of commercial applications (e.g., cellophane, too). Often used with polyolefins.	Flexible.	Bonds not very strong.	Generally applied by spray with bonding effected by light pressure.
Mechanical				
Mechanical fasteners (staples, screws, molded-in inserts, snap fits, and a variety of proprietary fasteners)	Typical mechanical fasteners are listed on the left. Devices are made of metal or plastic. Type selected depends on how strong the end product must be, appearance factors. Often used to join dissimilar plastics or plastics to nonplastics.	Adaptable to many materials; low to medium cost, can be used for parts that must be disassembled.	Some have limited pull-out strength; molded-in inserts may result in stresses.	Nails and staples are applied by simply hammering or stapling. Other fasteners may be inserted by drill press, ultrasonics, air or electric gun, hand tool. Special molding, i.e., molded-in-hole, may be required.

*Typical adhesives in each class are as follows. Liquids: 1. solvent—polyester, vinyl, phenolics, acrylics, rubbers, epoxies, polyamide; 2. water—acrylics, rubber-casein; 3. anaerobics—cyanoacrylate; mastics—rubbers, epoxies; hot melts—polyamides, PE, PS, PVA; film—epoxies, polyamide, phenolics; pressure sensitive—rubbers.

rapidly at their melt temperatures. The principal methods of direct heat welding are

- Heated tool
- Hot gas
- Resistance wire
- Laser
- Infrared

These methods are generally capable of joining thermoplastics to themselves and other thermoplastics and, in certain cases, they may also be used to weld thermoplastics to non-plastic substrates.

11.4.1 Heated Tool Welding

Fusion or heated tool welding is an excellent method of joining many thermoplastics. In this method, the surfaces to be fused are heated by holding them against a hot surface. When the plastic becomes molten, and a flash about half the thickness of the substrate is visible, the parts are removed from the hot surface. They are then immediately joined under slight pressure (5–15 psi) and allowed to cool and harden. The molten polymer acts as a hot melt adhesive, providing a bond between the substrates.

This method is often used in high-volume operations where adhesive bonding is objectionably long. It is also often used to join low-surface-energy materials, such as polypropylene, where the cost and complexity required for substrate treatment and adhesive bonding cannot be tolerated. Surface treatment, other than simple cleaning, is usually not required for thermal welding. Heated tool welding is a simple, economical technique in which high-strength joints can be achieved with large and small parts. Hermetic seals can also be achieved. Heated tool welding does not introduce foreign materials into the part and, as a result, plastic parts are more easily recycled.

Success in heated tool welding depends primarily on having the proper temperature at the heating surface and on the timing of the various steps in the process. These periods include time for application of heat, time between removal of heat and joining of parts, and time the parts are under pressure. The tool should be hot enough to produce sufficient flow for fusion within 10 s. The parts are generally pressed against the heated tool with a certain degree of pressure. However, to avoid strain, the pressure on the parts should be released for a period of time before they are removed from contact with the heated tool. While some rules of thumb can apply, the final process settings for temperature, duration of heating and cooling times, and pressures will depend on the polymer. Adjustments will be required until the desired bond quality is achieved. The thickness of the molten layer is an important determinant of weld strength. Dimensions are usually controlled through the incorporation of displacement stops at both the heating and mating steps in the process. If welds are wider than 1/4 in., the heated parts should be glided across each other during the mating step to prevent air entrapment in the joint.

Heated tool welding is suitable for almost any thermoplastic but is most often used for softer, semicrystalline thermoplastics. Common plastic substrates that are suitable for heated tool welding include polyethylene, polypropylene, polystyrene, ABS, PVC, and acetals. It is usually not suitable for nylon or other materials that have long molecular chains. Dissimilar yet chemically compatible materials that have different melting temperatures can be welded in heated tool welding by using two heated platens each heated to the melting temperature of the part to be welded.

Heated tool welding can be accomplished with either no surface treatment or very minor surface preparation (degreasing and removal of mold release), depending on the strength and reliability dictated by the application. Generally, surface degreasing to remove mold release or other organic contaminants is the only prebond treatment necessary. Mechanical roughening or chemical treatment of the surface provides no advantage since the surface will be melted and a new surface will be formed. Plastic parts that have significant degree of internal moisture may have to be dried before heated tool welding, or else the moisture will tend to escape the molten surface in the form of vapor bubbles.

Electric-strip heaters, soldering irons, hot plates, and resistance blades are common methods of providing heat locally. Usually, the heating platen is coated with a fluorocarbon such as PTFE for non sticking. A simple hot plate has been used extensively with many plastics. Figure 11.1 illustrates an arrangement for direct heat welding consisting of heated platens and fixturing. The parts are held on the hot plate until sufficient fusible material has been developed. Table 11.3 lists typical hot-plate temperatures for a variety of plastics. A similar technique involves butting flat plastic sheets on a flat table against a heated blade that runs the length of the sheet. Once the plastic begins to soften, the blade is raised and the sheets are pressed together and fused.

The direct heat welding operation can be completely manual, as in the case of producing a few prototypes, or it can be semiautomatic or fully automatic for fast, high-volume production. For automated assembly, rotary machines are often used where there is an independent

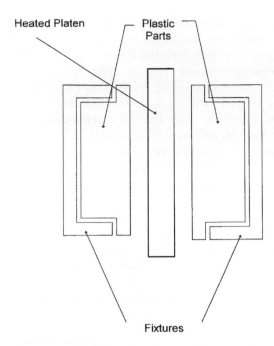

Heated Platen   **Plastic Parts**

Fixtures

Figure 11.1 In direct hot plate welding, two fixtures press components into a hot moving platen, causing the plastic to melt at the interface.

TABLE 11.3 Hot Plate Temperatures To Weld Plastics[3]

Plastic	Temp., °F
ABS	450
Acetal	500
Phenoxy	550
Polyethylene LD	360
Polyethylene HD	390
Polycarbonate	650
PPO	650
Noryl[*]	525
Polypropylene	400
Polystyrene	420
SAN	450
Nylon 6,6	475
PVC	450

[*]Trademark of General Electric Co.

station for each process: (1) clamping into fixtures, (2) heating, (3) joining and cooling, (4) unloading.

Heated wheels or continuously moving heated bands are tools commonly used to bond thin plastic sheet and film. This is commonly used for sealing purposes such as packaging of food. Care must be taken, especially with thin film, not to apply excessive pressure or heat. This could result in melting through the plastic. Table 11.4 provides heat-sealing temperature ranges for common plastic films.

TABLE 11.4 Heat Sealing Temperatures for Plastic Films[4]

Film	Temp., °F
Coated cellophane	200–350
Cellulose acetate	400–500
Coated polyester	490
Poly(chlorotrifluoroethylene)	415–450
Polyethylene	250–375
Polystyrene (oriented)	220–300
Poly(vinyl alcohol)	300–400
Poly(vinyl chloride) and copolymers (nonrigid)	200–400
Poly(vinyl chloride) and copolymers (rigid)	260–400
Poly(vinyl chloride)—nitrile rubber bland	220–350
Poly(vinylidene chloride)	285
Rubber hydrochloride	225–350
Fluorinated ethylene–propylene copolymer	600–750

Heated tool welding is commonly used in medium- to high-volume industries that can make use of the simplicity and speed of these joining processes. Industries that commonly use this fastening method include appliance and automotive. Welding times range from 10–20 s for small parts to up to 30 min for larger parts such as heavy-duty pipe. Typical cycle times are less than 60 s. Although heated tool welding is faster than adhesive bonding, it is not as fast as other welding methods such as ultrasonic or induction welding.

The heated tool process is extremely useful for pipe and duct work, rods and bars, and for continuous seals in films. However, irregular surfaces are difficult to heat unless complicated tools are provided. Special tooling configuration can be used for bonding any structural profile to a flat surface. In certain applications, the direct heating can also be used to shape the joint. With pipe, for example, a technique called *groove welding* is generally employed. Groove welding involves two heating elements. One element melts a groove in one substrate that is the exact shape of the mating part, and the other element heats the edge of the mating substrate. The heated part is quickly placed into the heated groove and allowed to cool.

11.4.2 Hot Gas Welding

An electrically heated welding gun can be used to bond many thermoplastic materials. An electrical heating element in the welding gun is capable of heating either compressed air or an inert gas to 425–700°F and forcing the heated gas onto the substrate surface. The pieces to be joined are beveled and positioned with a small gap between them. A welding rod made of the same plastic that is being bonded is laid in the joint with a steady pressure. The heat from the gun is directed to the tip of the rod, where it fills the gap, as shown in Fig. 11.2. Several passes may be necessary with the rod to fill the pocket. Thin sheets that are to be butt welded together, as in the case of tank linings, use a flat strip instead of a rod. The strip is laid over the joint and is welded in place in a single pass. Usually, the parts to be joined are held by fixtures so that they do not move during welding or while the weld is cooling. Alternatively, the parts can be first tacked together using a tool similar to a soldering iron.

Hot gas welding is usually a manual operation where the quality of the joint corresponds to the skill and experience of the operator. However, automatic welding machines are available and are used for overlap welding of seams or membranes. In either case, bond strength of approximately 85% of the strength of the bulk material can be achieved. Hot gas welding is a relatively fast operation. It can be used to weld a 1-in. wide tank seam at rates up to 60 in./min. It can also be used to do temporary tack work and to repair faulty or damaged joints that are made by gas welding or other joining processes.

Hot gas welding can be used to join most thermoplastics, including polypropylene, polyethylene, acrylonitrile butadiene styrene, polyvinyl chloride, thermoplastic polyurethane, high-density polyethylene, polyamide, polycarbonate, and polymethylmethacrylate. For polyolefins and other plastics that are easily oxidized, the heated gas must be inert (e.g., nitrogen or argon), since hot air will oxidize the surface of the plastic.

Process parameters that are responsible for the strength of hot gas weld include the type of plastic being welded, the temperature and type of gas, the pressure on the rod during welding, the preparation of the material before welding, and the skill of the welder. After welding, the joint should not be stressed for several hours. This is particularly true for polyolefins, nylons, and polyformaldehyde. Hot-gas welding is not recommended for filled materials or substrates that are less than 1/16 in. thick. Conventional hot gas welding joint designs are shown in Fig. 11.3.

Ideally, the welding rod should have a triangular cross section to match the bevel in the joint. A joint can be filled in one pass using tri-

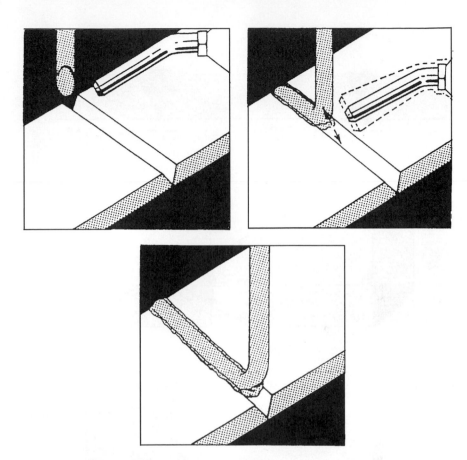

THERMOPLASTIC WELDING CHART

	PVC	H.D. Poly- ethylene	Poly- pro- pylene	Penton	ABS	Plexi- glass
Welding Temperature	525	550	575	600	500	575
Forming Temperature	300	300	350	350	300	350
Welding Gas	Air	WP* Nitro- gen	WP* Nitro- gen	Air	WP* Nitro- gen	Air
*W.P.—water pumped nitrogen						

Figure 11.2 Hot gas welding apparatus, method of application, and thermoplastic welding parameters.[5]

angular rod, saving time and material. Plastic welding rods of various types and cross sections are commercially available. However, it is also possible to cut welding rod from the sheet of plastic that is being joined. Although this may require multiple passes for filling, and the

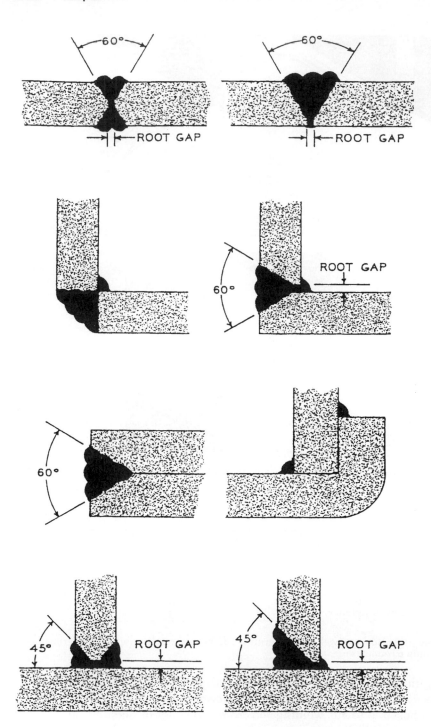

Figure 11.3 Conventional hot gas welding joint designs.[5]

chance of air pockets are greater, the welding rod is very low in cost, and the user is guaranteed of material compatibility between the rod and the plastic being joined.

Hot gas welding can be used in a wide variety of welding, sealing, and repair applications. Applications are usually large structural assemblies. Hot gas welding is used very often in industrial applications such as chemical storage tank repair, pipe fittings, etc. It is an ideal system for a small fabricator or anyone looking for an inexpensive welding system. Welders are available for several hundred dollars. The weld may not be as cosmetically attractive as with other joining methods, but fast processing and tensile strengths of 85 percent of the parent material can be obtained easily.

Another form of hot gas welding is extrusion welding. In this process, an extruder is used instead of a hot gas gun. The molten welding material is expelled continuously from the extruder and fills a groove in the preheated weld area. A welding shoe follows the application of the hot extrudate and actually molds the seam in place. The main advantage with extrusion welding is the pressure that can be applied to the joint. This adds to the quality and consistency of the joint.

11.4.3 Resistance Wire Welding

The resistance wire welding method of joining employs an electrical resistance heating element laid between mating substrates to generate the needed heat of fusion. Once the element is heated, the surrounding plastic melts and flows together. Heating elements can be anything that conducts current and can be heated through Joule heating. This includes nichrome wire, carbon fiber, woven graphite fabric, and stainless steel foil. Figure 11.4 shows an example of such a joint where a nichrome wire is used as the heating element. After the bond has been made, the resistance element that is exterior to the joint is cut off. Implant materials should be compatible with the intended application, since they will remain in the bond line for the life of the product.

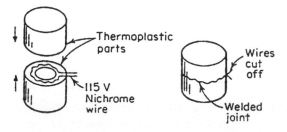

Figure 11.4   Resistance wire welding of thermoplastic joints.[6]

Like hot plate welding, resistance welding has three steps: heating, pressing, and maintaining contact pressure as the joint gels and cures. The entire cycle takes from 30 s to several minutes. Resistance welders can be automated or manually operated. Processing parameters include power (voltage and current), weld pressure, peak temperature, dwell time at temperature, and cooling time.

With resistance wire welding, surface preparation steps are necessary only when one of the substrates cannot be melted (e.g., thermosets and metals). Standard adhesive joining surface preparation processes such as those suggested in Chap. 12 can be used with these substrates.

The resistance heating process can be performed at either constant power or at constant temperature. When using constant power, a particular voltage and current is applied and held for a specified period of time. The actual temperatures are not controlled and are difficult to predict. In constant temperature resistance wire welding, temperature sensors monitor the temperature of the weld and automatically adjust the current and voltage to maintain a predefined temperature. Accurate control of heating and cooling rates is important when welding some plastics such as semicrystalline thermoplastics or when welding substrates having significantly different melt temperatures or thermal expansion coefficients. This heating and cooling control can be used to minimize internal stresses in the joint due to thermal effects.

Resistance wire welding can be used to weld dissimilar materials including thermoplastics, thermoplastic composites, thermosets, and metal in many combinations. When the substrate is not the source of the adhesive melt, such as when bonding two aluminum strips together, then a thermoplastic film with an embedded heating element can be used as the adhesive. Large parts can require considerable power requirements. Resistance welding has been applied to complex joints in automotive applications, including vehicle bumpers and panels, joints in plastic pipe, and in medical devices. Resistance wire welding is not restricted to flat surfaces. If access to the heating element is possible, repair of badly bonded joints is possible, and joints can be disassembled in a reverse process to which they were made. A similar type of process can be used to cure thermosetting adhesives when the heat generated by the resistance wire is used to advance the cure.

11.4.4 Laser Welding

Laser welding of plastic parts has been available for the last 30 years. However, only recently has the technology and cost allowed these joining techniques to be considered broadly.[7] Laser welders produce small beams of photons and electrons, respectively. The beams are focused onto the workpiece. Power density varies from a few to several thousand W/mm^2, but low-power lasers (less than 50 W/mm^2) generally are used for plastic parts.

Laser welding is a high-speed, noncontact process for welding thermoplastics. It is expected to find applications in the packaging and medical products industries.[8] Thermal radiation absorbed by the workpiece forms the weld. Sold state Nd:YAG and CO_2 lasers are most commonly used for welding. Laser radiation, in the normal mode of operation, is so intense and focused that it very quickly degrades thermoplastics. However, lasers have been used to butt weld polyethylene by pressing the unwelded parts together and tracking a defocused laser beam along the joint area. High-speed laser welding of polyethylene films has been demonstrated at weld speeds of 164 ft/min using carbon dioxide and Nd:YAG lasers. Weld strengths are very near to the strength of the parent substrate.

Processing parameters that have been studied in laser welding are the power level of the laser, shielding gas flow rate, offset of the laser beam from a focal point on the top surface of the weld interface, travel speed of the beam along the interface, and welding pressure.[9] Butt joint designs can be laser welded; lap joints can be welded by directing the beam at the edges of the joint.

Lasers have been used primarily for welding polyethylene and polypropylene. Usually, laser welding is applied only to films or thin-walled components. The least powerful beams, around 50 W, with the widest weld spots are used for fear of degrading the polymer substrate. The primary goal in laser welding is to reach a melt temperature at which the parts can be joined quickly before the plastic degrades. To avoid material degradation, accurate temperature measurement of the weld surfaces and temperature control by varying laser strength is essential.

Lasers have been primarily used for joining delicate components that cannot stand the pressure of heated tool or other thermal welding methods. Applications exist in the medical, automotive, and chemical industries. Perhaps the greatest opportunity for this process will be for the high-speed joining of films.

Laser welding has also been used for filament winding of fiber reinforced composite materials using a thermoplastic prepreg. A defocused laser beam is directed on the area where the prepreg meets the winding as it is being built up. With suitable control over the winding speed, applied pressure, and the temperature of the laser, excellent reinforced structures of relatively complex shape can be achieved.

Laser welding requires a substantial investment in equipment and the need for a ventilation system to remove hazardous gaseous and particulate materials resulting for the vaporization of polymer degradation products. Of course, suitable precautions must also be taken to protect the eyesight of anyone in the vicinity of a laser welding operation.

11.4.5 Infrared Welding

Infrared radiation is a noncontact alternative to hot plate welding. Infrared is particularly promising for higher melting polymers, since the parts do not contact and stick to the heat source. Infrared radiation can penetrate into a polymer and create a melt zone quickly. By contrast, hot plate welding involves heating the polymer surface and relying on conduction to create the required melt zone.

Infrared welding is at least 30% faster than heated tool welding. High reproducibility and bond quality can be obtained. Infrared welding can be easily automated, and it can be used for continuous joining. Often, heated tool welding equipment can be modified to accept infrared heating elements.

Infrared radiation can be supplied by high -intensity quartz heat lamps. The lamps are removed after melting the polymer, and the parts are forced together, as with hot plate welding. The depth of the melt zone depends on many factors, including minor changes in polymer formulation. For example, colorants and pigments will change a polymer's absorption properties and will affect the quality of the infrared welding process. Generally, the darker the polymer, the less infrared energy is transferred down through a melt zone, and the more likely will surface degradation occur through overheating.

11.5 Indirect Heating Methods

Many plastic parts may be joined by indirect heating. With these methods, the materials are heated by external energy sources. The heat is induced within the polymer or at the interface. The most popular indirect heating methods are *induction welding* and *dielectric welding*.

For induction welding, the energy source is an electromagnetic field; for dielectric welding, the energy source is an electric field of high frequency.

Indirect heat joining is possible for almost all thermoplastics; however, it is most often used with the newer engineering thermoplastics. The engineering thermoplastics generally have greater heat and chemical resistance than the more conventional plastics. In many applications, engineering plastics are reinforced to improve structural characteristics. They are generally stronger than other plastics and have excellent strength-to-weight ratios. However, many of the engineering plastics are not well suited to joining by direct heat because of the high melt temperatures. Indirect heating methods and frictional heating methods must be used to obtain fast, high-quality bonds with these useful plastic materials.

11.5.1   Induction Welding

The electromagnetic induction field can be used to heat a metal grid or insert placed between mating thermoplastic substrates. Radio frequency energy from the electromagnetic field induces eddy currents in the conductive material, and the material's resistance to these currents produces heat. When the joint is positioned between induction coils, the hot insert causes the plastic to melt and fuse together. Slight pressure is maintained as the induction field is turned off and the joint hardens. The main advantage is that heating occurs only where the electromagnetic insert is applied. The bulk substrate remains at room temperature, avoiding degradation or distortion.

Induction welding is very much like resistance wire welding. An implant is heated to melt the surrounding polymer. Rather than heating the implant restively, in induction welding the implant is heated with an electromagnetic field. More popular forms of induction welding have been developed that use a bonding agent consisting of a thermoplastic resin filled with metal particles. This bonding agent melts in the induction field and forms the adhesive joint. The advantage of this method is that stresses caused by large metal inserts are avoided.

The bonding agent should be similar to the substrates. When joining polyethylene, for example, the bonding agent may be a polyethylene resin containing 0.5–0.6% by volume magnetic iron oxide powder. Electromagnetic adhesives can be made from iron-oxide filled thermoplastics. These adhesives can be shaped into gaskets or film that will melt in the induction field. Induction welding of a nozzle to a thermoplastic hose with an electromagnetic adhesive gaskets is illustrated in Fig. 11.5.

Four basic components embody the electromagnetic welding process:

1. An induction generator that converts 60 Hz electrical supply to 3–40 MHz output frequency and output power from 1 to 5kW

2. The induction heating coil, consisting of water-cooled copper tubing, usually formed into hairpin-shaped loops

3. Fixturing, used to hold parts in place

4. The bonding material, in the form of molded or extruded preforms, which becomes an integral part of the welded product

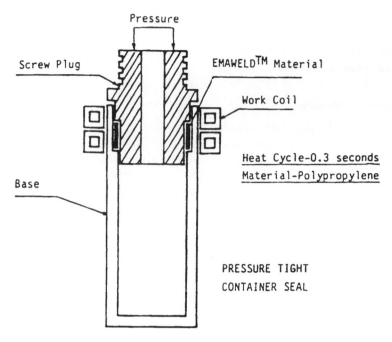

Figure 11.5 Schematic of Emaweld system of induction welding.[10]

Induction heating coils should be placed as close as possible to the joint. For complex designs, coils can be contoured to the joint. Electromagnetic welding systems can be designed for semiautomatic or completely automatic operation. With automated equipment, a sealing rate of up to 150 parts/min can be achieved. Equipment costs are generally in the $10,000 to hundreds of thousands of dollars range, depending on the degree of automation required.

The bonding agent is usually produced for the particular application to ensure compatibility with the materials being joined. However, induction welding equipment suppliers also offer proprietary compounds for joining dissimilar materials. The bonding agent is often shaped into a profile to match the joint design (i.e., gaskets, rings, ribbon). The fillers used in the bonding agents are micron sized ferromagnetic powders. They can be metallic, such as iron or stainless steel, or a ceramic ferrite material.

Quick bonding rates are generally obtainable, because heating occurs only at the interface. Heat does not have to flow from an outside source or through the substrate material to the point of need. Polyethylene joint fabrication can take as little as 3 s with electromagnetic welding. Depending on the weld area, most plastics can be joined by electromagnetic welding in 3–12 s cycle times.

Plastics that are readily bonded with induction methods include all grades of ABS, nylon, polyester, polyethylene, polypropylene, and polystyrene, as well as those materials often considered more difficult to bond such as acetals, modified polyphenylene oxide, and polycarbonate. Reinforced thermoplastics with filler levels up to 65% have been joined successfully.[10] Many combinations of dissimilar materials can be bonded with induction welding processes. Table 11.5 shows compatible plastic combinations for electromagnetic adhesives. Thermoset and other nonmetallic substrates can also be electromagnetically bonded. In these applications, the bonding agent acts as a hot melt adhesive.

Advantages of induction welding include the following:

TABLE 11.5   Compatible Plastic Combinations for Bonding with Electromagnetic Adhesives[12]

● = Compatible

	ABS	Acetals	Acrylic	Cellulosics	Ionomer (Surlyn)	Nylon 6,6 11, 12	Polybutylene	Polycarbonate	Polyethylene	Polyphenylene Oxide (Noryl)	Polypropylene	Polystyrene	Polysulfone	Polyvinyl Chloride	Polyurethane	SAN	Thermoplastic Polyester	Copolyester	Styrene Bl. Copolymer	Olefin Type
ABS	●		●					●								●				
Acetals		●																		
Acrylic	●		●					●				●				●				
Cellulosics				●																
Ionomer (Surlyn)					●															
Nylon 6.6, 11, 12						●														
Polybutylene							●													
Polycarbonate	●		●					●				●	●			●				
Polyethylene									●										●	
Polyphenylene Oxide (Noryl)										●		●								
Polypropylene											●									●
Polystyrene			●					●		●		●				●				
Polysulfone								●					●							
Polyvinyl Chloride														●						
Polyurethane															●					
SAN	●		●					●				●				●				
Thermoplastic Polyester																	●			
Thermoplastic Elastomers: Copolyester																		●	●	●
Thermoplastic Elastomers: Styrene Bl. Copolymer									●									●	●	
Thermoplastic Elastomers: Olefin Type											●							●		●

- Heat damage, distortion and, over-softening of the parts are reduced.
- Squeeze-out of fused material form the bond line is limited.
- Hermetic seals are possible.
- Control is easily maintained by adjusting the output of the power supply.
- No pretreatment of the substrates is required.
- Bonding agents have unlimited storage life.

The ability to produce hermetic seals is cited as one of the prime advantages in certain applications such as in medical equipment. Welds can also be disassembled by placing the bonded article in an electromagnetic field and remelting the joint. There are few limitations on part size or geometry. The only requirement is that the induction coils can be designed to apply a uniform field. The primary disadvantages of electromagnetic bonding are that the metal inserts remain in the finished product, and they represent an added cost. The cost of induction welding equipment is high. The weld is generally not as strong as those obtained by other welding methods.

Induction welding is frequently used for high-speed bonding of many plastic parts. Production cycles are generally faster than with other bonding methods. It is especially useful on plastics that have a high melt temperature such as the modern engineering plastics. Thus, induction welding is used in many under-the-hood automotive applications. It is also frequently used for welding large or irregularly shaped parts.

Electromagnetic induction methods have also been used to quickly cure thermosetting adhesives such as epoxies. Metal particle fillers or wire or mesh inserts are used to provide the heat source. These systems generally have to be formulated so that they cure with a low internal exotherm or else the joint will overheat and the adhesive will thermally degrade.

11.5.2   Dielectric Heating

Dielectric sealing can be used on most thermoplastics except those that are relatively transparent to high-frequency electric fields. This method is used mostly to seal vinyl sheeting such as automobile upholstery, swimming pool liners, and rainwear. An alternating electric field is imposed on the joint, which causes rapid reorientation of polar molecules. As a result, heat is generated within the polymer by molecular friction. The heat causes the polymer to melt and pressure is applied to the joint. The field is then removed, and the joint is held until the weld cools. The main difficulty in using dielectric heating as a bonding method is in directing the heat to the interface. Generally, heating occurs in the entire volume of the polymer that is exposed to the electric field.

Variables in the bonding operation are the frequency generated, dielectric loss of the plastic, the power applied, pressure, and time. The materials most suitable for dielectric welding are those that have strong dipoles. These can often be identified by their high electrical dissipation factors. Materials most commonly welded by this process include polyvinyl chloride, polyurethane, polyamide, and thermoplastic polyester. Since the field intensity decreases with distance from the source, this process is normally used with thin polymer films.

Dielectric heating can also be used to generate the heat necessary for curing polar, thermosetting adhesives, and it can be used to quickly evaporate water from a water based adhesive formulation. Dielectric processing water based adhesives are commonly used in the furniture industry for very fast drying of wood joints in furniture. Common white glues, such as polyvinyl acetate emulsions, can be dried in seconds using dielectric heating processes.

There are basically two forms of dielectric welding: radio frequency welding and microwave welding. Radio frequency welding uses high-frequency (13–100 MHz) to generate heat in polar materials, resulting in melting and weld formation after cooling. The electrodes are usually designed into the platens of a press. Microwave welding uses high frequency (2–20 GHz) electromagnetic radiation to heat a susceptor material located at the joint interface. The generated heat melts thermoplastic materials at the joint interface, producing a weld upon cooling. Heat generation occurs in microwave welding through absorption of electrical energy similar to radio frequency welding.

Polyaniline doped with an aqueous acid is used as a susceptor in microwave welding. This introduces polar groups and a degree of conductivity into the molecular structure. It is these polar groups that preferentially generate heat when exposed to microwave energy. These doped materials are used to produce gaskets which can be used as an adhesive in dielectric welding.

Dielectric welding is commonly used for sealing thin films such as polyvinyl chloride for lawn waste bags, inflatable articles, liners, and clothing. It is used to produce high-volume stationary items such as loose-leaf notebooks and checkbook covers. Because of the cost of the equipment and the nature of the process, industries of major importance for dielectric welding are the commodity industries.

11.6   Friction Welding

In friction welding, the joint interface alone is heated due to mechanical friction caused by one substrate surface contacting and sliding over another substrate surface. The frictional heat generated is sufficient to create a melt zone at the interface. Once a melt zone is created, the relative movement is stopped, and the parts are held together under slight pressure until the melt cools and sets. Common friction welding processes include

- Spin welding
- Ultrasonic welding
- Vibration welding

11.6.1 Spin Welding

Spin welding uses frictional forces to provide the heat of fusion at the interface. One substrate is rotated very rapidly while in touch with the other substrate, which is fixed in a stationary position. The surfaces melt by frictional heating without heating or otherwise damaging the areas outside of the joint. Sufficient pressure is applied during the process to force out a slight amount of resinous flash along with excess air bubbles. Once the rotation is stopped, position and pressure are maintained until the weld sets. The rotation speed and pressure are dependent on the type of thermoplastic being joined.

Spin welding is an old and uncomplicated technique. The equipment required can be as simple as lathes or modified drills. Spin welding has a lower capital cost than other welding methods. The base equipment required is comparatively inexpensive; however, auxiliary equipment such as fixtures, part feeders, and unloaders can drive up the cost of the system. Depending on the geometry and size of the part, the fixture that attaches the part to the rotating motor may be complex. A production rate of 300 parts/min is possible on simple circular joints with an automated system containing multiple heads.

The main advantages of spin welding are its simplicity, high weld quality, and the wide range of possible materials that can be joined with this method. Spin welding is capable of very high throughput. Heavy welds are possible with spin welding. Actual welding times for most parts are only several seconds. A strong hermetic seal can be obtained that is frequently stronger than the material substrate itself. No foreign materials are introduced into the weld, and no environmental considerations are necessary. The main disadvantage of this process is that spin welding is used primarily on parts where at least one substrate is circular.

When considering a part as a candidate for spin welding, there are three items that must be considered:

1. The type of material and the temperature at which it starts to become tacky

2. The diameter of the parts

3. How much flash will develop and what to do with the flash

The parts to be welded must be structurally stiff enough to resist the pressure required. Joint areas must be circular, and a shallow matching grove is desirable to index the two parts and provide a uniform bearing surface. In addition, the tongue-and-groove type joint is useful in hiding the flash that is generated during the welding process. However, a flash *trap* will usually lower the ultimate bond strength. It is generally more desirable to either remove the flash or to design the part so that the flash accumulates on the inside of the joint and hidden from view. Figure 11.6 shows conventional joint designs used in spin welding.

Since the heating that is generated at the interface depends on the relative surface velocity, the outside edges of circular components will see higher temperatures by virtue of their greater diameter and surface velocity. This will cause a thermal differential that could result in internal stress in the joint. To alleviate this affect, joints with hollow section and thin walls are preferred.

The larger the part, the larger the motor required to spin the part as more torque is required to spin the part and obtain sufficient friction. Parts with diameters of 1 to 15 in. have been spin welded using motors from 1/4 to 3 horsepower.[14] The weld can be controlled by the rpm of the motor and somewhat by the pressure on the piece being joined, the timing of the pressure during spin and during joining, and the cooling time and pressure. In commercial rotation welding machines, speed

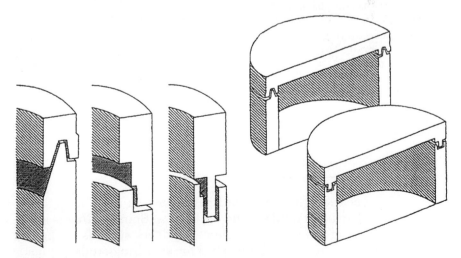

Figure 11.6 Common joints used in the spin welding process.[13]

can range from 200 to 14,000 rpm. Welding times range from tenths of a second to 20 s, and cool-down times are in the range of 1/2 s. A typical complete process time is two seconds.[13] Axial pressure on the part ranges from 150 to 1000 psi. A prototype appraisal is usually completed to determine the optimum parameters of the process for a particular material and joint design.

Table 11.6 shows the temperature at which tackiness starts for most thermoplastics that can be spin welded. This data is useful for all forms of thermal welding not only spin welding. The tackiness temperature can be used as a guide to determine rpm or sfm (surface feet per minute) required at the part. Rotational speeds from 200 to 14,000 rpm are often used. An unfilled 1-in. diameter polyethylene part may be spun at 1000 rpm (260 sfm) to reach the tackiness temperature of 280°F. As the amount of inert filler increases in the part, the rpm needs to be increased. Effects of increasing rpm are similar to those of increased pressure.

TABLE 11.6   Tack Temperatures of Common Thermoplastics (from Ref. 14)

Plastic	Tackiness temperature, °F
Ethylene, vinyl, acetate	150
PVC	170
Polystyrene, high-impact	180
ABS, high-impact	200
Acetal	240
Polyurethane, thermoplastic	245
SAN, CAB	250
Polypropylene, noryl	260
Cellulose acetate	270
Polycarbonate	275
Polyethylene	280
Acetal	290
Acrylic	320
Polysulfone	325
PET	350
PES	430
Fluorocarbon, melt-processible	630

Typical applications include small parts such as fuel filters, check valves, aerosol cylinders, tubes, and containers. Spin welding is also a popular method of joining large-volume products such as packaging and toys. Spin welding can also be used for attaching studs to plastic parts.

11.6.2 Ultrasonic Welding

Ultrasonic welding is another frictional process that can be used on many thermoplastic parts. Frictional heat in this form of welding is generated by high-frequency vibration. The basic parts of a standard ultrasonic welding device are shown in Fig. 11.7. During ultrasonic welding, a high-frequency electrodynamic field is generated that resonates a metal horn that is in contact with one substrate. The horn vibrates the substrate sufficiently fast relative to a fixed substrate that significant heat is generated at the interface. With pressure and subsequent cooling, a strong bond can be obtained. The stages of the ultrasonic welding process is shown in Fig. 11.8.

The frequency generally used in ultrasonic assembly is 20 kHz, because the vibration amplitude and power necessary to melt thermoplastics are easy to achieve. However, this power can produce a great deal of mechanical vibration, which is difficult to control, and tooling becomes large. Higher frequencies (40 kHz) that produce less vibration are possible and are generally used for welding the engineering thermoplastics and reinforced polymers. Higher frequencies are also more appropriate for smaller parts and for parts where less material degradation is required.

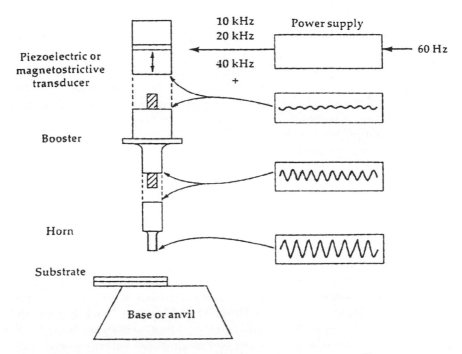

Figure 11.7 Equipment used in a standard ultrasonic welding process.[15]

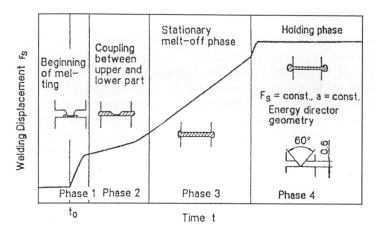

Figure 11.8 Stages in the ultrasonic welding process. In phase 1, the horn is placed in contact with the part, pressure is applied, and vibratory motion is started. Heat generation due to friction melts the energy director, and it flows into the joint interface. The weld displacement begins to increase as the distance between the parts decreases. In phase 2, the melting rate increases, resulting in increased weld displacement, and the part surfaces meet. Steady-state melting occurs in phase 3, as a constant melt layer thickness is maintained in the weld. In phase 4, the holding phase, vibrations cease. Maximum displacement is reached, and intermolecular diffusion occurs as the weld cools and solidifies.[16]

Ultrasonic welding is clean, fast (20–30 parts per minute), and usually results in a joint that is as strong as the parent material. The method can provide hermetically sealed components if the entire joint can be welded at one time. Large parts generally are too massive to be joined with one continuous bond, so spot welding is necessary. It is difficult to obtain a completely sealed joint with spot welding. Material handling equipment can be easily interfaced with the ultrasonic system to further improve rapid assembly.

Rigid plastics with a modulus of elasticity are best. Rigid plastics readily transmit the ultrasonic energy, whereas softer plastics tend to dampen the energy before it reaches the critical joint area. Excellent results generally are obtainable with polystyrene, SAN, ABS, polycarbonate, and acrylic plastics. PVC and the cellulosics tend to attenuate energy and deform or degrade at their surfaces. Figure 11.9 shows an index for the ultrasonic weldability of conventional thermoplastics. Dissimilar plastics may be joined if they have similar melt temperatures and are chemical compatible. The plastic compatibility chart for ultrasonic welding is shown in Table 11.7. Materials such a polycarbonate and nylon must be dried before welding, otherwise their high level of internal moisture will cause foaming and interfere with the joint.

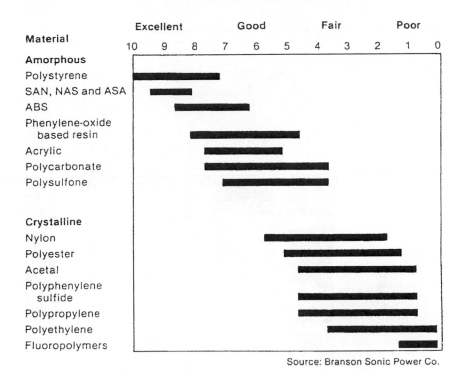

Figure 11.9 Ultrasonic weldability index for common thermoplastics.[17]

Common ultrasonic welding joint designs are shown in Fig. 11.10. The most common design is a butt joint that uses an *energy director*. This design is appropriate for most amorphous plastic materials. The wedge design concentrates the vibrational energy at the tip of the energy director. A uniform melt then develops where the volume of material formed by the energy director becomes the material that is consumed in the joint. Without the energy director, a butt joint would produce voids along the interface, resulting in stress and a low strength joint. Shear and scarf joints are employed for crystalline polymeric materials. They are usually formed by designing an interference fit.

Ultrasonic welding can also be used to stake plastics to other substrates and for inserting metal parts. It can also be used for spot welding two plastic components. Figure 11.11 illustrates ultrasonic insertion, swaging, stacking, and spot welding operations. In ultrasonic spot welding, the horn tip passes through the top sheet to be welded. The molten plastic form a neat raised ring on the surface that is shaped by the horn tip. Energy is also released at the interface of the two sheets, producing frictional heat. As the tip penetrates the bottom

TABLE 11.7 Compatibility of Plastics for Ultrasonic Welding[15]

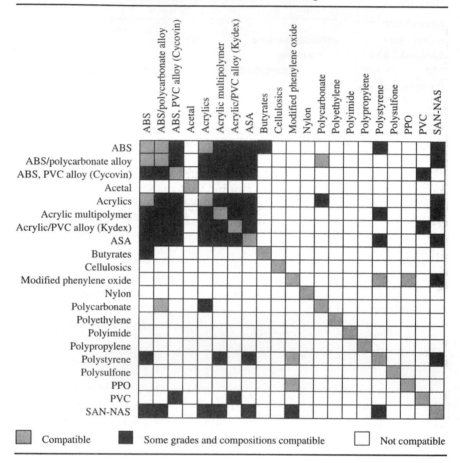

Compatible   Some grades and compositions compatible   Not compatible

substrate, displaced molten plastic flows between the sheets into the preheated are and forms a permanent bond.

Ultrasonic heating is also applicable to hot-melt and thermosetting adhesives.[20] In these cases, the frictional energy is generated by the substrate contacting an adhesive film between the two substrates. The frictional energy generated is sufficient either to melt the hot melt adhesive or to cure the thermosetting adhesive.

11.6.3 Vibration Welding

Vibration welding is similar to ultrasonic welding in that it uses the heat generated at the surface of two parts rubbing together. This frictional heading produces melting in the interfacial area of the joint.

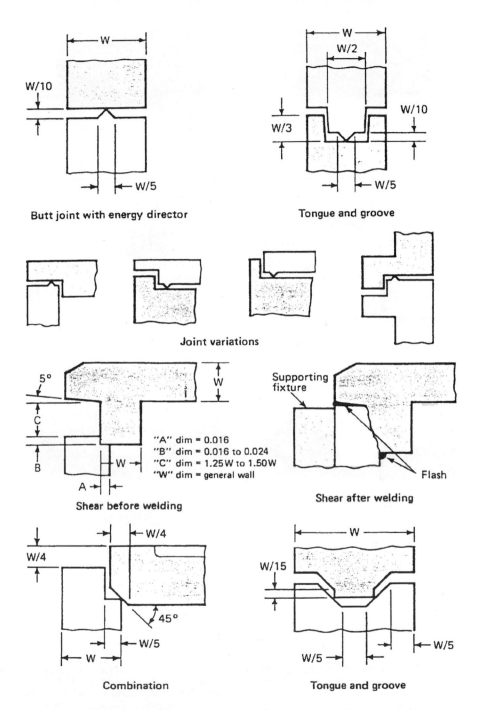

Figure 11.10 Ultrasonic welding joints for amorphous and crystalline polymeric materials.[18]

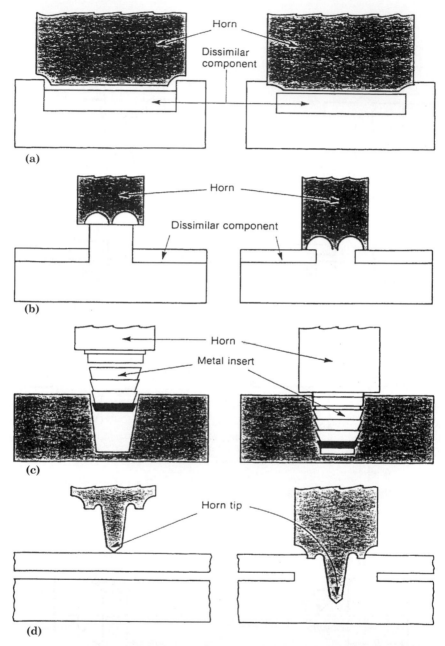

Figure 11.11 Ultrasonic joining operations. (a) Swaging: the plastic ridge is melted and reshaped (left) by ultrasonic vibration to lock another part into place. (b) Staking: ultrasonic vibrations melt and reform a plastic stud (left) to lock a dissimilar component into place (right). (c) Insertion: a metal insert (left) is embedded in a preformed hold in a plastic part by ultrasonic vibration (right). (d) Spot welding: two plastic components (left) are joined at localized points (right).[19]

Vibration welding is different from ultrasonic welding, however, in that it uses lower frequencies of vibration, 120–240 Hz rather than 20–40 kHz as used for ultrasonic welding. With lower frequencies, much larger parts can be bonded because of less reliance on the power supply. Figure 11.12 shows the joining and sealing of a two-part plastic tank design of different sizes using vibration welding.

There are two types of vibration welding: linear and axial. Linear vibration welding is most commonly used. Friction is generated by a linear, back-and-forth motion. Axial or orbital vibration welding allows irregularly shaped plastic parts to be vibration welded. In axial welding, one component is clamped to a stationary structure, and the other component is vibrated using an orbital motion.

Vibration welding fills a gap in the spectrum of thermoplastic welding in that it is suitable for large, irregularly shaped parts. Vibration welding has been used successfully on large thermoplastic parts such as canisters, pipe sections, and other parts that are too large to be excited with an ultrasonic generator and ultrasonically welded. Vibration welding is also capable of producing strong, pressure-tight joints at rapid rates. The major advantage is its application to large parts and to non-circular joints, provided that a small relative motion between the parts in the welding plane is possible.

Usually, the manufacturers of ultrasonic welding equipment will also provide vibration welding equipment. Vibration welding equipment can be either electrically driven (variable frequency) or hydraulically driven (constant frequency). Capital cost is generally higher than with ultrasonic welding.

Process parameters to control in vibration welding are the amplitude and frequency of motion, weld pressure, and weld time. Most in-

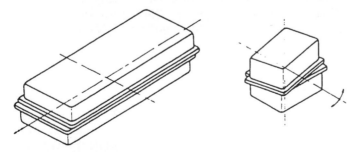

Linear vibration (left) is employed where the length-to-width ratio precludes the use of axial welding (right) where the axial shift is still within the width of the welded edge.

Figure 11.12 Linear and axial vibration welding of a two-part container.[21]

dustrial vibration welding machines operated at frequencies of 120 to 240 Hz. The amplitude of vibration is usually less than 0.2 in. Lower weld amplitudes are used with higher frequencies. Lower amplitudes are necessary when welding parts into recessed cavities. Lower amplitudes (0.020 in) are used for high-temperature thermoplastics. Joint pressure is held in the rage of 200 to 250 psi, although at times much higher pressures are required. High mechanical strength can usually be obtained at shorter weld times by decreasing the pressure during the welding cycle. Vibration welding equipment has been designed to vary the pressure during the welding cycle to improve weld quality and decrease cycle times. This also allows more of the melted polymer to remain in the bond area, producing a wider weld zone.

Vibration welding times depend on the melt temperature of the resin and range from 1 to 10 s with solidification times of less than 1 s. Total cycle times typically range form 6 to 15 s. This is slightly longer than typical spin welding and ultrasonic welding cycles but much shorter than hot plate welding and solvent cementing.

A number of factors must be considered when vibration welding larger parts. Clearances must be maintained between the parts to allow for movement between the halves. The fixture must support the entire joint area, and the parts must not flex during welding. Vibration welding is applicable to a variety of thermoplastic parts with planar or slightly curved surfaces. The basic joint is a butt joint but, unless parts have thick walls, a heavy flange is generally required to provide rigidity and an adequate welding surface. Typical joint designs for vibration welds are shown in Fig. 11.13.

Vibration welding is ideally suited to injection molded or extruded parts in engineering thermoplastics as well as acetal, nylon, polyethylene, ionomer, and acrylic resins. Almost any thermoplastic can be vibration welded. Unlike other welding methods, vibration welding is applicable to crystalline or amorphous or filled, reinforced, or pigmented materials. Vibration welding also can be utilized with fluoropolymers and polyester elastomers, none of which can be joined by ultrasonic welding. By optimizing welding parameters and glass fiber loadings, nylon 6 and nylon 6,6 butt joints can be produced having up to 17% higher strength than the base resin.[23] Any pair of dissimilar materials that can be ultrasonically joined can also be vibration welded.

Vibration welding techniques have found several applications in the automobile industry, including emission control canisters, fuel pumps and tanks, headlight and tail light assemblies, heater valves, air intake filters, water pump housings, and bumper assemblies. They have also been used for joining pressure vessels and for batteries, motor housings, and butane gas lighter tanks.

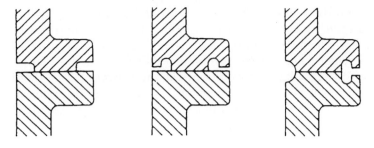

Unless the parts have thick walls, it is necessary to form a flange for the butt weld. In practice, melt traps are usually formed so that the molten resin does not create a flashing.

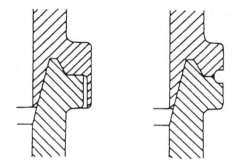

Circular parts can be joined more effectively with tongue-in-groove joints, which retain most of the melt and create a joint of high integrity.

Figure 11.13 Typical vibration welding joint designs.[21]

11.7   Solvent Cementing

Solvent cementing is the simplest and most economical method of joining noncrystalline thermoplastics. In solvent cementing, the application of the solvent softens and dissolves the substrate surfaces being bonded. The solvent diffuses into the surface allowing increased freedom of movement of the polymer chains. As the parts are then brought together under pressure, the solvent softened plastic flows. Van der Walls attractive forces are formed between molecules from each part, and polymer chains from each part intermingle and diffuse into one another. The parts then are held in place until the solvent evaporates from the joint area.

Solvent-cemented joints of like materials are less sensitive to thermal cycling than joints bonded with adhesives, because there is no

stress at the interface due to differences in thermal expansion between the adhesive and the substrate. When two dissimilar plastics
are to be joined, adhesive bonding is generally desirable because of
solvent and polymer compatibility problems. Solvent cemented joints
are as resistant to degrading environments as the parent plastic. Bond
strength greater than 85% of the parent plastic can generally be obtained. Solvents provide high strength bonds quickly due to rapid
evaporation rates.

Solvent bonding is suitable for all amorphous plastics. It is used primarily on ABS, acrylics, cellulosics, polycarbonates, polystyrene,
polyphenylene oxide, and vinyls. Solvent welding is not suitable for
crystalline thermoplastics. It is not affective on polyolefins, fluorocarbons, or other solvent resistant polymers. Solvent welding is moderately affective on nylon and acetal polymers. Solvent welding cannot
be used to bond thermosets. It can be used to bond soluble plastics to
unlike porous surfaces, including wood and paper, through impregnation and encapsulation of the fibrous surface.

The major disadvantage of solvent cementing is the possibility of
stress cracking in certain plastic substrates. Stress cracking or *crazing* is the formation of microcracks on the surface of a plastic part that
has residual internal stresses due to its molding process. The contact
with a solvent will cause the stresses to release uncontrollably, resulting in stress cracking of the part. When this is a problem, annealing of
the plastic part at a temperature slightly below its glass transition
temperature will usually relieve the internal stresses and reduce the
stress cracking probability. Annealing time must be sufficiently long to
allow the entire part to come up to the annealing temperature. Another disadvantage of solvent welding is that many solvents are flammable and/or toxic and must be handled accordingly. Proper
ventilation must be provided when bonding large areas or with high-
volume production.

Solvent cements should be chosen with approximately the same solubility parameter as the plastic to be bonded. Table 11.8 lists typical
solvents used to bond major plastics. Solvents used for bonding can be
a single pure solvent, a combination of solvents, or a solvent(s) mixed
with resin. It is common to use a mixture of a fast-drying solvent with
a less volatile solvent to prevent crazing. The solvent cement can be
bodied up to 25% by weight with the parent plastic to increase viscosity. These bodied solvent cements can fill gaps and provide less shrinkage and internal stress than if only pure solvent is used.

The parts to be bonded should be unstressed and annealed if necessary. For solvent bonding, surfaces should be clean and should fit together uniformly throughout the joint. Close-fitting edges are
necessary for good bonding. The solvent cement is generally applied to

TABLE 11.8   Typical Solvents for Solvent Cementing of Plastics[23]

Solvents for bonding plastics	Acetic acid (glacial)	Acetone	Acetone: ethyl acetate: R (40:40:20)	Acetone: ethyl lactate (90:10)	Acetone: methoxyethyl acetate (80:20)	Acetone: methyl acetate (70:30)	Butyl acetate: acetone: methyl acetate (50:30:20)	Butyl acetate: methyl methacrylate (40:60)	Chloroform	Ethyl acetate	Ethyl acetate: ethyl alcohol (80:20)	Ethylene dichloride	Ethylene dichloride: methylene chloride (50:50)	Methyl acetate	Methylene chloride	Methylene chloride: methyl methacrylate (60:40)	Methylene chloride: methyl methacrylate (50:50)	Methylene chloride: trichloroethylene (85:15)	Methyl ethyl ketone	Methyl isobutyl ketone	Methyl methacrylate	Tetrachloroethylene	Tetrachloroethane	Tetrahydrofuran: cyclohexanone (80:20)	Toluene	Toluene: ethyl alcohol (90:10)	Toluene: methyl ethyl ketone (50:50)	Trichloroethane	Tricholorethylene	Xylene	Xylene: methyl isobutyl ketone (25:75)
ABS																			●	●				●							●
Cellulose acetate		●	●							●					●																
Cellulose acetate butrate		●					●	●							●																
Cellulose propionate								●																							
Ethyl cellulose										●																		●			
Nylon	●																														
Polycarbonate									●			●												●							
Polymethyl methacrylate									●		●					●					●			●							
Polyphenylene oxide									●				●			●	●							●							
Polystyrene															●	●	●								●					●	
Polysulfone																	●	●													
Polyvinyl chloride																			●				●								●
SAN							●			●									●												
Styrene butadiene	●																		●												

the substrate with a syringe or brush. In some cases, the surface may be immersed in the solvent. However, solvent application generally must be carefully controlled, since a small difference in the amount of solvent applied to a substrate greatly affects joint strength. After the area to be bonded softens, the parts are mated and held under light pressure until dry. Pressure should be low and uniform so that the joint will not be stressed. After the joint hardens, the pressure is released, and an elevated-temperature cure may be necessary, depending on the plastic and desired joint strength. Exact processing parameters for solvent welding are usually determined by trial and error. They will depend on the exact polymer, ambient conditions, and type of solvent used.

The bonded part should not be packaged or stressed until the solvent has adequate time to escape form the joint. Complete evaporation of solvent may not occur for hours or days. Some solvent-joined parts may have to be "cured" at elevated temperatures to encourage the release of solvent prior to packaging.

11.8   Methods of Mechanical Joining

There are instances when adhesive bonding, thermal welding, or solvent cementing are not practical joining methods for plastic assembly. This usually occurs because the optimum joint design is not possible, the cost and complexity are too great, or the skill and resources are not present to attempt these forms of fastening. Another common reason for foregoing bonding or welding is when repeated disassembly of the product is required. Fortunately, when these situations occur, the designer can still turn to mechanical fastening as a possible solution.

There are basically two methods of mechanical assembly for plastic parts. The first uses fasteners, such as screws or bolts, and the second uses interference fit, such as press fit or snap fit, and is generally used in thermoplastic applications. This latter method of fastening is also called *design for assembly of self-fastening.* If possible, the designer should try to design the entire product as a one-part molding or with the capability of being press-fit or snap-fit together, because this will eliminate the need for a secondary assembly operation. However, mechanical limitations often will make it necessary to join one part to another using a fastening device. Fortunately, there are a number of mechanical fasteners designed for metals that are also generally suitable with plastics, and there are many other fasteners specifically designed for plastics. Typical of these are thread-forming screws, rivets, threaded inserts, and spring clips.

As in adhesive bonding or welding, special considerations must be given to mechanical fastening because of the nature of the plastic material. Care must be taken to avoid overstressing the parts. Mechanical creep can result in loss of preload in poorly designed systems. Reliable mechanically fastened plastic joints require

- A firm strong connection
- Materials that are stable in the environment
- Stable geometry
- Appropriate stresses in the parts, including the correct clamping force

In addition to joint strength, mechanically fastened joints should prevent slip, separation, vibration, misalignment, and wear of parts. Well designed joints provide the above without being excessively large or heavy, or burdening assemblers with bulky tools. Designing plastic parts for mechanical fastening will depend primarily on the particular plastic being joined and the functional requirements of the application.

11.8.1   Mechanical Fasteners

A large variety of mechanical fasteners can be used for joining plastic parts to themselves and to other materials. These include

- Machine screws and bolts
- Self-threading screws
- Rivets
- Spring fasteners and clips

In general, when repeated disassembly of the product is anticipated, mechanical fasteners are used. Metal fasteners of high strength can overstress plastic parts, so torque controlled tightening or special design provisions are required. Where torque cannot be controlled, various types of washers can be used to spread the compression force over larger areas.

11.8.1.1   Machine screws, bolts, etc.   Parts molded of thermoplastic resin are sometimes assembled with machine screws or with bolts, nuts, and washers, especially if it is a very strong plastic. Machine screws are generally used with threaded inserts, nuts, and clips. They rarely are used in pretapped holes. Figure 11.14 shows correct and incorrect methods of mechanical fastening of plastic parts using this hardware.

Inserts into the plastic part can be effectively used to provide the female part of the fastener. Inserts that are used for plastic assembly consist of molded-in inserts and post-molded inserts.

Molded-in inserts represent inserts that are placed in the mold before the plastic resin is injected. The resin provided is then shaped to the part geometry and locks the insert into its body. Molded-in inserts provide very high-strength assemblies and relatively low unit cost. However, molded-in inserts could increase part cycle time while the inserts are manually placed in the mold. When the application involves infrequent disassembly, molded-in threads can be used successfully. Coarse threads can also be molded into most materials. Threads

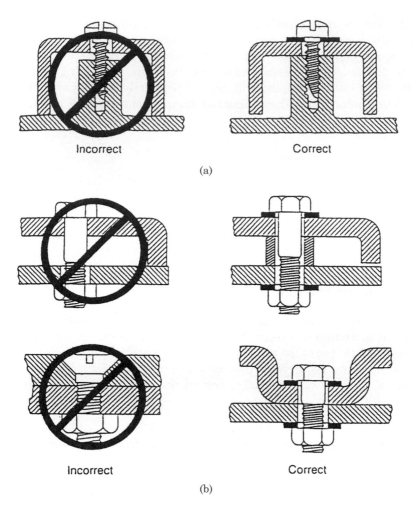

Figure 11.14 Mechanical fastening with (a) self-tapping screws and (b) bolts, nuts, and washers.[24]

of 32 or finer pitch should be avoided, along with tapered threads, because of excessive stress on the part. If the mating connector is metal, overtorque will result in part failure.

Post molded inserts come in four types: (a) press-in, (b) expansion, (c) self-tapping and thread-forming, and (d) inserts that are installed by some method of heating (e.g., ultrasonic). Metal inserts are available in a wide range of shapes and sizes for permanent installation. Inserts are typically installed in molded bosses, designed with holes to suit the insert to be used. Some inserts are pressed into place, and others are installed by methods designed to limit the stress and increase strength. Generally, the outside of the insert is provided with projections of various configurations that penetrate the plastic and prevent movement under normal forces exerted during assembly.

Whatever mechanical fastener is used, particular attention should be paid to the head of the fastener. Conical heads, called *flat heads,* produce undesirable tensile stresses and should not be used. Bolt or screw heads with a flat underside, such as pan heads, round heads, and so forth (Fig. 11.15) are preferred, because the stress produced is more compressive. Flat washers are also suggested and should be used under both the nut and the fastener head. Sufficient diametrical clearance for the body of the fastener should always be provided in the parts to be joined. This clearance can nominally be 0.25 mm (0.010 in.).

11.8.1.2 Self-threading screws. Self-threading screws can be either thread cutting of thread forming. To select the correct screw, the de-

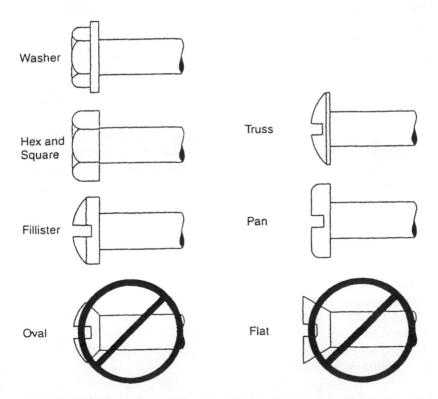

Figure 11.15 Common head systems of screws and bolts. Flat underside of head is preferred.[25]

signer must know which plastic will be used and its modulus of elasticity. The advantages of using these types of screws are

- They are generally off-the-shelf items.
- They are low in cost.
- They allow high production rates.
- Minimal tooling investment is required.

The principal disadvantage of these screws is limited reuse; after repeated disassembly and assembly, these screws will cut or form new threads in the hole, eventually destroying the integrity of the assembly.

Thread -forming screws are used in the softer, more ductile plastics with moduli below 1380 MPa (200,000 psi). There are a number of fasteners especially designed for use with plastics (Fig. 11.16). Thread-forming screws displace plastic material during the threading operation. This type of screw induces high stress levels in the part and is not recommended for parts made of weak resins.

Assembly strengths using thread-forming screws can be increased by reducing hole diameter in the more ductile plastics, by increasing screw thread engagement, or by going to a larger diameter screw when space permits. The most common problem encountered with these types of screws is boss cracking. This can be minimized or eliminated by increasing the size of the boss, increasing the diameter of the hole, decreasing the size of the screw, changing the thread configuration of the screw, or changing the part to a more ductile plastic.

Thread-cutting screws are used in harder, less-ductile plastics. Thread-cutting screws remove material as they are installed, thereby avoiding high stress. However, these screws should not be installed and removed repeatedly.

11.8.1.3 Rivets. Rivets provide permanent assembly at very low cost. Clamp load must be limited to low levels to prevent distortion of the part. To distribute the load, rivets with large heads should be used with washers under the flared end of the rivet. The heads should be three times the shank diameter.

Riveted composite joints should be designed to avoid loading the rivet in tension. Generally, a hole 1/64 in. (0.4 mm) larger than the rivet shank is satisfactory for composite joints. A number of patented rivet designs are commercially available for joining aircraft or aerospace structural composites.

11.8.1.4 Spring steel fasteners. Push-on spring steel fasteners (Fig. 11.17) can be used for holding light loads. Spring steel fasteners are

Barbs provide holding power.

BARBED

Pushtite fastener is pushed into place and can be screwed out.

PUSHTITE

Dual-height thread design boosts holding power by increasing the amount of plastic captured between threads.

HI-LO

Some specials have thread angles smaller than the 60° common on most standard screws. Included angles of 30 or 45° make sharper threads that can be forced into ductile plastics more readily, creating deeper mating threads and reducing stress. With smaller thread angles, boss size can sometimes be reduced.

SHARP THREAD

Reverse saw-tooth edges bite into the walls of the plastic.

MILFORD

Triangular configuration is another technique for capturing large amounts of plastic. After insertion, the plastic cold-flows or relaxes back into the area between lobes. The Trilobe design also creates a vent along the length of the fastener during insertion, eliminating the "ram" effect. In some ductile plastics, pressure builds up in the hole under the fastener as it is inserted, shattering or cracking the material.

TRILOBE

For rapid installation on lightly loaded joints, some fasteners have a thread configuration that allows the screws to be pushed into place. Typical is this design. Suitable for ductile plastics, this fastener relies on plastics relaxation around the shank to form threads. The thread is helical so that it can be unscrewed, but reuse is limited.

PUSH-IN THREAD

Blunt-tip fasteners are suitable for most commercial plastics. Harder plastics require a fastener with a cutting lip. Hardest plastics require both a piercing and drilling tip, as in these fasteners.

BLUNT

CUTTING

PIERCING

Twin lead fastener seats in two revolutions.

TWIN LEAD

Figure 11.16 Thread-forming fasteners for plastics.[25]

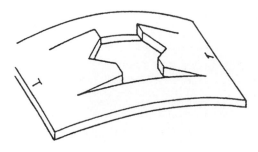

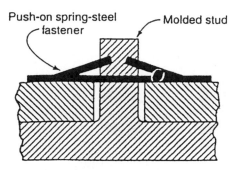

Push-on spring-steel
fastener

Molded stud

Figure 11.17   Push-on spring steel fasteners.[24]

simply pushed on over a molded stud. The stud should have a mini-mum 0.38 mm (0.015 in.) radius at its base. Too large a radius could create a thick section, resulting in sinks or voids in the plastic mold-ing.

11.8.2   Design for Self-Assembly

It is often possible and desirable to incorporate fastening mechanisms in the design of the molded part itself. The two most common methods of doing this are by interference fit (including press fit or shrink fit) and by snap-fit. Whether these methods can be used will depend heavily on the nature of the plastic material and the freedom one has in part design.

11.8.2.1   Press fit.   In press or interference fits, a shaft of one material is joined with the hub of another material by a dimensional interfer-ence between the shaft's outside diameter and the hub's inside diame-ter. This simple, fast assembly method provides joints with high strength and low cost. Press fitting is applicable to parts that must be joined to themselves or to other plastic and non-plastic parts. The advisability of its use will depend on the relative properties of the two materials being assembled. When two different materials are being assembled, the harder material should be forced into the softer. For example, a metal shaft can be press-fitted into plastic hubs. Press-fit joints can be made by simple application of force or by heating or cooling one part relative to the other.

Press fitting produces very high stresses in the plastic parts. With brittle plastics, such as thermosets, press-fit assembly may cause the plastic to crack if conditions are not carefully controlled.

Where press fits are used, the designer generally seeks the maximum pullout force using the greatest allowable interference between parts that is consistent with the strength of the plastic. Figure 11.18 provides general equations for interference fits (when the hub and shaft are made of the same materials and for when they are a metal shaft and a plastic hub). Safety factors of 1.5 to 2.0 are used in most applications.

For a press-fit joint, the effect of thermal cycling, stress relaxation, and environmental conditioning must be carefully evaluated. Testing of the factory assembled parts under expected temperature cycles, or under any condition that can cause changes to the dimensions or modulus of the parts, is obviously indicated. Differences in coefficient of thermal expansion can result in reduced interference due either to one material shrinking or expanding away from the other, or it can cause thermal stresses as the temperature changes. Since plastic materials will creep or stress-relieve under continued loading, loosening of the press fit, at least to some extent, can be expected during service. To counteract this, the designer can knurl or groove the parts. The plastic will then tend to flow into the grooves and retain the holding power of the joint.

11.8.2.2 Snap fit. In all types of snap-fit joints, a protruding part of one component, such as a hook, stud, or bead, is briefly deflected during the joining operation, and it is made to catch in a depression (undercut) in the mating component. This method of assembly is uniquely suited to thermoplastic materials due to their flexibility, high elongation, and ability to be molded into complex shapes. However, snap-fit joints cannot carry a load in excess of the force necessary to make or break the snap-fit. Snap-fit assemblies are usually employed to attach lids or covers that are meant to be disassembled or that will be lightly loaded. The design should be such that, after the assembly, the joint will return to a stress-free condition.

The two most common types of snap-fits are those with flexible cantilevered lugs (Fig. 11.19) and those with a full cylindrical undercut

General equation for interference

$$I = \frac{S_d D_s}{W}\left[\frac{W\mu_h}{E_h} + \frac{1-\mu_s}{E_s}\right]$$

in which

$$W = \frac{1+\left(\dfrac{D_s}{D_h}\right)^2}{1-\left(\dfrac{D_s}{D_h}\right)^2}$$

I =   Diametral interference, mm (in.)

S_d =   Design stress limit or yield strength of the polymer, generally in the hub, MPa (psi) (A typical design limit for an interference fit with thermoplastics is 0.5% strain at 73°C.)

D_h =   Outside diameter of hub, mm (in.)

D_s =   Diameter of shaft, mm (in.)

E_h =   Modulus of elasticity of hub, MPa (psi)

E_s =   Elasticity of shaft, MPa (psi)

μ_h =   Poisson's ratio of hub material

μ_s =   Poisson's ratio of shaft material

W =   Geometric factor

If the shaft and hub are of the same material, $E_h = E_s$ and $\mu_h = \mu_s$. The above equation simplifies to:

Shaft and hub of same material

$$I = \frac{S_d D_s}{W} \times \frac{W+1}{E_h}$$

If the shaft is a high modulus metal or other material, with $E_s > 34.4\times10^3$ MPa, the last term in the general interference equation is negligible, and the equation simplifies to:

Metal shaft, plastic hub

$$I = \frac{S_d D_s}{W} \times \frac{W+\mu_h}{E_h}$$

Figure 11.18 General calculation of interference fit between a shaft and hub.[26]

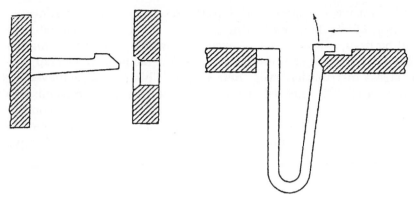

Figure 11.19 Snap-fitting cantilevered arms.[24]

and mating lip (Fig. 11.20). Cylindrical snap fits are generally stronger but require deformation for removal from the mold. Materials with good recovery characteristics are required.

To obtain satisfactory results, the undercut design must fulfill certain requirements, as follows:

- The wall thickness should be kept uniform.

- The snap fit must be placed in an area where the undercut section can expand freely.

- The ideal geometric shape is circular.

- Ejection of an undercut core from the mold is assisted by the fact that the resin is still at relatively high temperatures.

- Weld lines should be avoided in the area of the undercut.

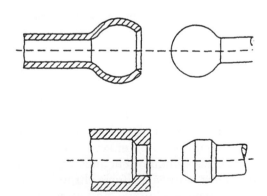

Figure 11.20 Undercuts for snap joints.[24]

In the cantilevered snap-fit design, the retaining force is essentially a function of the bending stiffness of the resin. Cantilevered lugs should be designed so as not to exceed allowable stresses during assembly. Cantilevered snap fits should be dimensioned to develop constant stress distribution over their length. This can be achieved by providing a slightly tapered section or by adding a rib. Special care must be taken to avoid sharp corners and other possible stress concentrations. Cantilever design equations have been recently developed to allow for both the part and the snap fit to flex.[28] Many more designs and configurations can be used with snap-fit configuration than only cantilever or snap fit joints. The individual plastic resin suppliers are suggested for design rules and guidance on specific applications.

11.9   Recommended Assembly Processes for Common Plastics

When decisions are to be made relative to assembly methods (mechanical fastening, adhesive bonding, thermal welding, or solvent cementing), special considerations must be taken because of the nature of the substrate and possible interactions with the adhesive or the environment. The following sections identify some of these considerations and offer an assembly guide to the various methods of assemblies that have been found appropriate for specific plastics.

11.9.1   Acetal Homopolymer and Acetal Copolymer

Parts made of acetal homopolymer and copolymer are generally strong and tough, with a surface finish that is the mirror image of the mold surface. Acetal parts are generally ready for end-use or further assembling with little or no postmold finishing.

Press fitting has been found to provide joints of high strength at minimum cost. Acetal copolymer can be used to provide snap-fit parts. Use of self-tapping screws may provide substantial cost savings by simplifying machined parts and reducing assembly costs.

Epoxies, isocyanate cured polyester, and cyanoacrylates are used to bond acetal copolymer. Generally, the surface is treated with a sulfuric-chromic acid treatment. Epoxies have shown 150–500 psi shear strength on sanded surfaces and 500–1000 psi on chemically treated surfaces. Plasma treatment has also shown to be effective on acetal substrates. Acetal homopolymer surfaces should be chemically treated prior to bonding. This is accomplished with a sulfuric-chromic acid treatment followed by a solvent wipe. Epoxies, nitrile, and nitrile-phenolics can be used as adhesives.

Thermal welding and solvent cementing are commonly used for bonding this material to itself. Heated tool welding produces exceptionally strong joints with acetal homopolymers and copolymers. With the homopolymer, a temperature of the heated surface near 550°F and a contact time of 2 to 10 s are recommended. The copolymer can be hot plate welded from 430 to 560°F. Annealing acetal copolymer joints is claimed to strengthen them further. Annealing can be done by immersing the part in 350°F oil. Acetal resin can be bonded by hot wire welding. Pressure on the joint, duration of the current, and wire type and size must be varied to achieve optimum results. Shear strength on the order of 150 to 300 lb/in. or more have been obtained with both varieties, depending on the wire size, energizing times (wire temperature), and clamping force.

Hot gas welding is used effectively on heavy acetal sections. Joints with 50% of the tensile strength of the acetal resin have been obtained. Conditions of joint design and rod placement are similar to those presented for ABS. A nitrogen blanket is suggested to avoid oxidation. The outlet temperature of the welding gun should be approximately 630°F for the homopolymer and 560°F for the copolymer. For maximum joint strength, both the welding rod and parts to be welded must be heated so that all surfaces are melted.

Acetal components can easily be joined by spin welding, which is a fast and generally economical method to obtain joints of good strength. Spin welded acetal joints can have straight 90° mating surfaces, or surfaces can be angles, molded in a V-shape, or flanged.

Although not common practice, acetal copolymer can be solvent welded at room temperature with full-strength hexafluoroacetone sesquihydrate (Allied Chemical Corp., Morristown, N.J.). The cement has been found to be an effective bonding agent for adhering to itself, nylon, or ABS. Bond strengths in shear are greater than 850 psi using *as-molded* surfaces. Hexafluoroacetone sesquihydrate is a severe eye and skin irritant. Specific handling recommendations and information on toxicity should be requested from Allied Chemical Corp. Because of its high solvent resistance, acetal homopolymer cannot be solvent cemented.

11.9.2   Acrylonitrile-Butadiene-Styrene (ABS)

ABS parts can be designed for snap-fit assembly using a general guideline of 5% allowable strain during the interference phase of the assembly. Thread-cutting screws are frequently recommended for non-foamed ABS, and thread-forming screws are recommended for foamed grades. Depending on the application, the use of bosses and boss caps may be advantageous.

The best adhesives for ABS are epoxies, urethanes, thermosetting acrylics, nitrile-phenolics, and cyanoacrylates. These adhesives have shown joint strength greater than that of the ABS substrates being bonded. ABS substrates do not require special surface treatments other than simple cleaning and removal of possible contaminants.

ABS can also be bonded to itself and to certain other thermoplastics by either solvent cementing or any of the heat welding methods. For bonding acrylonitrile butadiene styrene (ABS) to itself, it is recommended that the hot plate temperatures be between 430 to 550°F. Lower temperatures will result in sticking of the materials to the heated platens, while temperatures above 550°F will increase the possibility of thermal degradation of the surface. In joining ABS, the surfaces should be in contact with the heated tool until they are molten, then brought carefully and quickly together and held with minimum pressure. If too much pressure is applied, the molten material will be forced from the weld, resulting in poor appearance and reduced weld strength. Normally, if a weld flash greater than 1/8 in. occurs, too much joining pressure has been used.

Hot gas welding has been used to join ABS thermoplastic with much success. Joints with over 50% of the strength of the parent material have been obtained. The ABS welding rod should be held approximately at a 90° angle to the base material; the gun should be held at a 45° angle with the nozzle 1/4 to 1/2 in. from the rod. ABS parts to be hot gas welded should be bonded at 60° angles. The welding gun, capable of heating the gas to 500 to 600°F, must be moved continually in a fanning motion to heat both the welding rod and bed. Slight pressure must be maintained on the rod to ensure good adhesion.

Spin welded ABS joints can have straight 90° mating surfaces, or the surface can be angled, molded in a V-shape, or flanged. The most important factor in the quality of the weld is the joint design. The area of the spinning part should be as large as possible, but the difference in linear velocity between the maximum and minimum radii should be as small as feasible.

One of the fastest methods of bonding ABS and acetal thermoplastics is induction welding. This process usually takes 3–10 s but can be done in as little as 1 s. During welding a constant pressure of at least 100 psi should be applied on the joint to minimize the development of bubbles; this pressure should be maintained until the joint has sufficiently cooled. When used, metal inserts should be 0.02 to 0.04 in. thick. Joints should be designed to enclose completely the metal insert. Inserts made of carbon steel require less power for heating although other types of metal can be used. The insert should be located as close as possible to the electromagnetic generator coil and centered within the coil to assure uniform heating.

The solvents recommended for ABS are methyl ethyl ketone, methyl isobutyl ketone, tetrahydrofuran, and methylene chloride. The solvent used should be quick drying to prevent moisture absorption yet slow enough to allow assembly of the parts. The recommended cure time is 12 to 24 hr at room temperature. The time can be reduced by curing at 130 to 150°F. A cement can be made by dissolving ABS resin in solvent up to 25% solids. This type of cement is very effective in joining parts that have irregular surfaces or areas that are not readily accessible. Because of the rapid softening actions of the solvent, the pressure and amount of solvent applied should be minimal.

11.9.3 Cellulosics (Cellulose Acetate, Cellulose Acetate Butyrate, Cellulose Nitrate, Ethyl Cellulose, Etc.)

Cellulosic materials can be mechanically fastened by a number of methods. However, their rigidity and propensity to have internal molding stresses must be carefully considered.

Adhesives commonly used are epoxies, urethanes, isocyanate-cured polyesters, nitrile-phenolic, and cyanoacrylate. Only cleaning is required prior to applying the adhesive. A recommended surface cleaner is isopropyl alcohol. Cellulosic plastics may contain plasticizers. The extent of plasticizer migration and the compatibility with the adhesive must be evaluated. Cellulosics can be stress cracked by uncured cyanoacrylate and acrylic adhesives. Any excess adhesive should be removed from the surface immediately.

Cellulosic materials can also be solvent cemented. Where stress crazing is a problem, adhesives are a preferred method of assembly.

11.9.4 Fluorocarbons (PTFE, CTFE, FEP, Etc.)

Because of the lower ductility of the fluorocarbon materials, snap-fit and press-fit joints are seldom used. Rivets or studs can be used in forming permanent mechanical joints. These can be provided with thermal techniques on the melt-processible grades. Self-tapping screws and threaded inserts are used for many mechanical joining operations. In bolted connections, some stress relaxation may occur the first day after installation. In such cases, mechanical fasteners should be tightened; thereafter, stress relaxation is negligible.

The combination of properties that makes fluorocarbons highly desirable engineering plastics also makes them nearly impossible to heat or solvent weld and very difficult to bond with adhesives without proper surface treatment. The most common surface preparation for fluorocarbons is a sodium naphthalene etch, which is believed to remove fluorine atoms from the surface to provide better wetting properties. A formulation and description of the sodium naphthalene process can be found in Chap. 12. Commercial chemical products for etching fluorocarbons are also listed.

Another process for treating fluorocarbons as well as some other hard to bond plastics (notably polyolefins) is plasma treating. Plasma surface treatment has been shown to increase the tensile shear strength of Teflon bonded with epoxy adhesive from 50 to 1000 psi. The major disadvantage of plasma treating is that it is a batch process, which involves large capital equipment expense, and part size is often limited because of available plasma treating vessel volume. Epoxies and polyurethanes are commonly used for bonding treated fluorocarbon surfaces.

Melt processible fluorocarbon parts have been successfully heat welded, and certain grades have been spin welded and hermetically sealed with induction heating. However, because of the extremely high temperatures involved and the resulting weak bonds, these processes are seldom used for structural applications.

Fluorocarbon parts cannot be solvent welded because of their great resistance to all solvents.

11.9.5 Polyamide (Nylon)

Because of their toughness, abrasion resistance, and generally good chemical resistance, parts made from polyamide or resin (or nylon) are generally more difficult to finish and assemble than other plastic parts. However, nylons are used virtually in every industry and market. The number of chemical types and formulations of nylon available also provide difficulty in selecting fabrication and finishing processes.

Nylon parts can be mechanically fastened by most of the methods described in this chapter. Mechanical fastening is usually the preferred method of assembly, because adhesives bonding and welding often show variable results mainly due to the high internal moisture levels in nylon. Nylon parts can contain a high percentage of absorbed water. This water can create a weak boundary layer under certain conditions. Generally, parts are dried to less than 0.5% water before bonding.

Some epoxy, resorcinol formaldehyde, phenol resorcinol, and rubber based adhesives have been found to produce satisfactory joints between nylon and metal, wood, glass, and leather. The adhesive tensile shear strength is about 250–1000 psi. Adhesive bonding is usually considered inferior to heat welding or solvent cementing. However, priming of nylon adherends with compositions based on resorcinol formaldehyde, isocyanate modified rubber, and cationic surfactant have been reported to provide improved joint strength. Elastomeric (nitrile, urethane), hot melt (polyamide, polyester), and reactive (epoxy, urethane, acrylic, and cyanoacrylate) adhesives have been used for bonding nylon.

Induction welding has also been used for nylon and polycarbonate parts. Because of the variety of formulation available and their direct effect on heat welding parameters, the reader is referred to the resin manufacturer for starting parameter for use in these welding methods. Both nylon and polycarbonate resins should be predried before induction welding.

Recommended solvent systems for bonding nylon to nylon are aqueous phenol, solutions of resorcinol in alcohol, and solutions of calcium chloride in alcohol. These solvents are sometimes bodied by adding nylon resin.

11.9.6 Polycarbonate

Polycarbonate parts lend themselves to all mechanical assembly methods. Polycarbonate parts can be easily joined by solvents or thermal welding methods; they can also be joined by adhesives. However, polycarbonate is soluble in selected chlorinated hydrocarbons. It also exhibits crazing in acetone and is attacked by bases.

When adhesives are used, epoxies, urethanes, and cyanoacrylates are chosen. Adhesive bond strengths with polycarbonate are generally 1000–2000 psi. Cyanoacrylates, however, are claimed to provide over 3000 psi when bonding polycarbonate to itself. No special surface preparation is required of polycarbonate other than sanding and cleaning. Polycarbonates can stress crack in the presence of certain solvents. When cementing polycarbonate parts to metal parts, a room temperature curing adhesive is suggested to avoid stress in the interface caused by differences in thermal expansion.

Polycarbonate film is effectively heat sealed in the packaging industry. The sealing temperature is approximately 425°F. For maximum strength the film should be dried at 250°F to remove moisture before bonding. The drying time varies with the thickness of the film or sheet. A period of approximately 20 min is suggested for a 20-mil thick film and 6 hr for a 1/4-in. thick sheet. Predried films and sheets should be sealed within two hours after drying. Hot plate welding of thick sheets of polycarbonate is accomplished at about 650°F. The faces of the substrates should be butted against the heating element for 2 to 5 s or until molten. The surfaces are then immediately pressed together and held for several seconds to make the weld. Excessive pressure can cause localized strain and reduce the strength of the bond. Pressure during cooling should not be greater than 100 psi.

Polycarbonate parts having thickness of at least 40 mils can be successfully hot gas welded. Bond strengths in excess of 70% of the parent resin have been achieved. Equipment should be used capable of providing gas temperature of 600 to 1200°F. As prescribed for the heated tool process, it is important to adequately predry (250°F) both the polycarbonate parts and welding rods. The bonding process should occur within minutes of removing the parts from the predrying oven.

For spin welding, tip speeds of 30–50 ft/min create the most favorable conditions to get polycarbonate resin surfaces to their sealing temperature of 435°F. Contact times as short as 1/2 s are sufficient for small parts. Pressures of 300–400 psi are generally adequate. For the best bonds, parts should be heat treated for stress relief at 250°F for several hours after welding. However, this stress relief step is often unnecessary and may lead to degraded impact properties of the parent plastics.

Methylene chloride is a very fast solvent cement for polycarbonate. This solvent is recommended only for temperature climate zones and on small areas. A mixture of 60% methylene chloride and 40% ethylene chloride is slower drying and the most common solvent cement used. Ethylene chloride is recommended in very hot climate. These solvents can be bodied with 1 to 5% polycarbonate resin where gap filling properties are important. A pressure of 200 psi is recommended.

11.9.7   Polyethylene, Polypropylene, and Polymethyl Pentene

Because of their ductility, polyolefin parts must be carefully assembled using mechanical fasteners. These assembly methods are normally used on the materials having higher modulus such as high molecular weights of polyethylene and polypropylene.

Epoxy and nitrile-phenolic adhesives have been used to bond these plastics after surface preparation. The surface can be etched with a sodium sulfuric-dichromate acid solution at elevated temperature. Flame treatment and corona discharge have also been used. However, plasma treatment has proven to be the optimum surface process for these materials. Shear strengths in excess of 3000 psi have been reported on polyethylene treated for 10 min in an oxygen plasma and bonded with an epoxy adhesive. Polyolefin materials can also be thermally welded, but they cannot be solvent cemented.

Polyolefins can be thermally welded by almost any technique. However, they cannot be solvent welded because of their resistance to most solvents.

11.9.8   Polyethylene Terephthalate and Polybutylene Terephthalate

These materials can be joined by mechanical self-fastening methods or by mechanical fasteners.

Polyethylene terephthalate (PET) and polybutylene terephthalate (PBT) parts are generally joined by adhesives. Surface treatments recommended specifically for PBT include abrasion and solvent cleaning with toluene. Gas plasma surface treatments and chemical etch have been used where maximum strength is necessary. Solvent cleaning of PET surfaces is recommended. The linear film of polyethylene terephthalate (Mylar®) surface can be pretreated by alkaline etching or plasma for maximum adhesion, but often a special treatment is unnecessary. Commonly used adhesives for both PBT and PET substrates are isocyanate cured polyesters, epoxies, and urethanes. Polyethylene terephthalate cannot be solvent cemented or heat welded.

Ultrasonic welding is the most common thermal assembly process used with polybutylene terephthalate parts. However, heated tool welding and other welding methods have proven satisfactory joints when bonding PET and PBT to itself and to dissimilar materials.

Solvent cementing is generally not used to assemble PET or PBT parts because of their solvent resistance.

11.9.9   Polyetherimide (PEI), Polyamide-imide, Polyetheretherketone (PEEK), Polyaryl Sulfone, and Polyethersulfone (PES)

These high-temperature thermoplastic materials are generally joined mechanically or with adhesives. The high modulus, low creep strength, and superior fatigue resistance make these materials ideal for snap-fit joints.

They are easily bonded with epoxy or urethane adhesives; however, the temperature resistance of the adhesives do not match the temperature resistance of the plastic part. No special surface treatment is required other than abrasion and solvent cleaning. Polyetherimide (ULTEM®), polyamide-imide (TORLON®), and polyethersulfone can be solvent cemented, and ultrasonic welding is possible.

These plastics can also be welded using vibration and ultrasonic thermal processes. Solvent welding is also possible with selected solvents and processing conditions.

11.9.10   Polyimide

Polyimide parts can be joined with mechanical fasteners. Self-tapping screws must be strong enough to withstand distortion when they are inserted into the polyimide resin, which is very hard.

Polyimide parts can be bonded with epoxy adhesives. Only abrasion and solvent cleaning is necessary to treat the substrate prior to bonding. The plastic part will usually have higher thermal rating than the adhesive. Thermosetting polyimides cannot be heat welded or solvent cemented.

11.9.11   Polymethylmethacrylate (Acrylic)

Acrylics are commonly solvent cemented or heat welded. Because acrylics are noncrystalline materials, they can be welded with greater ease than semicrystalline parts. Ultrasonic welding is the most popular process for welding acrylic parts. However, because they are relatively brittle materials, mechanical fastening processes must be carefully chosen.

Epoxies, urethanes, cyanoacrylates, and thermosetting acrylics will result in bond strengths greater than the strength of the acrylic part. The surface needs only to be clean of contamination. Molded parts may stress crack when in contact with an adhesive containing solvent or monomer. If this is a problem, an anneal (slightly below the heat distortion temperature) is recommended prior to bonding.

11.9.12   Polyphenylene Oxide (PPO)

Polystyrene modified polyphenylene oxide can be joined with almost all techniques described in this chapter. Snap-fit and press-fit assemblies can be easily made with this material. Maximum strain limit of 8% is commonly used in the flexing member of PPO parts. Metal screws and bolts are commonly used to assemble PPO parts or for attaching various components.

Epoxy, polyester, polyurethane and thermosetting acrylic have been used to bond modified PPO to itself and other materials. Bond strengths are approximately 600–1500 psi on sanded surfaces and 1000–2200 psi on chromic acid etched surfaces.

Polystyrene modified polyphenylene oxide (PPO) or Noryl® can be hot plate welded at 500 to 550°F and 20 to 30 s contact time. Unmodified PPO can be welded at hot plate temperatures of 650°F. Excellent spin welded bonds are possible with modified polyphenylene oxide (PPO), because the low thermal conductivity of the resin prevents head dissipation from the bonding surfaces. Typical spin welding specifications are rotational speed of 40–50 ft per min and a pressure of 300–400 psi. Spin time should be sufficient to ensure molten surfaces.

Polyphenylene oxide joints must mate almost perfectly; otherwise, solvent welding provides a weak bond. Very little solvent cement is needed. Best results are obtained by applying the solvent cement to only one substrate. Optimum holding time has been found to be 4 min at approximately 400 psi. A mixture of 95% chloroform and 5% carbon tetrachloride is the best solvent system for general-purpose bonding, but very good ventilation is necessary. Ethylene dichloride offers a slower rate of evaporation for large structures or hot climates.

11.9.13 Polyphenylene Sulfide (PPS)

Being a semicrystalline thermoplastic, PPS is not ideally suited to ultrasonic welding. Because of its excellent solvent resistance, PPS cannot be solvent cemented. PPS assemblies can be made by a variety of mechanical fastening methods as well as by adhesives bonding.

Adhesives recommended for polyphenylene sulfide include epoxies and urethanes. Joint strengths in excess of 1000 psi have been reported for abraded and solvent cleaned surfaces. Somewhat better adhesion has been reported for machined surfaces over as-molded surfaces. The high heat and chemical resistance of polyphenylene sulfide plastics make them inappropriate for solvent cementing or heat welding.

Polyimide and polyphenylene sulfide (PPS) resins present a problem in that their high temperature resistance generally requires that the adhesive have similar thermal properties. Thus, high-temperature epoxy adhesives are most often used with polyimide and PPS parts. Joint strength is superior (>1000 psi), but thermal resistance is not better than the best epoxy systems (300–400°F continuous).

11.9.14 Polystyrene

Polystyrene parts are conventionally solvent cemented or heat welded. However, urethanes, epoxies, unsaturated polyesters, and cyanoacrylates will provide good adhesion to abraded and solvent cleaned surfaces. Hot melt adhesives are used in the furniture industry. Polystyrene foams will collapse when in contact with certain solvents. For polystyrene foams, a 100% solids adhesive or a water-based contact adhesive is recommended.

Polystyrene can be joined by either thermal or solvent welding techniques. Preference is generally given to ultrasonic methods because of its speed and simplicity. However, heated tool welding and spin welding are also commonly used.

11.9.15 Polysulfone

Polysulfone parts can be joined with all the processes described in this chapter. Because of their inherent dimensional stability and creep resistance, polysulfone parts can be press fitted with ease. Generally, the amount of interference will be less than that required for other thermoplastics. Self-tapping screws and threaded inserts have also been used.

Urethane and epoxy adhesives are recommended for bonding polysulfone substrates. No special surface treatment is necessary. Polysulfones can also be easily joined by solvent cementing or thermal welding methods.

Direct thermal welding of polysulfone requires a heated tool capable of attaining 700°F. Contact time should be approximately 10 s, and then the parts must be joined immediately. Polysulfone parts should be dried 3 to 6 hr at 250°F before attempting to heat seal. Polysulfone can also be joined to metal, since polysulfone resins have good adhesive characteristics. Bonding to aluminum requires 700°F. With cold rolled steel, the surface of the metal first must be primed with 5 to 10% solution of polysulfone and baked for 10 min at 500°F. The primed piece then can be heat welded to the polysulfone part at 500 to 600°F.

A special tool has been developed for hot gas welding of polysulfone. The welding process is similar to standard hot gas welding methods but requires greater elevated temperature control. At the welding temperature, great care must be taken to avoid excessive application of heat, which will result in degradation of the polysulfone resin.

For polysulfone, a 5% solution of polysulfone resin in methylene chloride is recommended as a solvent cement. A minimum amount of cement should be used. The assembled pieces should be held for 5 min under 500 psi. The strength of the joint will improve over a period of several weeks as the residual solvent evaporate.

11.9.16 Polyvinyl Chloride (PVC)

Rigid polyvinyl chloride can be easily bonded with epoxies, urethanes, cyanoacrylates, and thermosetting acrylics. Flexible polyvinyl chloride parts present a problem because of plasticizer migration over time. Nitrile adhesives are recommended for bonding flexible polyvinyl chloride because of compatibility with the plasticizers used. Adhesives that are found to be compatible with one particular polyvinyl chloride plasticizer may not work with another formulation. Solvent cementing and thermal welding methods are also commonly used to bond both rigid and flexible polyvinyl chloride parts.

11.9.17 Thermoplastic Polyesters

These materials may be bonded with epoxy, thermosetting acrylic, urethane, and nitrile-phenolic adhesives. Special surface treatment is not necessary for adequate bonds. However, plasma treatment has been reported to provide enhanced adhesion. Solvent cementing and certain thermal welding methods can also be used with thermoplastic polyester.

Thermoplastic polyester resin can be solvent cemented using hexafluoroisopropanol or hexafluoroacetone sesquihydrate. The solvent should be applied to both surfaces and the parts assembled as quickly as possible. Moderate pressure should be applied as soon as the parts are assembled. Pressure should be maintained for at least 1 to 2 min; maximum bond strength will not develop until at least 18 hr at room temperature. Bond strengths of thermoplastic polyester bonded to itself will be in the 800 to 1500 psi range.

11.9.18 Thermosetting Plastics (Epoxies; Diallyl Phthalate; Polyesters; Melamine, Phenol, and Urea Formaldehyde; Polyurethanes; Etc.)

Thermosetting plastics are joined either mechanically or by adhesives. Their thermosetting nature prohibits the use of solvent or thermal welding processes; however they are easily bonded with many adhesives.

Abrasion and solvent cleaning are generally recommended as the surface treatment. Surface preparation is generally necessary to remove contaminant, mold release, or gloss from the part surface. Simple solvent washing and abrasion is a satisfactory surface treatment for bonds approaching the strength of the parent plastic. An adhesive should be selected that has a similar coefficient of expansion and modulus as the part being bonded. Rigid parts are best bonded with rigid adhesives based on epoxy formulations. More flexible parts should be bonded with adhesives that are flexible in nature after curing. Epoxies, thermosetting acrylics, and urethanes are the best adhesives for the purpose.

11.10 More Information on Joining Plastics

Additional details on joining plastics by adhesive bonding, direct heat welding, indirect heat welding, frictional welding, solvent cementing, or mechanical fastening can be found in numerous places. The best source of information is often the plastic resin manufacturers themselves. They often have recipes and processes that they will freely offer, because it is in their interest to encourage manufacturers to incorporate their materials in joined components.

Another source of information is the equipment manufacturers. The manufacturers of induction bonding, ultrasonic bonding, spin welding, vibration welding, and other related equipment will often provide guidance on the correct parameter to be used for specific materials and joint designs. Many of these equipment suppliers will have customer service laboratories where prototype parts can be tried and guidance provided regarding optimum processing parameters.

Of course, the adhesive supplier and the mechanical fastener supplier can provide detailed information on their products and advice about the substrate for which it is most appropriate. They can generally provide complete processes and specification relative to the assembly operation. They usually also have moderate amounts of test data to provide an indication of strength and durability.

Finally, a very useful source of information is the technical literature, conference publications, books and handbooks relative to the subject of joining plastics. The following works are especially recommended for anyone requiring detailed information in this area:

- *Handbook of Plastics Joining*, Plastics Design Laboratory, 1999

- *Designing Plastic Parts for Assembly* (Paul A. Tres), Hanser/Gardner, 1998

- *Decorating and Assembly of Plastic Parts* (E. A. Muccio), ASM Publications, 1999

- *Handbook of Adhesives and Sealants* (Edward M. Petrie), McGraw-Hill, 1999

References

1. "Engineer's Guide to Plastics," *Materials Engineering,* May 1972.
2. Trauenicht, J. O., "Bonding and Joining, Weigh the Alternatives; Part 1: Solvent Cement, Thermal Welding," *Plastics Technology,* August 1970.
3. Gentle, D. F., "Bonding Systems for Plastics," *Aspects of Adhesion,* Vol. 5, D. J. Almer, ed., University of London Press, London, 1969.
4. Mark, H. F., Gaylord, N. G., and Bihales, N. M., eds., *Encyclopedia of Polymer Science and Technology,* Vol.1, Wiley, New York, 1964, p. 536.
5. "All About Welding of Plastics," Seelyte Inc., Minneapolis, MN.
6. "How to Fasten and Join Plastics," *Materials Engineering,* March 1971.
7. Spooner, S. A., "Designing for Electron Beam and Laser Welding," *Design News,* September 23, 1985.
8. Troughton, M., "Lasers and Other New Processes Promise Future Welding Benefits," *Modern Plastics,* Mid-November, 1997.
9. "Laser Welding," Chapter 13, *Handbook of Plastics Joining, Plastics Design Library,* Norwich, NY, 1997.
10. Chookazian, M., "Design Criteria for Electromagnetic Welding of Thermoplastics," Emabond Corporation, Norwood, NJ.
11. Leatherman, A., "Induction Bonding Finds a Niche in an Evolving Plastics Industry," *Plastics Engineering,* April 1981.
12. "Electromagnetic Welding System for Assembling Thermoplastic Parts," Emabond Corporation, Norwood, NJ.
13. "Spin Welding," Chapter 4, *Handbook of Plastics Joining,* Plastics Design Library, Norwich, NY, 1997.

14. LaBounty, T. J., "Spin Welding Up-Dating and Old Technique," *SPE ANTEC,* 1985, pp. 855–856.
15. Grimm, R. A., "Welding Process for Plastics," *Advanced Materials and Processes,* March 1995.
16. "Ultrasonic Welding," Chapter 5, *Handbook of Plastics Joining,* Plastics Design Library, Norwich, NY, 1997.
17. Branson Sonic Power Company.
18. "Ultrasonic Joining Gains Favor With Better Equipment and Knowhow," *Product Engineer,* January 1977.
19. Mainolfi, S. J., "Designing Component Parts for Ultrasonic Assembly," *Plastics Engineering,* December 1984.
20. Hauser, R. L., "Ultra Adhesives for Ultrasonic Bonding," *Adhesives Age,* 1969.
21. Scherer, R., "Vibration Welding Could Make the Impossible Design Possible," *Plastics World,* September 1976.
22. Kagan, V. A., et. al., "Optimizing the Vibration Welding of Glass Reinforced Nylon Joints," *Plastics Engineering,* September 1996.
23. Raia, D. C., "Adhesives—the King of Fasteners," *Plastics World,* June 17, 1975.
24. "Engineering Plastics," *Engineered Materials Handbook,* vol. 2, ASM International, Metals Park, OH, 1988.
25. *Machine Design,* November 17, 1988.
26. "Mechanical Fastening," Chapter 14, *Handbook of Plastics Joining,* Product Design Library, Norwich, NY, 1997.
27. McMaster, W., and Lee, C., "New Equations Make Fastening Plastic Components a Snap," *Machine Design,* September 10, 1998.

12

Plastics and Elastomers as Adhesives

Edward M. Petrie
ABB Power T & D Company, Inc.
Raleigh, North Carolina

12.1 Introduction to Adhesives

Adhesives were first used many thousands of years ago, and most were derived from naturally occurring vegetable, animal, or mineral substances. Synthetic polymeric adhesives displaced many of these early products due to stronger adhesion and greater resistance to operating environments. These modern plastic- and elastomer-based adhesives are the principal subject of this chapter.

An adhesive is a substance capable of holding substrates (adherends) together by surface attachment. A material merely conforming to this definition does not necessarily ensure success in an assembly process. For an adhesive to be useful, it must not only hold materials together but also withstand operating loads and last the life of the product.

The successful application of an adhesive depends on many factors. Anyone using an adhesive faces a complex task of selecting the proper adhesive and the correct processing conditions that allow a bond to form. One must also determine which substrate-surface treatment will permit an acceptable degree of permanence and bond strength. The adhesive joint must be correctly designed to avoid stresses within the joint that could cause premature failure. Also, the physical and chemical stability of the bond must be forecast with relation to its service environment. This chapter is intended to guide the adhesives user through these numerous considerations.

12.1.1   Advantages and Disadvantages of Adhesive Bonding

Adhesive bonding presents several distinct advantages over conventional mechanical methods of fastening. There are also some disadvantages that may make adhesive bonding impractical. These are summarized in Table 12.1.

TABLE 12.1   Advantages and Disadvantages of Adhesive Bonding

Advantages	Disadvantages
1. Provides large stress-bearing area.	1. Surfaces must be carefully cleaned.
2. Provides excellent fatigue strength.	2. Long cure times may be needed.
3. Damps vibration and absorbs shock.	3. Limitation on upper continuous operating temperature (generally 350°F).
4. Minimizes or prevents galvanic corrosion between dissimilar metals.	
5. Joins all shapes and thicknesses.	4. Heat and pressure may be required.
6. Provides smooth contours.	5. Jigs and fixtures may be needed.
7. Seals joints.	6. Rigid process control usually necessary.
8. Joins any combination of similar of dissimilar materials.	7. Inspection of finished joint difficult.
9. Often less expensive and faster than mechanical fastening.	8. Useful life depends on environment.
10. Heat, if required, is too low to affect metal parts.	9. Environmental, health, and safety considerations are necessary.
11. Provides attractive strength-to-weight ratio.	10. Special training sometimes required.

The design engineer must consider and weigh these factors before deciding on a method of fastening. However, in many applications, adhesive bonding is the only practical method for assembly. In the aircraft industry, for example, adhesives make the use of thin metal and honeycomb structures feasible, because stresses are transmitted more effectively by adhesives than by rivets or welds. Plastics and elastomers can also be more reliably joined with adhesives than by other methods.

12.1.1.1   Mechanical advantages. The most common methods of structural fastening are shown in Fig. 12.1. Because of the uniformity of an adhesive bond, certain mechanical advantages can be provided as shown. The stress-distribution characteristics and inherent toughness of polymeric adhesives provide bonds with superior fatigue resistance, as shown in Fig. 12.2. Generally, in well designed joints, the adherends will fail in fatigue before the adhesive.

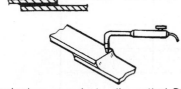

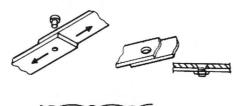

Brazing is an expensive bonding method. Requiring excessive heat, it often results in irregular, distorted parts. Adhesive bonds are always uniform.

When joining a material with mechanical fasteners, holes must be drilled through the assembly. These holes weaken the material and allow concentration of stress.

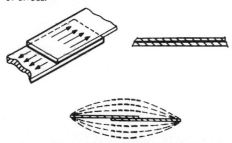

A high strength adhesive bond withstands stress more effectively than either welds or mechanical fasteners.

Figure 12.1 Common methods of structural fastening.[1]

12.1.1.2 Design advantages. Adhesives offer certain design advantages that are often valuable.

- Unlike rivets or bolts, adhesives produce smooth contours that are aerodynamically and cosmetically beneficial.

- Adhesives also offer a better strength-to-weight ratio than mechanical fasteners.

- Adhesives can join any combination of solid materials, regardless of shape or thickness. Materials such as plastics, elastomers, ceramics, and wood can be joined more economically and efficiently by adhesive bonding than by other methods.

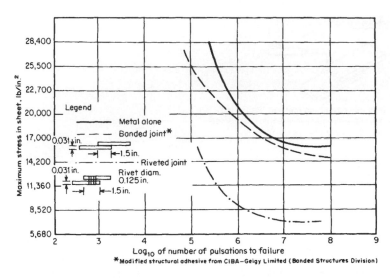

Figure 12.2   Fatigue strengths of aluminum-alloy specimens under pulsating tensile load.[2]

- Adhesive bonding is frequently faster and less expensive than conventional fastening methods. As the size of the area to be joined increases, the time and labor saved by using adhesives instead of mechanical fasteners become progressively greater, because the entire joint area can be assembled in one operation.

12.1.1.3 Other advantages.
Adhesives can be made to function as electrical and thermal insulators. The degree of insulation can be varied with different adhesive formulations and fillers. Adhesives can even be made electrically and thermally conductive with silver and boron nitride fillers, respectively. Adhesives can also perform sealing functions, offering a barrier to the passage of fluids and gases. Adhesives may also act as vibration dampers to reduce the noise and oscillation encountered in assemblies.

Frequently, adhesives may be called upon to do multiple functions. In addition to being a mechanical fastener, an adhesive may also be used as a sealant, vibration damper, insulator, and gap filler in the same application.

12.1.1.4 Mechanical limitations.
The most serious limitation to the use of modern polymeric adhesives is their time-dependent strength in degrading service environments such as moisture, high temperatures, or chemicals.

There are polymeric adhesives that perform well at temperatures between –60 and 350°F. But only a few adhesives can withstand operating temperatures outside that range. Adhesives can also be degraded by chemical environments and outdoor weathering. The rate of strength degradation may be accelerated by continuous stress or elevated temperatures.

12.1.1.5 Design limitations. The adhesive joint must be carefully designed for optimum performance. Design factors must include the type of stress, environmental influences, and production methods that will be used. The strength of the adhesive joint depends on the type and direction of stress. Generally, adhesives perform better when stressed in shear or tension than when exposed to cleavage or peel forces.

Since nearly every adhesive application is somewhat unique, the adhesive manufacturers often do not have data concerning the aging characteristics of their adhesives in specific environments. Thus, before any adhesive is incorporated into production, a thorough evaluation should be made in a simulated operating environment. Time must also be allowed to train personnel in what can be a rather complex and critical manufacturing process.

12.1.1.6 Production limitations. All adhesives require clean surfaces to attain optimum results. Depending on the type and condition of the substrate and the bond strength desired, surface preparations ranging from a simple solvent wipe to chemical etching are necessary.

If the adhesive has multiple components, the parts must be carefully weighed and mixed. The setting operation often requires heat and pressure. Lengthy set time could make assembly jigs and fixtures necessary.

Adhesives may be composed of materials that present personnel hazards, including flammability and dermatitis, in which case necessary precautions must be considered. Finally, the inspection of finished joints for quality control is very difficult. This requires strict control over the entire bonding process to ensure uniform bond quality.

Although the material cost is relatively low, some adhesive systems may require metering, mixing, and dispensing equipment as well as curing fixtures, ovens, and presses. Capital equipment investment must be included in any economic evaluation. The following items contribute to a "hidden cost" of using adhesives, and they also could lead to serious production difficulties:

1. The storage life (shelf life) of the adhesive may be unrealistically short; some adhesives require refrigerated storage.

2. The adhesive may begin to solidify or gel too early in the bonding process.

3. Waste, safety, and environmental concerns can be essential cost factors.

4. Cleanup is a cost factor, especially where misapplied adhesive may ruin the appearance of a product.

5. Once bonded, samples cannot easily be disassembled; if misalignment occurs and the adhesive cures, usually the part must be scrapped.

12.1.2 Theories of Adhesion

Various theories attempt to describe the phenomena of adhesion. No single theory explains adhesion in a general way. However, knowledge of adhesion theories can assist in understanding the basic requirements for a good bond.

12.1.2.1 Mechanical theory. The surface of a solid material is never truly smooth but consists of a maze of microscopic peaks and valleys. According to the mechanical theory of adhesion, the adhesive must penetrate the cavities on the surface and displace the trapped air at the interface.

Such mechanical anchoring appears to be a prime factor in bonding many porous substrates. Adhesives also frequently bond better to abraded surfaces than to natural surfaces. This beneficial effect may be due to:

1. Mechanical interlocking

2. Formation of a clean surface

3. Formation of a more reactive surface

4. Formation of a larger surface area

12.1.2.2 Adsorption theory. The adsorption theory states that adhesion results from molecular contact between two materials and the surface forces that develop. The process of establishing intimate contact between an adhesive and the adherend is known as *wetting*. Figure 12.3 illustrates good and poor wetting of a liquid spreading over a surface.

For an adhesive to wet a solid surface, the adhesive should have a lower surface tension than the solid's critical surface tension. Tables

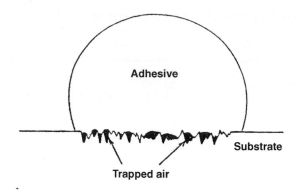

Trapped air

Poor wetting and rough surface. Adhesive has not
flowed into surface irregularities, and air is trapped
at the interface.

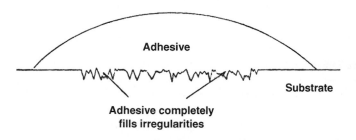

**Adhesive completely
fills irregularities**

Good wetting and rough surface. Adhesive is in
intimate contact with the substrate.

Figure 12.3 Illustration of good and poor wetting by adhesive
spreading over a surface.

12.2 and 12.3 list the surface tensions of common adherends and
liquids.

Most organic adhesives easily wet metallic solids. But many solid
organic substrates have surface tensions less than those of common
adhesives. From Tables 12.2 and 12.3, it is apparent that epoxy adhesives will wet clean aluminum or copper surface. However, epoxy resin
will not wet a substrate having a critical surface tension significantly
less than 47 dynes/cm. Epoxies will not, for example, wet either a
metal surface contaminated with silicone oil or a clean polyethylene
substrate.

After intimate contact is achieved between adhesive and adherend
through wetting, it is believed that adhesion results primarily through

TABLE 12.2   Critical Surface Tension of Common Plastics and Metals

Materials	Critical surface tension, dyn/cm
Acetal	47
Acrylonitrile-butadiene-styrene	35
Cellulose	45
Epoxy	47
Fluoroethylene propylene	16
Polyamide	46
Polycarbonate	46
Polyethylene	31
Polyethylene terephthalate	43
Polyimide	40
Polymethylmethacrylate	39
Polyphenylene sulfide	38
Polystyrene	33
Polysulfone	41
Polytetrafluoroethylene	18
Polyvinyl chloride	39
Silicone	24
Aluminum	≈500
Copper	≈1000

TABLE 12.3   Surface Tension of Common Adhesives Bonding

Material	Surface tension, dyn/cm
Epoxy resin	47
Fluorinated epoxy resin[*]	33
Glycerol	63
Petroleum lubricating oil	29
Silicone oils	21
Water	73

[*]Experimental resin; developed to wet low–energy surfaces. (Note low surface tension relative to most plastics.)

forces of molecular attraction. Four general types of chemical bonds are recognized: electrostatic, covalent, and metallic (which are referred to as *primary bonds*), and van der Walls forces (which are referred to as *secondary bonds*). The adhesion between adhesive and adherend is thought to be primarily due to van der Walls forces of attraction.

12.1.2.3   Electrostatic and diffusion theories.   The electrostatic theory states that electrostatic forces in the form of an electrical double layer are formed at the adhesive–adherend interface. These forces account for resistance to separation. The theory gathers support from the fact that electrical discharges have been noticed when an adhesive is peeled from a substrate.

The fundamental concept of the diffusion theory is that adhesion arises through the interdiffusion of molecules in the adhesive and adherend. The diffusion theory is primarily applicable when both the adhesive and adherend are polymeric, having long-chain molecules capable of movement. Bonds formed by solvent or heat welding thermoplastics result from the diffusion of molecules.

12.1.2.4   Weak boundary layer theory.   According to the weak boundary layer theory, when bond failure seems to be at the interface, usually a cohesive break of a weak boundary layer is the real event.[3] Weak boundary layers can originate from the adhesive, the adherend, the environment, or any combination of the three.

Weak boundary layers can occur on the adhesive or adherend if an impurity concentrates near the bonding surface and forms a weak attachment to the substrate. When bond failure occurs, it is the weak boundary layer that fails, although failure seems to occur at the adhesive-adherend interface.

Two examples of a weak boundary layer effect are polyethylene and metal oxides. Conventional grades of polyethylene have weak, low-molecular-weight constituents evenly distributed throughout the polymer. These weak elements are present at the interface and contribute to low failing stress when polyethylene is used as an adhesive or adherend. Certain metal oxides are weakly attached to their base metals.

Failure of adhesive joints made with these adherends will occur cohesively within the weak oxide. Weak boundary layers can be removed or strengthened by various surface treatments.

Weak boundary layers formed from the shop environment are very common. When the adhesive does not wet the substrate as shown in Fig. 12.3, a weak boundary layer of air is trapped at the interface, causing lowered joint strength. Moisture from the air may also form a weak boundary layer on hydrophilic adherends.

12.1.3   Requirements of a Good Bond

The basic requirements for a good adhesive bond are cleanliness, wetting, solidification, and proper selection of adhesive and joint design.

12.1.3.1   Cleanliness.   To achieve an effective adhesive bond, one must start with a clean surface. Foreign materials such as dirt, oil, moisture, and weak oxide layers must be removed from the substrate surface, or else the adhesive will bond to these weak boundary layers rather than the actual substrate. There are various surface preparations that remove or strengthen the weak boundary layer. These treatments generally involve physical or chemical processes or a combination of both. Surface-preparation methods for specific substrates will be discussed in a later section.

12.1.3.2   Wetting.   While it is in the liquid state, the adhesives must *wet* the substrate. Examples of good and poor wetting have been explained. The result of good wetting is greater contact area between adherend and adhesive over which the forces of adhesion may act.

12.1.3.3   Solidification.   The liquid adhesive, once applied, must be capable of conversion into a solid. The process of solidifying can be completed in different ways (e.g., chemical reaction by any combination of heat, pressure, and curing agents; cooling from a molten liquid to a solid; and drying due to solvent evaporation). The method by which solidification occurs depends on the adhesive.

12.1.3.4   Adhesive choice.   Factors most likely to influence adhesive selection are listed in Table 12.4. With regard to the controlling factors involved, the many adhesives available can usually be narrowed to a few candidates that are most likely to be successful. The general areas of concern to the design engineer when selecting adhesives should be the material to be bonded, service requirements, production requirements, and overall cost.

12.1.3.5   Joint design.   The adhesive joint should be designed to optimize the forces of adhesion. Such design considerations will be discussed in the next section. Although adequate adhesive-bonded assemblies have been made from joints designed for mechanical fastening, maximum benefit can be obtained only in assemblies specifically designed for adhesive bonding.

12.1.4   Mechanism of Bond Degradation

Adhesive joints may fail adhesively or cohesively. *Adhesive* failure is interfacial bond failure between the adhesive and adherend. *Cohesive* failure occurs when the adhesive fractures, allowing a layer of adhesive to remain on both substrates. When the adherend fails before the adhesive, it is known as a *cohesive failure of the adherend.* The various modes of possible bond failures are shown in Fig. 12.4.

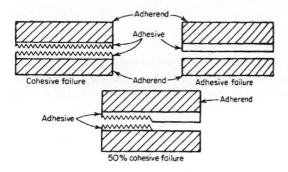

Figure 12.4   Cohesive and adhesive bond failures.

TABLE 12.4   Factors Influencing Adhesive Selection

Stress	Tension
	Shear
	Impact
	Peel
	Cleavage
	Fatigue
Chemical factors	External (service-related)
	Internal (effect of adhered on adhesives)
Exposure	Weathering
	Light
	Oxidation
	Moisture
	Salt spray
Temperature	High
	Low
	Cycling
Biological factors	Mold
	Rodents or vermin
Working properties	Application
	Bonding time and temperature range
	Tackiness
	Curing rate
	Storage stability
	Coverage

Cohesive failure within the adhesive or one of the adherends is the ideal type of failure, because the maximum strength of the materials in the joint has been reached. However, failure mode should not be used as a criterion for a useful joint. Some adhesive-adherend combinations may fail in adhesion but provide sufficient strength margin to be practical. An analysis of failure mode can be an extremely useful tool to determine if a failure is due to a weak boundary layer or improper surface preparation.

The exact cause of adhesive failure is very hard to determine, because so many factors in adhesive bonding are interrelated. However, there are certain common factors at work when an adhesive bond is made that contribute to the weakening of all bonds. The influences of these factors are qualitatively summarized in Fig. 12.5.

If the adhesive does not wet the surface of the substrate, the joint will be inferior. It is also important to allow the adhesive enough time to wet the substrate effectively before gelation occurs.

Internal stresses occur in the adhesive joint because of a natural tendency of the adhesive to shrink during solidification and because of differences in physical properties between adhesive and substrate. The coefficient of thermal expansion of adhesive and adherend should be as close as possible to limit stresses that may develop during ther-

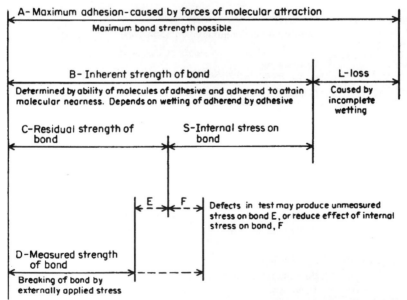

Figure 12.5 Relations between factors involved in adhesion.[4]

mal cycling or after cooling from an elevated-temperature cure. Polymeric adhesives generally have a thermal-expansion coefficient an order of magnitude greater than metals. Adhesives can be formulated with various fillers to modify their thermal-expansion characteristics and limit internal stresses. A relatively elastic adhesive capable of accommodating internal stress may also be useful when thermal-expansion differences are of concern.

Once an adhesive bond is placed in service, other forces are at work weakening the bond. The type of stress, its orientation to the joint, and the rate in which the stress is applied are important. Sustained loads can cause premature failure in service, even though similar unloaded joints may exhibit adequate strength when tested after aging. Most adhesives have poor strength when stresses are acting to peel or cleave the adhesive from the substrate. Many adhesives are sensitive to the rate in which the joint is stressed. Rigid, brittle adhesives sometimes have excellent tensile or shear strength but stand up very poorly under an impact test.

Operating environments are capable of degrading an adhesive joint in various ways. The adhesive may have to withstand temperature variation, weathering, oxidation, moisture, and other exposure conditions. If more than one of these factors are present in the operating environment, their synergistic effect could cause a rapid decline in adhesive strength.

12.1.5   Adhesive Classification

Adhesives may be classified by many methods. The most common methods are by function, chemical composition, mode of application and setting, and end use.

The functional classification defines adhesives as being structural or nonstructural. Structural adhesives are materials of high strength and permanence. Their primary function is to hold structures together and be capable of resisting high loads. Nonstructural adhesives are not required to support substantial loads. They merely hold lightweight materials in place or provide a seal without having a high degree of strength.

The chemical composition classification broadly describes adhesives as thermosetting, thermoplastic, elastomeric, or combinations of these. There are then many chemical types within each classification. They are described in Table 12.5.

Adhesives are often classified by their mode of application. Depending on viscosity, adhesives are sprayable, brushable, or trowelable. Heavily bodied adhesive pastes and mastics are considered extrudable; they are applied by syringe, caulking gun, or pneumatic pumping equipment.

TABLE 12.5 Adhesives Classified by Chemical Composition (from Ref. 5)

Classification	Thermoplastic	Thermosetting	Elastomeric	Alloys
Types within group...	Cellulose acetate, cellulose acetate butyrate, cellulose nitrate, polyvinyl acetate, vinyl vinylidene, polyvinyl acetals, polyvinyl alcohol, polyamide, acrylic, phenoxy	Cyanoacrylate, polyester, polyacrylate, urea formaldehyde, melamine formaldehyde, resorcinol and phenol-resorcinol formaldehyde, epoxy, polyimide, polybenzimidazole, acrylic, acrylate acid diester	Natural rubber, reclaimed rubber, butyl, polyisobutylene, nitrile, styrene-butadiene, polyurethane, polysulfide, silicone, neoprene	Epoxy-phenolic, epoxy-polysulfide, epoxy-nylon, nitrilephenolic, neoprene-phenolic, vinyl-phenolic
Most used form...	Liquid, some dry film	Liquid, but all forms common	Liquid, some film	Liquid, paste, film
Common further classifications...	By vehicle (most are solvent dispersions or water emulsions)	By cure requirements (heat and/or pressure most common but some are catalyst types)	By cure requirements (all are common); also by vehicle (most are solvent dispersions or water emulsions)	By cure requirements (usually heat and pressure except some epoxy types); by vehicle (most are solvent dispersions or 100% solids); and by type of adherends or end-service conditions
Bond characteristics...	Good to 150–200°F; poor creep strength; fair peel strength	Good to 200–500°F; good creep strength; fair peel strength	Good to 150–400°F; never melt completely; low strength; high flexibility	Balanced combination of properties of other chemical groups depending on formulation; generally higher strength over wider temp range
Major type of use...	Unstressed joints; designs with caps, overlaps, stiffeners	Stressed joints at slightly elevated temp	Unstressed joints on light-weight materials; joints in flexure	Where highest and strictest end-service conditions must be met; sometimes regardless of cost, as military uses
Materials most commonly bonded...	Formulation range covers all materials, but emphasis on nonmetallics—esp. wood, leather, cork, paper, etc.	For structural uses of most materials	Few used "straight" for rubber, fabric, foil, paper, leather, plastics films; also as tapes. Most modified with synthetic resins	Metals, ceramics, glass, thermosetting plastics; nature of adherends often not as vital as design or end-service conditions (i.e., high strength, temp)

Another distinction between adhesives is the manner in which they flow or solidify. Some adhesives solidify simply by losing solvent, whereas others harden as a result of heat activation or chemical reaction. Pressure-sensitive adhesives flow under pressure and are stable when the pressure is removed.

Adhesives may also be classified according to their end use. Thus, metal adhesives, wood adhesives, and vinyl adhesives refer to the substrates they bond; and acid-resistant adhesives, heat-resistant adhesives, and weatherable adhesives indicate the environments for which each is suited.

12.1.6 Adhesive Bonding Process

A typical flow chart for the adhesive bonding process is shown in Fig. 12.6. The elements of the bonding process are as important as the adhesive itself for a successful end product.

Many of the adhesive problems that develop are not due to a poor choice of material or joint design but are directly related to faulty production techniques. The adhesive user must obtain the proper processing instructions from the manufacturer and follow them closely and consistently to ensure acceptable results. Adhesive production involves four basic steps:

1. Design of joints and selection of adhesive

2. Preparation of adherends

3. Applying and curing the adhesive

4. Inspection of bonded parts

The remainder of this chapter will discuss these steps in detail

12.2 Design and Test of Adhesive Joints

12.2.1 Types of Stress

Four basic types of loading stress are common to adhesive joints: tensile, shear, cleavage, and peel. Any combination of these stresses, illustrated in Fig. 12.7, may be encountered in an adhesive application.

Figure 12.6 Basic steps in bonding process.

Tensile stress develops when forces acting perpendicular to the plane of the joint are distributed uniformly over the entire bonded area. Adhesive joints show good resistance to tensile loading, because all the adhesive contributes to the strength of the joint. In practical applications, however, loads are rarely axial, and unwanted cleavage or peel stresses tend to develop.

Shear stress results when forces acting in the plane of the adhesive try to separate the adherends. Joints that depend on the adhesive's shear strength are relatively easy to design and offer favorable properties. Adhesive joints are strong when stressed in shear, because all the bonded area contributes to the strength.

Cleavage and peel stresses are undesirable. Cleavage occurs when forces at one end of a rigid bonded assembly act to split the adherends apart. Peel stress is similar to cleavage but applies to a joint where one or both of the adherends are flexible. Joints loaded in peel or cleavage offer much lower strength then joints loaded in shear because the stress is concentrated on only a very small area of the total bond. The remainder of the bonded area makes no contribution to the strength of the joint. Peel and cleavage stresses should be avoided where possible.

12.2.2   Joint Efficiency

To avoid concentration of stress, the joint designer should take into consideration the following rules:

1. Keep the stress on the bond line to a minimum.

2. Design the joint so that the operating loads will stress the adhesive in shear.

3. Peel and cleavage stresses should be minimized.

4. Distribute the stress as uniformly as possible over the entire bonded area.

5. Adhesive strength is directly proportional to bond width. Increasing width will always increase bond strength; increasing the depth does not always increase strength.

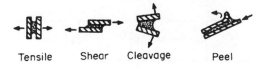

Tensile   Shear   Cleavage   Peel

Figure 12.7   Four basic types of adhesive stress.

6. Generally, rigid adhesives are better in shear, and flexible adhesives are better in peel.

Brittle adhesives are particularly weak in peel, because the stress is localized at only a very thin line at the edge of the bond, as shown in Fig. 12.8. Tough, flexible adhesives distribute the peeling stress over a wider bond area and show greater resistance to peel.

For a given adhesive and adherend, the strength of a joint stressed in shear depends primarily on the width and depth of the overlap and the thickness of the adherend. Adhesive shear strength is directly proportional to the width of the joint. Strength can sometimes be increased by increasing the overlap depth, but the relationship is not linear. Since the ends of the bonded joint carry a higher proportion of the load than the interior area, the most efficient way of increasing joint strength is by increasing the width of the bonded area.

In a shear joint made from thin, relatively flexible adherends, there is a tendency for the bonded area to distort because of eccentricity of the applied load. This distortion, illustrated in Fig. 12.9, causes cleavage stress on the ends of the joint, and the joint strength may be considerably impaired. Thicker adherends are more rigid, and the distortion is not as much a problem as with thin-gage adherends. Figure 12.10 shows the general interrelationship between failure load, depth of overlap, and adherend thickness for a specific metallic adhesive joint.

Since the stress distribution across the bonded area is not uniform and depends on joint geometry, the failure load of one specimen cannot be used to predict the failure load of another specimen with different joint geometry. The results of a particular shear test pertain only to joints that are exact duplicates. To characterize overlap joints more closely, the ratio of overlap length to adherend thickness l/t can be

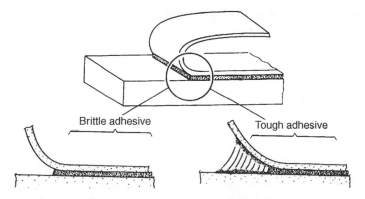

Figure 12.8 Tough, flexible adhesives distribute peel stress over a larger area.[6]

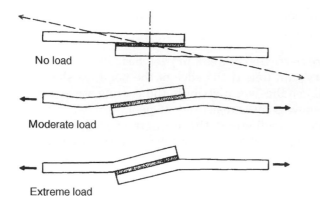

No load

Moderate load

Extreme load

Figure 12.9 Distortion caused by loading can introduce cleavage stresses and must be considered in joint design.[5]

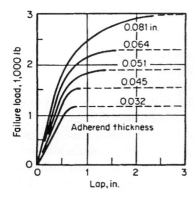

Figure 12.10 Interrelation of failure loads, depth of lap, and adherend thickness for lap joints with a specific adhesive and adherend.[7]

plotted against shear strength. A set of l/t curves for aluminum bonded with a nitrile-rubber adhesive is shown in Fig. 12.11.

The strength of an adhesive joint also depends on the thickness of the adhesive. Thin adhesive films offer the highest strength provided that the bonded area does not have "starved" areas where all the adhesive has been forced out. Excessively heavy adhesive-film thicknesses cause greater internal stresses during cure and concentration of stress under load at the ends of a joint. Optimum adhesive thickness for maximum shear strength is generally considered to be between 2 and 10 mils. Strength does not vary significantly with bondline thickness in this range.

12.2.3 Joint Design

A favorable stress can be applied by using proper joint design. However, joint designs may be impractical, expensive to make, or hard to

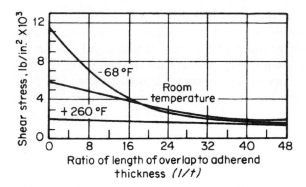

Figure 12.11 l/t curves at three test temperatures for aluminum joints bonded with nitrile-rubber adhesive.[8]

align. The design engineer will often have to weigh these factors against optimum joint performance.

12.2.3.1 Flat adherends.   The simplest joint to make is the plain butt joint. However, butt joints cannot withstand bending forces, because the adhesive would experience cleavage stress. The butt joint can be improved by redesigning in a number of ways, as shown in Fig. 12.12.

Lap joints are commonly used because they are simple to make, are applicable to thin adherends, and stress the adhesive in its strongest direction. Tensile loading of a lap joint causes the adhesive to be stressed in shear. However, the simple lap joint is offset, and the shear forces are not in line, as was illustrated in Fig. 12.9. Modifications of lap-joint design include:

1. Redesigning the joint to bring the load on the adherends in line

2. Making the adherends more rigid (thicker) near the bond area (see Fig. 12.10)

3. Making the edges of the bonded area more flexible for better conformance, thus minimizing peel

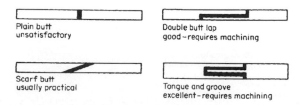

Figure 12.12   Butt connections.

Modifications of lap joints are shown in Fig. 12.13.

Strap joints keep the operating loads aligned and are generally used where overlap joints are impractical because of adherend thickness. Strap-joint designs are shown in Fig. 12.14. Like the lap joint, the single strap is subjected to cleavage stress under bending forces.

When thin members are bonded to thicker sheets, operating loads generally tend to peel the thin member from its base, as shown in Fig. 12.15a. The subsequent illustrations show what can be done to decrease peeling tendencies in simple joints.

12.2.3.2   Cylindrical adherends.   Several recommended designs for rod and tube joints are illustrated in Fig. 12.16. These designs should be used instead of the simpler butt joint. Their resistance to bending forces and subsequent cleavage is much better, and the bonded area is larger. Unfortunately, most of these joint designs require a machining operation.

12.2.3.3   Angle and corner joints.   A butt joint is the simplest method of bonding two surfaces that meet at an angle. Although the butt joint has good resistance to pure tension and compression, its bending strength is very poor. Dado, L, and T angle joints, shown in Fig. 12.17,

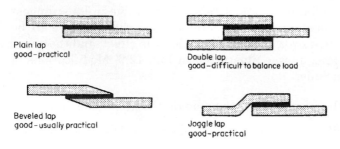

Plain lap
good – practical

Double lap
good – difficult to balance load

Beveled lap
good – usually practical

Joggle lap
good – practical

Figure 12.13   Lap connections.

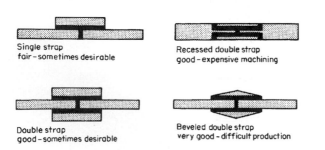

Single strap
fair – sometimes desirable

Recessed double strap
good – expensive machining

Double strap
good – sometimes desirable

Beveled double strap
very good – difficult production

Figure 12.14   Strap connections.

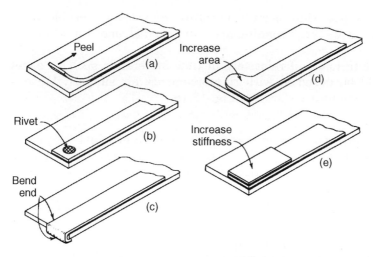

Figure 12.15 Minimizing peel in adhesive joints.[9]

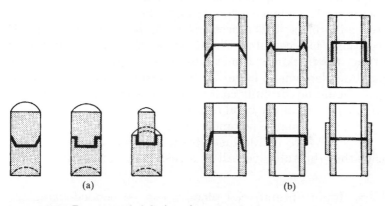

Figure 12.16 Recommended designs for rod and tube joints: (a) three joint designs for adhesive bonding of round bars and (b) six joint configurations that are useful in adhesive-bonding cylinders or tubes.[10]

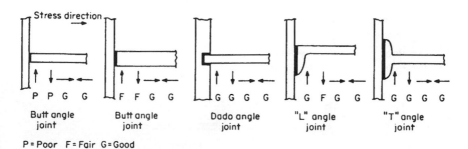

P = Poor   F = Fair   G = Good

Figure 12.17 Types of angle joints and methods of reducing cleavage.[9]

offer greatly improved properties. The T design is the preferable angle joint because of its large bonding area and good strength in all directions.

Corner joints made of relatively flexible adherends such as sheet metal should be designed with reinforcements for support. Various corner-joint designs are shown in Fig. 12.18.

12.2.3.4 Plastic and elastomeric joints. The design of joints for plastic and elastomeric substrates follows the same practice as for metal. However, certain characteristics of these materials require special consideration.

Flexible plastics and elastomers. Thin or flexible polymeric substrates may be joined using a simple or modified lap joint. The double strap joint is best, but it is also the most time consuming to fabricate. The strap material should be made out of the same material as the parts to be joined, or at least have approximately equivalent strength, flexibility, and thickness. The adhesive should have the same degree of flexibility as the adherends. If the sections to be bonded are relatively thick, a scarf joint is acceptable. The length of the scarf should be at least four times the thickness; sometimes larger scarfs may be needed.

When bonding elastomers, forces on the substrate during setting of the adhesive should be carefully controlled, since excess pressure will cause residual stresses at the bond interface.

As with all joint designs, polymeric joints should be designed to avoid peel stress. Figure 12.19 illustrates methods of bonding flexible substrates so that the adhesive will be stressed in its strongest direction.

Rigid plastics. Rigid unreinforced plastics can be bonded using the joint design principles for metals. However, reinforced plastics are often anisotropic materials. This means that their strength properties are directional. Joints made from anisotropic substrates should be de-

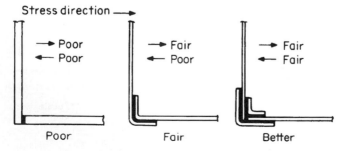

Figure 12.18 Reinforcement of bonded corners.[9]

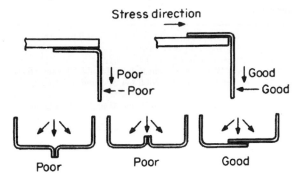

Figure 12.19 Methods of joining flexible plastic or rubber.[9]

signed to stress both the adhesive and adherend in the direction of greatest strength. Laminates, for example, should be stressed parallel to the laminations. Stresses normal to the laminate may cause the substrate to delaminate. Single and joggle lap joints are more likely to cause delamination than scarf or beveled lap joints. The strap-joint variations are useful when bending loads are expected.

12.2.4   Test Methods

12.2.4.1   Standard ASTM test methods. A number of standard tests for adhesive bonds have been specified by the American Society for Testing and Materials (ASTM). Selected ASTM standards are presented in Table 12.6. The properties usually reported by adhesive suppliers are ASTM tensile-shear and peel strength.

Lap-shear tests. The lap-shear or tensile-shear test measures the strength of the adhesive in shear. It is the most common adhesive test, because the specimens are inexpensive, easy to fabricate, and simple to test. This method is described in ASTM D 1002, and the standard test specimen is shown in Fig. 12.20a. The specimen is loaded in tension, causing the adhesive to be stressed in shear until failure occurs. Since the test calls for a sample population of five, specimens can be made and cut from large test panels, illustrated in Fig. 12.20b.

Tensile tests. The tensile strength of an adhesive joint is seldom reported in the adhesive suppliers' literature, because pure tensile stress is not often encountered in actual production. Tensile test specimens also require considerable machining to ensure parallel surfaces.

ASTM tension tests are described in D 897 and D 2095 and employ bar- or rod-shaped butt joints. The maximum load at which failure oc-

TABLE 12.6   ASTM Adhesive Standards

Aging	Resistance of Adhesives to Cyclic Aging Conditions, Test for (D1183) Bonding Permanency of Water- or Solvent-Soluble Liquid Adhesives for Labeling Glass Bottles, Test for (D 1581) Bonding Permanency of Water- or Solvent-Soluble Liquid Adhesives for Automatic Machine Sealing Top Flaps of Fiber Specimens, Test for (D1713) Permanence of Adhesive-Bonded Joints in Plywood under Mold Conditions, Test for (D 1877) Accelerated Aging of Adhesive Joints by the Oxygen-Pressure Method, Practice for (D 3632)
Amylaceous matter	Amylaceous Matter in Adhesives, Test for (D 1488)
Biodeterioration	Susceptibility of Dry Adhesive Film to Attack by Roaches, Test for (D 1382) Susceptibility of Dry Adhesive Film to Attack by Laboratory Rats, Test for (D 1383) Permanence of Adhesive-Bonded Joints in Plywood under Mold Conditions, Test for (D 1877) Effect of Bacterial Contamination of Adhesive Preparations and Adhesive Films, Test for (D 4299) Effect of Mold Contamination on Permanence of Adhesive Preparation and Adhesive Films, Test for (D 4300)
Blocking point	Blocking Point of Potentially Adhesive Layers, Test for (D1146)
Bonding permanency	(See Aging.)
Chemical reagents	Resistance of Adhesive Bonds to Chemical Reagents, Test for (D 896)
Cleavage	Cleavage Strength of Metal-to-Metal Adhesive Bonds, Test for (D 1062)
Cleavage/peel strength	Strength Properties of Adhesives in Cleavage Peel by Tension Loading (Engineering Plastics-to-Engineering Plastics), Test for (D 3807)
Corrosivity	Determining Corrosivity if Adhesive Materials, Practice for (D 3310)
Creep	Conducting Creep Tests of Metal-to-Metal Adhesives, Practice for (D 1780) Creep Properties of Adhesives in Shear by Compression Loading (Metal-to-Metal), Test for (D 2293) Creep Properties of Adhesives in Shear by Tension Loading, Test for (D 2294)
Cryogenic temperatures	Strength Properties of Adhesives in Shear by Tension Loading in the Temperature Range from −267.8 to −55°C (−450 to −67°F), Test for (D 2557)
Density	Density of Adhesives in Fluid Form, Test for (D 1875)
Durability (including weathering)	Effect of Moisture and Temperature on Adhesive Bonds, Test for (D 1151) Atmospheric Exposure of Adhesive-Bonded Joints and Structures, Practice for (D 1828) Determining Durability of Adhesive Joints Stressed in Peel, Practice for (D 2918) Determining Durability of Adhesive Joints Stressed in Shear by Tension Loading, Practice for (D 2919) (See also Wedge Test)

TABLE 12.6   ASTM Adhesive Standards *(Continued)*

Electrical properties	Adhesives Relative to Their Use as Electrical Insulation, Testing (D 1304)
Electrolytic corrosion	Determining Electrolytic Corrosion of Copper by Adhesives, Practice for (D 3482)
Fatigue	Fatigue Properties of Adhesives in Shear by Tension Loading (Metal/Metal), Test for (D 3166)
Filler content	Filler content of Phenol, Resorcinol, and Melamine Adhesives, Test for (D 3166)
Flexibility	(See Flexural strength)
Flexural strength	Flexural Strength of Adhesive Bonded Laminated Assemblies, Test for (D 1184) Flexibility Determination of Hot Melt Adhesives by Mandrel Bend Test Methods, Practice for (D 3111)
Flow properties	Flow Properties of Adhesives, Test for (D 2183)
Fracture strength in cleavage	Fracture Strength in Cleavage of Adhesives in Bonded Joints, Practice for (D 3433)
Gap-filling adhesive bonds	Strength of Gap Filling Adhesive Bonds in Shear by Compression Loading, Practice for (D 3931)
High-temperature effects	Strength Properties of Adhesives in Shear by Tension Loading at Elevated Temperatures (Metal-to-Metal), Test for (D 2295)
Hydrogen-ion concentration	Hydrogen Ion Concentration, est for (D 1583)
Impact strength	Impact Strength of Adhesive Bonds, Test for (D 950)
Light exposure	(See Radiation Exposure)
Low and cryogenic temperatures	Strength Properties of Adhesives in Shear by Tension Loading in the Temperature Range from -267.8 to $-55°C$ (-450 to $-67°F$), Test for (D 2557)
Nonvolatile content	Nonvolatile Content of Aqueous Adhesives, Test for (D 1489) Nonvolatile Content of Urea-Formaldehyde Resin Solutions, Test for (D 1490) Nonvolatile Content of Phenol, Resorcinol, and Melamine Adhesives, Test for (D 1582)
Odor	Determination of the Odor of Adhesives, Test for (D 4339)
Peel strength (stripping strength)	Peel or Stripping Strength of Adhesive Bonds, Test for (D 903) Climbing Drum Peel Test for Adhesives, Method for (D 1781) Peel Resistance of Adhesives (T-Peel Test), Test for (D 1876) Evaluating Peel Strength of Shoe Sole Attaching Adhesives, Test for (D 2558) Determining Durability of Adhesive Joints Stressed in Peel, Practice for (D 2918) Floating Roller Peel Resistance, Test for (D 3167)

TABLE 12.6   ASTM Adhesive Standards (Continued)

Penetration	Penetration of Adhesives, Test for (D 1916)
pH	(See Hydrogen-Ion Concentration)
Radiation exposure (including light)	Exposure of Adhesive Specimens to Artificial (Carbon-Arc Type) and Natural Light, Practice for (D 904) Exposure of Adhesive Specimens to High-Energy Radiation, Practice for (D 1879)
Rubber cement tests	Rubber Cements, Testing of (D 816)
Salt spray (fog) testing	Salt Spray (Fog) Testing, Method of (B 117) Modified Salt Spray (Fog) Testing, Practice for (G 85)
Shear Strength (Tensile Shear Strength)	Shear Strength and Shear Modulus of Structural Adhesives, Test for (E 229) Strength Properties of Adhesive Bonds in Shear by Compression Loading, Test for (D 905) Strength Properties of Adhesives in Plywood Type Construction in Shear by Tension Loading, Test for (D 906) Strength Properties of Adhesives in Shear by Tension Loading (Metal-to-Metal), Test for (D 1002) Determining Strength Development of Adhesive Bonds, Practice for (D 1144) Strength Properties of Metal-to-Metal Adhesives by Compression Loading (Disk Shear), Test for (D 2181) Strength Properties of Adhesives in Shear by Tension Loading at Elevated Temperatures (Metal-to-Metal), Test for (D 2295) Strength Properties of Adhesives in Two-Ply Wood construction in Shear by Tension Loading, Test for (D 2339) Strength Properties of Adhesives in Shear by Tension Loading in the Temperature Range from −267.8 to −55°C (−450 to −67°F), Test for (D 2557) Determining Durability of Adhesive Joints Stressed in Shear by Tension Loading, Practice for (D 2919) Determining the Strength of Adhesively Bonded Rigid Plastic Lap-Shear Joints in Shear by Tension Loading, Practice for (D 3163) Determining the Strength of Adhesively Bonded Plastic Lap-Shear Sandwich Joints in Shear by Tension Loading, Practice for (D 3164) Strength Properties of Adhesives in Shear by Tension Loading of Laminated Assemblies, Test for (D 3165) Fatigue Properties of Adhesives in Shear by Tension Loading (Metal/Metal), Test for (D 3166) Strength Properties of Double Lap Shear Adhesive Joints by Tension Loading, Test for (3528) Strength of Gap-Filling Adhesive Bonds in Shear by Compression Loading, Practice for (D 3931) Measuring Strength and Shear Modulus of Nonrigid Adhesives by the Thick Adherend Tensile Lap Specimen, Practice for (D 3983) Measuring Shear Properties of Structural Adhesives by the Modified-Rail Test, Practice for (D 4027)
Specimen Preparation	Preparation of Bar and Rod Specimens for Adhesion Tests, Practice for (D 2094)
Spot-Adhesion Test	Qualitative Determination of Adhesion of Adhesives to Substrates by Spot Adhesion Test Method, Practice for (D 3808)

TABLE 12.6   ASTM Adhesive Standards *(Continued)*

Spread (Coverage)	Applied Weight per Unit ARea of Dried Adhesive Solids, Test for (D 898) Applied Weight per Unit Area of Liquid Adhesive, Test for (D 899)
Storage Life	Storage Life of Adhesives by Consistency and Bond Strength, Test for (D 1337)
Strength Development	Determining Strength Development of Adhesive Bonds, Practice for (D 1144)
Stress-Cracking Resistance	Evaluating the Stress Cracking of Plastics by Adhesives Using the Bent Beam Method, Practice for (D 3929)
Stripping Strength	(See Peel Strength)
Surface Preparation	Preparation of Surfaces of Plastics Prior to Adhesive Bonding, Practice for (D 2093) Preparation of Metal Surfaces for Adhesive Bonding, Practice for (D 2651) Analysis of Sulfochromate Etch Solution Used in Surface Preparation of Aluminum, Methods of (D 2674) Preparation of Aluminum Surfaces for Structural Adhesive Bonding (Phosphoric Acid Anodizing), Practice for (D 3933)
Tack	Pressure Sensitive Tack of Adhesives Using an Inverted Probe Machine, Test for (D 2979) Tack of Pressure-Sensitive Adhesives by Rolling Ball, Test for (D 3121)
Tensile Strength	Tensile Properties of Adhesive Bonds, Test for (D 897) Determining Strength Development of Adhesive Bonds, Practice for (D 1144) Cross-Lap Specimens for Tensile Properties of Adhesives, Testing of (D 1344) Tensile Strength of Adhesives by Means of Bar and Rod Specimens, Method for (D 2095)
Torque Strength	Determining the Torque Strength of Ultraviolet (UV) LIght-Cured Glass/Metal Adhesive Joints, Practice for (D 3658)
Viscosity	Viscosity of Adhesives, Test for (D 1084) Apparent Viscosity of Adhesives Having Shear-Rate-Dependent Flow Properties, Test for (D 2556) Viscosity of Hot Melt Adhesives and Coating Materials, Test for (D 3236)
Volume Resistivity	Volume Resistivity of Conductive Adhesives, Test for (D 2739)
Water Absorptiveness (of Paper Labels)	Water Absorptiveness of Paper Labels, Test for (D 1584)
Weathering	(See Durability)
Wedge Test	Adhesive Bonded Surface Durability of Aluminum (Wedge Test) (D 3762)
Working Life	Working Life of Liquid or Paste Adhesive by Consistency and Bond Strength, Test for (D 1338)

*The latest revisions of ASTM standards can be obtained from the American Society for Testing and Materials, 1916 Race Street, Philadelphia, PA 19103.

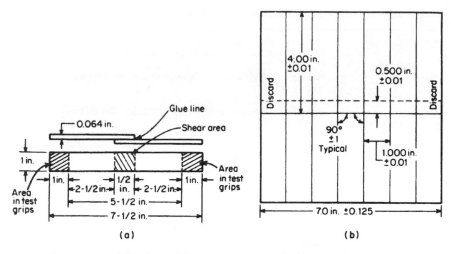

Figure 12.20 Standard lap-shear test specimen design: (a) form and dimensions of lap-shear test specimen and (b) standard test panel of five lap-shear specimens. *(From ASTM D 1002)*

curs is recorded in pounds per square inch of bonded area. Test environment, joint geometry, and type of failure should also be recorded.

A simple cross-lap specimen to determine tensile strength is described in ASTM D 1344. This specimen has the advantage of being easy to make, but grip alignment and adherend deflection during loading can cause irreproducibility. A sample population of at least 10 is recommended for this test method.

Peel test. Because adhesives are notoriously weak in peel, tests to measure peel resistance are very important. Peel tests involve stripping away a flexible adherend from another adherend that may be flexible or rigid. The specimen is usually peeled at an angle of 90 or 180°.

The most common types of peel test are the T-peel, Bell, and climbing-drum methods. Representative test specimens are shown in Fig. 12.21. The values resulting from each test method can be substantially different; hence, it is important to specify the test method employed.

Peel values are recorded in pounds per inch of width of the bonded specimen. They tend to fluctuate more than other adhesive test results because of the extremely small area at which the stress is localized during loading.

The T-peel test is described in ASTM D 1876 and is the most common of all peel tests. The T-peel specimen is shown in Fig. 12.22. Generally, this test method is used when both adherends are flexible.

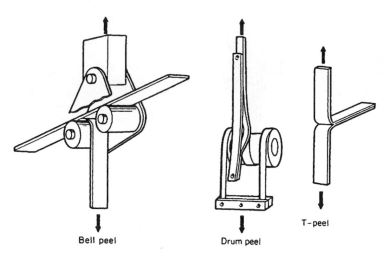

Figure 12.21 Common types of adhesive peel tests.[11]

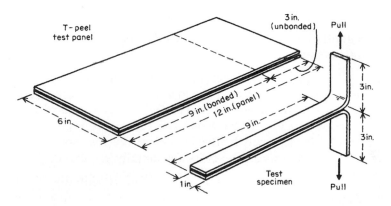

Figure 12.22 Test panel and specimen for T peel. *(From ASTM D 1876)*

A 90° peel test, such as the Bell peel (ASTM D 3167), is used when one adherend is flexible and the other is rigid. The flexible member is peeled at a constant 90° angle through a spool arrangement. Thus, the values obtained are generally more reproducible.

The climbing-drum peel specimen is described in ASTM D 1781. This test method is intended for determining peel strength of thin metal facings on honeycomb cores, although it can be generally used for joints where at least one member is flexible.

A variation of the T-peel test is a 180° stripping test illustrated in Fig. 12.23 and described in ASTM D 903. It is commonly used when one adherend is flexible enough to permit a 180° turn near the point of

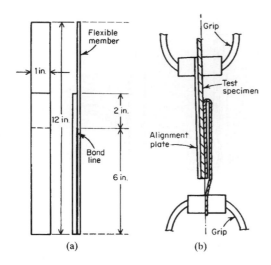

Figure 12.23   180° peel test specimens: (a) specimen design and (b) specimen under test. *(From ASTM D 903)*

loading. This test offers more reproducible results than the T-peel test, because the angle of peel is maintained constant.

Cleavage test.   Cleavage tests are conducted by prying apart one end of a rigid bonded joint and measuring the load necessary to cause rupture. The test method is described in ASTM D 1062. A standard test specimen is illustrated in Fig. 12.24. Cleavage values are reported in pounds per inch of adhesive width. Because cleavage test specimens involve considerable machining, peel tests are usually preferred.

Fatigue test.   Fatigue testing places a given load repeatedly on a bonded joint. Standard lap-shear specimens are tested on a fatiguing machine capable of inducing cyclic loading (usually in tension) on the

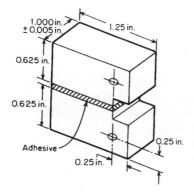

Figure 12.24 Cleavage test specimen. *(From ASTM D 1062)*

joint. The fatigue strength of an adhesive is reported as the number of cycles of a known load necessary to cause failure.

Fatigue strength is dependent on adhesive, curing conditions, joint geometry, mode of stressing, magnitude of stress, and duration and frequency of load cycling.

Impact test. The resistance of an adhesive to impact can be determined by ASTM D 950. The specimen is mounted in a grip as shown in Fig. 12.25 and placed in a standard impact machine. One adherend is struck with a pendulum hammer traveling at 11 ft/s, and the energy of impact is reported in pounds per square inch of bonded area.

Creep test. The dimensional change occurring in a stressed adhesive over a long time period is called *creep*. Creep data are seldom reported in the adhesive suppliers' literature, because the tests are time consuming and expensive. This is very unfortunate, since sustained loading is a common occurrence in adhesive applications. All adhesives tend to creep—some much more than others. With weak adhesives, creep may be so extensive that bond failure occurs prematurely. Certain adhesives have also been found to degrade more rapidly when aged in a stressed rather than an unstressed condition.

Creep tests are made by loading a specimen with a predetermined stress and measuring the total deformation as a function of time or measuring the time necessary for complete failure of the specimen. Depending on the adhesive, loads, and testing conditions, the time required for a measurable deformation may be extremely long. ASTM D 2294 defines a test for creep properties of adhesives utilizing a spring-loaded apparatus to maintain constant stress.

Environmental tests. Strength values determined by short-term tests do not give an adequate indication of an adhesive's permanence dur-

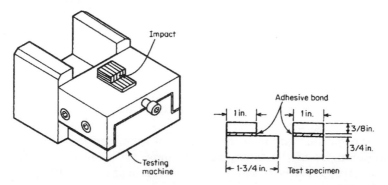

Figure 12.25 Impact test specimen and holding fixture. *(From ASTM D 950)*

ing continuous environmental exposure. Laboratory-controlled aging tests seldom last longer than a few thousand hours. To predict the permanence of an adhesive over a 20-year product life requires accelerated test procedures and extrapolation of data. Such extrapolations are extremely risky, because the causes of adhesive-bond deterioration are many and not well understood.

12.2.4.2 Federal test specifications. A variety of federal and military specifications describing adhesives and test methods have been prepared. Selected government specifications are described in Table 12.7.

12.3 Surface Preparation

12.3.1 Importance of Surface Preparation

Surface preparation of adherends prior to bonding is one of the most important factors in the adhesive-bonding process. Initial bond strength and joint permanence are greatly dependent on the quality of surface that is in contact with the adhesive. Prebond treatments are intended to remove weak boundary layers and provide easily wettable surfaces. As a general rule, all adherends must be treated in some manner prior to bonding.

Surface preparations can range from simple solvent wiping to a combination of mechanical abrading, chemical cleaning, and acid etching. In many low- to medium-strength applications, extensive surface preparation may be unnecessary. But, where maximum bond strength, permanence, and reliability are required, carefully controlled surface-treating processes are necessary. The following factors should be considered in the selection of a surface preparation process:

1. The ultimate initial bond strength required

2. The degree of permanence necessary and the service environment

3. The degree and type of contamination on the adherend

4. The type of adherend and adhesive

Table 12.8 shows the effect of various metallic-surface preparations on adhesive-joint strength.

Surface preparations enhance the quality of a bonded joint by performing one or more of the following functions: (1) remove contaminants, (2) control adsorbed water, (3) control oxide formation, (4) poison surface atoms that catalyze adhesive breakdown, (5) protect the adhesive from the adherend and vice versa, (6) match the adher-

TABLE 12.7   Government Adhesive Specifications

Military Specifications

MIL-A-928	Adhesive; metal to wood, structural
MIL-A-1154	Adhesive; bonding, vulcanized synthetic rubber to steel
MIL-C-1219	Cement, iron, and steel
MIL-C-3316	Adhesive; fire resistant, thermal insulation
MIL-C-4003	Cement; general purpose, synthetic base
MIL-C-5092	Adhesive; rubber (synthetic and reclaimed rubber base)
MIL-A-5534	Adhesive; high temperature setting resin (phenol, melamine and resorcinol base)
MIL-C-5339	Cement; natural rubber
MIL-A-5540	Adhesive; polychloroprene
MIL-A-8576	Adhesive; acrylic monomer base, for acrylic plastic
MIL-A-8623	Adhesive; epoxy resin, metal-to-metal structural bonding
MIL-A-9117	Adhesive; sealing, for aromatic fuel cells and general repair
MIL-C-10523	Cement, gasket, for automobile applications
MIL-S-11030	Sealing compound, noncuring polysulfide base
MIL-S-11031	Sealing compound, adhesive; curing, polysulfide base
MIL-A-11238	Adhesive; cellulose nitrate
MIL-C-12850	Cement, rubber
MIL--13554	Adhesive for cellulose nitrate film on metals
MIL-C-13792	Cement, vinyl acetate base solvent type
MIL-A-13883	Adhesive, synthetic rubber (hot or cold bonding)
MIL-A-14042	Adhesive, epoxy
MIL-C-14064	Cement; grinding disk
MIL-P-14536	Polyisobutylene binder
MIL-I-15126	Insulation tape, electrical, pressure-sensitive adhesive and pressure-sensitive thermosetting adhesive
MIL-C-18726	Cement, vinyl alcohol-acetate
MIL-A-22010	Adhesive, solvent type, polyvinyl chloride
MIL-A-22397	Adhesive, phenol and resorcinol resin base
MIL-A-22434	Adhesive, polyester, thixotropic
MIL-C-22608	Compound insulating, high temperature

TABLE 12.7   Government Adhesive Specifications *(Continued)*

Military Specifications

MIL-A-22895	Adhesive, metal identification plate
MIL-C-23092	Cement, natural rubber
MIL-A-25055	Adhesive; acrylic monomer base, for acrylic plastics
MIL-A-25457	Adhesive, air-drying silicone rubber
MIL-A-25463	Adhesive, metallic structural honeycomb construction
MIL-A-46050	Adhesive, special; rapid room temperature curing, solventless
MIL-A-46051	Adhesive, room-temperature and intermediate-temperature setting resin (phenol, resorcinol, and melamine base)
MIL-A-52194	Adhesive, epoxy (for bonding glass reinforced polyester)
MIL-A-9067C	Adhesive bonding, process and inspection requirement for
MIL-C-7438	Core material, aluminum, for sandwich construction
MIL-C-8073	Core material, plastic honeycomb, laminated glass fabric base, for aircraft structural applications
MIL-C-21275	Core material, metallic, heat-resisting, for structural sandwich construction
MIL-H-9884	Honeycomb material, cushioning, paper
MIL-S-9041A	Sandwich construction, plastic resin, glass fabric base, laminated facings for aircraft structural applications

Federal Specifications

MMM-A-181	Adhesive, room-temperature and intermediate-temperature setting resin (phenol, resorcinol and melamine base)
MMM-A-00185	Adhesive, rubber
MMM-A-00187	Adhesive, synthetic, epoxy resin base, paste form, general purpose
MMM-A-132	Adhesives, heat resistant, airframe structural, metal-to-metal
Federal Test Method 175	Adhesives: methods of testing
MIL-STD-401	Sandwich construction and core materials; general test methods

end crystal structure to the adhesive molecular structure, and (7) control surface roughness. Thus, surface preparations can affect the permanence of the joint as well as its initial strength.

Plastic and elastomeric adherends are even more dependent than metals on surface preparation. Many of these surfaces are contaminated with mold-release agents or processing additives. Such contami-

TABLE 12.8   Effect of Substrate Pretreatment on Strength of Adhesive-Bonded Joints

Adherend	Treatment	Adhesive	Shear strength, lb/in.^{2}	Ref.
Aluminum	As received	Epoxy	444	12
	Vapor degreased		837	
	Grit blast		1751	
	Acid etch		2756	
Aluminum	As received	Vinyl-phenolic	2442	13
	Degreased		2741	
	Acid etch		5173	
Stainless steel	As received	Vinyl-phenolic	5215	13
	Degreased		6306	
	Acid etch		7056	
Cold-rolled steel	As received	Epoxy	2900	14
	Vapor degreased		2910	
	Grit blast		4260	
	Acid etch		4470	
Copper	Vapor degreased	Epoxy	1790	15
	Acid etch		2330	
Titanium	As received	Vinyl-phenolic	1356	13
	Degreased		3180	
	Acid etch		6743	
Titanium	Acid etch	Epoxy	3183	16
	Liquid pickle		3317	
	Liquid hone		3900	
	Hydrofluorosilicic acid etch		4005	

nants must be removed before bonding. Because of their low surface energy, polytetrafluoroethylene, polyethylene, and certain other polymeric materials are completely unsuitable for adhesive bonding in their natural state. The surfaces of these materials must be chemically or physically altered prior to bonding to improve wetting.

12.3.2   General Surface Preparation Methods

Listed here are several methods of preparing both metal and polymer substrates for adhesive bonding. The chosen method will ultimately be the process that yields the necessary strength and permanence with the least cost.

12.3.2.1   Solvent wiping.   Where loosely held dirt, grease, and oil are the only contaminants, simple solvent wiping will provide surfaces for weak- to medium-strength bonds. Solvent wiping is widely used, but it is the least effective substrate treatment. Volatile solvents such as acetone and trichloroethylene are acceptable. Trichloroethylene is often favored because of its nonflammability. A clean cloth should be saturated with the solvent and wiped across the area to be bonded until no signs of residue are evident on the cloth or substrate. Special precautions are necessary to prevent the solvent from becoming contaminated. For example, the wiping cloth should never touch the solvent container, and new wiping cloths must be used often. After cleaning, the parts should be air dried in a clean, dry environment before being bonded.

12.3.2.2 Vapor degreasing. Vapor degreasing is a reproducible form of solvent cleaning that is attractive when many parts must be prepared. It consists of suspending the adherends in a container of solvent vapor such as trichloroethylene or perchloroethylene. When the hot vapors come into contact with the relatively cool substrate, solvent condensation occurs on the surface of the part, which dissolves the organic contaminants. Vapor degreasing is preferred to solvent wiping, because the surfaces are continuously being washed in distilled, uncontaminated solvent. The vapor degreaser must be kept clean, and a fresh supply of solvent is used when the contaminants in the solvent trough lower the boiling point significantly.

Modern vapor degreasing equipment is available with ultrasonic transducers built into the solvent rinse tank. The parts are initially cleaned by vapor and then subjected to ultrasonic scrubbing. The cleaning solutions and processing parameters must be optimized by test.

12.3.2.3 Abrasive cleaning. Mechanical methods for surface preparation include sandblasting, wire brushing, and abrasion with sandpaper, emery cloth, or metal wool. These methods are most effective for removing heavy, loose particles such as dirt, scale, tarnish, and oxide layers.

The parts should always be degreased prior to abrasive treatment to prevent contaminants from being rubbed into the surface. Solid particles left on the surfaces after abrading can be removed by blasts of clean, oil free, dry air or solvent wiping.

Each metal reacts favorably with a specific range of abrasive sizes. It has been found that joint strength generally increases with the degree of surface roughness, provided that the adhesive wets the adherend (Table 12.9). However, excessively rough surfaces increase the probability that voids will be left at the interface, causing stress risers detrimental to the joint.

TABLE 12.9   Effect of Surface Roughness in Butt Tensile Strength of DER 332-Versamid 140 (60/40)[*] Epoxy Joints (from Ref. 17)

	Adherend	Adherend surface[†]	Butt tensile strength, lb/in.^{2}
6061	A1	Polished	4720 ± 1000
6061	A1	0.005-in. grooves	6420 ± 500
6061	A1	0.005-in. grooves, sandblasted	7020 ± 1120
6061	A1	Sandblasted (40–50 grit)	7920 ± 530
6061	A1	Sandblasted (10–20 grit)	7680 ± 360
304	SS	Polished	4030 ± 840
304	SS	0.010-in. grooves	5110 ± 1020
304	SS	0.010-in. grooves, sandblasted	5510 ± 770
304	SS	Sandblasted (40–50 grit)	7750 ± 840
304	SS	Sandblasted (10–20 grit)	9120 ± 470

[*]74°C/16 hr cure.
[†]Adherend surfaces were chromate etched.

12.3.2.4   Chemical cleaning. Strong detergent solutions are used to emulsify surface contaminants on both metallic and nonmetallic substrates. These solutions are usually heated. Parts for cleaning are generally immersed in a well agitated solution maintained at 150 to 210°F for approximately 10 min. The surfaces are then rinsed immediately with deionized water and dried. Chemical cleaning is often used in combination with other surface treatments. Chemical cleaning by itself will not remove heavy or strongly attached contaminants such as rust or scale.

Alkaline detergents recommended for prebond cleaning are combinations of alkaline salts such as sodium metasilicate and tetrasodium pyrophosphate with surfactants included. Many commercial detergents are available.

12.3.2.5   Other cleaning methods. Vapor-honing and ultrasonic cleaning are efficient treating methods for small, delicate parts. In cases where the substrate is so delicate that usual abrasive treatments may be too rough, contaminants can be removed by vapor honing. This method is similar to grit blasting except that very fine abrasive particles are suspended in a high-velocity water or steam spray. Thorough rinsing after vapor honing is usually not required.

Ultrasonic cleaning employs a bath of cleaning liquid or solvent that is ultrasonically activated by a high-frequency transducer. The part to be cleaned is immersed in the liquid, which carries the sonic waves to the surface of the part. High-frequency vibrations then dislodge the contaminants. Commercial ultrasonic cleaning units are available from a number of manufacturers.

12.3.2.6 Alteration of surfaces. Certain treatments of substrate surfaces change the physical and chemical properties of the surface to produce greater wettability and/or a stronger surface. Specific processes are required for each substrate material. The part or area to be bonded is usually immersed in an active solution for a matter of minutes. The parts are then immediately rinsed with deionized water and dried.

Chemical solutions must be changed regularly to prevent contamination and ensure repeatable concentration. Tank temperature and agitation must also be controlled. Personnel need to be trained in the safe handling and use of chemical solutions and must wear the proper clothing.

Some paste-type acid etching products are available that simultaneously clean and chemically treat surfaces. They react at room temperature and need only be applied to the specific area to be bonded. However, these paste etchants generally require much longer treatment time than hot acid-bath processes.

Treatment of certain polymeric surfaces with excited inert gases greatly improves the bond strength of adhesive joints prepared from these materials. With this technique, called *plasma treatment,* a low-pressure inert gas is activated by an electrodeless radio-frequency discharge or microwave excitation to produce metastable species that react with the polymeric surface. The type of plasma gas can be selected to initiate a wide assortment of chemical reactions. In the case of polyethylene, plasma treatment produces a strong, wettable, cross-linked skin. Commercial instruments are available that can treat polymeric materials in this manner. Table 12.10 presents bond strength of various plastic joints pretreated with activated gas and bonded with an epoxy adhesive.

12.3.2.7 Combined methods. More than one cleaning method is usually required for optimum adhesive properties. A three-step process that is recommended for most substrates consists of (1) degreasing,

TABLE 12.10   Typical Adhesive-Strength Improvement with Plasma Treatment;
Aluminum-Plastic Shear Specimen Bonded with Epon 828-Versamide 140 (70/30)
Epoxy Adhesive (from Ref. 18)

Material	Strength of bond, lb/in.^{2}	
	Control	After plasma treatment
High-density polyethylene	315 + 38	>3125 + 68
Low-density polyethylene	372 +52	>1466 ± 106
Nylon 6	846 ± 166	>3956 ± 195
Polystyrene	566 ± 17	>4015 ± 85
Mylar*A	530 ± 51	1660 ± 40
Mylar D	618 ± 25	1185 ± 83
Polyvinylfluoride (Tedar)†	278 ± 2	>1280 ± 73
Lexan*	410 ± 10	928 ± 66
Polypropylene	370 ±	3080 ± 180
Teflon* TFE	75	750

*Trademark of E.I. du Pont de Nemours & Company, Inc.
†Trademark of General Electric Co.

(2) mechanical abrasion, and (3) chemical treatment. Table 12.11
shows the effect of various combinations of aluminum-surface preparations on lap-shear strength.

12.3.3   Surface Treatment of Polymers

Treating of plastic surfaces usually consists of one or a combination of
the following processes: solvent cleaning, abrasive treatment and
etching, flame, hot air, electric discharge, or plasma treatments. The
purpose of the treatment is to either remove or strengthen the weak
boundary layer or to increase the critical surface tension.

Abrasive treatments consist of scouring, machining, hand sanding,
and dry and wet abrasive blasting. The choice is generally determined
by available production facilities and cost. Laminates can be prepared
by either abrasion or the tear-ply technique. In the tear-ply design,
the laminate is manufactured so that one ply of heavy fabric, such as
Dacron or equivalent, is attached at the bonding surface. Just prior to
bonding the tear-ply is stripped away, and a fresh, clean, bondable
surface is exposed.

TABLE 12.11   Surface Preparation of Aluminum Substrates vs. Lap-Shear Strength[12]

Group treatment	X, lb/in.^{2}	s, lb/in.^{2}	Cv, %
1. Vapor degrease, grit blast 90-mesh grit, alkaline clean, $Na_2Cr_2O_7$-H_2SO_4, distilled water	3091	103	3.5
2. Vapor degrease, grit blast 90-mesh grit, alkaline clean, $Na_2Cr_2O_7$-H_2SO_4, tap water	2929	215	7.3
3. Vapor degrease, alkaline clean, $Na_2Cr_2O_7$-H_2SO_4, distilled water	2800	307	10.96
4. Vapor degrease, alkaline clean, $Na_2Cr_2O_7$-H_2SO_4, tap water	2826	115	4.1
5. Vapor degrease, alkaline clean, chromic-H_2SO_4, deionized water	2874	163	5.6
6. Vapor degrease, $Na_2Cr_2O_7$-H2SO4, tap water	2756	363	1.3
7. Unsealed anodized	1935	209	10.8
8. Vapor degrease, grit blast 90-mesh grit	1751	138	7.9
9. Vapor degrease, wet and dry sand, 100 + 240 mesh grit, N_2 blown	1758	160	9.1
10. Vapor degrease, wet and dry sand, wipe off with sandpaper	1726	60	3.4
11. Solvent wipe, wet and dry sand, wipe off with sandpaper (done rapidly)	1540	68	4
12. Solvent wipe, sand (not wet and dry), 120 grit	1329	135	1.0
13. Solvent wipe, wet and dry sand, 240 grit only	1345	205	15.2
14. Vapor degrease, aluminum wood	1478		
15. Vapor degrease, 15% NaOH	1671		
16. Vapor degrease	837	72	8.5
17. Solvent wipe (benzene)	353		
18. As received	444	232	52.2

X = average value. s = standard deviation. C_v = coefficient of variation. Resin employed is EA 934 Hysol Division, Dexter Corp.; cured 16 hr at 75°F plus 1 hr at 180°F.

Chemical etching treatments vary with the type of plastic surface. The plastic resin supplier is the best guide to the appropriate etching chemicals and process. Etching processes can involve the use of corrosive and hazardous materials. The most common processes are sulfuric-dichromate etch (polyethylene and polypropylene) and sodium-naphthalene etch (fluorocarbons).

Flame, hot air, electrical discharge, and plasma treatments physically and chemically change the nature of polymeric surfaces. The plasma treating process has been found to be very successful on most hard-to-bond plastic surfaces. Table 12.10 shows that plasma treatment results in improved plastic joint strength with common epoxy adhesives. Plasma treatment requires high vacuum and special processing equipment.

Better adhesion can be obtained if parts are formed, treated, and coated with an adhesive in a continuous operation. The sooner an article can be bonded after surface treatment, the better will be the adhesion. After the part is treated, handling and exposure to shop environments should be kept to a minimum. Some surface treatments, such as plasma, have a long effective shelf life. However, some treating processes, such as electrical discharge and flame treating, will become less effective the longer the time between surface preparation and bonding.

12.3.4 Evaluation of Surface Preparation before and after Bonding

On many nonporous surfaces, a useful and quick method for testing the effectiveness of the surface preparation is the *water-break test.* If distilled water beads up when sprayed on the surface and does not wet the substrate, the surface-preparation steps should be repeated. If the water wets the surface in a uniform film, an effective surface operation has been achieved.

The objective of surface treating is to obtain a joint where the weakest link is the adhesive layer and not the interface. Thus, destructively tested joints should be examined for mode of failure. If failure is cohesive (within the adhesive layer or adherend), the surface treatment is the optimum for that particular combination of adherend, adhesive, and testing condition. However, it must be realized that specimens could exhibit cohesive failure initially and interfacial failure after aging. Both adhesive and surface preparations need to be tested with respect to the intended service environment.

12.3.5 Substrate Equilibrium

After the surface-preparation process has been completed, the substrates may have to be stored before bonding. During storage, the surface could regain its oxide or weak boundary layer and become exposed to contaminating environments. Typical storage life for various metals subjected to different treatments is shown in Table 12.12. Because of the relatively short storage life of many treated materials, bonding should be conducted as soon after the surface preparation as possible.

TABLE 12.12   Maximum Allowable Time between Surface
Preparation and Bonding or Priming (from Ref. 19)

Metal	Surface	Time
Aluminum	Wet-abrasive-blasted	72 hr
Aluminum	Sulfuric-chromic acid etched	6 days
Aluminum	Anodized	30 days
Stainless steel	Sulfuric acid etched	30 days
Steel	Sandblasted	4 hr
Brass	Wet-abrasive-blasted	8 hr

If prolonged storage is necessary, a compatible primer may be used to coat the treated substrates after surface preparation. The primer will protect the surface during storage and interact with the adhesive during bonding. Many primer systems are sold with adhesives for this purpose.

12.3.6   Primers and Adhesion Promoters

Primers are applied and cured onto the adherend prior to adhesive bonding. They serve four primary functions singly or in combination:

1. Protection of surfaces after treatment (primers can be used to extend the time between preparing the adherend surface and bonding)

2. Developing tack for holding or positioning parts to be bonded

3. Inhibiting corrosion during service

4. Serving as an intermediate layer to enhance the physical properties of the joint and improve bond strength

Some primers have been found to provide corrosion resistance for the joint during service. The primer protects the adhesive-adherend interface and lengthens the service life of the bonded joint. Representative data are shown in Fig. 12.26.

Primers may also be used to modify the characteristics of the joint. For example, elastomeric primers are used with rigid adhesives to provide greater peel or cleavage resistance.

Primers can also chemically react with the adhesive and adherend to provide greater joint strengths. This type of primer is referred to as an *adhesion promoter*. The use of reactive silane to improve the adhesion of resin to glass fibers in polymeric laminates is well known in the plastic industry.

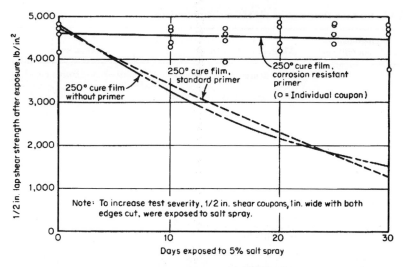

Figure 12.26 Effect of primer on lap-shear strength of aluminum joints exposed to 5% salt spray.[20]

12.3.7 Specific Surface Treatments and Characteristics

12.3.7.1 Metallic adherends. Table 12.13 lists common surface-treating procedures for metallic adherends. The general methods previ-

TABLE 12.13 Surface Preparation for Metals

Adherend	Degreasing solvent	Method of treatment	Remarks
Aluminum and aluminum alloys	Trichloroethylene	1. Sandblast or 100-grit emery cloth followed by solvent degreasing	Medium- to high-strength bonds, suitable for noncritical applications
		2. Immerse for 10 min at 77 ± 6°C in a commercial alkaline cleaner or	Optimum bond strength, specified in ASTM D 2651 and MIL-A-9067. Solvent degrease may replace alkaline cleaning

Parts by wt.

Sodium metasilicate..	30.0
Sodium hydroxide....	1.5
Sodium pyrophosphate	1.5
Nacconol NR (Allied Chemical Co.)......	0.5
Water (distilled)......	128.0

Wash in water below 65°C and etch for 10 min at 68±3°C in

Parts by wt.

Sodium dichromate...	1
Sulfuric acid (96%, sp. gr. 1.84).........	10
Water (distilled)......	30

Rinse in distilled water after washing in tap water, and dry in air

TABLE 12.13   Surface Preparation for Metals *(Continued)*

Adherend	Degreasing solvent	Method of treatment	Remarks
		3. Vapor degrease or solvent wipe. Immerse for 5 min at 71–82°C in Sulfuric acid (96%, sp. gr. 1.84)......... 1 gal Chromic acid........ 45 oz Distilled water....... 9 gal Rinse in distilled water and dry in air	Alternative to sulfuric-dichromate etch

		4. Degrease with 50/50 solution of methyl ethyl ketone and chlorothene. Abrade lightly with mildly abrasive cleaner. Rinse in deionized water; wipe; or air dry. Etch 20 min at RT in
		
		               Parts by wt.
		Sodium dichromate... 2
		Sulfuric acid (96%,
		  sp. gr. 1.84)......... 7
		Rinse thoroughly in deionized water; dry at 70°C for 30 min
			Room-temp etch

		5. Form a paste using sulfuric–dichromate solution and finely divided silica or fuller's earth. Apply; do not permit paste to dry. Time depends on degree of contamination (usually greater than 10 min at RT). Wash very thoroughly with deionized water, and air-dry	Paste form of acid etch useful when part cannot be immersed. ASTM D 2651
Beryllium	Trichloroethylene	1. Scrub with a nonchlorinated cleaner. Rinse with deionized water. Force dry at 250–300°F. Check for water break and repeat procedure if necessary	
		2. Immerse for 5–10 min at 20°C in *Parts by wt.* Sodium hydroxide..... 20–30 Water (distilled)...... 170–180 Rinse in distilled water after washing in tap water, and oven-dry for 10 min at 121–177°C	
		3. Same procedure as 2, except immerse for 3–4 min at 75–80°C	

Brass and bronze (see also Copper and copper alloys)	Trichloroethylene	1. Etch for 5 min at 20°C in
		
		               Parts by wt.
		Zinc oxide.......... 20
		Sulfuric acid (96%,
		  sp. gr. 1.84)......... 460
			Temperatures must not exceed 65°C when washing and drying

TABLE 12.13   Surface Preparation for Metals *(Continued)*

Adherend	Degreasing solvent	Method of treatment	Remarks
		Nitric acid (69%, sp. gr. 1.41)............ 360 Rinse in water, below 65°C, and re-etch in the acid solution for 5 min at 49°C. Rinse in distilled water after washing, and dry in air	
Cadmium	Trichloro-ethylene	1. Abrasion. Grit or vapor blast, or 100-grit emery cloth, followed by solvent degreasing	Suitable for general-purpose bonding
		2. Electroplate with nickel or silver	For maximum bond strength
Chromium	Trichloro-ethylene	1. Abrasion. Grit or vapor blast, or 100-grit emery cloth, followed by solvent degreasing	Suitable for general-purpose bonding
		2. Etch for 1–5 min at 90–95°C in	
		Parts by wt. Hydrochloric acid (37%).............. 17 Water............... 20 Rinse in distilled water after cold/hot-water washing, and dry in hot air	For maximum bond strength
Copper and copper alloys	Trichloro-ethylene	1. Abrasion. Sanding, wire brushing, or 100-grit emery cloth, followed by vapor or solvent degreasing	Suitable for general-purpose bonding. Use 320-grit emery cloth for foil
		2. Etch for 10 min at 66°C in *Parts by wt.* Ferric sulfate........ 1.0 Sulfuric acid (96%)... 0.75 Water.............. 8.0 Wash in water at 20°C, and etch in cold solution of *Parts by wt.* Sodium dichromate... 5 Sulfuric acid (96%)... 10 Water.............. 85 Etch until a bright clean surface has been obtained. Rinse in water, dip in ammonium hydroxide (sp. gr. 0.88), and wash in tap water. Rinse in distilled water, and dry in warm air	For maximum bond strength. Suitable for brass and bronze. ASTM D 2651
		3. Etch for 1–2 min at 20°C in *Parts by wt.* Ferric chloride (42% w/w solution)....... 0.75 Nitric acid (sp. gr. 1.41).............. 1.5 Water.............. 10.0 Rinse in distilled water after cold-water wash and dry in air stream at 20°C	Room-temp etch. ASTM D 2651

TABLE 12.13   Surface Preparation for Metals *(Continued)*

Adherend	Degreasing solvent	Method of treatment	Remarks
		4. Etch for 30 s at 20°C in *Parts by wt.* Ammonium persulfate 1 Water.............. 4 Rinse in distilled water after cold-water wash, and dry in air stream at 20°C	Alternative etching solution to above where fast processing is required
		5. Solvent degrease. Immerse 30 s at 20°C in *Parts by vol.* Nitric acid (69%).... 10 Deionized water..... 90 Rinse in running water and transfer immediately to next solution; immerse for 1–2 min at 98°C in Ebonol C (Ethone, Inc., New Haven, Conn.) 24 oz and equivalent water to make 1 gal Rinse in deionized water, and air-dry	For copper alloys containing over 95% copper. Stable surface for hot bonding. ASTM D 2651
Germanium	Trichloro-ethylene	Abrasion. Grit or vapor blast followed by solvent degreasing	
Gold	Trichloro-ethylene	Solvent or vapor degrease after light abrasion with a fine emery cloth	
Iron		See Steel (mild)	
Lead and solders	Trichloro-ethylene	Abrasion. Grit or vapor blast, or 100-grit emery cloth, followed by solvent degreasing	
Magnesium and magnesium alloys	Trichloro-ethylene	1. Abrasion with 100-grit emery cloth followed by solvent degreasing	Apply the adhesive immediately after abrasion
		2. Vapor degrease. Immerse for 10 min at 60–70°C in *Parts by wt.* Deionized water...... 95 Sodium metasilicate.. 2.5 Trisodium pyrophosphate.............. 1.1 Sodium hydroxide.... 1.1 Nacconal NR (Allied Chemical Co.)...... 0.3 Rinse in water and dry below 60°C	Medium to high bond strength. ASTM D 2651
		3. Vapor degrease. Immerse for 10 min at 71–88°C *Parts by wt.* Water.............. 4 Chromic acid........ 1 Rinse in water, and dry below 60°C	High bond strength. ASTM D 2651

TABLE 12.13   Surface Preparation for Metals *(Continued)*

Adherend	Degreasing solvent	Method of treatment	Remarks
		4. Vapor degrease. Immerse for 5–10 min at 63–80°C in <div align="right">*Parts by wt.*</div>Water.............. 12 Sodium hydroxide.... 1 Rinse in water. Immerse for 5–15 min at RT in <div align="right">*Parts by wt.*</div>Water.............. 123 Chromic acid........ 24 Calcium nitrate...... 1.8 Rinse in water, and dry below 60°C	Alternative to procedure 2
		5. Light anodic treatment and various corrosion-preventive treatments have been developed by magnesium producers (Dow 17 and Dow 7, Dow Chemical Co.)	Dow 17 preferred under extreme environmental conditions
Nickel	Trichloro-ethylene	1. Abrasion with 100-grit emery cloth followed by solvent degreasing	For general-purpose bonding
		2. Etch for 5 s at 20°C in nitric acid (69%, sp.gr.1.41). Wash in cold and hot water followed by a distilled-water rinse, and air-dry at 40°C	For general-purpose bonding
Platinum	Trichloro-ethylene	Solvent or vapor degrease after light abrasion with a 320-grit emery cloth	
Silver	Trichloro-ethylene	Abrasion with 320-grit emery cloth followed by solvent degreasing	
Steel (stainless)	Trichloro-ethylene	1. Abrasion with 100-grit emery cloth, grit or vapor blast, followed by solvent degreasing	

		2. Solvent degrease and abrade with grit paper. Degrease again. Immerse for 10 min at 71–82°C in
		<div align="right">*Parts by wt.*</div>Sodium metasilicate.. 3
		Tetrasodium pyro-
		  phosphate......... 1.5
		Sodium hydroxide.... 1.5
		Nacconol NR (Allied
		  Chemical Co.)...... 0.5
		Distilled water...... 138.0
		Rinse in deionized water; dry in air, at 93°C. Immerse for 10 min at 85–91°C in
		<div align="right">*Parts by wt.*</div>Oxalic acid......... 1
		Sulfuric acid (sp. gr.
		  1.84).............. 1
		Distilled water...... 8
		Rinse in deionized water; dry at 93°C for 10–15 min
			Heat-resistant bond. Alkaline clean alone sufficient for general bonding. Commercial alkaline cleaners (Prebond 700, American Cyanamid) available

TABLE 12.13   Surface Preparation for Metals *(Continued)*

Adherend	Degreasing solvent	Method of treatment	Remarks
		3. Etch for 15 min at 63°C in *Parts by wt.* Sodium dichromate (saturated solution).. 0.30 Sulfuric acid......... 10.0 Remove carbon residue with nylon brush while rinsing. Rinse in distilled water and dry in warm air, at 93°C	ASTM D 2651
		4. Etch for 2 min at 93°C in *Parts by wt.* Hydrochloric acid (37%)............. 20 Orthophosphoric acid (85%)............. 3 Hydrofluoric acid (35%)............. 1 Rinse in warm water with a final rinse in distilled water. Dry in air below 93°C	For maximum resistance to heat and environment. ASTM D 2651
		5. Vapor degrease for 10 min and pickle for 10 min at 20°C in *Parts by vol.* Nitric acid (69%, sp. gr. 1.41)............ 10 Hydrofluoric acid (48%)............. 2 Water.............. 88 Dry in air under 70°C	Room-temperature etch. Treatment may be followed by passivation for 20 min in 5–10% w/v chromic acid (CrO_3) solution
Steel (mild), iron, and ferrous metals other than stainless	Trichloroethylene	1. Abrasion. Grit or vapor blast followed by solvent degreasing with water-free solvents	Xylene or toluene is preferred to acetone and ketone, which may be moist enough to cause rusting
		2. Etch for 5–10 min at 20°C in *Parts by wt.* Hydrochloric acid (37%)............. 1 Water.............. 1 Rinse in distilled water after cold-water wash, and dry in warm air for 10 min at 93°C	Bonding should follow immediately after etching treatment since ferrous metals are prone to rusting. Abrasion is more suitable for procedure where bonding is delayed
		3. Etch for 10 min at 60°C in *Parts by wt.* Orthophosphoric acid (85%)............. 1 Ethyl alcohol (denatured)........ 2 Brush off carbon residue with nylon brush while washing in running water. Rinse with distilled water, and heat for 1 h at 120°C	For maximum strength
Tin	Trichloroethylene	Solvent or vapor degrease after light abrasion with a fine emery cloth (320-grit)	

TABLE 12.13   Surface Preparation for Metals *(Continued)*

Adherend	Degreasing solvent	Method of treatment	Remarks
Titanium and titanium alloys	Trichloro-ethylene	1. Abrasion. Grit or vapor blast, or 100-grit emery cloth, followed by solvent degrease; or scour with a nonchlorinated cleaner, rinse, and dry 2. Etch for 5–10 min at 20°C in *Parts by wt.* Sodium fluoride...... 2 Chromium trioxide... 1 Sulfuric acid (96%, sp. gr. 1.84)........ 10 Water.............. 50 Rinse in water and distilled water. Dry in air at 93°C	For general-purpose bonding
		3. Etch for 2 min at 20°C in *Parts by vol.* Hydrofluoric acid (60%)............. 63 Hydrochloric acid (37%)............. 841 Orthophosphoric acid (85%)............. 89 Rinse in water and distilled water. Dry in air at 93°C	Suitable for alloys to be bonded with polybenzimidazole adhesives. Bond within 10 min of treatment. ASTM D 2651
		4. Etch for 10–15 min at 38–52°C in *Parts by vol.* Nitric acid (69%).... 6 Hydrofluoric acid (60%)............. 1 Water.............. 20 Rinse with water and distilled water. Dry in oven at 71–82°C for 15 min	Alternative etch for alloys to be bonded with polyimide adhesives is nitric: hydrofluoric: water in ratio 5:1:27 by wt. Etch 30 s at 20°C
		5. Commercial etching liquids and pastes (Plasa-Jell 107C available from Semco, South Hoover, Los Angeles, Calif. 18881)	
Tungsten and tungsten alloys	Trichloro-ethylene	Etch for 1–5 min at 20°C in *Parts by wt.* Nitric acid (69%, sp. gr. 1.41)........... 6 Hydrofluoric acid (60%)............. 1 Sulfuric acid (96%, sp. gr. 1.84)........ 10 Water.............. 3 Add a few drops of hydrogen peroxide (20%). Rinse with water and distilled water. Dry in air at 71–82°C for 15 min	
Uranium	Trichloro-ethylene	Abrasion of the metal in a pool of liquid adhesive (epoxy resins have been used)	Prevents oxidation of metal

TABLE 12.13   Surface Preparation for Meta s *(Continued)*

Adherend	Degreasing solvent	Method of treatment	Remarks
Zinc and zinc alloys	Trichloro-ethylene	1. Abrasion. Grit or vapor blast, or 100-grit emery cloth, followed by solvent degreasing	For general-purpose bonding

		2. Etch for 2–4 min at 20°C in
		
		               Parts
		               by vol.
		Hydrochloric acid
		  (37%)............... 10–20
		Water............... 90–80
		Rinse with warm water and distilled water. Dry in air at 66–71°C for 30 min
			Glacial acetic acid is an alternative to hydrochloric acid

		3. Etch for 3–6 min at 38°C in
		
		               Parts
		               by wt.
		Sulfuric acid (96%,
		  sp. gr. 1.84)......... 2
		Sodium dichromate
		  (crystalline)......... 1
		Water............... 8
			Suitable for freshly galvanized metal

Source: Based on Refs. 11, 21–24, 26

ously described are all applicable to metallic surfaces, but the processes listed in Table 12.13 have been specifically found to provide reproducible structural bonds and fit easily into the bonding operation. The metals most commonly used in bonded structures and their respective surface treatments are described more fully in the following sections.

Aluminum and aluminum alloys .   The effects of various aluminum surface treatments have been studied extensively. The most widely used process for high-strength, environment-resistant adhesive joints is the sodium dichromate-sulfuric acid etch, developed by Forest Product Laboratories and known as the FPL etch process. Abrasion or solvent degreasing treatments result in lower bond strengths, but these simpler processes are more easily placed into production. Table 12.14 qualitatively lists the bond strengths which can be realized with various aluminum treatments.

Copper and copper alloys.   Surface preparation of copper alloys is necessary to remove weak oxide layers attached to the copper surface. This oxide layer is especially troublesome, because it forms very rapidly. Copper specimens must be bonded or primed as quickly as possible after surface preparation. Copper also has a tendency to form brittle surface compounds when used with certain adhesives that are corrosive to copper.

One of the better surface treatments for copper, utilizing a commercial product named Ebonol C (Enthane, Inc., New Haven, CT), does not remove the oxide layer but creates a deeper and stronger oxide formation. This process, called *black oxide,* is commonly used when bond-

TABLE 12.14   Surface Treatment for Adhesive Bonding Aluminum (from Ref. 25)

Surface treatment	Type of bond
Solvent wipe (MEK, MIBK, trichloroethylene)	Low to medium strength
Abrasion of surface, plus solvent wipe (sandblasting, coarse sandpaper, etc.)	Medium to high strength
Hot-vapor degrease (trichloroethylene)	Medium strength
Abrasion of surface, plus vapor degrease	Medium to high strength
Alodine treatment	Low strength
Anodize	Medium strength
Caustic etch[*]	High strength
Chromic acid etch (sodium dichromate-surface acid)[†]	

[*]A good caustic etch is Oakite 164 (Oakite Products, Inc., 19 Rector Street, New York).
[†]Recommended pretreatment for aluminum to achieve maximum bond strength and weatherability:
1. Degrease in hot trichloroethylene vapor (160°F).
2. Dip in the following chromic acid solution for 10 min at 160°F:

Sodium dischromate ($Na_2Cr_2)H \cdot 2H_2O$	1 part/wt
Cone, sulfuric acid (sp. gr. 1.86)	10 parts/wt.
Distilled water	30 parts/wt.

3. Rinse thoroughly in cold, running, distilled, or deionized water.
4. Air dry for 30 min, followed by 10 min at 150°F.

ing requires elevated temperatures, for example, for laminating copper foil. Chromate conversion coatings are also used for high-strength copper joints.

Magnesium and magnesium alloys. Magnesium is one of the lightest metals. The surface is very sensitive to corrosion, and chemical products are often formed at the adhesive–metal interface during bonding. Preferred surface preparations for magnesium develop a strong surface coating to prevent corrosion. Proprietary methods of producing such coatings have been developed by magnesium producers.

Steel and stainless steel. Steels are generally easy to bond, provided that all rust, scale, and organic contaminants are removed. This may be accomplished easily by a combination of mechanical abrasion and solvent cleaning. Table 12.15 shows the effect of various surface treatments on the tensile shear strength of steel joints bonded with a vinyl-phenolic adhesive.

Prepared steel surfaces are easily oxidized. Once processed, they should be kept free of moisture and primed or bonded within 4 hr. Stainless surfaces are not as sensitive to oxidation as carbon steels, and a slightly longer time between surface preparation and bonding is acceptable.

TABLE 12.15   Effect of Pretreatment on the Shear Strength of Steel Joints Bonded with a Polyvinyl Formal Phenolic Adhesive (from Ref. 26)

Pretreatment	Martensitic steel		Austenitic steel		Mild steel	
	M	C	M	C	M	C
Grit blast + vapor degreasing	5120	13.1	4100	4.7	4360	7.8
Vapor blast + vapor degreasing	6150	5.6	4940	7.1	4800	5.9
The following treatments were preceded by vapor degreasing:						
Cleaning in metasilicate solution	4360	5.7	3550	7.8	4540	6.1
Acid-dichromate etch	5780	5.8	2150	22.5	4070	4.0
Vapor blast + acid dichromate etch	6180	4.1				
Hydrochloric acid etch + phosphoric acid etch	3700	17.9	950	20.2	3090	20.7
Nitric/hydrofluoric acid etch	6570	7.5	3210	15.2	4050	8.4

M = mean failing load, lb/in^2. C = coefficient of variation, %.

Titanium alloys.   Because of the usual use of titanium at high temperatures, most surface preparations are directed at improving the thermal resistance of titanium joints. Like magnesium, titanium can also react with the adhesive during cure and create a weak boundary layer.

12.3.7.2   Plastic adherends.   Many plastics and plastic composites can be treated by simple mechanical abrasion or alkaline cleaning to remove surface contaminants. In some cases, it is necessary that the polymeric surface be physically or chemically modified to achieve acceptable bonding. This applies particularly to crystalline thermoplastics such as the polyolefins, linear polyesters, and fluorocarbons. Methods used to improve the bonding characteristics of these surfaces include:

1. Oxidation via chemical treatment or flame treatment

2. Electrical discharge to leave a more reactive surface

3. Ionized inert gas, which strengthens the surface by cross-linking and leaves it more reactive

4. Metal-ion treatment

Table 12.16 lists common recommended surface treatments for plastic adherends. These treatments are necessary when plastics are to be joined with adhesives. Solvent and heat welding are other methods of fastening plastics that do not require chemical alteration of the surface. Welding procedures were discussed in the previous chapter.

As with metallic substrates, the effects of plastic surface treatments decrease with time. It is necessary to prime or bond soon after the surfaces are treated. Listed below are some common plastic materials that require special physical or chemical treatments to achieve adequate surfaces for adhesive bonding.

Fluorocarbons. Fluorocarbons such as polytetrafluoroethylene (TFE), polyfluoroethylene propylene (FEP), polychlorotrifluoroethylene (CFE), and polymonochlorotrifluoroethylene (Kel-F) are notoriously difficult to bond because of their low surface tension. However, epoxy and polyurethane adhesives offer moderate strength if the fluorocarbon is treated prior to bonding.

The fluorocarbon surface may be made more *wettable* by exposing it for a brief moment to a hot flame to oxidize the surface. The most satisfactory surface treatment is achieved by immersing the plastic in a bath consisting of sodium-naphthalene dispersion in tetrahydrofuran. This process is believed to remove fluorine atoms, leaving a carbonized surface that can be wet easily. Fluorocarbon films pretreated for adhesive bonding are available from most suppliers. A formulation and description of the sodium-naphthalene process may be found in Table 12.16. Commercial chemical products for etching fluorocarbons are also listed.

Polyethylene terephthalate (mylar). A medium-strength bond can be obtained with polyethylene terephthalate plastics and films by abrasion and solvent cleaning. However, a stronger bond can be achieved by immersing the surface in a warm solution of sodium hydroxide or in an alkaline cleaning solution for 2 to 10 min.

Polyolefins, polyformaldehyde, polyether. These materials can be effectively bonded only if the surface is first located. Polyethylene and polypropylene can be prepared for bonding by holding the flame of an oxyacetylene torch over the plastic until it becomes glossy, or else by heating the surface momentarily with a blast of hot air. It is important not to overheat the plastic, thereby causing deformation. The treated plastic must be bonded as quickly as possible after surface preparation.

Polyolefins, such as polyethylene, polypropylene, and polymethylpentene, as well as polyformaldehyde and polyether, may be more effectively treated with a sodium dichromate-sulfuric acid solution. This treatment oxidizes the surface, allowing better wetting. Activated gas plasma treatment, described in the general section on surface treatments is also an effective treatment for these plastics. Table 12.17

TABLE 12.16   Surface Preparations for Plastics

Adherend	Degreasing solvent	Method of treatment	Remarks
Acetal (co-polymer)	Acetone	1. Abrasion. Grit or vapor blast, or medium-grit emery cloth followed by solvent degreasing	For general-purpose bonding
		2. Etch in the following acid solution: *Parts by wt.* Potassium dichromate 75 Distilled water...... 120 Concentrated sulfuric acid (96%, sp. gr. 1.84).............. 1,500 for 10 s at 25°C. Rinse in distilled water, and dry in air at RT	For maximum bond strength. ASTM D 2093
Acetal (homo-polymer)	Acetone	1. Abrasion. Sand with 280A-grit emery cloth followed by solvent degreasing	For general-purpose bonding
		2. "Satinizing" technique. Immerse the part in *Parts by wt.* Perchloroethylene.... 96.85 1,4-Dioxane.......... 3.00 *p*-Toluenesulfonic acid.............. 0.05 Cab-o-Sil (Cabot Corp.)............. 0.10 for 5–30 s at 80–120°C. Transfer the part immediately to an oven at 120°C for 1 min. Wash in hot water. Dry in air at 120°C	For maximum bond strength. Recommended by du Pont
Acrylonitrile butadiene styrene	Acetone	1. Abrasion. Grit or vapor blast, or 220-grit emery cloth, followed by solvent degreasing	
		2. Etch in chromic acid solution for 20 min at 60°C	Recipe 2 for methyl pentane
Cellulosics: Cellulose, cellulose acetate, cellulose acetate butyrate, cellulose nitrate, cellulose propionate, ethyl cellulose	Methanol, isopropanol	1. Abrasion. Grit or vapor blast, or 220-grit emery cloth, followed by solvent degreasing	For general bonding purposes
		2. After procedure 1, dry the plastic at 100°C for 1 h, and apply adhesive before the plastic cools to room temperature	
Diallyl phthalate, diallyl isophthalate	Acetone, methyl ethyl ketone	Abrasion. Grit or vapor blast, or 100-grit emery cloth, followed by solvent degreasing	Steel wool may be used for abrasion

TABLE 12.16   Surface Preparations for Plastics *(Continued)*

Adherend	Degreasing solvent	Method of treatment	Remarks
Epoxy resins	Acetone, methyl ethyl ketone	Abrasion. Grit or vapor blast, or 100-grit emery cloth, followed by solvent degreasing	Sand or steel shot are suitable abrasives
Ethylene vinyl acetate	Methanol	Prime with epoxy adhesive and fuse into the surface by heating for 30 min at 100°C	
Furane	Acetone, methyl ethyl ketone	Abrasion. Grit or vapor blast, or 100-grit emery cloth, followed by solvent degreasing	
Ionomer	Acetone, methyl ethyl ketone	Abrasion. Grit or vapor blast, or 100-grit emery cloth, followed by solvent degreasing	Alumina (180-grit) is a suitable abrasive
Melamine resins	Acetone, methyl ethyl ketone	Abrasion. Grit or vapor blast, or 100-grit emery cloth, followed by solvent degreasing	
Methyl pentene	Acetone	1. Abrasion. Grit or vapor blast, or 100-grit emery cloth, followed by solvent degreasing 2. Immerse for 1 h at 60°C in *Parts by wt.* Sulfuric acid (96%, sp. gr. 1.84)......... 26 Potassium chromate.. 3 Water............... 11 Rinse in water and distilled water. Dry in warm air 3. Immerse for 5–10 min at 90°C in potassium permanganate (saturated solution), acidified with sulfuric acid (96%, sp. gr. 1.84). Rinse in water and distilled water. Dry in warm air 4. Prime surface with lacquer *based on urea-formaldehyde* resin diluted with carbon tetrachloride	For general-purpose bonding Coatings (dried) offer excellent *bonding surfaces without further* pretreatment
Phenolic resins, phenolic melamine resins	Acetone, methyl ethyl ketone, detergent	1. Abrasion. Grit or vapor blast, or abrade with 100-grit emery cloth, followed by solvent degreasing 2. Removal of surface layer of one ply of fabric previously placed on surface before curing. Expose fresh bonding surface by tearing off the ply prior to bonding	Steel wool may be used for abrasion. Sand or steel shot are suitable abrasives. Glass-fabric decorative laminates may be degreased with detergent solution

TABLE 12.16   Surface Preparations for Plastics *(Continued)*

Adherend	Degreasing solvent	Method of treatment	Remarks
Polyamide (nylon)	Acetone, methyl ethyl ketone, detergent	1. Abrasion. Grit or vapor blast, or abrade with 100-grit emery cloth, followed by solvent degreasing	Sand or steel shot are suitable abrasives
		2. Prime with a spreading dough based on the type of rubber to be bonded in admixture with isocyanate	Suitable for bonding polyamide textiles to natural and synthetic rubbers
		3. Prime with resorcinol-formaldehyde adhesive	Good adhesion to primer coat with epoxy adhesives in metal-plastic joints
Polycarbonate, allyl diglycol carbonate	Methanol, isopropanol, detergent	Abrasion. Grit or vapor blast, or 100-grit emery cloth, followed by solvent degreasing	Sand or steel shot are suitable abrasives
Fluorocarbons: Polychlorotrifluoroethylene, polytetrafluoroethylene, polyvinyl fluoride, polymonochlorotrifluoroethylene	Trichloroethylene	1. Wipe with solvent and treat with the following for 15 min at RT: Naphthalene (128 g) dissolved in tetrahydrofuran (1 liter) to which is added sodium (23 g) during a stirring period of 2 h. Rinse in deionized water, and dry in warm air	Sodium-treated surfaces must not be abraded before use. Hazardous etching solutions requiring skillful handling. Proprietary etching solutions are commercially available (see 2). PTFE available in etched tape. ASTM D 2093
		2. Wipe with solvent and treat as recommended in one of the following commercial etchants: Bond aid.....W. S. Shamban and Co. 11617 W. Jefferson Blvd. Culver City, Calif. Fluorobond...Joclin Mfg. Co. 15 Lufbery Ave. Wallingford, Conn. Fluoroetch....Action Associates 1180 Raymond Blvd. Newark, N.J. Tetraetch....W. L. Gore Associates 487 Paper Mill Rd. Newark, Del.	
		3. Prime with epoxy adhesive, and fuse into the surface by heating for 10 min at 370°C followed by 5 min at 400°C	
		4. Expose to one of the following gases activated by corona discharge: Air (dry) for 5 min Air (wet) for 5 min Nitrous oxide for 10 min Nitrogen for 5 min	Bond within 15 min of pretreatment
		5. Expose to electric discharge from a tesla coil (50,000 V ac) for 4 min	Bond within 15 min of pretreatment

TABLE 12.16   Surface Preparations for Plastics (Continued)

Adherend	Degreasing solvent	Method of treatment	Remarks
Polyesters, polyethylene terephthalate (Mylar)	Detergent, acetone, methyl ethyl ketone	1. Abrasion. Grit or vapor blast, or 100-grit emery cloth, followed by solvent degreasing	For general-purpose bonding
		2. Immerse for 10 min at 70–95°C in Parts by wt. Sodium hydroxide.... 2 Water.............. 8 Rinse in hot water and dry in hot air	For maximum bond strength. Suitable for linear polyester films (Mylar)
Chlorinated polyether	Acetone, methyl ethyl ketone	Etch for 5–10 min at 66–71°C in Parts by wt. Sodium dichromate...... 5 Water................. 8 Sulfuric acid (96%, sp. gr. 1.84).............. 100 Rinse in water and distilled water. Dry in air	Suitable for film materials such as Penton. ASTM D 2093
Polyethylene, polyethylene (chlorinated), polyethylene terephthalate (see polyesters), polypropylene, polyformaldehyde	Acetone, methyl ethyl ketone	1. Solvent degreasing	Low-bond-strength applications
		2. Expose surface to gas-burner flame (or oxyacetylene oxidizing flame) until the substrate is glossy	
		3. Etch in the following: Parts by wt. Sodium dichromate... 5 Water.............. 8 Sulfuric acid (96%, sp. gr. 1.84)......... 100 Polyethylene 60 min at 25°C or and polypropylene 1 min at 71°C Polyformaldehyde......10 s at 25°C	For maximum bond strength. ASTM D 2093
		4. Expose to following gases activated by corona discharge: Air (dry)...........For 15 min Air (wet)...........For 5 min Nitrous Oxide......For 10 min Nitrogen...........For 15 min	Bond within 15 min of pretreatment. Suitable for polyolefins.
		5. Expose to electric discharge from a tesla coil (50,000 V ac) for 1 min	Bond within 15 min of pretreatment. Suitable for polyolefins.
Polymethyl methacrylate, methacrylate butadiene styrene	Acetone, methyl ethyl ketone, detergent, methanol, trichloroethylene, isopropanol	Abrasion. Grit or vapor blast, or 100-grit emery cloth, followed by solvent degreasing	For maximum strength relieve stresses by heating plastic for 5 h at 100°C

TABLE 12.16   Surface Preparations for P astics *(Continued)*

Adherend	Degreasing solvent	Method of treatment	Remarks
Poly-phenylene	Trichloro-ethylene	Abrasion. Grit or vapor blast, or 100-grit emery cloth, followed by solvent degreasing	
Poly-phenylene oxide	Methanol	Solvent degrease	Plastic is soluble in xylene and may be primed wjth adhesive in xylene solvent
Polysty-rene	Methanol, isopro-panol, deter-gent	Abrasion, Grit or vapor blast, or 100-grit emery cloth, followed by solvent degreasing	Suitable for rigid plastic
Poly-sulfone	Methanol	Vapor degrease	
Polyure-thane	Acetone, methyl ethyl ketone	Abrade with 100-grit emery cloth and solvent degrease	
Polyvinyl chloride, polyvinyl-idene chloride polyvinyl fluoride	Trichloro-ethylene, methyl ethyl ketone	1. Abrasion. Grit or vapor blast, or 100-grit emery cloth followed by solvent degreasing 2. Solvent wipe with ketone	Suitable for rigid plastic. For maximum strength, prime with nitrile-phenolic adhesive Suitable for plasticized material
Styrene acrylo-nitrile	Trichloro-ethylene	Solvent degrease	
Urea for-malde-hyde	Acetone, methyl ethyl ketone	Abrasion. Grit or vapor blast, or 100-grit emery cloth, followed by solvent degreasing	

Source: Based on Refs. 11, 21–24, 26.

shows the tensile-shear strength of bonded polyethylene pretreated by these various methods.

12.3.7.3 Elastomeric adherends. Vulcanized-rubber joints are often contaminated with mold release and plasticizers or extenders that can migrate to the surface. As shown in Table 12.18, solvent washing and abrading are common treatments for most elastomers, but chemical treatment may be required for maximum properties. Synthetic and natural rubbers may require *cyclizing* with concentrated sulfuric acid until hairline fractures are evident on the surface.

12.3.7.4 Other adherends. Table 12.19 provides surface treatments for a variety of materials not covered in the preceding tables. Bonding to painted or plated parts requires special consideration. The resulting adhesive bond is only as strong as the adhesion of the paint or plating to the base material.

TABLE 12.17 Effects of Surface Treatments on Bonding to Polyethylene with Various Types of Adhesives (from Ref. 27)

Specimen No.	Control	Flame treated	Sanded	Acid treated, oven-dried at 90°C	Acid treated, oven-dried at 71°C	Acid treated, wiped, air-dried at 22°C	Acid treated, acetone-dried	Plasma treatment			
								Helium (30 s)	Helium (30 min)	Oxygen (30 s)	Oxygen (30 min)
Epoxy											
1	40	464*	186	454	428*	500*	516*	468*	...	423*	490*
2	48	480*	166	480*	440*	524*	490*	470*	...	463*	424*
3	24	452*	182	472*	532*	500*	502*	470*	...	484*	439*
4	58	486*	220	462*	460*	524*	500*	450*	...	495*	424*
5	58	520*	216	506*	424*	448*	476*				
Avg lb/in.²	46	480	195	475	457	499	497	464	...	466	445
Polyester											
1	74	502*	214	290	300	294	452*	196	284	264	480*
2	102	472*	146	290	288	416*	462*	230	396*	246	514*
3	70	430*	170	230	322	412	392	214	380	320	300
4	70	364*	188	268	318	426*	464*	160	400*	240	370
5	108	400*	178	308	256	236	200	148	372	244	484*
Avg lb/in.²	85	434	175	277	297	357	394	190	346	263	430
Nitrile–Rubber–Phenolic											
1	42	196	54	102	100	124	106	...	166	...	210
2	38	120	54	100	92	128	136	...	110	...	170
3	46	88	52	88	64	120	102	...	276	...	220
4	46	120	64	96	158	88	54	...	112	...	110
5	48	166	56	96	124	88	164	...	224	...	170
Avg lb/in.²	44	138	56	96	108	110	112	...	178	...	176

* Adherend failed rather than bond.   All values in this table are based upon lap-shear strength calculated as lb/in.².

TABLE 12.18 Surface Preparation for E astomers

Adherend	Degreasing solvent	Method of treatment	Remarks
Natural rubber	Methanol, isopropanol	1. Abrasion followed by brushing. Grit or vapor blast, or 280-grit emery cloth, followed by solvent wipe	For general-purpose bonding
		2. Treat the surface for 2–10 min with sulfuric acid (sp. gr. 1.84) at RT. Rinse thoroughly with cold water/hot water. Dry after rinsing in distilled water. (Residual acid may be neutralized by soaking for 10 min in 10% ammonium hydroxide after hot-water washing)	Adequate pretreatment is indicated by the appearance of hairline surface cracks on flexing the rubber. Suitable for many synthetic rubbers when given 10–15 min etch at room temperature. Unsuitable for use on butyl, polysulfide, silicone, chlorinated polyethylene, and polyurethane rubbers
		3. Treat surface for 2–10 min with paste made from sulfuric acid and barium sulfate. Apply paste with stainless-steel spatula, and follow procedure 2, above	
		4. Treat surface for 2–10 min in *Parts by vol.* Sodium hypochlorite.. 6 Hydrochloric acid (37%)............. 1 Water............... 200 Rinse with cold water and dry	Suitable for those rubbers amenable to treatments 2 and 3
Butadiene styrene	Toluene	1. Abrasion followed by brushing. Grit or vapor blast, or 280-grit emery cloth, followed by solvent wipe	Excess toluene results in swollen rubber. A 20-min drying time will restore the part to its original dimensions
		2. Prime with butadiene styrene adhesive in an aliphatic solvent.	
		3. Etch surface for 1–5 min at RT, following method 2 for natural rubber.	
Butadiene nitrile	Methanol	1. Abrasion followed by brushing. Grit or vapor blast, or 280-grit emery cloth, followed by solvent wipe	
		2. Etch surface for 10–45 s at RT, following method 2 for natural rubber	
Butyl	Toluene	1. Solvent wipe	For general-purpose bonding
		2. Prime with butyl-rubber adhesive in an aliphatic solvent	For maximum strength
Chlorosulfonated polyethylene	Acetone or methyl ethyl ketone	Abrasion followed by brushing. Grit or vapor blast, or 280-grit emery cloth, followed by solvent wipe	General-purpose bonding
Ethylene propylene	Acetone or methyl ethyl ketone	Abrasion followed by brushing. Grit or vapor blast, or 280-grit emery cloth, followed by solvent wipe	General-purpose bonding

12.4 Types of Adhesives

12.4.1 Adhesive Composition

Modern-day adhesives are often fairly complex formulations of components that perform specialty functions. The adhesive base or binder is

TABLE 12.18   Surface Preparation for E astomers *(Continued)*

Adherend	Degreasing solvent	Method of treatment	Remarks
Fluoro-silicone	Methanol	Application of fluorosilicone primer (A 4040) to metal where intention is to bond unvulcanized rubber	Primer available from Dow Corning
Polyacrylic	Methanol	Abrasion followed by brushing. Grit or vapor blast, or 100-grit emery cloth followed by solvent wipe	General-purpose bonding
Polybutadiene	Methanol	Solvent wipe	General-purpose bonding
Polychloroprene	Toluene, methanol, isopropanol	1. Abrasion followed by brushing. Grit or vapor blast, or 100-grit emery cloth, followed by solvent wipe 2. Etch surface for 5–30 min at RT, following method 2 for natural rubber	Adhesion improved by abrasion with 280-grit emery cloth followed by acetone wipe
Polysulfide	Methanol	Immerse overnight in strong chlorine water, wash and dry	
Polyurethane	Methanol	1. Abrasion followed by brushing. Grit or vapor blast, or 280-grit emery cloth followed by solvent wipe 2. Incorporation of a chlorosilane into the adhesive-elastomer system. 1% w/w is usually sufficient	Chlorosilane is available commercially. Addition to adhesive eliminates need for priming and improves adhesion to glass, metals. Silane may be used as a surface primer
Silicone	Acetone or methanol	1. Application of primer, Chemlok 607, in solvent (dries 10–15 min) 2. Expose to oxygen gas activated by corona discharge for 10 min	Primer available from Hughson Chemical Company

Source: Based on Refs. 11, 21–24, 26.

TABLE 12.19   Surface Preparations for Materials Other than Metals, Plastics, and Elastomers

Adherend	Degreasing solvent	Method of treatment	Remarks
Asbestos (rigid)	Acetone	1. Abrasion. Abrade with 100-grit emery cloth, remove dust, and solvent degrease 2. Prime with diluted adhesive or low-viscosity rosin ester	Allow the board to stand for sufficient time to allow solvent to evaporate off
Brick and fired non-glazed building materals	Methyl ethyl ketone	Abrade surface with a wire brush; remove all dust and contaminants	
Carbon graphite	Acetone	Abrasion. Abrade with 220-grit emery cloth and solvent degrease after dust removal	For general-purpose bonding

the primary component of an adhesive. The binder is generally the resinous component from which the name of the adhesive is derived. For example, an epoxy adhesive may have many components, but the primary material is epoxy resin.

TABLE 12.19   Surface Preparations for Materials Other than Metals, Plastics, and Elastomers (Continued)

Adherend	Degreasing solvent	Method of treatment	Remarks
Glass and quartz (nonoptical)	Acetone, detergent	1. Abrasion. Grit blast with carborundum and water slurry, and solvent degrease. Dry for 30 min at 100°C. Apply the adhesive before the glass cools to RT	For general-purpose bonding. Drying process improves bond strength
		2. Immerse for 10–15 min at 20°C in Parts by wt. Sodium dichromate... 7 Water.............. 7 Sulfuric acid (96%, sp. gr. 1.84)........ 400 Rinse in water and distilled water. Dry thoroughly	For maximum strength
Glass (optical)	Acetone, detergent	Clean in an ultrasonically agitated detergent bath. Rinse; dry below 38°C	
Ceramics and porcelain	Acetone	1. Abrasion. Grit blast with carborundum and water slurry, and solvent degrease	Suitable for unglazed ceramics such as alumina, silica
		2. Solvent degrease or wash in warm aqueous detergent. Rinse and dry	For glazed ceramics such as porcelain
		3. Immerse for 15 min at 20°C in Parts by wt. Sodium dichromate... 7 Water.............. 7 Sulfuric acid (96%, sp. gr. 1.84)........ 400 Rinse in water and distilled water. Oven-dry at 66°C	For maximum strength bonding of small ceramic (glazed) artefacts
Concrete, granite, stone	Perchloroethylene, detergent	1. Abrasion. Abrade with a wire brush, degrease with detergent, and rinse with hot water before drying	For general-purpose bonding
		2. Etch with 15% hydrochloric acid until effervescence ceases. Wash with water until surface is litmus-neutral. Rinse with 1% ammonia and water. Dry thoroughly before bonding	Applied by stiff-bristle brush. Acid should be prepared in a polyethylene pail 10–12% hydrochloric or sulfuric acids are alternative etchants. 10% w/w sodium bicarbonate may be used instead of ammonia for acid neutralization
Wood, plywood		Abrasion. Dry wood is smoothed with a suitable emery paper. Sand plywood along the direction of the grain	For general-purpose bonding
Painted surface	Detergent	1. Clean with detergent solution, abrade with a medium emery cloth, final wash with detergent	Bond generally as strong as the paint
		2. Remove paint by solvent or abrasion, and pretreat exposed base	For maximum adhesion

Source: Based on Refs. 11, 21–24, 26.

A hardener is a substance added to an adhesive formulation to initiate the curing reaction and take part in it. Two-component adhesive systems have one component, which is the base, and a second component, which is the hardener. Upon mixing, a chemical reaction ensues that causes the adhesive to solidify. A catalyst is sometimes incorporated into an adhesive formulation to speed the reaction between base and hardener.

Solvents are sometimes needed to lower viscosity or to disperse the adhesive to a spreadable consistency. Often, a mixture of solvents is required to achieve the desired properties.

A reactive ingredient added to an adhesive to reduce the concentration of binder is called a *diluent*. Diluents are principally used to lower viscosity and modify processing conditions of some adhesives. Diluents react with the binder during cure, become part of the product, and do not evaporate as does a solvent.

Fillers are generally inorganic particulates added to the adhesive to improve working properties, strength, permanence, or other qualities. Fillers are also used to reduce material cost. By selective use of fillers, the properties of an adhesive can be changed tremendously. Thermal expansion, electrical and thermal conduction, shrinkage, viscosity, and thermal resistance are only a few properties that can be modified by use of selective fillers.

A carrier or reinforcement is usually a thin fabric used to support a semicured (B-staged) adhesive to provide a product that can be used as a tape or film. The carrier can also serve as a spacer between the adherends and reinforcement for the adhesive.

Adhesives can be broadly classified as being thermoplastic, thermosetting, elastomeric, or an alloy blend. These four adhesive classifications can be further subdivided by specific chemical composition as described in Tables 12.20 through 12.23. The types of resins that go into the thermosetting and alloy adhesive classes are noted for high strength, creep resistance, and resistance to environments such as heat, moisture, solvents, and oils. Their physical properties are well suited for structural-adhesive applications.

Elastomeric and thermoplastic adhesive classes are not used in applications requiring continuous load, because of their tendency to creep under stress. They are also degraded by many common service environments. These adhesives find greatest use in low-strength applications such as pressure-sensitive tape, sealants, and hot melt products.

12.4.2 Structural Adhesives

12.4.2.1 Epoxy. Epoxy adhesives offer a high degree of adhesion to all substrates except some untreated plastics and elastomers. Cured epoxies have thermosetting molecular structures. They exhibit excellent tensile-shear strength but poor peel strength unless modified with a more resilient polymer. Epoxy adhesives offer excellent resistance to

TABLE 12.20 Thermosetting Adhesives

Adhesive	Description	Curing method	Special characteristics	Usual adherends	Price range
Cyanoacrylate	One-part liquid	Rapidly at RT in absence of air	Fast setting; good bond strength; low viscosity; high cost; poor heat and shock resistance; will not bond to acidic surfaces	Metals, plastics, glass	Very high
Polyester	Two-part liquid or paste	RT or higher	Resistant to chemicals, moisture, heat, weathering; good electrical properties; wide range of strengths; some resins do not fully cure in presence of air; isocyanate-cured system bonds well to many plastic films	Metals, foils, plastics, plastic laminates, glass	Low–med
Urea formaldehyde	Usually supplied as two-part resin and hardening agent; extenders and fillers used	Under pressure	Not as durable as others but suitable for fair range of service conditions; generally low cost and ease of application and cure; pot life limited to 1 to 24 hr	Plywood	Low
Melamine formaldehyde	Powder to be mixed with hardening agent	Heat and pressure	Equivalent in durability and water resistance (including boiling water) to phenolics and resorcinols; often combined with ureas to lower cost; higher service temp than ureas	Plywood, other wood products	Med
Resorcinol and phenol-resorcinol formaldehyde	Usually alcohol-water solutions to which formaldehyde must be added	RT or higher with moderate pressure	Suitable for exterior use; unaffected by boiling water, mold, fungus, grease, oil, most solvents; bond strength equals or betters strength of wood; do not bond directly to metal	Wood, plastics, paper, fiberboard, plywood	Med

Type	Form	Cure	Properties	Adherends	Cost
Epoxy	Two-part liquid or paste; one-part liquid, paste or solid; solutions	RT or higher	Most versatile adhesive available; excellent tensile-shear strength; poor peel strength; excellent resistance to moisture and solvents; low cure shrinkage; variety of curing agents/hardeners results in many variations	Metals, plastic, glass, rubber, wood, ceramics	Med
Polyimide	Supported film, solvent solution	High temp	Excellent thermal and oxidation resistance; suitable for continuous use at 550°F and short-term use to 900°F; expensive	Metals, metal foil, honeycomb core	Very high
Polybenzimidazole	Supported film	Long, high-temp cure	Good strength at high temperatures; suitable for continuous use at 450°F and short-term use at 1000°F; volatiles released during cure; deteriorates at high temperatures on exposure to air; expensive	Metals, metal foil, honeycomb core	Very high
Acrylic	Two-part liquid or paste	RT	Excellent bond to many plastics, good weather resistance, fast cure, catalyst can be used as a substrate primer; poor peel and impact strength	Metals, many plastics, wood	Very high
Acrylate acid diester	One-part liquid or paste	RT or higher in absence of air	Chemically blocked, anaerobic type; excellent wetting ability; useful temperature range of −65 to 300°F; withstands rapid thermal cycling; high-tensile-strength grade requires cure at 250°F; cures in minutes at 280°F	Metals, plastics, glass, wood	Very high

TABLE 12.21 Thermoplastic Adhesives

Adhesive	Description	Curing method	Special characteristics	Usual adherends	Price range
Cellulose acetate, cellulose acetate butyrate	Solvent solutions	Solvent evaporation	Water-clear, more heat resistant but less water resistant than cellulose nitrate; cellulose acetate butyrate has better heat and water resistance than cellulose acetate and is compatible with a wider range of plasticizers	Plastics, leather, paper, wood, glass, fabrics	Low
Cellulose nitrate	Solvent solutions	Evaporation of solvent	Tough, develops strength rapidly, water resistant; bonds to many surfaces; discolors in sunlight; dried adhesive is flammable	Glass, metal, cloth, plastics	Low
Polyvinyl acetate	Solvent solutions and water emulsions, plasticized or unplasticized, often containing fillers and pigments; also dried film that is light stable, water-white, transparent	On evaporation of solvent or water; film by heat and pressure	Bond strength of several thousand lb/in.^{2} but not under continuous loading; the most versatile in terms of formulations and uses; tasteless, odorless; good resistance to oil, grease, acid; fair water resistance	Emulsions particularly useful with porous materials like wood and paper; solutions used with plastic films, mica, glass, metal, ceramics	Low
Vinyl vinylidene	Solutions in solvents like methyl ethyl ketone	Evaporation of solvent	Tough, strong, transparent, and colorless; resistant to hydrocarbon solvents, greases, oils	Particularly useful with textiles; also porous materials, plastics	Med

Polyvinyl acetals	Solvent solutions, film, and solids	Evaporation of solvent; film and solid by heat and pressure	Flexible bond; modified with phenolics for structural use; good resistance to chemicals and oils; includes polyvinyl formal and polyvinyl butyral types	Metals, mica, glass, rubber, wood, paper	Med
Polyvinyl alcohol	Water solutions, often extended with starch or clay	Evaporation of water	Odorless, tasteless, and fungus-resistant (if desired); excellent resistance to grease and oils; water soluble	Porous materials such as fiberboard, paper, cloth	Low
Polyamide	Solid hot-melt, film, solvent solutions	Heat and pressure	Good film flexibility; resistant to oil and water; used for heat-sealing compounds	Metals, paper, plastic films	Med
Acrylic	Solvent solutions, emulsions, and mixtures requiring added catalysts	Evaporation of solvent; RT or elevated temp (two-part)	Good low-temperature bonds; poor heat resistance; excellent resistance to ultraviolet; clear; colorless	Glass, metals, paper, textiles, metallic foils, plastics	Med
Phenoxy	Solvent solutions, film, solid hot-melt	Heat and pressure	Retains high strength from 40 to 180°F; resists creep up to 180°F; suitable for structural use	Metals, wood, paper, plastic film	Med

TABLE 12.22 Elastomeric Adhesives

Adhesive	Description	Curing method	Special characteristics	Usual adherends	Price range
Natural rubber	Solvent solutions, latexes, and vulcanizing type	Solvent evaporation, vulcanizing type by heat or RT (two-part)	Excellent tack, good strength; shear strength 30–180 lb/in.^{2}; peel strength 0.56 lb/in. width; surface can be tack-free to touch and yet bond to similarly coated surface	Natural rubber, masonite, wood, felt, fabric, paper, metal	Med
Reclaimed rubber	Solvent solutions, some water dispersions; most are black, some gray and red	Evaporation of solvent	Low cost, widely used; peel strength higher than natural rubber; failure occurs under relatively low constant loads	Rubber, sponge rubber, fabric, leather, wood, metal, painted metal, building materials	Low
Butyl	Solvent system, latex	Solvent evaporation, chemical cross-linking with curing agents and heat	Low permeability to gases, good resistance to water and chemicals, poor resistance to oils, low strength	Rubber, metals	Med
Polyisobutylene	Solvent solution	Evaporation of solvent	Sticky, low-strength bonds; copolymers can be cured to improve adhesion, environmental resistance, and elasticity; good aging; poor thermal resistance; attacked by solvents	Plastic film, rubber, metal foil, paper	Low
Nitrile	Latexes and solvent solutions compounded with resins, metallic oxides, fillers, etc.	Evaporation of solvent and/or heat and pressure	Most versatile rubber adhesive; superior resistance to oil and hydrocarbon solvents; inferior in tack range, but most dry tack-free, an advantage in precoated assemblies; shear strength of 150–2000 lb/in.^{2}, higher than neoprene, if cured	Rubber (particularly nitrile), metal, vinyl plastics	Med

Styrene butadiene	Solvent solutions and latexes; because tack is low, rubber is compounded with tackifiers and plasticizing oils	Evaporation of solvent	Usually better aging properties than natural or reclaimed; low dead load strength; bond strength similar to reclaimed; useful temp range from −40 to 160°F	Fabrics, foils, plastic film laminates, rubber and sponge rubber, wood	Low
Polyurethane	Two-part liquid or paste	RT or higher	Excellent tensile-shear strength from −400 to 200°F; poor resistance to moisture before and after cure; good adhesion to plastics	Plastics, metals, rubber	Med
Polysulfide	Two-part liquid or paste	RT or higher	Resistant to wide range of solvents, oils, and greases; good gas impermeability; resistant to weather, sunlight, ozone; retains flexibility over wide temperature range; not suitable for permanent load-bearing applications	Metals, wood, plastics	High
Silicone	Solvent solution; heat or RT curing and pressure sensitive; and RT vulcanizing pastes	Solvent evaporation, RT or elevated temp	Of primary interest is pressure-sensitive type used for tape; high strengths for other forms are reported from −100 to 500°F; limited service to 700°F; excellent dielectric properties	Metals; glass; paper; plastics and rubber, including silicone and butyl rubber and fluorocarbons	High–very high
Neoprene	Latexes and solvent solutions, often compounded with resins, metallic oxides, fillers, etc.	Evaporation of solvent	Superior to other rubber adhesives in most respects—quickness; strength; max temp (to 200°F, sometimes 350°F); aging; resistant to light, weathering, mild acids, oils	Metals, leather, fabric, plastics, rubber (particularly neoprene), wood, building materials	Med

TABLE 12.23 Alloy Adhesives

Adhesive	Description	Curing method	Special characteristics	Usual adherends	Price range
Epoxy-phenolic	Two-part paste, supported film	Heat and pressure	Good properties at moderate cures; volatiles released during cure; retains 50% of bond strength at 500°F; limited shelf life; low peel strength and shock resistance	Metals, honeycomb core, plastic laminates, ceramics	Med
Epoxy-polysulfide	Two-part liquid or paste	RT or higher	Useful temperature range −70 to 200°F; greater resistance to impact, higher elongation, and less brittleness than epoxies	Metals, plastic, wood, concrete	Med
Epoxy-nylon	Solvent solutions, supported and unsupported film	Heat and pressure	Excellent tensile-shear strength at cryogenic temperature; useful temperature range −423 to 180°F; limited shelf life	Metals, honeycomb core, plastics	Med
Nitrile-phenolic	Solvent solutions, unsupported and supported film	Heat and pressure	Excellent shear strength; good peel strength; superior to vinyl and neoprene—phenolics; good adhesion	Metals, plastics, glass, rubber	Med
Neoprene-phenolic	Solvent solutions, unsupported and supported film	Heat and pressure	Good bonds to a variety of substrates; useful temp range −70 to 200°F, excellent fatigue and impact strength	Metals, glass, plastics	Med
Vinyl-phenolic	Solvent solutions and emulsions, tape, liquid, and coreacting powder	Heat and pressure	Good shear and peel strength; good heat resistance; good resistance to weathering, humidity, oil, water, and solvents; vinyl formal and vinyl butyral forms available, vinyl formal–phenolic is strongest	Metals, paper, honeycomb core	Low –med

oil, moisture, and many solvents. Low cure shrinkage and high resistance to creep under prolonged stress are characteristic of epoxy resins.

Epoxy adhesives are commercially available as liquids, pastes, and semicured (B-staged) film and solids. Epoxy adhesives are generally supplied as a 100% solids (nonsolvent) formulation, but some sprayable epoxy adhesives are available in solvent systems. Epoxy resins have no evolution of volatiles during cure and are useful in gap-filling applications.

Depending on the type of curing agent, epoxy adhesives can cure at room or elevated temperatures. Higher strengths and better heat resistance are usually obtained with the heat-curing types. Room-temperature-curing epoxies can harden in as little as 1 min at room temperature, but most systems require from 18 to 72 hr. The curing time is greatly temperature-dependent, as shown in Fig. 12.27.

Epoxy resins are the most versatile of structural adhesives, because they can be cured and co-reacted with many different resins to provide widely varying properties. Table 12.24 describes the influence of curing agents on the bond strength of epoxy to various adherends. The type of epoxy resin used in most adhesives is derived from the reaction of bisphenol A and epichlorohydrin. This resin can be cured with amines or polyamides for room-temperature-setting systems; anhydrides for elevated-temperature cure; or latent curing agents such as boron trifluoride complexes for use in one-component, heat-curing adhesives. Polyamide curing agents are used in most "general-purpose" epoxy adhesives. They provide a room-temperature cure and bond well to many substrates including plastics, glass, and elastomers. The polyamide-cured epoxy also offers a relatively flexible adhesive with fair peel and thermal-cycling properties.

12.4.2.2 Epoxy alloy or hybrids. A variety of polymers can be blended and co-reacted with epoxy resins to provide certain desired properties. The most common of these are phenolic, nylon, and polysulfide resins.

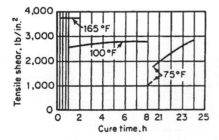

Figure 12.27 Characteristics of particular epoxy adhesive under different curing time and temperature relationships.[28]

TABLE 12.24 Influence of Epoxy Curing Agent on Bond Strength Obtained with Various Base Materials[29]

Curing agent	Amount*	Cure cycle, h at °F	Tensile-shear strength, lb/in.²					
			Polyester glass-mat laminate	Polyester glass-cloth laminate	Cold-rolled steel	Aluminum	Brass	Copper
Triethylamine.............	6	24 at 75, 4 at 150	1,850	2,100	2,456	1,810	1,765	655
Trimethylamine.............	6	24 at 75, 4 at 150	1,054	1,453	1,385	1,543	1,524	1,745
Triethylenetetramine.............	12	24 at 75, 4 at 150	1,150	1,632	1,423	1,675	1,625	1,325
Pyrrolidine.............	5	24 at 75, 4 at 150	1,250	1,694	1,295	1,733	1,632	1,420
Polyamid amine equivalent 210–230.....	35–65	24 at 75, 4 at 150	1,200	1,450	2,340	3,120	2,005	1,876
Metaphenylenidiamine.............	12.5	4 at 350	780	640	2,150	2,258	2,150	1,650
Diethylenetriamine.............	11	24 at 75, 4 at 150	1,010	1,126	1,350	1,420	1,135	1,236
Boron trifluoride monoethylamine.....	3	3 at 375	.....	.....	1,732	1,876	1,525	1,635
Dicyandiamide.............		4 at 350	530	432	2,680	2,785	2,635	2,550
Methyl nadic anhydride.............	85	6 at 350	600	756	2,280	2,165	1,955	1,835

Epoxy resin used was derived from bisphenol A and epichlorohydrin and had an epoxide equivalent of 180 to 195; the adhesives contained no filler
* Per 100 parts by weight of resin.

Epoxy-phenolic. Adhesives based on epoxy-phenolic blends are good for continuous high-temperature service in the 350°F range or intermittent service as high as 500°F. They retain their properties over a very wide temperature range, as shown in Fig. 12.28. Shear strengths of up to 3,000 lb/in.² at room temperature and 1,000 to 2,000 lb/in.² at 500°F are available. Resistance to oil, solvents, and moisture is very good. Because of their rigid nature, epoxy-phenolic adhesives have low peel strength and limited thermal-shock resistance.

These adhesives are available as pastes, solvent solutions, and B-staged film supported on glass fabric. Cure generally requires 350°F for 1 hr under moderate pressure. Epoxy-phenolic adhesives were developed primarily for bonding metal joints in high-temperature applications.

Epoxy-nylon. Epoxy-nylon adhesives offer both excellent shear and peel strength. They maintain physical properties at cryogenic temperatures but are limited to a maximum service temperature of 180°F.

Epoxy-nylon adhesives are available as unsupported B-staged film or in solvent solutions. A moderate pressure of 25 lb/in.² and temperature of 350°F are generally required for 1 hr to cure the adhesive. Because of their excellent filleting properties and high peel strength, epoxy-nylon adhesives are used to bond aluminum skins to honeycomb core in aircraft structures.

Epoxy-polysulfide. Polysulfide resins combine with epoxy resins to provide adhesives with excellent flexibility and chemical resistance. These adhesives bond well to many different substrates. Shear strength and elevated temperature properties are poor, but resistance to peel forces and low temperatures is very good. The epoxy-polysulfide alloy is supplied as a two-part, flowable paste that cures to a rubbery solid at room temperature. A common application for epoxy-polysulfide adhesives is as a sealant.

12.4.2.3 Resorcinol and phenol resorcinol. Resorcinol adhesives are primarily used for bonding wood, plastic skins to wood cores, and

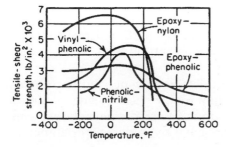

Figure 12.28 Effect of temperature on tensile-shear strength of adhesive alloys (substrate material is aluminum).[28]

primed metal to wood. Adhesive bonds as strong as wood itself are obtainable. Resorcinol adhesives are resistant to boiling water, oil, many solvents, and mold growth. Their service temperature ranges from −300 to 350°F. Because of the high cost, resorcinol resins are often modified by the addition of phenolic resins to form phenol resorcinol.

Resorcinol and phenol resorcinol adhesives are available in liquid form and are mixed with a powder hardener before application. These adhesives are cold-setting, but they can also be pressed at elevated temperatures for faster production.

12.4.2.4 Melamine formaldehyde and urea formaldehydes.

Melamine formaldehyde resins are colorless adhesives for wood. Because of the high cost, they are sometimes blended with urea formaldehyde. Melamine formaldehyde is usually supplied in powder form and reconstituted with water; a hardener is added at the time of use. Temperatures of about 200°F are necessary for cure. Adhesive strength is greater than the strength of the wood substrate.

Urea formaldehyde adhesives are not as strong or as moisture-resistant as resorcinol adhesives. However, they are inexpensive, and both hot- and cold-setting types are available. Maximum service temperature of urea adhesive is approximately 140°F. Cold water resistance is good, but boiling-water resistance may be improved by the addition of melamine formaldehyde or phenol resorcinol resins. Urea-based adhesives are used in plywood manufacture.

12.4.2.5 Modified phenolics.

Phenolic or phenol formaldehyde is also used as an adhesive for bonding wood. However, because of its brittle nature, this resin is unsuitable alone for more extensive adhesive applications. By modifying phenolic resin with various synthetic rubbers and thermoplastic materials, flexibility is greatly improved. The modified adhesive is well suited for structural bonding of many materials.

Nitrile-phenolic. Certain blends of phenolic resins with nitrile rubber produce adhesives useful to 300°F. On metals, nitrile phenolics offer shear strength in excess of 4,000 lb/in.^{2} and excellent peel properties. Good bond strengths can also be achieved on rubber, plastics, and glass. These adhesives have high impact strength and resistance to creep and fatigue. Their resistance to solvent, oil, and water is also good.

Nitrile-phenolic adhesives are available as solvent solutions and as supported and unsupported film. They require heat curing at 300 to 500°F under pressure of up to 200 lb/in.^{2}. The nitrile-phenolic systems with the highest curing temperature have the greatest resistance to elevated temperatures during service. Because of good peel strength and elevated-temperature properties, nitrile-phenolic adhesives are commonly used for bonding linings to automobile brake shoes.

Vinyl-phenolic. Vinyl-phenolic adhesives are based on a combination of phenolic resin with polyvinyl formal or polyvinyl butyral resins. They have excellent shear and peel strength. Room-temperature shear strength as high as 5,000 lb/in.^{2} is available. Maximum operating temperature, however, is only 200°F, because the thermoplastic constituent softens at elevated temperatures. Chemical resistance and impact strength are excellent.

Vinyl-phenolic adhesives are supplied in solvent solutions and as supported and unsupported film. The adhesive cures rapidly at elevated temperatures under pressure. They are used to bond metals, elastomers, and plastics to themselves or each other. A major application of vinyl-phenolic adhesive is the bonding of copper sheet to plastic laminate in printed circuit board manufacture.

Neoprene-phenolic. Neoprene-phenolic alloys are used to bond a variety of substrates. Normal service temperature is −70 to 200°F. Because of high resistance to creep and most service environments, neoprene-phenolic joints can withstand prolonged stress. Fatigue and impact strengths are also excellent. Shear strength, however, is lower than that of other modified phenolic adhesives.

Temperatures over 300°F and pressure greater than 50 lb/in.^{2} are needed for cure. Neoprene-phenolic adhesives are available as solvent solutions and film. During cure, these adhesives are quite sensitive to atmospheric moisture, surface contamination, and other processing variables.

12.4.2.6 Polyaromatics. Polyimide and polybenzimidazole resins belong to the aromatic heterocycle polymer family, which is noted for its outstanding thermal resistance. These two highly cross-linked adhesives are the most thermally stable systems commercially available. The polybenzimidazole (PBI) adhesive has shear strength on steel of 3000 lb/in.^{2} at room temperature and 2500 lb/in.^{2} at 700°F. The polyimide adhesive offers a shear strength of approximately 3,000 lb/in.^{2} at room temperature, but it does not have the excellent strength at 700 to 1,000°F, which is characteristic of PBI. Polyimide adhesives offer better elevated-temperature aging properties than PBI. The maximum continuous operating temperature for a polyimide adhesive is 600 to 650°F, whereas PBI adhesives oxidize rapidly at temperatures over 500°F.

Both adhesives are available as supported film, and polyimide resins are also available in solvent solution. During cure, temperatures of

550 to 650°F and high pressure are required. Volatiles are released during cure which contribute to a porous, brittle bond line with relatively low peel strength.

12.4.2.7 Polyester. Polyesters are a large class of synthetic resins having widely varying properties. They may be divided into two distinct groups: saturated and unsaturated.

Unsaturated polyesters are fast-curing, two-part systems that harden by the addition of catalysts, usually peroxides. Styrene monomer is generally used as a reactive diluent for polyester resins. Cure can occur at room or elevated temperature, depending on the type of catalyst. Accelerators such as cobalt naphthalene are sometimes incorporated into the resin to speed cure. Unsaturated polyester adhesives exhibit greater shrinkage during cure and poorer chemical resistance than epoxy adhesives. Certain types of polyesters are inhibited from curing by the presence of air, but they cure fully when enclosed between two substrates. Depending on the type of resin, polyester adhesives can be quite flexible or very rigid. Uses include patching kits for the repair of automobile bodies and concrete flooring. Polyester adhesives also have strong bond strength to glass-reinforced polyester laminates.

Saturated polyester resins exhibit high peel strength and are used to laminate plastic films such as polyethylene terephthalate (Mylar). They also offer excellent clarity and color stability. These polyester types, in both solution and solid form, can be chemically cross-linked with curing agents such as the isocyanates for improved thermal and chemical stability.

12.4.2.8 Polyurethane. Polyurethane-based adhesives form tough bonds with high peel strength. Generally supplied as a two-part liquid, polyurethane adhesives can be cured at room or elevated temperatures. They have exceptionally high strength at cryogenic temperatures, but only a few formulations offer operating temperatures greater than 250°F. Like epoxies, urethane adhesives can be applied by a variety of methods and form strong bonds to most surfaces. Some polyurethane adhesives degrade substantially when exposed to high-humidity environments.

Polyurethane adhesives bond well to many substrates, including hard-to-bond plastics. Since they are very flexible, polyurethane adhesives are often used to bond films, foils, and elastomers. Moisture curing one-part urethanes are also available. These adhesives utilize the humidity in the air to activate their curing mechanism.

12.4.2.9 Anaerobic adhesives. Acrylate acid diester and cyanoacrylate resins are called *anaerobic* adhesives because they cure when air is excluded from the resin. Anaerobic resins are noted for being simple to use, one-part adhesives, having fast cure at room temperature and high cost. However, the cost is moderate when considering a bonded-area basis, because only a small volume of adhesive is required. Most anaerobic adhesives do not cure when gaps between adherend surfaces are greater than 10 mils, although some monomers have been developed to provide for thicker bond-lines.

The acrylate acid diester adhesives are available in various viscosities. They cure in minutes at room temperature when a special primer is used or in 3 to 10 min at 250°F without the primer. Without the primer, the adhesive requires 3 to 4 hr at room temperature to cure.

The cyanoacrylate adhesives are more rigid and less resistant to moisture than acrylate acid diester adhesives. They are available only as low-viscosity liquids that cure in seconds at room temperature without the need of a primer. The cyanoacrylate adhesives bond well to a variety of substrates, as shown in Table 12.25, but have relatively poor thermal resistance. Modifications of the original cyanoacrylate resins have been introduced to provide faster cures, higher strengths with some plastics, and greater thermal resistance.

12.4.2.10 Thermosetting acrylics. Thermosetting acrylic adhesives are newly developed two-part systems that provide high shear strength to many metals and plastics, as shown in Table 12.26. These acrylics retain their strength to 200°F. They are relatively rigid adhesives with poor peel strength. These adhesives are particularly noted for their weather and moisture resistance as well as fast cure at room temperature.

One manufacturer has developed an acrylic adhesive system in which the hardener is applied to the substrate as a primer solution. The substrate can then be dried and stored for up to six months. When the parts are to be bonded, only the acrylic resin need be applied between the already primed substrates. Cure can occur in minutes at room temperature, depending on the type of acrylic resin used. Thus, this system offers the user a fast-reacting, one-part adhesive (with primer) with long shelf life.

12.4.3 Nonstructural Adhesives

Nonstructural adhesives are characterized by low shear strength, (usually less than 1000 psi) and poor creep resistance at slightly elevated temperatures. The most common nonstructural adhesives are

TABLE 12.25   Performance of Cyanoacrylate Adhesives on Various Substrates (from Ref. 30)

Substrate	Age of bond	Shear strength, lb/in.^{2} of adhesive bonds
Steel–steel	10 min	1920
	48 hr	3300
Aluminum–aluminum	10 min	1480
	48 hr	2270
Butyl rubber–butyl rubber	10 min	150[*]
SBR rubber–SBR rubber	10 min	130
Neoprene rubber–neoprene rubber	10 min	100[*]
SBR rubber–phenolic	10 min	110[*]
Phenolic–phenolic	10 min	930[*]
	48 hr	940[*]
Phenolic–aluminum	10 min	650
	48 hr	920[*]
Aluminum–nylon	10 min	500
	48 hr	950
Nylon–nylon	10 min	330
	48 hr	600
Neoprene rubber–polyester glass	10 min	110[*]
Polyester glass–polyester glass	10 min	680
Acrylic–acrylic	10 min	810[*]
	48 hr	790[*]
ABS–ABS	10 min	640[*]
	48 hr	710[*]
Polystyrene–polystyrene	10 min	330[*]
Polycarbonate–polycarbonate	10 min	790
	48 hr	850[*]

[*]Substrate failure.

based on elastomers and thermoplastics. Although these systems have low strength, they usually are easy to use and fast setting. Most nonstructural adhesives are used in assembly-line fastening operations or as sealants and pressure-sensitive tapes.

12.4.3.1 Elastomer-based adhesives. Natural- or synthetic-rubber-based adhesives usually have excellent peel strength but low shear strength. Their resiliency provides good fatigue and impact properties. Temperature resistance is generally limited to 150 to 200°F, and creep under load occurs at room temperature.

The basic types of rubber-based adhesives used for nonstructural applications are shown in Table 12.27. These systems are generally supplied as solvent solutions, latex cements, and pressure-sensitive tapes. The first two forms require driving the solvent or water vehicle from the adhesive before bonding. This is accomplished by either sim-

TABLE 12.26   Tensile-Shear Strength of Various Substrates Bonded with Thermosetting Acrylic Adhesives (from Ref. 31)

Substrate[*]	Average lap shear, lb/in.^{2} at 77°F		
	Adhesive A	Adhesive B	Adhesive C
Alclad aluminum, etched	4430	4235	5420
Bare aluminum, etched	4305	3985	5015
Bare aluminum, blasted	3375	3695	4375
Brass, blasted	4015	3150	4075
302 stainless steel, blasted	4645	4700	5170
302 stainless steel, etched	2840	4275	2650
Cold-rolled steel, blasted	2050	3385	2135
Copper, blasted	2915	2740	3255
Polyvinyl chloride, solvent wiped	1375[†]	1250[†]	1250[†]
Polymethyl methacrylate, solvent wiped	1550[†]	1160[†]	865[†]
Polycarbonate, solvent wiped	2570[†]	960	2570[†]
ABS, solvent wiped	1610[†]	1635	1280[†]
Alclad aluminum–PVC	1180[†]		
Plywood, 5/8-in. exterior glued (lb/in.)	802[†]	978[†]	
AFG-01 gap fill (1/16-in.) (lb/in.)		1083[†]	

[*]Metals solvent cleaned and degreased before etching or blasting.
[†]Substrate failure.

ple ambient air evaporation or forced heating. Some of the stronger and more environmental-resistant rubber-based adhesives require an elevated-temperature cure. Generally, only slight pressure is required to achieve a substantial bond.

Pressure-sensitive adhesives are permanently tacky and flow under pressure to provide intimate contact with the adherend surface. Pressure-sensitive tapes are made by placing these adhesives on a backing material such as rubber, vinyl, canvas, or cotton cloth. After pressure is applied, the adhesive tightly grips the part being mounted as well as the surface to which it is affixed. The ease of application and the many different properties that can be obtained from elastomeric adhesives account for their wide use.

In addition to pressure-sensitive adhesives, elastomers go into mastic compounds that find wide use in the construction industry. Neoprene and reclaimed-rubber mastics are used to bond gypsum board and plywood flooring to wood-framing members. Often, the adhesive bond is much stronger than the substrate. These mastic systems cure by evaporation of solvent through the porous substrates.

12.4.3.2 Silicone. Silicone pressure-sensitive adhesives have low shear strength but excellent peel strength and heat resistance. Silicone adhesives can be supplied as solvent solutions for pressure-sensi-

TABLE 12.27   Properties of Elastomeric Adhesives Used in Nonstructural Applications (from Ref. 29)

Adhesive	Application	Advantages	Limitations
Reclaimed rubber	Bonding paper, rubber, plastic and ceramic tile, plastic films, fibrous sound insulation and weather-stripping; also used for the adhesive on surgical and electrical tape	Low cost, applied very easily with roller coating, spraying, dipping, or brushing, gains strength very rapidly after joining, excellent moisture and water resistance	Becomes quite brittle with age, poor resistance to organic solvents
Natural rubber	Same as reclaimed rubber; also used for bonding leather and rubber sides to shoes	Excellent resilience, moisture and water resistance	Becomes quite brittle with age; poor resistance to organic solvents; does not bond well to metals
Neoprene rubber	Bonding weather stripping and fibrous sound-proofing materials to metal; used extensively in industry; bonding synthetic fibers, i.e., Dacron	Good strength to 150°F, fair resistance to creep	Poor storage life, high cost; small amounts of hydrochloric acid evolved during aging that may cause corrosion in closed systems; poor resistance to sunlight
Nitrile rubber	Bonding plastic films to metals, and fibrous materials such as wood and fabrics to aluminum, brass, and steel; also, bonding nylon to nylon and other materials	Most stable synthetic-rubber adhesive, excellent oil resistance, easily modified by addition of thermosetting resins	Does not bond well to natural rubber or butyl rubber
Polyiso-butylene	Bonding rubber to itself and plastic materials; also, bonding polyethylene terephthalate film to itself, aluminum foil and other plastic films	Good aging characteristics	Attacked by hydrocarbons; poor thermal resistance
Butyl	Bonding rubber to itself and metals; forms good bonding with most plastic films such as polyethylene terephthalate and polyvinylidene chloride	Excellent aging characteristics; chemically cross-linked materials have good thermal properties	Metals should be treated with an appropriate primer before bonding; attacked by hydrocarbons

tive application. The adhesive reaches maximum physical properties after being cured at elevated temperature with an organic peroxide catalyst. A lesser degree of adhesion can also be developed at room temperature. Silicone adhesives retain their qualities over a wide temperature range and after extended exposure to elevated temperature. Table 12.28 shows typical adhesive-strength properties of a silicone pressure-sensitive tape prepared with aluminum-foil backing.

Room-temperature-vulcanizing (RTV) silicone-rubber adhesives and sealants form flexible bonds with high peel strength to many substrates. These resins are one-component pastes that cure by reacting with moisture in the air. Because of this unique curing mechanism, nonporous substrates should not overlap by more than 1 in.

TABLE 12.28   Effect of Temperature and Aging on Si icone Pressure-Sensitive Tape (Aluminum Foil Backing) (from Ref. 32)

Testing temperature		Adhesive strength, oz/in.	
°C	°F	Uncatalyzed adhesive	Catalyzed adhesive (1% benzoyl peroxide)
−70	−94	Over 100*	Over 100*
−20	−4	Over 100*	Over 100*
100	212	60	48
150	302	60	45
200	392	Cohesive failure	40
250	482	Cohesive failure	35
Heat-aging cycle prior to testing (tested at 25°C, 77°F):			
No aging............................		60	52
1 h at 150°C (302°F)................		55	50
24 h at 150°C (302°F)...............		60	50
7 days at 150°C (302°F).............		70	50
1 h at 250°C (482°F)................		60	50
24 h at 250°C (482°F)...............		65	65
7 days at 250°C (482°F).............		Cohesive failure	65

* maximum limit of testing equipment.

RTV silicone materials cure at room temperature in about 24 hr. Fully cured adhesives can be used for extended periods up to 450°F and for shorter periods up to 500°F. Figure 12.29 illustrates the peel strength of an RTV adhesive on aluminum as a function of heat aging. With most RTV silicone formulations, acetic acid is released during cure. Consequently, corrosion of metals such as copper and brass in the bonding area may be a problem. However, special formulations are available that liberate methanol instead of acetic acid during cure. Silicone rubber bonds to clean metal, glass, wood, silicone resin, vulcanized silicone rubber, ceramic, and many plastic surfaces.

12.4.3.3 Thermoplastic adhesives. Table 12.21 describes the most common types of thermoplastic adhesives. These adhesives are useful in the −20° to 150°F temperature range. Their physical properties vary with chemical type. Some resins like the polyamides are quite tacky and flexible, while others are very rigid.

Thermoplastic adhesives are generally available as solvent solutions, water-based emulsions, and hot melts. The first two systems are

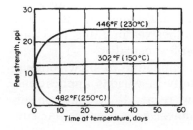

Figure 12.29   Peel strength of RTV silicone rubber bonded to aluminum as a function of heat aging.[33]

useful in bonding porous materials such as wood, plastic foam, and paper. Water-based systems are especially useful for bonding foams that could be affected adversely by solvents. When hardened, thermoplastic adhesives are very nonresistant to the solvent in which they are originally supplied.

Hot-melt systems are usually flexible and tough. They are used extensively for sealing applications involving paper, plastic films, and metal foil. Table 12.29 offers a general comparison of hot-melt adhesives. Hot melts can be supplied as (1) tapes or ribbons, (2) films, (3) granules, (4) pellets, (5) blocks, or (6) cards, which are melted and pressed between the substrate. The rate at which the adhesive cools and sets depends on the type of substrate and whether it is preheated. Table 12.30 lists the advantages and disadvantages associated with the use of water-based, solvent-based, and hot-melt thermoplastic adhesives.

12.5   Selecting an Adhesive

12.5.1   Factors Affecting Selection

There is no general-purpose adhesive. The best adhesive for a particular application will depend on the materials to be bonded, the service and assembly requirements, and the economics of the bonding operation. By using these factors as criteria for selection, the many commercially available adhesives can be narrowed down to a few possible candidates. One can seldom achieve all of the desired properties in a single adhesive system. However, a compromise adhesive can usually be chosen by deciding which properties are of major and minor importance.

12.5.1.1 Adhesive substrates. The materials to be bonded are a prime factor in determining which adhesive to use. Some adherends such as stainless steel or wood can be successfully bonded with a great many adhesive types; other adherends such as nylon can be bonded with only a few. Typical adhesive-adherend combinations are listed in Table 12.31. A number in a given column indicates the particular adhesive that will bond to a cross-referenced substrate. If two different materials are to be bonded, the recommended adhesives in Table 12.31 are those showing identical numbers under both substrates.

This information is intended only as a guideline to show common adhesives that have been used successfully in various applications. The adhesive selections are listed without regard to strength or service requirements. Lack of a suggested adhesive for a particular sub-

TABLE 12.29 General Comparison of Common Hot-Melt Adhesives (from Ref. 34)

Property	Ethylene vinyl acetate	Ethylene ethyl acetate	Ethylene acrylic acid	Ionomer	Phenoxy	Polyamide	Polyester	Polyethylene	Polyvinyl acetate	Polyvinyl butyral
Softening point, °C	40	60	70	75	100	100	.....	.....	65–195	.....
Melting point, °C	95	90 (80–100)	.....	.....	.....	.....	267	137	.....	130
Crystallinity	L	L	M	L	L	L	H	H or L	L	L
Melt index	6	3 (2–20)	.3 (0.5–400)	2	2.5	2	5	5 (0.5–20)	.....	5
Tensile strength, lb/in.^{2}	2,750	2,000	3,300	4,000	9,500	2,000	4,500	2,000	5,000	6,500
% elongation	800	700	600	450	75	300	500	150	10	100
Lap shear on metal, lb/in.^{2}	.....	.....	1,700	.....	3,500	1,050	500	.....	.....	.....
Film peel, lb/in.	1	4.5	12	12	.....	.....	5	2	.....	.....
Cost	M	M	M	M	H	M	H	L	M–L	M
Usage	H	M	L	L	L	H	H	L	M	H

H = high. M = medium. L = low.

TABLE 12.30   Advantages and Disadvantages of Thermop astic Adhesive Forms (from Ref. 35)

Form	Advantages	Disadvantages
Water base.......	Lower cost Nonflammable Nontoxic solvent Wide range of solids content Wide range of viscosity High concentration of high-molecular-weight material can be used Penetration and wetting can be varied	Poorer water resistance Subject to freezing Shrinks fabrics Wrinkles or curls paper Can be contaminated by some metals used for storage and application Corrosive to some metals Slow drying Poorer electrical properties
Solvent base......	Water-resistant Wide range of drying rates and open times Develop high early bond strength Easily wets some difficult surfaces	Fire and explosive hazard Health hazard Special explosion proof and ventilating equipment needed
Hot melt.........	Lower package and shipping cost per pound of solid material Will not freeze Drying and equipment for drying are unnecessary Impervious surfaces easily bonded Fast bond-strength development Good storage stability Provides continuous water-vapor impermeable and water-resistant adhesive film	Special applicating equipment needed Limited strength because of viscosity and temperature limitations Degrades on continuous heating Poorer coating weight control Preheating of adherends may be necessary

strate does not necessarily mean that a poor bond will result, only that information is not commonly available concerning that particular combination.

12.5.1.2 Service requirements.   Adhesives must also be selected with regard to the type of stress and environmental conditions to which they will be exposed. These factors will further limit the number of candidate adhesives to be tested. Information on the environmental resistance of various adhesive classifications will appear in the next section.

The chosen adhesive should have strength great enough to resist the maximum stress during any time in service with reasonable safety factor. Overspecifying could result in certain adhesives being overlooked that can do the job at lower cost and with less demanding curing conditions.

12.5.1.3 Production requirements.   Adhesives require time, pressure, and heat, singly or in combination, to harden or set. Production requirements often constitute a severe restricting factor in the selection of an adhesive. Typical factors involved in assembly are the equipment available, allowable cure time, pressure required, necessary bonding temperature, degree of substrate preparation required, and physical form of the adhesive. Table 12.32 lists available forms and processing requirements for various types of adhesives.

TABLE 12.31 Selecting Adhesives for Use with Various Substrates

Adhesives (number key given in Table 12.32)

Adherend	Adhesives
Metals, Alloys	
Aluminum	1 2 3 5 9 10 11 14 15 21 23 24 25 27 28 29 30 31 32 33 34 35 36 37 38 39 40 41 42 43
Brass	1 5 9 11 14 21 23 24 30 32 35 36 39 40 41 42 43
Bronze	1 5 9 11 21 23 24 30 32 35 36
Cadmium	9 10
Chromium	1 2 3 9 10 11 21 24 29 32 34 35 36 39 40 41 42 43
Copper	1 2 5 9 10 11 14 15 21 23 24 30 32 34 35 36 39 40 41 42 43
Germanium	9 10 11
Gold	2 3 5 10 11 29 30 32 37
Lead	2 3 9 11
Magnesium	1 5 9 10 11 19 21 23 24 27 28 29 30 31 32 33 34 35 36 37 39 40 41 42 43
Nickel	9 10 11 29 36
Platinum	9 11 23 38
Silver	5 9 10 11 36
Steel (mild), iron	1 2 3 5 9 10 11 12 13 14 15 16 17 18 19 21 23 24 25 27 28 29 30 31 32 33 34 35 36 37 39 40 41 43
Steel (stainless)	1 2 3 5 9 10 11 12 13 14 15 16 17 18 19 21 23 24 25 27 28 29 30 31 32 33 34 35 36 37 39 40 41 43
Tin	1 2 5 9 10 11 12 13 23
Titanium	1 2 5 9 10 11 18 19 23 29 30 31 32 33 34 35 36 37 39 40 41 42 43
Tungsten	9 10
Zinc	1 2 3 5 10 11 19 23 24 27 28 29 30 31 32 33 34 35 36 40
Uranium	2 11
Plastics	
Acetals	1 2 11 30 36 40
Acrylonitrile butadiene styrene	1 2 3 9 11 14 15 24 27 29 30 32 36 40
Allyl diglycol carbonate	24
Cellulose	16 18
Cellulose acetate	1 2 16 17 18 19 23 26 29 30 36 40
Cellulose acetate—butyrate	1 2 16 17 18 19 26 29 30 36 40
Cellulose nitrate	18 23
Cellulose propionate	1 2 16 17 18 19 23 26 29 30 36 40
Diallyl isophthalate	4 9 31 36 40 42
Diallyl phthalate	4 9 31 36 40 42
Epoxy resins	2 5 9 11 27 28 29 30 32 35 36 37 40
Ethyl cellulose	18
Methacrylate butadiene styrene	31
Melamine formaldehyde	4 7 9 11 19 30 36 40 43
Methyl pentene	9
Phenol formaldehyde	4 7 9 11 19 36 40 43
Phenolic—melamine	2 5 7 9 10 11 14 23 27 28 29 30 31 32 35 36 37 39 41 43

TABLE 12.31 Selecting Adhesives for Use with Various Substrates (Continued)

Adherend	Adhesives (number key given in Table 12.32)
Plastics (Continued)	
Phenoxy	11, 40
Polyamide (nylon)	1, 2, 7, 8, 11, 15, 23, 30, 32, 40
Polycarbonate	1, 2, 3, 11, 14, 15, 32, 40
Polychlorotrifluoroethylene	9, 10, 11, 29, 34, 35, 40
Polyester (fiber composite)	2, 3, 9, 14, 24, 27, 28, 29, 32, 35, 36, 40
Polyether (chlorinated)	30, 36
Polyethylene	9, 11, 23, 30, 32, 36, 40
Polyethylene (film)	2, 23, 29, 30, 35, 36, 40
Polyethylene (chlorinated)	2, 5, 30, 35, 36, 40
Polyethylene terephthalate (Mylar)	2, 5, 10, 19, 23, 29, 30, 32, 35, 40
Polyformaldehyde	2, 9, 10, 11, 24, 30, 40
Polyimide	9, 10, 11, 12, 30, 34, 35, 36, 37, 39, 40
Polymethyl methacrylate	1, 5, 11, 14, 15, 19, 24, 29, 30, 32, 34, 35, 36, 40
Polyphenylene	11
Polyphenylene oxide	2, 3, 9, 11, 14, 29, 30, 32, 36, 40
Polypropylene	2, 11
Polypropylene (film)	23, 30, 35
Polystyrene	1, 3, 14, 15, 24
Polystyrene (film)	1, 15, 19
Polysulfone	14
Polytetrafluoroethylene	9, 10, 11, 29, 30, 32, 34, 35, 40
Polyurethane	1, 2, 9, 11, 24, 29, 30, 32, 36, 40
Polyvinyl chloride	2, 14, 15, 24, 30, 32, 35, 40
Polyvinyl chloride (film)	2, 3, 30, 32, 40
Polyvinyl fluoride	2, 11, 33, 40
Polyvinylidene chloride	2, 23, 30, 32
Silicone	14, 34, 35
Styrene acrylonitrile	11, 30
Urea formaldehyde	4
Foams	
Epoxy	9, 11, 26, 27
Latex	1, 2, 9, 11, 19, 30, 32
Phenol formaldehyde	2, 5, 9, 11, 27, 28, 30, 32, 36, 41
Polyethylene	9, 11, 29, 30
Polyethylene–cellulose acetate	9, 21
Polyphenylene oxide	2, 9, 11, 14, 28, 30
Polystyrene	2, 9, 11, 14, 19, 30, 31, 36
Polyurethane	2, 9, 11, 15, 28, 30, 32, 35
Polyvinylchloride	2, 3, 9, 11, 19, 24, 28, 30, 32, 33, 36
Silicone	4, 34
Urea formaldehyde	7

Rubbers

Material	Reference numbers
Butyl	1 2 · 7 · 9 · 15 · 28 · 30 · 32 · 35 · 40 · 43
Butadiene nitrile	1 2 · 7 · 9 10 · 15 · 30 31 32 · 35 · 40 · 43
Butadiene styrene	1 2 · 9 10 · 15 · 30 32 · 35 · 40 · 43
Chlorosulfonated polyethylene	
Ethylene propylene	11
Fluorocarbon	1 · 9 10 11 · 34 35 · 40 41
Fluorosilicone	34
Polyacrylic	24
Polybutadiene	2 · 8 10 11 · 31
Polychloroprene (neoprene)	1 2 · 7 8 · 10 11 · 15 · 28 · 30 31 32 33 · 34 35 36 · 40 41 · 43
Polyisoprene (natural)	1 2 · 7 · 9 10 11 · 15 · 30 31 32 · 34 35 · 40 · 43
Polysulfide	33
Polyurethane	1 2 · 11 · 26 27 · 28 29 30 · 32 · 34 35 · 40
Silicone	34 35 · 35 36

Wood, Allied Materials

Material	Reference numbers
Cork	2 3 4 5 6 7 8 9 · 11 · 16 17 18 19 · 21 · 23 · 26 27 28 · 30 31 32 33 34 · 36 · 40 41
Hardboard, chipboard	4 7 · 14 15 · 25 26 · 31
Wood	2 3 4 5 6 7 8 9 · 11 · 16 17 18 19 · 21 · 23 · 25 26 27 28 29 · 30 31 32 33 34 · 36 · 40
Wood (laminates)	2 3 4 5 6 7 8 9 · 11 · 16 17 18 19 · 21 · 23 · 26 · 28 29 · 30 31 32 33 34 · 36 · 40

Fiber Products

Material	Reference numbers
Cardboard	2 3 4 5 · 7 8 9 · 11 · 16 17 18 19 · 21 22 23 24 · 26 27 28 29 · 30 31 32 33 · 36 37 · 40 41
Cotton	2 3 4 5 · 7 · 9 · 11 · 16 17 18 19 · 23 24 · 26 27 · 30 31 32 33 34 35 36 · 40 41
Felt	3 4 5 · 7 · 9 · 11 · 16 17 18 19 · 23 24 · 26 27 · 30 31 32 33 34 35 36 · 40 41
Jute	2 3 4 5 · 7 · 9 · 11 · 16 17 · 19 · 23 24 · 26 27 · 30 31 32 33 34 · 36 · 40 41
Leather	2 3 · 5 · 7 · 9 · 11 · 16 17 18 19 · 21 · 23 24 · 26 27 28 29 · 30 31 32 33 34 · 36 · 40 41
Paper (bookbinding)	16 17 18 19 · 21 22 23 · 26 27 28 29 · 30 31 32 33 · 36
Paper (labels)	16 17 18 19 · 21 22 23 24 · 26 27 28 29 · 30 31 32 33 34 · 36
Paper (packaging)	2 · 16 17 18 19 · 21 22 · 26 27 28 29 · 30 31 32 33 · 36
Rayon	2 3 4 5 · 7 · 9 · 11 · 16 17 18 19 · 23 24 · 26 27 28 · 30 31 32 33 34 35 36 · 40 41
Silk	2 3 4 5 · 7 · 9 · 11 · 16 17 18 19 · 23 24 · 26 27 28 · 30 31 32 33 34 35 36 · 40 41
Wool	2 3 4 5 · 7 · 9 · 11 · 16 17 18 19 · 23 24 · 26 27 28 · 30 31 32 33 34 35 36 · 40 41

Inorganic Materials

Material	Reference numbers
Asbestos	2 3 · 5 · 7 8 9 10 11 · 14 · 16 · 18 19 · 21 · 27 28 29 · 30 31 32 33 34 35 · 37 · 40 · 43
Carbon	9 · 11
Carborundum	9 · 11
Ceramics (porcelain, vitreous)	1 2 3 4 · 7 8 9 10 11 · 14 15 16 17 18 19 · 21 · 27 28 29 · 30 31 32 33 34 35 36 37 · 40 41
Concrete, stone, granite	2 3 4 · 7 8 · 14 · 35
Ferrite	9 · 11
Glass	1 2 3 · 5 · 9 10 11 · 14 15 16 17 18 · 28 · 30 31 32 · 34 35 · 37 · 40
Magnesium fluoride	35
Mica	14
Quartz	9 · 11
Sodium chloride	11
Tungsten carbide	9 · 11

Source: Based on Schields.[26]

TABLE 12.32 Selecting Adhesives with Respect to Form and Processing Factors

Adhesive type	Common forms available: Solid	Film	Paste	Liquid	Solvent sol, emulsion	Method of cure: Solvent release	Fusion on heating	Pressure-sensitive	Chemical reaction	Processing conditions: Room temp	Elevated temp	Bonding pressure required	Bonding pressure not required
Thermosetting Adhesives													
1. Cyanoacrylate				X					X	X		X	
2. Polyester + isocyanate		X			X	X			X	X	X	X	
3. Polyester + monomer			X	X					X	X	X		X
4. Urea formaldehyde	X			X					X	X	X	X	
5. Melamine formaldehyde	X								X		X	X	
6. Urea–melamine formaldehyde	X								X		X	X	
7. Resorcinol formaldehyde				X					X	X	X	X	
8. Phenol-resorcinol formaldehyde				X					X	X	X	X	
9. Epoxy (+ polyamine)		X	X	X					X	X	X		X
10. Epoxy (+ polyanhydride)	X	X	X	X					X		X		X
11. Epoxy (+ polyamide)			X	X					X	X	X		X
12. Polyimide		X			X				X		X	X	
13. Polybenzimidazole		X							X		X	X	
14. Acrylic			X	X					X	X	X		X
15. Acrylate acid diester			X	X					X	X	X	X	
Thermoplastic Adhesives													
16. Cellulose acetate					X	X				X		X	
17. Cellulose acetate butyrate					X	X				X			X
18. Cellulose nitrate					X	X				X			X

19. Polyvinyl acetate		X				X*	X*	X		X*	X
20. Vinyl vinylidene	X	X				X	X	X		X	X
21. Polyvinyl acetal		X				X	X	X		X*	X*
22. Polyvinyl alcohol	X	X				X		X	X	X	X
23. Polyamide	X	X					X			X	X
24. Acrylic		X	X						X	X	X
25. Phenoxy	X	X				X	X	X	X	X	X

Elastomer Adhesives

26. Natural rubber					X	X	X		X	X	X
27. Reclaimed rubber					X	X	X	X	X	X	X
28. Butyl					X	X	X	X	X	X	X
29. Polyisobutylene					X	X	X	X	X	X	X
30. Nitrile					X	X	X	X	X	X	X
31. Styrene butadiene	X				X	X	X	X*	X	X*	X*
32. Polyurethane		X	X								
33. Polysulfide		X	X								
34. Silicone (RTV)					X	X	X		X	X	X
35. Silicone resin					X	X	X	X	X	X	X
36. Neoprene								X			

Alloy Adhesives

37. Epoxy-phenolic			X	X			X	X		X	X
38. Epoxy-polysulfide			X	X			X	X		X	X
39. Epoxy-nylon		X	X				X	X		X	X
40. Phenolic-nitrile		X	X					X		X	X
41. Phenolic-neoprene		X	X					X		X	X
42. Phenolic-polyvinyl butyral		X	X					X		X	X
43. Phenolic-polyvinyl formal			X					X		X	X

* Heat and pressure required for hot-melt types.

12.5.1.4 Cost requirements. Economic analysis of the bonding operation should consider not only raw-material cost but also the processing equipment necessary, time and labor required, and cost incurred by wasted adhesive and rejected parts. Final cost per bonded part is a realistic criterion for selection of an adhesive.

12.5.2 Adhesives for Metal

The chemical types of structural adhesives for metal bonding were described in the preceding section. Since organic adhesives readily wet most metallic surfaces, the adhesive selection does not depend as much on the type of metal substrate as on other bonding requirements.

Selecting a specific adhesive from a table of general properties is difficult, because formulations within one class of adhesive may vary widely in physical properties. General physical data for structural metal adhesives are presented in Table 12.33. This table may prove useful in making preliminary selections or eliminating obviously unsuitable adhesives. Nonstructural adhesives for metals include elastomeric and thermoplastic resins. These are generally used as pressure-sensitive or hot-melt adhesives. They are noted for fast production, low cost, and low to medium strength. Typical adhesives for nonstructural-bonding applications were previously described. Most pressure-sensitive and hot-melt cements can be used on any clean metal surface and on many plastics and elastomers.

12.5.3 Adhesives for Plastics

The physical and chemical properties of both the solidified adhesive and the plastic substrate affect the quality of the bonded joint. Major elements of concern are the thermal expansion coefficient and glass transition temperature of the substrate relative to the adhesive. Special consideration is also required of polymeric surfaces that can change during normal aging or exposure to operating environments.

Significant differences in thermal expansion coefficient between substrates and the adhesive can cause serious stress at the plastic joint's interface. These stresses are compounded by thermal cycling and low-temperature service requirements. Selection of a resilient adhesive or adjustments in the adhesive's thermal expansion coefficient via fillers or additives can reduce such stress.

Structural adhesives must have a glass transition temperature higher than the operating temperature to avoid a cohesively weak bond and possible creep problems. Modern engineering plastics, such as polyimide or polyphenylene sulfide, have very high glass transition

TABLE 12.33 Properties of Structural Adhesives Used to Bond Metals

Adhesive	Service temp, °F		Shear strength, lb/in.^{2}	Peel strength	Impact strength	Creep resistance	Solvent resistance	Moisture resistance	Type of bond
	Max	Min							
Epoxy-amine	150	−50	3,000–5,000	Poor	Poor	Good	Good	Good	Rigid
Epoxy-polyamide	150	−60	2,000–4,000	Medium	Good	Good	Good	Medium	Tough and moderately flexible
Epoxy-anhydride	300	−60	3,000–5,000	Poor	Medium	Good	Good	Good	Rigid
Epoxy-phenolic	350	−423	3,200	Poor	Poor	Good	Good	Good	Rigid
Epoxy-nylon	180	−423	6,500	Very good	Good	Medium	Good	Poor	Tough
Epoxy-polysulfide	150	−100	3,000	Good	Medium	Medium	Good	Good	Flexible
Nitrile-phenolic	300	−100	3,000	Good	Good	Good	Good	Good	Tough and moderately flexible
Vinyl-phenolic	225	−60	2,000–5,000	Very good	Good	Medium	Medium	Good	Tough and moderately flexible
Neoprene-phenolic	200	−70	3,000	Good	Good	Good	Good	Good	Tough and moderately flexible
Polyimide	600	−423	3,000	Poor	Poor	Good	Good	Medium	Rigid
Polybenzimidazole	500	−423	2,000–3,000	Poor	Poor	Good	Good	Good	Rigid
Polyurethane	150	−423	5,000	Good	Good	Good	Medium	Poor	Flexible
Acrylate acid diester	200	−60	2,000–4,000	Poor	Medium	Good	Poor	Poor	Rigid
Cyanoacrylate	150	−60	2,000	Poor	Poor	Good	Poor	Poor	Rigid
Phenoxy	180	−70	2,500	Medium	Good	Good	Poor	Good	Tough and moderately flexible
Thermosetting acrylic	250	−60	3,000–4,000	Poor	Poor	Good	Good	Good	Rigid

temperatures. Most common adhesives have a relatively low glass transition temperature, so the weakest thermal link in the joint may often be the adhesive.

Use of an adhesive too far below the glass transition temperature could result in low peel or cleavage strength. Brittleness of the adhesive at very low temperatures could also manifest itself in poor impact strength.

Plastic substrates could be chemically active, even when isolated from the operating environment. Many polymeric surfaces slowly undergo chemical and physical change. The plastic surface, at the time of bonding, may be well suited to the adhesive process. However, after aging, undesirable surface conditions may present themselves at the interface, displace the adhesive, and result in bond failure. These weak boundary layers may come from the environment or from within the plastic substrate itself.

Moisture, solvent, plasticizers, and various gases and ions can compete with the cured adhesive for bonding sites. The process where a weak boundary layer preferentially displaces the adhesive at the interface is called *desorption*. Moisture is the most common desorbing substance, being present both in the environment and within many polymeric substrates.

Solutions to the desorption problem consist of eliminating the source of the weak boundary layer or selecting an adhesive that is compatible with the desorbing material. Excessive moisture can be eliminated from a plastic part by postcuring or drying the part before bonding. Additives that can migrate to the surface can possibly be eliminated by reformulating the plastic resin. Also, certain adhesives are more compatible with oils and plasticizers than others. For example, the migration of plasticizer from flexible polyvinyl chloride can be counteracted by using a nitrile based adhesive. Nitrile adhesive resins are capable of absorbing the plasticizer without degrading.

12.5.3.1 Thermoplastics. Many thermoplastics can be joined by solvent or heat welding as well as with adhesives. These alternative joining processes are discussed in detail in the previous chapter. The plastic manufacturer is generally the leading source of information on the proper methods of joining a particular plastic.

12.5.3.2 Thermosetting plastics. Thermosetting plastics cannot be heat or solvent welded. They are easily bonded with many adhesives, some of which have been listed in Table 12.31. Abrasion is generally recommended as a surface treatment.

12.5.3.3 Reinforced plastics. Adhesives that give satisfactory results on the resin matrix alone may also be used to bond reinforced plastics. Surface preparation of reinforced thermosetting plastics consists of abrasion and solvent cleaning. A degree of abrasion is desired so that the reinforcing material is exposed to the adhesive.

Reinforced thermoplastic parts are generally abraded and cleaned prior to adhesive bonding. However, special surface treatment such as used on the thermoplastic resin matrix may be necessary for optimal strength. Care must be taken so that the treatment chemicals do not wick into the substrate and cause degradation. Certain reinforced thermoplastics may also be solvent cemented or heat welded. However, the percentage of filler in the substrate must be limited, or the bond will be starved of resin.

12.5.3.4 Plastic foams. Some solvent cements and solvent-containing pressure-sensitive adhesives will collapse thermoplastic foams. Water-based adhesives, based on styrene butadiene rubber (SBR) or polyvinyl acetate, and 100% solid adhesives are often used. Butyl, nitrile, and polyurethene adhesives are often used for flexible polyurethane foam. Epoxy adhesives offer excellent properties on rigid polyurethane foam.

12.5.4 Adhesives for Elastomers

12.5.4.1 Vulcanized elastomers. Bonding of vulcanized elastomers to themselves and other materials is generally completed by using a pressure-sensitive adhesive derived from an elastomer similar to the one being bonded. Flexible thermosetting adhesives such as epoxy-polyamide or polyurethane also offer excellent bond strength to most elastomers. Surface treatment consists of washing with a solvent, abrading, or acid cyclizing as described in Table 12.18.

Elastomers vary greatly in formulation from one manufacturer to another. Fillers, plasticizers, antioxidants, etc., may affect the adhesive bond. Adhesives should be thoroughly tested on a specific elastomer and then reevaluated if the elastomer manufacturer or formulation is changed.

12.5.4.2 Unvulcanized elastomers. Unvulcanized elastomers may be bonded to metals and other rigid adherends by priming the adherend with a suitable air- or heat-drying adhesive before the elastomer is molded against the adherend. The most common elastomers to be bonded in this way include nitrile, neoprene, urethane, natural rubber, SBR, and butyl rubber. Less common unvulcanized elastomers such as the silicones, fluorocarbons, chlorosulfonated polyethylene, and polyacrylate are more difficult to bond. However, recently developed adhesive primers improve the bond of these elastomers to metal. Surface treatment of the adherend before priming should be according to good standards.

12.5.5   Adhesives for Wood

Resorcinol-formaldehyde resins are cold-setting adhesives for wood structures. Urea-formaldehyde adhesives, commonly modified with melamine formaldehyde, are used in the production of plywood and in wood veneering for interior applications. Phenol-formaldehyde and resorcinol-formaldehyde adhesive systems have the best heat and weather resistance.

Polyvinyl acetates are quick-drying, water-based adhesives commonly used for the assembly of furniture. This adhesive produces bonds stronger than the wood itself, but it is not resistant to moisture or high temperature. Table 12.34 describes common adhesives used for bonding wood.

TABLE 12.34   Properties of Common Wood Adhesives (from Ref. 36)

Resin type used	Resin solids in glue mix, %	Principal use	Method of application	Principal property	Principal limitation
Urea formaldehyde	23–30	Wood to wood interior	Spreader rolls	Bleed-through-free; good adhesion	Poor durability
Phenol formaldehyde	23–27	Plywood exterior	Spreader rolls	Durability	Comparatively long cure times
Melamine formaldehyde	68–72	Wood to wood, splicing, patching, scarfing	Sprayed, combed	Adhesion, color, durability	Relative cost; poor washability; needs heat to cure
Melamine urea 2/1	55–60	End and edge gluing exterior	Applicator	Colorless, durability and speed	Cost
Resorcinol formaldehyde	50–56	Exterior wood to wood (laminating)	Spreader rolls	Cold sets durability	Cost, odor
Phenol-resorcinol 10/90	50–56	Wood to wood exterior (laminating)	Spreader rolls	Warm-set durability	Cost, odor
Polyvinylacetate emulsion	45–55	Wood to wood interior	Brushed, sprayed, spreader rolls	Handy	Lack of H_2O and heat resistance

12.5.6 Adhesives for Glass

Glass adhesives are generally transparent, heat-setting resins that are water-resistant to meet the requirements of outdoor applications. Adhesives generally used to bond glass, and their physical characteristics, are presented in Table 12.35.

TABLE 12.35   Commercial Adhesives Most Desirable for Glass (from Ref. 37)

		Bond characteristics		
Trade name	Chemical type	Strength, lb/in.^{2}	Type of failure	Weathering quality
Butacite, Butvar........	Polyvinyl butyral	2,000–4,000	Adhesive	Fair
Bostik 7026, FM-45, FM-46..............	Phenolic butyral	2,000–5,500	Glass	Excellent
EC826, EC776..........	....	....	Adhesion and glass	
N-199, Scotchweld.......	Phenolic nitrile	1,000–1,200	....	Excellent
Pliobond M-20, EC847 ..	Vinyl nitrile	1,200–3,000	Adhesion and glass	Fair to good
EC711, EC882..........	....	....	....	....
EC870................	Neoprene	800–1,200	Adhesion and cohesive	Fair
EC801, EC612.........	Polysulfide	200– 400	Cohesive	Excellent
EC526, R660T, EC669...	Rubber base	200– 800	Adhesive	Fair to poor
Silastic................	Silicone	200– 300	Cohesive	Excellent
Rez-N-Glue, du Pont 5459................	Cellulose vinyl	1,000–1,200	Adhesive	Fair
Vinylite AYAF, 28-18....	Vinyl acetate	1,500–2,000	Adhesive	Poor
Araldite, Epon L-1372, ERL-2774, R-313, C-14, SH-1, J-1152..........	Epoxy	600–2,000	Adhesive	Fair to good

12.6   Effect of the Environment

For an adhesive bond to be useful, it not only must withstand the mechanical forces acting on it, it must also resist the service environment. Adhesive strength is influenced by many common environments including temperature, moisture, chemical fluids, and outdoor weathering. Table 12.36 summarizes the relative resistance of various adhesive types to common environments.

12.6.1   High Temperature

All polymeric materials are degraded to some extent by exposure to elevated temperatures. Not only are physical properties lowered at high temperatures, they also degrade due to thermal aging. Newly developed polymeric adhesives have been found to withstand 500 to 600°F continuously. To use these materials, the designer must pay a premium in adhesive cost and also be capable of providing long, high-temperature cures.

TABLE 12.36 Relative Resistance of Synthetic Adhesives to Common Service Environments (from Ref. 38)

Adhesive type	Shear	Peel	Heat	Cold	Water	Hot water	Acid	Alkali	Oil, grease	Fuels	Alcohols	Ketones	Esters	Aromatics	Chlorinated solvents
Thermosetting Adhesives															
1. Cyanoacrylate	2	6	5		6	6	6	6	3	3	5	5	5	4	4
2. Polyester + isocyanate	2	2	3	2	1	3	3	2	2	2	3	2	2	6	2
3. Polyester + monomer	2	6	5	3	3	6	3	6	2	2	2	6	6	6	6
4. Urea formaldehyde	2	6	3	3	2	6	2	2	2	2	2	2	2	2	2
5. Melamine formaldehyde	2	6	2	2	2	5	2	2	2	2	2	2	2	2	2
6. Urea–melamine formaldehyde	2	6	2	2	2	2	1	1	2	2	2	2	2	2	2
7. Resorcinol formaldehyde	2	6	2	2	2	2	2	2	2	2	2	2	2	2	2
8. Phenol-resorcinol formaldehyde	2	6	2	2	2	2	2	2	2	3	1	2	6	1	
9. Epoxy (+ polyamine)	2	5	3	5	2	3	2	2	2	2	1	6	6	2	
10. Epoxy (+ polyanhydride)	2	5	1	4	3	3	2	2		2	1	6	6	3	
11. Epoxy (+ polyamide)	2	2	6	2	2	6	3	6	2	2	1	6	6	2	2
12. Polyimide	2	4	1	1	2	4	2	2	2	2	2	2	2	2	2
13. Polybenzimidazole	2	4	1	1	2	4	2	2	2	2	2	2	2	2	2
14. Acrylic	2	6	5	3	1	3	2	2	2	2	2	2	2	2	2
15. Acrylate acid diester	2	5	3	3	4	4	6	6	3	3	5	5	5	4	4
Thermoplastic Adhesives															
16. Cellulose acetate	2	6	2	3	1	6	1	2	3	2	4	6	6	6	6
17. Cellulose acetate butyrate	2	3	3	3	2		3	2			6	6	6	6	6
18. Cellulose nitrate	2	6	3	3	3	3	3	6	2	2	6	6	6	6	6
19. Polyvinyl acetate	2	6	6		3	6	3	3	2	2	6	6	6	6	6
20. Vinyl vinylidene	2	3	3	3	3	3	3	3	2	2	2	2	2		
21. Polyvinyl acetal	2	6	5	2	2		6	3	2	2	3	3	6	3	2
22. Polyvinyl alcohol		2	3		6	6	5	5	2	1	3	1	1	1	1
23. Polyamide	2	3	5	3	5	6	6	2	2	2	6	2	2	2	6
24. Acrylic	2	2	4	3	3	3						4	4		4
25. Phenoxy	2	3	4	3	3	4	3	2	3	5	5			6	

Elastomer Adhesives

	1	2	3	4	5	6	7	8	9	10	11	12	13	14	15
26. Natural rubber	2	3	3	:	3	:	3	3	6	6	2	2	4	4	6	6
27. Reclaimed rubber	2	3	3	:	2	:	3	3	6	6	2	2	4	4	6	6
28. Butyl	3	6	6	3	2	6	1	2	6	6	2	2	2	2	6	6
29. Polyisobutylene	6	6	6	3	2	6	2	2	6	6	3	2	2	2	6	6
30. Nitrile	2	3	3	3	2	5	5	6	2	2	3	6	6	6	3	6
31. Styrene butadiene	3	6	3	3	1	:	3	2	5	2	2	6	6	5	6	5
32. Polyurethane	2	3	2	2	2	3	3	3	2	2	2	5	5	:	6
33. Polysulfide	3	2	6	2	1	6	2	2	2	2	2	6	6	2	6
34. Silicone (RTV)	3	5	1	1	2	2	3	2	3	2	3	3	3	3	3
35. Silicone resin	2	2	1	2	2	2	2	3	2	2	3	3	4	3	3
36. Neoprene	2	3	3	3	2	:	2	2	2	2	3	6	6	6	6

Alloy Adhesives

	1	2	3	4	5	6	7	8	9	10	11	12	13	14	15
37. Epoxy-phenolic	1	6	1	3	2	2	2	2	3	3	2	6	6	2	6
38. Epoxy-polysulfide	2	2	6	2	1	6	2	2	2	2	3	6	6	2	6
39. Epoxy-nylon	1	1	6	2	2	6	:	:	2	3	2	6	6	6	6
40. Phenolic-nitrile	2	2	2	3	2	2	2	2	2	2	2	6	6	6	6
41. Phenolic-neoprene	2	3	3	2	2	:	3	3	2	3	4	6	6	6	6
42. Phenolic-polyvinyl butyral	2	3	3	3	2	3	3	2	2	2	4	6	6	6	6
43. Phenolic-polyvinyl formal	2	3	6	6	2	6	4	2	2	2	4	6	6	6	6

Key: 1. Excellent   2. Good   3. Fair   4. Poor   5. Very poor   6. Extremely poor

For an adhesive to withstand elevated-temperature exposure, it must have a high melting or softening point and resistance to oxidation. Materials with a low melting point, such as many of the thermoplastic adhesives, may prove excellent adhesives at room temperature. However, once the service temperature approaches the glass transition temperature of these adhesives, plastic flow results in deformation of the bond and degradation in cohesive strength. Thermosetting materials, exhibiting no melting point, consist of highly cross-linked networks of macromolecules. Many of these materials are suitable for high-temperature applications. When considering thermoset adhesives, the critical factor is the rate of strength reduction due to thermal oxidation and pyrolysis.

Thermal oxidation initiates progressive chain scission of molecules, resulting in losses of weight, strength, elongation, and toughness within the adhesive. Figure 12.30 illustrates the effect of oxidation by comparing adhesive joints aged in both high-temperature air and inert-gas environments. The rate of strength degradation in air depends on the temperature, the adhesive, the rate of airflow, and even the type of adherend. Certain metal-adhesive interfaces are capable of accelerating the rate of oxidation. For example, many structural adhesives exhibit better thermal stability when bonded to aluminum than when bonded to stainless steel or titanium (Fig. 12.30).

High-temperature adhesives are usually characterized by a rigid polymeric structure, high softening temperature, and stable chemical groups. The same factors also make these adhesives very difficult to process. Only epoxy-phenolic-, polyimide-, and polybenzimidazole-based adhesives can withstand long-term service temperatures greater than 350°F.

12.6.1.1 Epoxy. Epoxy adhesives are generally limited to continuous applications below 300°F. Figure 12.31 illustrates the aging character-

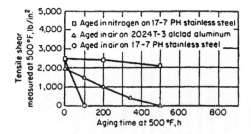

Figure 12.30 The effect of 500°F aging in air and nitrogen on an epoxy-phenolic adhesive (HT-424).[39]

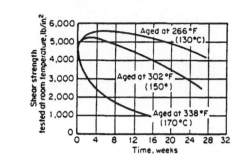

Figure 12.31 Effect of temperature aging on typical epoxy adhesive in air. Strength measured at room temperature.[40]

istics of a typical epoxy adhesive at elevated temperatures. Certain epoxy adhesives are able to withstand short terms at 500°F and long-term service at 300 to 350°F. These systems were formulated especially for thermal environments by incorporation of stable epoxy co-reactants, high-temperature curing agents, and antioxidants into the adhesive.

One successful epoxy co-reactant system is an epoxy-phenolic alloy. The excellent thermal stability of the phenolic resins is coupled with the adhesion properties of epoxies to provide an adhesive capable of 700°F short-term operation and continuous use at 350°F. The heat-resistance and thermal-aging properties of an epoxy-phenolic adhesive are compared with those of other high-temperature adhesives in Fig. 12.32.

Anhydride curing agents give unmodified epoxy adhesives greater thermal stability than most other epoxy curing agents. Phthalic anhydride, pyromellitic dianhydride, and chlorendic anhydride allow greater cross-linking and result in short term-heat resistance to 450°F. Long-term thermal endurance, however, is limited to 300°F. Typical epoxy formulations cured with pyromellitic dianhydride offer 1,200-2,600 lb/in.^{2} shear strength at 300°F and 1,000 lb/in.^{2} at 450°F.

12.6.1.2 Modified phenolics. Of the common modified phenolic adhesives, the nitrile-phenolic blend has the best resistance to shear at elevated temperatures. Nitrile phenolic adhesives have high shear strength up to 250 to 350°F, and the strength retention on aging at these temperatures is very good. The nitrile phenolic adhesives are also extremely tough and provide high peel strength.

12.6.1.3 Silicone. Silicone adhesives have very good thermal stability but low strength. Their chief application is in nonstructural applications such as high-temperature pressure-sensitive tape.

Attempts have been made to incorporate silicones with other resins such as epoxies and phenolics, but long cure times and low strength have limited their use.

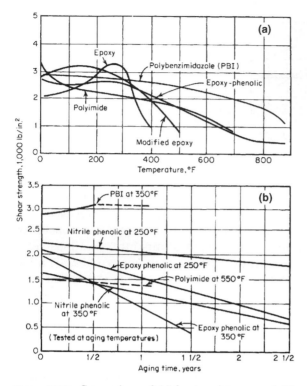

Figure 12.32 Comparison of (a) heat resistance and (b) thermal aging of high-temperature structural adhesives.[41]

12.6.1.4 Polyaromatics. The most common polyaromatic resins, polyimide and polybenzimidazole, offer greater thermal resistance than any other commercially available adhesive. The rigidity of their molecular chains decreases the possibility of chain scission caused by high temperatures. The aromaticity of these structures provides high bond-dissociation energy and acts as an *energy sink* to the thermal environment.

Polyimide. The strength retention of polyimide adhesives for short exposures to 1000°F is slightly better than that of an epoxy-phenolic alloy. However, the thermal endurance of polyimides at temperatures greater than 500°F is unmatched by other commercially available adhesives.

Polyimide adhesives are usually supplied as a glass-fabric-reinforced film having a limited shelf life. A cure of 90 min at 500 to 600°F and 15 to 200 lb/in.^{2} pressure is usually necessary for optimal properties. High-boiling volatiles can be released during cure, which causes a somewhat porous adhesive layer. Because of the inherent rigidity of this material, peel strength is low.

Polybenzimidazole. As illustrated in Fig. 12.32, polybenzimidazole (PBI) adhesives offer the best short-term performance at elevated temperatures. However, PBI resins oxidize rapidly and are not recommended for continuous use at temperatures over 450°F.

PBI adhesives require a cure at 600°F. Release of volatiles during cure contributes to a porous adhesive bond. Supplied as a very stiff, glass-fabric-reinforced film, this adhesive is expensive, and applications are limited by a long, high-temperature curing cycle.

12.6.2 Low Temperature

The factors that determine the strength of an adhesive at very low temperatures are (1) the difference in coefficient of thermal expansion between adhesive and adherend, (2) the elastic modulus, and (3) the thermal conductivity of the adhesive. The difference in thermal expansion is very important, especially since the elastic modulus of the adhesive generally decreases with falling temperature. It is necessary that the adhesive retain some resiliency if the thermal-expansion coefficients of adhesive and adherend cannot be closely matched. The adhesive's coefficient of thermal conductivity is important in minimizing transient stresses during cooling. This is why thinner bonds have better cryogenic properties than thicker ones.

Low-temperature properties of common structural adhesives used for cryogenic applications are illustrated in Fig. 12.33.

Epoxy-polyamide adhesives can be made serviceable at very low temperatures by the addition of appropriate fillers to control thermal expansion. However, the epoxy-based systems are not as attractive as some others because of brittleness and corresponding low peel and impact strength at cryogenic temperatures.

Epoxy-phenolic adhesives are exceptional in that they have good adhesive properties at both elevated and low temperatures. Vinyl-phenolic adhesives maintain fair shear and peel strength at −423°F, but strength decreases with decreasing temperature. Nitrile-phenolic adhesives do not have high strength at low service temperatures because of rigidity.

Polyurethane and epoxy-nylon systems offer outstanding cryogenic properties. Polyurethane adhesives are easily processible and bond well to many substrates. Peel strength ranges from 22 lb/in. at 75°F to 26 lb/in. at −423°F, and the increase in shear strength at −423°F is even more dramatic. Epoxy-nylon adhesives also retain flexibility and yield 5,000 lb/in.^{2} shear strength in the cryogenic temperature range.

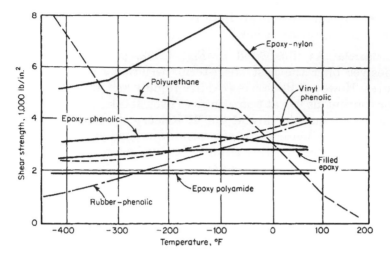

Figure 12.33 Properties of cryogenic structural adhesive systems.[41]

Heat-resistant polyaromatic adhesives also have shown promising low-temperature properties. The shear strength of a polybenzimidazole adhesive on stainless-steel substrates is 5,690 lb/in.[2] at a test temperature of −423°F, and polyimide adhesives have exhibited shear strength of 4,100 lb/in.[2] at −320°F. These unique properties show the applicability of polyaromatic adhesives on structures seeing both very high and low temperatures.

12.6.3 Humidity and Water Immersion

Moisture can affect adhesive strength in two significant ways. Some polymeric materials, notably ester-based polyurethanes, will *revert,* i.e., lose hardness, strength, and (in the worst cases) turn fluid during exposure to warm, humid air. Water can also permeate the adhesive and preferentially displace the adhesive at the bond interface. This later mechanism is the most common cause of adhesive-strength reduction in moist environments.

The rate of reversion or hydrolytic instability depends on the chemical structure of the base adhesive, the type and amount of catalyst used, and the flexibility of the adhesive. Certain chemical linkages such as ester, urethane, amide, and urea can be hydrolyzed. The rate of attack is fastest for ester-based linkages. Ester linkages are present in certain types of polyurethanes and anhydride-cured epoxies. Generally amine-cured epoxies offer better hydrolytic stability than anhydride-cured types. Figure 12.34 illustrates the hydrolytic stability of various polymeric materials determined by a hardness measurement.

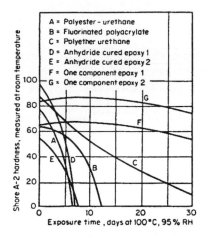

A = Polyester - urethane
B = Fluorinated polyacrylate
C = Polyether urethane
D = Anhydride cured epoxy 1
E = Anhydride cured epoxy 2
F = One component epoxy 1
G = One component epoxy 2

Shore A-2 hardness, measured at room temperature

Exposure time, days at 100°C, 95% RH

Figure 12.34 Hydrolytic stability of potting compounds. Materials showing rapid hardness loss will soften similarly after two to four years at ambient temperatures in a high-humidity, tropical climate.[42]

Reversion is usually much faster in flexible materials, because water permeates more easily.

Structural adhesives not susceptible to the reversion phenomenon are also likely to lose adhesive strength when exposed to moisture. The degradation curves shown in Fig. 12.35 are typical for an adhesive exposed to moist, high-temperature environments. The mode of failure in the initial stages of aging is usually truly cohesive. After aging, the failure becomes one of adhesion. It is expected that water vapor permeates the adhesive through its exposed edges and concentrates in weak boundary layers at the interface. This effect is greatly dependent on the type of adhesive and substrate.

Stress accelerates the effect of environments on the adhesive joint. Little data are available on this phenomenon because of the time and expense associated with stress-aging tests. However, it is known that moisture, as an environmental burden, markedly decreases the ability of an adhesive to bear prolonged stress. Figure 12.36 illustrates the effect of stress aging on specimens exposed to relative humidity cycling from 90 to 100% and simultaneous temperature cycling from 80 to

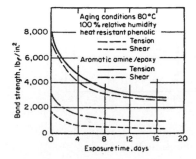

Aging conditions 80°C
100% relative humidity
heat resistant phenolic
————— Tension
————— Shear
Aromatic amine/epoxy
————— Tension
————— Shear

Bond strength, lbf/in²

Exposure time, days

Figure 12.35 Effect of humidity on adhesion of two structural adhesives to stainless steel.[43]

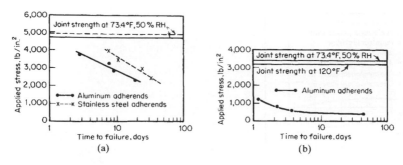

Figure 12.36 Time to failure vs. stress for two adhesives in a warm, high-humidity environment. (a) Adhesive = one-part, heat curing, modified epoxy. (b) Adhesive = flexibilized, amine-cured epoxy.[44]

120°F. The loss of load-bearing ability of a certain flexibilized epoxy adhesive (Fig. 12.36) is exceptional. The stress on this particular adhesive had to be reduced to 13% of its original strength for the joint to last a little more than 44 days in the test environment.

12.6.4 Outdoor Weathering

The most detrimental factors influencing adhesives aged outdoors are heat and humidity. Thermal cycling, ultraviolet radiation, and cold are relatively minor factors. The reasons why warm, moist climates degrade adhesive joints were presented in the last section.

When exposed to weather, structural adhesives rapidly lose strength during the first six months to one year. After two to three years, the rate of decline usually levels off, depending on the climate zone, adherend, adhesive, and stress level. Figure 12.37 shows the

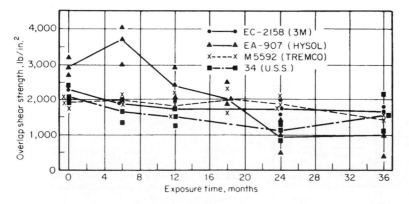

Figure 12.37 Effect of outdoor weathering on typical aluminum joints made with four different two-part epoxies cured at room temperature.[25]

weathering characteristics of unstressed epoxy adhesives to the Richmond, Virginia, climate.

The following generalizations are of importance in designing an adhesive joint for outdoor service:

1. The most severe locations are those with high humidity and warm temperatures.

2. Stressed panels deteriorate more rapidly than unstressed panels.

3. Stainless-steel panels are more resistant than aluminum panels because of corrosion.

4. Heat-cured adhesive systems are generally more resistant than room-temperature-cured systems.

5. With the better adhesives, unstressed bonds are relatively resistant to severe outdoor weathering, although all joints will eventually exhibit some strength loss.

MIL-STD-304 is a commonly used accelerated-exposure technique to determine the effect of weathering and high humidity on adhesive specimens. Adhesive comparisons can be made with this type of test. In this procedure, bonded panels are exposed to alternating cold ($-65°F$), dry heat ($160°F$), and heat and humidity ($160°F$, 95% RH) for 30 days. The effect of MIL-STD-304 conditioning on the joint strength of common structural adhesives is presented in Table 12.37.

12.6.5 Chemicals and Solvents

Most organic adhesives tend to be susceptible to chemicals and solvents, especially at elevated temperatures. Standard test fluids and

TABLE 12.37   Effect of MIL-STD-304 Aging on Bonded Aluminum Joints (from Ref. 45)

Adhesive	Shear, lb/in.², 73°F		Shear, lb/in.², 160°F	
	Control	Aged	Control	Aged
Room-temp. cured:				
Epoxy–polyamide	1,800	2,100	2,700	1,800
Epoxy–polysulfide	1,900	1,640	1,700	6,070
Epoxy–aromatic amine	2,000	Failed	720	Failed
Epoxy–nylon	2,600	1,730	220	80
Resorcinol epoxy–polyamide	3,500	3,120	3,300	2,720
Epoxy–anhydride	3,000	920	3,300	1,330
Polyurethane	2,600	1,970	1,600	1,560
Cured 45 min at 330°F, epoxy–phenolic	2,900	2,350	2,900	2,190
Cured 1 h at 350°F:				
Modified epoxy	4,900	3,400	4,100	3,200
Nylon–phenolic	4,600	3,900	3,070	2,900

immersion conditions are used by adhesive suppliers and are defined in MMM-A-132. Unfortunately, exposure tests lasting less than 30 days are not applicable to many requirements. Practically all adhesives are resistant to these fluids over short time periods and at room temperatures. Some epoxy adhesives even show an increase in strength during aging in fuel or oil. This effect is possibly due to a postcuring or plasticizing of the epoxy by oil.

Epoxy adhesives are generally more resistant to a wide variety of liquid environments than other structural adhesives. However, the resistance to a specific environment is greatly dependent on the type of epoxy curing agent used. Aromatic amine, such as metaphenylene diamine, cured systems are frequently preferred for long-term chemical resistance.

There is no "best adhesive" for universal chemical environments. As an example, maximum resistance to bases almost axiomatically means poor resistance to acids. It is relatively easy to find an adhesive that is resistant to one particular chemical environment. It becomes more difficult to find an adhesive that will not degrade in two widely differing chemical environments. Generally, adhesives that are most resistant to high temperatures have the best resistance to chemicals and solvents.

The temperature of the immersion medium is a significant factor in the aging properties of the adhesive. As the temperature increases, more fluid is generally absorbed by the adhesive, and the degradation rate increases.

From the rather limited information reported in the literature, it may be summarized that

1. Chemical-resistance tests are not uniform in concentrations, temperature, time, properties measured.

2. Generally, chlorinated solvents and ketones are severe environments.

3. High-boiling solvents, such as dimethylformamide and dimethyl sulfoxide, are severe environments.

4. Acetic acid is a severe environment.

5. Amine curing agents for epoxies are poor in oxidizing acids.

6. Anhydride curing agents are poor in caustics.

12.6.6 Vacuum

The ability of an adhesive to withstand long periods of exposure to a vacuum is of primary importance for certain applications. Loss of low-molecular-weight constituents such as plasticizers or diluents could result in hardening and porosity of cured adhesives or sealants.

Since most structural adhesives are relatively high-molecular-weight polymers, exposure to pressures as low as 10^{-9} torr is not harmful. However, high temperatures, radiation, or other degrading environments may cause the formation of low-molecular-weight fragments, which tend to bleed out of the adhesive in a vacuum.

Epoxy and polyurethane adhesives are not appreciably affected by 10^{-9} torr for seven days at room temperature. However, polyurethane adhesives exhibit significant outgassing when aged under 10^{-9} torr at 225°F.

12.6.7   Radiation

High-energy particulate and electromagnetic radiation including neutron, electron, and gamma radiation have similar effects on organic adhesives. Radiation causes molecular-chain scission of the adhesive, which results in weakening and embrittlement of the bond. This degradation is worsened when the adhesive is simultaneously exposed to elevated temperatures and radiation.

Figure 12.38 illustrates the effect of radiation dosage on the tensile-shear strength of structural adhesives. Generally, heat-resistant adhesives have been found to resist radiation better than less thermally stable systems. Fibrous reinforcement, fillers, curing agents, and reactive diluents affect the radiation resistance of adhesive systems. In epoxy-based adhesives, aromatic curing agents offer greater radiation resistance than aliphatic-type curing agents.

12.7   Processing and Quality Control of Adhesive Joints

Processing and quality control are usually the final considerations in the design of an adhesive-bonding system. These decisions are very important, however, because they alone may (1) restrict the degrees of freedom in designing the end product, (2) determine the types and number of adhesives that can be considered, (3) affect the quality and reproducibility of the joint, and (4) affect the total assembly cost.

12.7.1   Measuring and Mixing

When a multiple-part adhesive is used, the concentration ratios have a significant effect on the quality of the joint. Strength differences caused by varying curing-agent concentration are most noticeable when the joints are tested at elevated temperatures or after exposure to water or solvents. Exact proportions of resin and hardener must be

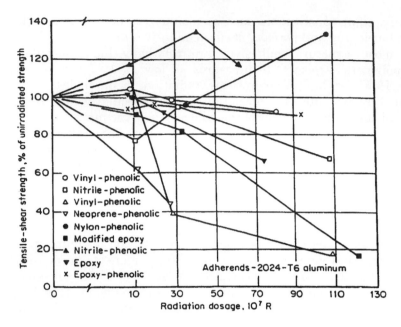

Figure 12.38 Percent change of initial tensile-shear strength caused by nuclear radiation dosage.[46]

weighed out on an accurate balance or in a measuring container for best adhesive quality and reproducibility.

The weighed-out components must be mixed thoroughly. Mixing should be continued until no color streaks or density stratifications are noticeable. Caution should be taken to prevent air from being mixed into the adhesive through overagitation. This can cause foaming of the adhesive during heat cure, resulting in porous bonds. If air does become mixed into the adhesive, vacuum degassing may be necessary before application.

Only enough adhesive should be mixed to work with before the adhesive begins to cure. Working life of an adhesive is defined as *the period of time during which an adhesive remains suitable for use after mixing with catalyst.* Working life is decreased as the ambient temperature increases and as the batch size becomes larger. One-part adhesives, and some heat-curing, two-part adhesives, have very long working lives at room temperature, and application and assembly speed or batch size are not critical.

For a large-scale bonding operation, hand mixing is costly, messy, and slow, and repeatability is entirely dependent on the operator. Equipment is available that can meter, mix, and dispense multicomponent adhesives on a continuous or shot basis.

12.7.2   Application of Adhesives

The selection of an application method depends primarily on the form of the adhesive: liquid, paste, powder, or film. Table 12.38 describes the advantages and limitations realized in using each of the four basic forms. Other factors influencing the application method are the size and shape of parts to be bonded, the total area where the adhesive is to be applied, and production volume and rate.

12.7.2.1   Liquids.   Liquids, the most common form of adhesive, can be applied by a variety of methods. Brush, simple rollers, and glue guns are manual methods that provide simplicity, low cost, and versatility. Spray, dipping, and mechanical roll coaters are generally used on large production runs. Mechanical-roller methods are commonly used to apply a uniform layer of adhesive to a continuous roll or coil. Such automated systems are used with adhesives that have a long working life and low viscosity. Spray methods can be used on both small and large production runs. The spray adhesive is generally in solvent solution, and sizable amounts of adhesive may be lost from overspray. Two-component adhesives are usually mixed prior to placing in the spray-gun reservoir. Application systems are available, however, that meter and mix the adhesive in the spray-gun barrel. This is ideal for fast-reacting systems.

12.7.2.2   Pastes.   Bulk adhesives such as pastes and mastics are the simplest and most reproducible adhesive to apply. These systems can be troweled or extruded through a caulking gun. Little operator skill is required. Since the thixotropic nature of the paste prevents it from flowing excessively, application is usually clean, and little waste is generated.

12.7.2.3   Powders.   Powder adhesives can be applied in three ways. They may be sifted onto a preheated substrate. The powder that falls onto the substrate melts and adheres. The assembly is then mated and cured according to recommended processes. A preheated substrate could also be dipped into a fluidized bed of the powder and then extracted with an attached coating of adhesive. This method helps to ensure an even distribution of powder. Finally, the powder can be melted into a paste or liquid and applied by conventional means.

Powder adhesives are generally one-part, epoxy-based systems that require heat and pressure to cure. They do not require metering and mixing but often must be refrigerated for extended shelf life.

TABLE 12.38 Characteristics of Various Adhesive Application Methods.[47]

Application method	Viscosity	Operator skill	Production rate	Equipment cost	Coating uniformity	Material loss
Liquid						
Manual, brush or roller	Low to medium	Little	Low	Low	Poor	Low
Roll coating, reverse, gravure	Low	Moderate	High	High	Good	Low
Spray, manual, automatic, airless, or external mix	Low to high	Moderate to high	Moderate to high	Moderate to high	Good	Low to high
Curtain coating	Low	Moderate	High	High	Good to excellent	Low
Bulk						
Paste and mastic	High	Little	Low to moderate	Low	Fair	Low to high
Powder						
Dry or liquid primed	....	Moderate	Low	High	Poor to fair	Low
Dry Film	....	Moderate to high	Low to high	Low to high	Excellent	Lowest

12.7.2.4 Films. Dry adhesive films have the following advantages:

1. High repeatability—no mixing or metering, constant thickness

2. Easy to handle—low equipment cost, relatively hazard-free, clean operating

3. Very little waste—preforms can be cut to size

4. Excellent physical properties—wide variety of adhesive types available

Films are limited to flat surfaces or simple contours. Application requires a relatively high degree of care to ensure nonwrinkling and removal of separator sheets. Characteristics of available film adhesives vary widely, depending on the type of adhesive used.

Film adhesives are made as both unsupported and supported types. The carrier for supported films is generally fibrous fabric or mat. Film adhesives are supplied in heat-activated, pressure-sensitive, or solvent-activated forms. Solvent-activated adhesives are made tacky and pressure-sensitive by wiping with solvent. They are not as strong as the other types but are well suited for contoured, curved, or irregularly shaped parts. Manual solvent-reactivation methods should be closely monitored so that excessive solvent is not used. Solvent-activated films include neoprene, nitrile, and butyral phenolics. Decorative trim and nameplates are usually fastened onto a product with solvent-activated adhesives.

12.7.3 Bonding Equipment

After the adhesive is applied, the assembly must be mated as quickly as possible to prevent contamination of the adhesive surface. The substrates are held together under pressure and heated if necessary until cure is achieved. The equipment required to perform these functions must provide adequate heat and pressure, maintain the prescribed pressure during the entire cure cycle, and distribute pressure uniformly over the bond area. Of course, many adhesives require only simple contact pressure at room temperature, and extensive bonding equipment is not necessary.

12.7.3.1 Pressure equipment. Pressure devices should be designed to maintain constant pressure on the bond during the entire cure cycle. They must compensate for thickness reduction from adhesive flow or thermal expansion of assembly parts. Thus, screw-actuated devices like C-clamps and bolted fixtures are not acceptable when a constant pressure is important. Spring pressure can often be used to supplement clamps and compensate for thickness variations. Dead-weight loading may be applied in many instances; however, this method is sometimes impractical, especially when heat cure is necessary.

Pneumatic and hydraulic presses are excellent tools for applying constant pressure. Steam or electrically heated platen presses with hydraulic rams are often used for adhesive bonding. Some units have multiple platens, thereby permitting the bonding of several assemblies at one time.

Large bonded areas such as on aircraft parts are usually cured in an autoclave. The parts are mated first and covered with a rubber blanket to provide uniform pressure distribution. The assembly is then placed in an autoclave, which can be pressurized and heated. This method requires heavy capital equipment investment.

Vacuum-bagging techniques can be an inexpensive method of applying pressure to large parts. A film or plastic bag is used to enclose the assembly, and the edges of the film are sealed airtight. A vacuum is drawn on the bag, enabling atmospheric pressure to force the adherends together. Vacuum bags are especially effective on large areas because size is not limited by equipment.

12.7.3.2 Heating equipment. Many structural adhesives require heat as well as pressure. Most often, the strongest bonds are achieved by an elevated-temperature cure. With many adhesives, trade-offs between cure times and temperature are permissible. But, generally, the manufacturer will recommend a certain curing schedule for optimal properties.

If, for example, a cure of 60 min at 300°F is recommended, this does not mean that the assembly should be placed in a 300°F oven for 60 min. It is the bond line that should be at 300°F for 60 min. Total oven time would be 60 min plus whatever time is required to bring the assembly up to 300°F. Large parts act as a heat sink and may require substantial time for an adhesive in the bond line to reach the necessary temperature. Bond-line temperatures are best measured by thermocouples placed very close to the adhesive. In some cases, it may be desirable to place the thermocouple in the adhesive joint for the first few assemblies being cured.

Oven heating is the most common source of heat for bonded parts, even though it involves long curing cycles because of the heat-sink action of large assemblies. Ovens may be heated with gas, oil, electricity, or infrared units. Good air circulation within the oven is mandatory to prevent nonuniform heating.

Heated-platen presses are good for bonding flat or moderately contoured panels when faster cure cycles are desired. Platens are heated with steam, hot oil, or electricity and are easily adapted with cooling-water connections to further speed the bonding cycle.

12.7.3.3 Adhesive-thickness control.

It is highly desirable to have a uniformly thin (2- to 10-mil) adhesive bond line. Starved adhesive joints, however, will yield exceptionally poor properties. Three basic methods are used to control adhesive thickness. The first method is to use mechanical shims or stops that can be removed after the curing operation. Sometimes it is possible to design stops into the joint.

The second method is to employ a film adhesive that becomes highly viscous during the cure cycle preventing excessive adhesive flow-out. With supported films, the adhesive carrier itself can act as the *shims*. Generally, the cured bond-line thickness will be determined by the original thickness of the adhesive film. The third method of controlling adhesive thickness is to use trial and error to determine the correct pressure-adhesive viscosity factors that will yield the desired bond thickness.

12.7.4 Quality Control

A flowchart of a quality-control system for a major aircraft company is illustrated in Fig. 12.39. This system is designed to ensure reproducible bonds and, if a substandard bond is detected, to make suitable corrections. Quality control should cover all phases of the bonding cycle from inspection of incoming material to the inspection of the completed assembly. In fact, good quality control will start even before receipt of materials.

12.7.4.1 Prebonding conditions.

The human element enters the adhesive-bonding process more than in other fabrication techniques. An extremely high percentage of defects can be traced to poor workmanship. This generally prevails in the surface-preparation steps but may also arise in any of the other bonding steps. This problem can be largely overcome by proper motivation and education. All employees from design engineer to laborer to quality-control inspector should be somewhat familiar with adhesive-bonding technology and be aware of the circumstances that can lead to poor joints.

The plant's bonding area should be as clean as possible prior to receipt of materials. The basic approach to keeping the assembly area clean is to segregate it from the other manufacturing operations either in a corner of the plant or in isolated rooms. The air should be dry and filtered to prevent moisture or other contaminants from gathering at a possible interface. The cleaning and bonding operations should be sep-

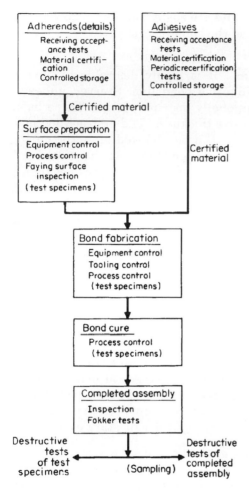

Figure 12.39 Flowchart of a quality-control system for adhesive bonding.[48]

arated from each other. If mold release is used to prevent adhesive flash from sticking to bonding equipment, it is advisable that great care be taken to ensure that the release does not contaminate the adherends. Spray mold releases, especially silicone release agents, have a tendency to migrate to undesirable areas.

12.7.4.2 Quality control of adhesive and surface treatment. Acceptance tests on adhesives should be directed toward assurance that incoming materials are identical from lot to lot. The tests should be those that can quickly and accurately detect deficiencies in the adhesive's physical or chemical properties. ASTM lists various test methods that are commonly used for adhesive acceptance. Actual test specimens should also be made to verify strength of the adhesive. These specimens should be stressed in directions that are representative of the forces that the bond will see in service, i.e., shear, peel, tension, or cleavage. If possible, the specimens should be prepared and cured in the same manner as actual production assemblies. If time permits, specimens should also be tested in simulated service environments, e.g., high temperature, humidity.

Surface preparations must be carefully controlled for reliable production of adhesive-bonded parts. If a chemical surface treatment is required, the process must be monitored for proper sequence, bath temperature, solution concentration, and contaminants. If sand or grit blasting is employed, the abrasive must be changed regularly. An adequate supply of clean wiping cloths for solvent cleaning is also mandatory. Checks should be made to determine if cloths or solvent containers have become contaminated.

The specific surface preparation can be checked for effectiveness by the water-break free test. After the final treating step, the substrate surface is checked for a continuous film of water that should form when deionized water drops are placed on the surface.

After the adequacy of the surface treatment has been determined, precautions must be taken to ensure that the substrates are kept clean and dry until bonding. The adhesive or primer should be applied to the treated surface as quickly as possible.

12.7.4.3 Quality control of the bonding process. The adhesive metering and mixing operation should be monitored by periodically sampling the mixed adhesive and testing it for adhesive properties. A visual inspection can also be made for air entrapment and degree of mixing. The quality-control engineer should be sure that the oldest adhesive is used first and that the specified shelf life has not been exceeded.

During the actual assembly operation, the cleanliness of the shop and tools should be verified. The shop atmosphere should be controlled as closely as possible. Temperature in the range of 65 to 90°F and relative humidity from 20 to 65 percent is best for almost all bonding operations.

The amount of the applied adhesive and the final bond-line thickness must also be monitored, because they can have a significant effect on joint strength. Curing conditions should be monitored for heat-up rate, maximum and minimum temperature during cure, time at the required temperature, and cool-down rate.

12.7.4.4   Bond inspection.   After the adhesive is cured, the joint can be inspected to detect gross flaws or defects. This inspection procedure can be either destructive or nondestructive in nature. Destructive testing generally involves placing samples in simulated or accelerated service and determining if they have similar properties to a specimen that is known to have a good bond and adequate service performance. The causes and remedies for faults revealed by such mechanical tests are described in Table 12.39. Nondestructive testing (NDT) is far more economical, and every assembly can be tested if desired. Great amounts of energy are now being devoted to improve NDT techniques.

TABLE 12.39   Faults Revealed by Mechanical Tests

Fault	Cause	Remedy
Thick, uneven glue line	Clamping pressure too low	Increase pressure. Check that clamps are seating properly
	No follow-up pressure	Modify clamps or check for freedom of moving parts
	Curing temperature too low	Use higher curing temperature. Check that temperature is above the minimum specified throughout the curing cycle
	Adhesive exceeded its shelf life, resulting in increased viscosity	Use fresh adhesive
Adhesive residue has spongy appearance or contains bubbles	Excess air stirred into adhesive	Vacuum-degas adhesive before application
	Solvents not completely dried out before bonding	Increase drying time or temperature. Make sure drying area is properly ventilated
	Adhesive material contains volatile constituent	Seek advice from manufacturers
	A low-boiling constituent boiled away	Curing temperature is too high
Voids in bond (i.e., areas that are not bonded), clean bare metal exposed, adhesive failure at interface	Joint surfaces not properly treated	Check treating procedure; use clean solvents and wiping rags. Wiping rags must not be made from synthetic fiber. Make sure cleaned parts are not touched before bonding. Cover stored parts to prevent dust from settling on them
	Resin may be contaminated	Replace resin. Check solids content. Clean resin tank
	Uneven clamping pressure	Check clamps for distortion
	Substrates distorted	Check for distortion; correct or discard distorted components. If distorted components must be used, try adhesive with better gap-filling ability
Adhesive can be softened by heating or wiping with solvent	Adhesive not properly cured	Use higher curing temperature or extend curing time. Temperature and time must be above the minimum specified throughout the curing cycle. Check mixing ratios and thoroughness of mixing. Large parts act as a heat sink, necessitating larger cure times

12.7.4.5 Nondestructive-testing procedures

Visual inspection. A trained eye can detect a surprising number of faulty joints by close inspection of the adhesive around the bonded area. Table 12.40 lists the characteristics of faulty joints that can be detected visually. The most difficult defect to be found by any way are those related to improper curing and surface treatment. Therefore, great care and control must be given to surface-preparation procedures and shop cleanliness.

Sonic inspection. Sonic and ultrasonic methods are at present the most popular NDT technique for use on adhesive joints. Simple tapping of a bonded joint with a coin or light hammer can indicate an unbonded area. Sharp, clear tones indicate that adhesive is present and adhering to the substrate in some degree; dull, hollow tones indicate a void or unattached area. Ultrasonic testing basically measures the response of the bonded joint to loading by low-power ultrasonic energy.

Other NDT methods. Radiography (X-ray) inspection can be used to detect voids or discontinuities in the adhesive bond. This method is more expensive and requires more skilled experience than ultrasonic methods.

Thermal-transmission methods are relatively new techniques for adhesive inspection. Liquid crystals applied to the joint can make voids visible if the substrate is heated. This test is simple and inex-

TABLE 12.40 Visual Inspection for Faulty Bonds

Fault	Cause	Remedy
No appearance of adhesive around edges of joint or adhesive bond line too thick	Clamping pressure roo low	Increase pressure. Check that clamps are seating properly
	Starved joint	Apply more adhesive
	Curing temperature too low	Use higher curing temperature. Check that temperature is above the minimum specified
Adhesive bond line too thin	Clamping pressure too high	Lessen pressure
	Curing temperature too high	Use lower curing temperature
	Starved joint	Apply more adhesive
Adhesive flash breaks easily away from substrate	Improper surface treatment	Check treating procedure; use clean solvents and wiping rags. Make sure cleaned parts are not touched before bonding
Adhesive flash is excessively porous	Excess air stirred into adhesive	Vacuum-degas adhesive before application
	Solvent not completely dried out before bonding	Increase drying time or temperature
	Adhesive material contains volatile constituent	Seek advice from manufacturers
Adhesive flash can be softened by heating or wiping with solvent	Adhesive not properly cured	Use higher curing temperature or extend curing time. Temperature and time must be above minimum specified. Check mixing

pensive, although materials with poor heat-transfer properties are difficult to test, and the joint must be accessible from both sides. An infrared inspection technique has also been developed for detection of internal voids and nonbonds. This technique is somewhat expensive, but it can accurately determine the size and depth of the flaw.

The science of holography has also been used for NDT of adhesive bonds. Holography is a method of producing photographic images of flaws and voids using coherent light such as that produced by a laser. The major advantage of holography is that it photographs successive "slices" through the scene volume. A true three-dimensional image of a defect or void can then be reconstructed.

12.7.5   Environmental and Safety Concerns

Four primary safety factors must be considered in all adhesive bonding operations: toxicity, flammability, hazardous incompatibility, and equipment.

All adhesives, solvents, chemical treatments, and so forth must be handled in a manner preventing toxic exposure to the work force. Methods and facilities must be provided to ensure that the maximum acceptable concentrations of hazardous materials are never exceeded. These values are prominently displayed on the material's Material Safety Data Sheet (MSDA), which must be maintained and available for the workforce.

Where flammable solvents and adhesives are used, they must be stored, handled, and used in a manner that prevents any possibility of ignition. Proper safety containers, storage areas, and well ventilated workplaces are required.

Certain adhesive materials are hazardous when mixed together. Epoxy and polyester catalysts, especially, must be well understood prior to departing from the manufacturers' recommended procedure for mixing. Certain unstabilized solvents, such as trichloroethylene and perchloroethylene, are subject to chemical reaction on contact with oxygen or moisture. Only stabilized grades of solvents should be used.

Certain adhesive systems, such as heat-curing epoxy and room-temperature-curing polyesters, can develop very large exothermic reactions on mixing. The temperature generated during this exotherm depends on the mass of the material being mixed. Exotherm temperatures can get so high that the adhesive will catch fire and burn. Adhesive products should always be applied in thin bond lines to minimize the exotherm until the chemistry of the product is well understood.

Safe equipment and proper operation is, of course, crucial to a workplace. Sufficient training and safety precautions must be installed in the factory before the bonding process is established.

References

1. Structural Adhesives, Minnesota Mining and Manufacturing Co., Adhesives Coatings and Sealers Division, *Technical Bulletin.*
2. Powis, C. N., Some Applications of Structural Adhesives, in D. J. Alner, ed., *Aspects of Adhesion,* vol. 4, University of London Press, Ltd., London 1968.
3. Bikerman, J. J., Causes of Poor Adhesion, *Ind. Eng. Chem.*, September, 1967.
4. Reinhart, F. W., Survey of Adhesion and Types of Bonds Involved, in J. E. Rutzler and R. L. Savage, eds., *Adhesion and Adhesives Fundamentals and Practices,* Society of Chemical Industry, London, 1954.
5. Merriam, J. C., Adhesive Bonding, *Mater. Des. Eng.*, September, 1959.
6. Rider, D. K., Which Adhesives for Bonded Metal Assembly, *Prod. Eng.*, May 25, 1964.
7. Perry, H. A., Room Temperature Setting Adhesives for Metals and Plastics, in J. E. Rutzler and R. L. Savage, eds., *Adhesion and Adhesives Fundamentals and Practices,* Society of Chemical Industry, London, 1954.
8. Lunsford, L. R., Design of Bonded Joints, in M. J. Bodnar, ed., *Symposium on Adhesives for Structural Applications,* Interscience, New York, 1962.
9. Koehn, G. W., Design Manual on Adhesives, *Machine Design,* April, 1954.
10. "Adhesive Bonding Alcoa Aluminum," Aluminum Co. of America, 1967.
11. DeLollis, N. J., "Adhesives for Metals Theory and Technology," Industrial Press, New York, 1970.
12. Chessin, N., and V. Curran, Preparation of Aluminum Surface for Bonding, in M. J. Bodnar, ed., *Structural Adhesive Bonding,* Interscience, New York, 1966.
13. Muchnick, S. N., Adhesive Bonding of Metals, *Mech. Eng.*, January, 1956.
14. Vazirani, H. N., Surface Preparation of Steel and Its Alloys for Adhesive Bonding and Organic Coatings, *J. Adhesion,* July, 1969.
15. Vazirani, H. N., Surface Preparation of Copper and Its Alloys for Adhesive Bonding and Organic Coatings, *J. Adhesion,* July, 1969.
16. Walter, R. E., D. L. Voss, and M. S. Hochberg, Structural Bonding of Titanium for Advanced Aircraft, *Nat. SAMPE Tech. Conf. Proc.*, vol 2, Aerospace Adhesives and Elastomers, 1970.
17. Jennings, C. W., Surface Roughness and Bond Strength of Adhesives, *J. Adhesion,* May, 1972.
18. Bersin, R.L., How to Obtain Strong Adhesive Bonds via Plasma Treatment, *Adhesive Age,* March, 1972.
19. Rogers, N. L., Surface Preparation of Metals, in M. J. Bodnar, ed., *Structural Adhesive Bonding,* Interscience, New York, 1966.
20. Krieger, R. B., Advances in Corrosion Resistance of Bonded Structures, *Nat. SAMPE Tech. Conf. Proc.*, vol 2, Aerospace Adhesives and Elastomers, 1970.
21. Cagle, C. V., *Adhesive Bonding Techniques and Applications,* McGraw-Hill, New York, 1968.
22. Landrock, A. H., *Adhesives Technology Handbook,* Noyes Publications, New Jersey, 1985.
23. Guttmann, W. H., *Concise Guide to Structural Adhesives,* Reinhold, New York, 1961.
24. "Preparing the Surface for Adhesive Bonding," Hysol Div., Dexter Corp., Bull. G1-600.
25. *Adhesive Bonding Aluminum,* Reynolds Metal Co., 1966.
26. Schields, J., *Adhesives Handbook,* CRC Press, 1970.
27. Devine, A. T., and M. J. Bodnar, Effects of Various Surface Treatments on Adhesive Bonding of Polyethylene, *Adhesives Age,* May, 1969.
28. Austin, J. E., and L. C. Jackson, Management: Teach Your Engineers to Design Better With Adhesives, *SAE J.*, October, 1961.
29. Burgman, H. A., Selecting Structural Adhesive Materials, *Electrotechnol.*, June, 1965.
30. Three New Cyanoacrylate Adhesives, Eastman Chemical Co., Leaflet R-206A.

31. TAME, A New Concept in Structural Adhesives, B. Goodrich General Products Co. Bull. GPC-72-AD-3.
32. Information about Pressure Sensitive Adhesives, Dow Corning, Bull. 02-032.
33. Information about Silastic RTV Silicone Rubber, Dow Corning, Bull. 61-015a.
34. Bruno, E. J., ed., *Adhesives in Modern Manufacturing*, p. 29, Society of Manufacturing Engineers, 1970.
35. Lichman, J., Water-based and Solvent-based Adhesives, in I. Skeist, ed., *Handbook of Adhesives,* Reinhold, New York, 1977.
36. Hemming, C. B., Wood Gluing, in I. Skeist, ed., *Handbook of Adhesives*, 1/e, Reinhold, New York, 1962.
37. Moser, F., Bonding Glass, in I. Skeist, ed., *Handbook of Adhesives*, 1/e, Reinhold, New York, 1962.
38. Weggemans, D. M., Adhesive Charts, in *Adhesion and Adhesives,* vol. 2, Elsevier, Amsterdam, 1967.
39. Krieger, R. B. and R. E. Politi, High Temperature Structural Adhesives, in D. J. Alner, ed., *Aspects of Adhesion,* vol. 3, University of London Press, London, 1967.
40. Burgman, H. A., The Trend in Structural Adhesives, *Machine Design*, November 21, 1963.
41. Kausen, R. C., Adhesives for High and Low Temperature, *Mater. Eng.*, August-September, 1964.
42. Bolger, J. C., New One-Part Epoxies are Flexible and Reversion Resistant, *Insulation*, October, 1969.
43. Falconer, D. J., et. al., The Effect of High Humidity Environments on the Strength of Adhesive Joints, *Chem. Ind.*, July 4, 1964.
44. Sharpe, L. H., Aspects of the Permanence of Adhesive Joints, in M. J. Bodnar, ed., *Structural Adhesive Bonding,* Interscience, New York, 1966.
45. Tanner, W. C., Adhesives and Adhesion in Structural Bonding for Military Material, in M. J. Bodnar, ed., *Structural Adhesive Bonding,* Interscience, New York, 1966.
46. Arlook, R. S., and D. G. Harvey, Effects of Nuclear Radiation on Structural Adhesive Bonds, Wright Air Development Center, Report WADC-TR-46-467.
47. Carroll, K. W., How to Apply Adhesive, *Prod. Eng.*, November 22, 1965.
48. Smith, D. F., and C. V. Cagle, A Quality Control System for Adhesive Bonding Utilizing Ultrasonic Testing, in M. J. Bodnar, ed., *Structural Adhesive Bonding,* Interscience, New York, 1966.

13

Testing of Materials

Perry L. Martin
Martin Testing Laboratories, Inc.
Yuba City, California

13.1 Introduction

Any product is an assemblage of various components, packaging and interconnection materials, and technologies. The reliability of even a "well designed" product is usually reduced by some finite number of defects in materials imparted by "normal manufacturing processes." To achieve the level of reliability inherent in the design, these defects must be detected and fixed before the product leaves the factory. If not corrected, the defects will show up as fielded product failures, which increase warranty costs and reduce the perceived quality and value to the customer. Additionally, manufacturing production costs will increase due to rework and repair if corrective actions are not taken. It is for these reasons that manufacturers have a requisite interest in the most effective means for the earliest elimination of defects.

Most manufacturers traditionally have depended on the final acceptance testing to catch manufacturing defects. They have learned over time to add intermediate material tests, electrical tests, screens, and inspections as part of the continuing course of manufacturing process improvement. As parts fail, failure analysis can provide valuable feedback as to the type of test, inspection, or screen to use; duration or resolution required; and placement within the production process. The optimization of the overall reliability program will effectively cover the material, component, and assembly fault spectrum while preventing redundant coverage and maximizing the resulting cost savings.

It must be emphasized that, for the purposes of testing, the materials cannot be considered independent of the product or the production processes used to produce the product. Given the almost infinite number of material/product combinations, we cannot hope to cover all of the possible materials tests. ASTM alone produces multiple volumes of test procedures covering thousands of pages. Materials testing will be the focus of this chapter, we will introduce some of the thought processes that go into designing a test program and review some of the more prevalent tests.

For example, when trying to determine which test or inspection technique is best, and most cost-effective, for identifying a particular type of defect, the following points must be considered:

- The complexity of the technology or material being tested/inspected
- The volume to be tested/inspected
- The per piece cost
- The required accuracy for defect location

These ideas will be developed in the following section.

13.1.1   Designing a Test Regime[1]

13.1.1.1   General considerations. Paint and coating materials will be used as an example of the materials and process considerations that go into designing a test regimen. Painting and coating is a process involving materials, surface preparation, and how the coating materials are applied. All three of these are linked; a defect in one causes the entire process to fail.

Some reasons for the development of a new test program might include the following:

- Development of a new process to apply a coating and selection of appropriate materials
- Problems with an existing process and/or existing materials
- Changes in existing processes and existing materials because of new environmental restrictions or unavailability of desired materials
- Qualification or evaluation of a new product

Essentially, all four of these will require consultation with the process personnel, end users, and testing personnel. The laboratory facilities required can be in-house or outside sources but, either way, communication is essential. There are situations in which the problems can be solved by consultation without doing any laboratory work. However, a true solution usually involves both.

There are three basic steps:

1. Define the problem or area of interest so that both you and the test personnel clearly understand what is generally required. We want to be able to clearly understand the objectives of the testing. This involves discussions among the "experts" on all sides. We cannot propose the best effort and minimize the costs until we have a mutual understanding at this point. This may be very difficult initially, because of the backgrounds of the parties and the need of the "customer" to not divulge essential information at this time. There may be a need to sign nondisclosure agreements before the free flow of information needed to attain this understanding can begin.

2. Establish a general plan of effort. This may involve experimental designs or some of the newer methods to gain the objectives at the least cost. There can be substantial cost savings by utilizing one "sample" for many tests and evaluations. If multiple test facilities are being used, determine who will be prime, and manage the effort.

3. Cost out the plan, prepare proposals, describe a general overall plan of effort, and submit it. The proposal should define what is intended to be delivered.

If the "customer" has a shopping list of ASTM tests, there still needs to be an understanding on these issues:

- *Just what does the customer provide in materials and test items?* If the customer provides a set of coated panels, then there may be problems. Cutting up panels to meet ASTM test requirements is poor practice, since the cutting up damages the panel. Some of the ASTM tests are valid only when they are done on a certain type and size of panel, and on panels that have been through a very specific preparation process. The preferred method should always be to prepare the panels in the test laboratory, where quality control can be maintained on the whole process.

- *Does the customer want strict adherence to the tests or just general adherence?* For certification, strict adherence is required. It should be noted that there are some ASTM tests that are equipment and equipment-procedure specific. If the customer is having problems with an existing process or existing coating material, then "ad hoc" tests are a better choice. Here, the intent is to use testing to solve the problem, and conformance to a specific ASTM test is not important.

- *What does the test lab have to provide in terms of materials? What do they have to buy? What do they have to deliver? Is it just a series of simple test result sheets, or is it a comprehensive evaluation report?*

Example: The Painting/Coating Process

Surface pretreatment and preparation

    Removal of old coatings if required

    Surface cleaning

    Masking, barrier treatments

    Prepaint substrate treatments

Application

    Application equipment and tools

    Cleanup and waste disposal

    Mixing

    Personal protective equipment

    Area preparation

    Environment control

    Film thickness and coverage requirements

    Personnel skills and qualifications

    Application information, specifications, MSDSs, etc.

    Inspection requirements

    Curing, drying process requirements

Coating materials

    Sealers, fillers, undercoaters

    Primers

    Topcoats

    Thinners, reducers

13.1.1.2 Tests on coating materials. There are three areas in which tests on coating materials are made. Coating materials start off as liquids and, after application, end up as hard cured films on some substrate. You can't do liquid tests on coated panels, and you can't do cured film tests such as corrosion resistance on the liquid paint. Also, the *reverse engineering* process of trying to infer the process and the coating materials used (e.g., formulations) from the final cured coating is very expensive and, in most cases, is not successful.

Tests on coating materials are to some extent specific to a given industry. For example, coating materials used on automobiles and trucks are tested in accord with many automobile manufacturers' peculiar test methods. There are two general groups of coating material test methods—those put out by the ASTM under part 6, and the Federal test methods put out under Federal Test Method Standard 141. Most paint manufacturers have their own peculiar tests, which are not often known outside the company.

Certain classes of coatings have their own set of such tests. For example, tests on paints used on roadways have one set of tests, and paints used for the outside of houses have different sets of tests. In some cases, the tests may have similar names, but they are different. That is why it is difficult to generalize a proposed test program.

In the next section, a long list of test capabilities are presented. They are based on the ASTM and FTMS test sets. Not all of the ASTM or FTMS tests are given. This is only a subset, representing some of the more common ones. The list is divided into specific properties for which we are testing.

- FTMS is Federal Test Method Standard 141.

- A (T) after an ASTM document listing indicates that it is a *test method*.

- A (P) after an ASTM document listing indicates that it is a *recommended practice*.

- A (G) after an ASTM document listing indicates that it is only *guidance*.

The list below is based primarily on architectural and industrial maintenance coatings and the processes involved in their application. The list *does not* comprehensively cover ASTM tests on the following; however, the listed ASTM test may still be applicable in the special area:

- Materials used in graphic art applications
- Cleaning applications
- Exterior pavement and roadway coatings
- Factory applied surface coatings of miscellaneous metal parts, furniture, and wood products
- Aerospace component manufacturing and assembly
- Automotive, truck, and heavy equipment manufacturing and refinishing operations
- Pipeline coatings and applications

As needs develop in these other areas, the list can be expanded.

Typical Paint and Coating Test Program

Liquid Coating Properties

This generally requires a quart to a gallon of the coating material, depending on the extent of the testing.

Composition
   ASTM D2369 Volatile content of coatings (T)
   ASTM D4017 Water in paints by Karl Fisher method (T)
   EPA Method 24 Volatile organic compound content
   VOC content by headspace absorption and GC analysis

Physical properties
   ASTM D5201 Calculating physical constants of coatings (P)
   ASTM D1475 Density (T)
   ASTM D3278 D1310, D3941, D3278, D3934 Flash point (T)
   ASTM D6165 Odor (G)

Storage and aging
   ASTM D869 Settling (T)
   ASTM D1849 Package stability (T)
   ASTM D2243 Freeze-thaw for waterborne (T)
   ASTM D2337 Freeze-thaw for lacquers (T)
   ASTM D154 Skinning (G)
   ASTM D185 Coarse particles and foreign matter (T)
   ASTM D2574 Emulsion paints attack by microorganisms (T)
   ASTM D5588 Microbial conditions of coatings (T)
   FTMS Method 3011 Condition in container
   FTMS Method 3019 Storage stability at extreme temperatures
   FTMS Method 3021 Skinning
   FTMS Method 3022 Storage stability

Application and Film Formation Properties

This generally requires a quart to several gallons, depending on the number of tests that are required. Spray and roller application takes a lot of paint due to the size of the panels required.

Brushing properties
   FTMS Method 4321 Brushing properties
   ASTM D4958 Comparison of brush drag, latex paints (T)
   ASTM D344 Visual evaluation of brushouts (T)

Dry time

This can be done at any number of set environmental conditions of temperature and relative humidity.

ASTM D1640 Drying, curing and film formation (T)

ASTM D5895 Times of drying using mechanical recorders (T)

FTMS Method 4061 Drying time

Flow/consistency characteristics

ASTM D562 Consistency by Stormer viscometer (T)

ASTM D1200 Ford cup (T)

ASTM D2196 Brookfield viscometer (T)

ASTM D4212 Dip-type viscosity cups (T)

ASTM D4287 ICI cone/plate viscometer (T)

ASTM D5125 ISO flow cups (T)

ASTM D5478 Falling needle viscometer (T)

FTMS Method 4287 Brookfield viscosity

Hiding power and spreading (wet) rate

We generally combine hiding power testing with spreading rate, because it can all be done on one chart. This determines a characteristic S parameter of the coating and provides data on wet-film thickness required to achieve a specified dry-film thickness on coated panels.

ASTM D5007 Wet-to-dry hiding change (T)

ASTM D2805 Hiding power of paints by reflectometry (T)

Applicator height-wet-dry thickness relationships

FTMS Method 4122 Hiding power (contrast ratio)

Leveling

ASTM D4062 Leveling by draw-down (Leneta test) (T)

NYPC Leveling test (blade)

Pot life

This can be a simple flow/consistency measurement or a complete set of application tests at periodic intervals during the pot life time period.

Roller coating properties

FTMS Method 4335 Roller coating properties

Sag resistance

ASTM D4400 Using a multi-notch applicator (T)

FTMS Method 4493 Sag resistance, Baker method

FTMS Method 4494 Sag resistance, multinotch blade

Spatter resistance
ASTM D4707 Paint spatter resistance from roller application (T)

Spraying properties
FTMS Method 4331 Spraying properties

Wet film thickness
ASTM D1212 Measurement (T)
ASTM D4414 By notch-type gages (P)

Dried/Cured Film Properties

These are all tests done on flat panels. The number, the size, and the type of panel substrate are frequently dictated by the selected ASTM test. The preferred method is to do the coating of the panels in the lab, where the uniformity of film thickness can be controlled. In many cases, a number of tests can be done on the same panel. This allows for certain economies when tests are combined.

Adhesion to substrate
ASTM D2197 Scrape adhesion (T)
ASTM D3359 Tape test (wet) (T)
ASTM D4541 PATTI test (T)
ASTM D5179 Adhesion to plastic by direct tensile testing (T)
FTMS Method 6251 Lacquer lifting test
FTMS Method 6252 Self-lifting test
FTMS Method 6301 Wet tape test

Appearance and finish
These are done before and after any test that is likely to alter appearance, when appearance following the test is an important criterion. Outside of gloss and color, the tests are all subjective, using standards.
ASTM D523 Specular gloss (T)
ASTM D610 Degree of rusting (T)
ASTM D660 Degree of checking (P)
ASTM D661 Degree of cracking (T)
ASTM D662 Degree of erosion (T)
ASTM D714 Degree of blistering (T)
ASTM D772 Degree of flaking (scaling) (T)
ASTM D1654 Evaluation of specimens subject to corrosion (P)
ASTM D1729 Color differences (P)
ASTM D1848 Reporting film failures (C)
ASTM D2244 Color differences (T)

ASTM D2616 Color differences with a gray scale (T)

ASTM D3134 Color and gloss tolerances (P)

ASTM D3928 Gloss or sheen uniformity (T)

ASTM D4039 Reflection-haze of high gloss surfaces (T)

ASTM D4214 Degree of chalking (T) {Non-instrument Method}

ASTM D4449 Gloss differences from surfaces of similar appearance (T)

ASTM D5065 Condition of aged coatings on steel (G)

ASTM E284 Standard terminology of appearance (A)

ASTM E308 Computation of colors by CIE (P)

ASTM E313 Yellowness index (P)

ASTM E805 Identification of instrumental methods for color (P)

ASTM E1345 Reducing the effect of color measurement variability (P)

ASTM E1347 Color and color difference measurements (T)

ASTM E1349 Reflectance factor and color by spectrophotometry (T)

FTMS Method 4251 Color specification from spectrophotometric data

FTMS Method 4252 Color specification from tristimulus data

FTMS Method 6101 60 degree specular gloss

FTMS Method 6103 85 degree specular gloss

FTMS Method 6104 20 degree specular gloss

FTMS Method 6122 Lightness index difference

FTMS Method 6123 Color difference of opaque materials

FTMS Method 6131 Yellowness index

Chemical resistance, industrial

In general, there has to be an agreement with the customer on just what chemicals that are to be used, the temperature of contact, and the duration. A preferred test method is one that gives a large enough area of contact so as to be able to make a good appearance evaluation. Chemical resistance is to assess the ability of the coating to withstand the chemical, not to protect the substrate against corrosion from the chemical.

ASTM D2792 Solvent and fuel resistance (T)

ASTM D3023 Stain and reagent resistance (P)

ASTM D3260 Acid and mortar resistance (T)

ASTM D3912 Coatings used in light water nuclear plants (T)

ASTM D5402 Solvent rubs (P)

FTMS Method 6011 Hydrocarbon resistance (T)

Modified Tnemec test for industrial chemical resistance

Corrosion resistance. These are all laboratory tests done on test specified panels, inside specified special pieces of equipment and under specified conditions. Customer has to specify the test duration in hours, and any special test environments.

ASTM B117 Salt spray (T)

ASTM B287 Acetic acid-salt spray (T)

ASTM D2803 Filiform corrosion (G)

ASTM D5894 Cyclic salt fog/UV exposure, coated metal (P)

ASTM G85 Modified salt spray testing, annex Al, acetic acid-salt spray (P)

ASTM G85 Modified salt spray testing, annex A2, cyclical acidified salt fog (P)

ASTM G85 Modified salt spray testing, annex A3, acidified synthetic sea water fog (P)

ASTM G85 Modified salt spray testing, annex Al, salt/sulfur dioxide spray (fog) testing (P)

ASTM G85 Modified salt spray testing, annex Al, dilute electrolyte cyclic fog/dry test (P), prohesion test

Dirt resistance. This is basically an outdoor test. It can be run in conjunction with outdoor weathering tests.

ASTM D3719 Quantifying dirt collection (T)

Dirt removal ability (washability)

ASTM D3450 Washability properties (T)

ASTM D4828 Practical washability (T)

FTMS Method 6141 Washability

ASTM D2198 Stain removal from multicolor lacquers (T)

Environmental (atmosphere) resistance

ASTM D1211 Temperature change resistance (T)

ASTM D2246 Cracking resistance (T)

ASTM D2247 100% humidity (T)

ASTM D3459 Humid-dry cycling (T)

FTMS Method 6201 Humidity test

Film flexibility

ASTM D522 Mandrel bend test, method A (T)

ASTM D522 Mandrel bend test, method B (T)

ASTM D2370 Tensile properties (T)

ASTM D2794 Impact test (T)

ASTM D4146 Formability of primers on steel (T)

FTMS Method 6221 Flexibility

FTMS Method 6304 Knife test

Film thickness

ASTM D1005 Measurement (T)

ASTM D1186 Measurement over ferrous substrate (T)

ASTM D1400 Nonconductive coatings over a nonferrous metal base (T)

ASTM D4138 Protective coatings by destructive methods (T)

ASTM D5235 Microscopic measurements on wood substrates (T)

ASTM D5796 By destructive means using a boring device (T)

ASTM D6132 Over concrete using an ultrasonic gauge (T)

Fire retardancy

ASTM D1360 Cabinet method (T)

ASTM D3806 Small scale evaluation of retardancy by two-foot tunnel (T)

Hardness

ASTM D1474 Indentation hardness (T)

ASTM D2134 By Sward type hardness rocker (T)

ASTM D3363 Pencil hardness (T)

ASTM D4366 Pendulum damping tests (T)

Heat resistance

ASTM D2485 High temperature surface coatings (T)

ASTM D5499 Heat resistance of polymer linings (T)

FTMS Method 6051 Heat resistance

Hiding of substrate surface

These are special tests to evaluate in a practical sense the ability of the coating to hide the underlying surface.

ASTM D344 Visual evaluation of brushouts (T)

ASTM D2064 Print resistance of architectural paints (T)

ASTM D2091 Substrate print resistance (T)

ASTM D5150 Visual evaluation of roller applied coating (T)

FTMS Method 4121 Dry opacity

FTMS Method 6262 Primer absorption and topcoat holdout (requires modification)

Household chemical resistance

In general, there has to be an agreement with the customer on just what detergents and chemicals that are to be used, the temperature of contact, and the duration. The preferred test method is one that gives a large enough area of contact so as to be able to make a good appearance evaluation. Drops do not give this.

ASTM D1308 Household chemicals (T)
ASTM D2248 Detergent resistance (T)

Mildew and fungus resistance

ASTM D3273 Growth of mold resistance (T)
ASTM D3456 Microbiological attack (P)
ASTM D3623 Antifouling in shallow submergence (T)
ASTM D4610 Presence of microbial growth on coatings (G)
ASTM D5589 Resistance to algal defacement (T)
ASTM D5590 Resistance to fungal defacement (T)
MIL-STD-810 Method 508, Fungus
FTMS Method 6271 Mildew resistance

Penetration of water through coating

ASTM D5401 Clear water repellent coatings on wood (T)
ASTM D5860 Effect of water repellent treatments on mortar specimens (T)

Permeability of cured film

ASTM D1653 Water vapor transmission (T)
ASTM D2354 Minimum film formation temperature (T)
ASTM D3258 Porosity (hydrocarbon) (T)
ASTM D3793 Porosity (hydrocarbon, 40 F and 70 F) (T)

Sanding properties

FTMS Method 6321 Sanding characteristics

Stain transfer blocking/staining resistance

ASTM D1546 Evaluation of clear wood sealers (P)

Surface contact transfer effects (blocking)

ASTM D2199 Plasticizer migration (T)
ASTM D2793 Block resistance (T)
ASTM D4946 Blocking resistance of architectural paints (T)
ASTM D3003 Pressure mottling and blocking resistance (T)

Traction properties

ASTM D5859 Traction of footwear on surfaces (T)

Water resistance

ASTM D870 Immersion (P)
ASTM D1735 Water fog water resistance (P)

ASTM D4585 Controlled condensation (P)

ASTM D5860 Freeze-thaw resistance of water repellent treated mortar (T)

Wear, mar, and abrasion resistance

ASTM D968 Falling abrasive (T)

ASTM D2486 Scrub resistance (T)

ASTM D3170 Chipping resistance (Gravelometer) (T)

ASTM D4060 Tabor abrader (T)

ASTM D4213 Scrub resistance by weight loss (T)

ASTM D5178 Mar resistance (T)

ASTM D6037 Dry abrasion mar resistance of high gloss coatings (T)

FTMS Method 6142 Scrub resistance (T)

FTMS Method 6192 Abrasion resistance (Tabor)

Weathering resistance, accelerated, laboratory

ASTM D822 Filtered open-flame carbon arc exposure (P)

ASTM D2620 Light stability of clear coatings (T)

ASTM D3361 Unfiltered open-flame carbon arc exposure (P)

ASTM D4587 Using the QUV apparatus (P)

ASTM D5031 Enclosed carbon arc exposure (P)

FTMS Method 4561 Light fastness of pigments

Weathering resistance, outdoor, normal, and accelerated
Florida; Washington, DC; Eastern states; and Arizona are typical locations.

ASTM D1006 Exterior exposure, coatings on wood (P)

ASTM D1014 Exterior exposure, coatings on steel (P)

ASTM D1641 Outdoor exposure (P)

ASTM D2830 Exterior durability (T)

ASTM D4141 Accelerated outdoor exposure tests (G)

13.1.2   A Note About Regulatory Testing for Equipment Safety

A variety of products require accredited testing to ensure that a manufacturer can make products that meet specified safety requirements according to nationally recognized test standards. In the United States, OSHA now runs the Nationally Recognized Testing Laboratories (NRTL) program for accreditation of independent test laboratories. The most widely known is Underwriters Laboratories (UL). Original equipment manufacturers should note that having agency approvals on a product's components or subassemblies can make the process of obtaining approval for the entire system much easier. This is usually done under the UL-recognized component program for use in a particular type of equipment or application. The main point to keep in mind is that the cost of obtaining all these certifications is relatively high, and it may be prohibitive to production if the product volume is relatively low. An off-the-shelf product that already carries the approvals may be the only viable alternative.

There may also be product test requirements derived from where the product is to be marketed (an example is the IEC standards) or based on the type of equipment (FCC standards). An approval that must be sought if the product is to be marketed in Canada is that of the Canadian Standards Association (CSA), which is the Canadian equivalent of the U.S. National Bureau of Standards. CSA product requirements are usually quite similar to UL, and CSA may request a copy of applicable UL reports. CSA has at least one facility that is a member of the NRTL. It is important for managers to realize that these tests are performed to meet government regulations concerning product safety and only indicate a valid design. These tests usually provide no insight into production anomalies, offer no protection from product liability claims, and have no affect on warranty costs.[2]

13.2   Chemical Characterization[3]

13.2.1   Introduction

The materials engineer will deal with four major classifications of chemical analysis: bulk analysis, microanalysis, thermal analysis, and surface analysis. Bulk analysis techniques utilize a relatively large volume of the sample and are used to identify the elements or compounds present and verify conformance to applicable specifications. Microanalysis techniques explore a much smaller volume of the sample and are typically used to identify the elements or compounds present for studies of particles, contamination, or material segregation. Thermal analysis techniques are used to obtain thermomechanical information on sample materials to identify the coefficient of thermal expansion and other properties relevant to the failure analyst. Finally, surface analysis examines only the top few atomic layers of a material. These techniques are used in microcontamination, adhesion, and microelectronic studies. Surface analysis will not be covered in this chapter, so the interested reader is advised to use the materials in the recommended reading list.

Before deciding which technique(s) to use, the materials engineer must ask a number of questions:

- What type of information is required—quantitative, qualitative, or a mixture of both?
- What analytical accuracy and precision* are required?
- What is the physical state of the material? Is the material a powder, pellet, paste, foam, thin film, fiber, liquid, bar, gel, irregular chunk, or tubing?
- What is known about the material or samples?
- What are the important properties of the material?
- Is the sample a single component or a complex mixture?
- What is the material's future?
- How much material is available for analysis, and is there a limitation on sample size?
- How many samples must be run?
- What is the required analysis turnaround time?
- Are there any safety hazards to be concerned about?

The ability to answer these questions, and the use of the answers, will depend on the experience of the materials engineer and analyst and the equipment available. The importance of chemical analysis for raw material characterization is illustrated using the example of steel and the effect of alloying elements presented in Table 13.1.

13.2.2 Techniques and Applications of Atomic Spectroscopy

Atomic spectroscopy is actually not one technique but three: atomic absorption, atomic emission, and atomic fluorescence. Of these, atomic absorption (AA) and atomic emission are the most widely used. Our discussion will deal with them and an affiliated technique, ICP-mass spectrometry.

13.2.2.1 Atomic absorption.
Atomic absorption is the process that occurs when a ground-state atom absorbs energy in the form of light of a specific wavelength and is elevated to an excited state. The amount of light energy absorbed at this wavelength will increase as the number of atoms of the selected element in the light path increases. The relationship between the amount of light absorbed and the concentration of an-

---

* *Accuracy* is the extent to which the results of a measurement approach the true values, while *precision* is the measure of the range of values of a set of measurements.

TABLE 13.1   The Effect of Alloying Elements in Steel
If incoming material does not have the specified amount of alloying element, the desired material properties will not be attained after processing.

Alloying element	Effect on steel
Aluminum	Deoxidation, ease of nitriding
Boron	Hardenability
Carbon	Hardness, strength, wear
Chromium	Corrosion resistance, strength
Cobalt	Hardness, wear
Columbium	Reduction/elimination of carbide precipitation
Copper	Corrosion resistance, strength
Lead	Machinability
Manganese	Strength, hardenability, more response to heat treatment
Molybdenum	High-temperature strength, hardenability
Nickel	Toughness, strength, hardenability
Phosphorus	Strength
Silicon	Deoxidation, hardenability
Sulfur	Machinability
Tellurium	Machinability
Titanium	Reduction/elimination of carbide precipitation
Vanadium	Fine grain, toughness

alyte present in known standards can be used to determine unknown concentrations by measuring the amount of light they absorb. Instrument readouts can be calibrated to display concentrations directly.

The basic instrumentation for atomic absorption requires a primary light source, an atom source, a monochromator to isolate the specific wavelength of light to be used, a detector to measure the light accurately, electronics to treat the signal, and a data display or logging device to show the results. The light source normally used is either a hollow cathode lamp or an electrodeless discharge lamp.

The atom source used must produce free analyte atoms from the sample. The source of energy for free atom production is heat, most commonly in the form of an air-acetylene or nitrous oxide-acetylene flame. The sample is introduced as an aerosol into the flame. The flame burner head is aligned so that the light beam passes through the flame, where the light is absorbed.

13.2.2.2 Graphite furnace atomic absorption. The major limitation of atomic absorption using flame sampling (flame AA) is that the burner-nebulizer system is a relatively inefficient sampling device. Only a small fraction of the sample reaches the flame, and the atomized sample passes quickly through the light path. An improved sampling device would atomize the entire sample and retain the atomized sample in the light path for an extended period to enhance the sensitivity of the technique. Electrothermal vaporization using a graphite furnace provides those features.

With graphite furnace atomic absorption (GFAA), the flame is replaced by an electrically heated graphite tube. The sample is introduced directly into the tube, which is then heated in a programmed series of steps to remove the solvent and major matrix components and then to atomize the remaining sample. All of the analyte is atomized, and the atoms are retained within the tube (and the light path, which passes through the tube) for an extended period. As a result, sensitivity and detection limits are significantly improved.

Graphite furnace analysis times are longer than those for flame sampling, and fewer elements can be determined using GFAA. However, the enhanced sensitivity of GFAA and the ability of GFAA to analyze very small samples and directly analyze certain types of solid samples significantly expand the capabilities of atomic absorption.

13.2.2.3 Atomic emission. Atomic emission spectroscopy is a process in which the light emitted by excited atoms or ions is measured. The emission occurs when sufficient thermal or electrical energy is available to excite a free atom or ion to an unstable energy state. Light is emitted when the atom or ion returns to a more stable configuration or the ground state. The wavelengths of light emitted are specific to the elements that are present in the sample.

The basic instrument used for atomic emission is very similar to that used for atomic absorption, with the difference that no primary light source is used for atomic emission. One of the more critical components for atomic emission instruments is the atomization source, because it must also provide sufficient energy to excite the atoms as well as atomize them.

The earliest energy sources for excitation were simple flames, but these often lacked sufficient thermal energy to be truly effective sources. Later, electrothermal sources such as arc/spark systems were used, particularly when analyzing solid samples. These sources are useful for doing qualitative and quantitative work with solid samples but are expensive, difficult to use, and have limited applications.

Due to the limitations of the early sources, atomic emission initially did not enjoy the universal popularity of atomic absorption. This changed dramatically with the development of the inductively coupled plasma (ICP) as a source for atomic emission. The ICP eliminates many of the problems associated with past emission sources and has caused a dramatic increase in the utility and use of emission spectroscopy.

13.2.2.4 ICP.   The ICP is an argon plasma maintained by the interaction of an RF field and ionized argon gas. The ICP is reported to reach temperatures as high as 10,000 K, with the sample experiencing useful temperatures between 5,500 K and 8,000 K. These temperatures allow complete atomization of elements, minimizing chemical interference effects.

The plasma is formed by a tangential stream of argon gas flowing between two quartz tubes. Radio frequency (RF) power is applied through the coil, and an oscillating magnetic field is formed. The plasma is created when the argon is made conductive by exposing it to an electrical discharge, which creates seed electrons and ions. Inside the induced magnetic field, the charged particles (electrons and ions) are forced to flow in a closed annular path. As they meet resistance to their flow, heating takes place, and additional ionization occurs. The process occurs almost instantaneously, and the plasma expands to its full dimensions.

As viewed from the top, the plasma has a circular, "doughnut" shape. The sample is injected as an aerosol through the center of the doughnut. This characteristic of the ICP confines the sample to a narrow region and provides an optically thin emission source and a chemically inert atmosphere. This results in a wide dynamic range and minimal chemical interactions in an analysis. Argon is also used as a carrier gas for the sample.

13.2.2.5 ICP-mass spectrometry.   As its name implies, ICP-mass spectrometry (ICP-MS) is the synergistic combination of an inductively coupled plasma with a quadrupole mass spectrometer. ICP-MS uses the ability of the argon ICP to efficiently generate singly charged ions from the elemental species within a sample. These ions are then directed into a quadrupole mass spectrometer.

The function of the mass spectrometer is similar to that of the monochromator in an AA or ICP emission system. However, rather than separating light according to its wavelength, the mass spectrometer separates the ions introduced from the ICP according to their mass-to-charge ratio. Ions of the selected mass/charge are directed to a detector that quantifies the number of ions present. Due to the similarity of the sample introduction and data handling techniques, using an ICP-MS is very much like using an ICP emission spectrometer.

ICP-MS combines the multielement capabilities and broad linear working range of ICP emission with the exceptional detection limits of graphite furnace AA. It is also one of the few analytical techniques that permit the quantization of elemental isotopic concentrations and ratios.

13.2.2.6 Overview of FT-IR microspectroscopy.

Typical applications of the technique include bulk composition, surface contamination, and inclusions. The overall physical size of sample is restricted to what can be accommodated by the stage of an optical microscope. Sample thickness for transmission is generally limited to <0.5 mm for mid IR, and a few millimeters for near IR.

Detection is by fingerprints of individual chemical compounds or characteristic absorptions of chemical functional groups organic or inorganic. This is an absorption technique where the limit is determined by a concentration × path length product. In absolute terms, this puts the smallest amount detectable in the picogram range. Monomolecular layers are readily detectable on metal surfaces.

The measured signal is a ratio of the infrared energy detected in the presence of a sample to that with no sample present. Indirectly, this gives the amount of IR absorbed by the sample. Absorption arises from molecular vibrations and can be associated with specific chemical bonds that absorb at different frequencies (wavelengths). The measurement can be by transmission (thin samples), reflection (reflective surface), or by attenuated total reflectance (ATR). ATR is a near-field technique that uses a crystal of a high refractive index material brought into contact with the surface. It is used for opaque samples.

The presence or absence of particular chemical functional groups is detected. Using the identity of specific compounds by comparison with libraries of spectra, the IR spectrum giving a unique fingerprint. Libraries with tens of thousands of spectra are available. Quantitative information can be extracted on the amount of material present and the composition of mixtures.

Reflection from a homogeneous material is that of the surface molecules. Layers on a reflective substrate give transmission/reflection spectra. These correspond to transmission through a layer of double the actual thickness. It is useful from monolayers to hundreds of micrometers. ATR spectra come from a thickness of less than the wavelength, the effective depth being proportional to the wavelength. In the mid IR, this depth typically would be a few micrometers.

For transmission this is limited by diffraction, giving a typical limit of 10×10 micrometers. The normal arrangement is to mask the region to be measured by using an aperture at an image of the sample rather than at the sample itself. Physical masking of the sample can give spectra from smaller regions. Spectra can also be obtained from smaller samples if they can be physically isolated.

Using an x-y stage, samples can be mapped automatically, pixel by pixel. A full spectrum is obtained at each point. Maps can then be generated based on specific spectral features. Multiple maps can be created from a single experiment to show the distribution of different species.

IR is very versatile, being capable of obtaining spectra from almost any surface. It is a very simple technique, operating in air. IR is the best technique for identifying specific compounds. It provides very good specificity for different chemical types. It has limited spatial resolution because of the wavelengths utilized. Because maps are generated sequentially, pixel-by-pixel mapping is fairly slow, occurring at a rate of several seconds per pixel.

13.2.3 Selecting the Proper Atomic Spectroscopy Technique

With the availability of a variety of atomic spectroscopy techniques such as flame atomic absorption, graphite furnace atomic absorption, inductively coupled plasma emission, and ICP-mass spectrometry, laboratory managers must decide which technique is best suited for the analytical problems of their laboratory. Because atomic spectroscopy techniques complement each other so well, it may not always be clear which technique is optimal for a particular laboratory. A clear understanding of the analytical problem in the laboratory and the capabilities provided by the different techniques is necessary.

Important criteria for selecting an analytical technique include detection limits, analytical working range, sample throughput, interferences, ease of use, and the availability of proven methodology. These criteria are discussed below for flame AA, graphite furnace AA (GFAA), ICP emission, and ICP-mass spectrometry (ICP-MS).

13.2.3.1 Atomic spectroscopy detection limits. The detection limits achievable for individual elements represent a significant criterion for the usefulness of an analytical technique for a given analytical problem. Without adequate detection limit capabilities, lengthy analyte concentration procedures may be required prior to analysis.

Typical detection limit ranges for the major atomic spectroscopy techniques are shown in Fig. 13.1 for six atomic spectroscopic techniques: flame AA, hydride generation AA, graphite furnace AA (GFAA), ICP emission with radial and axial torch configurations and ICP-mass spectrometry.

Generally, the best detection limits are attained using ICP-MS or graphite furnace AA. For mercury and elements that form hydrides, the cold vapor mercury or hydride generation techniques offer exceptional detection limits.

Detection limits should be defined very conservatively, with a 98% confidence level, based on established conventions for the analytical technique. This means that, if a concentration at the detection limit were measured many times, it could be distinguished from a zero or baseline reading in 98% (3σ) of the determinations.

13.2.3.2 Analytical working range. The analytical working range can be viewed as the concentration range over which quantitative results can be obtained without having to recalibrate the system. Selecting a technique with an analytical working range (and detection limits) based on the expected analyte concentrations minimizes analysis times by allowing samples with varying analyte concentrations to be analyzed together. A wide analytical working range also can reduce sample handling requirements, minimizing potential errors.

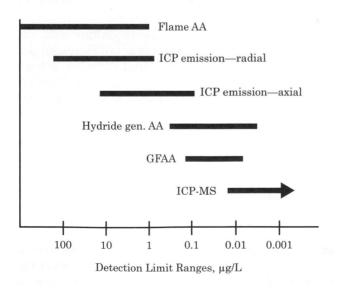

Figure 13.1 Typical detection ranges for the major atomic spectroscopy techniques.

13.2.3.3 Sample throughput. Sample throughput is the number of samples that can be analyzed or elements that can be determined per unit time. For most techniques, analyses performed at the limits of detection or where the best precision is required will be more time consuming than less-demanding analyses. Where these factors are not limiting, however, the number of elements to be determined per sample and the analytical technique will determine the sample throughput.

Flame AA. Flame AA provides exceptional sample throughput when analyzing a large number of samples for a limited number of elements. A typical determination of a single element requires only 3 to 10 s. However, flame AA requires specific light sources and determination of optical parameters for each element, and it may require different flame gases for different elements. In automated multielement flame AA systems, all samples normally are analyzed for one element, the system is then automatically adjusted for the next element, and so on. As a result, even though it is frequently used for multielement analysis, flame AA is generally considered to be a single-element technique.

Graphite furnace AA. As with flame AA, GFAA is basically a single-element technique. Because of the need to thermally program the system to remove solvent and matrix components prior to atomization, GFAA has a relatively low sample throughput. A typical graphite furnace determination normally requires 2 to 3 minutes.

ICP emission. ICP emission is a true multielement technique with exceptional sample throughput. ICP emission systems typically can determine 10 to 40 elements per minute in individual samples. Where only a few elements are to be determined, however, ICP is limited by the time required for equilibration of the plasma with each new sample, typically about 15 to 30 seconds.

ICP-MS. ICP-MS is also a true multielement technique with the same advantages and limitations of ICP emission. The sample throughput for ICP-MS is typically 20 to 30 element determinations per minute, depending on such factors as the concentration levels and required precision.

13.2.3.4 Interferences. Few, if any, analytical techniques are free of interferences. With atomic spectroscopy techniques, however, most interferences have been studied and documented, and methods exist to correct or compensate for those that may occur. A summary of the types of interferences seen with atomic spectroscopy techniques, all of which are controllable, and the corresponding methods of compensation are shown in Table 13.2.

TABLE 13.2 Atomic Spectroscopy Interferences

Technique	Interference type	Compensation method
Flame AA	Ionization	Ionization buffer
		Releasing agent or nitrous oxide-acetylene
	Chemical	flame
	Physical	Dilution, matrix matching, or method of additions
GFAA	Physical and chemical	STPF conditions
	Molecular absorption	Zeeman or continuum source background correction
	Spectral	Zeeman background correction
ICP emission	Spectral	Background correction or the use of alternate analytical lines
	Matrix	Internal standardization
ICP-MS	Mass overlap	Interelement correction, use of alternate mass values, or higher mass resolution
	Matrix	Internal standardization

SOURCE: *Electronic Failure Analysis Handbook, Table 9.2, McGraw-Hill.*

13.2.3.5 Other comparison criteria. Other comparison criteria for analytical techniques include the ease of use, required operator skill levels, and availability of documented methodology.

Flame AA. Flame AA is very easy to use. Extensive applications information is available. Excellent precision makes it a preferred technique for the determination of major constituents and higher concentration analytes.

GFAA. GFAA applications are well documented, although not as completely as with flame AA. GFAA has exceptional detection limit capabilities but within a limited analytical working range. Sample throughput is less than that of other atomic spectroscopy techniques. Operator skill requirements are somewhat more extensive than for flame AA.

ICP emission. ICP emission is the best overall multielement atomic spectroscopy technique, with excellent sample throughput and very wide analytical range. Good documentation is available for applications. Operator skill requirements are intermediate between flame AA and GFAA.

ICP-MS. ICP-MS is a relatively new technique with exceptional multi-element capabilities at trace and ultra-trace concentration levels and the ability to perform isotopic analyses. Good basic documentation for interferences exists. Applications documentation is limited but growing rapidly. ICP-MS requires operator skills similar to those for ICP emission and GFAA.

13.2.4 Comparison Summary

The main selection criteria for atomic spectroscopy techniques, concentration range, and analytical throughput are summarized in Fig. 13.2. Where the selection is based on analyte detection limits, flame AA and ICP emission are favored for moderate to high levels, while graphite furnace AA and ICP-MS are favored for lower levels. ICP emission and ICP-MS are multielement techniques, favored where large numbers of samples are to be analyzed.

13.3 Thermal Analysis

The material physical properties that can be analyzed by thermal analysis include:

- Softening point
- Effects of plasticizer
- Glass transition temperature
- Melting point
- Heat of fusion
- Specific heat
- Purity

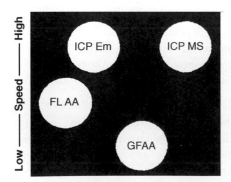

Figure 13.2 General atomic spectroscopy instrumentation selection guide based on sample throughput and concentration range. *(From Electronic Failure Analysis Handbook, Fig. 9.4, McGraw-Hill.)*

- Oxidative stability
- Rates of reaction, curing, and cross-linking
- Degree of cure
- Decomposition
- Crystallization temperature
- Pyrolysis
- Weight percent filler
- Weight loss vs. time
- Weight loss vs. temperature
- Loss of water, solvent, or plasticizer
- Impact properties
- Viscoelastic behavior as a function of stress, strain, frequency, and temperature
- Mechanical behavior
- Long-term behavior such as creep and creep recovery
- Modulus vs. temperature
- Damping vs. temperature
- Chain branching
- Molecular weight
- Molecular weight distribution
- Coefficient of thermal expansion
- Liquidus temperature
- Solidus temperature

Many times, a combination of techniques is required to solve the problem, as enumerated below.

Multiple thermal analysis techniques

ASTM methods

D1525 Test Method for VICAT Softening Temperature of Plastics

D0256 Test Method for Impact Resistance of Plastics and Electrical Insulating Materials

D0789 Test Method for Determination of Relative Viscosity, Melting Point, and Moisture Content of Polyamide

Tables 13.3, 13.4, and 13.5 summarize the techniques of choice for cured and uncured material identification and degree of cure analysis.

TABLE 13.3   Techniques for Degree-of-Cure Analysis *(from Ref. 4)*

Technique	Description	Use	Value
Thermomechanical analysis (TMA)	Measure probe displacement as a function of temperature	Determines glass transition temperature	Indication of degree of cure or environmental effects
Differential scanning calorimetry (DSC)	Performs enthalpy measurements	Determines glass transition temperature Determines residual heat of reaction	Indication of degree of cure or environmental effects
Dynamic mechanical analysis (DMA)	Measures mechanical response to oscillating dynamic loading	Determines glass transition temperature Observes mechanical transition due to additional cross-linking	Indication of degree of cure or environmental effects Indication of under-cured condition
Infrared spectroscopy	Measures IR spectrum	Distinguishes between reacted and unreacted functional groups	Indicates amount of unreacted functional groups to determine extent of cure
Solvent extraction	Exposure to an organic solvent	Removes unreacted material leaving reacted network behind	Indication of the degree of cure

TABLE 13.4   Techniques for Uncured Material Identification *(from Ref. 4)*

Technique	Description	Use	Value
High-pressure liquid chromatography (HPLC)	Produces liquid chromatograms of any soluble liquid	Identifies individual components of differing solubilities or size	Formulation verification
Infrared spectroscopy	Measures IR spectra	Identifies functional groups attached to carbon backbone	Formulation verification
Differential scanning calorimetry (DSC)	Performs enthalpy measurements	Determines heat of reaction	Formulation verification
X-ray fluorescence	Measures X-ray fluorescence spectra	Determines sulfur content	Hardener content

13.3.1   Differential scanning calorimetry (DSC)

Differential scanning calorimetry (DSC) is the workhorse of the thermal analysis laboratory. The DSC can examine materials between -170 and $+750°C$. DSC is a technique upon which the heat flow to or from a sample specimen is measured as a function of temperature or time as it is subjected to a controlled temperature program in a controlled atmosphere.

TABLE 13.5   Techniques for Cured Material Identification *(from Ref. 4)*

Technique	Description	Use	Value
Pyrolysis—gas chromatography (PGC)	Determines gas chromatograms formed from nonvolatile organics by thermal decomposition	Qualitative and quantitative analysis of cured epoxy	Formulation/impurity verification
Pyrolysis—gas chromatography/ mass spectroscopy (PGC/MS)	Allows mass spectrometer to act as a detector for the gas chromatograph	Qualitative and quantitative analysis of cured epoxy	Formulation/impurity verification
Infrared spectroscopy	Measures IR spectra	Functional group analysis	Formulation verification
Thermomechanical analysis (TMA)	Measures material thermal-mechanical response	Determines glass transition	Identify general resin system
X-ray fluorescence	Measures X-ray fluorescence spectra	Determines sulfur content	Hardener content

Some of the applications of DSC are listed below:

- Identify the softening point of the material (glass transition).
- Compare additive effects on a material.
- Identify the glass transition temperature.
- Identify the material's minimum process temperature.
- Identify the amount of energy required to melt the material.
- Quantify the material's specific heat.
- Perform oxidative stability testing (OST).
- Understand the reaction kinetics of a thermoset material as it cures.
- Compare the degree of cure of one material to another.
- Characterize a material as it cures under ultraviolet light.
- Characterize a material as it is thermally cured.
- Determine the crystallization temperature upon cooling.

Differential scanning calorimetry (DSC) methods

ASTM methods

D2471 Test Method for Gel Time and Peak Exothermic Temperature of Reacting Thermosetting Resins

D5028 Test Method for Curing Properties of Pultrusion Resins by Thermal Analysis

D4816 Test Method for Determining the Specific Heat Capacity of Materials by DSC

D4565 Test Method for Determining the Physical/Environmental Performance Properties of Insulation and Jackets for Telecommunications Wire and Cable

D4591 Test Method for Determining Temperatures and Heats of Transitions of Fluoropolymers by DSC

D3012 Test Method for Thermal Oxidative Stability of Polypropylene Plastics Using a Biaxial Rotator

D4803 Test Method for Predicting Heat Buildup in PVC Building Products

D2117 Test Method for Melting of Semicrystalline Polymers by the Hot Stage Microscopy Method

D3417 Test Method for Heats of Fusion and Crystallization of Polymers by Thermal Analysis

D3418 Test Method for Transition Temperature of Polymers by Thermal Analysis

D3895 Test Method for Oxidative Induction Time (OIT) of Polyolefins by Differential Scanning Calorimetry

D4419 Test Method for Determining the Transition Temperatures of Petroleum Waxes by DSC

E698 Standard Test Method for Arrhenius Kinetic Constants (of thermally unstable materials) Using DSC

E1559 Standard Test Method for Contamination Outgassing Characteristics of Space Craft Materials by DSC

E537 Standard Test Method for Determining the Thermal Stability of Chemicals by DSC

E793 Standard Test Method for Determining the Heat of Crystallization (of solid samples in granular form) by DSC

E1269 Standard Test Method for Specific Heat Capacity by DSC

E1356 Standard Test Method for Glass Transition Temperature by DSC

13.3.2 Thermogravimetric Analysis (TGA)

Thermogravimetric analysis (TGA) examines materials between ambient and +1500°C. TGA is a technique in which the mass of a substance is monitored as a function of temperature or time as the sample specimen is subjected to a controlled temperature program in a controlled atmosphere.

Some plastics applications of TGA are listed below:

- Identify the filler content of a material by weight percent.
- Identify the ash content of a material by weight percent.
- Characterize the materials weight loss within a certain temperature range.
- Characterize the material's weight loss vs. time at a given temperature.
- Quantify the material's loss of water, solvent, or plasticizer within a certain temperature range.
- Examine flame retardant properties of a material.
- Examine the combustion properties of a material.

Thermogravimetric analysis (TGA) methods

ASTM Methods

> D2288 Test Method for Weight Loss of Plasticizers on Heating
>
> D4202 Test Method for Thermal Stability of PVC Resin
>
> D2115 Test Method for Volatile Matter (including water) of Vinyl Chloride Resins
>
> D2126 Test Method for Response of Rigid Cellular Plastics to Thermal and Humid Aging
>
> D3045 Recommended Practice for Heat Aging of Plastics Without Load
>
> D1870 Practice for Elevated Temperature Aging Using a Tubular Oven
>
> D4218 Test Method for Determination of Carbon Black Content in Polyethylene Compounds by a Muffle-Furnace
>
> D1603 Test Method for Carbon Black in Olefin Plastics
>
> D5510 Practice for Heat Aging of Oxidatively Degradable Plastics
>
> E1131 Standard Test Method for Compositional Analysis by TGA
>
> E1641 Standard Test Method for Decomposition Kinetics by TGA

13.3.3 Dynamic Mechanical Analysis (DMA)

Dynamic mechanical analysis (DMA) examines materials between −170 and +1000°C.

DMA is a technique in which a substance under an oscillating load is measured as a function of temperature or time as the substance is subjected to a controlled temperature program in a controlled atmosphere.

Some plastics applications of DMA are listed below:

- Quantify the impact properties of a material.
- Examine the viscoelastic behavior of a material as a function of stress, strain, frequency, time, or temperature.
- Examine a material's mechanical behavior.
- Examine a material's long term behavior with respect to creep or creep recovery.
- Identify the material's modulus vs. temperature.
- Identify the material's damping qualities vs. temperature.
- Examine effects of temperature on molecular chain branching.
- Compare material's molecular weight and molecular weight distribution.
- Examine additive effects on a material's mechanical properties.

Dynamic mechanical analysis (DMA) methods

ASTM Methods

D4440 Practice for Rheological Measurement of Polymer Melts Using Dynamic Mechanical Procedures

D4473 Practice for Measuring the Cure Behavior of Thermosetting Resins Using Dynamic Mechanical Properties

D5023 Test Method for Measuring the Dynamic Mechanical Properties of Plastics in Three Point Bending

D5024 Test Method for Measuring the Dynamic Mechanical Properties of Plastics in Compression

D5026 Test Method for Measuring the Dynamic Mechanical Properties of Plastics in Tension

D5418 Test Method for Measuring the Dynamic Mechanical Properties of Plastics Using a Dual Cantilever Beam

D0638 Test Method for Tensile Properties of Plastics

D0695 Test Method for Compressive Properties of Rigid Plastics

D1708 Test Method for Tensile Properties of Plastics by Use of Microtensile Specimens

D5296 Test Method for Molecular Weight Averages and Molecular Weight Distribution of Polystyrene by DMA

D3029 Test Method for Impact Resistance of Flat, Rigid, Plastic Specimens by Means of a Tup (falling weight)

D4508 Test Method for Chip Impact Strength of Plastics

D4812 Test Method for Unnotched Cantilever Beam Impact Strength of Plastics

D5083 Test Method for Tensile Properties of Reinforced Thermosetting Plastics Using Straight-sided Specimens

D0790 Test Method for Flexural Properties of Plastics

D0882 Test Method for Tensile Properties of Thin Plastics

D0952 Test Method for Bond or Cohesive Strength of Sheet Plastics and Electrical Insulating Materials

D0953 Test Method for Bearing Strength of Plastics

D1043 Test Method for Stiffness Properties of Plastics as a Function of Temperature by Means of a Torsion Test

D5420 Test Method for Impact Resistance of Flat, Rigid, Plastic Specimen by Means of a Striker Impacted by a Falling Weight

D5628 Test Method for Impact Resistance of Flat, Rigid, Plastic Specimen by Means of a Falling Dart (tup or falling weight)

D0671 Test Method for Flexural Fatigue of Plastics by Constant Amplitude of Force

D0747 Test Method for Apparent Bending Modulus Plastics by Means of a Cantilever Beam

D0785 Test Method for Rockwell Hardness of Plastics and Electrical Insulating Materials

D2990 Test Method for Tensile, Compressive, and Flexural Creep and Creep Rupture of Plastics

D2765 Test Method for Determination of Gel Content and Swell Ratio of Cross-linked Ethylene Plastics

D4476 Test Method for Flexural Properties of Fiber Reinforced Pultruded Plastic Rods

D2343 Test Method for Tensile Properties of Glass Fiber Strands, Yams, and Rovings Used in Reinforced Plastics

D1939 Practice for Determining Residual Stresses in Extruded or Molded ABS

E1640 Standard Test Method for Glass Transition by DMA

13.3.4 Thermomechanical Analysis (TMA)

Thermomechanical analysis (TMA) examines materials between −170 and +1000°C.

TMA is a technique in which the deformation of a substance under a nonoscillating load is measured as a function of temperature or time as the substance is subjected to a controlled temperature program in a controlled atmosphere.

Some applications of TMA are listed below:

- Determine a material's coefficient of thermal expansion.

- Determine the volumetric growth of a material, dilatometry measurements.

- Identify a material's glass transition temperature.

- Identify the material's VICAT softening point.
- Measure a material's heat deflection temperature (HDT).

Thermomechanical analysis (TMA) methods

ASTM Methods

D2566 Test Method for Linear Shrinkage of Cured Thermosetting Casting Resin During Cure

D0789 Test Method for Determination of Relative Viscosity, Melting Point, and Moisture Content of Polyamide

D2732 Test Method for Unrestrictive Linear Thermal Shrinkage of Plastic Film and Sheeting

D0696 Test Method for Determination of Coefficient of Linear Thermal Expansion of Plastics

D3386 Test Method for Determination of Coefficient of Linear Thermal Expansion of Electrical Insulating Materials by TMA

E1545 Test Method for Determination of Glass Transition Temperatures Under Normal Range (−100 to +600°C)

E831 Test Method for Determination of Glass Transition Temperatures of Solid Materials by TMA

13.4   Thermal Testing[5]

13.4.1   Vicat Softening Temperature, ASTM D1525 (ISO 306)

This test gives a measure of the temperature at which a plastic starts to soften rapidly. A round, flat-ended needle of 1 mm^2 cross section penetrates the surface of a plastic test specimen under a predefined load, and the temperature is raised at a uniform rate. The Vicat softening temperature, or VST, is the temperature at which the penetration reaches 1 mm.

ISO 306 describes two methods:

- Method A—load of 10 N
- Method B—load of 50 N with two possible rates of temperature rise: 50°C/h and 120°C/h

This results in ISO values quoted as A50, A120, B50, or B120. The test assembly is immersed in a heating bath with a starting temperature of 23°C (73°F). After 5 min, the load is applied: 10 or 50 N. The temperature of the bath at which the indenting tip has penetrated by 1 ± 0.01 mm is reported as the VST of the material at the chosen load and temperature rise.

13.4.2   Ball Pressure EC335-1

This is a softening temperature test, similar to the Vicat method. The sample is horizontally positioned on a support in a heating cabinet, and a steel ball of 5 mm diameter is pressed onto it with a force of 20 Newtons. After 1 h, the ball is removed, the sample is cooled in water for 10 s, and if the remaining impression diameter is <2 mm, the material is reported to meet the ball pressure test at the applied temperature.

Depending on the application, the test temperature can be varied:

- 75°C (167°F) for non-current-carrying parts
- 125°C (257°F) for live parts

13.4.3   Heat Deflection Temperature and Deflection, Temperature under Load, ASTM D648 (ISO 75)

Heat deflection temperature (HDT) is a relative measure of a material's ability to perform for a short time at elevated temperatures while supporting a load. The test measures the effect of temperature on stiffness; a standard test specimen is given a defined surface stress, and the temperature is raised at a uniform rate.

In both ASTM and ISO standards, a loaded test bar is placed in a silicone-oil-filled heating bath.

The surface stress on the specimen is:

- Low—for ASTM and ISO 64 psi (0.45 MPa)
- High—for ASTM 264 psi (1.82 MPa) and for ISO 1.80 MPa

The force is allowed to act for five minutes; this waiting period may be omitted with testing materials that show no appreciable creep during the initial five minutes. After five minutes, the original bath temperature of 23°C (73°F) is raised at a uniform rate of 2°C/min.

The deflection of the test bar is continuously observed; the temperature at which the deflection reaches 0.32 mm (ISO) or 0.25 mm (ASTM) is reported as rejection temperature under load, or heat deflection temperature.

Two acronyms are commonly used:

DTUL—deflection temperature under load

HDT—heat distortion temperature or heat deflection temperature

It is common practice to use the acronym DTUL for ASTM values and HDT for ISO values.

Depending on the applied surface stress, the letters A or B are added to HDT—HDT/A for a load of 1.80 Mpa, and HDT/B for a load of 0.45 MPa.

13.4.3.1 HDT—Amorphous vs. semicrystalline polymers. In amorphous polymers, HDT is nearly the same as the glass transition temperature (Tg) of the material.

Because amorphous polymers have no defined melting temperature, they are processed in their rubbery state above Tg. Crystalline polymers may show low HDT values and still have structural utility at higher temperatures. The HDT test method is more reproducible with amorphous polymers than with crystalline polymers. With some polymers, it may be necessary to anneal the test specimens to obtain reliable results.

Addition of glass fibers to the polymer will increase its modulus. Since the HDT represents a temperature where the material exhibits a defined modulus, increasing the modulus will also increase the HDT. Glass fibers have a more significant effect on the HDT of crystalline polymers than on amorphous polymers.

Although widely used to indicate high-temperature performance, the HDT test simulates only a narrow range of conditions. Many high-temperature applications involve higher temperatures, more loading, and unsupported conditions. Therefore, the results obtained by this test method do not represent maximum use temperatures, because in real life, essential factors such as time, loading, and nominal surface stress may differ from the test conditions.

13.4.4 Thermal Conductivity, ASTM E1530

The thermal insulating capacity of plastics is rated by measuring the thermal conductivity. Plaques of plastic are placed on both sides of a small, heated platen, and heat sinks are put against the free surfaces of the plaques. Insulators positioned around the test cell prevent radial heat loss. The axial flow of heat through the plastic plaques can then be measured. The results are reported in W/m°C.

13.4.5 Relative Thermal Index, RTI UL 746B

Formerly named the *continuous use temperature rating* or CUTR, the RTI is the maximum service temperature at which the critical properties of a material will remain within acceptable limits over a long period of time. As established by UL 746B, there can be up to three independent RTI ratings assigned to a material:

1. Electrical—by measuring dielectric strength

2. Mechanical with impact—by measuring tensile impact strength

3. Mechanical without impact—by measuring tensile strength

These three properties were selected as critical indicators due to their sensitivity to the high temperatures as used in the test.

The long-term thermal performance of a material is tested relative to a second control material, which already has an established RTI and exhibits good performance. Hence, the term *relative thermal index*. The control material is used because thermal degradation characteristics are inherently sensitive to variables in the testing program itself. The control material will be affected by the same unique combination of these factors during the tests, thus providing a valid basis for comparison with the subject material.

Ideally, long-term thermal performance would be evaluated by aging the subject material at normal operating temperatures for a long time. However, this is impractical for most applications. Therefore, accelerated aging is conducted at much higher temperatures. In the aging process, samples of the subject and control materials are placed in ovens maintained at set constant temperatures. Samples of both materials are removed at predetermined times and then tested for retention of key properties. By measuring the three mentioned properties as functions of time and temperature, the theoretical end of useful service may be mathematically determined for each temperature. This end of service life is defined as the time at which a material property has degraded to 50% of its original value. Through the Arrhenius representation of the test data, the maximum temperature at which the subject material can be expected to have a satisfactory service life may be determined. This predicted temperature is RTI for each property.

An understanding of how the RTI is determined enables engineers to use the temperature index to help predict how parts molded from a given material will work in elevated temperature end-use environments.

13.4.6   Coefficient of Linear Thermal Expansion, ASTM E831

Every material will expand when heated. Injection-molded polymer parts will expand and change dimensions in proportion to the increase in temperature. To characterize this expansion, engineers rely on the *coefficient of linear thermal expansion* or CLTE to describe the changes in length, width, or thickness of a molded part. Amorphous polymers will generally show consistent expansion rates over their useful temperature range. Crystalline polymers generally have increased rates of expansion above their glass transition temperature.

The addition of fillers, causing anisotropy, significantly alters the CLTE of a polymer. Glass fibers will generally align in the direction of the flow front; when the polymer is heated, the fibers restrict expansion along their axis and reduce the CLTE. In directions perpendicular to flow direction and thickness, the CLTE will be higher.

Polymers may be formulated with CLTEs to match those of metal or other materials used in complex constructions.

13.5 Mechanical Testing[5]

13.5.1 Hardness Testing

Just some of the ASTM specifications for hardness testing are listed below:

E140-97e2 Standard Hardness Conversion Tables for Metals (Relationship Among Brinell Hardness, Vickers Hardness, Rockwell Hardness, Rockwell Superficial Hardness, Knoop Hardness, and Scleroscope Hardness)

F1957-99 Standard Test Method for Composite Foam Hardness-Durometer Hardness

E110-82(1997)e2 Standard Test Method for Indentation Hardness of Metallic Materials by Portable Hardness Testers

E18-00 Standard Test Methods for Rockwell Hardness and Rockwell Superficial Hardness of Metallic Materials

D2134-93 Test Method for Determining the Hardness of Organic Coatings with a Sward-Type Hardness Rocker

B647-84(1994) Standard Test Method for Indentation Hardness of Aluminum Alloys by Means of a Webster Hardness Gage

A833-84(1996) Standard Practice for Indentation Hardness of Metallic Materials by Comparison Hardness Testers

B277 Standard Test Method for Hardness of Electrical Contact Materials (Discontinued 2000), Replaced By No Replacement

F1151-88(1998) Standard Test Method for Determining Variations in Hardness of Film Ribbon Pancakes

E1842-96 Standard Test Method for Macro-Rockwell Hardness Testing of Metallic Materials

E448-82(1997)e1 Standard Practice for Scleroscope Hardness Testing of Metallic Materials

E384-99 Standard Test Method for Microindentation Hardness of Materials

E103-84(1996)e1 Standard Test Method for Rapid Indentation Hardness Testing of Metallic Materials

E92-82(1997)e3 Standard Test Method for Vickers Hardness of Metallic Materials

E10-00a Standard Test Method for Brinell Hardness of Metallic Materials

D5873-95 Standard Test Method for Determination of Rock Hardness by Rebound Hammer Method

D5230-00 Standard Test Method for Carbon Black-Automated Individual Pellet Hardness

D4366-95 Standard Test Methods for Hardness of Organic Coatings by Pendulum Damping Tests

D3802-79(1999) Standard Test Method for Ball-Pan Hardness of Activated Carbon

D3363-00 Standard Test Method for Film Hardness by Pencil Test

D3313-99 Standard Test Method for Carbon Black—Individual Pellet Hardness

D2583-95 Standard Test Method for Indentation Hardness of Rigid Plastics by Means of a Barcol Impressor

D2240-97e1 Standard Test Method for Rubber Property-Durometer Hardness

D1865-89(1999)e1 Standard Test Method for Hardness of Mineral Aggregate Used on Built-Up Roofs

D1474-98 Standard Test Methods for Indentation Hardness of Organic Coatings

D1415-88(1999) Standard Test Method for Rubber Property-International Hardness

D785-98 Standard Test Method for Rockwell Hardness of Plastics and Electrical Insulating Materials

C1327-99 Standard Test Method for Vickers Indentation Hardness of Advanced Ceramics

C1326-99 Standard Test Method for Knoop Indentation Hardness of Advanced Ceramics

C886-98 Standard Test Method for Scleroscope Hardness Testing of Carbon and Graphite Materials

C849-88(1999) Standard Test Method for Knoop Indentation Hardness of Ceramic Whitewares

C748-98 Standard Test Method for Rockwell Hardness of Graphite Materials

C730-98 Standard Test Method for Knoop Indentation Hardness of Glass

C661-98 Standard Test Method for Indentation Hardness of Elasto-meric-Type Sealants by Means of a Durometer

B724-83(1991)e1 Standard Test Method for Indentation Hardness of Aluminum Alloys by Means of a Newage Portable Non-Caliper-Type Instrument

B648-78(1994) Standard Test Method for Indentation Hardness of Aluminum Alloys by Means of a Barcol Impressor

B294-92(1997) Standard Test Method for Hardness Testing of Cemented Carbides

As can be seen, there are a variety of methods and procedures for hardness testing. An array of hardness testers is shown in Fig. 13.3.

13.5.1.1 Hardness testing of plastics. Since hardness testing covers a large number of materials and methodologies, the area of hardness testing of plastics will be used to illustrate this field of testing. For plastics, the Rockwell hardness test determines the hardness after allowing for elastic recovery of the test specimen. For both Ball and

Figure 13.3 From left to right, Rockwell C tester, Rockwell B tester, Superficial Rockwell tester, and Vickers tester (on floor). *(Courtesy of Martin Testing Laboratories.)*

Shore hardness, hardness is derived from the depth of penetration under load, thus excluding any elastic recovery of the material. Rockwell values cannot, therefore, be directly related to Ball or Shore values. Although the ranges for Shore A and D values can be compared to ranges for Ball indentation hardness values, a linear correlation does not exist.

Ball hardness. A polished hardened steel ball with a diameter of 5 mm is pressed into the surface of a test specimen (at least 4 mm thick) with a force of 358 Newtons (ISO 2039-1). After 30 s of load application, the depth of the impression is measured, from which the surface area of the impression is calculated. Ball hardness H358/30 is calculated as applied load divided by surface area of impression. Results are reported in Newtons per mm^2.

Rockwell hardness, ASTM D785 (ISO 2039-2). A Rockwell hardness number is directly related to the indentation hardness of a plastic material: the higher the number, the harder the material. Due to a short overlap of Rockwell hardness scales, two different numbers on two different scales may be obtained for the same material, both of which may be technically correct.

The indentor, a polished hardened steel ball, is pressed into the surface of a test specimen. The diameter of the ball depends on the Rockwell scale in use. The specimen is loaded by a minor load, followed by a major load, and then again by the same minor load. The actual measurement is based on the total depth of penetration: this penetration is calculated from the total penetration, minus the elastic recovery after removal of the major load, and minus the penetration resulting from the minor load. The Rockwell hardness number is derived from 130 minus depth of penetration in units of 0.002 mm.

Rockwell hardness numbers should be between 50 and 115. Values above this range are inaccurate: the determination should be repeated using the next severest scale. The scales increase in severeness from R over L to M (increasing material hardness). If a less severe scale than the R scale is needed (for a softer material), the Rockwell test is not suitable.

Shore hardness, ASTM D2240 (ISO 868). Shore hardness values are scale numbers resulting from the indentation of a plastic material with a defined steel rod. It is measured with durometers of two types, both with calibrated springs for applying force to the indentor. Durometer A is used for softer materials and durometer D for harder materials.

The specimen is placed in the durometer, the pressure base is applied to the specimen, and the scale of the indenting device is read after 15 s. Scale numbers are shown in terms of units, ranging from 0 for full protrusion of 2.5 mm, to 100 for nil protrusion (obtained by using a flat piece of glass).

Shore values vary from 10 to 90 for Shore A—soft materials to 20 to 90 for Shore D—hard materials. No simple relationship exists between indentation hardness determined by this test method and any fundamental property of the material tested.

13.5.2 Tensile Strength, Strain, and Modulus, ASTM D638 (ISO 527)

Fundamental to the understanding of a material's performance is knowledge of how the material will respond to any load. By knowing the amount of deformation (strain) introduced by a given load (stress), the designer can begin to predict the response of the application under its working conditions. Stress/strain relationships under tension are the most widely reported mechanical property for comparing materials or designing an application (see Fig. 13.4).

Tensile stress/strain relationships are determined as follows. A dogbone-shaped specimen is elongated at a constant rate, and the load applied and elongation are recorded. Stress and strain are then calculated as follows:

$$\text{Stress} = \frac{\text{load}}{\text{original cross-sectional area}} \text{ Mpa (psi)}$$

Figure 13.4 Automated tensile testing equipment with a thermal chamber for hot/cold testing of materials. Source: Courtesy of Martin Testing Laboratories, Inc.

$$\text{Strain} = \left(\frac{\text{elongation}}{\text{original percent length}}\right) \times 100\%$$

Other mechanical properties determined from the stress/strain relationship are:

Modulus stress/strain	MPa (psi)
Stress at yield	initial max. stress MPa (psi)
Stress at break	stress at failure MPa (psi)
Strain at break	strain at failure or percent maximum elongation
Proportional limit	point of first nonlinearity
Elastic modulus	modulus below the proportional limit MPa (psi)

13.5.3 Flexural Strength and Modulus, ASTM D790 (ISO 178)

Flexural strength is the measure of how well a material resists bending, or the stiffness of the material. Unlike tensile loading, in flexural testing, all force is applied in one direction. A simple, freely supported beam is loaded at mid-span, thereby producing three-point loading. On a standard testing machine, the loading nose pushes the specimen at a constant rate of 2 mm/min.

To calculate the flexural modulus, a load deflection curve is plotted using the recorded data. This is taken from the initial linear portion of the curve by using at least five values of load and deflection.

The flexural modulus (ratio of stress to strain) is most often quoted when citing flexural properties. Flexural modulus is equivalent to the slope of the line tangential to the stress/strain curve, for the portion of the curve where the plastic has not yet deformed. Values for flexural stress and flexural modulus are reported in MPa (psi).

13.5.4 Impact Testing

In standard testing, such as tensile and flexural testing, the material absorbs energy slowly. In real life, materials often absorb applied forces very quickly: falling objects, blows, collisions, drops, etc. The purpose of impact testing is to simulate these conditions. Izod and Charpy methods are used to investigate the behavior of specified specimens under specified impact stresses and to estimate the brittleness or toughness of specimens. They should not be used as a source of data for design calculations on components. Information on the typical behavior of a material can be obtained by testing different types of test specimens prepared under different conditions, varying notch radius and test temperatures.

Both tests are performed on a pendulum impact machine. The specimen is clamped in a vice;, and the pendulum hammer—with a hardened steel striking edge with specified radius—is released from a predefined height, causing the specimen to shear from the sudden load. The residual energy in the pendulum hammer carries it upward, and the difference in the drop height and return height represents the energy to break the test bar. The test can be carried out at room temperature, or at lower temperatures to test cold-temperature embrittlement. Test bars can vary in type and in dimensions of the notches.

The test results of falling-weight impact, such as Gardner and flexed plate, are dependent of the geometry of both the falling weight and support. They should be used only to obtain relative rankings of materials.

Impact values cannot be considered absolute unless the geometry of the test equipment and specimen conform to the end-use requirement. The relative ranking of materials may be expected to be the same between two test methods if the mode of failure and impact velocities are the same.

Impact properties can be very sensitive to test specimen thickness and molecular orientation. The differences in specimen thickness as used in ASTM and ISO methods may affect impact values strongly. A change from 3 to 4 mm thickness can even provide a transition in the failure mode from ductile to brittle behavior, through the influence of molecular weight and specimen thickness on Izod notched impact. Materials already showing a brittle fracture mode in 3 mm thickness, such as mineral and glass filled grades, will not be affected. Neither will impact modified materials.

It must be realized, however, that materials have not changed—only test methods. The ductile/brittle transition mentioned rarely plays a role in real life; part thicknesses are mostly designed as 3 mm or lower.

13.5.4.1 Izod impact strength, ASTM D256 (ISO 180). The notched Izod impact test has become the standard for comparing the impact resistance of plastic materials. However, this test has little relevance to the response of a molded part to an actual environmental impact. Because of the varying notch sensitivity of materials, this test will penalize some materials more than others. Although they have often been questioned as a meaningful measure of impact resistance (the test tends to measure notch sensitivity rather than the ability of plastic to withstand impact), the values are widely accepted as a guide for comparison of toughness among materials. The notched Izod test is best applied in determining the impact resistance for parts with many sharp corners such as ribs, intersecting walls, and other stress risers. The unnotched Izod test uses the same loading geometry with the exception that there is no notch cut into the specimen (or the specimen is clamped in a reversed way). This type of testing always provides superior values over notched Izod because of the lack of a stress concentrator.

The ISO designation reflects type of specimen and type of notch:

- ISO 180/IA means specimen type I and notch type A. The dimensions of specimen type I are 80 mm long, 10 mm high, and 4 mm thick.

- ISO 180/IU means the same type I specimen, but it is clamped in a reversed way (indicating unnotched). The specimens as used in the ASTM method have similar dimensions, same notch radius, and same height, but they differ in length (63.5 mm) and, more importantly, in thickness (3.2 mm).

The ISO results are defined as the impact energy in joules used to break the test specimen, divided by the specimen area at the notch. Results are reported in kJ/m^2.

The ASTM results are defined as the impact energy in joules, divided by the length of the notch (or thickness of the specimen). They are reported in J/m.

The difference in specimen thickness may result in different interpretations of impact strength, as already discussed.

13.5.4.2 Charpy impact strength, ASTM D256 (ISO 179). The main difference between Charpy and Izod tests is the way the test bar is held. In Charpy testing, the specimen is not clamped but lies freely on the support in a horizontal position. The ISO designation reflects type of specimen and type of notch:

- ISO 179/2C means specimen type 2 and notch type C.

- ISO 179/2D means specimen type 2, but unnotched.

ISO results are defined as the impact energy in joules absorbed by the test specimen, divided by the surface area of the specimen at the notch. Results arc reported in kJ/m^2.

13.6   Miscellaneous Testing

13.6.1   Taber Abrasion Resistance, ASTM D1044 (ISO 3537)

This test measures the quantity of abrasion loss by abrading a test specimen with a Taber Abrasion machine (see Fig. 13.5). The specimen is mounted on a rotating disc, turning at a speed of 60 rpm. A load is applied by means of weights, which push the abrasive wheels against the specimen. The abrasive wheels are available in different types, depending on the test to be performed. After a specified number of cycles, the test is stopped. The mass of abrasion loss is defined as the total of the masses of test piece fragments that have dropped off. It is reported in mg/1000 cycles.

13.6.2   Flammability Testing

13.6.2.1   UL 94 flammability.   The most widely accepted flammability performance standards for plastic materials are UL 94 ratings. These are intended to provide an indication of a material's ability to extinguish a flame after it is ignited. Several ratings can be applied, based on the rate of burning, time to extinguish, ability to resist dripping, and whether drips are burning.

Each material tested may receive several ratings, based on color and/or thickness. When specifying a material for an application, the UL rating should be applicable for the thickness used in the wall section in the plastic part. The UL rating should always be reported with the thickness; just reporting the UL rating without mentioning thickness is insufficient. This test is not intended to reflect hazards presented by any material under actual fire conditions. Table 13.6 summarizes UL 94 rating categories.

Figure 13.5 Taber abrasion apparatus. *(Courtesy of Martin Testing Laboratories, Inc.)*

TABLE 13.6   Summary of the UL 94 Rating Categories

Category	Description
HB	Slow burning on a horizontal specimen, burning rate < 76 mm/min for thickness < 3 mm
V-0	Burning stops within 10 s on a vertical specimen; no drips allowed
V-1	Burning stops within 30 s on a vertical specimen; no drips allowed
V-2	Burning stops within 30 s on a vertical specimen; drips of flaming particles are allowed
5V	Burning stops within 60 s after five applications of a flame—larger than used in V-testing each of 5 s, to a test bar
5VB	Plaque specimens may have a burn-through (hole)
5VA	Plaque specimens may not have a burn-through (hole)—highest UL rating

13.6.2.2   UL 94 HB horizontal testing procedure.   Where flammability is a safety requirement, HB materials are normally not permitted. In general, HB classified materials are not recommended for electrical applications except for mechanical and/or decorative purposes. Sometimes this is misunderstood: non-FR materials (or materials that are not meant to be FR materials) do not automatically meet HB requirements. UL 94 HB is (although the least severe) a flammability classification and has to be checked by testing. This test is not intended to reflect hazards presented by any material under actual fire conditions.

13.6.2.3   UL 94 VO, V1, and V2 vertical testing procedure.   The vertical tests take the same specimens as are used for the HB test. Burning times, glowing times, when dripping occurs, and whether the cotton beneath ignites are all noted. Flaming drips, widely recognized as a main source for the spread of fire or flames, distinguish V1 from V2.

13.6.2.4   UL 94-5V vertical testing procedure.   UL 94-5V is the most severe of all UL tests. It involves two steps:

Step I.   A standard flammability bar is mounted vertically and subjected to each of five applications of a 127-mm flame, 5 s duration. To pass, no bar specimen may burn with flaming or glowing combustion for more than 60 s after the fifth flame application. Also, no burning drips are allowed that ignite cotton placed beneath the samples. The total procedure is repeated with five bars.

Step 2. A plaque with the same thickness as the bars is tested in a horizontal position with the same flame. The total procedure is repeated with three plaques. Two classifications result from this horizontal test: 5VB and 5VA. 5VB allows holes (burn-through). 5VA does not allow holes.

UL94-5VA is the most stringent of all UL tests, specified for fire enclosures on larger office machines. For those applications with expected wall thickness of less than 1.5 mm, glass-filled material grades should be used, These tests are not intended to reflect hazards presented by any material under actual fire conditions.

13.6.2.5 CSA flammability, CSA C22.2 no. 0.6, test A*.

This Canadian Standard Association flammability test is carried out in a similar way to the UL 94 5V test. However, the test is more severe: each flame application is of 15 s duration. In addition, during the first four flame applications, the sample must extinguish within 30 s and, after the fifth application, within 60 s (compared to UL 94-5V with five flame applications of 5 s each).

13.6.2.6 Limited oxygen index, ASTM D2863 (ISO 4589).

The purpose of the oxygen index test is to measure the relative flammability of materials by burning them in a controlled environment. The oxygen index represents the minimum level of oxygen in the atmosphere that can sustain flame on a thermoplastic material.

The test atmosphere is an externally controlled mixture of nitrogen and oxygen. A supported specimen is ignited with a pilot flame, which is then removed. In successive test runs, the oxygen concentration is reduced to a point where the sample can no longer support combustion. *Limited oxygen index* or LOI is defined as the minimum oxygen concentration in which the material will burn for 3 min or can keep the sample burning over a distance of 50 mm. The higher the LOI value, the less the likelihood of combustion. These tests are not intended to reflect hazards presented by any material under actual fire conditions.

13.6.2.7 Glow wire, IEC 695-2-1.

The glow wire test simulates thermal stresses that may be produced by sources of heat or ignition, such as overloaded resistors or glowing elements. A sample of the insulating

---

* Test results complying with this CSA test should be considered to be in accordance with UL 94-5V as well.

material is held vertically for 30 s with a 1 Newton force against the tip of an electrically heated glowing wire. The travel of the glow wire tip through the sample is limited. After withdrawing the sample, the time for extinguishing flames and the presence of any burning drops are noted. The specimen is considered to have withstood the glow wire test if one of the following situations applies:

1. There is no flame and no glowing.

2. Flames or glowing of the specimen, or the surroundings and the layer below, extinguish within 30 s after removal of the glow wire, and the surrounding parts and the layer below have not burned away completely. When a layer of tissue paper is used, there shall be no ignition of this paper and no scorching of the pinewood board.

Actual live parts or enclosures are tested in a similar way. The temperature level of the glow wire tip is dependent on how the finished part is used:

Attended or unattended

Continuously loaded or not

Used near or away from a central supply point

In contact with a current-carrying (live) part or used as an enclosure or cover

Under less or more stringent conditions

Depending on the required level of severity for the final part environment, the following test temperatures are preferred: 550, 650, 750, 850 or 960°C (1020, 1200, 1380, 1560, or 1760°F). Estimating the risk of failure due to abnormal heat, ignition, and spread of fire should determine the appropriate test temperature.

13.6.2.8 Needle flame, IEC 695-2-2. The needle flame test simulates the effect of small flames that may result from faulty conditions within electrical equipment. To evaluate the likely spread of fire (burning or glowing particles), either a layer of the subject material or components normally surrounding the specimen, or a single layer of tissue paper, is positioned underneath the specimen. The test flame is applied to the sample for a certain time period, usually 5, 10, 20, 30, 60 or 120 s. Other levels of severity can be adopted for specific requirements. This test is not intended to reflect hazards presented by any material under actual fire conditions.

Unless otherwise specified in the relevant specification, the specimen is considered to have withstood the needle flame test if one of the following four situations apply:

1. The specimen does not ignite.

2. Flames or burning or glowing particles falling from the specimen do not spread fire to the surrounding parts or to the layer placed below the specimen, and there is no flame or glowing of the specimen at the end of application of the test flame.

3. The duration of the burning is less than 30 s.

4. The extent of burning specified in the relevant specification has not been exceeded.

13.6.3   Electrical Testing

13.6.3.1   Dielectric strength, ASTM D149 (IEC 243-1).   Dielectric strength reflects the electric strength of insulating materials at power frequencies (48 Hz to 62 Hz) or the measure of dielectric breakdown resistance of a material under an applied voltage. The applied voltage just before breakdown is divided by the specimen thickness to give the value in kV/mm. The surrounding medium can be air or oil. The thickness dependence can be significant; all values are reported at specimen thickness.

Many factors influence the values:

- Thickness, homogeneity, and moisture content of the test specimen
- Dimensions and thermal conductivity of the test electrodes
- Frequency and waveform of the applied voltage
- Ambient temperature, pressure, and humidity
- Electrical and thermal characteristics of the ambient medium

13.6.3.2   Surface resistivity, ASTM D257 (IEC 93).   When an insulating plastic is subjected to a voltage, some portion of the resultant current will flow along the surface of the plastic molding if there is another conductor or ground attached to the same surface. Surface resistivity is a measure of the ability to resist that surface current. It is measured as the resistance when a direct voltage is applied between surface mounted electrodes of unit width and unit spacing. It is reported in ohms (Ω) (sometimes referenced to ohms per square, Ω/sq).

13.6.3.3 Volume resistivity, ASTM D257 (IEC 93). When an electric potential is applied across an insulator, the current flow will be limited by the resistance capabilities of the material. Volume resistivity is the electrical resistance when an electric potential is applied between opposite faces of a unit cube. It is measured in Ω-cm. Volume resistivity will be affected by environmental conditions imposed on the material. It varies inversely with temperature and decreases slightly in moist environments. Materials with volume resistivity values above 1088 Ω-cm are considered insulators. Partial conductors have values of 1033 to 1088 Ω-cm.

13.6.3.4 Relative permittivity, ASTM D150 (IEC 250). The relative permittivity of an insulating material is the ratio of capacitance of a capacitor, in which the space between and around the electrodes is entirely and exclusively filled with the insulating material in question, to the capacitance of the same configuration of electrodes in vacuum (see Fig. 13.6).

In ac dielectric applications, good resistivity as well as low energy dissipation are desirable characteristics. The dissipation of electrical energy results in inefficiencies in an electronic component and causes heat buildup in the plastic part that acts as a dielectric.

In an ideal dielectric material, such as a vacuum, there is no energy loss to dipole motion of the molecules. In solid materials, such as plas-

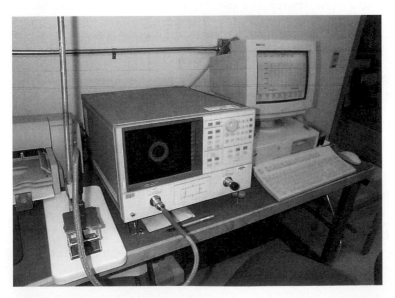

Figure 13.6 Automated, probe-based test equipment for measuring dielectric constant. *Courtesy of Martin Testing Laboratories, Inc.)*

tics, the dipole motion becomes a factor. A measure of this inefficiency is the relative permittivity (formerly called *dielectric constant*).

It is a dimensionless factor derived by dividing the parallel capacitance of the system with a plastic material by that of an equivalent system with a vacuum as dielectric. The lower the number, the better the performance of the material as an insulator.

13.6.3.5   Dissipation factor, ASTM D150 (IEC 250). The dielectric loss angle of an insulating material is the angle by which the phase difference between applied voltage and resulting current deviates from $\pi/2$ rad, when the dielectric of the capacitor consists exclusively of the dielectric material. The dielectric dissipation factor tan d of an insulating material is the tangent of the loss angle d.

In a perfect dielectric, the voltage wave and the current are exactly 90° out of phase. As the dielectric becomes less than 100% efficient, the current wave begins to lag the voltage in direct proportion. The amount the current wave deviates from being 90° out of phase with the voltage is defined as the *dielectric loss angle*. The tangent of this angle is known as the *loss tangent* or *dissipation factor*.

A low dissipation factor is important for plastic insulators in high-frequency applications such as radar equipment and microwave parts; smaller values mean better dielectric materials. A high dissipation factor is important for welding capabilities.

Both relative permittivity and dissipation factor are measured using the same test equipment. Test values obtained are highly dependent on temperature, moisture levels, frequency, and voltage.

13.6.3.6   Arc resistance, ASTM D495. When an electric current is allowed to travel across an insulator's surface, this surface will damage over time and become conductive. Arc resistance is a measure of the time in seconds required to make an insulating surface conductive under a high-voltage, low-current arc. Or the arc resistance is the elapsed time during which the surface of a plastic material will resist the formation of a continuous conducting path when subjected to a high-voltage, low-current arc under specific conditions.

13.6.3.7   Comparative tracking index, ASTM D3638 (IEC 112). The tracking index is the relative resistance of electrical insulating materials to tracking when the surface is exposed, under electrical stress, to contaminants containing water. Comparative tracking index, or CTI, test is undertaken to evaluate the safety of components carrying live parts; insulating material between live parts must be resistant to tracking.

CTI is defined as the maximum voltage at which no failure occurs at 50 drops of ammonium chloride in water. High values of CTI are desirable. Materials meeting the CTI requirements at 600 V are called *high-tracking resins.*

The CTI test procedure is complex. Influencing factors are the condition of the electrodes, electrolyte, and sample surface, and of the applied voltage. Values can be lowered by additives such as

- Pigments, particularly carbon black
- Flame retardants
- Glass fibers

Thus, black, FR, and GF materials in general, are not recommended when tracking resistance is a key requirement. Minerals (TiO_2) tend to raise CTI values.

The CTI test is carried out with two platinum electrodes of given dimensions placed with the slightly rounded chisel edge flat on the test sample. A potential difference is applied between the electrodes, usually a minimum of 175 V. Where parts will be exposed to high electrical stress, 250 V is specified. The voltage is applied in steps of 25 V, with the maximum voltage being 600 V. The surface of the test specimen is wetted by 50 drops of a solution of 0.1% ammonium chloride in distilled water (the so-called *solution A*) falling centrally between the two electrodes. Both the size and frequency of the electrolyte drops are specified. When no current is flowing at the chosen voltage, the test is repeated with a 25-V higher voltage until there is a flow of current. This voltage, decreased with one step of 25 V, is called the CTI. The test is then repeated with a voltage of 25 V below the CTI value and with 100 drops instead of 50. The voltage is determined at which 100 drops do not create a flow of current. This 100-drop value can be reported in addition to the 50-drop CTI values between brackets ().

13.6.4   Optical Testing

13.6.4.1   Haze and light transmission, ASTM D1003.
Haze is caused by the scattering of light within a material, and it can be affected by a molecular structure, degree of crystallinity, or impurities at the surface or interior of the polymer. Haze is appropriate only for translucent or transparent materials—not for opaque ones. Haze is sometimes thought of as the opposite of gloss, which would properly be absorption of an incident beam. However, the haze test method actually measures absorption, transmittance, and deviation of a direct beam by a translucent material.

A specimen is placed in the path of a narrow beam of bright light so that some of the light passes through the specimen and some continues unimpeded. Both parts of the beam pass into a sphere equipped with a photodetector. Two quantities can be determined:

- The total strength of the light beam
- The amount of light deviated by more than 2.5° from the original beam

From these two quantities, two values are calculated:

- Haze, or the percentage of incident light scattered more than 2.5°
- Luminous transmittance, or the percentage of incident light that is transmitted through the specimen

13.6.4.2 Gloss, ASTM D523.   Gloss is associated with the capacity of a surface to reflect more light in some directions than in others. Gloss can be measured in a glossmeter. A bright light is reflected off a specimen at an angle, and the luminance or brightness of the reflected beam is measured by a photodetector. Most commonly, a 60° angle is used. Shinier materials can be measured at 20° and matte surfaces at 85°. The glossmeter is calibrated by using a black glass standard with a gloss value of 100. Plastics show lower values; they strongly depend on the method of molding.

13.6.4.3 Haze and gloss.   Haze and gloss test methods measure how well a material reflects and transmits light. They quantify qualifications such as clear and shiny. While haze is only appropriate for transparent or translucent materials, gloss can be measured for any material. Both gloss and haze tests are precise, but they are often used to measure appearance, which is more subjective. The correlation between haze and gloss values and how people rate the clarity or shininess of a plastic is uncertain.

13.6.4.4 Refractive index, ASTM D542.   A light beam is transmitted through a transparent specimen under a certain angle. The deviation from the beam, caused by the material when passing the specimen, is the index of refraction found by dividing $\sin x / \sin y$.

Test specimens are predried at 122°F (55°C) for 24 hr, cooled to room temperature, and weighed before immersion in water for a specified time and at a specified temperature. Water absorption may be measured as follows:

- *At 23°C (73°F).*   The specimens are placed in a container with distilled water at 23°C (73°F). After 24 hr, the specimens are dried and weighed.

- *At 100°C (212°F).*   The specimens are immersed for 30 min in boiling water, cooled for 15 min in water at 23°C (73°F), and reweighed.

- *Until saturation.*   The specimens are immersed in water at 23°C (73°F) until full saturation is reached.

Water absorption may be expressed as

- The mass of water absorbed
- The mass of water absorbed per unit of surface area
- A percentage of water absorbed with respect to the mass of the test specimen

13.6.5   Mold Shrinkage ASTM D955 (ISO 2577)

Mold shrinkage is the difference between the dimensions of the mold and of the molded article produced therein. It is reported in percent or millimeters per millimeter.

Values for mold shrinkage are reported as parallel to the flow (inflow direction) and perpendicular to the flow (crossflow direction). These values may differ significantly for glass-filled materials. Mold shrinkage may also vary with other parameters such as part design, mold design, mold temperature, injection pressure, and cycle time.

Mold shrinkage figures, as measured on simple parts such as a tensile bar or disc, are typical data only for material selection purposes. They should not be used for part or tool design.

13.6.6   Melt Mass-Flow Rate/Melt Volume-Flow Rate, ASTM D1238 (ISO 1133)

The melt-flow rate (MFR) or melt volume-flow rate (MVR) test measures the flow of a molten polymer through an extrusion plastometer under specific temperature and load conditions. The extrusion plastometer consists of a vertical cylinder with a small die of 2 mm at the bottom and a removable piston at the top. A charge of material is placed in the cylinder and preheated for several minutes. The piston is placed on top of the molten polymer, and its weight forces the polymer through the die and on to a collecting plate. The time interval for the test ranges from 15 s to 6 min to accommodate the different viscosities of plastics. Temperatures used are 220, 250, and 300°C (428, 482, and 572°F). Loads used are 1.2, 5, and 10 kg.

The amount of polymer collected after a specific interval is weighed and normalized to the number of grams that would have been extruded in 10 min; melt flow rate is expressed in grams per reference time.

For example, "MFR (220/10) = xx g/10 min" refers to the melt flow rate for a test temperature of 220°C (428°F) and a nominal load of 10 kg.

The melt flow rate of polymers is dependent on the rate of shear. The shear rates in this test are much smaller than those used under normal conditions of processing. Therefore, data obtained by this method may not always correlate with polymers' behavior in actual use.

Recommended Readings

D. Briggs and M. P. Seah, *Practical Surface Analysis*, John Wiley & Sons, 1983.
Materials Characterization, Metals Handbook, Vol. 10, 9th ed., American Society for Metals, 1986.

References

In addition to the numerous ASTM, ISO, IPC, IEC, CSA, UL, EC, and other specifications discussed in the text, the following references were consulted:

1. Private communication with David A. Heiser, P.E.
2. See 29 CFR 1910.7 for OSHA regulations.
3. P. Martin, *Electronic Failure Analysis Handbook,* Chapter 9, McGraw-Hill, 1999.
4. Kar, R.J., WL-TR-91-4032 *Composite Failure Analysis Handbook,* Vol. II, Part I, pp. 6–4, Northrop Corporation, Hawthorne, California, February 1992.
5. Shastri, Ranganath, "Plastics Testing," *Modern Plastics Handbook,* McGraw-Hill, New York, NY, 2000.

14

Materials Recycling*

Susan E. Selke, Ph.D.
School of Packaging
Michigan State University
East Lansing, Michigan

14.1 Introduction

Recycling of materials has an extremely long history in the United States and all over the world. Whenever raw materials were scarce or expensive, it was common sense to use the materials already on hand as many times as possible. Clothes, for example, were handed down from one family member to another, dresses became aprons, scraps of fabric became quilts, cloth too worn to be used in clothing became a rag rug, or, in a complete change of identity, paper. During World War II, recycling was essential to the war effort. After that time, however, in the United States, increased affluence led to a large decline in recycling, as the U.S. became characterized as a "throw-away society." Glimmerings of increased interest in conservation and recycling emerged around 1970, with the first Earth Day signaling a renewed interest in environmental issues. Concern about wasting resources was fueled by the oil crisis of 1973. Recycling was recognized as a way of conserving both resources and, at least in many cases, energy as well. Recycling programs were started in a number of communities, though most remained small and had limited impact. In the mid-1980s, however, recycling entered a period of dramatic growth with the emergence of a new issue—lack of disposal capacity for municipal solid waste. By the late 1990s, concern about solid waste disposal had declined in the U.S., though it was still a major issue in many other

---

* With credits to *Modern Plastics Handbook,* Chapter 12, Dr. Susan E. Selke, McGraw-Hill, New York, N.Y., February 2000.

parts of the world. As we enter the twenty-first century, solid waste disposal capacity remains a concern in much of the world, oil prices are rising again, the threat of global warming is fueling new calls for energy conservation, and recycling rates are rising for some materials in some places and falling for others.

This chapter will examine the current status of recycling, including recycling rates, recycling technology, properties of recycled materials, and markets for them, along with a discussion of legislative requirements and consumer pressure. The emphasis will be on recycling of post-consumer materials—those that have served their intended purpose and are destined for disposal. The emphasis will also be on recycling in the United States, though there will be some discussion of the situation in other countries.

14.1.1 Solid Waste Disposal Concerns

As mentioned above, concerns about disposal of municipal solid waste (MSW) were the primary driver behind the very large growth in material recycling that has taken place since 1980. Municipal solid waste is defined as residential, commercial, institutional, and industrial wastes such as durable goods, nondurable goods, containers and packaging, food scraps, yard trimmings, and similar materials. It does not include construction and demolition debris, automobiles (though it does include tires), municipal sludge, combustion ash, and industrial process waste.

In the mid-1980s, many major metropolitan areas, particularly on the East Coast, were very close to being out of MSW disposal capacity. Disposal costs were rising astronomically, reaching over $100 per ton in New Jersey for *tipping fees* (the amount charged by the disposal facility for accepting the waste) alone. The voyage of the garbage scow Mobro, which sailed from Long Island, New York, around a good part of the Western Hemisphere, turned away from port after port as it searched for a home for its cargo, caught the attention of the media and the American public and came to epitomize the "garbage crisis."

Governments at various levels, from individual communities to whole states, were struggling to find ways to deal with ensuring continuation of the necessary public service of garbage disposal while containing the costs that were threatening to ruin their budgets—and the chances of reelection for the responsible officials. Acronyms such as NIMBY (not in my backyard), NIMTO (not in my term of office), and PITBY (put it in their backyard) were coined.

Some communities and states solved at least their immediate problem by shipping their garbage to adjoining communities or states—or even farther. Garbage from Long Island, New York, reached landfills as far away as Michigan. Predictably, "host" communities were not always happy with their role. Many states tried to write laws to prohibit the import of "outsiders'" waste, only to have them struck down based on the free trade between states provisions in the U.S. Constitution. A number of incineration facilities for municipal solid waste were built, but public resistance to these facilities soon became even greater than resistance to landfills, and their costs were typically much higher than landfills. Recycling programs were started up around the country, first in the hundreds and then in the thousands. In contrast to incineration, recycling proved to be very politically popular.

At the same time, slowly but surely, new landfills were sited and built. Due to new regulations, these landfills were constructed much differently from the old landfills that were being shut down. They contained liners—often double liners—to protect against groundwater pollution, and caps to help prevent ingress of water. More care was given to locating them in geologically appropriate areas as well. The cost of these new landfills was also higher but, with increase in capacity and decrease in demand (as recycling increased), the average tipping fees in landfills actually declined in many areas from the record highs set in the early 1990s. For example, in New Jersey, the average landfill tipping fee in 1998 was down to $60 per ton.[1] While the absolute number of landfills in the U.S. continued to decline, to 2,314 in 1998, capacity increased. In 1988, 14 states reported having less than 5 years of disposal capacity remaining. In 1998, no states reported less than 5 years capacity, and only 7 had less than 10 years capacity.[1] The average landfill tipping fee in the U.S. was $33.60 per ton in 1998, up $2 per ton from 1997.[1]

Incineration increased in the last half of the 1980s but then leveled off in the face of growing public resistance. New York City, for example, at one time planned to build five large incineration facilities but found its plans tied up for years because of public opposition, and eventually scrapped the idea. EPA reports indicate that incineration rates have been relatively steady at about 16–17% since 1990.[2] BioCycle, with a somewhat different definition of MSW, reported an incineration rate of only 7.5% in 1998, down from 9% in 1997, and in the 10–11% range from 1990 through 1996. BioCycle draws its figures from state reports, and some states include other types of waste, especially construction and demolition debris, in their definition of municipal solid waste.[1]

Recycling rates have increased steadily in the U.S. since the mid 1980s, as many new recycling programs were begun. In 1998, the number of curbside recycling programs in the U.S. reached 9,349. The proportion of municipal solid waste that was recycled and composted reached 31.5%.[1]

During the mid to late 1990s, another factor also began to moderate the amount of MSW destined for landfill. The overall generation rate for MSW began to fall. Initially, the decline could be seen on a per capita basis only, as the rising population made overall MSW generation go up, even when the amount generated per person declined slightly. By 1995, declines were seen in total tonnage as well. However, in 1997, the strong economy led to generation of a record 217 million tons of municipal solid waste, back to the 1995 level of 4.4 pounds per person per day, after falling to 4.3 pounds in 1996.[2] BioCycle reported just under 375 million tons of MSW in 1998, up from 340 million tons in 1997.[1] Discards to landfill and other disposal increased to 119.6 million tons from 115.8 million tons in 1996, though remaining well below the 140.1 million tons in 1990, according to EPA.[2] Historical trends in generation and disposal of MSW in the U.S. are shown in Fig. 14.1. Recycling rates for various categories of materials are shown in Fig. 14.2.

In much of Europe, the lack of landfill capacity was more real than in the U.S. Many countries had been heavily dependent on incineration for a long time, since space for landfill was very hard to find. However, public resistance to incineration was increasing. These ongoing problems led to increased reliance on composting and recycling as alternatives to incineration and landfill. However, increasing waste generation has undermined efforts to reduce reliance on landfilling. According to a recent report, landfilling in 1995 accounted for 67% of

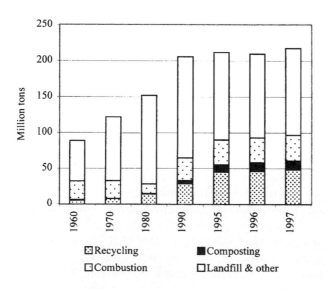

Figure 14.1 Trends in generation and disposal of municipal solid waste in the U.S.[2]

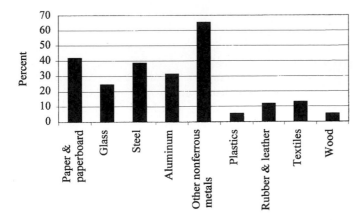

Figure 14.2 Recycling rates for materials in U.S. municipal solid waste, 1997.[2]

municipal solid waste, up from 64% in the 1985–1995 period (Fig. 14.3).[3]

Other parts of the world, too, have experienced problems with continuing to dispose of materials as they have done in the past. In much of the developing world, the usual location of waste disposal is in open dumps. A considerable amount of unorganized recycling is common in these societies, with individuals scavenging reusable materials from the dump sites. As more modern forms of waste disposal are implemented in efforts to curtail the problems resulting from open dumping, recycling in a more organized fashion is becoming part of the solid waste management strategy.

Thus, around the world, there is increasing reliance on recycling, not as the only method for handling solid waste but as an important part of what has become known as *integrated solid waste management*—the mix of strategies used to handle disposal of the wastes we generate.

14.1.2 Composition of Municipal Solid Waste

The U.S. Environmental Protection Agency classifies municipal solid waste by product type and by material. Categorization by material shows that the largest single component of MSW, by weight, is paper and paperboard at 38.6%, for a total of 83.8 million tons in 1997 (Fig. 14.4).[2] Yard trimmings are the next largest category, followed by food waste and then by plastics. The relative contributions of paper and paperboard and of plastics have increased over the years, while the contributions of food and yard wastes have declined (Fig. 14.5).

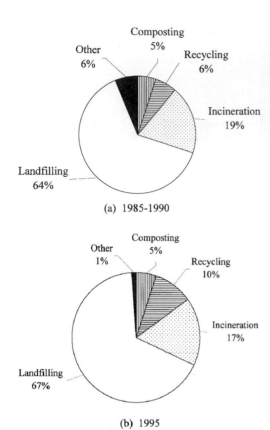

(a) 1985-1990

(b) 1995

Figure 14.3 Municipal solid waste disposal trends in Europe.

EPA categorizes products as durable goods, nondurable goods, containers and packaging, and other wastes, including food and yard wastes. The proportion of materials in MSW classified as containers and packaging has stayed relatively constant, at about one-third of the waste stream. The durable goods category has increased somewhat, and the nondurable goods category has increased substantially (Figs. 14.6 and 14.7), while food and yard wastes have declined.

14.1.3 Legislative Requirements

In the United States, legislative requirements related to recycling are in effect predominantly at the state, rather than federal, level. A number of states have some kind of requirement that recycling opportunities be available to residents. Some require residents to participate in recycling, requiring that the target recyclable materials be kept out of

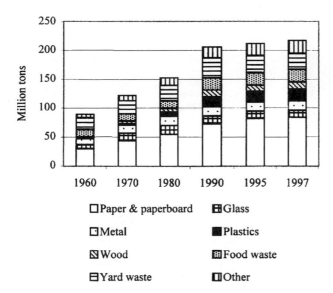

Figure 14.4 Materials in U.S. municipal waste.[2]

the disposal stream and instead diverted to recycling (source separation). Some do not mandate recycling per se but prohibit the disposal, by landfill or incineration, of the target recyclables. Others require that communities incorporate recycling as part of their solid waste management strategy. Still others simply require that consideration be given to recycling as an option (Tables 14.1 and 14.2).[4,5]

Several states have considered the establishment of taxes or fees to promote recycling. Bottle deposit legislation can be put in this category. States with mandatory deposits on certain containers, typically carbonated beverages (Table 14.3), achieve high rates of return, typically 90% or more, facilitating the recycling of the containers. Hawaii currently has an advance disposal fee on glass containers, and a bill that would impose such a fee on plastic containers has been introduced into the state legislature.[4] In 1993, Florida instituted an advance disposal fee on containers that did not meet a minimum recycling rate or satisfy recycled content options. The fee had a 1995 sunset date and was not extended. Many states have given favorable treatment to recycling activities in the tax code (Table 14.4).

A number of states have instituted grant or loan programs to assist in the establishment of recycling. Funds from such programs have been used in a variety of ways, from developing educational materials for children to convince them of the value of recycling, to buying equipment for processing recyclable materials or for manufacturing products from these materials.

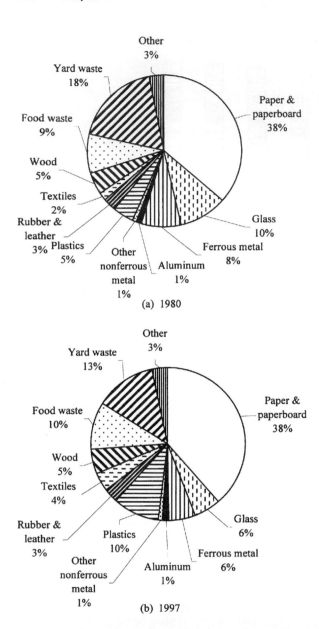

Figure 14.5 Proportions of materials in U.S. municipal solid waste: (a) 1980 and (b) 1987.[2]

In the late 1980s, when curbside recycling programs were growing at a very rapid pace, problems were encountered with supply of recyclables outstripping processing capacity and demand, resulting in dramatic decreases in the value of materials collected for recycling. This

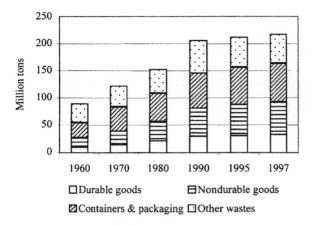

Figure 14.6   Products in U.S. municipal solid waste.[2]

TABLE 14.1   Summary of Recycling Requirements in the U.S.[4]

Recycling requirement	State
Goals only	Idaho, Louisiana, Mississippi, Nebraska, North Dakota
Local governments required to write recycling plans	Alabama, California, Hawaii, Iowa, Kentucky, Maryland, Massachusetts, Michigan, Missouri, New Hampshire, New Mexico, Ohio, South Dakota, Tennessee, Texas, Vermont, Virginia
Local governments required to implement recycling	Arizona, Arkansas, Delaware, Georgia, Illinois, Minnesota, Nevada (certain counties only), North Carolina, Oklahoma (municipalities), Oregon, South Carolina, Washington
Local governments required to meet recycling goals	Florida
Source separation of recyclables required	Connecticut, District of Columbia, Maine (businesses only), New Jersey, New York, Pennsylvania, Rhode Island, West Virginia (urban areas)
Volume-based disposal fees mandated if recycling goals not met	Wisconsin

was especially the case for newsprint. Community programs that depended on the revenue from selling old newspapers (ONP) to support
their activities found themselves, in a significant number of cases, not
only no longer receiving the counted-on revenue, but actually being

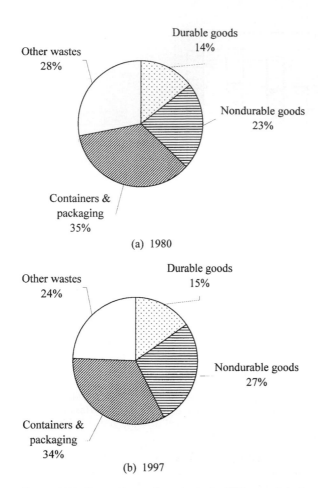

Figure 14.7 Proportions of products in U.S. municipal solid waste: (a) 1980 and (b) 1997.[2]

forced to pay to get rid of the collected newspapers. To bolster the markets and processing capacity for the material, a number of states considered the imposition of mandatory recycled content requirements on newspaper publishers. In several states, such laws were enacted. In a number of other states, newspaper publishers and state governments reached voluntary agreements for use of recycled content to avoid imposition of mandatory requirements (Table 14.5). These laws and agreements did result not only in creating expanded markets for recycled paper but, significantly, in stimulating investment by the paper industry in processing capacity for recycled paper. Because a significant amount of newsprint in the U.S. is imported from Canada, they affected paper manufacturing in Canada as well as in the U.S.

TABLE 14.2   Selected Disposal Bans in the U.S.[4]

State	Material
Alabama	Used oil, lead-acid batteries
Arizona	Used oil, lead-acid batteries, tires
Arkansas	Used oil, lead-acid batteries, white goods
California	Used oil, lead-acid batteries, tires, white goods
Colorado	Tires, construction waste
Connecticut	Designated recyclables, batteries, used oil
District of Columbia	Used oil
Florida	Used oil, lead-acid batteries, white goods, C/D waste
Georgia	Lead-acid batteries, tires
Hawaii	Used oil, lead-acid batteries
Idaho	Lead-acid batteries, tires
Illinois	Used oil, lead-acid batteries, tires, white goods
Iowa	Used oil, batteries, tires
Kansas	Tires
Kentucky	Lead-acid batteries, tires
Louisiana	Recyclables, used oil, lead-acid batteries, tires, white goods
Maine	Lead-acid batteries, tires, white goods
Maryland	Batteries, tires, white goods
Massachusetts	Recyclables, lead-acid batteries, tires, white goods
Michigan	Used oil, lead-acid batteries
Minnesota	Source-separated recyclables, used oil, batteries, tires, oil filters, white goods
Mississippi	Used oil, lead-acid batteries, tires
Missouri	Used oil, lead-acid batteries, tires, white goods
Nebraska	Used oil, lead-acid batteries, tires, white goods
New Hampshire	Recyclables, used oil, lead-acid batteries
New Jersey	Lead-acid batteries, white goods

TABLE 14.2   Selected Disposal Bans in the U.S.[4] *(Continued)*

State	Material
New Mexico	Lead-acid batteries, white goods
New York	Lead-acid batteries, tires
North Carolina	Used oil, lead-acid batteries, white goods
North Dakota	Used oil, lead-acid batteries, white goods
Ohio	Lead-acid batteries, tires
Oregon	Used oil, lead-acid batteries, tires, white goods
Pennsylvania	Lead-acid batteries
Rhode Island	Recyclables
South Carolina	Used oil, lead-acid batteries, tires, white goods
South Dakota	Used oil, lead-acid batteries, tires
Tennessee	Used oil, lead-acid batteries, tires
Texas	Used oil, lead-acid batteries
Utah	Used oil, lead-acid batteries, tires
Vermont	Used oil, batteries, tires, white goods
Virginia	Used oil, lead-acid batteries, white goods
Washington	Used oil, lead-acid batteries
West Virginia	Used oil, lead-acid batteries, tires
Wisconsin	Recyclables, used oil, lead-acid batteries, white goods
Wyoming	Lead-acid batteries

The success of mandatory recycled content requirements for newspapers in increasing markets for recycled paper set the stage for consideration of similar requirements for other materials. Oregon requires a minimum of 35% recycled content in glass containers, effective Jan. 1, 1999, which will increase to 50% on Jan. 1, 2002. California requires 35% minimum "postfilled" recycled content in glass containers. The term "postfilled" refers to "any container which had been previously filled with a beverage or food." The requirement is reduced to 25% for manufacturers who can demonstrate that their use of recycled glass consisted of at least 75% mixed color cullet.[4] California

TABLE 14.3   Bottle Deposit Legislation in the U.S.[4]

State	Containers Covered	Characteristics
Connecticut	Beer, malt beverages, carbonated soft drinks, soda water, mineral water	5 cent deposit
California	Beer, malt beverages, carbonated soft drinks, wine coolers; effective 1/1/00 water, sports drinks, fruit drinks, coffee, tea	Refund system rather than deposit, 2.5 cents on most containers
Delaware	Non-alcoholic carbonated beverages, beer and other malt beverages	5 cent deposit, aluminum cans exempt
Iowa	Beer, soft drinks, wine, liquor	5 cent deposit
Maine	Beer, soft drinks, distilled spirits, wine, juice, water and other noncarbonated beverages	5 cent deposit, 15 cents on wine and liquor, no deposit on milk
Massachusetts	Carbonated soft drinks, mineral water, beer and other malt beverages	5 cent deposit; containers 2 gallons or larger exempt
Michigan	Beer, soda, canned cocktails, carbonated water, mineral water, wine coolers	10 cent deposit; 5 cents on some refillable bottles
New York	Beer, soda, wine coolers, carbonated mineral water and soda water	5 cent deposit
Oregon	Beer, malt beverages, soft drinks, carbonated and mineral water	5 cent deposit; 3 cents on standard refillable bottles
Vermont	Beer and soft drinks, liquor	5 cent deposit; 15 cents on liquor bottles; all glass bottles must be refillable

also requires that a minimum of 30% average recycled content from glass containers be used in fiberglass sold or manufactured in the state.

Three states have passed laws affecting rigid plastic packaging, requiring minimum recycled content, minimum recycling rates, or source reduction. Wisconsin requires plastic containers, except for food, beverages, drugs, and cosmetics, effective in 1995, to consist of at least 10% recycled or remanufactured material by weight.[5] Reportedly, there is little enforcement of this legislation. Oregon requires rigid plastic containers, except food and medical packaging, to contain 25% recycled content, meet target 25% recycling rates (defined in a va-

TABLE 14.4   States Providing Favorable Tax Treatment for Recycling[4]

Arizona	Iowa	Missouri	Oklahoma
Arkansas	Kansas	Montana	South Carolina
California	Kentucky	New Jersey	Texas
Delaware	Louisiana	New Mexico	Virginia
Hawaii	Maine	North Carolina	West Virginia
Illinois	Maryland	North Dakota	Wisconsin
Indiana	Minnesota	Ohio	

TABLE 14.5   States with Minimum Recycled Content Requirements for Newspapers[4]

Mandatory	Voluntary
Arizona	Colorado
California	Hawaii
Connecticut	Indiana
District of Columbia	Iowa
Illinois	Kentucky
Maryland	Louisiana
Missouri	Maine
North Carolina	Massachusetts
Oregon	New Hampshire
Rhode Island	New York
Texas	Ohio
West Virginia	Pennsylvania
Wisconsin	South Dakota
	Vermont
	Virginia

riety of ways), or be source-reduced by 10%.[4] Since this law went into effect in 1995, the aggregate recycling rate for all plastic containers in Oregon has been above the required 25%, so all plastic containers satisfy the law's requirements automatically. In 1996, the estimated recycling rate was 33.3% in the state.[6] However, in 1998, the rate was down to 28.4% and, in early 2000, the state warned that its projections show that the plastics recycling rate may fall below 25% as early as 2002 if the current decline in rates continues.

California has a rigid plastic recycling law very similar to that of Oregon. In contrast to Oregon, which projects a recycling rate for the current year, California certifies the recycling rate for a past year, typically two years after the fact. In 1998, the California Waste Management Board determined that the aggregate recycling rate for rigid plastic containers in California fell below the required 25% in 1996, and it surveyed a selected group of manufacturers, requiring them to certify to the board how they were meeting the requirements of the law.[7] The previous year, the board had, after considerable controversy, adopted a recycling rate range for 1995 that spanned the required 25%, so no enforcement of the law was needed. The 1998 action took manufacturers by surprise, as there was widespread disbelief that the recycling rate had fallen as much as the state determined. Companies that had expected to be in automatic compliance were now being asked to demonstrate compliance retroactively. The state reached agreements with several large companies that admitted noncompliance, waiving penalties in exchange for the companies working to come into compliance and documenting those efforts. In early 2000, the California Waste Management Board levied a fine against one company (out of business at that point) that not only failed to show compliance but also refused to cooperate, stating that it did not feel it came under the jurisdiction of the law. At last report, the company planned to appeal the fine. The California rigid plastic container recycling rate for 1997, released in 1999, was also below 25%, and a wider spectrum of companies was surveyed for compliance. The 1998 rate will be released in 2000 and is also expected to be below 25%, since recycling rates for rigid plastic containers have continued to fall.

The majority of U.S. states require coding of plastic to facilitate plastics recycling by identifying the type of resin used (Table 14.6). The regulations specify use of the coding system developed by the Society of the Plastics Industry (SPI), consisting of a triangle formed from chasing arrows, with a number code inside the triangle and a letter code underneath. This system has been controversial since its inception, with complaints by many environmental groups that consumers misinterpret the identification symbol as an indication that the container is recyclable, or even that it contains recycled content. The problem is aggravated by the use of the symbols on a variety of objects other than rigid plastic containers. At the same time, the identification symbol has been criticized by recyclers as not providing enough information. For example, it does not differentiate between

TABLE 14.6   States Requiring the SPI Code on Rigid Plastic Containers[4]

Alaska	Illinois	Minnesota	Oregon
Arizona	Indiana	Mississippi	Rhode Island
Arkansas	Iowa	Missouri	South Carolina
California	Kansas	Nebraska	South Dakota
Colorado	Kentucky	Nevada	Tennessee
Connecticut	Louisiana	New Jersey	Texas
Delaware	Maine	North Carolina	Virginia
Florida	Maryland	North Dakota	Washington
Georgia	Massachusetts	Ohio	Wisconsin
Hawaii	Michigan	Oklahoma	

high- and low-melt flow grades of HDPE, even though the two are incompatible in a recycling system, and blending can result in a product that no end-users find appropriate for their needs. There was a long series of meetings between representatives of environmental groups, recyclers, and plastics industry representatives to try to develop a solution to these problems, but the effort eventually failed.

The approach to MSW management and recycling is very different in Europe as compared to the United States. Germany took the lead in Europe in implementing the producers' responsibility philosophy, later expanded to the entire European Union, which makes companies responsible for the proper disposal of the packages for their products, with requirements that certain percentages of such packaging be collected and that a certain percentage of collected materials be recycled. In most cases, industry responded by forming industry organizations to collectively handle the collection and recovery of the packaging so that they did not have to do it individually. The first such organization, in Germany, was Duales System Deutschland (DSD), commonly known as the Green Dot system.

It was not long before the idea of extending producers responsibility to products, not just to packaging, emerged. There are now requirements that limit to five percent the amount of new cars that can be landfilled. Automobile manufacturers have responded by changing the design of their products to make them more recyclable and are also using more recycled materials in their construction. The same philosophy is expected to be extended to other materials, such as construction and demolition debris, electronics, and household appliances.

Canada adopted a National Packaging Protocol, with a requirement to reduce the amount of packaging waste reaching disposal to 50% of 1988 levels by 2000.[8] To the surprise of many, this target was reached by 1996, when packaging waste disposal was reported to be 51.2% less than in 1988.[9] Various Canadian provinces have their own regulations in support of this goal, including deposits and fees, landfill bans, and requirements for the use of refillable containers.[10]

Japan has had a deposit system for beer and sake bottles for many years. In 1995, a law was passed to require businesses to recycle designated packaging wastes beginning in 1997. Polyethylene terephthalate (PET) bottles and other containers are covered, and non-PET plastic containers are included as of 2000. Industry responded by creating the Japan Container and Package Recycling Association, a third-party organization, similar to the Green Dot system, which collects a fee in exchange for handling the recycling of the packaging waste.[5]

South Korea has a deposit system on most containers, and it is designed to encourage the use of reusable packaging and promote recycling of nonrefillable containers. It has also adopted guidelines intended to reduce the volume of polystyrene cushioning used in packaging.[5]

One municipal council in Malaysia began in July 1997 to restrict the use of plastic packaging because of concern about disposal of plastics and about adverse effects on wildlife from littered plastics, especially those that enter bodies of water.[11]

Israel passed a law in 1997 that requires local councils to recycle at least 15% of their solid waste by the year 2000 and to recycle 25% by 2008. The recycling rate in 1996 was slightly over 25%, nationwide.[12]

Switzerland requires the take-back and recycling or environmentally sound disposal of electronic entertainment, office, information, communications, and household equipment.[13]

A variety of other countries around the world have adopted, or are adopting, policies aimed at promoting the recycling of packaging materials and thus reducing their disposal burdens. In many cases, they are following the European producer responsibility approach.

14.1.4   Other Pressures for Recycling

Legislation is not the only approach that can be used to push recycling. Various other incentives include consumer pressure, purchasing preferences, and pay-as-you-throw schemes.

Consumer pressure is often exhibited as preference for the purchase of products containing recycled materials, or as protest against goods or companies that fail to use recycled materials. The "consumer" may be an individual, an institution, or a business. For example, many businesses have made a commitment to purchase more recycled materials (sometimes under pressure from their own consumers) and therefore, as consumers, give preferential treatment to products containing recycled materials in making their own purchasing decisions. The recent action by Coca Cola to again begin using recycled content in soft drink bottles in the U.S.[14] is in response, at least in part, to high-profile activity by environmental groups criticizing the company for failing to do so. The campaign included full-page advertisements in the New York Times and Wall Street Journal and encouragement for individuals to mail empty soft drink bottles to the company.[15]

Purchasing preferences can be seen as one manifestation of consumer pressure. The term is more often associated with governments rather than businesses, and it entails some type of formal commitment to buy products made from recycled material when costs and performance are competitive. In many cases, formal provision is made for paying somewhat more for recycled materials than for competitive products from virgin materials. The largest such program in the U.S. is the federal purchasing preference program, which applies to government agencies and contractors (above a minimum size), requiring them to give preference to products made from recycled materials when purchasing listed products. EPA issues the *Comprehensive Guideline for Procurement of Products Containing Recovered Materials*,[16] which contains the list of covered products, and updates it periodically to reflect changes in technology and availability of recycled products.

Pay-as-you-throw schemes provide an indirect way of promoting recycling. If individuals or businesses are charged for their waste according to the amount they produce, and if the cost of getting rid of recyclables is less than the cost of getting rid of garbage, there is a built-in financial incentive to minimize generation of waste, in part by maximizing the diversion of potential recyclables from the garbage stream to recycling. Charging for waste disposal by the amount generated has been reasonably standard practice for businesses. In recent years, many communities have incorporated this idea into residential waste collection.

14.2   Collection of Materials for Recycling

For recycling to occur, three basic elements must be in place. First, there must be a system for collecting the targeted materials. Second, there must be a facility capable of processing the collected recyclables into a form that can be utilized by manufacturers to make a new product. Third, new products made in whole or part from the recycled material must be manufactured and sold. For recycling to be self-sustaining, the system as a whole must be economically feasible as well.

While processing and end uses differ substantially for different materials, there are similarities in collection that can usefully be discussed in a generic fashion. There are three main approaches: go out and get the material, create conditions such that the material will be brought to you, or use some combined approach. For all of these approaches, there is a relationship between the convenience the system provides to the generator of the materials and the motivation required to initiate and sustain the collection activity.

14.2.1 Deposit Systems

Deposit systems work by imposing a monetary deposit on the targeted items, paid by the consumer, which is refunded to the consumer when the item is properly returned. In some cases, the enabling legislation requires that the collected deposit items be recycled. In other cases, the economic incentives provided by the assemblage of materials are sufficient to ensure that recycling usually occurs.

The first widespread deposit system in the United States was not government mandated at all. Refillable glass beverage bottles (milk, soft drinks, and beer) had a deposit designed to ensure that the bottles got back to the distributor for refilling. The value of the bottles was high enough that the bottlers could not afford to have too many people fail to return them. No such deposits were charged on disposable plastic, glass, or aluminum containers, since the bottlers had no interest in their return. The "bottle bills" discussed in Sec. 14.1.3, which imposed deposits on most carbonated beverage containers in these states, were targeted initially at reducing litter but also served to greatly increase recycling of covered containers. The deposit of 5 or 10 cents per container proved to be a sufficient incentive to get consumers to bring in 90% or more of the covered containers to centralized collection points (retail stores). When there was a desire to increase recycling beyond PET soft drink bottles to other types of plastic containers, this was one model that could be followed. Maine was the first state to significantly expand its deposit law for the purpose of promoting recycling. In 1990, Maine extended deposits to containers for most beverages, excluding milk. In 1999, California expanded its refund value law (effective Jan. 1, 2000), bringing it closer to a deposit law format and extending coverage to water, juices, and sports drinks.

The beverage and retail industries have been adamantly opposed to deposit laws, citing high costs, sanitary concerns, and other objections. The primary advantages of deposit systems are their high rates of return of the targeted containers and relatively low levels of contamination. Handling of returned containers is often done by employees at retail stores. However, use of reverse vending machines, which can identify the container, check for contamination, and issue either money or a refund slip, is increasing.

14.2.2 Buyback Centers

For many years, the aluminum beverage can industry has depended on recycled aluminum to reduce the cost of manufacturing aluminum cans. The large amount of energy savings from using recycled aluminum, compared to ore, is very important to the overall economic viability of the industry. While deposit programs are very successful, the industry needed an effective alternative for states without such systems. Therefore, it established a network of buyback centers across much of the U.S. Individuals can bring collected aluminum cans to these facilities and are paid for the material based on its weight and the current value of the recycled metal. The industry also encourages can drives as fund-raising activities for local organizations. While collection rates at such centers are considerably lower than in deposit states, they did serve to increase the overall U.S. aluminum beverage can recycling rate to over 50%, at a time when rates for plastic and glass beverage containers hovered around 20%. Somewhat ironically, the success of such buyback centers has declined with the increased availability of curbside recycling across the U.S. The convenience of curbside recycling, for many consumers, outweighs the monetary reward offered by the buyback centers.

The idea of buyback centers is, of course, applicable to materials other than aluminum cans. Many glass plants will buy back glass containers from consumers. Many manufacturers of recycled paper will buy paper. At one time, there was a widespread network of Beverage Industry Recycling Program (BIRP) buyback centers, most of which accepted newspapers along with beverage containers, paying consumers for the material they brought in. These were established largely as part of efforts to ward off passage of bottle deposit legislation, and they, too, have declined as curbside recycling has become more commonplace.

14.2.3 Drop-Off Facilities

Drop-off recycling centers, like buyback centers, depend on consumers to deliver the targeted materials to a designated collection point. The difference between these facilities and buyback centers is that there is no monetary reward for the consumer. Therefore, they require the consumer to have higher intrinsic motivation for the behavior to continue.

Participation rates in such systems are often quite low, even when they are reasonably convenient.

The major advantage of drop-off facilities is that they are reasonably low in cost, especially if they are unattended. Their primary disadvantages are relatively low rates of participation and relatively high rates of contamination with undesired materials. Drop-off facilities are the primary means of collecting recyclables in much of Europe. In the U.S., BioCycle counted 12,699 drop-off recycling programs in 1997. No number was reported for 1998.[1,17]

14.2.4 Curbside Collection

In the U.S., most recycling of post-consumer materials is done through curbside collection. A BioCycle survey counted 9,349 curbside recycling programs in 1998, up from 8,937 in 1997, serving almost 139.5 million people.[1] In these systems, households set their recyclables out for collection in much the same way as they do their garbage, often in the same place and on the same day. Many of these systems provide a bin (usually colored blue) to consumers as a collection container. In most systems, the consumer places a variety of recyclables in the bin, perhaps with others bundled alongside, and the materials are sorted in a material recovery facility (MRF). Sometimes the sorting is done at truckside instead. In other systems, the consumers must sort the materials into designated categories before they are picked up. Both the latter systems rely on use of a compartmented recycling vehicle to keep the materials from intermingling. Some curbside systems use a bag (also usually blue) rather than a bin. In some of these, garbage and recyclables are collected at the same time, in the same vehicle, and the recyclables are sorted out after the load is dumped.

Because collection systems enhance convenience for the generator of the waste materials, participation in recycling is typically higher in these systems than in drop-off systems, where individuals must make more of an effort to feed the materials into the recycling system. Deposit systems are an exception; here, the added incentive of the monetary reward, plus the fact that the redemption center is typically located in a retail establishment where the consumers will be going anyway to buy their groceries, more than makes up for the little extra effort involved.

14.2.5 Recovery of Recyclables from Mixed Waste

One way to avoid the issue of how to convince individuals (or businesses) to participate in recycling is to separate recyclable materials from mixed municipal solid waste streams. The U.S. Bureau of Mines supported extensive work in the 1970s on the processing of garbage streams, using techniques derived from mineral processing. While most such efforts were unsuccessful because of high costs and unacceptable levels of contamination, the use of electromagnets to separate ferrous metals from mixed refuse is quite effective. In fact, ferrous metals can even be recovered from incinerator ash.

It is fairly common for waste disposal facilities to do at least some recovery of materials from municipal solid waste. Sometimes, the diversion is simple, involving, for example, separate collection of large items such as appliances. In other cases, loads of MSW that are known to contain higher than average quantities of recyclables are targeted for processing, often by hand-picking of desired materials. Loads from retail stores, for example, are often good sources of corrugated. Similarly, construction and demolition debris, though not classified as MSW, is often processed to remove steel and wood.

In some countries, recycling collection occurs primarily through the activities of scavengers. In Brazil, India, and Egypt, for example, groups of people make their living by sorting through dumps, recovering materials that can be sold. In other cases, the scavengers go door to door, collecting materials before they reach the disposal facilities.

14.3   Ferrous Metal Recycling

Recycling of ferrous metals (iron and steel) has a very long history. Most steel products manufactured contain at least 25% recycled content, and many are 100% recycled material. Worldwide, the recycling rate for the greater than 750 million tons per year of steel produced is over 50%.[13] Use of recycled metals saves energy, reduces pollution, and lowers costs. Scrap generated in manufacturing iron and steel, or in manufacturing products from ferrous metals, is routinely recovered and recycled. Most post-consumer recycled steel is recovered from items such as construction materials, automobiles, ships, and other such materials, which are not classified as part of the municipal solid waste stream. In recent years, however, the value of recovering steel from MSW has been recognized, and recycling of such materials has grown dramatically, even though the proportion of ferrous metals in MSW in the U.S. has steadily declined from 11.7% in 1960 to 5.7% in 1997.[2]

The majority of ferrous metal in MSW is found in durable goods, 9.23 million tons of the total 12.33 million tons. In 1997, recovery of ferrous metal from durable goods amounted to 2.84 million tons, for a recycling rate of 30.8%. The other major source of ferrous metal in MSW is containers and packaging. Food and other cans amounted to

2.86 million tons in 1997; 1.73 million tons were recycled, a rate of 60.5%. Other steel packaging, mostly barrels and drums, amounted to 240,000 tons, of which 160,000 tons were recovered, for a 66.7% recycling rate (Fig. 14.8). The overall recycling rate for ferrous metals in MSW was 38.4%.[2]

Overall steel recycling rates are significantly higher than the rate for steel in municipal solid waste, 65.2% in 1997 and 63.8% in 1998, for a total of over 67 million tons in 1998.[18] Automobiles and construction beams and plates had the highest recycling rates of the major categories reported by the Steel Recycling Institute (SRI) (Fig. 14.9).[18] The industry voiced concern about the decline in recycling rates in most categories of steel in 1998, compared to 1997, attributing the decline to increased competition from imported steel, and calling for the U.S. government to reconsider unrestricted steel imports, in light of the "poor environmental records of foreign steel producers."[19] U.S. steel production was down more than 20% in 1998.[20]

Steel recycling rates vary considerably around the world. Japan has shown dramatic improvement in recycling rates for steel cans, reaching 82.5% in 1998 (Fig. 14.10).[21] The steel packaging recycling rate in the European Union was 51% in 1998, down slightly from 52% in 1997, but still well above the 1996 rate of 45%. There were large differences between various EU countries, with Germany highest at 81%, and Luxembourg at the bottom at 10% (Fig. 14.11).[22]

At one time, recovery of tin from tin-plated steel was a dominant factor in recycling of steel cans. However, the increased use of tin-free steel in packaging applications, coupled with the decrease in average tin weight in tinplate, has led to a decline in the importance of this ac-

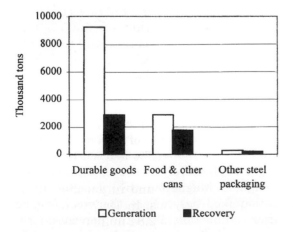

Figure 14.8 Steel generation and recovery in U.S. municipal solid waste, 1997.[2]

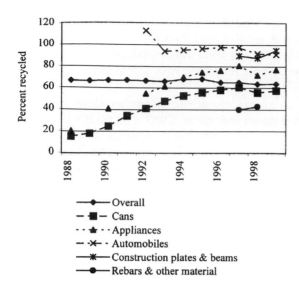

Figure 14.9 Recycling rates for selected steel products in the U.S.[18]

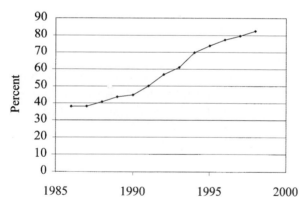

Figure 14.10 Steel can recycling rates in Japan.[21]

tivity. In 1996, a ton of steel cans contained only about 3.8 lb of tin. Some detinning facilities have closed in recent years because of these factors. In 1997, there were nine detinners in operation in the U.S.

As mentioned earlier, steel can be effectively recovered from municipal solid waste using electromagnets. Magnetic recovery of steel is especially common at incineration facilities, where the steel can be recovered either before or after combustion. Both the percentage recovered and the quality of the recovered material are somewhat higher when recovery precedes combustion. Recovery from MSW has

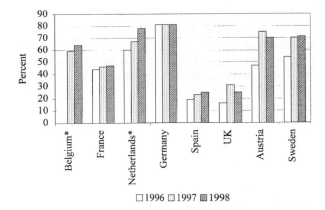

Figure 14.11 Steel packaging recycling rates in Europe.[22]
(*Rates for Belgium and the Netherlands are for steel and
aluminum cans combined.)

been a significant factor in increasing the recovery of steel from MSW,
and particularly from packaging wastes.

In the U.S., recovery of steel cans from curbside and drop-off recycling programs is an important factor in recovery of packaging steel. It
was estimated that steel cans were included in 6,900 of the 7,200 curbside collection programs in the U.S. in 1998, and in about 10,000 drop-
off programs. Steel cans were recovered "dockside" from a large number of commercial and institutional kitchens as well. In addition, 96 of
the 114 waste-to-energy facilities in the U.S. recovered steel magneti-
cally, for an annual recovery of about 775,000 tons.[23]

In the last several years, SRI has worked to encourage the recycling
of empty paint and aerosol cans. Michigan became the first U.S. state
to have statewide inclusion of empty aerosol cans in recycling programs. As of 1998, about 3,500 recycling programs were collecting
paint cans, and more than 4,700 were collecting empty aerosol cans,
mixed with other steel cans.[23]

Recycling of oil filters is increasing, reaching nearly 30% in 1997, in
large part due to the used oil filter hotline program sponsored by the
Filter Manufacturers Council, which supplies information to consumers about companies that accept used filters for recycling. There are
about 10,000 appliance recycling programs in the U.S., and the total
number of appliances recycled in 1997 was more than 46 million.[23]

Recycling of steel by the steel industry differs significantly in the
two major types of steel manufacturing processes. The basic oxygen
furnace (BOF) method of manufacturing steel uses 25 to 30% recycled
steel in the mix. In 1998, over 18 million tons of ferrous scrap were
used in the production of 59.7 million tons of liquid steel in BOF facili-

ties. Just under 6 million tons of scrap used was classified as pre-consumer "home scrap," unsalable material produced during steel manufacture. Of purchased scrap, 83.4% was post-consumer, and the remainder was pre-consumer "prompt scrap" produced by manufacturing processes for items made with steel. The overall post-consumer recycled content was 16.9%.[24] BOF steel is used for flat-rolled products such as sheet for cans, appliances, and automobiles.

The electric arc furnace (EAF) method uses virtually 100% recycled steel. In 1998, 44.3 million tons of ferrous scrap were used in the production of 45.8 million tons of liquid steel in EAF facilities. Just over 9 million tons was pre-consumer "home scrap" generated within the facilities. As for BOF facilities, outside purchased scrap was 83.4% post-consumer material, and the remainder pre-consumer "prompt scrap." The overall post-consumer recycled content was 64.3%.[24] EAF steel is used for long or heavy shapes, such as I-beams, rebar, wire, and nails. The percentage of U.S. steel manufactured in electric arc furnaces has grown from about 30% in the early 1980s to about 40% in 1997. It is expected to increase to 50% by 2020.[23] Electric arc furnaces are found primarily in mini-mills. In 1999, 89 mini-mills were in operation in 31 states and 6 Canadian provinces, with five U.S. states (Pennsylvania, Texas, Arkansas, Illinois, and Indiana) accounting for half of U.S. mini-mill production, and Ontario accounting for over a third of Canadian production.[25]

Steel recycling is primarily an open-loop process. Rather than going back into the same or similar applications, available scrap steel typically goes to the closest steel manufacturing facility. Thus, a steel can may become an automobile, which may become an appliance, which may become a can again.

In addition to reducing landfill space, steel recycling conserves natural resources, including energy. Each ton of recycled steel is estimated to conserve 2,500 pounds of iron ore, 1,400 pounds of coal, and 120 pounds of limestone. Overall, recycling steel is estimated to save enough energy each year to power about 18 million North American households.[26]

14.4 Aluminum Recycling

Historically, aluminum has had a higher recycling rate than most materials found in municipal solid waste. Much of this was due to recycling of aluminum beverage cans, which have had a recycling rate in the U.S. of over 50% every year since 1981. Beverage cans are the largest source of aluminum in the U.S. municipal solid waste stream, accounting for 1.53 million tons in 1997. Of this, 970,000 tons were recycled, for a 59.5% rate. About 50,000 tons of non-beverage aluminum cans were discarded, along with 360,000 tons of foil and closures. Recovery of foil and closures was about 30,000 tons, for a recycling rate of 8.3%. Recovery of non-beverage cans was less than 5,000 tons. The overall recycling rate for aluminum packaging was 48.5%. About 180,000 tons of aluminum in nondurable goods and 890,000 tons in durable goods were discarded, with no substantial recovery in either of these categories. The overall recycling rate for aluminum in MSW was 31.2%, and for aluminum packaging was 48.5%.[2]

Recycling rates for aluminum beverage cans in the U.S. have fallen somewhat in recent years from their high of 67.9% in 1992 (Fig. 14.12).[27] Drops in recycling rates for used beverage cans (UBC) have been attributed to decline in UBC prices, the demise of buy-back centers, and decreased use of UBCs in fund-raising activities by community groups. These, in turn, are associated with a decline in the value of virgin aluminum, increased curbside collection, and higher costs for buyback centers.

In Europe, where the aluminum can has had a smaller share of the beverage container market, recycling rates have been lower than in the U.S., though they have increased rapidly in the last several years. In 1994, the average European recycling rate for aluminum beverage cans was 30%. By 1998, it reached 41%, and a 50% rate is predicted for 2000.[28] Rates differ significantly among various European countries, as can be seen in Fig. 14.13.

Non-MSW sources of recycled aluminum include construction materials, sign posts, fencing, airplanes, etc. In 1998, the total amount of aluminum recycled in the U.S. was 9.1 billion pounds, of which almost

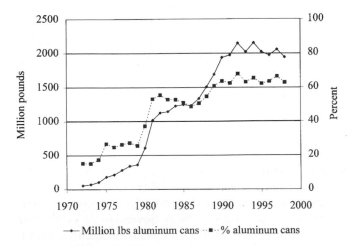

Figure 14.12 Aluminum beverage can recycling rates in the United States.[26]

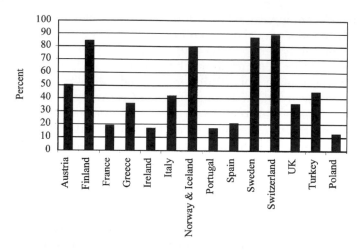

Figure 14.13 Aluminum can recycling rates in Europe.[28]

2.0 billion pounds was from used beverage cans.[29] In Europe, the recycling rate for aluminum in building and construction was estimated at 85% in 1998, and the recycling rate for transportation at 95%.[30] The overall recycling rate for aluminum in the U.S. is about 33%, the same as the overall global recycling rate.[30,31]

A significant motivating factor for aluminum recycling is the large energy savings provided by using recycled aluminum in place of aluminum from ore. Making aluminum ingot requires, on average, 186,262 MJ of energy if the aluminum is made from ore, but only 11,690 MJ, 6.3%, if the ingot is made from recycled aluminum (not counting energy required for transportation).[32] Generation of pollutants is also reduced substantially.

The processing of recycled aluminum from cans (UBCs) begins with shredding the collected material to a size of about 1/4 to 1 in. The primary reason for shredding is to avoid carrying large amounts of moisture into the melting furnace, where it could cause a serious explosion. The fines and aluminum dust created during shredding are collected and removed using high-efficiency cyclones to eliminate the hazard of dust explosions. Magnetic separation is used to remove any steel contained in the material, and pneumatic processing is used to remove paper and other contaminants. The material is then sent through a delacquering furnace where the metal is baked bare and paint or coating materials are removed. Next, the aluminum is charged into the melting furnace, where temperatures reach about 1450°F. Salt is added at this stage to remove some silicon and calcium. Some contaminants rise to the surface of the melt, forming the *dross* or *skim,* which must be removed and sent for further processing. The melted aluminum is next sent to the ingot plant, where alloying ingredients are added as needed to obtain the desired composition, and where the metal is filtered and fluxed to remove remaining contaminants. Finally, the ingots are rolled into can sheet.[33] The average recycled content of aluminum beverage cans in the U.S. in 1998 was 51.4%.[29] The dross is commonly recycled by toll processors, who also handle a variety of other types of scrap aluminum.

Recycling of other types of aluminum scrap proceeds similarly to recycling of beverage cans. The material is shredded, passed by magnetic separators, and screened to remove small particles of dust and dirt. Driers are used to remove moisture, oil, and organics before the scrap is melted.

14.5 Recycling of Other Nonferrous Metals

The primary nonferrous metals (other than aluminum) found in MSW are lead, copper, and zinc. Lead, primarily in lead-acid batteries, is the most common, amounting to 880,000 tons in U.S. MSW in 1997. Of this, 830,000 tons was recovered for recycling, for a rate of 94.3%. This very high recovery rate is attributable to deposits on lead-acid batteries in many states, coupled with bans on landfill or incineration of lead-acid batteries in most states (see Table 14.2).[2]

Other nonferrous metals in MSW amounted to about 390,000 tons in 1997, nearly all in durable goods. Recovery was insignificant, in terms of total tonnage. There is recovery of some silver from X-ray and photographic film, and there is some recovery of copper from materials such as pipe and wire (which are not classified as MSW).[2]

Recovery of metals from sources other than municipal solid waste is quite significant for some materials. Copper, for example, can be recovered from wiring in building and construction debris. Such types of metal recovery are not covered in this chapter.

14.6 Glass Recycling

Containers are the largest source of glass in MSW, amounting to 10.61 million tons in 1997 in the U.S. About 2.92 million tons were recovered, for a recycling rate of 27.5%. Figure 14.14 shows the breakdown by container category, for both generation and recovery. An additional 1.4 million tons of glass were found in durable goods, such as appliances, furniture, and electronics, with no significant recovery in this category. The overall recycling rate for glass in MSW was 24.3%.[2] The Glass Packaging Institute (GPI) claims considerably higher recycling rates for glass containers but has differences in calculation methodology, including an allowance for reuse of refillable glass containers. The

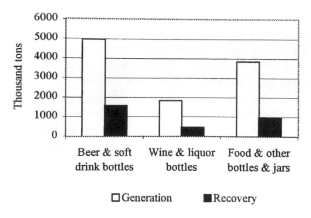

Figure 14.14 Glass container generation and recovery in the U.S., 1997.[2]

GPI reported a rate of 35.2% for 1997 (30.8% if refillables are removed), compared to EPA's 27.5% rate. The 1998 glass recycling rate reported by the Glass Packaging Institute fell to 34.8%, 30.1% if the allowance for refillables is removed.[34]

In the European Union, the average recycling rate for glass containers was about 58% in 1997.[35] Recycling rates differ substantially from country to country, as shown in Fig. 14.15.[36]

The primary market for recycled glass is use in the manufacture of new glass containers. The average recycled content for glass containers in 1995 was 25.35%.[37] Dividing the GPI's total for purchased cullet

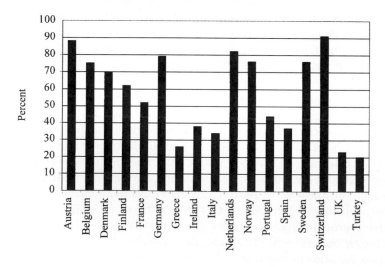

Figure 14.15 Glass recycling rates in Europe, 1997.[36]

by domestic production gives an estimate of 25.33% recycled content for 1998.[34] Successful use in manufacture of new containers requires that the glass be free of contaminants, including glass of nonstandard chemical composition, ceramics, stones, metal, and others. Crushed glass, ready for recycling, is called *cullet*. Every one percent increase in cullet use is reported to save 0.25% of the energy required to produce glass containers, primarily by lowering the required processing temperature. Use of cullet also prolongs furnace life; it is estimated that use of 50% cullet can double furnace life. Emissions of nitrogen oxides, particulates, and other pollutants are also decreased, as is water usage.[38]

One major issue in recycling glass is the need to separate glass of different chemical compositions. Most packaging glass is soda-lime glass, containing approximately 70% silicon oxide, 15% sodium oxide, 12% calcium oxide, 2% aluminum oxide, and 1% minor ingredients. Glass used in non-packaging applications may have a considerably different chemical makeup. Since the primary market for recycled glass is containers, most recycling programs will not accept non-container glass. A related issue is color. The color in glass is imparted through the addition of small amounts of minerals, generally iron and sulfur for amber (brown) glass, and chrome oxides for emerald (green) glass. In 1996, glass containers in the U.S. MSW stream were 58% flint, 33% amber, and 9% green.[39] If colors are mixed together indiscriminately, the resulting glass will have an undesirable greenish-gray color, and even very small levels of colored glass can impart an undesirable tint to clear (flint) glass, which makes up the majority of production. Therefore, if recycled glass is to be used in containers, it is necessary to pay careful attention to color separation.

Most sorting of glass by color is done with unbroken containers, either at the source by the consumer, during collection, or at a material recovery facility (MRF). Even in MRFs, the most common sorting method is hand-picking. For obvious reasons, broken containers present a hazard, so containers that break before color sorting result in, at best, a stream of low-quality mixed-color cullet, or at worst represent a disposal burden. Some studies have shown that it is common for more than 50% of glass collected through commingled container collection and processing at a MRF to be lost due to contamination and color mixing.[37] The broken glass can also contaminate other streams, especially paper, adding to waste and quality problems. Technology to automatically color-sort glass containers has been developed and is reportedly effective on broken glass as well as on intact containers. If this technology comes into widespread use, it could be of significant benefit. At present, however, broken glass is a major part of the waste stream in many MRFs.

There are non-packaging applications for cullet, some of which are suitable for mixed color and even contaminated materials. At the high-value end is production of fiberglass. California requires by law that 30% of the content of fiberglass be recycled glass containers. At the low-value end, cullet can be substituted for some of the aggregate in a variety of construction applications, such as concrete, road beds, pavement (glasphalt), drainage medium, and backfill. Other applications include use in sand blasting, manufacture of glass pellets and beads, frictionators for lighting matches and similar applications, and decorative uses such as manufacture of ceramic tiles, picture frames, costume jewelry, and household items.[34]

14.7   Paper Recycling

Recycled paper has long been an important source of material for paper and paperboard packaging. In recent years, packaging, especially the corrugated variety, has also become an important source of recycled paper. Paper and paperboard account for the largest fraction of municipal solid waste in the U.S., and also the largest component of the material recovered from MSW (Figs. 14.4 and 14.16). Containers and packaging account for about 47% of all the paper and paperboard

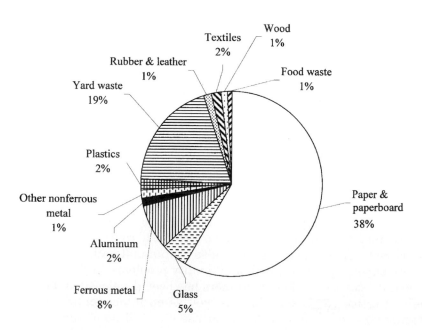

Figure 14.16 Materials recovered from municipal solid waste in the U.S., 1997.[2]

products in MSW and for over 60% of all the paper and paperboard recovered from MSW (Fig. 14.17).

14.7.1 Paper Recycling Rates

In 1996, the average recovery rate for paper and paperboard around the world was 42%.[40] Both the tonnage of paper and paperboard recycled and the recycling rate have grown substantially in the U.S. in recent years (Fig. 14.18), increasing more than 50% since 1990 for a total of over 45 million tons in 1998, a recovery rate of 44.7%.[41] Recovery of corrugated boxes (old corrugated, OCC) accounts for the largest fraction, followed by old newspapers (ONP) and mixed papers (Fig. 14.19). (It should be noted that, in calculating these statistics, the American Forest and Paper Association defined "corrugated" to include Kraft grocery bags, multi-wall shipping sacks, and similar mate-

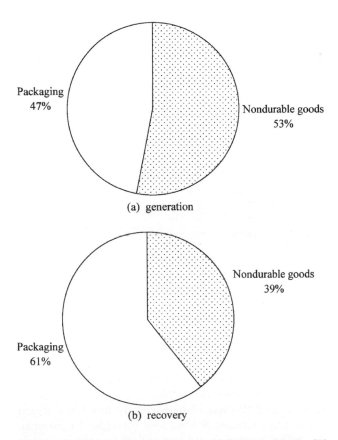

Figure 14.17 Paper and paperboard (a) in U.S. municipal solid waste and (b) recovered from municipal solid waste, 1997.[2]

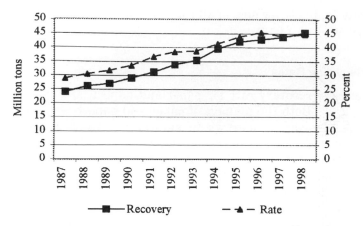

Figure 14.18 Paper and paperboard recycling in the U.S.[41]

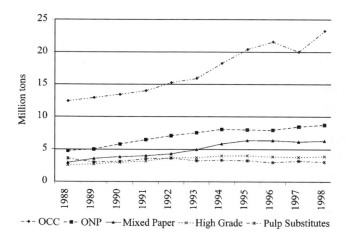

Figure 14.19 Paper and paperboard recycling in the U.S., by category.[41]

rials in addition to corrugated boxes.) Canada has shown similar growth in recycling, reaching a recovery rate of 45.1% in 1997, up from 19.6% in 1981 and 27.6% in 1990.[42] Manufacture of containerboard is the largest tonnage use for recovered paper in the U.S. (see Fig. 14.20).[43] The term *containerboard* refers to both linerboard and medium used in manufacture of corrugated board.

Many other countries around the world historically have had much higher recycling rates than the United States and Canada. In general, countries with abundant forest resources have relied less on recycling than countries without such resources, as would be expected. The

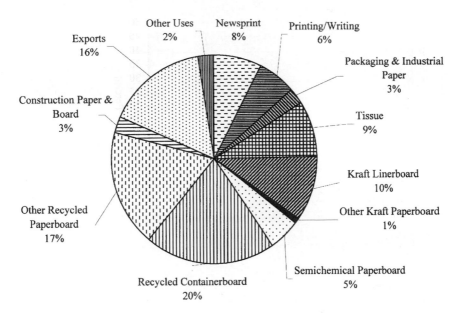

Figure 14.20 Uses for recycled paper and paperboard in the U.S., 1996.[43]

1997 paper recycling rate in Mexico was close to 50%.[44] Recycling rates for Canada and Japan are summarized in Fig. 14.21.[21,42] Paper recovery rates for 1995 for some other Asian countries are shown in Fig. 14.22, and rates for some European countries are shown in Fig. 14.23. It should be noted that not all of this material is recycled; rather, some is disposed by incineration with energy recovery.[45] The

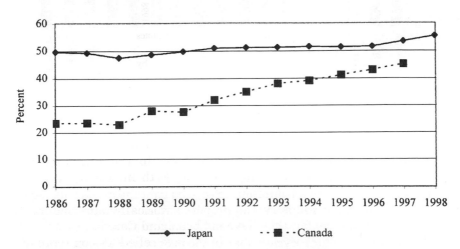

Figure 14.21 Paper and paperboard recycling rates in Japan and Canada.[42,45]

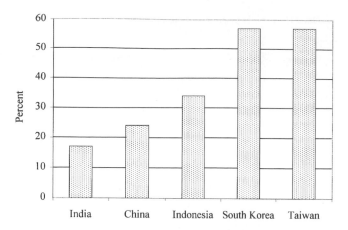

Figure 14.22 Paper and paperboard recovery rates in Asia, 1996.[40]

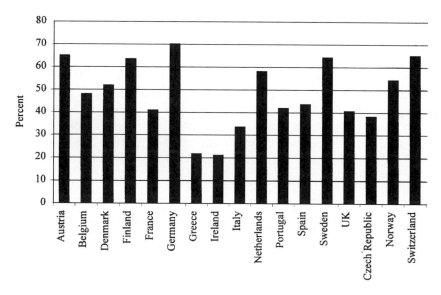

Figure 14.23 Paper and paperboard recovery rates in Europe, 1998.[46]

recycling rates for European countries as a percentage of collected material are shown in Fig. 14.24. Recently, the EU has considered imposition of requirements for minimum recycled content in various paper products. The paper industry has proposed voluntary agreements as an alternative but, so far, they have not been accepted.[46]

Utilization rates for recycled fiber in the production of paper and paperboard materials also vary significantly around the world. In the

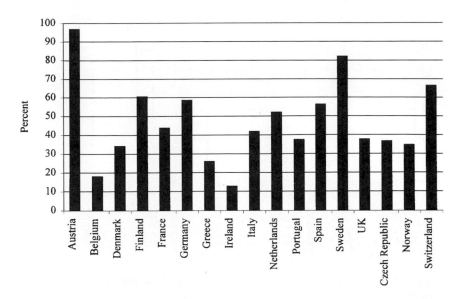

Figure 14.24   Paper and paperboard recycling as a fraction of recovery, Europe, 1998.[46]

U.S., about 80% of all papermakers use some recovered fiber in manufacturing, with the average recycled content exceeding 37%, up from 25% in 1988.[41] The historical trend in utilization is shown in Fig. 14.25. The largest category of use by U.S. paper and paperboard mills is the manufacture of recycled corrugated (Fig. 14.26).[49]

In Canada, about 71% of the fiber used in papermaking comes from recovered paper plus sawmill residues.[42] Regional waste paper utiliza-

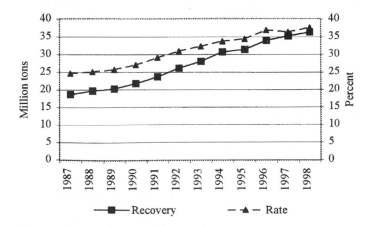

Figure 14.25   Utilization of recovered fiber in U.S. paper and paperboard mills.[41]

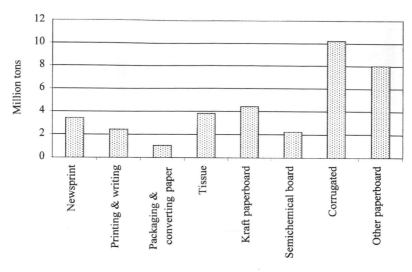

Figure 14.26 Products made from recovered fiber in U.S. paper and paperboard mills, 1998.[48]

tion rates for about 1995 are summarized in Fig. 14.27.[48] Figure 14.28 shows the historical pattern in use of domestic and imported pulp and waste paper in the U.K.[49]

Recycled paper, and paper and paperboard, are significant commodities in international trade. Table 14.7 categorizes regions of the world as net importers or exporters of recovered paper, pulp, and paper products. Of course, there are differences within regions as well as between regions. In North America, Canada imported about 45% of its

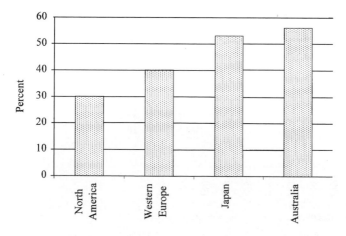

Figure 14.27 Regional recycled fiber utilization rates in papermaking.[49]

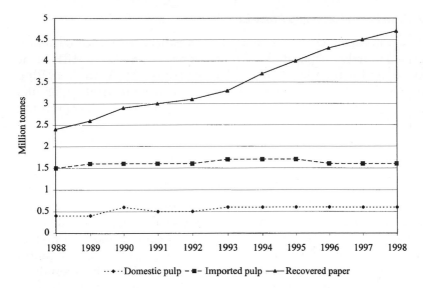

Figure 14.28 Domestic and imported pulp and waste paper utilization in the U.K.[50]

TABLE 14.7   Net Importers and Exporters of Pulp and Paper[43]

Region	Paper products	Pulp	Recovered paper
North America	Export	Export	Export
Western Europe	Export	Import	Export
Eastern Europe and Russia	Export	Export	Import
Oceania and Africa	Import		Export
Latin America	Import	Export	Import
Asia	Import	Import	Import

recovered paper in 1997, almost exclusively from the United States, and exported about 75% of the paper and paperboard it produced, mostly to the U.S.[42] The U.S. exported 16% of the paper it recovered for recycling in 1998, mostly to Canada and the Far East.[41] In Western Europe, Sweden, Austria, Spain, France, and Italy were importers of recovered paper; Finland, Norway, Portugal, Ireland, Greece, Denmark, Belgium, the Netherlands, Germany, and the UK were net exporters in 1997.[43]

14.7.2   The Paper Recycling Process

Recycling of paper begins, of course, with collection. Collected material may be further separated by grade at the source, at a MRF, or at truckside, and it then is usually baled for transport. At the recycling facility, the baled material is deposited in a hydropulper, which breaks apart the bale and resuspends the fibers using a large tank containing a blade that provides intense agitation. The removal of contaminants starts at this point. A ragger hangs into the hydropulper to remove long stringy objects such as baling wire. A junk remover removes heavy materials. From the hydropulper, the suspended fibers pass through a variety of cleaning mechanisms, typically including centrifugal cleaners and various types of screening devices. Centrifugal cleaners are designed to separate materials by density, removing fractions that are either too light, such as plastic film, or too heavy, such as staples and stones. Screens or filters are designed to separate materials by size, removing materials that are too small (such as small fiber fragments and dirt) or too large, (such as large contaminants or fiber bundles that have not been broken up sufficiently by the repulping action of the hydropulper and need to be recirculated). Systems even exist that can separate the long fibers in corrugated originating in the liners from the short fibers originating in the medium.

High-grade papers intended for applications where white paper is desired must generally be de-inked to remove previous printing. Newsprint intended for recycling into new newsprint must also be de-inked. For corrugated boxes and for other recycled paper streams intended for use in packaging paperboard, de-inking is not usually required. The predominant de-inking technology is flotation de-inking, in which chemical treatment combined with generation of air bubbles is used to remove the ink from the paper fibers and attach it to bubbles of air, which convey the ink to the top of the flotation cell, where it is removed in the form of scum. The presence of clay in the cell facilitates the attachment of the ink to the bubbles. This has resulted, in the U.S., in old magazines (OMG) changing from an undesirable material to one that is typically in higher demand than is available. In newspaper recycling, between 5 and 30% OMG is desired, with the remainder ONP, depending on the mill technology. An 80/20 ratio of ONP to OMG is about average.[50]

After processing, the recovered fiber is usually made into paper immediately, either alone or mixed with virgin fibers, although it can also be dewatered and baled for shipment to another paper manufacturer. The recycling process results in significant shortening and weakening of the paper fibers. Consequently, the properties of paper containing recycled fiber are generally somewhat inferior to equivalent paper made from virgin pulp. There is also significant loss of material during the recycling and paper-making operation. A rule of thumb is about 10% of the weight of recovered material delivered to the papermaker for recycling will be lost if de-inking is not required. If de-inking is involved, the loss may be about 30%.

The decrease in properties as a consequence of recycling, along with other considerations, means that paper recycling is largely an open-loop process, with a significant fraction of the recovered fiber going to somewhat downgraded applications rather than back to the same use. Typical applications of various types of recycled paper and paperboard will be discussed in more detail in the following subsections.

14.7.3 Recycling of Packaging Paper and Paperboard

As mentioned, nearly half of all paper and paperboard found in U.S. municipal solid waste originates in packaging, and more than half of all the paper and paperboard recycled is packaging. Figure 14.29 shows the proportion of types of packaging paper and paperboard found in the U.S. waste stream in 1997, and Fig. 14.30 shows the proportion of packaging paper and paperboard recycled. As can be seen, corrugated dominates recovery of paper-based packaging. In fact, slightly over one-half of all paper materials recovered for recycling in the U.S. fell into the general category of corrugated. The recycling rate for corrugated (Fig. 14.31) is significantly higher than that for most other paper or paperboard materials.

Most corrugated is collected for recycling through retail stores and businesses, which receive large quantities of goods in corrugated

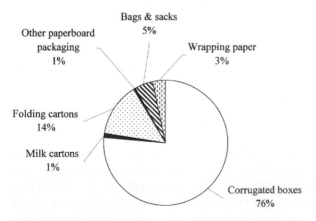

Figure 14.29 Packaging paper and paperboard in U.S. municipal solid waste, 1997.[2]

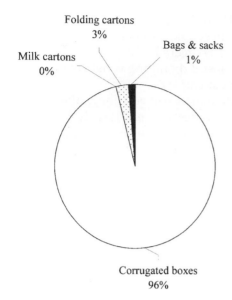

Folding cartons
3%

Bags & sacks
1%

Milk cartons
0%

Corrugated boxes
96%

Figure 14.30 Recovery of packaging paper and paperboard in the U.S., 1997.[2]

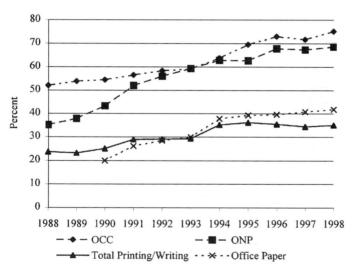

Figure 14.31 Recycling rates in the U.S. by category of paper materials.[41]

boxes. Some is collected from individuals through drop-off centers or through curbside recycling. However, residential corrugated amounts to only about 13% of all OCC available for recovery in the U.S. (Fig. 14.32).[51] The recovery rate for OCC from manufacturing, retail, and commercial facilities was 70–81% in 1995, compared to only about 5% for residential OCC (Fig. 14.33).[51] Corrugated generated by busi-

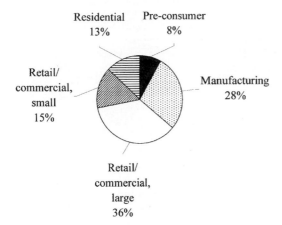

Figure 14.32 Supply of old corrugated containers in the U.S.[52]

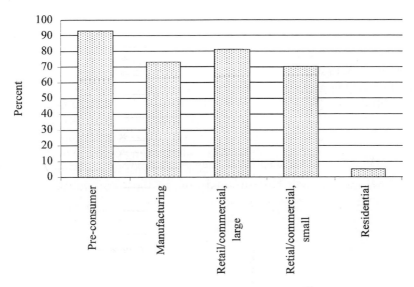

Figure 14.33 Recovery rates for OCC in the U.S. by sector.[452]

nesses is often baled on-site for economy in transporting the material. The business may or may not receive payment from the recycler or other collector for the material generated.

Corrugated is produced primarily from high-quality Kraft pulp, which has excellent strength properties. Thus, even after recycling, corrugated is a valuable source of fiber. The largest use of recycled corrugated is back in the production of containerboard (defined as corrugated and related materials) as shown in Fig. 14.34, where it may be

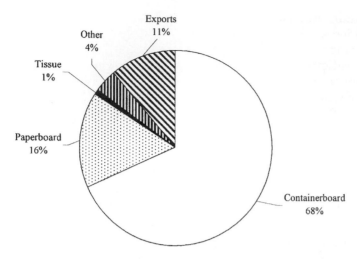

Figure 14.34 Uses of OCC in the U.S.

used alone or blended with virgin fiber. In 1999, over one-fifth of the total industry capacity for linerboard manufacture was in facilities using 100% recycled fiber.[47]

Kraft bags, such as grocery sacks and multi-wall bags are, as mentioned, included in AFPA's definition of corrugated. Kraft bags are often recycled along with corrugated, although their recovery rate is considerably lower than that of corrugated boxes (Fig. 14.35). Recycling of multi-wall bags is more complex, since many of these bags contain additional materials such as plastic film to enhance strength and barrier. This makes such bags undesirable in most corrugated recycling programs.

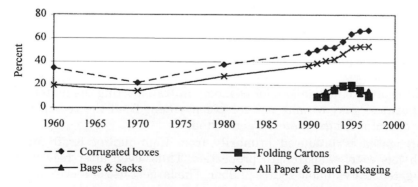

Figure 14.35 Recycling rates for packaging paper and paperboard in U.S. municipal solid waste, 1997.[2]

Another undesirable material is corrugated boxes that have been modified for added strength in humid conditions, such as by wax impregnation, coating with plastic, or modification of the paper structure through incorporation of thermosetting plastic resins. These materials present problems in paper recycling operations, either through failing to repulp properly, adding contaminants to the pulp, or both. Recently, there have been efforts to press for coding of such boxes to help keep them out of the recycling stream, while preventing desirable boxes from being discarded because the generators were fearful that they contained wax or other undesirable constituents. It has been estimated that as much as 15% of non-waxed boxes generated in grocery stores were not being recycled because they could not be distinguished from waxed boxes.[53]

The U.S. corrugated industry produces about 1.5 million tons of waxed boxes per year, about five percent of all corrugated boxes.[54] A voluntary coding system for wax-impregnated boxes that has been adopted in North America involves marking them on the box flaps with the word "wax" in English, Spanish, and French *(WAX/CERA/CIRE)*. A longer-term solution is the development of repulpable wax, and considerable research is devoted to this end. Other research is directed at developing recycling mill processes that can separate waxed fibers from non-waxed fibers. Inland Paperboard and Packaging, Inc., of Indiana, and Thermo Black Clawson, of Ohio, have developed a special recycling process for waxed boxes called *Xtrax*. The process uses reverse screening along with other fiber cleaning techniques to reduce the wax content in the material from over 30% initially to less than 1% in the end product. Average recycled OCC is reported to contain about 3% wax, which is acceptable for most grades.[55] The American Forest and Paper Association maintains an online directory of waxed corrugated recovery facilities on its web site, www.afandpa.org.

Wet-strength grades of paper, which incorporate thermoset resins, can be repulped if they are properly identified and sorted by type. However, few if any recycling opportunities are available for post-consumer wet-strength materials.

Folding cartons and other paperboard packaging are recovered at significantly lower rates than corrugated. Many of these materials are manufactured from 100% recycled fiber. Thus, the overall quality of the potentially recoverable fiber is lower than for corrugated, which contains, on average, a substantial percentage of virgin fiber. In addition, because these materials are dispersed in households rather than being concentrated in businesses, recovery is more expensive. Fewer curbside and drop-off recycling programs accept these materials, compared to programs accepting corrugated boxes, newspaper, and even mixed paper "junk mail."

14.7.4 Newspaper and Magazine Recycling

Newspapers are the largest category of non-packaging paper and paperboard found in municipal solid waste (Fig. 14.36) and also the largest category of recovered non-packaging paper and paperboard (Fig. 14.37). In contrast to collection of corrugated, recycling of newsprint is largely accomplished through collection from individuals, predominantly through curbside collection. Newsprint is a relatively low-quality paper, produced primarily through mechanical pulping rather than through Kraft or other chemical pulping methods. (The American Forest and Paper Association uses the term *groundwood* to refer both to pulp produced by the old stone groundwood process and to that produced by the more modern thermal mechanical pulping, both of which are classified as mechanical pulping processes.) These methods result in a higher than average yield of pulp of lower than average quality. Consequently, ONP is lower in quality than OCC and more limited in use. U.S. recycling rates for newsprint are shown in Fig. 14.38.

The largest use of ONP is in the production of new newsprint, usually in a blend with virgin fiber (Fig. 14.39). As discussed in Sec. 14.1.3, a number of U.S. states have either mandatory requirements for recycled content in newspapers or have reached voluntary agreements with the newspaper industry, which has had a significant effect

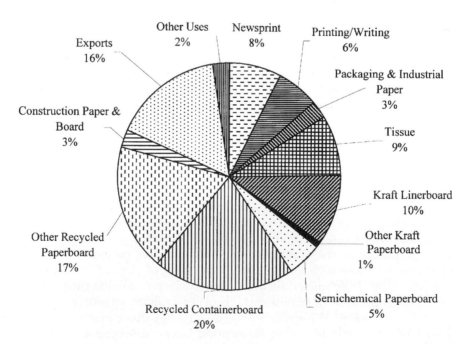

Figure 14.36   Non-packaging paper and paperboard in U.S. municipal solid waste, 1997.[2]

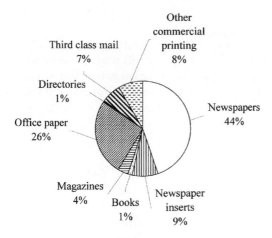

Figure 14.37   Recovery of non-packaging paper and paperboard in the U.S., 1997.[2]

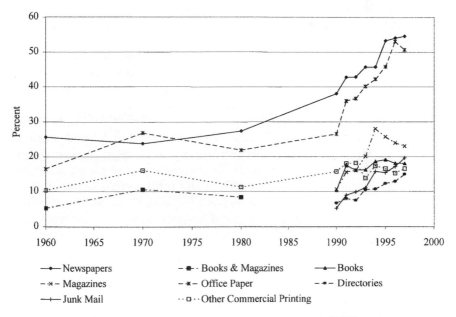

Figure 14.38   Recycling rates for non-packaging paper in the U.S.[2,43,52]

on utilization rates. In 1997, use of recycled fiber amounted to about 30% of newsprint production in North America. Many states are pushing for 40%, a goal that is not likely to be attained.[51] In the U.K., the proportion of waste paper used in newsprint reached 52.42% in 1998, up from 47.43% in 1997. However, because of the effects of imports,

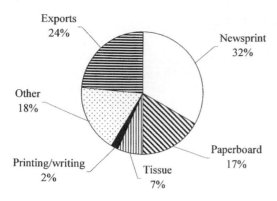

Figure 14.39 Uses of recycled newsprint in the U.S.[41]

the overall recycled content of newsprint consumed by UK newspapers was only 46.54%.[56] In 2000, the UK was considering legislation that could impose a mandatory recycled content of 80% on newsprint.[57]

Newsprint insets are usually also manufactured from mechanical pulp but often contain clay coating to enhance their printing surface. These inserts are usually recycled right along with the newsprint. In the U.S., in 1997, newsprint accounted for 81.2% of the total newspaper stream, with inserts making up the remainder. The recycling rate for newsprint was 56.3%, and for groundwood inserts it was 47.4%.[2]

Magazine paper is primarily made from a high-quality grade of mechanical pulp (though some is made from chemical pulp) and contains substantial amounts of clay coating to produce a glossy printing surface. The clay coating, which actually consists of clay plus other minerals, may make up as much as 40% of the total weight of the sheet.[58] As discussed in Sec. 14.7.2, this clay is now seen as an asset in de-inking, so magazines are in high demand, particularly by mills producing recycled newsprint. Most magazines are collected through curbside collection and drop-off programs, though there are also office recycling programs for magazines.

The paper in telephone and similar directories is a low-quality groundwood fiber, similar to newsprint. However, because of the binding and glue, it is not generally desirable to mix these materials with newsprint for recycling. These materials tend to be generated in a highly seasonal fashion, with large numbers of directories available for recycling at the time that new directories are distributed, and the number then falling to very low levels. Therefore, most of such materials that are recycled are collected during seasonal campaigns as a temporary addition to curbside, drop-off, or office paper recycling programs.

14.7.5   Recycling of Printing and Writing Paper

The primary collection source of printing and writing paper for recycling is office paper recycling programs. Many offices, in various sectors of the U.S. economy, have a two-tiered collection system for paper: one high-quality white paper stream and a lower-quality colored or mixed paper stream. In some cases, newspapers make up a third recycling stream, and occasionally the collection of magazines and telephone directories may be added.

Office paper is predominantly made from high-grade chemical pulp, so it can be suitable for uses for which Kraft paper and corrugated are not appropriate, such as the manufacture of new high-grade paper, providing the paper is properly sorted to remove undesirable materials. Sorting typically begins with source separation. Often, the generators of office paper are asked.to put the highest-quality white paper with no adhesives, groundwood paper, or other contaminants, into one collection bin, and other types of paper with greater tolerance for contamination in another. This stream, or in some cases a single mixed office paper stream, is sometimes further sorted in an MRF. However, automated equipment to do this effectively has not yet been developed, so such sorting is labor intensive and expensive.

Other sources of high-grade paper for recycling include stationary, books, reports, etc. In many cases, some materials in the category are manufactured from high-grade chemical pulp, while others are manufactured from lower-grade groundwood pulp. Often, the materials are collected as part of a mixed paper recycling stream and are diverted to low-end uses. Some are collected as part of office paper recycling programs, and others through curbside or drop-off programs. For example, in the UK, the post office ran a collection program for old Christmas cards, targeting children at elementary schools.[59]

Recycling rates for various high-grade papers are shown in Fig. 14.38, and typical uses for recycled printing and writing paper in Fig. 14.40. The average recycled content in printing and writing paper in the U.S. was only about 6 to 7% in 1997, down from a high of about 10% earlier in the decade.[60]

14.7.6   Contamination Issues

Several different types of contamination are issues for the use of recycled paper. The presence of inks, dyes, or pigments changes the appearance of recycled paper, generally in undesirable ways. Mixing of fibers produced by mechanical pulping with fibers produced by chemical pulping is a concern where the better strength and permanence of chemically pulped fibers is important. The presence of added materi-

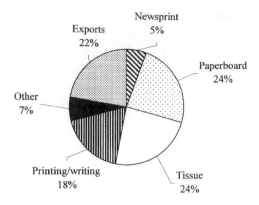

Figure 14.40 Uses for recycled printing and writing paper in the U.S., 1998.[41]

als such as waxes and adhesives can detract from the appearance and from the strength of paper. Adhesives are a particular problem, since they are so widely used with paper products. In general, pressure-sensitive adhesives present more problems than other types. Efforts are continuing to develop pressure-sensitive adhesives for applications such as postage stamps and labels that are more compatible with recycling systems.

14.8   Plastics Recycling

The origins of plastics in MSW in the U.S. are shown in Fig. 14.41. Figure 14.42 describes the relative proportions of various plastic res-

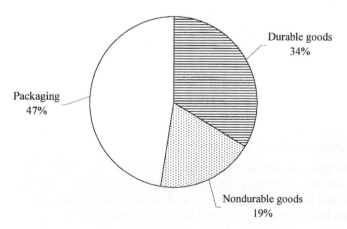

Figure 14.41 Plastics in U.S. municipal solid waste, by product category, 1997.[2]

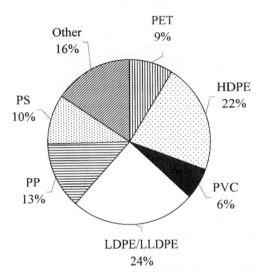

Figure 14.42 Plastics in U.S. municipal solid waste, by resin, 1997.[2]

ins found in MSW. The overall recycling rate for plastics is generally considerably smaller than the rates for metal, glass, and paper. In part, this is due to the shorter history of plastics recycling. It is also due in part to the greater complexity in recycling plastics, since the category contains a large variety of different materials, each with its own characteristics. Since polymers are generally incompatible with polymers of different chemical type, mixing plastics together indiscriminately usually produces a material with very poor performance characteristics. Thus, production of high-value products from recycled plastics is intimately tied to the ability to separate materials by resin type. Some markets exist for commingled recycled plastics, as will be discussed in Sec. 14.8.13, but these are generally considerably lower in value than the potential markets for single resin recycled materials. Historical trends in recycling rates for plastics in MSW in the U.S. and in Western Europe are shown in Fig. 14.43. The recycling rates shown for plastics in Western Europe exclude incineration with energy recovery. If this option is included, recovery rates increase, ranging from 20% in 1994 to 30% in 1998.[61] In most countries, including the U.S., recovery of energy is not classified as recycling.

Recycling rates are significantly higher for some plastic materials than for others, and for some types of plastic products, as will be discussed in more detail in the following subsections. Many recycling programs for plastics focus on plastic containers, or even more narrowly on plastic bottles. The American Plastics Council (APC) calculated the 1996 recycling rate for rigid plastic containers as 21.2%. The rate for

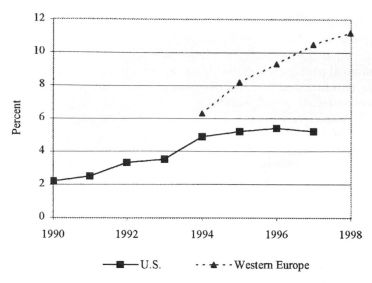

Figure 14.43 Plastics recycling rates in the U.S. and Western Europe.[2,61]

1997 fell to 20.2%, although the tonnage of plastic collected for recycling increased from 1.321 billion to 1.375 billion pounds. Use of virgin plastic increased at a higher rate, from 6.221 billion to 6.800 billion pounds during the same period. The APC reported the recycling rate for plastic bottles as 23.5% in 1998, down from 23.7% in 1997 and 24.5% in 1996. The 1996 recycling rate for flexible plastic packaging was only 2.8%.[62,63]

RECOUP reported that 7,000 tonnes of plastic bottles were collected in the UK in 1997, a total of 140 million bottles. More than 1 in 3 local solid waste authorities ran a plastic bottle recycling scheme, including over 3,000 plastic bottle banks and nearly 2 million homes served by curbside collection.[64]

In Germany, under the DSD Green Dot system, more than 5 million metric tons of post-consumer plastic packaging was collected in 1995, 79% of the total amount of plastic packaging generated by households and small businesses. Almost 90% of that came from curbside recycling collection systems, with the remainder coming from drop-off systems at supermarkets, gas stations, and public buildings. Slightly less than 55% of the collected material was recycled, with the remainder used for energy recovery.[65]

How recycling rates are calculated is itself a source of controversy. There have been charges in the past that surveys that ask recyclers for data receive inflated figures and thus inflate recycling rates. Surveying organizations take various steps to avoid this problem but cannot completely eliminate it. On the other hand, some organizations that do recycling may be missed in the survey, thus decreasing recycling rates. Even if the accuracy of the data could be guaranteed, a more fundamental problem remains. What should be counted as being recycled? The two fundamental options are (a) measuring the amount of material collected for recycling or (b) measuring the amount of material actually reused. Since 5 to 15% of collected material is lost during processing of the material, mostly because it is some type of contaminant such as paper labels, product residues, undesired types of plastic, etc., there is a significant difference in recycling rates between the two approaches. The American Plastics Council (APC) is the major source of information about plastics recycling rates in the U.S. In 1997, APC switched from using the amount of cleaned material ready for use to the amount of material collected for processing. Their justification was that the latter method is more in keeping with the way recycling rates are calculated for other materials. Especially since this resulted in inflating recycling rates at a time when recycling rates, if calculated by comparable measures, were declining, this move brought considerable criticism. For instance, the PET bottle recycling rate in 1997 was 27.1% if based on material collected but only 22.7% based on clean material ready for reuse.[66] Criticism of APC was further heightened by their decision to restrict distribution of their annual plastic recycling report to APC members. The Environmental Defense Fund, in response, issued a report titled "Something to Hide: The Sorry State of Plastics Recycling," in which they used APC's numbers to highlight the decline in plastic recycling rate that was evident when 1996 data was compared to 1995 on the same basis. They also noted that polystyrene food service items were deleted from APC's definition of plastic packaging, beginning in 1995—a move that further shored up plastics recycling rates. EDF calculated that the recycling rate for plastic packaging in 1996 was only 9.5% and would have been only 8.5% if polystyrene food service items had been included.[62]

14.8.1 Processing of Plastics

Three major categories of recycling processes for plastics exist: mechanical recycling, chemical recycling, and thermal recycling.

Mechanical, or physical, recycling involves changing the size and shape of the plastic materials, removing contaminants, blending in additives if desired, and similar activities that change the appearance of the recycled material but do not alter (at least not to a large extent) its basic chemical structure. Common processing steps include grinding, air classification to remove light contaminants, washing, gravity-based separation of resins that are heavier than water from those that are lighter than water, screening, rinsing, drying, and often melting and pelletization, perhaps with the addition of colorants, heat stabilizers, or other ingredients. Mechanical recycling is by far the most common type of plastics recycling.

Chemical recycling involves using chemical reactions to break down the molecular structure of the plastic. The products of the reaction then can be purified and used again to produce either the same or a related polymer. An example is the glycolysis process sometimes used to recycle PET, in which the PET is broken down into monomers, crystallized, and repolymerized. Condensation polymers such as PET, nylon, and polyurethane are typically much more amenable to chemical recycling than are addition polymers such as polyolefins, polystyrene, and PVC. Most commercial processes for depolymerization and repolymerization are restricted to a single polymer, usually PET, nylon 6, or polyurethane.

Thermal recycling uses heat to break down the chemical structure of the polymer. In pyrolysis, for example, the polymer (or mixture of polymers) is subjected to high temperature in the absence of sufficient oxygen for combustion, causing the polymer structure to break down. Thermal recycling can be applied to all types of plastics, addition polymers as well as condensation polymers. The typical yield is a complex mixture of products, even when the feedstock is a single polymer resin. If reasonably pure compounds can be recovered, products of thermal recycling can be used as feedstock for new materials. When the products are a complex mixture that is not easily separated, the products are most often used as fuel. There are relatively few commercial operations today that involve thermal recycling of plastics, though research continues. Germany has the largest number of such facilities, due to its requirements for recycling of plastics packaging. A consortium of European plastic resin companies, the Plastics to Feedstock Recycling Consortium, has a pilot plant for thermal recycling in Grangemouth, Scotland, and hopes to use the technology in a full-scale commercial plant by late 2000. The system uses fluidized bed cracking to produce a waxlike material from mixed plastic waste. The product, when mixed with naphtha, can be used as a raw material in a cracker or refinery to produce feedstocks such as ethylene and propylene.[67]

Sometimes, chemical and thermal recycling are, together, termed *feedstock recycling*. Figure 14.44 shows the growth in mechanical and feedstock recycling in Western Europe between 1994 and 1998.[61]

14.8.1.1 Separation and contamination issues. When plastics are collected for recycling, they are virtually never in a pure homogeneous

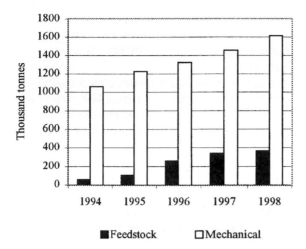

Figure 14.44 Mechanical and feedstock recycling in Europe.[61]

form. The collected materials will contain product residues, dirt, labels, and other materials. Often, the material will contain more than one base polymer and resins with a variety of additives, including coloring agents. Usefulness of the material is enhanced if it can be cleaned and purified. Therefore, technologies for cleaning and separating the materials are an important part of most plastics recycling systems.

It is useful to differentiate between separation of plastics from non-plastic contaminants, and separation of plastics of one type from those of another type. Separation of plastics from non-plastics typically relies on a variety of fairly conventional processing techniques. Typically, the plastic is granulated, sent through an air classifier to remove light materials such as label fragments, washed with hot water and detergent to remove product residues and remove or soften adhesives, and screened to remove small heavy contaminants such as dirt. If necessary, magnetic separation can be used to remove ferrous metals, and techniques such as eddy current separators or electrostatic separators can be used to remove other metals. Many of these techniques were developed in the minerals processing industries and have been adapted to use with plastics.

A particular concern for recycled plastics that are to be used in food contact applications is the potential presence of materials that may be dissolved in the recycled plastic and later migrate out into a product. Special care is needed in design of recycling processes to ensure that potentially hazardous substances do not migrate from recycled plastic into food products in amounts that might adversely impact human health. Companies desiring to produce recycled resins suitable for food contact generally challenge the process with known amounts of contaminant simulants and then determine whether the processing is able to adequately remove the contaminants. The U.S. Food and Drug Administration, while it does not formally approve recycled resins for food contact, has issued "letters of nonobjection" to a few processes that have demonstrated, to the satisfaction of the FDA, the ability to reduce contamination levels below the "threshold of regulation" of 0.5 ppb dietary concentration, which the FDA regards as an acceptable level of protection for most potential contaminants.[68] Another approach that has been accepted by the FDA is to interpose a barrier layer of virgin polymer between the recycled polymer and the food product. The amount of barrier that is sufficient depends on the mass transfer characteristics of the polymer and the intended use of the resin, among other factors.

14.8.1.2 Sorting. Separation of different types of polymers from each other is often a required or a desired part of plastics recycling processes. Such separation procedures can be classified as *macrosorting, microsorting,* or *molecular sorting.* Macrosorting refers to the sorting of whole or nearly whole objects. Microsorting refers to sorting of chipped or granulated plastics. Molecular sorting refers to sorting of materials whose physical form has been wholly disrupted, such as by dissolving the plastics.

Examples of macrosorting processes include separation of PVC bottles from PET bottles, separation of polyester carpet from nylon carpet, and sorting of automobile components by resin type. Much of this sorting is still done by hand, using people who pick materials off a conveyor belt and place them in the appropriate receptacle. However, a lot of effort has gone into development of more mechanized means of sorting to make this process both more economical and more reliable, and the use of such mechanized systems is increasing.

Various devices are now commercially available to separate plastics by resin type. They typically rely on differences in the absorption or transmission of certain wavelengths of electromagnetic radiation. Many of these systems can be used to separate plastics by color as well as by resin type. For example, the process used at the plastics recovery facility in Salem, Oregon, which was developed by Magnetic Separation Systems (MSS) of Nashville, Tennessee, sorts two to three bottles per second, using four sensors and seven computers to separate plastic bottles according to resin and color. X-ray transmission is used to detect PVC, an infrared light high-density array to separate clear from translucent or opaque plastics, a machine vision color sensor to identify bottle color (ignoring the label), and a near-infrared spectrum detector to identify resin type.[69]

Frankel Industries, of Edison, New Jersey, developed a system that combines manual sorting with differences in optical dispersion and refraction for separating PET and PVC from each other and from PETG and polystyrene. In this system, a special light shines on containers on a conveyor, and workers wear special goggles that give the different resins a distinctive appearance.[70]

Particularly for recycling of appliances, carpet, and automobile plastics, several companies have developed equipment to scan the plastic, usually with infrared light, and compare its spectrum, using a computer, to known types of plastic, resulting in identification of the plastic resin. One such device is the Portasort, developed jointly by Ford and the University of Southampton, Highfield, Southampton, U.K. It compares the spectrum of the unknown plastic against a library of 200 or more different polymers. A larger version, called the PolyAna system, can identify nearly 1000 different plastics, including blends and fillers. The same group developed the Tribo-pen, which uses triboelectric technology for plastics identification. This equipment, which has a sensing device about the size of a small flashlight, was developed for sorting automotive plastics. It comes in two basic types, the first to identify nylon, polypropylene, ABS and polyacrylite, and the second designed for more limited sorting, such as differentiating between PE and PVC.[71,72,73]

The first step in microsorting is a size reduction process, like chipping or grinding, to reduce the plastic articles to small pieces that will then be separated by resin type and perhaps also by color. One of the oldest examples is separation of high-density polyethylene base cups from PET soft drink bottles using a sink-float tank. More modern separation processes, such as the use of hydrocyclones, also rely primarily on differences in the density of the materials for the separation.

A number of other attributes have also been used as the basis for microsorting systems, including differences in melting point and in triboelectric behavior.[74] In many of these systems, proper control over the size of the plastic flakes is important in being able to reliably separate the resins. Some systems rely on differences in grinding behavior of the plastics combined with sieving or other size-based separation mechanisms for sorting. Sometimes cryogenic grinding is used to facilitate grinding and to generate size differences.

Systems that use electromagnetic radiation are under development and have had limited commercial application. SRC Vision, Inc., of Medford, Oregon, has an optical-based technology, originally developed for sorting of foods, which is used primarily by large processors for color-sorting single resins, such as in separating green from clear

PET. Union Carbide has used an SRC Vision system to separate colored HDPE flake into red, yellow, blue, green, black, and white product streams. The full SRC system uses X-rays, ultraviolet light, visible light, infrared light, reflectance, a monochromatic camera, and a color camera that is reported to be able to separate 16 million colors of red, green, and blue combinations.[74,75] ESM International, Inc. of Houston, Texas, also has developed an optical-based system.[74]

A novel European process being used for separation of plastics from durable goods separates the materials, including laminated structures, by blowing them apart at supersonic speeds. Various materials deform differently, permitting the use of sieving and classifying based on differences in size, geometry, specific gravity and ballistic behavior, using fluid bed separators and other equipment.[76] The Multi-Products Recycling Facility operated by wTe Corporation is designed to recover engineering plastics (as well as metals) from durable goods. It uses air classifiers to remove light materials, including foam and fiber, and a series of sink/float classifiers operating with water solutions at different specific gravities to separate chipped plastics by density, as well as using infrared technology to identify plastics before grinding.[77] KHD Humboldt Wedag AG in Cologne, Germany, has designed a system for separation of plastics by density using centrifuges and water or salt solutions. The intense turbulence in the centrifuges also helps clean the flaked plastics, as well as dewatering them.[78] Recovery Processes International, of Salt Lake City, Utah, has a froth flotation system designed to separate PET from PVC.[74]

Molecular sorting involves complete destruction of the physical structure of the plastic article prior to separation of the resins. Such systems typically use dissolution in solvents and reprecipitation, using either a single solvent at multiple temperatures or combinations of solvents. Because of the use of organic solvents, and consequently the need to control emissions and to recover the solvents, costs of such systems tend to be high. There is also a concern about residual solvent in the recovered polymer and its tendency to leach into products. There are, at present, no commercial systems using this approach.

Some research effort has focused on facilitating plastics separation by incorporating chemical tracers into plastics, particularly packaging materials, so that they can be more easily identified and separated. One such effort, funded by the European Union, has resulted in a pilot plant for separating PVC, PET, and HDPE bottles using fluorescent trace compounds that have been incorporated into the bottles.[79]

14.8.2 Uses for Recycled Plastic

Recycled plastics are used in a variety of applications, including automobiles, housewares, packaging, and construction. More information about their uses is found in the sections on recycling of individual types of plastics. Recycled materials, including plastics, also are an important segment of world trade activities. For example, in 1995, recycled plastic exports from the U.S. alone amounted to 652.8 million pounds, for a value of about $205 million. Most of these exports went to Hong Kong, and much of that material probably went on to China.[80] The Far East is also an important market for other countries.

14.8.3 Polyethylene Terephthalate (PET)

PET soft drink bottles were the first post-consumer plastic containers to be recycled on a large scale. In the U.S., the existence of bottle deposit legislation caused large numbers of these containers to become available in reasonably centralized locations, creating an opportunity for the development of systems to take advantage of the value of this material. One of the first companies to successfully develop systems for recycling PET soft drink bottles was Wellman, which began processing clear PET bottles in 1979 and is still the largest PET recycler in the U.S. St. Jude was another early entrant into PET recycling, beginning about the same time as Wellman but on a smaller scale, and concentrating on the green bottles while Wellman concentrated on clear bottles.[81]

PET beverage bottles are the largest single source of PET in municipal solid waste, and packaging accounts for more PET in MSW than does either durable or nondurable goods, as shown in Figs. 14.45 and 14.46.

The existence of bottle deposit legislation continues to be an important factor in PET recycling. It was estimated that, in 1997, 54% of the

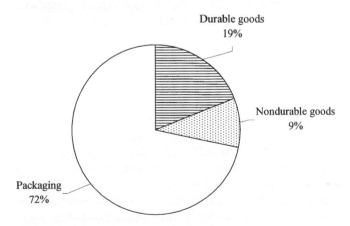

Durable goods
19%

Nondurable goods
9%

Packaging
72%

Figure 14.45 PET in U.S. municipal solid waste, 1997.[2]

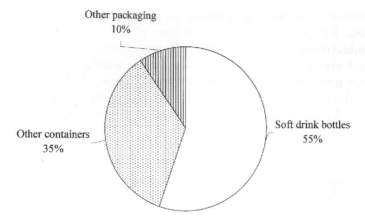

Figure 14.46   PET in packaging in U.S., 1997.[2]

PET soft drink bottles recycled came from bottle-bill states, while these states accounted for only 29% of the population. Recycling rates for soft drink bottles in deposit states range from 76 to 90%.[82] It should be noted that the low end represents California, which has a bottle refund system rather than a true deposit. The monetary incentive in California is lower than in true deposit states, and the refund system is less convenient.

14.8.3.1   PET recycling rates.   PET recycling in the U.S. grew rapidly from its beginnings in 1979 but was confined almost exclusively to deposit states until the mid 1990s. When concerns about solid waste disposal led to the creation of a large number of new recycling programs, many of them providing curbside collection, many of these programs included PET soft drink bottles and HDPE milk bottles in the mix of materials they accepted for recycling. This significantly increased the available amount of PET. At the same time, uses of PET bottles began to expand significantly outside the soft drink bottle market. These "custom" bottles began to be included as accepted materials, along with the soft drink bottles. In the deposit states, where PET soft drink bottles were not included in curbside collection programs because they were collected through the deposit system, programs began to add PET to the collected materials. The result was a significant increase in both the amount of PET bottles potentially available for recycling and the amount that was actually collected and recycled. During the late 1980s and early 1990s, both the overall tonnage of PET recycled and the recycling rate continued to grow, with the soft drink bottle recycling rate higher than the rate for custom bottles, and the rates for bottles very much higher than the rates for other forms of PET (Figs 14.47 and 14.48).

During the mid-1990s, the growth in use of PET, both in packaging applications and elsewhere, led a number of companies to invest in new facilities for production of virgin PET worldwide. As these facilities entered production, the supply of PET increased at a faster rate than markets, and prices fell. Additionally, during start-up, these facilities produced large amounts of off-spec resin, which was sold at very low prices. At the same time, in the U.S., there was a decrease in legislative pressure to use recycled plastic, particularly in Florida and California, and export markets decreased. The result, in mid-1996, was a drastic fall in the price at which recycled PET could be sold. Some PET recyclers shut down, because their costs for processing the

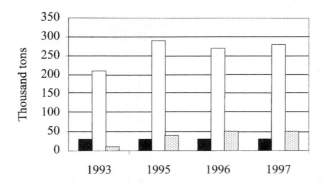

■ Durables □ Soft drink bottles ▨ Other containers

Figure 14.47   Trends in U.S. sources of recycled PET.[2,43,52,83]

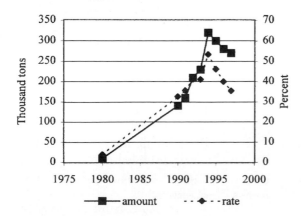

——■—— amount   - - ◆ - - rate

Figure 14.48   PET soft drink bottle recycling rates.[2,43,52]

material were higher than the price they could obtain for it. A few re-cycling collection programs stopped accepting plastics. During this same time period, there was increasing use of PET in small single-serve beverage bottles, and it became evident that the willingness of consumers to divert these containers for recycling was less, on aver-age, than with the larger size bottles. Much of this probably is because these bottles are more likely to be consumed away from home, where they may be tossed into the trash instead of taken home to the recy-cling bin. The result of this combination of factors in the U.S. was a decrease in both the total tonnage of PET recycled and, of course, in the recycling rate. Late 1997 brought a small increase in value of recy-cled PET and other signs of recovery, but the rate remained below the highs reached earlier in the decade (Fig. 14.49). The overall PET bot-tle recycling rate, according to the APC, was 25.4% in 1997, down from 27.8% in 1996.[82] In 1998, the tonnage of PET bottles recycled in-creased by more than 9% to 710 million pounds, but the recycling rate fell to 24.4%.[63] In Europe, where PET recycling is driven by govern-ment mandates, recycling rates and amounts continued to increase during this period, despite the low prices.

In addition to recycling of PET bottles, there is some recycling of PET strapping. In non-packaging applications, some PET photo-graphic film, including X-ray film, is recycled. In that case, PET is ob-tained as a by-product of silver recovery. Recovery of PET from durable goods was estimated at 30,000 tons in 1997, 8.3% of the amount discarded. Recovery of PET in nondurable goods was insignif-icant. Recovery of PET soft drink bottles was 280,000 tons, 37.3%, and

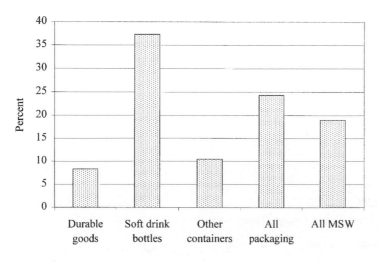

Figure 14.49 Tonnage of PET recycled in U.S. by sector, 1997.[2]

of other PET containers 50,000 tons, 10.4%. Recovery of other PET packaging was negligible. Overall, 330,000 tons of PET packaging were recovered, 24.3%; total recovery of PET from MSW was 360,000 tons, 18.9%.[2]

Because of differences in data and methodology, different sources sometimes quote significantly different rates for PET recycling. The American Plastics Council reported a PET soft drink bottle recycling rate of 35.8% in 1997, down from 38.6% in 1996, and slightly lower than the 37.3% rate in 1997 reported by the EPA. The National Association for PET Container Resources (NAPCOR) calculated still lower rates, 31.7% in 1996 and 27.1% in 1998.[84]

In an effort to increase recycling of PET, the National Association for PET Container Resources (NAPCOR) is sponsoring the placement of "Big Bin" collection containers in locations such as stadiums, convenience stores, and amusement parks, in an effort to capture more of the single-serving PET bottles that are consumed away from home.[84]

14.8.3.2 Mechanical recycling of PET.   Most PET is processed by mechanical recycling. In these systems, the PET is typically first color-sorted to separate clear from green and other bottles, since the clear PET has a higher value. Next, the PET is chipped, washed, and purified in various ways so that a pure resin can be obtained. One of the major issues is separation of PET from PVC. Because both are transparent plastics, they are difficult to separate reliably by manual sorting. Furthermore, their densities overlap, so they cannot be separated by conventional float-sink methods. To complicate the matter, PVC may be present in the recycled stream in the form of labels, or as inner liners in caps, in addition to bottles. This presents major problems to recyclers, since very small amounts of PVC contamination, e.g., 4 to 10 ppm, can cause significant adverse effects on PET properties.[81] At PET melt temperatures, PVC decomposes, generating HCl, which can catalyze PET decomposition, as well as leaving black specs in the recovered material. Thus, both performance and appearance can be significantly damaged.

Another contamination issue stems from the adhesives that may be used to attach labels or base cups. Often, not all of the adhesive residue can be removed by washing. These residues can cause color changes in the PET. Furthermore, the ethylene vinyl acetate can decompose, releasing acetic acid that, along with the rosin acids in some adhesives, can catalyze PET decomposition. Thus, these contaminants also can detract from both performance and appearance of the recycled material.

PET is also sensitive to degradation from the additional heat history and exposure to moisture during recycling. This commonly shows up as a decrease in intrinsic viscosity (IV). It is possible to subject the material to solid-stating, much as is done in resin manufacture, to increase the molecular weight (and consequently IV) back to the desired level.

Physically recycled PET from certain operations that add additional intensive cleaning steps, perhaps along with controls over the source of the material, has been approved (in the form of a letter of nonobjection) by FDA for unrestricted food contact applications, either alone or in a blend with virgin PET. The companies involved have released very little information about the details of the cleaning procedures. They are believed to involve intensive high-temperature washing along with limitation of the incoming material to soft drink bottles from deposit states, which are known to provide a cleaner recycled stream than does curbside collection. Less intensively cleaned PET has been approved for use as a buried inner layer in food packaging, with virgin PET used as a barrier to prevent migration of contaminants from the recycled layer.

On the whole, recycled PET retains very good properties and can be used for a variety of applications. Markets will be discussed further in Sec. 14.8.3.4.

14.8.3.3 Chemical recycling of PET. Chemical recycling of PET depends on chemical reactions that break down the PET into small molecules, which can then be used as chemical feedstocks, either for repolymerizing PET or for manufacturing related polymers. Two procedures, glycolysis and methanolysis, are in commercial use. Both can be used to produce PET that is essentially chemically identical to virgin polymer, and have been approved for food contact use.[85,86]

The first of these processes to receive a "letter of nonobjection" from the FDA, in 1991, was Goodyear's glycolysis process (later sold to Shell). Later that same year, Eastman Chemical and Hoechst-Celanese received approval for their methanolysis processes. The glycolysis processes typically produce partial depolymerization, which is followed by purification and repolymerization. Methanolysis processes provide full depolymerization, followed by purification by crystallization, and repolymerization. Glycolysis cannot remove colorants and certain impurities that can be removed by methanolysis. DuPont operated a methanolysis facility for recycling PET but discontinued the operation for economic reasons.

14.8.3.4 Markets for recycled PET. Historically, the first large market for recycled PET was in fiber applications, in particular polyester fiberfill for use in ski jackets, sleeping bags, pillows, and similar products. While there are now many additional markets for recycled PET bottles, fiber markets still dominate (Fig. 14.50). These fiber uses now include substantial use in carpet and even in clothing. A contest held by the Toronto-based Environment and Plastics Industry Council (EPIC) featured wedding dresses made from recycled plastic, with the average entry requiring 80 soft drink bottles to make. Half the polyester carpet manufactured in the U.S. now contains recycled PET.

Some PET is used in manufacture of new bottles. For a time, PET soft drink bottles made from 25% repolymerized PET were being used in parts of the U.S. However, the higher cost of the repolymerized (chemically recycled) PET that was being used caused such applications to disappear when legislative and consumer pressure to use packages with recycled content declined. In 1998, the GrassRoots Recycling Network began a campaign asking consumers to mail empty PET bottles back to the Coca Cola Co. in an effort to convince them to use recycled resin in soft drink bottles, as well as urging a boycott of Coke products until the company began using recycled PET.[15] In early 2000, Coca Cola issued a press release stating that, in 1999, it began using significant amounts of mechanically recycled PET in beverage containers.[14] The company had already been using recycled material in some overseas markets such as Australia, New Zealand, Saudi Arabia, and parts of Europe.

Veryfine, headquartered in Westford, Massachusetts, is one of only a few other U.S. users of recycled PET in food or beverage bottles. Veryfine packages all their juice and juice drinks in bottles containing recycled PET in a middle layer, surrounded by ethylene vinyl alcohol for oxygen barrier and containing layers of virgin PET on the inner and outer surfaces. Recycled PET makes up about 35% of the container. Heinz USA uses essentially the same structure for ketchup bottles.[87]

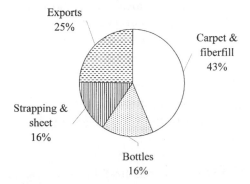

Figure 14.50 Markets for recycled PET in the U.S., 1996.[43]

In 1999 and 2000, Miller's introduction in the U.S. of five-layer PET beer bottles with two nylon barrier layers incited considerable concern from recyclers and environmental groups, particularly since many of the bottles are amber in color, and all used aluminum caps. In announcing the company's decision to go to nationwide distribution of the containers, Miller also committed to changing from aluminum to plastic closures, modifying the labels for ease of recycling, and also to using recycled content in the middle PET layer.[88]

Recycled PET is also found in thermoformed trays for uses such as packaging eggs, fresh produce, and pastries. In these applications, purity standards are less stringent, since there is less tendency for migration of contaminants to the food product. In fact, egg cartons were the earliest food-contact application for recycled PET, using essentially the same grade of recyclate as non-food packaging applications.

The use of physically reprocessed PET in non-food containers is more common than in food packaging. Up to 100% recycled PET can be used, or the recycled material can be blended with virgin. For example, Clorox uses about 50% post-consumer recycled PET in bottles for its Pine-Sol cleaner, after having tried and abandoned use of 100% recycled content due to processing problems.[87] Most often, the recycled material in these containers is blended with virgin, since this is less costly than using multilayer technology.

Recycled PET is also used in sheet and strapping. For example, blister packages made with recycled PET have been used for products ranging from pet supples to electronics to personal care products. Physically recycled PET is sometimes used in a buried inner layer, in either sheet or bottles, for food contact applications. It is also used either alone or blended with virgin PET for a variety of non-food applications. PET films with recycled content are also available.

The automotive industry is increasing its use of recycled PET. Several companies, including Lear-Donnelly, Johnson Controls, and United Technologies, now manufacture headliners that incorporate recycled soft drink bottles as an alternative to polyurethane. Eventually, old headliners will be a source of recycled material.[89]

Use of the products of chemical recycling of PET in the production of new PET resin has already been mentioned. In addition, the products from chemical recycling can be used as a feedstock in manufacturing of unsaturated polyesters, often for glass-fiber reinforced applications such as bath tubs, shower stalls, and boat hulls. Unsaturated polyesters have also found uses in polymer concrete.

14.8.4 High-Density Polyethylene (HDPE)

Sources of high-density polyethylene (HDPE) in U.S. municipal solid waste are shown in Fig. 14.51. As is the case for PET, packaging is the

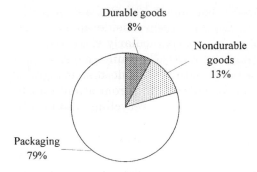

Figure 14.51 HDPE in U.S. municipal solid waste, 1997.[2]

largest source of HDPE in MSW. The single largest type of HDPE packaging (Fig. 14.52) is milk and water bottles, formed by blow molding from unpigmented homopolymer HDPE with a fractional melt index.

Recycling of high-density polyethylene milk bottles has about as long a history as recycling of PET soft drink bottles. For a long time, HDPE recycling rates were very much lower than PET recycling rates, but they have continued to increase in the late 1990s while PET recycling rates were falling and, in 1998, the HDPE bottle recycling rate exceeded the PET bottle recycling rate for the first time.

While deposit programs provided the impetus for soft drink bottle recycling, no such programs existed for milk bottles. Therefore, milk bottle recycling got its start with drop-off programs, relying on the willingness of individuals to deliver the bottles for recycling. In the

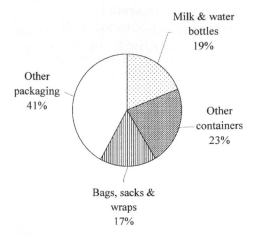

Figure 14.52 HDPE in packaging in the U.S., 1997.[2]

early years, the presence of paper labels on the bottles was a major problem, since many recyclers did not have technology that could successfully remove the paper. Many recycling programs requested that participants remove the labels from the bottles, some even suggesting placing a small amount of water in the bottles and heating them in a microwave to soften the adhesive so the labels could be peeled off. Such requests met with little success. In one of the early HDPE milk bottle recycling programs in Grand Rapids, Michigan, an employee cut out the label-bearing part of each bottle with a utility knife and discarded it before feeding the rest of the bottle into the shredder. Such solutions obviously entailed high labor costs, as well as loss of potentially recyclable materials, and kept most HDPE milk bottle recycling programs on the borderline of profitability, at best.

As technology developed to better handle this and other contamination issues, and as pressure to recycle plastics mounted, HDPE milk bottle recycling expanded and many programs began to include non-milk bottle HDPE containers. Now the majority of curbside and drop-off collection programs for recyclables include blow-molded HDPE bottles as one of the materials collected. The recycling rate for HDPE milk and water bottles in the U.S. in 1997 was 31.3%, according to the EPA. The recycling rate for other HDPE containers was 18.5%. Overall, the HDPE packaging recycling rate was 10.1%. In the durable goods category, the HDPE recycling rate was 12.2%. There was no significant recycling of HDPE from nondurable goods. The overall recycling rate for HDPE in MSW was 9.1%.[2] Figure 14.53 illustrates the sources of recycled HDPE, and Fig. 14.54 shows trends in HDPE milk bottle recovery. The American Plastics Council calculated that nearly 734 million pounds of HDPE bottles were recycled in 1998, for a rate of 25.2%, up from 24.7% in 1997 and 24.4% in 1996.[63,82]

As can be seen, the majority of HDPE recycling in the U.S., as for PET, is from bottles. There is some recovery of bags, sacks, and wraps,

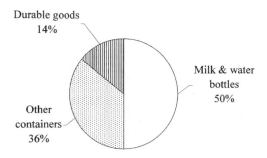

Figure 14.53 Sources of recycled HDPE in the U.S., 1997.[2]

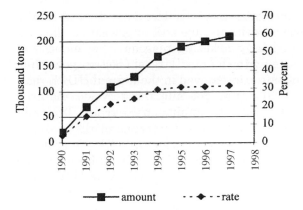

Figure 14.54 HDPE milk bottle recycling in the U.S.[2,43,52]

although this has declined considerably in the last several years. The HDPE collected in this category is mostly merchandise sacks, usually collected through drop-off bins located in retail stores that accept plastic bags of all types. There is also some recycling of HDPE envelopes. Recycling of HDPE base cups from PET bottles has largely disappeared with the phasing out of this style of container.

Collected HDPE is, typically, first sorted to separate the natural (unpigmented) containers, which have higher value, from pigmented containers. Separation is usually manual, although automated systems have been developed. It is also possible to further separate the pigmented HDPE into various color categories, either automatically or manually, but this is still relatively uncommon. Sorting out of the natural HDPE is often done prior to baling the materials for delivery to a processor, though it can be done at a later stage.

At the processor, the HDPE containers are typically shredded, washed, and sent through either a float/sink tank or a hydrocyclone to separate out heavy contaminants. Air classification may be employed prior to the washing step as well. The clean materials are dried and then usually pelletized in an extruder equipped with a melt filter to remove any residual non-plastic contaminants. When mixed colors are processed, the result is usually a grayish-green color, which is most often combined with a black color concentrate for use in producing black products. Natural bottles are of more value, because they can be used to produce products having a variety of colors.

Four major types of contamination are of concern in HDPE recycling. The first is contaminants that add undesired color to a natural HDPE stream. The primary source of this unwanted color is caps on bottles. While nearly all recycling programs ask consumers to remove the caps before placing the bottles in the collection system, a significant number of bottles arrive with the caps still in place, and the caps are usually brightly colored for marketing reasons. The majority of these caps are polypropylene, with the next largest fraction polyethylene. Neither of these materials are removed in the normal HDPE recycling systems. Thus, any caps that get into the recycled material stream will remain and discolor the unpigmented resin. Typically, amounts are low enough that mechanical properties of the material are not adversely affected, but they are sufficient to impart a gray coloration to what would otherwise be white HDPE pellets. Recently, the introduction of pigmented high-density polyethylene milk bottles has concerned recyclers, who fear these materials will cut into the use of the more profitable natural bottles. Pigmented HDPE resin typically sells for only 60% of the price of natural HDPE.[90]

The second type of contamination that is of concern is the mixing of injection molding (high melt flow) grades of HDPE with blow molding (low melt flow) grades. The result can be a resin that does not have flow properties desired for either of these types of processing, rendering it nearly unusable. The coding system for plastic bottles does not differentiate between these two types of polyethylene, so it is difficult to convey to consumers in any simple fashion which bottles are desired (the extrusion blow molded ones) and which are not (the injection blow molded ones). Some collection programs attempt to instruct consumers to place for collection the bottles "with a seam" and not the ones that do not have this characteristic. Other programs ignore the issue and simply accept the resulting contamination and its adverse effects on properties. Fortunately, the vast majority of HDPE bottles, particularly in larger sizes, are extrusion blow molded. A few years ago, however, when extrusion blow molded base cups were introduced as an alternative to injection molded base cups, some recyclers found themselves with HDPE resins that they could not sell, because the materials were not suitable for processing into new base cups or desired for other applications, due to the mixing of the different grades of resin.

A third significant contamination issue is the mixing of polypropylene into the HDPE stream. The polypropylene arises primarily from caps which, as discussed above, are included in the recycling stream despite requests that consumers remove them. Some PP also arises from fitments on detergent bottles and from inclusion of PP bottles with HDPE bottles when materials are collected. The density-based separation systems commonly employed in HDPE recycling do not separate PP from HDPE, since both are lighter than water. Fortunately, in most applications, a certain level of PP contamination can be tolerated. However, particularly in the pigmented HDPE stream, lev-

els of PP contamination are often sufficient to limit the amount of recycled HDPE that can be used, forcing manufacturers to blend the
post-consumer materials with other scrap that is free of PP, or with
virgin (often off-grade) HDPE. Commercially viable systems for separating PP from HDPE are, at least for the most part, not yet available.

The fourth type of contamination that is an issue is contamination
of the HDPE with chemical substances that may later migrate from
an HDPE container into the product. This is a more serious issue for
HDPE than for PET, for two reasons. First, the solubility of foreign
substances of many types is greater in HDPE than in PET. Therefore,
there is often more potential for migration. Second, the diffusivity of
many substances is greater in HDPE than in PET. Consequently, the
ability of substances to move through the HDPE and reach a contained product is greater. The strategies for dealing with this potential
problem are essentially the same as for PET.

First, a combination of selection of materials and processing steps
can be used to minimize the contamination levels in the HDPE. The
FDA has issued letters of nonobjection for recycling systems for HDPE
that permit those material to be used in some food contact applications. The first company to obtain a letter of nonobjection from FDA
for such purposes was Union Carbide. Their technology was later sold
to Ecoplast, which also received a letter of nonobjection.[91]

Second, the recycled HDPE can be used in a multilayer structure
that provides a layer of virgin polymer as the product contact phase.
This approach was first used for laundry products when problems
were encountered with migration of odorous substances from recycled
plastic to the products. The inner layer of virgin polymer provided a
sufficient barrier to solve the problem. In these same applications,
problems were also encountered with the appearance of the bottle.
This was solved by incorporating a thin layer of virgin polymer on the
outside of the polymer to carry the pigment. One added benefit was
that this minimized the amount of (often expensive) pigment required
to achieve the desired marketing image. The layer of virgin polymer
on the inside of the container also provided an added benefit by reducing the tendency to environmental stress cracking in these containers.
Since the recycled layer being incorporated was most often homopolymer HDPE from milk bottles, it did not have the stress crack resistance of the copolymer HDPE typically used for detergents. Later,
with the development of better washing technology, it was found to be
possible to package such products in single-layer bottles formed from a
blend of virgin and recycled HDPE. Nonetheless, such three-layer bottles, with the inner layer containing a combination of recycled milk
bottles and regrind from bottle manufacture, remain standard for
laundry detergents and similar products.

There are a variety of markets for recycled HDPE bottles. In the early days, the major market was agricultural drainage pipe. Today, this market accounts for only about 18% of recycled HDPE, with containers the largest market, followed by pallets and plastic lumber (Fig. 14.55). Film, mostly merchandise sacks and trash bags, is also a significant market.

Proctor & Gamble, which pioneered the use of three-layer bottles with an inner layer of recycled HDPE between outer layers of virgin material for its fabric softener and liquid detergent, is now the largest user of recycled plastic in the U.S. P&G packaging typically contains between 25 and 100% recycled HDPE, depending on product requirements. Clorox is another major user of recycled HDPE in bottles, as is DowBrands.[87]

DuPont uses 25% recycled HDPE in its Tyvek envelopes. The company also operates a program for recycling used envelopes. For small users, the system involves selecting one envelope to be filled with other used envelopes, and mailing them back to the company. For large users, other systems can be put in place.[92]

14.8.5   Low-Density Polyethylene (LDPE)

Because of the similarity in properties and uses between low-density polyethylene (LDPE) and linear low-density polyethylene (LLDPE), and because they are often blended in a variety of applications, use and recycling of LDPE and LLDPE are often both reported and carried out together. Therefore, in the remainder of this discussion, we will use the term *low-density polyethylene,* or *LDPE,* to refer to both LDPE and LLDPE. About half of the LDPE found in municipal solid waste comes from packaging. Another sizable fraction comes from nondurable goods, especially trash bags (Fig. 14.56).

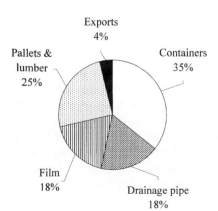

Exports
4%

Pallets &
lumber
25%

Containers
35%

Film
18%

Drainage pipe
18%

Figure 14.55 Markets for recycled HDPE in the U.S., 1996.[43]

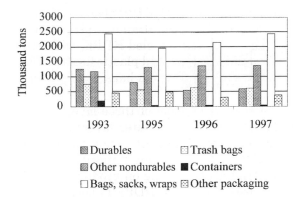

Figure 14.56   LDPE and LLDPE in U.S. municipal solid waste.[2,43,52,83]

Recycled LDPE comes from two main sources, stretch wrap and merchandise bags. In contrast to PET and HDPE, curbside collection does not play a significant role in LDPE recycling systems in the U.S. Stretch wrap is collected primarily from warehouses, retailers, and similar establishments where large quantities of goods arrive on pallets, with the loads stabilized by use of stretch wrap. These materials must be disposed of, so separating the stretch wrap and sending it for recycling avoids the disposal costs that would otherwise be incurred. In this way, recycling of stretch wrap is much like recycling of corrugated boxes.

Merchandise sacks are collected primarily through drop-off locations. Many retailers maintain a bin or barrel near the front of the store, where customers can bring plastic bags for recycling. The majority of these bags are LDPE, though a significant amount of HDPE is usually present as well. A few communities have experimented with adding plastic bags to curbside collection programs, but this remains very rare in the U.S. Most multimaterial drop-off facilities do not include plastic bags in the materials they accept, either. In recent years, there appears to have been some decrease in the availability of merchandise bag recycling. Some merchants have discontinued programs because of contamination of the stream with undesired materials, unfavorable economics, or for other reasons. Another source of recycled LDPE is garment bags, collected from department stores in a similar manner to collection of stretch wrap.

Recovery of LDPE and LLDPE bags, sacks, and wraps in the U.S. in 1997 was 100,000 tons, 4.1%, according to EPA. Recovery of LDPE in other categories of packaging and in durable and nondurable goods was negligible, for an overall recovery rate for LDPE of 1.9% (Fig. 14.57).[2]

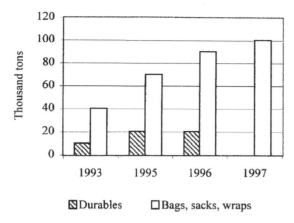

Figure 14.57   Recycled LDPE in the U.S.[2,43,52,83]

In Canada, the Plastic Film Manufacturers Association of Canada and the Environment and Plastics Institute of Canada (EPIC) have sponsored curbside recycling programs for plastic film of all types. In 1994, more than 100 communities in Ontario and 5 in Quebec were participating.[93] By 1996, the program had grown to 19 municipalities in the Montreal area and 146 in Ontario.[94] In 1998, EPIC published "The Best Practices Guide for the Collection and Handling of Polyethylene Plastic Bags and Film in Municipal Curbside Recycling Programs." This 27-page guide offers step-by-step instructions for "best practices" for successful curbside recycling of plastic bags, intended for use by Canadian municipalities that already have curbside film collection in place.[95]

Processing of film plastic is, in general, more difficult than processing of containers. The lower bulk density of the film leads to difficulty in handling the material. Contaminants such as paper from labels or from sales slips left in plastic bags are also harder to remove. In processing of containers, air separation is commonly employed to remove much of the light material, mostly paper and film plastic, from the heavier containers. Obviously, this will not be successful if the feedstock is plastic film. Historically, a significant fraction of the collected merchandise bags have been shipped to the Far East, where low labor costs permitted hand-sorting to be employed. For pallet stretch wrap, recyclers have worked with product manufacturers, distributors, and retailers to avoid contamination of the recovered wrap with paper labels.

A major market for recycled plastic film and bags is manufacture of trash bags. Recycled plastic has also been used in the manufacture of new bags, bubble wrap, plastic lumber, housewares, and other applications.

14.8.6   Polyvinyl Chloride (PVC)

Polyvinyl chloride (PVC) in U.S. municipal solid waste originates most often in nondurable goods, followed closely by durable goods and packaging (Fig. 14.58). In addition, a substantial amount of PVC is found in building and construction debris, which is not categorized as municipal solid waste. Materials found in this category include vinyl siding, pipe, roofing, and floor tile among others. A substantial majority of vinyl production goes into such long-term uses. According to the Vinyl Institute, 13.3 billion pounds of vinyl were sold by the U.S. in 1996 while, according to EPA, only 1,230,000 tons, 2.4 billion pounds, entered the municipal solid waste stream that year.[43,96] Similarly, estimates are that only about 18% of the 15 million metric tons of PVC produced in Europe each year is used in packaging.[97]

The EPA reported negligible recycling of PVC from MSW in the U.S. in 1997, with *negligible* defined as less than 5,000 tons.[2] A study by Principia Partners, funded by the Vinyl Institute and the Chlorine Chemistry Council, reported about 18 million pounds of post-consumer vinyl (of all types) recycled in the U.S. and Canada combined in 1997 (Fig. 14.59).[98] Many of these materials, such as bottles, binders, and medical products, would be considered part of the municipal solid waste stream, while others, such as carpet, windows, and siding, would not. The report states that the demand for rigid post-consumer vinyl was smaller than the supply due to the low price for virgin vinyl resin. Flexible materials were recovered primarily for their plasticizer content, which is higher in value than the resin.

In Europe, where PVC has had wider use in packaging, particularly in water bottles, PVC recycling has a longer history and has been more successful. About half of the 4 billion PVC mineral water bottles used in France each year are reported to be recycled, for example.[96]

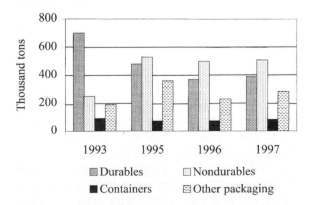

Figure 14.58   PVC in U.S. municipal solid waste.[2,43,52,83]

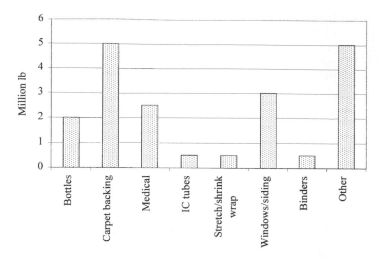

Figure 14.59 Post-consumer recycled PVC in North America, 1997.[98]

One of the issues facing PVC recycling is separation of PVC from a stream of mixed plastics. This is particularly important in the U.S. for PVC packaging, since PVC, where it is collected at all, is generally collected in a program that targets all types of plastic bottles. Several systems have been developed to automatically identify PVC, generally by picking up the chlorine presence using some type of radiative signal. For example, the vinyl bottle sorter developed by ASOMA Instruments uses X-ray fluorescence to detect the presence of chlorine. Several such systems are commercially available, including those made by National Recovery Technologies, ASOMA Instruments, and Magnetic Separation Systems.[96] A number of efforts have been made to permit separation of PVC from other plastics after chipping or grinding, but there are apparently no such systems in full commercial use at present.

In the U.S., there was an effort in the early 1990s, much of it spearheaded by the Vinyl Institute, to establish recycling for PVC packaging. None of these efforts were very successful in substantially increasing the PVC recycling rate and, in recent years, there appears to have been a decline in the availability of recycling. During 1998, there was substantial criticism of PVC recycling efforts, in particular those of the Vinyl Institute, by the Association for Post-consumer Plastic Recyclers (APR). APR requested assistance from the Vinyl Institute in 1996, when PVC markets began to dry up significantly and many of their members were faced with landfilling recovered PVC bottles due to lack of markets.[99] The APR, in 1998, raised the issue publicly again, as they were not pleased with the Vinyl Institute's response to their problems.[100] The recycling rate for PVC bottles fell to 0.1% in 1997, from 2% in 1996, after Occidental Chemical Corp. ended their subsidization of a program to buy back PVC bottles.

Use of PVC in packaging is declining in some areas, due in part to competition from PET, which can provide similar properties of clarity, strength, and rigidity. PET has a better environmental image than PVC, and recent declines in PET prices have increased its economic competitiveness. In some cases, some of the new very clear polypropylene bottles are being chosen as an alternative to PVC as well. In 1998, three Japanese firms announced that they will discontinue all use of PVC containers, switching to PET and PP. The firms are facing a government requirement for recycling their packaging beginning in April 2000.[101]

Recycling of vinyl siding and other pre-consumer scrap, on the other hand, may have better success. A number of pilot projects have been carried out for recycling of vinyl siding, mostly focusing on scrap from building construction or remodeling, and often with financial support from the Vinyl Institute. As a result of a pilot project in Grand Rapids, Michigan, recycling of vinyl siding waste is included in the "Residential Construction Waste Management: A Builders' Field Guide," which was funded by the EPA.[102] Among the most high profile of these pilot projects are those involving Habitat for Humanity, which builds housing for low-income families. The Vinyl Institute and other PVC-related industry organizations have donated money and materials to some of these projects in addition to supporting recycling efforts for the vinyl scrap generated during construction.

Polymer Reclaim and Exchange, in Burlington, North Carolina, recycles about 300,000 pounds per month of vinyl siding from construction debris. Drop-off sites are located at landfills and near manufacturers of mobile and manufactured homes, and material is collected from as far as 500 miles away. The collected materials are cleaned and flaked and then sold to molders, extruders, and compounders.[103]

In addition to vinyl siding, recycling efforts for building-related PVC wastes have focused on window profiles, carpet backing, pipe, and automotive scrap. The Vinyl Institute estimates that about 300 million pounds of pre-consumer vinyl scrap is recycled each year in the U.S., far exceeding the 9.5 million pounds of post-consumer vinyl. It further estimates that recycling of post-industrial vinyl of all types amounts to over 500 million pounds annually in North America.[96] In France, the Autovinyle recycling program for PVC automotive scrap recycled 1,740 metric tons of PVC in its first year of operation, 1997–1998. Its goal was 5,000 metric tons by the end of 1999.[104]

Several firms recycle PVC (as well as PE) wire and cable insulation, primarily from the telecommunications industry. A primary focus is the recovery of the copper and aluminum wire and cable, with the plastic insulation being recovered as a by-product. Since the recovered plastic typically contains small amounts of metal, it is suitable only for applications where high purity is not required. Uses include truck mud flaps, flower pots, traffic stops, and reflective bibs for construction workers. Most of the wire and cable originate in phone and business equipment wiring that is being replaced by fiber optic cable.[105]

Some recycling of PVC intravenous bags from hospitals is going on. One participating hospital is Beth Israel Medical Center in New York City, which is one of the pioneers in hospital recycling.[106]

A number of uses are possible for recycled PVC, depending on its source and purity. Most often, the recycled material is blended with virgin PVC. Applications include packaging, both bottles and blister packages; siding; pipe; floor tiles; and many others. Rhovyl, a French clothing manufacturer, is producing sweaters from post-consumer PVC mineral water bottles, combined with wool in a 70/20% vinyl/wool blend. Collins & Aikman Floorcoverings uses discarded carpeting for parking stops and industrial flooring. Crane Plastics uses scrap from vinyl windows and siding to make retaining walls and bulkheads. In the United Kingdom, IBM has achieved closed-loop recycling of PVC monitor housings into 100% recycled content PVC computer keyboard backs. Philip Environmental, in Hamilton, Ontario, Canada, is recycling about 125 million pounds per year of wire and cable scrap into products such as sound-deadening panels for cars, truck mud flaps, and floor mats. Conigliaro Industries of Massachusetts recycles 500,000 pounds per year of post-consumer PVC medical plastics, along with roofing membrane and other PVC scrap, into checkbook covers, plastic binders, and other products.[96] In India, PVC shoes are often recycled into new PVC shoes.[107]

14.8.7 Polystyrene (PS)

Slightly more than half of the polystyrene (PS) in municipal solid waste originates in packaging materials (Fig. 14.60). Nondurable goods, particularly plastic plates and cups, are the next largest category. As was true for PVC, a substantial amount of PS is used in the building and construction industry, mostly for insulation materials.

According to the U.S. EPA, recovery of PS in 1997 from durable and nondurable goods was negligible. Recovery from packaging materials totaled about 10,000 tons, for a packaging recycling rate of 10.0% and an overall recycling rate for PS in MSW of 0.5%. The Polystyrene Packaging Council reported a total of 54 million pounds of polystyrene recycled in the U.S. in 1996, with 10 million pounds of food service polystyrene, 23 million pounds of transport and protective packaging,

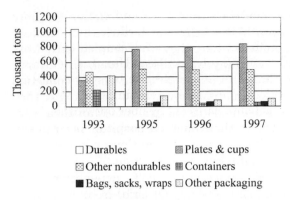

Figure 14.60 PS in U.S. municipal solid waste.[2,43,52,83]

and 21 million pounds of other non-packaging polystyrene applications including audio and video cassettes, CD jewel cases, insulation board, and other products.[108]

During the mid to late 1980s, PS was under attack on a variety of fronts, including ozone depletion and litter as well as the perception that it contributed a great deal to solid waste problems. In response, eight polystyrene resin suppliers formed the National Polystyrene Recycling Company (NPRC, later named the Polystyrene Recycling Company of America) in 1989 to concentrate on recycling of food-service polystyrene. Plans were to operate six plants around the U.S. to recycle these materials, with a goal of achieving a 25% recycling rate by 1995. High levels of contamination with food wastes and an inability to operate profitably plagued the facilities. In 1990, a highly publicized decision by McDonalds, which had instituted pilot recycling programs in several of its facilities, to abandon the PS clamshells and discontinue PS recycling dealt a further blow to recycling of food-service PS. By 1997, the NPRC was down to two facilities, one located in Chicago and the other in Corona, California, and to five PS resin company owners, and it had lost money through almost all of its history.[108,109] The economic woes were attributed in part to a worldwide oversupply of virgin general purpose polystyrene, which had driven down prices. With virgin resin selling at $0.40 to 0.50 per pound, recycled resin was selling for $0.38 to 0.45 per pound. Since it cost $0.10 to $0.50 per pound to sort, clean, and remanufacture the recycled polystyrene, depending on its quality and cleanliness, it was difficult for recycled PS to compete with virgin. Processing costs for food-service PS tend to be at the high end of this range, so the Polystyrene Packaging Council stated that it was not economical to recycle at that time.[108] In the U.S., collection of food-service PS focused on large generators of material, such as school, office, and other institutional cafeterias. Little attempt was made to collect PS at curbside, where collection costs would be even higher. Drop-off facilities for PS were initiated, often located at retailers where they were coupled with drop-offs for merchandise bags, but many were discontinued. In Ontario, Canada, however, PS is included in about 20% of curbside collection programs.[110] According to the Canadian Polystyrene Recycling Association, more than a million households in Ontario and Manitoba were expected to recycle polystyrene in their blue box curbside programs in 2000.[111]

Recycling of PS from non-food-service applications has been more successful than that from food-service ones. PS recycling from audio and video cassettes, CD jewel cases, and insulation board increased almost 70% between 1994 and 1996.[108] Another source of recycled polystyrene is clothing hangers used by department stores.[112]

Recycling of EPS protective foam packaging has also been considerably more effective than the recycling of food-service PS. The Alliance of Foam Packaging Recyclers was formed in 1991 to promote recycling of protective foam forms and peanuts. In 2000, they reported more than 200 collection locations for such materials, in the U.S. and Canada.[113] AFPR has also worked with members to provide recycling opportunities for those without direct access to a PS recycling facility, sometimes by arranging for prepaid UPS shipment of cushioning materials back to the manufacturer. Reuse is also part of the strategy for dealing with EPS protective foam. Some molded cushions are reused by manufacturers. Molded shapes are collected for reuse, among other places, at a network of Mail Boxes Etc. facilities.

The 1998 recycling rate for EPS protective foam was 9.0% in the U.S., up from 7.4% in 1997 (Fig. 14.61). AFPR attributes the increase mostly to receipt of more accurate information from certain recyclers

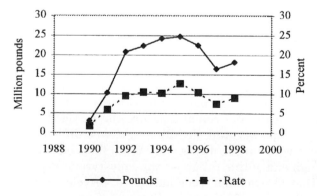

Figure 14.61 Recycling of EPS protective cushioning in the U.S.[2113]

along with improved methodology for calculating the amount of EPS protective packaging produced annually. About 55% of the EPS packaging recycled was made into new packaging.[113]

AFPR also reports that nearly 30% of all loose fill EPS is reused. For mailing services, the reuse rate is as high as 50%. Source reduction activities are also significant. AFPR says that the amount of polystyrene source reduced in 1994 amounted to an energy savings equivalent to recycling 24% of the polystyrene packaging and disposables produced in that year.[114]

End uses for recycled PS vary. For cushioning materials, the most frequent use is back into cushioning materials. Other applications include products such as bean bag filler, and construction applications such as ceiling texture. Studies of recycled cushioning showed that 25% recycled content EPS (more in some cases) can be used without adversely affecting cushion performance. In one study, foam containing 25% recycled content actually outperformed virgin EPS.[115] Insulation board containing recycled PS is also available. However, Amoco Foam Products, which had sold such material since 1991, discontinued use of recycled PS in insulation in 1997, stating that recycled PS was more expensive than virgin resin, and buyers were not willing to pay more for the recycled product.[116] Other applications for recycled PS include other types of packaging, housewares, and durable goods such as cameras and video cassette casings.

EPS recycling also occurs around the world. In many countries, however, recovery for combustion as fuel is included in the reported rates. In Japan, the Japan Expanded Polystyrene Recycling Association reported a recycling rate of 33.2% in 1995, 107.8 million pounds, which were recycled into pellets, soil improvers, and fuel.[117] In 1999, the recycling rate reached 33.2% (Fig. 14.62).[45] The recycling rate in

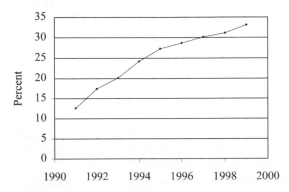

Figure 14.62 Recovery of EPS in Japan (includes energy recovery).[45]

Korea was reported to be 21% in 1994. Taiwan has government regulations that require recycling of polystyrene, and the country reported a rate of over 56%, 8.6 million pounds, in 1995.[117] Australia has programs for recycling of EPS boxes for fruit and vegetables, with a reported collection rate of 2.2 million pounds, 10% of all boxes, in 1996.[118]

In the United Kingdom, recycling of polystyrene packaging has increased faster than most forecasts, from 11.8% in 1993 to 28% in 1996, for a total of 6,000 tonnes. Forecasts for mechanical recycling of EPS by 2010 are now set at 35 to 40%, 16,600 tonnes. The faster than expected growth was attributed to more rapid than anticipated growth in virgin EPS markets, along with advances in technology that enabled recyclers to handle contaminated EPS material such as fish boxes and horticultural trays. End uses include EPS foam cushioning materials, non-foam applications such as CD and video cases, and extruded EPS applications such as hardwood and slate replacements. In addition to mechanical recycling, recovered EPS is used as a fuel for energy generation.[119]

One of the problems faced by recyclers of foamed PS, the kind most commonly used, is the very low bulk density of the materials. Compaction and baling are commonly employed to increase the bulk density, but the material is still expensive to ship for long distances. Recently, several companies have focused their efforts on transforming the PS foam into a gel, with a much higher density, to improve the transport economics. This method also can allow removal of contaminants by filtering the liquid material. One such company, International Foam Solutions, Inc., of Delray, Florida, has developed STYRO SOLVE, a citrus-based biodegradable solvent for EPS foam and other PS products. The gel is diluted after it is received at the processing facility so that it can be filtered. Resource Recovery Technologies, located near Philadelphia, PA, uses a nonflammable chemical solvent to dissolve the PS and separate it from contaminants. The solvent is recovered for reuse.[120]

The most common recycling of non-packaging, non-foam PS is recycling of disposable camera bodies. The recovered camera bodies are ground, mixed with virgin resin, and used in the production of new disposable cameras. The internal frame and chassis of the cameras, which are also polystyrene, are recovered intact and reused in new cameras.[121] Eastman Kodak reported in 1996 that 77% of its disposable cameras were being recycled under a program in which photofinishers are reimbursed for returning the cameras.[122]

Polystyrene can also been recovered from appliances. Philips, in Hamburg, Germany, studied recycling of non-flame-retardant PS from TV sets, using the recovered material to injection mold equipment housings, and concluded that there was no significant reduction in the properties of the recovered material.[123]

It is possible to use chemical methods to depolymerize PS into monomer, which could then be used for making new polystyrene. Some years ago, the Toyo Dynam company in Japan developed a prototype system in which polystyrene foam was ground and then sprayed with styrene monomer to dissolve the PS and separate it from contaminants, including other plastics and food scraps. The resultant solution was cracked and vaporized in a heated reflux vessel.[124]

14.8.8 Polypropylene (PP)

While packaging is the major source of polypropylene (PP) in municipal solid waste (Fig. 14.63), durable goods are the primary source of recycled polypropylene (Fig. 14.64). Much of this material comes from

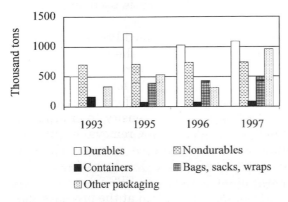

Figure 14.63   PP in U.S. municipal solid waste.[2,43,52,83]

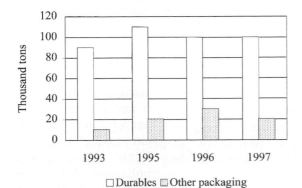

Figure 14.64   PP recycling in the U.S.[2,43,52,83]

recycling of polypropylene automotive battery cases. Due to concerns about the effect of lead emissions from disposal of such batteries, many states have prohibited their disposal in landfills or by incineration, and several states have instituted deposits to help ensure that batteries are collected for recycling rather than going to disposal, as was discussed in Sec. 14.5. The recovery of the polypropylene is a side benefit of lead recycling, which is the driver for battery recycling. PP makes up about seven percent of the battery, by weight. The largest market for the recovered material is, blended with virgin PP, new battery cases. The Battery Council International reported a 1996 recycling rate for lead-acid batteries of 96.5%.[125] The recovery of PP from battery cases was at about the same rate, since it is a routine part of battery reprocessing.

Recycling of PP from durable goods, most of which were battery cases, was 100,000 tons in 1997. No recovery of PP from nondurable goods was reported. Recycling of PP from packaging was about 20,000 tons, all from the "other plastics packaging" EPA category. Recovery of PP from containers was listed as negligible (under 5,000 tons), and no recycling of PP from the bags, sacks, and wraps category was reported. The total PP packaging recycling rate was 2.1%, and the overall recycling rate for PP in MSW was 4.3%.[2]

Recycling of PP battery cases also takes place outside the U.S. Germany began such recycling in 1984, yielding about 3,000 tonnes per year of polypropylene. The process involves crushing the batteries and then separating the light fraction, which is predominantly PP, from the heavy fraction, containing lead and other components. The light fraction, which is about 97% PP, is further size reduced, dried, and sent through a cyclone separator, which increases its purity to 99.5%. The resulting material is compounded to user specifications and then pelletized.[126]

Polypropylene can also be recovered from appliances. A recycling plant in Frankfurt reports recovery of about a third of the weight of discarded coffee machines as polypropylene, with greater than 99% purity and excellent property retention.[123]

Polypropylene spools and wheel counters in disposable cameras are removed when the returned cameras are disassembled, and used again in new cameras.[121] Polypropylene hangers from department stores are also sometimes recycled.[92]

Some recycling of polypropylene packaging is carried out, most often targeting bottles collected as part of an "all plastic bottle" collection system. Recycled resin from collected containers has been used in soap bottles.[127]

Packaging uses of PP are approximately evenly divided between film, container, and closure applications. Little deliberate recycling of

PP film or closures is done. However, PP closures and fitments are sometimes inadvertently recycled with HDPE containers, where they are an undesired contaminant. Since both PP and HDPE have densities lower than water, the gravity-based systems (float/sink tanks or hydrocyclones) commonly used to separate HDPE from other plastics do not separate out PP. Currently, there are no commercially available systems for separating HDPE from PP once the materials are chipped or ground. While the presence of a small amount of PP contamination does not result in significant performance problems, larger amounts do make the recycled resin unsuitable for some uses or require it to be blended with other resin to dilute the PP.

One use of polypropylene by industry is for the production of dye tubes, which carry yarn as it passes through the dye process. These dye tubes can be used only once, because the textile dye penetrates the polypropylene. At least one company, Wellmark Inc. of Asheboro, North Carolina, is recycling these tubes. First, residual scrap yarn is removed, and then the tubes are separated by type of polypropylene and granulated. Air separation is used to remove paper labels and other contaminants. The resulting PP is used in injection molding processes. Wellmark recycles 3,500 tons of scrap tubes per year. Because of the high volume of the hollow tubes, the company estimates that the annual amount of tubes they process is enough to fill five football fields stacked 10 ft high.[128]

14.8.9 Nylon

Nylon recycling has increased substantially in the last several years. Most recycling efforts have focused on recovery of carpet. According to the U.S. Department of Energy, about 3.5 billion pounds of waste carpet are discarded each year in the U.S., with about 30% of them made from nylon 6. (For more on carpet recycling, see Sec. 14.8.16.) Recycling systems for condensation polymers such as nylon and PET can more effectively use chemical depolymerization techniques than can systems for addition polymers such as polyolefins and PVC. Most of the efforts directed at nylon recycling have taken this route.

The U.S. Department of Energy provided support to AlliedSignal, as well as carrying out research at the National Renewable Energy Laboratory, for the development of a process for chemical conversion of nylon 6 to caprolactam, which could then be repolymerized to nylon 6 for a variety of applications.[129,130] The economics of this process were particularly promising. A commercial-size plant recycling 100 million pounds of waste carpet could produce high-grade caprolactam for about 15 to 20 cents per pound, while caprolactam from virgin materials currently sells for 90 cents to $1 per pound.[130]

Several companies are now engaged in commercial recycling of nylon 6 carpet. BASF Corp. collects only its own nylon 6 carpet, depolymerizes and purifies it, and then uses it in the manufacture of new carpet.[131] In 1999, Evergreen Nylon Recycling LLC, a joint venture between AlliedSignal and DSM Chemicals North America, opened a new facility to recycle more than 200 million pounds of nylon carpet each year. The process involves depolymerization of the nylon to caprolactam, which is then repolymerized and used for a variety of applications, including carpet and engineering plastics for automobiles. In addition to permitting closed-loop recycling, at full capacity, the facility saves 4.9 trillion BTUs each year compared with conventional caprolactam production. Evergreen Recycling has also developed a laser device for resin identification to facilitate the sorting of nylon from other types of products. The company plans to adapt its process in the near future to recycle nylon automobile parts as well.[132,133]

DuPont is recycling mineral-reinforced nylon 66 carpet, including that generated by Ford Motor Co., and containing 25% post consumer recycled content, to manufacture engine air cleaner housings. A proprietary process is used to reclaim the face fiber, which is melt-recycled and compounded with virgin nylon.[134]

An exception to the general method of using depolymerization for carpet is the Lear Corp. system for recycling scrap carpet pieces. Relatively clean scrap carpet made from nylon 6, nylon 6/6, and other resins is baled and then pelletized, producing a black-colored resin that they hope to sell to automobile manufacturers for nonvisible applications such as acoustical parts and the backs of floor mats. The plant is currently recycling more than 1 million pounds of scrap per year.[88]

Recycling of nylon 6 is not limited to carpets and automobile parts. Toray, a Japanese company, in 1995 began recycling apparel made from nylon 6 by depolymerization and repolymerization.[135] Used U.S. Postal Service nylon mailbags that are not repairable are being recycled into pellets that are used, among other applications, for automobile parts.[136]

An unusual recycling system is that of *Preserve* brand toothbrushes, manufactured by Recycline, Inc., of Somerville, Massachusetts. Buyers of the toothbrushes, which have handles made from a blend of recycled and virgin polypropylene, and nylon bristles, can send the brushes back to the company in a postage-paid return envelope. The brushes are then recycled into plastic lumber, with the claim that the blend of PP and nylon actually strengthens the lumber.[137]

14.8.10   Polyurethane (PU)

Polyurethane recycling, like that for nylon, has often taken a chemical approach. This is particularly important to polyurethane, since most of its applications are for cross-linked material that cannot, therefore, be melted and reformed. The first North American facility for chemical recycling of molded polyurethanes opened in September, 1997, in Detroit, Michigan. The 10 million pound per year facility is owned and operated by BASF Corporation and Philip Services Corp. While targeted first at automotive waste streams, it could eventually include appliances and construction materials as well as recreational products such as bowling balls and pins. The facility uses BASF's patented glycolysis process to chemically reduce PU to polyol, which can be used like a virgin polyol, combined with isocyanate to produce polyurethane products. Various types of PU can be mixed together in the process, including polyurethane paint on polyurethane parts. BASF will use the recovered polyol for manufacture of new polyurethanes for use in automotive, construction, and other applications.[138]

ICI Polyurethanes is working on development of a split-phase glycolysis chemical recycling technology for polyurethane and plans to build a full-scale plant in Britain to recycle polyurethane foam from mattresses, furniture, and automotive seat cushioning.[139]

Other approaches to recycling polyurethane have focused on chopping up the flexible varieties and using them as filler in new polyurethane products. Rebonded seat foam materials have been studied for use in floor carpet underlayment in vehicles and found to be comparable to currently used materials.[140]

Bayer Corp. of Pittsburgh, through its Hennecke Machinery division, along with the German company Greiner Schaumstofftechnik, has developed a process for producing molded parts from scrap polyurethane foam. The foam is shredded to flakes and then mixed with a prepolymer adhesive. The mixture is filled into molds using a system that precisely and individually meters the amount fed into each mold to allow all molds to fill completely. It is reported that at least one European company is using the system commercially. Another process developed by Hennecke grinds polyurethane foam cuttings into a fine powder, 2 micron size, which is then added to the polyol in quantities up to 30% during polyurethane production.[141]

14.8.11   Polycarbonate (PC)

Some recycling of polycarbonate from products such as automobile bumpers, compact discs, computer housings, and telephones is being carried out. General Electric began buying back polycarbonate from five-gallon water bottles several years ago and tried for a considerable amount of time to introduce its idea of a cascade of uses for polycarbonate, beginning with reusable packaging and ending in various product applications such as automobile parts. GE also initiated a pilot buyback recycling program in 1995 for polycarbonate scrap and other resins generated by its customers, including recovered PC parts from used automobiles, if they were manufactured from GE resins.[142]

Bayer, in Leverkusen, Germany, in 1995 built Europe's first polycarbonate CD recycling facility. The PC is separated from aluminum coatings, protective layers, and imprinting, and the pulverized material is combined with virgin resin and sold for various uses.[143]

Polycarbonate winding levers and outer covers on disposable cameras are recovered intact during processing of the cameras and reused in new cameras.[121]

14.8.12   Acrylonitrile/Butadiene/Styrene Copolymers (ABS)

Recycling of ABS is primarily focused on appliances. British Telecommunications plc recovers and recycles about 2.5 million telephones every year. About 300,000 are refurbished for reuse, and the remainder are recycled. The recovered ABS from the telephone housing is typically used in molded products such as printer ribbon cassettes and car wheel trims.[144] Similarly, AT&T Bell Laboratories recovers ABS phone housings and recycles them primarily into mounting panels for business telephone systems.[145] GE Specialty Chemicals recycles ABS from refrigerator liners.[146]

Hewlett-Packard recycles computer equipment using some of the recovered ABS in manufacture of printers, blended with virgin ABS. The recycled content in the printer cases is reported to be at least 25%.[147]

HM Gesellschaft für Wertstoff-Recycling recovers ABS from vacuum cleaners, reporting recovery of more than 25% of the weight as ABS with better than 99% purity and properties comparable to those of impact-resistant polystyrene.[123]

Blends of ABS and polycarbonate (PC) are also recycled, with the usual source again being used appliances. Siemens Nixdorf Informationsysteme (SNI) in Munich, Germany, recycles a variety of equipment made by the company, such as computers and peripherals and automatic teller machines. Customers are charged a fee for the reprocessing. Much of the recovered material, which is manually disassembled and resin type identified by molded-in markings or by IR spectroscopy, is PC/ABS blends, which are ground and shipped to Bayer AG in Leverkusen, Germany, where they are blended with virgin resin and used by SNI to mold equipment housings.[123]

14.8.13   Commingled Plastics and Plastic Lumber

In some cases, separation of collected plastics by resin type is either not feasible with current technology or is economically unattractive.

For such streams, use of the plastics in a commingled form may be preferable. A number of applications have been developed for commingled plastics. Many fall in the general category of plastic lumber, while others are replacements for concrete or other materials. Some of these processes are also able to handle substantial concentrations of non-plastic contaminants, reducing the need for cleaning the collected material. Some plastic lumber products use single-resin recycled plastics or single-resin plastics combined with fillers or reinforcers.

Most processes for recycling of commingled plastics depend on a predominance of polyolefin in the mix to serve as a matrix, within which the other components are dispersed. One of the early processes was the Mitsubishi Reverzer, developed in Japan. This process used a short-time high-temperature high-shear machine that could handle up to 50% filler. The plastics were first softened in a hopper and then mixed in an extruder. Products were formed in a variety of shapes by flow molding, extrusion, or compression molding. However, the equipment was never successfully marketed.[148]

The Klobbie process, developed in the Netherlands in the 1970s, consisted of an extruder coupled to several long linear molds mounted on a rotating turret. The molds were filled directly from the extruder, without any extrusion nozzle or filter pack, and then rotated into a tank of cooling water. After solidification, the products were removed using air pressure. This process is often called an *intrusion process,* since it is a cross between injection molding and extrusion. Klobbie also patented the use of foaming agents in the equipment.[148] The Klobbie process is the father of several commercial extrusion processes in use today. Like the Reverzer, these processes can generally incorporate substantial amounts of non-plastic contaminants, which tend to migrate toward the middle of the thick cross-section items being produced.

Another early process was the Recycloplast process, developed in Germany, in which materials are heated in a "plastificator," using kneading to produce a paste. The paste is then extruded into roll-shaped loaves, which are either immediately compression-molded or granulated for further use. The British Regal Converter system is another compression molding process, in which commingled plastics are granulated, distributed on a steel belt, melted in an oven, and then compacted into a continuous board.[148]

The ET/1 system developed by Advanced Recycling Technology of Brakel, Belgium, is a more modern variant of the Klobbie process, which has been used by a number of operations in several different countries. The feed must contain a minimum of 50 to 60% polyolefin. The short, high-speed adiabatic screw extruder melts the polyolefin and fills molds up to 12 ft long, mounted on a revolving turret that is cooled in a water bath. Products formed are in the plastic lumber category. Properties of lumber formed from mixed plastic bottles can be improved by adding 10 to 30% recycled polystyrene.[148]

The Superwood Process, developed in Ireland by Superwood International, is another Klobbie variant in which fillers or blowing agents are commonly added to vary the product properties. Hammer's Plastic Recycling in Iowa developed another Klobbie variant in which closed molds, a heated nozzle, and a screen pack are used to increase molding pressure.[148,149] Other processes include those of WormserKunststoffe Recycling GmbH in Germany and C. A. Greiner and Sohne in Austria.[148]

Electrolux, in Stockholm, Sweden, is reportedly using 40 to 80% commingled recycled plastics from sources such as used telephones, car parts, and other appliances in the manufacture of vacuum cleaner housings. The company is also moving toward closed-loop reuse of in-plant scrap in future product designs.[150]

Tipco Industries Ltd, in Bombay, India, produces Tipwood, a composite material from mixed plastic waste and 20% filler, for use as a wood substitute for applications such as pallets, benches, fencing, and road markers.[151]

Hettinga Technologies of Des Moines, Iowa, has used commingled plastics to make injection molded panels, using a controlled-density molding process that accommodates a range of plastics. The panels are reported to have a thin, smooth outer skin around an integral core that is foamed in situ. Costs are reported to be about 60% of the cost of plywood, for panels that handle like pine, hold nails and screws like hardwood, and resist rot and mildew. The process involves incorporating a dissolved blowing agent into the mold, which expands the part once it is removed from the mold.[152]

The Plastic Lumber Trade Association, PLTA, was formed in the mid 1990s to promote the growth of recycled plastic lumber. One of the concerns of the PLTA and other interested parties was the diversity of properties of "plastic lumber" produced from different mixes of materials by different processes. As a result, the American Society of Testing and Materials (ASTM) began developing standards for the material. In late 1997, ASTM issued five new test methods for plastic lumber, intended to help establish a benchmark for minimum plastic lumber performance. The standards include tests for compressive properties, flexural properties, bulk density and specific gravity, compressive and flexural creep and creep-rupture, and tests for performance of mechanical fasteners.[153] Two additional test methods have subsequently been developed, and several additional ones are expected to be released by the end of 2000.[154]

Plastic lumber is used for fencing, decks, benches, picnic tables, and other applications as an alternative to wood lumber. The plastic products generally are more expensive to purchase than wood but have a considerably longer life. Furthermore, they do not need to be painted or varnished. Since wood products in these applications generally are made of treated lumber, interest in them is expected to grow as concern about the environmental effects of the compounds used to treat the lumber increases. Some plastic lumber products are used in applications such as parking stops and machinery bases, where they replace concrete. While not all such applications use commingled recycled plastics, many do. Some use single-resin recycled plastics, usually HDPE or LDPE, while others use virgin plastic. In some cases, composites of plastic (virgin or recycled) are used with fibrous reinforcements, usually wood or glass fiber. The fibers may be recycled material as well.

Another application of plastics lumber is in pallets. A number of designs have emerged, most using high-density polyethylene, and some using fibrous reinforcements or steel inserts. Recently, researchers at Battelle developed a pallet made from recycled plastic, including both HDPE and commingled plastics, that can hold up to 10 tons. It is intended to replace wood and steel for storing 55 gallon drums of hazardous and radioactive materials. In these applications, the plastic pallets are less likely to absorb contaminants and easier to clean than wood and metal alternatives.[155]

Most recently, plastic lumber has broken into two potentially large-scale applications: plastic railroad ties and bridges. Plastic rail ties have been used for some time in Japan, where wood is scarce, but these ties are made from virgin foamed polyurethane with a continuous glass fiber reinforcement. The ties now being tested in the U.S. are made from recycled plastics, which are pre-dimpled to sit better on the rocks used as ballast in the rail lines and have the same size and appearance as traditional wood ties.[156] One of the first commercial purchases of plastic lumber railroad ties in the U.S. was by the Chicago Transit Authority, which purchased 250 ties for its elevated train line.[157]

The U.S. Army Corps of Engineers used 13,000 pounds of commingled recycled plastics to build a bridge at Fort Leonard Wood in St. Robert, Missouri. The bridge is primarily used for pedestrian traffic but can hold up to 30 tons, allowing it to also support light trucks. The Army expects the bridge to last 50 years without maintenance—significantly longer than the 15 years for treated wood or 5 years for untreated wood. Even the joists are made of plastic, opening the door to structural applications that once were considered unsuitable for plastic lumber. The material used for the joists is a polystyrene-modified lumber, formed from recycled post-consumer polystyrene foam along with recycled high-density polyethylene, which was produced by Poly-wood Inc. of South Plainfield, New Jersey. It has a modulus 2.5 times that of conventional plastic lumber and therefore can withstand higher loading. The deck planks were produced by Plastic Lumber Co. of Akron, Ohio, and the railing by Hammer's Plastic Recycling Corp. of Iowa Falls, Iowa, and Renew Plastics Inc. of Luxemburg, Wisconsin. One unique property of the plastic lumber is that it melts around screws when they are driven into the boards. When the plastic cools, it hardens around the screw, locking it into place and ensuring a long-lasting bond.[158]

Research on composite materials made of wood fiber or flour or recovered paper fiber, along with recycled plastics, has been ongoing in several localities for a number of years. Investigators include CSIRO and the Cooperative Research Centre for Polymers in Australia, the University of Toronto and University of Quebec in Canada, the Forest Products Research Laboratory and the Risk Reduction Engineering Laboratory in Cincinnati, and Michigan State University, among others. Polymers used include low- and high-density polyethylene, polypropylene, polystyrene, and PVC, along with commingled plastics. For example, the USDA Forest Products Laboratory has studied the use of recovered paper fiber with a mix of PP, LDPE, HDPE, and PVC. This mix has been investigated by a plastic lumber manufacturer for use as a core layer in plastic lumber for pallets.[159] Boise Cascade is researching the use of plastic film, recovered from residents in the Seattle area, and wood waste to make a composite that they intend as a replacement for virgin wood material in building construction.[160]

On a commercial scale, Advanced Environmental Recycling Technologies Inc. (AERT) of Rogers, Arkansas, for several years has been producing window frames from a composite of recycled polyethylene film and wood fiber.[161] Atma Plastics Pvt. Ltd., of Chandigarh, India, produces a recycled PE/PP/wood composite for use as a wood substitute or as a filler in cast polyester for furniture.[151] Natural Fiber Composites Inc. of Baraboo, Wisconsin, has commercialized a pelletized wood fiber-filled plastic, using recovered paper or wood fibers along with PP, HDPE, or PS.[162,163] Comptrusion Corp., of Richmond Hill, Ontario, makes polyethylene wood flour composites and is working on a PVC/wood composite. Formtech Enterprises Inc. of Stow, Ohio, manufactures a PVC wood-fiber composite.[164] Whether these materials use recycled plastic was not clear. Mikron Industries is manufacturing a wood-plastic composite for window frames that is based on an undisclosed mixture of plastics, and that uses material that otherwise would be headed for landfill disposal.[165]

While many processes for using commingled plastics look at collection of bottles from curbside as the primary source of materials, they frequently combine these materials with industrial waste stream plastics, including coextruded scrap and other examples of multi-resin, perhaps contaminated, materials. Other companies focus entirely on these types of waste streams. For example, Northern Telecom Ltd, based in Toronto, has a recycling facility for plastic reclaimed from wire and cable, along with materials from phone, fax, business machine, and pager equipment. The material, which contains small amounts of residual copper and aluminum, is used for truck mud flaps, flower pots, traffic stops, reflective bibs for construction workers, and other applications where high purity is not required.[166]

14.8.14 Compatibilization of Commingled Plastics

As discussed above, one approach to the use of commingled plastics is to use them in wood-substitute and similar applications where the incompatibility between the various ingredients of the mix does not present severe problems. A different approach is to compatibilize the resins in some way, hence improving the performance characteristics of the blend. The compatibilization can be done by the way the blend is processed, perhaps including some degree of control over the resins in the mix or by additives.

Northwestern University has developed a process they refer to as solid-state shear extrusion pulverization. In this process, a twin-screw extruder is used to convert mixed plastics and scrap rubber into a uniform fine powder, which can then be used in a variety of products.[167]

Manas Laminations, of New Delhi, India, has developed compatibilizers that enable PET/PE waste streams to be processed into useful products, including flooring, office partitions, corrugated roofing, and slates for benches.[151]

A number of other companies are also marketing or developing compatibilizing additives. For example, Dexco Polymers, of Houston, Texas, sells a line of styrenic block copolymers that can enhance properties of mixed streams of polyolefins. BASF is developing compatibilizers for mixtures of polystyrene with polyethylene or polypropylene, and for mixtures of ABS with polypropylene. DuPont sells compatibilizers based on maleic anhydride-grafted polyolefins for compatibilization of mixed polyolefins.[146]

14.8.15 Automotive Plastics Recycling

The automotive industry is under pressure, in various countries, to make automobiles more recyclable and to use more recycled materials in automobile manufacture.[168] For example, European Union regulations require that only 5% of automobiles, by weight, be landfilled beginning in 2015. Many countries are incorporating, either by regulation or voluntarily, the philosophy of extended product responsibility on automobile manufacturers. With the increasing use of plastics in automobiles, this has forced attention to the problem of recycling these automotive plastics.

While some recycling of auto parts has been covered in the sections on recycling of individual plastic resins, it is also useful to look at automotive plastics recycling in a more unified way. Recycling of automobiles has a very long history, but most of the effort in the past has focused on recovery of metals, and steel in particular. The plastics in cars remained in the "auto shredder residue" or fluff and have routinely been destined for disposal rather than recycling. Recycling of rubber, both from tires and from car parts, is an important part of automotive recycling efforts, and is discussed in Sec. 14.9.

Recovery of plastics from fluff, which contains a variety of plastic and non-plastic components, has not been achieved commercially. Much as recovery of plastics from household waste is more successful when the targeted materials are diverted from the garbage rather than trying to separate them out from a mixed waste stream, recycling of plastics from automobiles is easier when parts are disassembled than when the whole car is ground up. However, in the case of automobiles, even separating functional units such as bumpers or dashboards does not always result in a single type of plastic, since many of these components use a combination of materials. Furthermore, the cost of disassembly is significantly higher than the cost of processing the auto body in the usual way. Automobile manufacturers are engaged in design changes to simplify the recovery of car parts for recycling. Efforts include improving the ease of removal of automobile subassemblies, reducing the number of different resins used in auto parts, and ensuring that compatible resins are used in parts where multiple resins are necessary, so that they need not be separated for recycling. Design of hand-held systems to reliably identify resin types is an important adjunct to these efforts. While newer cars have plastic parts with molded-in resin identification, most cars being scrapped now were manufactured before those systems were in place.

Automakers have made substantial strides in use of parts containing recycled plastic in building of automobiles. As mentioned previously, recycled nylon is being used in air-cleaner housing as well as in fan assemblies and other automotive parts.[169] Recycled ABS and polyester/polycarbonate alloys are being used for brackets to hold radio antennae, splash shields, and small under-the-hood parts.[170] Recycled polyurethane is being used in reaction injection molded bumper fascias.[171] Recycled polycarbonate/polybutylene terephthalate bumpers are being incorporated into new bumpers.[172] Thermoplastic olefin

(TPO) bumpers are being recycled into bumper fascias, splash shields, air dams, and claddings.[172] Recycled polypropylene is used in power-train applications, fender liners, air conditioning evaporative housings, vents, and other applications.[169,172,173] Recycled PET is being used in a variety of parts, including headliners and engine covers.[173,174] Recycled polycarbonate has found use in instrument panel covers.[172] Other examples can be found, and recycled plastics are also being used in packaging for parts used in manufacturing automobiles. Cost remains a key issue for the automobile industry, with most manufacturers insisting that parts made from recycled material deliver comparable performance at no more cost than virgin materials.

14.8.16 Carpet Recycling

In the past several years, interest in recycling of carpeting has grown substantially. Most efforts have focused on commercial carpeting. It is estimated that about 1.7 million tons of waste carpet are landfilled in the U.S. each year, most of it during construction or remodeling of office space. Carpet is typically formed by bonding a face fiber onto a backing fiber, most commonly using a styrene-butadiene adhesive that incorporates fillers such as calcium carbonate. The face fibers are generally nylon, polypropylene, polyester, wool, or acrylic. The backing is usually polypropylene, PVC, nylon, or jute. In addition to the complexity of the structures themselves, used carpet is generally heavily contaminated with dirt, staples, food, and other materials.[175] Many efforts to recycle carpeting have focused on nylon carpet, which makes up about two-thirds of the face fiber market, and particularly on nylon 6 because of the ease of chemical depolymerization of this material (see Sec. 14.8.9). The next most commonly used face fiber is polypropylene.[175]

One company, Interface Flooring Systems, of LaGrange, Georgia, has embraced the producer responsibility concept by leasing rather than selling its carpet, retaining responsibility for carpet care and ultimate disposal. The company works with Custom Cryogenic Grinding Corp., of Simcoe, Ontario, to process the returned carpet using a cryogenic grinding process to make the carpet brittle, facilitating its separation into nylon face fibers and PVC backing.[175]

Monsanto will recycle all types of carpeting that its customers replace with Monsanto nylon carpet. Nylon carpet face fiber, latex, and backing are recycled into thermoplastic pellets, which are reused industrially. The company is also exploring opportunities for reuse of carpet in fuel recovery systems as well as for respinning post-consumer nylon carpet face fiber.[176]

In Minnesota, more than 60 businesses, along with government groups and the University of Minnesota, participate in the Minnesota

CA-RE (carpet recycling) Program. Some of the collected material is used by resin and fiber companies, and some is used by United Recycling, a subsidiary of Environmental Technologies USA, to make "grey felt," which is used for padding installed under commercial floor coverings as well as for sound insulation in automobiles.[177]

Collins & Aikman Floorcoverings, of Dalton, Georgia, uses vinyl-backed carpet to make solid commingled plastic products such as car stops and highway sound-wall barriers. They are now using up to 75% reclaimed carpet materials to make a nylon-reinforced backing for new carpet for modular tile products.[175]

DuPont is investigating the use of ammonolysis to depolymerize mixed nylon 6 and nylon 6/6 from used carpet. The company is also using reclaimed fiberized material from nylon carpet to make nylon building products for use in wet environments such as kitchens and bathrooms.[175]

Researchers in Georgia are investigating an unconventional use of carpet fibers, incorporating them into the surface of unpaved roads to improve road performance.[178]

To simplify the task of carpet material identification, the Carpet and Rug Institute has developed a seven-part universal coding system whereby a code on the carpet backing can be used to describe the components of the carpet, including facing, backing, adhesive, and fillers. As of 1997, it was estimated that 85% of the carpet now being made in the U.S. uses this code. However, the average 10-year life span of carpet means that, for the next several years, most carpet entering the recycling stream will not be so labeled.[175] Thus, as for automobile parts, equipment for identifying carpet materials will continue to be needed.

14.8.17 Other Plastics

While the major types of plastics recycling have been addressed, a variety of other types of plastics are being recycled, often on a small-scale or experimental basis. For example, Arco Chemical Co. has a process for recycling glass-reinforced styrene maleic anhydride from industrial scrap.[179] The University of Nottingham has a project for developing recycling techniques for thermoset materials, including polyesters, vinyl esters, epoxies, phenolics, and amino resins, along with glass and carbon-fiber reinforced resins.[180] The Fraunhofer Institute in Teltow is developing a process for recycling thermosets using an amine-based reagent in a one-step process that requires little added heat. The process is said to be applicable to almost all thermosets.[181] Imperial Chemical Industries plc and Mitsubishi Rayon Co. Ltd. are developing technology for recycling of acrylics by chemical depolymerization and repolymerization.[182] The introduction of polyethylene naphthalate (PEN) in U.S. packaging markets was delayed by the perceived need to develop processes for automatic separation of PEN from PET, as well as technology for recycling of PEN. Other examples could be cited as well. The field of plastics recycling is constantly evolving in response to changing demands and opportunities as well as the emergence of new resins and new applications.

14.9 Recycling of Rubber

Recycling of rubber from tires in the U.S. was estimated at 770,000 tons in 1997, 22.3% of the 3,450,000 tons of rubber in tires that entered the municipal solid waste stream.[2] EPA's information on trends in generation and recovery of rubber tires is shown in Fig. 14.65, and recycling rates in Fig. 14.66.

The Scrap Tire Management Council of the Rubber Manufacturers Association claims that markets for scrap tires consumed approximately 66% of the total number of tires scrapped in 1998, down from nearly 76% in 1996. The decrease in use as fuel, which accounted for 42% of the market, was responsible for the decline. Markets other than fuel included agricultural and civil engineering applications and manufacture of rubber products. Exports accounted for about 5.5%. Use of scrap tires in civil engineering application increased from 10 million tires in 1996 to 20 million in 1998. The tire material replaces material such as soil, clean fill, drainage aggregate, and other fill material.[183]

Products manufactured from ground rubber accounted for about 460 million pounds of scrap tires. Of this, 45.6% originated in tire buffings from retreading operations, and the remainder from whole scrap tires. One significant market is blends with asphalt, in rubber modified as-

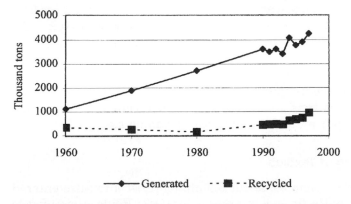

Figure 14.65 Generation and recovery of tires in U.S. municipal solid waste.[2,52]

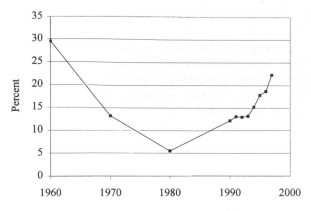

Figure 14.66 Tire recycling rates in the U.S.[2,52]

phalt, RMA. Another market is bound rubber products, where the rubber particles are held together by an adhesive, such as carpet underlay, dock bumpers, and patio flooring. Powdered scrap tire rubber can be used in the manufacture of new tires. Michelin North America is producing an original equipment tire for Ford that contains five percent recycled rubber content. Ground rubber is also used in athletic and recreational applications such as in running track material, for playground surfaces, and as a turf top dressing. Other applications include the manufacture of friction brake material, addition to thermoplastic polymers to modify their properties, and the manufacture of auto parts. Tires can also be used in the manufacture of cut, stamped, or punched rubber products. However, tires with steel belts or body plies are not suitable for these products.[183]

In Europe, the goal is to reach a recycling rate of 65% of scrap tires by 2000. Japan recycled 91% of the 102 million scrap tires generated in 1997. Energy recovery was the largest market, accounting for 51% of the total. Export for reuse consumed 17%, 12% were recycled, 8% were retreaded, and 3% were used in miscellaneous ways.[184]

Research on novel applications for scrap tires is ongoing. University of Illinois researchers have investigated the recycle of tires into activated carbon absorbents.[185] Primix Corp. is manufacturing composite railroad ties formed from a concrete base with a shell of polyethylene, rubber from scrap tires, and a proprietary blending agent.[186]

14.10   Recycling of Textiles

According to EPA estimates, about 8.2 million tons of textiles entered the U.S. municipal solid waste stream in 1997. While a significant amount of textiles are recovered for reuse, they still enter the waste stream eventually. The amount of recycling, including export, was estimated at 12.9%, 1.1 million tons.[2]

The Council for Textile Recycling claims that its 350 members remove 2.5 billion pounds of post-consumer textile product waste from the U.S. municipal solid waste stream each year. The majority of the collected material, 61%, is exported, with about half for use as used clothing and the other half used as fiber. The primary domestic market for recycled textiles is in the manufacture of wipers, which consumes 25% of the collected material. Domestic fiber reprocessing uses another seven percent, and seven percent of the collected material is landfilled as waste.[187]

14.11   Recycling of Wood

The EPA estimates that about 590,000 tons of wood was recycled from U.S. municipal solid waste in 1997, out of 11.6 million tons of wood discarded, for a recycling rate of 5.1%. Important sources of wood in the waste stream include furniture and other durable goods and wood packaging such as crates and pallets. Most of the recycling involved chipping for uses such as mulch or bedding material. Refurbishing and reuse of objects such as pallets is classified as *source reduction* and not as recycling. This is a significant activity, since EPA estimates that over 5 million tons of wooden pallets were refurbished and returned to service rather than being discarded.[2]

Wood packaging accounted for over 7 million tons of material entering the municipal solid waste stream, about 61% of all wood. All recycling of wood from MSW tabulated by EPA originated in packaging. Patterns of generation and recovery of wood in municipal solid waste are illustrated in Fig. 14.67. Recycling rates for wood packaging are shown in Fig. 14.68.

Construction and demolition debris, suburban land-clearing debris, and forestry waste, including mill waste, are important sources of wood waste that are not included in the definition of municipal solid waste. It was estimated that the volume of waste wood in the U.S. was 194 million tons in 1996. About 63% of this, primarily mill residuals, was recovered and used, mostly as mulch and fuel.[188] A recent study in North Carolina estimated that 30% of construction and demolition debris generated in the state was wood, of which less than five percent was recovered.[189] Recycling of construction and demolition debris is becoming more common with the growth of the deconstruction industry. Reclaimed lumber salvaged by these companies often originated in old-growth forests and has a number of desirable characteristics, including larger sizes than are currently available and dense, tight-grain wood that is often quite free of defects. The U.S. Forest Products

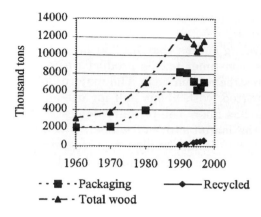

Figure 14.67 Generation and recovery of wood in U.S. municipal solid waste.[2,43]

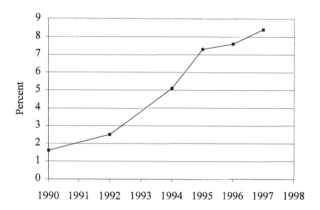

Figure 14.68 Wood packaging recycling rates in the U.S.[2,43]

Laboratory has been involved in testing reclaimed wood for structural integrity in support of the development of reclamation protocols and grading rules for reclaimed wood.[190]

Wooden pallets are already recycled at a relatively high rate, 71% in 1998.[191] One company even manufactures fine furniture from old pallets.[192] Mulch, animal bedding, and playground cover are more common uses.

References

1. Glenn, J., "The State of Garbage in America" *BioCycle*, Apr. 1999, pp. 60–71.
2. U.S. EPA, *Characterization of Municipal Solid Waste in the United States: 1998 Update*, EPA 530-R-99-021, September 1999.

3. *Food Packaging Bulletin*, Vol. 8, No. 9, 1999, pp. 7–8.
4. Thompson Publishing Group, *Environmental Packaging: U.S. Guide to Green Labeling, Packaging and Recycling*, Washington, D.C., 1999.
5. Raymond Communications, *State Recycling Laws Update*, College Park, Maryland, 1999.
6. "Oregon," *Resource Recycling*, Feb. 1996, p. 10.
7. "Plastic Recycling Rate in California," *Reuse/Recycle*, April 1998, p. 30.
8. Canadian Council of Ministers of the Environment, *National Packaging Protocol*, CCME-TS/WM-FS 020, 1990.
9. "Packaging Waste in Canada Cut by Half," *Plastics News*, March 2, 1998.
10. Gies, G., "The State of Garbage in Canada," *BioCycle*, Feb. 1996, pp. 46–52.
11. Majid, M., "Restricting the Use of Plastic Packaging," *Professional Bulletin of the National Poison Centre*, Malaysia, June 1997, pp. 1–4, available online, prn.usm.my/ bulletin/1997/prn13.html.
12. "Mandatory Recycling Coming to Israel," *BioCycle*, Feb. 1998, p. 71.
13. "Switzerland: Recovery of Electronic Scrap Now Obligatory," *Resources Report*, July 1998, pp. 6–7.
14. Coca-Cola, "Consumer Advisory: Coca-Cola System Using Recycled Plastic," press release, March 1, 2000.
15. Toloken, S., "Campaign Asks Coke to Use Recycled PET," *Plastics News*, September 14, 1998, p. 6.
16. U.S. Environmental Protection Agency, *Comprehensive Guideline for Procurement of Products Containing Recovered Materials*, 40 CFR Part 247, 2000.
17. Glenn, J., "The State of Garbage in America, Part I" *BioCycle*, Apr. 1998, pp. 32–43.
18. Steel Recycling Institute, "Fact Sheet: a Few Facts about Steel—North America's #1 Recycled Material," available online, www.recycle-steel.org/fact/main.html.
19. Steel Recycling Institute, "1998 Steel Recycling Rates Fall after 10 Year Gain. Foreign Imports Dramatically Impact Value of Steel Scrap," press release, April 14, 1999.
20. Steel Recycling Institute, "Unrestricted Foreign Steel Imports Impact U.S. Recycling Efforts," press release, Nov. 23, 1998.
21. Japan Information Network, "Recycling Rates (1986–1999)," available online jin.jcic.or.jp/stat/stats/19ENV51.html.
22. Association of European Producers of Steel for Packaging (APEAL), "Recycling Achievements," available online, www.apeal.org/Contents/Enviroment/Envi03.html. (sic)
23. Crawford, G. L., "Steeling for Major Recycling Gains," *Resource Recycling*, June 1998, pp. 43–47.
24. Steel Recycling Institute, "The Inherent Recycled Content of Today's Steel," available online, www.recycle-steel.org/buy/BuyInherent.html.
25. Powell, J., "Directory of U.S. and Canadian Steel Mini-mills," *Resource Recycling*, March 1999, pp. 30–31.
26. The Steel Alliance, "Recycle," available online, www.thenewsteel.org/recycle/environ.htm.
27. The Aluminum Association, "Recycling: Reclamation Statistics," available online, www.aluminum.org/default2.cfm/4/36.
28. aluNet, "European Aluminium Can Recycling Continued to Increase in 1998," Alupress S.A., Greece, available online, www.alunet.gr/market/english/recycling.htm.
29. The Aluminum Association, "Recycling Facts," available online, www.aluminum.org/default2.cfm/4/34.
30. European Aluminium Association, available online, www.eaa.net/pages/material/recycled.html.
31. The Aluminum Association, "Aluminum Facts at a Glance," Washington, D.C., June 1999.
32. Roy F. Weston, Inc., *Life Cycle Inventory Report for the North American Aluminum Industry*, The Aluminum Association, Washington, D.C., 1998.
33. Cobb, G., "Keeping Contaminants to a Minimum," *BioCycle*, Jan. 1993, p. 75.

34. Glass Packaging Institute, available online, www.gpi.org.
35. *Recycling World*, Oct. 8, 1999, available online, www.tecweb.com.
36. FEVE, *Glass Gazette*, No. 24, Sept. 1998, p. 5., available online www.brit-glass.co.uk/publications/glassgazette1998/page5.html.
37. Tyler, N. M., "Glass Recycling: Cause and Effect," *Resource Recycling*, Aug. 1996, pp. 39–42.
38. Walsh, P. and P. O'Leary, "Recycling Offers Benefits, Opportunities…And Challenges," *Waste Age*, 19(1): 54–60, 1988.
39. U.S. Environmental Protection Agency, *MSW Factbook*, Version 4.0, Office of Solid Waste, Washington, D.C., 1997.
40. Cesar, M., "Asian Currency Crisis Affects U.S. Recovered Paper Markets," *Resource Recycling*, June, 1998, pp. 30–36.
41. American Forest and Paper Association, available online, www.afandpa.org/recycling.
42. Canadian Pulp and Paper Association, available online, www.open.doors.cppa.ca/english/wood/guide.
43. U.S. EPA, *Characterization of Municipal Solid Waste in the United States: 1997 Update*, EPA 530-R-98-007, May 1998.
44. Rovelo, C., "An Opportunity for Recovered Fiber and Allied Paper Products," *Resource Recycling*, April, 1998, pp. 10–17.
45. Confederation of European Paper Industries, "The Growth in Paper Recycling Has Largely Outset Increased Paper Consumption During the Nineties," CEPI Newsletter, No. 13, Aug. 1999.
46. *Recycling World Magazine*, "Recycled Content Targets for Paper Not Ruled Out," Feb. 2, 2000
47. Iannazzi, F., "A Decade of Progress in U.S. Paper Recovery, 1989–1998," *Resource Recycling*, June, 1999, pp. 19–25.
48. Australian Paper, "Waste Paper Utilisation," available online, www.apg.com.au/paperr/prwpu.htm.
49. The Paper Federation of Great Britain, "Raw Materials Usage," available online, www.paper.org.uk/htdocs/Statistics/raw.html.
50. Iannazzi, F. and R. Clarke, "Can We Reach 40 Percent Recovered Paper Use in Newsprint by 2000?" *Resource Recycling*, Nov. 1998, pp. 12–17.
51. Iannazzi, F. and R. Clarke, "When Will OCC Prices improve?" *Resource Recycling*, Jan. 1998, pp. 26–33.
52. U.S. EPA, *Characterization of Municipal Solid Waste in the United States: 1996 Update*, EPA 530-R-97-015, May 1997.
53. *BioCycle*, "Marking Waxed Corrugated Boxes," May, 1998, p. 6.
54. Kunzler, C., "What's What with Waxed Corrugated," *BioCycle*, Aug., 1998, pp. 64–66.
55. Curley, J., "Waxed Boxes Still a Thorny Issue for US Boxmakers," *International Paper Board Industry*, April, 1998, pp. 21–22.
56. *Recycling World Magazine*, "Recycled Content of Newsprint Reaches 46.54%," July 16, 1999.
57. *Recycling World Magazine*, "Newsprint Bill Progress," April 7, 2000.
58. Moore, B., "Magazines: Where Recycling Demand Exceeds Supply," *Resource Recycling*, June, 1999, pp. 26–30.
59. *Recycling World Magazine*, "Christmas Card Recycling," Jan. 15, 1999.
60. Kinsella, S., "Recycled Paper Buyers, Where Are You?" *Resource Recycling*, Nov. 1998, pp. 18–21.
61. APME, *Plastics: An Analysis of Plastics Consumption and Recovery in Western Europe 1998,*"Association of Plastics Manufacturers in Europe, 2000.
62. Denison, R. A., *Something to Hide: The Sorry State of Plastics Recycling*, Environmental Defense Fund, Washington, D.C., 1997.
63. American Plastics Council, "1998 Recycling Rate Study," available online, www.plasticsresource.com/recycling_rate_study.
64. RECOUP, "Plastic Bottle Recycling Statistics 1997," available online, www.tecweb.com/ recoup.

65. Bilitewski, B. and C. Copeland, "Packaging Take-back in Germany: The Plastics Recycling Picture," *Resource Recycling*, Feb. 1997, pp. 46–52.
66. Toloken, S., "Contaminants Seep into Rates for Recycling," *Plastics News*, Aug. 31, 1998, p. 1.
67. Higgs, R., "New Technology to Boost European Recycling," *Plastics News*, June 16, 1997, p. 56.
68. Food and Drug Administration, *Recommendations for Chemistry Data for Indirect Food Additive Petitions*, Center for Food Safety & Applied Nutrition, Washington, D.C., 1995.
69. Powell, J., "The PRFect Solution to Plastic Bottle Recycling," *Resource Recycling*, Feb. 1995, pp. 25–27.
70. "Plastics Recyclers Stay on the Cutting Edge," *BioCycle*, May 1996, pp. 42–46.
71. Colvin, R., "Sorting Mixed Polymers Eased by Hand-held Unit," *Modern Plastics*, April, 1995.
72. Stambler, I., "Plastic Identifiers Groomed to Cut Recycling Roadblocks," *R&D Magazine*, Oct. 1996, pp. 29–30.
73. Smith, S., "PolyAna System Identifies Array of Plastics," *Plastics News*, Dec. 15, 1997, p. 14.
74. Apotheker, S., "Flake and Shake, Then Separate," *Resource Recycling*, June 1996, pp. 20–27.
75. Smith, S., "Fry-sorting Process Promising in Post-consumer Plastics Use," *Plastics News*, Nov. 11, 1996, p. 27.
76. Schut, J. H., "Process for Reclaiming Durables Takes off in U.S.," *Modern Plastics*, March 1998, pp. 56–57.
77. Maten, A., "Recovering Plastics from Durable Goods: Improving the Technology," *Resource Recycling*, Sept. 1995, pp. 38–43.
78. Schut, J. H., "Centrifugal Force Puts New Spin on Separation," *Plastics World*, Jan. 1996, p. 15.
79. "Packaging Research Case Study," *Management & Technology*, June 1996, p. 23.
80. King, R., "U.S. Exports of Waste Plastics Climb," *Plastics News*, March 25, 1996, p. 1.
81. Babinchak, S., "Current Problems in PET Recycling Need to Be Resolved," *Resource Recycling*, Oct. 1997, pp. 29–31.
82. Toloken, S., "Plastic Bottle Recycling Rate Keeps Sliding," *Plastics News*, Aug. 24, 1998, p. 1.
83. U.S. EPA, *Characterization of Municipal Solid Waste in the United States: 1994 Update*, EPA 530-R-94-042, Nov. 1994.
84. National Association for PET Container Resources, available online, www.napcor.com
85. Barham, V. F., "Closing the Loop for P.E.T. Soft Drink Containers," *Journal of Packaging Technology*, Jan./Feb. 1991, pp. 28–29.
86. Bakker, M., "Using Recycled Plastics in Food Bottles: The Technical Barriers," *Resource Recycling*, May 1994, pp. 59–64.
87. Newcorn, D., "Plastics' Broken Loop," *Packaging World*, June 1997, pp. 22–24.
88. Toloken, S., "Miller Uncaps PET Bottles," *Plastics News*, March 13, 2000, p. 1, 67.
89. Pryweller, J., "Lear Corp. Offers Automakers Recycled Plastic for Vehicles," *Plastics News*, Sept. 21, 1998, p. 21.
90. Toloken, S., "Recyclers Worried over Opaque Milk Bottles," *Plastics News*, Oct. 27, 1997, p. 7.
91. "Ecoplast Gets FDA Nod," *Plastics News*, April 6, 1998.
92. DuPont, available online, www.dupont.com/tyvek.
93. Apotheker, S., "Film at 11; A Picture of Curbside Recovery Efforts for Plastic Bags," *Resource Recycling*, May 1995, pp. 35–40.
94. "Montreal-area Citizens Begin Film Program," *Plastics News*, Feb. 12, 1996.
95. Environment and Plastics Industry Council, "EPIC Moving Ahead on Several Initiatives," available online, www.plastics.ca.
96. Vinyl Institute, available online, www.vinylinfo.org, 1998.

97. The PVC Centre, European Vinyl Corp., available online, www.ramsay.co.uk/pvc/pvcenvir. htm, 1998.
98. Principia Partners, *Post-industrial and Post-consumer Vinyl Reclaim: Material Flow and Uses in North America*, July, 1999, available online, www.principiaconsulting.com/reports_vinyl.pdf.
99. Smith, S. S., "APR Urges Vinyl Institute to Find Markets," *Plastics News*, July 14, 1997, p. 6.
100. Toloken, S., "PVC Bottle Recyclers Chastise Vinyl Institute," *Plastics News*, August 18, 1997.
101. "Japanese Packagers to Phase out PVC Bottles," *Modern Plastics*, March, 1998, p. 20.
102. Wisner, D., "Recycling Post-consumer Durable Vinyl Products," presented at The World Vinyl Forum, Sept. 7–9, 1997, Akron, Ohio.
103. Alaisa, C, "Giving a Second Life to Plastic Scrap," *Resource Recycling*, Feb. 1997, pp. 29–31.
104. Vinyl Institute, "In France, Vinyl Automotive Recycling Proving Successful," *EnVIronmental Briefs*, Aug. 1998, available online, www.vinylinfo.org.
105. Ford, T., "Nortel May Add to its Recycling Process," *Plastics News*, Dec. 18, 1995, p. 10.
106. Kiser, J., "Hospital Recycling Moves Ahead," *BioCycle*, Nov. 1995, pp. 30–33.
107. Moore, S., "Plastics Recycling Profit Soars in India," *Modern Plastics*, June 1995, pp. 19–21.
108. Ehrlich, R. J., "The Economic Realities of Recycling," *Polystyrene News*, Fall 1997, p. 1, 3–4.
109. Toloken, S., "NPRC to Shut Failing PS Recycling Plant," *Plastics News*, August 4, 1997, p. 1, 39.
110. Canadian Polystyrene Recycling Association, "CPRA Collection Program Update," *News from Canada's First Polystyrene Recycling Facility*, Spring 1994.
111. Canadian Polystyrene Recycling Association, "CPRA Facts and Figures," available online, www.cpra-canada.com/facts.html.
112. "A&E Starts Program to Rescue Hangers," *Plastics News*, Dec. 2, 1996.
113. Alliance of Foam Packaging Recyclers, "EPS Meets Environmental Expectations," available online, www.epspackaging.org/accolades.html.
114. Alliance of Foam Packaging Recyclers, available online, www.presstar.com/afpr.
115. Hornberger, L. and T. Hight, "How Recycled Content Affects Foam Cushioning Performance," *Packaging Technology & Engineering*, April, 1998, pp. 48–51.
116. "Recycled PS Insulation Is No More," *Resource Recycling*, April 1997, p. 59.
117. "Malaysia Establishes EPS Recycling Group," *Modern Plastics*, June, 1996, p. 25.
118. Tilley, K., "Aussies Upgrade EPS Program," *Plastics News*, Nov. 11, 1996, p. 25.
119. UK EPS Information Service, available online, www1.mailbox.co.uk/www.eps.co.uk, 1998.
120. Alliance of Foam Packaging Recyclers, "EPS Recycling—What's Next," *Molding the Future*, October, 1998.
121. "Eastman Kodak Recycles 50 Million Cameras," *Plastics News*, Aug. 14, 1995, p. 10.
122. "Snapshot of Recycling," *BioCycle*, June 1996, p. 88.
123. Myers, S., "Recyclers of Appliances, Durables Looking to Germany's Proposals," *Modern Plastics*, March 1995, pp. 14–15.
124. "A Noncatalytic Process Reverts Polystyrene to its Monomer," *Chemical Engineering*, Feb. 1993, p. 19.
125. Battery Council International, "Lead-acid Batteries Head List of Recycled Products," press release, Nov. 12, 1998.
126. Heil, K., and R. Pfaff, "Quality Assurance in Plastics Recycling by the Example of Polypropylene; Report on the Experience Gathered with a Scrap Battery Recycling Plant," in F. LaMantia (ed.), *Recycling of Plastic Materials*, ChemTec Pub., Ontario, 1993, pp. 171–185.
127. "Murphy Scrubs Virgin PP bottle in Favor of PCR," *Packaging Digest*, March 1995.
128. "Recovering Dye Tubes," *BioCycle*, May 1996, p. 25.

129. U.S. Dept. of Energy, "Waste Carpet Recycling," available online, www.oit.doe.gov/ chemicals/ citar96/ CITAR96p31.htm, 1998.
130. Texas Society of Professional Engineers, "Research into Chemical Recycling Could Open New Opportunities," available online, www.tspe.org/recycle-6.htm.
131. BASF, "6ix Again®: Technology That Sustains the Earth," available online, www.basf.com/ businesses/fibers/ sixagain/index2.html, 1998.
132. AlliedSignal, Inc., press release, June 25, 1998.
133. Evergreen Nylon Recycling, available online, www.n6recycling.com.
134. "Ford Parts Incorporate Reclaimed Nylon Carpet," *Modern Plastics*, March 1997, p. 14.
135. Toray, available online, www.toray.co.jp/e/kankyou/risai.html, 1998.
136. Federal Prison Industries, Inc., available online, www.unicor.gov/_vti_bin/ shtml.exe/unicor/ environtextiles.html/map, 1998.
137. Recycline, available online, www.recycline.com/recinfo.html, 1998.
138. BASF Corp. Philip Services Corp., press release, September 16, 1997.
139. Higgs, R., "ICI to Recycle PU Foam Waste in U.K.," *Plastics News*, Oct. 20, 1997, p. 24.
140. Duranceau, C. M., G. R. Winslow and P. Saha, "Recycling of Automotive Seat Foam: Acoustics of Post Consumer Rebond Seat Foam for Carpet Underlayment Application," Society of Automotive Engineers, 1998, available online, www.polyurethan.org/PURRC/REPORT1/ index.html.
141. "Bayer Introduces Recycling Processes," *Plastics News*, Jan. 13, 1997.
142. White, K., "GE Plastics Begins Buyback Program Aimed at Plastic Auto Scrap," *Waste Age's Recycling Times*, March 7, 1995, p. 8.
143. "CD Recycling Plant Is Europe's First," *Modern Plastics*, Sept. 1995, p. 13.
144. British Telecommunications plc, available online, www.bt.co.uk/corpinfo/enviro/ phones.htm, 1998.
145. Texas Society of Professional Engineers, available online, www.tspe.org/recycle-4.HTM, 1998.
146. Graff, G., "Additive Makers Vie for Reclaimed Resin Markets," *Modern Plastics*, Feb. 1996, pp. 51–53.
147. Ford, T., "Hewlett-Packard Printers Use Recycled ABS," *Plastics News*, Aug. 7, 1995, p. 40.
148. Bisio, A. L., and M. Xanthos (eds.), *How to Manage Plastics Waste: Technology and Market Opportunities*, Hanser Pub., Munich, 1994.
149. Van Ness, K. E., and T. J. Nosker, "Commingled Plastics," in R. Ehrig (ed.), *Plastics Recycling: Products and Processes*, Hanser Pub., Munich, 1992, pp. 187–229.
150. "Electrolux is Putting Recyclate in Vacuums," *Modern Plastics*, Jan. 1997, p. 22.
151. Moore, S., "Plastics Recycling Profit Soars in India," *Modern Plastics*, June 1995, pp. 19–21.
152. Mapleston, P., "Housing May Be Built from Scrap in Low-Pressure Process," *Modern Plastics*, Sept. 1995, p. 21.
153. "Five Standard Test Methods on Plastic Lumber Approved," *ASTM Standardization News*, Nov. 1997, p. 11.
154. Krishnaswamy, P. and D. Stusek, "Structural Applications for Recycled Plastic Lumber," *Resource Recycling*, Oct. 1999, pp. 27–30.
155. "Plastic Pallets Rival Wood, Steel Models," *Plastics News*, Nov. 17, 1997.
156. Bregar, B., "Plastic Rail Ties Gaining Favor," *Plastics News*, May 11, 1998.
157. Buwalda, T. and B. Halpin, "What's New with Plastic Lumber?" *Resource Recycling*, Oct. 1998, pp. 22–26.
158. Urey, C., "Uncle Sam Recruits Recycled Plastic Lumber," *Plastics News*, July 13, 1998.
159. Solomon-Hess, J., "Process for Plastic-Paper Mix Tested," *Plastics News*, Dec. 4, 1995, p. 13.
160. "Boise Cascade Eyes Film-Wood Composite." *Plastics News*, Oct. 5, 1998.
161. Bregar, B., "AERT Wins Major Contract for its Recycled Material," *Plastics News*, May 20, 1991, p. 1.
162. "Wood-fiber Composite Is Now Commercial," *Modern Plastics*, May 1997, p. 16.

163. Lavendel, B., "Recycled Wood and Plastic Composites Find Markets," *BioCycle*, Dec. 1996, pp. 39–43.
164. Urey, C., "Wood Composites Make Show at Meeting," *Plastics News*, Sept. 29, 1997.
165. Urey, C., "Mikron Invests in Wood-Plastic Composite," *Plastics News*, Aug. 17, 1998, p. 8.
166. Ford, T., "Nortel May Add to its Recycling Process," *Plastics News*, Dec. 19, 1995, p. 10.
167. White, K., "Research Center to Demonstrate New Commingled Plastic Processing System," *Waste Age's Recycling Times*, Oct. 4, 1994, p. 9.
168. Wilt, C. and L. Kincaid, "There Auto Be a Law: End-of-life Vehicle Recycling Policies in 21 Countries," *Resource Recycling*, March 1997, pp. 42–50.
169. Pryweller, J., "Ford Driving Recycled Nylon Applications," *Plastics News*, Feb. 24, 1997, pp. 7, 9.
170. Pryweller, J., "Cost is King in Auto-related Recycling," *Plastics News*, Sept. 8, 1997, pp. 1, 8.
171. Pryweller, J., "Recycling Center Pitching PU to Carmakers," *Plastics News*, Sept. 22, 1997, p. 35.
172. Sherman, L. M., "Compounders Take the Lead in Post-use Bumper Recycling," *Plastics Technology*, March 1996, pp. 27–29.
173. Grande, J. A., "Ford Is Targeting 50% Use of Recycle-content Resin by 2002," *Modern Plastics*, July 1996, pp. 32–33.
174. Pryweller, J., "Projects Could Turn Plastics into a Recycling 'Headliner'," *Plastics News*, Feb. 23, 1998, pp. 2, 9.
175. Powell, J., "Magic Carpet Ride: the Coming of a New Recyclable," *Resource Recycling*, April 1997, pp. 42–46.
176. Monsanto, available online, www.floorspecs.com/reference_Library/Monsanto_ Recycling/ index.htm, 1998.
177. United Recycling, available online, www.cais.net/publish/stories/0596haz9.htm, 1998.
178. "Rolling Out the Red Carpet for Recycling," *BioCycle*, Feb. 1998, pp. 12–13.
179. "Arco Chemical Devises SMA Recycling Process," *Plastics News*, March 11, 1996.
180. "British Project for Recycling Polymer Composites Launched," *C&EN*, June 19, 1995, p. 11.
181. "Thermoset Reclaim Bets on Chemical Process," *Modern Plastics*, April 1995, p. 13.
182. Higgs, R., "ICI, Mitsubishi Research Acrylic Recycling," *Plastics News*, May 26, 1997.
183. Scrap Tire Management Council, "Scrap Tire Use/Disposal Study, 1998/1999 Update, Executive Summary," Rubber Manufacturers Association, available online, www.rma.org/exsumn.html.
184. Moore, M., "Officials from Around the Globe Discuss Ways to Recycle Scrap Tires," *Rubber & Plastics News*, Nov. 2, 1998, p. 6.
185. "Worn-out Tires Get New Life," *Chemical Engineering Progress*, Jan. 1999, p. 15.
186. Wenger, R., "Primix Develops a Composite Railroad Tie," *Plastics News*, Aug. 23, 1999, p. 43.
187. Council for Textile Recycling, "Textile Recycling Fact Sheet," available online, www.textilerecycle.org/ctrfacts.html.
188. Horne-Brine, P., "Recycling Wood for Value and Volume: Work at the Cutting Edge," *Resource Recycling*, March 1998, pp. 44–52.
189. Gray, K., "The Many Routes to Recycling Wood," *BioCycle*, March 1999, pp. 64–66.
190. Horne-Brine, P. and R. Falk, "Knock on Wood: Real Recycling Opportunities are Opening Up," *Resource Recycling*, Aug. 1999, pp. 42–46.
191. "Pallets Top 'Most Recycled' List," *Packaging Technology & Engineering*, Sept. 1999, p. 10.
192. "Fine Furniture is Crafted from Used Pallets," *BioCycle*, May 1999, p. 27.

Index

ABOUT THE EDITOR

Charles A. Harper is President of Technology Seminars, Inc., of Lutherville, Maryland. He is widely recognized as a leader in materials for product design, having worked and taught extensively in this area. Mr. Harper is also Series Editor for the Materials Science and Technology Series and the Electronic Packaging and Interconnection Series, both published by McGraw-Hill. He has been active in many professional societies, including the Society of Plastics Engineers, American Society for Materials, and the Society for the Advancement of Materials Engineering, in which he holds the honorary level of Fellow of the Society. He is a past President and Fellow of the International Microelectronics and Packaging Society. Mr. Harper is a graduate of the Johns Hopkins University, Baltimore, Maryland, where he has also served as Adjunct Professor.